Computational Medicinal Chemistry for Drug Discovery

Computational Medicinal Chemistry for Drug Discovery

edited by

Patrick Bultinck

Ghent University
Ghent, Belgium

Hans De Winter
Wilfried Langenaeker

Johnson & Johnson Pharmaceutical Research and Development
A Division of Janssen Pharmaceutica N.V.
Beerse, Belgium

Jan P. Tollenaere

Utrecht University
Utrecht, The Netherlands

CRC Press
Taylor & Francis Group
Boca Raton   London   New York

CRC Press is an imprint of the
Taylor & Francis Group, an **informa** business

FIRST INDIAN EDITION, 2009

Although great care has been taken to provide accurate and current information, neither the author(s) nor the publisher, nor anyone else associated with this publication, shall be liable for any loss, damage, or liability directly or indirectly caused or alleged to be caused by this book. The material contained herein is not intended to provide specific advice or recommendations for any specific situation.

Trademark notice: Product or corporate names may be trademarks or registered trademarks and are used only for identification and explanation without intent to infringe.

Library of Congress Cataloging-in-Publication Data
A catalog record for this book is available from the Library of Congress.

ISBN: 0-8247-4774-7

Printed and Bound in India by Saurabh Printers Pvt. Ltd.

Headquarters
Marcel Dekker, Inc., 270 Madison Avenue, New York, NY 10016, U.S.A.
tel: 212-696-9000; fax: 212-685-4540

Distribution and Customer Service
Marcel Dekker, Inc., Cimarron Road, Monticello, New York 12701, U.S.A.
tel: 800-228-1160; fax: 845-796-1772

Eastern Hemisphere Distribution
Marcel Dekker AG, Hutgasse 4, Postfach 812, Ch-4001 Basel, Switzerland
tel: 41-61-260-6300; fax: 41-61-260-6333

World Wide Web
http://www.dekker.com

The publisher offers discounts on this book when ordered in bulk quantities. For more information, write to Special Sales/Professional Marketing at the headquarters address above.

Preface

Computational approaches to medicinal chemical problems have developed rapidly over the last 40 years or so. In the late 1950s and early 1960s, gigantic mainframe computers were used to perform simple HMO (Huckel molecular orbital) and PPP (Pariser-Parr-Pople) calculations on aromatic compounds such as substituted benzenes, naphthalenes, anthracenes, etc., to explain their UV spectral properties. In the early 1960s, stand-alone programs became available to simulate NMR spectra. With the advent of Hansch-type analysis of structure-activity relationships (SAR), computers were used to solve multiple regression equations. In 1963 the Quantum Chemistry Program Exchange (QCPE) started distribution of programs such as Extended Huckel Theory (EHT) and early versions of Complete Neglect of Differential Overlap (CNDO), which to the delight of theoretical chemists eventually made it possible to perform conformational analyses on nonaromatic molecules. However scientifically exciting, all these computations involved quite some expertise in mastering the computer's operating system as well as manual labor punching cards and hauling boxes of punched cards to and from the mainframe computer center. Of greater concern, however, was the fact that real-life molecules such as those routinely synthesized by medicinal chemists were most often too big to be treated theoretically using the computers of those days. This resulted in a situation in which the contribution of a theoretical chemist was, at best, politely tolerated but in general considered irrelevant to the work of a classically trained medicinal chemist.

All this changed, although slowly, in the 1970s, with improvements in the speed, manageability, and availability of computer technology. A considerable impediment in the late 1970s and early 1980s was the lack of proper visualization of the theoretical results. Indeed, it was discouraging to discuss theoretical results with a suspicious chemist on the basis of pages and pages of computer output. This obstacle was dramatically removed with the advent of graphics computers able to depict HOMOs, LUMOs, MEPs (molecular electrostatic potential), dipole moment vectors, etc, superimposed on a 3D representation of the molecule(s) of interest. By the early 1990s graphics workstations linked to multiprocessor machines were powerful enough to perform reliable calculations on real-life molecules in a time frame sufficiently small to keep the interest of the medicinal chemist alive and to show the results in an understandable and appealing way.

Nowadays, one can safely state that the computational chemist has become a respectable member of a drug (ligand) design team, standing on an equal footing with the synthetic chemists, pharmacologists, and others at the beginning of the long and arduous path of ligand creation aimed toward bringing a medicine to the market.

The title of this book refers to two topics, namely, Computational Medicinal Chemistry and Drug Design. It unites these topics by giving an overview of the main methods at the disposal of the computational chemist and to highlight some applications of these methods in drug design. Although drug and ligand appear to be synonymous in this volume, they most definitely are not. Notwithstanding "drug design" in the title, this volume essentially deals with methods that can be applied to molecules that may possibly become drugs. Whether, when, and how a molecule may acquire the status of a drug or a medicine is investigated and decided by, among others, toxicologists, pharmacists, and clinicians and is therefore explicitly outside the scope of this volume.

Similarly, a choice had to be made regarding the topics covered in this volume. For example, molecular dynamics (MD) based free-energy changes in solution calculations are not treated, because these are not yet a day-to-day practice in actual ligand design due to the very high computational demands for the long MD simulations required.

This book starts with seven chapters devoted to methods for the computation of molecular structure: molecular mechanics, semiempirical methods, wave function–based quantum chemistry, density-functional theory methods, hybrid methods, an assessment of the accuracy and applicability of these methods, and finally 3D structure generation and conformational analysis.

In the next chapters, one or several of those formalisms are used to describe some aspects of molecular behavior toward other molecules in terms of properties such as electrostatic potential, nonbonded interactions, behavior in solvents, reactivity and behavior during interaction with other molecules, and finally similarity on the basis of nonquantum and quantum properties.

Before addressing some aspects of, broadly speaking, ligand-receptor interactions, a critical evaluation of protein structure determination was felt in order. This is then followed by accounts of docking and scoring, pharmacophore identification 3D searching, substructure searching, and molecular descriptors.

The following chapters address 2D and 3D models using classical molecular and quantum-based descriptors and models derived from data mining techniques as well as library design.

Given the increasing demand for enantiomerically pure drugs, vibrational circular dichroism (VCD) will become a standard technique in the medicinal chemical laboratory. The VCD chapter illustrates the use of high-level quantum chemical calculations and conformational analysis discussed in previous chapters. Similarly, the chapter on neuraminidase highlights the combined use of protein crystallography, ligand receptor interaction theory, and computational methods. Finally, this volume ends with a concise glossary.

Thanks are due to Anita Lekhwani, who initially suggested this project, and to Lila Harris, who helped in realizing the project. Each individual chapter was reviewed by at least three editors. During monthly editorial meetings reviews were critically compared.

The editors are grateful to those authors who strictly adhered to the time schedule.

Finally, it is hoped that this volume may give the reader a useful overview of the main computational techniques that are currently in use on a day-to-day basis in modern ligand (drug) design, both in academia and in an industrial pharmaceutical environment.

Johnson & Johnson Pharmaceutical Research and Development–Beerse (Belgium) is gratefully acknowledged for financial and logistic support for this project.

Patrick Bultinck
Hans De Winter
Wilfried Langenaeker
Jan P. Tollenaere

Contents

Contributors

Orlando Acevedo   Center for Computational Studies and Department of Chemistry and Biochemistry, Duquesne University, Pittsburgh, Pennsylvania, U.S.A.

Paul W. Ayers   Department of Chemistry, McMaster University, Hamilton, Ontario, Canada

Christopher J. Barden   Department of Chemistry, Dalhousie University, Halifax, Nova Scotia, Canada

John M. Barnard   Barnard Chemical Information Ltd., Stannington, Sheffield, S. Yorks, United Kingdom

Andrey A. Bliznynk   ANU Supercomputer Facility, Australian National University, Canberra, Australian Capital Territory, Australia

Thomas Bredow   Theoretical Chemistry, University of Hannover, Hannover, Germany

Ramon Carbó-Dorca   Institute of Computational Chemistry, University of Girona, Campus Montilivi, Catalonia, Spain

P. K. Chattaraj   Department of Chemistry, Indian Institute of Technology, Kharagpur, India

Lingran Chen   MDL Information Systems, Inc., San Leandro, California, U.S.A.

Peter L. Cummins   Division of Molecular Bioscience, John Curtin School of Medical Research, Australian National University, Canberra, Australian Capital Territory, Australia

Geoff M. Downs   Barnard Chemical Information Ltd., Stannington, Sheffield, United Kingdom

Jeffrey C. Dyason   Griffith University (Gold Coast), Bundall, Queensland, Australia

Michael F. M. Engels   Johnson & Johnson Pharmaceutical Research and Development, A Division of Janssen Pharmaceutica N.V., Beerse, Belgium

Istvan Enyedy   Bayer Research Center, West Haven, Connecticut, U.S.A.

Jeffrey D. Evanseck   Department of Chemistry and Biochemistry, Duquesne University, Pittsburgh, Pennsylvania, U.S.A.

Johann Gasteiger   Computer-Chemie-Centrum, Institute for Organic Chemistry, Erlangen-Nuernberg University, Erlangen, Germany

Valerie J. Gillet   Department of Information Studies, University of Sheffield, Sheffield, United Kingdom

Xavier Gironés   Institute of Computational Chemistry, University of Girona, Campus Montilivi, Catalonia, Spain

Jill E. Gready   Division of Molecular Bioscience, John Curtin School of Medical Research, Australian National University, Canberra, Australian Capital Territory, Australia

Klaus Gundertofte   Department of Computational Chemistry, H. Lundbeck A/S Copenhagen-Valby, Denmark

Trygve Helgaker   Department of Chemistry, University of Oslo, Oslo, Norway

Rémy D. Hoffmann   Accelrys SARL, Parc Club Orsay Université, Orsay, France

Rob W. W. Hooft   Bruker Nonius BV, Delft, The Netherlands

Mark von Itzstein   Institute for Glycomics, Griffith University (Gold Coast Campus), Queensland, Australia

Poul Jørgensen   Department of Chemistry, University of Aarhus, Aarhus, Denmark

Mati Karelson   Centre of Strategic Competence, University of Tartu, Tartu, Estonia

Wim Klopper   Institute of Physical Chemistry, University of Karlsruhe (TH), Karlsruhe, Germany

Elmar Krieger   Centre for Molecular and Biomolecular Informatics, University of Nijmegen, Nijmegen, The Netherlands

Hugo Kubinyi   Molecular Modelling and Combinatorial Chemistry, BASF AG, Ludwigshafen, Germany (retired)

Thierry Langer Department of Pharmaceutical Chemistry, University of Innsbruck, Innsbruck, Austria

Tommy Liljefors Department of Medicinal Chemistry, The Danish University of Pharmaceutical Sciences, Copenhagen, Denmark

B. Maiti Department of Chemistry, Indian Institute of Technology, Kharagpur, India

Sonja Meddeb Accelrys SARL, Parc Club Orsay Université, Orsay, France

Paul G. Mezey Scientific Modeling and Simulation Laboratory, Memorial University of Newfoundland, St. John's, Newfoundland, Canada

Ed E. Moret Department of Medicinal Chemistry, Utrecht Institute for Pharmaceutical Sciences, Utrecht University, Utrecht, The Netherlands

Ingo Muegge Boehringer Ingelheim Pharmaceuticals, Inc., Ridgefield, Connecticut, U.S.A.

Jane S. Murray Department of Chemistry, University of New Orleans, New Orleans, Louisiana, U.S.A.

Sander B. Nabuurs Centre for Molecular and Biomolecular Informatics, University of Nijmegen, Nijmegen, The Netherlands

S. Nath Chemistry Department, Indian Institute of Technology, Kharagpur, India

Per-Ola Norrby Department of Chemistry, Technical University of Denmark, Lyngby, Denmark

Jeppe Olsen Department of Chemistry, University of Aarhus, Aarhus, Denmark

Tudor I. Oprea EST Chemical Computing, AstraZeneca R&D Mölndal, Mölndal, Sweden

Ingrid Pettersson Novo Nordisk A/S, Måløv, Denmark

Peter Politzer Department of Chemistry, University of New Orleans, New Orleans, Louisiana, U.S.A.

Theo H. Reijmers Johnson & Johnson Pharmaceutical Research and Development, A Division of Janssen Pharmaceutica N.V., Beerse, Belgium

Jean-Louis Rivail Groupe de Chimie théorique, "Structure et Réactivité des Systèmes Moléculaires Complexes," Henri Poincaré University, Nancy-Vandoeuvre, France

Jens Sadowski Structural Chemistry Laboratory, AstraZeneca R&D Mölndal, Mölndal, Sweden

Henry F. Schaefer III Center for Computational Quantum Chemistry, University of Georgia, Athens, Georgia, U.S.A.

Steve Scheiner Department of Chemistry and Biochemistry, Utah State University, Logan, Utah, U.S.A.

Christof H. Schwab Molecular Networks GmBH, Erlangen, Germany

Chris A. E. M. Spronk Centre for Molecular and Biomolecular Informatics, University of Nijmegen, Nijmegen, The Netherlands

Philip J. Stephens Department of Chemistry, University of Southern California, Los Angeles, California, U.S.A.

Jan P. Tollenaere Department of Medicinal Chemistry, Utrecht Institute for Pharmaceutical Sciences, Utrecht University, Utrecht, The Netherlands

John H. Van Drie Vertex Pharmaceuticals, Cambridge, Massachusetts, U.S.A.

Gert Vriend Centre for Molecular and Biomolecular Informatics, University of Nijmegen, Nijmegen, The Netherlands

Jennifer C. Wilson Griffith University (Gold Coast), Bundall, Queensland, Australia

Weitao Yang Department of Chemistry, Duke University, Durham, North Carolina, U.S.A.

Computational Medicinal Chemistry for Drug Discovery

1

Molecular Mechanics and Comparison of Force Fields

TOMMY LILJEFORS

The Danish University of Pharmaceutical Sciences, Copenhagen, Denmark

KLAUS GUNDERTOFTE

H. Lundbeck A/S, Copenhagen-Valby, Denmark

PER-OLA NORRBY

Technical University of Denmark, Lyngby, Denmark

INGRID PETTERSSON

Novo Nordisk A/S, Måløv, Denmark

1. INTRODUCTION

Molecular mechanics (force field) calculation is the most commonly used type of calculation in computational medicinal chemistry, and a large number of different force fields have been developed over the years. The results of a molecular mechanics (MM) calculation are highly dependent on the functional forms of the potential energy functions of the force field and of the quality of their parameterization. Thus in order to obtain reliable computational results it is crucial that the merits and limitations of the various available force fields are taken into account. In this chapter, the basic principles of force-field calculations are reviewed, and a comparison of calculated and experimental conformational energies for a wide range of commonly used force fields is presented. As quantum mechanical (QM) methods have undergone a rapid development in the last decade, we have also undertaken a comparison of these force fields with some commonly employed QM methods. The chapter also includes a review of force fields with respect to their abilities to calculate intermolecular interactions.

Finally, as solvent effects play an important role in computational medicinal chemistry, a discussion of force-field calculations including solvation is also included in this chapter.

2. BASIC PRINCIPLES OF MOLECULAR MECHANICS

Empirical force-field methodology is based on classical mechanics and on the fundamental assumption that the total "steric" energy of a structure can be expressed as a sum of contributions from many interaction types [1–3]. Another important assumption is that the force field and its parameters, which have been determined from a set of molecules, are transferable to other molecules.

Molecular mechanics methods are several orders of magnitude faster than QM methods, and for problems where MM methods are well defined, the accuracy may be as good as or better than QM calculations at a relatively high level (see Sec. 4). The main drawback of MM is that the method and the quality of the calculations are extremely dependent on empirical parameters. Such parameters are generally determined by experimental studies or high-level ab initio calculations, and the parameterization is often based on a small number of model systems.

2.1. Atom Types, Bonds, and Angles

The fundamental unit of most force fields is the *atom type*, determining what parameters to apply for all interactions involving the same constituent atom types. The various interaction types include bond lengths, angles, distances, etc. (see Fig. 1). In theory, every combination of atom types needs to be specifically parameterized. In practice, however, only the relevant combinations of these will ever be determined. For

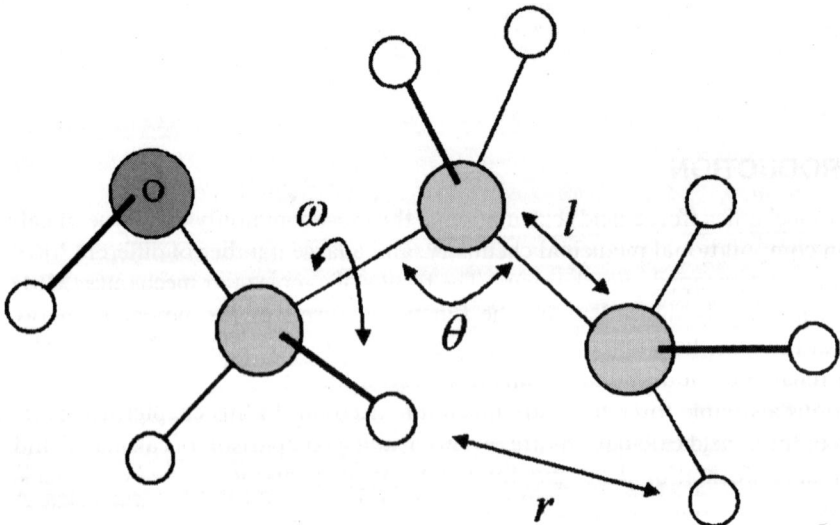

Figure 1 Definition of basic parameters in force fields. Bond lengths (*l*), angles (*θ*), torsion angles (*ω*), and nonbonded distances (*r*) are exemplified in *n*-propanol.

example, force fields with a carbonyl oxygen atom type will include bonds from this to carbon, but rarely to anything else. Thus the number of bond types in most force fields is only a few times higher than the number of atom types. In most force fields the parameters are further differentiated, based on the particular structural surroundings such as bond orders or the like.

Each bond in a structure will contribute a stretch term to the total energy. Bonds are normally described as harmonic bonds, and like springs, are characterized by a preferred length. The resistance to change from the optimum value is then defined by a "force constant," and each bond type is thus described by at least two parameters and the energies calculated by Hooke's law (Eq. (1)). Here the reference bond length is l_0 (Fig. 1).

$$E_s = k_s(l - l_0)^2 \tag{1}$$

Hooke's law can represent the energy increase on small distortions from the reference value and is applied in the CHARMm force field [4] and is default in the Dreiding [5] and UFF force fields [6]. However, for larger distortions, the energy of a true bond is normally represented by a Morse function (Eq. (2)) that can describe the process of dissociation energy correctly. In CVFF [7], a Morse potential is default, but a Hooke potential may be applied. The Morse potential requires one more parameter and, therefore, a wider range of reference data is needed for the parameterization. The potential is given in Eq. (2), where D is the dissociation energy and α is a parameter which, together with D, determines the curvature at the minimum.

$$E_s = D(1 - e^{-\alpha(l-l_0)})^2 \tag{2}$$

This representation is normally not needed for organic structures of a reasonable input quality with small distortions and the difference between the two functions is then negligible. A harmonic potential or a higher-order derivative of such is normally used in the initial optimization phase. Additional accuracy gained from a well-determined Morse function, at the cost of increase in complexity, may be important when studying more complex systems.

Modified Hooke's law corrected with cubic (as in the MM2-based force fields [8]) and further extensions to quartic terms (as in MM3 [9], CFF [10], and MMFF [11] force fields; see Eq. (3) [9]) or other expansions [12] have been developed to mimic the Morse potential and are used to speed up convergence in very distorted starting geometries, while keeping a proper description of the potential energy.

$$E_s - k_s(l - l_0)^2[1 + c_s(l - l_0) + q_s(l - l)^2] \tag{3}$$

The simplest approach to obtaining optimized bond angles close to the reference value θ_0 (Fig. 1) is to introduce a quadratic energy penalty, the harmonic approximation, similar to the representation of bond energies (Eq. (4)), although some methods use nonbonded interactions to model angle forces [3].

$$E_b = k_b(\theta - \theta_0)^2 \tag{4}$$

This simple representation is used in, e.g., the CVFF and CHARMm force fields. As for bonds, two parameters are needed, a reference angle and a force constant, and only a fraction of all the possible combinations of atom types are represented in real chemical structures. In certain cases, generalized parameters are used because of lack of accurate reference data, e.g., using a reference value close to 109.5° for all unknown angles around an sp3 carbon. To avoid losing the convergence properties for very large distortions, expansions to higher order terms, similar to those in bond energies discussed above, are applied in most force fields. Expansions to the power of four (MMFF) and even six (MM2 and MM3) are used.

Special care has to be taken in the representations of angles of 180°, which are wrongly represented as a cusp. To correct this problem with the slope going to zero, trigonometric functions as exemplified in Eq. (5) can be applied [13–15]. Close to a maximum this correction may lead to convergence problems, but this price is worth paying in most cases.

$$E_b = k[1 + \cos(n\theta + \psi)] \tag{5}$$

2.2. Nonbonded Interactions

Interactions between atoms that are not transmitted through bonds are referred to as nonbonded interactions. Most interactions are between centers of atoms, while some force fields use through-space interactions between points that are not centered on nuclei, such as lone pairs and bond-center dipoles. Interactions between atoms separated by only one or two bonds are normally not calculated, whereas atoms in the 1, 4-position with three intervening bonds interact both via torsional and nonbonded potentials. Thus these interactions become partially dependent. Introduction of scalable parameters for nonbonded 1,4-interactions can reduce this interdependence.

2.3. Electrostatic Interactions

Calculation of electrostatic interaction energies can be done simply by using Coulomb's law, Eq. (6), providing charges q centered on each nucleus.

$$E_{el} = \frac{q_i q_j}{\varepsilon r} \tag{6}$$

Most force fields except those derived from the native MM2 and MM3 implementations apply the Coulomb potentials. Charge assignment can be done using a variety of schemes, including fragment matching [16] and contributions through bonds [10, 17]. Furthermore, there is currently an increasing interest in polarizable force fields incorporating electrostatics dependent on the surroundings [18]. A major problem lies in the fact that atomic charges are statistical properties rather than observable items, and it is not always possible to find one set of charges that will reproduce all properties of interest. For most major force fields, one charge determination scheme has been adopted and used in the further development of new parameter sets securing internal consistency. Quantum mechanical calculations are generally a good source of data for electrostatic parameters and derived charges. Inclusion of the dielectric constant ε in Eq. (6) opens the possibility of developing simple solvation models by raising the value from 1 in the gas phase. More elaborate models are described in Sec. 6.

Eq. (7) describes a charge model primarily based on bond-center dipoles as applied by Allinger in MM2 and MM3 [3]. Such parameterization requires dipoles to be determined for each bond type independent of the surroundings. χ and α_i, α_j are the angle between the dipoles and the angles between each dipole and the connecting vector, respectively.

$$E_{el} = \frac{\mu_i \mu_j}{\varepsilon r^3} (\cos \chi - 3 \cos \alpha_i \cos \alpha_j) \tag{7}$$

2.4. Van der Waals Interactions

Short-range repulsions and London dispersion attractions are balanced by a shallow energy minimum at the van der Waals distance (Eq. (8)), describing the Lennard–Jones' potential, used by most force fields. Here the parameters A and B are calculated based on atomic radii and the minimum found at the sum of the two radii.

$$E_{vdW} = \frac{A}{r^{12}} - \frac{B}{r^6} \tag{8}$$

Most force fields use the Lennard–Jones functional form or close derivatives (9-6 or 14-7 functional forms as opposed to the standard 12-6 form). To compensate for the too hard repulsive component, MM2 and MM3 use the Buckingham potential shown in Eq. (9).

$$E_{vdW} = A e^{-\alpha r} - \frac{B}{r^6} \tag{9}$$

2.5. Hydrogen Bonding

The simplest way to handle hydrogen bonding is to rely on the other nonbonded potentials to reproduce hydrogen bonds. Some methods include specific pair parameters [19] while others use special potentials for the nonbonded interactions between hydrogen bond donors and acceptors [20,21].

2.6. Torsional Angles

Four consecutive atoms define the torsional bond (see Fig. 1). A large number of different torsional types therefore exist, and general parameters for the central bond are often used. Whereas certain preferred values for bond lengths and angles exist, torsions are even softer than bond angles and all possible values can be found in real structures. Thus the energy function must be valid over the entire range and, furthermore, be periodic. For symmetry reasons, the function should have stationary points at $0°$ and $180°$. A simple cosine function as exemplified in Eq. (10) has been used in the CVFF, CHARMm, and Dreiding force fields.

$$E_t = v \cos n\omega \tag{10}$$

where the periodicity n is the number of minima for the potential, usually 3 for an sp^3 sp^3 bond and 2 for a conjugated bond, and v is proportional to the rotational barrier. The Fourier expansion described in Eq. (11) allows the flexibility to model more complex torsional profiles and is used in most force fields today, including the MM2 and MM3 suite of programs. The form depicted in Eq. (11) also allows setting the minimum contribution to zero.

$$E_t = v_1(1 + \cos \omega) + v_2(1 - \cos 2\omega) + v_3(1 + \cos 3\omega) \tag{11}$$

2.7. Out-of-Plane Bending

Special parameterization is needed to prevent atoms bound to sp^2 carbons with three substituents to deviate from planarity. Many implementations apply an energy term E_{oop} that increases the energy when one of the atoms deviates from the plane defined by the three others. Several functions have been implemented, e.g., improper torsions or Hooke's law functions [22,23].

2.8. Modifications

Several force fields apply various modifiers and additional terms to address specific problems with the reduced set of standard terms. Allinger's electronegativity effect corrects the problem with substituents reducing the preferred bond lengths [24]. Adaptation of bond orders in conjugated systems is done by a simplified QM interpolation scheme [25–27], and cross terms can be used to, e.g., correct for the elongation of bonds when angles are compressed as shown in Eq. (12) [23].

$$E_{sb} = k_{sb}(\theta - \theta_0)(l - l_0) \tag{12}$$

3. COMPARISON OF CALCULATED CONFORMATIONAL ENERGIES

The comparison of force fields presented in this section focuses on the ability of different force fields to reproduce conformational energies. The relative performance of the ability to reproduce geometries is not included as this is done reasonably well by most force fields. The force fields included in the comparison are AMBER* [20, 28], CFF91 [10,29], CFF99 [10,29], CHARMm2.3 [4,29,30], CVFF [7,29], Dreiding 2.21 [5,29], MM2* [28], MM3*[28], MMFF [11,28], OPLS_AA [28,31], Sybyl5.21 [32, 33], and UFF1.1 [6,29]. These force fields have been selected as they are widely distributed as summarized in Table 1 and commonly used by computational and medic-

Table 1 A Summary of Different Force Fields Native to and Available in Different Software Packages[a]

	Cerius 2 (Accelrys Inc.)	InsightII (Accelrys Inc.)	MacroModel (Schrödinger Inc.)	Quanta (Accelrys Inc.)	Sybyl5.21 (Tripos Inc.)
AMBER*/AMBER		×	◆		×
CFF91	×	◆			
CFF99	◆				
CHARMm2.3		×		◆	
CVFF	×	◆			
Dreiding2.21	×				
MM2*			◆		×
MM3*			◆		
MMFF	×		◆		×
OPLS_AA			◆		
Sybyl5.21					◆
UFF1.1	×				

[a] ◆ = Native force field, × = available force field.

inal chemists. MM2(91) [8] and MM3(92) [9] are also included in the comparison. The comparison is an update of previously reported evaluations [34–36]. The data set used in the evaluation is given in Appendix A and is the same as previously employed. For further information on the dataset and the selection of experimental values, see Refs. [34–36].

Fig. 2 summarizes the overall results obtained by the different force fields and for different structural classes of compounds in terms of mean absolute errors. The performance of the force fields for particular classes of compounds is discussed in the following sections. Fig. 2 also includes the overall results for three QM methods (PM3, HF/6-31G* and B3LYP/6-31G*). These results will be discussed in Sec. 4.

3.1. Acyclic Hydrocarbons

As can be seen in Fig. 2, the calculated errors for the hydrocarbons in the data set are rather small for all tested force fields. The simplest hydrocarbon that can adopt two conformers is butane. As butane represents a fragment that can occur several times in a molecule and thus adds up errors, it is of importance that the force field can reproduce the experimental gauche-anti energy difference. Different experimental values for this energy difference have been reported [37–40]. The smallest reported experimental energy difference is 0.67 and the largest 1.09 kcal/mol. The experimental value 0.97 kcal/mol [37] has been used in the calculations of mean absolute errors in Fig. 2. Fig. 3 shows that all force fields correctly calculate the anti-conformer to be the most stable conformer and that most of the force fields can reproduce the experimental value within the variation of the experimental data. The force fields showing the largest errors are UFF 1.1, AMBER*, Sybyl5.21, and CVFF.

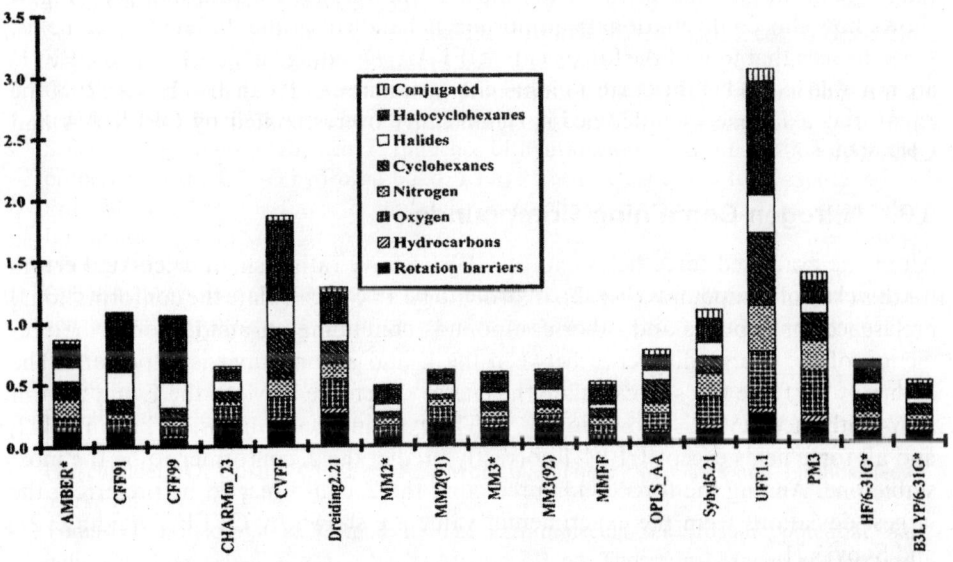

Figure 2 Comparison of mean absolute errors (in kcal/mol) for different structural classes of organic compounds obtained in calculations of conformational energy differences by using different commonly used force fields.

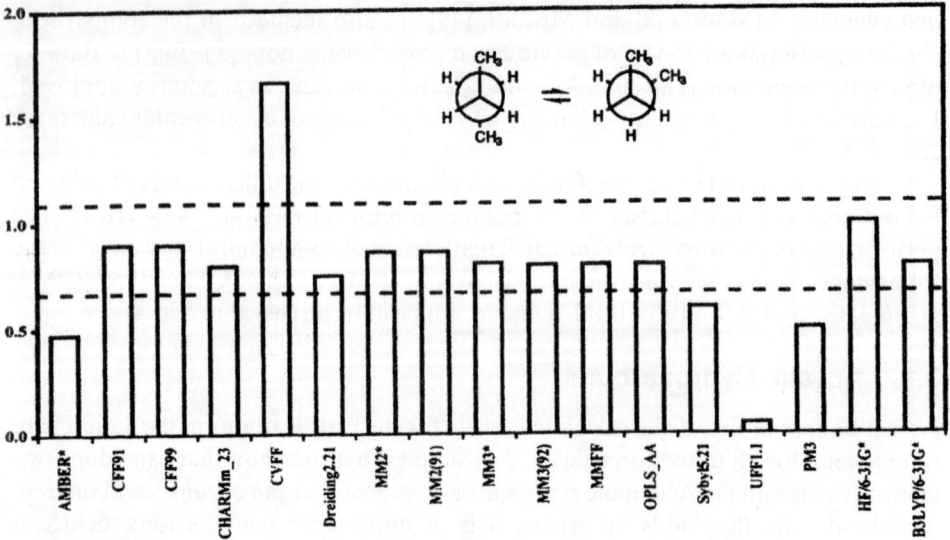

Figure 3 Calculated gauche-anti energy differences for butane in kcal/mol. The dashed horizontal lines show the range of reported experimental values.

3.2. Oxygen-Containing Compounds

Fig. 2 shows that the class of oxygen-containing compounds may give rise to larger errors than the hydrocarbons (Dreiding2.21, Sybyl5.21, and UFF1.1). For 2-methoxy-tetrahydropyrane (Fig. 4), the anomeric effect makes the conformer with the methoxy group in an axial position the most stable one by 1.0 kcal/mol [41]. Fig. 4 shows how this conformational equilibrium is handled by the different force fields. It can be seen that four of the force fields (UFF 1.1, Dreiding2.21, CVFF, and CFF91) are not able to predict the correct global energy minimum. It can also be seen that the equatorial–axial energy difference is significantly overestimated by OPLS_AA and CHARMm 2.3.

3.3. Nitrogen-Containing Compounds

All of the evaluated force fields except UFF1.1 have rather small calculated errors for this class of compounds (Fig. 2). In order to be able to calculate the conformational preference for peptides and other compounds containing an amide bond, the prediction of the energy difference between the E and Z conformer is important. The ability of the force fields to calculate the energy difference between the E and Z form in N-methylacetamide is shown in Fig. 5. The experimental value is 2.3 kcal/mol [42] and all force fields except UFF1.1 correctly predict the Z conformer to be the most stable one. Among the force fields predicting the Z conformer to be preferred, the largest deviations from the experimental value are shown by CVFF, Dreiding2.21, and Sybyl5.21.

Another common fragment in medicinal chemistry is N-methylpiperidine [43]. Fig. 6 shows the calculated energy difference between the axial and equatorial conformers for the different force fields. All force fields correctly predict the equatorial

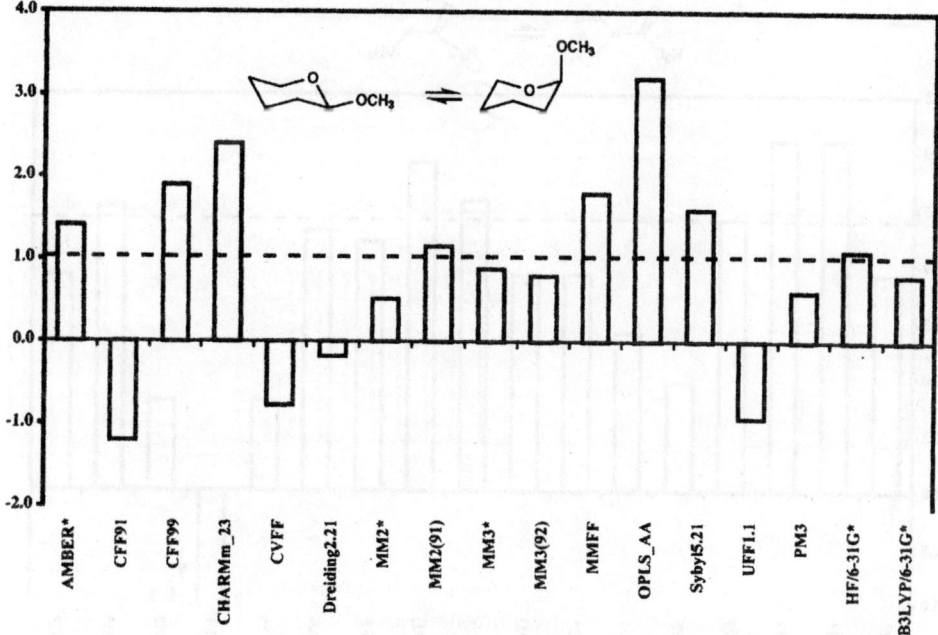

Figure 4 Calculated equatorial–axial conformational energy differences in kcal/mol for 2-methoxy-tetrahydropyran. The dashed line indicates the experimental value.

conformer to be the most stable one. However, the energy difference is significantly overestimated by UFF1.1 and underestimated by more than 1 kcal/mol by AMBER*, CVFF, Dreiding2.21, OPLS_AA, and Sybyl5.21.

3.4. Cyclohexanes

For substituted cyclohexanes, two conformational properties are of fundamental importance. A force field should be able to predict both the correct conformation of the ring system and the position (axial or equatorial) of a substituent. Fig. 7 shows the ability of the different force fields to predict the energy difference between the twist-boat and chair conformation of cyclohexane [44]. As can be seen in the figure most of the force fields reproduce this well. However, the energy difference is overestimated by several of the force fields, in particular by CVFF and UFF1.1.

For testing the ability of the force fields to reproduce the energy difference between an axial and equatorial substituent, methylcyclohexane and aminocyclohexane have been chosen as examples. The experimental value for the energy difference between the two chair conformers in methylcyclohexane is 1.75 kcal/mol [45]. All force fields correctly calculate the equatorial conformer to be the most stable one as displayed in Fig. 8. Again, the energy difference is strongly overestimated by CVFF and UFF1.1.

For aminocyclohexane, the experimental value for the energy difference between the axial and equatorial conformer is 1.49 kcal/mol with the equatorial conformer as the most stable one [46]. In Fig. 9 it is shown that AMBER* predicts the axial

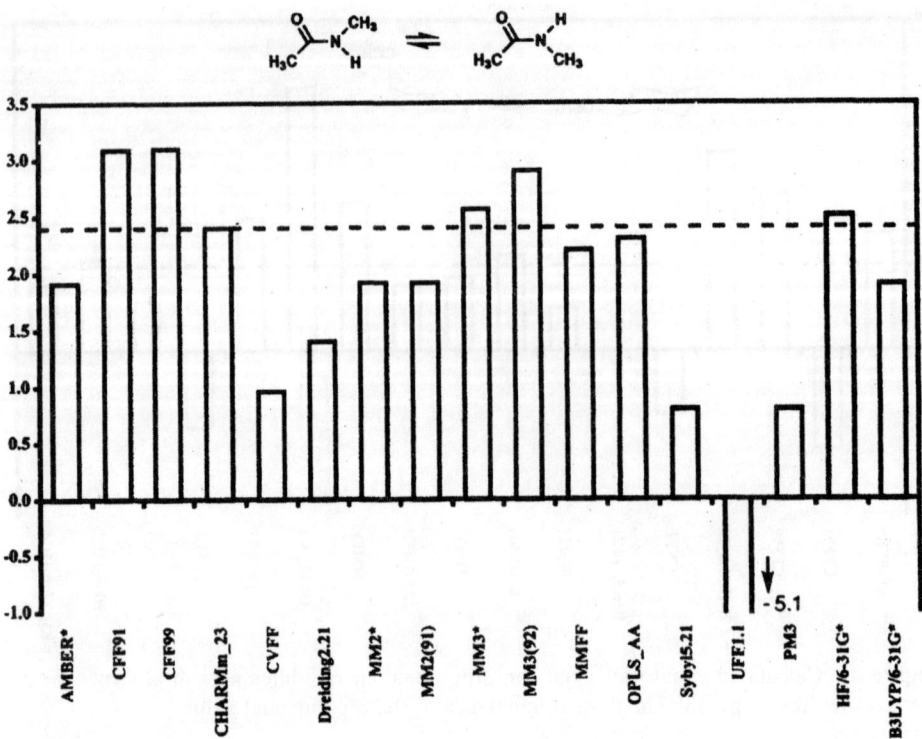

Figure 5 Calculated energy differences in kcal/mol between the E and Z conformer of *N*-methylacetamide. The dashed line indicates the experimental value.

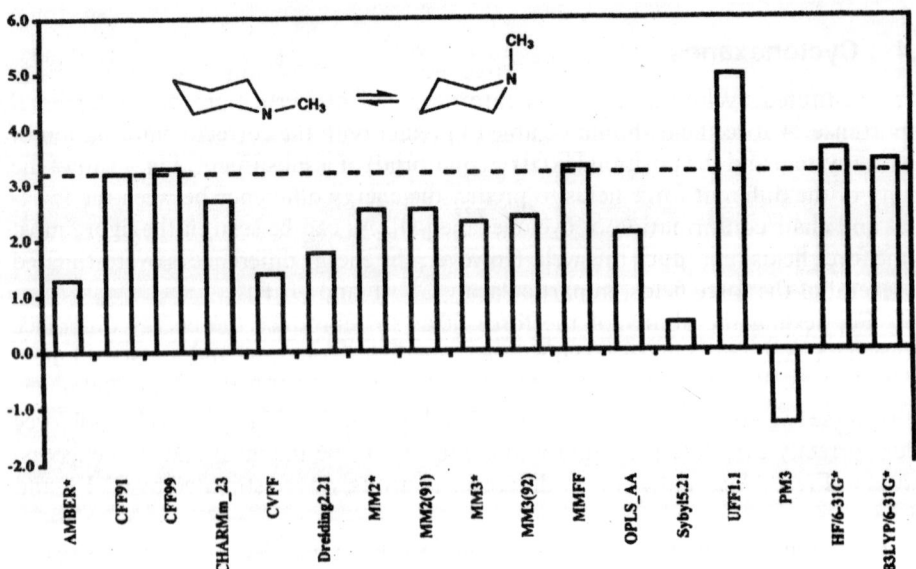

Figure 6 Calculated conformational energy differences (axial–equatorial) in kcal/mol for *N*-methylpiperidine. The dashed line shows the experimental value.

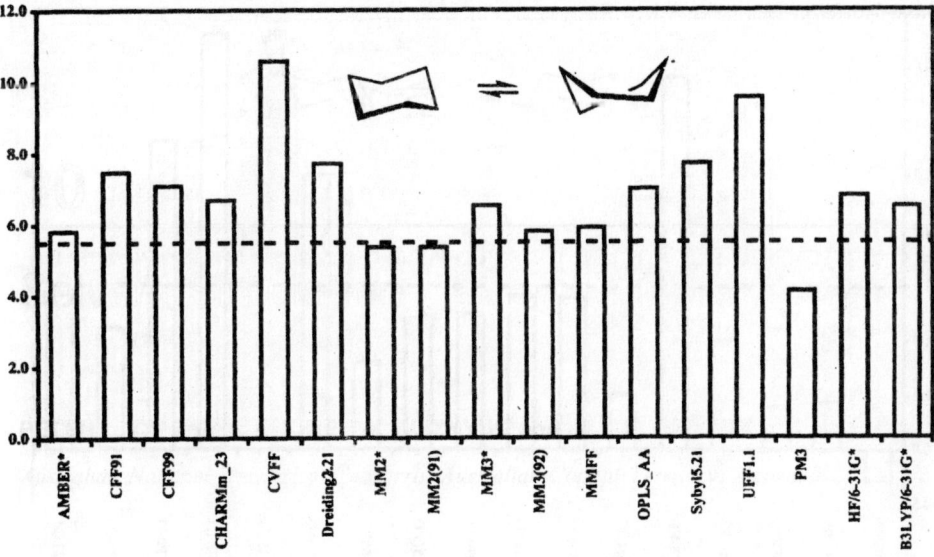

Figure 7 Calculated energy differences in kcal/mol between the twist-boat and chair conformers of cyclohexane. The dashed line indicates the experimental value.

conformer to be the most stable one and that Sybyl5.21 predicts the two conformers to be essentially equally stable. It can also be seen that the energy difference is significantly underestimated by CFF91, CF99, Dreiding2.21, and MMFF and overestimated by CVFF, OPLS_AA, and UFF1.1.

In conclusion, the overall results displayed in Fig. 2 show that for the data set employed in this comparison of force fields the best results are obtained by MM2*, MM2(91), MM3*, MM3(92), MMFF, and CHARMm. The least successful results are clearly obtained by CVFF, Dreiding 2.21, and UFF1.1.

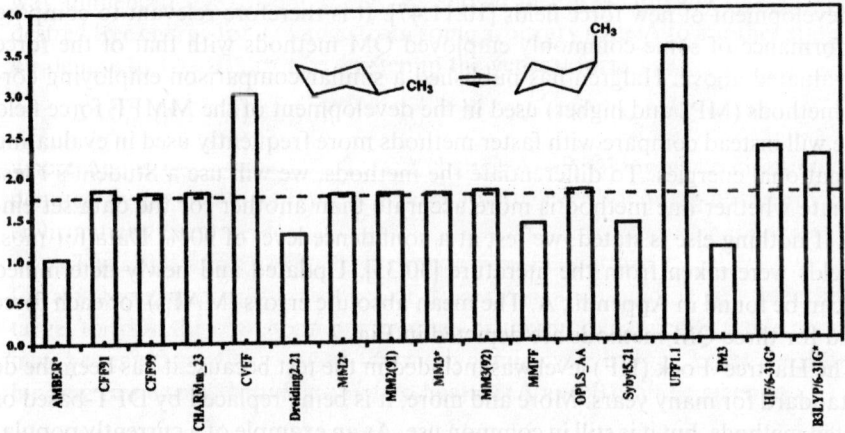

Figure 8 Calculated conformational energy differences between axial and equatorial methyl-cyclohexane in kcal/mol. The dashed line shows the experimental value.

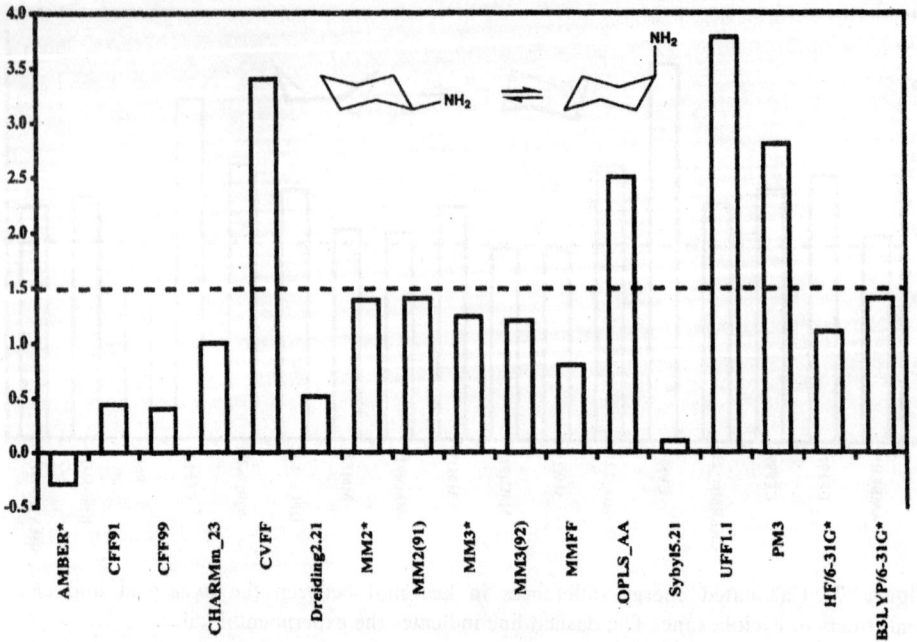

Figure 9 Calculated energy differences in kcal/mol between axial and equatorial amino-cyclohexane. The dashed line indicates the experimental value.

4. COMPARISON OF QUANTUM MECHANICS AND MOLECULAR MECHANICS

Quantum mechanical methods have been undergoing an explosive development in the last decade, in performance but even more in accessibility [1]. At present, several QM methods can routinely be applied to geometry optimization and evaluation of conformational energies for small organic molecules. This has traditionally been the domain of force-field methods. Furthermore, QM results are increasingly being used in the development of new force fields [10,11,47]. It is therefore relevant to compare the performance of some commonly employed QM methods with that of the force fields evaluated above. Halgren has published a similar comparison employing correlated methods (MP2 and higher) used in the development of the MMFF force field [11]. We will instead compare with faster methods more frequently used in evaluating conformational energies. To differentiate the methods, we will use a Student's t test to evaluate whether one method is more accurate than another for the data set employed. If nothing else is stated, we test at a confidence level of 90%. Data for most force fields were taken from the literature [30,35]. Updated and newly determined results can be found in Appendix A. The mean absolute errors (MAEs) for each force field and for three QM methods are depicted in Fig. 2.

The Hartree–Fock (HF) level was included in the test because it has been the de facto standard for many years. More and more, it is being replaced by DFT-based or correlated methods, but it is still in common use. As an example of a currently popular DFT method we have chosen B3LYP [48], a hybrid functional employing three empirical parameters to weigh the contributions from HF exchange and different DFT functional components. This method has been shown to be a good alternative to high-level ab initio methods for many types of energy comparisons [49]. Both methods have been employed with the 6-31G* basis set. A frequently employed method is to calculate the energies at a correlated level using geometries from a simpler calculation. This has been done here using the MP2/6-31 + G** method with either HF or B3LYP geometries (see Appendix A) in the Jaguar [50] and Gaussian98 [51] programs. Finally, we have tested two popular semi-empirical methods, AM1 [52] and PM3 [53]. For conformational energies of molecules with around 1000 atoms, semiempirical methods are still the only feasible QM alternatives. The difference between these two methods was not statistically significant, but PM3 gave a slightly lower total error in the test and was therefore used in all comparisons.

Comparing the QM methods to each other (see Fig. 2 and Appendix A), we can see that the MAE over the entire set of conformational energies is 0.49 kcal/mol for B3LYP, 0.66 kcal/mol for HF, and 1.37 kcal/mol for PM3. We can say with 98% confidence that B3LYP is more accurate than HF for this type of comparison, and with more than 99.9% confidence that both methods are better than PM3. Thus the three methods form a convenient scale for grading the force-field methods. The two MP2 methods are not shown in Fig. 2, as both overall appearance and MAE are similar to B3LYP (MAE 0.48 and 0.50 kcal/mol, respectively; see Appendix A). An interesting corollary of the MP2 results is that the geometries from HF and B3LYP are of similar quality for this type of comparison. They are not identical, but differences are obviously systematic and thus cancel in a comparison of conformational energies.

Looking at the force fields, we can see that most of them fall in about the same accuracy range as the QM methods (Fig. 2). The two best force fields, MM2* and MMFF, are significantly better than HF (> 95%) and are not significantly different from B3LYP. It should be noted that force fields are limited compared to QM methods in that they are only applicable to molecules with identical connectivity (e.g., conformations and possibly stereoisomers), and then only for systems where parameters have been well determined. However, within this limitation, it is noteworthy that the best force fields are as accurate as any affordable QM method and certainly many orders of magnitude more cost effective. This also makes clear that parameterization of force fields requires methods that are significantly better than HF [11a], because the best possible result in parameterizing a force field is to reproduce the reference data exactly.

Following this star group, we find a set of force fields which are not significantly better than HF, but nor are these significantly worse than B3LYP, with mean errors up to 0.67 kcal/mol. These include most of the MM2 and MM3 implementations, as well as CHARMm. It is also quite probable that with a complete set of parameters, CFF91 and CFF99 would fall within this group. However, detailed halogen parameters are unavailable for the CFF methods, which causes the program to automatically supply rule-based parameters of a lower quality. For this reason, we cannot grade CFF with certainty. Closely following this group is a single force field, the newly implemented OPLS_AA force field, with a performance probably worse than B3LYP (MAE = 0.75 kcal/mol).

It is significant that the group of force fields with an accuracy at least equal to HF all contain some well-parameterized cross terms. Obviously, a few such terms are necessary for an accurate calculation of conformational energies. The best parameterized diagonal (i.e., lacking cross-terms) force field in the study, AMBER*, yielded an MAE of 0.87 kcal/mol. This is significantly worse than HF (and indeed AMBER is extensively parameterized from HF), but it is still better than the Tripos force field in Sybyl (MAE = 1.07 kcal/mol). However, both are significantly more accurate than the semi-empirical method PM3.

In the next group, we find early diagonal force fields included in previous studies [35] but not in Fig. 2, such as an earlier version of the Tripos force field and ChemX, but also a rule-based force field, Dreiding (included in Fig. 2). All of these have about the same accuracy as the PM3 method. The early diagonal force fields are still being used to some extent but are slowly being replaced by more modern force fields. However, it is interesting to note that for conformational energies of large systems beyond the scope of the HF method and if the presence of unknown groups make application of specifically parameterized force fields impossible, a rule-based force field is preferable to a semiempirical calculation. It has about the same accuracy and is still many orders of magnitude faster than PM3.

A few force fields have an accuracy worse than that of semiempirical methods. CVFF was developed from an initially diagonal force field by adding a large number of cross terms, with insufficient reparameterization. It is obvious that this resulted in a force field with low predictivity, and its use cannot be recommended for any application. UFF was intended to cover the entire periodic table and is still the only published force field that can accomplish this task. However, the accuracy for organic molecules was sacrificed in the process: the MAE for UFF is ca. 3 kcal/mol.

The force fields can also be compared to a "blank" result, the mean absolute of all conformational energies to be predicted. This is the performance that would be expected by any random-number generator symmetrically centered around 0 kcal/mol. Most force fields yield a performance substantially better than this random guess, but of the methods considered here, CVFF is not significantly different from the blank, and UFF is actually worse.

5. INTERMOLECULAR INTERACTIONS

Calculations of intermolecular interactions are extremely important in many aspects of modern medicinal chemistry. Docking of ligands into cavities in targets is used in structure-based design and precise estimation of ligand–target energetics is required to predict binding affinities. A prerequisite to do reasonable qualitative docking is to have a well-defined target and a good parameterized method for calculating intermolecular interactions. These are quite difficult to calculate and quantification of binding energies requires even better methods. Clearly, the best understood experimental cases are crystal structures, and parameterization is often based on such studies. In the absence of experimental data, high-level ab initio calculations can be used for computations of intermolecular energies and geometries, and the results may be employed in the parameterization process.

Electrostatic and van der Waals intermolecular interactions are involved in the binding process. Apart from a valid description of the conformational energies, accurate description of these interactions is crucial for the determination of intermolecular energies. The energy functions in the MM methods are normally parameterized against standard models, which involve interactions between atoms as in hydrogen bonding. Other important interactions include those between aromatic moieties in receptor–ligand interactions. It has been known for a long time that charge–transfer interactions between electron-rich and electron-deficient rings occur. Weaker interactions from the edge to the face of rings are also important [54].

An extensive comparative study on intermolecular interactions has been made by Halgren [11b,11g]. Interaction energies in small model systems calculated by MMFF94, MM2, MM2X, MM3, OPLS, and CHARMm [11b] were compared with high-level ab initio energies. Important differences between these methods stem from differences in their charge models. The neglect of polarizability is a limitation of all the models. In general terms, MMFF94 and MM3 performed well in nonbonded aliphatic systems. Considering hydrogen bonding, MMFF94 and OPLS in most cases correctly predict interaction energies within 10% and also manage to correctly classify the strengths of hydrogen bond acceptors and donors. Predictions of geometries of the complexes are good with the largest discrepancies in weakly bound sulfur-containing complexes.

In an extended comparison of intermolecular interaction energies and distances based on scaled HF/6-31G* data for 66 hydrogen-bonded complexes and including the MMFF94, MMFF94s, CFF95, CVFF, CHARMm, CHARMM22, OPLS*, AMBER*, MM2*, and MM3* force fields, the MMFF94/MMFF94s and CHARM22 force fields clearly performed best [11g]. The next best performance was displayed by the AMBER* force field followed by OPLS* and CFF95.

6. FORCE-FIELD CALCULATIONS INCLUDING SOLVATION

As described above, force fields are generally developed and validated for gas phase properties on the basis of gas-phase experimental data or from data obtained in low dielectric solvents. More recently, data from high-level ab initio calculations have been employed. Thus straightforward force-field calculations refer to molecules in vacuo. However, solvation plays an important role in many aspects of chemistry in general and of medicinal chemistry in particular. For instance, most compounds of relevance for medicinal chemistry are flexible and very often have polar functional groups. As the most important solvent in medicinal chemistry is water, conformational properties in this highly polar solvent may be drastically different from the properties in vacuo and only calculations including the solvent may yield meaningful results. Calculations on various aspects of ligand–enzyme/receptor interactions and partitioning between phases require the consideration of solvation effects. Thus the accurate estimation of solvent effects is a key problem in computational medicinal chemistry.

6.1. Explicit vs. Implicit (Dielectric Continuum) Solvation Models

The most straightforward way to account for solvation effects is to explicitly include a large number of solvent molecules in the calculations. However, this requires the explicit consideration of hundreds or thousands of solvent molecules around the solute. In addition, the need for the generation of a large number of water configurations requires Monte Carlo or molecular dynamics methodologies to be employed, resulting in a very high computational cost. This problem has led to the development of implicit solvation models in which the solvent is treated as a polarizable continuous medium surrounding the solute beginning at or near its van der Waals surface. The solvent is then characterized by its bulk dielectric constant. Such methods are orders of magnitude faster to use in calculations of solvent effects compared to explicit solvation models and have therefore received much attention in computational medicinal chemistry. A disadvantage of implicit solvation models is that no structural information on specific solvent–solute interactions can be obtained.

The most rigorous dielectric continuum methods employ numerical solutions to the Poisson–Boltzmann equation [55]. As these methods are computationally quite expensive, in particular in connection with calculations of derivatives, much work has been concentrated on the development of computationally less expensive approximate continuum models of sufficient accuracy. One of the most widely used of these is the Generalized Born Solvent Accessible Surface Area (GB/SA) model developed by Still and coworkers [56,57]. The model is implemented in the MacroModel program [17,28] and parameterized for water and chloroform. It may be used in conjunction with the force fields available in MacroModel, e.g., AMBER*, MM2*, MM3*, MMFF, OPLS*. It should be noted that the original parameterization of the GB/SA model is based on the OPLS force field.

Dielectric continuum models have also been developed to be used in conjunction with ab initio as well as semiempirical quantum chemical methods. For a comprehensive discussion on dielectric continuum models in general and on its use in connection with quantum chemistry calculations in particular, the reviews by Cramer and Truhlar are highly recommended [58].

As the present chapter is restricted to force-field calculations, only the GB/SA dielectric continuum model and similar models will be discussed. The aim is not to give an exhaustive review of the rapidly increasing literature in this area but to describe the basic properties of the GB/SA model and to discuss some aspects of the model and its use that are of particular interest in computational medicinal chemistry.

6.2. The GB/SA Model

In the GB/SA model, the solvation free energy (G_{solv}) is calculated as a sum of three terms

$$G_{solv} = G_{cav} + G_{vdW} + G_{pol} \tag{13}$$

where G_{cav} is the cavitation energy, i.e., the free energy required to form a cavity in the solvent in which the solute is embedded. G_{vdW} is the solute–solvent van der Waals energy (first hydration shell effects) and G_{pol} the solute–solvent electrostatic polarization energy. The sum of $G_{cav} + G_{vdW}$ is taken to be proportional to the solvent accessible surface (SA) of the solute and is calculated as the sum of atomic surface area contributions SA_k multiplied by an empirical atomic solvation parameter σ_k for atoms of type k as shown in Eq. (14).

$$G_{cav} + G_{vdW} = \sum \sigma_k SA_k \tag{14}$$

The $G_{\rm pol}$ term is calculated by the generalized Born equation (Eq. (15))

$$\Delta G_{\rm pol} = -332\left(1 - \frac{1}{\varepsilon}\right)\sum_{i=1}^{n-1}\sum_{j=i+1}^{n}\frac{q_i q_j}{r_{ij}} - 166\left(1 - \frac{1}{\varepsilon}\right)\sum_{i}^{n}\frac{q_i^2}{\alpha_i} \tag{15}$$

modified in Eq. (16) to allow for irregularly shaped solutes.

$$\Delta G_{\rm pol} = -166\left(1 - \frac{1}{\varepsilon}\right)\sum_{i=1}^{n}\sum_{j=1}^{n}\frac{q_i q_j}{\left(r_{ij}^2 + \alpha_{ij}^2 e^{-D_{ij}}\right)^{0.5}} \tag{16}$$

where $\alpha_{ij} = (\alpha_i \alpha_j)^{0.5}$ and $D_{ij} = r_{ij}^2/(2\alpha_{ij})^2$.

A computationally efficient analytical method has been developed for the crucial calculation of Born radii, which is required for each atom of the solute that carries a (partial) charge, and the $G_{\rm pol}$ term has been parameterized to fit atomic polarization energies obtained by Poisson–Boltzmann equation [57]. The GB/SA model is thus fully analytical and affords first and second derivatives allowing for solvation effects to be included in energy minimizations, molecular dynamics, etc. The $G_{\rm pol}$ term is most important for polar molecules and describes the polarization of the solvent by the solute. As force fields in general are not polarizable, it does not account for the polarization of the solute by the solvent. This is clearly an important limitation of this type of calculations.

Qui et al. have compared experimental and calculated hydration free energies for a set of 35 small organic molecules with diverse functional groups by using the OPLS force field and the GB/SA hydration model [57]. These calculations resulted in a mean absolute error of 0.9 kcal/mol. It is of interest to note that the results obtained with the GB/SA model were very similar to those obtained by the corresponding calculations using the full Poisson–Boltzmann equation.

6.3. Comparisons of Calculations Employing Explicit and Implicit Solvation Models

In the first paper reporting the GB/SA model, Still and coworkers compared the calculated $G_{\rm pol}$ values obtained by the GB/SA algorithm with the corresponding values obtained by free-energy perturbation (FEP) calculations using explicit solvation with TIP4P water molecules [56]. For the neutral compounds in the dataset, a linear correlation between the two sets of calculated values was obtained with a slope of 1.1 and a correlation coefficient of 0.98. Thus the results obtained by two different types of solvation models are very similar. The same conclusion was drawn by Reddy et al. on comparing the results of Monte Carlo-free-energy perturbation calculations with explicit TIP4P waters with the corresponding results from GB/SA calculations [59]. Thus at least for simple neutral organic compounds there is no need for computationally expensive explicit solvent simulations for estimating free energies of hydration.

As an example of the relative performance of explicit and implicit solvation models in calculations of conformational equilibria, Scarsi et al. [60] compared the calculated conformational properties obtained by the CHARMM force field of liquid 1,2-dichloroethane and of terminally blocked alanine dipeptide in aqueous solution. They employed (i) a systematic conformational search with solvation energies calculated by a dielectric continuum generalized Born model similar to the G_{pol} part in GB/SA, and (ii) molecular dynamics simulations including 216 1,2-dichloroethane molecules and 207 water molecules, respectively. Good agreement between the results obtained by the two computational methods were shown for both cases. The increase in the gauche/trans ratio for 1,2-dichloroethane on going from gas phase to the liquid phase as experimentally observed was reproduced by both methods.

6.4. The Dependence of Calculated Hydration Energies on Different Charge Sets

As mentioned above, the accurate calculation of the electrostatic contribution (G_{pol}) is crucial for the calculation of hydration free energies for polar molecules. This implies that the quality of the atom-centered partial charges used by the force field to describe electrostatic interactions is of decisive importance for the results. The atomic partial charges in the various force fields are assigned in different ways. For instance, the basic OPLS* charge set is based on liquid-phase simulations in explicit solvents, but more recent versions also employ fitting of partial charges to the electrostatic potential surface calculated by ab initio calculations [31]. The AMBER* force field also uses charges derived from fitting to molecular electrostatic potentials. In contrast, the partial charges in MM2* and MM3* are derived from the empirically determined bond dipoles in the "authentic" parent programs MM2 [3,8] and MM3 [9]. Charges for MMFF are basically calculated to mimic electrostatic potential derived charges calculated by using the HF/6-31G* basis set and formulated as "bond charge incre-ments" to be added to full or fractional "formal atomic charges" [11b]. It is important to note that the assigned partial charges in all force fields are an integral part of the force field and should not be modified by the user.

Reddy et al. have systematically studied the sensitivity of hydration free ener-gies calculated using the GB/SA model on different charge sets and force fields [59]. Using a small database of 11 monofunctional compounds with standard geometries and using single-point energy calculations, they compared the calculated free ener-gies of hydration for force fields available in Macromodel (MM2*, MM3*, OPLS*, AMBER* and MMFF). The charge sets of OPLS* and AMBER* clearly performed best with mean absolute errors (MAE) of 1.02 and 1.38 kcal/mol, respectively, whereas those of MM3* (MAE = 1.82 kcal/mol) and in particular MM2* (MAE = 2.65 kcal/mol) display significantly inferior performance. The results obtained by MMFF were of similar quality as MM3* (MAE = 1.97 kcal/mol). The good performance of the OPLS* charge set is not surprising as the original GB/SA parameterization was based on the OPLS force field. As noted by Reddy et al., it is likely that reparam-eterization of GB/SA for a particular force field may improve the results for the force field. This has been demonstrated by Cheng et al. [61], who partly reparameterized the GB/SA model for the MMFF force field with a resulting significant improvement in performance. An average unsigned error of 0.74 kcal/mol for 129 neutral com-pounds was obtained to be compared with an error of 1.43 kcal/mol from using the original GB/SA model with MMFF. A generalized Born model with parameters spe-cifically tailored to the AMBER force field has been reported by Jayaram et al. [62].

The force field dependence on calculated conformational equilibria in aqueous solution has been demonstrated in a study of the strongly polar ionotropic glutamate receptor agonist kainate [63]. Conformational analyses of kainate in aqueous solution were performed using the MM3*, AMBER*, and MMFF94 force fields in conjunction with the GB/SA hydration model. The conformational properties of kainate in aqueous solution have been studied by Todeschi et al. using ^{13}C and ^1H NMR spectroscopy [64]. The experimental data indicate that the predominating conformation of kainate is of type A (Fig. 10), with no significant contribution of the internally hydrogen bonded type B conformation.

AMBER* + GB/SA and MMFF94 + GB/SA predict that the conformational ensemble consist of 72% and 83%, respectively, of type A conformers, whereas 96% of the MM3* conformational ensemble consists of type B conformers with one strongly dominating conformer. This study indicates that MM3* + GB/SA strongly overestimates the stability of the hydrogen bonded ion-pair in aqueous solution as shown by conformer B, in comparison with the separated ions as in conformer A.

6.5. Calculations of Conformational Energy Penalties for Ligand–Protein Binding

Calculated conformation energy penalties for ligand binding are useful in, e.g., pharmacophore modeling and structure-based ligand design. In pharmacophore modeling, such energies may be employed to find suitable candidates for the bioactive conformations of a set of molecules. In structure-based ligand design it is necessary to ensure that the designed ligand does not require a prohibitively high conformational energy for binding to the receptor. This is important as the equation $\Delta G = RT\ln K_i$ implies that each 1.4 kcal/mol ($T = 300$ K) of increased conformational energy of the bound conformation leads to a decrease in the affinity by a factor of 10.

High conformational energy penalties have often been reported in the literature. For instance, Nicklaus et al. studied 27 flexible ligands extracted from experimentally determined ligand–protein complexes and obtained calculated energy differences between the protein-bound and unbound conformations between 0 and 39.7 kcal/mol with an average of 15.9 kcal/mol [65]. The most important reason for these high energies is that the calculations were performed for gas phase. Such calculations are not meaningful in connection with structure–activity studies and ligand design, as it is

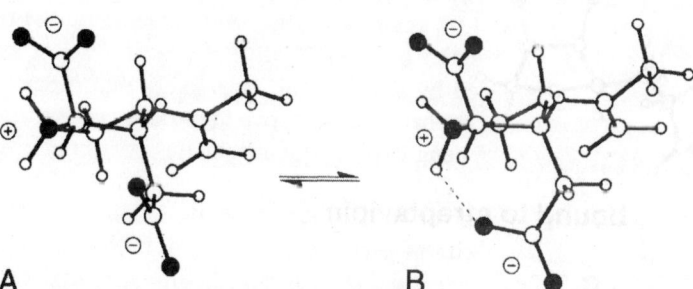

Figure 10 Conformational equilibrium for kainate involving separated ions (A) and intermolecular ion-pair hydrogen bonding (B).

clear that the aqueous conformational ensemble for the unbound ligand must be used as the reference state in this type of calculations [66].

This has been demonstrated in a study of 33 ligand–protein complexes including 28 different ligands using the MM3* and AMBER* force fields with the GB/SA hydration model [66]. By using the aqueous conformational ensemble for the unbound ligand as the reference state, the great majority of conformational energy penalties for binding were calculated to be smaller than 3 kcal/mol. As an example of the strong influence of solvent effects in this type of calculations, the preferred conformation of biotin in vacuo displays a strong hydrogen bond between the carboxylate group and the NH group (Fig. 11a). However, in aqueous solution biotin strongly prefers an extended conformation (Fig. 11b) according to MM3* as well as AMBER*. The conformation of biotin bound to the enzyme streptavidin is shown in Fig. 11c. Using the in vacuo conformation as the reference state for calculating the conformational energy penalty gives a calculated energy penalty of 12.8 kcal/mol (MM3*) and 6.4 kcal/mol (AMBER*). These high energies are clearly not compatible with the very high affinity of biotin to streptavidin ($K_a = 2.5 \times 10^{13}$, $\Delta G = -18.3$ kcal/mol, $\Delta H = -32$ kcal/mol). When using the predominating conformation in aqueous

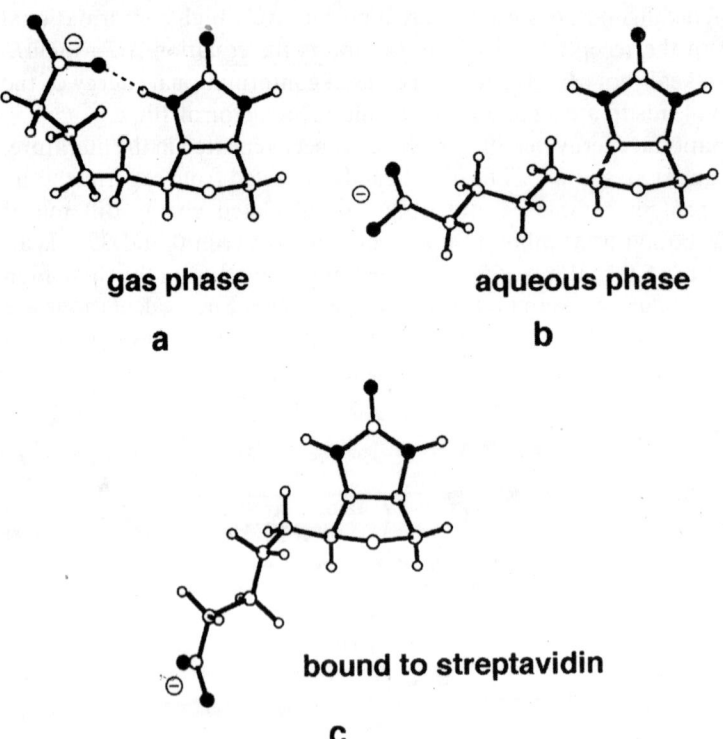

gas phase

a

aqueous phase

b

bound to streptavidin

c

Figure 11 Calculated lowest energy conformations for biotin in (a) gas phase and (b) aqueous phase. The conformation observed in the biotin–streptavidine ligand–protein complex (pdb-code: 1stp) is shown in (c).

phase as the reference conformer, the corresponding energies are calculated to be small, less than 1.6 kcal/mol.

Dielectric continuum models are excellent tools for fast and reliable calculations of hydration energies and solvent effects on, e.g., conformational equilibria, ligand–receptor interactions, and partitioning phenomena. For neutral solutes, the performance of such solvation models is already very good, whereas calculations on ionic compounds still pose significant problems. Force fields that include polarization effects may be required for accurate calculations on strongly polar molecules. A problem in the further development and validation of solvation models is the lack of experimental data for, e.g., conformational equilibria in aqueous solution. For optimal accuracy of calculations using a dielectric continuum model, it would be an advantage if the model is parameterized for the particular force field to be used.

7. CONCLUSION

The ability of 14 widely distributed and commonly used force fields to reproduce experimental conformational energies for a data set of 44 conformational energy differences or rotational barriers has been compared. The results show that the best results are obtained by the MM2*, MM2(91), MM3*, MM3(92), MMFF, and CHARMm force fields, whereas the least successful results are obtained by the CVFF, Dreiding2.21, and UFF1.1 force fields. AMBER*, CFF91, CFF99, OPLS_AA, and Sybyl5.21 display results of intermediate quality. A further comparison was made with results obtained by the semiempirical PM3, ab initio HF/6-31G* and density functional B3LYP/6-31G* calculations. (Among the quantum chemical methods themselves B3LYP is, as expected, more accurate than HF, and both methods are better than PM3.) A significant result of this comparison is that the best force fields are as accurate, for the data set used, as the QM methods and certainly many orders of magnitude more cost effective. The two best force fields, MM2* and MMFF, are significantly better than HF/6-31G* and are not significantly different from B3LYP/6-31G*. It is concluded that parameterization of force fields from data obtained by quantum mechanics methods requires methods that are significantly better than HF. The group of force fields with an accuracy at least equal to HF all contain some well-parameterized cross terms. All force fields tested, except CVFF and UFF1.1, perform better than the semiempirical PM3 method.

For calculations on intermolecular interactions, the extensive comparisons reported by Halgren clearly show that for hydrogen-bonded complexes the MMFF94/MMFF94s and CHARM22 force fields perform best, followed by AMBER*, OPLS*, and CFF95. For nonbonded aliphatic systems, MMFF94 and MM3 are the best performers. The neglect of polarizability is a limitation for all current force fields.

Dielectric continuum models such as the Generalized Born Solvent Accessible Surface Area (GB/SA) model are, in conjunction with force fields, excellent tools for fast and reliable calculations of hydration energies and solvent effects on, e.g., conformational equilibria and ligand–receptor interactions. The performance for neutral solutes is very good, whereas calculations on ionic compounds are currently more problematic. A solution to these problems most probably requires force fields that include polarization effects. For optimal accuracy of calculations using a dielectric continuum model, it is a clear advantage if the model is parameterized for the particular force field used.

Appendix Updated MM and newly determined QM energies, in kcal/mol. Conformational equilibria and selection of experimental values are described in the main text and Ref. [35].

	Macromodel 7.2					Cerius2	Jaguar v. 4.1		Gaussian 98, rev. A.7				
	MM2*	MM3*	Amber*	MMFF	OPLS_AA	CFF99	HF/6-31G*	B3LYP/6-31G*	HF/6-31G*	B3LYP/6-31G*	MP2a//HF	MP2a//B3LYP	Exp.
Ethane, TS-GS	2.7	2.4	2.8	3.2	2.8	2.6	3.0	2.8	3.0	2.8	3.2	3.2	2.9
Propene, TS-GS	2.1	1.7	1.6	2.0	1.9	1.7	2.1	2.1	2.1	2.1	1.9	1.9	2.0
Isoprene, TS-GS	2.9	0.8	1.4	3.1	1.0	2.6	2.9	2.5	2.9	2.5	2.7	2.7	2.7
Ethylbenzene, TS-GS	1.5	1.8	0.4	1.2	1.4	1.7	1.4	1.1	1.4	1.1	1.5	1.6	1.7
Trimethyl isopropyl benzene, TS-GS	11.0	10.9	10.4	12.9	13.3	11.1	14.8	12.8	14.6	12.7	12.2	12.3	12.8
Styrene, TS-GS	1.7	3.3	2.4	1.0	3.0	1.1	2.9	4.4	2.9	4.4	2.2	1.8	1.8
Butane, g-a	0.9	0.8	0.5	0.8	0.8	0.9	1.0	0.8	1.0	0.8	0.7	0.7	1.0
2,3-Dimethylbutane, g-a	0.1	0.4	-0.1	-0.2	-0.2	0.5	-0.1	0.1	-0.1	0.0	0.1	0.1	0.1
1,3,5-Trineopentylbenzene, twosyn–allsyn	0.8	0.4	0.3	0.5	0.7	0.5	-0.2	-0.1	-0.2	-0.1	0.2	0.1	1.0
Methyl acetate, cis–trans (C=O)	5.6	7.0	6.4	8.3	9.0	8.7	9.4	7.8	9.4	7.8	8.6	8.6	8.0
2-Butanone, skew-ecl (C=O)	1.6	1.6	1.6	0.8	0.0	1.4	1.5	1.5	1.5	-0.0	1.0	0.0	2.0
Ethyl methyl ether, g-a	1.7	1.5	1.4	1.5	1.5	1.5	1.7	1.4	1.7	1.4	1.4	1.4	1.5
2-Methoxy-THP, eq–ax[b]	0.5	0.9	1.4	1.8	3.2	1.9	1.0	0.8	1.1	0.9	1.3	1.4	1.0
Ethanol (C–O), g-a	0.6	0.4	0.2	0.2	0.1	0.4	0.1	-0.3	0.1	-0.3	0.2	0.1	0.7
Propanol (C–C), g-a	0.3	0.4	-0.2	0.3	-0.1	0.3	-0.1	-0.1	-0.1	-0.1	-0.3	-0.3	-0.3
Ethyl amine (C–N), g-a	-0.1	-0.1	-0.1	-0.4	0.0	-0.2	-0.1	0.4	-0.1	0.4	-0.2	-0.2	0.7
N-Methylacetamide, E–Z	1.9	2.6	1.9	2.2	2.3	3.1	2.5	1.9	2.5	2.5	2.4	2.1	2.4
N-Methylpiperidine, ax–eq	2.5	2.3	1.3	3.3	2.1	3.3	3.6	3.4	3.6	3.4	3.9	3.9	3.2
2-Methylpiperidine, ax–eq	2.1	2.3	1.2	2.4	2.3	2.4	3.1	3.0	3.1	2.9	3.1	3.1	2.5
3-Methylpiperidine, ax–eq	1.6	1.5	0.5	1.1	2.3	1.4	1.6	1.6	1.6	1.6	1.3	1.3	1.6
4-Methylpiperidine, ax–eq	1.7	1.7	1.1	1.4	2.5	1.8	2.5	2.4	2.4	2.4	2.3	2.3	1.9

Cyclohexane, twist-chair	5.4	6.5	5.8	5.9	7.0	7.1	6.8	6.5	6.8	6.5	6.6	6.6	5.5
Phenylcyclohexane, ax–eq[b]	3.9	4.3	2.0	2.3	4.0	3.9	4.2	3.6	4.2	3.6	3.6	3.5	2.9
Methylcyclohexane, ax–eq	1.8	1.8	1.0	1.4	1.8	1.8	2.3	2.2	2.3	2.2	2.0	2.0	1.8
Aminocyclohexane, ax–eq[b]	1.4	1.2	-0.3	0.8	2.5	0.4	1.1	1.4	1.1	1.4	1.1	1.0	1.5
N,N-Dimethylamino-cyclohexane, ax–eq[b]	1.0	1.2	1.2	1.4	5.5	1.0	2.1	1.8	2.1	1.8	0.4	0.4	1.3
trans-1,2-Dimethylcyclohexane, ax–eq,eq	2.4	2.6	1.3	1.8	2.3	2.7	3.2	3.1	3.2	3.1	2.8	2.7	2.6
cis-1,3-Dimethylcyclohexane, ax,ax–eq,eq	5.3	5.7	4.4	5.1	5.4	5.4	6.6	6.0	6.5	5.9	5.8	5.7	5.5
FCH₂CH₂F, g–a	-0.6	-0.6	0.8	-0.6	-0.6	1.6	0.5	-0.4	0.5	-0.4	-0.5	-0.5	-0.8
PrCl, g–a	0.2	0.3	0.2	0.0	-0.3	1.0	0.4	0.1	0.4	0.2	0.1	0.1	-0.2
ClCH₂CH₂Cl, g–a	0.4	0.4	0.7	1.2	0.8	0.8	1.9	1.7	1.9	1.7	1.5	1.4	1.1
ClCH₂CH₂Cl, g–a-g,g	0.3	0.0	-0.2	0.4	-0.2	-0.9	0.4	0.7	0.4	0.7	0.7	0.7	1.1
ClCH₂CH₂Cl, a,a-g,g	0.8	0.2	-0.1	1.1	-0.1	-1.9	0.8	1.3	0.8	1.3	1.3	1.3	1.5
F, ax–eq	0.2	0.2	0.3	-0.4	0.3	1.6	-0.3	-0.2	-0.3	-0.2	-0.2	-0.1	0.2
Cl, ax–eq	0.4	3.6	0.5	-0.3	-0.2	1.7	1.0	0.8	1.0	0.8	0.9	1.0	0.5
Br, ax–eq	0.5	0.6	0.8	0.0	0.3	1.7	0.9	0.7	0.0	0.2	0.2	0.2	0.7
trans-1,2-diF, ax,ax–eq,eq	0.8	-0.3	-0.3	-0.2	1.4	2.1	-1.4	-0.8	-1.4	-0.8	-0.8	-0.8	0.6
trans-1,2-diCl, ax,ax–eq,eq	0.9	-0.4	0.1	-2.0	-2.2	2.2	-0.4	-0.6	-0.4	-0.5	-0.2	-0.1	-0.9
trans-1,2-diBr, ax,ax–eq,eq	-0.7	-0.4	0.7	-1.7	-2.3	1.7	-2.0	-2.1	-2.5	-2.5	-3.2	-3.0	-1.5
trans-1,4-diF, ax,ax–eq,eq	-0.5	-0.3	-0.2	-2.6	-0.5	2.7	-1.7	-1.5	-1.7	-1.5	-1.9	-1.9	-1.1
trans-1,4-diCl, ax,ax–eq,eq	0.3	0.7	0.1	-2.0	-0.5	3.4	0.8	0.3	0.6	0.3	0.6	0.6	-0.8
trans-1,4-diBr, ax,ax–eq,eq	0.6	0.9	1.2	-0.8	-0.6	3.1	0.4	-0.1	-1.4	-1.8	-1.0	-1.0	-0.9
Butadiene, s-cis-s-trans	2.7	1.7	3.2	3.4	3.6	2.9	3.1	3.9	3.0	3.5	2.6	2.8	2.5
Acrolein, s-cis-s-trans	1.6	1.9	2.4	2.0	2.6	1.8	1.7	1.7	1.7	1.7	1.9	2.0	1.7
Root mean square error (RMS)	0.72	0.80	1.07	0.61	1.07	1.58	0.84	0.69	0.82	0.75	0.62	0.67	3.00[c]
Mean absolute error (MAE)	0.48	0.58	0.87	0.50	0.75	1.06	0.66	0.49	0.66	0.54	0.48	0.50	2.02[c]

[a] MP2/6-31+G** single-point energies at the indicated G98 geometries.
[b] Total energy calculated by including contributions from up to three side chain rotamers.
[c] Deviation from zero, the expected error from a random-number generator centered around zero.

REFERENCES

1. Jensen F. Introduction to Computational Chemistry. Chichester: Wiley, 1999.
2. Goodman J. Chemical Applications of Molecular Modelling. Royal Society of Chemistry, Cambridge, 1998.
3. Burkert U, Allinger NL. Molecular Mechanics. ACS Monogr 1982; vol. 177. Washington, DC: ACS Monogr 177, ACS, 1982.
4. Momany FA, Rone R. Validation of the general purpose QUANTA®3.2/CHARMm® force field. J Comput Chem 1992; 13:888–900.
5. Mayo SL, Olafson BD, Goddard WA III. DREIDING: A generic force field for molecular simulations. J Phys Chem 1990; 94:8897–8909.
6a. Rappé AK, Casewit CJ, Colwell KS, Goddard WA III, Skiff WM. UFF, A full periodic table force field for molecular mechanics and molecular dynamics simulations. J Am Chem Soc 1992; 114:10024–10035.
6b. Casewit CJ, Colwell KS, Rappé AK. Application of a universal force field to organic molecules. J Am Chem Soc 1992; 114: 10035–10046.
6c. Casewit CJ, Colwell KS, Rappé AK. Application of a universal force field to main group compounds. J Am Chem Soc 1992; 114:10046–10053.
7a. Lifson S, Warshel A. Consistent force field for calculations of conformations, vibrational spectra and enthalpies of cycloalkane and n-alkane molecules. J Chem Phys 1968; 49:5116–5129.
7b. Hagler AT, Huler E, Lifson S. Energy functions for peptides and proteins: I. Derivation of a consistent force field including the hydrogen bond from amide crystals. J Am Chem Soc 1974; 96:5319–5327.
7c. Warshel A, Lifson S. Consistent force field calculations: II. Crystal structures, sublimation energies, molecular and lattice vibrations, molecular conformations, and enthalpies of alkanes. J Chem Phys 1970; 53: 582–594.
8. Allinger NL. Conformational analysis: 130. MM2. A hydrocarbon force field utilizing V1 and V2 torsional terms. J Am Chem Soc 1977; 89:8127–8134.
9a. Allinger NL, Yuh HY, Lii JH. Molecular Mechanics. The MM3 force field for hydrocarbons. J Am Chem Soc 1989; 111:8551–8566.
9b. Lii JH, Allinger NL. Molecular mechanics. The MM3 force field for hydrocarbons: 3. The van der Waals potentials and crystal data for aliphatic and aromatic hydrocarbons. J Am Chem Soc 1989; 111:8576–8582.
9c. Lii JH, Allinger NL. Molecular mechanics. The MM3 force field for hydrocarbons: 2. Vibrational frequencies and thermodynamics. J Am Chem Soc 1989; 111:8566–8575.
10a. Maple JR, Hwang M-J, Stockfish TP, Dinur U, Waldman M, Ewig CS, Hagler AT. Derivation of class II force fields: I. Methodology and quantum force field for the alkyl functional group and alkane molecules. J Comput Chem 1994; 15:162–182.
10b. Hwang MJ, Stockfisch TP, Hagler AT. Derivation of class II force fields: 2. Derivation and characterization of a class II force field, CFF93, for the alkyl functional group and alkane molecules. J Am Chem Soc 1994; 116:2515–2525.
11a. Halgren TA. Merck molecular force field: I. Basis, form, scope, parameterization, and performance of MMFF94. J Comput Chem 1996; 17:490–519.
11b. Halgren TA, Merck molecular force field: II. MMFF94 van der Waals and electrostatic parameters for intermolecular interactions. J Comput Chem 1996; 17:520–552.
11c. Halgren TA. Merck molecular force field: III. Molecular geometries and vibrational frequencies for MMFF94. J Comput Chem 1996; 17:553–586.
11d. Halgren TA, Nachbar RB. Merck molecular force field: IV. Conformational energies and geometries for MMFF94. J Comput Chem 1996; 17:587–615.
11e. Halgren TA. Merck molecular force field: V. Extension of MMFF94 using experimental

data, additional computational data, and empirical rules. J Comput Chem 1996; 17:616–641.

11f. Halgren TA. MMFF: VI. MMFF94s option for energy minimization studies. J Comput Chem 1999; 20:720–729.

11g. Halgren TA. MMFF: VII. Characterization of MMFF94, MMFF94s, and other widely available force fields for conformational energies and for intermolecular-interaction energies and geometries. J Comput Chem 1999; 20:730–748.

12. Dinur U, Hagler AT. On the functional representation of bond energy functions. J Comput Chem 1994; 15:919–924.

13. Comba P, Hambley TW. Molecular Modeling of Inorganic Compounds. New York: VCH Publishers, 1995.

14. Allured VS, Kelly CM, Landis CR. SHAPES empirical force field: new treatment of angular potentials and its application to square-planar transition-metal complexes. J Am Chem Soc 1991; 113:1–12.

15. Comba P, Hambley TW, Ströhle M. The directionality of d-orbitals and molecular-mechanics calculations of octahedral transition metal compounds. Helv Chim Acta 1995; 78:2042–2047.

16. Cornell WD, Cieplak P, Bayly CI, Gould IR, Merz KM Jr., Ferguson DM, Spellmeyer DC, Fox T, Caldwell JW, Kollman PA. A second generation force field for the simulation of proteins, nucleic acids, and organic molecules. J Am Chem Soc 1995; 117:5179–5197.

17. Mohamadi F, Richards NGJ, Guida WC, Liskamp R, Lipton M, Caulfield C, Chang G, Hendrickson T, Still WC. MacroModel—an integrated software system for modeling organic and bioorganic molecules using molecular mechanics. J Comput Chem 1990; 11:440–467.

18a. Rappé AK, Goddard WA III. Charge equilibration for molecular dynamics simulations. J Phys Chem 1991; 95:3358–3363.

18b. Halgren TA, Dam W. Polarizable force fields. Curr Opinion Struct Biol 2001; 11:236–242.

19. Allinger NL, Kok RA, Imam MR. Hydrogen bonding in MM2. J Comput Chem 1988; 9:591–595.

20a. Weiner SJ, Kollman PA, Nguyen DT, Case DA. An all atom force field for simulations of proteins and nucleic acids. J Comput Chem 1986; 7:230–252.

20b. Weiner SJ, Kollman PA, Case DA, Singh UC, Ghio C, Alagona G, Profeta S Jr, Weiner P. A new force field for molecular mechanical simulation of nucleic acids and proteins. J Am Chem Soc 1984; 106:765–784.

20c. Weiner SJ, Kollman PA. AMBER: assisted model building with energy refinement. A general program for modeling molecules and their interactions. J Comput Chem 1981; 2:287–303.

21. Lii JH, Allinger NL. Directional hydrogen bonding in the MM3 force field: II. J Comput Chem 1998; 19:1001–1016.

22. Hay BP. Methods for molecular mechanics modeling of coordination compounds. Coord Chem Rev 1993; 126:177–236.

23. Dinur U, Hagler AT. In: Lipkowitz KB, Boyd DB, eds. Reviews in Computational Chemistry. New York: VCH Publishers, Inc., 1991:99.

24. Thomas HD, Chen K, Allinger NL. Toward a better understanding of covalent bonds: the molecular mechanics calculation of C–H bond lengths and stretching frequencies. J Am Chem Soc 1994; 116:5887–5897.

25. Sprague JT, Tai JC, Yuh Y, Allinger NL. The MMP2 calculational method. J Comput Chem 1987; 8:581–603.

26. Tai JC, Allinger NL. Effect of inclusion of electron correlation in MM3 studies of cyclic conjugated compounds. J Comput Chem 1998; 19:475–487.

27. Tai J, Nevins N. In: Schleyer PvR, ed. Encyclopedia of Computational Chemistry. Vol 3. 1998:1013.

28. Schrödinger Inc. www.schrodinger.com.

29. Accelrys Inc. www.accelrys.com.

30. Nicklaus MC. Conformational energies calculated by the molecular mechanics program CHARMm. J Comput Chem 1997; 18:1056–1060.

31. Jorgensen WL, Maxwell DS, Tirado-Rives J. Development and testing of the OPLS all-atom force field on conformational energetics and properties of organic liquids. J Am Chem Soc 1996; 118:11225–11236.

32. Tripos Inc. www.tripos.com.

33. Clark M, Cramer RD III., Van Opdenbosch N. Validation of the general purpose Tripos 5.2 force field. J Comput Chem 1989; 10:982–1012.

34. Gundertofte K, Palm J, Pettersson I, Stamvik A. A comparison of conformational energies calculated by molecular mechanics (MM2(85), Sybyl 5.1, Sybyl 5.21, and ChemX) and semiempirical (AM1 and PM3) methods. J Comp Chem 1991; 12:200–208.

35. Gundertofte K, Liljefors T, Norrby P-O, Pettersson I. A comparison of conformational energies calculated by several molecular mechanics methods. J Comp Chem 1996; 17:429–449.

36. Pettersson I, Liljefors T. Molecular mechanics calculated conformational energies of organic molecules: a comparison of force fields. Reviews in computational chemistry. In: Lipkowitz KB, Boyd DB, eds. New York: VCH, 1996:167–189.

37. Verma AL, Murphy WF, Bernstein HJ. Rotational isomerism: XI. Raman spectra of n-butane, 2-methylbutane, and 2,3-dimethylbutane. J Chem Phys 1974; 60:1540–1544.

38. Herrebout WA, van der Veken BJ, Wang A, Durig JR. Enthalpy difference between conformers of n-butane and the potential function governing conformational interchange. J Phys Chem 1995; 99:578–585.

39. Durig JR, Wang A, Beshir W, Little TS. Barrier to asymmetric internal-rotation, conformational stability, vibrational-spectra and assignments, and ab initio calculations of normal-butane-D0, normal-butane-D5 and normal-butane-D10. J Raman Spectrosc 1991; 22:683–704.

40. Heenan RK, Bartell LS. Electron-diffraction studies of supersonic jets: 4. Conformational cooling of normal-butane. J Chem Phys 1983; 78:1270–1274.

41. de Hoog AJ, Buys HR, Altona C, Havinga E. Conformation of non-aromatic ring compounds—LII. NMR spectra and dipole moments of 2-alkoxytetrahydropyrans. Tetrahedron 1969; 25:3365–3375.

42. Kitano M, Fukuyama T, Kuchitsu K. Molecular structure of N-methylacetamide as studied by gas electron diffraction. Bull Chem Soc Jpn 1973; 46:384–387.

43. Crowley PJ, Robinson MJT, Ward MG. Conformational effects in compounds with 6-membered rings—XII. The conformational equilibrium in N-methylpiperidine. Tetrahedron 1977; 33:915–925.

44. Squillacote M, Sheridan RS, Chapman OL, Anet FAL. Spectroscopic detection of the twist-boat conformation of cyclohexane. A direct measurement of the free energy difference between the chair and the twist-boat. J Am Chem Soc 1975; 97:3244–3246.

45. Booth H, Everett JR. The experimental determination of the conformational free energy, enthalpy, and entropy differences for alkyl groups in alkylcyclohexanes by low temperature carbon-13 magnetic resonance spectroscopy. J Chem Soc Perkin II 1980; 255–259.

46. Booth H, Jozefowicz ML. The application of low temperature ^{13}C nuclear magnetic resonance spectroscopy to the determination of the A values of amino-, methyl-amino, and dimethylamino-substituents in cyclohexane. J Chem Soc Perkins Trans II 1976; 895–901.

47. Norrby P-O, Brandt P. Deriving force field parameters for coordination complexes. Coord Chem Rev 2001; 212:79–109.

48a. Becke AD. Density-functional thermochemistry: III. The role of exact exchange. J Chem Phys 1993; 98:5648–5652.

48b. Lee C. Yang W. Parr RG. Development of the Colle–Salvetti correlation-energy formula into a functional of the electron-density. Phys. Rev. B 1988; 37:785–789.

49. Foresman JB, Frisch Æ. Exploring Chemistry with Electronic Structure Methods. 2d ed. Pittsburgh, PA: Gaussian, Inc., 1996.

50. Jaguar v. 4.1 from Schrödinger Inc., www.schrodinger.com.

51. Frisch MJ, Trucks GW, Schlegel HB, Scuseria GE, Robb MA, Cheeseman JR, Zakrzewski VG, Montgomery JA, Stratmann RE Jr., Burrant JC, Dapprich S, Millam JM, Daniels AD, Kudin KN, Strain MC, Farkas O, Tomasi J, Barone V, Cossi M, Cammi R, Mennucci B, Pomelli C, Adamo C, Clifford S, Ochterski J, Petersson GA, Ayala PY, Cui Q, Morokuma K, Malick DK, Rabuck AD, Raghavachari K, Foresman JB, Cioslowski J, Ortiz JV, Baboul AG, Stefanov BB, Liu G, Liashenko A, Piskorz P, Komaromi I, Gomperts R, Martin RL, Fox DJ, Keith T, Al-Laham MA, Peng CY, Nanayakkara A, Gonzalez C, Challacombe M, Gill PMW, Johnson B, Chen W, Wong MW, Andres JL, Gonzalez C, Head-Gordon M, Replogle ES, Pople JA. Gaussian 98, Revision A7. Pittsburgh, PA: Gaussian, Inc., 1998.

52. Dewar MJS, Zoebisch EG, Healy EF, Stewart JJP. AM1: A new general purpose quantum mechanical molecular model. J Am Chem Soc 1985; 107:3902–3909.

53a. Stewart JJP. Optimization of parameters for semiempirical methods: I. Method. J Comput Chem 1989; 10:209–220.

53b. Stewart JJP. Optimization of parameters for semiempirical methods: II. Applications. J Comput Chem 1989; 10:221–264.

54. Jennings WB, Farrell BM, Malone JF. Attractive intramolecular edge-to-face aromatic interactions in flexible organic molecules. Acc Chem Res 2001; 34:885–894.

55. Davis ME, McCammon JA. Electrostatics in biomolecular structure and dynamics. Chem Rev 1990; 90:509–524.

56. Still WC, Tempczyk A, Hawley RC, Hendrickson T. Semianalytical treatment of solvation for molecular mechanics and molecular dynamics. J Am Chem Soc 1990; 112:6127–6129.

57. Qiu D, Shenkin PS, Hollinger FP, Still WC. The GB/SA continuum model for solvation. A fast analytical method for the calculations of approximate Born radii. J Phys Chem A 1997; 101:3005–3014.

58a. Cramer CJ, Truhlar DG. Continuum solvation models: classical and quantum mechanical implementations. In: Lipkowitz KB, Boyd DB, eds. Reviews in Computational Chemistry. Vol. 6. New York: VCH Publishers, Inc, 1995:1–71.

58b. Cramer CJ, Truhlar DG. Implicit solvation models: equilibria, structure, spectra, and dynamics. Chem Rev 1999; 99:2161–2200.

59. Reddy MR, Erion MD, Agarwal A, Viswabnadhan VN, McDonald DQ, Still WC. Solvation Free energies calculated using the GB/SA model: sensitivity of results on charge sets, protocols and force fields. J Comput Chem 1998; 19:769–780.

60. Scarsi M, Apostolakis J, Caflisch A. Comparison of a GB Solvation model with explicit solvent simulations: potentials of mean force and conformational preferences of alanine dipeptide and 1,2-dichloroethane. J Phys Chem B 1998; 102:3637–3641.

61. Cheng A, Best SA, Merz KM Jr., Reynolds CH. GB/SA water model for the Merck molecular force field (MMFF). J Mol Graphics Modell 2000; 18:273–282.

62. Jayaram B, Sprous D, Beveridge DL. Solvation free energy of biomacromolecules: parameters for a modified generalized Born model consistent with the AMBER force field. J Phys Chem B 1998; 102:9571–9576.

63. Nielsen PAa, Liljefors T. Conformational analysis of kainate in aqueous solution in relation to its binding to AMPA and kainic acid receptors. J Comput-Aided Mol Des 2001; 15:753–763.

64. Todeschi N, Gharbi-Benarous J, Acher F, Larue V, Pin J-P, Bockaert J, Azerad R,

Girault J-P. Conformational analysis of glutamic acid analogues as probes of glutamate receptors using molecular modelling and NMR methods. Comparison with specific agonists. Bioorg Med Chem 1997; 5:335–352.

65. Nicklaus MC, Wang S, Driscoll JS, Milne GWA. Conformational changes of small molecules binding to proteins. Bioorg Med Chem 1995; 3:411–428.

66. Boström J, Norrby P-O, Liljefors T. Conformational energy penalties of protein-bound ligands. J Comput-Aided Mol Des 1998; 12:383–396.

2

Semiempirical Methods

THOMAS BREDOW

University of Hannover, Hannover, Germany

1. INTRODUCTION

Like ab initio methods, semiempirical approaches are based on the electronic Schrödinger equation (1) obtained after separation of nuclear and electronic motion (Born–
Oppenheimer approximation) [1].

$$\hat{H}(\mathbf{r}, \mathbf{R})\Psi(\mathbf{r}, \mathbf{R}) = E^{el}(\mathbf{r}, \mathbf{R})\Psi(\mathbf{r}, \mathbf{R}) \tag{1}$$

Here $\mathbf{r}$ and $\mathbf{R}$ denote the coordinates of electrons and nuclei, respectively. The Hamilton operator H contains operators for the kinetic energy of the electrons, $\hat{T}$, their
electrostatic interaction with the nuclei $\hat{V}_{en}$, and among themselves $\hat{V}_{ee}$.

$$\hat{H}(\mathbf{r}, \mathbf{R}) = \hat{T} + \hat{V}_{en} + \hat{V}_{ee} \tag{2}$$

The total energy E of the system is obtained by adding the nuclear repulsion V_{nn} (a
classical sum of Coulomb interactions) to the electronic energy.

$$E(\mathbf{r}, \mathbf{R}) = E^{el}(\mathbf{r}, \mathbf{R}) + V_{nn} \tag{3}$$

The binding energy E_B of a molecule is the energy gain due to bond formation with
respect to the isolated atoms I in the gas phase.

$$E_B = E - \sum_I E_I \tag{4}$$

Two fundamental methodologies have been developed to solve the electronic Schrödinger equation (1) from first principles. The Molecular Orbital (MO) theory aims at
finding an expression for the wave function Ψ, while in Density Functional Theory
(DFT), the electron density distribution ρ plays this role. These methods are described
in detail in other chapters of this book. Their advantage is that they can, in principle,

provide an exact numerical solution to Eq. (1) provided that the convergence with respect to the inherent technical parameters (i.e., the completeness of the active space in multireference-determinant MO theory approaches, the quality of the exchange-correlation functional in DFT methods, and the basis set size for both) has been achieved. A disadvantage of first-principles methods is that they are—despite of the simplifications already introduced by the Born–Oppenheimer approximation and the neglect of relativistic effects in Eq. (1)—computationally demanding and therefore have been traditionally restricted to relatively small molecules with less that 10 atoms. Due to the advances in computer hardware and software efficiency in the last decade, the applicability of first-principles methods has been extended to systems with up to a few hundred atoms (see other chapters of this book). Nevertheless, there is still a need for simpler methods that are able to treat problems that are beyond the capabilities of the more correct theories [2]. This is particularly true for the treatment of large biomolecules with many hundreds to thousands of atoms, computer-aided drug design, or the development of quantitative structure activity relationships (QSAR), where a routine treatment of a large number of molecules is necessary. In some cases, approximate methods have been shown to provide similar accuracy as the more sophisticated approaches with a very small fraction of the computational effort [3]. In addition, even if the quantitative accuracy of semiempirical methods is usually limited, they can give insights into qualitative trends that are sometimes lost with high-level methods (for a more philosophical discussion of this point, see a recent article by Hoffmann [4]).

For these and other reasons, semiempirical methods have been developed as early as in the 1930s, starting with the famous Hückel method [1]. From the mid-1960s and so on, a vast variety of different semiempirical methods have been developed [1,5–10]. They have been widely used for the prediction of structural, energetic, and spectroscopic properties of molecular and solid-state systems in chemistry, biochemistry, biology, and pharmaceutics. The development of new semiempirical methods and their improvement and extension of application to larger and more complex systems have been continued until present [2,3,11].

This review gives a brief summary of the theoretical background of some of the most popular semiempirical methods together with recent examples of their use in modern chemical, biological, and pharmaceutical research.

2. APPROXIMATIONS

The basis for all semiempirical methods described in this chapter is the ab initio Hartree–Fock (HF) theory [1]. Here the wave function Ψ of an N-electron system is approximated as a single Slater-determinant Ψ_0 of spin orbitals ϕ. The spin orbitals are a product of a molecular orbital (MO) φ and a spin function (α or β), $\phi_a = \varphi_a\alpha$, $\overline{\phi}_a = \varphi_a\beta$.

$$\Psi_0 = |\phi_1\overline{\phi}_1\phi_2\overline{\phi}_2\ldots\phi_M\overline{\phi}_M > \tag{5}$$

In a closed-shell system, N is an even number and all M MOs are occupied with two electrons with α and β spin. This implies $M = N/2$. The molecular orbitals are expressed as a linear combination of $m \geq M$ atomic orbitals χ (LCAO).

$$\varphi_a = \sum_{\mu}^{m} c_{\mu a}\chi_\mu \tag{6}$$

By substituting Ψ with Eq. (5), integrating, and applying the variation principle, the Schrödinger equation (1) is converted into a system of linear equations, the Roothaan equation [12] (recently reviewed by Zerner [13]).

$$\mathbf{FC} = \mathbf{SC}\varepsilon \tag{7}$$

where the Fock matrix $\mathbf{F}$ is the effective Hamilton operator, $\mathbf{C}$ is the matrix of the MO coefficients $c_{\mu a}$, $\mathbf{S}$ is the overlap matrix, and ε is a diagonal matrix including the orbital energies. A similar equation can be constructed for open-shell systems [14]. The Fock matrix includes all information about the quantum-chemical system, i.e., all interactions that are included in the calculation. Its general ab initio formulation is the following:

$$
\begin{aligned}
F_{\mu\nu} &= H_{\mu\nu} + J_{\mu\nu} - \frac{1}{2}K_{\mu\nu} \\
&= H_{\mu\nu} + \sum_{\rho}^{m}\sum_{\sigma}^{m} P_{\rho\sigma}\left[(\mu\nu|\rho\sigma) - \frac{1}{2}(\mu\sigma|\rho\nu)\right]
\end{aligned} \tag{8}
$$

with

$$H_{\mu\nu} = \int \chi_{\mu}^{*}(1)\hat{h}\chi_{\nu}(1)\mathrm{d}q_1 \tag{9}$$

$$(\mu\nu|\rho\sigma) = \iint \chi_{\mu}^{*}(1)\chi_{\nu}(1)\frac{1}{r_{12}}\chi_{\rho}^{*}(2)\chi_{\sigma}\mathrm{d}q_1\mathrm{d}q_2 \tag{10}$$

$$P_{\rho\sigma} = 2\sum_{a}^{m} c_{\rho a}^{*}c_{\sigma a} \tag{11}$$

where μ, ν, ρ, and σ denote AOs and $H_{\mu\nu}$ are one-electron integrals representing the expectation values of the kinetic energy operator and the electron-nuclear potential energy operator $\hat{V}_{en}$ of Eq. (2). The $(\mu\nu|\rho\sigma)$ are two-electron repulsion integrals representing $\hat{V}_{ee}$, and the $P_{\rho\sigma}$ are elements of the density matrix $\mathbf{P}$. $J_{\mu\nu}$ and $K_{\mu\nu}$ are the matrix representations of the so-called *Coulomb* and *Exchange* operator $\hat{J}$ and $\hat{K}$, respectively. The electronic energy [Eq. (1)] can be expressed by the eigenvalues ε_a

$$E_{el} = 2\sum_{a}^{M}\varepsilon_a - \frac{1}{2}\sum_{\mu\nu}^{m} P_{\mu\nu}\left(J_{\mu\nu} - \frac{1}{2}K_{\mu\nu}\right) \tag{12}$$

Since the Fock matrix is dependent on the orbital coefficients, the Roothaan equations have to be repeatedly solved in an iterative process, the self-consistent field (SCF) procedure. One important step in the SCF procedure is the conversion of the general eigenvalue equation (7) into an ordinary one by an orthogonalization transformation [15].

$$\mathbf{F}^{\lambda}\mathbf{C}^{\lambda} = \mathbf{C}^{\lambda}\varepsilon \tag{13}$$

with

$$\mathbf{F}^{\lambda} = \mathbf{S}^{-1/2}\mathbf{FS}^{-1/2} \tag{14}$$

and

$$\mathbf{C}^{\lambda} = \mathbf{S}^{1/2}\mathbf{C} \tag{15}$$

$\mathbf{S}^{-1/2}$ is constructed from the overlap matrix $\mathbf{S}$. Solving Eq. (13) is equivalent to finding the matrix of the eigenvectors $\mathbf{C}^{\lambda}$, which transforms $\mathbf{F}^{\lambda}$ into diagonal form.

$$(\mathbf{C}^{\lambda}) + \mathbf{F}^{\lambda}\mathbf{C}^{\lambda} = \varepsilon \tag{16}$$

The orthogonalization transformation [Eq. (14)] is also a very important consideration in semiempirical calculations as will be discussed in later sections. In semiempirical methods, the Hartree–Fock–Roothaan approach is simplified.

1. In the construction of Ψ_0: usually, only valence electrons are treated explicitly. In some cases (Hückel, PPP), only π electrons of certain systems are taken into account. Only a minimal atomic basis set is taken into account. This means that H atoms are described by a 1s function, the elements Li–F by a {2s,2p} set, the elements Na–Cl by a {3s, 3p} set, Ca, K, and Zn–Br with {4s, 4p}, Sc–Cu with {4s, 4p, 3d}, and so on. In some cases, additional sets of shells have been used on selected elements.

2. In the construction of F^λ: a large part of the interactions is neglected, particularly in the two-electron part $(\mu\nu|\rho\sigma)$. All integrals involving AOs centered at more than two centers are neglected. Certain classes of integrals are replaced by parameterized functions. This is mainly the case for the two-center one-electron integrals $H_{\mu\nu}$ that are, to a large extent, responsible for the chemical bonding.

The above simplifications are responsible for the reduction in computational cost of semiempirical methods compared to ab initio (HF and post-HF) or density-functional methods which makes them a useful tool for the investigation of large systems or the routine treatment of large numbers of systems. This statement is, to a large extent, correct even today, despite the considerable technical improvements in ab initio and DFT methodology of the recent years. On the same computer, the computational time for an electronic structure calculation of a relatively small molecule, tetracene $C_{18}H_{12}$, was compared for a semiempirical method and several standard DFT, HF, and post-HF methods [3]. It turned out that HF and DFT methods require about 1000 to 10000 times more CPU time, while the factor increases to 36000 with a post-HF method. Future technical developments of first-principles methods might reduce these factors, but also the efficiency of semiempirical methods has been improved as will be shown in the next sections. The price to pay when approximate methods are used is of course the accuracy of the calculated properties which is, in general (but not always), inferior to that of first-principles calculations. In this respect, it is interesting to note that recent analyses of ab initio effective Hamilton operators [3,16] showed that the approximations and parametric forms of semiempirical methods can at least be partially justified based on fundamental theory.

The various semiempirical methods can be distinguished by the way in which these simplifications are introduced in the model. This will be shown in detail in the following sections.

3. METHODS

3.1. Hückel Method

The most serious approximations are made in the Hückel MO (HMO) method developed for conjugated planar hydrocarbons [1]. In the original method, the elements of the effective Fock matrix [Eq. (8)] are completely parameterized and no molecular integral has to be calculated. Only one $2p\pi$ AO per C atom is considered in the LCAO [Eq. (6)]. For a system with N_C carbon atoms, one gets

$$\phi_a^{HMO} = \sum_{\mu}^{N_C} c_{\mu a} C_\mu(2p\pi) \tag{17}$$

The Hückel MOs ϕ_a^{HMO} are normalized.

$$\sum_{\mu}^{N_C} c_{\mu a}^2 = 1 \tag{18}$$

The diagonal elements of the HMO "Fock" matrix $F_{\mu\mu}^{HMO}$, the so-called *Coulomb integrals*, are assumed to have the same value α for every carbon atom in the molecule, and the only nonzero off-diagonal elements $F_{\mu\sigma}^{HMO}$, the *resonance integrals*, are those between neighboring C atoms μ and σ. They are also set to the same value, β, irrespective of the molecular structure. This parameter can be obtained from a fitting to experimental optical spectra. Both parameters α and β have the dimension energy and are usually given in *atomic units, a.u.*, or *Hartree, E_h*. In order to convert a.u. into the more familiar caloric or SI units, the following conversion factors have to be used [17]: 1 a.u. $\equiv$ 627.5 kcal/mol $\equiv$ 2625 kJ/mol.

The hydrogen atoms are not taken into account. For this reason, and since only nearest-neighbor interactions are considered, the Hückel matrices for *cis*- and *trans*-butadiene C_4H_6 are identical (Fig. 1).

$$\mathbf{F}^{HMO}(\text{butadiene}) = \begin{pmatrix} \alpha & \beta & 0 & 0 \\ \beta & \alpha & \beta & 0 \\ 0 & \beta & \alpha & \beta \\ 0 & 0 & \beta & \alpha \end{pmatrix} \tag{19}$$

In general, the corresponding Hückel orbitals ϕ_a^{HMO} and eigenvalues ε_a can be obtained by standard diagonalization techniques for which a large number of programmed routines are available [18–20], in correspondence to the ab initio procedure [Eq. (16)]. Due to the simple structure of the Hückel matrix, it is, however, possible to obtain analytical expressions for the eigenvalues. In the case of conjugated chains with N_C carbon atoms, solving the secular equation $|\mathbf{F}^{HMO} - \varepsilon| = 0$ gives [21]

$$\varepsilon_a = \alpha - x_a \beta$$

$$x_a = 2\cos\left(\frac{a\pi}{N_C + 1}\right) \tag{20}$$

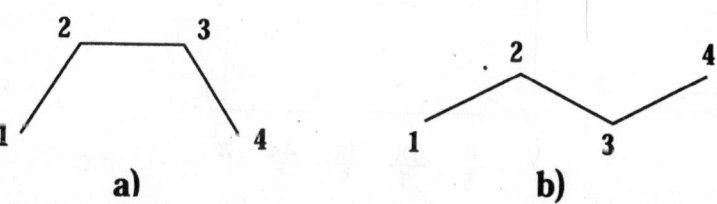

Figure 1 Numbering of C atoms in (a) *cis*- and (b) *trans*-butadiene.

Since the HMO Fock matrix is not dependent on the orbital coefficients, no SCF procedure as for ab initio methods (Sec. 2) has to be performed. In the present example with $N_C = 4$, the four orbital energies are $x_1 = 1.618\beta$, $x_2 = 0.618\beta$, $x_3 = -0.618\beta$, and $x_4 = -1.618\beta$. Since the parameter β is chosen to be negative, the first energy x_1 has the lowest value, and the fourth x_4 has the highest value. This corresponds to the usual convention in quantum chemistry. From these values, the *orbital diagram* of butadiene can be drawn.

In Fig. 2, the two lowest MOs were occupied with two electrons of opposite spin according to the *aufbau principle*. Two results can be obtained from an HMO calculation: the excitation energies comparable to optical transitions in absorption spectroscopy [22] and the total energy E^{el}. The lowest HMO excitation energy of butadiene corresponding to the transition of one electron from the highest occupied MO (HOMO), No. 2, to the lowest unoccupied MO (LUMO), No. 3 (Fig. 2), is $\varepsilon_3-\varepsilon_2$ = $1.236|\beta|$. In spectroscopy, excitation energies are usually given as reciprocal wavelength (*wave numbers*, unit cm^{-1}).

$$\frac{1}{\lambda}(\text{butadiene}) = \frac{\varepsilon_{LUMO} - \varepsilon_{HOMO}}{hc} = 1.236|\beta| \tag{21}$$

Here, h and c are Planck's constant and the speed of light, respectively [23]. On the other hand, Eq. (21) can be used to determine the semiempirical parameter β of the HMO method if the reciprocal wavelength $1/\lambda$ is taken from an accurate measurement. In practice, β is chosen to approximate the excitation energies of a large variety of hydrocarbons [1] as close as possible. For excitation energies, the parameter α does not have to be specified at all and could be set to zero. It only plays a role for the calculation of the total energy. Since the HMO theory does not contain Coulomb or Exchange operators, the calculation of the total energy [Eq. (12)] simplifies to

$$E^{el,HMO} = \sum_a^M n_a \varepsilon_a \tag{22}$$

where n_a is the occupation number of the ath MO. n_a is, in general, equal to 2 but is 1 for the HOMO of conjugated polyenes with an odd number of carbon atoms. For

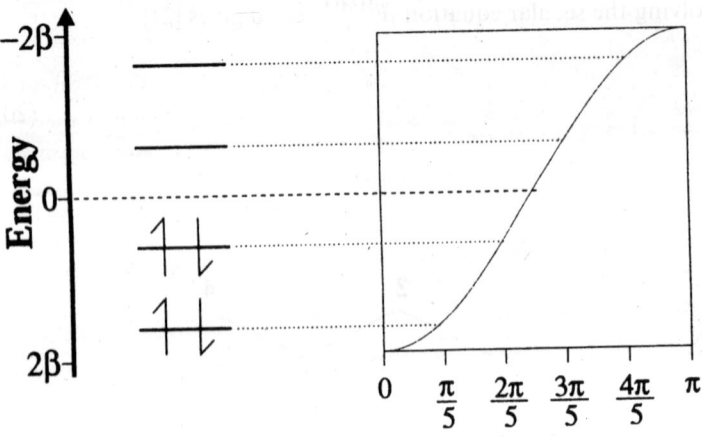

Figure 2   Hückel MO diagram for butadiene derived from analytical expression (21).

butadiene, Eq. (22) gives $E^{el,HMO}$ (butadiene) $= 4\alpha + 4.472\beta$. A famous example of the use of HMO total energies is the calculation of *delocalization energies* or *resonance energies* for aromatic compounds. For the prototype aromatic molecule benzene C_6H_6, the HMO energy is $6\alpha + 8\beta$. If the molecule would consist of three separated π systems (or, alternatively, three C_2H_2 molecules), the HMO energy would be $6\alpha + 6\beta$. Thus the stabilization energy due to delocalization of the three π orbitals is 2β (Fig. 3).

The resonance energy per π electron (REPE) has been used as a measure for the aromaticity of molecules [24]. For a recent and comprehensive overview of the aromaticity concept, see, for example, Refs. [25,26].

A generalization of the Hückel method to nonplanar systems comprised of carbon and heteroatoms is the Extended Hückel Theory (EHT) [27–30]. It takes explicitly into account all valence electrons, i.e., {1s} for H and {2s,2p} for C, N, O, and F. Similar to the HMO method, the "Fock" matrix in EHT $\mathbf{F}^{EHT}$ does not contain two-electron integrals. The diagonal elements $F_{\mu\mu}^{EHT}$ are obtained from experimental ionization potentials (IPs) where the Koopmans theorem [31] has been used.

$$IP_a(\text{Koopmans}) = -\varepsilon_a \tag{23}$$

The off-diagonal elements are approximated by the *Wolfsberg–Helmholz* formula [27]

$$F_{\mu\sigma}^{EHT} = \frac{1}{2} K S_{\mu\sigma} \left(F_{\mu\mu}^{EHT} + F_{\sigma\sigma}^{EHT} \right) \tag{24}$$

This expression takes into account the overlap $S_{\mu\sigma}$ between two AOs χ_μ and χ_σ centered at different atoms. The atomic basis functions χ are represented by Slater functions [32],

$$\chi^{Slater} = \frac{[2\zeta]^{n+1/2}}{[(2n)!]^{1/2}} r_A^{n-1} e^{-\zeta r_A} Y_l^m(\theta, \varphi) \tag{25}$$

where ζ is the *orbital exponent* and Y is a spherical harmonic. The two-center integrals $S_{\mu\sigma}$ over Slater functions can be easily evaluated [32].

The overlap matrix $\mathbf{S}$ is also taken into account in the EHT version of the general eigenvalue equation (7) which can be solved by applying the mentioned orthogonalization transformation and matrix diagonalization techniques. Due to the independence of $\mathbf{F}^{EHT}$ from the orbital coefficients, no SCF procedure has to be performed. This is similar to HMO.

Since the overlap depends on the interatomic distance, the EHT distinguishes between molecular conformations and it is possible to calculate equilibrium structures

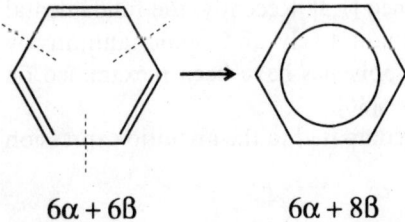

$6\alpha + 6\beta$ $6\alpha + 8\beta$

Figure 3 Resonance energy of benzene: transition from a fictitious system with three separated π bonds to the delocalized ground state.

using standard minimization techniques [33]. Thus the EHT represents the most simple all-valence electron semiempirical method. It has been successfully used by Hoffmann et al. in applications to a vast variety of systems, and it is still frequently used nowadays (see Sec. 6).

3.2. Pariser–Parr–Pople Method

The Hückel Theory in general gives reliable values only for the lowest excitation energies of the HOMO $\rightarrow$ LUMO transition of aromatic or conjugated hydrocarbons. It cannot also distinguish between singlet and triplet excited states that are experimentally known to have different luminescence characteristics. The reason for this deficiency is the neglect of electron–electron interactions in the Hückel method. A semiempirical π electron method that explicitly takes into account electron–electron repulsion $\hat{V}_{ee}$ in the effective Hamilton operator [Eq. (2)] is the Pariser–Parr–Pople (PPP) method [34,35]. It has been designed for the calculation of optical absorption spectra of aromatic hydrocarbons. Similar to the HMO theory, only C 2p π functions are taken into account and all other atoms in the molecule are ignored.

In the evaluation of the effective PPP Fock matrix elements, $F_{\mu\nu}^{\text{PPP}}$, the *zero differential overlap* (ZDO) approximation is used.

$$\chi_\mu(1)\chi_\nu(1)dq_1 = 0 \quad \text{for} \quad \mu \neq \nu \tag{26}$$

This is a fundamental assumption that is used more or less strictly in all semiempirical SCF MO methods. It has several important consequences.

1. The overlap matrix S becomes the unit matrix E (or, by integrating Eq. (26),∫ $\chi_\mu(1)\,\chi_\nu(1)\,dq_1 = S_{\mu\nu} = \delta_{\mu\nu}$ is obtained.)
2. From all two-electron integrals $(\mu\nu|\rho\sigma)$, only the "diagonal" terms of type $(\mu\mu|\rho\rho)$ remain. This reduces the size dependence of semiempirical methods from formally N^4 (as is the case for HF methods) to only N^2. Here lies the main reason why semiempirical methods are able to treat very large systems with up to $N = 10\,000$ electrons. (It has to be kept in mind that for the PPP method, the number of electrons and the number of atomic basis functions are identical. This is, in general, not the case for all-valence semiempirical methods and certainly not for ab initio methods using extended basis sets.)

At first sight, these are rather drastic approximations. However, it has been proven that they are at least partly justified in an orthogonalized basis after a transformation according to Eq. (14). In the orthogonal basis, most of the integrals neglected in the ZDO approximation become smaller in absolute magnitude, and their relevance for the total energy and excitation energy is diminished [1,3]. Recently, the fundamental reasons for the PPP model Hamiltonian to qualitatively and semiquantitatively reproduce spectroscopic features of conjugated polyenes have been reexamined on the basis of high-level coupled cluster calculations [36].

The explicit form of the PPP Fock matrix [compared to the ab initio expression (9)] is

$$F_{\mu\mu}^{\text{PPP}} = H_{\mu\mu}^{\text{PPP}} + \sum_B \sum_\sigma^B P_{\sigma\sigma}\gamma_{AB} - \frac{1}{2}P_{\mu\mu}\gamma_{AA} \tag{27}$$

$$F_{\mu\sigma}^{PPP} = H_{\mu\sigma}^{PPP} - \frac{1}{2} P_{\mu\sigma} \gamma_{AB} \quad \mu \text{ at atom } A, \sigma \text{ at atom } B \tag{28}$$

The integrals $H_{\mu\sigma}^{PPP}$ and γ_{AB} are calculated from semiempirical formulas that contain adjustable parameters. These are optimized in order to reproduce properties of a given class of molecules in an optimal manner (see Sec. 4). The intra-atomic electron repulsion integral γ_{AA} is estimated from experimentally measured ionization energies and electron affinities (see, for example, Ref. [23]).

The PPP method is the first semiempirical method presented here where the Fock matrix does depend on the MO coefficients **C** [via the density matrix elements **P**, see Eq. (11)]. Therefore, the Roothaan equations (by definition due to the ZDO approximation) in the orthogonal basis, Eq. (13), have to be solved in an iterative process until self-convergence is achieved [self-consistent field (SCF) procedure]. As starting coefficients $\mathbf{C}^0$, usually the orbitals of an HMO calculation are used.

Within the framework of PPP theory, it is possible to obtain excitation energies with the *Configuration Interaction* (CI) method. This method is described in greater detail in other chapters of this book. Here only a brief description is given. The PPP ground-state wave function is a single Slater determinant [Eq. (5)]. For closed-shell molecules with an even number N of electrons, $M = N/2$ MOs are doubly occupied and also M orbitals remain unoccupied. It is now possible to construct "excited" determinants by exchanging one or more occupied spin orbitals [i.e., either with α spin (ϕ) or with β spin ($\bar{\phi}$)] with unoccupied orbitals. For example, a determinant where the occupied α spin orbital a has been substituted by the unoccupied α spin orbital p is denoted as Ψ_a^p. An improved ground-state wave function Ψ^{CI} can then be obtained by a linear combination of the original ground-state Slater determinant Ψ_0 and the modified determinants.

$$\Psi^{CI} = C_0 \Psi_0 + \sum_{ap}^{M} (C_a^p \Psi_a^p + C_{\bar{a}}^{\bar{p}} \Psi_{\bar{a}}^{\bar{p}}) + \sum_{abpq}^{M} (C_{ab}^{pq} \Psi_{ab}^{pq} + \cdots) \tag{29}$$

The optimal coefficients C of the CI ground state are obtained by a variational procedure. At the same time also, CI states of higher energy are obtained. In the simplest CI expansion, only single substitutions of the type $a \leftrightarrow b$ and $\bar{a} \leftrightarrow \bar{b}$ are taken into account in Eq. (29). This procedure is called single-excitation CI (SCI). Only singlet (multiplicity of 1) and triplet (multiplicity 3) states are obtained in this way, while the inclusion of multiple substitutions gives rise to states of multiplicity 5 and higher. The lowest singlet state $^1\Psi_0^{CI}$ is the CI ground state, and the higher CI singlet states are denoted by $^1\Psi_1^{CI}, ^1\Psi_2^{CI}, \cdots$. The energy difference $E(^1\Psi_1^{CI}) - E(^1\Psi_0^{CI})$ is then comparable to the lowest excitation energy of an experimental absorption spectrum.

The PPP method is little used nowadays and has been largely superseded by more general semiempirical methods (see the following sections). However, it is still worth to mention since the approximations of the PPP approach are the prototype model of all methods that were developed later.

3.3. CNDO Method

The *Complete Neglect of Differential Overlap* method (CNDO) of Pople et al. [8,37–39] makes use of the ZDO approximation [Eq. (26)] for all pairs of atomic basis functions. It treats explicitly all valence electrons (e.g., C 2s, C 2p), but neglects completely the effect of inner (core) electrons (e.g., C 1s). The CNDO Fock matrix elements therefore reduce to

$$F_{\mu\mu}^{\mathrm{CNDO}} = H_{\mu\mu}^{\mathrm{CNDO}} + \sum_{\nu}^{\mathrm{A}} P_{\nu\nu}\gamma_{\mathrm{AA}} + \sum_{\mathrm{B}\neq\mathrm{A}}^{\mathrm{B}}\sum_{\sigma} P_{\sigma\sigma}\gamma_{\mathrm{AB}}$$

$$F_{\mu\nu}^{\mathrm{CNDO}} = -\frac{1}{2}P_{\mu\nu}\gamma_{\mathrm{AA}}$$

(30)

$$F_{\mu}^{\mathrm{CNDO}} = H_{\mu}^{\mathrm{CNDO}} - \frac{1}{2}P_{\mu\sigma}\gamma_{\mathrm{AB}} \quad (\mu, \nu \text{ at atom A}, \sigma \text{ at atom B})$$

(31)

For symmetry reasons, the intra-atomic one-electron integrals $H_{\mu\nu}^{\mathrm{CNDO}}$ vanish. The one-electron matrix elements $H_{\mu\mu}$ are subdivided into intra-atomic contributions U and interatomic contributions due to the electron attraction $V_{\mathrm{en}}^{\mathrm{B}}$ with other nuclei B.

$$H_{\mu\mu}^{\mathrm{CNDO}} = U_{\mu\mu} - \sum_{\mathrm{B}\neq\mathrm{A}} V_{\mathrm{en}}^{\mathrm{B}}$$

$$H_{\mu\sigma}^{\mathrm{CNDO}} = \beta_{\mathrm{AB}}S_{\mu\sigma}$$

(32)

$$\gamma_{\mathrm{AB}} = (s_{\mathrm{A}}s_{\mathrm{A}}|s_{\mathrm{B}}s_{\mathrm{B}})$$

In fact, the nuclear attraction integrals include an *effective core charge Z^** which is the nuclear charge Z reduced by the number of core electrons. The intra-atomic integral $U_{\mu\mu}$ contains the kinetic energy and the nuclear attraction with the nucleus where the AO μ is centered. The two-electron repulsion integrals $(\mu\mu|\sigma\sigma)$ are calculated over s-type functions invariably, and also when μ or σ are p-type atomic functions. These approximate integrals are called γ_{AB}. The replacement of p functions by angular-independent s functions is necessary in order to fulfill the requirement of *rotational invariance* as discussed in Refs. [37,40]. Calculated molecular properties must be independent from the orientation in a global coordinate system (in the absence of an external field). For the same reason, the nuclear attraction integrals $V_{\mathrm{en}}^{\mathrm{B}}$ are calculated using s-type AOs only.

All two electron one-center integrals $(\mu\mu|\nu\nu)$ are reduced to a single γ_{AA}. The two-center one-electron integrals $H_{\mu\sigma}^{\mathrm{CNDO}}$ are set proportional to the overlap integral $S_{\mu\sigma}$ which is calculated over Slater-type functions (and only here not neglected according to the ZDO assumption). The coefficient β_{AB} is calculated either from atomic parameters β_A or orbital-dependent parameters β_μ.

$$\beta_{\mathrm{AB}} = \tfrac{1}{2}(\beta_{\mathrm{A}} + \beta_{\mathrm{B}})$$

or

(33)

$$\beta_{\mu_{\mathrm{A}}\nu_{\mathrm{B}}} = \tfrac{1}{2}(\beta_{\mu_{\mathrm{A}}} + \beta_{\nu_{\mathrm{B}}})$$

According to the Wolfsberg–Helmholz formula (24), the β parameters can be regarded as diagonal terms of the CNDO Fock matrix.

There are two versions of CNDO, CNDO/1 and CNDO/2, which differ in the evaluation of the intra-atomic integral $U_{\mu\mu}$. In CNDO/1, it is taken from experimental IP of the free atom [8]. In CNDO/2, a modified procedure is used [41]. In order to compensate for deficiencies of CNDO/1 in describing long-range intermolecular electrostatic interactions, the evaluation of V_{en}^B was modified in CNDO/2 (see Ref. [42]).

$$V_{en}^B(\text{CNDO}/1) = \left\langle s_A \left| \frac{Z_B^*}{R_A} \right| s_A \right\rangle \tag{34}$$

$$V_{en}^B(\text{CNDO}/2) = Z_B^* \gamma_{AB} \tag{35}$$

The CNDO method is used to a lesser extent than EHT or less approximate semi-empirical methods (see Sec. 6). The accuracy of energetic and electronic properties obtained with CNDO is, in general, inferior to that of the methods described in the next sections, while the computational effort of the SCF calculation is comparable.

3.4. INDO Method

The *Intermediate Neglect of Differential Overlap* (INDO) method, originally developed by Pople and Beveridge [8] and Pople et al. [44], uses the ZDO approximation [Eq. (26)] only for two-center integrals. The elements of the INDO Fock operator are therefore modified with respect to CNDO mainly by the inclusion of one-center exchange-type integrals $(\mu v | \mu v)$.

$$F_{\mu\mu}^{\text{INDO}} = H_{\mu\mu}^{\text{CNDO}} + \sum_v^A P_{vv}\left[(\mu\mu | vv) - \frac{1}{2}(\mu v | \mu v)\right] + \sum_{B \neq A}^B \sum_v P_{vv}\gamma_{AB} \tag{36}$$

$$F_{\mu v}^{\text{INDO}} = P_{\mu v}\left[\frac{3}{2}(\mu v | \mu v) - \frac{1}{2}(\mu\mu | vv)\right] \tag{37}$$

$$F_{\mu\sigma}^{\text{INDO}} = H_{\mu\sigma}^{\text{CNDO}} - \frac{1}{2}P_{\mu\sigma}\gamma_{AB} \quad (\mu, v \text{ at atom } A, \sigma \text{ at atom } B) \tag{38}$$

The one-center two-electron integrals $(\mu\mu | vv)$ and $(\mu v | \mu v)$ are partly calculated analytically and partly derived from atomic spectra [8]. The original INDO method gives unsatisfying results for geometries and dissociation energies and was soon replaced by several improved versions which are still in use nowadays, namely, MINDO/3, INDO/S, and SINDO1. The first of these newer INDO methods is MINDO/3 (third version of the modified INDO) by Bingham et al. [45], the successor of MINDO/1 [46] and MINDO/2 [47]. The new idea of MINDO was to replace the time-consuming analytical calculation of two-electron integrals γ_{AB} (e.g., using the Harris algorithm [48]; recent algorithms for two-electron integrals over Slater orbitals can be found in Ref. [49]) by a simple multipole expansion, suggested by Ohno [50] and Klopman [51].

$$\gamma_{AB}(\text{MINDO}) = \frac{1}{\sqrt{R_{AB}^2 + \frac{1}{4}(\gamma_{AA} + \gamma_{BB})^2}} \tag{39}$$

The calculation of two-center one-electron integrals has been modified compared to INDO.

$$H_{\mu\sigma}^{\text{MINDO}} = \beta_{AB}S_{\mu\sigma}(H_{\mu\mu}^{\text{CNDO}} + H_{\sigma\sigma}^{\text{CNDO}}) \tag{40}$$

Here β_{AB} are empirical interatomic parameters that have been optimized to minimize the errors in heats of formation with respect to experiment of some reference compounds (see Sec. 4). The Pauli repulsion, which is not included in the CNDO or INDO method, has been incorporated into the core–core potential V_{nn} in the total energy calculation [Eq. (3)]. In order to obtain a balance between attractive and repulsive terms, the analytical $1/R_{AB}$ dependence has been replaced by γ_{AB} plus a correction term. This also in part accounts for the neglected interactions of inner orbitals.

$$V_{nn}^{MINDO} = \sum_{A>B} Z_A^* Z_B^* \gamma_{AB} + f_{AB} \tag{41}$$

$$f_{AB}^{MINDO} = \left(\frac{1}{R_{AB}} - \gamma_{AB}\right) e^{-\alpha_{AB} R_{AB}} \tag{42}$$

with the additional bond parameter α_{AB}. The one-center integrals are calculated by a method due to Oleari et al. [52]. The MINDO/3 version improved in general geometries and dissociation energies compared to the original INDO methods, but also had several failures for specific compounds.

The most frequently used INDO-type method nowadays, known as INDO/S or ZINDO, has been developed by Zerner and Ridley [54] and Bacon and Zerner [55]. From the very beginning, it has been designed for the calculation of molecular spectra of organic molecules and complexes containing transition metals, while its results for structural properties are less accurate. INDO/S starts from the original INDO Fock matrix terms [Eq. (38)] together with the analytic expression for the internuclear repulsion V_{nn}. Special attention has been given to the calculation of one-center two-electron integrals from Slater–Condon factors F and G and the evaluation of $U_{\mu\mu}$ from experimental ionization energies [55]. For example, the ionization process that removes an s electron of an atom with a $\{s^l p^m d^n\}$ electronic configuration can be expressed in terms of $U_{\mu\mu}$, F, and G.

$$IP_s = E(s^{l-1} p^m d^n) - E(s^l p^m d^n)$$
$$= -U_{ss} - (l-1)F_{ss}^0 - m\left[F_{sp}^0 - \frac{1}{6}G_{sp}^1\right] - n\left[F_{sd}^0 - \frac{1}{10}G_{sd}^2\right] \tag{43}$$

The one-center Coulomb integrals $(\mu\mu|\nu\nu) = F_{\mu\nu}^0$ are calculated analytically while the G integrals are taken as parameters. The IP are taken from atomic spectra. After rearranging Eq. (43), the U_{ss} are obtained. A special feature of INDO/S is the use of distance-dependent Slater exponents for the calculation of two-center integrals [55].

$$\zeta(R) = a + b/R \quad \text{for } \zeta(R) < \zeta(0) \tag{44}$$

$$\zeta(R) = \zeta(0) \quad \text{elsewhere} \tag{45}$$

This procedure to some extent mimics the use of multiple-zeta basis sets in high-quality ab initio calculations. Later, also charge-dependent orbital exponents that are more flexible with respect to the chemical environment have been implemented in INDO/S [56]. For transition metal atoms, one-center integrals of the general form $(\mu\nu|\rho\lambda)$ are taken into account which do not appear in the original INDO method.

For the two-center one-electron integrals, a conventional INDO formalism is applied. There are two sets of optimized parameter sets for INDO/S. One has been optimized for electronic spectra and the other for molecular geometries. Therefore, in principle, two successive INDO/S calculations would be necessary: first, a geometry optimization using the second parameter set, and then a calculation of spectroscopic properties at fixed geometry. Since it is known that the INDO/S geometries are not very accurate, usually another semiempirical method is used for the structure optimization in practice. Recently, a modification of the original INDO/S method has been reported [57].

One year after INDO/S, the method SINDO1 (symmetrically orthogonalized INDO/1) by Nanda and Jug [58] was introduced. Originally developed for organic compounds of first-row elements, it was later extended to elements of the second and third row [59,60]. This method has several distinct features. The most important is that the orthogonalization transformation [Eq. (14)] is taken into account by a Taylor expansion. The matrix $S^{-1/2}$ is approximated as

$$S = E + \sigma$$
$$S^{-1/2} = E - \frac{1}{2}\sigma + \frac{3}{8}\sigma^2 - \frac{5}{16}\sigma^3 + \cdots \qquad (46)$$

and expansion (46) is truncated after the second order. Only the one-electron integral matrix H is transformed (for a discussion of the consequences of finite-order expansion on molecular integrals, see, e.g., Ref. [61]). Another special feature of SINDO1 is the explicit treatment of inner orbitals by a pseudo-potential proposed by Zerner [62]. The calculation of one-center integrals is similar to that in INDO/S. Two-center one-electron integrals $H_{\mu\sigma}$ are calculated by the following empirical formula:

$$H_{\mu\sigma}^{\text{SINDO1}} = L_{\mu\sigma} + \Delta H_{\mu\sigma} \quad \mu \text{ at atom A}, \ \sigma \text{ at atom B} \qquad (47)$$

Here L is a correction of the Mulliken approximation for the kinetic energy and ΔH is entirely empirical and contains adjustable bond parameters. These are optimized in order to minimize the deviation from experiment for a set of reference compounds. In a way similar to INDO/S [Eq. (45)], two sets of Slater orbital exponents are used: one [$\zeta(0)$] for intra-atomic integrals and the other (ζ) for molecular integrals. For comparison with experimental heats of formation, the calculated binding energies E_B [Eq. (4)] are corrected by the zero-point energies obtained from vibration analyses. Later, a substantially modified version of SINDO1, MSINDO, was developed and re-parameterized for the elements H, C–F, Na–Cl, Sc–Zn, and Ga–Br [63–65].

A semiempirical method with a similar acronym, SINDO, but with completely different features has been developed by Golebiewski at el. [66].

An early review about the performance of these and other semiempirical methods has been given by Jug [67].

3.5. NDDO Method

The method *Neglect of Diatomic Differential Overlap* (NDDO) was originally developed by Pople and Beveridge [8] and Pople et al. [37]. The ZDO approximation [Eq. (26)] is only applied for orbital pairs centered at different atoms. Consequently, new types of two-center integrals appear compared to the INDO method, $(\mu\nu|\rho\lambda)$ and $(\mu|V_B|\nu)$. This means that not only monopole–monopole interactions are taken into account, but also dipole and quadrupole terms. Thus, in principle, NDDO-based methods should give an improved description of long-range intra- and intermolecular forces as they become important in large biomolecules. The NDDO Fock matrix becomes

$$F_{\mu\mu}^{NDDO} = H_{\mu\mu}^{CNDO} + \sum_{\rho}^{A}\sum_{\lambda}^{A} P_{\rho\lambda}\left[(\mu\mu|\rho\lambda) - \frac{1}{2}(\mu\rho|\mu\lambda)\right] + \sum_{B\neq A}^{B}\sum_{\tau}^{B}\sum_{\sigma}^{B} P_{\tau\sigma}(\mu\mu|\tau\sigma)$$

$$F_{\mu\nu}^{NDDO} = H_{\mu\nu}^{NDDO} + \sum_{\rho}^{A}\sum_{\lambda}^{A} P_{\rho\lambda}\left[(\mu\nu|\rho\lambda) - \frac{1}{2}(\mu\rho|\nu\lambda)\right] + \sum_{B\neq A}^{B}\sum_{\tau}^{B}\sum_{\sigma}^{B} P_{\tau\sigma}(\mu\nu|\tau\sigma)$$

$$F_{\mu\sigma}^{NDDO} = H_{\mu\sigma}^{CNDO} - \frac{1}{2}\sum_{\nu}^{A}\sum_{\tau}^{B} P_{\nu\tau}(\mu\nu|\sigma\tau) \quad (\mu, \nu, \rho, \text{ at atom } A, \tau, \sigma \text{ at atom } B)$$

$$(48)$$

with $H_{\mu\nu}^{NDDO}$ including the additional interactions $(\mu|V_B|\nu)$. In this original version, NDDO represented no consistent improvement over the then existing INDO methods. Only after considerable modification, the *Modified Neglect of Diatomic Overlap* (MNDO) method introduced by Dewar and Thiel [68] gave an improved agreement with experimental heats of formation ionization potentials, dipole moments, and geometries. MNDO was originally developed for first-row elements (H, C, N, O, and F). Later, it was extended by Thiel and Voityuk [69,70] to second-row elements and transition metals after inclusion of *d* functions (MNDO/d). At present, the elements H, Li, Be, B, C, N, O, F, Al, Si, Ge, Sn, Pb, P, S, Cl, Br, I, Zn, and Hg are parameterized.

The one-electron integrals U and $H_{\mu\sigma}$ and the core–core repulsion V_{nn} are calculated and parameterized in a similar way as in MINDO/3 (see the last section). For two-center two-electron integrals $(\mu\nu|\sigma\tau)$, the Ohno–Klopman approximation is used, initially with up to dipole–dipole terms [68] and later up to quadrupole–quadrupole terms [69]. The value of these approximate integrals is considerably smaller than the exact value in the bond regime. This has been interpreted as an intrinsic inclusion of electron correlation in the MNDO method [68].

The MNDO method has been continually modified and improved by Thiel et al. The most important aspects of these modifications are the use of Zerner's [62] effective core potential and the inclusion of orthogonalization corrections in a way similar to the INDO method SINDO1 (see the last section) leading to the two models OM1 and OM2 (orthogonalization models 1 and 2), respectively [3,71,72]. These corrections have been found to be important for the description of torsion angles in organic compounds. So far, it has been parameterized for elements H, C, N, and O.

Many popular semiempirical methods are based on the original MNDO method. The most prominent of these are *Austin Model 1* (AM1) by Dewar et al. [73] and *Parametric Method 3* (PM3) by Stewart [74]. These three methods represent the "semiempirical standard" for the calculation of organic molecules and are included in popular program packages such as Gaussian [78], CERIUS [79], SPARTAN [80], MOPAC [81], and AMPAC [82]. Recently, AM1 and PM3 have also been extended for the treatment of transition metal compounds [75–77]. In principle, they only differ in the parameterization and in the empirical function f_{AB} [Eq. (42)].

The newest semiempirical method is *Semi-Ab initio Model 1* (SAM1) by Dewar et al. [83]. As the acronym suggests, it is based on AM1. The two-electron integrals $(\mu\nu|\sigma\tau)$, however, are calculated analytically over Gaussian-type functions and scaled empirically. Up to the present, no comprehensive list of SAM1 parameters and error statistics is available [3]. Recently, several NDDO-type methods have been modified for the calculation of optical spectra [84].

4. PARAMETERIZATION

As described in the previous sections, all semiempiricals contain parameters. They either replace integrals that are calculated analytically in ab initio approaches, or they are part of empirical formulas that describe the chemical bonding, usually in the two-center one-electron part. These parametric formulas are designed to compensate for the neglect of a large part of the interatomic, three-, and four-center terms that have to be taken into account in first-principles methods. The quality of a semiempirical method therefore strongly depends not only on the formulation of the Fock operator, but also on the choice of the parameter sets.

4.1. Classification of Parameters

The parameters discussed here can be classified into two groups

1. Experimentally derived fixed parameters
2. Adjustable parameters.

All quantum-chemical methods, even if they are considered as derived from first-principles, make use of at least the first group of parameters. Examples are the atomic masses and atomic heats of formation. Atomic orbital exponents used in all LCAO-based methods are an example of the second group.

The experimentally derived parameters used in semiempirical methods are the orbital energies of valence and inner orbitals (i.e., the corresponding ionization potentials). The adjustable parameters can be further classified into atomic and bond parameters. The most important adjustable atomic parameters are the exponents of the atomic basis functions (usually Slater-type orbitals). In some methods (INDO/S, MSINDO, and AM1), different values of exponents are used for the evaluation of intra- and interatomic integrals. Other methods use atomic parameters (Slater–Condon factors) for the one-center terms U or use this integral directly as adjustable parameter. For the two-center one-electron integrals, all methods use atomic parameters, either called β (EHT, MINDO/3, MNDO, AM1, PM3, and SAM1) or K (MSINDO, with a slightly different functionality). Bond parameters α_{AB} that depend on the atomic number of two atoms appear in some semiempirical methods. They are either used in the correction term f_{AB} of the internuclear repulsion [Eq. (42)] or in the parameterized function for $H_{\mu\sigma}$ [Eq. (47)]. Generally, an element is described with 10–20 parameters, depending on the main quantum number and the specific method.

4.2. Optimization Methods

A semiempirical method only gives accurate results for molecular properties when the parameters for all elements and combinations of elements in a specific compound have been optimized before. A parameter optimization is performed on a set $\{i\}$ of test molecules for which reliable experimental reference values f_i^{ref} are available for the properties of interest. In general, these are heats of formation $\Delta_f H$, bond lengths R, angles θ, torsion angles Φ, ionization potentials IP, and dipole moments μ. For this reference set, the adjustable parameters are varied so that the total quadratic error s of the calculated properties f_i^{calc}

$$s = \sum_i s_i^2 = \sum_i \left[(f_i^{\text{ref}} - f_i^{\text{calc}})\omega_i \right]^2 \tag{49}$$

is minimized. The weighting factor ω_i is chosen in order to balance the different kinds of observables. It can also be used to give a lower or higher priority to certain reference properties according to the reliability of the corresponding experiment. In some cases, in particular, for transition metal compounds where the number of reliable experimental reference values is limited, results of accurate first-principles methods have been used [65]. The optimization procedure is generally based on gradient-driven methods [3] where the Jacobian matrix {the partial derivatives of s_i [Eq. (49)] with respect to the parameters} is calculated either numerically with finite differences or analytically [74]. Recently, also genetic algorithms have been used for the optimization of semiempirical parameters [86–88]. They have the advantage that they always find the global minimum for a given parameter set, while gradient-driven algorithms generally only find the minimum closest to the starting values.

Great care has to be taken that the optimization leads to a "reasonable" set of parameters where there is no overemphasis on the accuracy of specific properties [3,89]. For example, the orbital exponents for the elements within a period should follow the general trends of the corresponding values of ab initio methods [90]. The ordering $\zeta_s > \zeta_p$ for main-group elements or $\zeta_s > \zeta_p > \zeta_d$ for transition metals must hold. Only if the parameters show this *internal consistency* the quality of the reference set used for parameter optimization can be transferred to other compounds [65]. The reference set $\{i\}$, on the other hand, must be selected to contain a broad spectrum of different bonding situations for each parameterized element.

4.3. Comparison of Some Methods

All semiempirical methods are parameterized for the elements H, C, N, F, and O which form the basis of most organic molecules that are of importance in pharmaceutics, biochemistry, and organic chemistry. Here it is possible to compare the statistical errors for some of the more recent methods. In Table 1, the statistics for energetic, structural, and electronic properties for first-row elements are compared for MSINDO, MNDO, AM1, and PM3.

Apparently, there is no significant difference in the average errors for these four methods. It has to be noted that the reference set of MNDO, AM1, and PM3 is almost twice as large as the MSINDO reference set. These methods are therefore expected to give reliable results for a larger variety of organic molecules, and nowadays, applications of semiempirical methods on ground-state properties of biomolecules are restricted to these three methods while MSINDO has specialized in problems of solid-state chemistry (see Sec. 5). INDO/S is more in use for the calculation of spectroscopic features for organic systems with and without transition metals.

Another example for error statistics is given in Tables 2 and 3 for elements of the second row. Complete statistics for all second-row elements Na–Cl have so far only been published for two methods, MSINDO and MNDO/d. Both methods perform similar for the calculation of heats of formation. The agreement with experimental bond lengths seems to be slightly better with MSINDO than with MNDO/d, but as for

Table 1 Mean Absolute Errors for Ground-State Properties of First-Row Elements with Number of Values in Parentheses [63]

	Group	MSINDO	MNDO	AM1	PM3
$\Delta_f H$ (kcal/mol)	HCNO	5.12 (64)	6.26 (133)	5.52 (133)	4.23 (133)
	F	5.59 (25)	10.51 (43)	6.76 (43)	6.45 (43)
R (Å)	HCNO	0.011 (164)	0.015 (228)	0.017 (228)	0.011 (228)
	F	0.022 (46)	0.037 (124)	0.027 (124)	0.022 (124)
$\theta(°)$	HCNO	1.84 (72)	2.69 (92)	2.01 (92)	2.22 (92)
	F	1.20 (22)	3.04 (68)	3.11 (68)	2.72 (68)
IP (eV)	HCNO	0.44 (67)	0.47 (51)	0.36 (51)	0.43 (51)
	F	0.37 (16)	0.34 (40)	0.54 (40)	0.40 (40)
μ (D)	HCNO	0.34 (32)	0.32 (57)	0.25 (57)	0.27 (57)
	F	0.33 (17)	0.38 (40)	0.31 (40)	0.29 (40)

Comparison of MSINDO, MNDO, AM1, and PM3.

first-row elements, the size of the reference molecule set is considerably larger for MNDO/d so that a direct comparison is not possible. The average errors for both methods are slightly larger for second-row elements compared to the first row. This is due to the occurrence of more complex binding situations (ionic complexes and hypervalent compounds) with elements S and Cl. It was necessary in both semi-empirical methods to augment the minimal {3s, 3p} basis with 3d functions to describe these compounds with reasonable accuracy.

Until now, no complete error statistics has been published for third-row transition metals Sc–Zn except for MSINDO [65]. MNDO/d, AM1/d, and PM3d/PM3tm have so far only been parameterized for a small number of transition elements.

5. APPLICATIONS

The following subsections will present a selection of applications of the most frequently used semiempirical methods INDO/S (or ZINDO), MSINDO (or SINDO1), MNDO and MNDO/d, AM1, PM3, and SAM1 and modifications of these methods

Table 2 Mean Absolute Errors for Heats of Formation (kcal/mol) of Second-Row Elements with Number of Values in Parentheses [63]

Element	MSINDO	MNDO/d
Na	8.37 (13)	7.57 (23)
Mg	5.80 (25)	9.61 (46)
Al	6.76 (16)	4.93 (29)
Si	6.69 (41)	6.33 (84)
P	4.83 (21)	7.62 (43)
S	7.49 (24)	5.57 (99)
Cl	7.13 (37)	3.76 (178)

Comparison of MSINDO and MNDO/d.

Table 3 Mean Absolute Errors for Bond Lengths R (Å) of Second-Row Elements with Number of Values in Parentheses [63]

Element	MSINDO	MNDO/d
Na	0.051 (12)	0.120 (16)
Mg	0.030 (32)	0.120 (55)
Al	0.031 (13)	0.067 (20)
Si	0.018 (57)	0.047 (68)
P	0.019 (37)	0.048 (58)
S	0.022 (37)	0.040 (77)
Cl	0.037 (45)	0.038 (117)

Comparison of MSINDO and MNDO/d.

to problems in current chemistry. This choice cannot be expected to be complete, given the fact that every year, several hundred studies are published in this area (see Sec. 6). Several recent reviews about the use of semiempirical methods in the field of organic chemistry, biochemistry, and pharmacology can be found, e.g., in Refs. [3,11,91–93].

5.1. INDO/s

A theoretical model for the active site of an enzyme (*Azotobacter vinelandii*) FeMo cofactor for the fixation of nitrogen has been investigated by Stavrev and Zerner [94]. A small subsystem of the cofactor with the Fe and Mo atoms (Fig. 4) has been selected for ZINDO and DFT calculations of possible reaction pathways.The electronic excitations in monomers and aggregates of bacteriochlorophylls were calculated by means of INDO/S-CI calculations [95] as a model for photosynthetic processes in organisms, and the results were generalized by means of an effective Hamilton

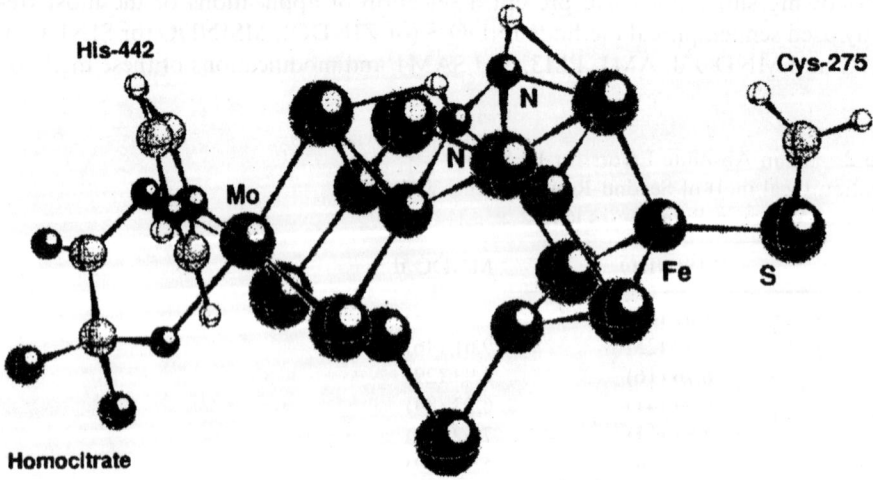

Figure 4 FeMo cofactor model system used for ZINDO calculations of nitrogen fixation [94].

operator. ZINDO together with HF calculations have been employed for a study of electronic properties of the DNA base guanine [97]. Emphasis was given to the sequence-specific regions of lowest ionization potentials that were calculated using Koopmans' theorem. The structures, stabilities, and electronic spectra of the hetero-fullerenes $C_{59}N$ and $C_{69}N$ and the formation of dimers with N–N bonds were examined with INDO/S [98]. An example of the strength of INDO/S for the calculation of optical spectra is Ref. [96], where spectroscopic red shifts due to dissolution of benzene in liquid cyclohexane are obtained in excellent agreement with the experimentally observed shifts, and where the spectroscopy of 4-hydroxy-1-methylstilbazolium betaine in dependence of the solvent polarity has been examined. The effect of conformation, nature of substituents, and endgroups on the molecular hyperpolarizability of dicyanomethylene (hetero)aromatic dyes has been investigated with AM1 (for geometries) and ZINDO (for electronic properties). The spectroscopy of 4-hydroxy-1-methylstilbazolium betaine including solvent effects has been studied with INDO/S, and good agreement with experimental values in water and methanol has been obtained [99].

5.2. MSINDO

As mentioned in Sec. 4, the method MSINDO has been designed for applications to solid-state problems such as adsorption phenomena, surface reactions, and properties of ionic materials [100]. Nevertheless, it has been found to give results of similar or even better quality as NDDO-type methods also for organic systems such as amines [101]. Special techniques for the description of solids and surfaces (embedding procedures and cyclic boundary conditions) have been developed and incorporated into SINDO1 and later into MSINDO [102,103]. MSINDO adsorption studies comprise water and other small molecules at surfaces of oxides Cr_2O_3 [104], MgO [105,106], and chalkogenides like NaCl [107]. In general, qualitative or even semi-quantitative agreement with experimental results has been obtained. Large siloxane clusters have been studied [108] in order to examine the experimentally observed growth pattern in the gas phase reaction of $SiCl_4$ with O_2. Embedded cluster models have been used to calculate the surface structure of MgO(100) [109]. A combination of semiempirical and DFT techniques has been applied in a study of Cu deposition on magnesium oxide surfaces [110]. The most recent MSINDO application to transition metal compounds is the study of Al doping of anatase and rutile TiO_2 particles [111].

5.3. MNDO/d

MNDO/d and the more recent orthogonalization models have been successfully used for the structure optimization of large organic molecules, particularly where high accuracy is required for the description of weak interactions that determine the conformation of long carbon chains. They have also been applied to calculations of very large systems like the geometrical optimization of a C_{960} fullerene [112]. Some of the most recent applications are a conformation analysis of cyclic ADP-ribose in connection with an experimental NMR study [113], the binding of methylguanidinium to a methylphosphate entity in a combined ab initio and semiempirical model study of the thymidylate synthase G52S mutation [114], and the potential surface for the approach of the carcinogen N-2-acetylamino-fluorene to the carbon $C_{(8)}$ of deoxy-guanosine [115]. A comparison between the new MNDO versions OM1 and OM2 with

AM1 and PM3 for the description of the secondary structure in peptides and proteins has been performed recently [116], and it was shown that the description of the peptide conformers is considerably improved by OM1 and OM2 compared with AM1 and PM3, although in some cases, there still were discrepancies with available ab initio data. MNDO-PSDCI molecular orbital theory has recently been used to calculate the spectroscopic properties of sensory rhodopsin from *Natronobacterium pharaonis* [117], demonstrating that MNDO is also a reliable tool for the calculation of optical spectra.

5.4. AM1

This method is widely applied for the structure determination and electronic structure calculation of large and very large organic systems. Some of the most impressing calculations in recent semiempirical applications are the AM1 energy calculations on a 19995-atom polymer of glycine and a 6304-atom RNA molecule [118] where the Millam–Scuseria conjugate-gradient density-matrix search was applied to replace conventional matrix diagonalization. The molecular electron density calculated with AM1 has been used as a basis for the parameterization of simple electrostatic field models that are used to obtain quantitative structure activity relationships (QSAR) for a series of singly substituted amines as well as para-substituted benzoic acids and other bioactive substances [119,120]. The conformational flexibility of the Ibuprofen molecule has been analyzed by crystallographic database searching and AM1 potential energy calculations [121]. A combined AM1 and HF study has been carried out for the interaction of anhydrotetracycline, the major toxic decomposition product of the antibiotic tetracycline, with aluminum [122]. In this study, a perfect agreement of semiempirical and ab initio calculations was observed demonstrating the reliability of the AM1 method for structural and energetic properties of organic molecules. A systematic conformational analysis of anhydrotetracycline, a toxic decomposition product of the widely used antibiotic tetracycline, has been carried out with AM1 [123]. The results were used to explain the toxic effects of the anhydrous derivative.

5.5. PM3

PM3 has been used in drug design, for example, on the effect of dinitrosubstitution on the methylation reaction of catechol and endogenous catechol derivatives catalyzed by catechol *O*-methyltransferase (COMT) [124], where the implications of the derived reaction mechanism to the design of COMT inhibitors are discussed. PM3 has been modified for the calculation of very large systems. Examples are the geometry optimization on the 1226-atom Kringle 1 of plasminogen [125] with a conjugate-gradient technique replacing matrix diagonalization. A series of epibatidine analogs and their positional isomers bearing an 8-azabicyclo[3.2.1]octane moiety has been described in a combined experimental and PM3 study [126,127]. The oxidation decomposition of 2,4,6-tri-*tert*-butylphenol and related compounds was investigated by calculating the reaction enthalpies with PM3 [128] in a study of the cytotoxic activity of these compounds. Recently, the electronic structure of a mutagen, chloroimide 3,3-dichloro-4-(dichloromethylene)-2,5-pyrrolidinedione, has been calculated for a correlation with its known bacterial mutagenicity value [129].

5.6. SAM1

A comparison between AM1 and SAM1 for the calculation of vibrational frequencies was carried out for 41 organic molecules by Holder and Dennington [130]. Both methods showed reasonable agreement with the experimental values, and SAM1 performed slightly better than AM1. Three systems of isomeric fullerenes, C_{88}, C_{36}, and C_{72}, have been optimized at semiempirical SAM1 and ab initio levels [131], and the calculated structural data are related to the observed data. A combined DFT and SAM1 study was performed to describe the interaction between nitric oxide (NO) and the active site of ferric cytochrome $P450$ [132]. Reactions of the same system, cytochrome $P450$, with alkanes were studied with SAM1 using a model system consisting of unsubstituted porphyrin, iron, and methylthiolate [133].

5.7. Combinations of Methods

Since the methods MNDO, AM1, PM3, and SAM1 are available together in many program packages (e.g., MOPAC, AMPAC, Gaussian, and SPARTAN), they are frequently used together in combined and comparative studies. An important issue in the theoretical treatment of biomolecules and the reactivity of enzymes is intra- and intermolecular hydrogen-bonding interaction. The calculated H-bonding interactions calculated with AM1, PM3, and SAM1 were compared with accurate ab initio results and experiments [134]. It has been found that AM1 performs better than the other two methods, but still is not satisfactory for O–HO interactions. A similar comparison of the same methods was performed for the normal modes in several local anesthetics of amino-ester type [135]. MOPAC calculations were used in a computer-aided conformational analysis in order to characterize the pharmacophore for the intestinal peptide carrier [136]. The gastrin CCK antagonist activity of 67 benzodiazepines has been studied by molecular modeling using MOPAC [137]. Conformationally constrained analogs of the potent muscarinic agonist 3-(4-(methylthio)-1,2,5-thiadiazol-3-yl)-1,2,5,6-tetrahydro-1-methylpyridine (methylthio-TZTP, 17) were designed and synthesized in a combined experimental and theoretical study using MOPAC 6.0 [138]. AM1, PM3, and MNDO electrostatic-potential-derived atomic charges have been compared in correlations with solvatochromic hydrogen-bonding acidity for QSAR studies [139]. Here the best correlation has been obtained with the AM1 and MNDO methods. The AMPAC program package has been used to obtain quantitative structure activity relationships (QSARs) in dental monomers that influence their mutagenicity [140]. A theoretical investigation into the possibility of designing bioreductive analogs of cyclophosphamides as anti-cancer drugs has been undertaken with AM1 and PM3 included in MOPAC93 [141] and gave results in agreement with the experiment. These methods were also used to calculate molecular vibration modes as a basis for a three-dimensional quantitative structure activity relationship using the eigenvalue analysis (EVA) paradigm applied to 41 HIV-1 integrase inhibitors [142].

6. PERSPECTIVES

Due to the rapid development of accurate first-principles method and their extension to larger systems in the past years, semiempirical methods are sometimes considered to be obsolete. The large number of recent applications reported in the last sections demonstrates that this is presently not the case and that approximate methods still are used for exploratory studies on complex systems. However, it is interesting to note some of the recent developments and trends in connection with semiempirical methods that will be presented in this section.

One might consider the number of published scientific articles on a given subject as a quantitative measure of its relevance. This has been investigated for the Hückel-, CNDO-, INDO-, and NDDO-type methods by an analysis of the Web of Science database, Institute of Scientific Information [143]. The search keywords were chosen carefully in order to eliminate false positive hits. Surprisingly, the number of investigations using the most accurate NDDO family of methods (AM1, PM3, and MNDO) seems to decrease gradually, although these still represent the most widely used approach. This result is completely different from that of an earlier investigation for the interval 1989 to 1993 [144]. A similar downward trend can be found for the extended Hückel method and for the CNDO method, while the number of researchers that make use of INDO-type methods (INDO/S, MINDO, and SINDO) remains almost constant during the last decade (Fig. 5). Of course, this analysis can only be regarded as approximate since most studies use more than one method at a time. It is also of course only about the quantity and not the quality of the studies.

A possible explanation of the trends shown in Fig. 5 is that wherever high accuracy is required, semiempirical methods have recently been replaced by DFT approaches, unless they are designed for the description of certain properties (like INDO/S) or classes of materials (like MSINDO).

6.1. New Techniques and Properties

Due to the ZDO approximation [Eq. (26)], the evaluation of molecular integrals is an N^2 process (i.e., the number of floating point operations to solve this equation is

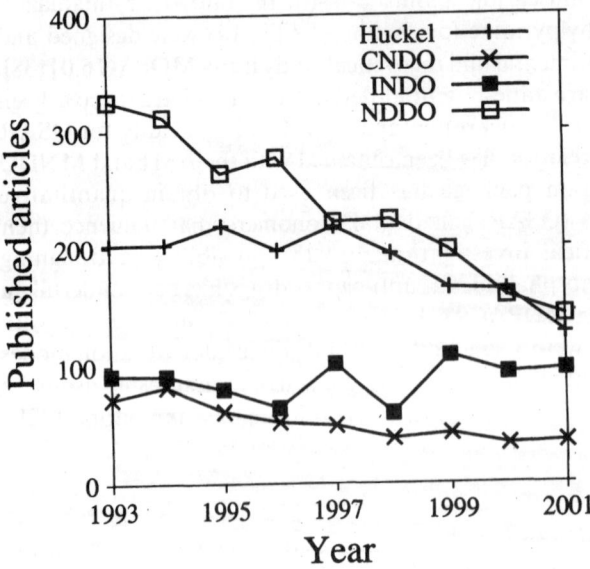

Figure 5 Number of scientific studies published in the past years that employed semiempirical methods.

proportional to N^2) as compared to N^3 in DFT and N^4 in HF. The bottleneck for the calculation of larger systems with $N \geq 10\,000$ electrons is the linear algebra connected with the Roothaan equation (13) which is an N^3 process. Several methods to reduce this effort have been developed for semiempirical methods (since they experienced this difficulty much earlier than first-principles methods). Stewart et al. [145] proposed a simplified diagonalization procedure based on the Jacobi method that still scales as N^3 but is one or two orders of magnitude faster than conventional diagonalizers. As mentioned in Sec. 5, Daniels et al. [118] introduced a conjugate-gradient density-matrix search for the AM1 treatment of huge polymers. They also used a divide-and-conquer technique which separates the large system into smaller subsystems that can be treated at much lower cost. A similar technique based on localized MOs has been developed by Stewart for biomolecules [146]. A technical solution for the problem of large systems is the development of parallelized program codes that can be used on modern parallel computers. Efficient parallel implementations exist for MNDO and MOPAC [147,148]. Semiempirical methods at CNDO, INDO, and NDDO level have been extended for the calculation of solid-state properties [103,149–152]. Electronic excitations can be now calculated with higher accuracy using the Green's function technique [153]. AM1 and PM3 have also been extended for the calculation of polarizabilities and hyperpolarizabilities [154]. It is possible to evaluate NMR and ESR spectra from semiempirical wave functions [1]. Thus semiempirical methods offer the same variety of properties as first-principles methods.

6.2. Hybrid Approaches

Semiempirical methods are the middle ground between highly accurate ab initio methods and completely empirical molecular mechanical (MM) methods [155]. For the treatment of very large biomolecules, hybrid approaches have been developed where the reactive center is described by a semiempirical method and the "inert" rest of the molecule by a classical force field [3,156,157]. This technique can also be applied for the description of solvent effects. The solvent molecules are then described by the MM method. If an even higher accuracy is required for the reactive center of the system, a hybrid approach of three different methods can be applied, e.g., in the ONIOM model by Vreven and Morokuma [158]. Here the center is described at DFT or post-HF level, the nearest-neighbor atoms at semiempirical level, and the outer surrounding at MM level. There also exist hybrid schemes between semiempirical and DFT methods only [159].

This coexistence with higher-level methods represents one of the most promising outlooks for the future of semiempirical methods. At the present and in the near future, there still exists a wide field in which approximate methods can aid the researcher to gain an understanding of the fundamental principles in chemistry and biology.

REFERENCES

1. Levine IN. Quantum Chemistry. 5th ed. New Jersey: Prentice-Hall, 2000.
2. Clark T. J Mol Struct, Theochem 2000; 530:1.
3. Thiel W. Adv Chem Phys 1996; 93:703.
4. Hoffmann R. J Mol Struct, Theochem 1998; 424:1.

5. Parr RG. The Quantum Theory of Molecular Electronic Structure. New York: WA Benjamin, 1963.
6. Salem L. The Molecular Orbital Theory of Conjugated Systems. New York: WA Benjamin, 1966.
7. Dewar MJS. The Molecular Orbital Theory of Organic Chemistry. New York: McGraw-Hill, 1969.
8. Pople JA, Beveridge DL. Approximate Molecular Orbital Theory. New York: McGraw-Hill, 1970.
9. Scholz M, Köhler HJ. Quantenchemie Bd 3. Heidelberg: Hüthig, 1981.
10. Clark T. A Handbook of Computational Chemistry. New York: Wiley, 1985.
11. Thiel W. J Mol Struct, Theochem 1997; 1:398–399.
12. Roothaan CCJ. Rev Mod Phys 1951; 23:69.
13. Zerner MC. Theor Chem Acc 2000; 103:217.
14. Roothaan CCJ. Rev Mod Phys 1960; 32:179.
15. Löwdin P-O. J Chem Phys 1950; 18:365.
16. Ruette F, Gonzalez C, Octavio A. J Mol Struct, Theochem 2001; 537:17.
17. Homann K-H. International Union of Pure and Applied Chemistry (IUPAC), Grössen. Einheiten und Symbole in der Physikalischen Chemie. Weinheim: Verlag Chemie, 1996.
18. Wilkinson JH, Reinsch C. Handbook for Automatic Computation, Linear Algebra. Berlin: Springer, 1971.
19. Press WH. Numerical Recipes in FORTRAN: the Art of Scientific Computing. Cambridge: Cambridge University Press, 1995.
20. LAPACK (Linear Algebra PACKage) User's Guide. 3rd ed. http://www.netlib.org/lapack/lug/lapack_lug.html, 1999.
21. Muir T. The Theory of Determinants in the Historical Order of Development. New York: Dover, 1960.
22. Huber KP, Herzberg G. Molecular Spectra and Molecular Structure. New York: Van Nostrand Reinhold, 1979.
23. Lide DR, ed. CRC Handbook of Chemistry and Physics. 81st ed. Boca Raton: CRC Press, 2000.
24. Hess, BA, Schaad, LJ. J Am Chem Soc. 1971; 93:305, 2413.
25. Katritzky AR, Jug K, Oniciu DC. Chem Rev 2001; 101:1421.
26. Jug K, Hiberty PC, Shaik S. Chem Rev 2001; 101:1477.
27. Wolfsberg M, Helmholz L. J Chem Phys 1952; 20:837.
28. Hoffmann R. J Chem Phys 1963; 39:1397.
29. Hoffmann, R. J Chem Phys. 1964; 40:2445, 2474, 2480.
30. Hoffmann R. Tetrahedron 1966; 22:521.
31. Koopmans T. Physica 1934; 1:104.
32. Glaeske H-J, Reinhold J, Volkmer P. Quantenchemie Band 5. Berlin: VEB Deutscher Verlag der Wissenschaften, 1987.
33. Fletcher R. Practical Methods of Optimization. 2nd ed. Chichester: John Wiley & Sons, 1987.
34. Pariser, R, Parr, R. J Chem Phys. 1953; 21: 466, 767.
35. Pople JA. Trans Faraday Soc 1953; 49:1375.
36. Podeszwa R, Kucharski SA, Stolarczyk LZ. J Chem Phys 2002; 116:480.
37. Pople JA, Santry DP, Segal GA. J Chem Phys 1965; 43:S129.
38. Pople JA, Segal GA. J Chem Phys 1965; 43:S136.
39. Pople JA, Segal GA. J Chem Phys 1966; 44:3289.
40. Jug K. Int J Quantum Chem 1969; 3S:241.
41. Murrell JM, Harget AJ. Semi-Empirical Self-Consistent-Field Molecular Orbital Theory of Molecules. London: Wiley, 1972.
42. Coffey P. Int J Quantum Chem 1974; 8:263.

43. Deleted in Proof.
44. Pople JA, Beveridge DL, Dobosh PA. J Chem Phys 1967; 47:2026.
45. Bingham, RC, Dewar, MJS, Lo, DII. J Am Chem Soc. 1975; 97:1285, 1294, 1302, 1307.
46. Baird NC, Dewar MJS. J Chem Phys 1969; 50:1262.
47. Dewar MJS, Haselbach E. J Am Chem Soc 1970; 92:590.
48. Harris FE. J Chem Phys 1969; 51:4770.
49. Guseinov II, Mamedov BA. Int J Quantum Chem 2001; 81:117.
50. Ohno K. Theor Chim Acta 1964; 2:219.
51. Klopman G. J Am Chem Soc 1964; 86:4550.
52. Oleari L, DiSipio L, DeMichelis G. Mol Phys 1966; 10:97.
53. Deleted in Proof.
54. Zerner MC, Ridley J. Theor Chim Acta 1973; 32:111.
55. Bacon AD, Zerner MC. Theor Chim Acta 1979; 53:21.
56. Longo RL. Int J Quantum Chem 1999; 75:585.
57. da Motta JD, Zener MC. Int J Quantum Chem 2001; 81:187.
58. Nanda DN, Jug K. Theor Chim Acta 1980; 57:95.
59. Jug K, Iffert R, Schulz J. Int J Quantum Chem 1987; 32:265.
60. Li J, Correa de Mello P, Jug K. J Comput Chem 1992; 13:85.
61. Brown RD, Roby KR. Theor Chim Acta 1970; 16:175.
62. Zerner MC. Mol Phys 1972; 23:963.
63. Ahlswede B, Jug K. J Comput Chem. 1999; 20:563, 572.
64. Jug K, Geudtner G, Homann T. J Comput Chem 2000; 21:974.
65. Bredow T, Geudtner G, Jug K. J Comput Chem 2001; 22:861.
66. Golebiewski, A, Nalewajski, R, Witko, M. Acta Phys Pol A. 1977; 51:617, 629.
67. Jug K. Theor Chim Acta 1980; 54:263.
68. Dewar, MJS, Thiel, W. J Am Chem Soc. 1977; 99:4899, 4907.
69. Thiel W, Voityuk AA. Theor Chim Acta 1992; 81:391.
70. Thiel W, Voityuk AA. J Phys Chem 1996; 100:616.
71. Kolb M, Thiel W. J Comput Chem 1993; 14:775.
72. Weber W, Thiel W. Theor Chem Acc 2000; 103:495.
73. Dewar MJS, Zoebisch EG, Healy EF, Stewart JJP. J Am Chem Soc 1985; 107:3902.
74. Stewart JJP. J Comput Chem 1989; 10:209.
75. Voityuk AA, Rösch N. J Phys Chem A 2000; 104:4089.
76. Ignatov SK, Razuvaev AG, Kokorev VN, Alexandrov YA. J Phys Chem 1996; 100:6354.
77. Bosque R, Maseras F. J Comput Chem 2000; 21:562.
78. Gaussian 98, Frisch MJ, Trucks GW, Schlegel HB, Scuseria GE, Robb MA, Cheeseman JR, Zakrzewski VG, Montgomery JA, Stratmann RE Jr, JC Burant, Dapprich S, Millam JM, Daniels AD, Kudin KN, Strain MC, Farkas O, Tomasi J, Barone V, Cossi M, Cammi R, Mennucci B, Pomelli C, Adamo C, Clifford S, Ochterski J, Petersson GA, Ayala PY, Cui Q, Morokuma K, Malick DK, Rabuck AD, Raghavachari K, Foresman JB, Cioslowski J, Ortiz JV, Stefanov BB, Liu G, Liashenko A, Piskorz P, Komaromi I, Gomperts R, Martin RL, Fox DJ, Keith T, Al-Laham MA, Peng CY, Nanayakkara A, Gonzalez C, Challacombe M, Gill PMW, Johnson B, Chen W, Wong MW, Andres JL, Gonzalez C, Head-Gordon M, Replogle ES, Pople JA. Pittsburgh, PA: Gaussian, Inc, 1998.
79. CERIUS(2). J Mol Graph Model 1997; 15:63.
80. SPARTAN. http://www.wavefun.com
81. Stewart JJP. J Comput-Aided Mol Des 1990; 4:1.
82. AMPAC with Graphic User Interface, Version 655 Semichem; http://www. semichem.-com, Shawnee Mission, KS.
83. Dewar MJS, Jie C, Yu G. Tetrahedron 1993; 49:5003.
84. Voityuk AA, Zerner MC, Rösch N. J Phys Chem A 1999; 103:4553.

85. Deleted in Proof.
86. Goldberg DE. Genetic Algorithms in Search, Optimization, and Machine Learning. Reading, Mass: Addison-Wesley, 1985.
87. Rossi I, Truhlar DG. Chem Phys Lett 1995; 233:231.
88. Cundari TR, Deng J, Fu W. Int J Quantum Chem 2000; 77:421.
89. Jug K, Krack M. Int J Quantum Chem 1992; 44:517.
90. Clementi E, Raimondi DL. J Chem Phys 1963; 38:2686.
91. Ford GP. J Mol Struct, Theochem 1997; 401:253.
92. Ohrn NY, Sabin JR, Zerner MC, eds. Proceedings of the International Symposium on the Application of Fundamental Theory to Problems of Biology and Pharmacology, Held at Ponce de Leon Resort, St Augustine, Florida, February 27–March 5, 1999, Int J Quantum Chem 1999; 75(6).
93. Gogonea V, Suarez D, van der Vaart A. Curr Opin Struct Biol 2001; 11:217.
94. Stavrev KK, Zerner MC. Int J Quantum Chem 1998; 70:1159.
95. Cory MG, Zerner MC, Xu XC, et al. J Phys Chem B 1998; 102:7640.
96. Coutinho K, Canuto S, Zerner MC. J Chem Phys 2000; 112:9874.
97. Zhu QQ, LeBreton PR. J Am Chem Soc 2000; 122:12824.
98. Ren AM, Feng JK, Sun XY, Li W, Tian WQ, Sun CC, Zheng XH, Zerner MC. Int J Quantum Chem 2000; 78:422.
99. de Alencastro RB, da Motta Neto JD. Int J Quantum Chem 2001; 85:529.
100. Jug K, Bredow T. Ragué Schleyer Pv, Allinger NL, Clark T, Gasteiger J, Kollman PA, Schaefer HF III., Schreiner PR, eds. Encyclopedia of Computational Chemistry. Vol. 4. 1998. New York: Wiley, 1998:2599.
101. Raabe G, Wang YK, Fleischhauer J. Z Naturforsch A 2000; 55:687.
102. Bredow T, Geudtner G, Jug K. J Chem Phys 1996; 105:6395.
103. Bredow T, Geudtner G, Jug K. J Comput Chem 2001; 22:89.
104. Bredow T. Surf Sci 1998; 401:82.
105. Ahlswede B, Homann T, Jug K. Surf Sci 2000; 445:49.
106. Tikhomirov VA, Jug K. J Phys Chem B 2000; 104:7619.
107. Jug K, Geudtner G. J Mol Catal A 1997; 119:143.
108. Jug K, Wichmann D. J Comput Chem 2000; 21:1549.
109. Gerson AR, Bredow T. Phys Chem Chem Phys 1999; 1:4889.
110. Geudtner G, Jug K, Köster AM. Surf Sci 2000; 467:98.
111. Steveson M, Bredow T, Gerson AR. Phys Chem Chem Phys 2002; 4:358.
112. Bakowies D, Buhl M, Thiel W. J Am Chem Soc 1995; 117:10113.
113. Rutherford TJ, Wilkie J, Vu CQ, Schnackerz KD, Jacobson MK, Gani D. Nucleosides Nucleotides Nucleic Acids 2001; 20:1485.
114. Sapse AM, Capiaux GM, Bertino JR. Molecular Medicine 2001; 7:200.
115. Besson M, Mihalek CL. Mutat Res, Fundam Mol Mech Mutagen 2001; 473:211.
116. Möhle K, Hofmann H-J, Thiel W. J Comput Chem 2001; 22:509.
117. Ren L, Martin CH, Wise KJ, Gillespie NB, Luecke H, Lanyi JK, Spudich JL, Birge RR. Biochem 2001; 40:13906.
118. Daniels AD, Millam JM, Scuseria GE. J Chem Phys 1997; 107:425.
119. Vaz RJ. Quant Struct-Act Relatsh 1997; 16:303.
120. Sulea T, Kurunczi L, Oprea TI, Simon Z. J Comput-Aided Mol Des 1998; 12:133.
121. Shankland N, Florence AJ, Cox PJ, Wilson CC, Shanklan K. Int J Pharm 1998; 165:107.
122. De Almeida WB, Dos Santos HF, Zerner MC. J Pharm Sci 1998; 87:1101.
123. Dos Santos HF, De Almeida WB, Zerner MC. J Pharm Sci 1998; 87:190.
124. Ovaska M, Yliniemelä A. J Comput-Aided Mol Des 1998; 12:301.
125. Daniels AD, Scuseria GE, Farkas Ö, Schlegel HB. Int J Quantum Chem 2000; 77:82.
126. Radl S, Hafner W, Budesinsky M, Hejnova L, Krejci I. Arch Pharm 2000; 333:167.
127. Radl S, Herzky P, Proska J, Hejnova L, Krejci I. Arch Pharm 2000; 333:107.

128. Saito M, Atsumi T, Satoh K, Ishihara M, Iwakura I, Sakagami H, Yokoe I, Fujisawa S. In Vitro Mol Toxicol 2001; 14:53.

129. Freeman BA, Wilson RE, Binder RG, Haddon WF. Mutat Res 2001; 490:89.

130. Holder AJ, Dennington RD II. J Mol Struct Theochem 1997; 401:207.

131. Slanina Z, Zhao X, Osawa E. Mol Mater 2000; 13:13.

132. Scherlis DA, Cymeryng CB, Estrin DA. Inorg Chem 2000; 39:2352.

133. Goller AH, Clark T. J Mol Struct (Theochem) 2001; 541:263.

134. Dannenberg JJ. J Mol Struct (Theochem) 1997; 401:279.

135. Palafox MA, Melendez FJ. J Mol Struct (Theochem) 1999; 459:239.

136. Swaan PW, Tukker JJ. J Pharm Sci 1997; 86:596.

137. Huche M, Legendre JJ. Chemometr Intell Lab Syst 1998; 41:43.

138. Sauerberg P, Olesen PH, Sheardown MJ, Rimvall K, Thogersen H, Shannon HE, Sawyer BD, Ward JS, Bymaster FP, DeLapp NW, Calligaro DO, Swedberg MDB. J Med Chem 1998; 41:109.

139. Ghafourian T, Dearden JC. J Pharm Pharmacol 2000; 52:603.

140. Yourtee D, Holder AJ, Smith R, Morrill JA, Kostoryz E, Brockmann W, Glaros A, Chappelow C, Eick D. J Biomater Sci Polym Edn 2001; 12:89.

141. Wu JH, Reynolds CA. J Comput-Aided Mol Des 2000; 14:307.

142. Makhija MT, Kulkarni VM. J Chem Inf Comput Sci 2001; 41:1569.

143. ISI Web of Science.

144. Boyd DB. J Mol Struct (Theochem) 1997; 401:219.

145. Stewart JJP, Császár P, Pulay P. J Comput Chem 1982; 3:227.

146. Stewart JJP. Int J Quantum Chem 1996; 58:133.

147. Thiel W, Green DG. Clementi E, Corongiu G, eds. Methods and Techniques in Computational Chemistry, METECC95 STEF Cagliari, Italy, 1995:141.

148. Früchtl HA, Nobesa RH, Bliznyuk A. J Mol Struct, (Theochem) 2000; 506:87.

149. Evarestov RA, Lovchikov VA. Phys Status Solidi B 1977; 79:743.

150. Smith PV, Szymanski JE, Matthews JAD. J Phys C 1985; 18:3157.

151. Stefanovich EV, Shidlovskaya EK, Shluger AL, Zakharov MK. Phys Status Solidi B 1990; 160:529.

152. Stewart JIP. J Mol Struct (Theochem) 2000; 556:59.

153. Danovich D. J Mol Struct (Theochem) 1997; 401:235.

154. Martin B, Gedeck P, Clark T. Int J Quantum Chem 2000; 77:473.

155. Reynolds CH. J Mol Struct (Theochem) 1997; 401:267.

156. Antes I, Thiel W. J Phys Chem A 1999; 103:9290.

157. Berweger CD, Thiel W, van Gunsteren WF. Proteins 2000; 41:299.

158. Vreven T, Morokuma K. J Comput Chem 2000; 21:1419.

159. Ohno K, Kamiya N, Asakawa N, Inoue Y, Sakurai M. Chem Phys Lett 2001; 341:387.

3

Wave Function–Based Quantum Chemistry

TRYGVE HELGAKER

University of Oslo, Oslo, Norway

POUL JØRGENSEN and JEPPE OLSEN

University of Aarhus, Aarhus, Denmark

WIM KLOPPER

University of Karlsruhe (TH), Karlsruhe, Germany

1. INTRODUCTION

The field of molecular electronic structure theory has developed rapidly during the last decades, allowing chemists to study theoretically systems of increasing size and complexity, often with an accuracy that rivals or even surpasses that of experimental measurements [1–4]. This situation has come about partly as a result of new developments in computational techniques, partly as a result of spectacular advances in computer technology. Consequently, practicing chemists now have at their disposal a wide range of powerful techniques of varying cost and accuracy, all of which may be applied to solve problems at the microscopic and molecular levels.

In molecular electronic structure theory, we study the properties of molecular systems as functions of the nuclear geometrical configuration, generating hypersurfaces of the potential energy and other molecular properties. Information about equilibrium structures, transition states, charge distributions, and electrostatic potentials provides insight into molecular structure and reactivity. Calculations of excitation energies and various constants of rotational, vibrational, electronic, and magnetic spectroscopies help unravel experimental observations and characterize new molecular species.

Electronic structure calculations may be carried out at many levels, differing in cost, accuracy, and reliability. At the simplest level, *molecular mechanics* (this volume, Chapter 1) may be used to model a wide range of systems at low cost, relying on large sets of adjustable parameters. Next, at the *semiempirical level* (this volume, Chapter 2), the techniques of quantum mechanics are used, but the computational cost is reduced by extensive use of empirical parameters. Finally, at the most complex level, a rigorous quantum mechanical treatment of electronic structure is provided by *nonempirical, wave function-based quantum chemical methods* [1] and by *density functional theory* (DFT) (this volume, Chapter 4). Although not treated here, other less standard techniques such as *quantum Monte Carlo* (QMC) have also been developed for the electronic structure problem (for these, we refer to the specialist literature, Refs. 5–7).

In the present chapter, we discuss wave function–based quantum chemical methods for the rigorous calculation of molecular electronic structure. In short, we are concerned with obtaining approximate solutions to the (nonrelativistic) time-dependent electronic Schrödinger equation [8]:

$$\hbar \frac{\partial \psi}{\partial t} = \hat{H}\psi, \tag{1}$$

where $\hat{H}$ is the Hamiltonian of the molecular electronic system and ψ is the wave function, which is a time-dependent, antisymmetrical function of the electronic coordinates. In the absence of external fields, the Hamiltonian becomes time-independent, and we then obtain the stationary states Ψ_n by solving the time-independent electronic Schrödinger equation:

$$\hat{H}\Psi_n = E_n\Psi_n, \tag{2}$$

where, in the Born–Oppenheimer approximation, the eigenvalue E_n is the total energy of the electronic system for a given nuclear configuration. The field-free molecular electronic Hamiltonian takes the form:

$$\hat{H} = -\frac{\hbar^2}{2m_e}\sum_i \nabla_i^2 + \frac{e^2}{4\pi\varepsilon_0}\left(\sum_{i>j}\frac{1}{|\vec{r}_i - \vec{r}_j|} - \sum_{iI}\frac{Z_I}{|\vec{r}_i - \vec{R}_I|} + \sum_{I>J}\frac{Z_I Z_J}{|\vec{R}_I - \vec{R}_J|}\right), \tag{3}$$

where $\vec{r}_i$ are the coordinates of electron i with mass m_e and charge $-e$, and $\vec{R}_I$ are the coordinates of nucleus I of atomic number Z_I. In atomic units, the prefactor $-\hbar^2/2m_e$ in the Hamiltonian becomes $-1/2$, whereas $e^2/4\pi\varepsilon_0$ becomes 1. In the following, atomic units are used except as noted.

Our task is to find approximate solutions to the time-independent Schrödinger equation (Eq. (2)) subject to the Pauli antisymmetry constraints of many-electron wave functions. Once such an approximate solution has been obtained, we may extract from it information about the electronic system and go on to compute different molecular properties related to experimental observations. Usually, we must explore a range of nuclear configurations in our calculations to determine critical points of the potential energy surface, or to include the effects of vibrational and rotational motions on the calculated properties. For properties related to time-dependent perturbations (e.g., all interactions with radiation), we must determine the time development of the wave function. In such cases, the solutions to the time-independent Schrödinger equation (Eq. (2)) are used as zero-order approximate solutions in a perturbational treatment of the time-dependent Schrödinger equation (Eq. (1)).

For systems containing heavy atoms, the Schrödinger equation becomes inadequate and the calculations must instead be based directly or indirectly on Dirac's relativistic equation [9], although in many cases, the relativistic corrections may be sufficiently well accounted for by effective potentials [10] or by low-order perturbation theory [11].

Before beginning our discussion of wave function-based electronic structure theory, we note that an alternative, rigorous approach to electronic structure is provided by DFT (this volume, chapter by Ayers and Yang). DFT is based on the premise that all information about the electronic system can be extracted from the electron density, rather than from the electronic wave function. The attraction of DFT is that the electron density is a much simpler entity than the wave function, depending on just three spatial coordinates rather than on the $4n$ spatial and spin coordinates of n electrons. However, a difficulty of DFT is that no accurate, nonempirical method has yet been devised to extract the necessary information from the electron density. Current DFT calculations are therefore, to a large extent, based on semiempirical functionals [12], in which a set of parameters is fitted to experimental data. Nevertheless, the fitted parameters are universal in the sense that they are not atom-dependent or molecule-dependent. Also, the accuracy achieved in this manner is often high, surpassed only by the most elaborate wave function methods [13].

Wave function methods, by contrast, make no use of adjustable parameters, are more generally applicable (to excited states, different spin states, etc.), and are often capable of considerably higher accuracy. Most important, wave function methods form hierarchies of increasing sophistication, allowing the user to approach the exact solution in a systematic manner, restricted only by computational resources [4,14–17]. In this sense, it constitutes the most satisfactory and useful theory that has been developed for the study of molecular electronic structure.

2. ORBITALS AND SLATER DETERMINANTS

We begin our discussion of wave function–based quantum chemistry by introducing the concepts of n-electron and one-electron expansions. First, in Sec. 2.1, we consider the expansion of the approximate wave function in Slater determinants of spin orbitals. Next, we introduce in Sec. 2.2 the one-electron Gaussian functions (basis functions) in terms of which the molecular spin orbitals are usually constructed; the standard basis sets of Gaussian functions are finally briefly reviewed in Sec. 2.3.

2.1. Slater Determinants and *n*-Electron Expansions

The construction of approximate electronic wave functions is a difficult many-body problem. The source of the difficulties is the presence of the two-body electron–electron repulsion term in the Hamiltonian equation [Eq. (3)]. In the absence of this term, there would be no interactions among the electrons and it would be sufficient to consider one electron at the time, independently of the others. Indeed, in this case, the many-electron Hamiltonian equation [Eq. (3)] becomes, apart from a constant nuclear–nuclear repulsion term, a sum of independent one-electron Hamiltonians:

$$\hat{H}^{\text{nonint}} = \sum_i \hat{h}_i + \sum_{I > J} \frac{Z_I Z_J}{|\vec{R}_I - \vec{R}_J|} \tag{4}$$

$$\hat{h}_i = -\frac{1}{2}\nabla_i^2 - \sum_I \frac{Z_I}{|\vec{r}_i - \vec{R}_I|} \tag{5}$$

whose eigenfunctions:

$$\hat{h}_i \psi_k(\vec{r}_i) = \varepsilon_k \psi_k(\vec{r}_i) \tag{6}$$

are called *orbitals*. To account for the two observed spin states of the electron, we introduce *spin orbitals* as products of such orbitals with one of two possible spin states:

$$\varphi_{k\alpha}(\mathbf{x}) \equiv \varphi_{k\alpha}(\vec{r}, s) = \psi_k(\vec{r})\alpha(s), \tag{7}$$

$$\varphi_{k\beta}(\mathbf{x}) \equiv \varphi_{k\beta}(\vec{r}, s) = \psi_k(\vec{r})\beta(s), \tag{8}$$

where the spin coordinate s takes on the values $1/2$ and $-1/2$. The spin functions are given by:

$$\alpha\left(\frac{1}{2}\right) = 1, \quad \alpha\left(-\frac{1}{2}\right) = 0, \tag{9}$$

$$\beta\left(\frac{1}{2}\right) = 0, \quad \beta\left(-\frac{1}{2}\right) = 1. \tag{10}$$

For brevity of notation, we shall in the following include the spin part of the spin orbitals in the spin orbital label:

$$\varphi_j(\mathbf{x}) = \varphi_{kj}(\vec{r})\sigma_j(s), \tag{11}$$

where $\sigma_j(s)$ is either $\alpha(s)$ or $\beta(s)$, depending on the spin orbital label j. For a given set of orbitals, there are twice as many spin orbitals.

A many-electron wave function may now be written as a product of these spin orbitals, properly antisymmetrized to comply with the Pauli principle. For two electrons, for example, we obtain:

$$\Phi_{k,l}(\mathbf{x}_1, \mathbf{x}_2) = \frac{1}{\sqrt{2}}[\varphi_k(\mathbf{x}_1)\varphi_l(\mathbf{x}_2) - \varphi_k(\mathbf{x}_2)\varphi_l(\mathbf{x}_1)], \tag{12}$$

which represents a state of noninteracting electrons. Note that this state vanishes when the two spin orbitals are identical or when $\mathbf{x}_1 = \mathbf{x}_2$. More generally, for a system of n independent electrons, the wave function may be written as a *Slater determinant*; that is, as the determinant of a square matrix whose elements are spin orbitals, with the electron labels as row indices and spin orbital labels as column indices:

$$\Phi_\mu(\mathbf{x}_1, \mathbf{x}_2, \ldots, \mathbf{x}_n) = \frac{1}{\sqrt{n!}} \begin{vmatrix} \varphi_{\mu_1}(\mathbf{x}_1) & \varphi_{\mu_2}(\mathbf{x}_1) & \cdots & \varphi_{\mu_n}(\mathbf{x}_1) \\ \varphi_{\mu_1}(\mathbf{x}_2) & \varphi_{\mu_2}(\mathbf{x}_2) & \cdots & \varphi_{\mu_n}(\mathbf{x}_2) \\ \vdots & \vdots & \ddots & \vdots \\ \varphi_{\mu_1}(\mathbf{x}_n) & \varphi_{\mu_2}(\mathbf{x}_n) & \cdots & \varphi_{\mu_n}(\mathbf{x}_n) \end{vmatrix}, \tag{13}$$

sometimes abbreviated as $\Phi_\mu = |\varphi_{\mu_1}, \varphi_{\mu_2}, \ldots, \varphi_{\mu_n}|$. In passing, we note that, in the following, we shall often use the word "configuration" for "Slater determinant." The term "configuration" is short for "configuration state function," which is a spin-symmetrized and space-symmetrized linear combination of Slater determinants.

A Slater determinant gives an exact representation of the n-electron wave function only in the (fictitious) limit of no interactions among the electrons (i.e., for a system of electrons described by the Hamiltonian in Eq. (4)). For a real system of interacting electrons, described by the Hamiltonian in Eq. (3), the Slater determinant can only serve as an approximate wave function. Nevertheless, in this case, we may still represent the true n-electron wave function Ψ exactly as a linear combination of Slater determinants (Eq. (13)):

$$\Psi(\mathbf{x}_1, \mathbf{x}_2, \ldots, \mathbf{x}_n) = \sum_{\mu=1}^{N_{\text{det}}} c_\mu \Phi_\mu(\mathbf{x}_1, \mathbf{x}_2, \ldots, \mathbf{x}_n), \tag{14}$$

where N_{det} is the number of unique n-electron Slater determinants that may be constructed from N spin orbitals:

$$N_{\text{det}} = \binom{N}{n}. \tag{15}$$

This representation of the wave function is exact provided that a complete set of exact eigenfunctions (spin orbitals) of $\hat{h}_i$ in Eq. (5) is used to construct the Slater determinants. In practical calculations, we do not have access to a complete set of orbitals. Moreover, even if we had these orbitals, we would not be able to construct the full set of Slater determinants that they give rise to, noting that the number of determinants (Eq. (15)) increases very steeply with the number of orbitals.

In practice, therefore, we must work with finite-dimensional orbital spaces, properly optimized so as to yield the best representation of the n-electron wave function. In addition, we must, for a given finite orbital basis, find a way to determine the coefficients in the expansion (Eq. (14)). The "best" wave function (i.e., the wave function with the lowest energy) is obtained by optimizing all expansion coefficients c_μ variationally, as done in the *full configuration interaction* (FCI) method [18–20]:

$$E_{\text{FCI}} = \min_{\{c_\mu\}} \frac{\langle \text{FCI}|\hat{H}|\text{FCI} \rangle}{\langle \text{FCI}|\text{FCI} \rangle} \geq E_{\text{exact}}. \tag{16}$$

However, this approach is too expensive because, even for very small molecules such as HF in moderately small orbital spaces, N_{det} becomes very large—several billions or more. Thus, we must instead be content to work with *approximate FCI wave functions*. Fortunately, several useful hierarchies of approximations to the FCI solution have been developed, enabling us to approach the FCI wave function closely, even for rather large systems.

From our discussion, it should be clear that the quality of our approximate solution to the Schrödinger equation depends not only on the particular n-electron model by which we choose to approximate the FCI solution, but also on the set of orbitals from which the FCI solution or its approximations are constructed. Before we consider in Secs. 3 and 4 the various techniques that have been developed to approximate the FCI wave function (Eq. (14)), we shall in the remainder of this section focus our attention on the spin orbitals.

2.2. Atomic Orbitals

In systems of high symmetry such as atoms, it is possible to represent the orbitals (i.e., the spatial part of the spin orbitals) numerically on a spatial grid. For polyatomic systems, however, it is more common to represent the spin orbitals as linear expansions of a set of N simple, analytical one-electron basis functions $\chi_\kappa(\mathbf{x})$, mostly centered on the atoms in the system. These linear expansions may then be written as:

$$\varphi_k(\mathbf{x}) = \sum_\kappa C_{\kappa k} \chi_\kappa(\mathbf{x}), \tag{17}$$

where, as before, the spin part has been included in the basis functions:

$$\chi_\kappa(\mathbf{x}) = \chi_{\lambda\kappa}(\vec{r})\sigma_\kappa(s). \tag{18}$$

In the following, the functions $\chi_\kappa(\mathbf{x})$ with $1 \leq \kappa \leq 2N$ are understood to be spin-dependent, whereas the functions $\chi_\lambda(\vec{r})$ with $1 \leq \kappa \leq N$ are pure spatial functions.

The nonorthogonal basis functions $\chi_\lambda(\vec{r})$ are referred to as *atomic orbitals* (AOs) and are often taken to be *Cartesian Gaussian-type orbitals* (GTOs) of the (unnormalized) form:

$$G_{ijk}\left(\vec{r}, a, \vec{A}\right) = x_A^i y_A^j z_A^k \exp(-ar_A^2), \tag{19}$$

where $a > 0$ is the orbital exponent and where:

$$r_A = |\vec{r} - \vec{A}| = \sqrt{x_A^2 + y_A^2 + z_A^2}. \tag{20}$$

The nonnegative integers $i, j,$ and k in Eq. (19) are related to the "angular momentum" of an electron in this AO as $l = i + j + k$. Gaussian functions are nearly always added in full shells (i.e., for a given orbital exponent a and a given l, all components $i + j + k = l$ are included in the basis simultaneously, thereby treating all Cartesian directions equivalently.

In the Cartesian scheme (Eq. (19)), there are $(l+1)(l+2)/2$ components of a given l, whereas the number of independent spherical harmonics is only $2l+1$. Usually, therefore, the Cartesian GTOs are not used individually but instead are combined linearly to give real solid harmonics (see Ref. 1). In addition, for a more compact and accurate description of the electronic structure, the GTOs (Eq. (19)) are not used individually as *primitive GTOs* but mostly as *contracted GTOs* (i.e., as fixed, linear combinations of primitive GTOs with different exponents a).

Although most molecular calculations are carried out using GTOs, in some cases (in particular for atoms and diatoms), *Slater-type orbitals* (STOs) are used instead. The STOs have a different radial form than the GTOs, proportional to $\exp(-\zeta r_A)$ rather than to $\exp(-ar_A^2)$. The GTOs are used in preference to the STOs because the evaluation of many-center integrals is much easier for GTOs than for STOs.

2.3. Gaussian Basis Sets

Over the years, a variety of standard *Gaussian basis sets* have been developed for virtually all atoms of the periodic table [21,22]. For qualitative or exploratory work, *minimal basis sets*, which contain only one shell of AOs for each (fully or partially) occupied shell in the parent atom, may be used. A popular minimal basis set is the STO-3G basis, where each AO is a linear combination of three primitive GTOs.

For semiquantitative work, *double-zeta* or *triple-zeta basis sets* (in which there are two or three shells of AOs for each fully or partially occupied atomic shell) are needed, at least for the valence shell. For first-row atoms, for example, the popular 6-31G basis has a minimal representation of the $1s$ core orbital and a double-zeta representation of the valence orbitals. Moreover, each AO is represented by a fixed linear combination of primitive GTOs: the $1s$ core orbital contains six primitive AOs, and each $2s$ and $2p$ valence orbital is represented by two contracted orbitals, containing three and one primitive functions.

In addition, *polarization functions* are needed. Such functions, which are of different symmetry than the AOs in the parent atom, are needed to describe the polarization of the atomic charge in the molecular environment. In the 6-31G* basis, for example, the 6-31G basis is augmented with a set of d-type polarization functions on the first-row atoms; in the 6-31G** basis, polarization functions (p-type) are also added to the hydrogens.

The basis sets described above are small and intended for qualitative or semi-quantitative, rather than quantitative, work. They are used mostly for simple wave functions consisting of one or a few Slater determinants such as the Hartree–Fock wave function, as discussed in Sec. 3. For the more advanced wave functions discussed in Sec. 4, it has been proven important to introduce hierarchies of basis sets. New AOs are introduced in a systematic manner, generating not only more accurate Hartree–Fock orbitals but also a suitable orbital space for including more and more Slater determinants in the n-electron expansion. In terms of these basis sets, determinant expansions (Eq. (14)) that systematically approach the exact wave function can be constructed . The *atomic natural orbital* (ANO) basis sets of Almlöf and Taylor [23] were among the first examples of such systematic sequences of basis sets. The ANO sets have later been modified and extended by Widmark et al. [24].

The *correlation-consistent basis sets* of Dunning [25], Kendall et al. [26], and Woon and Dunning [27] provide a particularly popular hierarchy of basis sets, which has been extensively used to extrapolate toward the FCI limit of a complete AO basis. For calculations correlating only the valence electrons, these basis sets are denoted cc-pVXZ, where $2 \leq X \leq 6$ is the *cardinal number* [25]. For first-row atoms, the smallest double-zeta basis cc-pVDZ with $X = 2$ contains three s-type contracted GTOs, two sets of p-type contracted GTOs, and one set of d-type GTOs (in total, 14 contracted AOs). At the next level, the triple-zeta cc-pVTZ basis with $X = 3$ contains 30 AOs, followed by the quadruple-zeta cc-pVQZ basis with 55 AOs, and the cc-pV5Z basis with 91 AOs. For first-row atoms, the number of AOs in the cc-pVXZ basis is given by $1/3(X+1)(X+3/2)(X+2)$. The largest correlation-consistent basis, cc-pV6Z, represents a $7s6p5d4f3g2h1i$ basis of 140 contracted GTOs. We note, however, that the cc-pVXZ basis sets are constructed for correlating only the valence electrons. For correlating all electrons (core as well as valence electrons), the correlation-consistent core valence basis sets cc-pCVXZ with $2 \leq X \leq 5$ are used [27].

3. SINGLE-CONFIGURATIONAL AND MULTICONFIGURATIONAL HARTREE–FOCK THEORY

In the present section, we discuss how the exact wave function may be approximately described by a few important configurations, constructed from a set of variationally optimized orbitals. These wave function models are often used on their own, for a crude but qualitatively correct description of the electronic system. In addition, they are important as starting points for the more advanced, quantitatively correct treatments discussed in Sec. 4.

3.1. The Hartree–Fock Model

The *Hartree–Fock model* is the simplest, most basic model in ab initio electronic structure theory [28]. In this model, the wave function is approximated by a single Slater determinant constructed from a set of orthonormal spin orbitals:

$$|\text{HF}\rangle = |\varphi_1, \varphi_2, \dots, \varphi_n|. \tag{21}$$

The spin orbitals are determined by invoking the variation principle (8) (i.e., by minimizing the energy with respect to variations in the spin orbitals):

$$E_{\text{HF}} = \min_{\{\varphi_i\}} \frac{\left\langle \text{HF}|\hat{H}|\text{HF} \right\rangle}{\langle \text{HF}|\text{HF}\rangle}. \tag{22}$$

The Hartree–Fock energy therefore constitutes a rigorous upper bound to the exact energy, $E_{\text{HF}} \geq E_{\text{exact}}$. By expanding each spin orbital in AOs according to Eq. (17), the minimization is achieved by varying the AO expansion coefficients.

In a variational sense, the Hartree–Fock model represents the best one-determinant approximation to the exact electronic state. It typically recovers 99% or more of the total electronic energy and it yields, for most molecular properties, results within 5%–10% of the exact values. For many purposes, therefore, the Hartree–Fock model represents an adequate model by itself. Just as important, it constitutes a natural starting point for the more elaborate treatments of electronic structure discussed in Sec. 4.

The optimization of the Hartree–Fock spin orbitals in Eq. (21) is a nonlinear minimization problem. By recasting Eq. (22) as a generalized eigenvalue problem, the optimization may be accomplished by repeated solution of the pseudo-eigenvalue equations:

$$\mathbf{F}\mathbf{C}_i = \varepsilon_i \mathbf{S}\mathbf{C}_i, \qquad i = 1, 2, \dots, n, \tag{23}$$

whose eigenvectors $\mathbf{C}_i$ represent the *molecular orbitals* (MOs) and whose eigenvalues ε_i are the orbital energies [29]. Assuming real AOs, the elements of the *overlap matrix* $\mathbf{S}$ are given by:

$$S_{\kappa\lambda} = \int \chi_\kappa(\mathbf{x})\chi_\lambda(\mathbf{x})d\mathbf{x}, \tag{24}$$

and the elements of the *Fock matrix* $\mathbf{F}$ are calculated as:

$$F_{\kappa\lambda} = h_{\kappa\lambda} + \sum_{\mu\nu} P_{\mu\nu}[(\mu\nu|\kappa\lambda) - (\mu\lambda|\kappa\nu)]. \tag{25}$$

Here the one-electron and two-electron Hamiltonian integrals are given by:

$$h_{\kappa\lambda} = \int \chi_\kappa(\mathbf{x}_1)\hat{h}_1\chi_\lambda(\mathbf{x}_1)d\mathbf{x}_1, \tag{26}$$

$$(\mu\nu|\kappa\lambda) = \iint \chi_\mu(\mathbf{x}_1)\chi_\kappa(\mathbf{x}_2)\frac{1}{r_{12}}\chi_\nu(\mathbf{x}_1)\chi_\lambda(\mathbf{x}_2)d\mathbf{x}_1 d\mathbf{x}_2, \tag{27}$$

and the AO *density matrix* elements are given by:

$$P_{\mu\nu} = \sum_{i=1}^{n} C_{\mu i} C_{\nu i}. \tag{28}$$

In Eq. (26), $\hat{h}_1$ is the one-electron Hamiltonian equation (Eq. (5)).

The generalized eigenvalue problem (Eq. (23)) is a pseudo-eigenvalue problem in the sense that the Fock matrix equation (Eq. (25)) depends (through $P_{\mu\nu}$) on its own eigenvectors. The eigenvalue problem (Eq. (23)) must therefore be iterated until the orbitals that are generated by the diagonalization are the same as those used in the construction of the Fock matrix. A *self-consistent field* (SCF) solution has then been established, and the resulting Fock matrix constitutes the AO representation of an effective one-electron operator called the *Fock operator*:

$$\hat{F}_1 = -\frac{1}{2}\nabla_1^2 + \hat{V}_1^{SCF}. \tag{29}$$

Apart from the attractive nuclear potential, the effective one-electron potential

$$\hat{V}_1^{SCF} = -\sum_I \frac{Z_I}{|\vec{r}_1 - \vec{R}_I|} + \hat{J}_1 - \hat{K}_1, \tag{30}$$

contains a repulsive potential, where the *Coulomb operator* J_1 and the *exchange operator* $\hat{K}_1$ are defined as:

$$\hat{J}_1\varphi_i(\mathbf{x}_1) = \sum_j \varphi_i(\mathbf{x}_1) \int \frac{\varphi_j(\mathbf{x}_2)\varphi_j(\mathbf{x}_2)}{r_{12}} d\mathbf{x}_2, \tag{31}$$

$$\hat{K}_1\varphi_i(\mathbf{x}_1) = \sum_j \varphi_j(\mathbf{x}_1) \int \frac{\varphi_j(\mathbf{x}_2)\varphi_i(\mathbf{x}_2)}{r_{12}} d\mathbf{x}_2. \tag{32}$$

In the limit of a complete basis, the pseudo-eigenvalue problem (Eq. (23)) may be expressed in the form:

$$\hat{F}_1\varphi_i(\mathbf{x}_1) = \varepsilon_i\varphi_i(\mathbf{x}_1) \tag{33}$$

showing that the MOs are eigenfunctions of the Fock operator.

The structure of Eq. (33) is similar to that of Eq. (5), indicating that, in the Hartree–Fock model, the electrons experience an average potential as described by the Coulomb and exchange operators. In Kohn–Sham DFT (this volume, chapter by Ayers and Yang), the exchange operator $\hat{K}_1$ is omitted and exchange is instead accounted for via an additional contribution to the effective potential from the exchange–correlation functional; in hybrid DFT, some proportion of the Hartree–Fock exchange operator $\hat{K}_1$ is retained.

The eigenvalues of the Fock eigenvalue problem—the orbital energies—satisfy *Koopmans' theorem*, which states that the orbital energy ε_i is equal to minus the ionization potential (IP) associated with the removal of an electron from orbital φ_i in the Hartree–Fock state without modifying the remaining orbitals. The agreement with the observed IPs is crude but useful for qualitative discussions.

Hartree–Fock calculations carried out without restrictions on the spatial parts of the alpha and beta spin orbitals are referred to as *unrestricted Hartree–Fock* (UHF) calculations. Often, it is useful to impose the condition that the alpha and beta spin orbitals occur in pairs, with the same spatial parts. Such calculations are referred to as *restricted Hartree–Fock* (RHF) calculations. Unlike UHF wave functions, RHF wave functions are pure spin states. On the other hand, because of the variation principle, the UHF energy is always equal to, or lower than, the RHF energy; when the two energies differ, the RHF model is said to be *unstable* [30].

The difference between the RHF and UHF models is illustrated for water in Fig. 1, where, for a fixed HOH bond angle, the UHF and RHF potential energy curves are plotted as functions of the OH bond distance, with the FCI curve included for comparison. The RHF instability sets in at $2.64a_0$, beyond which the UHF curve lies below the RHF curve.

3.2. Hartree–Fock Methods for Large Systems: Linear Scaling Methods

Nowadays, the Hartree–Fock method can be applied to systems containing several hundred atoms. In this section, we briefly review those aspects of Hartree–Fock theory that are important for large systems [32].

As described in Sec. 3.1, each Hartree–Fock iteration involves the construction of the Fock matrix for a given density matrix, followed by the diagonalization of the Fock matrix to generate a set of improved spin orbitals and thus an improved density matrix. Formally, the construction of the Fock matrix requires a number of operations proportional to K^4, where K is the number of atoms (because the number of two-electron integrals scales as K^4). For large systems, however, this quartic scaling with K (i.e., with system size) can be reduced to linear by special techniques, as will now be discussed.

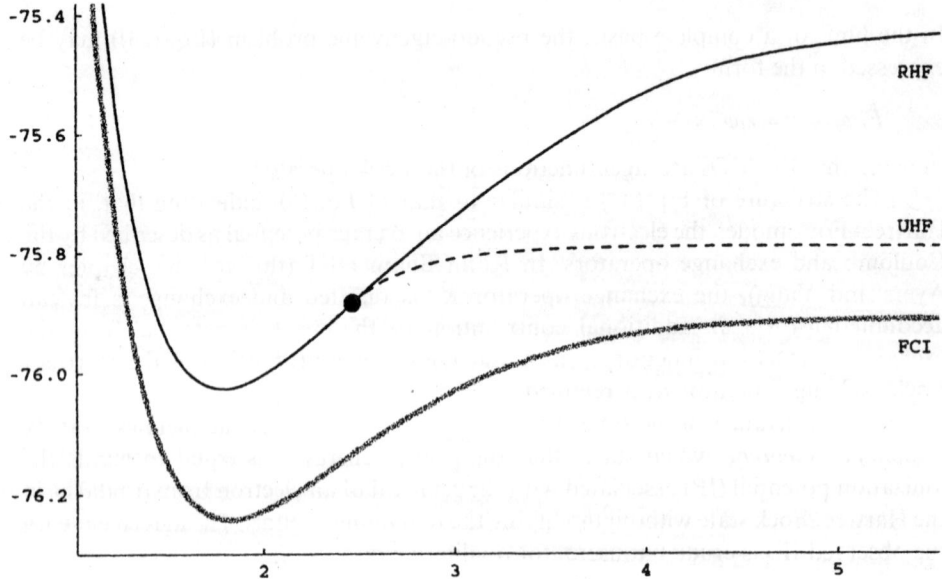

Figure 1 RHF and UHF dissociation of H_2O (atomic units).

A first reduction in cost is achieved by recognizing that the AOs are localized in space and that, for insulating electronic systems at least, the density matrix **P** is sparse. Therefore, many of the two-electron integrals that formally contribute to the Fock matrix need not be computed. In the construction of the Fock matrix, prescreening techniques are used to identify and calculate only those integrals that make a significant contribution (i.e., a contribution greater than some prescribed threshold) [28]. All other integrals are neglected, resulting in a dramatical reduction in computational cost for all but the smallest systems. Indeed, for large systems, the cost of this *direct Hartree–Fock method* scales only quadratically with system size.

By further rearranging the calculations, it is possible to reduce the scaling of the Fock matrix construction to linear [33]. This may be achieved by treating the classical, long-range (one-electron and two-electron) Coulomb interactions by special multipole methods, organized in such a manner that the total cost of the Fock matrix construction scales linearly with the size of the system. Because, for systems containing up to several hundred atoms, the Fock matrix construction is the time-critical step, such *fast multipole methods* (FMMs) have significantly extended the range of systems that can be treated by the Hartree–Fock method [34]. In passing, we note that all steps in the construction of the Fock matrix are ideally suited to modern parallel computer architectures.

Having reduced the cost of the Fock matrix construction to linear, another computational bottleneck arises for large systems—the diagonalization of the Fock matrix, whose cost scales cubically with system size. By developing schemes for directly optimizing the AO density matrix **P** in Eq. (28) without introducing MOs, linear scaling has been achieved also for this step [36,37]. Although promising, these experimental techniques cannot yet be applied in a routine manner.

3.3. Calculation of Molecular Properties

As discussed hitherto, the Hartree–Fock method allows for the calculation of the electronic energy at a given nuclear configuration. From the density matrix **P**, we obtain:

$$\rho(\mathbf{x}) = \sum_{\mu\nu} P_{\mu\nu} \chi_\mu(\mathbf{x}) \chi_\nu(\mathbf{x}) \tag{34}$$

from which we may extract the *electron density* $\rho(\vec{r})$ and the *spin density* $\rho_s(\vec{r})$:

$$\rho(\vec{r}) = \rho(\vec{r}, \tfrac{1}{2}) + \rho(\vec{r}, -\tfrac{1}{2}), \tag{35}$$

$$\rho_s(\vec{r}) = \rho\left(\vec{r}, \tfrac{1}{2}\right) - \rho\left(\vec{r}, -\tfrac{1}{2}\right), \tag{36}$$

as well as various one-electron properties such as dipole and quadrupole moments. Moreover, a *molecular electrostatic potential* (MEP) (this volume, chapter by Politzer and Murray) can be derived by computing the Coulomb interaction between a charged particle and the electronic charge given by $\rho(\vec{r})$. Furthermore, by splitting the summation over μ and ν in Eq. (34) into sums over atoms and their respective basis functions, the (spin) densities can be partitioned into atomic contributions known as *Mulliken charges*.

For a quantum chemical method to be useful for the general chemist, algorithms for calculating other properties must also be developed. For example, to determine the equilibrium structure, the change in the energy induced by a nuclear displacement must be known. The theoretical prediction of harmonic frequencies involves the second derivative of the electronic energy with respect to changes in the nuclear coordinates. Similarly, electrical and magnetic properties such as polarizabilities and magnetizabilities as well as NMR parameters may be calculated as second derivatives of the energy with respect to various time-independent perturbations. Efficient schemes for calculating first and higher derivatives of the Hartree–Fock energy have been developed, applicable to small and large systems [38,39].

The response to frequency-dependent external fields may be obtained from Hartree–Fock response theory, yielding dynamical polarizabilities and hyperpolarizabilities. The identification of excitation energies as the poles of the dynamical polarizability tensor may be invoked to calculate excitation energies as well as one-photon and two-photon transition moments from the time development of the ground state [40–42].

The performance of the Hartree–Fock model is illustrated in Table 1, where we have listed the electronic dissociation energy (D_e), the equilibrium bond distance (r_e), and the harmonic (ω_e) and fundamental (v) frequencies calculated at the Hartree–Fock/cc-pVXZ levels. Basis set convergence is in all cases rapid. Compared with the

Table 1 Calculations of the Electronic Dissociation Energy D_e (kJ/mol), the Equilibrium Geometry r_e (pm), and the Harmonic ω_e (cm^{-1}) and Fundamental v (cm^{-1}) Vibrational Frequencies of the N_2 Molecule

Method	Basis	D_e	r_e	ω_e	v
RHF	cc-pVDZ	469.3	107.73	2758.3	2735.7
	cc-pVTZ	503.7	106.71	2731.7	2710.3
	cc-pVQZ	509.7	106.56	2729.7	2708.1
	cc-pV5Z	510.6	106.54	2730.3	2708.5
CASSCF	cc-pVDZ	857.8	111.62	2354.3	2325.6
	cc-pVTZ	885.3	110.56	2339.4	2312.1
	cc-pVQZ	890.9	110.39	2339.5	2312.1
	cc-pV5Z	891.9	110.37	2340.4	2313.0
MP2	cc-pCVDZ	897.0	112.84	2175.8	2135.7
	cc-pCVTZ	962.8	111.01	2207.6	2169.9
	cc-pCVQZ	988.1	110.78	2218.1	2180.7
	cc-pCV5Z	998.8	110.70	2221.8	2184.4
CCSD	cc-pCVDZ	813.4	111.12	2411.8	2384.9
	cc-pCVTZ	873.7	109.35	2434.3	2408.4
	cc-pCVQZ	896.9	109.08	2446.8	2421.0
	cc-pCV5Z	905.6	108.99	2451.6	2425.7
CCSD(T)	cc-pCVDZ	843.8	111.74	2341.3	2312.4
	cc-pCVTZ	911.8	110.06	2354.7	2326.8
	cc-pCVQZ	936.3	109.81	2365.8	2338.0
	cc-pCV5Z	945.6	109.72	2370.1	2342.2
Experiment		956.3	109.77	2358.6	2329.9

In the correlated calculations, all electrons are correlated.

experiment, the dissociation energy is strongly underestimated, the bond distance is too short, and the vibrational frequencies are too high. This behavior is typical of the Hartree–Fock model, reflecting the inadequacy of the mean field description, which ignores the instantaneous interaction among the electrons.

3.4. Limitations of the Hartree–Fock Method

Although the Hartree–Fock model is applicable in many situations, providing a useful qualitative description of a wide variety of molecular systems and processes, it is important to realize that it fails in certain cases. In particular, the Hartree–Fock model fails to provide a reasonable approximation to the exact state whenever there are several Slater determinants with large weights in the FCI wave function (Eq. (14)). This often happens for excited electronic states and for molecules far away from their equilibrium geometry, particularly in regions of bond breaking and spin recoupling. Moreover, in molecules with more than one resonance structure or in molecules containing transition metal atoms, several determinants may be important even for the electronic ground state at equilibrium.

To illustrate the incorrect behavior of the RHF model upon bond breaking, we return to Fig. 1, which contains the potential energy curve for the symmetrical dissociation of the water molecule (i.e., for a fixed HOH bond angle). For OH bond distances far from equilibrium, the RHF curve is qualitatively different from the FCI curve, grossly overestimating the energy required for dissociation. By contrast, the UHF model dissociates correctly, at least in a qualitative—if not a quantitative—sense, due to the mixing of several states of different multiplicities in the dissociation limit.

From the optimization itself, it may often be difficult to judge whether the Hartree–Fock wave function is a good approximation to the exact wave function—in particular, whether or not the FCI wave function is dominated by one Slater determinant. However, the presence of several important determinants often gives rise to negative eigenvalues in the Hartree–Fock electronic Hessian (i.e., the second derivative of the Hartree–Fock energy with respect to changes in the MOs) [30,31]. For systems whose one-determinant dominance is in doubt, one should therefore inspect the electronic Hessian for negative eigenvalues (instabilities). However, even the absence of such instabilities does not ensure the correctness of the Hartree–Fock model. In difficult cases, therefore, it may be necessary to perform exploratory calculations using multiconfigurational methods, as discussed in Sec. 3.5.

3.5. Multiconfigurational Self-Consistent Field Theory

In Sec. 3.1, we saw that the Hartree–Fock model often gives results in qualitative agreement with experiment. It fails, however, in situations where *static or near-degeneracy* correlation becomes important (i.e., when several electronic configurations have the same or nearly the same energy). Such situations typically arise in the course of molecular reactions, when bonds are broken or formed. Sometimes, near-degeneracies may also be present at the equilibrium ground state geometry. Because only one of the nearly degenerate configurations can be occupied at the single-configuration Hartree–Fock level, the Hartree–Fock model breaks down and cannot be applied. Instead, even for qualitative agreement, we must adopt a multiconfigurational description of the electronic state.

The *multiconfigurational SCF* (MCSCF) model [43,44] is a generalization of the Hartree–Fock model to several configurations:

$$|\text{MCSCF}\rangle = \sum_{\mu} c_{\mu} \Phi_{\mu}, \tag{37}$$

whose expansion coefficients and MOs are simultaneously determined by optimizing the energy with respect to variations in both the MOs and the configuration coefficients:

$$E_{\text{MCSCF}} = \min_{\{\varphi_p\,;\,c_{\mu}\}} \frac{\left\langle \text{MCSCF}|\hat{H}|\text{MCSCF}\right\rangle}{\left\langle \text{MCSCF}|\text{MCSCF}\right\rangle}. \tag{38}$$

The MCSCF procedure may be applied to excited states as well as to the ground state. For the ground state, $E_{\text{MCSCF}} \geq E_{\text{exact}}$; for excited states, by contrast, the MCSCF energy may sometimes be lower than the corresponding exact energy (unless the calculated state is required to be orthogonal to all lower-lying states). This behavior of MCSCF theory occurs because the MCSCF energies of different electronic states are obtained not by the diagonalization of a single Hamiltonian but instead by separate nonlinear optimizations of the energy function.

For the optimization of Hartree–Fock wave functions, it is usually sufficient to apply the SCF scheme described in Sec. 3.1. By contrast, the optimization of MCSCF wave functions requires more advanced methods (e.g., the quasi-Newton method or some globally convergent modification of Newton's method, which involves, directly or indirectly, the calculation of the electronic Hessian as well as the electronic gradient at each iteration) [45].

3.6. Complete Active Space MCSCF Theory

When carrying out an MCSCF calculation, we must first decide which configurations to include in the wave function. Although the configurations may be selected individually, it is more convenient to proceed by dividing the orbital space into subspaces and then to generate configurations by distributing electrons among these subspaces. In the popular *complete active space SCF* (CASSCF) method, for example, the orbital space is divided into inactive, active, and secondary (external) subspaces [43,44]. The CASSCF model is now completely defined: the inactive orbitals are doubly occupied in all configurations, the secondary orbitals are unoccupied in all configurations, whereas the remaining electrons are distributed in all possible ways among the active orbitals. In a sense, we are carrying out an FCI calculation in the configuration space spanned by the active orbitals except that, during the optimization of the FCI wave function, not only the configuration coefficients but also the orbitals are optimized so as to yield the best possible wave function in the chosen configuration space. However, it is not always necessary to optimize all orbitals during the MCSCF optimization [e.g., the core orbitals are usually described well at the Hartree–Fock level and are therefore often kept "frozen" (i.e., unchanged) during the MCSCF optimization].

Let us consider how we may go about setting up an active space for an MCSCF calculation. For the study of reactive systems, we would preferably include in the active space all valence orbitals (at least of all atoms involved in the reactions), leaving the core inactive. Thus, for first-row atoms, all orbitals belonging to the L shell are active and those in the K shell are inactive. In this manner, we ensure a balanced description of the reactive system, no matter what reaction path is followed.

Unfortunately, at present, it is not possible to treat active spaces containing more than, say, 16 orbitals and the same number of valence electrons, confining this full-valence approach to rather small systems. To treat larger systems at the CASSCF level, we must exclude from the active space all orbitals that are deemed unimportant in a given chemical reaction, guided by our chemical intuition.

As an example, consider the symmetrical dissociation of H_2O. In H_2O, bonding arises from the combination of the two $1s$ orbitals on the hydrogens with two sp^3 hybrid orbitals on oxygen. A minimal active space consists of four active orbitals with four electrons, excluding the remaining two sp^3 hybrids, which do not participate in the dissociation. In general, a minimal active space of $2n$ orbitals is required to dissociate n single bonds (each with two paired electrons) into $2n$ unpaired electrons. The disadvantage of this scheme is that it introduces a bias toward the reaction path under study, making comparisons with other reactions difficult. In most cases, however, an unbiased full-valence CASSCF description of the reactive system will be prohibitively expensive and not applicable.

Fig. 2 shows the CASSCF potential energy curve for the symmetrical dissociation of H_2O, using a full-valence active space of six orbitals and eight electrons. For comparison, the figure also contains the FCI and RHF energy curves. Around equilibrium, the differences of the CASSCF and RHF curves from the FCI curve are similar. However, as the bonds are stretched, the RHF model dissociates incorrectly, whereas the CASSCF curve remains parallel to the FCI curve. Thus, the qualitative agreement that the RHF model exhibits around equilibrium has, in the CASSCF model, been extended to all bond distances, making it an ideal method for studies of reactions, at least in a qualitative sense.

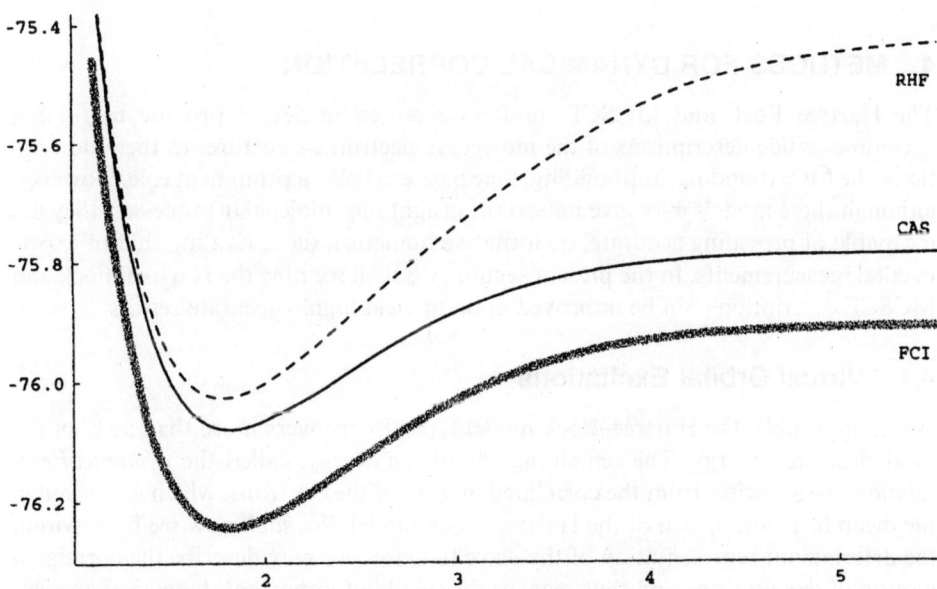

Figure 2 CASSCF dissociation of H_2O (atomic units).

More generally, properties such as vibrational frequencies and reaction energies, which depend on the form of the potential curve, are better predicted with CASSCF theory than with RHF theory, provided the active orbital space has been properly defined. In Table 1, we compare the N_2 full-valence CASSCF and RHF results for D_e, r_e, ω_e, and ν in various correlation-consistent basis sets. At the CASSCF level, D_e is in much better agreement with the experiment than at the RHF level. Moreover, the CASSCF r_e and ω_e are both close to the experimental values, although the RHF errors are slightly overcorrected because the CASSCF method overemphasizes the role of the antibonding orbitals.

The deviation of the CASSCF curve from the FCI curve in Fig. 2 is caused by *nonstatic or dynamical correlation* [1]. Although dynamical correlation is usually less geometry-dependent than static correlation, it must be included for high accuracy (see Sec. 4). One might think that it is possible to include the effects of dynamical correlation simply by extending the active space. For small molecules, this is, to some extent true, in particular when using the techniques of *restricted active space SCF* (RASSCF) theory [46]. Nevertheless, because of the enormous number of determinants needed to recover dynamical correlation, the simultaneous optimization of orbitals and configuration coefficients as done in MCSCF theory is not a practical approach to the accurate description of electronic systems.

For large molecules, the application of the CASSCF method is possible only by a careful selection of active orbitals, and by including only those orbitals that are necessary to describe the static correlation for a given molecule or reaction. The application of the CASSCF method to large molecules thus requires a good knowledge both of the CASSCF method and of the electronic structure of the molecules under study. In conclusion, the CASSCF method is not a black box method but rather a highly flexible method that allows the description of all types of electronic systems. In the hands of an experienced computational chemist, it can be a powerful and versatile tool.

4. METHODS FOR DYNAMICAL CORRELATION

The Hartree–Fock and MCSCF models presented in Sec. 3 provide useful but sometimes crude descriptions of the molecular electronic structure. In these descriptions, the MOs (bonding, antibonding, lone pair, etc.) play a prominent role. However, although these models may give important insight into molecular processes, they are incapable of providing accurate, quantitative numerical data, rivaling that of experimental measurements. In the present section, we shall see how the Hartree–Fock and MCSCF descriptions can be improved upon to yield highly accurate results.

4.1. Virtual Orbital Excitations

When applicable, the Hartree–Fock model typically recovers more than 99% of the total electronic energy. The remaining 1% of the energy, called the *dynamical correlation energy*, arises from the correlated motion of the electrons, which is ignored in the mean field description of the Hartree–Fock model. We shall now see how, within the determinant representation of the wave function, we may describe the correlated motion of the electrons and thus recover the small but important dynamical correlation energy.

The Hartree–Fock determinant describes a situation where the electrons move independently of one another and where the probability of finding one electron at some point in space is independent of the positions of the other electrons. To introduce correlation among the electrons, we must allow the electrons to interact among one another beyond the mean field approximation. In the orbital picture, such interactions manifest themselves through *virtual excitations* from one set of orbitals to another. The most important class of interactions are the pairwise interactions of two electrons, resulting in the simultaneous excitations of two electrons from one pair of spin orbitals to another pair (consistent with the Pauli principle that no more than two electrons may occupy the same spatial orbital). Such virtual excitations are called *double excitations*. With each possible double excitation in the molecule, we associate a unique amplitude, which represents the probability of this virtual excitation happening. The final, correlated wave function is obtained by allowing all such virtual excitations to happen, in all possible combinations.

Mathematically, we may describe this approach to electron correlation in the following manner. Let the Hartree–Fock state be represented by $|\mathrm{HF}\rangle$. The virtual excitation of two electrons from spin orbitals i and j to the virtual orbitals a and b may now be expressed as:

$$\hat{X}_{ij}^{ab}|\mathrm{HF}\rangle = t_{ij}^{ab} a_b^\dagger a_a^\dagger a_i a_j |\mathrm{HF}\rangle. \tag{39}$$

We have here introduced the *annihilation operators* a_i and a_j, which remove electrons from spin orbitals i and j, respectively, and the *creation operators* $a_a^\dagger$ and $a_b^\dagger$, which put electrons into the spin orbitals a and b. To agree with the Pauli principle, the creation and annihilation operators obey the commutation relations:

$$a_p a_q + a_q a_p = 0, \quad a_p^\dagger a_q^\dagger + a_q^\dagger a_p^\dagger = 0, \quad a_p^\dagger a_q + a_q a_p^\dagger = \delta_{pq} \tag{40}$$

where δ_{pq} is zero when $p \neq q$ and one when $p = q$. By applying $1 + \hat{X}_{ij}^{ab}$ to the Hartree–Fock state, we produce a new state:

$$\left(1 + \hat{X}_{ij}^{ab}\right)|\mathrm{HF}\rangle = |\mathrm{HF}\rangle + t_{ij}^{ab} a_b^\dagger a_a^\dagger a_i a_j |\mathrm{HF}\rangle \tag{41}$$

as a superposition of the original Hartree–Fock state with an excited state, which represents a different spatial distribution of the electrons. The amplitude t_{ij}^{ab} in Eq. (41) may be determined variationally to yield the lowest energy, although in practice other techniques are often used, as discussed in Secs. 4.2 and 4.3.

To illustrate the effect of virtual excitations and the superposition of determinants, we have in Fig. 3 plotted the one-electron and two-electron densities of the hydrogen molecule in the uncorrelated Hartree–Fock (upper plots) and correlated FCI (lower plots) ground states:

$$|\mathrm{HF}\rangle = \left|1\sigma_g^2\right\rangle, \tag{42}$$

$$|\mathrm{FCI}\rangle = 0.9939\left|1\sigma_g^2\right\rangle - 0.1106\left|1\sigma_u^2\right\rangle, \tag{43}$$

evaluated in a minimal basis of STOs centered on the two nuclei A and B:

$$1s_\mathrm{A} = \frac{1}{\sqrt{\pi}}\exp(-r_\mathrm{A}), \quad 1s_\mathrm{B} = \frac{1}{\sqrt{\pi}}\exp(-r_\mathrm{B}). \tag{44}$$

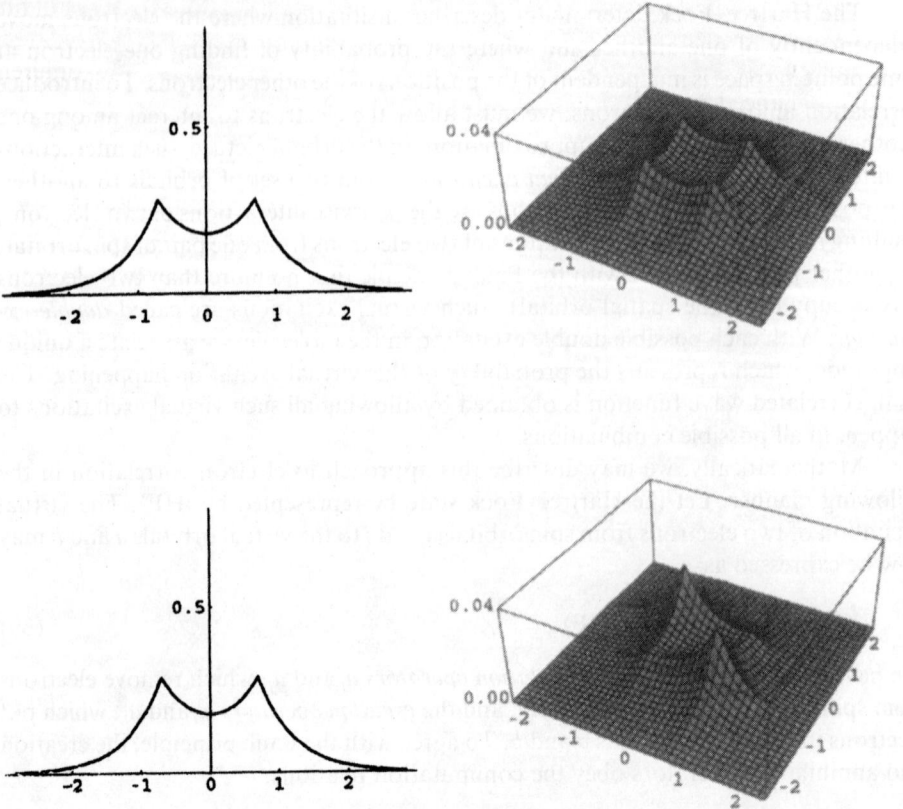

Figure 3 The one-electron and two-electron density functions of the $^1\Sigma_g^+$ ground state of the H_2 molecule. The upper plots contain the one-electron and two-electron densities of the uncorrelated Hartree–Fock description in a minimal basis; the lower plots contain the corresponding densities of the two-configuration correlated FCI description in the same basis. In all cases, the electron density has been plotted on the molecular axis (one axis for the one-electron densities, two axes for the two-electron densities).

The FCI state (Eq. (43)) has been generated from the Hartree–Fock state (Eq. (42)) by application of the double excitation operator as in Eq. (41), followed by a variational optimization. Whereas the one-electron density $\rho(\vec{r})$ represents the overall probability of finding an electron at a given point $\vec{r}$ in space, the two-electron density $\rho(\vec{r}_1, \vec{r}_2)$ represents the probability of finding one electron at position $\vec{r}_1$ when the other electron is known to be at $\vec{r}_2$.

We first note that the difference between the one-electron Hartree–Fock and FCI densities in Fig. 3 is very small. This is understandable as the wave function changes only little upon correlation (compare Eqs. (42) and (43)). By contrast, the two-electron density in Fig. 3 is strongly affected by correlation. In the uncorrelated Hartree–Fock state, the two-electron density is essentially a product of two separate one-electron densities because the instantaneous position of one electron does not affect the probability distribution of the second electron. In the FCI state, on the other hand, there is a strong correlation between the electrons in the sense that the presence of one electron at one nucleus greatly reduces the chance of finding the second electron at the same nucleus; at the same time, the probability of locating the electrons on different nuclei is enhanced by correlation.

4.2. Coupled-Cluster Theory

By applying all possible excitations to the Hartree–Fock state, we arrive at the following *coupled-cluster representation* of the FCI wave function:

$$|CC\rangle = \left[\prod_{ai} \left(1 + \hat{X}_i^a \right) \right] \left[\prod_{abij} \left(1 + \hat{X}_{ij}^{ab} \right) \right] \left[\prod_{abcijk} \left(1 + \hat{X}_{ijk}^{abc} \right) \right] \dots |HF\rangle, \qquad (45)$$

where the excitation operators are abbreviated as:

$$\hat{X}_i^a = t_i^a a_a^\dagger a_i, \quad \hat{X}_{ij}^{ab} = t_{ij}^{ab} a_b^\dagger a_a^\dagger a_i a_j, \quad \hat{X}_{ijk}^{abc} = t_{ijk}^{abc} a_c^\dagger a_b^\dagger a_a^\dagger a_i a_j a_k, \qquad (46)$$

and so on. Thus, in coupled-cluster theory, the wave function is parameterized in terms of the *coupled-cluster amplitudes* t_i^a, t_{ij}^{ab}, t_{ijk}^{abc}, ..., representing the probability that a given virtual excitation may happen. More commonly, the coupled-cluster wave function in Eq. (45) is expressed by means of an exponential operator working on the Hartree–Fock state [47,48]:

$$|CC\rangle = \exp\left(\hat{T} \right) |HF\rangle, \qquad (47)$$

where the *cluster operator* $\hat{T} = \hat{T}_1 + \hat{T}_2 + \dots$ contains the single-excitation and double-excitation operators:

$$\hat{T}_1 = \sum_{ai} \hat{X}_i^a, \quad \hat{T}_2 = \sum_{abij} \hat{X}_{ij}^{ab}, \quad \hat{T}_3 = \sum_{abcijk} \hat{X}_{ijk}^{abc}, \qquad (48)$$

as well as all higher-order excitation operators. Note that, in Eqs. (45) and (47), the order of the excitation operators is unimportant because, according to Eq. (40), the creation operators of the virtual spin orbitals always anticommute with the annihilation operators of the occupied spin orbitals.

To gain a better understanding of the structure of the coupled-cluster state, we expand the product state (Eq. (45)) in the following manner:

$$|CC\rangle = |HF\rangle + \sum_{ai} \hat{X}_i^a |HF\rangle + \sum_{abij} \left(\hat{X}_{ij}^{ab} + \hat{X}_i^a \hat{X}_j^b \right) |HF\rangle + \dots \qquad (49)$$

Clearly, the resulting wave function has contributions from all Slater determinants, whose expansion coefficients are determined by the cluster amplitudes. The doubly excited determinants, for example, have contributions both from pure double excitations $\hat{X}_{ij}^{ab}$ and from products of two independent single excitations $\hat{X}_i^a \hat{X}_j^b$. The former excitations are known as *connected*, the latter excitations are known as *disconnected*. In this manner, the amplitudes of different excitation processes contribute to the same expansion coefficients of the FCI wave function in Eq. (14).

The purpose of introducing the coupled-cluster expansion (Eq. (45)) is that it provides us with a convenient way of approximating the FCI coefficients, at least when the Hartree–Fock model is a good one. For example, we may base our approximate FCI description on the single and double excitations only, writing the wave function as [49]:

$$|CCSD\rangle = \left[\prod_{ai}\left(1 + \hat{X}_i^a\right)\right]\left[\prod_{abij}\left(1 + \hat{X}_{ij}^{ab}\right)\right]|HF\rangle. \tag{50}$$

In this *coupled-cluster singles and doubles* (CCSD) approximation, none of the coefficients in the FCI expansion is ignored. Instead, they are approximated by a much smaller set of singles and doubles amplitudes.

Alternatively, we could approximate the FCI wave function (Eq. (14)) directly by omitting, for example, all determinants that differ from the Hartree–Fock determinant by more than two spin orbitals, computing the remaining coefficients variationally. This *CI singles and doubles* (CISD) wave function has the same number of parameters as the CCSD wave function. However, the CISD model is less useful than the CCSD model, which allows independent (disconnected) excitations to occur throughout the molecule. In CISD theory, only connected double excitations are allowed, no matter how large the system is. Consequently, the CISD description deteriorates as the number of electrons increases. The CCSD wave function, by contrast, is *size-extensive* (i.e., it works equally well for small and large systems, providing a consistent description of the electronic structure) [1].

The CCSD model constitutes a particularly important correlated level as it includes the most important class of virtual excitations: the *connected doubles*. Higher excitations such as triples are less important because they represent the less probable, simultaneous interaction among three or more electrons. Single excitations, on the other hand, are also unimportant as they do not correspond to a physical interaction among electrons but instead represent a relaxation of the orbitals in response to the changes introduced by the virtual excitations. For total energies, therefore, the *coupled-cluster singles* (CCS) method represents no improvement on the Hartree–Fock description, as the Hartree–Fock orbitals are already fully optimized. Indeed, the single excitations come into play only at the CCSD level, in response to the double excitations. In passing, we note that, in the course of the coupled-cluster amplitude optimization, the coupled-cluster orbitals can be reoptimized such that there are no relaxation effects, as done in the *Brueckner doubles* (BD) method [50,51].

For technical reasons, the CCSD (and other coupled-cluster) wave functions are not optimized variationally (by minimizing the expectation value of the Hamiltonian). Instead, the CCSD amplitudes are obtained by projecting the Schrödinger equation onto the manifold of all singly and doubly excited determinants, thereby establishing as many equations as there are unknowns (i.e., amplitudes) in the wave function. The solution of these nonlinear equations is not much more complicated or expensive than the solution of the CISD equations, making the CCSD method preferable. It should be understood, however, that the resulting CCSD energy does not represent a strict upper bound to the exact energy. In practice, this does not matter much as the energies are anyway very accurate (at least when the Hartree–Fock approximation is a good one) and because we are primarily interested in energy differences rather than in total energies.

Coupled-cluster theory provides the most important hierarchy of models in ab initio quantum chemistry. At each new excitation level of this hierarchy, a significant improvement is observed in the calculated energies and properties. Typically, the CCSD model reduces the error in the calculated properties by a factor of three or four relative to the Hartree–Fock model. A similar improvement is observed at the *coupled-cluster singles, doubles, and triples* (CCSDT) level [52,53], which, for many properties, gives errors on the order of 1% or less, sometimes surpassing the accuracy of experimental measurements. Further reductions are achieved at higher levels such as the *coupled-cluster singles, doubles, triples, and quadruples* (CCSDTQ) level, although this is rarely possible because of the steep increase in cost with each new excitation level. Thus, whereas the cost of the Hartree–Fock model formally scales as K^4, where K is the number of atoms, the costs of the CCSD, CCSDT, and CCSDTQ models scale, formally at least, as K^6, K^8, and K^{10}, severely restricting the applicability of the coupled-cluster hierarchy for large systems (see, however, the discussion in Sec. 4.6).

4.3. Møller–Plesset Perturbation Theory

Whenever the Hartree–Fock wave function provides a good zero-order description of the electronic system, it is natural to investigate the possibility of treating dynamical correlation by perturbation theory rather than by coupled-cluster theory. In this manner, we may hope to recover the most important effects of dynamical correlation at a cost lower than that of coupled-cluster theory.

Indeed, this approach has been rather successful in quantum chemistry, at least to low orders in the perturbation. It begins with the separation of the electronic Hamiltonian into a zero-order operator and a perturbation operator called the *fluctuation potential*:

$$\hat{H} = \hat{H}_0 + \hat{V}. \tag{51}$$

The zero-order Hamiltonian $\hat{H}_0$ corresponds to the Fock operator, whereas the fluctuation potential $\hat{V}$ represents the difference between the full, instantaneous two-electron potential and the averaged SCF potential of the Hartree–Fock model:

$$\hat{H}_0 = \sum_i \hat{F}_i, \quad \hat{V} = \sum_{i>j} \frac{1}{|\vec{r}_i - \vec{r}_j|} - \sum_i \left(\hat{J}_i - \hat{K}_i \right). \tag{52}$$

We now apply standard Rayleigh–Schrödinger perturbation theory, using the Hartree–Fock determinant as the zero-order state, and expand the perturbed states in the set of excited determinants. This approach gives rise to *Møller–Plesset perturbation theory* [1]. To first order, we recover the Hartree–Fock energy E_{HF} and, to second order, we obtain the second-order Møller–Plesset (MP2) energy:

$$E_{MP2} = E_{HF} + \frac{1}{4} \sum_{abij} \frac{[(ia|jb) - (ib|ja)]^2}{\varepsilon_i + \varepsilon_j - \varepsilon_a - \varepsilon_b}. \tag{53}$$

The MP2 energy is always lower than the Hartree–Fock energy and usually represents a rather good approximation to the total electronic energy. The MP2 model usually works well whenever the Hartree–Fock wave function is a reasonable one, typically recovering about 90% of the total correlation energy, at a cost that scales formally as K^5. Still, it is less robust and somewhat less generally applicable than the CCSD model.

It is possible to extend the Møller–Plesset perturbation treatment to higher orders. However, the perturbation expansion often oscillates and diverges to higher orders, in particular in large AO basis sets. In general, therefore, we advocate only the MP2 model, followed if necessary by a more elaborate treatment at the coupled-cluster level.

4.4. Perturbative Corrections to Coupled-Cluster Theory: The CCSD(T) Model

As discussed in Sec. 4.2, the coupled-cluster hierarchy converges rapidly, the error in the total energy (and other properties) being reduced significantly with each new level of excitations. Unfortunately, the inclusion of higher-order connected excitations increases the computational cost enormously. In practice, although it is possible to carry out CCSD calculations for fairly large systems and basis sets (more than 10 atoms and 500 AOs), the full CCSDT model is presently too expensive for routine calculations. However, because we are anyway forced to neglect the connected quadruples (CCSDTQ) in our calculations, the overall quality of our results will not be adversely affected if, in the treatment of the connected triples, we make an error that is no larger than that incurred by neglecting the quadruples. In practice, therefore, any inexpensive, approximate treatment of the triples that gives an error of the order of 10% or less is welcome.

Among the various approximate methods for including the connected triple excitations, the *CCSD with perturbative triples* [CCSD(T)] method is the most popular [54]. In this approach, the CCSD calculation is followed by the calculation of a perturbational estimate of the triple excitations, reducing the overall scaling with respect to the size of the system from K^8 in CCSDT to K^7 in CCSD(T).

Of all the methods currently used in molecular electronic structure theory, the CCSD(T) model is probably the most successful, highly accurate level, at least for closed-shell molecular systems. For many properties of interest to chemists such as molecular structure, atomization energies, and vibrational frequencies, it provides numerical data of consistently high quality, sometimes surpassing that of experiment. Nevertheless, it does fail in certain cases, in particular for systems characterized by several important Slater determinants and also for certain properties such as indirect nuclear spin–spin couplings of magnetic resonance spectroscopy.

4.5. Dynamical Correlation in Multireference Systems

The theory for the calculation of dynamical correlation effects in systems dominated by a single determinant (RHF or UHF) is well developed and routinely applied in a black box manner. By contrast, the treatment of dynamical correlation in systems containing several important configurations is much more difficult and less amenable to routine calculations.

There are essentially two ways by means of which dynamical correlation can be treated for multireference systems: *multireference configuration interaction* (MRCI) methods [55] and *CAS perturbation theory* (CASPT), or more generally *multireference perturbation theory* (MRPT) [56]. All methods begin by setting up a reference space of important electronic configurations, usually by carrying out an MCSCF optimization to obtain the inactive and active orbitals for the reference configurations. In MRCI, the wave function is then generated by allowing all excitations up to a given order from the reference determinants, and the amplitudes of the included determinants are determined by the variation principle. Typically, all single and double excitations from the reference determinants are included, leading to the *multireference singles-and-doubles configuration interaction* (MRSDCI) method.

Except for small molecules, the number of MRSDCI reference determinants may become very large, rendering the calculations time-consuming or impossible. However, the steep increase in the number of MRSDCI amplitudes with the number of reference determinants may be significantly reduced by means of *internal contraction*. In internally contracted theory, the wave function is generated by combining the MCSCF state $|MC\rangle$ linearly with excitations such as $a_a^\dagger a_b^\dagger a_i a_j | MC >$. For systems that are reasonably well described by a small number of reference determinants, the MRSDCI method provides accurate results. However, because the MRSDCI theory is not size-extensive, it is ill suited for correlating the electrons in large molecules. A variety of schemes have been proposed to reduce the size extensivity error of MRSDCI, none of which has found widespread use.

In CASPT, a perturbation expansion is constructed with a CASSCF state as the zero-order state, thereby relaxing the single-determinant constraint of Møller–Plesset perturbation theory. The zero-order Hamiltonian is usually nondiagonal in the basis of the zero-order states, and the configurations of the first-order wave function are conveniently taken as the internally contracted singly excited and doubly excited configurations discussed above. To obtain the first-order wave function correction and the second-order energy correction, it is (unlike in Møller–Plesset theory, as presented in Sec. 4.3) necessary to solve a set of linear equations. Furthermore, as the formalism is significantly more complicated than that of Møller–Plesset theory, CASPT has only been developed to second and third orders, yielding the CASPT2 and CASPT3 corrections, respectively.

Although not rigorously size-extensive, CASPT behaves better than MRSDCI theory for systems of many electrons. The CASPT and other MRPT methods are important in the sense that they are the only generally applicable methods for ab initio calculations of dynamical correlation of open-shell and closed-shell multiconfigurational electronic systems.

4.6. Dynamical Correlation in Large Systems

The correlated methods discussed up to this point provide a delocalized description of the electronic system. The delocalized nature of these methods arises from their use of canonical orbitals (i.e., the eigenvectors of the Hartree–Fock equations) of Eq. (33). To treat large systems, it is better to express the theory in terms of orbitals that are localized in space, extending over only a few atoms. The virtual excitations then occur predominantly locally in the molecule (among localized occupied and virtual orbitals). As a result, the number of excitation amplitudes increases only linearly with system size.

A difficulty with this local approach to dynamical correlation is that, in Møller–Plesset theory, for example, the zero-order Fock operator is no longer diagonal in the space of the Slater determinants, making the application of such theories slightly more complicated than theories based on canonical orbitals. Currently, the development of local correlation methods is an active area of research [57–63]. The *diatomics-in-molecules* (DIM) method and the *triatomics-in-molecules* (TRIM) method, for instance, recover typically 95% and 99.7%, respectively, of the full MP2 correlation energy [63]. By means of a linear scaling local variant of the CCSDT method, calculations have been carried out on the Indinavir molecule (an HIV protease inhibitor) in Fig. 4, illustrating the potential of rigorous ab initio theory for biological systems.

A different approach to dynamical correlation in large systems is to approximate the four-center two-electron integrals in, for example, Eq. (53) by sums of products of three-center integrals, as done in the *resolution-of-identity MP2* (RI-MP2) method [64]. As the number of three-center integrals is much smaller than the number of four-center integrals, this approach reduces the computational cost of MP2 calculations dramatically. To ensure high accuracy in the calculations, special auxiliary basis sets have been developed for the RI expansion. Although the RI expansion

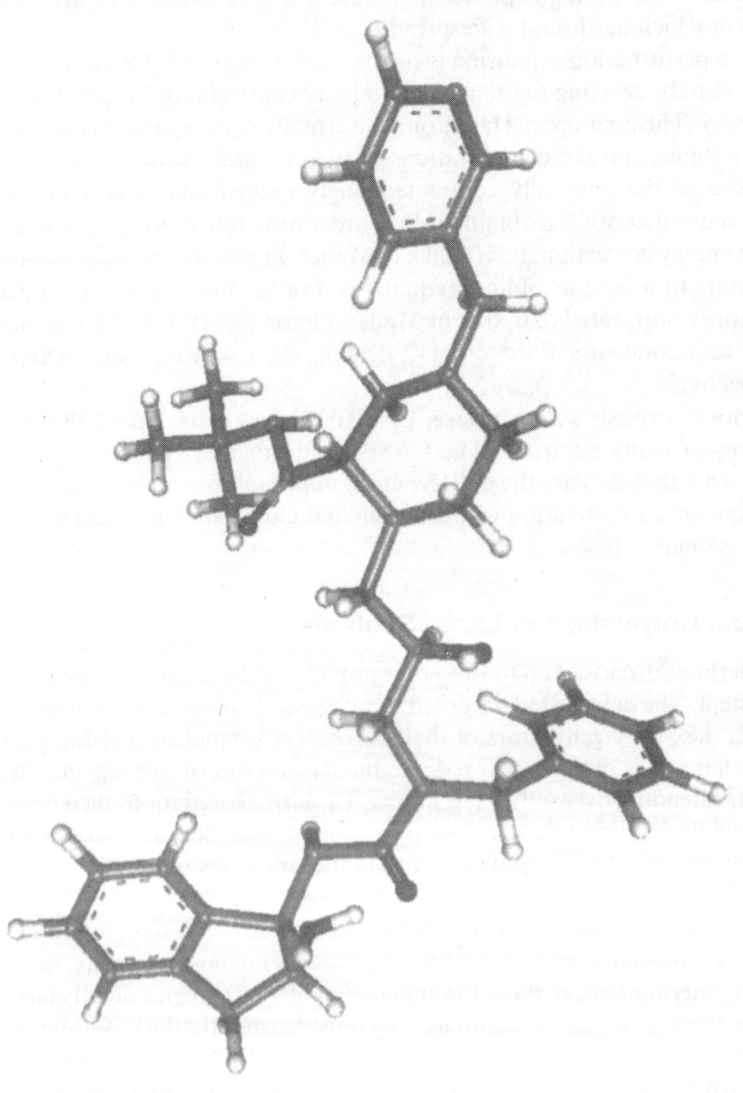

Figure 4   The electronic structure of the Indinavir molecule, an HIV protease inhibitor, has been calculated using a linear scaling local version of CCSDT by Schütz. (From Ref. 62.)

introduces a small error in the MP2 energy, the error is usually much smaller than the intrinsic MP2 error. By means of the RI-MP2 method, very large molecular systems have been treated rigorously. A related method for large systems is the *pseudospectral method* [65]. In this method, a numerical three-dimensional spatial grid is set up to reduce the four-center integrals to sums of products of three-center integrals. Moreover, efficient parallel computer implementations have been developed for various MP2 methods [66, 67].

5. CONVERGENCE TOWARD THE EXACT SOLUTION

As pointed out in Secs. 2.3 and 4.2, wave function-based quantum chemical calculations can be set up such that a smooth and often monotonic convergence is established toward the exact result (i.e., toward the exact solution of the nonrelativistic electronic Schrödinger equation) (Eq. (2)). From an understanding of this convergence, we may improve upon our calculations in a systematic manner. Moreover, we may provide reliable estimates of the exact solution by means of extrapolation.

The convergence of the quantum chemical calculations can be studied in terms of two types of hierarchies. First, the quality of the calculations depends on the flexibility of the MO space; the AOs that are used to expand the MOs may be extended in a well-defined and systematic manner, thereby establishing *a one-electron hierarchy*. Second, we can increase the excitation level in coupled-cluster theory or the order of perturbation expansion, thus setting up an *n-electron hierarchy* of approximate electronic wave functions. In Fig. 5, the roles of the one-electron and the *n*-electron hierarchies are illustrated.

Along the ordinate, a sequence of correlation-consistent cc-pVXZ basis sets with $X \geq 2$ is depicted. Along the abscissa, the FCI limit is approached—beginning with Hartree–Fock theory and followed by the first correlated level, at which the single and double excitations are described by MP2 perturbation theory. The same excitations are subsequently treated by coupled-cluster theory at the CCSD level, which is then further improved upon by a perturbation treatment of the triple excitations at the CCSD(T) level. At the CCSDT level, the triple excitations are fully treated by coupled-cluster theory, and so on. In this manner, the hierarchy Hartree–Fock $\rightarrow$ MP2 $\rightarrow$ CCSD $\rightarrow$ CCSD(T) $\rightarrow$ CCSDT $\rightarrow \cdots \rightarrow$ FCI is established.

In Table 1, the dissociation energy and some spectroscopic constants of N_2 are listed for correlation-consistent basis sets with $X \leq 5$ and for the *n*-electron models Hartree–Fock, full-valence MCSCF, MP2, CCSD, and CCSD(T). A clear (but not necessarily rapid) convergence is observed within the established hierarchies.

Concerning the convergence with the cardinal number X, it has been found that, to leading order, the errors arising from our approximate treatment of dynamical correlation vanish as X^{-3} [68]. This must be viewed as slow convergence; because the number of AOs N increases as X^3 (see Sec. 2.3), it implies that the error is inversely proportional to N. As the calculations involve the evaluation of N^4 two-electron integrals, it can be difficult to increase the AO basis to the size that is required to obtain a prescribed (high) precision, in particular for large molecular systems. In such cases, it can be very helpful to perform calculations in two or more (small) basis sets and to study the corresponding basis set convergence. Moreover, when systematic sequences such as cc-p(C)VXZ are used, the X^{-3} convergence of the basis set may be exploited to estimate the complete-basis limit by extrapolation.

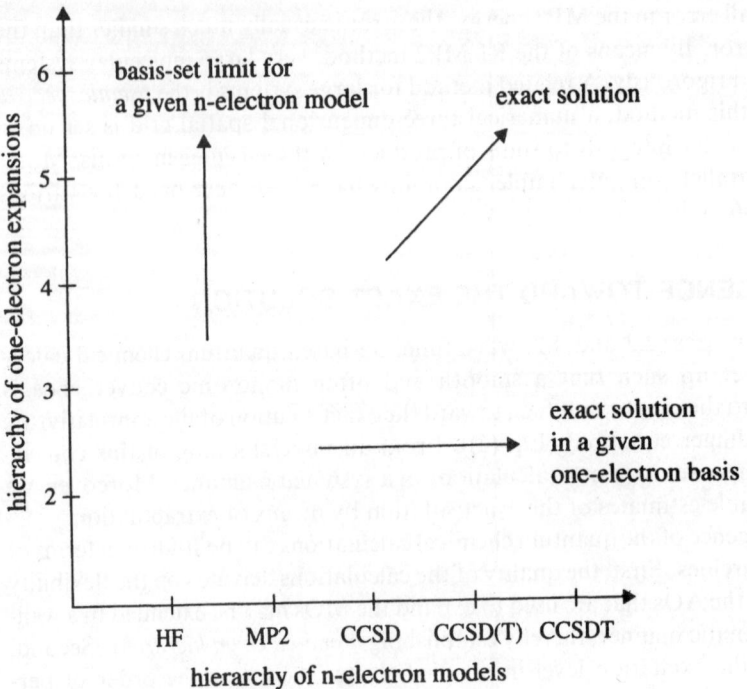

Figure 5   The one- and n-electron hierarchies of ab initio theory.

The slow X^{-3} convergence cannot easily be avoided. To improve on it, one can try to describe electron correlation not only by means of virtual orbital excitations but also by means of spatial two-electron basis functions that depend explicitly on the electron–electron distances. This is the idea behind the *explicitly correlated* methods [69]. Indeed, in this manner, it is possible to accelerate the convergence from X^{-3} to X^{-7}, greatly reducing the basis set requirements for high accuracy. Nevertheless, such calculations are complex and cannot yet be used routinely on large molecules.

6.   CALIBRATION OF THE STANDARD METHODS

For the one-electron and n-electron hierarchies of ab initio theory to be useful, it is necessary to carry out a careful and extensive calibration of their performance. Such a calibration is best carried out by comparing, in a statistical manner, calculated values of different properties with experimental measurements. In the present section, we give two examples of such comparisons, for bond lengths and reaction enthalpies. In Figs. 6 and 7, we have plotted the normal distributions of the errors in calculations of bond distances and reaction enthalpies, respectively. The statistics underlying the normal distributions are based on calculations on 19 closed-shell first-row molecules (e.g., HF and C_2H_4) and 13 reactions involving these molecules (for more details, see Refs. 1 and 17).

For both bond distances and reaction enthalpies, there is a marked improvement in the performance of the calculations as we improve the n-electron description.

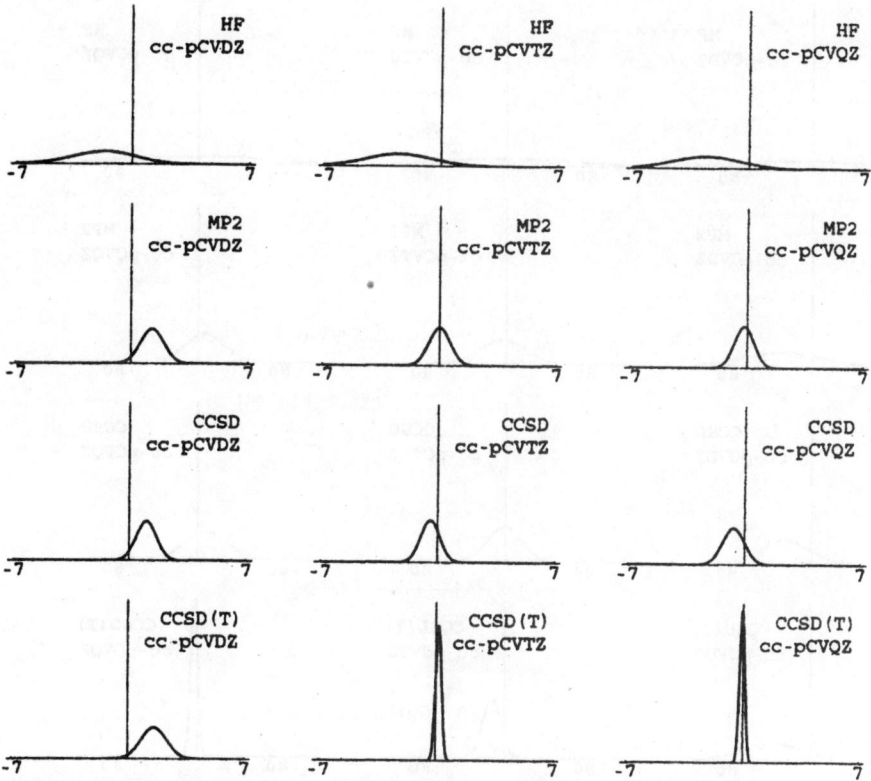

Figure 6 Normal distributions of the error in calculated bond distances (pm).

In the largest basis (cc-pCVQZ), the distribution of errors is very broad for Hartree–Fock but becomes sharply peaked at the CCSD(T) level. The MP2 and CCSD distributions are similar to each other and intermediate between those of Hartree–Fock and CCSD(T). It is important to note, however, that this progression of the n-electron models is not observed in the small cc-pCVDZ basis. Clearly, this basis is not sufficiently flexible for correlated calculations, providing too small virtual excitation spaces for these models to work properly—in particular, for reaction enthalpies. In the cc-pCVTZ basis, convergence is more satisfactory, especially for bond distances.

From Figs. 6 and 7, it may appear that there is little to be gained by going from MP2 (whose cost scales as K^5) to CCSD (with cost K^6). However, although the performance of MP2 and CCSD is quite similar for bond distances and reaction enthalpies of closed-shell systems of light atoms, for other properties and other systems, the more expensive and robust CCSD method outperforms MP2.

Finally, we note that, at the MP2/cc-pCVTZ level, there is (on average) a fortuitous cancellation of the one-electron and n-electron errors. Such cancellations are frequently encountered in ab initio quantum chemistry. Although in many cases unavoidable, these cancellations should not be relied on but instead treated with great caution.

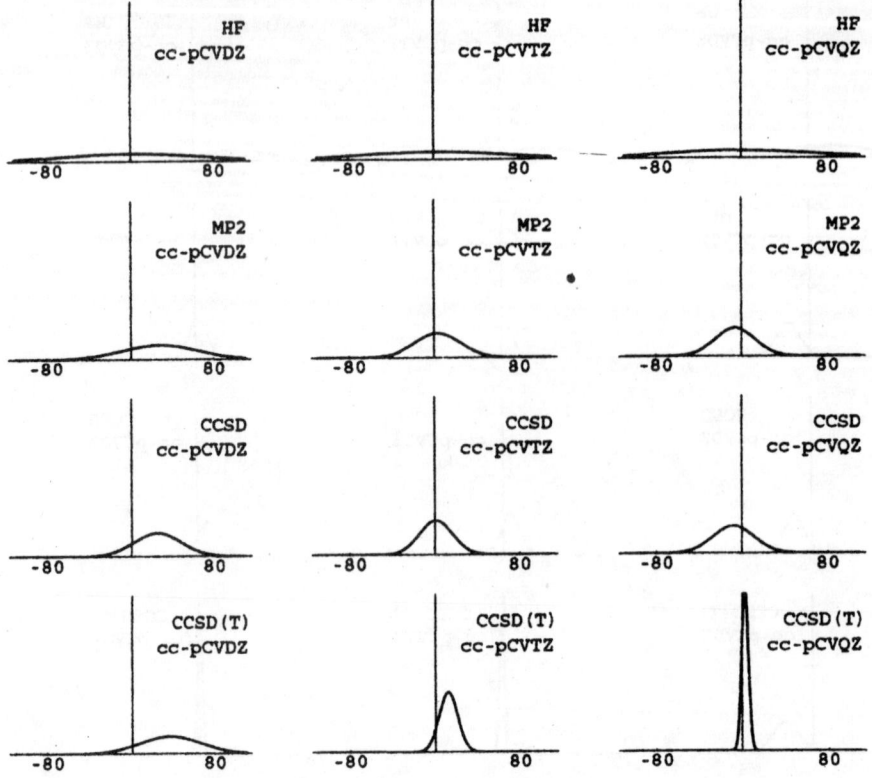

Figure 7   Normal distributions of the error in calculated reaction enthalpies (kJ/mol).

7.   CONCLUDING REMARKS

In this chapter, we have discussed the rigorous calculation of molecular electronic structure by means of n-electron wave functions. At present, such methods are capable of treating fairly large systems, including systems of biological and pharmaceutical interest, in an approximate but nonempirical manner, yielding results of qualitative accuracy. Because of their nonempirical nature, such calculations represent a useful, independent source of information about chemical systems, complementary to that obtained by experiment. For high accuracy, hierarchies of methods have been developed, allowing the exact solution to be approached in a systematic manner. At the highest level, wave function-based quantum chemistry provides very accurate information about molecular systems, rivaling that of many experimental measurements.

Although the most accurate calculations are usually not possible on systems of biological interest, a careful calibration of each level in the hierarchy ensures that even the results of the simpler calculations may be used with confidence. Indeed, the presence of such systematic, universally adopted hierarchies is probably the most distinctive feature of modern wave function-based quantum chemistry, setting it apart from other computational techniques of electronic structure.

ACKNOWLEDGMENTS

We thank Dr. M. Schütz (Stuttgart) for permission to reproduce Fig. 4. The research of W. K. is supported by the DFG Research Center for Functional Nanostructures (CFN) under project number C2.3. T. H. gratefully acknowledges support by MOLPROP. P. J. and J. O. acknowledge the support from the Danish National Research Council (grant no. 9901973).

REFERENCES

1. Helgaker T, Jørgensen P, Olsen J. Molecular Electronic-Structure Theory. Chichester: Wiley, 2000.
2. Jensen F. Introduction to Computational Chemistry. Chichester: Wiley, 1999.
3. Cioslowski J, ed. Quantum-Mechanical Prediction of Thermochemical Data. Dordrecht: Kluwer Academic Publishers, 2001.
4. Dunning TH Jr. A road map for the calculation of molecular binding energies. J Phys Chem A 2000; 104:9062–9080.
5. Hammond BL, Lester WA Jr, Reynolds PJ. Monte Carlo Methods in Ab Initio Quantum Chemistry. Singapore: World Scientific, 1994.
6. Lester WA Jr, ed. Recent Advances in Quantum Monte Carlo Methods. Singapore: World Scientific, 1997.
7. Lester WA Jr, Rothstein SM, Tanaka S, eds. Recent Advances in Quantum Monte Carlo Methods—Part II. Singapore: World Scientific, 2001.
8. Cohen-Tannoudji C, Diu B, Laloë F. Quantum Mechanics. New York: Wiley, 1977.
9. Pyykkö P. Relativistic Theory of Atoms and Molecules: A Bibliography. Berlin: Springer, 1986–2000.
10. Stoll H, Metz B, Dolg M. Relativistic energy-consistent pseudopotentials—recent developments. J Comput Chem 2002; 23:767–778.
11. Cowan RD, Griffin DC. Approximate relativistic corrections to atomic radial wave functions. J Opt Soc Am 1976; 66:1010–1014.
12. Tuma C, Boese AD, Handy NC. Predicting the binding energies of H-bonded complexes: a comparative DFT study. Phys Chem Chem Phys 1999; 1:3939–3947.
13. Curtiss LA, Raghavachari K, Redfern PC, Pople JA. Assessment of Gaussian-3 and density functional theories for a larger experimental test set. J Chem Phys 2000; 112:7374–7383.
14. Császár AG, Allen WD, Schaefer HF III. In pursuit of the ab initio limit for conformational energy prototypes. J Chem Phys 1998; 108:9751–9764.
15. Dunning TH Jr, Peterson KA, Woon DE. Basis sets: correlation consistent sets. In: Schleyer PvR, Allinger NL, Clark T, Gasteiger J, Kollman PA, Schaefer HF III, Scheiner PR, eds. Encyclopedia of Computational Chemistry. Vol. 1. Chichester: Wiley, 1998:88–115.
16. Bak KL, Jørgensen P, Olsen J, Helgaker T, Klopper W. Accuracy of atomization energies and reaction enthalpies in standard and extrapolated electronic wave function/basis set calculations. J Chem Phys 2000; 112:9229–9242.
17. Bak KL, Gauss J, Jørgensen P, Olsen J, Helgaker T, Stanton JF. The accurate determination of molecular equilibrium structures. J Chem Phys 2001; 114:6548–6556.
18. Shavitt I. The method of configuration interaction. In: Schaefer HF III, ed. Methods in Electronic Structure Theory, Modern Theoretical Chemistry III. New York: Plenum, 1977:189–275.
19. Sherrill CD, Schaefer HF III. The configuration interaction method: advances in highly correlated approaches. Adv Quantum Chem 1999; 34:143–269.

20. Bauschlicher CW Jr, Langhoff SR, Taylor PR. Accurate quantum chemical calculations. Adv Chem Phys 1990; 77:103–161.

21. Feller D, Davidson ER. Basis sets for ab initio molecular orbital calculations and inter-molecular interactions. In: Lipkowitz KP, Boyd DB, eds. Review in Computational Chemistry. New York: VCH Publishers, 1990:1–37.

22. Helgaker T, Taylor PR. Gaussian basis sets and molecular integrals. In: Yarkony DR, ed. Modern Electronic Structure Theory. Singapore: World Scientific, 1995:725–856.

23. Almlöf J, Taylor PR. General contraction of Gaussian basis sets. I. Atomic natural orbitals for first- and second-row atoms. J Chem Phys 1987; 86:4070–4077.

24. Widmark P-O, Malmqvist P-Å, Roos BO. Density matrix averaged atomic natural orbital (ANO) basis sets for correlated molecular wave functions. Theor Chim Acta 1990; 77:291–306.

25. Dunning TH Jr. Gaussian basis sets for use in correlated molecular calculations. J Chem Phys 1989; 90:1007–1023.

26. Kendall RA, Dunning TH Jr, Harrison RJ. Electron affinities of the first-row atoms revisited. Systematic basis sets and wave functions. J Chem Phys 1992; 96:6796–6806.

27. Woon DE, Dunning TH Jr. Gaussian basis sets for use in correlated molecular calculations. V. Core-valence basis sets for boron through neon. J Chem Phys 1995; 103:4572–4585.

28. Almlöf J. Direct methods in electronic structure theory. In: Yarkony DR, ed. Modern Electronic Structure Theory. Singapore: World Scientific, 1995:110–151.

29. Roothaan CCJ. New developments in molecular orbital theory. Rev Mod Phys 1951; 23:69–89.

30. Čiček J, Paldus J. Stability conditions for the solutions of the Hartree–Fock equations for atomic and molecular systems. Application to the pi-electron model of cyclic poly-enes. J Chem Phys 1967; 47:3976–3985.

31. Löwdin PO, Mayer I. Some studies of the general Hartree–Fock method. Adv Quantum Chem 1992; 24:79–114.

32. Frenking G, ed. Special issue: quantum chemical methods for large molecules. J Comp Chem 2000; 21:1419–1588.

33. Strain MC, Scuseria GE, Frisch MJ. Achieving linear scaling for the electronic quantum Coulomb problem. Science 1996; 271:51–53.

34. Greengard L. Fast algorithms for classical physics. Science 1994; 265:909–914.

35. Greengard L, Rokhlin V. A fast algorithm for particle simulations. J Comp Phys 1987; 73:325–348.

36. Li X-P, Nunes RW, Vanderbilt D. Density-matrix electronic-structure method with linear system-size scaling. Phys Rev B 1993; 47:10891–10894.

37. Helgaker T, Larsen H, Olsen J, Jørgensen P. Direct optimization of the AO density matrix in Hartree–Fock and Kohn–Sham theories. Chem Phys Lett 2000; 327:397–403.

38. Pulay P. Analytical derivative methods in quantum chemistry. Adv Chem Phys 1987; 69:241–286.

39. Ochsenfeld C, Head-Gordon M. A reformulation of the coupled perturbed self-consistent field equations entirely within a local atomic orbital density matrix-based scheme. Chem Phys Lett 1997; 270:399–405.

40. Olsen J, Jørgensen P. Time-dependent response theory with applications to self-consistent field and multiconfigurational self-consistent field wave functions. In: Yarkony DR, ed. Modern Electronic Structure Theory. Singapore: World Scientific, 1995:857–990.

41. Ochsenfeld C, Gauss J, Ahlrichs R. An ab initio treatment of the electronic absorption spectra of excess-electron alkali halide clusters $Na_{n+1}Cl_n$ up to $Na_{18}Cl_{17}$. J Chem Phys 1995; 103:7401–7407.

42. Larsen H, Jørgensen P, Olsen J, Helgaker T. Hartree–Fock and Kohn–Sham atomic-orbital based time-dependent response theory. J Chem Phys 2000; 113:8908–8917.

43. Roos BO. The complete active space self-consistent field method and its applications in electronic structure calculations. Adv Chem Phys 1987; 69:399–445.

44. Shepard R. The multiconfigurational self-consistent field method. Adv Chem Phys 1987; 69:63–200.

45. Olsen J, Yeager DL, Jørgensen P. Optimization and characterization of a multiconfigurational self-consistent field (MCSCF) state. Adv Chem Phys 1983; 54:1–176.

46. Olsen J, Roos BO, Jørgensen P, Jensen HJAA. Determinant based algorithms for complete and restricted configuration interaction spaces. J Chem Phys 1988; 89:2185–2192.

47. Bartlett RJ. Coupled-cluster theory: an overview of recent developments. In: Yarkony DR, ed. Modern Electronic Structure Theory. Singapore: World Scientific, 1995:1047–1131.

48. Bartlett RJ, ed. Recent Advances in Coupled-Cluster Methods [Recent Advances in Computational Chemistry]. Vol. 3. Singapore: World Scientific, 1997.

49. Purvis GD III, Bartlett RJ. A full coupled-cluster singles and doubles model: the inclusion of disconnected triples. J Chem Phys 1982; 76:1910–1918.

50. Chiles RA, Dykstra CE. An electron pair operator approach to coupled cluster wave functions. Applications to He_2, Be_2 and Mg_2 and comparison with CEPA methods. J Chem Phys 1981; 74:4544–4556.

51. Handy NC, Pople JA, Head-Gordon M, Raghavachari K, Trucks GW. Size-consistent Brueckner theory limited to double substitutions. Chem Phys Lett 1989; 164:185–192.

52. Noga J, Bartlett RJ. The full CCSDT model for molecular electronic structure. J Chem Phys 1987; 86:7041–7050; J Chem Phys 1988; 89:3401–3401 [erratum].

53. Scuseria GE, Schaefer HF III. A new implementation of the full CCSDT model for molecular electronic structure. Chem Phys Lett 1988; 152:382–386.

54. Raghavachari K, Trucks GW, Pople JA, Head-Gordon M. A fifth-order perturbation comparison of electronic correlation theories. Chem Phys Lett 1989; 157:479–483.

55. Werner H-J. Matrix-formulated direct multiconfiguration self-consistent field and multiconfiguration reference configuration-interaction methods. Adv Chem Phys 1987; 69:1–62.

56. Merchán M, Serrano-Andrés L, Fülscher MP, Roos BO. Multiconfigurational perturbation theory applied to excited states of organic compounds. In: Hirao K, ed. Recent Advances in Multireference Theory. Vol. 4. Singapore: World Scientific, 1999:161–195.

57. Saebø S, Pulay P. A low-scaling method for second order Møller–Plesset calculations. J Chem Phys 2001; 115:3975–3983.

58. Ayala PY, Scuseria GE. Linear scaling second-order Møller–Plesset theory in the atomic orbital basis for large molecular systems. J Chem Phys 1999; 110:3660–3671.

59. Scuseria GE, Ayala PY. Linear scaling coupled cluster and perturbation theories in the atomic orbital basis. J Chem Phys 1999; 111:8330–8343.

60. Schütz M, Hetzer G, Werner H-J. Low-order scaling local electron correlation methods. I. Linear scaling local MP2. J Chem Phys 1999; 111:5691–5705.

61. Schütz M, Werner H-J. Low-order scaling local electron correlation methods. IV. Linear scaling local coupled-cluster (LCCSD). J Chem Phys 2001; 114:661–681.

62. Schütz M. Low-order scaling local electron correlation methods. V. Connected triples beyond (T): linear scaling local CCSDT-1b. J Chem Phys 2002; 116:8772–8785.

63. Lee MS, Maslen PE, Head-Gordon M. Closely approximating second-order Møller–Plesset perturbation theory with a local triatomics in molecules model. J Chem Phys 2000; 112:3592–3601.

64. Weigend F, Häser M. RI-MP2: first derivatives and global consistency. Theor Chem Acc 1997; 97:331–340.

65. Murphy RB, Pollard WT, Friesner RA. Pseudospectral localized generalized Møller–Plesset methods with a generalized valence bond reference wave function: theory and calculation of conformational energies. J Chem Phys 1997; 106:5073–5084.

66. Schütz M, Lindh R. An integral direct, distributed-data, parallel MP2 algorithm. Theor Chim Acta 1997; 95:13–34.

67. Nielsen IMB, Janssen CL. Multi-threading: a new dimension to massively parallel scientific computation. Comp Phys Commun 2000; 128:238–244.

68. Helgaker T, Klopper W, Koch H, Noga J. Basis-set convergence of correlated calculations on water. J Chem Phys 1997; 106:9639–9646.

69. Klopper W. r_{12}-Dependent wavefunctions. In: Schleyer PvR, Allinger NL, Clark T, Gasteiger J, Kollman PA, Schaefer HF III, Scheiner PR, eds. Encyclopedia of Computational Chemistry. Vol. 4. Chichester: Wiley, 1998:2351–2375.

4

Density-Functional Theory

PAUL W. AYERS

McMaster University, Hamilton, Ontario, Canada

WEITAO YANG

Duke University, Durham, North Carolina, U.S.A.

1. MOTIVATION

Traditionally, quantum mechanical calculations for molecules and materials have been concerned with computing an accurate approximation to the electronic wave function, $\Psi(r_1, s_1, \ldots, r_N, s_N)$, for the electronic ground state (and/or a few of the more important excited states) of the system. From the wave function, one may compute all of the observable properties of the system by straightforward integration

$$Q[\Psi] \equiv \sum_{s_i} \int \int \cdots \int \left\{ \Psi^*(r_1, s_1, \ldots, r_N, s_N) \hat{Q} \Psi(r_1, s_1, \ldots, r_N, s_N) \right\} dr_1 \ldots dr_N,$$

(1)

where the Hermitian operator $\hat{Q}$ corresponding to the property Q may be determined using the quantum-classical correspondence principle. Because the observed value of a property depends upon the particular state being probed, Q is a function of the wave function. A function of a function is termed a *functional*, whence the suggestive notation on the left-hand side of Eq. (1).

Of paramount importance is the energy operator, called the Hamiltonian, which, for an N-electron M-atom molecule treated within the Born–Oppenheimer approximation, may be written as

$$\hat{H}_{\text{molecule}} \equiv \sum_{i=1}^{N} \left(-\frac{\hbar^2}{2m_e} \nabla_i^2 + \sum_{\alpha=1}^{M} -\frac{Z_\alpha e^2}{|r_i - R_\alpha|} + \sum_{j>i}^{N} \frac{e^2}{|r_i - r_j|} \right),$$

(2)

89

where m_e, e, and $\{r_i\}_{i=1}^N$ represent the mass, charge, and positions of the electrons, while $\{Z_\alpha\}_{\alpha=1}^M$ and $\{R_\alpha\}_{\alpha=1}^M$ represent the charges (atomic numbers) and positions of the nuclear centers, respectively. The energy of a molecule is then

$$W_{\text{molecule}}[\Psi] \equiv \sum_{\alpha=1}^M \sum_{\beta=\alpha+1}^M \frac{Z_\alpha Z_\beta e^2}{|R_\alpha - R_\beta|} + \left\langle \Psi | \hat{H}_{\text{molecule}} | \Psi \right\rangle, \tag{3}$$

where we have introduced the standard bracket shorthand for integrals of the form of Eq. (1). In passing, we note that the wave functions under consideration are normalized,

$$1 \equiv \langle \Psi | \Psi \rangle, \tag{4}$$

and antisymmetric with respect to exchange of spatial and spin coordinates,

$$\Psi(\ldots r_i, s_i, \ldots r_j, s_j, \ldots) = -\Psi(\ldots r_j, s_j, \ldots r_i, s_i, \ldots). \tag{5}$$

Given the importance of the electronic wave function to the description of molecular systems, a brief word on how one determines the electronic wave function is justified. The key idea is that the "best" wave function for the ground state is the one with the lowest energy. That is, no "trial" wave function, $\tilde{\Psi}$, has a lower energy than the exact ground-state wave function, $\Psi_{\text{g.s.}}$,

$$\left\langle \Psi_{\text{g.s.}} | \hat{H}_{\text{molecule}} | \Psi_{\text{g.s.}} \right\rangle \leq \left\langle \tilde{\Psi} | \hat{H}_{\text{molecule}} | \tilde{\Psi} \right\rangle \tag{6}$$

This leads to the computationally useful variational principle: minimizing $E[\Psi] \equiv \langle \Psi | \hat{H}_{\text{molecule}} | \Psi \rangle$ with respect to wave functions which satisfy the constraints from Eqs. (4) and (5) yields the exact ground-state energy. Performing this minimization by introducing a Lagrange multiplier for the normalization constraint of Eq. (4), one obtains the famous Schrödinger equation

$$\hat{H}_{\text{molecule}} \Psi = E\Psi. \tag{7}$$

The problem arises not with the elegant simplicity of Eqs. (1)–(7), but rather when we actually try to find an accurate numerical approximation to the ground-state wave function: the Hamiltonian operator is a second-order partial differential operator in $3N$ real-valued coordinates (r_i) and N dichotomic ($s_i = \pm 1/2$) spin coordinates. Unless the number of electrons is very small or the Hamiltonian is of a very special form (so that the electrons are not coupled together), finding the wave function of lowest energy is a very difficult task, with the computational cost growing exponentially as the number of electrons increases. While the computational expense of practical approaches for approximating the ground-state wave function tends to grow more slowly, the computational requirements of such methods (both in terms of computer time and memory) still tend to grow much faster than the size of the molecule under consideration,

$$\cos t \propto (\text{size})^k, \tag{8}$$

where $k > 1$ ($k = 7$ is not uncommon). This nonphysical scaling severely limits the size of systems that can be treated. For instance, models for enzymatic catalysis routinely require quantum mechanical treatment of an active site region that might contain 400 atoms, of which ca. 100 are non-hydrogen atoms (metal centers, carbon, nitrogen, oxygen, etc.); this is expected to be several orders of magnitude more difficult than the "diatomic + atom" reactions often used as benchmarks for computational methods in chemical dynamics.

Clearly, then, accurate treatments of most chemical processes of biological interest are well outside the scope of traditional methods for determining the ground-state wave function. Recalling that the source of the problem can be traced to the prohibitive dimensionality of the wave function, we speculate that if we could express a molecule's properties as functionals of a function with fewer coordinates than the wave function, then some of the problems encountered when treating large systems might be surmounted. This is the motivation for density-functional theory, in which the ground-state electron density replaces the electronic wave function as the basic descriptor of molecular systems. In the following, we review the theoretical background of density-functional theory (Sec. 2), review computational aspects of the theory with emphasis on its simplicity relative to wave function-based approaches (Sec. 3), and review the intuitive picture of chemical reactivity that emerges naturally from the density-functional theory (Sec. 4). As no attempt at full mathematical rigor will be made, Sec. 5 reviews the literature with special emphasis on the issues (none of which are serious) that are glossed over in the preceding sections.

2. DENSITY-FUNCTIONAL THEORY

2.1. The External Potential and the Electron Density

The reader familiar with the historical development of quantum mechanics can be forgiven if they greet with skepticism the notion that one can extract all the information contained in a ground-state electronic wave function from the probability distribution function for observing an electron at the point r (that is, the ground-state electron density). The first hint that such a construction might be possible follows directly from the form of the molecular Hamiltonian, Eq. (2). Consider that the form of the kinetic energy operator

$$\hat{T} \equiv \sum_{i=1}^{N} \left(-\frac{\hbar^2}{2m_c} \nabla_i^2 \right), \tag{9}$$

and the electron–electron repulsion energy operator

$$\hat{V}_{ee} = \sum_{i=1}^{N} \left(\sum_{j=i+1}^{N} \frac{e^2}{|r_i - r_j|} \right), \tag{10}$$

are determined by the fact that we are interested in an N-electron system. The only parts of the Hamiltonian that change when the electronic system changes are the number of electrons, N, and the potential the electrons feel due to their "external" environment—that is, those particles/fields that are not due to other electrons. For a system of N electrons bound by an inhomogeneous and non-isotropic electrostatic potential, $v(r)$, one may write

$$\hat{H}_{\text{molecule}} \equiv \hat{T} + \hat{V}_{ee} + \sum_{i=1}^{N} v(r_i)$$

$$= \hat{F} + \sum_{i=1}^{N} v(r_i). \tag{11}$$

Note what we have gained: because the Hamiltonian operator determines the wave function for the system from the variational principle, it follows that any property, Q, of the ground state of an electronic system may be written as a function of the number of electrons, N, and a functional of a real-valued trivariate function, $v(r)$, which we call the *external potential*. We denote this functional $Q[v(r); N]$. Unfortunately, no expression for $Q[v(r); N]$ with computational utility comparable to Eq. (1) is known. However, the fact that properties of a system can be expressed as a function of N and a functional of a single trivariate function, $v(r)$, does suggest that there might be a computational useful theory in terms of a trivariate function.

To motivate the subsequent development, recall that the wave function of a system possesses no direct physical significance. Rather, the most informative observable property of the wave function is its complex square, $|\Psi(r_1, s_1, \ldots, r_N, s_N)|^2$, which, according to the Born postulate, represents the probability that an electron has spin s_1 and is located at r_1, another electron has spin s_2 and is located at r_2, etc. Thus

$$\rho_N(r_1, r_2, \ldots r_N) \equiv \sum_{s_i} |\Psi(r_1, s_1, \ldots, r_N, s_N)|^2 \tag{12}$$

is the probability distribution function for the electrons in the system. A related trivariate function is the probability of observing an electron at the point r; this defines the electron density, $\rho(r)$, at any point in space. From the fact the electron density at the point r is the sum of the probabilities of any of the N electrons being at r (the other $N-1$ electrons can be anywhere in space),

$$\rho(r) \equiv \sum_{i=1}^{N} \int \int \cdots \int \delta(r_i - r) \rho_N(r_1, r_2, \ldots, r_N) dr_1 \ldots dr_N$$

$$= \left\langle \Psi \left| \sum_{i=1}^{N} \delta(r_i - r) \right| \Psi \right\rangle. \tag{13}$$

Insofar as the electron density is the probability of observing an electron at a point, it is clearly nonnegative; it is also observable experimentally. It is also clear from Eq. (13) that

$$N[\rho] \equiv \int \rho(r) dr; \tag{14}$$

that is, the number of electrons is a functional of the electron density.

2.2. The Ground-State Electron Density as a Descriptor of Electronic Systems

The sweeping theorem of Hohenberg and Kohn is that, like the wave function, "the ground state's electron density determines all the properties of an electronic system" [1]. The result is proved in three steps. First, one recalls that the number of electrons is determined from the electron density using Eq. (14). Next, one demonstrates that the external potential can be determined from the ground-state electron density. From N and $v(r)$, we may determine the electronic Hamiltonian and solve Schrödinger's equation for the wave function, subsequently determining all observable properties of the system.

A key to this development is the assertion that the ground-state electron density determines the electronic external potential. Stated mathematically, we must demonstrate that the external potential is a functional of the electron density. Just as f is a function of x if and only if no argument of f, x_0, corresponds to more than one value of f (which is to say that $f(x)$ is single-valued for all x in the domain of f), the external potential, $v(r)$ is a functional of the ground-state electron density, $\rho(r)$, if and only if no two external potentials correspond to the same ground-state electron density.

That this is true follows directly from the variational principle for the wave function. Consider two different N-electron systems; these systems have different ground-state wave functions, Ψ_0 and Ψ_1, and external potentials, $v_0(r)$ and $v_1(r)$, which differ by more than an additive constant. From the variational principle for the energy, Eq. (6),

$$\left\langle \Psi_0 \middle| \hat{H}_0 \middle| \Psi_0 \right\rangle < \left\langle \Psi_1 \middle| \hat{H}_0 \middle| \Psi_1 \right\rangle$$

$$\left\langle \Psi_1 \middle| \hat{H}_1 \middle| \Psi_1 \right\rangle < \left\langle \Psi_0 \middle| \hat{H}_1 \middle| \Psi_0 \right\rangle \tag{15}$$

Adding these equations [with substitution of the defining Eq. (11)] yields

$$\left\{ \left\langle \Psi_0 \middle| \hat{F} \middle| \Psi_0 \right\rangle + \left\langle \Psi_0 \middle| \sum_{i=1}^{N} v_0(r_i) \middle| \Psi_0 \right\rangle \right\}$$

$$+ \left\{ \left\langle \Psi_1 \middle| \hat{F} \middle| \Psi_1 \right\rangle + \left\langle \Psi_1 \middle| \sum_{i=1}^{N} v_1(r_i) \middle| \Psi_1 \right\rangle \right\}$$

$$< \left\{ \left\langle \Psi_0 \middle| \hat{F} \middle| \Psi_0 \right\rangle + \left\langle \Psi_0 \middle| \sum_{i=1}^{N} v_1(r_i) \middle| \Psi_0 \right\rangle \right\} \tag{16}$$

$$+ \left\{ \left\langle \Psi_1 \middle| \hat{F} \middle| \Psi_1 \right\rangle + \left\langle \Psi_1 \middle| \sum_{i=1}^{N} v_0(r_i) \middle| \Psi_1 \right\rangle \right\},$$

which simplifies to

$$\left\langle \Psi_0 \middle| \sum_{i=1}^{N} (v_0(r_i) - v_1(r_i)) \middle| \Psi_0 \right\rangle - \left\langle \Psi_1 \middle| \sum_{i=1}^{N} (v_0(r_i) - v_1(r_i)) \middle| \Psi_1 \right\rangle < 0. \tag{17}$$

From the definition of the electron density, Eq. (13), we obtain the key relation

$$\int (\rho_0(r) - \rho_1(r))(v_0(r) - v_1(r))dr < 0, \tag{18}$$

where $\rho_0(r)$ and $\rho_1(r)$ are the ground-state densities for the N-electron systems with external potentials $v_0(r)$ and $v_1(r)$, respectively. Since $v_0(r) \neq v_1(r)$ by assumption, Eq. (18) implies that $\rho_0(r) \neq \rho_1(r)$. Thus no two external potentials correspond to the same ground-state electron density. This result, first obtained by Hohenberg and Kohn [1] in 1964, is generally called the (first) Hohenberg–Kohn theorem. The present proof is modeled after the more general considerations of Levy [2] and Englisch and Englisch [3].

2.3. The Variational Principle for the Ground-State Electron Density

The first Hohenberg–Kohn theorem is an existence theorem: it indicates that we can, in principle, determine the ground-state wave function, $\Psi[\rho]$, the "purely electronic" contribution to the total energy, $F[\rho] = \langle \Psi[\rho] | \hat{F} | \Psi[\rho] \rangle$, the total energy $E[\rho]$, and all other properties of an electronic system directly from the ground-state electron density. Leveraging the Hohenberg–Kohn theorem to practical applications requires a method for accurately determining the ground-state electron density. Recalling the utility of the variational principle for determining the wave function, we now wish to derive a variational principle for the ground-state electron density.

To do this, consider a system consisting of N-electrons, with electron density $\rho_0(r)$, confined by external potential, $v_1(r)$, for which $\rho_0(r)$ is not a ground-state electron density. The purely electronic contribution to the energy,

$$F[\rho_0] = \langle \Psi_0 | \hat{F} | \Psi_0 \rangle, \tag{19}$$

does not depend on the external potential; such *universal functionals* play a key role in density-functional theory. The interaction energy between the external forces on the electrons and the electrons is, from classical electrostatics,

$$V_{\text{ext}}[v_1, \rho_0] = \int \rho_0(r) v_1(r) dr, \tag{20}$$

and so the total energy of this system can be written as

$$E_{v_1}[\rho_0] \equiv F[\rho_0] + \int \rho_0(r) v_1(r) dr. \tag{21}$$

Here we adopt the standard notation, which explicitly denotes the fact that the external potential, $v_1(r)$, is a parameter that is constructed at the beginning of the variational calculation from the molecular geometry and external fields of interest.

The comparison between the energies of the "wrong" electron density $\rho_0(r)$ and a ground-state electron density for $v_1(r)$ follows directly from Eq. (16) and the defining Eq. (21):

$$E_{\text{g.s.}}[\rho_1] = E_{v_1}[\rho_1] \equiv F[\rho_1] + \int \rho_1(r) v_1(r) dr$$

$$\leq E_{v_1}[\rho_0] \equiv F[\rho_0] + \int \rho_0(r) v_1(r) dr; \tag{22}$$

the equality holds only when $\rho_0(r)$ is an electron density for this system (which, by assumption, it is not). Equation (22), which is often referred to as the second Hohenberg–Kohn theorem [1], is the foundation of all practical procedures for finding the ground-state electron density:

$$E_{\text{g.s.}} = E_v[\rho_{\text{g.s.}}] = \underbrace{\min}_{\text{all } N-\text{electron } \rho(r)} E_v[\rho]. \tag{23}$$

2.4. Recapitulation: Analogies to the Wave Function Theory

It is instructive to compare the Hohenberg–Kohn theorems to their counterparts in conventional wave function–based quantum theory. The wave function provides a complete quantum mechanical description of any state of any system. The first Hohenberg–Kohn theorem indicates that the ground-state electron density determines all the properties of any electronic system, by which we mean a system whose Hamiltonian operator assumes the specific form of Eq. (11). The practical utility of quantum mechanics depends upon efficient computational methods for determining the wave function; the most fundamental of these is the variational principle, Eq. (6). Similarly useful in the density-functional theory context is the second Hohenberg–Kohn theorem, Eq. (22).

Instead of direct implementation of the variational principle, one often seeks to solve the associated Schrödinger equation, Eq. (7). A similarly useful equation in density-functional theory is derived from Eq. (23): given the N-electron ground-state electron density, $\rho_N(r)$, for a system, $v(r)$, with a nondegenerate ground state, all other N-electron densities have greater energy, $E_v[\rho] > E_v[\rho_N]$. Subject to sufficient smoothness in the energy functional, this indicates that the energy functional is stationary with respect to small normalization-preserving perturbations of the ground-state electron density. That is, given a function, $\Delta(r)$, for which $\int \Delta(r) dr = 0$, then

$$E_v[\rho_0(r) + \varepsilon\Delta(r)] - E_v[\rho_0(r)] \propto \varepsilon^2 \tag{24}$$

in the limit of small ε. Thus,

$$\left.\frac{dE_v[\rho_0(r) + \varepsilon\Delta(r)]}{d\varepsilon}\right|_{\varepsilon=0} = 0. \tag{25}$$

The restriction in Eq. (25) to variations for which $\int \Delta(r) dr = 0$ is computationally inconvenient. To avoid this, we introduce the notion of a functional derivative, $\left.\frac{\delta E_v[\rho]}{\delta\rho(r)}\right|_{\rho=\rho_0}$. Just as the gradient of a function at a point is *defined* as that vector, $\nabla f(x)|_{x=x_0}$, which maps small changes in x about x_0 to the resulting changes in the value of the function according to

$$\lim_{\varepsilon\to0}\left[f(x_0 + \varepsilon d) - f(x_0)\right] = \lim_{\varepsilon\to0}\nabla f(x)|_{x=x_0} \cdot \varepsilon d = \lim_{\varepsilon\to0}\varepsilon\sum_{i=1}^{\dim(x)}\left\{\nabla f(x)|_{x=x_0}\right\}_i d_i, \tag{26}$$

the functional derivative is *defined* as the function that maps small changes in the electron density to small changes in the energy according to the equation:

$$\lim_{\varepsilon\to0} E_v[\rho_0 + \varepsilon\Delta] - E_v[\rho_0] \equiv \lim_{\varepsilon\to0}\int \left.\frac{\delta E_v[\rho]}{\delta\rho(r)}\right|_{\rho=\rho_0} \varepsilon\Delta(r)\,dr$$

$$\lim_{\varepsilon\to0}\frac{E_v[\rho_0 + \varepsilon\Delta] - E_v[\rho_0]}{\varepsilon} = \int \left.\frac{\delta E_v[\rho]}{\delta\rho(r)}\right|_{\rho=\rho_0}\Delta(r)\,dr. \tag{27}$$

Choosing $\Delta(r) = \delta(r - r_0)$ in Eq. (27), we obtain the computationally useful formula

$$\left.\frac{\delta E_v[\rho]}{\delta\rho(r_0)}\right|_{\rho=\rho_0} = \lim_{\varepsilon\to0}\frac{E_v[\rho_0(r) + \varepsilon\delta(r - r_0)] - E_v[\rho_0(r)]}{\varepsilon}. \tag{28}$$

We can now express the density-functional variational principle in terms of functional derivatives, namely,

$$\left.\frac{\delta(E_v[\rho] - \mu N[\rho])}{\delta\rho(r)}\right|_{\rho=\rho_N} = 0, \tag{29}$$

where $\rho_N(r)$ is the ground-state N-electron density for the external potential $v(r)$ and the Lagrange multiplier, $\mu[\rho_N]$, constrains the variation to N-electron densities. To find the value of the Lagrange multiplier, consider that the stationary condition holds also for small changes in the electron density associated with changing the number of electron in the system; thus:

$$\frac{\partial\left[E_v[\rho] - \mu N[\rho]\right]}{\partial N} = 0. \tag{30}$$

There results

$$\mu = \left(\frac{\partial E}{\partial N}\right)_{v(r)}, \tag{31}$$

allowing us to identify μ as the chemical potential for the electrons in this system [4]. The introduction of the chemical potential in the density-functional variational principle, Eq. (30), is analogous to the transformation to the grand canonical ensemble in statistical mechanics [5].

Simplifying Eq. (29) yields a functional differential equation for the ground-state density,

$$\frac{\delta E_v[\rho]}{\delta\rho(r)} = \mu = \text{constant}, \tag{32}$$

that is directly analogous to Schrödinger's partial differential equation for the wave function. This is most apparent when the Schrödinger equation is written in the form

$$\frac{\hat{H}\Psi}{\Psi} = E = \text{constant}. \tag{33}$$

Equations (32) and (33) possess comparable importance and similar utility in the density-functional and wave-functional approaches to quantum mechanical systems, respectively.

3.   FURTHER DEVELOPMENTS IN DENSITY FUNCTIONAL THEORY (DFT)

3.1.   The Kohn–Sham Equations

In principle, solving Eq. (32) provides a straightforward approach to the ground-state electronic energy, ground-state electron density, and from the density, certain other properties of the ground-state system. Unlike the analogous Eq. (33), however, where the energy functional is known from the Hamiltonian operator, the exact expression for $E_v[\rho]$ is not known, or, more accurately, it is not known in an explicit and computational tractable form. What we have done is to trade an extremely difficult computational problem (solving the Schrödinger equation) for an extremely challenging theoretical problem (finding accurate approximations for $E_v[\rho]$).

The core problem is finding adequate approximations for the Hohenberg–Kohn functional [cf. Eq. (21)]

$$F[\rho] = E[\rho] - \int \rho(r)v[\rho;r]\,dr. \tag{34}$$

Indeed, in the formative years of quantum mechanics, there was substantial interest in determining the properties of systems directly from the electron density, with key developments being the Thomas–Fermi theory (1927–1928) [6,7], the Dirac [8] exchange correction (1930), the Wigner [9] correction for the correlation energy (1934), and the Weizsacker [10] functional for the kinetic energy (1935). In 1964, Hohenberg and Kohn proved that there exists a functional such that the solution to Eq. (32) gives the exact ground-state electron density and the exact ground-state energy of any system of electrons. Decomposing the Hohenberg–Kohn functional into its kinetic energy and electron–electron repulsion energy components,

$$F[\rho] \equiv T[\rho] + V_{ee}[\rho], \tag{35}$$

one observes that the primary problem is not the inadequate approximations for $V_{ee}[\rho]$ (even the most primitive of models, combining Coulomb repulsion, Dirac exchange, and Wigner correlation, gives a rather reasonable result), but the inadequate approximation of the kinetic energy. Most notably, the kinetic energy functionals of Thomas–Fermi and Weizsacker fail to adequately account for the influence of the Pauli exclusion principle on electrons [11,12]. Still, as the simplest explicit density-functional theory methods, Thomas–Fermi type models possess substantial theoretical importance and hence have been studied extensively [13,14].

In the pursuit of quantitative accuracy from density-functional theory, one typically abandons the idea of expressing the kinetic energy directly in terms of the electron density. Instead, one introduces a set of auxiliary functions, the Kohn–Sham orbitals, $\{\psi_i[\rho;r]\}_{i=1}^{\infty}$, themselves functionals of the ground-state electron density, and computes an accurate approximation to the kinetic energy from these orbitals according to the formula [15],

$$T[\rho] \equiv T_s[\rho] + T_c[\rho] \equiv \sum_{i=1}^{N/2} 2 \left\langle \psi_i[\rho;r] \left| -\frac{\hbar^2}{2m_e} \nabla^2 \right| \psi_i[\rho;r] \right\rangle + T_c[\rho]. \tag{36}$$

The limits on the summation indicate that the sum only runs over those orbitals which are occupied (according to the aufbau rule); moreover, in this section, we restrict ourselves to closed shell systems and so each orbital is doubly occupied. Fortunately, the correction to the Kohn–Sham kinetic energy, $T_s[\rho]$, is quite small. For notational simplicity and compactness, we shall henceforth employ atomic units.

To motivate the Kohn–Sham method, we return to molecular Hamiltonian [Eq. (2)] and note that, were it not for the electron–electron repulsion terms coupling the electrons, we could write the Hamiltonian operator as a sum of one-electron operators and solve Schrödinger equation by separation of variables. This motivates the idea of replacing the electron–electron repulsion operator by an average local representation thereof, $w(r)$, which we may term the "internal potential." The Hamiltonian operator becomes

$$\hat{H} \equiv \sum_{i=1}^{N} \left(-\frac{\nabla_i^2}{2} + v(r_i) + w(r_i) \right), \tag{37}$$

and, upon separation of variables, solving the Schrödinger equation is equivalent to solving the one-electron eigenproblems

$$\left(-\frac{\nabla^2}{2} + v(r) + w(r) \right) \psi_i(r) = \varepsilon_i \psi_i(r). \tag{38}$$

The associated approximation to the system's ground-state wave function is obtained as the antisymmetric product of lowest-energy orbitals, $\{\psi_i(r)\}_{i=1}^{N/2}$, with appropriate spin factors; we denote this as Slater determinate Φ. Excepting certain limiting cases, Φ is an inadequate approximation to the true wave function, as obtained by solving the Schrödinger equation.

Before proceeding further, it is necessary that we choose a method for constructing the internal potential, $w(r)$. The insight of Kohn and Sham was to choose the internal potential so that the systems defined by the true Hamiltonian [Eq. (11)] and the model Hamiltonian [Eq. (37)] have the same ground-state electron density [15], This can occur only if the ground-state energy density functionals, $E_v[\rho]$ [cf. Eq. (21)], and

$$E_v^{KS}[\rho] \equiv T_s[\rho] + \int \rho(r)(v(r) + w(r))\mathrm{d}r$$
$$= \sum_{i=1}^{N/2} 2\left\langle \psi_i \left| -\frac{\nabla^2}{2} \right| \psi_i \right\rangle + \int \rho(r)(v(r) + w(r))\mathrm{d}r, \tag{39}$$

are minimized by the same electron density. Referring to Eq. (32), we may express this condition as:

$$\left.\frac{\delta E_v[\rho]}{\delta\rho(r)}\right|_{\rho=\rho_0} = \mu$$
$$\left.\frac{\delta E_v^{KS}[\rho]}{\delta\rho(r)}\right|_{\rho=\rho_0} = \mu_{KS}. \tag{40}$$

Simplifying, one obtains

$$\left.\frac{\delta T[\rho]}{\delta\rho(r)}\right|_{\rho=\rho_0} + \left.\frac{\delta V_{ee}[\rho]}{\delta\rho(r)}\right|_{\rho=\rho_0} + v(r) + \mu - \mu_{KS} = \left.\frac{\delta T_s[\rho]}{\delta\rho(r)}\right|_{\rho=\rho_0} + v(r) + w(r)$$
$$\left.\frac{\delta V_{ee}[\rho]}{\delta\rho(r)}\right|_{\rho=\rho_0} + \left.\frac{\delta T_c[\rho]}{\delta\rho(r)}\right|_{\rho=\rho_0} + \mu - \mu_{KS} = w(r), \tag{41}$$

where we have used definitions (35) and (36). Insofar as the zero of energy is arbitrary, the Kohn–Sham chemical potential, μ_{KS}, is usually (but not always [16]) taken to be the same as the true chemical potential. For historical reasons and computational facility, the classical electrostatic repulsion energy functional,

$$J[\rho] \equiv \frac{1}{2} \int \int \frac{\rho(r)\rho(r')}{|r - r'|} \mathrm{d}r\mathrm{d}r' \tag{42}$$

is usually separated from the "non-electrostatic" terms in $V_{ee}[\rho]$, which are combined with $T_c[\rho]$ to form the exchange-correlation energy,

$$E_{xc}[\rho] \equiv V_{ee}[\rho] - J[\rho] + T_c[\rho]. \tag{43}$$

The equation for the internal potential then becomes

$$w(r) = \left.\frac{\delta J[\rho]}{\delta\rho(r)}\right|_{\rho=\rho_0} + \left.\frac{\delta E_{xc}[\rho]}{\delta\rho(r)}\right|_{\rho=\rho_0} \tag{44}$$

For notational simplicity, this is generally simplified further by defining the electro-static potential, $v_J[\rho; r] \equiv \frac{\delta J[\rho]}{\delta \rho(r)}$, and the exchange-correlation potential, $v_{xc}[\rho; r] \equiv \frac{\delta E_{xc}[\rho]}{\delta \rho(r)}$. Substitution of these results into Eq. (37) yields the celebrated Kohn–Sham equations [15]

$$\left(-\frac{\nabla^2}{2} + v(r) + v_J[\rho; r] + v_{xc}[\rho; r]\right)\psi_i(r) = \varepsilon_i \psi_i(r). \tag{45}$$

$$\rho(r) \equiv \left\langle \Phi_{KS} \left| \sum_{i=1}^{N} \delta(r_i - r) \right| \Phi_{KS} \right\rangle = \sum_{i=1}^{N/2} 2|\psi_i(r)|^2 \tag{46}$$

Because Eq. (45) depends on the electron density and Eq. (46) depends on the Kohn–Sham orbitals, these two equations must be solved self-consistently. The procedure for solving the Kohn–Sham system, then, is to guess an electron density, construct $v_J[\rho; r] + v_{xc}[\rho; r]$, and solve Eq. (45), subsequently obtaining a new electron density from Eq. (46). Unless the electron density from Eq. (46) equals the "guess density," one proceeds to construct a (suitably improved) guess for the electron density and repeats the process until the input density and the output density are the same.

The Kohn–Sham wave function, Φ_{KS}, is not expected to be a good approximation to the exact wave function; indeed, it is a worse approximation to the exact wave function than the Hartree–Fock wave function. However, unlike the electron density obtained from the Hartree–Fock equations, the Kohn–Sham method yields, in principle, the *exact* electron density. Thus we do not need to use the Kohn–Sham wave function to compute the properties of chemical systems. Rather, motivated by the first Hohenberg–Kohn theorem, we compute properties directly from the Kohn–Sham electron density. How one does this, for any given system and for any property of interest, is an active topic of research.

Because of its critical role in constructing the potential energy surface [cf. Eq. (3)] for a molecule, thence in the prediction of molecular structure and chemical reactivity, we mention how one may compute the electronic energy of a system using the Kohn–Sham method. In particular, one has

$$
\begin{aligned}
E[v; N] &\equiv E_v[\rho_{g.s.}] \\
&= T_s[\rho_{g.s.}] + J[\rho_{g.s.}] + \int \rho_{g.s.}(r)\, v(r)\,dr + E_{xc}[\rho_{g.s.}] \\
&= 2\sum_{i=1}^{N/2} \varepsilon_i - J[\rho_{g.s.}] + E_{xc}[\rho_{g.s.}] - \int \rho_{g.s.}(r) v_{xc}(r)\,dr.
\end{aligned}
\tag{47}
$$

This general form, in which the value of a property is computed expressed in terms of its value for the Kohn–Sham system plus a correction dependent on the exchange-correlation energy, recurs throughout Kohn–Sham density-functional theory.

3.2. Spin Density-Functional Theory

The Kohn–Sham equations as presented in the previous section are most useful for systems in which all electrons are paired. For systems with nonvanishing total spin, it is more convenient computationally (but by no means essential theoretically) to, taking a cue from unrestricted Hartree–Fock theory, construct spin-dependent Kohn–Sham equations. Briefly, then, the key elements of spin density-functional theory are

(a) The spin density for the α and β spin electrons, $\rho_\alpha(r) + \rho_\beta(r) = \rho(r)$.

(b) The exchange-correlation spin density functional, $E_{xc}[\rho_\alpha,\rho_\beta]$, and its functional derivatives, $v_{xc,\sigma}[\rho_\alpha,\rho_\beta; r] \equiv \frac{\delta E_{xc}[\rho_\alpha,\rho_\beta]}{\delta\rho_\sigma(r)}$ $(\sigma = \alpha,\beta)$.

(c) The unrestricted Kohn–Sham equations [17,18].

$$\left\{\left(-\frac{\nabla^2}{2} + v(r) + v_J[\rho; r] + v_{xc;\sigma}[\rho_\alpha,\rho_\beta; r]\right)\psi_{k;\sigma}(r) = \varepsilon_{k;\sigma}\psi_{k;\sigma}(r).\right\}_{\sigma=\alpha,\beta} \tag{48}$$

$$\rho_\sigma(r) \equiv \sum_{i=1}^{N_\alpha} |\psi_{i,\sigma}(r)|^2; \ \sigma = \alpha,\beta \tag{49}$$

Similar to the spin-compensated case, the solution of the unrestricted Kohn–Sham equations starts with the external potential and the number of spin α and spin β electrons in the state of interest (denoted N_α and N_β, respectively); then Eqs. (48) and (49) are solved until consistency is achieved. Using the Kohn–Sham orbitals and orbital energies, one then computes the total energy of the system using the spin-dependent generalization of Eq. (47),

$$E[v; N_\alpha, N_\beta] = \sum_{k=1}^{N_\alpha} \varepsilon_{k;\alpha} + \sum_{k=1}^{N_\beta} \varepsilon_{k;\beta} - J[\rho] + E_{xc}[\rho_\alpha,\rho_\beta]$$

$$- \int \rho_\alpha(r)v_{xc;\alpha}[\rho_\alpha,\rho_\beta; r]dr - \int \rho_\beta(r)v_{xc;\beta}[\rho_\alpha,\rho_\beta; r]dr. \tag{50}$$

For simplicity, we shall, throughout the remainder of this document, treat only the original Kohn–Sham equations, Eqs. (45) and (46).

3.3. Exchange-Correlation Energy Functionals

After solving the Kohn–Sham system, one may evaluate the total electronic energy of the system using Eq. (47) or Eq. (50), as appropriate. From these expressions for the energy and the dependence of the Kohn–Sham potential upon the functional derivative of the exchange-correlation energy functional, it is clear that the accuracy of a density-functional method is entirely dependent upon choosing an appropriate exchange-correlation energy functional, $E_{xc}[\rho]$. Indeed, if a practical and exact form for $E_{xc}[\rho]$ was known, then Kohn–Sham calculations employing this functional would give the exact energy and the exact ground-state electron density.

No useful explicit form for $E_{xc}[\rho]$ is known, but approximate exchange-correlation energy functionals often provide an excellent approximation to the energetic properties of the molecule. To explain how approximate exchange-correlation functionals work (and when they fail), recall the essence of the Kohn–Sham method: the Kohn–Sham method constructs the model system with the energy functional

$$E_{v_{KS}}^{KS}[\rho] \equiv \langle\Phi|\hat{T}|\Phi\rangle + \int \rho(r) v_{KS}(r)dr, \tag{51}$$

that has the exact same ground-state electron density as the real system of interest, which is associated with the energy expression,

$$E_v[\rho] = \left\langle \Psi \middle| \hat{T} + \hat{V}_{ee} \middle| \Psi \right\rangle + \int \rho(r)v(r)\mathrm{d}r. \tag{52}$$

Starting from these disparate systems, we construct a whole range of intermediate systems, all with identical electron density but with incrementally increasing strengths for the electron–electron repulsion term:

$$E_v^\lambda[\rho] \equiv \left\langle \Psi^\lambda \middle| \hat{T} + \lambda\hat{V}_{ee} \middle| \Psi^\lambda \right\rangle + \int \rho(r)v_\lambda(r)\mathrm{d}r; \tag{53}$$

clearly $\lambda = 0$ corresponds to the Kohn–Sham model and $\lambda = 1$ corresponds to the system of interest. From the fundamental theorem of calculus,

$$
\begin{aligned}
E_{xc}[\rho] &= T[\rho] + V_{ee}[\rho] - T_s[\rho] - J[\rho] \\
&= \left\langle \Psi^\lambda \middle| \hat{T} + \lambda\hat{V}_{ee} \middle| \Psi^\lambda \right\rangle \bigg|_{\lambda=0}^{\lambda=1} - J[\rho] \\
&= \int_0^1 \frac{\mathrm{d}}{\mathrm{d}\lambda} \left\langle \Psi^\lambda \middle| \hat{T} + \lambda\hat{V}_{ee} \middle| \Psi^\lambda \right\rangle \bigg|_{\lambda=0}^{\lambda=1} \mathrm{d}\lambda - J[\rho].
\end{aligned}
\tag{54}
$$

Application of the Hellmann–Feynman theorem yields a simple and useful expression for the exchange-correlation energy [19,20]

$$E_{xc}[\rho] = \int_0^1 \left\langle \Psi^\lambda \middle| \hat{V}_{ee} \middle| \Psi^\lambda \right\rangle \mathrm{d}\lambda - J[\rho]. \tag{55}$$

This approach to the exchange-correlation energy is known as the adiabatic connection formalism [21–24].

Defining $\rho_2^\lambda(x,x')$ to be the probability of observing a pair of electrons, one at x and one at x',

$$\rho_2^\lambda(x, x') \equiv \left\langle \Psi^\lambda \middle| \sum_{i=1}^N \sum_{j \neq 1} \delta(r_i - x)\,\delta(r_j - x') \middle| \Psi^\lambda \right\rangle, \tag{56}$$

we obtain from Eq. (55) the working expression

$$E_{xc}[\rho] - \frac{1}{2} \int \int \frac{\bar{\rho}_2(x,x')}{|x-x'|}\,\mathrm{d}x\mathrm{d}x' - \frac{1}{2} \int \int \frac{\rho(x)\rho(x')}{|x-x'|}\,\mathrm{d}x\mathrm{d}x' \tag{57}$$

where $\bar{\rho}_2(x,x')$ is defined as the value of $\rho_2^\lambda(x,x')$ averaged over the adiabatic connection path. If, in analogy to the classical theory of liquids, we define the exchange-correlation hole as

$$\bar{h}(x, x') \equiv \bar{\rho}_2(x, x') - \rho(x)\rho(x') \tag{58}$$

then we may write the exchange-correlation energy in the compact form

$$E_{xc}[\rho] \equiv \frac{1}{2} \int \rho(x) \int \frac{\rho(x')\bar{h}(x,x')}{|x-x'|}\,\mathrm{d}x'\mathrm{d}x. \tag{59}$$

Approximate exchange-correlation density functionals differ in the approximations they make to the innermost integral in Eq. (59). A common assumption is that the exchange-correlation charge, $\bar{\rho}_{xc}(x,x') \equiv \rho(x')\bar{h}(x,x')$, when spherically averaged about $x' = x$:

$$\bar{\rho}_{xc}(x, |x - x'|) \equiv \frac{1}{4\pi} \int_0^{2\pi} \int_0^1 \rho_{xc}(x, |x - x'|, \theta_{xx'}, \phi_{xx'}) d(\cos\theta_{xx'}) d\phi_{xx'}, \quad (60)$$

is strongly localized near x. This is often a good assumption: it is often true that $\bar{\rho}_{xc}(x,|x-x'|)$ takes its minimum value at $|x-x'| = 0$ [where $\bar{\rho}_{xc}(x,0)$ is slightly larger than $-\rho(x)$] and decays rapidly and monotonically to its limiting asymptotic value

$$\lim_{|x-x'| \to \infty} \bar{\rho}_{xc}(x, |x - x'|) = 0 \quad (61)$$

as $|x-x'|$ increases [25]. This suggests that information about the electron density at and near the point x may be used to form an effective approximation to $\bar{\rho}_{xc}(x,|x-x'|)$.

The simplest choice, of course, is to express $\bar{\rho}_{xc}(x,|x-x'|)$ as a function of the electron density at x; this yields the class of models for the exchange-correlation energy known as local-density approximations (LDAs) [15],

$$E_{xc}^{LDA}[\rho] \equiv \int \rho(x) f(\rho(x)) dx. \quad (62)$$

Recognizing that information about how the electron density changes in the vicinity of the point x is also relevant; numerous other approximations express $\bar{\rho}_{xc}(x,|x-x'|)$ as a function of not only the density at the point x, but also the derivatives thereof. These generalized gradient approximations (GGAs) to the exchange-correlation energy take the general form [26–28]

$$E_{xc}^{CCA}[\rho] \equiv \int \rho(x) f(\rho(x), \nabla\rho(x), \nabla^2\rho(x), \ldots) dx. \quad (63)$$

Most recently developed functionals use either the GGA for or a generalization thereof. Of particular importance are hybrid functionals, which express the total exchange-correlation as a sum of an "exact" exchange (Hartree–Fock) term and a GGA term [29,30].

From this argument, one expects that neither LDAs nor GGAs are reliable when the exchange correlation charge is delocalized over several atomic regions. Such behavior is in fact observed: modern generalized gradient approximations are reliable and accurate when $\bar{\rho}_{xc}(x,|x-x'|)$ is localized and centered on x; errors are typically no more than a few kilocalories per mole. In some cases, however, $\bar{\rho}_{xc}(x,|x-x'|)$ does not increase more or less monotonically as one moves away from x, instead having additional minima in other nearby portions of the molecule. In such cases, errors are often an order of magnitude larger, frequently tens of kilocalories per mole. Even when $\bar{\rho}_{xc}(x,|x-x'|)$ is rather delocalized, however, generalized gradient approximations *sometimes* give reasonable results, owing mostly to the cancellation of errors [and helped by the fact that the factor of inverse distance in Eq. (59) helps to reduce the energetic importance of the "far away portion" of the exchange-correlation charge].

Given the importance of localized exchange-correlation charges to the accuracy of approximate density functionals, a rule of thumb for deciding when $\bar{\rho}_{xc}(x,|x-x'|)$ is likely to be localized is of great utility. One such rule, due to Zhang and Yang [31], is that a dissociating molecule will tend to have a nonlocalized hole when the ionization potential of one of the dissociating fragments resembles the electron affinity of the other. Another rule was proposed by Gritsenko, Ensing, Schipper, and Baerends (GESB) [32,33], who present a semiempirical argument that when the number of bonding electrons divided by the number of atomic orbitals composing a bond is an integer, then $\overline{\rho}_{xc}(x,|x-x'|)$ is usually localized, increasing more or less monotonically as one moves away from the reference point x. By contrast, systems with 1-electron 2-center bonds (as H_2^+) [31], 2-electron 3-center bonds (as the bridging bonds in diborane), 4-electron 3-center bonds (as the transition state in an S_N2 reaction) [33], and 3-electron, 2-center bonds (as F_2^-) tend to have exchange correlation charges with two or more significant minima and are problematic for every known density-functional theoretic technique [32], including many much more elaborate than the simple generalized gradient approximations. Indeed, systems of these types usually have substantial multireference character; such systems, then, are often problematic for conventional wave function-based approaches also.

Fortunately, for a wide range of molecular structure, including "normal" covalent bonds, ionic bonds, and closed-shell/closed-shell interactions, and for the most common reaction pathways (simple bond formation and cleavage), the GESB rule indicates that $\overline{\rho}_{xc}(x,|x-x'|)$ is relatively well localized. Modern GGAs are accurate and reliable predictors of molecular structure and reactivity for such systems. On the other hand, when the GESB rule suggests that $\overline{\rho}_{xc}(x,|x-x'|)$ is not localized, then it is essential that one carefully check the results of one's calculations, preferably by recourse to experiment or to a more conventional wave function-based technique. Density-functional theory can still be useful in these cases, especially since the energy obtained from density-functional calculations on these systems tends to be systematically too low: in assuming the exchange-correlation charge to be more highly localized than it actually is, GGAs typically overestimate the attraction between the electron and its hole in Eq. (59), thereby overestimating the magnitude of the exchange-correlation energy.

3.4. Linear-Scaling Methods for Solving the Kohn–Sham System

The Kohn–Sham construction is a pragmatic one, justified by computational utility. Of special computational utility is the fact that each Kohn–Sham orbital experiences the same potential and that this potential, in turn, is a functional of the electron density alone. This allows us to rewrite the Kohn–Sham energy in terms of the *first-order density matrix*,

$$\gamma_{KS}(r,r') \equiv \sum_{i=1}^{N/2} \left[\psi_i^*(r)\psi_i(r') + \text{c.c.} \right] \tag{64}$$

Specifically,

$$E_{v_{KS}}^{KS}[\gamma] \equiv \int\int \delta(r-r')\left(-\frac{\nabla_r^2}{2}\right)\gamma(r,r')drdr' + \int \rho(r)v_{KS}(r)dr, \tag{65}$$

where the electron density is given by

$$\rho(\mathbf{r}) \equiv \gamma(\mathbf{r}, \mathbf{r}) \tag{66}$$

and we have defined the Kohn–Sham potential according to

$$v_{KS}(\mathbf{r}) \equiv v(\mathbf{r}) + v_J[\rho; \mathbf{r}] + v_{xc}[\rho; \mathbf{r}]. \tag{67}$$

To see what may be gained by such a construction, consider that the normal Kohn–Sham procedure requires finding the eigenvalues and eigenvectors of the Kohn–Sham equations; computational cost thus increases as the cube of the size of the system. However, due to the particular structure of the Kohn–Sham equations, we do not need to find the Kohn–Sham eigenvalues and eigenvectors: computing the internal potential only requires the electron density, while computing the kinetic energy only requires the first-order density matrix. Because $\gamma(\mathbf{r}, \mathbf{r}')$ depends on two spatial coordinates instead of just one (like the electron density), the computational cost inherent in finding $\gamma(\mathbf{r}, \mathbf{r}')$ formally increases only as the square of the system's size.

One may be tempted then to find the density matrix by minimizing the Kohn–Sham energy, Eq. (65), with respect to the density matrix subject to the constraint that

$$N = \int \rho(\mathbf{r}) \mathrm{d}\mathbf{r} = \int \gamma(\mathbf{r}, \mathbf{r}) \mathrm{d}\mathbf{r}. \tag{68}$$

Unfortunately, this does not give the correct answer, giving instead the state where all the electrons are in the lowest energy Kohn–Sham orbital; this violates the Pauli exclusion principle. Satisfying the Pauli exclusion principle requires that every state of the system be occupied by no fewer than zero and no more than two electrons (one with spin α and one with spin β). This indicates that the eigenvalues of the first-order density matrix [it follows from the defining Eq. (64) that the eigenvectors of $\gamma(\mathbf{r}, \mathbf{r}')$ are the Kohn–Sham orbitals]

$$\int \gamma(\mathbf{r}, \mathbf{r}') \psi_i(\mathbf{r}') \mathrm{d}\mathbf{r}' = n_i \psi_i(\mathbf{r}) \tag{69}$$

must be between zero and two,

$$0 \leq n_i \leq 2. \tag{70}$$

The eigenvalues of the first-order density matrix are identified with the occupation numbers for their associated Kohn–Sham orbitals.

For the minimum energy state of a closed shell system, all the Kohn–Sham orbital occupation numbers will be either 0 or 2, which allows us to replace the constraint (70) with the more compact idempotency condition

$$\gamma(\mathbf{r}, \mathbf{r}') \equiv \frac{1}{2} \int \gamma(\mathbf{r}, \mathbf{x}) \gamma(\mathbf{x}, \mathbf{r}') \mathrm{d}\mathbf{x}. \tag{71}$$

The "divide-and-conquer" method [34,35] was the first Kohn–Sham algorithm which delivered linear scaling: computational costs that grow linearly with the size of the system. In this technique, one first projects the density matrix onto a basis set, typically a set of Gaussian functions, $\{e^{-\zeta_i |r - R_\alpha|^2}\}$, centered at each atomic center, α. We may project the density matrix

$$\gamma(\boldsymbol{r}, \boldsymbol{r}') \rightarrow \int \int \chi_{\alpha i}(\boldsymbol{r}) \gamma(\boldsymbol{r}, \boldsymbol{r}') \chi_{\beta j}(\boldsymbol{r}') d\boldsymbol{r} d\boldsymbol{r}' \equiv \sum_{\gamma k} \sum_{\delta l} S_{\alpha i, \gamma k} \gamma_{\gamma k, \delta l} S_{\delta l, \beta j} \tag{72}$$

and the Kohn–Sham Hamiltonian operator

$$\left(-\frac{\nabla^2}{2} + v_{KS}(\boldsymbol{r}) \right) \rightarrow \int \int \delta(\boldsymbol{r} - \boldsymbol{r}') \left(-\frac{\nabla^2}{2} + v_{KS}(\boldsymbol{r}) \right) \xi_{\alpha i}(\boldsymbol{r}) \xi_{\beta j}(\boldsymbol{r}') d\boldsymbol{r} d\boldsymbol{r}' \tag{73}$$
$$= h_{\alpha i, \beta j}$$

onto this basis set; here S is the overlap matrix,

$$S_{\alpha i, \beta j} \equiv \int \chi_{\alpha i}(\boldsymbol{r}) \chi_{\beta j}(\boldsymbol{r}) d\boldsymbol{r}. \tag{74}$$

Because the basis functions decay strongly as one moves away from the atom on which they are centered, $h_{\alpha i; \beta j} \approx 0$ when $|\boldsymbol{R}_\alpha - \boldsymbol{R}_\beta|$ is large. In particular, this means that the $h_{\alpha i; \beta j}$ can be neglected whenever $|\boldsymbol{R}_\alpha - \boldsymbol{R}_\beta|$ is greater than some threshold, R_h. Typically $R_h \sim 7$ Å $= 13.5$ Bohr.

Computationally, one proceeds as follows. Starting at subsystem α (which could be an atom, but is more generally a molecular fragment), one constructs the subsystem Hamiltonian matrix $h_{\gamma i; \delta j}^{(\alpha)}$ where

$$h_{\gamma i, \delta j}^{(\alpha)} = \begin{cases} \left\langle \chi_{\gamma i} \left| -\frac{\nabla^2}{2} + v_{KS}(\boldsymbol{r}) \right| \chi_{\delta j} \right\rangle & \begin{cases} |\boldsymbol{R}_\delta - \boldsymbol{R}_\gamma| < R_h \\ \max \left(d(\boldsymbol{R}_\gamma, \boldsymbol{R}_{\text{system}}), d(\boldsymbol{R}_\delta, \boldsymbol{R}_{\text{system}}) \right) < R_b \end{cases} \\ 0 & \text{otherwise} \end{cases} \tag{75}$$

In addition to the aforementioned cutoff on the off-diagonal elements (R_h), there is an additional system-dependent cutoff on $h_{\gamma i; \delta j}^{(\alpha)}$: we do not calculate matrix elements in which either of the two centers, γ or δ, are further than some "buffer distance," R_b, from an atom in the subsystem α. Typically, $R_b \sim 6$ Å $= 11.5$ Bohr is sufficient to ensure that the interactions between the subsystem of interest and the neighboring systems are accurately modeled [36,37].

Given the subsystem Hamiltonian, $h_{\gamma i; \delta j}^{(\alpha)}$, one may solve the Kohn–Sham equations directly. Projecting the Kohn–Sham equations onto the nonorthogonal basis set yields a generalized eigenvalue problem for the molecular fragment.

$$\sum_{\delta j} h_{\gamma i; \delta j}^{(\alpha)} C_{\delta j}^{(\alpha), k} = \varepsilon^{(\alpha)k} \sum_{\delta j} S_{\gamma i; \delta j}^{(\alpha)} C_{\delta j}^{(\alpha), k} \tag{76}$$

where

$$C_{\delta j}^{(\alpha)k} \equiv \int \chi_{\delta j}(\boldsymbol{r}) \psi^{(\alpha)k}(\boldsymbol{r}) d\boldsymbol{r}, \tag{77}$$

$\psi_k^{(\alpha)}(\boldsymbol{r})$ is the kth Kohn–Sham orbital for subsystem α and $S_{\gamma i; \delta j}^{(\alpha)}$ is using the cutoff scheme from Eq. (75). We then define the density matrix of the subsystem as

$$\gamma_{\gamma i; \delta j}^{(\alpha)} = \sum_k f_T \left(\mu - \varepsilon^{(\alpha)k} \right) p_{\gamma \delta}^{(\alpha)} \left(C_{\gamma i}^{(\alpha)k} \right)^* C_{\delta j}^{(\alpha)k} \tag{78}$$

where $f_T(\mu-\varepsilon^{(\alpha)k})$ is the Fermi distribution function at temperature T and

$$p^{\alpha}_{\gamma\delta} \equiv \begin{cases} 1 & \gamma = \delta = \alpha \\ \dfrac{1}{2} & \left\{ \begin{matrix} \gamma = \alpha; \delta \neq \alpha \\ \delta = \alpha; \gamma \neq \alpha \end{matrix} \right\} \\ 0 & \text{otherwise.} \end{cases} \tag{79}$$

The Mulliken-like partitioning function [38] defined by Eq. (79) ensures that we include in the density matrix associated with fragment α only those elements which have at least one basis function centered in the subsystem of interest, and we fully include only those elements that have both indices in the subsystem of interest. (The weighting factor of one-half is motivated by symmetry, $p^{(\alpha)}_{\alpha\beta} = p^{(\beta)}_{\alpha\beta}$, and the requirement that we avoid "double counting" of the "cross terms" in the density matrix.) The density matrix of the entire system is then represented as the sum of the subsystems' density matrices

$$\gamma_{\gamma i; \delta j} \equiv \sum_{\alpha} \gamma^{(\alpha)}_{\gamma i; \delta j}. \tag{80}$$

It is clear that this construction has given the desired "nearsightedness" for the density matrix: $\gamma_{\gamma i;\delta j}$ is zero whenever $|R_\gamma - R_\delta| > R_h$ [39]. Moreover, because the cost of the method is proportional to the number of subsystems, and the number of subsystems may be chosen to be proportional to the size of the molecule, the cost of the divide-and-conquer method is proportional to the size of the molecule.

The divide-and-conquer relies upon our ability to write the Kohn–Sham Hamiltonian operator for a subsystem of a larger system. Alternatively, one may try to construct Kohn–Sham orbitals for molecular subsystems directly without recourse to a localized version of the Kohn–Sham equations. The idea, which is rooted in a long tradition of orbital localization transformations, is to write the exact Kohn–Sham density matrix as [40]

$$\gamma(r, r') \equiv \sum_{i=1}^{M} 2(\psi_i(r))^* S^-_{ij} \psi_j(r') \tag{81}$$

where S^- is the generalized inverse of S,

$$SS^- S = S, \tag{82}$$

and S is the overlap matrix of the localized Kohn–Sham orbitals,

$$S_{ij} \equiv \langle \psi_i | \psi_j \rangle. \tag{83}$$

The Kohn–Sham orbitals are chosen so that the orbitals are "as localized as possible" in some well-defined sense; this leads to S_{ij} being effectively zero for orbitals centered far from one another. Among the numerous advantages of methods based upon Eq. (81) is that the Pauli principle (idempotency) is automatically satisfied. Also, note that in Eq. (81), the number of localized orbitals, M, can exceed the number of occupied Kohn–Sham orbitals, $N/2$.

Direct minimization of the Kohn–Sham energy using Eq. (81) is infeasible: direct methods for computing S^- grow as the third power of the size of the system. However, suppose that the localized orbitals are nearly orthogonal to one another, so that S resembles the identity matrix, I. Then the generalized inverse of S resembles the identity matrix and to second order [41],

$$S^- \approx 2I - ISI \tag{84}$$

$$\approx 2I - S.$$

Inserting Eq. (84) in Eq. (82) and then inserting this expression for the density matrix into the variational principle yield the linear-scaling method proposed by Kim et al. [42]:

$$E_{KS}[v; N] \equiv \min_{\{\psi_i\}} Tr\left[2H(2I - S) + \mu(N - 2 \cdot (2I - S))\right] \tag{85}$$

where

$$H_{ij} \equiv \left\langle \psi_i \left| -\frac{\nabla^2}{2} + v_{KS}(\boldsymbol{r}) \right| \psi_j \right\rangle \tag{86}$$

and the chemical potential, $\varepsilon_{HOMO} < \mu < \varepsilon_{LUMO}$, are used to force the normalization constraint, Eq. (68). The Kohn–Sham Hamiltonian ensures that the number of nonnegligible elements in H grows only linearly with the size of the system, and so the cost of implementing Eq. (85) rises only linearly with the size of the system. At convergence, the minimizing set of orbitals in Eq. (85) are orthogonal localized orbitals [as it is only then that Eq. (84) is exact].

Yang [40] proposed several methods which may be viewed as refinements to Eq. (85). The simplest of these methods starts by, for each set of orbitals, computing the rank of the overlap matrix [Eq. (83)] using the procedure:

$$\text{rank}(S) \equiv \text{Tr}(SS^-) = \min_{X = X^T} \left[S(2X - XSX)\right]. \tag{87}$$

The matrix, X_{max}, that maximizes $\text{Tr}[S(2X - 2XSX)]$ is the generalized inverse, allowing the variational procedure for the Kohn–Sham energy to be written as [40]

$$\min_{\{\psi_i\}} \text{Tr}\left[H(X_{max}) + \mu(N - 2 \cdot \text{rank}[S])\right]. \tag{88}$$

Because X_{max} is an accurate inverse [instead of the approximate form from Eq. (84)], the desired minimum in Eq. (88) is the global minimum. On the other hand, Eq. (85) yields only a local minimum for the exact Kohn–Sham density matrix, which puts a premium on making a good initial guess. Furthermore, unlike Eq. (85), where the solution orbitals are orthogonal localized orbitals, in Yang's method, the orthogonality constraint is relaxed, resulting in a more localized orbital basis [43]. This reduces the number of nonzero matrix elements in S and H. Yang's construction subsumes several previous methods [42,44–46].

Yet another technique is to minimize the Kohn–Sham energy as a functional of the density matrix, Eq. (65). In the first method of this type, due to Li et al. [47], the ground state of the system was only a local minimum of the energy functional. To surmount this difficulty, one may choose to explicitly impose the idempotency constraint on the density matrix, Eq. (71), with a Lagrange multiplier. For example, forcing the normalization constraint, Eq. (68), and the idempotency constraint [48],

$$
0 \equiv \int \int \int \int \gamma(r, x_1) \big(2\delta(x_1 - x_2) - \gamma(x_1 - x_2)\big)
$$
$$
\times \big(2\delta(x_2 - x_3) - \gamma(x_2, x_3)\big) \gamma(x_3, r) dr dx_1 dx_2 dx_3
$$

(89)

with Lagrange multipliers gives the functional

$$
\Lambda_{v_{KS}, \lambda, \mu}[\gamma] \equiv E_{v_{KS}}^{KS}[\gamma] + \mu \left(\int \gamma(r, r) dr - N \right) + \lambda \mathrm{Tr}\left[\gamma^2 (2 - \gamma)^2 \right].
$$

(90)

[The final term in Eq. (90) introduces a standard shorthand notation for Eq. (89)]. The computational expense is significantly enhanced because the idempotency constraint is not satisfied except in the limit as $\lambda \to \infty$ (but see the alternative construction of Adhikari and Baer [49]). With these caveats, however, minimization of $\Lambda_{v_{KS}, \mu, \lambda}$ replaces solving the Kohn–Sham partial differential equations in a self-consistent procedure with the equation for the Kohn–Sham potential in terms of the electron density, Eq. (67).

Minimization of Eq. (90) with respect to the density matrix still has a cost that grows with the square of the size of the system of interest. Note, however, that, owing to the presence of the delta function in Eq. (65), evaluating the kinetic energy contribution to the energy only requires the value of $\gamma(r, r')$ in the immediate neighborhood of $|r - r'| = 0$; evaluation of the energy functional may thus be reduced to the pseudo-three-dimensional integral:

$$
E_{v_{KS}}^{KS}[\gamma] \equiv \int \int_{|r - r'| < d} \delta(r - r') \left(-\frac{\nabla_r^2}{2} + v_{KS}(r) \right) \gamma(r, r') dr' dr.
$$

(91)

Because the computational cost required to evaluate Eq. (91) grows only *linearly* with the size of the system, the cost of minimizing $\Lambda_{v_{KS}, \mu, \lambda}$ would reduce to linear scaling with respect to system size if we could similarly restrict integrations over the coordinates in Eq. (89) to

$$
0 \equiv \int \int \int \int_{|x_i - x_j| < D} \gamma(x_1, x_2) \big(2\delta(x_2 - x_3) - \gamma(x_2, x_3)\big)
$$
$$
\times \big(2\delta(x_3 - x_4) - \gamma(x_3, x_4)\big) \gamma(x_4, x_1) dx_1 dx_2 dx_3 dx_4
$$

(92)

Owing to the "nearsightedness" of the density matrix, $\gamma(r, r')$ falls off rapidly (typically exponentially) as $|r - r'|$ increases [39]; this justifies the restrictions on the range of integration imposed in Eq. (92). Although Eq. (92) is not inconsistent with linear scaling, the sequential four-coordinate integration and the fact that the cutoff in Eq. (92) is longer than that in Eq. (91) ($D \gg d$) indicate that the bottleneck in the direct minimization of $\Lambda_{v_{KS}, \mu, \lambda}$ is associated with the idempotency constraint.

3.5. Linear-Scaling Methods for the Kohn–Sham Potential and for Numerical Integration

The Kohn–Sham method involves two coupled equations which must be solved sequentially; in the first step, one solves the partial differential equations (45) for a given Kohn–Sham potential (which is constructed from a guessed electron density). In the next step, one uses the solution to the Kohn–Sham partial differential equations to construct a new and improved electron density, and with this, one infers a new Kohn–Sham potential

$$v_{KS}(r) \equiv v(r) + \int \frac{\rho(r')}{|r - r'|} dr' + v_{xc}[\rho; r]. \tag{93}$$

It follows that a linear-scaling solution to the Kohn–Sham density-functional problem requires not only the linear-scaling methods for constructing the Kohn–Sham density matrix from the Kohn–Sham potential explored in the previous section, but also a linear-scaling method for constructing the Kohn–Sham potential for any given density. For the types of density functionals commonly in use, constructing $v_{xc}[\rho; r]$ is relatively unproblematic. By contrast, because evaluating the Coulomb potential,

$$v_J[\rho; r] \equiv \int \frac{\rho(r')}{|r - r'|} dr', \tag{94}$$

at each point, r, requires an integration over the entire system, direct methods for evaluating $v_J[\rho; r]$ incur computational costs growing as the square of the size of the system. Several methods for solving the Coulomb problem with cost proportional to the size of the system are available; each method relies upon the use of a hierarchical expansion technique to reduce computational costs. Of especial importance are fast multipole techniques; these methods attack the Coulomb problem directly by replacing the detailed local behavior of $\rho(r')$ with a multipole expansion whenever $|r-r'|$ is large [50–54]. (Although in its more naive implementations, the fast multipole method was not a linear-scaling method [55,56]; recent improvements achieve this goal [57].) An alternative technique is to solve the Poisson's equation

$$\nabla^2 v_J[\rho; r] = -4\pi\rho(r). \tag{95}$$

Linear-scaling techniques for solving Poisson's equation include multigrid approaches [58–61] and fast Poisson solvers [62].

Often, the bottleneck in linear-scaling density-functional theory is the evaluation of the Coulomb potential; the trade off between the simple and direct method of integrating Eq. (94) and the more sophisticated linear-scaling approaches is evidenced by the fact that, for moderately large systems, linear-scaling density-functional techniques are often less efficient than direct solution to the Kohn–Sham system. As the size of the system increases beyond 10 to 20 Å, however, linear-scaling techniques become essential.

After constructing the Kohn–Sham potential, one must construct the electron density, $\rho(r')$, the Hamiltonian matrix, Eq. (86), and the overlap matrix, Eq. (83). Because the basis functions are localized and the Kohn–Sham Hamiltonian is a local operator [cf. Eq. (91)], most of the matrix elements

$$\rho_{ij}(r_0) \equiv \langle \psi_j | \delta(r - r_0) | \psi_i \rangle$$

$$H_{ij} \equiv \left\langle \psi_j \left| -\frac{\nabla^2}{2} + v_{KS}(r) \right| \psi_i \right\rangle \tag{96}$$

$$S_{ij} \equiv \langle \psi_j | \psi_i \rangle$$

are very small. Replacing all the matrix elements that are smaller than some threshold, τ, with zero, one finds that the number of nonzero elements in H and S increases only linearly with the size of the system.

It follows that if we do not compute the integrals that would be subsequently set to zero, then we will achieve a linear-scaling technique for evaluating the matrix elements [63]. The necessary idea for achieving this goal is already implicit in the divide-and-conquer approach: one neglects integrals when $\psi_i(r)$ and $\psi_j(r)$ are centered in distant locations in the molecule (according to some suitable criterion). Usually, the only matrix elements that cannot be evaluated analytically are those containing the exchange-correlation potential,

$$W_{ij}^{\mathrm{xc}} \equiv \int \psi_i^*(r) v_{\mathrm{xc}}(r) \psi_j(r) \mathrm{d}r. \tag{97}$$

To efficiently evaluate this integral, one decomposes the system's volume into a union of "atomic regions," Ω_A, and then evaluates the integral over each atomic region. Thus

$$W_{ij}^{\mathrm{xc}} \approx \sum_A{}' \int_{\Omega_A} \psi_i^*(r) v_{\mathrm{xc}}(r) \psi_j(r) \mathrm{d}r \tag{98}$$

where the prime on the summation indicates that the regions where the integrand is negligible are neglected [64,65].

4. DENSITY-FUNCTIONAL DESCRIPTORS FOR CHEMICAL CHANGE

4.1. The Electronic Energy as the State Function for a Chemical System

In the previous section, we explored how Kohn–Sham density-functional theory can be used to provide accurate quantitative characterizations of chemical systems. In the present section, we focus instead on density-functional theoretic tools for qualitative descriptions of chemical reactivity, with particular focus on describing the driving forces of chemical reactions.

Insofar as the energy may be expressed as a functional of the number of electrons and the external potential, we may regard the electronic energy, $E[v(r);N]$, as the state function for the system with N electrons bound by the external potential $v(r)$. Insofar as chemical changes are driven by changes in the electronic energy, we need to consider how the energy of a molecule changes as its number of electrons changes (as from electron transfer) and its external potential changes (as from the presence of an attacking reagent). To do this, we employ the total differential [4]

$$\mathrm{d}E[v;N] = \left(\frac{\partial E[v;N]}{\partial N}\right)_{v(r)} \mathrm{d}N + \int \left(\frac{\delta E[v;N]}{\delta v(r)}\right)_N \delta v(r) \mathrm{d}r. \tag{99}$$

The first term in Eq. (99) has been previously identified as the chemical potential for electrons, $\mu \equiv \left(\frac{\partial E[v;N]}{\partial N}\right)_{v(r)}$. That the functional derivative in the second term is simply the electron density follows directly from the definition of the functional derivative [Eq. (28)] and the Hellmann–Feynman theorem [19,20]:

$$\left(\frac{\delta E[v_0(r), N]}{\delta v(r_0)}\right)_N = \left(\frac{\partial E[v_0(r) + \varepsilon\delta(r - r_0); N]}{\partial \varepsilon}\right)\Bigg|_{\varepsilon=0} = \rho(r_0). \tag{100}$$

Thus [4]

$$dE[v; N] = \mu dN + \int \rho(r)\delta v(r)dr. \tag{101}$$

The resemblance between Eq. (101) and the total differential of the internal energy with respect to the number of moles of a chemical species, n, and the volume of the container, V, at absolute zero:

$$dU(n, V) = \left(\frac{\partial U(n, V)}{\partial n}\right)_V dn + \left(\frac{\partial U(n, V)}{\partial V}\right)_n dV = \mu dn - PdV \tag{102}$$

is not purely coincidental. Both n and N represent the number of particles in the system, while both V and $v(r)$ model the "container" that confines the particles. In particular, just as $dV > 0$ represents an increase in the size of a classical system, $\delta v(r) > 0$ increases the repulsiveness of the external potential and ordinarily prompts the electron cloud to expand.

4.2. The Electronic Chemical Potential

By analogy to Eq. (102), we deduce that the electronic chemical potential represents how energetically favorable it is for an electronic system to accept electrons. This leads us to expect that, in analogy to the thermodynamic treatment of multicomponent systems, electrons will transfer from molecules (or molecular fragments) with high chemical potential to molecules with low chemical potential, with equilibrium being established only when the chemical potential is uniform throughout the system [4,66]. Recalling that electrons flow from molecular fragments with low electronegativity to molecular fragments with high electronegativity until the electronegativity is every-where equalized [67,68], we see that the concepts of electronic chemical potential and electronegativity must be closely related. Indeed, taking the finite difference approx-imation to the electronic chemical potential, we obtain [4]:

$$\begin{aligned} \mu &\approx \frac{E[v; N + 1] - E[v; N - 1]}{2} \\ &\approx \frac{(E[v; N + 1] - E[v; N]) + (E[v; N] - E[v; N - 1])}{2} \\ &\approx -\frac{A + I}{2} \\ &\approx -\chi_{\text{Mulliken}}, \end{aligned} \tag{103}$$

where I is the ionization potential, A is the electron affinity, and χ_{Mulliken} is Mulliken's definition of electronegativity [69]. Consequently, we may *define* chemical electro-negativity with

$$\chi \equiv -\mu. \tag{104}$$

The equalization principle for the electronic chemical potential (equivalently, the electronegativity equalization principle) may be couched in a form reminiscent of the argument from classical thermodynamics [4]. However, the chemical potential equalization principle follows most directly from the variational principle and, in particular, Eq. (32). First, define the local chemical potential by

$$\mu(r) \equiv \frac{\delta E_v[\rho]}{\delta \rho(r)}. \tag{105}$$

Suppose the chemical potential is not equalized; then there are two points, r_1 and r_2, with $\mu(r_1) < \mu(r_2)$. Consider these two points to be separate systems with $N_1 = \rho(r_1)dr$ and $N_2 = \rho(r_2)dr$ electrons, respectively. From the definition of the local chemical potential [Eq. (105)] and the assumption that $\mu(r_1) < \mu(r_2)$, it follows that the state with $N_1 + dN = [\rho(r_1) + \delta\rho(r)]dr$ and $N_2 - dN = [\rho(r_2) - \delta\rho(r)]dr$ electrons has lower energy than the initial state. Electronic states, then, are stable only if the chemical potential is constant throughout the system.

4.3. Second-Order Description of Chemical Change

Proceeding further, we now explore how the chemical potential and the electron density change with respect to N and $v(r)$:

$$d\mu = \left(\frac{\partial \mu[v; N]}{\partial N}\right)_{v(r)} dN + \int \left(\frac{\delta \mu[v, N]}{\delta v(r)}\right)_N \delta v(r) dr \tag{106}$$

$$d\rho(r) = \left(\frac{\partial \rho[v; N, r]}{\partial N}\right)_{v(r)} dN + \int \left(\frac{\delta \rho[v; N, r]}{\delta v(r')}\right)_N \delta v(r') dr'. \tag{107}$$

In thermodynamics, the equations for the second-order response of the energy are linked by a Maxwell relation. Likewise, Eqs. (106) and (107) are linked by the Maxwell relation [70]:

$$f(r) = \left(\frac{\partial \rho[v; N, r]}{\partial N}\right)_{v(r)} = \left(\frac{\partial}{\partial N}\left(\frac{\delta E[v; N]}{\delta v(r)}\right)_N\right)_{v(r)} = \left(\frac{\delta}{\delta v(r)}\left(\frac{\partial E[v; N]}{\partial N}\right)_{v(r)}\right)_N$$

$$= \left(\frac{\delta \mu[v; N]}{\delta v(r)}\right)_N. \tag{108}$$

The *Fukui function, f(r)*, measures the response of a reactant to changes in both N and $v(r)$. That the Fukui function is normalized to unity follows from Leibniz's rule:

$$\int f(r) dr = \int \left(\frac{\partial \rho(r)}{\partial N}\right)_{v(r)} dr = \left(\frac{\partial \int \rho(r) dr}{\partial N}\right)_{v(r)} = \left(\frac{\partial N}{\partial N}\right)_{v(r)} = 1. \tag{109}$$

By definition, the Fukui function represents the change in electron density due to addition or removal of electrons from the system. Recalling results from the frontier molecular orbital theory, where the magnitude of the highest-occupied and lowest-unoccupied molecular orbitals are used to discern the propensity of a molecular site to attack by electron acceptors and electron donors, respectively, we deduce that the

Fukui function is a key density-functional index for predicting the regioselectivity of reactants [71,72]. The classical analogs of Eqs. (106) and (107) are [73]:

$$d\mu(n, V) = \left(\frac{\partial \mu(n, V)}{\partial n}\right)_V dn + \left(\frac{\partial \mu(n, V)}{\partial V}\right)_n dV$$

$$= \eta \cdot dn + \left(\frac{\partial \mu(n, V)}{\partial V}\right)_n dV \tag{110}$$

$$dP(n, V) = \left(\frac{\partial P(n, V)}{\partial n}\right)_V dn + \left(\frac{\partial P(n, V)}{\partial V}\right)_n dV$$

$$= \left(\frac{\partial P(n, V)}{\partial n}\right)_n dn - \frac{B}{V} dV, \tag{111}$$

where $B = -V\left(\frac{\partial P}{\partial V}\right)_n$ is the bulk modulus and $\eta = \frac{B}{n} \cdot \frac{V}{n}$ is the molar crystal hardness proposed by Yang et al. [74]

In particular, we note that $\left(\frac{\partial P}{\partial V}\right)_n = -\frac{B}{V}$ represents the "deformability" of a system. Similarly, $\left(\frac{\delta\rho[v;N,r]}{\delta v(r')}\right)_N$ represents the ease with which the molecular electron density is "deformed" by an external field; we refer to $\left(\frac{\delta\rho[v;N,r]}{\delta v(r')}\right)_N$ as the *polarizability kernel*.

In analogy to Eq. (110), we identify

$$\eta[v; N] \equiv \left(\frac{\partial \mu[v; N]}{\partial N}\right)_{v(r)} = \left(\frac{\partial^2 E[v; N]}{\partial N^2}\right)_{v(r)} \approx I - A \tag{112}$$

as the *chemical hardness*. [The reader is cautioned that this definition for the chemical hardness differs from the original definition of Parr and Pearson [75] by a factor of 2; while both definitions are prevalent in the literature, Eq. (112) is now preferred [76].] The chemical hardness of a system is associated with its band gap, which is a key chemical reactivity indicator for extended systems. The chemical hardness also plays a pivotal role in molecular reactivity, most famously in conjunction with the theory of hard and soft acids and bases [4,75,77–79].

Further insight into the chemical hardness may be obtained by considering relationship between the macroscopic $\left(\frac{\partial \mu}{\partial n}\right)_V$ and thermodynamic equilibrium. For simplicity, consider a system in liquid–vapor equilibrium at some fixed temperature,

$$A_l \rightleftarrows A_v. \tag{113}$$

Assume that in the initial state, there are n_l moles of A in the liquid phase and n_v moles of A in the vapor phase; since the system is in equilibrium, $\mu_l[n_l, V] = \mu_v[n_v, V]$. If we increase the volume of the container from V to $V + \Delta V$ at constant temperature, then $\mu_l[n_l, V + \Delta V] > \mu_v[n_v, V + \Delta V]$. Once equilibrium is reestablished, there will be $n_l - \Delta n$ and $n_v + \Delta n$ moles of A in the liquid and vapor phases, respectively. To first order in Δn, we have

$$\Delta n = \frac{\mu_l - \mu_v}{\eta_l(n_l, V) + \eta_v(n_v, V)}. \tag{114}$$

We observe that the equilibrium in Eq. (113) is most stable (i.e., the change in n_l and n_v needed to reestablish equilibrium is minimal) when η_l and η_v are large. Similarly, large values of the chemical hardness are associated with the stability of electronic systems [80].

4.4. Thermodynamic Analogies to the Hohenberg–Kohn Theorems [81,82]

The fundamental stability conditions in thermodynamics are formulated as variational principles. Within the zero-temperature canonical ensemble, the quantities n and V are used to specify the state of interest. Suppose one chooses a nonoptimum pressure, $P(r)$. For example, we can divide the system with a partition and place Maxwell's demon at the "door" between the partitions to ensure that the pressure on one side of the partition is greater than that on the other side of the partition. Then, we have that:

$$A_{n,V}(P(r), \mu) > A_{n,V}(P_0, \mu),\tag{115}$$

where P_0 is the pressure the system reverts to if we "kill" Maxwell's demon.

Similarly, when we choose to specify an electronic system with the number of electrons, N, and the external potential, $v(r)$, we say that we are working in the *electronic canonical ensemble* [70,73]. The second Hohenberg–Kohn theorem is the analog of Eq. (115): if we choose a nonoptimal density, we get too large an energy:

$$E_v[\bar{\rho}_N] > E_v[\rho_N].\tag{116}$$

From this "thermodynamic" perspective, the first Hohenberg–Kohn theorem is merely an assertion that that the Legendre transform from the electronic "ensemble" specified by N and $v(r)$ to that specified by $\rho(r)$ exists [73]. In analogy to classical thermodynamics, the state function for the ρ-ensemble is:

$$
\begin{aligned}
F_\rho[N, v] &\equiv E[v; N] - \int \left(\frac{\delta E[v; N]}{\delta v(r)} \right)_N v(r)\mathrm{d}r \\
&= E[v; N] - \int \rho(r)v(r)\mathrm{d}r.
\end{aligned}\tag{117}
$$

The ρ-ensemble is most directly analogous to the isothermal–isobaric ensemble in classical statistical thermodynamics.

5. MATHEMATICAL CONSIDERATIONS

In the preceding discussion, several subtle mathematical points have been overlooked. Of these, the most important is the v-representability problem: the Hohenberg–Kohn theorems indicate that a ground-state electron density uniquely determines its associated external potential and thus all the properties of the system, including the exchange-correlation energy. However, the Hohenberg–Kohn treatment does not address how one can tell whether a given density is a ground-state electron density for some system, that is, whether a given electron density is v-representable. For a long time, it was suspected that every reasonable electron density might be v-representable; however, Levy [2], Lieb [83], and Englisch and Englisch [3] have demonstrated that this is not the case. Fortunately, no essential difficulties arise; one may define the exchange-correlation energy (as well as all of the other properties of a system) for non-v-representable densities in such a way as to preserve the variational principle, Eq. (23) [83–86].

Related to the v-representability problem is the Kohn–Sham v-representability problem. That is, given a system of interest, can one always find an internal potential, $w(r)$, such that the ground-state electron density of the Kohn–Sham model system is the same as that of the state of interest? Again, the answer seems to be no [87], but if one allows fractional occupation numbers of the Kohn–Sham orbitals, then no essential difficulties arise [3,88,89]. We note that in this case, the idempotency constraint, Eq. (71), is no longer appropriate, and the less stringent Eq. (70) should be used instead.

Finally, there is the matter of fractional numbers of electrons; in Eqs. (30) and (31) and throughout Sec. 4, we found it convenient to consider systems with noninteger numbers of electrons. Since no such systems exist in nature, the properties of these systems must be defined in an appropriate way. Several different arguments converge on the same result: the properties of a system with $N + \varepsilon$ electrons ($0 \leq \varepsilon \leq 1$) should be taken as the appropriate weighted average of the properties of the systems with integer numbers of electrons [5,90]:

$$Q[v(r); N + \varepsilon] \equiv Q[v(r); N] + \varepsilon(Q[v(r); N + 1] - Q[v(r); N]). \qquad (118)$$

Among the unpleasant consequences of this result is that changes in properties of a system due to changes in the number of electrons (or nonnumber conserving changes in the electron density) cannot be computed without first explicitly specifying whether the number of electrons is increasing or decreasing. Again, however, this difficulty is not insurmountable and merely requires that derivatives with respect to the number of electrons and functional derivatives with respect to the electron density carry an additional notation as to whether the change in question increases or decreases the number of electrons in the system.

6. SUMMARY

Over the course of the last decade, density-functional theory and, in particular, the Kohn–Sham method have become the methods of choice for modeling the electronic structure and chemical reactivity of large systems. This is due, in large part, to the development of accurate density functionals for the exchange-correlation energy and availability of efficient computational implementations; progress in both directions has been facilitated by the simplifications obtained by using the ground-state electron density, instead of the many-electron wave function, as the descriptor of molecular states. In addition to providing accurate quantitative descriptions of most chemical systems, density-functional theory provides qualitative descriptors that elucidate the factors driving chemical reactions.

The present review has been very selective, stressing the rationale behind density-functional methods above their applications and excluding many important topics (both theoretical and computational). The interested reader may refer to anyone of the many books [91–93] or review articles [94–101] on density-functional theory for more details. Of special importance is the extension of density-functional theory to time-dependent external potentials [102–105], as this enables the dynamical behavior of molecules, including electronic excitation, to be addressed in the context of DFT [106–108]. As they are particularly relevant to the present discussion, we cite several articles related to the formal foundations of density-functional theory [85,100,109–111], linear-scaling methods [63,112–116], exchange-correlation energy functionals [25, 117–122], and qualitative tools for describing chemical reactions [123–126,126–132].

ACKNOWLEDGMENTS

P.W.A. acknowledges financial support from a NIH postdoctoral fellowship and W. Y. acknowledges support from the National Science Foundation and the National Institutes of Health.

REFERENCES

1. Hohenberg P, Kohn W. Phys Rev 1964; 136:B864–B871.
2. Levy M. Phys Rev A 1982; 26:1200.
3. Englisch H, Englisch R. Physica, A (Amsterdam) 1983; 121A:253.
4. Parr RG, Donnelly RA, Levy M, Palke WE. J Chem Phys 1978; 68:3801.
5. Perdew JP, Parr RG, Levy M, Balduz JL Jr.. Phys Rev Lett 1982; 49:1691.
6. Thomas LH. Proc Camb Philol Soc 1927; 23:542.
7. Fermi E. Z Phys 1928; 48:73.
8. Dirac PAM. Proc Camb Philol Soc 1930; 26:376.
9. Wigner E. Phys Rev 1934; 46:1002.
10. Weizsacker CFv. Z Phys 1935; 96:431.
11. March NH. Phys Lett A 1986; 113:476.
12. March NH. Phys Lett A 1985; 113:66.
13. Parr RG, Yang W. Density-Functional Theory of Atoms and Molecules. New York: Oxford UP.
14. Lieb EH. Rev Mod Phys 1981; 53:603.
15. Kohn W, Sham LJ. Phys Rev 1965; 140:A1133–A1138.
16. Tozer DJ, Handy NC. J Chem Phys 1998; 108:2545.
17. Von Barth U, Hedin L. J Phys C 1972; 5:1629.
18. Rajagopal AK, Callaway J. Phys Rev B 1973; 7:1912.
19. Feynman RP. Phys Rev 1939; 56:340.
20. Hellmann H. Einfuehrung in Die Quantenchemie. Leipzig: Deuticke.
21. Harris J. Phys Rev A 1984; 29:1648.
22. Harris J, Jones RO. J Phys F 1974; 4:1170.
23. Gunnarsson O, Lundqvist BI. Phys Rev B 1976; 13:4274.
24. Langreth DC, Perdew JP. Phys Rev B 1977; 15:2847.
25. Baerends EJ, Gritsenko OV. J Phys Chem A 1997; 101:5383.
26. Lee C, Yang W, Parr RG. Phys Rev B 1988; 37:785.
27. Becke AD. Phys Rev A 1988; 38:3098.
28. Perdew JP, Yue W. Phys Rev B 1986; 33:8800.
29. Becke AD. J Chem Phys 1993; 98:1372.
30. Becke AD. J Chem Phys 1993; 98:5648.
31. Zhang Y, Yang W. J Chem Phys 1998; 109:2604.
32. Sodupe M, Bertran J, Rodriguez-Santiago L, Baerends EJ. J Phys Chem A 1999; 103:166.
33. Gritsenko OV, Ensing B, Schipper PRT, Baerends EJ. J Phys Chem A 2000; 104:8558.
34. Yang W. Phys Rev Lett 1991; 66:1438.
35. Horsfield AP, Bratkovsky AM, Fearn M, Pettifor DG, Aoki M. Phys Rev B 1996; 53: 12694.
36. Lee T-S, York DM, Yang W. J Chem Phys 1996; 105:2744.
37. Yang W. Phys Rev A 1991; 44:7823.
38. Mulliken RS. J Chem Phys 1962; 36.
39. Kohn W. Phys Rev Lett 1996; 76, 3168.
40. Yang W. Phys Rev B 1997; 56:9294.

41. Greenspan D. Am Math Mon 1955; 62:303.
42. Kim JN, Mauri F, Galli G. Phys Rev B 1995; 52:1640.
43. Liu S, Perez-Jorda M, Yang W. J Chem Phys 2000; 112:1634.
44. Hierse W, Stechel EB. Phys Rev B 1994; 50:17811.
45. Ordejon P, Drabold DA, Grumbach MP, Martin RM. Phys Rev B 1993; 48:14646.
46. Mauri F, Galli G, Car R. Phys Rev B 1993; 47:9973.
47. Li XP, Nunes W, Vanderbilt D. Phys Rev B 1993; 47:10891.
48. Haynes PD, Payne MC. Phys Rev B 1999; 59:12173.
49. Adhikari S, Baer R. J Chem Phys 2001; 115:11.
50. White CA, Johnson BG, Gill PMW, Headgordon M. Chem Phys Lett 1994; 230:8.
51. Strain MC, Scuseria GE, Frisch MJ. Science 1996; 271:51.
52. White CA, Headgordon M. J Chem Phys 1994; 101:6593.
53. Greengard L. Science 1994; 265:909.
54. Greengard L, Rokhlin V. J Comput Phys 1987; 73:325.
55. Perez-Jorda JM, Yang W. Chem Phys Lett 1998; 282:71.
56. Aluru S. Siam J Sci Comput 1996; 17:773.
57. Sun XB, Pitsianis NP. Siam Rev 2001; 43:289.
58. Sagui C, Darden T. J Chem Phys 2001; 114:6578.
59. Briggs WL, Henson VE, McCormic SF. A Multigrid Tutorial. 2d ed. Philadelphia: SIAM.
60. Trottenberg U, Oosterlee CW, Schuller A. Multigrid. San Diego: Academic Press, 2001.
61. Wang J, Beck TL. J Chem Phys 2000; 112:9223.
62. Schroder J, Trottenberg U. Witsch Lect Notes Math 1978; 631:153.
63. Yang W, Perez-Jorda JM. Schleyer PvR, ed. Encyclopedia of Computational Chemistry. New York: Wiley, 1998;1496–1513.
64. Becke AD. J Chem Phys 1988; 88:2547.
65. Stratmann RE, Scuseria GE, Frisch MJ. Chem Phys Lett 1996; 257:213.
66. Parr RG, Bartolotti LJ. J Am Chem Soc 1982; 104:3801.
67. Sanderson RT. Science 1951; 114:670.
68. Sanderson RT. Chemical Bonds and Bond Energy. 2d ed. New York: Academic, 1976.
69. Mulliken RS. J Chem Phys 1934; 2:782.
70. Nalewajski RF, Parr RG. J Chem Phys 1982; 77:399.
71. Parr RG, Yang W. J Am Chem Soc 1984; 106:4049.
72. Ayers PW, Levy M. Theor Chem Acc 2000; 103:353.
73. Nalewajski RF. J Chem Phys 1983; 78:6112.
74. Yang W, Parr RG, Uytterhoeven L. Phys Chem Miner 1987; 15:191.
75. Parr RG, Pearson RG. J Am Chem Soc 1983; 105:7512.
76. Yang W, Parr RG. Proc Natl Acad Sci USA 1985; 82:6723.
77. Pearson RG. J Am Chem Soc 1963; 85:3533.
78. Pearson RG. J Chem Educ 1968; 45:643.
79. Pearson RG. J Chem Educ 1968; 45:581.
80. Pearson RG. J Chem Educ 1999; 76:267.
81. Ayers PW. Theor Chem Acc 2001; 106:271.
82. Pearson RG. J Chem Educ 1987; 64:561.
83. Lieb EH. Int J Quantum Chem 1983; 24:243.
84. Levy M. Proc Natl Acad Sci USA 1979; 76:6062.
85. Levy M, Perdew JP. NATO ASI Ser Ser B 1985; 123:11.
86. Levy M, Perdew JP. Int J Quantum Chem 1985; 743.
87. Morrison RC. J Chem Phys 2002; 117:10506.
88. Englisch H, Englisch R. Phys Status Solidi B 1984; 123:711.
89. Englisch H, Englisch R. Phys Status Solidi B 1984;124373.
90. Yang W, Zhang Y, Ayers PW. Phys Rev Lett 2000; 84:5172.

91. Dreizler RM, Gross EKU. Density Functional Theory: An Approach to the Quantum Many-Body Problem. Berlin: Springer-Verlag, 1990.
92. Parr RG, Ayers PW. J Phys Chem A 2002; 106:5060.
93. Ayers PW, Morrison RC, Roy RK. J Chem Phys 2002; 116:8731.
94. Stillinger FH. J Chem Phys 2000; 112:9711.
95. Gresh N, Leboeuf M, Salahub D. ACS Symp Ser 1994; 569:82.
96. Velde GT, Bickelhaupt FM, Baerends EJ, Guerra CF, VanGisbergen SJA, Snijders JG, Ziegler T. J Comput Chem 2001; 22:931.
97. Berces A, Ziegler T. Top Curr Chem 1996; 182:41.
98. Kohn W. Rev Mod Phys 1999; 71:1253.
99. March NH, Parr RG. Proc Natl Acad Sci USA Phys Sci 1980; 77:6285.
100. Parr RG. Annu Rev Phys Chem 1983; 34:631.
101. Kohn W, Becke D, Parr RG. J Phys Chem 1996; 100:12974.
102. Runge E, Gross EKU. Phys Rev Lett 1984; 52:997.
103. Dhara AK, Ghosh SK. Phys Rev A 1987; 35:442.
104. Kohl H, Dreizler RM. Phys Rev Lett 1986; 56:1993.
105. Gross EKU, Kohn W. Phys Rev Lett 1985; 55:2850.
106. Casida ME, Jamorski C, Bohr F, Guan JG, Salahub DR. ACS Symp Ser 1996; 628:145.
107. Van Leeuwen R. Int J Mod Phys B 2001; 15:1969.
108. Casida ME. Chong DP, ed. Recent Advances in Density Functional Methods. Part1. Singapore: World Scientific, 1995:155–192.
109. Levy M. Proc Natl Acad Sci USA 1979; 76:6062.
110. Parr RG. Philos Mag B 1994; 69:737.
111. Lieb EH. NATO ASI Ser Ser B 1985; 123:31.
112. Beck TL. Rev Mod Phys 2000; 72:1041.
113. Scuseria GE. J Phys Chem A 1999; 103:4782.
114. Wu SY, Jayanthi CS. Phys Rep-Rev Sec Phys Lett 2002; 358:1.
115. Goedecker S. Rev Mod Phys 1999; 71:1085.
116. Challacombe M. Comput Phys Commun 2000; 128:93.
117. Handy NC, Tozer DJ. Mol Phys 1998; 94:707.
118. Hardy GH. Messenger Math 1917; 46:175.
119. Cramer H, Wold H. J Lond Math 1936; 11:290.
120. Neumann R, Nobes RH, Handy NC. Mol Phys 1996; 87:1.
121. Hilbert D. Bull Am Math Soc 2000; 37:407.
122. Becke AD. ACS Symp Ser 1989; 394:165.
123. Ayers PW, Parr RG. J Am Chem Soc 2000; 122:2010.
124. Geerlings P, De Proft F. Int J Quantum Chem 2000; 80:227.
125. Boon G, De Proft F, Langenaeker W, Geerlings P. Chem Phys Lett 1998; 295:122.
126. Vasilescu FH. Trans Am Math Soc 2002; 354:1265.
127. Parr RG. NATO ASI Ser, Ser B 1985; 123:141.
128. Parr RG. Aspects of density fünctional theory. In: Dahl JP, Avery J, eds. Local Density Approximations in Quantum Chemistry and Solid State Physics. New York: Plenum Press, 1984: 21–31.
129. Chermette H. J Comput Chem 1999; 20:129.
130. Ayers PW, Parr RG. J Am Chem Soc 2001; 123:2007.
131. Nalewajski RF, Korchowiec J, Michalak A. Top Curr Chem 1996; 183:25.
132. Chattaraj PK. J Indian Chem Soc 1992; 69:173.

5

Hybrid Quantum Mechanical/Molecular Mechanical Methods

JEAN-LOUIS RIVAIL

Henri Poincaré University, Nancy-Vandoeuvre, France

1. INTRODUCTION

Understanding, at the molecular level, any elementary biological process usually requires the consideration of a large system made of thousands of atoms. This is true for macromolecules such as proteins or nucleic acids, but also for smaller molecules because they can hardly be considered independently of their usually complex surroundings, made of a large number of water, solvent, or host molecules.

In such systems, entropy is an important factor and the statistical nature of this quantity requires the computation of a large number of different configurations in Monte Carlo (MC) or molecular dynamics (MD) approaches. Computer simulations based on first principles are still unrealistic for such systems, even though the performance of modern computers is increasing impressively. Therefore, some simplified approaches that allow us to simulate a system made of thousands of atoms at a reasonable computational cost are still necessary and the methods of molecular mechanics (MM) (this volume, Chapter 1) have now become important tools to simulate the energy variations resulting from configurational changes.

The classical force fields used in molecular mechanics are based upon the concept of transferable properties of standard chemical bonds. This intuitive concept is now supported by quantum mechanical computations on small reference systems, which are currently used to parameterize the force fields. It fails when these bonds differ strongly from their equilibrium, and this is particularly true in the course of chemical reactions in which some bonds are broken and some are formed. In such cases, it is necessary to go beyond the level of molecular mechanics and to consider the elementary components of the system, electrons and nuclei, the behavior of which requires the use of quantum mechanics. At this level, the number of interacting particles is increased enormously. This is the reason why the first principle approach of very large systems is still out of reach nowadays. However, the transferability of bond properties means that the structural modifications that occur in one region of a large molecule do not influence the properties of the rest of the system significantly and so a local approach is still valid. Therefore, it becomes clear that the study of a system that undergoes large electronic changes, in particular chemical reactions, can be modeled by a quantum mechanical treatment limited to a part of this system, small enough to be compatible with the computational resources. One must, however, keep in mind that the properties of the subsystem treated at the quantum mechanical level are not independent of the rest of the system. The neighboring atoms may constrain the subsystem to adopt a geometry that departs from the preferred geometry, free of any interactions. The long-range electrostatic interactions are expected to introduce a perturbation that can hardly be neglected. This is the reason why the interaction of the quantum subsystem with its surroundings has to be taken into account, and the molecular mechanics force fields can be used to model that part of the system that is assumed not to be greatly modified during the process of interest, provided that one models the interactions between the quantum mechanical and the molecular mechanical parts of the system in a realistic way. This is the basis of the now-popular quantum mechanical/molecular mechanical (QM/MM) methods [1].

2. QM/MM STUDIES OF SOLUTIONS

The study of a reaction in solution involving small molecules (i.e., molecules having a size compatible with a full quantum chemical computation) can be performed by means of a QM/MM approach. The subsystem requiring a quantum mechanical treatment consists of the molecules that take part in the reaction, and all the "spectator" solvent molecules of the sample are represented by the classical force field. Therefore, the energy of the system can be written as follows:

$$E = E_{QM} + E_{MM} + E_{QM/MM} \tag{1}$$

In this expression, E_{QM} and E_{MM} stand for the energy of the quantum subsystem and the classical one, respectively, and $E_{QM/MM}$ stands for the interaction energy between these two subsystems. This quantity corresponds to an interaction between nonbonded atoms and, in classical force fields, it is usually decomposed into an electrostatic term V_e, which depends on the electrostatic parameters of the force field (point charges, permanent moments, and occasionally induced moments) and on the nuclear charges and the electronic density in the quantum subsystem and an empirical van der Waals contribution V_v.

The difference with classical force fields comes from the fact that the charged particles in the quantum subsystem are the nuclei and the electrons whose density is known when the electronic wavefunction Ψ is defined. When the electrostatic properties of the classical part are expressed by point charges q_M on the atoms, the operator V_e has the simple expression (in atomic units):

$$V_e = \sum_M \left(-\sum_i \frac{q_M}{R_{iM}} + \sum_K \frac{Z_K}{R_{KM}} \right) \tag{2}$$

where the sums are extended to all classical atoms M (charge q_M), all nuclei K of the quantum subsystem (charge Z_K), and all electrons i. R_{iM} and R_{KM} represent the interparticle distances. This expression is easy to extend when more elaborated expressions of the classical electrostatic energy of interaction are used, in particular when higher permanent or induced electric moments are located on the classical atoms.

The van der Waals interaction V_v is introduced to account for the fact that the classical atoms are not point charges but have an attractive dispersion interaction with any other atom and a short-range repulsion. This contribution is included in the force fields and is used without any modification in the expression of $E_{QM/MM}$. It is assumed to be independent from the actual electronic structure of the interacting atoms.

Thus, the quantum computation requires the minimization of the quantity $\langle \Psi | \mathbf{H}_o + V_e | \Psi \rangle$, where $\mathbf{H}_o$ is the Hamiltonian of the quantum part in the absence of the solvent and Ψ is the wavefunction for the quantum part interacting with its surroundings.

The different energy contributions are then given by:

$$E_{QM} = \langle \Psi | \mathbf{H}_o | \Psi \rangle \quad \text{and} \quad E_{QM/MM} = V_v + \langle \Psi | V_e | \Psi \rangle. \tag{3}$$

E_{MM} is calculated using the expressions implemented in the force field used.

The expression for the total energy of a system allows Monte Carlo evaluations of the thermodynamic properties of the sample. The forces (i.e., the energy first derivatives) make molecular dynamics studies possible.

A large number of chemical reactions in the liquid phase have been studied by means of such methodologies [2–7].

3. QM/MM STUDIES OF MACROMOLECULES

The principle of the quantum computation applied to a fragment of a macromolecule, while the rest of the system is treated at the molecular mechanical level, is the same as in the case of solutions, except that the separation between the subsystems goes through chemical bonds. Cutting these bonds gives rise to dangling bonds that have to be saturated. This is not compatible with standard QM computations, and several solutions designed to enable quantum chemical computations on a fragment of a large, covalently bonded system have been proposed.

3.1. The Link Atoms

The simplest way to keep the electronic structure of the quantum subsystem as close as possible to what it would be in the entire macromolecule consists of saturating the dangling bonds with monovalent atoms called link atoms. Typically, hydrogen atoms are used. The computation now consists of a model molecule of the reactive part interacting with classical surroundings, similar to the case of solutions. This approach has been introduced by Singh and Kollman [8] and has been put in a operational form by Field et al. [9].

The only difference with the case of solutions arises from the treatment of the electrostatic interaction between the link atom and the classical atoms. In a first version implemented in the CHARMM package [10], this interaction was simply neglected (the "QQ" link). An alternative version in which the link atom interacts with all the atoms has been added (the "HQ" link). Due to the size of the reactive parts of interest in biomolecules, semiempirical quantum chemical methods are often used, especially the AM1 [11] or PM3 [12,13] parameterizations working at the NDDO level (this volume, Chapter 2).

Semiempirical methods have, as a common feature, the fact that they only describe the electrons of the atomic valence shells that interact with a core made of the nuclei surrounded by the electrons of the inner shells. The nuclear charge Z_K of nucleus K is replaced by the core charge Z'_K, which is equal to the number of valence electrons. The molecular orbitals are expanded in the set of Slater atomic orbitals $|\mu\rangle, |\nu\rangle$ of the atoms' valence shell and assumed to be orthogonal in semiempirical approximations. These molecular orbitals are eigenfunctions of a semiempirical Fock operator, which will not be developed here. The molecular energy is the sum of the electronic energy and the core–core interaction energy, which, for a pair of atoms K, L, takes the form:

$$Z'_K Z'_L (s_K s_K | s_L s_L)[1 + f(R_{KL})] \tag{4}$$

where R_{KL} is the distance between nuclei K and L, and s_K, s_L denote s orbitals centered on atoms K and L so that $(s_K s_K | s_L s_L)$ is the Coulomb integral involving this pair of orbitals. $f(R_{KL})$ is a rapidly decreasing empirical function that varies from one method to another. To be consistent with the semiempirical formalism, the interaction between the point charges of the classical subsystem and the nuclei of the quantum subsystem has to be replaced by an interaction between the point charges and the atomic cores, which is written in the form:

$$E_{\text{QM/MM}}^{\text{charge/core}} = \sum_K \sum_M Z'_K q_M (s_K s_K | s_M s_M)[1 + f(R_{KM})] \tag{5}$$

and the (μ, ν) element of the Fock matrix is written as:

$$F_{\mu\nu} = F_{\mu\nu}^\circ - \sum_M q_M (\mu\nu | s_M s_M) \tag{6}$$

where $F_{\mu\nu}^\circ$ is the Fock matrix element for the quantum part isolated from its surroundings.

This version of the QM/MM approach has been used to study many biochemical reactions [14,15]. An implementation of analytical second derivatives makes it useful to interpret vibrational spectra and to characterize stationary points on a potential energy surface [16].

Several related methods have been developed during the past years. The Achilles' heel of this approach is the treatment of the interaction between the link atom and classical part of the system, and the proper simulation of the bonds between the two kinds of subsystems [17]. Other shortcomings come from the nature of the link atom.

Hydrogen may be rather convenient to replace a carbon atom, although the carbon–hydrogen equilibrium bond length is shorter than a carbon–carbon single bond. It may be more controversial in the case of a more polar bond. This is the reason why some authors prefer to use, instead of a hydrogen atom, a "dummy group" [18] or a pseudo-halogen [19,20]. To avoid the introduction of nonphysical extra degrees of freedom, Antes and Thiel [21] have defined special quantum atoms or "adjusted connection atoms" with a special parameterization to mimic as well as possible the broken bond. All these parametrical methods are tailored for semiempirical approaches.

3.2. The Frozen Orbitals

Another strategy to separate the QM part from the MM one is to freeze the pair of electrons in the broken bond (assumed to be a single bond). This has been suggested first by Warshel and Levitt [22] and the method has been developed recently at the semiempirical [23,24] and ab initio levels [26–28] as the *local self-consistent field* (LSCF) method.

If one assumes that the electron pair of the broken bond is described by a localized orbital, the electronic structure of the quantum part can be expanded in a set of molecular orbitals that have to be orthogonal to this localized orbital. The assumption that the properties of the bonds are transferable from one equilibrium structure to another can be applied to this bond, which is assumed to remain unchanged even if the structure of the quantum part is modified. This means that this localized orbital may be approximated by a combination of atomic orbitals centered on the pair of atoms that defines this bond only [i.e., we use a strictly localized bond orbital (SLBO), which can be extracted from a molecular orbital study of a model molecule in which the bond of interest is present].

In semiempirical methods, the neglect of diatomic differential overlap makes the orthogonality condition easy to fulfill. If we denote by X the last quantum atom and if we denote by Y the first classical one, we only have to consider the combination of the atomic orbitals of X that enter the SLBO of the $X-Y$ bond. In the case of an atom with only s and p orbitals in the valence shell (denoted by $|s\rangle$, $|x\rangle$, $|y\rangle$, and $|z\rangle$), one may define the contribution of atom X to the SLBO by a hybrid orbital $|l\rangle$:

$$|l\rangle = a_{11}|s\rangle + a_{12}|x\rangle + a_{13}|y\rangle + a_{14}|z\rangle \tag{7}$$

and if this orbital is normalized, it enters the localized orbital with a coefficient C_l or, equivalently, a density matrix element:

$$P_{ll} = 2C_l^2 \tag{8}$$

Within the assumptions of semiempirical methods, this orbital is considered orthogonal to all the orbitals of the quantum subsystem belonging to atoms other than X.

If we define on X three combinations of $|s\rangle$, $|x\rangle$, $|y\rangle$, and $|z\rangle$ orthogonal to $|l\rangle$, namely $|i\rangle$, $|j\rangle$, and $|k\rangle$, these functions can be combined with the other atomic orbitals to build molecular orbitals of the reactive part, which are orthogonal to the SLBO (Fig. 1). The computation of these orbitals can be achieved by means of a simple linear transformation of the Fock operator. If the molecular orbitals are expanded in a set of N atomic orbitals, including those of the L atoms involved in L bonds separating the reactive part from the classical one, the Fock matrix is an $(N–L) \times (N–L)$ square matrix, but the interaction of the electrons with the 2L electrons in the localized orbitals implies that in the Hartree–Fock equations, the density matrix is an $N \times N$ matrix including the contributions of the $|l\rangle$ hybrid orbitals. Like in the case of the link atom approximation, the Fock operator must include the electrostatic interaction of the electrons with the classical charges q_M of the classical atoms and the total energy requires the computation of the interaction of the atomic cores in the quantum part with the classical charges.

The case of the first classical atom Y deserves special attention. Its charge has been optimized with the usual approximations of the classical force fields in which the

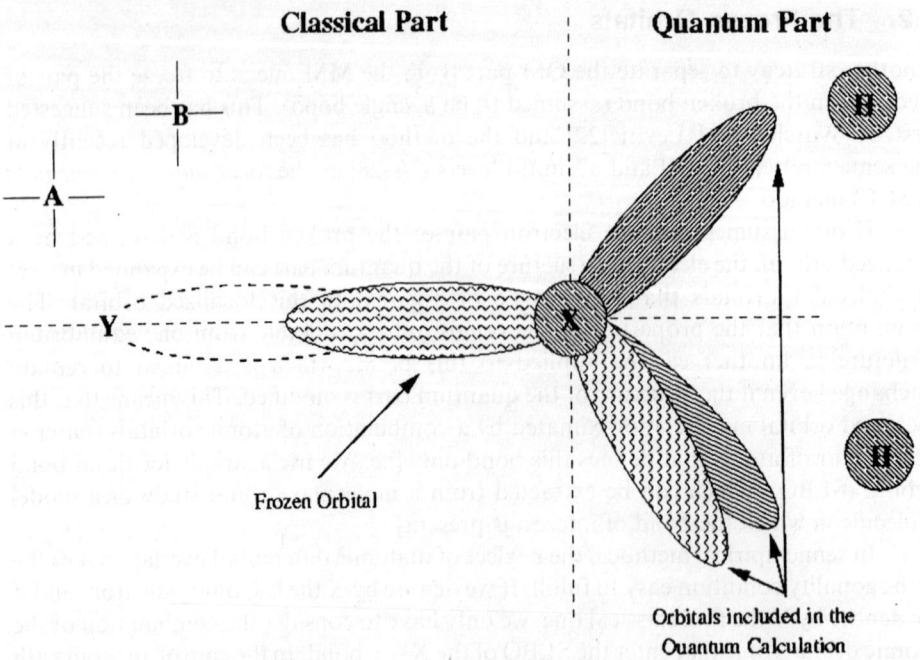

Figure 1   The hybrid frozen orbital and the three hybrid orbitals included in the quantum computation.

electrostatic interactions between nearest neighbors are included into the effective force constants and are, therefore, discarded. In the QM/MM scheme, this charge interacts with the electrons of the quantum subsystem. This means that the perturbation of the quantum subsystem requires a reparameterized effective charge $\mathring{q}_Y$. In addition, special attention has to be paid to the total charge, which must be equal to the actual charge of the system. For the same reasons, the force constant of the $X-Y$ bond must be reparameterized. This being done and the analytical derivatives of the energy of the quantum part having been calculated, a full classical quantum force field (CQFF) [24] is possible for multiple applications in biochemical reactivity and molecular dynamics [25].

A modification of this approach, still at the semiempirical level, has been proposed by Gao et al. [29] under the appellation of generalized hybrid orbital (GHO). In this method, the hybrid orbital of atom Y, which occurs in the SLBO, is explicitly considered and is included in the SCF procedure, which involves now all the orbitals of atom X. The other hybrid orbitals of Y, which would define the bonds with the other neighbors of this atom, are considered to define a core potential of Y, which is reparameterized in the semiempirical scheme to describe the $X-Y$ bond as correctly as possible. The parameterization of the Y atom and the $X-Y$ bond requires the same care as above.

A detailed comparison of LSCF, GHO, and link atom semiempirical methods has been done by Reuter et al. [30]. The advantages and disadvantages of the methods, as they were when the study was published, appear clearly.

3.3. The Ab Initio and DFT LSCF

The use of frozen orbitals, such as the bond orbitals connecting the quantum to the classical part of the system, can be extended to nonempirical quantum methods such as ab initio Hartree–Fock, post Hartree–Fock, or DFT. In these cases, the overlap between atomic orbitals is taken into account and the orthogonality conditions are more difficult to fulfill. The mathematical formulation of the method has been developed in the original papers [26–28] and the process can be summarized as follows.

The input requires the usual data of a quantum chemical treatment: the starting geometry of the system and the basis set to be used, plus the L strictly localized bond orbitals expanded in the basis set chosen for the computation.

The code performs the following operations:

1. Translate and rotate the L frozen orbitals to make them coincide with the broken bonds.
2. Symmetrically orthogonalize the L frozen orbitals and compute the corresponding density matrix.
3. Project the N atomic orbitals of the basis set out of the subspace defined by the frozen orbitals. The result is a set of N functions orthogonal to the frozen orbitals, but this set is not linearly independent because there exist L additional linear combinations orthogonal to them.
4. Perform a canonical orthogonalization [31] of these functions.
5. Compute the Fock (or Kohn Sham) matrix with all the occupied molecular orbitals including the frozen ones.
6. Perform the computation using the set of orthogonal basis functions.

The overall result of steps 2–4 is a transformation of a set of N atomic orbitals $\{\ldots\varphi_i\ldots\}$ into a set of N–L orthogonal functions $\{\ldots\varphi_j'\ldots\}$, which are orthogonal to the frozen orbitals. They can be combined to produce the required molecular orbitals. This transformation can be represented by a rectangular $N \times (N$–L$)$ matrix **B**. In the SCF or DFT calculation, this matrix replaces the usual orthogonalization matrix in the computation of the eigenvalues and molecular orbitals and reduces the $N \times N$ Fock matrix to an $(N$–L$) \times (N$–L$)$ one. The process consists of a slight modification of the standard codes. Another slight modification, similar to what happens in the semiempirical LSCF, regards the computations of the Fock matrix, which must include the interaction between the electrons in the molecular orbitals and the electrons of the frozen orbitals. This is achieved by adding to the SCF density matrix the density matrix of the frozen orbitals. Details about the equations, the computation of the energy derivatives, and the modification of the frontier bond potential can be found in Ref. 28. The performance of the method is illustrated by the geometry optimization of the structure of Crambin in which a RHF 6-31G* computation on THR1 is mixed with the AMBER [32] force field for the rest of the protein. The results are collected in Fig. 2. The same kind of computation on GLU23, which is separated from the MM subsystem by two bonds, gives the same level of agreement [28].

An alternative approach has been proposed by Philipp and Friesner [33] and Murphy et al. [34]. It differs from the previous one by the introduction of modified Roothaan equations to compute the electronic density and energy of the QM part, which avoids the orthogonalization process, and by the treatment of the interaction of

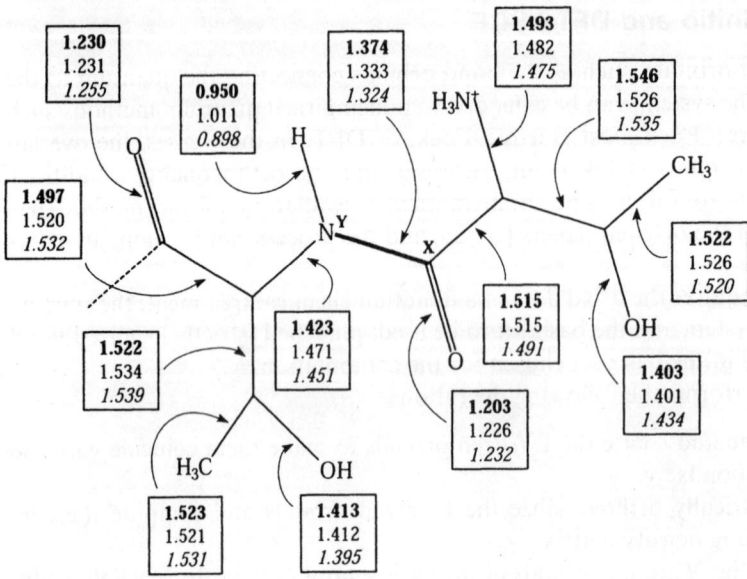

Figure 2 Optimized bond lengths (in Å) for THR1 and THR2 in Crambin. Bold roman corresponds to the LSCF (RHF 6-31G*/AMBER) calculation in which the quantum part is THR1. Light roman corresponds to the pure MM calculation and italic corresponds to crystallographic data. X is the QM frontier atom and Y is the MM frontier atom.

the electrons of the QM part with those of the frontier bonds. Another frozen localized molecular orbital-based approach, which differs from the LSCF one by the method utilized to orthogonalize the molecular orbitals to the frozen ones, has been implemented by Kairys and Jensen [35].

3.4. Other Ab Initio and DFT Methods

Although it has been first developed with semiempirical methods, the link atom approximation is working at the ab initio [36], post Hartree–Fock, or DFT level [37]. A related approach has been proposed by Maseras and Morokuma [38] as the integrated MO + MM (IMOMM) method. It has been further generalized and allows three different levels of computation including two QM levels (IMOMO), in addition to an MM one. It is now implemented in the GAUSSIAN98 package (39) under the name of ONIOM [40]. The principle is as follows. The system is divided into two or three parts. The real (large) system, which will be treated entirely at the low level of computation, yields an energy E_6 with the notations of Ref. 40, a possible intermediate model system to be computed both at a low level (energy E_3) and at a medium level (energy E_5), and finally a small model that requires a high level of computation (energy E_4) but is also computed at the medium level (energy E_2). Each model system is built by isolating the part of interest from the larger one and by saturating the dangling bonds by a link atom, or a link group of atoms. The ONIOM3 energy of the system is obtained by an extrapolation procedure, which consists of subtracting in a low level of computation the energy of the subsystem computed at this level and replacing it by the energy of the same subsystem computed at the next higher level. Therefore, it is obtained by the following equation:

$$E_{ONIOM3} = E_6 - E_3 + E_5 - E_2 + E_1 \tag{9}$$

The method allows geometry optimizations and the computation of vibrational frequencies by means of a procedure to compute the energy derivatives (at any order). The coordinates of the atoms of a subsystem common to two levels of computation are constrained to keep the values given by the higher level of computation, except for the bond length of the link atoms. If, as above, we denote by X—Y a broken bond between two layers in the ONIOM scheme, X belonging to the high level subsystem and L being the link atom (or group) used to define the model molecule, the X—Y bond length is related to the optimized X—L bond length by multiplying the latter by a scaling factor, which may be defined as the ratio of the standard corresponding bond lengths. This procedure allows the correction of the derivatives correspondingly.

This method proved to be quite efficient. It has been used to study the reaction mechanism of several reactions involving large molecular systems. It has recently been applied successfully to study the geometry of a model of the heme in hemoglobin [41]. Its shortcoming comes from the fact that the electrostatic perturbation that the classical part may exert on the electronic structure of the quantum model molecule is not fully considered, and the method, which is limited to a mechanical embedding of the model molecules in the whole system, proves to be very convenient to account for any kind of steric constrains but it may miss some important polarization effects.

An efficient method, which is an extension to enzymatic reactions of the approach to solvent effects by Warshel [42], consists of starting from ab initio computations on the reactants and products, or on model molecules in the case of a macromolecule, and incorporating them in the whole system by means of the Empirical Valence Bond mapping potential [43].

An original approach has been proposed recently by Poteau et al. [44,45]. Their effective group potentials (EGPs) are pseudopotentials fitted to represent the chemical groups bonded to the reactive part, which can then be computed with any standard ab initio method. Many potential applications of this methodology are expected. The same strategy has been implemented by Röthlisberger [46] in Car–Parrinello simulations of large systems.

4. APPLICATION OF QM/MM IN COMPUTATIONAL BIOCHEMISTRY

Detailed information on the mechanism of biochemical reactions may be of crucial importance in designing new molecules having a pharmacological activity. For example, the detailed mechanism of protein hydrolysis by thermolysin has been studied at the QM/MM semiempirical level [25]. The various steps of the reaction and their transition states have been characterized. Fig. 3 (see color plate) shows the structure of the transition state of the rate-determining step. The important consequence of this approach is the fact that it is possible to evaluate the influence of the whole macromolecular surroundings on the energetics of the process. It then becomes possible, for instance, to predict the influence of a mutation on the reaction kinetics.

Similarly, when the structure of transition states is known, it is possible to design inhibitors that act as blocking agents for this reaction and to test them, in silico, before

Figure 3 The rate-determining transition state in peptide hydrolysis by thermolysin from an AM1/AMBER QM/MM computation. The sticks and balls and sticks part correspond to the QM fragment. (See color plate at end of chapter.)

a rather long-lasting experimental study. These agents have a structure that mimics an important transition state. The active part of most of the biochemical reactions involves a large number of atoms, and using an ab initio or a DFT approach may be rather expensive from a computational point of view. This is the reason why most of the studies still use a semiempirical method for the quantum computation. Nevertheless, the results of a semiempirical computation may be refined by a more precise ab initio or DFT complementary investigation, provided that a limited number of structures are considered. This is already possible on large computers and the rapid improvement in the speed and capacity of modern computers will make this kind of computations more affordable in the near future.

5. CONCLUSION

The QM/MM methods are presently in a phase of rapid improvement. For the time being, each group has tried to develop its own method. Some of them are just a slight modification of a previously published idea, whereas some try to explore new original ideas. For instance, the use, at the semiempirical level, of an antisymmetrized product of strictly localized geminals (APSLG) [47] (i.e., orthogonal bond orbitals) to treat more accurately the junction between the quantum subsystem and the "spectator" one looks promising. Up to now, no standard method has really emerged from the number of various solutions proposed, in particular, to incorporate the QM part into the MM one. Some methods such as the link atoms or IMOMM are already in an accessible form, available in computational packages such as CHARMM [32] (link atom) or GAUSSIAN [39] (IMOMM). Many other codes are still being further developed or improved in their authors' laboratories. In many cases, it is possible to get a copy of these codes under request, but their use usually requires some expertise. It is easy to anticipate that several user-friendly QM/MM codes will soon become available. They are on the point of entering the list of the common tools of investigation into a large number of laboratories, which are concerned with molecular modeling of biochemical systems, computational medicinal chemistry, and drug discovery.

REFERENCES

1. Monard G, Merz KM Jr. Combined quantum mechanical/molecular mechanical methodologies applied to biomolecular systems. Acc Chem Res 1999; 32:904–911.
2. Gao J. Methods and applications of combined quantum mechanical and molecular mechanical potentials. In: Lipkowitz KB, Boyd DB, eds. Reviews in Computational Chemistry. Vol. 7. New York: VCH Publishers, Inc., 1996:119–185.
3. Tuñón I, Martins-Costa MTC, Millot C, Ruiz-López MF. Molecular dynamics simulations of elementary chemical processes in liquid water using combined density functional and molecular mechanics potential: I. Proton transfer in strongly H-bonded complexes. J Chem Phys 1997; 106:3633–3642.
4. Strnad M, Martins-Costa MTC, Millot C, Tuñón I, Ruiz-López MF, Rivail JL. Molecular dynamics simulations of elementary chemical processes in liquid water using combined density functional and molecular mechanics potential: II. Charge separation processes. J Chem Phys 1997; 106:3643–3657.
5. Castillo R, Andrés J, Moliner V. Quantum mechanical/molecular mechanical study of the Favorskii rearrangement in aqueous media. J Phys Chem B 2001; 105:2453–2460.
6. Byun Y, Mo YR, Gao JL. New insight on the origin of the unusual acidity of Meldrum's acid from ab initio and combined QM/MM simulation study. J Am Chem Soc 2001; 123:3974–3979.
7. Chaban GM, Gerber RB. Anharmonic vibrational spectroscopy of the glycine-water complex: calculations for ab initio, empirical and hybrid quantum mechanics/molecular mechanics potentials. J Chem Phys 2001; 115:1340–1348.
8. Singh UC, Kollman PA. A combined ab initio quantum mechanical and molecular mechanical method for carrying out simulations on complex molecular systems: applications to the $CH_3 Cl^+ Cl^-$ exchange reaction and gas phase protonation of polyethers. J Comput Chem 1986; 7:718–730.
9. Field MJ, Bash PA, Karplus M. A combined quantum mechanical and molecular mechanical potential for molecular dynamics simulations. J Comput Chem 1990; 11:700–733.
10. Brooks BR, Bruccoleri RE, Olafson BD, States DJ, Swaminathan S, Karplus M. CHARMM: A program for macromolecular energy, minimization and dynamics calculations. J Comput Chem 1983; 4:187–217.
11. Dewar MJS, Zoebisch EG, Healy EF, Stewart JJP. AM1: a new general purpose quantum mechanical molecular model. J Am Chem Soc 1985; 107:3902–3909.
12. Stewart JJP. Optimization of parameters for semiempirical methods. I. Method. J Comput Chem 1989; 10:209–220.
13. Stewart JJP. Optimization of parameters for semiempirical methods. II. Applications. J Comput Chem 1989; 10:221–264.
14. Proust de Martin F, Dumas R, Field MJ. A hybrid-potential free-energy study of the isomerization step of the acetohydroxy acid isomeroreductase reaction. J Am Chem Soc 2000; 122:7688–7697.

15. Guo H, Cui Q, Lipscomb WN, Karplus M. Substrate conformational transitions in the active site of chorismate mutase: their role in the catalytic mechanism. Proc Natl Acad Sci USA 2001; 98:9032–9037.

16. Cui Q, Karplus M. Molecular properties from combined QM/MM methods. I. Analytical second derivatives and vibrational calculations. J Chem Phys 2000; 112:1133–1149.

17. Hall RJ, Hindle SA, Burton NA, Hillier IH. Aspects of hybrid QM/MM calculations: the treatment of the QM/MM interface region and geometry optimization with an application to chorismate mutase. J Comput Chem 2000; 21:1433–1441.

18. Ranganathan S, Gready JE. Hybrid quantum and molecular mechanical (QM/MM) studies of the pyruvate to L-lactate interconversion in L-lactate dehydrogenase. J Phys Chem B 1997; 101:5614–5618.

19. Hyperchem™. Computational chemistry. Waterloo, Ontario: Hypercube, Inc., 1994:215–219.

20. Cummins PL, Gready JE. Combined quantum and molecular mechanics (QM/MM) study of the ionization state of 8-methylpteridin substrate bound to dihydrofolate reductase. J Phys Chem B 2000; 104:4503–4510.

21. Antes I, Thiel W. Adjusted connection atoms for combined quantum mechanical and molecular mechanical methods. J Phys Chem A 1999; 103:9290–9295.

22. Warshel A, Levitt M. Theoretical studies of enzymatic reactions: dielectric, electrostatic and steric stabilization of carbonium ion in the reaction of lysozyme. J Mol Biol 1976; 103:227–249.

23. Théry V, Rinaldi D, Rivail JL, Maigret B, Ferenczy GJ. Quantum mechanical computations on very large molecular systems: the local self-consistent field method. J Comput Chem 1994; 15:269–282.

24. Monard G, Loos M, Théry V, Baka K, Rivail JL. Hybrid classical quantum force field for modeling very large molecules. Int J Quantum Chem 1996; 58:153–159.

25. Antonczak S, Monard G, Ruiz-López MF, Rivail JL. Insights in peptide hydrolysis mechanism by thermolysin: a theoretical QM/MM study. J Mol Model 2000; 6:527–538.

26. Assfeld X, Rivail JL. Quantum chemical computations on parts of large molecules: the ab initio self consistent field method. Chem Phys Lett 1996; 263:100–106.

27. Assfeld X, Ferré N, Rivail JL. The local self consistent field. Principles and applications to combined QM/MM computations on biomacromolecular systems. In: Gao J, Thompson MA, eds. Combined Quantum Mechanical and Molecular Mechanical Methods. ACS Symposium Series 712. Washington, DC: American Chemical Society, 1998:234–239.

28. Ferré N, Assfeld X, Rivail JL. Specific force field parameters determination for the hybrid ab initio QM/MM LSCF method. J Comput Chem 2002; 23:610–624.

29. Gao J, Amara P, Alhambra C, Field MJ. A generalized hybrid orbital (GHO) method for the treatment of boundary atoms in combined QM/MM calculations. J Phys Chem 102, 4714–4721.

30. Reuter N, Dejaegere A, Maigret B, Karplus M. Frontier bonds in QM/MM methods: a comparison of different approaches. J Phys Chem A 2000; 104:1720–1735.

31. Szabo A, Ostlund NS. Modern Quantum Chemistry. New York: McGraw-Hill, 1986:144–145.

32. Cornell WD, Cieplak P, Bayly CI, Gould IR, Merz KM Jr, Ferguson DM, Spellmeyer DC, Fox T, Caldwell JW, Kollman P. A second generation force-field for the simulation of proteins, nucleic acids and organic molecules. J Am Chem Soc 1995; 117:5179–5197.

33. Philipp DM, Friesner RA. Mixed ab initio QM/MM modeling using frozen orbitals and tests with alanine dipeptide and tetrapeptide. J Comput Chem 1999; 20:1468–1494.

34. Murphy RB, Philipp DM, Friesner RA. Frozen orbital QM/MM methods for density functional theory. Chem Phys Lett 2000; 321:113–120.

35. Kairys V, Jensen JH. QM/MM boundaries across covalent bonds: a frozen localized

molecular orbital-based approach for the effective fragment potential method. J Phys Chem A 2000; 104:6656–6665.

36. Sheppard DW, Burton NA, Hillier IH. Ab initio hybrid quantum mechanical/molecular mechanical studies of the mechanisms of the enzymes protein kinase and thymidine phosphorylase. J Mol Struct Theochem 2000; 506:35–44.

37. Cui Q, Karplus M. Molecular properties from combined QM/MM methods. 2. Chemical shifts in large molecules. J Phys Chem B 2000; 104:3721–3743.

38. Maseras F, Morokuma K. IMOMM: a new integrated ab initio + molecular mechanics geometry optimization scheme of equilibrium structures and transition states. J Comput Chem 1995; 16:1170–1179.

39. Frisch M, Trucks G, Schlegel H, Scuseria G, Robb M, Cheeseman J, Zakrewski V, Montgomery J Jr, Stratsmann R, Burant J, Dapprich S, Millam J, Daniels A, Kudin K, Strain M, Farkas O, Tomasi J, Barone V, Cossi M, Cammi R, Mennucci B, Pomelli C, Adamo C, Clifford S, Ochterski J, Petersson G, Ayala P, Cui Q, Morokuma K, Malik D, Rabuck A, Raghavachari K, Foresman J, Cioslowski J, Ortiz J, Baboul A, Stefanov B, Liu G, Liashenko A, Piskorz P, Komaromi I, Gomperts R, Martin R, Fox D, Keith T, Al-Laham M, Peng C, Nanayakkara A, Gonzales C, Head-Gordon M, Repolge E, Pople J. Gaussian 98, Revision A.9. Pittsburg PA: Gaussian, Inc., 1998.

40. Dapprich S, Komáromi I, Byun KS, Morokuma K, Frisch MJ. A new ONIOM implementation in GAUSSIAN98: Part I. The calculation of energies, gradient, vibrational frequencies and electric field derivatives. J Mol Struct Theochem 1999; 461–462:1–21.

41. Maréchal JD, Maseras F, Lledós A, Mouawad L, Perahia D. Ab initio calculations predict a very low barrier for the rotation of the axial ligand in [Fe(P)(Im)]. Chem Phys Lett 2002; 353:379–382.

42. Warshel A. Computer Modeling of Chemical Reactions in Enzymes and Solutions. New York: John Wiley and Sons, 1991.

43. Bentzien J, Muller RP, Florian J, Warshel A. Hybrid ab initio quantum mechanics/molecular mechanics calculations of free energy surfaces for enzymatic reactions: the nucleophilic attack in subtilisin. J Phys Chem B 1998; 102:2293–2301.

44. Poteau R, Ortega I, Alary F, Ramirez-Solis A, Barthelat JC, Daudey JP. Effective group potentials. 1. Method. J Phys Chem A 2001; 105:198–205.

45. Poteau R, Alary F, El Makarim HA, Heully JL, Barthelat JC, Daudey JP. Effective group potentials. 2. Extraction and transferability for chemical groups involved in covalent or donor-acceptor bonds. J Phys Chem A 2001; 105:206–214.

46. Röthlisberger U, Carloni P, Doclo K, Parrinello M. A comparative study of galactose oxidase and active site analogs based on QM/MM Car Parrinello simulations. J Biol Inorg Chem 2000; 5:236–250.

47. Tokmachev AM, Tchougréeff AL, Misurkin IA. Effective electronic Hamiltonian for quantum subsystem in hybrid QM/MM method as derived from APSLG description of electronic structure of classical part of molecular system. J Mol Struct Theochem 2000; 506:17–34.

6

Accuracy and Applicability of Quantum Chemical Methods in Computational Medicinal Chemistry

CHRISTOPHER J. BARDEN

Dalhousie University, Halifax, Nova Scotia, Canada

HENRY F. SCHAEFER III

University of Georgia, Athens, Georgia, U.S.A.

1. INTRODUCTION

If scientists from an earlier age—even a mere hundred years past—were to somehow acquire this volume through a fantastic abrogation of temporal law, they would be most surprised to learn that enough is known about computational medicinal chemistry to warrant a text on the subject. After all, the early 20th century had scant knowledge of the basis of disease, with cures extant only in serendipitous cases, and the understanding of chemistry was equally dim. Imagine the amazement they would feel as they pored over this book, gleaning bits and pieces of a hundred years of science: that all things are composed of atoms combining with molecules to form chemical bonds; that these molecules can interact to redistribute their atoms in a new fashion; that some of these constructions can be quite large, forming mile-long chains of smaller building blocks; and that, in fact, life itself is based upon a bewildering array of systems not altogether understood, the blueprints to which are encoded in these chains!

The modern-day quantum chemist also regards this impressive edifice with admiration, but such wonder is tempered with frustration and a grim sense of purpose. Electronic structure theorists have been working at computing the structure and properties of chemical systems for some 75 years, and though it has not been easy, much has been accomplished. We owe a great deal of progress to Schrödinger:

$$H|\Psi\rangle = E|\Psi\rangle \tag{1}$$

This famous equation has always been a source of strength, fortifying us and strengthening our resolve to produce results in better agreement with experiment (and more and more often, results more accurate than experiment). Indeed, for those of us who are chiefly interested in ab initio quantum chemical methods, the serene exactitude of Schrödinger's equation provides a framework that cannot be improved; for in its deterministic purity, we know what we *do* know, and we know what we *do not* know. For example, we are certain that a full configuration interaction calculation (FCI) with a complete basis set will generate the "right" answer, subject to the Born–Oppenheimer approximation, neglecting relativity, and barring the possibility that the answer might be slightly different if we had the Grand Unified Theory. Regrettably, such accuracy comes at a computational price that cannot possibly be paid for an entire protein.

So what is to become of computational quantum chemistry under this new, unpleasant epistemology? The finite store of computer power is compelling: in the interest of speed, we must make assumptions, some of which will be unphysical (that is, unsupported by the Schrödinger equation). Some of these postulates, such as the apparent trivialities listed above, are largely upheld by experimental evidence. Other theories are more questionable, but the experiments that could incontrovertibly verify or refute them are not easily undertaken. In the end, many of the computational shortcuts taken today are known to be untrue, but it is hoped—*not* shown, albeit inductively supported—that the magnitude of the error will be within an acceptable margin. This final case is aesthetically distasteful but sometimes necessary (and, at times, surprisingly accurate). This chapter aims to demonstrate how each of these methods can be jointly and separately used to practically answer chemical questions in medicinal chemistry.

Before delving into the techniques, a semantic excursion seems necessary. First, "computational quantum chemistry" as used in this chapter reflects the broader definition, referring to any technique that uses computers to model a chemical system via the Schrödinger equation or some approximation thereof; this is a catch-all for every ab initio method, semiempirical scheme, and theoretical model chemistry. (Density functional theory also is included, although it does not stringently satisfy this definition, because it enjoys widespread identification with the ab initio methods.) Molecular mechanics, therefore, is not "computational quantum chemistry," but its application to hybrid QM/MM methods will be discussed regardless.

Second, while there are many other theories besides molecular mechanics for finding minima in the large degrees of freedom of biological systems (genetic algorithms, molecular dynamics, etc.), they are promulgated in such a way that it would be difficult to discuss them in the context of computational quantum chemistry. They are thus beyond the scope of this chapter, but later chapters in this volume will highlight each of the most important.

Finally, even "computational medicinal chemistry" could be confusing in a broad discussion of accuracy and applicability of methods, so for our purposes it will only be used to denote computational investigations directly undertaken for the advancement of medicine. Unfortunately, this definition excludes a great deal of theoretical work (including our own) because it is more fundamental in nature. Obviously, we consider ab initio studies on such important structures as DNA base pairs [1] and metalloenzyme active sites [2] to be essential in better understanding biological systems; nevertheless, unless a drug is involved, it cannot properly be called "computational medicinal chemistry."

2. THEORETICAL BACKGROUND

An overview is appropriate at this point, although some of the following material may be a review for anyone who has been reading this book in sequential order (see previous chapters and Ref. [3]). Eq. (1), the Schrödinger equation, has the simple form of an eigenvalue problem. The theory of quantum mechanics stipulates that in such an eigenvalue problem, performing an experiment (operator) upon the system (wavefunction) will result in an observable (eigenvalue) that is an intrinsic property. Surely, the most intrinsic of properties is the system's energy, and the special operator that reveals the energy is called the Hamiltonian. In computational quantum chemistry, the form most often used is the nonrelativistic, fixed-nucleus molecular electronic Hamiltonian (these two simplifications, while not necessary, are expeditious and reasonable, as most chemical systems are not dramatically affected by their omission):

$$\left[-\sum_{i=1}^{N}\frac{1}{2}\nabla_i^2 - \sum_{i=1}^{N}\sum_{A=1}^{M}\frac{Z_A}{r_{1A}} + \sum_{i=1}^{N}\sum_{j=1}^{N}\frac{1}{r_{ij}}\right]|\Psi\rangle = E_{\text{elec}}|\Psi\rangle \tag{2}$$

In this equation, N is the number of electrons while M is the number of nuclei. The first term denotes the electronic kinetic energy. The second term accounts for potential energy between electrons and nuclei. Life as a quantum chemist would be much easier if the third term did not exist—although one might argue that life in such a universe would be unlikely—for it represents the electron–electron potential energy. The first two terms can be analytically solved, but the third generally cannot. It is here, then, that the first approximation is made, usually in the manner suggested by Hartree [4] and Fock [5]

$$H_{\text{HF}} = -\sum_{i=1}^{N}\frac{1}{2}\nabla_i^2 - \sum_{i=1}^{N}\sum_{A=1}^{M}\frac{Z_A}{r_{1A}} + v^{\text{HF}} \tag{3}$$

in which an effective potential v^{HF} replaces the true interelectronic description. This substitution means that any given electron feels the influence of its brethren equally, without regard to the various distances which separate them. For this reason, it is often said that the Hartree–Fock approximation (HF) does not include "electron correlation." Calculations utilizing HF are nonetheless ubiquitous, because the method is computationally inexpensive and the central field description proves to be qualitatively correct in most cases. Is this a forgivable abridgement? Yes, provided one is interested only in qualitative results, but otherwise one must attempt to recapture the electron correlation. The second major ansatz replaces the actual integrals with groupings of one-electron functions called "basis sets" designed to mimic the structure of orbitals. This simplification improves performance substantially, but it introduces two sources of error: the chosen basis set may be insufficiently large to accurately model a given chemical system ("basis set incompleteness"), and technical differences in the way the integrals are calculated for related but nonidentical systems lead to an artifactual discrepancy ("basis set superposition error" or BSSE). Fortunately, the two errors usually act in opposition to each other and formally vanish at the complete basis set limit, should it be necessary to go that far [6].

As electron correlation is necessary for best results, several techniques have been developed to recover it. Most of these methods start with Hartree–Fock as a convenient jumping-off point and add back some of the instantaneous interelectronic influence; there are three of these methods in general use. Configuration Interaction (CI) aims to improve the electronic description by partitioning the true wavefunction into a set of relevant electron configurations to be variationally optimized [7]. The number of configurations that are "relevant" depends on the accuracy desired (a full CI contains them all), but generally only the single and double excitations are included. Møller–Plesset Perturbation Theory (MPPT) utilizes the mathematics of perturbation theory to treat the correlated part of the solution as a perturbation on top of the unperturbed Hartree–Fock reference. Often, it is truncated to include only double excitations and denoted "MP2" [8]. Far more accurate is coupled cluster (CC) theory, which uses the rather unintuitive description of the exact wavefunction as the uncorrelated wavefunction acted upon by an exponential e^T where T represents all excitations:

$$T = T_1 + T_2 + \cdots + T_n \tag{4}$$

Truncate this operator to third order with "CCSD(T)" and it still reproduces an estimated 97% of the correlation description [9]. (It is worth noting that methods exist which explicitly include the interelectronic potential. Recent calculations on the helium atom using Hylleraas-type r_{12} methods were able to match the exact non-relativistic energy to an astounding 10^{-12} kcal/mol [10].)

The "almost fourth" correlated method, the newest addition to the quantum chemist's arsenal, is density functional theory (DFT). While it is often used in a manner akin to wavefunction-based methods, DFT does not construct a wavefunction. Instead, DFT works with the electron density to determine molecular properties under the Hohenberg–Kohn paradigm [11]. This theorem shows that the electron density and the electronic Hamiltonian have a functional relationship which allows for computation of all ground-state molecular properties without a wavefunction. The chief consequence of this work is quite staggering—molecular properties are accessible after the determination of only three coordinates, regardless of molecular size! But the primary difficulty with "true" DFT can be stated bluntly (yet no less distressingly): we do not know the nature of the functional relationship, and it is entirely possible we never will. The only thing we can do is build trial exchange-correlation functionals (chimeras with abstruse names such as B3LYP [12,13] and HCTH [14]) and weigh their efficacy. The current ad hoc approach to DFT is a far cry from its promise, but modern-day Kohn–Sham DFT still has computational advantages over ab initio methods and can be applied logically in the context of HF/SCF procedures and machinery [15].

If one wishes for near-coupled-cluster accuracy ("chemical accuracy") without the additional cost of CCSD(T), a theoretical model chemistry may be a good choice. Theoretical model chemistries perform a number of reduced-size calculations and apply a formula to guess the full-sized results. Perhaps the most mathematically rigorous of these fitting theories, CBS-Q, attempts to extrapolate to full basis set convergence by applying smaller basis sets in a uniform manner (it should be noted that coupled cluster, being a high-level method, is notoriously unforgiving of small basis sets) [6]. More empirical, but probably more appropriate to computational medicinal chemistry, are the much-vaunted Gaussian-N schemes [16]. The Gaussian-2 formulations (for there are several versions) have overall errors of approximately 1.4

kcal/mol in energies. G2 is customarily used as a quick way to verify thermodynamic quantities such as reaction energetics for lower-level geometries, and it generally works well for organic systems.

Delving into semiempirical methods, of course, requires a further commitment to the principles of empiricism. Historically, empirical schemes were developed in tandem with ab initio quantum chemical methods, having been fashioned to emulate them at a fraction of their cost. Sometimes they are all that is needed for decent results; consider, for example, the energy level calculations on numerous systems by extended Hückel theory [17]. In other cases, a semiempirical method has a performance so consistent with what it is lacking that it is called by its very shortcoming, i.e., Complete Neglect of Differential Overlap (CNDO) [18]. Modern semiempirical techniques such as AM1 [19] and PM3 [20] include those repulsion terms most important to chemical bonding but replace the explicit calculation of molecular integrals with parameters determined from ab initio calculations or experiment. They represent a valuable compromise between accuracy and computational efficiency.

Molecular mechanics bears little resemblance to any of the previous theories. Acting under the rationale that a chemical bond can be thought of as a spring between two spheres, molecular mechanics calculations build a potential model of the system using the principles of classical mechanics and some additional empirical corrections:

$$V = V_{\text{stretch}} + V_{\text{bend}} + V_{\text{torsion}} + V_{\text{van der Waals}} + \text{etc.} \tag{5}$$

Each of these potential functions has a straightforward (usually Newtonian) definition and contains parameters to be adjusted for best results. Denoted "force fields," these parameter sets are often fit to the optimum average error among the properties of interest (i.e., geometries, heats of formation, etc.) for specific molecular classes. Some force fields (such as the venerable MM2/MM3/MM4 series) [21] are meant to be generalized for large organic systems, but others (such as AMBER) [22] were created for certain classes of macromolecules. Molecular mechanics' accuracy depends on the species being studied and how closely it chemically resembles the molecules used to create the force field.

3. ACCURACY AND APPLICABILITY OF METHODS

So much for the theories. We now turn to their practicality and usefulness as regards computational medicinal chemistry. The current state of the field can be thought of as a sliding scale of accuracy for each quantum chemical method, along with a concomitant list of applicable system sizes due to limited computational power. (Considering the vast dissimilarity among the techniques, it is not at all surprising that computational quantum chemists find themselves continually arguing over what constitutes an "accurate" calculation on a "large molecule.") In fact, "scaling" is an altogether appropriate word, for in computational science it denotes how quickly the time required for a calculation increases with an increase in size. Scalings for ab initio methods vary widely, so we will consider each in turn.

3.1. Ab Initio

The fastest scaling for an ab initio method is Hartree–Fock theory utilizing the self-consistent-field procedure (HF/SCF), which for a given basis set scales as the number of electrons N^4. In other words, double the size of the calculation and it will take around 16 times as long. This scaling measures only the generation of the energy of a single point on the potential energy surface, yet geometry optimizations require multiple single points or analytic derivative computations to determine a minimum. Thus, as a practical matter, a method can be used as a primary, day-to-day tool only when molecular geometries can be computed (although when necessary to determine energy barriers, a few higher-level energy points at the lower-level geometries may provide a decent substitute). Today's computers make full geometry optimizations or energy points with HF/SCF realistic for systems of 150 or 500 heavy atoms, respectively. For organic molecules (and most drugs and receptor sites are in that category), such a calculation will undoubtedly provide structures within 0.1 Å of bond lengths and 5° of bond angles and relative energies often within 5–10 kcal/mol of experiment. Perturbation theory's workhorse, MP2, scales as N^5. So in 2002, it is constructive to use MP2 with full geometry optimization only on systems up to 50 heavy atoms, and energy points are useful up to 200 heavy atoms. The additional cost of electron correlation generally provides better results (0.05 Å, 2°, 3 kcal/mol) and is especially helpful when determining the relative energetics of reactants, products, and transition states. When the desired accuracy involves distinguishing systems separated by only 1 kcal/mol, coupled cluster theory is the only practical solution. Its practicality is tempered by its N^7 scaling, however, and as such it is really only useful at present for systems of 20 heavy atoms or less (or up to 70 as an energy point). CCSD(T) can usually provide geometries to 0.01 Å and 0.5° with relative energies within 0.2 kcal/mol of experiment [3,23].

Is there a place for the very highest-level ab initio techniques in computational medicinal chemistry? Sadly, not at this time. To acquire all of the electron correlation, theorists use the FCI method described above, but only for diatomic and triatomic systems that can justify the prohibitive $N!$ cost. The explicit Hylleraas methods are so expensive that their full implementation may never be useful for systems with more than a few electrons, although the related R12 techniques (especially MP2-R12) might be [24]. It is worth mentioning that in certain situations involving excited states (i.e., electronic spectroscopy) or metal-containing systems, none of the practical ab initio techniques can be exhaustively accurate, as they are all based on a single-reference description of the wavefunction that fails when excited states are close in energy to the ground state. As traditional multiconfiguration SCF and CI are out of the question for large systems, the best choice is probably a linear response method [25] such as Configuration Interaction Singles (CIS) coupled with whatever reference is affordable.

Overall, perhaps the best argument for the use of ab initio quantum chemistry is its versatility. Ab initio methods are unmatched in the area of molecular property computation. Energy, being a zeroth-order property, is easily acquired. A myriad of other properties can be understood using analytic derivative theory; routines exist for infrared, Raman, NMR chemical shifts, circular dichroism, magnetic susceptibility, dipole moments, and spin-orbit coupling, among others [26]. Properties that relate to the essential thermodynamics or kinetics of a chemical reaction can be computed with mature, robust techniques such as Variational Transition State Theory (VTST) [27] and emerging dynamical methods such as the Reaction Path Hamiltonian [28]. Even bulk effects such as solvation can be treated, either explicitly through the addition of solvent molecules to the calculation, or in an averaged fashion utilizing the Polarizable Continuum Model (PCM) [29] or the Self-Consistent Reaction Field (SCRF) [30,31]. In the case of truly exotic physical phenomena such as the Mössbauer effect, ab initio methods are the only choice, and one must simply build a model compound to effectively estimate the property on the largest systems of interest. Today, the use of ab initio quantum chemical methods in computational medicinal chemistry is widespread, albeit not so widespread as it could be if computers were faster. The consensus view is that such techniques are extremely valuable for smaller systems, but the additional accuracy is not worth the computational cost for larger systems. Hartree–Fock theory, however, seems fast enough to warrant general use for many studies. Indeed, HF/3-21G (Hartree–Fock with the split-valence Gaussian basis set 3-21G) has become a standard of sorts when a relatively large amount of accurate conformational information is desired, as in a recent conformational analysis of the glycoprotein model compounds N-formyl-L-asparaginamide and N-acetyl-L-asparagine N-methylamide [32]. Structural studies of the conventional variety can bear fruit also, especially if the chemical problem involves possible adverse reactions in DNA [energies determined using an impressive MP2/6-311G(2d,p) treatment to be positively sure] [33]. In the end, when chemical intuition fails us, ab initio quantum chemistry is the time-tested method of last resort: CCSD(T) on acetone easily explains the dipole moment Stark effect shifts in the photosynthetic reaction center of various *Rhodobacter sphaeroides* mutants [34].

Berg and co-workers [35] plainly have a great deal of confidence in ab initio methods for the advancement of medicine. They point out that it took only a few decades before it was possible to perform a full geometry optimization on a 126-atom, 372 degree-of-freedom chain of 12 alanines (see Fig. 1) [36], and they feel computational power will continue to increase. In suggesting an ambitious computational effort toward understanding peptide folding, they note that ab initio results, while expensive to obtain, will likely provide enough accuracy that they will not need to be recalculated for a long time. By contrast, they believe that the tremendous undertaking of computing millions of conformational parameters for all possible tripeptides requires rigor that semiempirical calculations cannot provide. Their initiative suggests that protein folding can be tackled three peptides at a time using methods parameterized from their computations. The platform they have chosen: HF/3-21G for

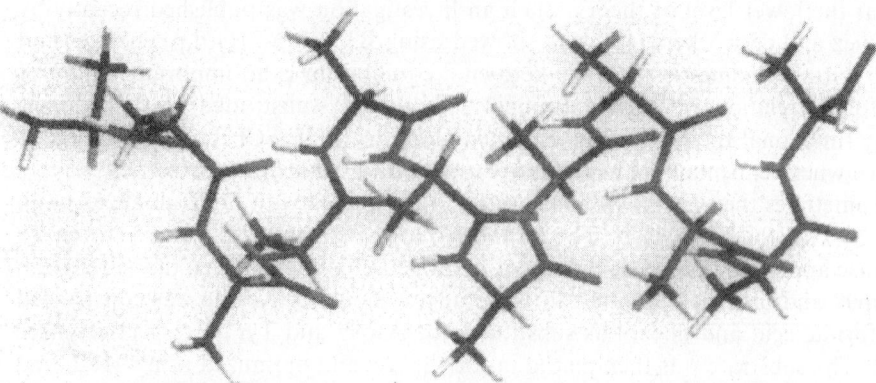

Figure 1   The helical alanine 12-mer, a recent landmark for full ab initio geometry optimizations of biomolecules. See Ref. 36. (See color plate at end of chapter.)

geometry optimizations and the DFT functional B3LYP/6-31G* for energies. Why DFT and not MP2 or CCSD(T)? Let us consider it.

3.2. Density Functional Theory

Modern density functional theory in its current Kohn–Sham formulation is still very much a method in development. Its patchwork of varied and often peculiar-looking functionals does little to simplify matters for the nonspecialists, yet these workers are increasingly expected to use DFT in support of their research. A few functionals such as B3LYP [12,13] and BP86 [37,38] have been deemed useful in describing most chemical systems; no doubt newer functionals perform the same or better than these two, but they have been extensively tested and are therefore recommended for general use. It is important to realize that none of these so-called "DFT functionals" can be shown to resemble the exact functional—the sooner that mythical creature is found, the better—but that their performance is roughly at the MP2 level with only an HF/SCF level of computational cost. They have also proven to be unusually adept at modeling metal-containing systems, even when the usual all-electron basis set for the metal is replaced with a simpler effective core potential (ECP) to model the inner electrons [39]. Many of these ECPs even account for important relativistic changes in the significantly larger cores of heavy nuclei. Most of the properties available in a HF/SCF calculation are available for DFT. Density functional theory is formally a ground state theory, but it has a linear-response formalism designated Time-Dependent DFT (TD-DFT) that can be used to produce an electronic spectrum and photochemical reaction data [40].

From a small molecule perspective, DFT functionals are still viewed with some suspicion due to their inconsistent record of accuracy for special cases such as annulenes [41]. However, for a large biomolecule, such problems are not likely to creep up at a significant rate, as unstrained organic chemistry is usually rather straightforward theoretically. Density functional theory is not very sensitive to basis set effects, so a medium-size set is already approaching the limit of its accuracy; this makes it ideal for geometries of systems up to 150 heavy atoms (DFT typically runs within a factor of two of the HF/SCF required time). Energies of larger systems can be computed using Hartree–Fock geometries in order to verify reaction intermediates and transition states at the lower level of theory. Such an investigation was published recently by Rodriguez and co-workers [42], who showed using B3LYP/6-31G(d,p) energies and HF/6-31G(d,p) geometries that the keto-enol equilibrium is an important figure of merit for correlation with the antifungal activity of α-substituted acetophenones. Density functional theory is often exploited along with HF/SCF for more energetic evidence when semiempirical methods are required to handle the geometries.

Sometimes the system is small enough to be studied with DFT alone, as in an article by Pan and McAllister. The tautomerization of steroids by Δ^5-3 ketosteroid isomerase apparently proceeds by way of hydrogen bonding between the steroid and the Asp99 and/or Tyr14 residues. In this study, a model active site was constructed using formic acid and phenol as substitutes for Asp99 and Tyr14, respectively (see Fig. 2). The substrate was then placed inside the site and optimized using MP2/ and B3LYP/6-31 + G(d,p) including solvation with the SCRF-SCIPCM [43] method. The results were unfortunately indeterminate, as the 1 kcal/mol difference between trial structures was not enough to determine which H-bond-mediated mechanism was favored (if either) [44]; although uses abound, modern DFT functionals' intermediate-level accuracy proves insufficient to describe bonding universally well.

3.3. Semiempirical Methods

Much of the important aspects of semiempirical methods has been previously discussed. Modern semiempirical methods include MNDO [45], AM1 [19], and PM3 [20], among others. MNDO is incorporated into the MOPAC software package [46], which is capable of computing many molecular properties including polarizibilities, IR, NMR, Raman, and nuclear quadrupole resonance parameters. MNDO is not as often used, however, as it does not adequately reproduce hydrogen bonding or heats of formation to better than 14 kcal/mol [47]. Each semiempirical method is built around a different effective Hamiltonian, and thus some are more useful than others in various circumstances. It has been suggested that the PM3 Hamiltonian is superior for modeling hydrogen bonds for precisely this reason [48].

All current semiempirical methods suffer from their valence-only implementation. They are also all parameterized for only the ground state for each nucleus of interest. As such, they are not very good choices for careful examinations of reactions. On the whole, however, they can be expected to perform within 8–10 kcal/mol of experiment for heats of formation [47]. Solvation effects may be easily included as well by way of the popular SMx technique [49]. Their most useful property is that they may be applied to systems of up to 500 atoms. Thus a full AM1 geometry optimization of the α-chymotrypsin (serine protease) active site is possible, along with its target *N*-acetyl-L-tryptophanamide [50]. In a far-looking viewpoint paper by Patel et al., "Will *ab initio* and DFT drug design be practical in the 21st century?", AM1 is used to probe

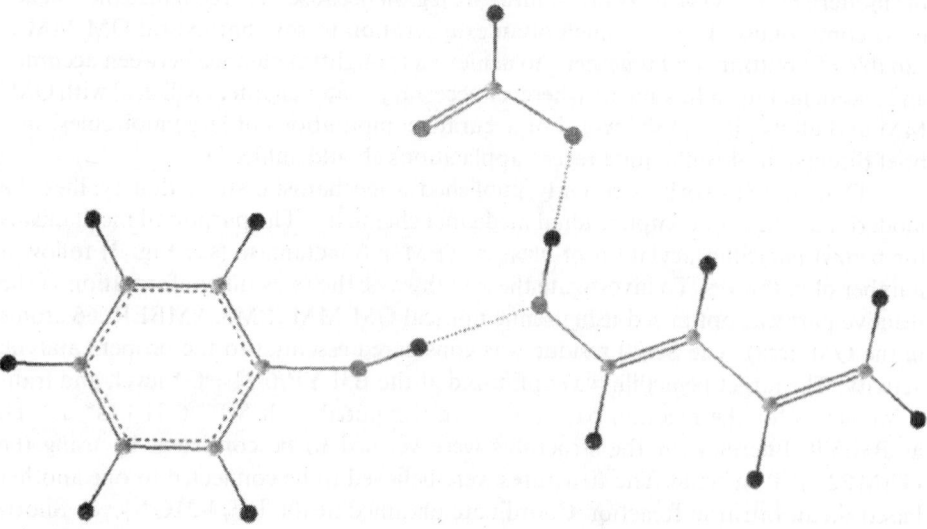

Figure 2   A model active site for Δ^5-3 ketosteroid isomerase studied using density functional theory. See Ref. 44.

the optimal (unknown experimentally) arrangement among helices in the seven-helix β2-adrenergic G-protein coupled receptor [51]. The most likely use of semiempirical methods, though, is as the quantum mechanical part of a hybrid calculation.

3.4. Hybrid QM/MM

Hybrid Quantum Mechanics/Molecular Mechanics (QM/MM) methods have computational symbiosis as their goal. Quantum mechanics methods are readily applicable to 15–500 atom systems with good-to-excellent accuracy depending on the specific method. Molecular mechanical methods can generate decent results (close to HF/SCF accuracy, sometimes better for conventional systems) [52] for many thousands of atoms, provided nothing in the molecule requires accurate modeling of bond breaking, polarization effects, etc. In hybrid QM/MM methods, the unparalleled speed of molecular mechanics may be applied to the parts of the molecular system that have a negligible chemical impact, while some quantum mechanical theory of higher accuracy may attack the difficult-to-model catalytic active site [53]. Such calculations are, in principle, capable of handling systems with several thousand atoms, which is why such studies hold a significant share of research in computational medicinal chemistry.

As hybrid QM/MM uses each tool for maximum practicality in a pragmatic fashion, it is perhaps not too surprising that the first application of QM/MM was reported before all the theory had been constructed. Vibrational structures and electronic transitions in conjugated polyenes and retinal were the subjects of the very first hybrid QM/MM study by Warshel and Karplus [54]. The pioneering work of Warshel and Levitt laid down the ground rules for a consistent QM/MM algorithm [55]. Further research solidified the theory and made it robust enough to handle such historically difficult aspects as reactivity [56] and solvation [57]. The practitioners of the modern hybrid QM/MM procedures are legion because they recognize the efficacy of its compromise. It is not much of an exaggeration to say that hybrid QM/MM is capable of "putting it all together" to achieve a thoughtful balance between accuracy and speed, making adjustments wherever necessary. Later chapters will deal with QM/MM and all its allies in the world of accurate computations of large molecules, so a brief discussion of some quite recent applications should suffice.

Díaz and co-workers recently published a mechanistic study that typifies the modern paradigm of computational medicinal chemistry. The purported mechanisms for benzyl penicillin acylation of class A TEM-1 β-lactamase (see Fig. 3) follow a number of pathways. To investigate these pathways, the relevant conformation of the reactive part was optimized using semiempirical QM/MM (PM3/AMBER, 66 atoms in the QM area). The Ser70 residue was considered essential to the proper catalytic activity. The target penicillin was optimized at the B3LYP/6-31 + G* level, and transition states for the reaction pathways were computed with MP2/6-31 + G* as well as B3LYP. Energies for the structures were verified to be consistent by using the G2(MP2, SVP) scheme. The structures were believed to be connected to one another based on an Intrinsic Reaction Coordinate obtained at the HF/3-21G* level. Short-range solvent effects were treated with explicit solvent molecules when practical, while the rest of the solvent effects were explicitly included in the QM/MM treatment but approximated using SCRF in the ab initio treatments. The complexation energy had

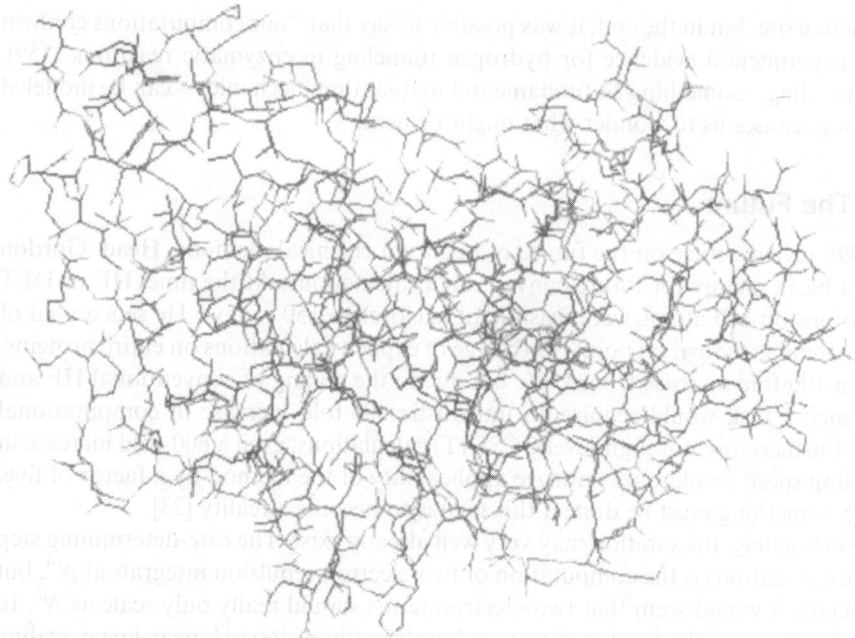

Figure 3 Class A TEM-1 β-lactamase (PDB ID: 1BTL), the subject of a study employing QM/MM techniques. See Ref. 58.

to be carefully derived using a formula that took all the various treatments into consideration:

$$\Delta E_{composite} \approx \Delta E_{B3LYP/6-31+G^*} \text{ (active site)} + \Delta\Delta G_{solvation} \text{ (protein-penicillin)}$$

$$+ [\Delta E_{PM3} \text{ (protein-penicillin)} - \Delta E_{PM3} \text{ (active site)}] \tag{6}$$

The authors made a determination after examining their diverse and voluminous results that "the acylation of class A β-lactamases by penicillin proceeds through a hydroxyl- and carboxylate-assisted mechanism" [58].

Another well-conceived research project was undertaken by Alhambra and co-workers which showcases the variety of problems that hybrid QM/MM can tackle with the help of conventional theory. The main interest of this work was the role of tunneling in the dynamics of the horse liver alcohol dehydrogenase (LADH) metalloenzyme. The specific kinetics had already been experimentally measured, so it was appropriate to compare those results to the best possible theory. Liver alcohol dehydrogenase transforms benzyl alcoholate into benzylaldehyde, and QM/MM (AM1/TIP3P, 9-31 atoms in the active site, depending on model) was used to explore the potential energy surface of that reaction to find stationary points. Those points were further refined with valence bond theory and the dynamics considered by VTST. In order to properly model the tunneling behavior, a three-stage approach was devised which treated the outer part with MM and the inner part with QM, SEVB, and VTST in such a way as to allow for an "equilibrium secondary zone" between the two parts. This rather difficult construction was no doubt complicated by the presence of a metal in the active site, but in the end, it was possible to say that "our computations confirm ... the experimental evidence for hydrogen tunneling in enzymatic reactions" [59]. That tunneling—something so fundamental to quantum mechanics—can be modeled in this way causes us to wonder what might be next.

3.5. The Future

In a 1996 review article on the future of quantum chemical methods, Head–Gordon paints a bleak picture for future conventional calculations. At the time, HF or DFT calculations on 100 atoms were feasible (it is closer to 150 today). He sets a goal of 10,000 atoms as the arrival point for the age of explicit calculations on entire proteins. This is a 100-fold increase from 1996, but due to the scaling of conventional HF and DFT, such a task would require an unrealistic 600-fold increase in computational power. Furthermore, for high-level CCSD(T) calculations, even a 600-fold increase in processing speed would only improve applicability of the method by a factor of five. Clearly, something must be done if this goal is to become a reality [23].

Fortunately, the solution may very well already exist. The rate-determining step in these calculations is the computation of two-electron repulsion integrals at N^4, but superficially it would seem that two-electron terms should really only scale as N^2. In fact, using the recently developed fast multipole methods [60,61], near-linear scaling can be achieved for Hartree–Fock and Density Functional theories. The consequences are staggering: the goal of 10,000 atoms could be reached with only a 100-fold increase in computer power! Additional work in the parallelization of quantum chemistry software [62] can further reduce this, because fully parallelized software can be run simultaneously on many machines and the work is additive. Linear scaling techniques utilizing sophisticated sparse matrix-multiply routines have demonstrated energy point HF and DFT calculations for, among other things, a 6304-atom nucleotide sequence and a nearly 20,000-atom polyglycine chain [63–65]. Even more impressive, semiempirical geometry optimizations of up to 3000 atoms including solvent effects have already been computed over a 10-day period using similar processes [66].

There are other exciting developments still to come. One may be the use of plane waves, a kind of basis set used on extended systems in physics which is fast and does not suffer from basis set superposition error. Recent work on applying plane waves to DFT suggests that plane waves might perform very well on systems containing hundreds of atoms [67]. In the vigorous area of hybrid QM/MM methods, new ways of interfacing quantum mechanical regions to molecular mechanical regions are reported to reproduce fully quantum mechanical DFT calculations to within 1 kcal/mol [68]. Most recently, the first all-electron density functional calculation of a metalloprotein was reported, made possible by massively parallel distributed computing [69]. All of these various improvements promise to make computational medicinal chemistry faster and more rigorous.

4. ACCURACY: A WILD GOOSE CHASE?

With all this discussion of varying accuracy, one might be inclined to question just how accurate the results need to be. After all, the position may well be moot if agreement with experiment is the goal, for the main experimental tool for protein structure, x-ray crystallography, provides at best 2.0-Å resolution for biomolecules. At that resolution, individual atoms are not resolved, although secondary structure is apparent and protein folding fairly clear. Side chains are hard to decipher, and the position of individual solvent molecules is an open question. The only way to gain explicit structural information is to resolve the system below 1.0 Å, but such precision is available only for small molecules, so it seems as though ab initio accuracy might be overkill (however, see Hargittai and Hargittai [70] to read the case for chemical accuracy structures in biology). Furthermore, a single geometry—such as a single x-ray structure—represents merely one conformation favored at some moment in time; there may be many others, and computing all of them is not only prohibitive but probably unnecessary.

The counterargument has both a practical and a philosophical thrust. Realistically speaking, if an experiment provides insufficient detail to elucidate the crucial structures, then it is up to theory to determine what might be occurring. The inner workings of a protein or receptor are infamously tricky to illuminate. Biological structures are a paradox: on one hand, they seem "fuzzy," as if every switch operates like a dimmer knob, everything being merely a matter of degree; on the other hand, there are instances where seemingly inconsequential changes in structure cause the activity to be completely negated. Lacking a sufficient experiment, how can one be confident about suspicions one way or the other? The answer must lie, we believe, in theory commensurate to the task before it. There is definitely some validity to taking issue with a method historically used to examine a single molecule in vacuo at absolute zero, but those limitations have been diminished considerably in recent years. No one can safely argue against the power of the high-level methods, and they are inherently superior to empiricism and inductive reasoning. In summary, if we adjust the throw of our light so that we might see further, we are still not seeing as well—regardless of how brightly it may have shined on previous occasions.

5. CONCLUSION

Our piece having been said, we present the following seven rules for the use of quantum chemical methods in medicinal chemistry:

1. Use the most accurate method that is practical for the system in question. However, if the system is homologous to hundreds of other systems, all of which must be computed, it would be best to err on the side of speed.
2. Organic chemistry has been largely explained by high-level ab initio quantum chemical methods. Thus it would be best to use such techniques or DFT whenever a reaction is involved. Semiempirical methods are, however, quite satisfactory for the most part.
3. For calculations on oligonucleotides and small polypeptides that form an active site, as well as any small molecules that will interface with the site, ab initio or DFT methods are good choices. For larger chains, semiempirical methods are preferred.
4. Metal-containing systems are problematic at present; DFT is probably the most effective compromise between speed and accuracy for such species. Avoid semiempirical methods for such cases, as they have typically been parameterized using data that is (owing to the nature of the experiments) far

less accurate than the organic data. In fact, more often than one would hope, the parameters simply do not exist for the desired atom.

5. For large biomolecules in which the conformations are numerous and only a few are chemically important, molecular mechanics, molecular dynamics, or other non-QM methods are good for paring down the possibilities.

6. Hybrid QM/MM methods are probably the best choice for mechanistic studies of enzymes. For best results, include coarse solvation effects in the MM part and explicit solvent molecules in the QM part. Any unusual characteristics of the system should be carefully considered using a conventional DFT or ab initio calculation.

7. Never completely abandon a more accurate method for a more computationally efficient one. Computers are always getting faster, and someday that extra power may produce dividends heretofore unanticipated in the world of computational medicinal chemistry.

ACKNOWLEDGMENTS

We would like to thank the National Science Foundation and the Department of Energy for their continuing support.

REFERENCES

1. Wesolowski SS, Leininger ML, Pentchev PN, Schaefer HF. Electron affinities of the DNA and RNA bases. J Am Chem Soc 2001; 123:4023–4028.
2. Clay M, Johnson MK, Barden CJ, Schaefer HF. Characterization of P. furiosa superoxide reductase (manuscript in preparation).
3. Barden CJ, Schaefer HF. Quantum chemistry in the twenty-first century. Pure Appl Chem 2000; 72:1405–1423.
4. Hartree DR. The wave mechanics of an atom with a non-coulomb central field. Proc Camb Philol Soc 1928; 24:89, 111, 426.
5. Fock VA. Naherungsmethode zur Losung des quantenmechanischen Mehrkorperproblems (Approximate methods for the solution of the quantum mechanical many-body problem). Z Phys 1930; 15:126–148.
6. Martin JML, Taylor PR. Benchmark *ab initio* thermochemistry of the isomers of diimide, N_2H_2, using accurate computed structures and anharmonic force fields. Mol Phys 1999; 96:681–692.
7. Meckler A. Electronic energy levels of molecular oxygen. J Chem Phys 1953; 21:1750–1761.
8. Møller C, Plesset MS. Note on an approximation treatment for many-electron systems. Phys Rev 1934; 46:618.
9. Raghavachari K, Trucks GW, Pople JA, Head-Gordon M. A 5th-order perturbation comparison of electron correlation theories. Chem Phys Lett 1989; 157:479–483.
10. Drake GWF, Yan ZC. Variational eigenvalues for the s-states of helium. Chem Phys Lett 1994; 229:486–490.
11. Hohenberg P, Kohn W. Inhomogeneous electron gas. Phys Rev 1964; 136:B864.
12. Lee C, Yang W, Parr RG. Development of the Colle–Salvetti correlation-energy formula into a functional of the electron-density. Phys Rev B 1988; 37:785–789.
13. Becke AD. Density-functional thermochemistry. 3. The role of exact exchange. J Chem Phys 1993; 98:5648–5652.

14. Chan GKL, Handy NC. An extensive study of gradient approximations to the exchange-correlation and kinetic energy functionals. J Chem Phys 2000; 112:5639–5653.
15. Kohn W, Sham LJ. Self-consistent equations including exchange and correlation effects. Phys Rev 1965; 140:A1133.
16. Curtiss LA, Raghavachari K, Trucks GW, Pople JA. Gaussian-2 theory for molecular-energies of 1st-row and 2nd-row compounds. J Chem Phys 1991; 94:7221–7230.
17. Wolfsberg M, Helmholz L. The spectra and electronic structure of the tetrahedral ions MnO_4^-, CrO_4^{--}, and ClO_4^-. J Chem Phys 1952; 20:837–843.
18. Pople JA, Santry DP, Segal GA. Approximate self-consistent molecular orbital theory. I. Invariant procedures. J Chem Phys 1965; 43:S129.
19. Dewar MJS, Zoebisch EG, Healy EF, Stewart JJP. The development and use of quantum-mechanical molecular-models. 76. AM1—a new general-purpose quantum-mechanical molecular-model. J Am Chem Soc 1985; 107:3902–3909.
20. Stewart JJP. Optimization of parameters for semiempirical methods. J Comput Chem 1989; 10:209–220, 221–264.
21. Allinger NL, Chen KS, Lii JH. An improved force field (MM4) for saturated hydrocarbons. J Comput Chem 1996; 17:642–668.
22. Cornell WD, Cieplak P, Bayly CI, Gould IR, Merz KM, Ferguson DM, Spellmeyer DC, Fox T, Caldwell JW, Kollman PA. A 2nd generation force-field for the simulation of proteins, nucleic-acids, and organic-molecules. J Am Chem Soc 1995; 117:5179–5197.
23. Head-Gordon M. Quantum chemistry and molecular processes. J Phys Chem 1996; 100:13213–13225.
24. Kutzelnigg W, Klopper W. Wave-functions with terms linear in the interelectronic coordinates to take care of the correlation cusp. J Chem Phys 1991; 94:1985–2001, 2002–2019, 2020–2030.
25. Stanton JF, Bartlett RJ. The equation of motion coupled-cluster method—a systematic biorthogonal approach to molecular-excitation energies, transition-probabilities, and excited-state properties. J Chem Phys 1993; 98:7029–7039.
26. Yamaguchi Y, Osamura Y, Goddard JD, Schaefer HF. A New Dimension to Quantum Chemistry: Analytic Derivative Methods in ab initio Molecular Electronic Structure Theory. New York: Oxford University Press, 1994.
27. Keck JC. Variational theory of chemical reaction rates applied to 3-body recombinations. J Chem Phys 1960; 32:1035–1050.
28. Miller WH, Handy NC, Adams JE. Reaction-path Hamiltonian for polyatomic-molecules. J Chem Phys 1960; 32:1035–1050.
29. Tomasi J, Persico M. Molecular-interactions in solution—an overview of methods based on continuous distributions of the solvent. Chem Rev 1994; 94:2027–2094.
30. Wong MW, Frisch MJ, Wiberg KB. Solvent effects. 1. The mediation of electrostatic effects by solvents. J Am Chem Soc 1991; 113:4776–4782.
31. Christiansen O, Mikkelsen KV. A coupled-cluster solvent reaction field method. J Chem Phys 1999; 110:1365–1375.
32. Berg MA, Salpietro SJ, Perczel A, Farkas Ö, Csizmadia IG. Side-chain conformational analysis of N-formyl-L-asparaginamide and N-acetyl-L-asparagine N-methylamide in their γ_L backbone conformation. J Mol Struct Theochem 2000; 504:127–139.
33. Harrison MJ, Burton NA, Hillier IH. Catalytic mechanism of the enzyme papain: predictions with a hybrid Quantum Mechanical/Molecular Mechanical potential. J Am Chem Soc 1997; 119:12285–12291.
34. Hughes JM, Hutter MC, Reimers JR, Hush NS. Modeling the bacterial photosynthetic reaction center. 4. The structural, electrochemical, and hydrogen-bonding properties of 22 mutants of Rhodobacter sphaeroides. J Am Chem Soc 2001; 123:8550–8563.
35. Berg MA, Chase GA, Deretey E, Fuzery AK, Fung BM, Fung DYK, Henry-Riyad H,

Lin AC, Mak ML, Mantas A, Patel M, Repyakh IV, Staikova M, Salpietro SJ, Tang TH, Vank JC, Perczel A, Csonka GI, Farkas O, Torday LL, Szekely Z, Csizmadia IG. Prospects in computational molecular medicine: a millennial mega-project on peptide folding. J Mol Struct Theochem 2000; 500:5–58.

36. Topol IA, Burt SK, Deretey E, Tang TH, Perczel A, Rashin A, Csizmadia IG. Alpha- and 3(10)-helix interconversion: a quantum-chemical study on polyalanine systems in the gas phase and in aqueous solvent. J Am Chem Soc 2001; 123:6054–6060.

37. Becke AD. Density-functional exchange-energy approximation with correct asymptotic behavior. Phys Rev A 1988; 38:3098–3100.

38. Perdew JP. Density-functional approximation for the correlation-energy of the inhomogeneous electron-gas. Phys Rev B 1986; 33:8822–8824.

39. Glukhovtsev MN, Pross A, McGrath MP, Radom L. Extension of Gaussian-2 (G2) theory to bromine-containing and iodine-containing molecules—use of effective core potentials. J Chem Phys 1995; 103:1878–1885.

40. Bauernschmitt R, Ahlrichs R. Treatment of electronic excitations within the adiabatic approximation of time dependent density functional theory. Chem Phys Lett 1996; 256: 454–464.

41. King RA, Crawford TD, Stanton JF, Schaefer HF. Conformations of [10]annulene: more bad news for density functional theory and second-order perturbation theory. J Am Chem Soc 1999; 121:10788–10793.

42. Rodríguez AM, Giannini FA, Suvire FD, Baldoni HA, Furlán R, Zacchino SA, Beke G, Mátyus P, Enriz RD, Csizmadia IG. Correlation of antifungal activity of selected alpha-substituted acetophenones with their keto-enol tautomerization energy. J Mol Struct Theochem 2000; 504:35–50.

43. Miertus S, Scrocco E, Tomasi J. Electrostatic interaction of a solute with a continuum—a direct utilization of ab initio molecular potentials for the prevision of solvent effects. Chem Phys 1981; 55:117–129.

44. Pan Y, McAllister MA. Theoretical investigation of the role of hydrogen bonding during ketosteroid isomerase catalysis. J Mol Struct Theochem 2000; 504:29–33.

45. Dewar MJS, Thiel W. Ground-states of molecules: 38. MNDO method—approximations and parameters. J Am Chem Soc 1977; 99:4899–4907.

46. Stewart JJP. Special Issue—MOPAC—A semiempirical molecular-orbital program. J Comput Aid Mol Des 1990; 4:1–45.

47. Stewart JJP. Comments on a comparison of AM1 with the recently developed PM3 method-reply. J Comput Chem 1990; 11:543–544.

48. Jurema MW, Shields GC. Ability of the PM3 quantum-mechanical method to model intermolecular hydrogen bonding between neutral molecules. J Comput Chem 1993; 14:89–104.

49. Cramer CJ, Truhlar DG. AM1-SM2 and PM3-SM3 parameterized SCF solvation models for free-energies in aqueous-solution. J Comput Aid Mol Des 1992; 6:629–666.

50. Dive G, Dehareng D, Ghuysen JM. Detail study of a molecule in a molecule—N-acetyl-L-tryptophanamide in an active-site model of α-chymotrypsin. J Am Chem Soc 1994; 116:2548–2556.

51. Patel MA, Deretey E, Csizmadia IG. Will *ab initio* and DFT drug design be practical in the 21st century? A case study involving a structural analysis of the β2-adrenergic G-protein coupled receptor. J Mol Struct Theochem 1999; 492:1–18.

52. Rappé AK, Casewit CJ, Colwell KS, Goddard WA, Skiff WM. UFF, a full periodic-table force-field for molecular mechanics and molecular-dynamics simulations. J Am Chem Soc 1992; 114:10024–10035.

53. Aqvist J, Warshel A. Simulation of enzyme-reactions using valence-bond force-fields and other hybrid quantum-classical approaches. Chem Rev 1993; 93:2523–2544.

54. Warshel A, Karplus M. Calculation of ground and excited-state potential surfaces of

conjugated molecules: 1. Formulation and parametrization. J Am Chem Soc 1972; 94: 5612.

55.   Warshel A, Levitt M. Theoretical studies of enzymic reactions—dielectric, electrostatic and steric stabilization of carbonium-ion in reaction of lysozyme. J Mol Biol 1976; 103:227–249.

56.   Singh UC, Kollman PA. A combined ab initio quantum-mechanical and molecular mechanical method for carrying out simulations on complex molecular systems—applications to the $CH_3Cl + Cl^-$ exchange reaction and gas-phase protonation of polyethers. J Comput Chem 1986; 7:718–730.

57.   Field MJ, Bash PA, Karplus M. A combined quantum-mechanical and molecular mechanical potential for molecular-dynamics simulations. J Comput Chem 1990; 11:700–733.

58.   Díaz N, Suárez D, Sordo T, Merz KM. Acylation of class A β-lactamases by penicillins: a theoretical examination of the role of serine 130 and the β-lactam carboxylate group. J Phys Chem B 2001; 105:11302–11313.

59.   Alhambra C, Corchado J, Sánchez M, Garcia-Viloca M, Gao J, Truhlar DG. Canonical variational theory for enzyme kinetics with the protein mean force and multidimensional quantum mechanical tunneling dynamics. Theory and application to liver alcohol dehydrogenase. J Phys Chem B 2001; 105:11326–11340.

60.   White CA, Johnson BG, Gill PMW, Head-Gordon M. The continuous fast multipole method. Chem Phys Lett 1994; 230:8–16.

61.   White CA, Johnson BG, Gill PMW, Head-Gordon M. Linear scaling density functional calculations via the continuous fast multipole method. Chem Phys Lett 1996; 253:268–278.

62.   Kendall RA, Harrison RJ, Littlefield RJ, Guest MF. High performance computing in computational chemistry: methods and machines. Rev Comput Chem 1995; 6:209–316.

63.   Daniels AD, Milliam JM, Scuseria GE. Semiempirical methods with conjugate gradient density matrix search to replace diagonalization for molecular systems containing thousands of atoms. J Chem Phys 1997; 107:425–431.

64.   Scuseria GE. Linear scaling density functional calculations with Gaussian orbitals. J Phys Chem A 1999; 103:4782.

65.   Scuseria GE, Ayala PY. Linear scaling coupled cluster and perturbation theories in the atomic orbital basis. J Chem Phys 1999; 111:8330–8343.

66.   Lee TS, York DM, Yang W. Linear-scaling semiempirical quantum calculations for macromolecules. J Chem Phys 1996; 105:2744–2750.

67.   Fellers RS, Barsky D, Gygi F, Colvin M. An ab initio study of DNA base pair hydrogen bonding: a comparison of plane-wave versus Gaussian-type function methods. Chem Phys Lett 1999; 312:548–555.

68.   Murphy RB, Philipp DM, Friesner RA. Frozen orbital QM/MM methods for density functional theory. Chem Phys Lett 2000; 321:113–120.

69.   Yoshihiro T, Sato F, Kashiwagi H. Distributed parallel processing by using the object-oriented technology in ProteinDF program for all-electron calculations on proteins. Chem Phys Lett 2001; 346:313–321.

70.   Hargittai M, Hargittai I. Aspects of structural chemistry in molecular biology. In: Domenicano A, Hargittai I, eds. Strength from Weakness: Structural Consequences of Weak Interactions in Molecules, Supermolecules, and Crystals. The Hague: Kluwer Academic, 2002.

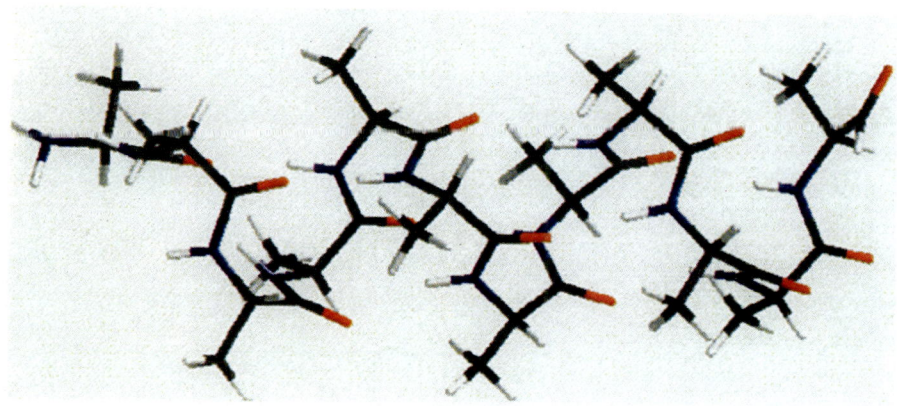

Figure 1 The helical alanine 12-mer, a recent landmark for full ab initio geometry optimizations of biomolecules. See Ref. 36.

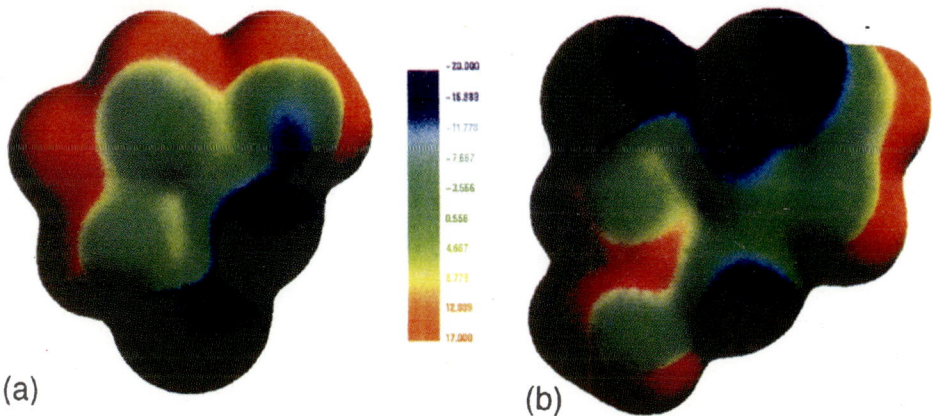

Figure 1 Electrostatic potentials on the molecular surfaces of (a) cytosine, **1**, and (b) guanine, **2**, computed at the Hartree–Fock 6-31G* level. Color ranges, in kcal/mol: *red*, more positive than 17; *blue*, more negative than −20 (see legend). The relative positions of the molecules are such that the portions that hydrogen bond are facing each other, showing how the extended positive and negative regions will interact.

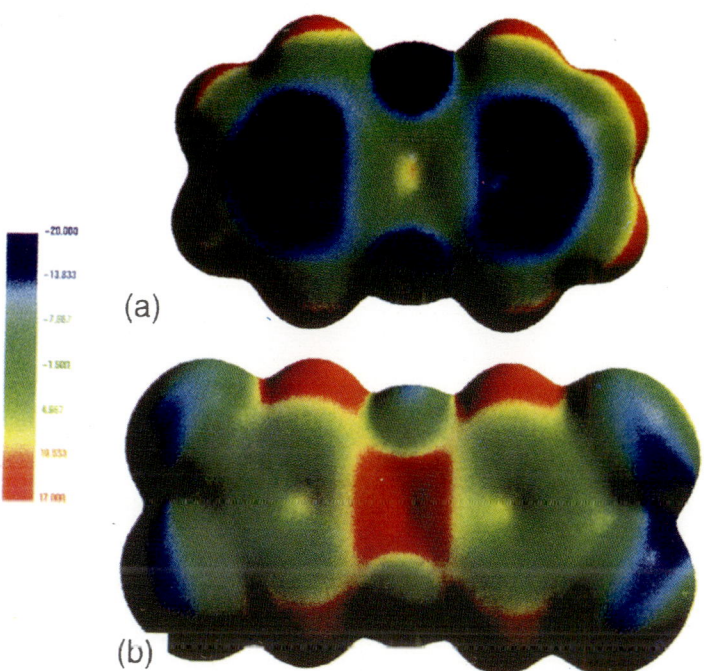

Figure 2 Electrostatic potentials on the molecular surfaces of (a) dibenzo-*p*-dioxin, **5**, and (b) 2,3,7,8-tetrachlorodibenzo-*p*-dioxin (TCDD), **6**, computed at the Hartree–Fock STO-5G* level. Color ranges, in kcal/mol: *red*, more positive than 17; *blue*, more negative than −20 (see legend).

7

3D Structure Generation and Conformational Searching

JENS SADOWSKI

AstraZeneca R&D Mölndal, Mölndal, Sweden

CHRISTOF H. SCHWAB

Molecular Networks GmbH, Erlangen, Germany

JOHANN GASTEIGER

Erlangen-Nuernberg University, Erlangen, Germany

1. INTRODUCTION

Many biological, physical, and chemical properties are clearly functions of the 3D structure of a molecule. Thus, the understanding of receptor–ligand interactions, molecular properties, or chemical reactivity requires not only information on how atoms are connected in a molecule (connection table) but also on their 3D structure. Experimental sources of 3D structure information are x-ray crystallography, microwave spectroscopy, electron diffraction, and nuclear magnetic resonance (NMR) spectroscopy. The largest source of experimentally determined molecular structures is the Cambridge Structural Database (CSD) [1], which contains at present about 250,000 x-ray structures. In addition, the Brookhaven Protein Data Bank (PDB) [2] contains about 17,000 structures of proteins and other biological macromolecules including several thousands of drug-sized molecules in their biologically active conformations docked into their receptors. For several reasons, the experimental sources of 3D structures are not sufficient and there is a real need for computer-generated models:

1. The number of compounds whose 3D structure has been determined (about 270,000) is small indeed when compared to the number of known compounds (more than 25 million).

2. Computational techniques in organic chemistry such as for drug design, structure elucidation, or synthesis planning quite often investigate enormous numbers of hypothetical molecules, which are not yet synthesized or even not stable as in the case of transition states of chemical reactions.

3. On the other hand, theoretical methods such as quantum mechanics or molecular mechanics (MM) can produce 3D molecular models of high quality and predict a number of molecular properties with high precision. Unfortunately, these techniques also require at least some reasonable 3D geometry of the molecule as starting point.

4. Very often, it is unknown which conformation of a flexible molecule is needed. For example, in drug design, we hunt often for the so-called bioactive conformation, which is the molecule in its receptor-bound state. In this case, any experimental structure of the isolated small molecule—in vacuum, in solution, or in a crystal—can be the wrong choice.

The missing link between the constitution of a molecule and its 3D structure in computational chemistry is a technique capable of automatically generating 3D models starting from the connectivity information of a given molecule. Because of its basic role, 3D structure generation is one of the fundamental problems in computational chemistry. As a consequence, in recent years, a number of automatic 3D model builders and conformer generators have become available [for two recent reviews, see Refs. 3 and 4].

In the following, we will discuss 2D-to-3D conversion in this context. However, it should be emphasized that we do so only for the sake of brevity. In reality, none of the conversion programs utilizes information of a 2D image of a chemical structure. Only the information on the atoms of a molecule and how they are connected is used (i.e., the starting information is the constitution of the molecule).

Most molecules of organic, biochemical, or pharmacological interest can adopt more than one conformation of nearly equal energy by rotation around single bonds. These molecular geometries correspond to the global and, in most cases, various local minima on the multidimensional molecular energy surface (also called potential surface). Which of these conformations is the preferred one may heavily depend on the interactions of the molecule with its environment. The conformations of one and the same molecule can significantly differ, if the molecule is observed isolated in the gas phase, if it is influenced by solvent effects in solution, if it is prior to reaction, or if it is exposed to directed electrostatic and steric forces caused by crystal packing in the solid state or by the amino acids at the binding site of a macromolecular biological receptor.

In structure-based drug design, the so-called *bioactive* conformation (the preferred conformation in the receptor-bound state) of potential drug molecules is of special interest. Its prediction is a challenging and demanding task, even if structural information on the biological receptor is available.

But in many cases, the biological target is not yet known or structurally not determined. Studies by superimposing sets of highly active compounds can provide a more detailed insight into the structural and electronic situation at the binding site of the receptor, if conformational degrees of freedom are allowed for the molecules during the superimposition procedure [5,6]. In addition, methods such as pharmacophore searches in 3D databases perform much more efficiently with higher hit rates and larger numbers of potential new drugs or lead structures if the molecules in the database are considered to be conformationally flexible [7].

Furthermore, even if the biological receptor is known and structurally determined, modern techniques such as virtual screening and docking experiments or de novo design systems have to take into account several alternative conformations of the small molecules under investigation to estimate and rank different binding modes and constants with locally optimized electrostatic and steric interactions between the ligand and its receptor [8].

Clearly, there is a need for computational tools that generate ensembles of conformations and a substantial step toward the understanding of the physical, chemical or biological, and pharmacological properties of a molecule to study or to analyze its possible conformations. Conformational analysis tries to correlate conformational changes of a molecule with the influence on its properties.

The major aim of conformational analysis is to identify the preferred conformations of a molecule under specific conditions. Therefore, conformational search techniques (i.e., methods that locate the global and local energy minima of a structure) play a crucial role in conformational analysis.

In the context of computer-aided drug discovery and design, the optimal search method would seek for and identify one single conformation, the biologically active one. The binding of a ligand to its receptor is a highly complex and multistep process and, thus, the protein-bound conformation depends on a large number of interconnected variables. In addition to this, typical druglike molecules can adopt quite a large number of low-energy conformations. These circumstances make it almost impossible, or at least rather difficult, to exactly predict the resulting binding mode in an absolute manner. Finally, experimental techniques such as high-resolution x-ray crystallography (which are quite time-consuming and cost-consuming and not always successful) have to be utilized to elucidate the biologically active conformation.

Therefore, computational methods have to rely on generating sets of conformations with the intention that these sets contain the geometry that explains the property under investigation (i.e., the biologically active conformation), or at least a conformation that is structurally quite similar.

2. PROBLEM DESCRIPTION

2.1. Computational Requirements

The main area of automatic structure generation is the 2D-to-3D conversion of large databases of druglike organic compounds. These databases often contain millions of structures, imposing some restrictions on the development of 3D structure generators. The decision to use a specific conversion program plays a crucial role because a change to another program can only be made with great difficulties. Firstly, the amount of computer resources for the conversion of hundreds of thousands of structures is quite large, and, secondly, much scientific work will be based on such a database and, thus, a change of these data makes a lot of the work already performed questionable or obsolete. Therefore, the choice to use a particular 3D structure generation program should be made only after a careful evaluation process. On the other hand, the task of generating 3D structures from connectivity information (the constitution of a molecule) is just too important and the problems to be solved are so diverse that it should always be open to new ideas and approaches. 3D database developers at Molecular Design Ltd. formulated the following criteria for a 2D-to-3D conversion program [9] and we will cover only those published approaches that fulfill more or less all of these criteria (the quotes are slightly abbreviated and modified):

Robustness. The program should run for a long time before failing and should indicate the actions taken on failure rather than simply crash.

Large files. The program should be able to handle large numbers of structures contained in a single file to minimize the number of conversion jobs.

Variety of chemical types. The program should be able to handle a wide variety of structural types.

Stereochemistry. The stereochemical information contained in the input data must be handled correctly.

Rapid and automated. The large size of the databases to be processed requires the conversion program to run in batch mode and to work with acceptable speed.

High-quality models. The generated models should be of high quality without further energy minimization and should represent at least one low-energy conformation. It should have internal diagnostics to validate the models generated

High conversion rate. As many 2D structures as possible should be converted.

For conformer generators, some specific additional criteria have to defined:

Coverage. The ensemble of conformations must include all relevant conformations and the method should be able to reproduce biologically active conformations.

Diversity. Because it is impossible to generate all conformations of reasonably large molecules in infinite resolution, the subset chosen from the whole conformational space has to be reasonably diverse.

Compactness. Given the size of today's databases and the requirement to store hundreds of conformers per molecule, compactness of storage becomes an issue with respect to both file size and retrieval efficiency.

2.2. General Problems

Each approach to automatic generation of 3D molecular models has to solve a number of general problems. The strategy for building a molecular model can be compared with the use of a mechanical molecular model building kit. Monocentric fragments that represent different hybridization states and provide the corresponding bond angles are connected using joints with a length corresponding to the required bond lengths. A basic assumption in this process of 3D structure generation is an allowed transfer of bond lengths and bond angles from one molecular environment to another (i.e., the usage of standard values for bond lengths and bond angles). However, this assumption requires to distinguish between a sufficiently large number of different atom types, hybridization states, and bond types with appropriate bond lengths and bond angles. Usually, the deviations from these standard values are rather small. A totally different situation is encountered for dihedral or torsional angles, which describe the twisting of a fragment of four atoms connected by a sequence of bonds. Because the steric energy may have multiple minima around a rotatable bond with similar energy content, this leads to more than one possibility for constructing a 3D molecular model for such molecules or, in other terms, to multiple conformations.

In acyclical molecules or substructures, the preferred torsional angles are those which simultaneously minimize torsional strain and the steric interactions between nonbonded atoms. The relatively large flexibility of such systems gives rise to multiple solutions (conformations) for the process of structure generation, which have quite similar energy. Account of this flexibility has to be taken and geometrically unacceptable situations as [e.g., the overlap of atoms ("clashes")] must strictly be avoided. With increasing numbers of possible conformations, it becomes less and less likely that the generated 3D structure corresponds to the experimentally determined geometry.

In cyclical structures, ring closure has to be taken into account as an additional geometrical constraint of the 3D structure generation process. Ring closure dramatically reduces the degrees of freedom as expressed in a reduction in the number of possible conformations compared to those in acyclical systems. In particular, the endocyclical torsional angles are mutually dependent. Due to this fact, many of the 3D structure generators use information on possible single-ring conformations. These conformations can be stored as 3D coordinate fragments or simply as lists of torsional angles. These so-called ring templates implicitly fulfill the condition of ring closure. Additional levels of sophistication are reached when the rings have exocyclical substituents, or when they are assembled in fused or bridged ring systems. Another challenge arises with increasing ring size. Large rings are apart from the requirement to ring closure, as flexible as acyclical systems. Fig. 1 shows the increase in the number of known conformations of cycloalkanes with dependence on ring size.

The conformational flexibility and thus the number of valid 3D molecular models steeply increase from ring size 9 upward. An explicit use of potential ring conformations becomes more and more infeasible. Some of the programs discussed below therefore refrain from generating 3D structures for macrocyclical and polymacrocyclical structures such as the trimacrocyclical system in Fig. 2.

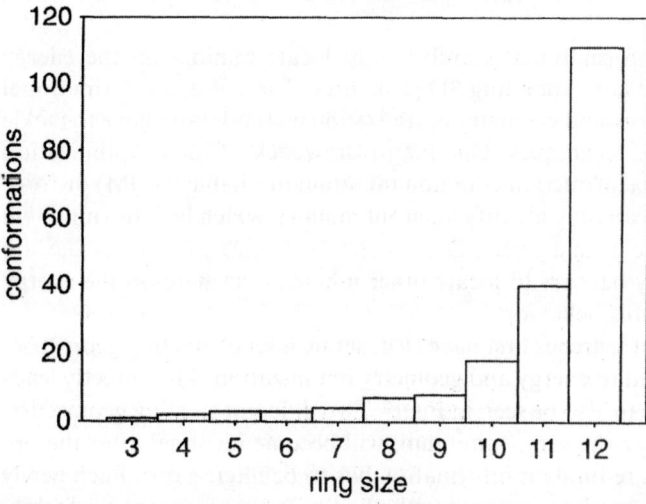

Figure 1 Increase of the number of known conformations of cycloalkanes with increasing size.

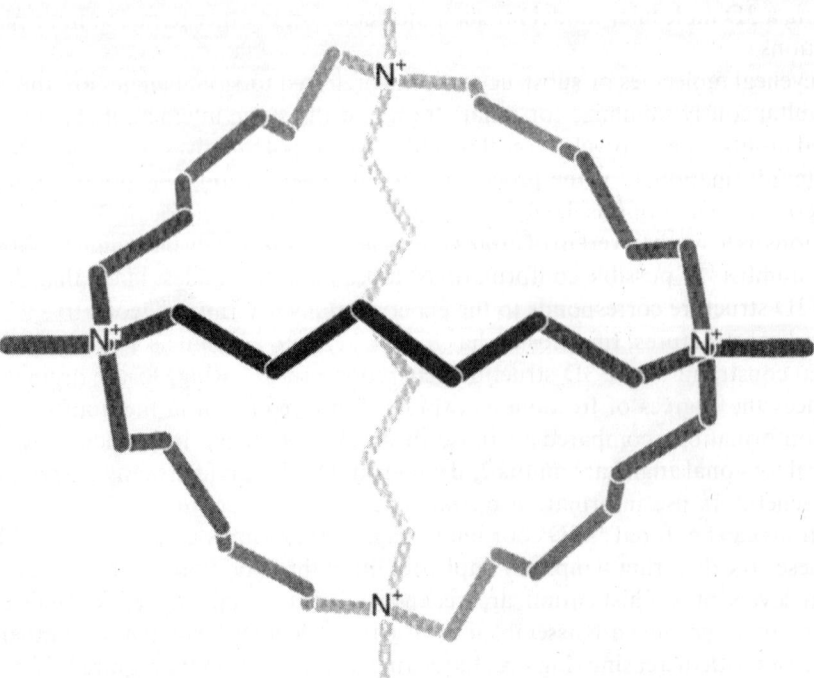

Figure 2   Trimacrocyclical bridged system.

Due to the specific complications when predicting the geometry of ring systems, many of the approaches to 3D structure generation dedicate most of the program intelligence to this part. Most often, the molecule under consideration is fragmented into acyclical and cyclical portions at the very beginning of the 3D generation process. The fragments are then handled separately and reassembled at the end of the whole process.

The objective of conformational searches is to locate minima on the energy surface and to generate the corresponding 3D structures. Therefore, conformational searches have to utilize energy and geometry optimization methods, and have to tackle problems inherent in these techniques. One major drawback of most optimization algorithms, as they are implemented in common quantum mechanical (QM) or force field packages, is that they can only identify adjacent minima, which lie "downhill" on the potential energy surface from a given 3D geometry as starting point (i.e., they are unable to overcome energy barriers to locate other minima elsewhere on the energy surface). Fig. 3 illustrates this behavior.

Thus, conformational searches first have to generate a set of starting geometries which then can be submitted to energy and geometry optimization. This directly leads to a third problem, which can also be seen in Fig. 3. Two different starting geometries, which both are located near the same minimum, will become identical after the optimization procedure. This redundant information has to be filtered out. Each newly generated conformation has to be compared with all previously generated conformations. It has to be stored if a new conformation, whose geometry relevantly differs from all previously generated conformations, has been found; otherwise, it has to be rejected. A common metric to perceive the similarity between two conformations is

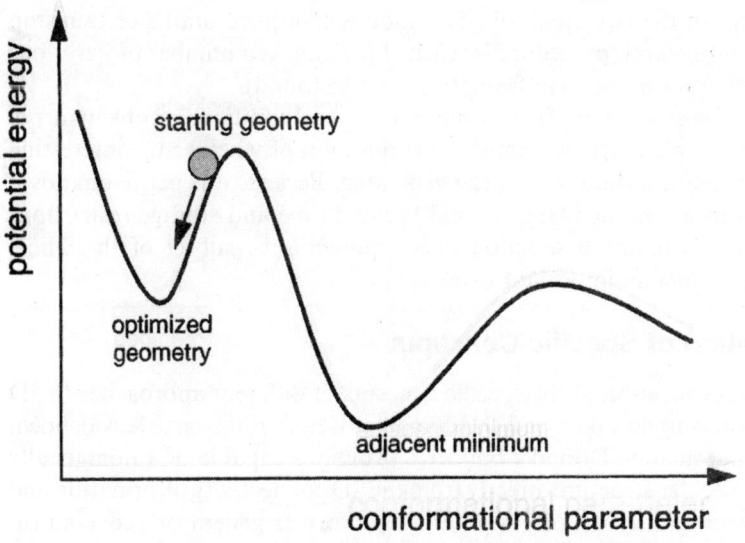

Figure 3 Identification of energy minima on the conformational energy surface (symbolic).

the RMS deviation of the positions of their atoms. The RMS deviation can either be measured in Cartesian space (RMS_{XYZ} [Å]), where the 3D Cartesian coordinates of all corresponding atom pairs are compared, or in torsion angle (TA) space (RMS_{TA} [°]), by calculating the deviation of all corresponding torsion angles of both conformations. As already discussed in the context of reproducing x-ray structures (see Sec. 4), two conformations can be regarded as identical if their RMS deviation is less than 0.3 Å or 15°, respectively.

Fig. 4 shows the general work flow of a conformational search. After generating an initial starting geometry, which is optimized in the subsequent step, the new structure is compared to all previously generated conformations (normally stored as a list of unique structures). If a substantially new geometry is found, it is added to this list of unique conformations; otherwise, it will be rejected. Then, a new starting structure

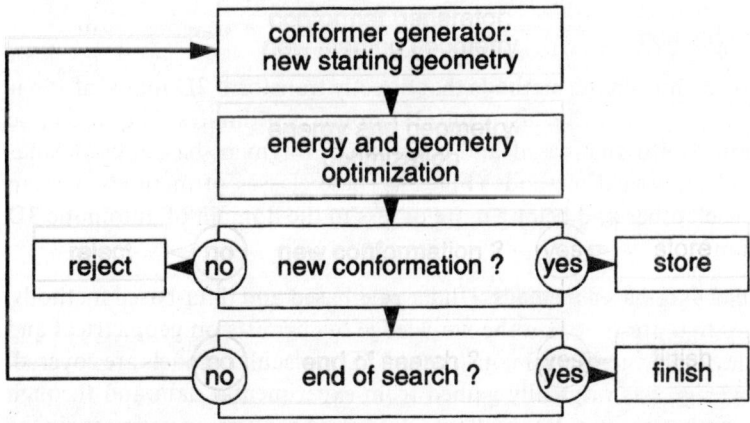

Figure 4 General work flow of conformational search techniques.

has to be generated for the next iteration. This loop is continued until a certain stop criterion for the entire search procedure is reached (i.e., a given number of iterations has been performed, or if no new conformations can be found).

When generating ensembles of conformations, several additional problems arise. First, the coverage problem arises, imposing the question of whether the interesting biologically active conformations have been generated. Because it is per se unknown which conformations are needed later on and because time and storage restrictions forbid to generate too many, a selection of a representative subset of the whole conformer space becomes an important issue.

2.3. Classification of Specific Concepts

In this chapter, a classification of the specific concepts of different approaches to 3D structure generation is undertaken and the domain covered in this article is defined. Under the term "automatic 3D model builder," programs capable of automatically predicting a 3D molecular structure directly from the 2D connectivity information and without user interaction are covered. The term "conformer generator" covers programs which, starting from the 2D structure or a single 3D model, generate sets of conformations. Most of the methods presented here are designed especially for small, druglike molecules. The prediction of the geometry of polymers, in particular of biopolymers, is a task of its own and not even attempted by the approaches discussed here.

2.3.1. Manual Methods

In the early beginning of thinking in three dimensions in organic chemistry, 3D molecular models were built by hand, using standard bond length and bond angle units from mechanical molecular model building kits. This technique, still useful today, found in the age of computational chemistry its modern expression in the well-known interactive 3D structure building options incorporated into nearly each program package for molecular modeling. The user may construct a 3D molecular geometry interactively, positioning atoms and bonds on a 3D graphics interface using standard bond lengths and angles, or connecting predefined fragments. All these methods are summarized under manual methods because all model building steps are performed by hand, irrespective of whether this is done in real space or with computer models.

2.3.2. Automatic Methods

Distinct from these are automatic methods that directly transform 2D input information on atoms, bonds, and stereochemistry into 3D atomic coordinates. The automatic methods are classified into rule-based and data-based, fragment-based, conformational analysis, and numerical methods (Fig. 5). These classes of methods overlap more or less with each other and belong more or less to the domain of automatic 3D structure generation:

> **Rule-based and data-based methods.** Under rule-based and data-based methods, approaches that are based on the knowledge of chemists on geometrical and energy rules and principles for constructing 3D molecular models are covered. This knowledge was originally gained from experimental data and through theoretical investigations. It is built into 2D-to-3D conversion programs in the

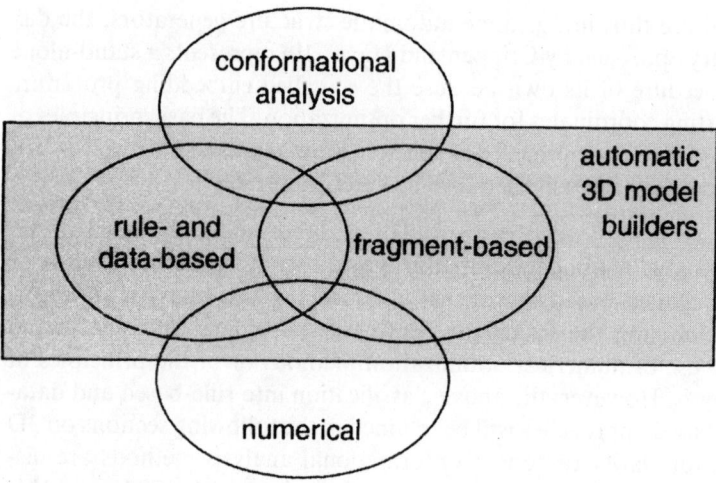

Figure 5 Classification of concepts.

form of chemical knowledge either in explicit (e.g., rules) or in implicit form (e.g., data on allowed ring conformations).

Fragment-based methods. At the far end of rule-based and data-based methods are approaches that are based almost exclusively on structural data. These methods are covered under a separate subdivision as fragment-based methods. These methods follow the concept of constructing molecular models from fragments that are as large and as similar as possible to the molecule to be built. The fragments are taken from a library of 3D structures. Fragment-based programs make extensive use of the implicit knowledge on model building represented by databases of 3D structures. Of course, fragment-based methods need also explicit rules on the fragmentation of the input structures, on finding closest analogs in the libraries, and on combining fragments to the entire molecular model.

Conformational analysis methods. In the field of conformational analysis, the 3D model builders and the conformer generators overlap. It is impossible to develop a 3D structure prediction program that does not implicitly look at several alternative conformations before settling down with the one written into the output file. The most common methods applied to conformational analysis and searching are systematic methods, random techniques, genetic algorithms (GAs), and simulation experiments. All these methods can be utilized either to identify the global minimum structure of a molecule under consideration, or to explore conformational space to generate an ensemble of low-energy conformations. Because pure conformer generation requires some additional issues to be addressed, this topic is described in another section.

Numerical methods. Quantum mechanical calculations, molecular mechanics, and distance geometry (DG) are summarized under numerical methods because they are based on extensive numerical optimization procedures often requiring substantial computation times (QMMM > DG). Although quantum mechanical or molecular mechanics programs need a reasonable starting geometry and are thus not genuine automatic structure generators, the distance geometry approach by Crippen and Havel [10] represents a stand-alone modeling procedure of its own because the so-called embedding procedure generates starting coordinates for further optimization. The basic principles of the distance geometry approach for 3D structure generation as well as for conformational searches will therefore be described briefly.

Clearly, there is no sharp border between all of the subdivisions discussed above. Rule-based and data-based methods use small fragments as at least bond lengths or ring templates, and fragment-based approaches of course use also rules for appropriately finding and combining the fragments. Both rule-based and fragment-based methods often make use of numerical optimization methods or of the principles of conformational analysis. However, the above classification into rule-based and data-based and fragment-based approaches will be retained in the following sections on 3D structure generation for clarity reasons. Conformational analysis methods are discussed in another section. The basic principles of numerical methods (QM and MM) are given elsewhere in this volume.

3. 3D STRUCTURE GENERATION: METHODS AND PROGRAMS

In this section, most of the currently available programs for automatic 3D structure generation will be discussed as far as they have been described in the literature. In addition, some early precursors of these methods are briefly presented due to their pioneering role in this field.

3.1. Early Precursors

3.1.1. Conformational Analysis for Six-Membered Rings in the LHASA Program

Corey and Feiner [11] assigned conformations of six-membered ring systems in a semiquantitative manner during the development of the synthesis design program, LHASA. The aim of this work was the prediction of the preferred conformations of synthetically important six-membered ring systems to evaluate the steric hindrance of different reaction sites in a molecule. In the first step, several possible geometries are assigned to the single rings (e.g., chair, half-chair, and boat) and the flexibility of these rings is evaluated (e.g., the possibility to distort them or to flip them into another conformation) using the 2D connection table and the stereochemical information. Secondly, the exocyclical substituents of the ring atoms are labelled to be either axial or equatorial. Thirdly, the relative energy differences between several possible conformations of flexible ring systems are calculated using empirical procedures based on energy increment schemes for the single-ring conformations, for intraring interactions (e.g., monoaxial substituents, 1,2-diequatorial, or 1,3-diaxial interactions in chair conformations), and inter-ring interactions between different rings of one ring system. Fig. 6 shows this increment scheme for intraring interactions in monoaxial, 1,2-diequatorial, and 1,3-diaxial substituted cyclohexane chair conformations. To predict destabilization energies E_D in monoaxial substituted cyclohexane chair conformations, energy increments A_R for a specific substituent, which describe the energy difference between the axial and equatorial configuration, are used. The interactions in

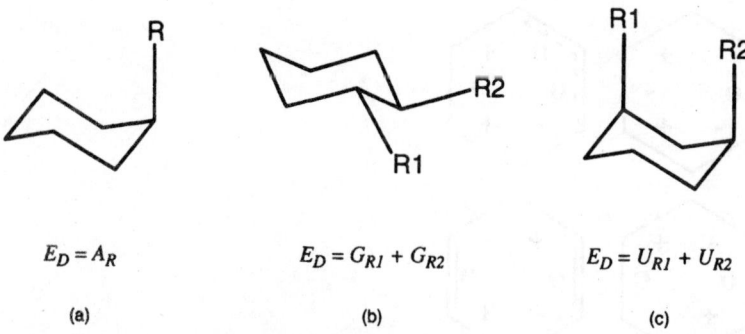

$$E_D = A_R \qquad\qquad E_D = G_{R1} + G_{R2} \qquad\qquad E_D = U_{R1} + U_{R2}$$

(a) (b) (c)

Figure 6 Incremental calculation scheme to predict destabilization energies E_D in monoaxial substituted (a), 1,2-diequatorial substituted (b), and 1,3-diaxial substituted (c) cyclohexane chair conformations in the LHASA program.

1,2-diequatorial or 1,3-diaxial substituted ring systems are calculated by separate increment schemes G_R and U_R, respectively. The increments for the substituents A_R, G_R, and U_R depend on the atom type, hydrogen attachment, and hybridization state of the atom directly connected to the ring.

Finally, the method is completed by using several rules to model the influence of endocyclical heteroatoms. In a series of examples, sufficient agreement was found with energies obtained by molecular mechanics and with geometries obtained by x-ray crystallography. The strength of the method was the use of symbolical logic (e.g., energy increments to calculate destabilization energies, rules to model the influence of endocyclical heteroatoms) for the geometry and energy prediction. However, the approach was limited to six-membered ring conformations and 3D structures were not generated explicitly.

3.1.2. The SCRIPT Program

Cohen et al. [12] presented in 1981 the SCRIPT program. A molecule is considered as an assembly of chain and ring fragments, possessing different conformations. The conformations are handled in an abstract form as "conformational diagrams" containing symbolical descriptions of the torsional angles of each bond. Chain fragments are treated as sequential four-atom fragments. Several possible low-energy conformations are given for the torsional angles of such a fragment that only depend on the nature of the central bond. Ring fragments are handled as templates that are joined. Possible conformers of rings of three to eight atoms are taken from a predefined table of templates that depend on the ring size and on the distribution of double bonds. These conformers are stored in the form of conformational diagrams as shown in Fig. 7 for the six-membered ring. The torsional angles of the ring bonds in these diagrams are represented only by their sign ($+/-$) for *gauche* angle types or zero (0) for a *cis* bond.

For ring fragments consisting of more than one ring, being either fused or bridged, a set of rules that restricts the allowed conformations of two adjacent rings is used. These rules consist of allowed combinations of torsional angles of the bond of fusion in the two regarded rings that depend on the stereochemistry of the bridgehead atoms.

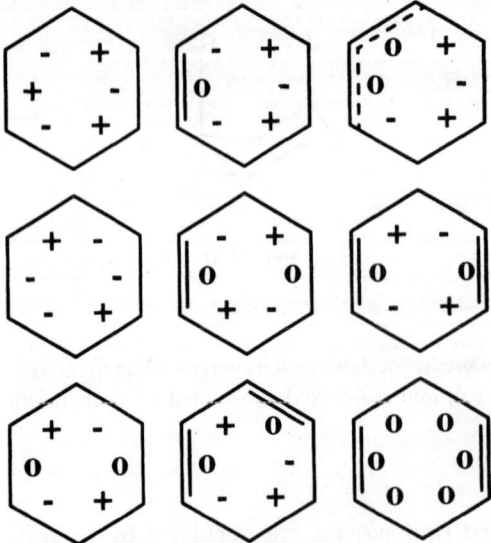

Figure 7 The nine possible conformational diagrams for a six-membered ring in the SCRIPT program. The torsional angles of the ring bonds are only defined by their sign ($+/-$) or zero (0) for a planar bond.

In a first step, the possible conformations are generated on a symbolical level of conformational diagrams. The combinatorial product of all conformational diagrams for rings and chains forms the conformational space of the molecule. In a second step, a set of rules and computational schemes allows the direct translation of the conformational diagrams into 3D atomic coordinates by using standard bond lengths, bond angles, and torsional angles calculated from the symbolical descriptions in the diagrams. This is achieved by computational schemes based on ring sizes. The 3D coordinates obtained are regarded to be rather crude. They may be evaluated by the calculation of the conformational energy based on molecular mechanics potentials. However, only the energies obtained after a geometry optimization are useful for a ranking of the conformers. In other words, to obtain a reasonable molecular model, a number of force field optimizations of different conformations are necessary.

The major strength of the SCRIPT method is the use of symbolical logic to construct possible ring conformations from a table of single-ring templates and the direct translation of these symbolical representations into 3D atomic coordinates, which makes these processing stages rather fast. The major weakness of this approach is the generation of rather crude 3D coordinates and the lack of an energy evaluation of the conformations at the symbolical level of conformational diagrams. The program was used with some benefit in reaction design studies.

3.1.3. SCA: Systematic Conformational Analysis for Cyclical Systems

De Clercq [13] has developed a program called SCA (systematic conformational analysis) for the construction of conformations of ring systems consisting of three-membered to seven-membered rings. Like the SCRIPT program (see above), it is based on lists of allowed conformations of single rings and a set of rules for determining torsion constraints in fused or bridged systems (i.e., the sign and the magnitude of the torsional angles common to two neighboring rings). The original procedures have been developed for a manual systematic conformational analysis starting from a two-dimensional structure with stereocenters indicated by a wedged/hashed bond notation.

After an interactive structure input via a 2D drawing of the structural formula augmented with stereodescriptors, the SCA program performs the following steps. Firstly, it analyses the input and assigns possible conformations to all five-membered, six-membered, and seven-membered single rings considering the torsion constraints introduced by unsaturated bonds and fused or bridged systems. These single-ring conformations are stored in the form of lists of torsional angles. An energy value is assigned to each conformation, calculated from the conformational energy of the un-substituted form, the influence of an exocyclical double bond, contributions from exocyclical substituents, and interactions of vicinal substituents. Secondly, the single-ring conformations are combined and the resulting abstract conformations of the entire ring system are ranked by the sum of the energies of the single-ring conformations. This energy ranking does not contain any information on long-range interactions as, for example, exerted by substituents of two different rings. Therefore, in a third step, the abstract representations are translated into 3D atomic coordinates using standard values for bond lengths, bond angles, and torsional angles. A special procedure is used to perfectly close the rings of strained systems by deforming some endocyclical bond angles. Then, a new energy ranking is calculated for these 3D structures using the above energy terms with the exception of the contributions of the substituents, which are replaced by separate nonbonded energy terms as functions of interatomic distance. The fine-tuning of the conformational energy by rather simple linear functions of the nonbonded distances was tested by calculating the energy differences between the axial and equatorial forms of the chair/chair conformations of several methyldecalins. The reported results compare rather favorably with the energy differences calculated by molecular mechanics.

The strength of the method is the rapid construction of reasonable 3D geometries of ring systems using symbolical logic and an energy ranking scheme that allows the derivation of best candidate conformations without having to invoke a geometry optimization. The weakness of the approach is the limitation to ring systems with up to seven members, although the handling of exocyclical chains is possible via the input of all necessary acyclical torsional angles.

Although somehow out of date, the program is still available from QCPE and has found use in at least two recent programs: MIMUMBA [14,15] for conformational analysis and FlexX [16] for the automatic and flexible docking of ligands into receptor sites of proteins.

3.2. Rule-Based and Data-Based Methods

3.2.1. WIZARD and COBRA

Dolata and Carter [17] and Leach and Prout [18] developed two programs, WIZARD and COBRA, for the systematic conformational analysis using symbolical logic and techniques of artificial intelligence (AI). The basic idea of this approach is to develop a set of rules for the construction of molecular models derived from the method of a human expert who recognizes conformational units with well-known optimum geometries (e.g., cyclohexane chair) and joins them to an entire system. The following steps are performed:

1. The molecule is analyzed and conformational units are recognized. A conformational unit is a connected substructure for which the AI system has some knowledge on its conformational behavior. Fig. 8 shows this fragmentation process for cyclazocine. The molecule contains four monocyclical and five acyclical conformational units. Cyclical units consist of one or more rings. Acyclical units consist of one to three bonds. Note that neighboring fragments overlap.

2. An abstract hierarchical representation of the molecule is generated in the form of a so-called unit graph. The conformational units are the nodes of this graph. The edges of the unit graph are formed by the type of junction between two neighboring units (i.e., acyclical join, fused rings, or bridged rings). Fig. 9 shows the unit graph for cyclazocine.

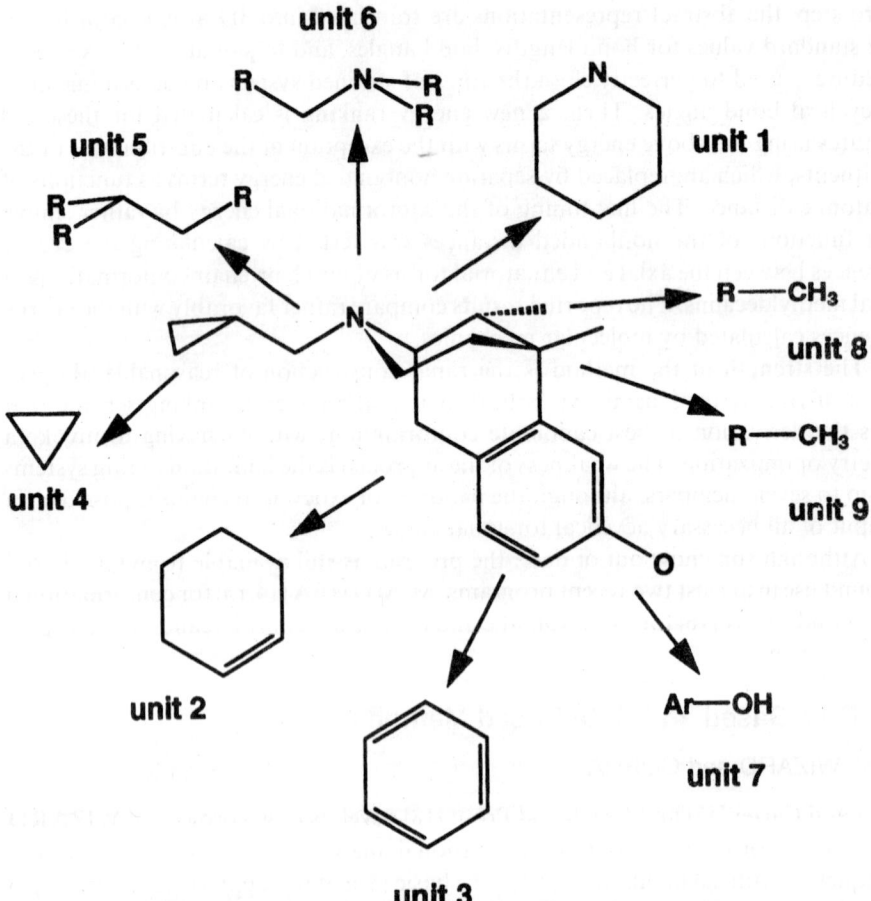

Figure 8   Recognition of nine conformational units in cyclazocine.

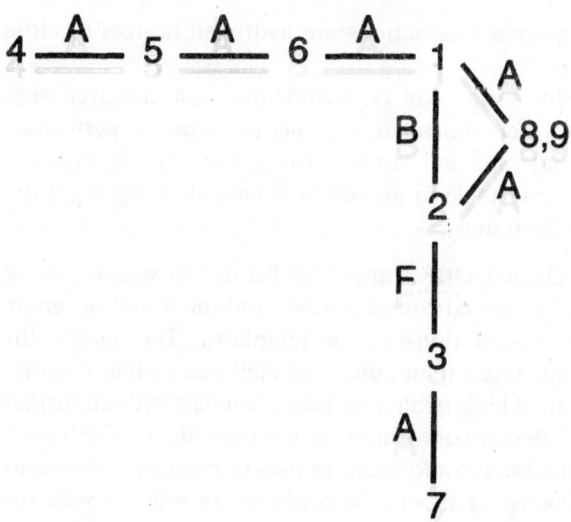

Figure 9 Abstract representation of cyclazocine. The conformational units are numbered in analogy to Fig. 8. The joins are marked by capital letters: A = acyclical join; F = fused rings; and B = bridged rings.

3. Lists of conformational templates are assigned to all conformational units, which are taken from a library. A template contains some knowledge on the fragment conformation (i.e., symbolical description of the conformation, strain energy, flexibility, and coordinates). If no exact expression of a specific unit can be found in the library, similar templates are searched on several levels of generalization. If, for example, no template for a heterocycle can be found, the corresponding carbocycle is taken. The templates are obtained either from molecular mechanics or x-ray crystallography.

4. Symbolical suggestions of conformations are built on the abstract level of the unit graph. The whole conformational space is formed by the combinatorial product of the templates assigned to the conformational units. The conformational space is searched by using a directed strategy, the A* algorithm [19]. The obtained symbolical suggestions are criticized using a set of predefined and self learned rules. The program looks for connections of units that are historically known to be bad (e.g., gauche$-$/gauche + pentane), or that have been found to be bad in an earlier stage of computation.

5. The symbolical suggestions are translated into coordinate representations combining the template coordinates. Because neighboring templates are overlapping, two templates can be joined by a least squares fit of the coordinates of the common atoms. The program has several weighting schemes for the common atoms. For instance, a substituent atom of a cyclical unit gets a lower weight than the atoms of the cycle itself. Different matching strategies are used for fused rings, for spiro rings, or for bridged systems. The coordinate representations are criticized after each combination step. Critics are the quality of the fit and problems arising from long-range interactions. The quality of the fit is characterized by the RMS value of the matching atom

positions. Types of long-range interactions are hydrogen bridges or close van der Waals contacts.

6. If no noncriticized conformation can be found, the least criticized suggestions are chosen for further refinement. Another tree search is performed to look for conformational units that can be deformed to solve the problem (i.e., changing one torsional angle in an acyclical unit, or assigning a deformed template to a cyclical unit).

The strength of the WIZARD and COBRA approach lies in the extensive use of symbolical representations for the suggested conformations and the use of optimum geometries for the coordinate representations of the templates. This makes the algorithm some orders of magnitude faster than numerical methods such as distance geometry. It allows the construction of high-quality molecular models without further optimization. In addition, when different conformations are possible, no conformation will be overlooked and a set of desired conformations may be produced. Problems may arise when templates are lacking, or fit only imperfectly. In other words, the quality of the result for a given problem strongly depends on whether suitable templates are contained in the library or not. On the other hand, the addition of a new template to the library requires database searches on x-ray structures and/or molecular mechanics calculations.

3.2.2. CONCORD

The popular program, CONCORD, of Pearlman [4,20] was for a long time the most widely used method for converting large databases of 2D structures to 3D representations. The program is based on rules and a simplified force field method. It performs the following steps for model building:

1. The input structure is analyzed and separated into ring systems and acyclical atoms. Two rings are regarded to belong to one and the same ring system if they both have at least two atoms in common with another ring of the same system. Thus, spiro-connected rings are handled separately.
2. Bond lengths and bond angles are taken from a table. They depend on atom type and bond order. The atom types are rather detailed and consider hybridization state and some first sphere neighbor atoms or small ring size. For carbon, for example, 21 atom types are considered.
3. Ring systems are processed by the assignment of a general conformation (e.g., chair, boat, etc.) to each ring. These general conformations reflect constraints from the conformations of the other rings of the entire ring system. Then, the rings are ordered according to a certain priority and are optimized in steps in this order by the minimization of a special strain function in internal coordinate space. The coordinates of rings already previously processed (on a higher level of priority) remain unchanged. Endocyclical bond angles and torsional angles are simultaneously changed to get perfectly closed rings and minimal steric energies.
4. Finally, the torsional angles of the acyclical parts are set to values that minimize the steric interactions of all 1–4, 1–5, and 1–6 interactions. Close contacts are relaxed by a limited energy minimization.

CONCORD considers the elements H, C, N, O, F, Si, P, S, Cl, Br, and I, and is able to process molecules with up to 200 nonhydrogen atoms. The maximum

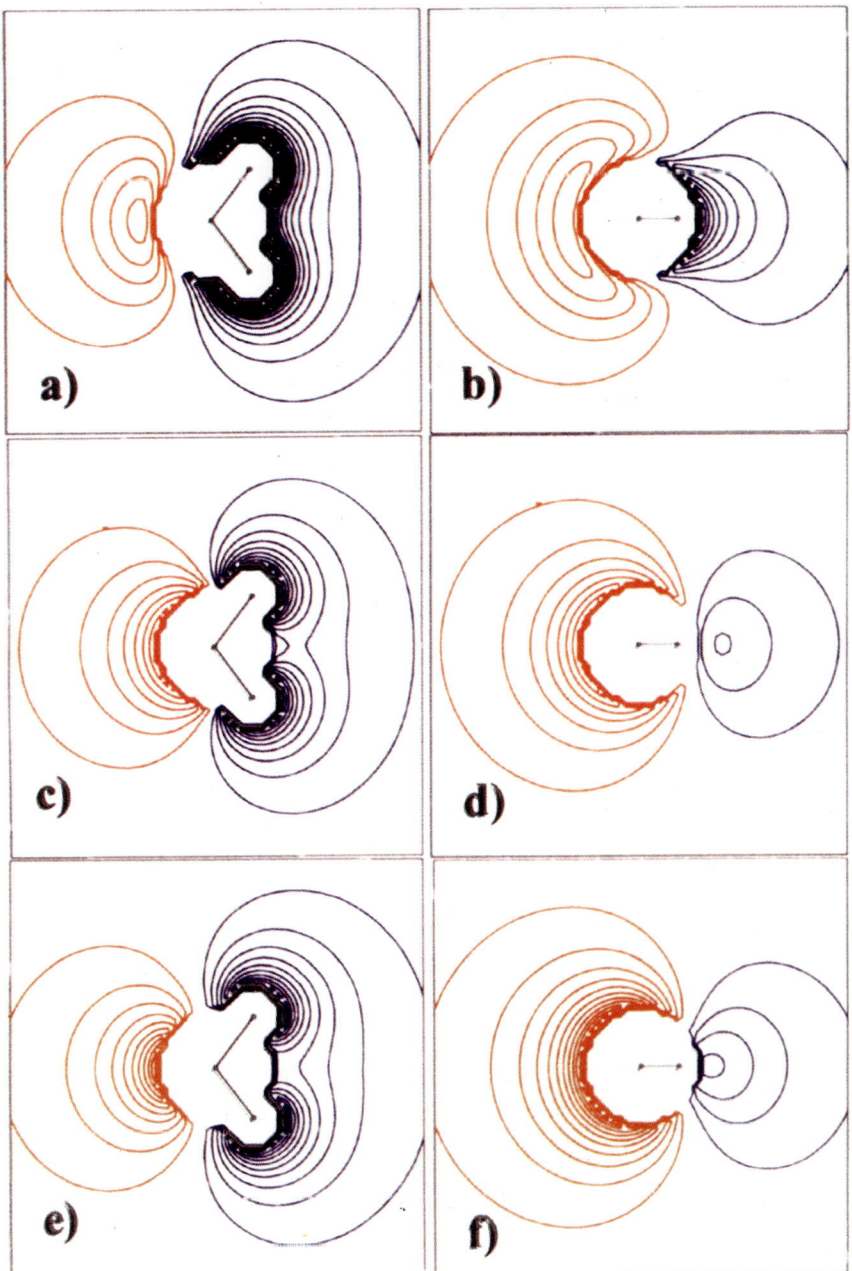

Figure 4 Molecular electrostatic potential of water molecule, represented as a contour plot with intervals of 0.025 au. Red contours indicate regions of negative potential and blue represents positive. (a–b) Potential generated from full electron density, in and perpendicular to the molecular plane, respectively; (c–d) potential generated from point charges situated at three atomic positions; (e–f) potential generated from point charges and dipoles situated at three atomic positions.

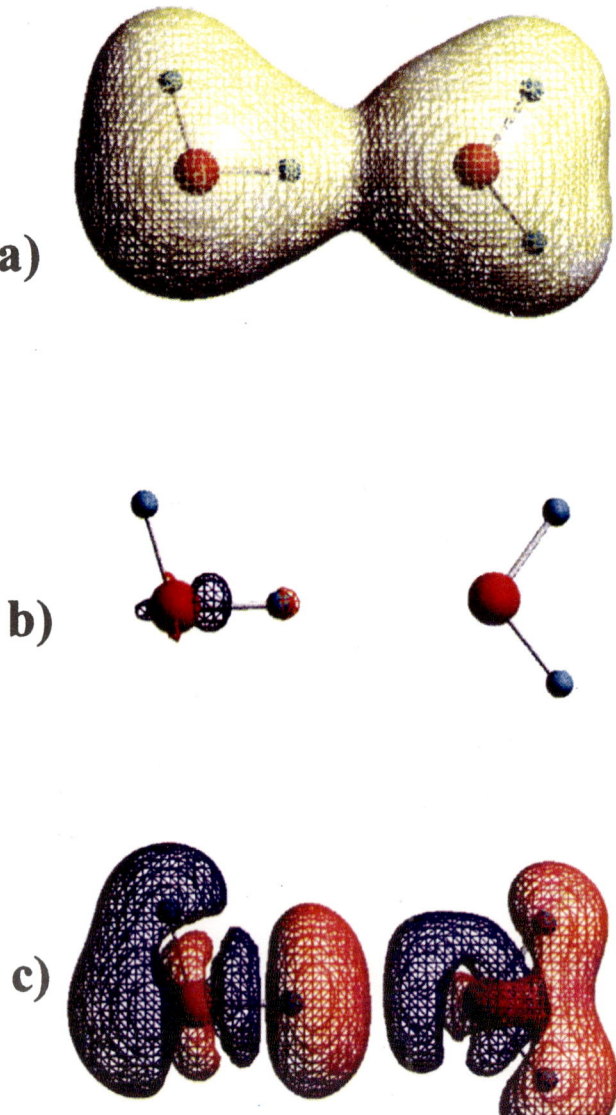

Figure 6 (a) Sum of the electron densities of the two water molecules, in the dimer configuration, illustrated as the 0.01-au contour. (b) Difference in density between the sum of monomers (a) and the dimer, again at the ±0.01 contour. Blue contour represents gain of density in the dimer vs. the pair of monomers, and loss of density is shown in red. (c) Same as (b) except that the ±0.0005 contour is illustrated.

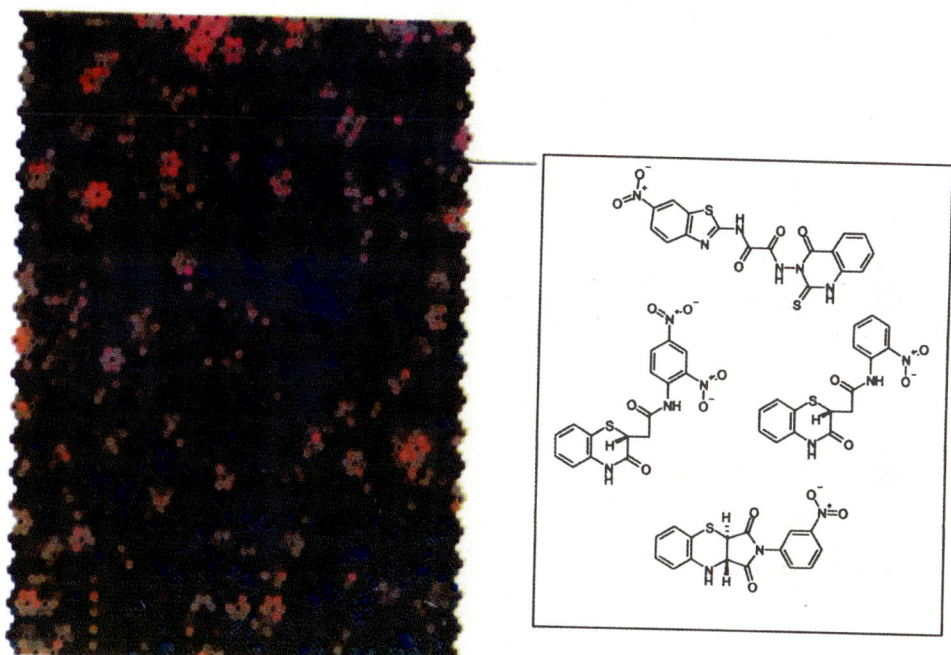

Figure 6 Example of a complex SOM visualizing the structural relationships of more than 40,000 chemical structures of the August 1999 release of the NCI anticancer database [52]. The SOM is partitioned into a hexagonal array of 966 clusters. Distances between clusters are indicated by the colors between clusters (red, close; black, intermediate; purple, far). Close and far neighbors are separated by dark and light blue colors, respectively. As an example, compounds in hexagon 9–23 are highlighted. (Courtesy of Drs. Rabow and Covell.)

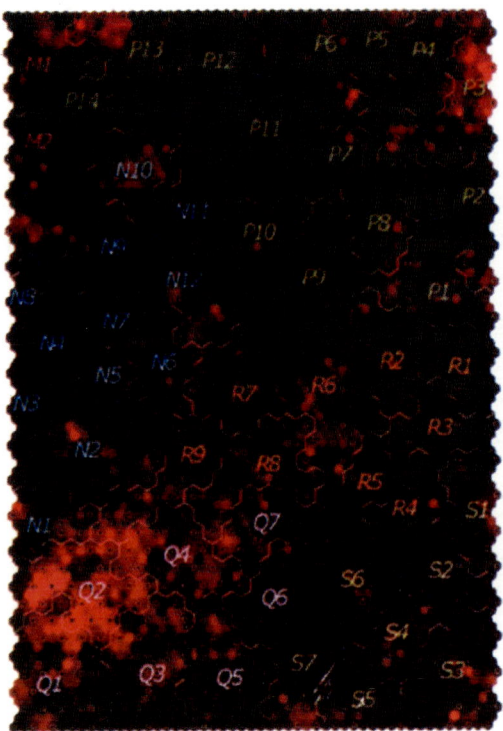

Figure 9 Self-organizing map of more than 20,000 compounds tested in the NCI's tumor cell screen [52]. The map consists of 966 clusters. Each compound in this data set is characterized by a set of more than 60 biological properties. Color bar at lower right indicates the distance between clusters (red, close; black, intermediate; purple, far). Fifty regions have been defined on this map that group together individual clusters with the most similar response profiles. These regions are assigned to six functional categories according to their apparent cellular activity: M, S, N, P. Regions Q and R have not been assigned to an activity class (see text for further clarification). (Courtesy of Drs. Rabow and Covell.)

connectivity (coordination number) of an atom is four. For multifragment compounds, CONCORD only models the largest fragment and passes the smaller fragments through. The produced structure is one single low-energy conformation.

3.2.3. CORINA

Extending an earlier work by Hiller and Gasteiger [21], Sadowski and Gasteiger [3] have developed the 3D structure generator, CORINA. The program was developed for the reaction prediction system, EROS, to model the influence of the spatial arrangement of the atoms in a molecule on its reactivity [22]. Therefore, the approach had to be applicable to the entire range of organic chemistry including reactive intermediates, macrocyclical, and organometallic compounds. To handle large amounts of hypothetical structures, it had to be automatic and rapid. The program performs the following steps in generating a 3D model:

1. Bond lengths and bond angles are set to standard values taken from a table. Bond lengths depend on the atom types, the atomic hybridization states, and the bond order of the regarded atom pair. For bond types not found in the table, reasonable values are calculated from covalent atomic radii and electronegativities. Bond lengths in conjugated systems are relaxed using a Hückel MO scheme. Bond angles only depend on the atom type and the hybridization state of the central atom. Atoms with up to six neighbors can be handled, using one of the following elementary geometry types: terminal, linear, planar, tetrahedral, trigonal bipyramidal, or octahedral. The tables of bond lengths and bond angles are parameterized for the entire periodic table.

2. The molecule is fragmented into ring systems and acyclical parts. Ring systems contain the ring atoms plus the exocyclic atoms directly bonded to ring atoms. The exocyclic atoms are included because their positions and their long-range interactions are strongly influenced by the ring conformation. Two rings belong to the same ring system if they have at least one atom in common with another ring of the same system. The ring systems are further classified into small ring systems that include rings with up to eight atoms, rigid macrocyclical systems that include large rings with low flexibility that is limited by bridges or fused rings, and flexible macrocyclical systems containing one flexible large ring that may be fused or bridged to a limited number of small rings.

3. Small-sized and medium-sized ring systems can be handled by using a table of allowed single-ring conformations because rings of three to eight members attain a limited number of conformations. These templates are stored as lists of torsional angles depending on the distribution of unsaturated bonds in the rings. They are characterized and ordered by a strain energy value describing the conformational energy. A backtracking algorithm is used for ring systems consisting of more than one ring being fused or bridged to find possible combinations of conformations of the single rings. Fig. 10 illustrates this procedure for cubane. After the smallest set of smallest rings (SSSR) has been determined (five four-membered rings in the case of cubane), the algorithm starts with the lowest-energy conformation (torsion angle sequence: $30°$, $-30°$, $30°$, and $-30°$; envelope form). Two of these conformations can be joined together, but a third one cannot be fused to this

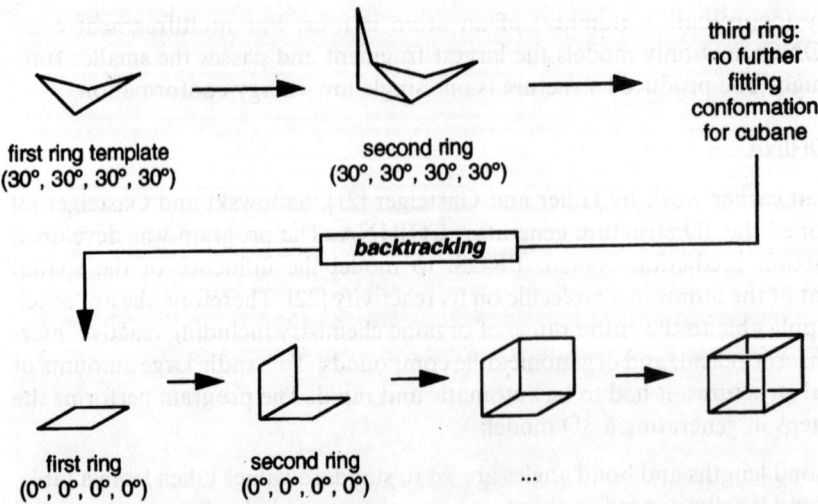

Figure 10 Backtracking procedure for the generation of a 3D model of cubane.

assembly. The backtracking algorithm then selects the next possible conformation from the list of ring templates ($0°$, $0°$, $0°$, and $0°$; planar form) to
fuse to the ring chosen first. Again, the fusion of more than one ring to the
envelope form is not possible. Thus, the planar geometry is assigned to the
first ring, too, which finally leads to the correct skeleton of cubane.

      Because heteroatoms and strained systems may cause imperfect ring
closure, a pseudo-force field calculation (see below) is performed to optimize
the ring geometries. Further details on the generation of a single conformation as well as sets of low-energy conformations of small-sized and
medium-sized ring systems with CORINA are given in Sec. 5.2.3.

4.   Rigid polymacrocyclical systems cannot be handled with the procedure
   described above for small ring systems. No conformations are available
   from the table of ring templates for rings with a size larger than nine. However, polymacrocyclical structures quite often show an overall general outline, a superstructure. For example, the polymacrocyclical molecule in the
   right-hand side of Fig. 11 has a cagelike superstructure as sketched at the
   left-hand side of Fig. 11. The procedure for generating a 3D structure for
   polymacrocyclical systems is based on the so-called principle of superstructure. It starts with a crude model of the hypothetical superstructure that
   retains the approximate shape and symmetry of the molecule and builds in
   several steps a model of the entire molecule.

5.   For flexible macrocyclical systems such as, for example, cyclononane, the
   above principle of superstructure cannot be used due to the conformational
   flexibility of these systems. A simple conformational analysis procedure for
   such large rings was developed, which was derived from the notation for the
   conformations of cycloalkanes. by Dale [23]. This notation is based on the
   assumption that low-energy conformations of large rings take a polygon
   shape. The notation consists of linear codes of the number of bonds between
   the corner atoms that define the polygon. Thus, possible conformations of

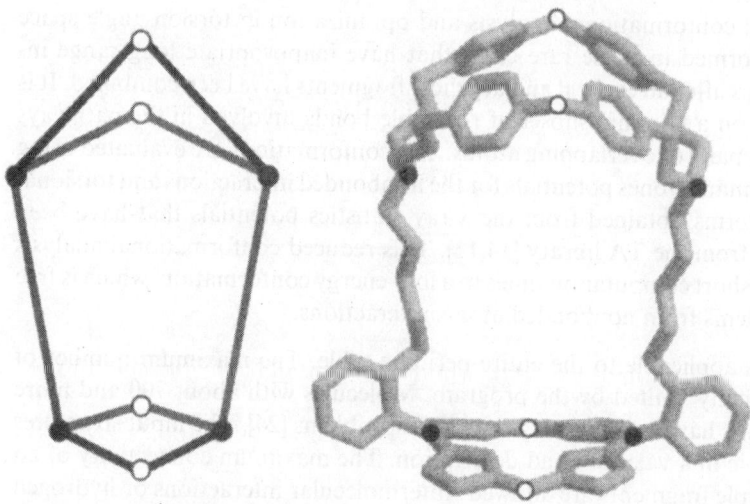

Figure 11 A macrocyclical molecule and the corresponding superstructure. Atoms in common are marked by black circles.

cyclononane have the codings [3,3,3], [2,3,4], [1,4,4,], [1,2,2,1,3], [1,2,1,2,3], and [1,2,2,2,2]. These 1D symbolical representations can quickly be generated and directly be translated into three-dimensional atomic coordinates, constructing the specified polygons. A simple linear combination of features calculated from the linear notations allows an energy ranking of the 1D conformations.

6. A pseudo-force field is used to optimize the geometries obtained by the above algorithms for ring systems. Two assumptions are made. Firstly, rather rigid ring systems will be optimized. Thus, torsional energies and nonbonded interactions can be regarded as second-order influences on the geometries. Secondly, the major aim is geometry optimization instead of energy calculation. Thus, no real energy values must be computed. These two assumptions lead to a rather simplified so-called pseudo-force field with a reduced number of energy terms and rather general parameters applicable to the entire range of organic chemistry. In addition, the energy functions are directly derived from geometrical considerations instead of physical functions. The pseudo-force field calculation is only applied to ring atoms. In this way, the adjustment of bond lengths, bond angles, and torsional angles in ring systems is rapidly achieved, converging after few iterations through the minimization procedure.

7. The torsional angles within acyclical chains are chosen according to a set of over 900 rules based on statistics on the conformational preferences of the x-ray structures that are contained in the Cambridge Structural Database [14,15]. These rules are stored in the so-called torsion angle library. For torsions with more than one possible low-energy conformation, angle values are chosen from the TA library, which result in the most extended conformation along the main chains to minimize repulsive nonbonded interactions.

8. Reduced conformational analysis and optimization in torsion angle space are performed in those rare cases that have inappropriate long-range interactions after all cyclical and acyclical fragments have been combined. It is focused on a minimal subset of rotatable bonds involved in the pathways between pairs of overlapping atoms. The conformations are evaluated using 12-6 Lennard–Jones potentials for the nonbonded interactions and torsional energy terms obtained from the x-ray statistics potentials that have been derived from the TA library [14,15]. This reduced conformational analysis leads in short computation times to a low-energy conformation, which is free of problems from nonbonded atoms interactions.

CORINA is applicable to the entire periodic table. The maximum number of atoms is not explicitly limited by the program. Molecules with about 700 and more nonhydrogen atoms have been processed without problems [24]. The input structures must be expressible in a valence bond description. The maximum connectivity of an atom is six. Multiple fragments are allowed. Intermolecular interactions or hydrogen bonds are not explicitly handled. Besides its general applicability to organic and organometallic compounds with no explicit limitations of the number of atoms or the ring sizes, the program offers three features not found in most of the other 3D structure generators:

Large rings. Large rings represent a special challenge and most of the other published 3D structure generators fail to process such systems. Following the "principle of superstructure," CORINA can process such systems. Fig. 12 compares the x-ray structures of three polymacrocyclic compounds with the corresponding CORINA models and the RMS deviations between them. Although rather large RMS values of 0.14–0.95 Å are measured, it can be seen that the program succeeded to predict correctly the overall shape and symmetry. Furthermore, the large RMS deviation for the molecule, CISZUZ (CSD refcode), results from the fact that it contains the anion I3$^-$ within the cage structure, thus pushing the bridging chains further to the outside in the experimental determined structure.

Metal complexes. Other types of structure commonly avoided by conventional structure generators are organometallic compounds. CORINA can process compounds containing atoms with up to six neighbors. Thus, in principle, metal complexes with up to octahedral centers can be modeled. The resulting structures often correspond quite well to the experimentally determined geometries (Fig. 13).

Multiple ring conformations. Methods dealing with molecular flexibility as, for example, conformer generation, flexible ligand docking, or 3D database searching, have to address also the problem of multiple ring conformations. The method of choice for flexible 3D database search is the use of 3D databases containing single low-energy conformations for each molecule to solve the flexibility problem on the fly instead of storing multiple conformations in the database—a rather time-consuming and disk space–consuming approach. For the purpose of flexible search there exist a number of methods as, for example, directed tweak [25]. These methods are rather efficient for chain portions of the molecules but run into problems when performed on ring systems [7]. A solution is to store 3D models with multiple ring conformations

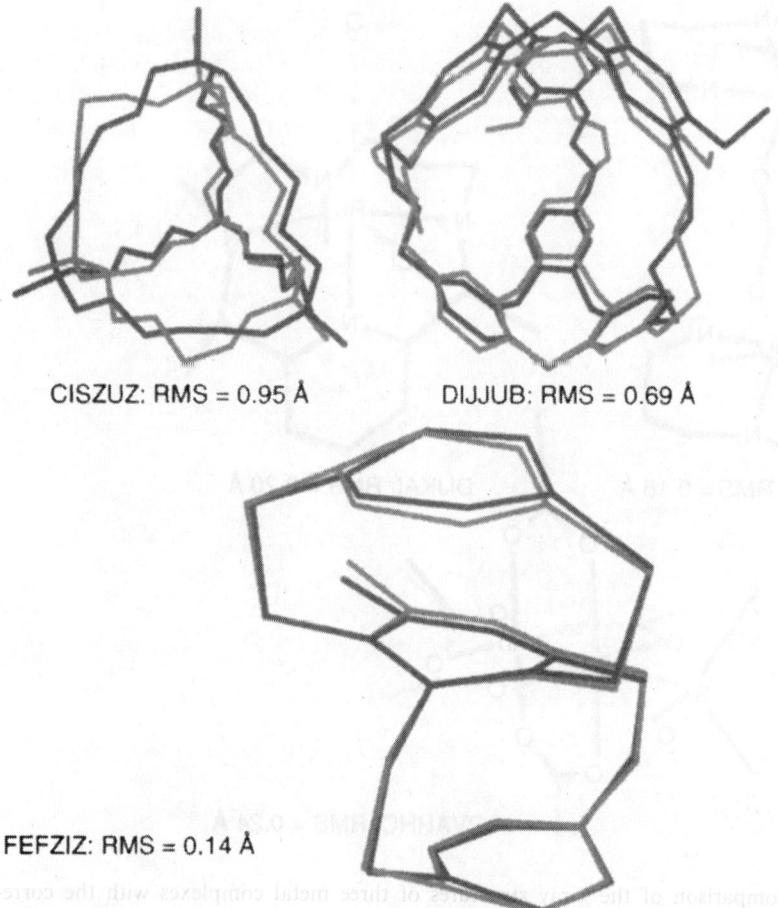

CISZUZ: RMS = 0.95 Å   DIJJUB: RMS = 0.69 Å

FEFZIZ: RMS = 0.14 Å

Figure 12   Comparison of the x-ray structures of three polymacrocyclical systems with the corresponding CORINA models and their RMS deviation.

and to apply the flexible search only to the chain portions. CORINA supports this technique by providing the option to generate multiple ring conformations and to write them to the output file (for further details, see Sec. 5.2.3). Additionally, CORINA has been interfaced to the conformer generator, MIMUMBA [14], and the docking program, FlexX16, to provide different ring geometries to these programs.

3.3.   Fragment-Based Methods

3.3.1.   AIMB

Wipke and Hahn [26] have developed a unique 3D model building technique that is based on finding near analogies for a molecule or substructures of it in the database of 3D molecular structures: AIMB (Analogy and Intelligence in Model Building). A human expert is able to construct a 3D model in a very efficient, non-numerical, and fast manner, reasoning by analogy based on one's knowledge on similar problems. The

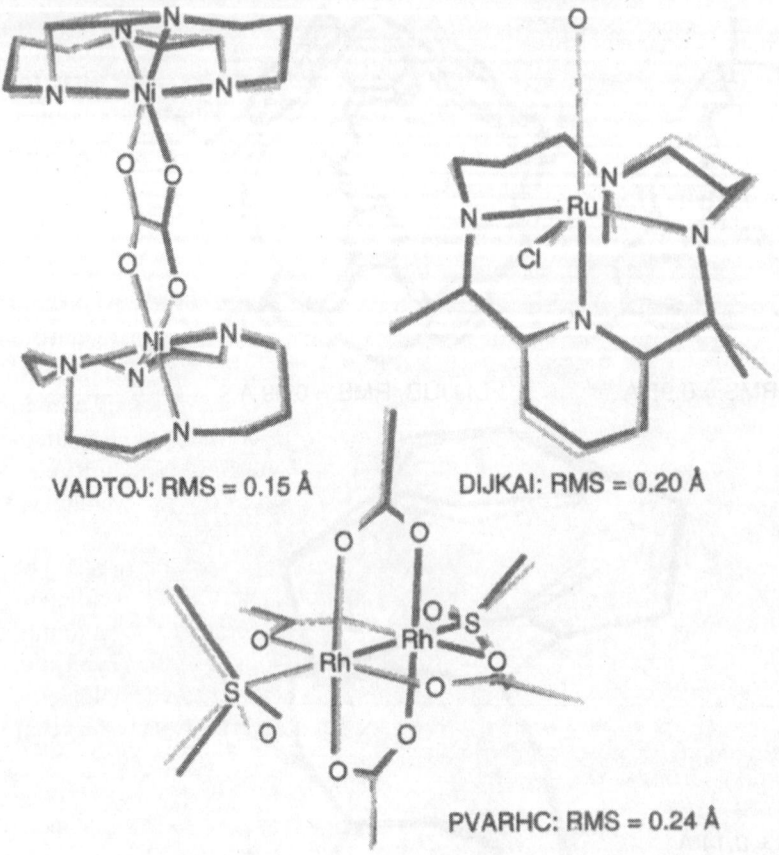

VADTOJ: RMS = 0.15 Å   DIJKAI: RMS = 0.20 Å

PVARHC: RMS = 0.24 Å

Figure 13 Comparison of the x-ray structures of three metal complexes with the corresponding CORINA models and their RMS deviation.

program tries to automate this method with knowledge already captured by crystallography and stored, for example, in the Cambridge Structural Database. The basic idea is that a large and widespread data collection of experimental molecular geometries implicitly contains "knowledge" on the molecules for model building. The following steps are performed by the different components of the method:

1. The knowledge base (KB) of AIMB was constructed from the Cambridge Structural Database, selecting organic molecules (C, N, O, P, S, Si, B, F, Cl, Br, and I) with less than 65 nonhydrogen atoms. Structures with atoms having a coordination number of more than five polymer structures and poor crystal structures were removed. Hydrogen atoms were removed because their positions are normally not determined experimentally. This subset of the Cambridge file was processed to generate abstractions that are hierarchically ordered for rapid access.

2. The problem analyzer perceives the target structure to identify rings, chains, aromaticity, and stereochemistry. If the target or a close analogy is not contained in the compound library, the decomposer uses a "divide and con-

quer" strategy to create substructures of the target and to treat them as new problems. The subdivision strategy follows the rule that there is maximum interaction within a unit and minimum interaction between units. First, the target is subdivided into ring assemblies and chains. If the program again fails in finding an analogy in the KB, the subproblems may be divided further. Ring assemblies are subdivided only once more into elemental single-ring units that cannot be divided further. A chain can be broken down into elemental chain fragments of simple bonded atom pairs. If an elemental subproblem cannot be solved, the model building process is aborted. In addition, the atoms of a subunit are weighted differently. These weights are assigned to atoms in descending order of priority: origin atoms, which form the join to another unit; α-origin atoms, which are nonorigin atoms in α-position to an origin atom; real atoms, all remaining atoms of a unit; and, in addition, dummy atoms, which are attached to origin atoms and contain some information about the chemical environment around the unit (e.g., rings, substituents, etc.). Fig. 14 shows the division of methyl cyclohexylketone into subunits.

3. The analogy finder searches for analogies of the subproblems in the KB. The hierarchical structure of the KB allows the probing of the file at different levels of abstraction. If no exact expressions of a subproblem can be found, the matching tolerances are increased until an analogy is found. Typically, this search is continued until a maximum search depth of 5–10 analogies is reached. This search strategy on several levels of abstraction guarantees that the best analogies are found first.

4. The analogy evaluator scores each found analogy to select the best analogies. The problem is twofold. Firstly, a similarity measure must be defined, which reasonably describes the distance between different analogies. Secondly, the mapping problem of projecting the target atoms onto the analogy atoms has to be solved (i.e., all possible mappings of the target and the analogy are to be explored) and it cannot be assumed that the target and the analogy are isomorphic. Because there are some constraints on atom and bond mapping (e.g., nondummy atoms must always be mapped onto nondummy atoms), not all possible permutations are to be checked. The similarity score of an analogy is calculated following a scoring function based on atomic attributes. The attributes include atom type, charge,

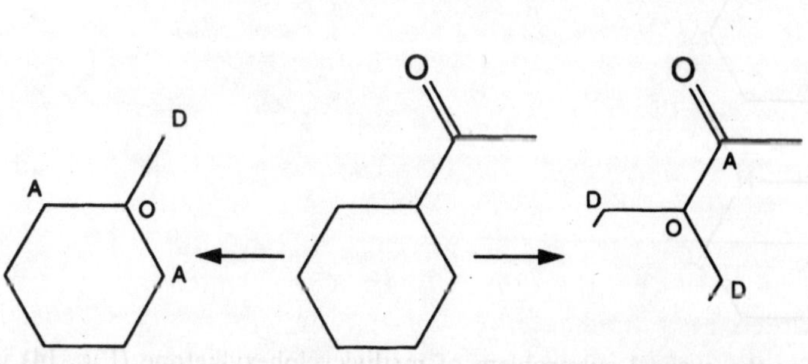

Figure 14 Two subproblems of methyl cyclohexylketone. Origin (O), α (A), real (R), and dummy (D) atoms are marked.

valence, hybridization, and stereochemistry. Weighting factors differentiate between origin, α-origin, real, and dummy atoms. Fig. 15 shows several analogies of the cyclical subunit of methyl cyclohexylketone in descending order of similarity.

5. The model assembler combines the analogies found for the subproblems to a coordinate representation of the original problem. The combination is performed in steps, superimposing the origin and dummy atoms of the subunits. The resulting differences in bond lengths and bond angles between both welded fragments are calculated as a measure of the quality of the fit.

The described algorithms rapidly build reasonable 3D molecular models that represent minimum energy conformations. The results are explained to the user using the information on the structures where the analogies are taken from. Because several analogies can be found for the subunits depending on the search depth, it is possible to perform a conformational search. Although the program does not contain any energy evaluation procedure and does not take into account long-range interactions, it was shown in several cases that the fragments used "know" something about these problems. A helical model of pentahelicene was built from single benzene rings because the best analogies found were taken from other helicenes. In this way, implicit knowledge on energy and long-range interactions can be extracted from the KB.

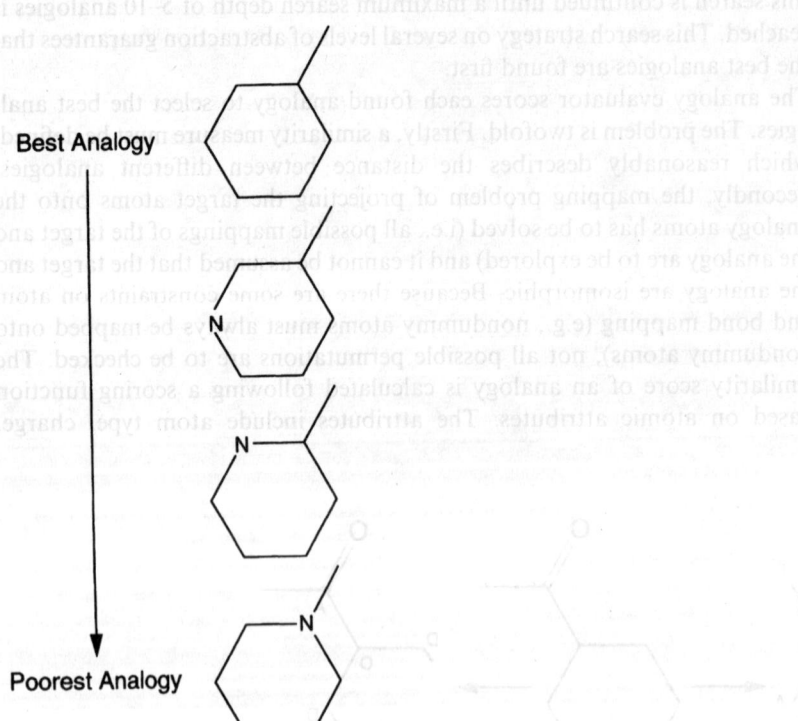

Figure 15 Analogies for the cyclical subproblem of methylcyclohexylketone (Fig. 14) in decreasing order of similarity: exact match, real atom mismatch, α-origin atom mismatch, and origin atom mismatch.

Several investigations have been performed to characterize the program's performance. Firstly, the problem-solving speed was studied as a function of the library size. It was shown for knowledge bases of 500, 1000, 5000, and 10,000 of 3D geometries that there is a substantial increase in speed with increasing size of the KB. The larger and more widespread the KB is, the earlier the AIMB finds good analogies of the subproblems. Secondly, the quality of the models was tested against the size of the KB. It was found that the better the models were constructed, the larger the KB was. Thirdly, the speed vs. the search depth was explored. The time needed for the model construction increased linearly with the search depth (i.e., the desired number of analogies for each unit). Finally, the speed as a function of the target complexity was studied. The time per molecule increased linearly with the number of atoms and with the number of subunits in the target.

The strength of the method is its speed and that it is exclusively based on experimental 3D structures and fast database searching techniques. The models built are as accurate as x-ray structures and can be explained by the parent structures where the subunits are taken from. One of the most interesting qualities of AIMB is its ability to build more accurate models more rapidly as the amount of knowledge present in the KB increases. The only limitation in the range of chemistry that can be handled is the content of the KB. Therefore, a possible difficulty of the program is that the quality of the models built strongly depends on the quality of the database of 3D structures available as KB. Another problem may be the use of redundant information because a lot of substructures with very similar geometries (e.g., benzene rings) are contained many times in the library. Unfortunately, the program has never been made publicly available.

3.3.2. Chem-X

Chemical Design Ltd. has developed a 2D-to-3D builder of its own [27], which assembles fragments retrieved from a database similar to the AIMB program by Wipke and Hahn (see above).

The heart of the 3D builder is a relatively small library of common ring substructure fragments containing specific carbocyclical and heterocyclical groups together with generalized fragments with unspecified atom types. Furthermore, the fragments are characterized by different patterns of unsaturation and by stereochemistry. The default library contains about 100 preoptimized cyclical structures. The model builder first tries to find exact matches of the cyclical substructures in the library. If no exact match can be found, generalized fragments are taken. Ring systems may be handled as whole fragments or as single-ring structures, which are fitted together. Acyclical parts of the molecule are constructed with torsional angles of the main chains in extended form. If more than one hit is found for a fragment, a conformational search can be performed. A special handling of stereochemistry allows the generation of different stereoisomers, which is useful in converting databases not containing stereoinformation. The range of validity of the model builder can be extended by updating of the library of ring fragments, but this slows down the program.

The program was used to convert large databases at Chemical Design Ltd. It seems to be more general than the AIMB approach because side chains are constructed straightforwardly instead of taking them from the library. Its major strength is the speed of the coordinate generation. Its major weakness is the rather simple construction scheme for side chains, which may result in problems from long-range interactions. The strategies used seem to be simpler than those used in AIMB and the models produced lack the explanation capabilities of AIMB. Especially the library search strategies seem to be less efficient because the addition of new fragments to the knowledge base slows the program down in contrast to the AIMB program, where the speed increases with the size of the database.

3.4. Distance Geometry

Although the distance geometry approach is assigned to numerical methods, a brief description of the distance geometry formalism is given here [10,28,29]. In addition, a distance geometry approach has been used in some of the programs discussed later in Sec. 4, such as MOLGEO [30] and CONVERTER [31]. A rather well-known program, which is based on distance geometry for conformational analysis purposes, is DGEOM [32] developed by Blaney and Dixon. The algorithm is built around a so-called distance matrix with upper and lower distance bounds for all atom pairs in a molecule or molecular ensemble. The 1–2 and 1–3 distances are simply derived from ideal values for bond lengths and bond angles. For 1–4 distances, upper and lower bounds are given referring to the minimum and maximum values allowed for a certain torsion angle. All other lower bounds are set to the minimal allowed distance between two nonbonded atoms—usually the sum of their van der Waals radii. The maximum bounds for these atom pairs can be estimated from the longest possible distance within the molecule based on the number of atoms and bonds. Additional distance constraints (e.g., from conformational restrictions or experimental data), such as those from 2D NMR spectra, are also used.

Given reasonable starting coordinates, a numerical minimization method called *triangle smoothing* is used along with a penalty function for violations of the distance bounds to optimize the geometry of the molecule. An additional algorithm called "embedding" can be used to derive starting coordinates directly from the distance matrix. Because embedding can produce a number of starting geometries, different conformations can be obtained and distance geometry is thus a conformational search method of its own.

Distance geometry methods tend to be faster than molecular mechanics methods and they are easier to parameterize. On the other hand, they are less accurate and the generated conformations can be rather crude because the distance matrix describes conformational properties only in a coarse manner as, for example, there is no possibility to describe multiple energy minima of torsional angles. Thus, the cyclohexane chair and boat conformations would be considered to be equally reasonable.

4. 3D STRUCTURE GENERATION: EVALUATION OF AVAILABLE PROGRAMS

The reliability of scientific work based on databases of generated 3D structures requires a careful evaluation of available 3D generators to find the program best suited for this purpose. This evaluation should cover the criteria given in Sec. 2.1 on computational requirements. In this section, a study comparing the results of seven currently available 3D structure generators with a set of 639 x-ray structures is presented. This is an updated version of a test published some years ago [33]. Of course, the comparison of computer-generated single low-energy conformations with x-ray structures will often not find any correspondence between them although the generated structures are reasonable. Thus, no absolute scoring of the results of a particular program can be expected. But such a comparison can reveal relative differences between the performances of a number of such programs and can figure out specific strengths and weaknesses.

4.1. Evaluation Procedure

A dataset of 639 x-ray structures was taken from the Cambridge Structural Database [1]. For all programs, a set of quality criteria was determined: the conversion rate, the number of program crashes, the number of stereo errors, the average computation time per molecule, the percentage of reproduced x-ray geometries, the percentage of reproduced ring geometries, the percentage of reproduced chain geometries, and the percentage of structures without crowded atoms.

An x-ray geometry is considered to be reasonably well reproduced if the RMS deviation of the atomic positions RMS_{XYZ} is less than 0.3 Å. A chain geometry is taken to be well reproduced if the RMS deviation of the torsion angles at rotatable bonds RMS_{TA} is less than $15°$. A 3D molecular model is considered to be free of nonbonded interactions if the close contact ratio CCR (the ratio of the smallest nonbonded distance to the smallest acceptable value for this distance) is greater than 0.8.

4.2. Programs

Table 1 gives information on the programs tested by the evaluation procedure. Four of them have been described in Secs. 3.2.1, 3.2.2, 3.2.3, and 3.3.2. The basic principles of the distance geometry approach, which is used in the programs MOLGEO and CONVERTER, are given in Sec. 3.4. An additional program called ALCOGEN has been included for which no information was available from the literature [34]. ALCOGEN was included because it clearly fulfills the criteria for automatic 3D converters and the two distance geometry programs were included to study the applicability of this approach to 3D database generation.

4.3. Results and Discussion

Table 2 shows the values for the quality criteria determined for the different model builders:

> **Conversion rate.** CORINA and CONVERTER came up with the largest conversion rate (98–100%). This indicates that these programs have the most widespread scope.

Table 1 3D Structure Generation: Programs Tested by the Evaluation Procedure

	CONCORD	ALCOGEN	Chem-X	MOLGEO	COBRA	CORINA	CONVERTER
Version	3.0.1	1.02		2.4		1.6	950
Year	1993	1993	1993	1993	1993	1994	1995
Reference	20	34	27	30	18	21	31

Table 2 3D Structure Generation: Summary of Results

	CONCORD	ALCOGEN	Chem-X	MOLGEO	COBRA	CORINA	CONVERTER
Conversion rate [%]	84	79	74	79	75	100	98
Program crashes	1	2	0	0	0	0	0
Stereo errors	0	1	23	1	0	0	0
CPU time [sec/molecule][a]	0.14	0.79	0.33	8.98	3.49	0.32	3.64
RMS_{XYZ} <0.3 Å [%][b]	38	40	33	19	38	46	37
RMS_{XYZ}^{rings} <0.3 Å [%][c]	89	88	89	69	89	90	87
RMS_{TA}^{chains} <15° [%][d]	49	55	45	41	49	58	53
CCR >0.8 [%][e]	91	94	71	86	87	97	100

The percentages refer to the total number of structures converted by each of the different programs and not to the total number of 639 structures in the original dataset.
[a] Scaled to a VAX 6000.
[b] Percentage of structures with an RMS deviation of the nonhydrogen atoms of less than 0.3 Å.
[c] Percentage of structures with an RMS deviation of the ring atoms of less than 0.3 Å.
[d] Percentage of structures with an RMS deviation of the torsional angles in open-chain portions of less than 15°.
[e] Percentage of structures with a close contact ratio of greater than 0.8.

Robustness. CONCORD and ALCOGEN encountered one or two program crashes, respectively, a rather high rate considering the rather limited size of the dataset.

Correctness of stereochemistry. All programs except Chem-X (23 failures) retained the stereochemistry of almost all stereocenters.

Computation time. CONCORD required extremely short computation times (0.14 sec/molecule), whereas MOLGEO, CONVERTER, and COBRA needed substantially larger times (3.49–8.98 sec/molecule). All other programs needed times of less than 1 sec/molecule. The computation times refer to the number of structures converted by the different programs.

Reproduction of x-ray geometries. CORINA reproduced the largest portion of x-ray structures (46%). Considering structural details such as ring systems as rigid, this rate becomes 87–90% for all programs except MOLGEO (69%). This is a hint that MOLGEO produces random conformations whereas the other programs try to find low-energy conformations. The highest rate of reproduced chain geometries generated ALCOGEN, CONVERTER, and CORINA (53–58%). Please note that the criterion for reproduced chain geometries has been redefined. In the original paper [33], all torsion angles at a rotatable bond have been taken into account. Because this overestimates some types of bonds, only one torsion angle per rotatable bond is counted. Thus, the percentages in the RMSTA row of Table 2 have slightly changed against Ref. 33.

Close contacts. The CONVERTER structures are completely free of close contacts. CONCORD, ALCOGEN, and CORINA generated between 91% and 97% structures without close contacts. The Chem-X builder produced only 71% of such overlap-free structures—an indication that the program does not perform any check for atom crowding.

Quantity–quality characteristics. The impression by the numbers in Table 2 is biased by the different conversion rates. As stated above, the percentages refer to the number of structures converted by the individual programs and not to the total number of 639 x-ray structures in the study. Thus, there is a sensitive relation between conversion rate and quality. Fig. 16 characterizes the relationship between quantity (conversion rate) and quality (the degree of reproduction of the x-ray structures) (i.e., the efficiency of the different programs) [35]. For each program, the ordered RMS values of the nonhydrogen atoms are plotted vs. the number of converted structures. Thus, the ends of the curves mark the number of totally converted structures and the ascents of the curves characterize the quality of the structures in terms of similarity to the x-ray structures. These quantity–quality characteristics show again the different suitability of the seven programs for automatic 2D-to-3D conversion.

4.4. Comparison of CONCORD and CORINA Using 25,017 X-ray Structures

4.4.1. Introduction

To address both the higher computational throughput of nowadays computers and the larger number of experimental 3D structures available now, the above evaluation study was repeated in 2001 using 25,017 x-ray structures. This evaluation was applied to the two now mostly used converters, CONCORD and CORINA. The new dataset

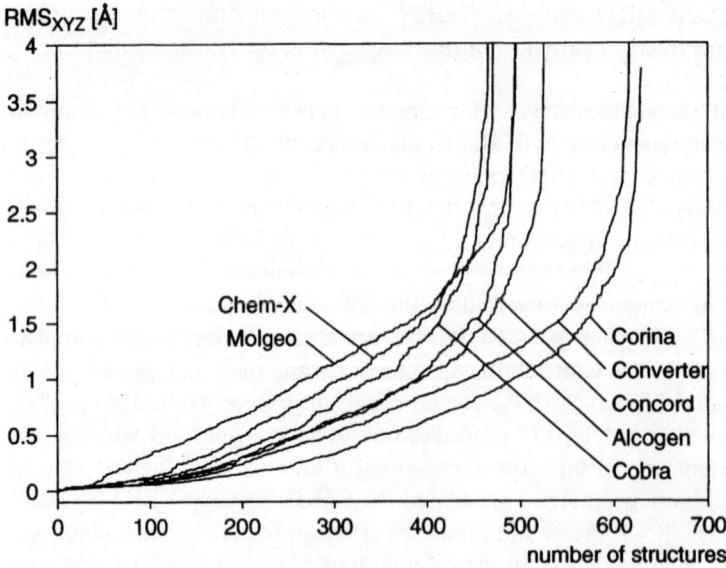

Figure 16 Quantity–quality characteristics of the seven 3D structure generators: conversion rate vs. RMS value of the nonhydrogen atoms. (From Ref. 35.)

should provide less bias and a more realistic impression of the performance of the programs under real-world conditions: both are designed to convert millions of structures as fast as possible while maintaining a good quality.

4.4.2. Dataset

The new dataset was obtained from the Cambridge Structural Database using the retrieval program, QUEST, in batch mode. The query was simply a combination of screens that selected error-free organic compounds that had been fully resolved, for which the connection table had been completely assigned, and which had an R factor of less than or equal to 5%. The compounds were exported in SYBYL MOL2 format. This initially gave 36,085 compounds. They were then converted into the MDL SDFile format and compounds with obvious errors in the connection tables were removed. This resulted in 35,556 compounds. From these, all purely inorganic compounds not containing any carbon atom, all compounds outside a molecular weight range between 100 and 750, compounds having more than six rotatable bonds, and compounds with rings larger than nine atoms were removed. These criteria should reduce the dataset to reasonably small and moderately flexible compounds, resulting in a total of 27,688 compounds. Finally, in cases with multiple species in the unit cell, all fragments but the largest one were removed (i.e., counterions, solvents, etc.).

In a last filtering step, all duplicate compounds were removed from the dataset. This finally gave 25,017 compounds. After calculating stereoparity values for stereo centers, this dataset was used for the new evaluation study.

4.4.3. Criteria

The same criteria were used as in the smaller evaluation study above with one minor change: The percentage of reproduced ring geometries (RMS < 0.3 Å) was restricted to flexible rings and was calculated relative to the number of compounds having flexible rings instead of the number of all compounds. This should provide a more realistic figure because it would exclude, for example, easy cases such as phenyl.

4.4.4. Programs

The program versions used for this study were CONCORD 4.0.4 and CORINA 2.6.

4.4.5. Results

Table 3 summarizes the results. The results are shown for both the complete dataset of 25,017 x-ray structures and for the subset of 22,768 compounds converted by both programs. None of the programs crashed or produced any stereo errors. Again, CORINA had a conversion rate near 100% whereas CONCORD converted only 91%. However, CONCORD was faster than CORINA with an average conversion time of 0.014 sec/compound compared to 0.049 sec/compound for CORINA. This relation changes if the smaller subset of 22,768 compounds converted by both programs is considered. Then, the timings for CONCORD and CORINA are 0.013 and 0.033 sec/molecule. Thus, the subset seems to include less time-consuming cases on average. Looking at the structure-related quality criteria, it becomes obvious that by using this dataset, the percentages of compounds fulfilling them are a bit lower for both programs compared with the smaller set of 639 compounds discussed above. This might have to do with a higher flexibility of the compounds in the larger set on average. Again, the relative differences of the percentages are much in favor of CORINA. Both programs seem to perform a robust and reasonably good 3D conversion. Whereas CONCORD performs 2.5–3.5 times faster, CORINA converts a significantly higher rate of structures with a better reproduction of the experimental geometries on average.

Table 3 3-D Structure Generation: Comparison of CONCORD and CORINA Using 25,017 X-ray Structures

	25,017 compounds		22,768 compounds[a]	
	CONCORD	CORINA	CONCORD	CORINA
Conversion rate [%]	91.2	99.7	100	100
Program crashes	0	0	0	0
Stereo errors	0	0	0	0
CPU time [sec/molecule][b]	0.014	0.049	0.013	0.033
$RMS_{XYZ} < 0.3$ Å [%][c]	20	28	20	28
$RMS_{XYZ}^{rings} < 0.3$ Å [%][d]	71	78	71	78
$RMS_{TA}^{chains} < 15°$ [%][e]	32	43	32	42
$CCR > 0.8$ [%][f]	95	98	95	98

[a] Subset converted by both programs.
[b] On an SGI R12000.
[c] Percentage of structures with an RMS deviation of the nonhydrogen atoms of less than 0.3 Å.
[d] Percentage of structures with an RMS deviation of the ring atoms of less than 0.3 Å (flexible rings only).
[e] Percentage of structures with an RMS deviation of the torsional angles in open-chain portions of less than 15°.
[f] Percentage of structures with a close contact ratio of greater than 0.8.

5. CONFORMATIONAL SEARCHING: METHODS AND PROGRAMS

In the following sections, common and frequently used approaches to generating initial starting geometries for a subsequent minimization will be the main focus and will be discussed briefly. Methods for energy and geometry optimization should be described elsewhere.

In general, five different approaches can be distinguished and are applied to explore the conformational space of a molecule: systematic searches, rule-based and data-based approaches (model building), random methods, genetic algorithms, distance geometry, and simulation methods. Some of the basic principles and ideas behind these concepts have already been described in the Secs. 2 and 3 of this article. In the following, the application of these concepts to conformational analysis and searches will be discussed.

5.1. Systematic Searches

Systematic methods generate conformational diversity by repetitively changing the molecular coordinates in a predefined, regular, and stepwise manner. One of the most widespread as well as oldest technique is the so-called grid search. The torsion angle of each rotatable bond of an input structure is rotated by a constant increment of $360°/n$ ($n = 1,2,3,\ldots,360$), whereas bond lengths and bond angles are kept fixed. This method is easily applicable to acyclical structures, but runs into problems when applied to ring systems because most of the generated conformations will not fulfill the ring closure restriction. Thus, cyclical and acyclical parts of a molecule are treated separately:

> **Acyclical portions.** Typically, systematic searches are using increments of 30° (n = 12) or 60° (n = 6) for rotating the torsion angles along the chains. The search problem can be represented by a so-called search tree, a method to graphically describe a system that can adopt different states. A tree consists of nodes that are connected by edges, and each node represents a certain state of the system. Fig. 17 shows such a search tree for a molecule with two rotatable bonds (e.g., n-pentane) and three discrete values for each torsion (n = 3). The search starts at the root node and each terminal node corresponds to a single conformation of the molecule. Starting with the root node and following the edges along the nodes marked with A and E in Fig. 17, the derived conformation would exhibit torsion angles of $\tau1 = 60°$ for the first and $\tau2 = 180°$ for the second rotatable bond.
>
> An advantage of a search tree is that several well-proven algorithms are available, which efficiently process this graph-theoretical representation to enumerate all possible conformations. The most commonly used strategy is a depth-first search (i.e., to move "downward" along the edges until a terminal node is reached, which represents a single conformation). A backtracking procedure is then applied to move "upward" the edges to find the next terminal node (e.g., back from node E to A in Fig. 17). After this operation, the depth-first search will move down to terminal node D, if not yet perceived, or F, respectively.
>
> Systematic grid searches represent a combinatorial problem and this is a major drawback of this technique. The number of possible conformations N

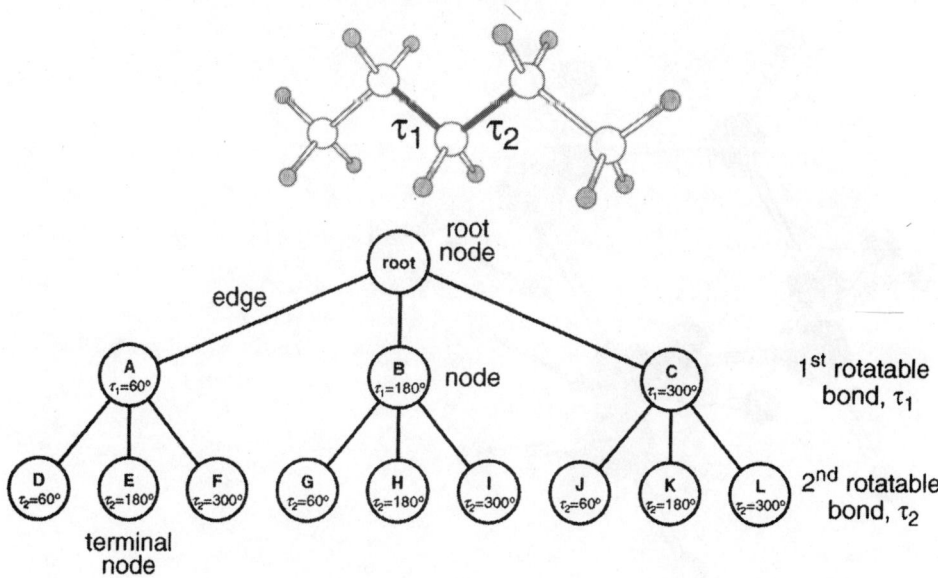

Figure 17 Tree representation of a systematic conformational search (grid search).

increases exponentially with the number of rotatable bonds, $k(N = (360°/n)^k$ ("combinatorial explosion"). Applying a grid of 30° ($n = 12$), a system with five rotors ($k = 5$) will result in a total number N of 248,832 conformations, which have to be minimized and compared in the following rather time-consuming steps. A coarser grid may lead to the loss of important geometries and many starting structures will fall into the same minimum after optimization.

To enhance the performance of grid searches, two advantages of the tree representation can be utilized. Firstly, parts of the tree that correspond to conformations with atom clashes or close contacts can be detected and pruned. Fig. 18 illustrates this for a conformation of n-heptane. Neither changes to the torsion τ_n nor to τ_{n-1} will avoid the steric clash exhibited at the marked atoms. Thus, the search tree can be pruned above the node of τ_{n-1}.

Secondly, the perception of symmetries within the molecule under investigation may also partly restrict the number of nodes and edges of the search tree that have to be processed.

Ring Systems. The methods described above can also be applied to ring systems by cutting one ring bond and treating the structure as "pseudoacyclical." Additionally, each generated conformation has to be checked for several intramolecular parameters before it is submitted to the optimization procedure. The most important parameter is the ring closure restriction (i.e., the distance between the two atoms, which are connected by the cut bond, has still to be within a range of a bond length). Fig. 19 shows two carbon skeletons of cyclohexane where the bond between the two marked atoms was cut. The left-hand conformation can be reconstructed into a real ring structure, whereas the distance between the marked atoms in the right-hand conformation is outside the range to form a ring bond.

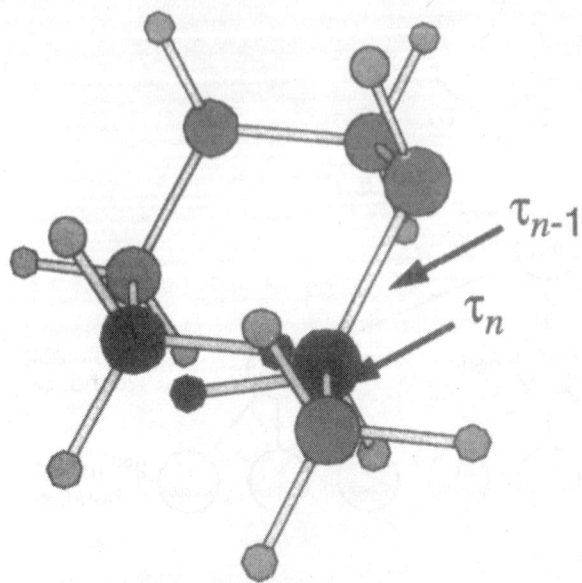

Figure 18   Close contacts in a conformation of n-heptane.

Clearly, in conjunction with the check for steric crowding, the consideration of the ring closure restriction results in much fewer numbers of conformations for geometry optimization compared to acyclical structure. Additionally, a fine grid and large distance ranges have to be chosen to generate enough starting geometries. Most of them are chemically unreasonable, or populate the same energy minimum after optimization.

A program system that strictly uses the methods described above is MULTIC, developed by Lipton and Still [36]. MULTIC generates starting geometries for a following geometry optimization procedure.

An interesting approach to treating ring systems in systematic searches was presented by Gotõ and Õsawa [37] already in 1989 and implemented in the program,

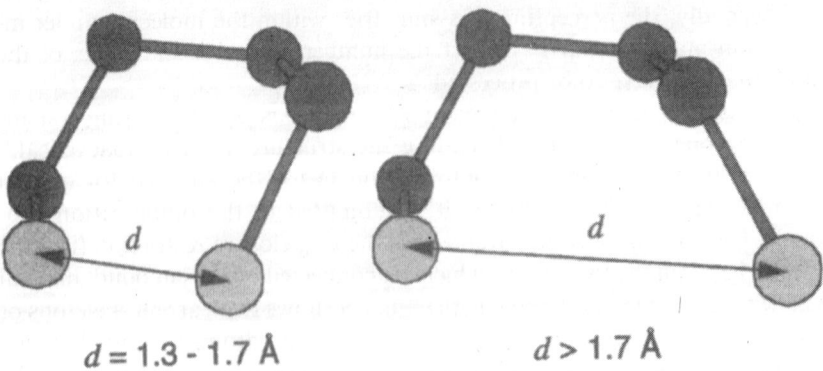

$d = 1.3 - 1.7 \text{ Å}$ $d > 1.7 \text{ Å}$

Figure 19   Consideration of the ring closure restriction in two conformations of cyclohexane.

CONFLEX. The so-called *corner flapping* algorithm does not need to cut ring bonds and therefore only generates real ring structures. The algorithm systematically inverts the geometry of each ring atom, which all are considered as corners of a polygon, by flapping them vertically to the mirrored position on the other side of a plane, which is defined by the adjacent ring atoms. Fig. 20 illustrates this process.

After a check for steric problems, the structures can directly be submitted to minimization.

Several commercially available program packages such as SPARTAN or the Cerius^2 system include routines or modules to systematically search the conformational space of small-sized and medium-sized molecules [38,39]. Additionally, some methods offer the possibility to generate the conformations under geometrical constraints to output only those structures that, for example, fulfill a given pharmacophore model within a certain range.

5.1.1. Confirm (Catalyst)

A widely used program for the generation of sets of conformation is ConFirm, which is integrated in the Cerius^2 and Catalyst software suites [39,40]. The basic algorithms implemented in ConFirm were developed by Smellie et al. [41]. The program provides conformational searches in a *best* and a *fast* mode. As the names already imply, the best search is exploring the conformational space in a more thorough fashion, whereas the fast mode is a more approximate technique requiring significantly less CPU time to process large datasets of molecules (e.g., 3D databases). Both methods significantly differ in the methods that are used to explore the conformational space. The fast mode applies a modified systematic search technique and, therefore, is briefly described in the following. The best search uses a distance geometry approach (see also Sec. 3.4). Thus, ConFirm's best mode is described in Sec. 5.5, where the basic principles of the application of distance geometry to conformational searching are discussed.

ConFirm's fast mode treats the ring systems and the acyclical parts of an input structure separately. Predefined ring conformations (templates) that are stored in a library are used to build appropriate ring geometries. For the open-chain portions, a modified systematic search technique is applied. A so-called *quasi-exhaustive* search is performed in torsion space by applying a maximum number of six predefined values of a torsion angle (grid points) depending on the hybridization states of the two atoms

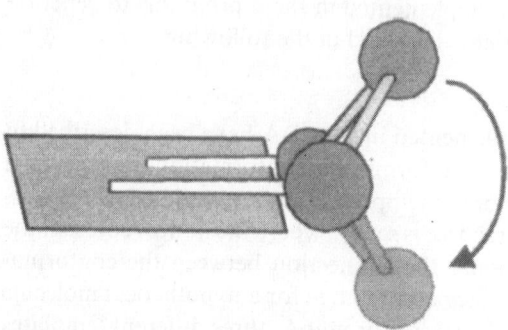

Figure 20 Corner flapping procedure for the treatment of ring structures in systematic searches.

that are connected by a rotatable bond. The major drawbacks of this relatively coarse grid (poor coverage of conformational space, loss of important geometries) are compensated by extending the fixed grid points to a permitted range of possible dihedral angle values ("fuzzy grid") as following. Each newly generated conformation is checked for atom clashes or close contacts. Usually, if such a situation is detected, grid search techniques reject this conformation and move to the next grid point. The quasi-exhaustive search now changes the torsion angle, which caused the steric interactions within the range between the two adjacent grid points ($\pm 60°$). A simple distance function with the torsion angle under consideration as variable is used to perform a minimization to resolve the van der Waals clashes. The minimization stops if all steric interactions have been eliminated, if the deviation to the original value of the changed torsion angle is larger than $\pm 60°$ (i.e., one of the adjacent grid points is reached), or if the distance function shows a global minimum. Thus, the number of conformations that are rejected due to van der Waals clashes in a "true" systematic search can be reduced and the accessible conformational space is explored more efficiently.

In the next step, the conformations obtained by the quasi-exhaustive search are refined in a restricted CHARMM force field. The restrictions are applied to the torsion angles, which are allowed to change only within a given range during the optimization procedure. This technique prevents the generation of duplicate conformations (i.e., that starting structures will populate the same minimum after the optimization). In addition, an estimate of the global energy minimum is calculated on the fly and only those conformations that have an internal energy less than a preset value (default value of 20 kcal/mol) after the optimization step are accepted. Furthermore, the number of the remaining conformations can be reduced by a simple heuristic [42]. The procedure is based on interconformational distances in Cartesian space and tries to maximize the conformational diversity in the generated subsets of conformations.

By default, ConFirm's fast search outputs a maximum number of 256 conformations per molecule within an energy range of 20 kcal/mol with respect to the estimated global minimum. Some results obtained with ConFirm are discussed in Sec. 6.

5.2. Rule-Based and Data-Based Approaches

The basic concepts of rule-based and data-based methods have already been described in Sec. 2.3 and several program systems utilizing this approach are discussed in Sec. 3.2. Some of these systems offer the possibility to output several geometries of one molecule. The extensions that have been implemented in these programs to generate reasonable sets of conformations are briefly discussed in the following.

5.2.1. COBRA

The basic principles and algorithms implemented in COBRA have been described in Sec. 3.2.1. As already mentioned, COBRA forms the conformational space of a molecule by the combinatorial product of the templates that are assigned to the conformational units. Thus, an internal search tree is set up, where the nodes represent the conformational units. The edges symbolize the connection between the conformational units and the templates. Fig. 21 shows a search tree for a hypothetical molecule consisting of two conformational units A and B. For unit A, three different templates (A1, A2, and A3) have been found in the library and two different templates (B1 and B2) are assigned to unit B.

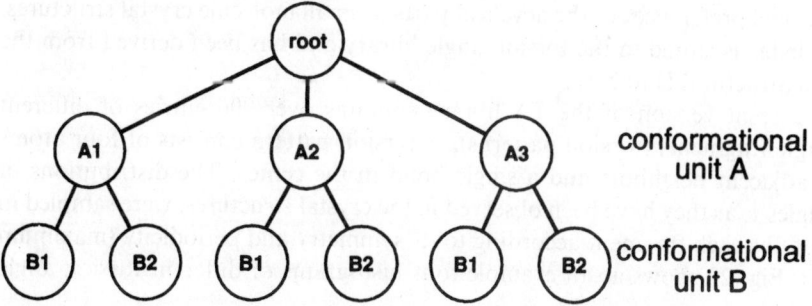

Figure 21 Conformational search tree of COBRA for a hypothetical molecule consisting of two conformational units (A) with three templates and (B) with two templates.

The so-called A* algorithm is used to identify the lowest-energy conformation by processing the search tree. The A* algorithm is based on a cost function and is able to determine the *least-cost* path (or *best-path-first* manner) from the root node to a terminal node. This requires that a cost is assigned to each edge in the search tree. A reasonable cost function for conformational analysis is the internal energy of the conformational template. Combining the conformational templates that lie on the *least-cost* path from the root to the terminal node results in the lowest-energy conformation of the molecule.

In principle, the internal search tree represents the conformational space that can be explored with COBRA. To generate a set of diverse conformations, COBRA uses the A* search method in combination with a clustering technique. The complete strategy is sometimes referred to as *direct clustering search* algorithm. The A* algorithm is used in a *maximum-cost* manner and, instead of conformational energies, the distances in torsion space of the conformations are taken as cost functions. Thus, the structure that differs most from the lowest-energy conformation can be identified and is stored as the second conformation. In the next step, the structure that is most dissimilar to the first two conformations is searched. This procedure is repeated until a user-defined number of conformations is found, or until no nodes of the tree are left. To output only conformations that have an energy within a certain range with respect to the lowest-energy conformation, an energy cutoff can be set.

5.2.2. MIMUMBA

Klebe and Mietzner [14] and Klebe et al. [15] developed the conformational analysis package, MIMUMBA, which generates ensembles of conformations for small-sized and medium-sized typical druglike compounds. MIMUMBA is a rule-based and data-based approach and treats the cyclical and acyclical parts of a molecule separately.

The program starts with a given 3D model. In the first step, the molecule under consideration is fragmented into ring systems and open-chain portions. A set of different ring geometries is obtained either by the algorithm implemented in CORINA (see Secs. 3.2.3 and 5.2.3) or by the program SCA (see Sec. 3.1.3). Both programs have been interfaced to MIMUMBA.

To explore the conformational space of the acyclical parts of a molecule, MIMUMBA uses a set of rules and data that results from a statistical analysis of the conformational preferences of the acyclical parts in small molecule crystal structures. This knowledge is stored in the torsion angle library and has been derived from the Cambridge Structural Database.

The current version of the TA library contains over 900 entries of different torsion angle fragments (torsion patterns). A torsion pattern consists of four atoms and their adjacent neighbors and a single bond in the center. The distributions of torsion angles τ, as they have been observed in the crystal structures, were sampled in steps of 10° for each fragment according to its symmetry and periodicity (maximum of 5–355°). Fig. 22 shows as an example four histograms of different torsion angle patterns.

Additionally, a set of tripeptide sequences was extracted from the PDB and analyzed to derive torsion histograms describing the conformational preferences of peptidic portions in ligands. Implicitly, these histograms contain the information about the conformational behavior of molecules in a structured molecular environment with varying intermolecular directional forces and dielectric conditions in the

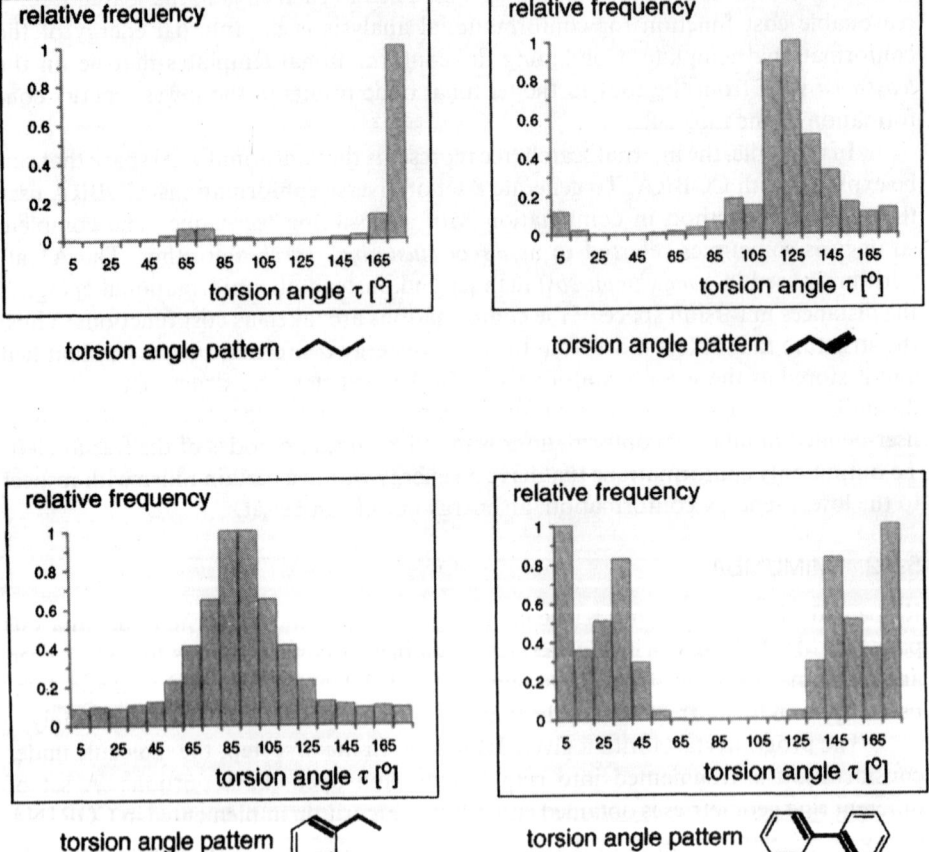

Figure 22   Histograms of angle distributions in torsion angle patterns as found in crystal structures contained in the CSD.

different crystal packings. Extensive studies have shown evidence for a correlation between ligand conformations at the binding site of a biological receptor (biologically active conformations) and their crystal structures [43]. Thus, MIMUMBA generates observed conformations likely to be of biological relevance, rather than sampling local energy minima or randomly enumerating conformations.

After determining the rotatable bonds of an input structure, a histogram taken from the TA library is assigned to each rotatable bond. If more than one entry of the TA library can be found for a rotor (maximum of nine), all matching distributions are consecutively superimposed and averaged. Thus, specific histograms can be obtained for a particular rotatable bond under consideration. Secondly, the histograms are transformed into empirical potential energy functions. The idea behind this is that a low frequency in the distribution of a certain angle value is correlated with a high-energy content for the structural fragment exhibiting this geometry and vice versa. This behavior can be described with the equation $E(\tau) = -A \ln f(\tau)$, where $E(\tau)$ is the energy value at torsion angle τ, A is an adjustable parameter, and $f(\tau)$ is the relative frequency of torsion angle τ [44]. The potential values $E(\tau)$ are calculated in steps of $10°$ and a smooth functional form is obtained by interpolating with a spline function between these steps. Fig. 23 illustrates the derivation of the empirical potential energy function for the fragment CN(H)C(H)(H)C.

The empirical energy function is used to select a set of preferred dihedral angles for the torsion fragment under consideration. In steps of $15°$, starting values are collected in ranges of τ where the potential exhibits a minimum below a preset energy threshold. However, two adjacent preferred torsion angles, which are used to build the conformations in the following step, have to have a minimum distance of $30°$. Prior to

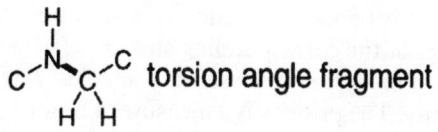

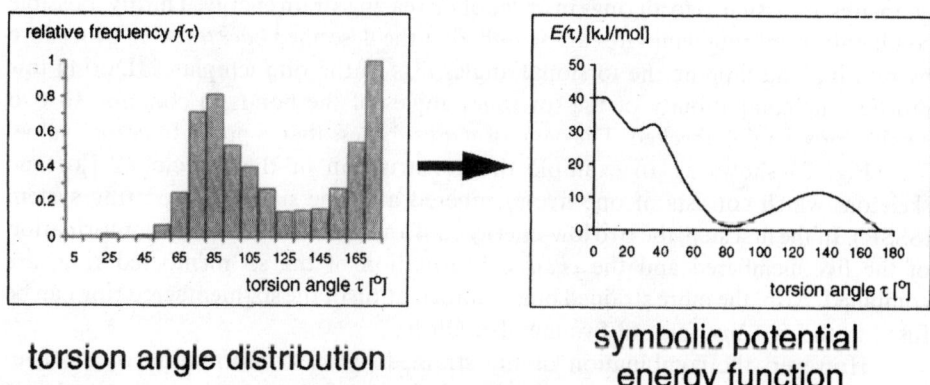

Figure 23 Derivation of a symbolical potential energy function from the torsion angle distribution of a torsion fragment.

the generation of the conformations, the number of possible conformations is enumerated, taking into account all possible combinations of different ring geometries and selected open-chain torsion angles. If this number exceeds a certain value, the selection of the dihedral angles is repeated with a coarser grid and a lower energy threshold.

Finally, the various cyclical and acyclical conformational fragments are combined. Those conformations that have an internal energy lower than a user-defined value are submitted to a geometry optimization step of the open-chain portions by applying the empirical energy function. Additionally, van der Waals potentials are used to avoid steric crowding or atom clashes. Furthermore, energetically and structurally identical and quite similar conformations are rejected.

MIMUMBA was tested on a set of 28 experimentally determined protein ligands (taken from the PDB) to evaluate how well the program can reproduce biologically active conformations [15]. The molecules in the test set contained from 3 to 13 rotatable bonds. For each generated conformation, the RMS deviation of the atom positions in Cartesian space (RMS_{XYZ} [Å]) to its bioactive geometry was calculated. The conformations generated by MIMUMBA with the smallest RMS_{XYZ} deviations were found within a range of 0.29–1.67 Å.

5.2.3.   CORINA and ROTATE

The basic methods of the 3D structure generator, CORINA, are described in Sec. 3.2.3. CORINA uses a table of allowed single-ring conformations (ring templates) for small-sized and medium-sized ring systems consisting of up to eight ring atoms. These templates are stored as lists of torsional angles for each ring size and each number of unsaturations in the ring, ordered by their conformational energy. Internally, CORINA performs a conformational analysis to find one low-energy conformation for single rings as well as for fused and bridged systems (see step 3 in Sec. 3.2.3). The results from this analysis can also be used to generate a set of conformations for cyclical systems containing rings of the corresponding size as detailed in the following.

Firstly, after determining the smallest set of smallest rings, all rings are ordered according to their priority. The priority is a measure of how central the ring lies within the complete ring system. Central rings are processed first because they have the highest number of interactions with neighboring rings. Secondly, lists of possible ring templates are assigned to all rings in order of increasing strain energy. Thirdly, possible combinations of ring templates are searched in the described *backtracking* procedure by rotating and flipping the torsional angles lists of the ring templates. During this process, the compatibility of the torsional angles of the bonds in common to two neighboring rings is checked. The deviation must be less than a preset tolerance value.

Fig. 24 shows as an example the construction of the bicyclo[3.2.1]octane skeleton, which consists of one five-membered and one six-membered ring system (SSSR). In the first step, the two low-energy conformations, the *envelope* conformation of the five-membered and the *chair* conformation of the six-membered ring, are combined. Also, the more strained boat conformation of the six-membered ring can be fused to the *envelope* form of five-membered ring.

However, the combination of the strained *planar* conformation of the five-membered ring to any of the six-membered ring conformations exceeds the preset threshold of torsional compatibility of the templates and the combination is rejected (see Fig. 25).

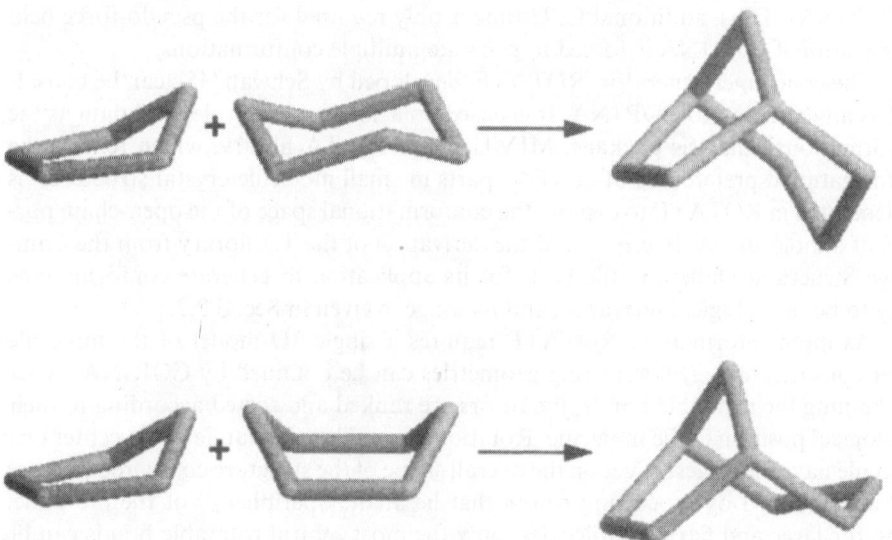

Figure 24 Combination of ring templates to generate conformations of a bicyclo[1.2.3]octane skeleton.

After all allowed combinations of the templates are found, the conformations are finally ordered by a simple symbolical energy function that takes into account the strain energies of the single-ring conformations; the strain contribution of exocyclical substituents in axial, 1,3-diaxial, or 1,2-diequatorial positions; and the deviation of the torsion angles at ring fusion bonds. Furthermore, this energy ranking allows the identification of the lowest-energy conformation, or the generation only of conformations within a certain energy range with respect to the lowest-energy conformation. In the last step of processing, the small-sized and medium-sized ring systems of all ring geometries are optimized by using the pseudo-force field procedure (see step 6 in Sec. 3.2.3).

Finally, the acyclical parts of the molecule are added to the optimized ring geometries and the complete conformations are checked for overlapping of atom or close contacts as described in Sec. 3.2.3.

As already mentioned, the conformational analysis procedure described above is also performed to identify and output the lowest-energy conformation (default mode

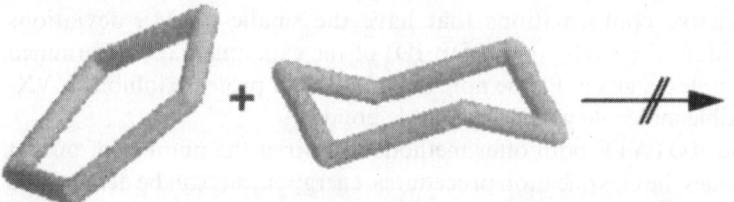

Figure 25 Unallowed combination of two-ring templates to generate a conformation of a bicyclo[1.2.3]octane skeleton.

of CORINA). Thus, additional CPU time is only required for the pseudo-force field optimization if CORINA is forced to generate multiple conformations.

The conformer generator, ROTATE, developed by Schwab [45], can be considered as an extension to CORINA. It is based on a similar set of rules and data as the conformational analysis package, MIMUMBA. The TA library, which reflects the conformational preferences of acyclical parts in small molecule crystal structures, is implemented in ROTATE to explore the conformational space of the open-chain portions of a molecule. A description of the derivation of the TA library from the Cambridge Structural Database, the basis for its application to generate conformations likely to be of biological relevance, and its usage is given in Sec. 5.2.2.

As input information, ROTATE requires a single 3D model of the molecule under consideration. Different ring geometries can be obtained by CORINA. After determining the rotatable bonds, the rotors are ranked and sorted according to their topological position in the molecule. Rotation around bonds that lie in the center of a molecule have the largest effect on the overall shape of the structure compared to those that are obtained by processing rotors that lie at the "periphery" of the molecule. Thus, for large and flexible molecules, only the most central rotatable bonds can be processed optionally. In addition, bonds to terminal groups and atoms are not considered to be rotatable.

After assigning appropriate torsion angle histograms from the TA library to each rotor, the empirical potential functions are derived as described above. The starting dihedral angles to build the conformations are selected as follows. A regular grid of 30° is applied and the potential energy values for the 12 angle values are calculated. Then, for each rotor, six torsion angles values that lie below a preset energy threshold are selected. These six values are regarded to be the most preferred torsion angles of the rotatable bond under consideration and are used as initial values to build the conformations (maximum of six per rotor). A *depth-first* search generates all possible combinations of the initial torsion angles of all rotatable bonds.

After a new conformation has been generated, each rotatable bond is geometry-optimized by applying the empirical energy function and only accepted if no steric problems can be detected. In addition, duplicate conformations are rejected. To restrict the number of geometries output, similar conformations can be combined to classes. The classification is based on the RMS deviation between the conformations, either in Cartesian or torsion space. The RMS threshold, which decides whether two conformations belong to the same class, is adjustable by the user and each class is finally represented by one conformation.

CORINA and ROTATE have been used to test whether the methods inherent are able to reproduce biologically active conformations. Fig. 26 shows three superimpositions of CORINA-generated and ROTATE-generated models (marked in grey) with biologically active conformations that have the smallest RMS deviations (RMS_{XYZ}). In addition, the PDB code (PDB ID) of the experimentally determined receptor–ligand complex is given. Please note that the HIV-1 protease inhibitor, VX-478, is a highly flexible molecule with 12 rotatable bonds.

CORINA and ROTATE both offer methods to restrict the number of output conformations. Besides the classification procedures, energy cutoffs can be defined and the number of rotors to be processed can be specified and weighted by the topological position within the molecule. Thus, the conformational space of a molecule can be explored to a degree required by the user.

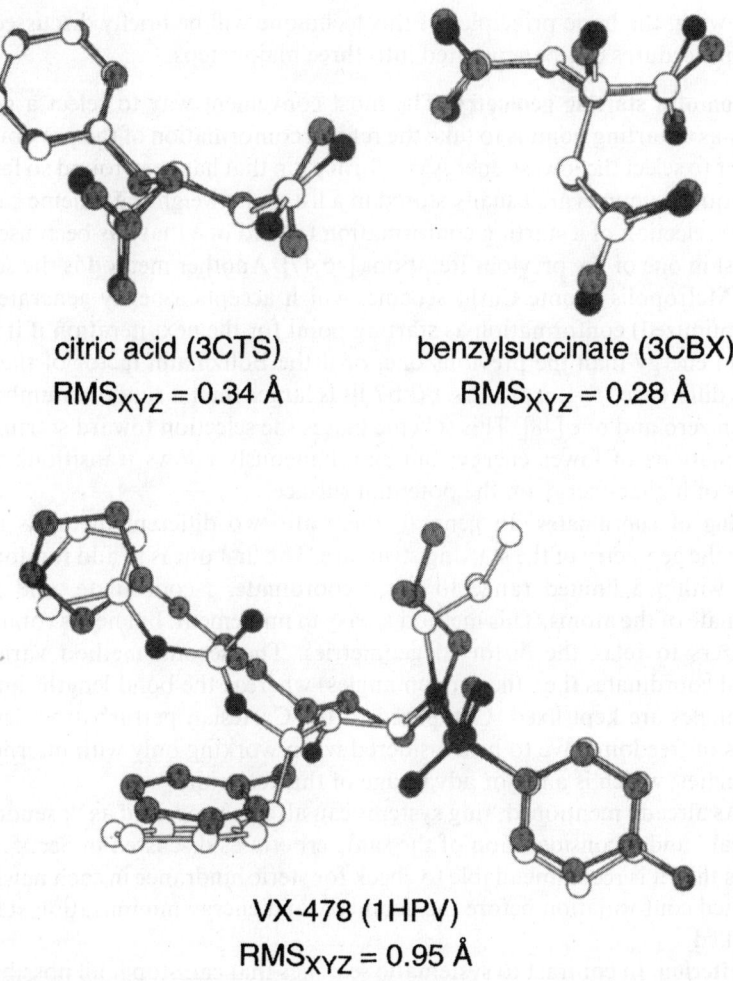

citric acid (3CTS)   benzylsuccinate (3CBX)
$RMS_{XYZ} = 0.34$ Å $RMS_{XYZ} = 0.28$ Å

VX-478 (1HPV)
$RMS_{XYZ} = 0.95$ Å

Figure 26 Superimposition of CORINA and ROTATE generated models (marked in grey) with biologically active conformations with the smallest RMS deviations.

5.3. Random Methods

As the name already implies, random methods generate a set of conformations by repetitively and randomly changing either the Cartesian coordinates or the internal coordinates of a starting geometry of the molecule under consideration. The changed structure is then geometry-optimized, compared to all previously generated conformations, and checked for its uniqueness. After choosing a new starting structure for the next iteration, the cycle starts again. Thus, random or stochastic techniques follow the general scheme of conformational searching, as illustrated in Fig. 4 in Sec. 2.2. In contrast to systematic searches, random methods explore the conformation space in an unpredictable fashion. A major advantage of random methods is that completely different regions of the energy surface may be investigated from one iteration step to the next. Nevertheless, some of the algorithms used for systematic methods can also be applied to stochastic searches such as the "pseudoacyclical" treatment of ring structures.

In the following, the basic principles of this technique will be briefly discussed. Random search procedures can be separated into three major steps:

1. **Selection of a starting geometry.** The most convenient way to select a geometry as a starting point is to take the refined conformation of the previous steps, or to select the lowest-energy conformation that has been found so far. All unique structures are usually stored in a list and a weighting scheme can bias the selection of a starting conformation toward one that has been used the least in one of the previous iterations [46,47]. Another method is the so-called Metropolis Monte Carlo scheme, which accepts a newly generated (and optimized) conformation as starting point for the next iteration if it is lower in energy than the previous one, or if the Boltzmann factor of their energy difference $(f = \exp[(-\Delta E)/(kbT)])$ is larger than a random number between zero and one [48]. This scheme biases the selection toward starting conformations of lower energy, but simultaneously allows transitions to regions of higher energy on the potential surface.

2. **Changing of coordinates.** In general, there are two different methods to change the geometry of the starting structure. The first one is to add random values within a limited range to the x-coordinate, y-coordinate, and z-coordinate of the atoms. This method is easy to implement, but needs robust optimizers to relax the distorted geometries. The second method varies internal coordinates (i.e., the torsion angles) whereas the bond lengths and bond angles are kept fixed. Compared to the Cartesian perturbation, less degrees of freedom have to be considered when working only with internal coordinates, which is a major advantage of this technique.

   As already mentioned, ring systems can also be processed as "pseudo-acyclical" under consideration of the same criteria as discussed in Sec. 5.1. Besides this, it is recommendable to check for steric hindrance in each newly generated conformation before the geometry and energy minimization step is invoked.

3. **Stop criterion.** In contrast to systematic searches that can stop if all possible combinations of different torsion angles have been enumerated, random methods do not have a "natural" endpoint. In most cases, the search is stopped if no new conformations are found (i.e., if several conformations have been generated several times).

A program that uses the Cartesian method to randomly change the starting geometries in each cycle is the RIPS (Random Incremental Pulse Search) system developed by Ferguson and Raber [49]. RIPS generates conformations that are relaxed in an MM2 force field. An internal coordinate random search method for finding low-energy conformation is implemented in the Monte Carlo Multiple Minimum (MCMM) package by Chang et al. [47] as well as in the Cerius^2 [39] conformer module (including the Metropolis selection criterion).

5.4. Genetic Algorithms

Genetic algorithms are a class of robust optimizers that simulate the process of evolution in nature [50]. Their objective is to find optimal solutions for a given problem. Before discussing their application to conformational search problems, the basic ideas and methods of genetic algorithms are briefly described. Fig. 27 illustrates one optimization step. Genetic algorithms do not start with a single individual, but with a population of individuals. Each member of the population is coded by a *chromosome* and represents a possible solution to the problem that has to be optimized. Some of these individuals provide a better solution to the problem under consideration than others (i.e., their *fitness* is higher). After the fitness of each member of the population has been calculated, a population is selected with a bias toward the fitter individuals (*selection*). Finally, the chromosomes are changed by genetic operators such as crossover and mutation to create a new population on the basis of the fittest members of the parent generation.

These steps comprise one optimization cycle. Such a cycle is usually repeated in a predefined number of iterations, which is called one experiment. In this way, a set of optimized solutions is obtained after one experiment.

In the application of genetic algorithms to conformational searching, the individuals of a population represent molecular conformations. The aim is to generate a set of diverse and low-energy geometries. Genetic algorithms can be applied to conformational searching in the following manner:

1. **Genetic code and operators.** Each individual of a population is coded by a chromosome. Chromosomes have to be decoded to obtain a proper phenotype and to calculate the fitness of each member. Usually, bitstrings are

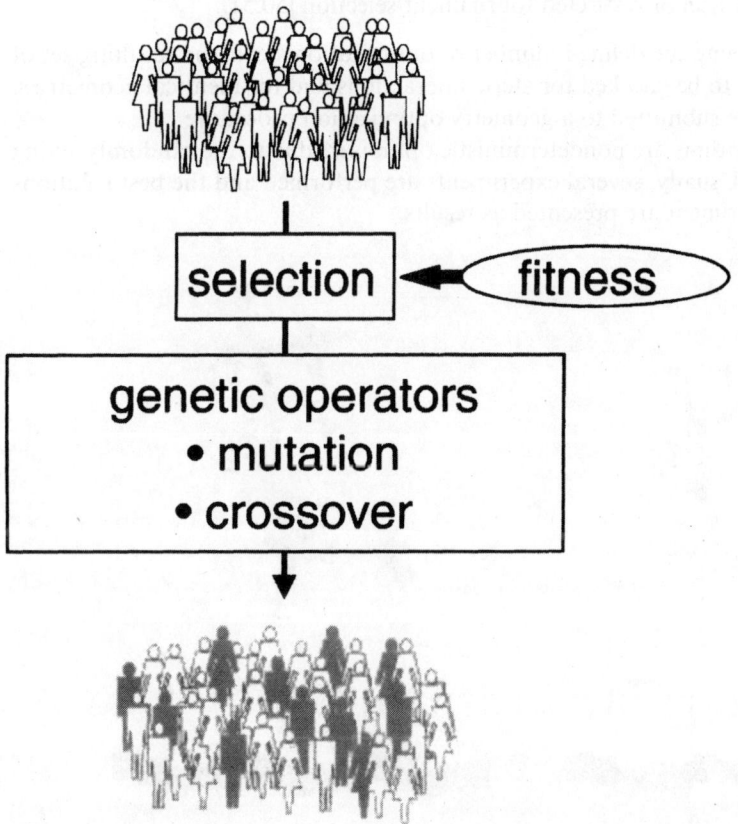

Figure 27 Basic principles of one optimization cycle of a genetic algorithm.

used to code the torsion angles of the rotatable bonds. Fig. 28 illustrates this for two conformations of n-hexane. In the first population (generation), the chromosomes can be randomly initialized with values of zero and one.

The most commonly used genetic operators are mutation and crossover. Although mutation randomly inverts bits of a chromosome of a single individual, crossover interchanges equally sized parts of chromosomes (bitstrings) between two different individuals. The crossover position is chosen in a random manner (see Fig. 29).

2. **Selection and fitness.** The driving force of the optimization is selection, which reflects Darwin's theorem of the "survival of the fittest." During several optimization steps and populations, the fittest individuals providing the best solutions for the given problem will survive. The fitness of each member of a population is calculated by the fitness function and the selection moves the individuals from one generation into the next generation based on their relative fitness. An appropriate fitness function for conformational search purposes is the conformational energy as obtained, for example, from a single point force field calculation. The lower the internal energy, the fitter is the conformation, and the higher should be the probability to select this conformation and to move it into the next generation. Several algorithms are described in the literature, which ensure that fitter individuals are selected with a higher probability as the parent for the following generation, such as roulette wheel or restricted tournament selection [50,51].

After performing the defined number of optimization steps, the resulting set of conformations has to be checked for steric interactions and for identical geometries. Finally, they can be submitted to a geometry optimization procedure.

Genetic algorithms are nondeterministic optimizers due to the randomly acting genetic operators. Usually, several experiments are performed and the best solutions of each single experiment are presented as results.

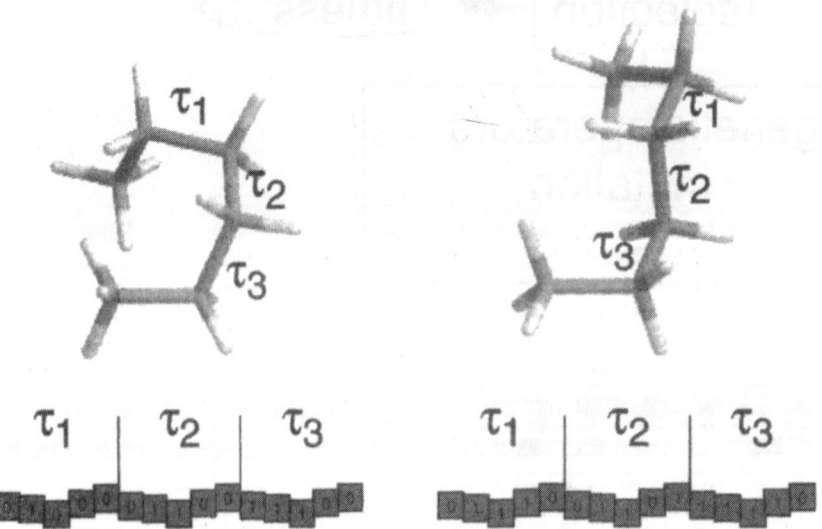

Figure 28   Two conformations of *n*-hexane and their coding of the torsion angles.

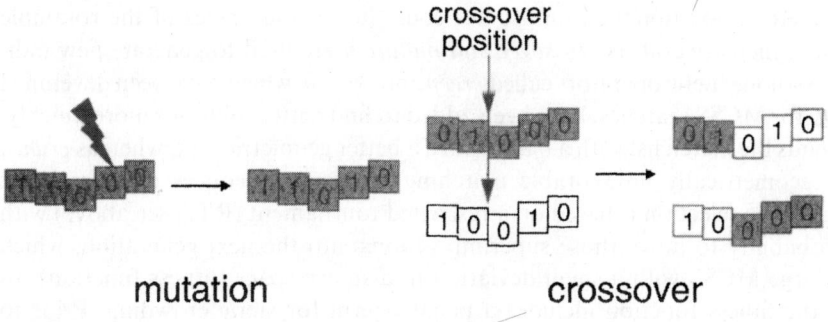

Figure 29 The genetic operators mutation and crossover in genetic algorithms.

An interesting application of genetic algorithms to conformational analysis is published by Nair and Goodman [52] and can be regarded as a hybrid method. In their approach, they combine the nondeterministic optimization of a genetic algorithm with the directed method of a force field relaxation (hybrid approach). Prior to selection, all conformations of the parent generation are optimized using an MM2 force field and the chromosomes are adapted accordingly. Then, the genetic operators are applied to form a second population. Both populations are used for selection by a roulette wheel procedure in the subsequent step. Therefore, individuals for the next generation are selected from a larger genetic pool, which increases the number of fitter solutions. The method was applied to find low-energy conformations of long-chain alkanes as well as for pathotoxic compounds.

The program package, SPARTAN, contains a module to explore the conformational space of small-sized and medium-sized molecules with a genetic algorithm [38]. After generating a user-defined number of conformations with a genetic optimization procedure, the structures can be submitted to geometrical refinement by quantum or molecular mechanics methods.

Furthermore, the program system, GAMMA (Genetic Algorithm for Multiple Molecule Alignment), explores the conformational space by applying a genetic algorithm [6,53]. GAMMA was designed to superimpose sets of different molecules (e.g., a series of ligands binding to the same biological receptor). The objective of the program is to determine structural similarities of the molecules under consideration (i.e., their 3D maximum common substructure, 3D MCSS). In addition, the program was extended to perceive physicochemical similarities of different compounds. This is an important task in structure-based drug design because the evaluation of such similarities can provide knowledge about the important steric and electronic features molecules have to expose for the binding to a certain protein whose structure is not yet known. In addition, pharmacophore models can be derived on the basis of these information. GAMMA combines a genetic algorithm with a numerical optimization procedure (hybrid method) to find optimal superimpositions.

One individual of a population represents a possible superimposition of the molecules under consideration. The chromosomes code the atoms of the different structures, which are mapped onto each other (match list) on the basis of either identical atom types or similar atomic properties (e.g., partial atomic charges) within a given range. The molecules are treated conformationally flexible during superimposition. Therefore, additional chromosomes code the torsion angles of the rotatable bonds. The genetic operators *crossover* and *mutation* are used to generate new individuals. Two nongenetic operators called *creep* and *crunch*, which have been developed specifically for MCSS searches, have been added to find better solutions more quickly. *Creep* extends the match list with a bias toward a better geometrical fit, whereas *crunch* seeks for geometrically unfavorable matching atoms and reduces the match list accordingly. The selection is based on a restricted tournament (RTS, see above) with a high probability to move those superimpositions into the next generation, which exhibit a large MCSS with a small deviation in distance space (fitness function). In addition, the fitness function includes a penalty term for steric crowding. Prior to selection, a directed tweak procedure (numerical component of the hybrid approach) improves the quality of the superimpositions by mapping the torsion angles of the rotatable bonds of the matched structures [25]. Thus, better solutions (geometrical fits) will have an advantage to get selected for the next generation.

After a defined number of optimization cycles and experiments are performed, a set of superimpositions with varying size of mapped substructures and different qualities of geometrical fits is obtained. The major aim of an MCSS search is to find a substructure size as large as possible with a deviation of the positions of the superimposed atoms as small as possible. As an increasing substructure size decreases the geometrical fit, two contradictory criteria have to be optimized. GAMMA solves this problem by applying a so-called Pareto optimization, which is able to identify for each possible substructure size the best geometrical fit [54]. Finally, this Pareto optimal set of superimpositions is presented as result.

In the context of conformational searching, GAMMA generates those geometries of a molecule that fit best in terms of geometrical or physicochemical properties to a set of different compounds. These geometries may not correspond to conformations in an energy minimum, but may reflect the spatial and electronic situations at the binding site of a biological receptor.

5.5. Distance Geometry

The basic principles of the distance geometry approach have been described in Sec. 3.4. There, it was mentioned that the embedding procedure can produce several molecular geometries that do not violate the preset distance restrictions. Therefore, distance geometry per se can be regarded as a conformational search method.

To overcome the lack of correspondence of a distance matrix to a conformation and its internal energy, Crippen [10] introduced the concept of *energy embedding*. This enables distance geometry approaches to generate initial distance matrices, which fulfill the preset distance restrictions and, when translated into Cartesian space, to obtain low-energy conformations. The quality of the generated conformations is improved by the inclusion of rotational degrees of freedom for open-chain portions during the embedding procedure and by repeated triangle smoothing.

The time-consuming step in distance geometry programs is the extensive numerical triangle smoothing, which has to be performed once for a given input structure. The following search in conformational space as well as the generation of conformations are quite fast, but the overall quality of the molecular models is rather crude. In addition, subsequent geometry optimization by force field methods is favorable, but often leads to identical or very similar conformations [55].

A strength of the distance geometry approach is that the conformation generation process can be augmented with experimental data (e.g., interatomic distances), which have been derived from 2D NMR spectroscopy experiments. The generated structures will then fulfill the observed conformational restrictions.

As already mentioned in Sec. 5.1.1, ConFirm's best mode uses a distance geometry approach to generate initial conformations for a subsequent geometry optimization and energy minimization step. Thus, also the cyclical portions are fully included in the search procedure and are not taken from a library of predefined ring geometries.

Again, a modified version of the CHARMM force field is used for geometry optimization and energy minimization. For a broad coverage of the low-energy conformational space, a method called *poling* is applied during the force field relaxation (only in the best mode of ConFirm) [42]. Poling biases the conformational sampling toward geometries, which are structurally far from a local minimum but are energetically quite near to each other and prevent the generation of duplicate conformations by geometry optimization. The algorithm adds huge artificial barriers (poles) at minima on the energy surface, which are already populated by a conformation. Fig. 30 illustrates this technique. In the upper diagram, conformation 1 (marked with a circle) is already optimized and populates a minimum on the energy surface. Conformation 2 would fall into the same minimum when relaxed in the force field. Therefore, a pole is added symmetrically at the place where conformation 1 is located, which can be seen in the lower diagram. Thus, conformation 2 will fall into the new, "artificial" minimum. In the next step, a pole is added starting at the minimum of conformation 2 (not shown in Fig. 30).

The poling algorithm forces the search procedure to explore the low-energy regions of conformational space more extensively. It generates conformations that may not correspond to an energy minimum but might be of interest in the context of structure-based drug design.

Some results obtained with ConFirm's best search are presented in Sec. 6.

5.6. Simulation Methods

Simulation is the modelling of a system with its dynamic processes to gain knowledge, which can be transferred into reality. The most important simulation methods in the field of conformational analysis are molecular dynamics (MD) and Monte Carlo simulations, as well as simulated annealing [56,57]. All these approaches are based on extensive numerical calculations and transformations. Therefore, only a brief introduction to these methods and their application to conformational searches is given in the following.

5.6.1. Molecular Dynamics and Monte Carlo Simulations

MD simulations are widely used for exploring conformational space. They are simulating the time-dependent movements and conformational changes of a molecular system by solving Newton's laws of motion. The result of an MD simulation is a so-called trajectory that specifies how the positions and velocities of the particles (atoms of a molecule) in the system vary with time.

The interactions between the atoms are usually described by a force field under the restriction that the forces between two atoms are not influenced by any other atom.

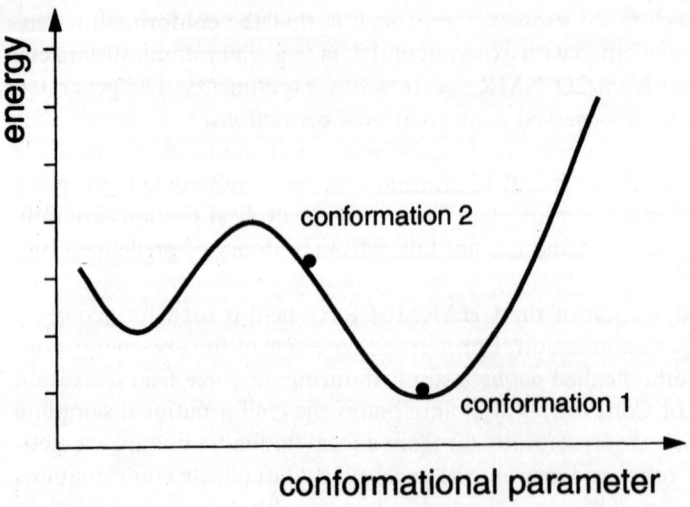

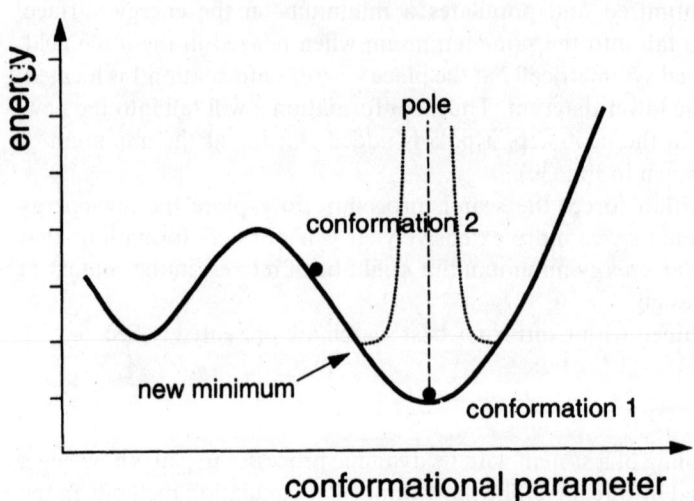

Figure 30   The application of poling for conformational sampling.

Furthermore, the temperature has to be set at which the system is be investigated. Additionally, solvent effects can be modelled by placing the molecule in a cage of solvent molecules. Starting with a single 3D structure, an initial velocity, which corresponds to the temperature of the system, is assigned to each atom. Usually, the velocities are chosen in such a way that the total momentum of the entire system equals to zero. By integrating Newton's laws, the forces that act on each atom are calculated and the atomic positions and velocities after a given time step can be derived. The rather fast processes on the atomic level require time steps of 1 fsec (1 fsec = 10^{-15} sec), which can then be repeated up to 10^7 times, depending on the number of atoms and available CPU power. Thus, the dynamic behavior of the system under investigation can be observed for several nanoseconds (1 nsec $= 10^{-9}$ sec), which is already quite enough to model the conformational changes of the side chains of a protein.

For conformational search purposes, molecular dynamics simulation can be used at very high—almost unrealistic—temperatures as bonds are not allowed to be broken. Thus, the system is able to surmount large energy barriers to explore wide regions of the conformational space. A set of conformations is obtained by sampling structures in regular intervals from the trajectory. Finally, an energy and geometry optimization provides the appropriate energy minimum conformations. Fig. 31 illustrates this schematically.

The principles of the Monte Carlo approach are briefly described in Sec. 5.3. It should be mentioned that a "true" Monte Carlo simulation does not involve any energy minimization step. Each randomly generated conformation is evaluated by applying the Metropolis criterion and is rejected or accepted without any further minimization. Furthermore, no time-dependent movements or conformational changes are investigated as is the case in MD simulations.

5.6.2. Simulated Annealing

Simulated annealing is a computer method that mimics the physical process of annealing matter to find a set of "best" solutions for a given problem. In reality, an important point of annealing is a very careful control of the temperature during the transition of the material from the liquid phase into the solid state, if, for example, single crystal structures should be obtained. The driving force is the free energy of the system, which reaches a minimum in the crystal state.

In simulated annealing, the system under consideration is heated up until a certain temperature is reached to occupy high-energy states and to overcome huge energy barriers. Then, it is thermally equilibrated for some time by using MD simulation techniques. Finally, the system is cooled down and as the temperature falls, states of lower energy become more probable according to the Boltzmann distribution.

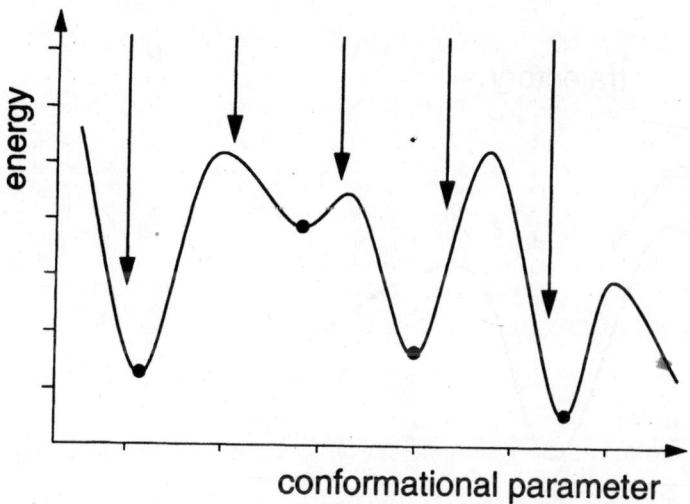

Figure 31   Sampling of conformations from the trajectory of a molecular dynamics simulation (arrows) followed by minimization to obtain minimum energy conformations (circles).

The driving force of the free energy is simulated by the internal energy of the system. At a temperature of 0 K, the system should have reached the global minimum (i.e., the lowest-energy conformation). This, of course, cannot be guaranteed because an infinite number of temperature steps and thermal equilibration would be required. Due to computational restrictions, this is not possible. Thus, repeated simulated annealing experiments can lead to different energy minimum conformations. Fig. 32 shows schematically the trajectory of an annealing process. Both minima on the energy surface can be reached when the system is cooled down, either the global minimum to the right-hand side or the local minimum to the left-hand side.

The software package, HyperChem, includes modules for molecular dynamics simulation and simulated annealing using the implemented MM + force field [58]. A well-known MD simulation software system is GROMOS (GROningen MOlecular Simulation), developed by Hermans et al. [59]. Some examples of simulated annealing experiments with large molecular systems are given in Ref. 24.

5.7. Peptide Structures

Peptides are ubiquitous in the living organism. Polypeptides and proteins are of overwhelming importance as biological receptors and as enzymes regulating biochemical pathways, whereas oligopeptides play a major role as messengers in signal transduction (e.g., as neuropeptides, hormones, or interleukins). Due to their poor bioavailability, the use of peptides in drug therapy is rather limited. Peptidomimetics try to overcome this drawback (e.g., with enzymatically noncleavable amide bonds, lower molecular weight, or less polar functional groups).

Peptides and peptidomimetics represent a class of compounds of its own for 3D structure generation and conformational analysis. Therefore, the programs developed for predicting spatial peptide models are rather restricted in their application to this specific group of molecules. In Sec. 6, approaches to the generation of low-energy conformations of peptide molecules will be briefly described. A detailed overview of

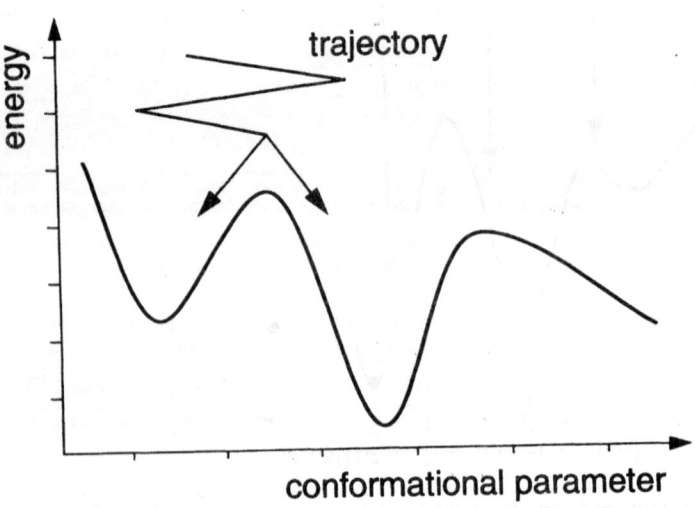

Figure 32 Trajectory of a simulated annealing experiment, which can end up in two different minima on the energy surface.

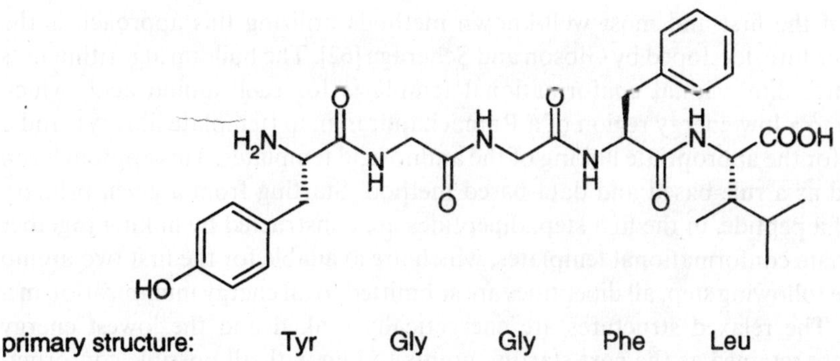

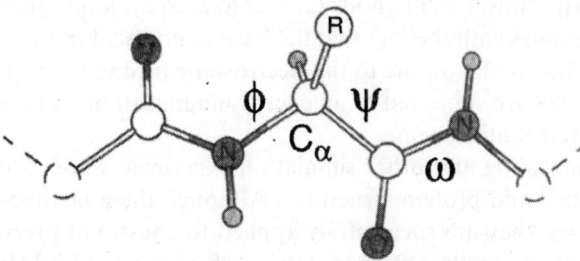

primary structure: Tyr Gly Gly Phe Leu

Figure 33 The pentapeptide Leu-enkephalin.

the various approaches to 3D structure generation and conformational analysis of oligopeptides is given in Ref. 60. The basic principles of protein modeling are given elsewhere in this volume.

The primary structure of peptides (oligopeptides and polypeptides) is described as a linear sequence consisting of the 20 naturally occurring L-amino acids. These amino acids are linked together by amide bonds. Fig. 33 illustrates this for the pentapeptide Leu-enkephalin.

A characteristic feature of peptides is the repetitive unit of the three directly bonded atoms N–C_α–$C_{carbonyl}$, which form the backbone of a peptide structure. Along the backbone, the geometry of the peptide is mainly determined by the torsion angles ω, ϕ, and ψ. In most cases of naturally occurring peptides, the dihedral angle of ω exhibits a value of 180° (see Fig. 34).

Studies of experimentally determined protein structures have shown that the torsion angles ϕ and ψ quite often appear in distinct combinations of values corresponding to certain frequently occurring structural motifs in proteins, such as the α-helix or the β-sheet. The most common way to visualize this behavior is a so-called Ramachandran plot of a protein [61]. These facts, together with the rather limited number of building blocks (20 amino acids in naturally occurring peptides), led to the idea to use a set of predefined low-energy conformations and most frequently occurring geometries to predict 3D structures of peptidic compounds. These templates can either be obtained from experimental data (i.e., from the low-energy regions of Ramachandran plots), or from computationally derived models of the amino acids.

Figure 34 Part of a peptide backbone with its three geometry-determining torsion angles ω, ϕ, and ψ.

One of the first and most well-known methods utilizing this approach is the *buildup* procedure developed by Gibson and Scheraga [62]. The buildup algorithm uses different three-dimensional conformational templates for each amino acid, which correspond to a low-energy region of a Ramachandran map (template library), and a set of rules for the appropriate linking of the amino acid templates. This approach can be classified as a rule-based and data-based method. Starting from a given primary structure of a peptide, in the first step, dipeptides are constructed by linking together all appropriate conformational templates, which are available for the first two amino acids. In the following step, all dipeptides are submitted to an energy minimization in a force field. The relaxed structures are energetically ranked and the lowest energy structures are retained as the next starting points to join with all possible conformations of the third amino acid. This procedure is repeated until the complete peptide is built up. Thus, the structure is gradually growing with an energy minimization step and selection of the lowest energy conformations so far obtained at each stage. In addition, Scheraga et al. developed algorithms to treat cyclical peptide structures to ensure a proper ring closure during the buildup procedure, also taking into account the symmetry involved in the ring system under consideration.

The major advantage of using experimental data (i.e., preferred conformational units of amino acid residues taken from protein crystal structures) is the implicit modeling of intramolecular as well as intermolecular interactions, which are contained in the data and become even more important with increasing size of the peptides. However, the conformational behavior of peptides and their larger analogues, proteins, is only comparable to a certain extent and the experimentally derived structure data on peptides are not sufficient enough for a statistical analysis (e.g., quite in contrast to the situation with small molecules; see Secs. 5.2.2 and 5.2.3). Furthermore, the force field relaxation should take into account terms for hydrogen bonding within the peptidic molecule and between the peptide and its surrounding aqueous solution.

Other approaches to generate 3D structures of peptides, mainly developed by Scheraga et al., involve Monte Carlo random search techniques. The Monte Carlo plus minimization method (MCM) allows random perturbations of one or more dihedral angels, giving a higher probability to changes in the torsion angles of the backbone rather than the dihedrals of the side chains and a subsequent minimization to overcome the inefficiency of MC searches [63]. After minimization, the resulting structure is retained for the next step, or rejected according to the Metropolis criterion [48]. The electrostatically driven Monte Carlo (EDMC) procedure combines the optimization of the alignment of the dipole moment of a randomly selected single amino acid unit with respect to the electrostatic field of the whole peptide with a random change of torsion angles [64]. Thus, this methods takes into account long-range electrostatic interactions and complies with the fact that the local amino acid residues quite often exhibit orientations that are favorable to the electrostatic field of the peptide. Again, the changed geometries are subjected to an energy minimization and are accepted or rejected using the Metropolis scheme.

Furthermore, simulated annealing and other simulation experiments are used to predict 3D structures of peptide and protein structures. Although these methods are still time-consuming processes, they are successfully applied to constraint problems (e.g., for refinement of experimentally obtained data, such as x-ray or NMR structures) [65].

An interesting approach to 3D structure generation and conformational analysis of peptides as well as of peptidomimetics was recently published by Bultinck et al. [66]. They developed a program called Generate, which is based on a modified buildup algorithm to generate three-dimensional peptidic structures. In contrast to Scheraga's original method, Generate uses a template library of computationally derived conformational units of amino acids. A major advantage of Generate is the open design of the library. A protocol was set up on the procedure to obtain the templates. Thus, future extensions of the library can provide the same level of quality as the current conformational templates. Furthermore, the program can be adapted to the specific needs of the user (e.g., 3D structure prediction of peptidomimetics) by simply adding new amino acid derivatives.

In essence, the conformational templates are derived from a series of Monte Carlo conformational searches and subsequent MMFF force field relaxation including a solvation model as implemented in the software package MacroModel [67]. To model the steric and electronic influences of neighboring amino acids, at least one alanine is attached to each amino acid during the template generation process, depending on whether it is a terminal residue or a peptidyl fragment in the chain. In addition, all calculations are carried out for both isomers with *cis* and *trans* amide bonds. Fig. 35 shows some examples of the models for peptide and peptidomimetic units, which are used to build the conformational library in Generate.

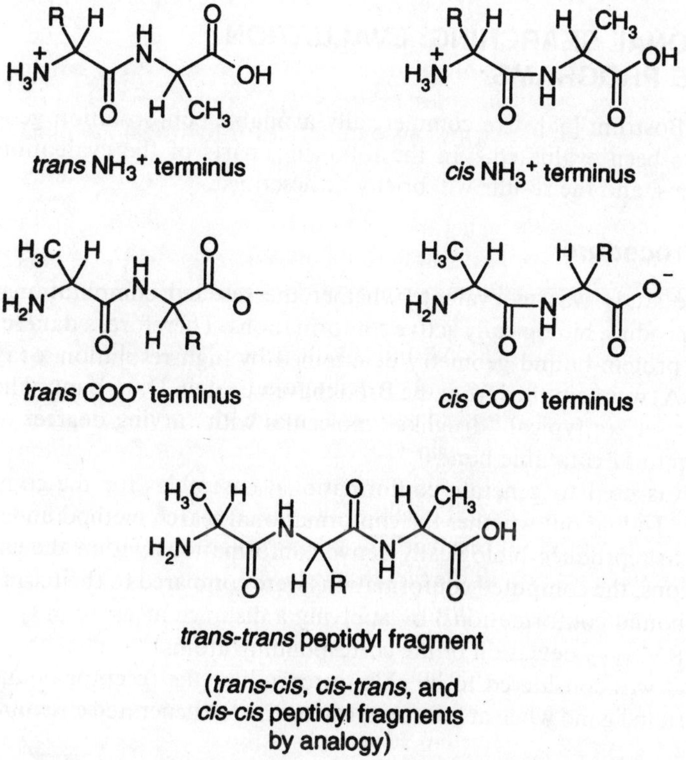

trans NH₃⁺ terminus

cis NH₃⁺ terminus

trans COO⁻ terminus

cis COO⁻ terminus

trans-trans peptidyl fragment

(*trans-cis*, *cis-trans*, and *cis-cis* peptidyl fragments by analogy)

Figure 35 Examples of models for peptide and peptidomimetic units used to build up the template library in Generate.

Starting from a given sequence of amino acids for a peptide or peptidomimetic, the user has to set each peptide bond either to *cis* or *trans*. In the first step of the buildup procedure, the program randomly selects one residue and an appropriate single conformational unit from the library as starting point. In the following step, the next amino acid is added by choosing the geometrically best fitting conformation of the second residue. This procedure continues in both directions of the chain until the complete structure is built up. To obtain an ensemble of conformations for the peptide structure under consideration, the buildup procedure is repeated until a user-defined number of geometries is obtained. Finally, duplicate structures may optionally be rejected and all remaining conformations are ranked energetically based on the calculation of van der Waals energies. This simplified energy ranking performs with a quite good correspondence to the total MMFF force field energy, but is far less time-consuming.

Generate usually builds up about 100 peptide conformations in less than 1 min on a common Linux workstation including the removal of duplicate geometries and the energy ranking procedure. By using preoptimized conformational templates of the amino acids, and thus avoiding time-consuming intermediate energy minimization steps, Generate performs much faster than the original buildup procedure described by Gibson and Scheraga [62]. In addition, the program can easily be extended to handle peptidomimetics. A major drawback is that the current version of Generate cannot be applied to cyclical structures.

6. CONFORMATIONAL SEARCHING: EVALUATION OF AVAILABLE PROGRAMS

In a recent paper by Boström [68], five commercially available conformation generation programs have been evaluated.* In the following, parts of the evaluation procedure, the programs, and the results will briefly be described.

6.1. Evaluation Procedure

The major aim of the study was to evaluate whether the selected computational methods are able to reproduce biologically active conformations. Therefore, a data set of 32 ligands in their protein-bound geometry determined by high-resolution x-ray crystallography (≤ 2.0 Å) was compiled from the Brookhaven Protein Data Bank. The compounds in the test set are typical "druglike" molecules with varying degrees of rotational freedom (up to 11 rotatable bonds).

Each program was used to generate conformational ensembles for the compounds in the data set. To find out whether the conformational search method under consideration is able to reproduce biologically active conformations within the ensembles of conformations, the computed conformations were compared to their x-ray geometries (receptor-bound conformations) by applying a distance measure in Cartesian space (i.e., the RMS_{XYZ} deviation of the corresponding atoms).

A search method was considered to be able to reproduce the receptor-bound conformation for a certain ligand when at least one structure in the generated ensemble

*The authors want to thank Dr. Jonas Boström (AstraZeneca R&D Mölndal, Sweden) for providing us with the details and the results of his study and for his kind permission to present them in this article.

of conformations has an RMS_{XYZ} deviation to the x-ray geometry below a certain value. The rather large value of 0.5 Å was chosen because the unmodified (unrefined) x-ray structures, as they are contained in the PDB, were used in the study.

The lowest-energy conformations of the ligands obtained by CORINA were taken as input structures for the different runs.

6.2. Programs

Table 4 summarizes some information on the programs that were evaluated in this study. Most of the methods inherent in the programs are only published to a certain extent. Thus, the programs could only be briefly described in the paper of Boström [68].

ConFirm (Catalyst) has already been described in the Secs. 5.1.1 and 5.5. Both search modes *best* and *fast* were used in this study to generate ensembles of conformations for the compounds in the test set. Furthermore, the default values of ConFirm to output a maximum of 256 conformations within an energy range of 20 kcal/mol with respect to the conformation with the lowest energy were applied.

CONFORT performs an exhaustive conformational analysis of a molecule [71]. Two different search modes either generate a user-defined number of conformations, or output a maximally diverse set of conformations, which was used in this study. The diversity metric is based on interconformational distances that circumvent the generation of duplicate structures. The conformations are relaxed and optimized by applying only internal coordinates and analytic gradients and by the Tripos force field package.

Flo99 relies on a random search technique and applies the Metropolis criterion to generate a set of 300 conformations within an energy window of 12 kcal/mol with respect to the lowest-energy conformation (default values) [72]. A modified version of the AMBER force field (QXP) is used for geometry optimization.

MacroModel offers a variety of methods for conformation generation. During the study, the so-called low-mode conformational search (LMCS) was used [67,73]. This efficient and accurate algorithm explores the conformational space by applying a (vibrational) mode-following or eigenvector-following technique. In addition, the program includes hydration and solvent-accessible surface area models. Geometry optimization was carried out with two different force field packages, MMFF and AMBER*, which are implemented in MacroModel.

Additionally, the program offers a variety of common force fields such as MM2* MM3*, Amber94, and OPLS-AA, which where not used in this study.

OMEGA is a fast rule-based method and generates the conformation in a systematic manner [74]. It cuts a structure into several torsional fragments and reconstructs the molecule according to different rules with varying torsion angles. Finally, duplicate conformations (which have an RMS_{XYZ} deviation less than or equal to 0.8 Å)

Table 4 Conformational Searching: Programs Tested by the Evaluation Procedure

	Catalyst	CONFORT	Flo99	MacroModel	Omega
Version	4.6	3.9		7.0	0.9.9
Reference	70	71	72	73	74

are discarded and the remaining structures are relaxed in a reduced force field, which only optimizes the torsion angles.

6.3. Results and Discussion

The programs described above were used to generate ensembles of conformations for each ligand in the test set. Finally, all conformations were compared to their corresponding bioactive geometry. A hit was defined when at least one conformation was found within an ensemble with an RMS_{XYZ} deviation to the x-ray geometry less than or equal to 0.5 Å.

The best result (i.e., the largest number of hits) was obtained by MacroModel by applying the hydration and solvent-accessible surface area model and the modified AMBER force field optimization with 22 reproduced biologically active conformations of the total set of 32 compounds. With a total CPU time of about 10 hr and an average number of 155 conformations per molecule, MacroModel needed the longest computation times. The next best program, Flo99, reproduced 20 x-ray geometries with an average number of 70 structures per molecule in approximately 50 min, followed by ConFirm. ConFirm's *best* search mode (15 reproduced x-ray geometries in 4.5 hr, 84 conformations per molecule on average) was slightly outperformed by its *fast* mode, which found 16 bioactive conformations within 1 min and 55 structures on average. The rule-based program OMEGA clearly performed fastest with only 18 sec for the complete data set and an average of 18 conformations per compound, but generated only 13 conformations within the required RMS range of 0.5 Å (17 conformations if the x-ray geometry was used as starting structure). CONFORT reproduced a maximum of 11 x-ray geometries in 3 hr and an average of 23 conformations per molecule.

Obviously, the computationally more demanding but accurate methods as implemented in MacroModel, which put more emphasis on intermolecular rather than intramolecular interactions (hydration model), lead to a better description of the problem under consideration. Also, the random search technique of Flo99 performed well without consideration of any intermolecular electrostatic interactions. Diverse sampling as performed by ConFirm and CONFORT gave only a low hit rate. It seems that in the context of searching for receptor-bound ligand geometries, important conformations get lost by these search methods. The OMEGA procedure performed rather impressively in terms of the high data-to-CPU ratio, although this approach is quite simplistic.

7. CONCLUSIONS

Automatic 3D structure generation, the conversion of a connection table into a 3D molecular model, has been pursued in the last years in both academic research and commercial software development and has become a standard technique routinely used in many fields of computational chemistry. Surveying the approaches developed since 1980, it was shown that a number of interesting solutions to this fundamental task of computational chemistry exist. The various methods derive their knowledge to various degrees from data of experimental or computed geometries, or from rules about the construction of molecular models. Great care has been devoted to making these 3D structure generators as rapid as possible to apply them to large datasets of molecules. The users of 3D structure generation programs, especially database developers, will increasingly become interested not only in the speed of the conversion programs but also in their accuracy and their conversion rate. They have to decide which of the criteria are the most important ones with respect to their application: conversion rate, quality of the 3D models, and/or computation time.

The in silico conformational search problem was initiated simultaneously with the demand for automatic 3D structure generation. As already stated, 3D model builders and conformation-generating systems overlap in the field of conformational analysis because it is almost impossible to predict a low-energy structure of a molecule without taking into account and evaluating several alternative conformations. Many approaches to conformational searching have been developed during the last 20 years and several of the most widely used and interesting methods were presented here. Exhaustive methods such as systematic searches can generate huge amounts of different geometries and explore the conformational space in quite a complete but unfeasible time-consuming manner. Any restrictions bias the search toward local regions on the energy surface and important minima may get lost. Random-based methods such as Monte Carlo and, to some extent, genetic algorithm approaches try to overcome this problem by generating conformations in an unpredictable (nondeterministic) fashion by applying large perturbations to obtain conformational diversity. Rule-based and data-based systems have the advantage that implicitly knowledge and information about the system under consideration can be included. In the context of structure-based drug design, the inclusion of rules and data on the conformational preferences of small molecules binding to a biological receptor might be an interesting further development as the number of experimentally determined receptor–ligand complexes will rise in the future. In addition, increasing CPU power will also arouse more interest in simulation experiments because already, nowadays, they are able to handle large systems such as proteins and to predict relative binding constants of sets of potential drug molecules.

The conformational space of a molecule can be regarded as a closed system. All conformations are interconnected simply by rotations around rotatable bonds. Nevertheless, conformational space also has a certain dimension of infinity and we are travelers on an endless journey in the land of the molecular energy hypersurfaces.

REFERENCES

1a. Allen FH, Davies JE, Galloy JJ, Johnson O, Kennard O, Macrae CF, Mitchell EM, Mitchell GF, Smith JM, Watson DG. J Chem Inf Comput Sci 1991; 31:187–204.
1b. Allen FH, Hoy VJ. Cambridge Structural Database. In: Schleyer, PvR, Allinger NL, Clark T, Gasteiger J, Kollman PA, Schaefer HF III, Schreiner PR, eds. Encyclopedia of Computational Chemistry. Chichester, UK: John Wiley and Sons, Inc., 1998:155–167.
2a. Berman HM, Westbrook J, Feng Z, Gilliland G, Bhat TN, Weissig H, Shindyalov IN, Bourne PE. Nucleic Acids Res 2000; 28:235–242.
2b. Sussman JL. Protein data bank: a database of 3D structural information of biological macromolecules. In: Schleyer PvR, Allinger NL, Clark T, Gasteiger J, Kollman PA, Schaefer HF III, Schreiner PR, eds. Encyclopedia of Computational Chemistry. Chichester, UK: John Wiley and Sons, Inc., 1998:2160–2168.
3a. Sadowski J, Gasteiger J. Chem Rev 1993; 7:2567–2581.
3b. Sadowski J. Three-dimensional structure generation: automation. In: Schleyer PvR, Allinger NL, Clark T, Gasteiger J, Kollman PA, Schaefer HF III, Schreiner PR, eds.

Encyclopedia of Computational Chemistry. Chichester, UK: John Wiley and Sons, Inc., 1998:2976–2988.

4. Pearlman RS. In: Kubinyi H, ed. 3D QSAR in Drug Design. Leiden: ESCOM, 1993:41–79.

5. Klebe G, Mietzner T, Weber FJ. Comput-Aided Mol Des 1994; 8:751–778.

6. Handschuh S, Gasteiger JJ. Mol Model 2000; 6:358–378.

7. Sadowski JJ. Comput-Aided Mol Des 1997; 11:53–60.

8. Bissantz C, Folkers G, Rognan D. J Med Chem 2000; 43:4759–4767.

9. Henry DR, McHale PJ, Christie BD, Hillman D. Tetrahedron Comput Methodol 1990; 3:531–536.

10. Crippen GM, Havel TF. Distance geometry and molecular conformations. In: Bawden D, ed. Chemometrics Research Studies Series 15. New York, NY, USA: Research Studies Press (Wiley), 1988.

11a. Corey EJ, Feiner NF. J Org Chem 1980; 45:757–764.

11b. Corey EJ, Feiner NF. J Org Chem 1980; 45:765–780.

12. Cohen NC, Colin P, Lemoine G. Tetrahedron 1981; 37:1711–1721.

13a. De Clercq PJ. J Org Chem 1981; 46:667–675.

13b. De Clercq PJ. Tetrahedron 1981; 37:4277–4286.

13c. De Clercq PJ. Tetrahedron 1984; 40:3717–3728.

13d. De Clercq PJ. Tetrahedron 3729–3739.

13e. Hoflack J, De Clercq PJ. Tetrahedron 1988; 44:6667–6676.

13f. SCA QCPE Program QCPE, No. QCMP079. Bloomington, IN, USA: Quantum Chemistry Program Exchange, Indiana University, 1989.

14. Klebe G, Mietzner T. J Comput-Aided Mol Des 1994; 8:583–606.

15. Klebe G, Mietzner T, Weber F. J Comput-Aided Mol Des 1999; 13:35–49.

16. Rarey M, Kramer B, Lengauer T, Klebe G. J Mol Biol 1996; 261:470–489.

17a. Dolata DP, Carter RE. J Chem Inf Comput Sci 1987; 27:36–47.

17b. Dolata DP, Leach AR, Prout K. J Comput-Aided Mol Des 1987; 1:73–85.

17c. Leach AR, Prout K, Dolata DP. J Comput-Aided Mol Des 1988; 2:107–123.

17d. Leach AR, Prout K, Dolata DP. J Comput-Aided Mol Des 1990; 4:271–282.

17e. Leach AR, Prout K, Dolata DP. J Comput Chem 1990; 11:680–693.

18a. Leach AR, Prout K. J Comput Chem 1990; 11:1193–1205.

18b. Leach AR, Smellie AS. J Chem Inf Comput Sci 1992; 32:379–385.

19. Leach AR. J Chem Inf Comput Sci 1994; 34:661–670.

20a. Pearlman RS. Chem Des Autom News 1987; 2:1/5–6.

20b. CONCORD, Tripos, Inc., St. Louis, MO, USA (http://www.tripos.com).

21a. Hiller C, Gasteiger J. In: Gasteiger J, ed. Software-Entwicklung in der Chemie. Berlin: Springer, 1987:53–66.

21b. Gasteiger J, Rudolph C, Sadowski J. Tetrahedron Comput Methodol 1990; 3:537–547.

21c. Sadowski J, Rudolph C, Gasteiger J. Anal Chim Acta 1992; 265:233–241.

21d. CORINA, Version 2.6, Molecular Networks GmbH, Nägelsbachstraße 25, Erlangen 91052, Germany (http://www.mol-net.de).

22a. Gasteiger J, Hutchings MG, Christoph B, Gann L, Hiller C, Löw P, Marsili M, Saller H, Yuki K. Top Curr Chem 1987; 137:19–73.

22b. Gasteiger J, Ihlenfeldt WD, Röse P. Recl Trav Chim Pays-Bas 1992; 111:270–290.

23. Dale J. Acta Chem Scand 1973; 27:1115–1129.

24. Schönberger H, Schwab CH, Hirsch A, Gasteiger J. J Mol Model 2000; 6:379–395.

25. Hurst T. J Chem Inf Comput Sci 1994; 34:190–196.

26a. Wipke WT, Hahn MA. In: Pierce T, Hohne B, eds. Applications of Artificial Intelligence in Chemistry. Symposium Series No. 306. Washington, DC: American Chemical Society, 1986:136–146.

26b. Wipke WT, Hahn MA. Tetrahedron Comput Methodol 1988; 2:141–167.
26c. Wipke WT, Hahn MA. In: Warr WE, ed. Chemical Structures. Vol. 1. Berlin: Springer, 1988:267–268,
26d. Hahn MA, Wipke WT. In: Warr WE, ed. Chemical Structures. Vol. 1. Berlin: Springer, 1988:269–278.
27. Davies K, Upton R. Tetrahedron Comput Methodol 1990; 3:665–671.
28. Crippen GM, Havel TF. J Chem Inf Comput Sci 1990; 30:222–227.
29. Wenger JC, Smith DH. J Chem Inf Comput Sci 1982; 22:29–34.
30. Goordeva EV, Katritzky AR, Shcherbukhin VV, Zefirov NS. J Chem Inf Comput Sci 1993; 33:102–111.
31. CONVERTER, Accelrys Inc., San Diego, CA, USA (http://www.accelrys.com).
32. DGEOM, QCPE Program QCPE, No. 590. Bloomington, IN, USA: Quantum Chemistry Program Exchange, Indiana University, 1995.
33. Sadowski J, Gasteiger J, Klebe G. J Chem Inf Comput Sci 1994; 34:1000–1008.
34. Hiller C. ALCOGEN. Weinheim, Germany: Chemical Concepts..
35. van Geerestein V, Grootenhuis P. Akzo Nobel. Oss, The Netherlands: NV Organon (unpublished results).
36. Lipton M, Still WC. J Comput Chem 1988; 9:343–355.
37a. Gotō H, Ōsawa E. J Am Chem Soc. 1989;111:8950–8951.
37b. CONFLEXQCPE Program QCPE, No. 592. Bloomington, IN, USA: Quantum Chemistry Program Exchange, Indiana University, 1989.
38. SPARTAN'O2, Wavefunction, Inc., Irvine, CA, USA (http://www.wavefun.com).
39. Cerius2, Version 4.6, Accelrys Inc., San Diego, CA, USA (http://www.accelrys.com).
40. Catalyst, Version 4.6, Accelrys Inc., San Diego, CA, USA (http://www.accelrys.com).
41a. Smellie A, Kahn SD, Teig SL. J Chem Inf Comput Sci 1995; 35:285–294.
41b. Smellie A, Kahn SD, Teig SL. J Chem Inf Comput Sci 1995; 35:295–304.
42. Smellie A, Teig SL, Towbin P. J Comput Chem 1995; 16:171–187.
43. Klebe G. Structure correlation and ligand/receptor interactions. In: Bürgi H-B, Dunitz JD, eds. Structure Correlation. Vol. 2. Weinheim: VCH Verlagsgesellschaft mbH, 1994:543–603.
44. Murray-Rust P. In: Griffin JF, Duax WL, eds. Molecular Structure and Biological Activity. New York, NY, USA: Elsevier Biomed, 1982:117–133.
45a. Schwab CH. Konformative Flexibilität von Liganden im Wirkstoffdesign. Ph.D Thesis, University of Erlangen-Nuremberg, Erlangen, 2001.
45b. ROTATE Version 1.1 α Molecular Networks GmbH, Nägelsbachstraße 25, Erlangen 91052, Germany (http://www.mol-net.de).
46a. Saunders M. J Am Chem Soc 1987; 109:3150–3152.
46b. Saunders M. J Comput Chem 1989; 10:203–208.
47. Chang G, Guida WC, Still WC. J Am Chem Soc 1989; 111:4379–4386.
48. Metropolis N, Rosenbluth AW, Rosenbluth MN, Teller AH, Teller E. J Chem Phys 1953; 21:1087–1092.
49. Ferguson DM, Raber DJ. J Am Chem Soc 1989; 111:4371–4378.
50. Goldberg DE. Genetic Algorithms in Search, Optimization and Machine Learning. New York, NY, USA: Addison-Wesley Publishing Company, 1989.
51. Harik GR. Finding multimodal solutions using restricted tournament selection. In: Eshelman LJ, ed. Proceedings of the Sixth International Conference on Genetic Algorithms. San Francisco, CA, USA: Morgan Kaufmann Publishers, Inc., 1995:24–31.
52. Nair N, Goodman JM. J Chem Inf Comput Sci 1998; 38:317–320.
53. Handschuh S, Wagener M, Gasteiger J. J Chem Inf Comput Sci 1998; 38:220–232.
54. Fonseca CM, Fleming PJ. In: Forrest S, ed. Proceedings of the 5th International Conference on Genetic Algorithms. San Mateo, CA, USA: Morgan Kaufmann Publishers, Inc., 1993:416–423.

55. Blaney JM, Dixon JS. Distance geometry in molecular modeling. In: Lipkowitz KB, Boyd DB, eds. Reviews in Computational Chemistry. Vol. 5. New York, NY, USA: VCH Verlagsgesellschaft mbH, 1994:299–335.

56. Lybrand TP. Computer simulations of biomolecular systems using molecular dynamics and free energy perturbation methods. In: Lipkowitz KB, Boyd DB, eds. Reviews in Computational Chemistry. Vol. 1. New York, NY, USA: VCH Verlagsgesellschaft mbH, 1990:295–320.

57. Kirkpatrick S, Gelatt CD, Vecchi MP. Science 1983; 220:671–680.

58. HyperChem, Version 6, Hypercube, Inc., Gainesville, FL, USA (http://www.hyper.com).

59a. Hermans J, Berendsen HJC, van Gunsteren WF, Postma JPM. Biopolymers 1984; 23: 1513–1518.

59b. GROMOS 96, BIOMOS BV, Zurich, Switzerland (http://www.igc.ethz.ch/cromos).

60. Scheraga HA. Distance predicting three-dimensional structures of oligopeptides. Lipkowitz KB, Boyd DB, eds. Reviews in Computational Chemistry. Vol. 3. New York, NY, USA: VCH Verlagsgesellschaft mbH, 1992:73–142.

61. Ramachandran GN, Ramakrishnan C, Sasiekharan V. Stereochemistry of polypeptide chain configurations. J Mol Biol 1963; 7:95–99.

62. Gibson KD, Scheraga HA. Revised algorithms for the build-up procedure for predicting protein conformations by energy minimization. J Comput Chem 1987; 8:826–834.

63. Li Z, Scheraga HA. Monte Carlo minimization approach to the multiple-minimum problem in protein folding. Proc Natl Acad Sci USA 1987; 84:6611–6615.

64a. Ripoll DR, Scheraga HA. On the multiple-minimum problem in the conformational analysis of polypeptides: II. An electrostatically driven Monte Carlo method: tests on poly(L-alanine). Biopolymers 1988; 27:1283–1303.

64b. Ripoll DR, Scheraga HA. On the multiple-minimum problem in the conformational analysis of polypeptides: III. An electrostatically driven Monte Carlo method: tests on met-enkephalin. J Protein Chem 1989; 8:263–287.

65. Daura X, Gademann K, Schäfer H, Jaun B, Seebach D, van Gunsteren WF. The β-peptide hairpin in solution: conformational study in methanol by NMR spectroscopy and MD simulation. J Am Chem Soc 2001; 123:2393–2404.

66. Bultinck P, Augustynen S, Hilbers HW, Moret EE, Tollenaere JP. Generate: a program for 3-D structure generation and conformational analysis of peptides and peptidomimetics. J Comput Chem 2002; 23:746–754.

67. MacroModel Version 7.0, Schrödinger, Inc., Portland, OR, USA (http://www.schrodinger.com).

68. Boström J. J Comput-Aided Mol Des 2001; 15(12):1137–1152.

70a. Spraque PW. Perspect Drug Discov Des 1995; 3:1.

70b. Catalyst, Version 4.6, Accelrys Inc., San Diego, CA, USA http://www.accelrys.com.

71a. Pearlman RS, Balducci R. Confort's User Manual, Version 3.9.

71b. CONFORT, Version 3.9, Tripos, Inc., St. Louis, MO, USA (http://www.tripos.com).

72. McMartin C, Bohacek R. J Comput-Aided Mol Des 1995; 11:333–342.

73. Kolossváry I, Guida WC. J Am Chem Soc 1996; 118:5011–5019.

74. Omega, Open Eyes Scientific Software, Santa Fe, NM, USA (http://www.eyesopen.com).

8

Molecular Electrostatic Potentials

PETER POLITZER and JANE S. MURRAY

University of New Orleans, New Orleans, Louisiana, U.S.A.

1. INTRODUCTION AND BACKGROUND

1.1. Fundamental Relationships

It follows from Coulomb's law that any distribution of electrical charge creates a potential $V(\mathbf{r})$ at each point $\mathbf{r}$ in the surrounding space. In the elementary example of a collection of point charges Q_i, which may be positive or negative,

$$V(\mathbf{r}) = \sum_i \frac{Q_i}{|\mathbf{r}_i - \mathbf{r}|} \tag{1}$$

in which $\mathbf{r}_i$ is the instantaneous position of Q_i. (All equations will be in atomic units and refer to a vacuum, so that the dielectric permittivity is $1/4\pi$.) If the charge distribution is continuous, with a density $D(\mathbf{r})$ (which again may be either positive or negative at any point $\mathbf{r}$), then integration replaces summation, and Eq. (1) becomes

$$V(\mathbf{r}) = \int \frac{D(\mathbf{r}')\mathrm{d}\mathbf{r}'}{|\mathbf{r}' - \mathbf{r}|} \tag{2}$$

$V(\mathbf{r})$ is also rigorously related to $D(\mathbf{r})$ through Poisson's equation [1],

$$\nabla^2 V(\mathbf{r}) = -4\pi D(\mathbf{r}) \tag{3}$$

A system of nuclei and electrons, such as a molecule, is generally viewed as having a continuous but static distribution of electronic charge around a rigid nuclear framework. Then,

$$D(\mathbf{r}) = \sum_A Z_A \delta(\mathbf{r} - \mathbf{R}_A) - \rho(\mathbf{r}) \tag{4}$$

where Z_A is the charge on nucleus A, located at $\mathbf{R}_A$, and $\rho(\mathbf{r})$ is the electronic density. Note that the latter is always positive; the electronic *charge* density is thus $-\rho(\mathbf{r})$. $V(\mathbf{r})$ is now called the "electrostatic" potential, and is given by Eq. (5),

$$V(\mathbf{r}) = \sum_A \frac{Z_A}{|\mathbf{R}_A - \mathbf{r}|} - \int \frac{\rho(\mathbf{r}')d\mathbf{r}'}{|\mathbf{r}' - \mathbf{r}|} \tag{5}$$

obtained by combining Eqs. (2) and (4). Poisson's equation for such systems takes the form

$$\nabla^2 V(\mathbf{r}) = 4\pi\rho(\mathbf{r}) - 4\pi \sum_A Z_A \delta(\mathbf{r} - \mathbf{R}_A) \tag{6}$$

From Eq. (5), it is seen that the sign of $V(\mathbf{r})$ in any given region of space depends on whether the positive contribution of the nuclei or the negative one of the electrons is dominant there.

In this chapter, $V(\mathbf{r})$ shall refer to the molecular electrostatic potential, which shall be the focus of our discussion. The applications that we will emphasize reflect the relationship of $V(\mathbf{r})$ to interaction energies. Thus the exact electrostatic interaction energy between the *unpolarized* molecule and a point charge Q (positive or negative) located at $\mathbf{r}$ is given by $QV(\mathbf{r})$. This assumes that the equilibrium charge distribution of the molecule has not been affected (i.e., polarized) by the presence of the point charge. If one wishes to find the *total* energy of the interaction, taking into account polarization and other effects, one approach is to use perturbation theory; then $QV(\mathbf{r})$ is the first-order term, the contributions of the other factors being represented by the higher-order terms. Accordingly, $V(\mathbf{r})$, although a potential, is customarily expressed in energy units (e.g., kcal/mol). This is really $QV(\mathbf{r})$, with $Q = +1$.

However, the significance of $V(\mathbf{r})$ goes beyond interaction energies; it is a fundamental quantity, a physical observable, which can be determined experimentally by diffraction methods [2,3] as well as computationally. It is rigorously linked to $\rho(\mathbf{r})$ by Poisson's equation, and $\rho(\mathbf{r})$ is known, through the Hohenberg–Kohn theorem [4], to be the ultimate determinant of a system's properties. Thus $V(\mathbf{r})$ should be, and is, related to the intrinsic properties of an isolated molecule. For example, the Hellmann–Feynman theorem [5,6] relates the total energy of an atom or molecule to the electrostatic potential $V_{0,A}$ that is created at any nucleus A by the electrons and remaining nuclei (in the case of a molecule):

$$V_{0,A} = \left(\frac{\partial E}{\partial Z_A}\right)_N \tag{7}$$

where N is the number of electrons. This indicates that atomic and molecular energies can be expressed in terms of the electrostatic potentials at the nuclei and the nuclear charges:

$$E = f(\{V_{0,A}, Z_A\}) \tag{8}$$

Indeed, a number of such formulas have been developed, both exact and approximate, and have been applied with significant success. This work has been extensively reviewed elsewhere [7–11], and shall not be further discussed here because it falls outside the scope of this chapter.

1.2. Some General Features of Atomic and Molecular Electrostatic Potentials

The electrostatic potential due to a neutral, spherically averaged free atom is positive everywhere and is monotonically decreasing [12,13]; the effect of the highly concentrated nucleus dominates over that of the dispersed electrons. When atoms combine to form a molecule, the accompanying rearrangement of electronic charge, while relatively minor, normally produces some regions of negative $V(\mathbf{r})$. These are typically associated with (1) lone pairs of electronegative atoms (e.g., N, O, F, Cl, etc.), (2) π regions of unsaturated hydrocarbons, and (3) strained C–C bonds [14–17]. Each negative region must have one or more local minima, V_{min}, i.e., points at which $V(\mathbf{r})$ reaches its locally most negative values. These are the positions at which a positive point charge would undergo the most attractive interaction with the unperturbed molecule, and accordingly the V_{min} have often been used, with some success, to identify and rank the sites most susceptible to electrophilic attack [14–17].

An analogous interpretation of local maxima, V_{max}, with regard to nucleophilic attack is not valid, because it has been shown by Pathak and Gadre [18] that V_{max} are associated only with the positions of the nuclei. They reflect the nuclear charges, not relative affinities for nucleophiles. However, this problem can often be overcome by looking for buildups of positive potential in two-dimensional or especially three-dimensional surfaces sufficiently removed from the nuclei [19–22]; such $V_{S,max}$ will be further discussed in a later section.

1.3. V(r) and Reactive Behavior

Since the pioneering contributions of Scrocco, Tomasi, and their collaborators [14,15], the evolution of which has been described in an excellent fashion by Tomasi et al. [23], the use and applications of molecular electrostatic potentials have dramatically expanded. (For recent overviews, see Refs. 3 and 24.) This is continuing, with $V(\mathbf{r})$ now involved in the development of molecular-scale electronic systems [25].

The most prominent role of $V(\mathbf{r})$ has been in the area of molecular reactive behavior [2,3,14–17,23,24]. This follows from the fact that it is through their electrostatic potentials that interacting molecules or other chemical species first "see" or "feel" each other. The use of local V_{min} to interpret and predict electrophilic attack has already been mentioned. This is sometimes quite successful [14–17], but it can also fail. Among the reasons for failure may be (1) that the most negative regions are not necessarily those most amenable to charge transfer [26–29], (2) the charge distribution may be significantly perturbed (i.e., polarized) by the proximity of the electrophile, so that the original $V(\mathbf{r})$ computed for the isolated system is no longer valid, and (3) the properties of the electrophile are not being taken into account.

The problem that polarization effects are being ignored has been addressed on a number of occasions [30–38], e.g., by applying second-order perturbation theory to appropriately correct $V(\mathbf{r})$. However, a much more popular option has been to examine $V(\mathbf{r})$ only beyond some minimum distance from the molecule, e.g., its van der Waals radius, where an approaching reactant would be too far away to significantly affect the molecule's charge distribution.

The preceding discussion leads to the conclusion that the use of $V(\mathbf{r})$ alone should be primarily limited to the analysis of noncovalent interactions. These are largely electrostatic in nature [3,6,39–41], and typically involve relatively large separations; thus polarization and charge transfer are minor or insignificant. For these purposes, it is reasonable to work with $V(\mathbf{r})$ computed on an "outer surface" of the molecule, i.e., $V_S(\mathbf{r})$, because this is what is "felt" by an interacting species. (The question of how to define an appropriate outer surface shall be addressed in Sec. 2.3.) There has indeed been a longstanding interest in relating the overall patterns of $V(\mathbf{r})$ in the outer regions of biologically active molecules to "recognition" processes, e.g., enzyme–substrate or drug–receptor [2,3,16,42–44], the basic idea being that the positive and negative areas on the two systems should complement each other. This work will be mentioned further in Secs. 3.2 and 3.4.

2. METHODOLOGY

The formula for $V(\mathbf{r})$, Eq. (5), is exact. However, $\rho(\mathbf{r})$ is normally obtained by an ab initio, density functional or semiempirical computational procedure, and hence is approximate, as is therefore $V(\mathbf{r})$. Given this basic limitation of quantum chemistry, there still remains a choice between using the inexact $\rho(\mathbf{r})$ to evaluate $V(\mathbf{r})$ rigorously or approximately. If $\rho(\mathbf{r})$ is expressed in terms of atomic basis functions, then the rigorous approach entails calculating a large number of two- and three-center nuclear–electronic attraction integrals, in which the point $\mathbf{r}$ replaces the position of the nucleus. Among approximate techniques, the most common involves rewriting $V(\mathbf{r})$ as a multipole expansion. These alternatives will be briefly discussed. For further details and references, see Refs. 3 and 16.

2.1. Rigorous Evaluation of V(r)

The molecular electrostatic potential is a one-electron property, and therefore, by the Moller–Plesset theorem, is correct through first order at the Hartree–Fock level [45]; any errors due to the neglect of electronic correlation are second- or higher-order effects. These may not be insignificant [46]. However, it has been shown that the inclusion of correlation has relatively little effect on Hartree–Fock electronic densities [47–51] and electrostatic potentials [15–17,52–54] in the outer regions of the molecule. There are differences in detail, e.g., in the exact magnitudes and locations of the V_{min}, but the general pattern of $V(\mathbf{r})$, which is usually what is important in regard to noncovalent interactions, tends to essentially remain the same. This is true as well in comparing Hartree–Fock $V(\mathbf{r})$ obtained with minimum vs. extended basis sets. Thus we repeat our earlier conclusion [16,17,55] that the overall qualitative picture of $V(\mathbf{r})$, particularly on an outer surface of a molecule, is not greatly affected by the level of an ab initio calculation; however, we do recommend that polarization functions be included for second-row atoms even in minimum basis sets, e.g., STO-5G* rather than STO-5G.

Similarly, a generally reliable representation of an outer surface $V(\mathbf{r})$, i.e., $V_S(\mathbf{r})$, can be obtained by some semiempirical methods [56–59] and by density functional procedures [60–63]. The latter do take account of electronic correlation, yet are comparable to Hartree–Fock in terms of demands on computer resources [64–66].

With the remarkable advances of the past decade in processor technology, it is now quite feasible to rigorously evaluate $V_S(\mathbf{r})$, using, e.g., the Gaussian 98 code [67], for relatively large molecules of very practical chemical or biological significance. Thus we have computed $V_S(\mathbf{r})$ at Hartree–Fock minimum-basis-set levels for some tetracyclines [68,69] and reverse transcriptase inhibitors [70,71], and are currently doing so for carbon and boron/nitrogen nanotube models with as many as 90 atoms. However, for very large systems, there continues to be interest in approximate techniques for determining $V_S(\mathbf{r})$, two of which will be mentioned in the next section.

2.2. Approximate Evaluation of V(r)

A popular approximate technique has been to expand $V(\mathbf{r})$ as a sum of contributions from the various multipoles of the molecular charge distribution, either taking the latter as a whole or dividing it into portions associated with various centers, e.g., the individual nuclei [3,16,23,72]. How well this reproduces the rigorously evaluated $V(\mathbf{r})$, for the same $\rho(\mathbf{r})$, depends on the number of terms in the expansion. A very elementary approach is to use a multicenter expansion but to limit it to monopoles corresponding to the constituent atoms, with net charges assigned by one of the methods that have been proposed for this purpose [16]. This amounts to applying Eq. (1), and has met with only limited success [3,16]. (In fact, what is more defensible is to utilize a rigorously evaluated $V(\mathbf{r})$ to determine a meaningful set of atomic charges, these being the ones that best reproduce the $V(\mathbf{r})$ in a least-squares sense [3,16,73–75].) An effective and very practical technique, which takes advantage of the Gaussian basis sets that are now so widely employed, is based on the fact that the resulting $\rho(\mathbf{r})$ can readily be represented as a finite multicenter expansion, the centers not being restricted to the nuclei [76]. When $V(\mathbf{r})$ is expressed in terms of the corresponding mono-, di-, and quadrupoles, it agrees well with that rigorously determined from the same $\rho(\mathbf{r})$, with a marked reduction in computing time [77].

Another approximation procedure that has long been of interest involves constructing $V(\mathbf{r})$ for a large system by superposing the contribution of its constituent portions, or fragments, taking these to be transferable [2,14,16,23,78–80]. The fragments can range from localized orbitals, lone pairs and bonds to atoms, functional groups, and even larger units. In a recent variation on this technique that has achieved good results, $V(\mathbf{r})$ is computed from the $\rho(\mathbf{r})$ obtained by combining "transferable atom equivalents" [81].

2.3. Molecular Surfaces

As stated earlier in this chapter, we feel that the most effective use of the electrostatic potential in relation to reactive behavior is in analyzing and predicting noncovalent interactions, with $V(\mathbf{r})$ being computed on an outer surface of the molecule. This poses the question of how the latter should be defined, because there is no rigorous basis for it. A common approach has been to use a set of intersecting spheres centered on the individual nuclei, with van der Waals or other appropriate radii

[82–85]. However, our preference is to follow the suggestion of Bader et al. [86] and take the surface to be an outer contour of the molecular electronic density. This has the important advantage that it reflects features specific to the particular molecule, such as lone pairs or strained bonds. We normally choose $\rho(\mathbf{r}) = 0.001$ electrons/bohr³, but we have confirmed that other low values of $\rho(\mathbf{r})$, e.g., 0.002 electrons/bohr³, would serve equally well [87].

For illustrative purposes, Fig. 1 shows our computed electrostatic potentials on the surfaces of cytosine (**1**) and guanine (**2**). Only three ranges of values are given; however, this could of course be increased by using various colors.

In addition to such three-dimensional surfaces, there are of course other ways in which $V(\mathbf{r})$ can be presented. One alternative involves two-dimensional planes through or above the molecule. This was the most widely used procedure in the early years of computing $V(\mathbf{r})$ [2,14–17,19,20,35–38]. Another technique is to plot the three-dimensional contour corresponding to a single selected value of $V(\mathbf{r})$ [3,88]. However, all of the examples to be discussed in the remainder of this chapter will be based on $V(\mathbf{r})$ calculated on molecular surfaces defined by $\rho = 0.001$ electron/bohr³. This is designated as $V_S(\mathbf{r})$.

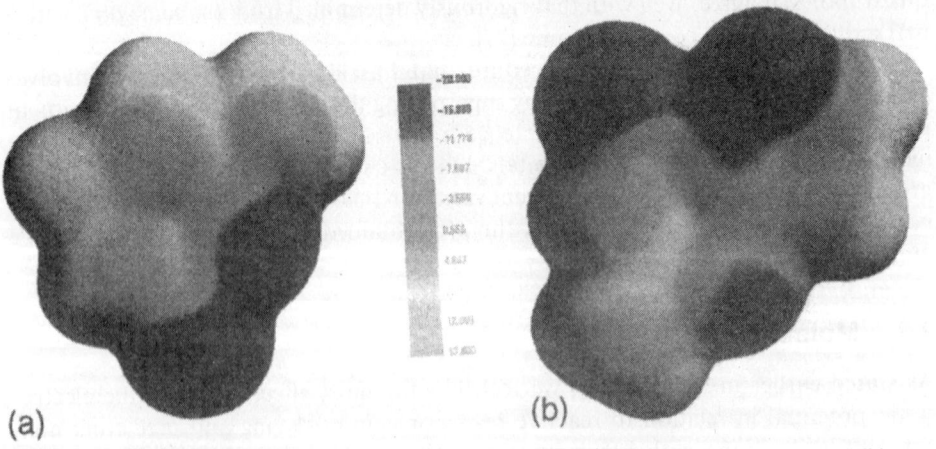

Figure 1 Electrostatic potentials on the molecular surfaces of (a) cytosine, **1**, and (b) guanine, **2**, computed at the Hartree–Fock 6-31G* level. Color ranges, in kcal/mol: *red*, more positive than 17; *blue*, more negative than −20 (see legend). The relative positions of the molecules are such that the portions that hydrogen bond are facing each other, showing how the extended positive and negative regions will interact. (See color plate at end of chapter.)

3. SOME APPLICATIONS

3.1. Hydrogen Bonding

One of the early applications of molecular electrostatic potentials was in relation to hydrogen bonding. The V_{min} (most negative potentials) were found to be effective in identifying sites at which hydrogen bonds would be accepted [16,89–91]. Furthermore, Kollman et al. [89] showed that the calculated energy of the interaction between hydrogen fluoride and a series of acceptors correlates well with the value of $V(\mathbf{r})$ at a fixed distance from the latter.

With the focus shifting to molecular surfaces, it became possible to address donating as well as accepting tendencies (termed hydrogen bond acidity and basicity, respectively). This is because the most positive potentials on a molecular surface (the $V_{S,max}$) do have meaning with respect to attracting nucleophiles [21], whereas the V_{max} in the three-dimensional molecular space do not (Sec. 1.2). Kamlet, Taft, and Abraham et al. [92–96] had developed experimentally based measures of hydrogen bond acidity and basicity for a series of compounds. We demonstrated that these properties can be quantitatively expressed in terms of just $V_{S,max}$ and $V_{S,min}$ [55,97]:

$$\text{H-bond acidity} = \alpha_1 V_{S,max} - \beta_1 \qquad \alpha_1, \beta_1 > 0 \tag{9}$$

$$\text{H-bond basicity} = \alpha_2 |V_{S,min}| - \beta_2 \qquad \alpha_2, \beta_2 > 0 \tag{10}$$

The correlation coefficients for Eqs. (9) and (10) were 0.991 and 0.920, respectively; the latter can be further improved by introducing an additional variable [98]. The reason that $V_{S,min}$ alone is not quite as effective for representing basicity as is $V_{S,max}$ for acidity may involve the fact that the latter is always near a hydrogen whereas the former is associated with various acceptor atoms in the molecules investigated. A pleasing feature of Eqs. (9) and (10) is that they satisfy different classes of compounds; that is, they are not family-dependent.

When a molecule has several possible donating or accepting sites in close proximity, the situation may become somewhat more complicated. There may now be several $V_{S,max}$ and/or $V_{S,min}$, the positions of which reflect not only specific atoms but also the overlapping of the various positive and negative regions. For example, cytosine (1) has three $V_{S,max}$, near its primary and secondary amino hydrogens [28]; it has a $V_{S,min}$ associated with its oxygen, but not with N_3 (as would have been anticipated), the negative surface potential of which has merged with that of the oxygen. For the same reason, N_7 of guanine (2) has no $V_{S,min}$, in contrast to the oxygen and N_3 [28]; guanine also has three $V_{S,max}$. To understand the multiple hydrogen bonding between cytosine and guanine, it is necessary to look for extended surface regions on the two molecules that complement each other and promote interaction. These can be seen in Fig. 1; the positive amino hydrogen and negative N_3 and oxygen potentials of cytosine exactly fit together with the negative oxygen and neighboring positive hydrogen regions of guanine.

An analogous example is provided by the diaryl ureas, of which the parent molecule is **3**. These are found to preferentially crystallize in homomeric fashion, i.e., with their own kind, as in **4**, rather than forming cocrystals with guest molecules [99]. This is true even for solutions containing strong hydrogen bond acceptors such as dimethylsulfoxide, which would be expected to interact with the amino hydrogens of the urea. We attribute this tendency for homomeric crystallization to the presence of extended negative potentials along the upper edges of the ureas, and positive ones along the lower edges [100]. The nonlocalized attraction between the complementary portions of adjoining molecules, e.g., **4**, is evidently sufficiently strong to preclude interaction with a guest molecule.

3

4

3.2. Interactions with Biological Receptors: A Qualitative Treatment of Dibenzo-*p*-dioxins

The interpretation of biological recognition interactions has continued, since the earliest days, to be one of the most active areas of application of electrostatic potentials [2,3,16,42–44,68–71,78,101–114]. The original studies of two-dimensional $V(\mathbf{r})$ plots have evolved into detailed quantitative characterization of the potentials on molecular surfaces and the investigation of factors such as shape, similarity, and flexibility. In this section, we shall focus on one specific example: a qualitative analysis of the molecular determinants of toxicity among the dibenzo-*p*-dioxins. We shall proceed to some more quantitative treatments in a later section, after establishing a basis for them.

Dibenzo-*p*-dioxin (**5**) is the parent molecule of a large family of halogenated derivatives, involving especially chlorine and/or bromine in various numbers and positions on the two outer rings. Depending on the extent and distribution of the substitution, these compounds display a wide range of toxicities [115,116], the two extremes being represented by **5** (virtually nontoxic) and the notorious 2,3,7,8-

tetrachlorodibenzo-*p*-dioxin (TCDD, **6**). The problems associated with TCDD and some of the others in this family include carcinogenesis, hepatoxicity, gastric lesions, loss of lymphoid tissue, urinary tract hyperplasia, chloracne, and acute loss of weight [116]. These effects involve an initial interaction with a cytosolic receptor, the structure of which is not known but has been envisioned as being porphine-like (**7**) [117,118].

5 TCDD, **6** **7**

8 **9**

Poland and Knutson [116] identified several structural features of the halogenated dibenzo-*p*-dioxins that are associated with high levels of both toxicity and receptor binding. The molecules should be essentially planar and rectangular in shape, with at least three of the four lateral positions (2,3,7,8; see **5**) bearing substituents; however, at least one ring position should be unsubstituted. Activity decreases from bromine to chlorine to fluorine.

In a series of studies [110,113,119–122], we have examined the computed electrostatic potentials of 12 variously halogenated dibenzo-*p*-dioxins, including **5** and **6**, as well as some related molecules, such as **8** and **9**. The toxicity of **8** is similar to that of TCDD (indicating that the oxygens of the latter are not needed), whereas **9** is much less toxic [116,123]. The electrostatic potentials of the nontoxic **5** and the very dangerous **6**, whether in planes above the molecules or on their surfaces, differ as markedly as do their biological activities, as is shown in Fig. 2. The $V_S(\mathbf{r})$ of **5** is positive above the hydrogens in the lateral regions, weakly negative above the aromatic rings, and more strongly negative over the oxygens. In complete contrast, TCDD (**6**) is negative above the chlorines (the lateral regions), very weakly so over the oxygens, and positive elsewhere; the electron-withdrawing effect of the chlorines has sufficed to almost eliminate, on the molecular surface, the relatively strong negative potentials normally observed near oxygens.

When we looked at the surface potentials of other halogenated dibenzo-*p*-dioxins, we found that their biological activities correlate with the extent to which their surface potentials resemble that of TCDD. Thus high levels of receptor binding and toxicity are associated with negative regions above all or most of the lateral posi-

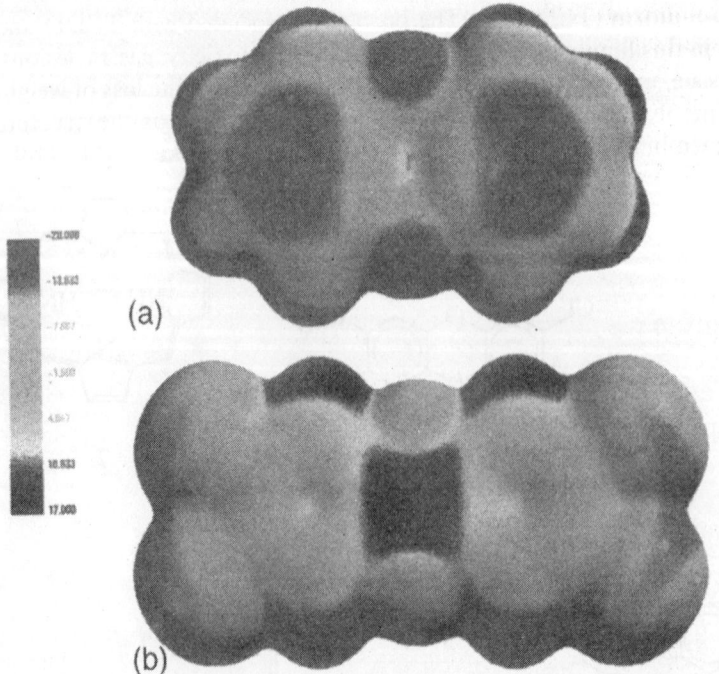

Figure 2 Electrostatic potentials on the molecular surfaces of (a) dibenzo-*p*-dioxin, **5**, and (b) 2,3,7,8-tetrachlorodibenzo-*p*-dioxin (TCDD), **6**, computed at the Hartree–Fock STO-5G* level. Color ranges, in kcal/mol: *red*, more positive than 17; *blue*, more negative than −20 (see legend). (See color plate at end of chapter.)

tions, separated by a large positive area over the three rings. The negative potentials of the oxygens must accordingly be small and quite weak, and nearly absent on the molecular surface. The $V_S(\mathbf{r})$ of **8** matches this pattern very well, and indeed its activity is similar to that of TCDD. On the other hand, **9** has protruding significant negative regions above the oxygens, and its activity is greatly diminished. Why is this so? An answer is provided by modeling studies of the interaction of TCDD with the proposed porphine receptor, **7** [117,118], which show the dioxin oxygens to be roughly above the doubly coordinated (unsubstituted) nitrogens of **7**. This means that if the negative regions of the oxygens are even somewhat prominent, as in the case of **9**, then there are likely to be repulsive interactions with the nitrogen lone pairs, which would inhibit binding to the receptor.

3.3. The General Interaction Properties Function (GIPF) Procedure

In analyzing the electrostatic potentials on molecular surfaces, our emphasis so far has been on the locations and magnitudes of the most positive and most negative values, $V_{S,max}$ and $V_{S,min}$, and on the qualitative general pattern of $V_S(\mathbf{r})$. However, already some years ago, it seemed desirable to develop the means for *quantitatively* characterizing the overall features of $V_S(\mathbf{r})$. For this purpose, we gradually introduced certain global statistically defined quantities, to complement the site-specific $V_{S,max}$ and $V_{S,min}$. The global ones include:

(a) the average positive and negative potentials over the entire surface, $\overline{V}_S^+$ and $\overline{V}_S^-$,

$$\overline{V}_S^+ = \frac{1}{\alpha}\sum_{j=1}^{\alpha} V_S^+(\mathbf{r}_j) \tag{11}$$

$$\overline{V}_S^- = \frac{1}{\beta}\sum_{k=1}^{\beta} V_S^-(\mathbf{r}_k) \tag{12}$$

(b) the average deviation, Π,

$$\Pi = \frac{1}{n}\sum_{i=1}^{n} |V_S(\mathbf{r}_i) - \overline{V}_S| \tag{13}$$

(c) the positive, negative, and total variances, σ_+^2, σ_-^2, and σ_{tot}^2,

$$\sigma_{tot} = \sigma_+^2 + \sigma_-^2 = \frac{1}{\alpha}\sum_{j=1}^{\alpha} [V_S^+(\mathbf{r}_j) - \overline{V}_S^+]^2 + \frac{1}{\beta}\sum_{k=1}^{\beta} [V_S^-(\mathbf{r}_k) - \overline{V}_S^-]^2 \tag{14}$$

(d) a balance parameter, ν,

$$\nu = \frac{\alpha_+^2 \sigma_-^2}{[\sigma_{tot}^2]^2} \tag{15}$$

In Eq. (13), $\overline{V}_S$ is the overall average of $V_S(\mathbf{r})$,

$$\overline{V}_S = (\alpha\overline{V}_S^+ + \beta\overline{V}_S^-)/(\alpha + \beta) \tag{16}$$

We interpret Π as a measure of internal charge separation, which is present even in molecules that have zero dipole moment, e.g., p-dinitrobenzene. It has been shown to correlate with various empirical indices of polarity [124,125]. The variances (σ_+^2, σ_-^2, and σ_{tot}^2) indicate the variability, or range, of the positive, negative, and overall surface potentials [126,127]. Because of the terms in Eq. (14) being squared, the variances are particularly sensitive to the extrema of $V_S(\mathbf{r})$, i.e., $V_{S,max}$ and $V_{S,min}$; as a result, σ_{tot}^2 tends to be much larger in magnitude than Π. The two quantities focus on different aspects of $V_S(\mathbf{r})$, and in fact often vary in opposite directions. Finally, ν is intended to quantify the degree of balance between the strengths of the positive and negative potentials on the surface [126,127]. When σ_+^2 and σ_-^2 are equal, whether large or small, then ν reaches its maximum value of 0.250. Thus the closer that ν is to 0.250, the better is the molecule able to interact through both its positive and negative regions (whether strongly or weakly).

The global quantities defined by Eqs. (11)–(16), plus the positive and negative areas, A_S^+ and A_S^-, and the site-specific $V_{S,max}$ and $V_{S,min}$, provide a detailed characterization of the electrostatic potential on a molecular surface. We have found that subsets of these quantities can be used to analytically represent, with good accuracy, a variety of liquid-, solution-, and solid-phase macroscopic properties that depend on noncovalent interactions. Among these are heats of vaporization, sublimation and fusion, boiling points and critical constants, solubilities and solvation energies, partition coefficients, surface tensions, diffusion constants, viscosities, and liquid and crystal densities. This work has been reviewed several times [125,128–130].

Our approach involves compiling an experimental database for the property of interest and then using a statistical analysis package to determine which subset of our computed global and site-specific quantities, in some combination, provides the best fit of the data. Because one of our objectives is to gain insight into the nature of the interactions, we try to represent each property in terms of as few variables as possible (typically three or four), to not obscure the key physical factors that are involved. We also make the relationships as general as possible, preferably covering a wide variety of compounds, although the correlations could probably be improved by separately treating different classes, e.g., halogenated alkanes, polycyclic aromatics, etc. We conceptually summarize our approach in terms of a general interaction properties function (GIPF),

$$\text{Property} = f[V_{S,\min}, V_{S,\max}, \overline{V}_S^+, \overline{V}_S^-, \Pi, \sigma_+^2, \sigma_-^2, \sigma, \nu, A_S^+, A_S^-] \tag{17}$$

again emphasizing that normally only three or four of the quantities on the right side of Eq. (17) are used in representing a given property.

For illustrative purposes, some of our GIPF expressions are shown in Eqs. (18)–(21):

Heat of vaporization [131] $= \alpha_1 A_S^{0.5} + \beta_1 (\nu \sigma_{\text{tot}}^2)^{0.5} - \gamma_1$ \hfill (18)

Compounds : 41	Correlation coefficient : 0.965
Range : 37.7 kJ/mol	Root mean square error : 2.4 kJ/mol

Liquid density [132] $= \alpha_2 (M/A_S) + \beta_2 \Pi + \gamma_2$ \hfill (19)

Compounds : 61	Correlation coefficient : 0.982
Range : 1.173 g/cm^3	Root mean square error : 0.055 kJ/mol

Diffusion constant in gelatin [133] $= \alpha_3 A_S^{-1} - \beta_3 \sigma_+^2 + \gamma_3 \sigma_-^2 - \delta_3$ \hfill (20)

Compounds : 10	Correlation coefficient : 0.990
Range : 1.53 × 10^{-7} cm^2/s	Root mean square error : 0.09 × 10^{-7} cm^2/s

Log(octanol/water partition coefficient) [134] $= \alpha_4 A_S - \beta_4 (\sigma_-^2) - \gamma_4 A_S \Pi - \delta_4$

\hfill (21)

Compounds : 70	Correlation coefficient : 0.961
Range : 6.56	Root mean square error : 0.44

In these equations, A_S is the total surface area, $A_S = A_S^+ + A_S^-$, M is the molecular mass, and the α_i, β_i, γ_i, and δ_i are all positive. More extensive compilations of GIPF relationships, with references to the original work, are given elsewhere [128–130, 135]. It should be noted that liquid-, solution-, and solid-phase properties are being expressed in terms of quantities calculated for a single (i.e., gas phase) molecule; no explicit account is being taken of the surroundings.

Once the GIPF representation of a particular property has been developed, no experimental data are needed to apply it, because all of the variables in Eq. (17)

can be computationally obtained (as can the molecular geometry). Thus the property can be predicted even for compounds that have not yet been synthesized, meaning that the GIPF procedure can be used in molecular design [135,136].

3.4. Interactions with Biological Receptors: Quantitative Treatments of Anti-HIV Drugs

In recent years, we have begun to analyze interactions in biological systems in terms of the GIPF approach. We began by computing the molecular surface electrostatic potentials and the global and site-specific GIPF quantities for a group of 19 anti-convulsants of various types, including hydantoins, barbiturates, carbamazepines, succinimides, etc. [137]. Most of them are derivatives of five-, six- or seven-membered heterocyclic rings, usually having ureide and/or amide linkages. A striking feature of their surface potentials is the similarity of their internal charge separations, as measured by Π. Our experience, having computed Π for roughly 150 organic mole-cules [125,138], is that it varies from about 2 kcal/mol for alkanes to mid-20's for molecules with several strongly electron-attracting substituents, e.g., some polynitro derivatives. However, for 16 of the 19 anticonvulsants, Π is between 10.0 and 13.0; the most extreme value is 14.4. The internal charge separations are accordingly intermediate in magnitude but quite restricted. We have found qualitatively the same feature, although with somewhat different ranges of Π, in other families of drugs: tetracyclines (antibiotics) [69], reverse transcriptase inhibitors [70], and cocaine ana-logs [139]. This suggests that each of these drug types needs to have a certain optimum balance between hydrophobicity and hydrophilicity, to permit the necessary move-ment between media of different polarities.

Among the drug systems just mentioned, only for the reverse transcriptase (RT) inhibitors did we have available sufficient experimental data to develop quantitative GIPF relationships for their activities. These are anti-HIV agents, which function by inhibiting the RT enzyme that promotes the reverse transcription of genomic RNA into double-stranded DNA, a key step in HIV replication [140,141]. We used data-bases compiled by Garg et al. [141] to obtain expressions for the anti-HIV potencies of three families of RT inhibitors [70,71]. The correlation coefficients are between 0.930 and 0.952. Two of these equations involve only global GIPF quantities and emphasize the positive surface potentials, suggesting that the RT inhibiting inter-actions of these molecules involve extended portions of their surfaces, whereas the third includes site-specific quantities, consistent with the interaction having some localized aspects.

In the context of RT inhibition, we are currently investigating whether better correlations might result from applying the GIPF approach to portions of the mole-cule rather than the whole. This might identify the source of the activity, or in other instances might indicate that it is quite delocalized.

3.5. Miscellaneous Interactions

It has been known for some time that certain organic halides can favorably interact through their C–X bonds (X = Cl, Br, I) with heteroatom lone pairs and with aro-matic π electrons [142–144]. This phenomenon, which appears to contradict the nor-mally negative nature of halogen substituents, has been termed "halogen bonding" [145]. An explanation can be given in terms of molecular surface electrostatic potentials. We have computed $V_S(\mathbf{r})$ for a series of halogenated methanes [146,147], and found that as the number of halogen atoms increases, there develops a significant positive potential on the outer tip of each atom X, centered on the C–X axis. The remainder of the surface near X is negative, as expected. This positive region (which is not observed on fluorines) can interact with negative potentials on other molecules. Halogen bonding may account for the fact that various non-hydrogen-containing halocarbons are able to disrupt hydrogen bonds, by displacing the donors [148]; this has been linked to the anesthetic potencies of the former.

Halogen bonding also manifests itself in the relative orientations of halogen derivatives in the crystalline state [149]. Indeed, the modes of interaction in many non-hydrogen-bonded noncovalent systems, ranging from gas phase complexes to molecular crystals, can be satisfactorily rationalized in terms of molecular surface electrostatic potentials [44,55,150]. In several instances, we have used this approach to explain anomalously high measured solid densities [151,152].

Another very important type of noncovalent interaction is that between solutes and solvents. We have developed GIPF relationships in which the free energies of solvation in seven different solvents, with various polarities, are expressed in terms of quantities characterizing the solute's molecular surface electrostatic potentials [153,154]. However, there have been many more elaborate treatments that explicitly evaluate the energy of the interaction between the solute and the solvent; the latter may be described, e.g., as a dielectric continuum, as a fixed lattice, or in terms of individual molecules. Detailed accounts can be found in several reviews [23,155–158].

4. SUMMARY

The electrostatic potential is a fundamental determinant of intrinsic atomic and molecular properties (e.g., energies, chemical potentials, covalent radii) as well as a guide to reactive behavior, especially in noncovalent interactions. In the present chapter, we have focused on the latter feature, in both qualitative and quantitative terms. In particular, we have sought to demonstrate how the electrostatic potentials on molecular surfaces, when effectively characterized, permit the correlation and prediction of a wide array of condensed-phase macroscopic physical properties, as well as interactions with biological receptors. As these procedures continue to evolve in scope and reliability, we can expect to gain increasing insight into chemical and biological processes, and an expanding capacity for designing compounds with specific desired features.

REFERENCES

1. Coalson RD, Beck TL. Poisson–Boltzmann type equations: numerical methods. In: Schleyer PvR, ed. Encyclopedia of Computational Chemistry. Vol. 3. New York: Wiley, 1998:2086–2100.
2. Politzer P, Truhlar DG, eds. Chemical Applications of Atomic and Molecular Electrostatic Potentials. New York: Plenum, 1981.
3. Naray-Szabo G, Ferenczy GG. Molecular Electrostatics. Chem Rev 1995; 95:829–847.
4. Hohenberg P, Kohn W. Inhomogeneous electron gas. Phys Rev, B 1964; 136:864–871.
5. Hellmann H. Einfuehrung in die Quantenchemie. Leipzig: Deuticke, 1937.

6. Feynman RP. Forces in molecules. Phys Rev 1939; 56:340–343.
7. Politzer P. Observations on the significance of the electrostatic potentials at the nuclei of atoms and molecules. Isr J Chem 1980; 19:224–232.
8. Politzer P. Relationships between the energies of atoms and molecules and the electrostatic potentials at their nuclei. In: Politzer P, Truhlar DG, eds. Chemical Applications of Atomic and Molecular Electrostatic Potentials. New York: Plenum, 1981:7–28.
9. Politzer P. Atomic and molecular energy and energy difference formulae based upon electrostatic potentials at nuclei. In: March NH, Deb BM, eds. The Single-Particle Density in Physics and Chemistry. London: Academic, 1987:59–72.
10. March NH. Electron Density Theory of Atoms and Molecules. London: Academic, 1992:245–256.
11. Politzer P, Lane P, Murray JS. The fundamental significance of electrostatic potentials at nuclei. In: Sen KD, ed. Reviews in Modern Quantum Chemistry: A Celebration of the Contributions of RG Parr. Vol. 1. Singapore: World Scientific, 2002:63–84.
12. Weinstein H, Politzer P, Srebrenik S. A misconception concerning the electronic density distribution of an atom. Theor Chim Acta 1975; 38:159–163.
13. Sen KD, Politzer P. Characteristic features of the electrostatic potentials of singly-negative monatomic ions. J Chem Phys 1989; 90:4370–4372.
14. Scrocco E, Tomasi J. The electrostatic molecular potential as a tool for the interpretation of molecular properties. Topics in Current Chemistry. No 42. Berlin: Springer-Verlag, 1973:95–170.
15. Scrocco E, Tomasi J. Electronic molecular structure, reactivity and intermolecular forces: an euristic interpretation by means of electrostatic molecular potentials. Adv Quantum Chem 1978; 11:115–193.
16. Politzer P, Daiker KC. Models for chemical reactivity. In: Deb BM, ed. The Force Concept in Chemistry. New York: Van Nostrand Reinhold, 1981:294–387.
17. Politzer P, Murray JS. Molecular electrostatic potentials and chemical reactivity. In Lipkowitz KB, Boyd DB, eds. Reviews in Computational Chemistry. Vol. 2. New York: VCH, 1991:273–312.
18. Pathak RK, Gadre SR. Maximal and minimal characteristics of molecular electrostatic potentials. J Chem Phys 1990; 93:1770–1773.
19. Politzer P, Laurence PR, Abrahmsen L, Zilles BA, Sjoberg P. The aromatic C–NO$_2$ bond as a site for nucleophilic attack. Chem Phys Lett 1984; 111:75–78.
20. Murray JS, Lane P, Politzer P. Electrostatic potential analysis of the π regions of some naphthalene derivatives. J Mol Struct (Theochem) 1990; 209:163–175.
21. Sjoberg P, Politzer P. Use of the electrostatic potential at the molecular surface to interpret and predict nucleophilic processes. J Phys Chem 1990; 94:3959–3961.
22. Murray JS, Lane P, Brinck T, Politzer P, Sjoberg P. Electrostatic potentials on the molecular surfaces of cyclic ureides. J Phys Chem 1991; 95:844–848.
23. Tomasi J, Mennucci B, Cammi R. MEP: a tool for interpretation and prediction. from molecular structure to solvation effects. In: Murray JS Sen K, eds. Molecular Electrostatic Potentials. Amsterdam: Elsevier, 1996:1–103.
24. Murray JS, Sen K, eds. Molecular Electrostatic Potentials. Amsterdam: Elsevier, 1996.
25. Tour JM, Kozaki M, Seminario JM. Molecular scale electronics: a synthetic/computational approach to digital computing. J Am Chem Soc 1998; 120:8486–8493.
26. Brinck T, Murray JS, Politzer P. Molecular surface electrostatic potentials and local ionization energies of group V–VII hydrides and their anions: relationships for aqueous and gas-phase acidities. Int J Quantum Chem 1993; 48:73–88.
27. Murray JS, Politzer P. Average local ionization energies: significance and applications. In: Parkanyi C, ed. Theoretical Organic Chemistry. Amsterdam: Elsevier, 1998:189–202.
28. Murray JS, Peralta-Inga Z, Politzer P, Ekanayake K, LeBreton P. Computational characterization of nucleotide bases: molecular surface electrostatic potentials and local

ionization energies, and local polarization energies. Int J Quantum Chem 2001; 83:245–254.

29. Politzer P, Murray JS, Concha MC. The complementary roles of molecular surface electrostatic potentials and average local ionization energies with respect to electrophilic processes. Int J Quantum Chem, 2002; 88:19–27.

30. Clark DT, Adams DB. Model potential energy surfaces for approach of an electrophile to acetylene and fluoroacetylene. Tetrahedron 1973; 29:1887–1889.

31. Bartlett RJ, Weinstein H. Theoretical treatment of multiple site reactivity in large molecules. Chem Phys Lett 1975; 30:441–447.

32. Bertran J, Silla E, Fernandez-Alonso JL. The van der Waals interactions as a tool for the interpretation of aromatic substitutions. Tetrahedron 1975; 31:1093–1096.

33. Bonaccorsi R, Scrocco E, Tomasi J. A representation of the polarization term in the interaction energy between a molecule and a point-like charge. Theor Chim Acta 1976; 43:63–73.

34. Chang S-Y, Weinstein H, Chou D. Perturbation treatment of multiple site reactivity: molecule–molecule interactions. Chem Phys Lett 1976; 42:145–150.

35. Moriishi H, Kikuchi O, Suzuki K, Klopman G. Reaction potential map analysis of chemical reactivity—III. Theor Chim Acta 1984; 64:319–338.

36. Francl MM. Polarization corrections to electrostatic potentials. J Phys Chem 1985; 89:428–433.

37. Alkorta I, Villar HO, Perez JJ. Effect of the basis set on the computation of molecular polarization. J Phys Chem 1993; 97:9113–9119.

38. Alkorta I, Perez JJ, Villar HO. Molecular polarization maps as a tool for studies of intermolecular interactions and chemical reactivity. J Mol Graph 1994; 12:3–13.

39. Hirschfelder JO, Curtiss CF, Bird RB. Molecular Theory of Gases and Liquids. New York: Wiley, 1954.

40. Hirschfelder JO. Intermolecular forces. Molecular Forces. Amsterdam: North-Holland, 1967:73–113.

41. Hunt KLC. Dispersion dipoles and dispersion forces: proof of Feynman's "conjecture" and generalization to interacting molecules of arbitrary symmetry. J Chem Phys 1990; 92:1180–1187.

42. Barnett G, Trsic M, Willette RE, eds. QuaSAR Quantitative Structure Activity Relationships of Analgesics, Narcotic Antagonists, and Hallucinogens. NIDA Research Monograph 22. Rockville, MD: National Institute on Drug Abuse, 1978.

43. Politzer P, Laurence PR, Jayasuriya K. Molecular electrostatic potentials: an effective tool for the elucidation of biochemical phenomena. Environ Health Perspect 1985; 61:191–202.

44. Price SL. Applications of realistic electrostatic modelling to molecules in complexes, solids and proteins. J Chem Soc, Faraday Trans 1996; 92:2997–3008.

45. Moller C, Plesset MS. Note on an approximation treatment for many-electron systems. Phys Rev 1934; 46:618–622.

46. Pople JA, Seeger R. Electron density in Moller–Plesset theory. J Chem Phys 1975; 62:4566.

47. Smith VH Jr. Theoretical determination and analysis of electronic charge distributions. Phys Scr 1977; 15:147–162.

48. Lauer G, Meyer H, Schulte K-W, Schweig A, Hase H-L. Correlated electron density of N_2. Chem Phys Lett 1979; 67:503–507.

49. Gatti C, MacDougall PJ, Bader RFW. Effect of electron correlation on the topological properties of molecular charge distributions. J Chem Phys 1988; 88:3792–3804.

50. Boyd RJ, Wang L-C. The effect of electron correlation on the topological and atomic properties of the electron density distributions of molecules. J Comp Chem 1989; 10:367–375.

51. Wang L-C, Boyd RJ. The effect of electron correlation on the electron density distri-

butions of molecules: comparison of perturbation and configuration interaction methods. J Chem Phys 1989; 90:1083–1090.

52. Daudel R, Leronzo H, Cimiraglia R, Tomasi J. Dependence of the electrostatic molecular potential upon the basis set and the method of calculation of the wave function. Case of the ground[3] $A_1(\pi\rightarrow\pi^*)$ and $^1A_1(\pi\rightarrow\pi^*)$ states of formaldehyde. Int J Quantum Chem 1978; 13:537–552.

53. Seminario JM, Murray JS, Politzer P. First-principles theoretical methods for the calulation of electronic charge densities and electrostatic potentials. In: Jeffrey GA, Piniella JF, eds. The Application of Charge Density Research to Chemistry and Drug Design. New York: Plenum, 1991:371–381.

54. Luque FJ, Orozco M, Illas F, Rubio J. Effect of electron correlation on the electrostatic potential distribution of molecules. J Am Chem Soc 1991; 113:5203–5211.

55. Murray JS, Politzer P. The molecular electrostatic potential: a tool for understanding and predicting molecular interactions. In: Sapse A-M, ed. Molecular Orbital Calculations for Biological Systems. New York: Oxford University Press, 1998:49–84.

56. Luque FJ, Illas F, Orozco M. Comparative study of the molecular electrostatic potential obtained from different wavefunctions. Reliability of the semiempirical MNDO wavefunction. J Comput Chem 1990; 11:416–430.

57. Luque FJ, Orozco M. Reliability of the AM1 wavefunction to compute molecular electrostatic potentials. Chem Phys Lett 1990; 168:269–275.

58. Ferenczy GG, Reynolds CA, Richards WG. Semiempirical AM1 electrostatic potentials and AM1 electrostatic potential derived charges: a comparison with ab initio values. J Comput Chem 1990; 11:159–169.

59. Krack M, Koster AM, Jug K. Approximate molecular electrostatic potentials from semiempirical wavefunctions. J Comput Chem 1997; 18:301–312.

60. Murray JS, Seminario JM, Concha MC, Politzer P. An analysis of molecular electrostatic potentials obtained by a local density functional approach. Int J Quantum Chem 1992; 44:113–122.

61. Geerlings P, DeProft F, Martin JML. Density functional theory concepts and techniques for studying molecular charge distributions and related properties. In: Seminario JM, ed. Recent Developments and Applications of Modern Density Functional Theory. Amsterdam: Elsevier, 1996:773–809.

62. Soliva R, Orozco M, Luque FJ. Suitability of density functional methods for calculation of electrostatic properties. J Comput Chem 1997; 18:980–991.

63. Leboeuf M, Koster AM, Jug K, Salahub DR. Topological analysis of the molecular electrostatic potential. J Chem Phys 1999; 111:4893–4905.

64. Seminario JM, Politzer P, eds. Modern Density Functional Theory: A Tool for Chemistry. Amsterdam: Elsevier, 1995.

65. Seminario JM, ed. Recent Developments and Applications of Modern Density Functional Theory. Amsterdam: Elsevier, 1996.

66. Springborg M, ed. Density–Functional Methods in Chemistry and Materials Science. New York: Wiley, 1997.

67. Frisch MJ, Trucks GW, Schlegel HB, Scuseria GE, Robb MA, Cheeseman JR, Zakrezewski VG, Montgomery JA, Stratmann RE, Burant JC, Dappich S, Millam JM, Daniels AD, Kudin KN, Strain MC, Farkas O, Tomasi J, Barone V, Cossi M, Cammi R, Mennucci B, Pomelli C, Adamo C, Clifford S, Ochterski J, Petersson G, Aayala PY, Cui Q, Morokuma K, Malick DK, Rubuck AD, Raghavachari K, Foresman JB, Cioslowski J, Ortiz JV, Stefanov BB, Liu G, Liashenko A, Piskorz P, Komaromi I, Gomperts R, Martin RL, Fox DJ, Keith T, Al-Laham MA, Peng CY, Nanayakkara A, Gonzalez C, Challacombe M, Gill PMW, Johnson BG, Chen W, Wong MW, Andres JL, Head-Gordon M, Replogle ES, Pople JA. Gaussian 98, Revision A.5. Pittsburgh, PA: Gaussian, Inc., 1998.

68. Murray JS, Peralta-Inga Z, Politzer P. Computed molecular surface electrostatic poten-

tials of the nonionic and zwitterionic forms of glycine, histidine, and tetracycline. Int J Quantum Chem 2000; 80:1216–1223.

69. Hussein W, Walker CG, Peralta-Inga Z, Murray JS. Computed electrostatic potentials and average local ionization energies on the molecular surfaces of some tetracyclines. Int J Quantum Chem 2001; 82:160–169.

70. Galvez Gonzalez O, Murray JS, Peralta-Inga Z, Politzer P. Computed molecular surface electrostatic potentials of two groups of reverse transcriptase inhibitors: relationships to anti-HIV-1 activities. Int J Quantum Chem 2001; 83:115–121.

71. Politzer P, Murray JS, Peralta-Inga Z. Molecular surface electrostatic potentials in relation to noncovalent interactions in biological systems. Int J Quantum Chem 2001; 85:676–684.

72. Rein R. On physical properties and interactions of polyatomic molecules: with applications to molecular recognition in biology. Adv Quantum Chem 1973; 7:335–396.

73. Chirlian LE, Francl MM. Atomic charges derived from electrostatic potentials: a detailed study. J Comput Chem 1987; 8:894–905.

74. Woods RJ, Khalil M, Pell W, Moffat SH, Smith VH Jr. Derivation of net atomic charges from molecular electrostatic potentials. J Comput Chem 1990; 11:297–310.

75. Williams DE. Net atomic charge and multipole models for the ab initio molecular electrostatic potential. Lipkowitz KB, Boyd DB, eds. Reviews in Computational Chemistry. Vol. 2. New York: VCH, 1991:219–271.

76. Rabinowitz JR, Namboodiri K, Weinstein H. A finite expansion method for the calculation and interpretation of molecular electrostatic potentials. Int J Quantum Chem 1986; 29:1697–1704.

77. Murray JS, Grice ME, Politzer P, Rabinowitz JR. Evaluation of a finite multipole expansion technique for the computation of electrostatic potentials of dibenzo-*p*-dioxins and related systems. J Comput Chem 1990; 11:112–120.

78. Naray-Szabo G. Electrostatic isopotential maps for large biomolecules. Int J Quantum Chem 1979; 16:265–272.

79. Bonaccorsi R, Ghio C, Scrocco E, Tomasi J. The effect of intramolecular interactions on the transferability properties of localized descriptions of chemical groups. Isr J Chem 1980; 19:109–126.

80. Lavery R, Pullman B. Molecular electrostatic potential on the surface envelopes of macromolecules: B-DNA. Int J Quantum Chem 1981; 20:259–272.

81. Breneman C, Martinov M. The use of the electrostatic potential field in QSAR and QSPR. In: Murray JS, Sen KD, eds. Molecular Electrostatic Potentials: Concepts and Applications. Amsterdam: Elsevier, 1996:143–179.

82. Connolly ML. Computation of molecular volume. J Am Chem Soc 1985; 107:1118–1124.

83. Du Q, Arteca GA. Derivation of fused-sphere molecular surfaces from properties of the electrostatic potential distribution. J Comput Chem 1996; 17:1258–1268.

84. Brickmann J, Exner T, Keil M, Marhofer R, Moeckel G. Molecular models: visualization. In: Schleyer PvR, ed. Encyclopedia of Computational Chemistry. Vol. 3. New York: Wiley, 1998:1678–1693.

85. Connolly ML. Molecular surface and volume. In: Schleyer PvR, ed. Encyclopedia of Computational Chemistry. Vol. 3. New York: Wiley, 1998:1698–1703.

86. Bader RFW, Carroll MT, Cheeseman JR, Chang C. Properties of atoms in molecules: atomic volumes. J Am Chem Soc 1987; 109:7968–7979.

87. Murray JS, Brinck T, Grice ME, Politzer P. Correlations between molecular electrostatic potentials and some experimentally-based indices of reactivity. J Mol Struct (Theochem) 1992; 256:29–45.

88. Matthew JB. Electrostatic and dynamic aspects of macromolecular recognition. In: Beveridge DL, Lavery R, eds. Theoretical Biochemistry & Molecular Biophysics. Vol. 2: Proteins. Schenectady, NY: Adenine, 1991:107–120.

89. Kollman PA, McKelvey J, Johansson A, Rothenberg S. Theoretical studies of hydrogen-bond dimers. J Am Chem Soc 1975; 97:955–965.

90. Leroy G, Louterman-Leloup G, Ruelle P. Contribution to the theoretical study of the hydrogen bond. I–III. Bull Soc Chim Belg 1976; 85:205–218, 219–228, 229–238.

91. Kollman PA. The role of the electrostatic potential in modeling hydrogen bonding and other non-covalent interactions. In: Politzer P, Truhlar DG, eds. Chemical Applications of Atomic and Molecular Electrostatic Potentials. New York: Plenum, 1981: 243–255.

92. Kamlet MJ, Taft RW. The solvatochromic comparison method.1. The β-scale of solvent hydrogen-bond acceptor (HBA) basicities. J Am Chem Soc 1976; 98:377–383.

93. Taft RW, Kamlet MJ. The solvatochromic comparison method: 2. The α-scale of solvent hydrogen-bond donor (HBD) acidities. J Am Chem Soc 1976; 98:2886–2894.

94. Kamlet MJ, Abboud J-LM, Abraham MH, Taft RW. Linear solvation energy relationships. 23. J Org Chem 1983; 48:2877–2887.

95. Abraham MH, Grellier PL, Prior DV, Duce PP, Morris JJ, Taylor PJ. A scale of solute hydrogen-bond acidity based on log K values for complexation in tetrachloromethane. J Chem Soc, Perkin Trans II, 699–711.

96. Abraham MH, Grellier PL, Prior DV, Morris JJ, Taylor PJ. A scale of solute hydrogen-bond basicity using log K values for complexation in tetrachloromethane. J Chem Soc, Perkin Trans II, 521–529.

97. Hagelin J, Brinck T, Berthelot M, Murray JS, Politzer P. Family-independent relationships between computed molecular surface quantities and solute hydrogen bond acidity/basicity and solute-induced methanol O–H infrared frequency shifts. Can J Chem 1995; 73:483–488.

98. Taft RW, Murray JS. Some effects of molecular structure on hydrogen-bonding interactions. some macroscopic and microscopic views from experimental and theoretical results. In: Politzer P, Murray JS, eds. Quantitative Treatments of Solute/Solvent Interactions. Amsterdam: Elsevier, 1994:55–82.

99. Etter MC, Urbanczyk-Lipowska Z, Zia-Ebrahimi M, Pananto TW. Hydrogen bond directed cocrystallization and molecular recognition properties of diarylureas. J Am Chem Soc 1990; 112:8415–8426.

100. Murray JS, Grice ME, Politzer P, Etter MC. A computational analysis of some diaryl ureas in relation to their observed crystalline hydrogen bonding patterns. Mol Eng 1991; 1:75–87.

101. Petrongolo C, Tomasi J. The use of the electrostatic molecular potential in quantum pharmacology. I. Ab initio results. Int J Quantum Chem, Quantum Biol Symp 1975; 2:181–190.

102. Loew GH, Berkowitz DS. Quantum chemical studies of morphinelike opiate narcotic analgesics: I. Effect of N-substituent variations. J Med Chem 1975; 18:656–662.

103. Hayes DM, Kollman PA. Role of electrostatics in a possible catalytic mechanism for carboxypeptidase A. J Am Chem Soc 1976; 98:7811–7816.

104. Petrongolo C, Preston HJT, Kaufman JJ. Ab nitio LCAO-MO-SCF calculation of the electrostatic molecular potential of chlorpromazine and promazone. Int J Quantum Chem 1978; 13:457–468.

105. Osman R, Weinstein H, Topiol S. Models for active sites of metalloenzymes. II. Ann NY Acad Sci 1981; 367:356–369.

106. Weinstein H, Osman R, Topiol S, Green JP. Quantum chemical studies on molecular determinants for drug action. Ann NY Acad Sci 1981; 367:434–451.

107. Cheney BV. Structural factors affecting aryl hydrocarbon hydroxylase induction by dibenzo-*p*-dioxins and dibenzofurans. Int J Quantum Chem 1982; 21:445–463.

108. Martin M, Sanz F, Campillo M, Parelo L, Perez J, Turino J. Quantum chemical study of the molecular patterns of mao inhibitors and substrates. Int J Quantum Chem 1983; 23:1627–1641.

109. Thomson C, Brandt R. Theoretical investigations of the structure of potential inhibitors of the enzyme glyoxalase—I. Int J Quantum Chem, Quantum Biol Symp 1983; 10:357–373.

110. Politzer P. Computational approaches to the identification of suspect toxic molecules. Toxicol Lett 1988; 43:257–276.

111. Arteca GA, Jammal VB, Mezey PG, Yadav JS, Hermsmeiers MA, Gund TM. Shape group studies of molecular similarity: relative shapes of van der Waals and electrostatic potential surfaces of nicotinic agonists. J Mol Graph 1988; 6:45–53.

112. Fisher CL, Tainer JA, Pique ME, Getzoff ED. Visualization of molecular flexibility and its effects on electrostatic recognition. J Mol Graph 1990; 8:125–145.

113. Politzer P, Murray JS. Electrostatic potential analysis of dibenzo-*p*-dioxins and structurally similar systems in relation to their biological activities. In: Beveridge DL, Lavery R, eds. Theoretical Biochemistry & Molecular Biophysics. Vol. 2: Proteins. Schenectady, NY: Adenine, 1991:165–191.

114. Platt DE, Silverman D. Registration, orientation and similarity of molecular electrostatic potentials through multipole matching. J Comput Chem 1996; 17:358–366.

115. Pitot HC, Goldsworthy T, Campbell HA, Poland A. Quantitative evaluation of the promotion by 2,3,7,8-tetrachlorodibenzo-*p*-dioxin of hepatocarcinogenesis from diethylnitrosamine. Cancer Res 1980; 40:3616–3620.

116. Poland A, Knutson JC. 2,3,7,8-Tetrachlorodibenzo-*p*-dioxin and related halogenated aromatic hydrocarbons: examination of the mechanism of toxicity. Annu Rev Pharmacol Toxicol 1982; 22:517–554.

117. McKinney JD, Long GA, Pedersen LG. PCB and dioxin binding to cytosol receptors: a theoretical model based on molecular parameters. Quant Struct-Act Relat 1984; 3:99–105.

118. McKinney JD, Darden T, Lyerly JA, Pedersen LG. Dioxin and related compound binding to the Ah receptor(s). Theoretical model based on molecular parameters and molecular mechanics. Quant Struct-Act Relat 1985; 4:166.

119. Murray JS, Zilles BA, Jayasuriya K, Politzer P. Comparative analysis of the electrostatic potentials of dibenzofuran and some dibenzo-*p*-dioxins. J Am Chem Soc 1986; 108:915–918.

120. Murray JS, Politzer P. Electrostatic potentials of some dibenzo-*p*-dioxins in relation to their biological activities. Theor Chim Acta 1987; 72:507–517.

121. Murray JS, Evans P, Politzer P. A comparative analysis of the electrostatic potentials of some structural analogues of 2,3,7,8-tetrachlorodibenzo-*p*-dioxin and of related aromatic systems. Int J Quantum Chem 1990; 37:271–289.

122. Sjoberg P, Murray JS, Brinck T, Evans P, Politzer P. The use of the electrostatic potential at the molecular surface in recognition interactions: dibenzo-*p*-dioxins and related systems. J Mol Graph 1990; 8:81–90.

123. Poland A, Greenlee WF, Kende AS. Studies on the mechanism of action of the chlorinated dibenzo-*p*-dioxins and related compounds. Ann NY Acad Sci 1979; 320:214–230.

124. Brinck T, Murray JS, Politzer P. Quantitative determination of the total local polarity (charge separation) in molecules. Mol Phys 1992; 76:609–617.

125. Murray JS, Brinck T, Lane P, Paulsen K, Politzer P. Statistically-based interaction indices derived from molecular surface electrostatic potentials; a general interaction properties function (GIPF). J Mol Struct (Theochem) 1994; 307:55–64.

126. Murray JS, Lane P, Brinck T, Politzer P. Relationships between computed molecular properties and solute/solvent interactions in supercritical solutions. J Phys Chem 1993; 97:5144–5148.

127. Murray JS, Lane P, Brinck T, Paulsen K, Grice ME, Politzer P. Relationships of critical constants and boiling points to computed molecular surface properties. J Phys Chem 1993; 97:9369–9373.

128. Murray JS, Politzer P. Statistical analysis of the molecular surface electrostatic potential: an approach to describing noncovalent interactions in condensed phases. J Mol Struct (Theochem) 1998; 425:107–114.

129. Politzer P, Murray JS. Representation of condensed phase properties in terms of molecular surface electrostatic potentials. Trends Chem Phys 1999; 7:157–168.

130. Politzer P, Murray JS. Computational prediction of condensed phase properties from statistical characterization of molecular surface electrostatic potentials. Fluid Phase Equilib 2001; 185:129–137.

131. Politzer P, Murray JS. A general interaction properties function (GIPF): an approach to understanding and predicting molecular interactions. In: Politzer P, Murray JS, eds. Quantitative Treatments of Solute/Solvent Interactions. Amsterdam: Elsevier, 1994: 243–289.

132. Murray JS, Brinck T, Politzer P. Relationships of molecular surface electrostatic potentials to some macroscopic properties. Chem Phys 1996; 204:289–299.

133. Politzer P, Murray JS, Flodmark P. Relationship between measured diffusion coefficients and calculated molecular surface properties. J Phys Chem 1996; 100:5538–5540.

134. Brinck T, Murray JS, Politzer P. Octanol/water partition coefficients expressed in terms of solute molecular surface areas and electrostatic potentials. J Org Chem 1993; 58: 7070–7073.

135. Politzer P, Murray JS, Brinck T, Lane P. Analytical representation and prediction of macroscopic properties: a general interaction properties function. In: Nelson JO, Karu AE, Wong RB, eds. Immunoanalysis of Agrochemicals: Emerging Technologies. Washington: American Chemical Society, 1995:109–118.

136. Politzer P, Murray JS, Concha MC, Brinck T. Some proposed criteria for simulants in supercritical systems. J Mol Struct (Theochem) 1993; 281:107–111.

137. Murray JS, Abu-Awwad F, Politzer P, Wilson LC, Troupin AS, Wall RE. Molecular surface electrostatic potentials of anticonvulsant drugs. Int J Quantum Chem 1998; 70: 1137–1143.

138. Murray JS, Lane P, Politzer P. Effects of strongly electron-attracting components on molecular surface electrostatic potentials; application to predicting impact sensitivities of energetic molecules. Mol Phys 1998; 93:187–194.

139. Ma Y, Murray JS, Politzer P. Unpublished work.

140. Fauci AS. The human immunodeficiency virus: infectivity and mechanisms of pathogenesis. Science 1988; 239:617–622.

141. Garg R, Gupta SP, Gao H, Suresh Babu M, Kumar Debnath A, Hansch C. Comparative quantitative structure–activity relationship studies on anti-HIV drugs. Chem Rev 1999; 99:3525–3601.

142. Dumas J-M, Peurichard H, Gomel M. C-X₄-base interactions as models of weak charge transfer interactions. J Chem Res (S) 1978;54–55.

143. Blackstock SC, Lorand JP, Kochi JK. Charge transfer interactions of amines with tetrahalomethanes. J Org Chem 1987; 52:1451–1460.

144. Gotch AJ, Garrett AW, Zwier TS. The ham bands revisited: spectroscopy and photophysics of the C_6H_6-CCl_4 complex. J Phys Chem 1991; 95:9699–9707.

145. Lorand JP. AL Spek, Private communication.

146. Brinck T, Murray JS, Politzer P. Surface electrostatic potentials of halogenated methanes as indicators of directional intermolecular interactions. Int J Quantum Chem, Quantum Biol Symp 1992; 19:57–64.

147. Politzer P, Murray JS. General and theoretical aspects of the C–X bonds (X = F, Cl, Br, I): Integration of Theory and Experiment. In: Patai S, Rappoport Z, eds. Supplement D2: The Chemistry of Halides, Pseudo-Halides and Azides. Part 1. New York: Wiley-Interscience, 1995:1–30.

148. DiPaulo T, Sandorfy C. On the hydrogen bond breaking ability of fluorocarbons containing higher halogens. Can J Chem 1974; 52:3612–3622.

149. Ramasubba N, Parthasarathy R, Murray-Rust P. Angular preferences of intermolecular forces around halogen centers. J Am Chem Soc 1986; 108:4308–4314.

150. Murray JS, Paulsen K, Politzer P. Molecular surface electrostatic potentials in the analysis of non-hydrogen-bonding noncovalent interactions. Proc Indian Acad Sci, Chem Sci 1994; 106:267–275.

151. Murray JS, Lane P, Brinck T, Politzer P. Electrostatic potentials on the molecular surfaces of cyclic ureides. J Phys Chem 1991; 95:844–848.

152. Murray JS, Gilardi R, Grice ME, Lane P, Politzer P. Structures and molecular surface electrostatic potentials of high-density C, N, H systems. Struct Chem 1996; 7:273–280.

153. Murray JS, Abu-Awwad F, Politzer P. Prediction of aqueous solvation free energies from properties of solute molecular surface electrostatic potentials. J Phys Chem, A 1999; 103:1853–1856.

154. Politzer P, Murray JS, Abu-Awwad F. Prediction of solvation free energies from computed properties of solute molecular surfaces. Int J Quantum Chem 2000; 76:643–647.

155. Kollman P. Free energy calculations: applications to chemical and biochemical phenomena. Chem Rev 1993; 93:2395–2417.

156. Davis ME. The inducible multipole solvation model: a new model for solvation effects on solute electrostatics. J Chem Phys 1994; 100:5149–5159.

157. Cramer CJ, Truhlar DG. Development and biological applications of quantum mechanical continuum solvation models. In: Politzer P, Murray JS, eds. Quantitative Treatments of Solute/Solvent Interactions. Amsterdam: Elsevier, 1994:9–54.

158. Bacskay GB, Reimers JR. Solvation: modeling. In: Schleyer PvR, ed. Encyclopedia of Computational Chemistry. Vol. 4. New York: Wiley, 1998:2620–2632.

9

Nonbonded Interactions

STEVE SCHEINER

Utah State University, Logan, Utah, U.S.A.

1. INTRODUCTION

Anyone who has taken a general chemistry class has some sense of what is meant by a bonding interaction. When one draws the structure of a molecule, whether in two dimensions, as in a simple Lewis dot structure, or three dimensions using VSEPR or some other representation, the lines that are drawn between the various atomic nuclei represent covalent bonds or "bonding interactions." This bond might be a single bond, as the O–H bonds in water, a double bond as in ethylene, or a triple bond for which acetylene serves as the most common example. In most cases, this covalent bond represents a shared pair of electrons and represents a good deal of binding energy holding the two atoms together, typically on the order of 50–100 kcal/mol.

In a general sense, the "noncovalent interaction" appellation is given to interactions between atoms that are *not* connected by a covalent bond. The simplest example would be the force between a pair of molecules. When two H_2 molecules, for instance, approach one another, there is clearly some sort of force between them, although no intermolecular covalent bond exists. This particular interaction is fairly weak and attractive at long distances. The nature of the force is typically characterized as "van der Waals" but is more complicated than one might expect. Even at the optimum separation of these two H_2 molecules, the interaction energy is less than 1 kcal/mol [1]. As the two molecules are jammed closer together than their equilibrium separation, a strong repulsive force emerges that pushes them further apart.

A higher level of attractive force is encountered if the two molecules happen to belong to a certain class that includes water as one of its members. The hydrogen bond that forms between the two molecules is responsible for many unusual characteristics of water, including its existence as a liquid at room temperature, and for the ability of ice to float on water, instead of sinking as would be expected of a solid and a liquid of the same substance. Hydrogen bonding is not limited to water or even to interactions between O atoms. It is commonly seen when the X and Y atoms of X–H $\cdots$ Y are electronegative, e.g., O, N, or F [2,3]. The NH $\cdots$ O H-bond is particularly important in protein molecules, for example, where it results in the now familiar α-helix and β-sheets of proteins. It appears that the electron pair of the acceptor molecule can be replaced by almost any rich source of electrons, e.g., the π-cloud of a CC multiple bond or a benzene molecule [4,5]. Nor does the donor have to be very electronegative: C–H groups appear to act as proton donors in H-bonds as well [6–8].

Nonbonding interactions are not limited to pairs of molecules, but can occur also within the confines of a single molecule. It frequently happens that as a molecule rotates around its various bonds, some atoms that are not covalently bonded to one another come into close proximity. This approach results in some sort of interaction between them. A standard example is associated with the energy difference between the eclipsed and staggered conformers of ethane. The higher energy of the former is usually attributed to the nonbonded repulsion between the hydrogen atoms that arises from their closer proximity in this arrangement. Hydrogen bonds, too, can occur in cases where both the donor and acceptor lie in the same molecule. A classic example of such an intramolecular H-bond is associated with malonaldehyde where the OH of one end of the HOCHCHCHO molecule forms a H-bond with the aldehydic O on the other end, as the two ends "curl up" toward one another. The aforementioned H-bonds in the α-helix of proteins represent another example of an intramolecular H-bond, although the donor and acceptor atoms are separated from one another by more than 10 covalent bonds.

Noncovalent interactions are not necessarily weak. If the two entities in question happen to be a cation and anion, for example, the coulombic force between them is very strong, particularly as the two ions more closely approach one another. The distance dependence of the interaction energy varies as $1/R$. If separated by a distance of 3 Å, the interaction energy would amount to some 110 kcal/mol, comparable in strength to a covalent bond, although the interatomic separation may be longer here. Another interaction which is strongly coulombic in origin arises when an ion is brought toward a neutral molecule. In particular, if the latter has a permanent dipole moment, the interaction energy is dominated by a $1/R^2$ dependence. Note, however, that the very sign of this interaction, whether attractive or repulsive, is reversed upon 180° rotation of the neutral, emphasizing the strong anisotropy of the interaction.

In summary, then, nonbonded forces vary over a wide spectrum. They include very strong coulombic interactions that typically involve charged species. Also, under this same rubric are very weak interactions between closed-shell molecules such as H_2. Somewhat stronger than the latter are hydrogen bonds, special cases in that only certain sorts of molecules participate in such interactions. However, whether very strong or nearly immeasurably weak, each sort of interaction is far more complicated than might appear at first sight. All behave differently as the two entities are rotated relative to one another or are pulled apart. Yet at the same time, much like all atoms can be shown to consist of protons, neutrons, and electrons, it would appear that all noncovalent forces are comprised of the same "elementary" forces, but in varying proportions. The purposes of this chapter are to enumerate and describe these elementary forces and to demonstrate how they fit together, like pieces of a jigsaw puzzle, in a fundamental and comprehensive understanding of noncovalent interactions.

It might be worthwhile to stress at this point that the following dissection of the interaction energy makes the assumption that the internal geometries of the two partner molecules, i.e., nuclear positions, are unchanged by the interaction. Of course, there will be some degree of geometric distortion within each subunit that is induced by the interaction, which will be accompanied by a certain amount of "nuclear distortion energy." This quantity is typically considered as an additional term, over and above the components described below. This term is separate and distinct from the "deformation energy" defined below, which refers to electronic and not nuclear rearrangements. (Unfortunately, one must be very cautious in reading the literature as the word "nuclear" is frequently omitted from the term "nuclear distortion energy." Worse, it is not uncommon for the nuclear distortion energy to be loosely referred to as deformation energy although it is quite different from the "true" deformation energy defined below.)

2. ELECTROSTATIC FORCES

One of the most important forces in any sort of interaction is electrostatic. The reader should be cautioned at the outset that this term is thrown around rather loosely in the literature, and one should be very careful to understand what precisely is meant in any given case. In a sense, *all* interactions are fully electrostatic to the exclusion of all other forces. After all, atomic nuclei are positively charged and the surrounding electrons are negative, and it is the coulombic interactions among these particles that give rise to all aspects of molecular interactions. Therefore in this sense, even the most fundamental covalent bond that binds two hydrogen atoms together in H_2 can be considered as a purely electrostatic interaction. This definition clearly defies the way in which people normally think of bonding forces in general and electrostatic forces in particular. It does underscore, however, the necessity of clarity in defining this term.

Perhaps the best and least ambiguous definition of what is normally meant by the electrostatic component of an interaction is illustrated in Fig. 1. The nuclear framework of a water molecule is illustrated in Fig. 1a, where each nucleus is of course positively charged. The electrons cannot be pinpointed due to the statistical nature of quantum chemistry as exemplified by the Heisenberg Uncertainty Principle, giving rise to the concept of probability density, a sort of time-lapse photograph which corresponds to a description of where the electrons spend their time. That is, regions of space where electrons spend a good deal of time are represented by "thick" density, which thins out proportionately as the electrons spend less and less time in other areas of space. The electron density can be extracted simply from a quantum mechanical calculation as the square of the wave function. As an example, the electron density of a water molecule is represented by the contour plot in Fig. 1b, which illustrates the high density around the nuclei, especially the O nucleus with its charge of $+8$.

Let us now consider the situation when two water molecules are brought toward one another, in a configuration that corresponds to the classical $O-H \cdots O$ hydrogen bond, as pictured in Fig. 1c. The electrostatic part of the interaction between these two molecules can be derived from Coulomb's equation that assigns the interaction between two particles as

$$E_C = (q_i q_j)/r_{ij} \tag{1}$$

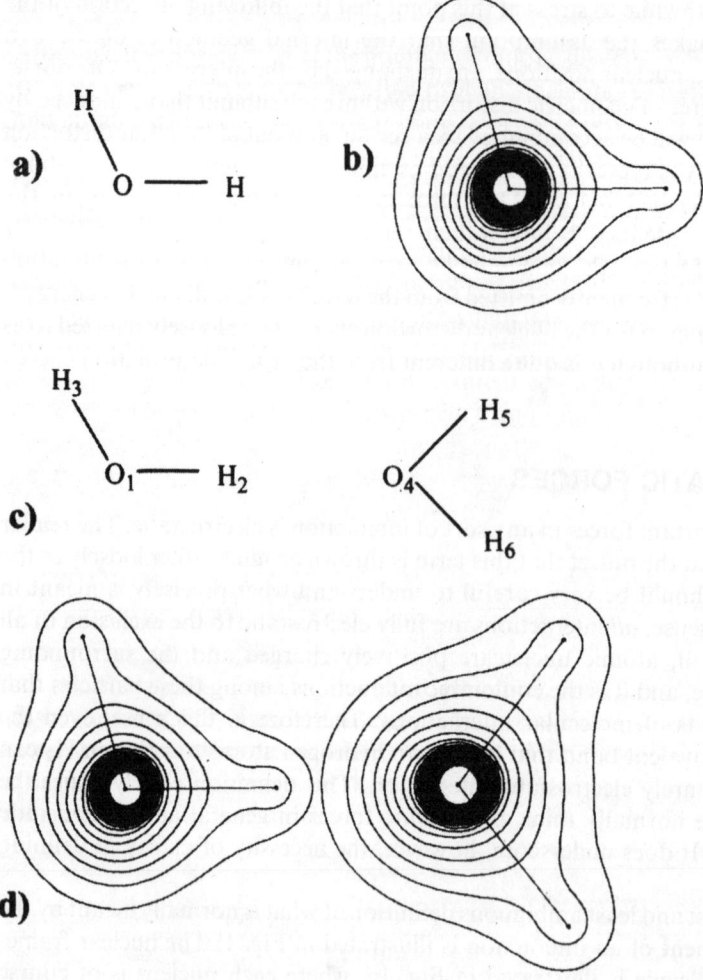

Figure 1   (a) Disposition of nuclei in water molecule, (b) contour plot of electron density, in intervals of 0.15 au, (c) nuclear positions in water dimer, (d) monomer electron densities in the dimer arrangement.

where q_i and q_j represent the charges of particles i and j and r_{ij} is the distance separating them.

One can easily compute the coulombic repulsion between the three nuclei ($i = 1$–3) on the first water molecule and nuclei $j = 4$–6 on the second as the sum:

$$E_C = \sum_i \sum_j (q_i q_j)/r_{ij} \tag{2}$$

But the situation for interactions involving electrons is more complex since one does not have single fixed locations for any of them. Instead, one uses the concept of electron density as a continuous function to compute the interactions of the electrons with the nuclei and with one another. Instead of an integral charge q at a fixed location, one makes use of infinitesimal volume elements, each with a charge proportional to the electron density function $\rho(x,y,z)$. The summation in Eq. (2) is replaced by an integral over all space [9]. In essence, since the charges of the electrons are "smeared" over space, their coulombic interactions with other particles must be integrated over all space.

The electrostatic interaction between the two water molecules would thus include first the internuclear repulsion term of Eq. (2). A second term represents the attraction between the nuclei on the first molecule (nuclei 1–3) and the electron cloud on the second molecule. To this would be added the conjugate attraction between nuclei on molecule 2 (4–6) and the electron cloud of molecule 1. Finally, one must take account of the interelectronic repulsion. For purposes of the intermolecular contact, this term would entail repulsion between the electron clouds of the two molecules.

One can hence consider the electrostatic interaction between the two water molecules in Fig. 1 in terms of their charge clouds, i.e., the distributions of electrons. It is essential to understand that this quantity encompasses *all* of space, not just the nuclei or regions close to the nuclei. This point is stressed in Fig. 1d, which places the charge clouds of the two molecules (as in Fig. 1b) in appropriate juxtaposition. In a particular calculation of this system [10], with the O atoms of the two waters separated by 3.25 Å, the electrostatic interaction was computed to be -5.73 kcal/mol, where the negative sign indicates an overall attraction.

In the absence of a quantum mechanical (or some other) elucidation of the electron density, one cannot precisely evaluate the electrostatic interaction. On the other hand, an accurate quantum calculation of the density of a large molecule can be problematic, and the ensuing evaluation of its interaction with the density of a partner molecule can be time-consuming. Consequently, there have been many attempts over the years to derive efficient means of approximating the full electrostatic interaction while minimizing any loss of accuracy.

2.1. Multipole Approximation

Many of the approximation methods are based on the principle of a multipole expansion [9,11,12]. We consider the water molecule for illustrative purposes once again. Since this molecule is neutral, its monopole moment, i.e., charge, is zero. [Had we been considering an ion such as $(H_3O)^+$, its monopole would be $+1$.] The next-highest-order moment is dipole and represents the separation between positive and negative centers in a single dimension. As is well known, the dipole moment of water is nonzero and is oriented along the HOH bisector, as illustrated in Fig. 2a. However, this of course does not convey the full picture of water's charge distribution. Water has a quadrupole moment as well, with components in each of the three principal coordinate directions. As illustrated in Fig. 2b, the component in the x-direction, perpendicular to the molecular plane, reflects the presence of two unshared O electron pairs above and below this plane. The presence of these lone pairs also contributes to the negative sign of the z-component in Fig. 2c, overshadowing the partial positive charges of the two H atoms. In the y-direction of Fig. 2d, however, the latter positive charges cause this component to be positive in sign. By continuing to higher and higher-order moments, octapole and so on, one obtains a progressively better description of the true charge distribution of water or any other molecule.

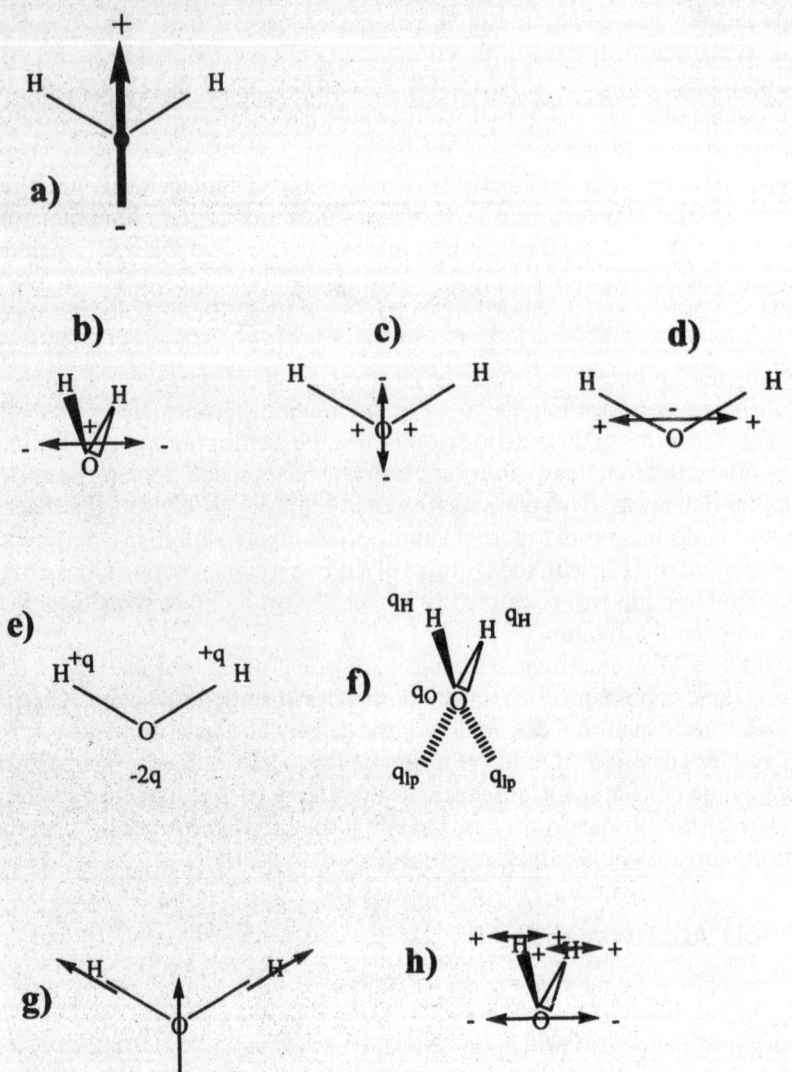

Figure 2 Various representations of the charge distribution in the water molecule: (a) dipole moment, (b–d) principal components of the quadrupole moment, (e) partial atomic charge approximation, augmented in (f) by partial charges assigned to O lone electron pairs, point dipoles (g), and quadrupoles (h) assigned to each atomic center.

Armed with a multipole approximation of the full charge distribution of the two molecules in question, one can evaluate the correct electrostatic interaction as a series of terms, in inverse powers of R, the intermolecular separation [9,11]. We will use the notation that i represents the order of any nonzero $2i$ moments of the first molecule ($i = 0$ for monopole, 1 for dipole, 2 for quadrupole, etc.). Terms expressing the interaction of the ith moment of one with the jth moment of the other are collected in the inverse nth power of R where $n = i + j + 1$ [13]. For example, the

R^{-1} term would be nonzero only if both partners were charged, with nonzero monopoles ($i = j = 0$). An R^{-2} term appears in the interaction between the monopole of one ($i = 0$) and the dipole of the other ($j = 1$). The first term that occurs in the case of a pair of neutral molecules, as in the water dimer, is R^{-3} which corresponds to the interaction between the dipole moments of the two molecules ($i = j = 1$). Any nonzero monopole–quadrupole terms would appear in this term as well. Dipole–quadrupole interactions die off as R^{-4}, which would also contain charge–octapole interactions, should they exist. Just as the continuation of the multipole expansion to higher orders progressively improves the approximation of the true charge distribution of each monomer, the continuation of the R^{-n} summation yields a progressively better approximation to the true electrostatic interaction energy.

An example of this summation process is provided in Table 1 which reports the various terms in the multipole approximation of the water dimer [10]. These data are provided for two different interoxygen separations of 3.25 and 2.75 Å. The first is longer than the optimized equilibrium distance of 2.84 Å, and the second is somewhat compressed. It may be first noted that the R^{-2} term is zero, consistent with the fact that neither of the partner molecules is charged. The first nonzero term corresponds to the dipole–dipole interaction in the R^{-3} column. This term is attractive due to the favorable arrangement of the two molecular dipoles. The next term is more complicated. It contains first the interaction between the dipole of the first molecule and the quadrupole of the second. This term represents a sum in that the quadrupole comprises three separate components (see Fig. 2). Also contained in the R^{-4} term is the interaction between the quadrupole of the first molecule and the dipole of the second. Altogether, one arrives at an attractive (negative) contribution. The complexity grows in the succeeding entry in Table 1. The R^{-5} term contains not only dipole–octapole (and the converse octapole–dipole), but also quadrupole–quadrupole elements.

There are several important points to be made about the data in Table 1. First, while all of the R^{-n} terms listed happen to be negative, as is their summation, this may not always be the case. There is no reason that the signs cannot alternate from term to term. It is not surprising to see that the terms are larger in magnitude for the closer approach of the two molecules, i.e., smaller $R(O \cdots O)$, which will typically be the case. One might also note that the magnitudes of the terms diminish as one

Table 1 Components (kcal/mol) of Multipole Expansion of Electrostatic Interaction and Other Components (from Ref. [10])

$R(O \cdots O)$	R^{-2}	R^{-3}	R^{-4}	R^{-5}	$\Sigma R^{-n, a}$	ES[b]	EX[c]	DEF[d]
HOH $\cdots$ OH$_2$								
3.25 Å	0.0	−2.92	−1.71	−0.51	−5.14	−5.73	1.16	−1.65
2.75 Å	0.0	−4.82	−3.33	−1.17	−9.32	−12.77	9.14	−3.74
H$_2$OH$^+ \cdots$ OH$_2$								
3.25 Å	−15.79	−0.51	−1.57	−1.91	−19.79	−20.14	0.58	−10.85
2.75 Å	−22.06	−0.84	−3.07	−4.40	−30.37	−32.04	5.95	−16.95

[a] Summation of previous terms, through R^{-5}.
[b] Full electrostatic interaction.
[c] Exchange repulsion.
[d] Deformation energy, resulting from relaxation of electron densities.

progresses through the series in either row. Again, this may not always be the case (see below for counterexample). If one follows the entries in each row, it might be expected that the succeeding terms (R^{-6} and so on) will continue to become smaller, but they may not become negligible until further out in the series. In other words, the series may not be rapidly convergent. This slow convergence becomes more of a problem as the distance between the two molecules shrinks. For example, while the series terminated at the R^{-5} term (-5.14 kcal/mol) is fairly close to the true electrostatic interaction of -5.73 kcal/mol (in the ES column of Table 1) for $R(O \cdot \cdot O) = 3.25$ Å, the discrepancy is much larger, -9.32 vs. -12.77 kcal/mol, for $R(O \cdot \cdot O) = 2.75$ Å. Indeed, there is no guarantee that the multipole series will converge at all, a caveat in the use of the multipole approximation at close intermolecular contact.

And even when the convergence is rapid, the multipole expansion is not a complete representation of the true electrostatic energy, as it neglects the effects of overlap of the charge distributions of the two molecules in question [12,14], sometimes referred to as a penetration term. Fortunately, the latter complicating effect dies off exponentially with intermolecular separation [15], adding to the validity of the multipole expansion at sufficient spacing.

It should last be pointed out that the electrostatic interaction does not pass through a minimum at the equilibrium separation (2.84 Å in this case), as does the total interaction energy. Indeed, the electrostatic interaction will tend to continue to become more and more negative, even when the two subunits have approached much closer than their equilibrium. It is not the electrostatic force which prevents an overly close approach, but rather the exchange repulsion, described below.

The same quantities are presented in the next two rows for the $H_2OH^+ \cdots OH_2$ system, pairing a charged subunit with a neutral molecule. The presence of an ionic component, with a nonzero monopole, leads to a large attractive R^{-2} term, corresponding to the monopole–dipole interaction. It is instructive to observe the small magnitude of the R^{-3} term, smaller than the succeeding terms, an illustration that the various terms of the series do not necessarily diminish steadily in magnitude. Indeed, there is every indication that the multipole approximation is far from convergent by the R^{-5} term since the latter is even larger than the preceding R^{-4} term. It is thus somewhat surprising that the sum through R^{-5} is fairly close to the full electrostatic interaction in the last column. The closeness of this approximation must be considered fortuitous and is likely the result of cancellation between higher terms in the series. As in the above neutral dimer pair, all terms grow in magnitude for closer approach of the two subunits.

In summary, the multipole series may provide an estimate of the full ES interaction with a certain measure of accuracy, particularly if the two subunits are well separated. This concept offers the opportunity to gain insight into the nature of the ES force, based on the direction and magnitudes of some of the lower moments of the individual molecules. It also has a predictive capability with regard to the anisotropy of the full ES interaction, i.e., its sensitivity to angular aspects of the intermolecular geometry.

2.2. Atom-Centered Means of Approximation

The multipole expansion described above rests on the idea that the full charge distribution of a molecule can be replaced by a series of progressively higher-order multipole moments, all originating at a single point, which is generally taken as the molecule's center of mass. While this approach may be sensible for a small molecule such as HOH, the characterization of the charge density of a large and complicated molecule, e.g., a polypeptide, by moments that all emanate from the molecule's center, is manifestly problematic. In such cases, it would make more sense to "distribute" the moments over the entire span of the molecule [9,14,16]. That is, one can more easily and efficiently simulate the molecule's charge distribution by centering moments at various locations along the molecule simultaneously.

Indeed, this idea lies at the heart of the venerable notion of assigning partial charges to the atoms in a molecule [17]. One may think of this practice as a representation of the true charge distribution of the molecule by a series of distributed multipoles (in this case, limited to monopoles) at various sites, namely, atomic centers. Even if limited to monopoles, the act of spreading them out over the entire molecule is equivalent in some sense to simulation of high orders of molecular center-based multipoles.

Once partial atomic charges are assigned to the various atoms, one can fall back on Eq. (2) to evaluate the electrostatic interaction energy. Whereas Eq. (2) represents the internuclear repulsion, using integral charges of $+8$ and $+1$ for the O and H nuclei, respectively, the same formulation will express the full intermolecular ES energy if the q_i and q_j nuclear charges are replaced by fractional atomic charges. One is then left with the question as to how to assign fractional charges to each atom. This is, unfortunately and most definitely, *not* a simple question.

The first order of business is to dispel the common misconception that atomic charges correspond to a real physical quantity. In order to understand why this is so, consider the electron density of the water molecule in Fig. 1b. This cloud covers the entire physical region around and between the three nuclei. The assignment of charge to any nucleus rests on some objective means of deciding where to draw the boundaries in this space; density on one side is assigned to one nucleus, and density on the other is assigned to the second nucleus. Several possible methods of doing so are indicated by dashed boundary lines in Fig. 3, any of which might seem quite reasonable. One might, for example, draw boundaries at the geometrical midpoint of the O–H bonds, as suggested in Fig. 3a. Or this same general planar separation could be taken but the boundaries moved toward or away from the central O. Another idea might be to draw spherical boundaries around the H atoms as in Fig. 3b, reflecting their approximate shape and assigning the remainder of the density to the heavier O atom. Still, another approach would draw a single straight line down along the two O–H bonds, as depicted in Fig. 3c. However, there is no reason to believe that any one method is fundamentally superior to any other. Indeed, *any* prescription for drawing these boundaries is arbitrary, none with more physical meaning than any other. In other words, while electron density is certainly a real physical phenomenon, which can be precisely determined in principle, the ambiguity arises in choosing atomic basins in which to wrap all this density.

As an example, the classic Mulliken [17] charge partitioning method would assign charges to the O and H atoms of -0.796 and $+0.398$, respectively, as indicated in the first row of Table 2. The corresponding "natural" atomic charges [18] for the same molecule are -0.961 and $+0.480$. Still, other methods would provide entirely different values. In addition, each would of course provide a correspondingly different estimate of the electrostatic energy. Hence the ES energy computed by plugging any

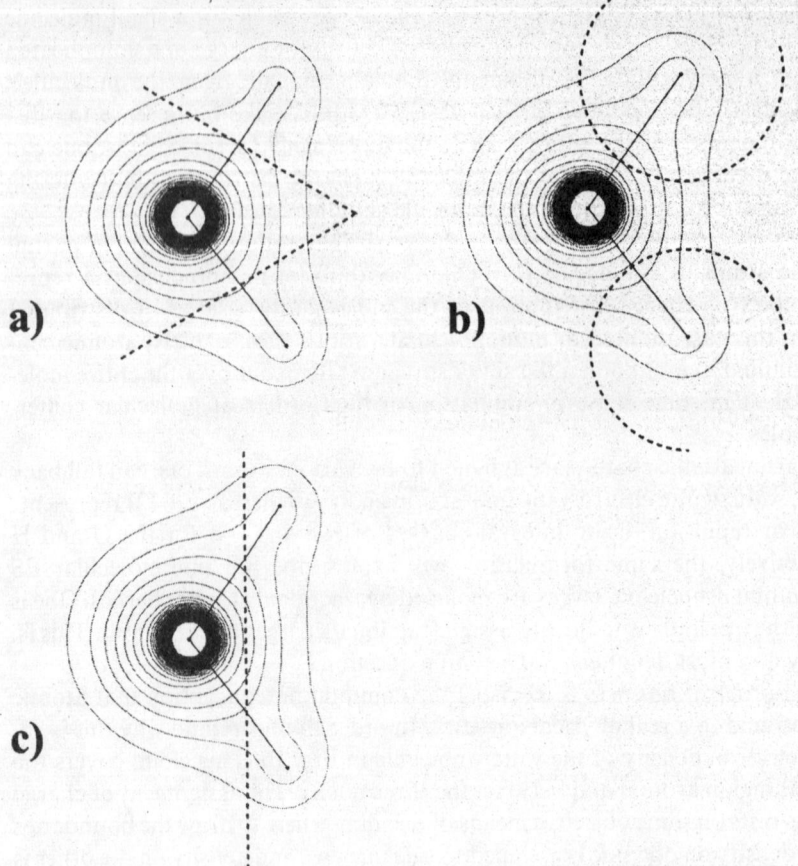

Figure 3 Possible means of drawing boundaries in space to assign electron density to one nucleus or another.

Table 2 Atomic Charge on O Atom of H_2O[a]

Mulliken	−0.796
Natural	−0.961
ESP q-fit[b]	−0.809
ESP qμ-fit[c]	−1.542

[a] Charge on H atom equal to −1/2 of this value.
[b] Best fit of atomic charges to electrostatic potential.
[c] Best fit of atomic charges and dipole moments to electrostatic potential.

set of atomic charges into Eq. (2) is no less arbitrary than is the means of assigning these charges.

Having said that, it must be acknowledged that the concept of atomic charges plays a historically important role in our understanding of intermolecular structure and reactivity. Although different means of computing atomic charges admittedly yield discrepant numerical values, they are consistent in a qualitative sense. That is, all methods agree that the greater electronegativity of O vs. H leads to a negative partial charge on the former and a positive charge on the latter. From that standpoint, electrostatic arguments will lead to the correct general geometry of the water dimer, wherein the positively charged H of one water molecule is attracted to the negative charge of the other molecule's O atom.

However, the representation of the full charge distribution of a molecule by monopoles centered on the nuclei can only take us so far toward an accurate reproduction of the true electrostatic energy. Taking water as an example again, the placement of fractional charges at the O and H centers, as indicated in Fig. 2e, does reproduce at least qualitatively the molecule's dipole moment. It is also consistent with the component of the quadrupole moment pictured in Fig. 2d. However, this scheme cannot reproduce even qualitatively the perpendicular component of the quadrupole in Fig. 2b, as all three centers lie in the molecular plane. Moreover, the signs of the charges on the atoms would lead to an incorrect sign for the quadrupole element in Fig. 2c. These incorrect quadrupole estimates are a symptom of a deeper problem. Any representation of even as simple a molecule as water based on atomic charges is incapable, in principle, of reproducing the nonplanar aspects of its interaction with another species as all three monopoles lie in the molecular plane.

The principal reason for the latter failings lies in the absence of any representation of the lone electron pairs on the O atom. It is the presence of these lone pairs, above and below the molecular plane, that result in the correct signs of all the multipole elements and that contribute heavily to the interaction of water with an approaching molecule. There are a number of ways in which the atomic charge model can be expanded to address this problem. One possibility is to add more centers to the representation, e.g., partial negative charges above and below the molecular plane, in the approximate centers of the two lone pairs. This concept is illustrated in Fig. 2f, where overall molecular neutrality requires that $2q_H + q_O + 2q_{lp} = 0$.

An alternative would retain the number of centers at three, but extend the expansion around each center beyond monopole, to include dipoles, quadrupoles, or perhaps even higher moments. One may, for example, place dipole moment vectors on each of the three atoms, as illustrated in Fig. 2g, to supplement the atomic charges in 2e. This scheme would not, however, provide for any charge distribution above or below the molecular plane. On the other hand, one can account for this aspect by adding a perpendicular component of a quadrupole moment to each atomic center, as indicated in Fig. 2h. This addition could add an important, and qualitatively correct, out-of-plane component to this molecule's interaction with another and help to model the O lone pairs.

2.2.1. Assignment of Parameters

Regardless of which philosophy is adopted, there remains the nontrivial question as to what values to assign to the various parameters, be they monopole, dipole, or higher moments. In the previous case of molecule-centered moments, these properties can be evaluated experimentally. For example, the dipole and higher moments of water have been measured and are tabulated, or, in the absence of reliable experimental data, one can calculate these quantities by fairly rigorous quantum mechanical calculations.

However, charges, dipoles, etc. do not correspond to real physical phenomena when assigned to atomic centers. Similarly, point-charge representations of lone pairs do not represent real physically observable quantities. With this understood, one can appreciate that atomic charges and so on can be looked upon not as physical quantities, but rather as adjustable parameters. These parameters can be fit to best reproduce certain desired quantities, either experimental or theoretical [14,19–21]. Of most relevance are those which are geared toward the electrostatic interactions of the molecule in question.

One philosophy might be to choose charges to reproduce as closely as possible certain multipole moments of the molecule [22,23]. As an example, one can choose the value of q in the atomic charge model of Fig. 2e such that the molecular dipole moment arising from this value matches the experimental dipole moment of water. Of course, this single q parameter could not be simultaneously chosen to accurately reproduce higher moments as well. The model in Fig. 2f, with its two linearly independent charges, coupled with two more additional parameters (the distance of q_{lp} from the O atom and the angle between the two $O \cdots q_{lp}$ axes) could, in principle, be fit to four physical phenomena. It would be natural to choose the latter as the dipole moment, plus the three components of the molecular quadrupole moment. This prescription would, of course, not be able to address the octapole and higher moments. By adding more degrees of freedom to the representation, it becomes possible to fit the parameters to higher and higher moments.

Another approach which has been taken is to abandon the individual molecular moments as the object of the fit. Instead, one can focus on all of the moments simultaneously, in a manner of speaking. To be more precise, the full electron density of a molecule such as water (see Fig. 1b) generates an electrical field in its vicinity, and this field (the product of all moments of the molecule) is associated with an electrostatic potential. This potential can be thought of as the electrostatic energy of interaction between the molecule's density and a point charge (whose presence does not perturb the density in any way). It is generated for a given point in space by integrating the density $\rho(x,y,z)$ of the molecule over all infinitesimal volume elements $d\tau(x,y,z)$ and dividing by the distance of this volume element from the point of interest [12,14,24,25]. (The reader is advised to consult the preceding chapter in this volume for an in-depth description of this property.)

This potential is illustrated as a contour plot in the molecular plane in Fig. 4a, whereas Fig. 4b corresponds to a plane perpendicular to the molecule, rotated 90° from the first. The red contours indicate regions of negative potential (attractive to a positive charge) and blue represents positive. The red and blue contours occur in the vicinity of the O and H atoms, respectively, as expected, based on their respective electronegativities. It is reiterated that this potential is a function of the full electron density and has no dependence on any arbitrary assignment of charges to atoms or other sites.

The values of the potential in a wide region of space offer an essentially infinite number of quantities to which the charges or other adjustable parameters may be fit. In summary, this approach provides the optimal fit of the electrostatic prop-

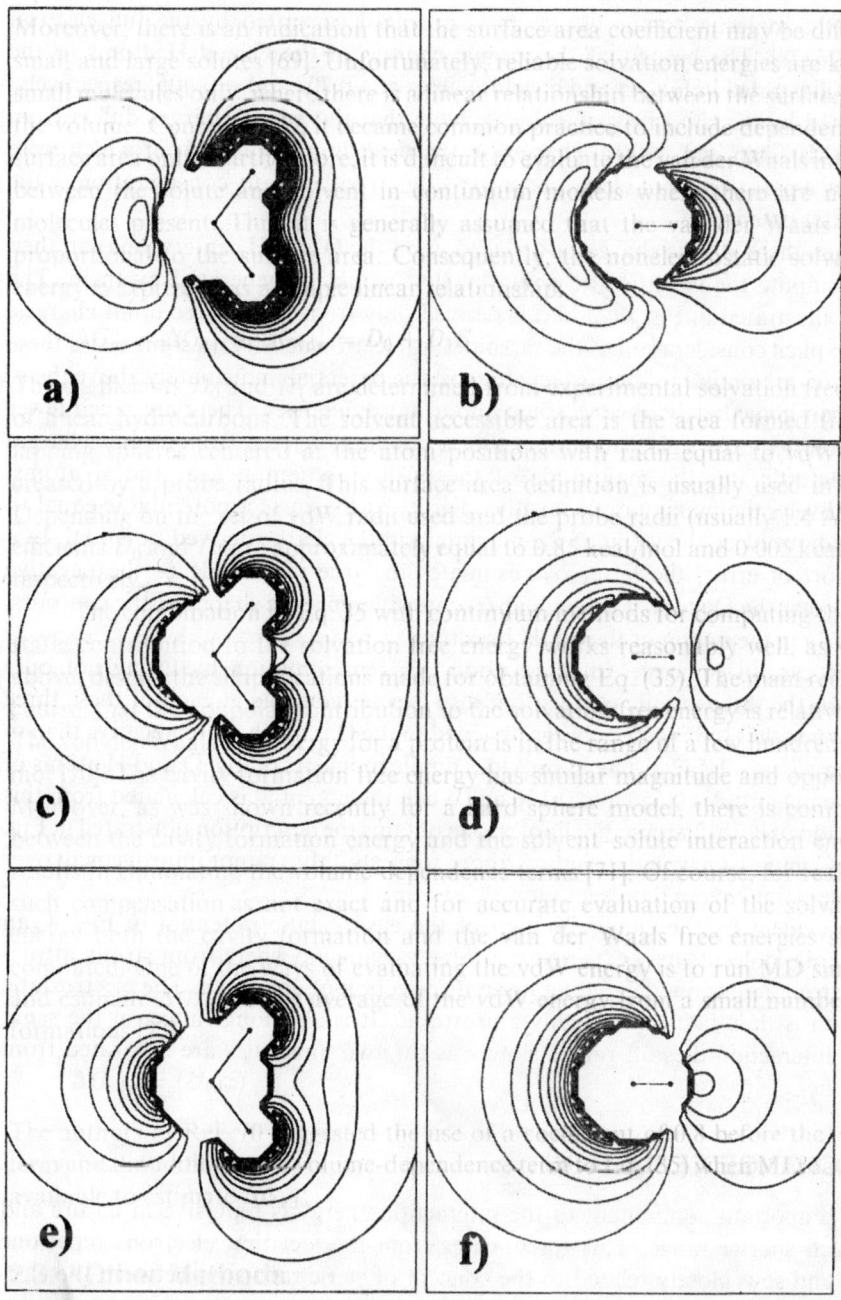

Figure 4 Molecular electrostatic potential of water molecule, represented as a contour plot with intervals of 0.025 au. Red contours indicate regions of negative potential and blue represents positive. (a–b) Potential generated from full electron density, in and perpendicular to the molecular plane, respectively; (c–d) potential generated from point charges situated at three atomic positions; (e–f) potential generated from point charges and dipoles situated at three atomic positions. (See color plate at end of chapter.)

erties of the molecule to the electrostatic potential generated by the full electron density [26–28]. The best fit of the atomic charges of the O and H atoms to the potential illustrated in Fig. 4a and b yields values of −0.809 and +0.405, respectively, as indicated by the third row of Table 2. Note that this set of charges differs from both the Mulliken and natural charges in the preceding rows, both of which were derived by a partitioning of electron density (Fig. 3) without any consideration of the electrostatic potential.

When these atomic charges are placed upon the O and H centers, one can then simply compute the electrostatic potential that would be generated by them. The potential illustrated in Fig. 4c and d is derived only from those three point charges, with no explicit consideration of the surrounding electron density. In some ways, these two pictures are similar to the correct electrostatic potential immediately above them, but one can immediately observe some important differences. There are much fewer blue contours on the right, suggesting that a potential derived from atomic charges only underestimates the magnitude of the positive potential near the H atoms. Quantitatively, the maximum of positive density in the correct potential is equal to 0.54 au, about twice the value of the maximum in the charge-derived potential. In a reverse sort of error, the latter overestimates the true magnitude of the negative potential near the O atom. There are also some differences in shapes of the contours, particularly in the region of the HOH bisector.

One can, in principle, improve upon this representation by fitting not only charges on the three atomic centers, but also dipole moments at each of these three points. When that is done, the charge assigned to the O is much more negative than in the earlier cases, −1.542 e. The fit provides dipole moments on the O and H atoms of respective magnitudes 0.556 and 0.136 au. When the potential is calculated from this formulation of both charges and dipoles, one obtains the description illustrated in Fig. 4e and f. Comparison with the pictures above suggests only a minor improvement over the simple point-charge representation.

As evident by the change in sign of the electrostatic potential in Fig. 4, an approaching molecule may experience a force varying anywhere from strong attraction to strong repulsion, depending upon its direction of approach. The electrostatic interaction is thus said to be highly anisotropic. It is also long range, in the sense that the interaction dies off rather slowly as the two molecules are separated from one another.

3. EXCHANGE REPULSION

Another important component to the interaction energy is repulsive in nature and is of much shorter range. This force arises from the fact that electrons repel one another and so is closely related to the concept of steric repulsion between the electron clouds of the partner molecules. This term is alternatively referred to as exchange energy or exchange repulsion since its formal mathematical origin lies in the Pauli exchange principle that keeps electrons of one molecule from occupying the orbital space of the other.

A simple exposition of this force begins with Fig. 1d, which illustrated the electron clouds of the two water molecules in the dimer. It must first be stressed that the density of each water molecule does not "end" at the outermost contour, but instead tapers off gradually and exponentially. The outermost contour of Fig. 1 was arbitrarily taken to be 0.15 au. If the outermost contour is drawn at a lower level, 0.05 au, one obtains the result shown in Fig. 5a. The larger area surrounded by the outer contours now makes it readily apparent that there is some overlap between the densities of the two molecules. It is this overlap which leads to an exchange repulsion between the two molecules which, in turn, prevents the water dimer from collapsing together into a single molecule with atoms unphysically close to one another.

The exchange repulsion energy is approximately proportional to the degree of overlap between the electron clouds, so it rises very quickly as the molecules approach. This rapid increase in the overlap is evident in Fig. 5b in which the two molecules have been brought slightly closer together. An energetic measure of this behavior can be gleaned from the EX column of Table 1. When $R(O \cdot \cdot O)$ in the water dimer decreases from 3.25 to 2.75 Å, the exchange repulsion increases by nearly an order of magnitude, as compared to the electrostatic attraction which merely doubles. This

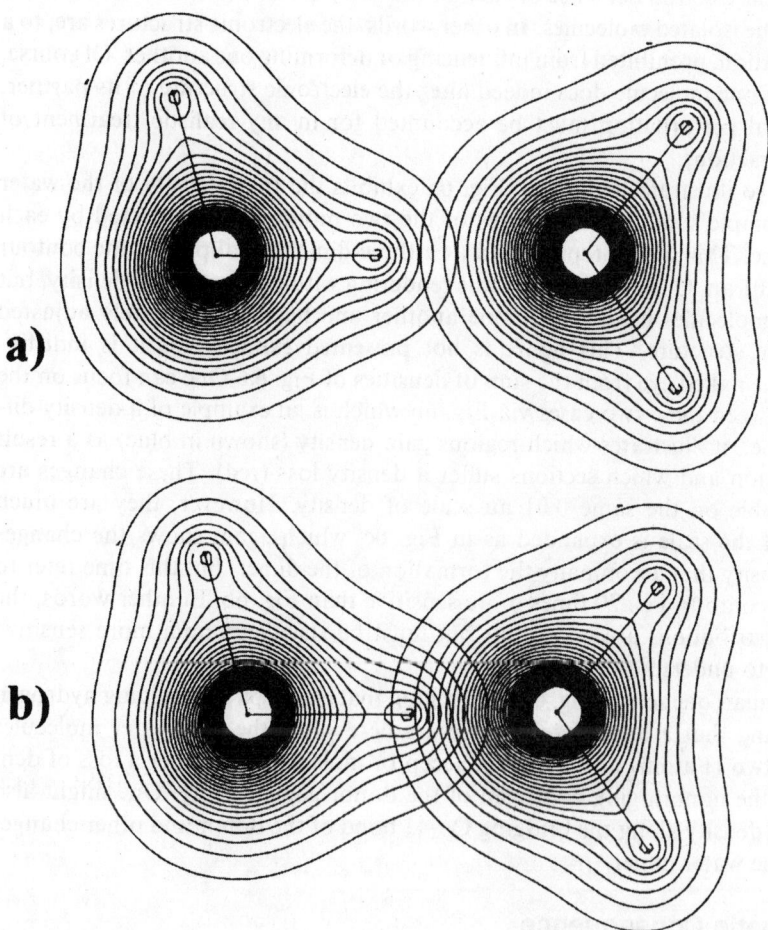

Figure 5 Electron densities of the two water monomers, oriented to coincide with the nuclear positions in the dimer arrangement (a) as in Fig. 1d and (b) with molecules brought closer together. Contour intervals are 0.5 au.

same sort of rapid increase is evident in the ionic $H_2OH^+ \cdots OH_2$ system, also shown in Table 1.

This very strong sensitivity to intermolecular distance makes exchange repulsion fairly straightforward to model. To a good approximation, one can simply center a highly distance-dependent function on each atomic center. Functions that have been used to good effect in the past have include aR^{-12} and $e^{-\alpha R}$, where a and α represent adjustable parameters. For many purposes, it has even been adequate to simply model the exchange repulsion by a set of "hard spheres," billiard balls if you will. Atoms on partner molecules are simply forbidden from approaching closer than a preassigned distance, and their mutual repulsion is assumed to be zero otherwise.

4. DEFORMATION

The electrostatic and exchange terms discussed above are evaluated within a framework wherein the electron densities of each of the two partner molecules are assumed to be those of the isolated molecules. In other words, the electronic structures are, to a first approximation, prohibited from influencing or deforming one another. Of course, the presence of one molecule does indeed alter the electronic structure of its partner, and this mutual perturbation must be accounted for in any realistic treatment of molecular interactions.

In order to illustrate this point, Fig. 6a exhibits the total density of the water dimer as the simple sum of the densities of the two monomers, unaffected by each other's presence. This sum is represented as a three-dimensional plot of the contour representing 0.01 au. One can compare this rendering with the same total density, but after the two molecules have "seen" one another and have appropriately adjusted their electronic structure. This figure is not presented explicitly as it is indistinguishable to the naked eye from the sum of densities of Fig. 6a. One can focus on the differences between these two cases via Fig. 6b which is an example of a density difference map, i.e., it illustrates which regions gain density (shown in blue) as a result of the interaction and which sections suffer a density loss (red). These changes are hardly noticeable on the same 0.01 au scale of density. However, they are much more visible if the scale is expanded as in Fig. 6c, which again shows the changes in electron density that accompany the formation of the dimer, but this time refer to the 0.0005-au contour, i.e., 20 times more sensitive than Fig. 6b. In other words, the electronic redistributions are important but must be visualized on a more sensitive scale in order to understand them.

In particular, one sees in Fig. 6c a red region that envelops the bridging hydrogen atom, indicating that a loss of density occurs here when the two water molecules interact. The two H atoms of the proton acceptor molecule also suffer a loss of density, whereas the nonbridging hydrogen of the donor gains density. One might also note a gain of density along the bridging O—H bond of the donor and other changes throughout the water dimer.

4.1. Energetic Consequence

The "relaxation" or "deformation" of the density cloud has an energetic consequence. The redistribution of electron density that accompanies the formation of the complex stabilizes the system, thus contributing to the interaction energy. In order to

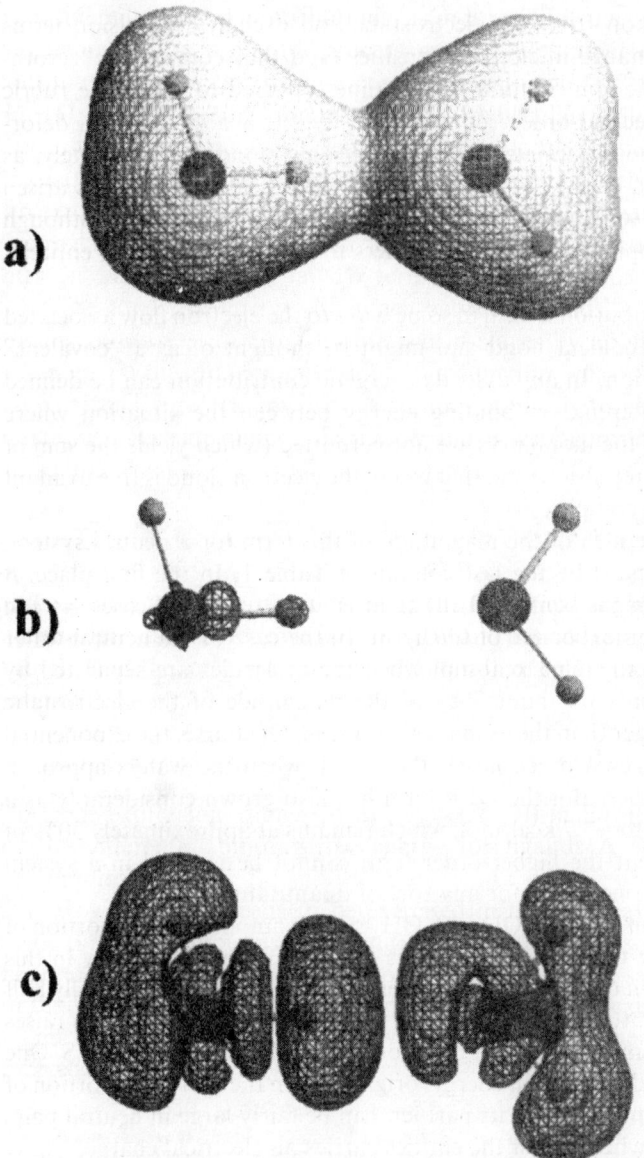

Figure 6 (a) Sum of the electron densities of the two water molecules, in the dimer configuration, illustrated as the 0.01-au contour. (b) Difference in density between the sum of monomers (a) and the dimer, again at the ±0.01 contour. Blue contour represents gain of density in the dimer vs. the pair of monomers, and loss of density is shown in red. (c) Same as (b) except that the ±0.0005 contour is illustrated. (See color plate at end of chapter.)

distinguish this phenomenon from the electrostatic and exchange repulsion terms which do not allow any change in electron densities (and thus comprise a "zeroth-order" term), this energetic contribution is sometimes referred to under the rubric of a "higher-order" or "second-order" contribution [29]. It is also termed a deformation energy (referring to deformation of the electronic cloud) or, alternately, as induction or delocalization energy. (The unfortunate historical situation has arisen that this same quantity is sometimes referred to as "polarization" energy, although the same designation is applied by some researchers to a different quantity entirely, vide infra.)

The electronic redistribution is akin in some ways to the electron flow associated with the formation of a covalent bond and might be thought of as a "covalent" contributor to the interaction. In any case, its energetic contribution can be defined simply enough as the difference in binding energy between the situation where electronic deformations of the monomers are not permitted (which yields the sum of electrostatic and exchange repulsion) and that where the electron cloud is free to adapt to the new situation of the complex.

One can obtain some idea of the magnitude of this term for a neutral system, and for an ion–molecule pair, by the last column of Table 1. In the first place, it should be noted that the higher-order (DEF) term is attractive in all cases, as one would expect from the physical origin of this term. In the case of the neutral water dimer, this term amounts to -1.65 kcal/mol when the molecules are separated by 3.25 Å. As such, it amounts to about 30% of the magnitude of the electrostatic attraction and is in fact larger than the exchange repulsion. Of course, the exponential rise of the latter term makes it much larger than DEF when the waters approach within 2.75 Å of one another. But the DEF term has also grown considerably as a result of the approach, up to -3.7 kcal/mol, which remains at approximately 30% of the ES term. It is clear that the higher-order term cannot be ignored in a system such as the water dimer if one hopes for any sort of quantitative accuracy.

The presence of an ion in the $H_2OH^+ \cdots OH_2$ system amplifies the distortion of the electron cloud, leading to much larger values of the deformation energy in this system. Even at a separation of 3.25 Å, the DEF term exceeds -10 kcal/mol, fully half as large as the electrostatic attraction itself. Compression of the system to 2.75 Å raises the DEF term to -17 kcal/mol, still on the order of 50% of the magnitude of ES. One can thus conclude that the deformation energy, originating in the mutual distortion of the electron cloud of each molecule by its partner, can be fairly large in neutral pairs and even more significant when one of the entities carries an electrical charge.

4.2. Further Partition of Deformation Energy

There have been attempts over the years to partition the deformation energy into a number of smaller factors. One of the earliest of these schemes was due to Morokuma and Kitaura (MK) [30,31] who defined "charge transfer" and "polarization" energy terms. The fundamental underpinning of their formulation rests on the concept that even in the complex, one molecule can be clearly distinguished from the other, with fully separate electron densities and associated molecular orbitals. Of course, this is incorrect since the molecular interaction by its very nature causes an intermingling of the two molecules. This assumption is thus reminiscent of the drawing of arbitrary borders around the nuclear centers to calculate atomic charges (vide supra).

Ignoring for the moment the arbitrariness of this definition, charge transfer is defined as the energetic consequence of electrons from molecule A drifting into molecule B's "airspace" and the analogous shifting of molecule B's electrons into vacant MOs that are part of the A subsystem. The shifting of density that occurs within the borders of molecule A, without crossing the boundary into molecule B, is associated with polarization energy, as is the internal redistribution of electron density within molecule B. Since these definitions do not encompass the entire interaction energy, Kitaura and Morokuma lumped the remaining effects into an umbrella "mixing" energy. In this scheme then, the DEF term is partitioned into charge transfer, polarization, and mixing energies.

Although perhaps not rigorously defensible, the MK scheme does offer physically sensible results in many cases. What is meant by this is that the mixing term is rather small, and the polarization and charge transfer energies both negative and smaller than the full DEF term. However, the intrinsic problem of attempting to "separate the inseparable" can lead to a breakdown of the approach, especially for short intermolecular contacts. An example of such a breakdown was illustrated for the system pairing NH_4^+ with NH_3 [32]. As the two entities began to approach one another, the various quantities behaved in a reasonable and expected fashion. But when the intermolecular separation became shorter than about 2.75 Å, the polarization energy plunged precipitously, exceeding -100 kcal/mol, clearly an unphysical finding. This large attractive component was opposed by a sharp rise in the mixing energy, which was positive in this case. Perhaps more disturbingly, this breakdown is more likely to occur for larger basis sets, of the sort that one would most want to apply to noncovalent interactions. The breakdown is not limited to this particular ionic H-bonded system, but occurs also in other complexes such as ScCO [33] and He_2 and $HeLi^+$ [34]; it has been attributed to the failure of the decomposition scheme to prevent the valence electrons of one fragment from trespassing into the core orbitals of its partner, despite the full occupation of the latter.

Of course, the MK scheme is only one of many approaches to partitioning the total interaction energy. A vast array of alternative methods has been described in the literature [18,29,35–38]. The reader is cautioned that different schemes may use the same name for two different formulations. Hence, a term that is attractive for a given interaction in one scheme may be repulsive within the context of another scheme.

5. DISPERSION

In addition to electrostatic, exchange, and induction (a.k.a. deformation) energy, the fourth principal contributor to the interaction energy is the so-called dispersion energy [39]. This quantity is closely related to the London forces that are well known from freshman chemistry texts that originate from instantaneous fluctuations of the electron density of one molecule, which cause a sympathetic series of instantaneous density fluctuations in its partner. Dispersion, by its very nature, is attractive. In terms of ab initio molecular orbital theory, the dispersion energy is not present at the SCF level, but is a byproduct of the inclusion of electron correlation into the calculation. The reader is hence alerted to the fact that calculations that do not include electron correlation (and there are many such, particularly in the early literature) cannot be expected to include this fourth, and sometimes very important, component of the noncovalent force.

In many ways, dispersion is the most demanding of the four forces to calculate accurately. The reason for this is not limited to the need to invoke electron correlation, typically much more computationally intensive than calculations limited to the SCF level. A chief difficulty is that the correlation contribution to the interaction energy includes not only dispersion energy, but also a host of other terms. For example, the addition of electron correlation modifies the charge distribution around a given molecule. This modification leads in turn to a change in this molecule's electrostatic interaction with its partner. Thus, the electrostatic interaction energy computed at a correlated level is different than that based on the SCF distribution. As a consequence, the correlation contribution to the interaction energy includes not only the dispersion energy, but also a "correction" to the ES term computed at the SCF level. This correction can be either positive or negative, regardless of whether the ES term itself is attractive or repulsive. Not only the ES term but the exchange and deformation terms too are prone to corrections when correlation is included. One can think of the dispersion energy as the intermolecular component of the correlation energy [40]; its separation from the intramolecular component is not a simple matter by any means. Further complicating the issue is the high sensitivity of dispersion energy to the quality of the basis set, necessitating the use of large and expensive basis sets to obtain dispersion energies of even reasonable accuracy.

There are theoretical approaches that compute the dispersion energy directly, so they do not suffer from the difficulty of extracting it from a jumble of other terms. Most of these are based on some form of perturbation theory. Szalewicz and Jeziorski [39] have summarized the basic equations and ideas of symmetry-adapted perturbation theory which has found wide use in studying noncovalent interactions.

As one might expect in a system like the helium dimer, where such forces as electrostatic and induction ought to be vanishingly small, dispersion contributes the lion's share of the attractive force (even if only 0.04 kcal/mol). As the other terms grow, as in the water dimer, the dispersion energy remains important. It contributes something in excess of 2 kcal/mol to the binding energy of the water dimer, nearly equal to the induction energy and about 1/3 the magnitude of the ES term [41]. In addition, in a tighter complex, the ion–molecule $NH_4^+ \cdots NH_3$ case, the dispersion can be significant as well. Calculations [32] of this system lead to an estimate of 3.5–5.0 kcal/mol for dispersion at an internuclear separation of 2.75 Å, depending upon the basis set used. Dispersion was shown to be of roughly 1/3 the magnitude of the induction energy, 1/6 of the ES, even in a system where the latter terms can be quite large.

The dispersion energy typically dies off as the inverse sixth power of the distance separating the electron clouds [11,13,40,42]. In that respect, it has shorter range than electrostatic forces, but has an effect at longer distances than does exchange repulsion. Indeed, dispersion and exchange energies are commonly tied together in "6–12" or "Lennard–Jones" empirical potentials that express part of the force between atoms on different molecules in the form $-a/R^6 + b/R^{12}$, where a and b are adjustable parameters [13,43].

6. SUMMARY

The various contributors to the noncovalent force each behaves differently with regard to the intermolecular distance and angular aspects of the geometry. The exchange repulsion is of very short range and becomes a factor only when the two units are in intimate contact. Moreover, this force is not very sensitive to different inter-molecular orientations. It can therefore be safely ignored in terms of steering molecules into an optimal alignment as they approach from some distance. Dispersion, too, is a short-range force, typically dying off as the inverse sixth power of the inter-molecular separation. It is always attractive, a sort of "general sticky force" between any pairs of atoms, and so is not very sensitive to misalignment of the molecules in question. The deformation energy is also attractive with a relatively small degree of anisotropy. It is generally of longer range than the exchange and dispersion. Longest range of all is the electrostatic interaction which dies off as the inverse third power of the distance (for neutral molecules). It can be strongly attractive or repulsive and so is highly dependent upon the mutual orientation. As a result, it is the electrostatic force which is dominant in the "steering" of a pair of molecules into an optimal alignment as they approach one another.

REFERENCES

1. Jeziorski B, Kolos W. Perturbation approach to the study of weak intermolecular inter-actions. In: Ratajczak H, Orville-Thomas WJ, eds. Molecular Interactions. New York: Wiley, 1982:1–46.
2. Scheiner S. Ab initio studies of hydrogen bonding. In: Maksic ZB, ed. Theoretical Models of Chemical Bonding. Berlin: Springer-Verlag, 1991:171–227.
3. Scheiner S. Hydrogen Bonding: A Theoretical Perspective. New York: Oxford University Press, 1997.
4. Alkorta I, Rozas I, Elguero J. Non-conventional hydrogen bonds. Chem Soc Rev 1998; 27:163–170.
5. Tarakeshwar P, Choi HS, Kim KS. Olefinic vs aromatic π–H interaction: A theoretical investigation of the nature of interaction of first-row hydrides with ethene and benzene. J Am Chem Soc 2001; 123:3323–3331.
6. Scheiner S. CH $\cdots$ O Hydrogen Bonding. In: Hargittai M, Hargittai I, eds. Advances in Molecular Structure Research. Stamford, CT: JAI Press, 2000:159–207.
7. Gu Y, Kar T, Scheiner S. Fundamental properties of the CH $\cdots$ O interaction: is it a true hydrogen bond? J Am Chem Soc 1999; 121:9411–9422.
8. Scheiner S, Grabowski SJ, Kar T. Influence of hybridization and substitution upon the properties of the CH$\cdots$O hydrogen bond. J Phys Chem A 2001; 105:10607–10612.
9. Price SL. Electrostatic forces in molecular interactions. In: Scheiner S, ed. Molecular Interactions: From van der Waals to Strongly Bound Complexes. Chichester: Wiley, 1997:297–333.
10. Cybulski SM, Scheiner S. Factors contributing to distortion energies of bent hydrogen bonds. Implications for proton-transfer potentials. J Phys Chem 1989; 93:6565–6574.
11. Stone AJ. Classical electrostatics in molecular interactions. In: Maksic ZB, ed. Theoretical Models of Chemical Bonding. Berlin: Springer-Verlag, 1991:103–131.
12. Náray-Szabó G, Ferenczy GG. Molecular electrostatics. Chem Rev 1995; 95:829–847.
13. Hirschfelder JO, Curtiss CF, Bird RB. Molecular Theory of Gases and Liquids. New York: John Wiley, 1954.
14. Tomasi J, Bonaccorsi R, Cammi R. The extramolecular electrostatic potential. An indicator of the chemical reactivity. In: Maksic ZB, ed. Theoretical Models of Chemical Bonding. Berlin: Springer-Verlag, 1991:229–268.
15. Beu TA, Buck U, Siebers JG, Wheatley RJ. A new intermolecular potential for hydrazine clusters: structures and spectra. J Chem Phys 2001; 106:6795–6805.

16. Stone AJ. Distributed multipole analysis, or how to describe a molecular charge distribution. Chem Phys Lett 1981; 83:233–239.

17. Mulliken RS. Electronic population analysis on LCAO-MO [linear combination of atomic orbital-molecular orbital] molecular wave functions. I. J Chem Phys 1955; 23:1833–1840.

18. Reed AE, Curtiss LA, Weinhold F. Intermolecular interactions from a natural bond orbital, donor–acceptor viewpoint. Chem Rev 1988; 88:899–926.

19. Farnum DG. Charge density-NMR chemical shift correlations in organic ions. Adv Phys Org Chem 1975; 11:123–175.

20. Coppens P, Hall MB. Electron distribution and the chemical bond. New York: Plenum, 1982.

21. Gussoni M. Role of vibrational intensities in the determination of molecular structure and charge distribution. J Mol Struct 1984; 113:323–340.

22. Ferenczy GG. Charges derived from distributed multipole series. J Comput Chem 1991; 12:913–917.

23. Chipot C, Angyan JG, Ferenczy GG. Transferable net atomic charges from a distributed multipole analysis for the description of electrostatic properties: a case study of saturated hydrocarbons. J Phys Chem 1993; 97:6628–6636.

24. Scrocco ET. J. Electronic molecular structure, reactivity and intermolecular forces: an heuristic interpretation by means of electrostatic molecular potentials. Adv Quantum Chem 1978; 11:115–196.

25. Politzer P, Murray JS, Peralta-Inga Z. Molecular surface electrostatic potentials in relation to noncovalent interactions in biological systems. Int J Quantum Chem 2001; 85:676–684.

26. Momany FA. Determination of partial atomic charges from ab initio molecular electrostatic potentials. Application to formamide, methanol, and formic acid. J Phys Chem 1978; 82:592–601.

27. Cox SR, Williams DE. Representation of the molecular electrostatic potential by a net atomic charge model. J Comput Chem 1981; 2:304–323.

28. Singh UC, Kollman PA. An approach to computing electrostatic charges for molecules. J Comput Chem 1984; 5:129–145.

29. van Lenthe JH, van Duijneveldt-van de Rigdt JGCM, van Duijneveldt FB. Weakly bonded systems. In: Lawley KP, ed. Ab Initio Methods in Quantum Chemistry. New York: Wiley, 1987:521–566.

30. Kitaura K, Morokuma K. A new energy decomposition scheme for molecular interactions within the Hartree–Fock approximation. Int J Quantum Chem 1976; 10:325–340.

31. Morokuma K, Kitaura K. Variational approach (SCF ab-initio calculations) to the study of molecular interactions: the origin of molecular interactions. In: Ratajczak H, Orville-Thomas WJ, eds. Molecular Interactions. New York: Wiley, 1980:21–87.

32. Cybulski SM, Scheiner S. Comparison of Morokuma and perturbation theory approaches to decomposition of interaction energy. $(NH_4)^+ \ldots NH_3$. Chem Phys Lett 1990; 166:57–64.

33. Frey RF, Davidson ER. Energy partitioning of the self-consistent field interaction energy of ScCO. J Chem Phys 1989; 90:5555–5562.

34. Gutowski M, Piela L. Interpretation of the Hartree–Fock interaction energy between closed-shell systems. Mol Phys 1988; 64:337–355.

35. Dreyfus M, Pullman A. Non-empirical study of the hydrogen bond between peptide units. Theor Chim Acta 1970; 19:20–37.

36. Fujimoto H, Kato S, Yamabe S, Fukui K. Molecular orbital calculations of the electronic structure of borazane. J Chem Phys 1974; 60:572–578.

37. Morokuma K, Kitaura K. Energy decomposition analysis of molecular interactions. In: Politzer P, Truhlar DG, eds. Chemical Applications of Atomic and Molecular Electrostatic Potentials. New York: Plenum, 1981:215–242.

38. Bonaccorsi R, Palla P, Cimiraglia R, Tomasi J. On the use of a MO polarized basis for the analysis of the interaction energy in molecular interactions: application to amine complexes. Int J Quantum Chem 1983; 24:307–316.

39. Szalewicz K, Jeziorski B. Symmetry-adapted perturbation theory of intermolecular interactions. In: Scheiner S, ed. Molecular Interactions. From van der Waals to Strongly Bound Complexes. New York: Wiley, 1997:3–43.

40. Hobza P, Zahradnik R. Weak Intermolecular Interactions in Chemistry and Biology. Amsterdam: Elsevier Scientific, 1980.

41. Mas EM, Szalewicz K. Effects of monomer geometry and basis set saturation on computed depth of water dimer potential. J Chem Phys 1996; 104:7606–7614.

42. Hunt KLC. Dispersion dipoles and dispersion forces: proof of Feynman's "conjecture" and generalization to interacting molecules of arbitrary symmetry. J Chem Phys 1990; 92:1180–1187.

43. Claverie P. Elaboration of approximate formulas for the interactions between large molecules: applications in organic chemistry. In: Pullman B, ed. Intermolecular Interactions: From Diatomics to Biopolymers. New York: Wiley, 1978:69–305.

10

Solvent Simulation

PETER L. CUMMINS, ANDREY A. BLIZNYUK and ,JILL E. GREADY

Australian National University, Canberra, Australian Capital Territory, Australia

1. INTRODUCTION

Water is ubiquitous. All biological processes must occur in an aqueous environment, thus making water an essential element of life [1]. Consequently, solvation processes play a crucial role in determining the strength of ligand binding to macromolecular targets, such as proteins and DNA, and in the solubility and transport of molecules across cell membranes. In the field of medicinal chemistry and drug design, researchers are frequently interested in questions of whether one ligand, or drug molecule, binds more tightly to its macromolecular target than another. The answers to these and related questions are given by free energy differences for changes occurring in different environments, taking into account that these environments consist of, or are in some way influenced by, solvent water. As shown in Fig. 1, for any type of system, the desired free energy for a change from some arbitrary state A to another state B, which we denote by (A→B), can be written in the general form

$$\Delta\Delta G_{j\to i}(A \to B) = \Delta G_i(A \to B) - \Delta G_j(A \to B) \tag{1}$$

where ΔG_i is the free energy for the change A→B taking place in an environment denoted by i, and ΔG_j is the free energy for the corresponding change in a different environment denoted j.

The application of Eq. (1) is probably best illustrated by the following example for a process involving a number of different ligands binding to a macromolecular target to form the corresponding number of thermodynamically stable complexes. It may be seen from the relevant thermodynamic cycle in Fig. 2 that the difference in binding free energy between any two ligands (A and B) can be given by

$$\Delta\Delta G_{bind}(A \to B) = \Delta G_{complex}(A \to B) - \Delta G_{sol}(A \to B) \tag{2}$$

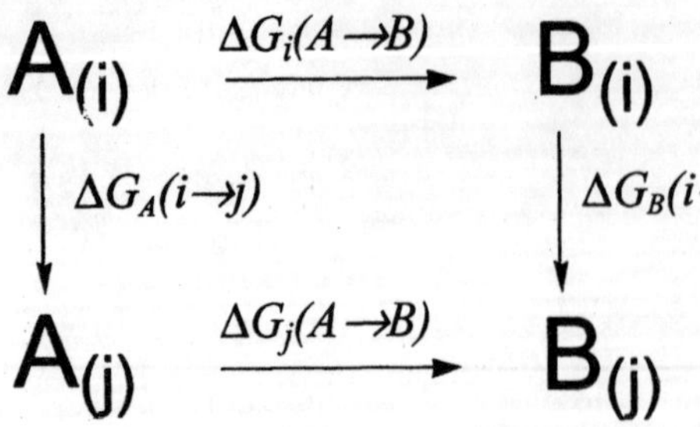

Figure 1 Generalized thermodynamic cycle for the change of state of a system (A→B) taking place in different environments denoted by i and j.

Thus the relative free energy of binding, ΔG_{bind}, is the difference between the free energy of complex formation, $\Delta G_{complex}$, i.e., a component that depends on the interactions between the ligands and the macromolecular binding site, and a solvation term, ΔG_{sol}, that relates to processes involved in solvation of the free (unbound) ligand. Note that the complex also resides in an aqueous environment, and, consequently, $\Delta G_{complex}$ may also be strongly influenced by the effects of solvent interactions with both the ligand and macromolecule at the binding site.

The other important application of Eq. (1) is in the partitioning of solutes between different phases, which has important implications for drug transport and solubility. The relative free energy on transferring any pair of solutes (A and B) from one solvent to a second solvent is given by

$$\Delta\Delta G_{transfer}(A \rightarrow B) = \Delta G_{sol.2}(A \rightarrow B) - \Delta G_{sol.1}(A \rightarrow B) \tag{3}$$

where $\Delta G_{sol.1}(A\rightarrow B)$ is the free energy difference between A and B in the first solvent and $\Delta G_{sol.2}(A\rightarrow B)$ is the corresponding free energy difference in the second solvent. The familiar partition coefficient P (or its log value) for a single solute (i.e., $A = B$) between two immiscible solvent phases is thus a particular case of Eq. (3):

$$\log(P) = -\frac{1}{2.303RT}[\Delta G_{sol.2}(A) - \Delta G_{sol.1}(A)] \tag{4}$$

where $\Delta G_{sol.2}(A)$ and $\Delta G_{sol.1}(A)$ are the solvation free energies of the solute A in solvents 2 and 1, respectively.

Clearly, solvent simulation is an essential element of the rational drug design process. The need for simulation methods to take proper account of solvation

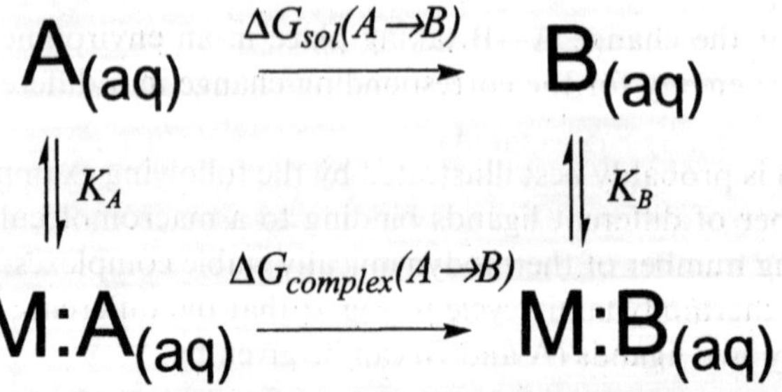

Figure 2 Thermodynamic cycle for the binding of ligands A and B to a macromolecular target M in aqueous solution, with equilibrium constants K_A and K_B, respectively.

processes in order to obtain free energy differences poses a considerable computational challenge. The computational approaches that are currently available can be broadly divided into implicit and explicit solvation models. In the explicit solvent approach, water molecules are treated as discrete entities necessitating a detailed description of interactions between solvent molecules at the atomic level. In contrast, implicit models dispense with this detail by considering the solvent to be a dielectric continuum, and are thus the more computationally efficient. The use of a continuum approximation can be justified by realizing that in order to obtain the desired information it is not necessary to know every detail about the system. It is only important to know how to model the solvent effects on the properties of interest. Specifically, it is unnecessary to quantify interactions between individual water molecules in the bulk solvent. At this simplest level, where the properties of the solvent are determined by a dielectric constant, only a knowledge of the noncovalent interactions (polar and nonpolar) between the solute and solvent molecules is required to compute solvation properties.

The use of explicit solvent models adds an extra level of complexity to the problem. The solvent is no longer a continuum. Both solute–solvent and solvent–solvent interaction terms must be considered. This additional complexity in the solvent is addressed using molecular dynamics (MD) and Monte Carlo (MC) simulation methods. These methods can be used to quantify solvation effects on ligand binding, including the calculation of free energy differences, but at greatly increased computational cost. A compromise solution to this computational bottleneck combines both explicit and implicit methods. In this mixed approach, at least part of the solvent—that in the immediate vicinity of the solute—requires the explicit treatment of molecules, while the remaining "bulk" solvent is treated as a continuum. Within the explicit solute and solvent region, there is also the choice of an appropriate methodology to consider. The solute, for example, may require some level of quantum chemical (semiempirical, density functional, or ab initio) rather than classical molecular mechanics treatment. Thus a combined quantum mechanical and molecular mechanical (QM/MM) method may be used in the overall description of the system force field.

The complexity of the solvation problem has led to a wide range of methodological developments. In practical applications, not only are solvated systems invariably divided into regions requiring different theoretical treatments, but also within each of these regions there is now a large number of optional methods that may be employed. Many of these have been extensively reviewed in a number of recent articles [2–8]. Despite much progress in the field, there remain considerable obstacles to the routine application of these methods in the field of medicinal chemistry and drug design, due to this inherent complexity. In this chapter we provide a theoretical background to those more recent techniques, in particular, that we judge are likely to become increasingly important in future simulation studies of biomolecules in aqueous environments. In Secs. 2 and 3 we deal with implicit and explicit solvation models, respectively. In Sec. 4, we also present a general summary of the major limitations of these various solvent simulation approaches.

2. IMPLICIT SOLVENT METHODS

Implicit solvation methodology is an actively developing field. Recent reviews of implicit solvent models [2,3,5] include more than 800 references each. Thus it is not possible to examine all the proposed methods here. In the present work we have restricted ourselves to reviewing well-established methods that are actively and successfully used in a variety of applications where solvation energy calculations are necessary. In our opinion, these methods will continue to be used in molecular modeling calculations. We have also restricted literature citation to papers published in the last few years, where possible, to give the reader an overview of the current developments in the field. More general reviews can be found elsewhere [2–7].

The total energy of solvation can be divided into three separate components, assuming that the electrostatic and the nonpolar or van der Waals energies do not depend upon each other. This assumption is true for all currently used molecular mechanics force fields. Thus the solvation free energy ΔG_{sol} is given by

$$\Delta G_{sol} = \Delta G_{cav} + \Delta G_{vdw} + \Delta G_{ele} \tag{5}$$

This equation corresponds to the following three-step process. First, the cavity inside a solvent is created and the molecule is inserted into the cavity. Next, nonpolar interactions between the solute and the solvent are switched on. Finally, the electrostatic interactions between the solute and the solvent are switched on. Of the three components in Eq. (5), the electrostatic component of the solvation energy (ΔG_{ele}) is by far the largest and is typically of an order of several thousands kcal/mol for an average protein. Consequently, it is convenient to start the examination of different approaches from the electrostatic solvation energy component.

The following list defines notations that are used below:

(i)   The electrostatic potential at point r, $\phi(r)$, which for a single-point charge is equal to $q/\varepsilon r$.
(ii) The electric field, $E(r)$, which for a single-point charge is equal to $\nabla\phi(r) = -q/\varepsilon r^2$, where ∇ is the gradient operator.
(iii) The electric displacement, $D(r)$, which is equal to $\varepsilon E(r)$ and is the electric field calculated in vacuum.

2.1. Electrostatic Component of the Solvation Energy

One of the simplest approximations would be to replace the solvent molecules by a grid of spherical dipoles. This would solve the problem of sampling possible conformations of solvent molecules. The interaction of the grid dipoles with the solvated molecule is then determined by a simple electrostatic term:

$$\Delta G_{ele} = -1/2 \sum_i E_{0i} \mu_0 \langle \cos(\varphi) \rangle \tag{6}$$

where E_{0i} is an electric field due to a solute molecule at point i of the grid, μ_0 is a dipole moment assigned to the grid point, and $\langle\cos(\varphi)\rangle$ is the average cosine between directions of the dipole moment and the electric field. The coefficient $1/2$ comes from the assumption of the linear response, i.e., half of the energy gained from the dipole field is spent reorienting the dipoles. Assuming that the dipoles are not polarizable, do not interact with each other, have only thermal movement, and are spherical, the average cosine can be easily evaluated assuming a Boltzmann distribution [9]

$$\langle\cos(\varphi)\rangle = \frac{\int\limits_{-1}^{1} \cos(\varphi)\exp(b_i\cos(\varphi))d(\cos(\varphi))}{\int\limits_{-1}^{1} \exp(b_i\cos(\varphi))d(\cos(\varphi))} = \coth(b_i) - \frac{1}{b_i} \tag{7}$$

where coth is the hyperbolic cotangent and b_i is a Boltzmann prefactor:

$$b_i = E_i \mu_0 / kT \tag{8}$$

In Eq. (8), T is the temperature and k is the Boltzmann constant. The electric field at the grid point, E_i, in Eq. (8) is different from the electric field E_{0i} in Eq. (6). It combines electric fields from the solute molecules and from other grid points (except the nearest); hence, iterations are required for its evaluation [10,11]. Test calculations [11] show that the electric field E_i in Eq. (8) can be substituted with the electric field due to solute atoms only (E_{0i}) This simplification speeds up considerably the computations, while giving reasonable solvation energies [11]. The function $(\coth(b)-1/b)$ in Eq. (7) is a well-known Langevin function [9], so the method of approximation of the solvent by grid dipoles became known as the Langevin dipoles (LD) method [10]. Together with induced protein dipoles (PD) it forms the basis of the PDLD and PDLD/S methods [10,11]. The grid forms a sphere around the solute molecule; the long-range solvation contribution due to the solvent outside the grid is estimated using Born and Onsager formulae (Eqs. (29) and (30)). The van der Waals (vdW) radii of the solute atoms were parameterized to reproduce solvation energies of small molecules in water. The values of the radii and detailed description of the method can be found in Ref. 11.

The validity of the model can be easily criticized on the basis of ignoring the interactions between solute dipoles [12]; indeed, it is well known that water molecules form strong directed hydrogen bonds. However, the model correctly reproduces solvation energies of small molecules [11], enzymatic reactions [11], ligand binding energies [11,13–15], and pK_a shifts [15]. Unfortunately, as only a very limited number of ligand association constants have been evaluated, the general applicability of the model to docking calculations is not clear. From the computational point of view, the solvation calculations with this model depend on the grid size and spacing, and solute molecule orientation. The LD model is best suited to accurate evaluation of the differences in ligand binding in combination with MD simulations [15] and is not practical for quick screening of many ligands.

Further simplifications can be achieved assuming that the solvent can be represented by a continuum with dielectric constant ε_w, which is different from the dielectric constant of the solute ε_s. When the solute molecule is immersed in the continuum, induced charges appear on the border, due to the difference between dielectric constants of the solute and solvent. According to Gauss's theorem, the total induced charge Q_s on a surface completely surrounding a number of charges q_i should be equal to

$$Q_s = (\varepsilon_s - \varepsilon_w)/\varepsilon_w \sum_i q_i \tag{9}$$

The density of the induced charge (σ) can be evaluated by solving the following equation:

$$\sigma 4\pi\varepsilon_s\varepsilon_w = -(\varepsilon_w - \varepsilon_w)\boldsymbol{Dn} \tag{10}$$

where $\boldsymbol{D}$ is the electrostatic displacement at the point with density σ and $\boldsymbol{n}$ is the normal to the cavity surface (both $\boldsymbol{D}$ and $\boldsymbol{n}$ are vectors). Equation (10) forms the basis of so-called boundary elements (BE) methods. A solute molecule is surrounded by some form of a cavity. The surface area of the cavity is divided by a number of small elements with surface area equal to ΔS. Assuming that the charge density is constant at each surface element so that $q_s = \sigma \Delta S$, and all charges are point charges, Eq. (10) then becomes

$$q_s = -(\varepsilon_w - \varepsilon_s)/4\pi\varepsilon_w\varepsilon_s \left[E_{self} + \Delta S \sum_{l \in solute} q_l/R_{sl}^2 \cos(\varphi_{sl}) + \Delta S \sum_{j \in \Delta S} q_j/R_{sj}^2 \cos(\varphi_{sj}) \right] \quad (11)$$

The three terms in brackets derive from the expansion of the $\boldsymbol{D} \boldsymbol{n}$ product in Eq. (10), i.e., the electric field due to the surface element itself, the electric field due to fixed charges inside the cavity denoted q_l, and the electric field due to induced charges q_j on the other surface elements, respectively. It is convenient to rewrite this equation in matrix form, to emphasize that the induced charges can be found by solving the system of linear equations:

$$AQ_s = B \quad (12)$$

Here Q_s is the vector of induced charges, the matrix A and the vector B have the following elements:

$$A_{ii} = -(\varepsilon_s - \varepsilon_w)/4\pi\varepsilon_s\varepsilon_w E_{self} - 1 \quad (13)$$

$$A_{ij} = -(\varepsilon_s - \varepsilon_w)/4\pi\varepsilon_s\varepsilon_w \Delta S_i \cos(\varphi_{ij})/R_{ij}^2 \quad (14)$$

$$B_i = -(\varepsilon_s - \varepsilon_w)/4\pi\varepsilon_s\varepsilon_w \Delta S_i \sum_l q_l/R_{il}^2 \cos(\varphi_{il}) \quad (15)$$

The summation in the last term is among all fixed charges q_l inside the cavity. The E_{self} value can be estimated using the following formula [17]:

$$E_{self} = -2\pi\left(1 - \sqrt{\Delta S_i/4R_i^2}\right) \quad (16)$$

This formula was derived assuming that the induced charge is located in a circular convex part of a sphere with radius R_i [17]. After the induced charges are found from Eq. (12), the solvation energy is calculated simply as the Coulomb interaction of fixed and induced charges, i.e.,

$$\Delta G_{ele} = 1/2 \sum_i \sum_s q_i q_s/R_{is} \quad (17)$$

Here, again, the coefficient 1/2 comes from the linear response approximation, i.e., that half of the solvation energy is spent on creating the induced charges.

The BE methods are widely used in quantum chemical calculations of solvent effects [2,3,5] with very good results. For historical reasons, the use of BE methods in molecular mechanics calculations is limited. Several versions of the BE method that differ in the cavity surface formation algorithms, evaluation of the E_{self} contribution, Eq. (16), and methods of solving the system of linear equations, Eq. (12), have been reported [18–22]. From a computational point of view, the BE methods are invariant to molecular rotations because the molecular surface is invariant to the rotation. The results will depend upon the method chosen for the formation and tessellation of the molecular surface. The time limiting step in BE methods is the solution of the system of linear equations (Eq. (12)). Even for relatively small proteins, the size of matrix A

(number of surface elements) is in the tens of thousands. Hence the iterative process together with multipole expansion of the electric field [20–22] is necessary to achieve good speed. In any case, calculations are typically of the order of minutes and are not suitable for fast ligand screening.

The continuum model of the solvent in the BE method ignores all the solvent-specific interactions and assumes that there is a sharp boundary between the dielectric constants of the solute and the solvent. These are very serious approximations, but the method works and gives accurate estimates of the total solvation energies [5,19,23]. Parameters of the method include the vdW radii of the solute atoms, which are necessary to compute the cavity surface, and the dielectric constant of the solute (ε_s).

It should be noted that the solvation energy is defined as the energy of transfer of a solute from vacuum, or an ideal gas phase, to a solvent [24]. Therefore, if the dielectric constant of the solute is different from 1 (the dielectric constant of vacuum or an ideal gas phase), two calculations are needed for the estimation of the solvation energy, one to find the energy of the solute in vacuum and another to find the energy of the solute in the solvent:

$$\Delta G_{ele} = \Delta G_{ele}(\varepsilon_w, \varepsilon_s) - \Delta G_{ele}(1, \varepsilon_s) \tag{18}$$

However, in many cases the last term of Eq. (18) is incorrectly ignored. Another interesting problem is what dielectric constant to choose for the description of a protein. Proteins are complex molecules containing polar groups inside, so it is not likely that there is a single dielectric constant that can be used for describing the Coulomb interactions inside a protein. Calculations of solvation energies, however, show that reasonable results are obtained when the ε_s constant is somewhere between 2 and 4.

Another approach that is similar to the BE continuum solvation model is the conductor-like screening model (COSMO) [22]. The total electrostatic energy, U_{ele}, of a cavity with some charges q_i inside a conductor is given by the following equation:

$$U_{ele} = 1/2 \sum_s \sum_k q_s q_k / r_{sk} + \sum_s \sum_i q_s q_i / r_{si} + 1/2 \sum_i \sum_{j \neq i} q_i q_j / r_{ij} \tag{19}$$

or in matrix form (q_s and q_i are now vectors of charges)

$$U_{ele} = 1/2\, q_s A q_s + q_s B q_i + 1/2\, q_i C q_i \tag{20}$$

Where the first term corresponds to the interaction of induced charges, the second term describes the interaction of the induced charges with point charges inside the cavity, and the third term is the Coulomb energy of the point charges inside the cavity. Here, as in the BE method, we assumed that the induced charge density is constant on some cavity segment ΔS and the charges are given by $q_s = \sigma \Delta S$. The total energy of the conductor is a minimum [25,28]. Therefore

$$\delta U_{ele}/\delta q_s = A q_s + B q_i = 0 \qquad \text{or} \qquad A q_s = -B q_i \tag{21}$$

Solving the system of linear equations (Eq. (21)) would give induced charges q_s for a solute in a conductor ($\varepsilon_w = \infty$). The COSMO method assumes that the induced charges of a solute inside a dielectric with dielectric constant ε_w are equal to

$$q_s(\varepsilon_w) = (\varepsilon_w - 1)/(\varepsilon_w + 1/2) q_s \tag{22}$$

After the q_s charges are computed using Eqs. (21) and (22), the solvation free energy is evaluated as in Eq. (23):

$$\Delta G_{\text{ele}} = 1/2 \, \boldsymbol{q}_s \boldsymbol{B} \boldsymbol{q}_i \tag{23}$$

It should be noted that, according to the Gauss theorem, Eq. (9), the total induced charge on the surface of a conductor is equal to the sum of all charges inside the cavity. Hence, the use of the scaling factor of Eq. (22) does not satisfy Gauss's theorem. For a solute with dielectric constant ε_s inside a dielectric with dielectric constant ε_w, the following scaling of the charges will satisfy Gauss's theorem [26]:

$$q_s(\varepsilon_s, \varepsilon_w) = (\varepsilon_w - \varepsilon_s)/(\varepsilon_s \varepsilon_w) q_s \tag{24}$$

For the most common solvation of a molecule in water ($\varepsilon_w = 80$ and $\varepsilon_s = 1$), the difference between the scaling factors, Eqs. (22) and (24), is small but may became important for different solvation processes. Note that the author of COSMO, A. Klamt, believes that the scaling factor of Eq. (22) is better than the scaling factor of Eq. (24) [27].

The COSMO method is also interesting as the basis of a very successful COSMO-RS method, which extends the treatment to solvents other than water [27,28]. The COSMO method is very popular in quantum chemical computations of solvation effects. For example, 29 papers using COSMO calculations were published in 2001. However, we are not aware of its use together with MM force fields. Compared with the BE method, COSMO introduces one more simplification, that of Eq. (22). On the other hand, the matrix A in Eq. (21) is positively defined [25], which makes solution of the system of linear equations simpler and faster. Also, because both A and B matrices contain only electrostatic potential terms, their computation in quantum chemistry is easier than calculation of the electric field terms in Eq. (12). Another potential benefit is that the long-range electrostatic potential contribution is easier to expand into multipoles than the electric field needed in BE methods, which may benefit linear-scaling approaches.

The most popular methods in MM computations of the solvation free energy are undoubtedly the methods based on the numerical solution of the Poisson equation:

$$\nabla \cdot [\varepsilon(r) \nabla \phi(r)] = -4\pi\rho(r) \tag{25}$$

Here $\varepsilon(r)$ is the dielectric constant. We use $\varepsilon(r)$ to emphasize that the constant is a function of position: for example, it is different inside a solvent and inside a solute. The $\phi(r)$ is the electrostatic potential to be determined and $\rho(r)$ is the charge distribution of the solute. Equation (25) is an exact equation of the electrostatic potential in a dielectric. If the solvent contains some dissolved salt (i.e., the concentration of positive and negative charges is the same) and we assume that the distribution of salt ions follows a Boltzmann distribution, then the electrostatic potential of the system is described by the Poisson–Boltzmann equation

$$\nabla \cdot [\varepsilon(r) \nabla \phi(r)] - \varepsilon(r) \kappa^2 \, \sinh[\phi(r) e/kT] = -4\pi\rho(r) \tag{26}$$

where $\kappa^2 = 8e^2 I/kT$ with I the ionic strength, e the unit charge, and sinh is the hyperbolic sine. This is a nonlinear equation, because sinh is a nonlinear function. When the electrostatic potential is small [$\phi(r)e \ll kT$], only the first term of the sinh expansion is significant, so Eq. (26) becomes the linear Poisson–Boltzmann equation

$$\nabla \cdot [\varepsilon(r) \nabla \phi(r)] - \varepsilon(r) \kappa^2 e \phi(r)/k_B T = -4\pi\rho(r) \tag{27}$$

In order to estimate the solvation free energy, two computations are necessary. The first calculation is performed to evaluate the electrostatic potential due to the solute itself (ϕ_0). The second computation is done to evaluate the electrostatic potential of the solute surrounded by continuum solvent (ϕ). The solvation free energy can then be computed as

$$\Delta G_{ele} = 1/2 \sum_i q_i(\phi - \phi_0) \tag{28}$$

where both ϕ and ϕ_0 are the electrostatic potentials at the point of the solute point charges q_i.

Analytical solution of the Poisson equation is possible only for a limited number of simple cavities. For example, for a single charge q_i in the center of a spherical cavity with radius R, the solvation energy is described by the Born formula:

$$\Delta G_{ele} = -1/2 \frac{(\varepsilon_w - \varepsilon_s)}{\varepsilon_w \varepsilon_s} \frac{q_i^2}{R} \tag{29}$$

Note that this formula can be easily obtained from Gauss's theorem of the induced charges, which, in this case, will be equal to $-q_i(\varepsilon_w - \varepsilon_s)/(\varepsilon_w \varepsilon_s)$. The coefficient $1/2$ comes from the linear-response hypothesis. Another useful formula is the solvation energy of the point dipole (μ) in the center of a spherical cavity [29] (Bell formula, sometimes also called the Onsager formula):

$$\Delta G_{ele} = -1/2 \frac{2(\varepsilon_w - \varepsilon_s)}{\varepsilon_s(2\varepsilon_w + \varepsilon_s)} \frac{\mu^2}{R^3} \tag{30}$$

Solvation energies for other multipoles inside a spherical cavity, including corrections due to salt effects, can be found, for example in Ref. 29. Analytical solutions of the Poisson equation for some other cavities, such as ellipse or cylinder, are also known [2] but are of little use in solvation calculations of biomolecules. For cavities of general shape only numerical solution of the Poisson and Poisson–Boltzmann equations is possible. There are two well-established approaches to the numerical solution of these equations: the finite difference and the finite element methods.

The finite difference method substitutes the whole system of solute and solvent by an equal-distance grid. The charges and dielectric constants of both solvent and solute are projected onto the grid points. The derivatives in the Poisson or Poisson–Boltzmann equation are substituted by their finite difference analog for each grid point [30,31]. The resulting system of linear equations, or nonlinear equations for the Poisson–Boltzmann Eq. (26), is solved iteratively. Formally, the size of the matrix of equations is N^3, where N is the number of grid points along one direction, but because the finite difference formula involves only six neighbor grid points, the matrix is sparse. In order to assign boundary conditions, the grid should cover a large enough space. The typical size of the grid should be about two times larger than the largest linear dimension of the solute molecule [31]. On the other hand, due to the N^3 dependence of the number of independent variables (electrostatic potentials at each grid point), the total size of the grid is quite constrained (N is around 100). Thus, for computations of large biomolecules, separation between grid points is usually about 1 Å. Clearly this grid is too coarse for the accurate evaluation of the electrostatic potential. Therefore a technique called focusing has been developed. Several calculations are performed with decreasing grid size, and results of the previous calculations are used to assign the electrostatic potential on the borders of the new grid. Of course, accurate computation of the electrostatic potential with this technique is possible only for a relatively small region of interest.

The finite difference solution results suffer from the necessity to project molecular quantities (charges and dielectric constants) onto the grid. As a result, the original charge distribution is distorted and the border between solvent and solute is not smooth. This leads to an error in the electrostatic potential and hence in the solvation free energy. For example, the deviation of the solvation free energy of a unit charge inside a sphere compared with the Born formula is about 1.5 kcal/mol for a 0.3-Å grid separation and 0.5 kcal/mol for a grid separation of 0.1 Å [32,33]. Also, the results of the finite difference calculations depend upon the grid position and are not invariant to molecular translation and rotation inside the grid. However, due to cancellation of errors, the error in ligand-binding energies is estimated to be of the order of 1% [34].

Considerable effort has been spent trying to eliminate the errors mentioned above. Error due to the nonsmooth solvent–solute interface can be reduced by computing the induced charges on the grid points and projecting them onto the smooth boundary. The solvation free energy is then computed using these projected charges [35]. Substantial reduction of the results dependence on the position and orientation of the grid was achieved using a smooth permittivity function [36]. Other major developments include the solution of the Poisson equation without charge projection [32], which may be beneficial for solutes with charge distributions described by dipoles and higher multipoles, not just point charges, and extension of the Poisson equation for nonuniform dielectric and multivalent ions [37].

As a consequence of these improvements, the finite difference solution of the Poisson equation may produce results as accurate as the results of the BE methods. The use of the Poisson equation has potential advantages over the BE method because different dielectric constants can be assigned to different parts of the solute molecule. Also, the dividing border between solute and solvent does not have to be sharp. Instead, some interchange region where the solute dielectric constant gradually changes to the solvent dielectric constant may be defined. Potentially, this may overcome some simplifications of the continuum model. However, these possible advantages have not been tested in calculations, so far.

Finite difference solution of the Poisson–Boltzmann equation (FDPD) was used successfully to compute pK_a shifts, solvation energies, and protein and ligand-binding energies [2–5,7,38]. Recent applications include accurate evaluation of ligand-binding energies [39–41], DNA-binding reactions [42], protein–protein interactions [43], and the stability of proteins [44].

An alternative way of solving the Poisson–Boltzmann equation is the finite element method, which uses nonuniform and not rectangular grids. For example, the grid may be made finer around an active site to accurately evaluate ligand binding, and coarser elsewhere. This achieves comparable accuracy with the finite difference methods, but with a smaller number of grid points. Unfortunately, the finite element method has not been used extensively in applications; only implementations of the method have been reported to date [45–47].

All the methods described so far are computationally quite expensive. CPU time on current computers for solvation energy evaluation ranges from several tens of minutes for finite difference solution of the Poisson–Boltzmann equation to approximately a minute for dipole lattice models. This is much larger than the molecular mechanics energy and gradients evaluation steps, which required about 1 sec of processor time. In order to increase the speed of computations, further simplifications are required.

Among the many approximate models for solvation free energy evaluation, the most frequently used is the generalized Born (GB) model. It evaluates the solvation energy using the following equation:

$$\Delta G_{ele} = -1/2\left(\frac{1}{\varepsilon_s} - \frac{1}{\varepsilon_w}\right)\sum_{ij} q_i q_j / f_{GB} \tag{31}$$

Here q_i and q_j are point atomic charges and f_{GB} is a distance-dependent function. The summation is over all pairs of atoms in the solute. The f_{GB} function is chosen to satisfy certain boundary conditions: for a distance $r_{ij} = 0$, $f_{GB} = R_i$, and for a long distance $r_{ij} = \infty$, $f_{GB} = r_{ij}$. These conditions turn Eq. (31) into the Born formula when $r_{ij} = 0$ and would recover the Coulomb interactions for $r_{ij} = \infty$. Indeed, the total electrostatic energy of interaction between charges at large distance is given by

$$U_{ele} = -1/2\sum_{ij}\left[\frac{q_i q_j}{\varepsilon_s r_{ij}} - \left(\frac{1}{\varepsilon_s} - \frac{1}{\varepsilon_w}\right)q_i q_j / f_{GB}\right] = 1/2\sum_{ij}\frac{q_i q_j}{\varepsilon_w r_{ij}} \tag{32}$$

Also, when the f_{GB} function is equal to

$$f_{GB} = \sqrt{r_{ij}^2 + R_i R_j \exp\left(\frac{-r_{ij}^2}{4R_i R_j}\right)} \tag{33}$$

the solvation energy of the point dipole inside a spherical cavity is within 10% of that from the Bell formula, Eq. (30) [48]. Initially, the effective Born radius (R_i) for each atom was determined by numerical integration using approximate calculations of the solvent accessible area [48]. Several area calculations have to be done for each atom, making the solvation free energy computation slow. Later, an analytical formula was proposed for evaluation of the effective Born radii [49]; this not only makes computations faster but also allows for the simple analytic evaluation of first and second derivatives of the solvation energy, which is not possible for the more accurate methods discussed above. This makes the new generation of GB methods very attractive in MD simulations of solvated molecules. Several versions of the GB model that differ mostly by different approaches to the evaluation of the effective Born radii and modifications of the f_{GB} function exist. A very good review of the variants and applications in molecular mechanics calculations was recently published [50]. A heavily parameterized GB version was also developed and successfully used with semiempirical QM methods [51]. Recent molecular mechanics applications are in protein–ligand binding [52], the binding of ions to RNA fragments [53,54], and in protein–protein interactions [55].

While in most applications the GB model works remarkably well, some problems were found when computing the binding of aliphatic cyclic ureas to HIV-1 protease [56]. The GB model failed to reproduce both experimental data and the results of FDPB calculations. This is not surprising. It is common for a parameterized model to fail sometimes. The problem is that it is impossible to predict for which systems this will happen. Thus, reference calculations with more accurate models may be needed. The GB model is fast enough to be used together with MD simulations and several such simulations were reported. It was found that GB models reproduce well the results for ligand binding obtained by explicit water simulations and by solution of the Poisson equation [57]. It also predicts correctly the minimum on the folding pathway of small peptides but overestimates the stabilization of transition structures [58]. Overestimation of transition structures by the GB model was also reported in MD simulations of small proteins [59]. In general, the GB models provide an attractive alternative to more accurate but slower models. However, we emphasize they should be used with some caution.

Two more models related to the GB formalism should be mentioned as they may provide some insights for future development. It was shown that a somewhat similar expression of the solvation energy can be obtained from the boundary element formalism [60]. The method was applied successfully to the docking of a variety of ligands to the DHFR protein [61]. Another approach for the estimation of the f_{GB} function was reported recently [62]. The method is based on the screened Coulomb potential and seems to reproduce the structure of small peptides correctly [63].

Several other models have been proposed for the evaluation of the electrostatic part of the solvent energy (see reviews [2–5,7]), but as they are very infrequently used, it is hard to estimate how well they perform in the estimation of solvation-related properties. It would be beyond our intentions to review them here.

Very few studies have tried to assess the validity of the assumptions used in continuum models of solvation. An attempt to verify the validity of linear response theory (the coefficient $1/2$ in continuum solvation free energy formulae) in polar solvents [64] showed that it holds reasonably well for monovalent ionic solutes but is less accurate for dipolar solutes. The concept of a fixed cavity was found to be inadequate in comparing MD simulation of a water molecule in liquid water and PB equation solutions [65]. The deviations obtained were attributed to hydrogen-bond formation and the decrease in solute cavity due to an increase in electrostatic interactions between the solute and solvent. It was also demonstrated that the optimal cavity size should be different for a charge and for a point dipole [66], thus making the concept of a cavity for a complex solute, such as a protein, problematic. The whole concept of using such macroscopic properties as dielectric constants in microscopic computations has been criticized repeatedly (see for example Refs. 16 and 67). Despite all these valid criticisms, continuum-based methods of solvation are used extensively and successfully in a variety of problems. However, the researchers using them should not forget that they are an approximate approach and should try to validate the results with more accurate explicit solvent simulations where possible.

2.2. Nonelectrostatic Contribution

The nonelectrostatic contribution to the solvation energy consists of two parts: the energy of creating a cavity in solvent and the energy of nonpolar interactions, or van der Waals energy, U_{vdw}. From theoretical considerations, the free energy of creating a cavity in a solvent should depend on the surface area (S) and on the volume (V) of a solute [68]:

$$\Delta G_{cav} = C_0 + C_1 S + C_2 V \tag{34}$$

Moreover, there is an indication that the surface area coefficient may be different for small and large solutes [69]. Unfortunately, reliable solvation energies are known for small molecules only, where there is a linear relationship between the surface area and the volume. Consequently, it became common practice to include dependence on the surface area only. Furthermore, it is difficult to evaluate the van der Waals interactions between the solute and solvent in continuum models where there are no solvent molecules present. Thus it is generally assumed that the van der Waals energy is proportional to the surface area. Consequently, the nonelectrostatic solvation free energy is expressed as a simple linear relationship:

$$\Delta G_{np} = \Delta G_{cav} + \Delta G_{vdw} = D_0 + D_1 S \tag{35}$$

The coefficients D_0 and D_1 are determined from experimental solvation free energies of linear hydrocarbons. The solvent accessible area is the area formed from overlapping spheres centered at the atom positions with radii equal to vdW radii increased by a probe radius. This surface area definition is usually used in Eq. (35). Depending on the set of vdW radii used and the probe radii (usually 1.4 Å), the coefficients D_0 and D_1 are approximately equal to 0.85 kcal/mol and 0.005 kcal/mol/Å^2, respectively.

The combination of Eq. 35 with continuum methods for computing the electrostatic contribution to the solvation free energy works reasonably well, as was cited above, despite the simplifications made for obtaining Eq. (35). The main reason is, of course, that the nonpolar contribution to the solvation free energy is relatively small. The van der Waals free energy for a protein is in the range of a few hundreds of kcal/mol [70]. The cavity formation free energy has similar magnitude and opposite sign. Moreover, as was shown recently for a hard-sphere model, there is compensation between the cavity formation energy and the solvent–solute interaction energy that results in eliminating the volume-dependence terms [71]. Of course, for real solvents such compensation is not exact and for accurate evaluation of the solvation free energy both the cavity formation and the van der Waals free energies should be computed. One of the ways of evaluating the vdW energy is to run MD simulations and estimate ΔG_{vdw} as the average of the vdW energy from a small number of conformations [23]:

$$\Delta G_{vdw} = \langle U_{vdw} \rangle \tag{36}$$

The authors of Ref. 70 suggested the use of a coefficient of 0.8 before the $< U_{vdw} >$ term and the addition of a volume-dependence term to Eq. (35) when MD data are not available to estimate ΔG_{vdw}.

2.3. Other Methods

Apart from methods based on continuum approaches, methods based on the division of the total solvation energy by atom or group contributions that are independent from each other are quite popular. The solvation free energy in these methods is computed as a sum of products of an empirical constant depending on the nature of atom or group (w_i), and a solvent accessible area of this atom or group (S_i):

$$\Delta G_{sol} = \sum_i w_i S_i \tag{37}$$

The coefficients w_i are usually determined by fitting the computed solvation free energies to experimental data. Equations of this type can be derived from Eq. (17). Indeed, if the cavity surface is constructed from elements depending on the nature of the atoms, for example, from overlapping spheres with radii depending on the atom type, then the induced charges on each part of the cavity surface can be classified as "belonging" to a particular atom. Consequently, the solvation energy can be expressed as follows:

$$\Delta G_{ele} = 1/2 \sum_i q_i \sum_{s \in i} \frac{q_s}{R_{is}} + 1/2 \sum_i q_i \sum_{s \notin i} \frac{q_s}{R_{is}} \tag{38}$$

The first term is analogous to the sum in Eq. (37). Comparing Eqs. (37) and (38) it may appear that Eq. (37) cannot reproduce the solvation free energy correctly, as the magnitudes of the induced charges q_s depend on all the other atoms of the solute and the second term in Eq. (38) is neglected. But in fact, with good parameterization, Eq. (37) may work very well. Indeed, recent parameterization of this scheme achieved 0.9 kcal/mol RMSD for a training set of 401 compounds and predicted the correct binding mode of a drug molecule (efavirenz) to HIV-1RT protein [72].

These types of models are often criticized for neglecting the solvation contribution from internal groups, the groups that are not exposed to solvent, i.e., $S_i = 0$ in Eq. (37). The parameterization in Ref. 72 is based on the atom types, and the atom types are in turn based on their chemical bonding, so the weighting coefficients w_i only indirectly include the influence of the neighboring atoms. The electrostatic contribution of the internal atoms to solvation is probably overestimated in simple continuum models based on atomic point charges. For example, much larger dielectric constants need to be used for evaluating interactions between groups inside a protein [16]. The vdW energy of buried groups is compensated, at least partially, by the cavity formation energy. Thus the main contribution to the solvation free energy should come from the exposed atoms and this is the basis for Eq. (37). Of course, the contribution of internal groups is not zero, so the methods based on atom or group contributions to the solvation free energy are not likely to compete with the continuum method approach or be applicable to a wide variety of solvent-related problems. But these methods may provide a useful tool for a quick estimation of the solvation free energy for molecules similar to those used in the parameterization.

3. EXPLICIT SOLVENT METHODS

In explicit solvent simulations, all the interactions in the system are described at the atomic level. Thus the expressions for the energy contain parameters characteristic of the atoms in the system. The nonbonded interaction energy (U_{nb}) between atoms in the system is usually given in terms of Coulomb (electrostatic) and van der Waals (vdW) contributions:

$$U_{nb} = U_{ele} + U_{vdw} \tag{39}$$

$$U_{ele} = \sum_{i<j} \frac{q_i q_j}{R_{ij}} \tag{40}$$

$$U_{vdw} = \sum_{i<j} \left[\frac{A_{ij}}{R_{ij}^{12}} - \frac{B_{ij}}{R_{ij}^6} \right] \tag{41}$$

where R_{ij} is the distance between atoms i and j, and the summation extends over solute and solvent atom pairs that are not covalently bonded. In the electrostatic term, Eq. (40), each atom i is assigned a partial charge q_i. The vdW term, Eq. (41), contains a short-range repulsion together with a longer-range attraction (dispersion) term, in which A_{ij} and B_{ij} are parameters characteristic of the interacting pair of atoms. These expressions for computing nonbonded interactions currently form the basis of the classical (molecular mechanics) force fields that are used in MD and MC simulations on biomolecules in aqueous solution. Molecular mechanics (MM) force fields are discussed in more detail elsewhere in this volume. In this section we begin by briefly outlining how these MM force fields have been used in biomolecular simulations to study ligand-binding and solvation problems. We then describe how free energies can be obtained by molecular simulation. The remainder of the discussion focuses on aspects of hybrid QM/MM methods for dealing with biomolecular solvation.

3.1. Molecular Simulations and Ligand Binding

Molecular dynamics simulations on biomolecular systems are now performed routinely using several classical force fields for macromolecules [73–76] and solvent water [77]. Together with generally available MD computer codes, these specialized biomolecular force fields have made possible the study of ligand–macromolecular interactions at the atomic level by an increasing number of researchers. Also, the thermodynamic cycle methods developed for determining free energy changes associated with ligand–protein interactions have had a considerable impact on the purposes for which MD simulations are employed. The use of thermodynamic cycle methods for studying ligand–macromolecular interactions was first suggested nearly 20 years ago by Tembe and McCammon [78]. Subsequent improvements in the speed of computers, and the development of MD methods incorporating these free energy techniques, have been accompanied by a steady rise in the theoretical study of drug–macromolecule interactions [79–93], and, more recently, solubility and transport properties [94–96].

In addition to free energies associated with ligand binding, MD simulations can yield a wealth of other information about the system under study. The MD trajectory gives the time evolution of the system and therefore can be used to study nonequilibrium properties, such as folding and conformational changes. Analysis of the MD trajectory provides average structural information about the macromolecules and their conformational flexibility in solution. The influence of solvent in mediating drug–protein interactions has been demonstrated in a number of studies [97–99]. As the solvent is modeled explicitly, its detailed molecular structure can be analyzed by calculating the radial distribution functions or 3-D distribution functions [5]. In fact, the radial distribution functions $g(R)$ can also be used to calculate the free energy as a function of distance of a water molecule from the macromolecule (i.e., the potential of mean force) [5,100]:

$$\Delta G(R) = -kT\ln g(R) \tag{42}$$

Thus there is a direct link between free energy of solvation and the structure of the solvent surrounding the solute.

3.2. Free Energy and Solvation Effects in Ligand Binding

3.2.1. Coupling Approaches

The solvation free energy plays a major role in rational drug design. This is illustrated in Fig. 2 for the drug-design problem. Note that this scheme does not yield the absolute free energy, or binding constants K_A and K_B of the ligands binding to the macromolecular target. Rather than calculating the free energy directly for the physical binding process, the relative free energies, $\Delta G_{sol}(A \rightarrow B)$ and $\Delta G_{complex}(A \rightarrow B)$, for the change in the chemical state of the system are calculated. One form of the ligand (A) is changed into another (B) during the MD simulation in order to obtain the relative free energy of binding or thermodynamic binding constants. This change or chemical "mutation" is achieved by coupling the ligands' force fields using a coupling parameter, λ, as described in more detail elsewhere [100]. Thus, for example, we may switch between A and B using the hybrid expression for the energy of the system ($0 \leq \lambda \leq 1$),

$$U(\lambda) = \lambda U^A + (1 - \lambda)U^B \tag{43}$$

where U^A is the total energy calculated using the MM parameters appropriate for ligand A, and U^B is the corresponding energy for ligand B. The free energy difference is usually computed using the free energy perturbation (FEP) or thermodynamic integration (TI) formula. Note that Eq. (43) defines any number of intermediate energies depending on the value of λ. In the FEP approach, λ is changed by finite increments, $\Delta\lambda$. The FEP formula are given by

$$\Delta G = 1/2[\Delta G(+) - \Delta G(-)] \tag{44}$$

$$\Delta G(+) = -\beta^{-1} \sum_{i=0}^{n-1} \ln\langle \exp[-\beta(U(\lambda_i + \Delta\lambda) - U(\lambda_i))]\rangle_{\lambda_i} \tag{45}$$

$$\Delta G(-) = -\beta^{-1} \sum_{i=1}^{n} \ln\langle \exp[-\beta(U(\lambda_i - \Delta\lambda) - U(\lambda_i))]\rangle_{\lambda_i} \tag{46}$$

where $\beta = (kT)^{-1}$, $n = (\Delta\lambda)^{-1} + 1$ is the total number of increments (windows), and $(+)$ and $(-)$ indicate statistics based on $+\Delta\lambda$ and $-\Delta\lambda$, respectively (double-wide sampling). Alternatively, differentiation of $G(\lambda)$ with respect to the coupling parameter λ leads to the following TI equation for the free energy difference:

$$\Delta G = \int_0^1 \left\langle \frac{\delta U}{\delta \lambda} \right\rangle_\lambda d\lambda \tag{47}$$

where $\langle\rangle_\lambda$ indicates an ensemble average taken at λ. The integrand in Eq. (47) is calculated numerically. Note that, whichever of the FEP or TI methods is used, the free energy must be calculated for the mutation of the free ligand in solution, and for the corresponding mutation of ligand bound to the macromolecules in the solvated complex. The effects of solvent are therefore rigorously treated for the ligand molecules in the different environments.

3.2.2. Free Energy Components

The FEP or TI methods outlined above may be applied to obtain the total free energy differences. However, it is often convenient to treat the solvation free energy as the sum over different types of contributions, as in Eq. (5) for the continuum approximation, which breaks the solvation process up into a number of physically meaningful contributions. The electrostatic contribution to the solvation free energy, ΔG_{ele} in Eq. (5), can be readily obtained from a molecular simulation in which only the electrostatic interactions between the solute molecule A and the solvent molecules are gradually switched on or off using the λ-coupling techniques as described, but otherwise A remains unchanged. In the switched-off state, these solute–solvent electrostatic interactions vanish and the solute is effectively nonpolar, i.e., only interacts with the solvent via vdW terms.

The total free energy difference between ligands A and B for solvation and binding processes can also be obtained in stages by application of the FEP or TI equations. As the total free energy is a state function, what exactly these stages are does not particularly matter. In FEP calculations on the hydrophobic hydration effect, the mutations between solute molecules A and B were conveniently carried out in two steps [101]:

$$A \overset{ele}{\to} A' \overset{vdw}{\to} B \tag{48}$$

In the above, the electrostatic terms in the force field are mutated first to obtain the intermediate state A′, followed by the vdW and bonded terms to obtain the final solute state B. Thus the total free energy for mutation of solute molecule A to B in solvent is given by

$$\Delta G_{sol}(A \to B) = \Delta G_{ele}(A \to A') + \Delta G_{vdw}(A' \to B) \tag{49}$$

The same expression applies to $\Delta G_{complex}(A \to B)$ in Eq. (2), and hence to $\Delta \Delta G_{bind}$ for the differential binding free energy for ligands A and B complexed with a macromolecule.

However, care needs to be exercised in the physical interpretation of the individual free energy terms for changes between different solutes. Although formally the total free energy is independent of the path A→B, the same is not necessarily true for the FEP or TI components. For example, consider the differences between species A and B in solution for the case where the order of the mutations in Eq. (48) is reversed, i.e.,

$$A \overset{vdw}{\to} A'' \overset{ele}{\to} B \tag{50}$$

In Eq. (48), the intermediate state A′ has effectively the same solute atomic charges as B, and the same vdw parameters as A. This is clearly not the same as in Eq. (50) where the intermediate state A″ has the charges belonging to A and the vdW parameters of B. Consequently, it follows that $\Delta G_{ele}(A \to A') \neq \Delta G_{ele}(A \to A'')$ and $\Delta G_{vdw}(A' \to B) \neq \Delta G_{vdw}(A'' \to B)$, i.e., electrostatic and vdW contributions to the differential free energy are not uniquely defined.

3.2.3. Linear Response Approximation

The success of the chemical "mutation" approach described above depends on the ability to transform the energy function, U, from one state to another during the MD simulations. This may not be applicable if the two states are very different, or for a QM treatment of the solute molecules. Consequently, in practice the scheme shown in Fig. 2 has its limitations and is not always useful. A possible solution to this problem lies in the linear response approximation (LRA), which has been used extensively in the implicit (dielectric continuum) solvation models. The coupling approach used to obtain free energies from molecular simulation can be thought of as a charging of the molecule in solution. If the system (solute plus solvent) response to this charging process is linear, the ratio of the free energy, ΔG_{ele} in Eq. (5), to the average electrostatic interaction energy between the solute and solvent molecules has been shown to be exactly one half [102,103], i.e., the free energy is given by

$$\Delta G_{ele} = 1/2 \langle U_{ele} \rangle \tag{51}$$

The LRA allows estimation of a free energy change without the need to mutate potential energy terms as in the FEP or TI methods. Consequently, the LRA is more generally applicable and well suited for obtaining the absolute binding free energy of a ligand, i.e., the free energy relative to the unbound ligand in solution, rather than just the free energy difference $\Delta\Delta G_{bind}$ between chemically similar ligands.

As discussed in Sec. 2 for the dielectric continuum models, this linear response assumption has been shown to be approximately true for the aqueous solvation free energy of ions, obtained by simulations using MM force fields with nonbonded interactions, as given by Eqs. (39) and (40) [64,102,103]. Consequently, a reliable estimate of the free energy of a solvation or binding process may often be obtained without the need to perform thermodynamic integration or perturbation calculations. This approach for obtaining the free energy may be especially useful in protein–ligand binding studies [103]. However, as originally formulated, the LRA does not provide a rigorous expression for the vdW and cavitation free energy contributions. Consequently, in practice its accuracy depends heavily on the availability of experimental data, e.g., known binding constants. A free energy of binding, ΔG_{bind}, incorporating the ligand desolvation term is given within the LRA by [102]

$$\Delta G_{bind} = 1/2 \langle \Delta U_{l-s}^{ele} \rangle + \alpha \langle \Delta U_{l-s}^{vdw} \rangle \tag{52}$$

where ΔU_{l-s}^{ele} and ΔU_{l-s}^{vdw} are, respectively, the electrostatic and van der Waals energy differences between the ligand in the solvated protein and the unbound ligand in solution. The parameter α is determined by fitting a set of calculated free energies of binding with experimental data. For example [104], to determine α for dihydrofolate reductase (DHFR)-binding ligands it was most suitable to choose compounds whose binding modes are expected to be close to that of a known ligand, biopterin. The 8-methyl-, 5,8-dimethyl-, 6,8-dimethyl-, and 7,8-dimethyl-N5-deazapterin cations were chosen as good candidates for this purpose as the substituents are small enough to avoid significant changes in the enzyme structure and they cover a range of free energy values from −6.4 to −8.7 kcal/mol. A value of $\alpha = -0.32$ produced a good correlation with the experimental data, resulting in an average absolute error of 0.31 kcal/mol for the set of ligands used in the calibration. This value was then used to estimate the binding free energy of a series of larger 8-substituted-N5-deazapterins in different binding pockets in the active site of DHFR that had been generated by simulated annealing and thus predict the most likely binding geometry by correlating with experimentally determined binding constants. These types of studies would be impossible using coupling approaches due to the large steric barriers separating the different conformations.

The LRA can also be used in an expression for the free energy of solvation in which the components are analogous to those in Eq. (5) for the implicit solvent model by introducing a term proportional to the solvent accessible surface area of the solute [105]:

$$\Delta G_{sol} = \beta \langle U_{ele} \rangle + \alpha \langle U_{vdw} \rangle + \gamma S \tag{53}$$

where U_{ele} and U_{vdw} are, respectively, the electrostatic and vdW energies for the interaction between solute and solvent molecules computed from an MD or MC simulation. Thus comparison with Eq. (5) suggests that the surface-area term, γS, would be interpreted as corresponding to the cavitation contribution to the solvation free energy. Notice also that the factor of 1/2 has been replaced by β to allow for possible deviations from linear behavior.

Chen and Tropsha [106] have developed a generalized linear response method for aqueous solvation which yields a general expression for the cavitation free energy and thus eliminates the need for the empirically determined parameters used in the other LRA-based methods. The generalized method is based on scaled particle theory which states that a cavity in solution can be created by a statistical fluctuation. The free energy of cavitation can be derived from the expression for the probability of finding the center of a water molecule in a spherical cavity created by such fluctuations. In general ΔG_{cav} at temperature T is given by

$$\Delta G_{cav} = -nkT \ln \left[1 - \frac{4}{3} \pi r_0^3 \rho \right] \tag{54}$$

where n is the number of solute atoms, r_0 the cavity radius, and ρ is the number density of the solvent. Substituting appropriate values for r_0 and ρ in water, and assuming the LRA is valid, the solvation free energy becomes [106]

$$\Delta G_{sol} = 1.49nkT + \frac{1}{2} \langle U_{ele} + U_{vdw} \rangle \tag{55}$$

where, as before, U_{ele} and U_{vdw} are, respectively, the electrostatic and vdW energies for the interaction between solute and solvent molecules computed from an MD or MC simulation. As the LRA is applied to both electrostatic and vdW interactions, there is no need for the empirically derived parameter α in Eqs. (52) and (53).

3.3. Quantum Mechanical/Molecular Mechanical Solvation Models for Biomolecules

Where quantum chemical methods have been used to study problems in medicinal chemistry and drug design, it has usually been combined with a continuum approximation [90,107–112], rather than explicit simulation, for the solvent effect. As noted, molecular simulations with an explicit solvent are traditionally performed using classical force fields. The reason for this is obvious: quantum mechanical calculations are too time consuming. The coupling of QM with continuum approximations has therefore become convenient. However, the so-called hybrid quantum mechanical and molecular mechanical (QM/MM) methods, which are discussed at length elsewhere in this book, can also offer a practical compromise between accuracy and efficiency. Here we discuss this QM/MM approach as it relates to the solvation problem.

The total potential energy of the system, U_T, partitioned into quantum (solute) and molecular mechanics (solvent) groups of atoms is given by

$$U_T = U_{QM}(\text{solute}) + U_{MM}(\text{solvent}) + U_{QM/MM} \tag{56}$$

where U_{QM} is the energy of the quantum system. In terms of the solute's normalized electronic wavefunction ψ_s, U_{QM} is given by

$$U_{QM} = \langle \psi_s | H_s | \psi_s \rangle \tag{57}$$

where H_s is the solute Hamiltonian operator. Note here that ψ_s is a general wavefunction that may or may not be perturbed by the solvent field. If ψ_s is unperturbed by the solvent field, i.e., the solute is "mechanically embedded" in the solvent, then it is an eigenfunction of H_s rather than the whole QM/MM system Hamiltonian. Also in Eq. (56), U_{MM} is the classical energy of the molecular mechanics part of the system, and $U_{QM/MM}$ is the interaction energy between the quantum and molecular mechanics parts of the system. Normally, the QM/MM interaction energy is the sum of polar (electrostatic and polarization) and nonpolar repulsion and dispersion (van der Waals) terms, analogous to U_{nb} given by Eq. (39) in the classical description of the force field, i.e.,

$$U_{QM/MM} = U_{ele} + U_{vdw} \tag{58}$$

The electrostatic term, U_{ele}, in Eq. (58) is given formally in terms of the solute wavefunction ψ_s, the sum over solvent atomic charges q_i, and the sum over solute nuclear charges Z_j by

$$U_{ele} = \sum_i q_i \left[\langle \psi_s | |r - r_i|^{-1} | \psi_s \rangle + \sum_j Z_j |r_i - r_j|^{-1} \right] \tag{59}$$

Clearly, the charge distribution $\langle \psi_s | \psi_s \rangle$ of the QM part can be polarized by the electrostatic potential produced by the atom-centered charges q_i of the MM system. Thus, QM/MM methods are capable of modeling polarization of the solute (QM region) by the solvent by relaxation of the wavefunction. Although U_{ele} explicitly includes polarization of the QM region by the field of atomic charges in the solvent, it is not usual to include explicit polarization terms in the calculation of U_{MM}. Also in Eq. (58), U_{vdw} is given by Eq. (41) for the classical force field, but the parameter values are not the same as in the classical force field due to the differences in the way U_{ele} is obtained. In order to obtain new vdW parameter values, a more precise definition of the QM/MM interaction energy is required. Consider, for example, a single solvent water molecule interacting with the solute molecule. Assuming that the supermolecule (solute plus water) system is treated quantum mechanically, U_{vdw} can be expressed as

$$U_{vdw} = \langle \psi_T | H_s + H_w + V_{sw} | \psi_T \rangle - U_{ele} - U_s - U_w \tag{60}$$

where H_w is the Hamiltonian for the water molecule, V_{sw} is the coulomb operator for the interaction between the solute and water, and ψ_T is the total system (supermolecule) wavefunction. U_s and U_w are the energies of the solute and water monomers, respectively, given in terms of the unperturbed monomer wavefunctions (ψ^0) by

$$U_s = \langle \psi_s^0 | H_s | \psi_s^0 \rangle \tag{61}$$

$$U_w = \langle \psi_w^0 | H_w | \psi_w^0 \rangle \tag{62}$$

For a given wavefunction, Eq. (60) is formally an exact definition of the interaction energy between the two molecules. Unfortunately, however, in practice it is rather cumbersome to work with. It necessitates a large number of ab initio calculations at various intermolecular separations and orientations to ensure the chosen parameters in U_{vdw} accurately reproduce the configurational energy of the system. Moreover, Eq. (60) should be solved for each new solvated system of interest, as there is no guarantee that the vdW parameters will be transferable.

It is also often convenient to decompose the interaction energy into different components, other than simply electrostatic and vdW as in Eq. (58). Given the wavefunction for a system, the total interaction energy between molecular fragments may be variously decomposed into electrostatic, exchange, polarization, and charge transfer contributions [113]. There is, of course, no unique way to describe intermolecular forces in terms of all of these different contributions. Consequently, the numerical values and relative importance of these contributions depend heavily on how they are defined within the various decomposition schemes. However, a useful first-order perturbation treatment of the interaction can be obtained if the electrostatic energy is defined on the basis of the noninteracting molecule wavefunctions ψ^0. The total wavefunction for the complex is then a product of the unperturbed molecular wavefunctions:

$$\psi_T^0 = \psi_s^0 \psi_w^0 \tag{63}$$

The electrostatic contribution U_{ele}^0 to the solute–solvent interaction energy in the absence of any polarization of the solute's charge density is obtained by substituting ψ_s^0 into Eq. (59). Although approximate, electrostatic models without explicit polarization terms are capable of predicting the binding energies, structures, and force constants of molecular complexes, provided that the short-range repulsions are also accurately described [114]. The combined effect of exchange repulsion, polarization, and charge transfer on the interaction energy is generally found to be of much lesser importance. Thus the electrostatic energy by itself can often provide a reasonable estimate of the binding energy.

The solute molecules can, in principle, be treated at any level of QM theory. However, in the majority of QM/MM studies of biologically important systems, U_{QM} is computed using one of the approximate semiempirical AM1, MNDO, and PM3 methods. The reason for this predominance of semiempirical methods is due solely to the computational cost of conventional ab initio or density functional methods. In fact, semiempirical methods are efficient enough to be used in MD simulations. In the following, we describe the most recent and significant advancements in the development of solvation models based on both semiempirical and ab initio QM/MM methods.

3.3.1. Semiempirical Methods

In ab initio methods the electrostatic term U_{ele} is the expectation value of the coulomb operator, as shown in Eq. (59). However, in the semiempirical MNDO, AM1, and

PM3 approximations, which lack a formal definition of the wavefunction, this energy is given by

$$U_{ele} = 1/2\, Tr\, VP + \sum_i q_i \sum_j Z_j(ss,ss) \tag{64}$$

where P is the SCF density matrix, q_i the atomic charges of the MM atoms, Z_j the core charges of the QM atoms, and (ss,ss) are two-center two-electron terms over s orbitals centered on the core of QM atom j. The matrix elements of V are given by

$$V_{\nu\mu} = \sum_i q_i(ss,\nu\mu) \tag{65}$$

where the sum is over all atomic charges q_i in the MM region. The two-center two-electron terms (ss,$\nu\mu$) depend on the distance between q_i and the QM atoms on which ν and μ are centered. This distance dependence is given by

$$(ss,\nu\mu) \rightarrow 1\big/\sqrt{R^2 + \alpha^2} \tag{66}$$

where R is the distance between the MM and QM atoms and α^2 is a fixed parameter depending on the types of atoms involved and the semiempirical method (MNDO, AM1, or PM3). The screened potential is required in semiempirical QM methods because the actual charge distribution is continuous. In molecular calculations the charge distributions overlap leading to a deviation from coulomb behavior. Values of atomic parameters are chosen so that in the limit of maximum overlap ($R = 0$), (ss,$\nu\mu$) yield the correct one-center integral, α^{-1}. In the opposite limit ($R \rightarrow \infty$), (ss,$\nu\mu$) approach the pure coulomb potentials.

In the semiempirical QM solvation model described here the electrostatic contribution to the solvation free energy was obtained using MD simulations [115,116]. The FEP or TI methods were used to compute the electrostatic solvation free energy as described in Sec. 3.2.2. The coupling parameter, λ, for scaling the QM/MM electrostatic terms in the MD simulations was introduced as follows:

$$U_{QM/MM}(\lambda) = (1 - \lambda)U_{ele} + U_{vdw} \tag{67}$$

A simple continuum approximation was also included to account for long-range contributions to the solvation free energy of charged solutes. The total solvation free energy (ΔG_{sol}) in terms of the contributions defined in Eq. (5) can be written in the general form:

$$\Delta G_{sol} = N^{-1} \sum_{i=1}^{N} [U_{ele}]_i + \frac{1 - \varepsilon}{2\varepsilon r_{cut}} Q^2 + \Delta G_{c/vdw} \tag{68}$$

The first term in Eq. (68) is readily derived from Eq. (47) for thermodynamic integration, although the perturbation Eqs. (44), (45), and (46) could also be applied. It is calculated during the MD simulation by the creation or annihilation of the solute–solvent interaction terms via the λ coupling parameter in Eq. (67). This corresponds to either a charging (solvation) or discharging (desolvation) of the solute in water. λ was coupled continuously with the MD time step during the simulation. Thus N is the number of MD integration time steps and $[U_{ele}]_i$ is the electrostatic/polarization part of the QM/MM interaction given by Eq. (64), evaluated using the configuration obtained at the ith time step in the MD simulation. The second (continuum) term is the Born correction in which ε is the dielectric constant of the solvent, Q is the total charge on the solute molecule, and r_{cut} is the cutoff radius for the neglect of solute–solvent interactions in the calculation of U_{ele}. The remaining nonpolar (i.e., cavity and van der Waals) free energy term

$$\Delta G_{c/vdw} = \Delta G_{vdw} + \Delta G_{cav} \tag{69}$$

represents a relatively small contribution (ca. 2 kcal/mol) and does not vary greatly between different solutes. Consequently, it can be obtained empirically, assuming a linear relationship between free energy and solvent-accessible surface area of the solute.

If the density matrix in Eq. (64) is obtained by performing the self-consistent field (SCF) calculation using the appropriately perturbed one-electron Fock matrix elements, the electrostatic and polarization contributions are contained in the first two terms of Eq. (68).

In terms of a decomposition analysis [113] of the wavefunction for the system, the effects of exchange repulsion and charge transfer are implicit in the remaining $\Delta G_{c/vdw}$ term. In a simplified nonpolarization QM/MM model in which only electrostatic terms are retained, the solvation free energy may be given by

$$\Delta G_{sol} = N^{-1} \sum_{i=1}^{N} [U_{ele}^0]_i + \frac{1-\varepsilon}{2\varepsilon r_{cut}} Q^2 \tag{70}$$

In Eq. (70) the neglect of polarization on the free energy is assumed to be partly offset by the neglect of $\Delta G_{c/vdw}$. Another equation can be derived using the LRA. The LRA typically underestimates, by a small amount, the electrostatic contribution to the free energy for classical force fields [102,103]. The mean electrostatic energy for the solvated state, $<U_{ele}>$, has also been calculated to test the validity of the linear response assumption for the QM/MM approach. For a range of ionic solutes $1/2 <U_{ele}>$ was found to be 88% to 99% of ΔG_{ele} calculated by TI [116]. Thus to a good approximation

$$\Delta G_{sol} = 1/2 \left[\langle U_{ele}^0 \rangle + \frac{1-\varepsilon}{\varepsilon r_{cut}} Q^2 \right] \tag{71}$$

can be used in place of Eq. (70).

The success of these semiempirical solvation models depends critically on careful parameterization. The importance of screening, Eq. (66), in determining accurate representations of the one-electron integrals depends on the extent of overlap of atomic charge densities. This overlap is simply the result of the strong electrostatic interaction between the ionic or polar solute molecules. Consequently, semiempirical methods may not adequately describe electrostatic solutes due to the overlap with the solvent charge densities, without some reparameterization. The importance of short-range repulsions is reflected in the choice of vdW parameters in the QM/MM solvation model [117]. The parameterization can be achieved by fitting to data obtained from quantum chemical calculations on small molecular clusters in the manner described above. These do not have to be semiempirical calculations. It may be preferable to use data from more accurate ab initio methods. Alternatively, simulation data can be used to fit the parameters to experimental solvation free energies as described in Refs. 115 and 116.

3.3.2. Ab Initio Methods—Effective Fragment Potentials

Ab initio methods offer a more general approach to the description of both the solute and solute–solvent interactions than do the less accurate semiempirical methods. They also present a more systematic way of improving the description of interactions between solvent molecules in the MM region. In Eq. (56), the interactions within the MM region are described by the classical force fields of the type given by Eqs. (40) and (41). These force fields then form part of the QM/MM interaction potential. For example, the atomic charges that are used in the classical MM potential also appear in Eq. (59) for the QM/MM interaction. It is possible, however, to construct a complete solvation model using ab initio quantum chemical methods only. This type of approach forms the basis of the effective fragment potential (EFP) method [118–120]. In this approach the MM potentials are replaced by so-called effective fragment potentials which are derived quantum mechanically. The total energy in the EFP method is given by the sum of three terms:

$$U_{efp} = U_{ele} + U_{pol} + U_{rep} \tag{72}$$

where U_{ele}, U_{pol}, and U_{rep} are the electrostatic, polarization, and repulsive interactions energies, respectively. The different terms in Eq. (72) are defined for interactions between the QM and MM regions (QM/MM), and for interactions between solvent molecules in the MM region.

In order to obtain the energy for the QM/MM interaction, the individual contributions are represented via one-electron terms in the ab initio Hamiltonian

$$V_{is}^{efp} = \sum_{k=1}^{K} V_{k}^{ele}(i,s) + \sum_{l=1}^{L} V_{l}^{pol}(i,s) + \sum_{m=1}^{M} V_{m}^{rep}(i,s) \tag{73}$$

where i represents a solvent molecule in the MM region and s the electronic coordinates in the QM solute region. For a water molecule, $K = 5$, $L = 4$, and $M = 2$. The total energy of interaction between the QM and MM regions is thus given by the expectation value of the sum of the EFP over all solvent molecules in the MM region:

$$U_{QM/MM} = \left\langle \psi_S \left| \sum_i V_{is}^{efp} \right| \psi_S \right\rangle \tag{74}$$

$$U_{MM}(\text{solvent}) = U_{efp} = \sum_{i<j} V_{ij}^{efp} \tag{75}$$

Note that a distinction is made between electrostatic and polarization energies. Thus the electrostatic term, U_{ele}, here refers to an interaction between monomer charge distributions as if they were infinitely separated (i.e., U_{ele}^0). A perturbative method is used to obtain polarization as a separate entity. The electrostatic and polarization contributions are expressed in terms of multipole expansions of the classical coulomb and induction energies. Electrostatic interactions are computed using a distributed multipole expansion up to and including octupoles at atom centers and bond midpoints. The polarization term is calculated from analytic dipole polarizability tensors for each localized molecular orbital (LMO) in the valence shell centered at the LMO charge centroid. These terms are derived from quantum calculations on the monomers and are, therefore, completely transferable between different interacting pairs of molecules.

The electrostatic and polarization contributions are uniquely defined within the EFP model in terms of the noninteracting monomer properties. By definition, the "repulsive" part of the energy is then given by

$$U_{\text{rep}} = U_{\text{tot}} - \left(U_{\text{ele}} + U_{\text{pol}}\right) \tag{76}$$

where U_{tot} is the total interaction energy between a pair of solute molecules. In the initial applications of the EFP methods, this total interaction energy was obtained from ab initio calculations on the supermolecule complex. As discussed in relation to Eq. (60), these calculations must be performed at a large number of intermolecular orientations. For the water dimer, approximately 200 Hartree–Fock calculations were performed [121]. The potentials V_m^{rep} were then fitted to these energies using a linear combination of two Gaussians ($M = 2$) for the QM/MM interactions and a single exponential function ($M = 1$) for the interactions between MM fragments. Calculations on water clusters containing up to 20 molecules [121,122], and on NaCl microsolvated with up to 10 water molecules [123], have been reported. The EFP calculations are quite competitive in terms of efficiency compared with traditional MM models. The EFP model calculations on water clusters were found to take approximately twice the CPU time taken by the TIP3P water model [121]. The EFP method has also been applied to study the solvation effects on a chemical reaction [124] and developed into a more general QM/MM scheme [120].

To address the problem of finite system size, the EFP method has also been combined with continuum models in order to model the effects of the neglected bulk solvent [125]. The Onsager equation was used to obtain the dipole polarization of the solute molecule (modeled quantum mechanically) and explicit water molecules (modeled by effective fragment potentials) due to the dielectric continuum. Thus the energy becomes

$$U = U_{\text{T}} - \frac{1 - \varepsilon}{2\varepsilon a} Q^2 - \frac{1}{2} g\mu^2 \tag{77}$$

$$g = [2(\varepsilon - 1)/(2\varepsilon + 1)]/a^3 \tag{78}$$

where Q and μ are the charge and dipole moment of the solute, respectively, and a is the radius of a spherical cavity containing the QM solute and EFP solvent molecules. The energy functional to be minimized with respect to the solute wavefunction parameters is given by

$$L = U_{\text{T}} - 1/2\, g\mu^2 - W(\langle \psi_s | \psi_s \rangle - 1) \tag{79}$$

where W is a Lagrange multiplier to ensure normalization of the wavefunction. Solving Eq. (79) gives the system polarization by the dielectric continuum representing the bulk solvent.

4. LIMITATIONS OF THE METHODS

When choosing a method for a particular application it is important to remember that they all have their limitations. Typically when optimizing the binding thermodynamics of a potential drug molecule, the free energy differences between ligand modifications are of interest. As we have noted [88], some of these differences may be approaching the limit of chemical accuracy, corresponding to free energies of 1 or 2 kcal/mol. Reliably computing such small free energy differences represents a major challenge, as a fully ab initio treatment of the thermodynamic problem remains impossible. While simplifying approximations are always necessary due to the complexity of the solvated systems under study, these approximations place limitations on the general applicability of the methods. In order to discuss the nature of these limitations, we have separated them into three main categories: force fields, long-range and boundary effects, and config-uration-space sampling.

4.1. Force Fields

Clearly, the quality of the force fields used to simulate the solvation process is critical to success. Much of the work on improving classical force fields focuses on the fine tuning of parameters [126]. Provided that parameterization is carried out with sufficient care, accurate solvation models can certainly be developed within most methodological approaches, including the implicit solvation models described in Sec. 2. The semiempirical QM/MM solvation models are a useful example of these parameterization issues [115–117]. However, as with all parameterizations there are limitations. Important effects are often only treated implicitly. Note that in the QM/MM solvation model, for example, polarization refers only to the electronic charge density of the QM solute: no explicit electronic polarization is included for the MM solvent. This treatment may therefore seem unbalanced, as polarization effects are not apparent in the MM region. The most widely used water models in biomolecular simulations, e.g., the TIP3P water model [77], include the effect of electronic polar-ization of solvent molecules in an averaged way. However, this average polarization is reflected in the static charges on the atoms of the water molecule and does not properly describe the microscopic variations in the local environment that take place over simulation time. The simplest way to include polarization in MM is via a separate polarization term. The electrostatic energy U_{ele} is written as

$$U_{ele} = U^0 + U_{pol} \tag{80}$$

$$U_{pol} = -1/2 \sum_i \Delta\mu_i E_i^0 \tag{81}$$

where E_i^0 and $\Delta\mu_i$ are the electric field and the induced atomic dipole, respectively. For each atom i, an atomic polarizability has then to be defined. Consequently, other force-field parameters need to be redefined, particularly the atomic charges used to calculate U^0. There is currently no generally accepted MM force field for biomolecular simulation that includes explicit polarization.

The EFP method attempts to overcome this parameterization problem by including electrostatics and polarization from first principles. This represents a considerable departure from the traditional MM potentials used in biomolecular simulations. The electrostatic energy remains an approximation, however, as the multipole expansions in U_{ele} are left uncorrected for the effects of charge penetration (U_{pen}). This penetration energy, U_{pen}, is then implicit in U_{rep}. Thus a limitation that is characteristic of all force-field methods, from the crudest MM to the most sophisti-cated QM/MM, is a certain lack of generality, i.e., there are arbitrary parameters that have to be refined and fitted for a particular application.

In the EFP approach, efforts are being made to improve this situation [120,127,128]. A general expression for the penetration energy, U_{pen}, for example, can be obtained by considering the Coulomb interaction between two identical spherical Gaussians [127]:

$$U_{pen} = -2 \sum_{ij} S_{ij} / \left(\alpha^{\frac{1}{2}} R_{ij}^2 \right) \tag{82}$$

where S_{ij} is the overlap of the Gaussians with exponent α. A value for the exponent is obtained on condition that the Gaussian overlap is equal to the LMO overlap. A general expression for the repulsions can also be obtained by realizing that the principal component is due to exchange effects, i.e., the Pauli repulsions between monomers. The exact zero-order exchange repulsion energy between monomers A and B is given by

$$U_{rep} = U_{ex} = \frac{\langle \psi_A^0 \psi_B^0 | \hat{A} H_{AB} | \psi_A^0 \psi_B^0 \rangle}{\langle \psi_A^0 \psi_B^0 | \hat{A} \psi_A^0 \psi_B^0 \rangle} - \langle \psi_A^0 \psi_B^0 | V_{AB} | \psi_A^0 \psi_B^0 \rangle - U_A - U_B \tag{83}$$

where $\hat{A}$ is the antisymmetrization operator. The EFP terms in U_{ex} consist of atomic basis set parameters, LMO coefficients, atom positions, and LMO centroids of charge [128]. Note that this treatment does not take into account higher-order exchange contributions $[E_{pol}(ex)]$ due to polarization of the solute wavefunction. Finally, the model requires the effects of charge transfer between molecules (U_{ct}) to be included. A more general expression for the EFP energy would then be given by

$$U_{efp} = U_{ele} + U_{pol} + U_{ex} + U_{pen} + U_{pol}(ex) + U_{ct} \tag{84}$$

U_{ct} will be difficult to quantify, as it is strictly a property of the supermolecule, rather than the noninteracting monomers.

Apart from the problems associated with the generality of the EFP equations for the nonbonded interactions, the model is restricted to rigid solvent molecules. The issue of intramolecular degrees of freedom has yet to be addressed. Currently, internal coordinates of the fragments are fixed at experimental values (e.g., OH bond = 0.944 Å and HOH angle = 106.7° for water), or at optimized ab initio values if experimental values are not available. Thus, although the QM atom positions can relax in the field of the EFP, the same is not true for EFPs themselves, i.e., the models lack an intramolecular force field in the MM region. This may not be a significantly large source of error in terms of changes in covalent bonding on complex formation but is critical for describing molecular conformations and therefore the vast majority of macromolecular systems.

Finally, we mention an approach that treats the whole (or a large fraction) of a protein by QM using linear-scaling methods. The ability to treat whole molecules the size of proteins using quantum chemical methods is a significant achievement and may obviate the need for the MM region. Thus, for a moderately sized solute molecule, hundreds of explicit water molecules could easily be treated quantum mechanically. However, at present these models have only limited application [108], as they remain too costly for the purpose of MD simulation and are restricted to semiempirical QM methods which are prone to unpredictable errors.

4.2. Long-Range Interaction and Boundary Effects

As computations with explicit solvent molecules are very time consuming in MD or MC simulations, spherical cutoffs are invariably applied to the list of nonbonded interactions. This leads to both unphysical discontinuities in the force field, which may lead to artefacts in the simulated structures, and the neglect of possibly important electrostatic interactions which decay slowly as q/r. Even in cases where it is practical to compute all of the nonbonded interactions, the total number of solvent molecules in a simulation is necessarily finite, so that the influence of the bulk has to be somehow modeled.

The discontinuities in the nonbonded energy, and, hence, forces on the atoms, may be addressed by scaling the nonbonded interaction using cubic spline functions (S). For the classical description of the force field, the nonbonded interaction energy is written as:

$$U_{nb} = \sum_{i<j} S(R_{ij}) \left[\frac{q_i q_j}{R_{ij}} + \frac{A_{ij}}{R_{ij}^{12}} - \frac{B_{ij}}{R_{ij}^{6}} \right] \tag{85}$$

$$S(R_{ij}) = \frac{\left[R_{cut}^2 - R_{ij}^2\right]^2 \left[R_{cut}^2 + 2R_{ij}^2 - 3R_{in}^2\right]}{\left[R_{cut}^2 - R_{in}^2\right]^3} \tag{86}$$

where R_{in} is a distance inside the cutoff, i.e., $R_{in} < R_{cut}$. For a given cutoff, it can be shown that the optimum size of the smoothing region, $\Delta = R_{cut} - R_{in}$, is given by $R_{in} \simeq 0.43 R_{cut}$ [129]. The use of smaller smoothing regions can result in artefacts due to the discontinuity. This result suggests that cutoffs (>20 Å) larger than those normally used in the past are required in order to effect a gentler transition across the cutoff region, while still retaining an accurate description of the electrostatic interaction.

The next issue to consider is the finite size of the simulated system. The proper treatment of long-range electrostatic interactions and boundary effects in simulations on systems of finite size remains problematic. These long-range effects are, of course, particularly important where ion solvation is involved [130]. Therefore, approximate but accurate models are a necessity in order to simplify the calculations. As yet there is insufficient comparative data to determine which of the proposed models are preferred for nonperiodic systems. The introduction of periodic boundary conditions for modeling the bulk in biomolecular simulations leads to an artificial symmetry. Thus, Ewald summations, which include the long-range electrostatic interactions and have been used in biomolecular simulations [131,132], may be criticized on the grounds that nonphysical periodic conditions are imposed. Alternatively, a continuum approach can be applied, as in Eqs. (70) and (77), which are examples of the use of a continuum model (Born and Onsager approximations, respectively), in a three-component (QM, MM, and continuum) model system.

The need to divide the system into multiple regions with quite different characteristics suggests that a flexible but unified theoretical formalism suitable for QM, MM, and continuum methods would be very useful. To this end, Boresch et al. [133] have introduced the dielectric field equation (DFE) for biomolecular solvation. The DFE is a general expression for the net electric field of the form:

$$E(r) = \int_V [\nabla \cdot \nabla \phi(r) P(r - r') - \nabla \phi(r) \rho(r - r')] dr' \tag{87}$$

where the integration is over V, the volume of the simulated system, and the dipole and charge densities, $P(r)$ and $\rho(r)$ respectively, can be defined for each of the three regions, QM, MM, and dielectric continuum. Boresch et al. [133] also suggest that an accurate solvation treatment requires calibration of a fourth, dielectric boundary, region surrounding the simulated system. Some of the flexibility of the DFE approach can be readily illustrated by considering the MM region. For classical MM potentials, the charge density and dipole density have particularly simple forms:

$$\rho_{MM}(r) = \sum_i q_i \delta(r - r_i) \tag{88}$$

$$P_{MM}(r) = \sum_i \mu_j \delta(r - r_i) \tag{89}$$

The dipole contribution to the field allows for electronic polarization in the MM region, i.e., the use of MM force fields including polarization terms. Moreover, charges may be grouped together, and their contribution to the field calculated more efficiently using the dipole, rather than the charge density. Thus, for example, some of the explicitly defined solvent molecules may be treated simply as dipoles.

4.3. Configuration-Space Sampling

With an accurate expression for the configurational energy, i.e., the energy as a function of the coordinates of all atoms in the system, it is, in principle, possible to compute with confidence any property of the system. For example, we can formally apply the FEP, TI, or LRA methods to compute the free energy of solvation. In order to obtain these thermodynamic quantities it is necessary to generate the canonical ensemble.

Unfortunately, this is often not a straightforward task for systems displaying high degrees of conformational flexibility. While MD or MC simulations can be used to study the conformational problem, sampling all the possible conformations of a macromolecule or ligand in solution represents a major challenge. On such complex potential energy (PE) surfaces there are a multitude of local minimum energy states corresponding to both solute and solvent configurations. In practice, it is impossible to scan the whole number of accessible configurations during the simulation. Consequently, it can be extremely difficult to obtain accurate canonical distributions at room temperature due to entrapment in one of these local minimum energy states. Of course, as the solvent degrees of freedom are removed, the continuum solvation methods we have described make this sampling much easier but at the expense of detailed accuracy.

Recently, generalized ensemble methods [134] have been proposed to sample the conformational space in MD and MC simulations of solvated systems. Whereas conventional simulations in the canonical ensemble may become trapped in these states of local energy minima, the generalized ensemble method works by performing a random walk in PE space during a single MD simulation trajectory. There are then techniques available that allow canonical-ensemble averages of any physical quantity to be obtained at any temperature. The method has been tested on a number of polypeptides in solution with encouraging results. In particular, such methods may prove useful for studying mobile loop conformations often associated with the active sites of enzymes. However, more testing needs to be done to determine the method's effectiveness for larger protein systems.

5.   CONCLUSION

Many of the problems that restrict the use of solvent simulation in medicinal chemistry and rational drug design are starting to be addressed. The methods used to compute free energy changes are well established and only awaiting refinements in force fields and sampling techniques in order to reach their full potential. There are now highly accurate first principles models for solvent simulations beginning to emerge, as in the EFP approach, which do not require calibration against empirical data and thus have a wider range of applicability. These models are based soundly on quantum chemical calculations of noncovalent interactions. In addition, generalized ensemble methods may prove to be useful for solving the sampling problem in highly solvated macromolecular systems. Implicit (dielectric continuum) solvation models will most likely continue to have application in the modeling of outer regions and system boundaries, i.e., the bulk solvent region or whenever an explicit description is not required. It remains to be seen whether EFP or related methods will lead to a new generation of high-quality force fields tailored for biomolecular simulations. It is clear, however, that such general force fields will be required if the application of solvent simulation is to advance to a stage where the chemist working in this field can use simulations with sufficient confidence.

REFERENCES

1.   Finney JL. The role of water perturbations in biological processes. In: Neilson GW, Enderby JE, eds. Water and Aqueous Solution. Bristol: Adam Hilger, 1986:227–244.
2.   Tomasi J, Perisco M. Molecular interactions in solutions: an overview of methods based on continuous distributions of solvent. Chem Rev 1994; 94:2027–2094.
3.   Cramer CJ, Truhlar DG. Implicit solvation models: equilibria, structure, spectra and dynamics. Chem Rev 1999; 99:2161–2200.
4.   Roux B, Simonson T. Implicit solvent models. Biophys Chem 1999; 78:1–20.
5.   Orozco M, Luque FJ. Theoretical methods for the description of the solvent effect in biomolecular systems. Chem Rev 2000; 100:4187–4225.
6.   Cramer CJ, Truhlar DG. Solvation thermodynamics and the treatment of equilibrium and nonequilibrium solvation effects by models based on collective solvent coordinates. In: Reddy MR, Erion MD, eds. Free Energy Calculations in Rational Drug Design. New York: Kluwer/Plenum, 2001:63–95.
7.   Simonson T. Macromolecular electrostatics: continuum models and their growing pains. Curr Opin Struct Biol 2001; 11:243–252.
8.   Agarwal A, Brown FK, Reddy MR. Relative solvation free energies calculated using explicit solvent. In: Reddy MR, Erion MD, eds. Free Energy Calculations in Rational Drug Design. New York: Kluwer/Plenum, 2001:95–117.
9.   Debye P. Polar Molecules. New York: Dover Publications, 1929.
10.   Warshel A. Computer Modeling of Chemical Reactions in Enzymes and Solutions. New York: Wiley, 1991.
11.   Lee FS, Chu Z-T, Warshel A. Microscopic and semimicroscopic calculations of electrostatic energies in proteins by POLARIS and ENZYMIX programs. J Comput Chem 1993; 14:161–185.
12.   Hill NE. Theoretical treatment of permittivity and loss. In: Hill NE, Vaughan WE, Price AH, Davies M, eds. Dielectric Properties and Molecular Behavior. London: Van Nostrand Reinhold Company, 1969:1–107.
13.   Lee FS, Chu Z-T, Bolger MB, Warshel A. Calculations of antibody–antigen interactions:

microscopic and semimicroscopic evaluation of free energies of binding of phosphorylcholine analogs to McPC603. Protein Eng 1992; 5:215–228.

14. Muegge I, Tao H, Warshel A. A fast estimate of electrostatic group contributions to the free energy of protein-inhibitor binding. Protein Eng 1997; 10:1363–1372.

15. Sham YY, Chu Z-T, Tao H, Warshel A. Examining methods for calculations of binding free energies: LRA, LRE, PDLD-LRA, and PDLD/S-LRA calculations of ligand binding to an HIV protease. Proteins 2000; 39:393–407.

16. Schultz CN, Warshel A. What are the dielectric "constants" of proteins and how to validate electrostatic models. Proteins 2001; 44:400–417.

17. Wang B, Ford GP. Molecular-orbital theory of a solute in a continuum with an arbitrary shaped boundary represented by finite surface element. J Chem Phys 1992; 97:4162–4169.

18. Rashin AA, Nambordi K. A simple method for the calculation of hydration enthalpies of polar molecules with arbitrary shapes. J Phys Chem 1987; 91:6003–6012.

19. Purisima EO, Nilar SH. A simple yet accurate boundary element method for continuum dielectric calculations. J Comput Chem 1995; 16:681–689.

20. Bharadwaj R, Windermuth A, Sridharan S, Honig B, Nichols A. The fast multipole boundary element method for molecular electrostatics: an optimal approach for large systems. J Comput Chem 1995; 16:898–913.

21. Zauhar RJ, Varnek A. A fast and space-efficient boundary element method for computing electrostatic and hydration effects in large molecules. J Comput Chem 1996; 17:864–877.

22. Vorobjev YN, Scheraga HA. A fast adaptive multigrid boundary element method for macromolecular electrostatic computations in a solvent. J Comput Chem 1997; 18:569–583.

23. Vorobjev YN, Hermanes J. ES/IS estimation of conformational free energy by combining dynamic simulations with explicit solvent with an implicit solvent continuum model. Biophys Chem 1999; 78:195–205.

24. Ben-Naim A. Solvation from small to macro molecules. Curr Opin Struct Biol 1994; 4:264–268.

25. Klamt A, Schuurmann GJ. COSMO: a new approach to dielectric screening in solvents with explicit expressions for the screening energy and its gradient. Chem Soc Perkin Trans 2 1993:799–805.

26. York DM, Karplus M. A smooth solvation potential based on the conductor-like screening model. J Phys Chem A 1999; 103:11060–11079.

27. Klamt A. COSMO and COSMO-RS. In: Schleyer PvR, ed. Encyclopedia of Computational Chemistry. Vol. 5. Chichester: John Wiley and Sons, 1998:604–614.

28. Klamt A, Eckert F. COSMO-RS: a novel and efficient method for the a priori predictions of thermophysical data of liquids. Fluid Phase Equilib 2000; 172:43–72.

29. Kirkwood JG. Theory of solutions of molecules containing widely separated charges with special applications to zwitterions. J Chem Phys 1934; 2:351–361.

30. Klapper I, Hagstrom R, Fine R, Sharp K, Honig B. Focusing of electric fields in the active site of Cu–Zn superoxide dismutase: effects of ionic strength and amino-acid modification. Proteins 1986; 1:47–59.

31. Nicholls A, Honig B. A rapid finite difference algorithm utilizing successive over-relaxation to solve the Poisson–Boltzmann equation. J Comput Chem 1991; 12:435–445.

32. Bruccoleri RE. Grid positioning independence and the reduction of self-energy in the solution of the Poisson–Boltzmann equation. J Comput Chem 1993; 14:1417–1422.

33. Zhou Z, Payne P, Vasquez M, Kuhn N, Levitt M. Finite-difference solution of the Poisson–Boltzmann equation: complete elimination of self-energy. J Comput Chem 1996; 17:1344–1351.

34. Shen J, Wendopolski J. Electrostatic binding energy calculation using the finite difference solution to the linearized Poisson–Boltzmann equation: assessment of its accuracy. J Comput Chem 1996; 17:350–357.

35. Rocchia W, Shridharan S, Nicholls A, Alexov E, Chiabbera A, Honig B. Rapid grid-based construction of the molecular surface and use of induced surface charge to calculate reaction field energies: application to molecular systems and geometrical objects. J Comput Chem 2002; 23:128–137.

36. Grant JA, Pickup BT, Nicholls A. A smooth permittivity function for Poisson–Boltzmann solvation methods. J Comput Chem 2001; 22:608–640.

37. Rocchia W, Alexov E, Honig B. Extending the applicability of the non-linear Poisson–Boltzmann equation: multiple dielectric constants and multivalent ions. J Phys Chem B 2001; 105:6507–6514.

38. Honig B, Sharp K, Yang A-S. Macroscopic models of aqueous solutions: biological and chemical applications. J Phys Chem 1993; 97:1101–1109.

39. Kuhn B, Kollman P. Binding of diverse set of ligands to avidin and streptavidin: an accurate quantitative prediction of their relative affinities by a combination of molecular mechanics and continuum solvent models. J Med Chem 2000; 43:3786–3791.

40. Eriksson MAL, Pitera J, Kollman P. Prediction of the binding free energies of new TIBO-like HIV-1 reverse transcriptase inhibitors using a combination of PROTEC, PB/SA, CMC/MD and free energy calculations. J Med Chem 1999; 42:868–881.

41. Woods CJ, King MA, Essex JW. The configurational dependence of binding free energies: a Poisson–Boltzmann study of neuraminidase inhibitors. J Comput-Aided Mol Des 2001; 15:129–144.

42. Chen SW, Honig B. Monovalent and divalent salt effects on electrostatic free energies defined by the nonlinear Poisson–Boltzmann equation: application to DNA binding reactions. J Phys Chem B 1997; 101:9113–9118.

43. Norel R, Sheinerman F, Petrey D, Honig B. Electrostatic contributions to protein-protein interactions: fast energetic filters for docking and their physical basis. Protein Sci 2001; 10:2147–2161.

44. Lee MR, Duan Y, Kollman PA. Use of MM-PB/SA in estimating the free energies of proteins: application to native, intermediates, and unfolded villin headpiece. Proteins 2000; 39:309–316.

45. Holst MJ, Saied F. Numerical solution of the nonlinear Poisson–Boltzmann equation: developing more robust and efficient methods. J Comput Chem 1995; 16:337–364.

46. Cortis CM, Friesner RA. Numerical solution of the Poisson–Boltzmann equation using tetrahedral finite-element meshes. J Comput Chem 1997; 18:1591–1608.

47. Friedrichs M, Zhou RH, Edinger SR, Friesner RA. Poisson–Boltzmann analytical gradients for molecular modeling calculations. J Phys Chem B 1999; 103:3057–3061.

48. Still WC, Tempczyk A, Hawley RC, Hendrickson T. Semianalytical treatment of solvation for molecular mechanics and dynamics. J Am Chem Soc 1990; 112:6127–6129.

49. Qui D, Shenkin PS, Hollinger FP, Still WC. The GB/SA continuum model for solvation. A fast analytical method for the calculation of approximate Born radii. J Phys Chem A 1997; 101:3005–3014.

50. Bashford D, Case DA. Generalized Born models of macromolecular solvation effects. Annu Rev Phys Chem 2000; 51:129–152.

51. Li JB, Zhu TH, Hawkins GD, Winget P, Liotard DA, Cramer CJ, Truhlar DG. Extension of the platform of applicability of the SM5.42R universal solvation model. Theor Chem Acc 1999; 103:9–63.

52. Zhou RH, Friesner RA, Ghosh A, Rizzo RC, Jorgensen WL, Levy RM. New linear interaction method for binding affinity calculations using a continuum solvent model. J Phys Chem B 2001; 105:10388–10397.

53. Tsui V, Case DA. Calculations of the absolute free energies of binding between RNA

and metal ions using molecular dynamics simulations and continuum electrostatics. J Phys Chem B 2001; 105:11314–11325.

54. Burkhardt C, Zacharias M. Modelling ion binding to AA platform motifs in RNA: a continuum solvent study including conformational adaptation. Nucleic Acids Res 2001; 29:3910–3918.

55. Noskov SY, Lim C. Free energy decomposition of protein–protein interactions. Biophys J 2001; 81:737–750.

56. Mardis KL, Luo R, Gilson MK. Interpreting trends in the binding of cyclic ureas to HIV-1 protease. J Mol Biol 2001; 309:507–517.

57. Bursulaya BD, Brooks CL. Comparative study of the folding free energy landscape of a three-stranded beta-sheet protein with explicit and implicit solvent models. J Phys Chem B 2000; 104:12378–12383.

58. Caliment N, Schaefer M, Simonson T. Protein molecular dynamics with the generalized/ACE solvent model. Proteins 2001; 45:144–158.

59. Zhang LY, Gallicchio E, Friesner RA, Levy RM. Solvent models for protein–ligand binding: comparison of implicit solvent Poisson and surface generalized Born models with explicit solvent simulations. J Comput Chem 2001; 22:591–607.

60. Bliznyuk AA, Gready JE. A new approach to estimation of the electrostatic component of the solvation energy in molecular mechanics calculations. J Phys Chem 1995; 99: 14506–14513.

61. Bliznyuk AA, Gready JE. Identification and energetic ranking of possible docking sites for pterin on dihydrofolate reductase. J Comput-Aided Mol Des 1998; 12:325–333.

62. Hassan SA, Guarnieri F, Mehler EL. A general treatment of solvent effects based on screened Coulomb potentials. J Phys Chem B 2000; 104:6478–6489.

63. Hassan SA, Mehler EL. A general screened Coulomb potential based implicit solvent model: calculation of secondary structure of small peptides. Int J Quant Chem 2001; 83:193–202.

64. Åqvist J, Hansson T. On the validity of electrostatic linear response in polar solvents. J Phys Chem 1996; 100:9512–9521.

65. Rick SW, Berne BJ. The aqueous solvation of water: a comparison of continuum methods with molecular dynamics. J Am Chem Soc 1994; 116:3949–3954.

66. Papazyan A, Warshel A. A stringent test of the cavity concept in continuum dielectrics. J Chem Phys 1997; 107:7975–7978.

67. van Duijnen PTh, De Vries AH. Utopia dielectrica. Int J Quant Chem Quant Biol Symp 1995; 29:523–531.

68. Ben-Naim A, Mazo RM. Size dependence of the solvation free energies of large solutes. J Phys Chem 1993; 97:10829–10834.

69. Huang DM, Chandler D. Temperature and length scale dependence of hydrophobic effects and their possible implications for protein folding. Proc Natl Acad Sci U S A 2000; 97:8324–8327.

70. Pitera JW, van Gunsteren WF. The importance of solute–solvent van der Waals interactions with interior atoms of biopolymers. J Am Chem Soc 2001; 123:3163–3164.

71. Shimizu S, Ikeguchi M, Nakamura S, Shimizu K. Size dependence of transfer free energies: a hard-sphere-chain-based formalism. J Chem Phys 1999; 110:2971–2982.

72. Wang J, Wang W, Huo S, Lee M, Kollman PA. Solvation model based on weighted solvent accessible surface area. J Phys Chem B 2001; 105:5055–5067.

73. Cornell WD, Cieplak P, Bayly CI, Gould IR, Merz K, Ferguson DM, Spellmeyer DC, Fox T, Caldwell JW, Kollman PA. A second generation force field for the simulation of proteins, nucleic acids, and organic molecules. J Am Chem Soc 1995; 117:5179–5197.

74. Jorgensen WL, Maxwell DS, Tirado-Rives J. Development and testing of the OPLS force field on conformational energetics and properties of organic liquids. J Am Chem Soc 1996; 118:11225–11236.

75. Scott WP, Hunenberger PE, Tironi HG, Mark AE, Billeter SR, Fennen J, Torda AE, Huber T, Kruger P, van Gunsteren WF. The GROMOS biomolecular simulation package. J Phys Chem A 1999; 103:3596–3607.

76. Foloppe N, MacKerell AD. All-atom empirical force field for nucleic acids: I. Parameter optimization based on small molecule and condensed phase macromolecular target data. J Comput Chem 2000; 21:86–104.

77. Jorgensen WL, Chandrasekhar J, Madura JD, Impey RW, Klein ML. Comparison of simple potential functions for simulating liquid water. J Chem Phys 1983; 79:926–935.

78. Tembe BL, McCammon JA. Ligand–receptor interactions. Comput Chem 1984; 8:281–283.

79. Singh UC, Benkovic SJ. A free energy perturbation study of the binding of methotrexate to dihydrofolate reductase. Proc Natl Acad Sci U S A 1988; 85:9519–9523.

80. Brooks CL. Thermodynamic calculations on biological molecules. Int J Quant Chem Quant Biol Symp 1988; 15:221–234.

81. van Gunsteren WF. Methods for calculations of free energies and binding constants: successes and problems. In: van Gunsteren WF, Weiner P, eds. Computer Simulation of Biomolecular Systems. Leiden: ESCOM, 1989:27–59.

82. Pearlman DA, Kollman PA. Free energy perturbation calculations: problems and pitfalls along the gilded road. In: van Gunsteren WF, Weiner PK, eds. Computer Simulation of Biomolecular Systems. Leiden: ESCOM, 1989:101–119.

83. Brooks CL, Fleischman SH. A theoretical approach to drug design: 1. Relative solvation thermodynamics for the antibacterial compound trimethoprim and ethyl derivatives substituted at the 3', 4' and 5' positions. J Am Chem Soc 1990; 112:3307–3312.

84. Fleischman SH, Brooks CL. Protein–drug interactions: Characterization of inhibitor binding in complexes of DHFR with trimethoprim and related derivatives. Proteins 1990; 7:52–61.

85. McDonald JJ, Brooks CL. Theoretical approach to drug design: 2. Relative thermodynamics of inhibitor binding by chicken dihydrofolate reductase to ethyl derivatives of trimethoprim substituted at 3'-, 4'- and 5'-positions. J Am Chem Soc 1991; 113:2295–2301.

86. Reynolds CA, King PM, Richards WG. Free energy calculations in molecular biophysics. Mol Phys 1992; 76:251–275.

87. Gerber PR, Mark AE, van Gunsteren WF. An approximate but efficient method to calculate free energy trends by computer simulation—application to dihydrofolate reductase inhibitor complexes. J Comput-Aided Mol Des 1993; 7:305–323.

88. Cummins PL, Gready JE. Computer-aided drug design: a free energy perturbation study on the binding of methyl-substituted pterins and N5-deazapterins to dihydrofolate reductase. J Comput-Aided Mol Des 1993; 7:535–555.

89. van Gunsteren WF. Molecular dynamics studies of proteins. Curr Opin Struct Biol 1993; 3:277–281.

90. Rao BG, Kim EE, Murcko MA. Calculation of solvation and binding free energy differences between VX-478 and its analogs by free energy perturbation and AMSOL methods. J Comput-Aided Mol Des 1996; 10:23–30.

91. McCarrick MA, Kollman PA. Predicting relative binding affinities of non-peptide HIV protease inhibitors with free energy perturbation calculations. J Comput-Aided Mol Des 1999; 13:109–121.

92. Dominy BN, Brooks CL. Methodology for protein–ligand binding studies: application to a model for drug resistance, the HIV/FIV protease system. Proteins 1999; 36:318–331.

93. Reddy MR, Erion MD. Calculation of relative binding free energy differences for

fructose 1,6-bisphosphatase inhibitors using the thermodynamic cycle perturbation approach. J Am Chem Soc 2001; 123:6246–6252.

94. Lombardo F, Blake JF, Curatolo WJ. Computation of brain–blood partitioning of organic solute via free energy calculations. J Med Chem 1996; 39:4750–4755.

95. Lyubartsev AP, Jacobsson SP, Sundholm G, Laaksonen A. Solubility of organic compounds in water/octanaol systems. An expanded ensemble molecular dynamics simulation study of log P parameters. J Phys Chem B 2001; 105:7775–7782.

96. Klamt A, Eckert F, Hornig M. COSMO-RS: a novel view to physiological solvation and partition questions. J Comput-Aided Mol Des 2001; 15:355–365.

97. Cummins PL, Gready JE. Solvent effects in active-site molecular dynamics simulations on the binding of 8-methyl-N5-deazapterin and 8-methylpterin to dihydrofolate reductase. J Comput Chem 1996; 17:1598–1611.

98. Marrone TJ, Resat H, Hodge CN, Chang CH, McCammon JA. Solvation studies of DMP323 and A76928 bound to HIV protease—analysis of water sites using grand canonical Monte Carlo simulations. Protein Sci 1998; 7:573–579.

99. Williams HEL, Searle MS. Structure dynamics and hydration of the nogalamycin-d(ATGCAT) (2) complex determined by NMR and molecular dynamics simulations in solution. J Mol Biol 1999; 290:699–716.

100. Beveridge DL, Jorgensen WL. Free energy simulations. Ann N Y Acad Sci 1986; 482:1–24.

101. Rao BG, Singh UC. Hydrophobic hydration: a free energy perturbation study. J Am Chem Soc 1989; 111:3125–3133.

102. Åqvist J, Medina C, Samuelsson J-E. New method for predicting binding affinity in computer-aided drug design. Protein Eng 1994; 7:385–391.

103. Åqvist J. Calculation of absolute binding free energies for charged ligands and effects of long-range electrostatic interactions. J Comput Chem 1996; 14:1587–1597.

104. Gorse A-D, Gready JE. Molecular dynamics simulations of the docking of substituted N5-deazapterins to dihydrofolate reductase. Protein Eng 1997; 10:23–30.

105. Carlson HA, Jorgensen WL. An extended linear response method for determining free energies of hydration. J Phys Chem 1995; 99:10667–10673.

106. Chen X, Tropsha A. Generalized linear response method: application to hydration free energy calculations. J Comput Chem 1999; 20:749–759.

107. Luque FJ, Barril X, Orozco M. Fractional description of free energies of solvation. J Comput-Aided Mol Des 1999; 13:139–152.

108. Gogonea V, Merz KM. Fully quantum mechanical description of proteins in solution. Combining linear scaling quantum mechanical methodologies with the Poisson–Boltzmann equation. J Phys Chem A 1999; 103:5171–5188.

109. Viswanadhan VN, Ghose AK, Wendoloski JJ. Estimating aqueous solvation and lipophilicity of small organic molecules: a comparative overview of atom/group contribution methods. Perspect Drug Discov Des 2000; 19:85–98.

110. Hoffmann M, Rychlcwski J. Effects of substituting a OH group by a F atom in D-glucose. Ab initio and DFT analysis. J Am Chem Soc 2001; 123:2308–2316.

111. Alagona G, Ghio C, Monti S. Ab initio modeling of competitive drug–drug interactions: 5-fluorouracil dimers in the gas phase and in solution. Int J Quant Chem 2001; 83:128–142.

112. Hoffmann M, Rychlewski J, Chrzanowska M, Hermann T. Mechanism of activation of an immunosuppressive drug: azathioprine. Quantum chemical study on the reaction of azathioprine with cysteine. J Am Chem Soc 2001; 123:6404–6409.

113. Gordon MS, Jensen JH. Wavefunctions and chemical bonding: interpretation. In: Schleyer PvR, ed. Encyclopedia of Computational Chemistry. Vol. 5. Chichester: John Wiley and Sons, 1998:3198–3214.

114. Dykstra CE. Molecular mechanics for weakly interacting assemblies of rare gas atoms and small molecules. J Am Chem Soc 1989; 111:6168–6174.

115. Cummins PL, Gready JE. Coupled semiempirical molecular orbital and molecular mechanics model (QM/MM) for organic molecules in aqueous solution. J Comput Chem 1997; 18:1496–1512.

116. Cummins PL, Gready JE. Coupled semiempirical quantum mechanics and molecular mechanics (QM/MM) calculations on the aqueous solvation free energies of ionized molecules. J Comput Chem 1999; 20:1028–1038.

117. Luque FJ, Reuter N, Cartier A, Ruiz-Lopez MF. Calibration of the quantum/classical Hamiltonian in semiempirical QM/MM AM1 and PM3 methods. J Phys Chem A 2000; 104:10923–10931.

118. Day PN, Jensen JH, Gordon MS, Webb SP, Stevens WJ, Krauss M, Garmer D, Basch H, Cohen D. An effective fragment method for modeling solvent effects in quantum mechanical calculations. J Chem Phys 1996; 105:1968–1986.

119. Chen W, Gordon MS. The effective fragment model for solvation: internal rotation in formamide. J Chem Phys 1996; 105:11081–11090.

120. Gordon MS, Freitag MA, Bandyopadhyay P, Jensen JH, Kairys V, Stevens WJ. The effective fragment potential method: a QM-based MM approach to modeling environmental effects in chemistry. J Phys Chem A 2001; 105:293–307.

121. Merrill GN, Gordon MS. Study of small water clusters using the effective fragment potential model. J Phys Chem A 1998; 102:2650–2657.

122. Day PN, Pachter R, Gordon MS, Merrill GN. A study of water clusters using the effective fragment potential and Monte Carlo simulated annealing. J Chem Phys 2000; 112:2063–2073.

123. Petersen CP, Gordon MS. Solvation of sodium chloride: an effective fragment study of $NaCl(H_2O)_n$. J Phys Chem A 1999; 103:4162–4166.

124. Webb SP, Gordon MS. Solvation of the Menshutkin reaction: a rigorous test of the effective fragment method. J Phys Chem A 1999; 103:1265–1273.

125. Bandyopadhyay P, Gordon MS. A combined discrete/continuum solvation model: application to glycine. J Chem Phys 2000; 113:1104–1109.

126. Halgren TA. Potential energy functions. Curr Opin Struct Biol 1995; 5:205–210.

127. Kairys V, Jensen JH. Evaluation of the charge penetration energy between non-orthogonal molecular orbitals using the spherical Gaussian overlap approximation. Chem Phys Lett 1999; 315:140–144.

128. Jensen JH, Gordon MS. An approximate formula for the intermolecular Pauli repulsion between closed shell molecules: II. Application to the effective fragment potential method. J Chem Phys 1998; 108:4772–4782.

129. Ding H-Q, Kaasawa N, Goddard WA III. Optimal spline cutoffs for Coulomb and van der Waals interactions. Chem Phys Lett 1992; 193:197–201.

130. Worth GA, King PM. Tautomerization and ionisation studies using free energy methods. In: Reddy MR, Erion MD, eds. Free Energy Calculations in Rational Drug Design. New York: Kluwer/Plenum, 2001:119–140.

131. Brooks CL III. Methodological advances in molecular dynamics simulations of biological systems. Curr Opin Struct Biol 1995; 5:211–215.

132. Essmann U, Perera L, Berkowitz ML, Darden T, Lee H, Pedersen LG. A smooth particle mesh Ewald method. J Chem Phys 1995; 103:8577–8593.

133. Boresch S, Ringhofer S, Hochtl P, Steinhauser P. Towards a better description and understanding of biomolecular solvation. Biophys Chem 1998; 78:43–68.

134. Mitsutake A, Sugita Y, Okamoto Y. Generalized-ensemble algorithms for molecular simulations of biopolymers. Biopolymers (Pept Sci) 2001; 60:96–123.

11

Reactivity Descriptors

P. K. CHATTARAJ, S. NATH, and B. MAITI
Indian Institute of Technology, Kharagpur, India

1. INTRODUCTION

Popular qualitative chemical concepts such as electronegativity [1] and hardness [2] have been widely used in understanding various aspects of chemical reactivity. A rigorous theoretical basis for these concepts has been provided by density functional theory (DFT). These reactivity indices are better appreciated in terms of the associated electronic structure principles such as electronegativity equalization principle (EEP), hard–soft acid–base principle, maximum hardness principle, minimum polarizability principle (MPP), etc. Local reactivity descriptors such as density, Fukui function, local softness, etc., have been used successfully in the studies of site selectivity in a molecule. Local variants of the structure principles have also been proposed. The importance of these structure principles in the study of different facets of medicinal chemistry has been highlighted. Because chemical reactions are actually dynamic processes, time-dependent profiles of these reactivity descriptors and the dynamic counterparts of the structure principles have been made use of in order to follow a chemical reaction from start to finish.

   In this chapter, we present different global and local reactivity descriptors vis-à-vis the associated electronic structure principles for analyzing structures, properties, reactivity, bonding, interactions, and dynamics in the contexts of various physico-chemical processes such as molecular vibrations, internal rotations, chemical reactions, aromaticity, stability of isomers, ion–atom collisions, atom-filled interactions, solvent effect, etc. Global reactivity descriptors are presented in Section 2, whereas Section 3 provides their theoretical basis. Section 4 delineates the local reactivity descriptors. Electronic structure principles associated with electronegativity and hardness are given in Sections 5 and 6, respectively. Section 7 presents the reactivity and selectivity analyses in terms of local quantities, and, finally, Section 8 contains some concluding remarks.

2. GLOBAL REACTIVITY DESCRIPTORS

Electronegativity [1] and hardness [2] are two important global reactivity descriptors. In order to understand the nature of the chemical bond, Pauling [3] first defined electronegativity as "the power of an atom in a molecule to attract electrons to itself." When there is a difference in the electronegativity values of atoms forming a molecule, there will be flow of electrons [3–5]. Because electronegativity is not an experimental observable, there are various definitions [1] of it having respective merits and demerits. The method of determination of electronegativity by Pauling was based on thermochemical data. Pauling's electronegativity values increase along horizontal rows of the periodic table and decrease down the vertical groups, which is in accordance with the general chemical intuition. Furthermore, the electronegativity calculated by Pauling's scale shows a linear correlation with the dipole moment, which is a measure of the ionic character of a bond.

Another electronegativity scale, which has versatile applications in chemistry, was defined by Mulliken [6]. This electronegativity is "absolute" [6] in the sense that it does not depend on the molecular environment and can be directly obtained in terms of two experimentally measurable quantities, ionization potential (I) and electron affinity (A), of any atom or molecule as:

$$\chi = \frac{I + A}{2} \tag{1}$$

Electronegativity depends on the hybridization of atoms in which the atom is present in the molecule. In order to calculate the power of an atom to attract electrons to itself, one has to consider the effect of charge on it. Mulliken's definition of electronegativity has been extended [7,8] to take care of these aspects. It was found [9] that in the valence state energy vs. atomic charge (net charge on an atom) plot, the atom with the higher slope at the origin will attract electrons from the atom with the lower slope, and the energy will be lowered in the process. This observation leads to the definition of electronegativity as the slope of the valence state energy (E) vs. atomic charge (q) plot [9,10]:

$$\chi = \left(\frac{\partial E}{\partial q}\right); \quad q = N - Z \tag{2}$$

where Z and N are the nuclear charge and the number of electrons, respectively. Equation (2) reduces to Mulliken's expression when one considers the valence state energies of neutral atoms and singly excited positive and negative ions. Electronegativity can be written as a linear function of charge as [11–13]:

$$\chi = \frac{\partial E}{\partial E} = \alpha + \beta q \tag{3}$$

in case the dependence of energy on charge is quadratic, that is,

$$E = a + bq + cq^2. \tag{4}$$

The parameter α is an inherent or neutral electronegativity, which is equivalent to the valence state electronegativity of Mulliken, and β is a charge coefficient, which measures the rate of change of electronegativity with charge [5,11]. Mulliken's definition can also be obtained from the concept of orbital electronegativity [8] and is defined as the electronegativity of singly occupied orbitals. Due to the charge flow from atoms A to B in the process of molecule formation, electronegativities of the valence orbitals get equalized [8,11–13]. The amount of charge flow leading to molecular stabilization can be calculated in terms of the fractional ionic character (FIC) of the bond as:

$$\text{FIC} = \frac{1}{2}(q_B - q_A) = \frac{1}{2}\frac{\chi_B^0 - \chi_A^0}{C_A + C_B} \tag{5}$$

where χ_A^0 and χ_B^0 are electronegativities of neutral atoms A and B, and C_A and C_B are the coefficients of q^2 terms in the corresponding energy expressions (Eq. (4)) for A and B, respectively.

In the scale of Allred and Rochow, electronegativity is measured as the electrostatic force exerted at the covalent radius (r_{cov}) of the atom [14,15]. This electronegativity can be written as a function of the screened nuclear charge (Z^*) and the atomic size as:

$$\chi = 3590\frac{Z^*}{r_{cov}^2} + 0.744. \tag{6}$$

The major advantage in the definition of Allred and Rochow is that it provides a direct physical interpretation of electronegativity as "the electron-attracting power of atoms."

There have been several other definitions [16–19] similar to that of Allred and Rochow which measure electronegativity as a function of covalent radius. Gordy [16] defined electronegativity as the potential exerted on a valence electron at the empirical covalent radius. The electronegativities calculated through Gordy's boundary covalent potential method show good correlations with the heats of formation [19], ionization potential [19], stabilization energies [18], and homopolar dissociation energies [19]. Sanderson's definition of electronegativity also depends on the size and the charge of the atoms, although in a somewhat different way [20]. Sanderson defined electronegativity as the "stability ratio" calculated as the average electron density of atoms expressed as a function of Z/r^3, where r is the ionic or covalent radius.

Recently, Allen [21] has introduced a method where electronegativity is calculated in terms of the ionization potentials of s and p orbitals. It has been considered [21] to be the first quantum mechanical realization of Pauling's electronegativity. The spectroscopic electronegativity is found to rationalize the different properties of the periodic table, such as the increase of the metallic character of elements as one goes down a vertical group, the separation of metals and nonmetals by a metalloid band, the formation of noble gas molecules, etc. This electronegativity scale is argued to be the third dimension of the periodic table where the first two dimensions are the number of orbitals and the atomic number, respectively. Allen's scale matches well with those of Pauling and Allred–Rochow but presents some discrepancies when compared to Mulliken's scale. The major discrepancy is that carbon has higher electronegativity than the hydrogen atom in Mulliken's scale, whereas χ_{spec} for hydrogen is higher. It has been argued by Kostyk and Whitehead [22] that the higher electronegativity value for carbon is quite justifiable as in alkenes, electrophiles are found to attack at the carbon atom rather than at hydrogen.

Electronegativities calculated using other definitions have been correlated with different properties of atoms and molecules, such as bond force constant of binary hydrides, ionization potential of atoms [23], polarizability [24,25], etc. Studies on the bond critical points of binary [26] and diatomic [26] hydrides provided correlation between the properties calculated at the bond critical points and the electron-attracting power of an atom [26].

A quantum thermodynamic definition of electronegativity has been provided by Gyftopoulos and Hatsopoulos [27] by considering the atom or the molecule as a member of a grand canonical ensemble where the energy (E) and the number of electrons (N) are continuous functions and all other properties of the ensemble are written in terms of these two independent variables. The chemical potential of the ensemble can be written as:

$$\mu = \frac{\partial E}{\partial N}, \text{ at constant entropy.} \tag{7}$$

As the electrochemical potential measures the escaping tendency of electrons, the electron-attracting power should be its negative. Hence, electronegativity is defined as:

$$\chi = -\mu = -\frac{\partial E}{\partial N}, \text{ at constant entropy,} \tag{8}$$

which is a continuous function of the number of electrons and temperature (θ). For ionic systems (e.g., for $N = Z + 1$ or $N \le Z - 1$), the chemical potential goes to $+\infty$ or $-\infty$ at all temperatures. For neutral atoms, this thermodynamic electronegativity takes the form of Mulliken's electronegativity at the limit of zero temperature.

Due to the electron-attracting property of atoms, electrons flow from the atoms with lower electronegativity to the atoms with higher electronegativity, leading to the equalization of electronegativity [20] in the molecule. The molecular electronegativity can be obtained as the geometrical mean [20] of the constituent free atoms' electronegativities, and the electronegativity equalization stabilizes the molecule.

However, in many cases, electronegativity difference alone cannot account for the stability of the molecule. For example, according to the electronegativity criterion, the CsF molecule should be very stable as the electronegativity difference between Cs and F is very large. But the reaction enthalpy data indicate that LiI and CsF will react to form CsI and LiF. In order to predict the direction of acid–base reactions and to account for the stability of the products, Pearson introduced two parameters "hardness" and "softness" in the vocabulary of chemistry. The qualitative definitions of hard and soft acids and bases are as follows [28–32]:

> Hard acids are acceptor atoms with small size, high positive charge, low polarizability, and the absence of easily excitable outer electrons (e.g., H^+, Li^+).
>
> Hard bases are donor atoms with small size, low polarizability, high electronegativity, having empty orbitals with large energy, and are hard to oxidize (e.g., NH_3, OH^-).

Soft acids are large, highly polarizable acceptor atoms with low positive charge having easily excitable outer electrons (e.g., I_2, Pd^{2+}).

Soft bases are large, highly polarizable donor atoms with low electronegativity, having low lying orbitals, and are easily oxidizable (e.g., H^-, CN^-).

The classification is purely empirical and based on the observations of bond energy, equilibrium constant, rate constant, and other experimental data [32]. These experimental observations finally lead to the prediction of a simple but important principle, which states that hard acids will prefer to coordinate with hard bases and soft acids will prefer to coordinate with soft bases for both kinetic and thermodynamic reasons. This is known as "hard–soft acid–base (HSAB)" principle [28–32]. It has also been argued [33] that hard–hard reactions are governed by charge-controlled interactions and that soft–soft interactions are of the covalent type. Different studies on reactivity suggest that soft molecules are more reactive compared to the corresponding harder counterparts. Hence, isomeric molecules having higher hardness are found to be more abundant in nature than that having lower hardness values. This leads to the principle of maximum hardness, which states [34] that "there seems to be a rule of nature that molecules arrange themselves so as to be as hard as possible." In an attempt to quantify the concepts of hardness and softness, Pearson proposed a relation that correlates the stability of the molecules with hardness and softness, as well as the inherent strengths of acids and bases. The stability constant of a reaction is given by [31]:

$$-pk = S_A S_B + \sigma_A \sigma_B \tag{9}$$

where S_A and S_B are the inherent strengths of acids and bases whereas σ_A and σ_B are softness factors. The HSAB principle has been criticized by Drago et al. [35], who pointed out that although the strengths of acids and bases are considered in Eq. (9), the HSAB principle explained molecular stability solely in terms of softness and hardness, and neglects the effect of acid–base strength in the molecule formation. Drago et al. [35] proposed a relation to measure the enthalpy change in terms of the parameters, which measures the strengths of hard and soft species as:

$$-\Delta H = C_A C_B + E_A E_B \tag{10}$$

where the first term measures the covalent contribution to the enthalpy change whereas the second term measures the corresponding electrostatic contribution. Here, C parameters are identified [35] with softness and E parameters are identified with hardness. However, it is suggested [5] that the contradictions in the theories of Pearson [28,30–32] and Drago et al. [35] are basically due to the difference in the approaches in understanding the acid–base reactions. Drago et al. [35] used E and C parameters in Eq. (10) to study the reaction of two species where the solvation effect is minimized or absent, whereas Pearson's theory considers the competition between forward and backward reactions in the acid–base equilibrium.

Although the qualitative concepts such as electronegativity and hardness have been found to be useful in understanding various chemical reactions, they were not taken very seriously until recently because they did not have legitimate theoretical genesis. Rigorous quantitative definitions and methods for calculations [36–38] of electronegativity, hardness, and related quantities such as chemical potential, local hardness, softness, Fukui function, etc., have been provided within density functional theory, where all quantities are expressed in terms of electron density. Various approximate forms for electronegativity and hardness have also been suggested in terms of ionization potential and electron affinity as well as the energies of the highest occupied molecular orbital (HOMO) and the lowest unoccupied molecular orbital (LUMO) wherein the computation of electronegativity, hardness, and related quantities becomes simpler.

3. THEORETICAL TREATMENT OF QUALITATIVE CONCEPTS

It has been discussed in Section 2 that concepts such as electronegativity and hardness could explain important aspects of chemical reactions and could be related to different physico-chemical properties. Density functional theory has been found to provide a rigorous theoretical background for electronegativity, hardness, and related concepts.

In density functional theory, the Lagrange multiplier associated with the normalization constraint is identified as chemical potential μ, maintaining the analogy with an ordinary thermodynamic system [39] viz.,

$$\mu = \left(\frac{\delta E}{\delta \rho}\right)_{v(\vec{r})} \tag{11}$$

where E is the total energy and $v(\vec{r})$ is the external potential. Chemical potential defined in Eq. (11) can be interpreted as the escaping tendency of electrons analogous to the chemical potential of macroscopical systems [39]. The Chemical potential of an N-electron system can as well be written as a partial derivative of energy with respect to the number of electrons because:

$$\int \rho d\vec{r} = N \tag{12}$$

The definition of Iczkowski and Margrave [9] identifies electronegativity (χ) as the slope of energy vs. N plot. Thus, chemical potential can be shown to be equivalent to the negative of electronegativity as [39]:

$$-\chi = \left(\frac{\partial E}{\partial N}\right)_v = \int \left(\frac{\delta E}{\delta \rho}\right)_v \left(\frac{\partial \rho}{\partial N}\right)_v d\vec{r} = \left(\frac{\delta E}{\delta \rho}\right)_v = \mu \tag{13}$$

The finite difference approximation of the partial derivative $(\partial E/\partial N)_v$ gives the equivalence of χ defined within DFT with that given by Mulliken [6], that is,

$$-\mu = \chi = \frac{I + A}{2} \tag{14}$$

The ground state energy curve as a function of N is continuous and shows a series of straight line segments [40]. The slope of energy vs. N plot shows discontinuity at integral numbers of N. Thus, at zero temperature limit, the chemical potential for neutral species is obtained by taking the average of limits of the $Z < N$ and $Z > N$ curves and is written as [40]:

$$\mu = -I \quad \text{for } Z - 1 < N < Z \tag{15a}$$

$$\mu = -\frac{I + A}{2} \quad \text{for } Z = N \tag{15b}$$

$$\mu = -A \quad \text{for } Z < N < Z + 1 \tag{15c}$$

where Z is the nuclear charge.

It should be noted that the correct definition for μ is difficult to evaluate and, for all practical purposes, μ is calculated simply as $\mu = (\delta E/\delta\rho)_v$ without any serious error as it has been shown that [41]:

$$\left(\frac{\delta E}{\delta\rho}\right)_{N,v} - \left(\frac{\delta E}{\delta\rho}\right)_v = C; \quad C \text{ being a constant.} \tag{16}$$

Electronegativity (χ) or chemical potential (μ), ionization potential (I), and electron affinity (A) can be computed for electronic systems from the Kohn–Sham (KS) equation, which has been extended by Janak [42] and others [43–45] using the $X\alpha$ method [46]. In this approach, one gets a meaning for orbital energy as:

$$\varepsilon_i = \frac{\partial E_i}{\partial n_i} \tag{17}$$

where n_i is the occupation number:

$$N = \sum_i n_i.$$

Now the integration of Eq. (17) between limits N and $N+1$ gives electron affinity for a species:

$$-A = E_{N+1} - E_N = \int_0^1 \varepsilon_{\text{LUMO}}(n)\mathrm{d}n \tag{18}$$

where $\varepsilon_{\text{LUMO}}$ is the energy of the lowest unoccupied molecular orbital. In Eq. (17), occupation number n has been assumed to vary continuously.

Equation (18) can be approximated within a transition state formulation [46,47] as the negative of $\varepsilon_{\text{LUMO}}$ of an intermediate transition state with $(N + 1/2)$ electrons. So,

$$-A \approx \varepsilon_{\text{LUMO}} \text{ for } n = \frac{1}{2}. \tag{19}$$

Similarly, ionization potential, which is written as:

$$-I = E_N - E_{N-1} = \int_0^1 \varepsilon_{\text{HOMO}}(n)\mathrm{d}n \tag{20}$$

may be approximated as $-\varepsilon_{\text{HOMO}}$ for transition state with $(N - 1/2)$ electrons:

$$-I \approx \varepsilon_{\text{HOMO}} \text{ for } n = \frac{1}{2}. \tag{21}$$

The computation of A from Eq. (19) often meets convergence problem [36]. Hence, it has been considered to be preferable to calculate electronegativity as the negative of the highest occupied orbital energy in the $X\alpha$ method with or without spin polarization [47–49] and with self-interaction correction [50]. Electron affinity (A) can be obtained from the following relation:

$$A = 2\chi - I \tag{22}$$

where χ is Mulliken's electronegativity.

In ab initio MO theory, I and A can be approximated as the negative of energies of HOMO and LUMO, respectively, using Koopmans' theorem. In this framework, the electronegativity is the negative of HOMO–LUMO energy average and can be written as [51]:

$$-\chi = \mu = \frac{1}{2}(\varepsilon_{HOMO} + \varepsilon_{LUMO}) \tag{23}$$

The curvature of E vs. N curve has been equated with hardness [36], another important parameter for understanding structure and reactivity. The absolute hardness is given as:

$$\eta = \frac{I - A}{2}. \tag{24}$$

Hardness can be equated to the second term in the Taylor series expansion of energy [52]:

$$\eta = \frac{1}{2}\left(\frac{\partial^2 E}{\partial N^2}\right)_v = \frac{1}{2}\left(\frac{\partial \mu}{\partial N}\right)_v \tag{25}$$

and would be always positive as the E vs. N curve is convex in nature. It can be interpreted as the resistance of the chemical potential of a system to change with the number of electrons [36,52]. The finite difference approximation of Eq. (25) leads to Eq. (24), which is the energy change of a species in a disproportionation reaction of the type:

$$\dot{A} + \dot{A} \rightarrow A^+ + \ddot{A}^- \tag{26}$$
$$\Delta E = I - A$$

In semiempirical Hückel-type model, the energy of one electron in $\dot{A}$ is taken to be the same as that of two electrons in $\ddot{A}^-$ and it results in zero energy change for disproportionation reaction of the type (Eq. (26)). However, in other semiempirical models such as PPP theory, which is related to Hubbard model in solid state physics, repulsion between two electrons in $\ddot{A}^-$ has been approximated as $(I-A)$ and, hence, in these type of models (PPP, CNDO, INDO, and MINDO), hardness may be thought of as electronic repulsion [36,53,54].

In ab initio wavefunction pictures using Koopmans' theorem, η becomes half of the energy gap between HOMO and LUMO as [51]:

$$\eta = \frac{1}{2}(\varepsilon_{LUMO} - \varepsilon_{HOMO}). \tag{27}$$

This definition, like previous ones, has a direct consequence on the reactivity theories as large HOMO–LUMO gap signifies reluctance of the system to take or give up electrons. For insulators and semiconductors, band gap is taken as the measure of η.

Hardness has been calculated in various other ways. For example, a five-point finite difference formula has been used [55] to approximate $(\partial^2 E/\partial N^2)$. The equality of chemical potential with the total electrostatic potential at the covalent radius [56–58] has been made use of in calculating η. The electron density required for this work [56] has been obtained from a self-consistent numerical solution of a quadratic Euler–Lagrange equation [59,60]. Orsky and Whitehead [61] have proposed another definition of η, which produces better hardness ordering and expected trends of bond energy values for a number of diatomic hard and soft acids and bases. This definition has given good hardness values [22] within density functional theory. Hardness has also been calculated as a time-dependent density functional [62,63].

The inverse of hardness is softness, which is given as [64]:

$$S = \frac{1}{2\eta} = \left(\frac{\partial N}{\partial \mu}\right)_v. \tag{28}$$

The concept of softness is associated with polarizability. The larger the chemical system is, the softer it will be. This correlation of softness with polarizability can be found directly from a bond charge model [65–68] where softness is found to be proportional to the internuclear distance of a molecule [69–72]. To extend this definition (Eq. (28)) to open systems, the system is considered as a member of a grand canonical ensemble with bath parameters μ, $v(\vec{r})$, and temperature θ. This definition of S in such an ensemble can be written in terms of a number fluctuation formula [64]:

$$S = \left(\frac{\partial \langle N \rangle}{\partial \mu}\right)_{v,\theta} = \frac{1}{k\theta}\left[\langle N^2 \rangle - \langle N \rangle^2\right] \tag{29}$$

where k is the Boltzmann constant. This statistical thermodynamic definition or charge fluctuation formula of softness relates it with bond index and valence [73].

Parr et al. [74] defined electrophilicity index (w) as:

$$w = \frac{\mu^2}{2\eta}. \tag{30}$$

This measures the propensity of electrophilic attack and is used [74] in understanding the reactivity of the human immunodeficiency virus type 1 (HIV-1) nucleocapsid protein p7 (NCp7) when reacted with a variety of electrophilic agents.

4. LOCAL REACTIVITY DESCRIPTORS

Although information about the overall reaction can be obtained from knowledge of global parameters such as electronegativity and hardness, the reactivity of a particular site of a molecular species can be explained by local quantities such as electron density ($\rho(\vec{r})$), Fukui function ($f(\vec{r})$) [75], local softness [64], or local hardness [76,77]. The dependence of these local quantities on reaction coordinate reflects the usefulness of these quantities in predicting the site selectivity of a chemical reaction. The most important local descriptor is the density $\rho(\vec{r})$ itself, the basic variable of DFT [78], given as:

$$\rho(\vec{r}) = \left(\frac{\delta E[\rho]}{\delta v(\vec{r})}\right)_N \tag{31}$$

The definition of Fukui function is given by [75]:

$$f(\vec{r}) = \left(\frac{\partial \rho}{\partial N}\right)_v = \left(\frac{\delta \mu}{\delta v(\vec{r})}\right)_N \tag{32}$$

such that $\int f(\vec{r}) d\vec{r} = 1$.

This definition of $f(\vec{r})$ is obtained by considering the change in energy and chemical potential when a system goes from one ground state to another, viz.,

$$dE = \mu dN + \int \rho(\vec{r})dv(\vec{r})d\vec{r} \tag{33}$$

$$d\mu = 2\eta dN + \int f(\vec{r})dv(\vec{r})d\vec{r} \tag{34}$$

and by application of a Maxwell relation in Eq. (34). The extent of a reaction can be given by $d\mu$ from Eq. (34). It can be predicted that the reaction would be favored in a direction of increasing $f(\vec{r})dv(\vec{r})(d\vec{r})$ at a particular site. As the slope of $\rho(\vec{r})$ vs. N curve has discontinuity for integral number of N, three types of Fukui functions can be defined, which separately account for electrophilic, nucleophilic or radical attack at a particular reaction site. Using finite difference and frozen core approximations, these three functions can be written as:

$$f^{+}(\vec{r}) = \left(\frac{\partial \rho}{\partial N}\right)_v^{+} \cong \rho_{N+1}(\vec{r}) - \rho_N(\vec{r}) \approx \rho_{\text{LUMO}}(\vec{r}) \quad \text{[for nucleophilic attack]} \tag{35a}$$

$$f^{-}(\vec{r}) = \left(\frac{\partial \rho}{\partial N}\right)_v^{-} \cong \rho_N(\vec{r}) - \rho_{N-1}(\vec{r}) \approx \rho_{\text{HOMO}}(\vec{r}) \quad \text{[for electrophilic attack]} \tag{35b}$$

$$f^{0}(\vec{r}) = \left(\frac{\partial \rho}{\partial N}\right)_v^{0} \cong \frac{1}{2}(\rho_{N+1}(\vec{r}) - \rho_{N-1}(\vec{r})) \approx \frac{1}{2}(\rho_{\text{HOMO}}(\vec{r})$$
$$+ \rho_{\text{LUMO}}(\vec{r})) \quad \text{[for neutral attack]}. \tag{35c}$$

The above equations provide a correspondence between this local parameter and the frontier orbital theory of chemical reactions [79] and thus justifies the nomenclature of Fukui (frontier) function. A large value of f^{-}, f^{+}, or f^{0} at a particular site denotes the high probability of electrophilic, nucleophilic, or radical attack to take place at that site.

The expression for condensed Fukui functions for the ith atom in a molecule can be obtained by considering finite difference approximation and Mulliken's population analysis scheme as [80,81]:

$$f_i^{+} = q_i(N+1) - q_i(N) \qquad \text{(for nucleophilic attack)} \tag{36a}$$

$$f_i^{-} = q_i(N) - q_i(N-1) \qquad \text{(for electrophilic attack)} \tag{36b}$$

$$f_i^{0} = \frac{1}{2}[q_i(N+1) - q_i(N-1)] \quad \text{(for radical attack)} \tag{36c}$$

Because electron number can be continuous in the extended version of Kohn–Sham theory [42], Fukui functions may be determined as derivatives (Eq. (32)). The explicit forms for f^{+} and f^{-} can be given in this formalism as [82]:

$$f^{+} = |\phi_{\text{LUMO}}(\vec{r})|^2 + \sum_{i=1}^{N} \frac{\partial}{\partial N}|\phi_i(\vec{r})|^2 \tag{37a}$$

$$f^{-} = |\phi_{\text{HOMO}}(\vec{r})|^2 + \sum_{i=1}^{N-1} \frac{\partial}{\partial N}|\phi_i(\vec{r})|^2 \tag{37b}$$

where ϕ is the spatial orbital of neutral atom. Prescriptions for calculation of the Fukui function using a variational technique [83] and a gradient expansion [84] are also provided.

The tendency of particular site to be involved in "frontier-controlled" [33] interactions, where frontier orbital densities play important roles, is given by a local softness parameter. Local softness $s(\vec{r})$ is defined as [64]:

$$s(\vec{r}) = -\left(\frac{\delta N}{\delta v(\vec{r})}\right)_\mu = \left(\frac{\partial \rho}{\partial \mu}\right)_v \tag{38}$$

and it integrates to global softness as:

$$S = \int s(\vec{r}) d\vec{r}. \tag{39}$$

Local softness is related to Fukui function, which may be defined as a normalized local softness by the following formula:

$$s(\vec{r}) = \left(\frac{\partial \rho(\vec{r})}{\partial \mu}\right)_v = \left(\frac{\partial \rho}{\partial N}\right)_v \left(\frac{\partial N}{\partial \mu}\right)_v = f(\vec{r}) S. \tag{40}$$

The information of Fukui function can be obtained from local softness although the reverse is not true [36].

The concept of local functions can be applied to the theory of metals. Local softness has been identified with the local density of states at the Fermi level $g(\varepsilon_F, \vec{r})$ at absolute zero [64]. So $s(\vec{r}) = g(\varepsilon_F, \vec{r})$ and $S = g(\varepsilon_F)$, where $g(\varepsilon_F)$ denotes the total energy of state at The Fermi level. Analogous to Eq. (29), a fluctuation formula for local softness can be written as:

$$s(\vec{r}) = \left[\frac{\partial \langle \rho(\vec{r}) \rangle}{\partial \mu}\right]_{v,\theta} = \frac{1}{k\theta}[\langle \rho(\vec{r})N \rangle - \langle \rho(\vec{r}) \rangle \langle N \rangle]. \tag{41}$$

To obtain the reciprocal relation between local quantities similar to Eq. (28), Berkowitz and Parr [85] defined two local kernels, which integrate to give local softness and local hardness. Softness kernel is defined as [85]:

$$-s(r - r') \equiv \frac{\delta \rho(\vec{r})}{\delta \mu(r'')} = \frac{\delta \rho(\vec{r}')}{\delta u(\vec{r})} \tag{42}$$

where the modified potential $u(\vec{r})$ has the form:

$$u(\vec{r}) = v(\vec{r}) - \mu = -\frac{\delta F}{\delta \rho(\vec{r})} \tag{43}$$

for which derivatives $\delta \rho(\vec{r})/\delta u(\vec{r}')$ as well as $\delta u(\vec{r})/\delta \rho(\vec{r}')$ exist. Local softness is obtained from softness kernel simply as:

$$s(\vec{r}) = \int s(\vec{r}, \vec{r}') dr' \tag{44}$$

Hardness kernel is defined as:

$$-2\eta(\vec{r}, \vec{r}') \equiv \frac{\delta u(\vec{r})}{\delta \rho(\vec{r}')} = \frac{\delta u(\vec{r}')}{\delta \rho(\vec{r})}. \tag{45}$$

Hardness kernel is the inverse of softness kernel in the sense:

$$2\int s(\vec{r}, \vec{r}')\eta(\vec{r}, \vec{r}''')dr' = \delta(\vec{r} - \vec{r}'''). \tag{46}$$

Local hardness cannot be obtained from hardness kernel by simple integration. But the relation exists as:

$$\eta(\vec{r}) = \frac{1}{N} \int \eta(\vec{r}, \vec{r}')\rho(\vec{r}')d\vec{r}'. \tag{47}$$

Inserting the expression for $\eta(\vec{r}, \vec{r}')$ from Eq. (45) into Eq. (47), we get $\eta(\vec{r})$ as [76]:

$$\eta(\vec{r}) = \frac{1}{2N} \int \frac{\delta^2 F}{\delta \rho(\vec{r})\delta \rho(\vec{r}')}\rho(\vec{r}')d\vec{r}'. \tag{48}$$

Local hardness can also be written in terms of Fukui functions [86,87] as:

$$\eta(\vec{r}) = \frac{1}{2} \int \frac{\delta^2 F}{\delta \rho(\vec{r})\delta \rho(\vec{r}')}f(\vec{r}')d\vec{r}' \tag{49}$$

which has been shown [87] to be its most unambiguous definition. The reciprocal relation at this level is:

$$2\int \eta(\vec{r})s(\vec{r})d\vec{r} = 1. \tag{50}$$

To obtain another definition of $\eta(\vec{r})$, we write $d\mu$ as [76]:

$$d\mu = 2\int \eta(\vec{r})d\rho(\vec{r}) + \frac{1}{N} \int \rho(\vec{r})dv(\vec{r})d\vec{r}' \tag{51}$$

which gives:

$$\eta(\vec{r}) = \frac{1}{2} \left(\frac{\delta\mu}{\delta\rho}\right)_v. \tag{52}$$

It should be noted that the definition of local hardness has inherent ambiguity in it [87]. The local hardness defined above requires a variation of $\rho(\vec{r})$, keeping $v(\vec{r})$ constant, which seems to be ambiguous because $v(\vec{r})$ and $\rho(\vec{r})$ are interdependent as has been proven by Hohenberg and Kohn [88].

Local hardness integrates to give global hardness [76] in a way similar to that of hardness kernel:

$$\eta = \int \eta(\vec{r})f(\vec{r})d\vec{r}. \tag{53}$$

5. ELECTRONEGATIVITY AND ASSOCIATED PRINCIPLES

The theoretical background of electronegativity and related concepts generated interest among chemists mainly because it can be evaluated in terms of experimental ionization potential and electron affinity [89] as well as through density functional calculation [22,47–50,90]. The systematic study on this topic revealed newer aspects such as relations between electronegativity and diamagnetic shielding [91] and high-temperature superconductivity [92–98], application of electronegativity difference in classifying the crystal structure of binary solids [99], explanation of alloy formation in terms of electronegativity [100], etc. Several other studies [101–103] included quantitative dependence of electronegativity on atomic number Z. It was shown [102] that in each group, electronegativity shows a periodic behavior and, at large Z, it is approximately $\sim Z^{1/3}$. The Z dependence of binding energy values of neutral atoms has also been studied [104].

The density functional theory not only provided a rigorous definition of electronegativity but also a basis for Sanderson's electronegativity equalization and geometrical mean principles [20]. The relation of electronegativity with the negative of chemical potential proved that for a system at equilibrium, electronegativity would be constant [39]. Moreover, in the chemical system of interest, electron will be distributed in such a way that the electronegativity of orbitals will be equal to the electronegativity of the system, that is,

$$\mu = \frac{\partial E}{\partial n_i} \text{ for all } i; \sum_i n_i = N \tag{54}$$

where n_i is the occupancy of the ith natural orbital. To find the dependence of charge flow on electronegativity difference, Parr et al. [39] proposed an "atom-in-molecule (AIM)" model [39,105–108]. Without remaining confined to any particular theoretical framework, Politzer and Weinstein [109] proved the validity of the principle of equalization of electronegativity for any arbitrary region of space in a molecule.

As the charge transfer leads to the formation of a new molecule, the molecular electronegativity, after electronegativity equalization, can be obtained from isolated atoms' electronegativity values by Sanderson's geometrical mean law [20] as:

$$\chi_{AB...N} = \left(\chi_A^0 \chi_B^0 \cdots \chi_N^0\right)^{1/N} \tag{55}$$

where $\chi_{AB...N}$ is the electronegativity of a polyatomic molecule and χ_i^0 ($i = A, B, ..., N$) are isolated atoms' electronegativities. The sufficient condition for the validity of Eq. (55) is that the energy of atoms should be an exponentially decaying function of the number of electrons [110]:

$$E(N) \approx E(Z)\exp[-\gamma(N - Z)] \tag{56}$$

or, equivalently:

$$\chi = \chi^0 \exp[-\gamma(N - Z)] \tag{57}$$

which agrees with the supposition that atomic energy is a quadratic function of the number of electrons [9]. In Eqs. (56) and (57), the decay parameter ($\gamma = I/A$) is found to be more or less constant for all atoms and has an approximate value of 2.2. However, it should be mentioned that Sanderson's geometrical mean principle is only an approximate scheme, which neglects the influence of external potential. The consideration of influence of another atom in AIM framework [111] leads to the evaluation of molecular electronegativity as the harmonic or geometrical mean of valence state electronegativities [111].

The concept of equalization of electronegativity can be applied to determine various properties of atoms and molecules. Using this concept, one can define the different atomic radii [57,112–116], which are the measures of the binding property of atoms. The electronegativity equalization principle gives a scheme for calculating the amount of charge transfer and partial charges on atoms in a molecule. For a diatomic molecule AB, the energy and chemical potential can be written as the following Taylor expansions:

$$E_A = E_A^0 - \chi_A \Delta N_A + \eta_A \Delta N_A^2 + \cdots \quad A \equiv A, B \tag{58}$$

and

$$-\chi_A = \left(\frac{\partial E_A}{\partial N_A}\right) = -\chi_A^0 + 2\eta_A \Delta N_A + \cdots A \equiv A, B \tag{59}$$

Truncation of Taylor expansion (Eq. (8)) after the second-order variation may be shown to be legitimate because the third-order derivative is often small [117], which is, however, not always true [56]. Application of EEP gives:

$$\chi_A = \chi_B$$

which implies

$$\Delta N = -\frac{\chi_B^0 - \chi_A^0}{2(\eta_A + \eta_B)} \tag{60}$$

and

$$\Delta E = -\frac{(\chi_B^0 - \chi_A^0)^2}{4(\eta_A + \eta_B)}. \tag{61}$$

From Eqs. (60) and (61), it can be seen that charge transfer is dependent on the first order of electronegativity difference, and stabilization energy has a second-order dependence on electronegativity difference. Equations (60) and (61) predict that the transfer process will be hindered by the hardness sum [52]. This model is only a crude model and Eq. (60) gives connectivity-independent charges for atoms in polyatomic molecules, which are not always acceptable. This model can be improved by taking into consideration change in the molecular environment. Considering electrostatic interaction between atoms in a molecule, an improved expression for the amount of charge transfer can be given, which depends on internuclear distance r as [118]:

$$\Delta N = (\chi_B^0 - \chi_A^0) \bigg/ \left[\left(\frac{\partial \chi_A}{\partial N_A}\right)_{N_A=Z_A} + \left(\frac{\partial \chi_B}{\partial N_B}\right)_{N_B=Z_B} + 2/r\right]. \tag{62}$$

The charges calculated from different orbital electronegativity equalization schemes were found to exhibit good correlations with ESCA or NMR shifts [119] and could differentiate between different isomers present in structurally different phases [119]. These methods are also used [119–121] to study electronegativity and charge distribution in solids. Connectivity-dependent charges [98] calculated this way [122,123] have been found to be adequate in explaining charge transfer in donor–acceptor atoms [124]. Concepts of bond electronegativity and bond hardness have also been introduced [125–128] in providing a model of covalent bonding in molecules.

Like atomic electronegativity, group electronegativity correlates with a number of theoretical as well as experimental quantities and is thus of extreme importance to chemists. However, calculation of group electronegativity is more complicated than that of atomic electronegativity. Several methods have been developed in this direction [128–146].

As the difference of electronegativity measures the charge transfer in a reaction (i.e., the extent of a reaction), it is always useful to study the nature of electronegativity change in chemical processes. It is known that when a molecule is formed, atomic electronegativity changes to "molecular electronegativity," which is constant everywhere in the molecule. Hence, necessary information about a process can be obtained in terms of a step-by-step change of atomic electronegativity. Different works on this aspect include ab initio and density functional studies of electronegativity profiles during various physico-chemical processes such as umbrella inversion [147,148], intramolecular atom transfer [147], internal rotation [149], and dissociation reactions of ordinary diatomic molecules [150], as well as hydrogen-bonded complexes [151]. Because chemical reactions are time-dependent processes, concepts of electronegativity and electronegativity equalization have been extended to dynamic situations [62].

It has been realized by Parr and Pearson [52] that electronegativity alone cannot properly account for all facets of a chemical process and another parameter, hardness, is necessary. Whereas electronegativity is the tangent to E vs. N curve, the corresponding curvature has been identified as hardness.

6. HARDNESS AND ASSOCIATED PRINCIPLES

The qualitative concepts of hardness and softness were first introduced by Pearson [30–32,34], which later culminated in enunciation of the famous hard–soft acid–base principle. Quantification of these concepts had been in order and was accomplished within density functional theory by Parr and Pearson [52]. The energy stabilization due to soft–soft interaction can be expressed by rearranging Eq. (61) as [52]:

$$\Delta E = -\frac{(\Delta \mu)^2}{2} \frac{s_A s_B}{s_A + s_B} \tag{63}$$

where $\Delta \mu = \mu_B - \mu_A$.

Equation (63), however, does not explain energy stabilization due to hard–hard interaction. Hence, a better approximation considering the effect of change of potential [111] gives the energy change of A up to second order as [36]:

$$\Delta E_A = \mu_A^0 \Delta N_A + \int \rho_A(\vec{r}) \Delta v_A(\vec{r}) d\vec{r} + \eta_A (\Delta N_A)^2 + \frac{1}{2} \int \Delta \rho_A(\vec{r}) \Delta v_A(\vec{r}) d\vec{r}. \tag{64}$$

The corresponding change in density is:

$$\Delta \rho_A(\vec{r}) = f_A(\vec{r}) \Delta N_A + \int \left[\frac{\delta \rho(\vec{r})}{\delta v_A(\vec{r}')} \right]_N \Delta v_A(\vec{r}) d\vec{r}' \tag{65}$$

Writing a similar set of equations for atom B, the total energy change can be obtained as $\Delta E_{AB} = \Delta E_A + \Delta E_B$, which can be written as Eq. (66) by separating covalent (ΔE_{cov}), electrostatic (ΔE_{el}), and polarization (ΔE_{pol}) energy contributions:

$$\Delta E = \Delta E_{cov} + \Delta E_{el} + \Delta E_{pol} \tag{66}$$

When a soft atom A reacts with another soft atom B, the energy stabilization due to covalent interaction will be high. For a hard–hard interaction, energy change will be governed by electrostatic interaction, which includes energy change due to electron–nuclear attraction (ΔE_{el}) as well as nuclear–nuclear repulsion. The corresponding charge transfer formula can be found by considering electronegativity equalization in a molecule as:

$$\Delta N = \frac{(\mu_B^0 - \mu_A^0) + \int f_B(\vec{r})\Delta v_B(\vec{r})d\vec{r} - \int f_A(\vec{r})\Delta v_A(\vec{r})d\vec{r}}{2(\eta_A + \eta_B)} \tag{67}$$

The HSAB principle has been used successfully in gaining insights into various aspects of medicinal chemistry especially the removal of toxic elements from the body as well as taking care of the deficiencies of various essential elements. It also helps in understanding the likely biochemical sites where the toxic metal ions would bind. A beautiful account of this subject is provided in Ref. 152. There are several hard acids such as Na^+, K^+, Mg^{2+}, and Ca^{2+}, which are essential for life because the body allows their hydrated forms to pass selectively across various barriers, resulting in the generation of an electric current, which in turn may trigger some response or transmit a nerve impulse. The double-helix structure of DNA is preserved by Mg^{2+}, whereas Ca^{2+} is essential in the formation of bones and teeth in the form of its phosphate or carbonate. Both beryllium and soluble barium salts are toxic and plutonium causes leukemia because of its radiotoxicity. Beryllium might be carcinogenic as well. Other hard acid metal ions are neither vital nor poisonous. Most of the acids and bases that are essential at low concentration are toxic at relatively larger concentrations. The hard base F^- is needed to prevent tooth decay but it causes damage to teeth and bones at larger concentrations. Lack of a borderline acid such as iron causes anemia, but an excess of it increases the risk of heart attacks. Copper and iron poisoning is treated using penicillamine and desferrioxamine, respectively.

Soft acid metal ions often rupture the SS single bond in cystine. Mercury, lead, cadmium, and thallium are known to be toxic. In the kidneys, soft acids such as cadmium bind to S-atoms in metallothionein. Chemotherapy using soft acids such as gold and platinum in the antiarthritic drug myocrisin and anticancer drug cisplatin, respectively, has well-known toxic side effects. Many of the soft bases are toxic. The hard base O_2 fails in the competition of binding with Fe^{2+} in heme, with soft carbon donors in carbon monoxide and cyanides. Also the toxicity of PH_3, AsH_3, H_2S, H_2Se, and their methylated variants can be attributed to the presence of the soft donor atoms. Soft bases such as S and P are present in amino and nucleic acids, whereas I is essential in thyroxin, which prevents goiter. Both the soft bases As and Se are essential to human metabolism in very low concentration, but their toxicity (possibly carcinogenicity also) in high dose is a matter of concern. It has been argued that [152] the HSAB principle may be applied in understanding this aspect in the way that an excess of these soft bases tries to bind to borderline acid metal ions important in the activity of various enzymes.

Although the meaning of the term "hardness" or "softness" does not necessarily identify the physical properties, using a thermodynamic approach this chemical hardness can be shown to be related to the compressibility factor, which in turn predicts mechanical hardness of minerals [153]. The definition of softness suggests that there would be correlation between softness and polarizability (α). Some workers [154]

(KD Sen, MC Bohm, and PC Schmidt in Ref. 1.) suggested that this relationship would be linear. Nagel [25] reported a good correlation between softness and (α/n) (n being the number of valence electrons) when both softness and polarizability were calculated by density functional methods. In another approach, assuming atom to be an electrodynamic system, charge capacitance has been identified with atomic softness [155], whereas a linear correlation between softness and atomic radius was found by many works [71,111,155]. It has been noted that the correlation of softness with these quantities exists even for molecular systems [156] when softness is approximated as the inverse of difference between I and A or ε_{LUMO} and ε_{HOMO}. Assuming spherical models for clusters and defining clusters' appropriate radii, it has been found that the cluster softness varies linearly [156] with cluster radius and cube root of polarizability. Furthermore, a good linear correlation between $\alpha^{1/3}$ and the cohesive energy of carbon clusters as well as a correlation between hardness multiplied with number of atoms and π resonance energy indicated [156] the existence of relation between hardness as well as softness parameters and binding energy. The relation of hardness or softness with binding energy has been studied in detail within semiempirical DFT [113,126–128].

When molecule formation takes place, like atomic electronegativity hardness parameter also undergoes change. There have been different propositions [71,122,157–161] on the nature of the dependence of molecular hardness on isolated atoms' values. The stabilization energy or charge transfer affinity corresponding to charge transfer process shows linear correlation with Hammett constant, which suggests that both electronegativity as well as hardness parameters are necessary to study substituent effects [161].

Experimental observations suggest that there should be a correlation between stability and hardness. Equilibrium states of atoms or molecules associated with maximum stability would have maximum hardness [34]. Initial study on this subject includes study of hardness of aromatic compounds that show low reactivity. Hardness values of alternant and nonalternant hydrocarbons calculated from ε_{HOMO} and ε_{LUMO} are found to be high and show a very good linear correlation [162] with the π resonance energy [163,164], which is a measure of aromaticity. It was argued [165] that in Hückel theory, minimum energy at constant chemical potential leads to the maximum π resonance integral and, consequently, the maximum hardness. Attempts have been made to obtain a formal proof for the maximum hardness principle [166–170]. It has been shown that a chemical system at equilibrium would evolve toward a state of maximum hardness in case the bath parameters do not change [166]. The maximum hardness principle has been found to be useful in predicting stability associated with various physico-chemical processes such as molecular vibrations, internal rotations, aromaticity, chemical reactions, HSAB principle, as well as the stability of closed-shell species and the isomers, statistical distributions, and dynamic situations [171–184].

The maximum hardness principle also demands that hardness will be minimum at the transition state. This has been found to be true for different processes including inversion of NH_3 [147] and PH_3 [148], intramolecular proton transfer [147], internal rotations [149], dissociation reactions for diatomics [150,151], and hydrogen-bonded complexes [152]. In all these processes, chemical potential remains either constant or passes through an extremum at the transition state. The maximum hardness principle has also been found to be true (a local maximum in hardness profile) for stable intermediate, which shows a local minimum on the potential energy surface [150]. The energy change in the dissociation reaction of diatomic molecules does not pass through a transition state. The corresponding hardness change has also been found to be monotonic and dependent on energies of neutral molecules as well as its ionic counterparts. The temporal evolution of local and global hardness during a chemical reaction provides [63] important insights into the associated molecular reaction dynamics.

A many-particle quantum system is completely characterized by N and $v(\vec{r})$. Whereas χ and η measure the response of the system when N changes at fixed $v(\vec{r})$, the polarizability (α) measures the response of the system for the variation of ($v(r)$) at fixed N when a weak electric field is the source of $v(\vec{r})$, in addition to that arising out of a set of nuclei. Based on the inverse relationship [185] between α and η, a minimum polarizability principle has been proposed: "the natural direction of evolution of any system is toward a state of minimum polarizability" [186].

The validity of both MHP and MPP has been tested in various physico-chemical processes [128,186–191]. The maximum hardness and minimum polarizability criteria complement the minimum energy criterion for stability. In general, a stable state (minimum energy configuration) or a favorable process is associated with the maximum hardness and minimum polarizability, and transition state is associated with the minimum hardness and maximum polarizability.

A molecule at equilibrium geometry possesses maximum hardness and minimum polarizability values when compared with the corresponding values for any other geometry obtained through a nontotally symmetric distortion. In the internal rotation process, the most stable isomer is associated with the maximum η and minimum α values, and the least stable isomer is associated with the minimum η and maximum α values. For several chemical reactions, it has been observed that the reaction proceeds in the direction that produces the hardest and least polarizable species [28,187,188]. It has been observed that a system is the hardest and the least polarizable in its ground state [189,190] and for the most stable species along the reaction path [191,192]. Chemical periodicity [118,193], improvement of basis set quality [194,195], and solvent effects [196] have also been studied in this connection.

Any theory of chemical reactivity requires a detailed knowledge of various sites in a molecule. Global quantities such as electronegativity and hardness appear to be inadequate in explaining site selectivity, and, as a natural consequence, concepts of local quantities started developing.

7. LOCAL QUANTITIES, REACTIVITY, AND SITE SELECTIVITY

Although a satisfactory theory of chemical reactivity is still awaited, earlier discussions indicate that local functions contain information about inherent reactivity of molecules and associated stereospecific control in a chemical reaction. Different studies on local functions reflect that reactivity generally follows a rule of thumb, viz., hard–hard interactions are charge-controlled whereas soft–soft interactions are frontier-controlled [33]. Studies on local softness show [64] that metals have high density of states at the Fermi level and are soft. Adsorption and heterogeneous catalysis can be understood in terms of soft–soft interactions at particular sites on metal surfaces having high values of $s(\vec{r})$. The fluctuation formula (Eq. (41)) for $s(\vec{r})$ strengthens [64] the argument of Falicov and Somorjai [197], who feel that low-density fluctuations can explain the catalytic activity of transition metals.

Density functional theory suggests that the HSAB principle, which was initially formulated to study global changes in a reaction, can also be applied to local interactions. The local version of HSAB principle [198] states that a reaction site that has large softness value would prefer to react with a soft species or a softer site of a species, and hard reaction sites will be involved in hard–hard interactions. Results from the calculation of local softness of silicon clusters [175] show that the softer site of silicon cluster will prefer nucleophilic attack from softer atoms such as Ga and the nucleophilic attack of harder atoms such as Si will be favored at the harder sites of the cluster where softness value is small. The concept of local HSAB principle has been extended to study impurity segregation process in solids [199]. It is possible to obtain information about the type of atoms that will prefer to segregate from the study of the hard–soft nature of the interface. The soft impurity atom will prefer to make a nucleophilic attack at the softer surface having larger s^+ $(\vec{r})$ values, whereas hard impurity will remain at the bulk. On the other hand, hard atom will segregate at the grain boundary if the boundary is harder than the bulk. As the atoms start segregating at the surface, the lattice starts relaxing and hardness increases. Hardness of the relaxed lattice will be maximum when the impurity segregates at the boundary, which can be thought of as a consequence of MHP. A detailed ab initio investigation on $\Sigma = 5$ tilts of [310] germanium [200] grain boundary has been carried out [199] and results have been found to be in conformity with HSAB prediction. Local HSAB principle has been made use of [201] in understanding the site selectivity in a molecule. It is observed that soft–soft interactions are preferred in the maximum Fukui function site and the minimum Fukui function site is preferred over the hard–hard interactions. Frontier-controlled soft–soft reactions are argued [202] to be favored at the maximum Fukui function site whereas the charge-controlled hard–hard reactions would prefer the site with the maximum net charge and not necessarily the minimum Fukui function.

The frontier orbital theory [79] of chemical reactivity can be justified in terms of three [75–80] types of Fukui functions, f^+, f^-, and f^0. Studies on different molecular systems reveal that the prediction made by these Fukui functions about the type of attacks (nucleophilic, electrophilic, and neutral) matches well with the experimental results [80,81,203,204]. For example, studies on formaldehyde show that nucleophilic attack from perpendicular direction will be favored on carbon atom as it has dense $f^+(\vec{r})$ contour lines [81]. Condensed Fukui functions calculated from the charges on neutral and ionic molecules provide information that is in agreement with the predictions made from local function contour maps. The highest value of f^- on the oxygen atom of formaldehyde predicts the preference of electrophilic attack of proton at oxygen during an acid-catalyzed hydration, whereas the largest f^+ on carbon suggests that the nucleophilic attack of OH^- will be preferred at carbon. The linkage isomerism can also be explained on a quantitative basis from the study of condensed Fukui functions. The high values of f^+ on sulphur explains the formation of $[Co(CN)_5SCN]^{3-}$-type complexes due to soft–soft interaction. But when the presence of hard ligands makes the metal atom harder, SCN^- prefers to bind metal atom through nitrogen, resulting in charge-controlled hard–hard interaction in $[Co(NH_3)NCS]^{2+}$ [81]. The chemical observation that carbon monoxide acts as a Lewis acid in neutral carbonyls is also found to be true in the studies on local quantities of carbon monoxide. Other studies on condensed Fukui function include predictions of stereoselectivity in nucleophilic addition reaction on maleimide [203], electrophilic substitution reactions [204], dissociation reactions [152], etc., which corroborate experimental results. Comparison of [204] contour maps of $f(\vec{r})$ and molecular electrostatic potential (MEP) shows that the information obtained from contour maps of $\cdot f(\vec{r})$ and MEP is complementary to each other as $f(\vec{r})$ indicates soft–soft interaction whereas MEP provides information about hard–hard interaction. It has been found that relaxation effect due to charge removal or addition is important in the prediction of stereoselectivity. Results of different studies [203,204] reveal that although $f^+(\vec{r})$ and $f^-(\vec{r})$ can be approximated as ρ_{LUMO} and ρ_{HOMO} (Eq. (5)), Fukui functions calculated from the densities of N, $(N+1)$, and $(N-1)$ electron systems give more accurate information than that obtained from the study of ρ_{LUMO} and ρ_{HOMO} alone, where orbital relaxation effects due to charge addition or removal are neglected. To study processes that involve a change in spin multiplicity, it is necessary to consider spin-polarized versions of local as well as global quantities [205]. It is expected that spin-polarized Fukui functions would be able to account for the photochemical reactions involving triplet transition state [206] and the catalytic behavior of para-magnetic substances [64].

Chemical reactivity associated with several typical organic reactions such as electrophilic substitution and nucleophilic addition is studied successfully using Fukui functions and local softness [207]. These local descriptors are found to be useful in following the reactivity of HIV-1 nucleocapsid protein p7 [208] as well as charybdo-toxin (ChTx), a 37-residue polypeptide acting as a K^+ channel blocker [209]. HIV-1 inhibition has been caused by various electrophiles through chemical modification of the NCp7 Cys_3His cores in the thiolate centers. Electrophilicity behavior dictates the reactivity of Cys_3His cores toward soft electrophiles [208]. It has also been demon-strated that Cys is the most labile site of NCp7 because the core of the C-terminal finger is much more reactive than the corresponding N-terminal finger, although the same retroviral zinc finger motif is shared by each of the two conserved NCp7 zinc fingers [208]. This analysis is important in understanding the nature of antiviral agents that selectively target retroviral nucleocapsid protein zinc fingers but do not affect the cellular zinc finger proteins. Local softness study reveals [209] that one of the major stabilization effects in the ChTx–K^+ channel complexation stems from the charge transfer to ChTx. A nuclear counterpart of a Fukui function is also defined [210] as the electrostatic force due to the electronic Fukui function. This quantity for several diatomics has been calculated recently [211]. Higher-order derivatives of both elec-tronic and nuclear Fukui functions are derived [212].

How the reactivity of a particular site changes during a chemical reaction has been studied [152]. Ab initio SCF calculations [152] of condensed Fukui functions at various sites at different stages of a dissociation reaction have been performed. The variation of these reactivity indices along the reaction path is consistent with chemical intuition.

In order to understand the importance of frontier orbitals in chemical reactivity, Berkowitz [213] studied the frontier-controlled reactions within the purview of density functional theory. It is evident that the directional characteristics of frontier orbitals determine the extent of charge transfer, and soft–soft interactions are frontier-con-trolled. A somewhat similar analysis showed that charge transfer would be facilitated at a place where the difference in local softness of two partners is large [87]. It may be noted that Fukui function is obtainable from local softness but the reverse is not true. On the other hand, local hardness suffers from the drawback of ambiguity [87], which allows one to even consider it to be equal to global hardness without disturbing their fundamental relations. Considering these aspects of the local quantities, it may be ascertained that local softness is the most appropriate index [87] for the chemical reactivity of a particular site in a molecule. A local softness map delineates the variations in reactivity in different sites in a molecule whereas global softness (obtained through a simple integration of $s(\vec{r})$) takes care of the relative reactivity, which varies from one molecule to other. It is hoped that a complete theory of chemical reactivity in terms of local softness, which would be able to account for stereospecificity in various chemical reactions including catalysis, will be developed one day.

8. CONCLUDING REMARKS

Among the various popular qualitative chemical concepts, electronegativity and hardness have been the most appreciated ones because they can account for a variety of important physico-chemical phenomena. Rigorous theoretical definitions of these quantities have been provided within density functional theory. Chemical potential (negative of electronegativity) and hardness, are respectively, the first-order and second-order derivatives of energy with respect to number of electrons. Therefore, the former measures the escaping tendency of electrons and latter gives the likelihood of that process. They are intimately connected with several other physical properties such as bond energy, covalent radius, polarizability, aromaticity, etc. There are several ways of calculating electronegativity and hardness in static and dynamic situations using density-based as well as wavefunction-based formalisms. There are two associated principles of electronic structure theory viz., electronegativity equalization and maximum hardness principles. During molecule formation, electronegativity of all constituent atoms equalizes to the molecular electronegativity defined as the geometrical mean of isolated atoms' electronegativity values. It has been observed in many occasions that an increase in hardness value (decrease in polarizability) is associated with an increase in stability. Another related concept important in rationalizing a generalized acid–base reaction is enunciated as HSAB principle, which states that hard likes hard and soft likes soft. This principle is helpful in understanding the nature of toxicity of a material and the possible ways through which it can be removed from the human body, as well as the importance of various essential elements in different metabolic activities. It may turn out to be a powerful tool in the overall strategy of drug design. The potential of different reactivity-related structure principles in analyzing the toxicity of various drugs and other essential elements needs to be explored. A satisfactory theory of chemical reactivity requires information of each site in a molecule. Local quantities such as local hardness and softness, respective kernels, and Fukui function are helpful in understanding the site selectivity during electrophilic, nucleophilic, or radical attack. Adsorption and catalysis may also be understood in terms of these quantities. It is difficult to get an unambiguous definition of local hardness. All information about Fukui function can be obtained from local softness, but the converse is not true. It appears that local softness contains maximum information about chemical reactivity. A molecular reaction dynamics can be envisaged in terms of temporal evolution of electronegativity and hardness during a chemical reaction. Therefore, it may be concluded that electronegativity and hardness are two cardinal indices of structure, properties, reactivity, and dynamics of many fermion systems encompassing atoms, molecules, and solids.

ACKNOWLEDGMENT

We are thankful to the CSIR, New Delhi for financial assistance.

REFERENCES

1. Sen KD, Jørgensen CK, eds. Electronegativity, Structure and Bonding. Vol. 66. Berlin: Springer-Verlag, 1987.
2. Sen KD, Mingos DMP, eds. Chemical Hardness, Structure and Bonding. Vol. 80. Berlin: Springer-Verlag, 1992.
3. Pauling L. The Nature of the Chemical Bond. 3rd ed. Ithaca, NY: Cornell University Press, 1960.
4. Mcweeny R. Coulson's Valence. 3rd ed. Oxford: Oxford University Press, 1979:161–171.
5. Huheey JE. Inorganic Chemistry. 2nd ed. New York: Harper and Row, 1979:159–173.
6. Mulliken RS. J Chem Phys 1934; 2:782; Mulliken RS. J Chem Phys 1935; 3:573.
7. Moffitt W. Proc R Soc A 1949; 196:510; Pritchard HO, Skinner HA. Chem Rev 1955; 55:745.
8. Hinze J, Jaffe HH. J Am Chem Soc 1962; 84:540; J Phys Chem 1963; 67:1501; Hinze J, Whitehead MA, Jaffe HH. J Am Chem Soc 1963; 85:148.
9. Iczkowski RP, Margrave JL. J Am Chem Soc 1961; 83:3547.
10. Pritchard HO, Sumner FH. Proc Roy Soc A 1956; 235:136.
11. Huheey JE. J Phys Chem 1965; 69:3284; J Org Chem 1971; 36:204.
12. Baird NC, Whitehead MA. Theor Chim Acta 1964; 2:259.
13. Whitehead MA, Baird NC, Kapllansky M. Theor Chim Acta 1965; 3:135.
14. Allred AL, Rochow EG. J Inorg Nucl Chem 1958; 5:264; Allred AL. J Inorg Nucl Chem 1961; 17:215.
15. Little EJ, Mark JM. J Chem Educ 1960; 37:231.
16. Gordy WE. Phys Rev 1946; 69:604; Gordy WE, Smith WV, Trambulo RF. Microwave Spectroscopy. New York: Wiley, 1953; Gordy W, Orville Thomas WJ. J Chem Phys 1956; 24:439.
17. Yuan HC. Huaxue Xuebao 1964; 30:341.
18. Walsh R. Organometallics 1989; 8:1973.
19. Luo YR, Benson SW. J Phys Chem 1988; 92:5255; J Am Chem Soc 1989; 111:2480; J Phys Chem 1990; 94:914; J Phys Chem 1989; 93:1674, 3791, 4643; J Phys Chem 1989; 93:7333; J Phys Chem 1989; 93:3304.
20. Sanderson RT. Science 1951; 114:670; 1952; 116:41; 1955; 121:207. J Chem Educ 1952; 29:539; 1954; 31:238; Inorganic Chemistry. New York: Van Nostrand-Reinhold, 1967; Chemical Bonds and Bond Energy. New York: Academic Press, 1976; J Am Chem Soc 1983; 105:2259; Polar Covalence. New York: Academic Press, 1983.
21. Allen LC. J Am Chem Soc 1989; 111:9003.
22. Kostyk RJ, Whitehead MA. J Mol Struct (Theochem) 1991; 203:83.
23. Walsh AD. Proc Roy Soc A 1951; 207:13.
24. Gorbunov AI, Kaganyuk DS. Russ J Phys Chem (Engl Transl) 1986; 60:1406.
25. Nagel JK. J Am Chem Soc 1990; 112:4741.
26. Edgecombe KE, Boyd RJ. Int J Quantum Chem 1986; 29:959; J Comput Chem 1988; 8:489; J Am Chem Soc 1988; 110:4182.
27. Gyftopoulos EP, Hatsopoulos GN. Proc Natl Acad Sci USA 1968; 60:786.
28. Pearson RG. J Am Chem Soc 1963; 85:3533; Science 1966; 151:172; Pearson RG, ed. Hard and Soft Acids and Bases, Hutchinson and Ross, Stroudsberg, Dowden: Pearson RG, Chemical Hardness: Applications from Molecules to Solids. Weinheim: Wiley-VCH, 1997.

29. Pearson RG, Songstad J. J Am Chem Soc 1967; 89:1827.
30. Pearson RG. J Chem Educ 1968; 45:581, 643.
31. Pearson RG. Inorg Chem 1972; 11:3146.
32. Pearson RG. Theoretical Models of Chemical Bonding: Part II. In: Maksic ZB, ed. Berlin: Sringer-Verlag, 1990:45–76.
33. Klopman G. J Am Chem Soc 1968; 90:223; Klopman G. In: Klopman G, ed. Chemical Reactivity and Reaction Paths. New York: Wiley, 1974, Chap. 4.
34. Pearson RG. J Chem Educ 1987; 64:561.
35. Drago RS, Vogel GC, Needham TE. J Am Chem Soc 1971; 93:6014; Drago RS, Kabler RA. Inorg Chem 1972; 11:3144; Drago RS. Inorg Chem 1973; 12:2211.
36. Parr RG, Yang W. Density Functional Theory of Atoms and Molecules. New York: Oxford University Press, 1989.
37. Chattaraj PK, Parr RG in Ref. 2.
38. Chattaraj PK. J Indian Chem Soc 1992; 69:173; Chermette H. J Comput Chem 1999; 20:129.
39. Parr RG, Donnelly RA, Levy M, Palke WE. J Chem Phys 1978; 68:3801.
40. Perdew JP, Parr RG, Levy M, Balduz JL Jr. Phys Rev Lett 1982; 49:1691; Levy M. Proc Natl Acad Sci USA 1982; 76:3946; Phys Rev A 1982; 26:1200.
41. Parr RG, Bartolotti LJ. J Phys Chem 1983; 87:2810.
42. Janak JF. Phys Rev B 1978; 18:7165.
43. Perdew JP, Zunger A. Phys Rev B 1981; 23:5048.
44. Harris J. Int J Quantum Chem 1979; 13:189; Phys Rev A 1984; 29:1684.
45. Gopinathan MS, Whitehead MA. Isr J Chem 1980; 19:209.
46. Slater JC. Phys Rev 1951; 81:385; Adv Quantum Chem 1972; 6:1.
47. Bartolotti LJ, Gadre SR, Parr RG. J Am Chem Soc 1980; 102:2945.
48. Robles J, Bartolotti LJ. J Am Chem Soc 1984; 106:3723.
49. Manoli S, Whitehead MA. J Chem Phys 1984; 81:841.
50. Goycoolea C, Barrera M, Zuloaga F. Int J Quantum Chem 1989; 36:455.
51. Pearson RG. Proc Natl Acad Sci USA 1986; 83:8440.
52. Parr RG, Pearson RG. J Am Chem Soc 1983; 105:7512; Chattaraj PK, Lee H, Parr RG. J Am Chem Soc 1991; 113:1855; Cedillo A, Chattaraj PK, Parr RG. Int J Quantum Chem 2000; 77:403.
53. Parr RG. The Quantum Theory of Molecular Electronic Structure. New York: Benjamin, 1963.
54. Segal GA, ed. Semiempirical Methods of Electronic Structure Calculation: Part A. Techniques; Part B. Applications. New York: Plenum, 1977.
55. Chattaraj PK, Nandi PK, Sannigrahi AB. Proc Indian Acad Sci (Chem Sci) 1991; 103: 583.
56. Chattaraj PK. J Indian Chem Soc 1993; 70:103.
57. Politzer P, Parr RG, Murphy DR. J Chem Phys 1983; 79:3859.
58. Harbola MK, Parr RG, Lee C. J Chem Phys 1991; 94:6055.
59. Haq S, Chattaraj PK, Deb BM. Chem Phys Lett 1984; 111:79; Chattaraj PK, Deb BM. Chem Phys Lett 1985; 121:143; Deb BM, Chattaraj PK. Phys Rev A 1988; 37:4030; Chattaraj PK. Phys Rev A 1990; 41:6505.
60. Deb BM, Chattaraj PK. J Indian Chem Soc 1989; 66:593; Phys Rev A 1992; 45:1412.
61. Orsky A, Whitehead MA. Can J Chem 1987; 65:1970.
62. Chattaraj PK. Int J Quantum Chem 1992; 41:854; Chattaraj PK, Nath S. Int J Quantum Chem 1994; 49:705; Chattaraj PK, Maiti B. J Phys Chem A 2001; 105:169–183.
63. Chattaraj PK, Nath S. Chem Phys Lett 1994; 217:342; Chattaraj PK, Sengupta S. J Phys Chem 1996; 100:16126; J Phys Chem A 1997; 101:7893.
64. Yang W, Parr RG. Proc Natl Acad Sci USA 1985; 82:6723.
65. Pasternak A. Chem Phys 1997; 26:101; J Chem Phys 1980; 73:593.

66. Parr RG, Borkman RF. J Chem Phys 1968; 49:1055.
67. Parr RG, Simons G. J Chem Phys 1971; 55:4197.
68. Politzer P. J Chem Phys 1970; 52:2157.
69. Roy NK, Samuels L, Parr RG. J Chem Phys 1979; 70:3680.
70. Nalewajski RF, Koninski M. Z Naturforsch 1987; 42a:451.
71. Komorowski L. Chem Phys Lett 1987; 134:536.
72. Gázquez JL, Ortiz E. J Chem Phys 1984; 81:2741.
73. Pitanga P, Giambiagi M, De Giambiagi MS. Chem Phys Lett 1986; 128:411.
74. Parr RG, Szentpaly LV, Liu S. J Am Chem Soc 1999; 121:1922.
75. Parr RG, Yang W. J Am Chem Soc 1984; 106:4049; Ayers PW, Levy M. Theor Chem Acc 2000; 103:353–360.
76. Berkowitz M, Ghosh SK, Parr RG. J Am Chem Soc 1985; 107:6811.
77. Ghosh SK, Berkowitz M. J Chem Phys 1985; 83:2976.
78. Parr RG, Yang W. Annu Rev Phys Chem 1995; 46:107.
79. Fukui K. Theory of Orientation and Stereoselection. Berlin: Springer-Verlag 1975; Science 1987; 218:747; Fukui K, Yonezawa T, Shingu H. J Chem Phys 1952; 20:722; Fukui K, Yonezawa T, Nagata C, Shingu H. J Chem Phys 1954; 22:1433.
80. Yang W, Mortier WJ. J Am Chem Soc 1986; 108:5708.
81. Lee C, Yang W, Parr RG. J Mol Struct (Theochem) 1988; 163:305.
82. Yang W, Parr RG, Pucci R. J Chem Phys 1984; 81:2862.
83. Chattaraj PK, Cedillo A, Parr RG. J Chem Phys 1995; 103:7645.
84. Chattaraj PK, Cedillo A, Parr RG. J Chem Phys 1995; 103:10620.
85. Berkowitz M, Parr RG. J Chem Phys 1988; 88:2554.
86. Ghosh SK. Chem Phys Lett 1990; 172:77.
87. Harbola MK, Chattaraj PK, Parr RG. Isr J Chem 1991; 321:395.
88. Hohenberg P, Kohn W. Phys Rev B 1964; 136:864.
89. Pearson RG. Inorg Chem 1988; 27:734.
90. Lackner KS, Zweig G. Phys Rev D 1983; 28:1671.
91. Ray NK, Parr RG. J Chem Phys 1980; 73:1334.
92. Balasubramanian S, Rao KJ. Solid State Commun 1989; 71:979.
93. Nepala DA, Mckay JM. Physica C (Amsterdam) 1989; 158:65.
94. Ichikawa S. J Phys Chem 1989; 93:7302.
95. Luo QG, Wang RY. J Phys Chem Solids 1987; 48:425.
96. Ichikawa S. J Phys Chem Solids 1989; 50:931.
97. Asokamani R, Manjula R. Phys Rev B 1989; 39:4217.
98. Ghanty TK, Ghosh SK. J Mol Struct (Theochem) 1992; 274:83.
99. Shankar S, Parr RG. Proc Natl Acad Sci USA 1984; 82:264.
100. Alonso JA, Girifalco LA. Phys Rev B 1979; 19:3889.
101. March NH, Parr RG. Proc Natl Acad Sci USA 1980; 77:6285.
102. Gázquez JL, Vela A, Galvan M. Phys Rev Lett 1986; 56:2606.
103. Gázquez JL, Galván M, Ortiz E, Vek A. In: Erdahl R, Smith VH Jr, eds. Density Matrices and Density Functionals. Dordrecht: Reidel, 643–662.
104. Chattaraj PK, Mukherjee A, Das MP, Deb BM. Proc Indian Acad Sci 1986; 96:231; Chattaraj PK, Nath S. J Indian Chem Soc 1994; 71:111.
105. Parr RG. Int J Quantum Chem 1984; 26:687.
106. Palke WE. J Chem Phys 1980; 72:2511.
107. Guse MP. J Chem Phys 1981; 75:828.
108. Reed JL. J Phys Chem 1981; 85:148.
109. Politzer P, Weinstein H. J Chem Phys 1979; 71:4218.
110. Parr RG, Bartolotti LJ. J Am Chem Soc 1982; 104:3801.
111. Nalewajski RF. J Am Chem Soc 1984; 106:944; J Chem Phys 1983; 78:6112; J Phys Chem 1985; 89:2831; Nalewajski RF, Koninski M. J Phys Chem 1984; 88:6234.

112. Politzer P, Parr RG, Murphy DR. Phys Rev B 1985; 31:6809.
113. Boyd RG, Markus GE. J Chem Phys 1981; 75:5385.
114. Balbas LC, Alonso JA, Vega LV. Z Phys 1986; D1:215.
115. Deb BM, Singh R, Sukumar N. J Mol Struct (Theochem) 1992; 259:121.
116. Nath S, Bhattacharjee S, Chattaraj PK. J Mol Struct (Theochem) 1995; 331:267.
117. Fuentealba P, Parr RG. J Chem Phys 1991; 94:5559.
118. Balbas LC, Alonso JA, Las Heras E. Mol Phys 1983; 48:981.
119. Mortier WJ, Van Genechten KA, Gasteiger J. J Am Chem Soc 1985; 107:829; Gasteiger J, Marsili M. Tetrahedron 1980; 36:3219; Hutchings MG, Gasteiger J. Tetrahedron Lett 1983; 24:2541; Marsili M, Gasteiger J. Stud Phys Theo Chem 1981; 16:56; Mortier WJ, Ghosh SK, Shankar S. J Am Chem Soc 1986; 108:4315; Van Genechten KA, Mortier WJ, Geerlings P. J Chem Soc Chem Commun 1986; 1278; Van Genechten KA, Mortier WJ, Geerlings P. J Chem Phys 1987; 86:5063.
120. Bertaut F. J Phys Radium 1952; 13:499.
121. Ewald PP. Ann Phys 1921; 64:253.
122. Yang W, Lee C, Ghosh SK. J Phys Chem 1985; 89:5412.
123. Ghanty TK, Ghosh SK. J Phys Chem 1991; 95:6512.
124. Gutmann V, Resch G, Linert W. Cord Chem Rev 1982; 43:133–164; Gutmann V. The Donor–Acceptor Approach to Molecular Interactions. New York: Plenum, 1978.
125. Bamzai AS, Deb BM. Rev Mod Phys 1981; 53:95.
126. Ghosh SK, Parr RG. Theor Chim Acta 1987; 72:379.
127. Ghanty TK, Ghosh SK. Inorg Chem 1992; 31:1951; J Chem Soc Chem Commun 1992; 1:1502.
128. Ghosh SK. Int J Quantum Chem 1994; 49:239.
129. Wilmshurst JK. J Chem Phys 1957; 27:1129.
130. Wells PR. Prog Phys Org Chem 1968; 6:111.
131. Inamoto N, Masuda S. Tetrahedron Lett 1977; 37:3287; Chem Lett 1982; 1003:1007; Inamoto N, Masuda S, Tori K, Yoshimura Y. Tetrahedron Lett 1978; 46:4547.
132. Mullay J. J Am Chem Soc 1984; 106:5842; 1985; 107:7271.
133. Huheey JE. J Phys Chem 1966; 70:2086.
134. Bratsch SG. J Chem Educ 1985; 62:101.
135. Reed LH, Allen LC. J Phys Chem 1992; 96:157.
136. Reynolds WF. Prog Phys Org Chem 1983; 13:165; Reynolds WF, Taft WR, Topsom RD. Tetrahedron Lett 1982; 23:1055; Marriott S, Reynolds WF, Taft RW, Topsom RD. J Org Chem 1984; 49:959.
137. Nath S, Nandi PK, Sannigrahi AB, Chattaraj PK. J Mol Struct (Theochem) 1993; 279: 207.
138. De Proft F, Langenaeker W, Geerlings P. J Phys Chem 1993; 97:1826.
139. Datta D. Proc Indian Acad Sci (Chem Sci) 1988; 100:549; Datta D, Majumdar D. Proc Indian Acad Sci (Chem Sci) 1991; 103.777.
140. Luo YR, Benson SW. J Phys Chem 1989; 93:3306; Luo YR, Pacey PD. J Am Chem Soc 1991; 113:1465.
141. Slee TS. J Am Chem Soc 1986; 108:606.
142. Hati S, Datta D. J Comput Chem 1992; 13:912.
143. Bader RFW. Acc Chem Res 1985; 18:9; Bader RFW, Nguyen-Dang TT. Adv Quantum Chem 1981; 14:63; Bader RFW, Nguyen-Dang TT, Tal Y. Rep Prog Phys 1981; 44:893; Bader RFW, Essen H. J Chem Phys 1984; 80:1943.
144. Slee TS. J Am Chem Soc 1986; 108:7541.
145. Wiberg KB, Breneman CM. J Am Chem Soc 1990; 112:8765.
146. Luo YR, Pacey PD. J Phys Chem 1991; 95:6745.
147. Datta D. J Phys Chem 1992; 96:4209.
148. Chattaraj PK, Nath S, Sannigrahi AB. Chem Phys Lett 1993; 212:223.

149. Chattaraj PK, Nath S, Sannigrahi AB. J Phys Chem 1994; 98:9143; Cárdenas-Jirón GI, Toro-Labbé A. J Phys Chem 1995; 99:5325; Cárdenas-Jirón GI, Toro-Labbé A. J Phys Chem 1995; 99:12730; Gutiérrez-Oliva S, Letelier JR, Toro-Labbé A. J Phys Chem A 2000; 104:1557.

150. Gázquez JL, Martinez A, Méndez F. J Phys Chem 1993; 97:4059; Pal S, Roy R, Chandra AK. J Phys Chem 1994; 98:2314.

151. Nath S, Sannigrahi AB, Chattaraj PK. J Mol Struct Theochem 1994; 309:65.

152. Wulfsberg G. Inorganic Chemistry. University Science Books, CA: Viva Books Private Limited, New Delhi Indian Edition, 2002. Chapter 5 and references therein.

153. Yang W, Parr RG, Uytterhoeven L. Phys Chem Miner 1987; 15:191.

154. Politzer P. J Chem Phys 1987; 86:1072.

155. Komorowski L. Chem Phys 1987; 114:55; Komorowski L. Z Naturforsch 1987; 42a:767.

156. Ghanty TK, Ghosh SK. J Phys Chem 1993; 97:4951; Fuentealba P, Reyes O. J Mol Struct (Theochem) 1993; 101:65.

157. Chattaraj PK. Curr Sci 1991; 61:391.

158. Nalewajski RF, Korchowiek J, Zhou Z. Int J Quantum Chem Symp 1988; 22:349.

159. Ohno K. Theor Chim Acta 1968; 10:111; Ohno K. Adv Quantum Chem 1967; 3:239.

160. Nalewajski RF in Ref. 2.

161. Komorowski L, Lipinski J, Pyka ML. J Phys Chem 1993; 97:3166; Komorowski L, Lipinski J. Chem Phys 1991; 157:45; Komorowski L in Ref. 2.

162. Zhou Z, Parr RG, Garst JF. Tetrahedron Lett 1988; 29:4843; Zhou Z, Parr RG. J Am Chem Soc 1989; 111:7371.

163. Hess BA Jr, Schaad LJ. J Am Chem Soc 1971; 93:305.

164. Aihara IJ. J Am Chem Soc 1976; 98:2750; Gutman I, Milun M, Trinajstic N. J Am Chem Soc 1977; 99:1692.

165. Zhou Z, Parr RG. J Am Chem Soc 1990; 112:5720.

166. Parr RG, Chattaraj PK. J Am Chem Soc 1991; 113:1854; Chattaraj PK. Proc Indian Natl Sci Acad Part A 1996; 62:513 (Review); Chattaraj PK, Liu GH, Parr RG. Chem Phys Lett 1995; 237:171; Liu S, Parr RG. J Chem Phys 1997; 106:5578; Ayers PW, Parr RG. J Am Chem Soc 2000; 122:2010.

167. Parr RG, Gázquez JL. J Phys Chem 1993; 97:3939.

168. Parr RG, Liu S, Kugler AA, Nagy A. Phys Rev A 1995; 52:969.

169. Chattaraj PK, Lee H, Parr RG. J Am Chem Soc 1991; 113:1955; Chattaraj PK, Maiti B. J Am Chem Soc 2003; 125:2705.

170. Chattaraj PK, Nath S. Indian J Chem 1992; 31A:954.

171. Pearson RG, Palke WE. J Phys Chem 1992; 96:3283.

172. Makov G. J Phys Chem 1995; 99:9337.

173. Pal S, Vaval N, Roy R. J Phys Chem 1993; 97:4404.

174. Gopinathan MS, Siddarth P, Ravimohan C. Theor Chim Acta 1986; 70:303; Siddarth P, Gopinathan MS. Proc Indian Acad Sci (Chem Sci) 1987; 99:91; J Am Chem Soc 1988; 110:96.

175. Galván M, Pino AD Jr, Joannopoulos JD. Phys Rev Lett 1993; 70:21.

176. Hertel IV, ed. Z Phys D At Mol Clust 1991; 19; Hertel IV, ed. Z Phys D At Mol Clust 1991; p. 20.

177. de Heer WA, Knight WD, Chou MY, Cohen ML. Solid State Phys 1987; 40:93; Knight WD, Clemenger K, de Heer WA, Saunders WA, Chou MY, Cohen ML. Phys Rev Lett 1984; 52:2141.

178. Martin TP, Bergmann T, Gohlich H. Z Phys D 1991; 19:25.

179. Harbola MK. Proc Natl Acad Sci USA 1992; 89:1036.

180. Alonso JA, Balbas LC in Ref. 2.

181. Parr RG, Zhou Z. Acc Chem Res 1993; 26:256.

182. Gunnarsson O, Lundqvist BI. Phys Rev B 1976; 13:4274; Gunnarsson O, Jonson M, Lundqvist BI. Phys Rev B 1979; 20:3136.

183. Datta D. Inorg Chem 1992; 31:2797; Hati S, Datta D. J Org Chem 1992; 57:6056.

184. Chattaraj PK, Schleyer PvR. J Am Chem Soc 1994; 116:1067.

185. Pearson RG. In: Structure and Bonding, Chemical Hardness. Vol. 80. Chapter 1, 1–10, Politzer P. J Chem Phys 1987; 86:1072; Fuentealba P, Reyes O. J Mol Struct (Theochem) 1993; 282:65; Ghanty TK, Ghosh SK. J Phys Chem 1996; 100:12295.

186. Chattaraj PK, Sengupta S. J Phys Chem 1996; 100:16126; J Phys Chem A 1997; 42: 7893; Chattaraj PK, Poddar A. J Phys Chem A 1998; 102:9944; Chattaraj PK, Poddar A. J Phys Chem A 1999 103:1274; Ghanty TK, Ghosh SK. J Phys Chem 1996; 100: 12295.

187. Parr RG, Chattaraj PK. J Am Chem Soc 1991; 113:1854.

188. Chattaraj PK, Fuentealba P, Jaque P, Toro-Labbé A. J Phys Chem A 1999; 103: 9307.

189. Chattaraj PK, Poddar A. J Phys Chem A 1999; 103:8691; Fuentealba P, Simon-Manso Y, Chattaraj PK. J Phys Chem A 2000; 104:3185.

190. Chattaraj PK, Sengupta S. J Phys Chem A 1999; 103:6122.

191. Chattaraj PK, Fuentealba P, Gomez B, Contreras R. J Am Chem Soc 2000; 122:348.

192. Chattaraj PK, Cedillo A, Parr RG, Arnett EM. J Org Chem 1995; 60:4707.

193. Chattaraj PK, Maiti B. J Chem Educ 2001; 78:811.

194. Pearson RG. Acc Chem Res 1993; 26:250.

195. Nath S, Sannigrahi AB, Chattaraj PK. J Mol Struct (Theochem) 1994; 306:87.

196. Chattaraj PK, Perez P, Zevallos J, Toro-Labbé A. J Phys Chem A 2001; 105:4272; Chattaraj PK, Gómez B, Chamorro E, Santos J, Fuentealba P. J Phys Chem A 2001; 105:8815; Perez P, Toro-Labbé A, Contreras R. J Am Chem Soc 2001; 123:5527.

197. Falicov LM, Somorjai GA. Proc Natl Acad Sci USA 1985; 82:2207.

198. Mendez F, Gazquez JL. J Am Chem Soc 1994; 116:9298; Gazquez JL, Mendez F. J Phys Chem 1994; 98:4591; Li Y, Evans JNS. J Am Chem Soc 1995; 117:7756.

199. Pino AD Jr, Galvan M, Arias TA, Joannopoulos JD. J Chem Phys 1993; 98:1606.

200. Bacmann JJ, Papan AM, Peit M, Silvestre G. Philos Mag A 1985; 51:697; Rouviere Bourret JL, Penisson JM. Acta Crystallogr A 1988; 44:838.

201. Nguyen LT, Le TN, De Proft F, Chandra AK, Langenaeker W, Nguyen MT, Geerlings P. J Am Chem Soc 1999; 121:5992 and references therein; Pal S, Chandrakumar KRS. J Am Chem Soc 2000; 122:4145 and references therein; Perez P, Simon-Manso Y, Aizman A, Fuentealba P, Contreras R. J Am Chem Soc 2000; 122:4756 and references therein.

202. Chattaraj PK. J Phys Chem A 2001; 105:511.

203. Mendez F, Galvan M, Garritz A, Vela A, Gazquez JL. J Mol Struct (Theochem) 1992; 277:81.

204. Langenaeker W, Demel K, Geerlings P. J Mol Struct (Theochem) 1991; 234:329.

205. Galvan M, Vela A, Gazquez JL. J Phys Chem 1988; 92:6470.

206. Cioslowski J, Martinov M, Mixon ST. J Phys Chem 1993; 97:10948.

207. Geerlings P, De Proft F, Langenaeker W. Adv Quantum Chem 1999; 33:303; Geerlings P, De Proft F, Martin JML. In: Seminario J, ed. Theoretical and Computational Chemistry: Vol. 5. Recent Developments in Density Functional Theory 1996:773–809; Geerlings P, De Proft F, Langenaeker W. In: Springborg M, ed. Density Functional Methods: Applications in Chemistry and Materials Science. New York: Wiley, 1997, Chapter 2; Geerlings P, Langenaeker W, De Proft F. Murray JS, Sen KD, eds. A Baeten in Theoretical and Computational Chemistry: Vol. 4. Molecular Electrostatic Potentials—Concepts and Applications. Elsevier, 1996:587–617.

208. Maynard AT, Huang M, Rice WG, Covell DG. Proc Natl Acad Sci USA 1998; 95: 11578–11583; Huang M, Maynard A, Turpin JA, Graham L, Janini GM, Covell DG, Rice WG. J Med Chem 1998; 41:1371–1381; Turpin JA, Song Y, Inman JK, Huang M, Wallqvist A, Maynard A, Covell DG, Rice WG, Appella E. J Med Chem 1999; 42:67–86; Maynard AT, Covell DG. J Am Chem Soc 2001; 123:1047–1058.

209. Ireta J, Galvan M, Cho K, Joannopoulos JD. J Am Chem Soc 1998; 120:9771.

210. Cohen MH, Ganduglia-Pirovano MV, Kudrnovský J. J Chem Phys 1995; 103:3543;
      Cohen MH, Ganduglia-Pirovano MV, Kudrnovský J. J Chem Phys 1994; 101:8988;
      Baekelandt BG. J Chem Phys 1996; 105:4664; Ayers PW, Parr RG. J Am Chem Soc
      2001; 123:2007.
211. DeProft F, Liu S, Geerlings P. J Chem Phys 1998; 108:7549.
212. Chamorro E, Contreras R, Fuentealba P. J Chem Phys 2000; 113:10861; Chamorro E,
      Fuentealba P, Contreras R. J Chem Phys 2001; 115:6822.
213. Berkowitz M. J Am Chem Soc 1987; 109:4823.

12

Transition States and Transition Structures

ORLANDO ACEVEDO and JEFFREY D. EVANSECK

Duquesne University, Pittsburgh, Pennsylvania, U.S.A.

1. INTRODUCTION

The *dynamics* of short-lived molecular species, such as transition states ($>10^{-13}$ sec), are now possible to observe. In fact, Ahmed H. Zewail received the 1999 Noble Prize for his contributions to the development and application of femtosecond spectroscopy [1,2]. Despite the significant progress in reaching femtosecond resolution and separate work using kinetic isotope effects [3], the *structure* of the transition state has not been directly observed by experiment. To date, only calculations can provide information about the geometric arrangement of constituent atoms for short-lived molecular species such as transition structures [4,5]. Although the transition state is central to understanding chemical reactivity and predicting thermochemical properties, it is not directly computed by commonly utilized quantum chemical methods. Instead, the transition structure is computed as a stationary point on the potential energy surface (PES), and is fundamentally different from a transition state [4,6]. It is critical to appreciate the differences between the transition structure and transition state, because different information regarding chemical reactivity can be concluded. Briefly, the transition structure is the saddle point on the vibrationless potential energy surface, whereas the transition state is the dividing plane at the free energy maximum separating two free energy minima [7]. It is important to note that the transition state is temperature-dependent and involves an ensemble of structures. In comparison, the transition structure is the single set of atomic coordinates that belong to the highest point on the potential energy surface along the reaction coordinate. A discussion on transition states in the context of medicinal chemistry naturally leads to dialog on enzymatic catalysis. In particular, the transition state has been implicated as a key element in understanding enzymatic catalysis [8–16]. Tight binding of the transition structure alone is insufficient to account for the rate enhancements achieved by enzymes. Other factors have been recently proposed that involve preorganization and orientational effects, and the role of the enzyme protecting the substrate from solvent. Regardless of the point of view taken, it is clear that to gain a deeper appreciation of enzymatic catalysis, it is essential to understand the properties and structure of the transition structure. How enzymatic catalysis is actually achieved continues to be a matter of debate. The focus of this work is to shed light on the role of transition structures and transition states in enzymatic catalysis, by describing our progress in formulating a better understanding of how the immediate molecular environment surrounding the transition structure affects the rate and selectivity of the renowned Diels–Alder reaction [17].

2. CRITICAL ISSUES AND CONCEPTIONS

2.1. Transition Structures

To begin the process of understanding transition structures and their properties, it is necessary to build a conceptual model of the potential energy surface. Briefly, the PES describes how the potential energy of a molecular system varies with structural change. For a large molecule, the corresponding PES has a high dimensionality resulting from a large number of degrees of freedom. Consequently, the full PES of large molecules cannot be plotted nor visualized. However, PESs are typically constructed through the structural variation from a few degrees of freedom within the molecular system. For example, consider the PES of butane, which has 36 degrees of freedom. From a chemical viewpoint, the interest is in the transformation of the different conformations of butane, which principally arise from changes in 1 degree of freedom defined by the torsional angle along the carbon atoms. The energy as a function of this single torsional angle results in the familiar one-dimensional PES identifying the *gauche* and *anti* energy minima, along with two energy maxima separating the minima. The typical two-dimensional PES involves structural variation in 2 degrees of freedom to create a map of the energetic response, or a surface of the PES in the region of interest. For simplicity, consider a two-dimensional PES that has only two minima, as shown in Fig. 1.

The underpinnings of a PES reside in the position of its stationary points and the variation of potential energy in between each of them. Geometry optimization techniques are used to locate energy extrema on the PES [18,19]. Stationary points, as either minima or maxima, ensue when the first derivatives of the potential energy with respect to coordinate variation are equal to zero. The gradient of the potential energy is the negative of the net molecular force; thus the force on the molecular system is zero at stationary points. The process of energy minimization also removes the kinetic energy from the system. Consequently, the PES does not reflect the effects of temperature, which gives the potential energy as a function of geometric variation of the molecule at 0 K. Knowledge of the PES enables chemists to make thermodynamic and kinetic interpretations of the processes of interest.

The two minima shown in Fig. 1 may correspond to the reactants and products resulting from a chemical reaction, two stable conformations interconverting through isomerization, the binding of a ligand and receptor, or any number of scenarios where one species is separated from another by an energy barrier. Note that there are an

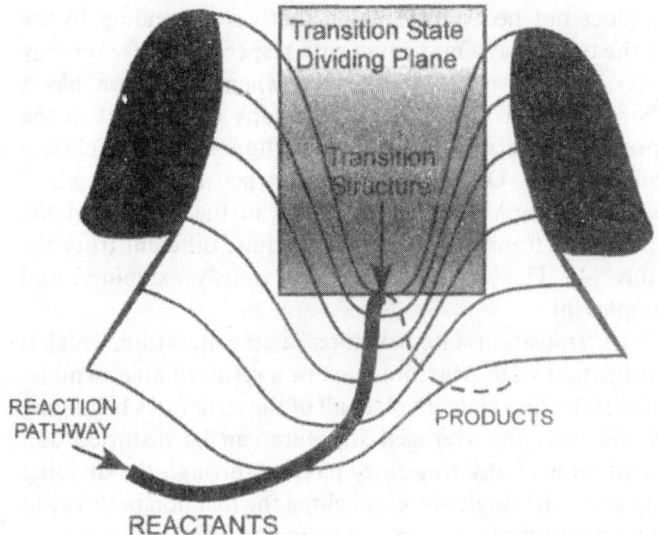

Figure 1 A schematic of a two-minimum potential energy surface.

infinite number of allowable and different pathways (also referred to as trajectories) that penetrate through the dividing surface in order to traverse between the reactants and products. However, there is only one lowest energy pathway interconnecting the two minima on the PES, which is known as the "reaction pathway." The "transition structure" is a single set of atomic coordinates that belong to the highest point on the PES along the reaction pathway. At the transition structure, the first derivatives of the potential energy with respect to the coordinates are zero. The energy is a maximum at the transition structure along the reaction pathway. However, for displacements in directions perpendicular to the reaction pathway, the energy is a minimum. This type of stationary point is known as a first-order saddle point, which is properly characterized by a vibrational analysis resulting in a single negative frequency or imaginary frequency. The vibration at the first-order saddle point corresponds to the nuclear displacements for interconversion between specific minimum.

2.2. Transition States

There is a fundamental difference between a transition state and transition structure, and the two continue to be misused in the literature. The transition state theory (TST) provides the necessary conceptual framework to understand the difference between the transition state and structure. A number of excellent reviews on TST are available and should be consulted for in-depth coverage of the material [2,6,7,20–23]. Briefly, classical TST replaces the rate constant by a one-way flux coefficient corresponding to the passage of trajectories through the free energy dividing plane that separates reactants from products. This dividing plane is the "transition state," which is also referred to as a hypersurface in phase space, as shown in Fig. 1. It is important to note that the transition state involves an ensemble of structures belonging to the dividing plane.

The transition state is situated at the free energy maximum along the reaction pathway that connects the two free energy minima. The position of the transition state along the reaction pathway does not necessarily match that corresponding to the transition structure, because the two are at a maximum with respect to the free energy and potential energy surfaces, respectively. Qualitatively, when a reaction has a relatively large activation barrier and a slowly varying entropy component in the region of the maximum potential energy, then the transition state will closely correspond to the transition structure. On the other hand, when a reaction has a relatively low activation barrier or rapidly varying entropy in the vicinity of the potential energy maximum, then the transition state may be quite different from the computed transition structure [4]. This issue has been previously examined and reported for illustrative examples [6].

Reference is often made to "transition state structures" in the literature, which is a source of confusion. The transition state structure must be a result of an ensemble-averaged quantity, because the transition state involves all of the structures belonging to the dividing plane. Consequently, the averaged structure can be distorted and potentially be a poor representation of any trajectory passing through the dividing plane. Alternatively, one may select the single structure along the reaction pathway at the free energy maximum to represent the transition state structure. This is also a poor selection, because only one structure is used to represent all of the structures from the dividing plane that are used to determine the free energy maximum. In either case, in our opinion, the term "transition state structure" is inappropriate and should not be used. Direct structural comparison between a transition structure and transition state cannot be made, because they are different entities.

2.3. Thermodynamic Quantities

Thermochemical quantities maybe extracted from a PES in terms of internal energy, E, enthalpy, H, or Gibbs energy free energy, G. The procedure described here is based on the implementation in the Gaussian software package [24]. By default, thermochemistry analysis is carried out at 298.15 K and 1 atm of pressure, which is easily modified [25,26]. The equations used to compute the thermochemical data from a PES can be found in most standard thermodynamic textbooks. In this specific implementation, there are two important assumptions that induce error into the computed thermochemical quantities. First, the equations used assume noninteracting particles, which is tantamount to an ideal gas treatment. Second, it is also assumed that the first and higher excited states are inaccessible. Both assumptions have the potential to introduce error into the computed thermochemical quantities, but, in general, do not impact the majority of molecular systems of interest.

The stationary points obtained by computational procedures on the PES are for vibrationless molecular systems. The electronic Hamiltonian used in ab initio calculations gives the total electronic energy, E_{elec}. A real molecule, however, has vibrational energy even at 0 K, which is the quantum mechanical (QM) zero-point energy (ZPE), $\frac{1}{2}h\nu$. At absolute zero, the internal energy, E_0, is defined as the computed electronic energy plus the zero-point energy.

$$E_0 = E_{\text{elec}} + \text{ZPE} \tag{1}$$

Thermal corrections to the computed internal energy are necessary to obtain energies at temperatures which are directly comparable to experimental conditions [27]. The thermal corrections to the internal energy are determined by contributions from the translational (q_t), electronic (q_e), rotational (q_r), and vibrational (q_v) partition functions. The internal thermal energy correction term is the sum of all four contributions. It is important to realize that the thermal energy corrections already include the zero-point energy through the vibrational partition fraction contribution, so it is not added redundantly.

$$E_{\text{thermal}}^{\text{temp}} = E_t + E_r + E_v + E_e \tag{2}$$

The internal energy at a specific temperature is then determined by the sum of the internal thermal energy correction term and the computed electronic energy. Take 298 K, as an example.

$$E_{298} = E_{\text{thermal}}^{298} + E_{\text{elec}} \tag{3}$$

Absolute thermodynamic quantities are difficult to compute accurately and are rarely reported in computational chemistry [28–31]. Rather, differences in the thermochemical quantities are used to improve the accuracy and agreement with experiment. Two thermodynamic quantities which are of common interest are defined in Fig. 2.

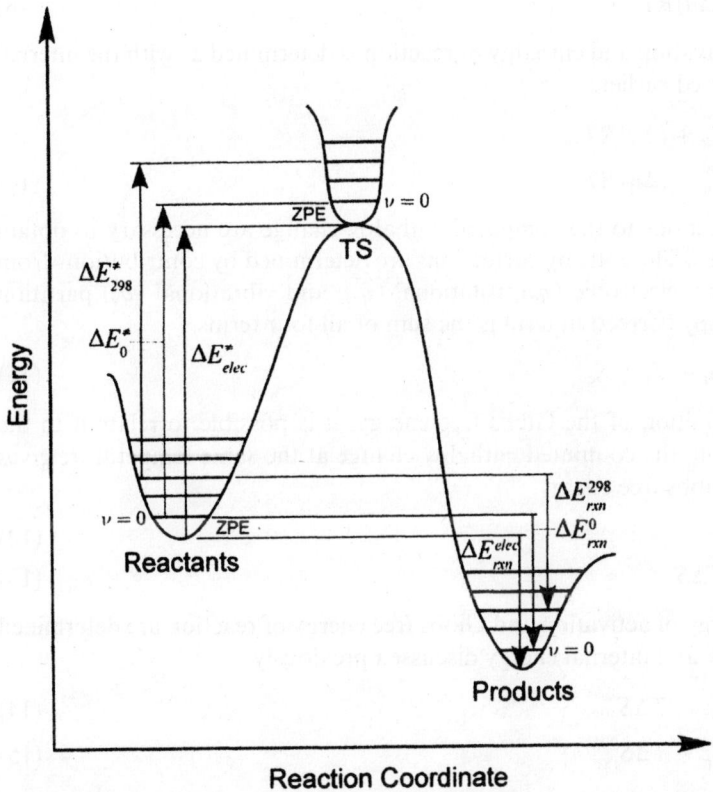

Figure 2 Compound thermodynamic quantities.

The first involves the activation energy, $\Delta E_{298}^{\neq}$, which is the internal energy difference between the transition structure and the ground state reactant at a specified temperature.

$$\Delta E_{298}^{\neq} = E_{298}^{TS} - E_{298}^{react} \tag{4}$$

The second quantity typically referred to is the energy of reaction ΔE_{rxn}^{298}, which is the energy of the products minus the energy of the reactants.

$$\Delta E_{rxn}^{298} = E_{298}^{product} - E_{298}^{react} \tag{5}$$

It is possible to use the standard thermodynamic equations to convert the thermally corrected internal energy into the enthalpies and Gibbs free energies at the same temperature. To determine the enthalpy, the thermodynamic definition of enthalpy is used.

$$H = E + PV \tag{6}$$

At constant pressure and temperature, and using the ideal gas approximation, the enthalpy difference can be related to the difference in internal energy. Therefore, the change in enthalpy will be computed correctly when the number of moles of gas changes during the course of the reaction.

$$\Delta H = \Delta E + P\Delta V \tag{7}$$

$$\Delta H = \Delta E + (\Delta n)RT \tag{8}$$

The enthalpy of activation and enthalpy of reaction is determined as with the internal energy terms discussed earlier.

$$\Delta H_{298}^{\neq} = \Delta E_{298}^{\neq} + (\Delta n)RT \tag{9}$$

$$\Delta H_{rxn}^{298} = \Delta E_{rxn}^{298} + (\Delta n)RT \tag{10}$$

Entropy corrections to the computed enthalpy change are necessary to obtain the Gibbs free energy. The entropy corrections are determined by contributions from the translational (q_t), electronic (q_e), rotational (q_r), and vibrational (q_v) partition fractions. The entropy correction term is the sum of all four terms.

$$S_{total} = S_t + S_r + S_v + S_e \tag{11}$$

By utilizing the definition of the Gibbs free energy, it is possible to relate it to the enthalpy. In addition, the computed enthalpy change at the same temperature gives the change in the Gibbs free energy.

$$G = H - TS \tag{12}$$

$$\Delta G = \Delta H - T\Delta S \tag{13}$$

The Gibbs free energy of activation and Gibbs free energy of reaction are determined as with the enthalpy and internal energy discussed previously.

$$\Delta G_{298}^{\neq} = \Delta H_{298}^{\neq} - T\Delta S_{298}^{\neq} \tag{14}$$

$$\Delta G_{rxn}^{298} = \Delta H_{rxn}^{298} - T\Delta S_{rxn}^{298} \tag{15}$$

In a Gaussian vibrational analysis output, as modified and shown in Table 1, several important pieces of information are printed to carry out the proper corrections for thermochemical quantities at stationary points. Gaussian provides absolute ther-

Table 1 Example of Gaussian Output. The Different
Thermodynamic Quantities

Sum of electronic and zero-point energies, $E_{elec} + ZPE = E_0$
Sum of electronic and thermal energies, $E_{elec} + E_{thermal}^{298} = E_{298}$
Sum of electronic and thermal enthalpies, $E_{elec} + H_{thermal}^{298} = H_{298}$
Sum of electronic and thermal Gibbs free energies, $E_{elec} + G_{thermal}^{298} = G_{298}$

mochemical quantities, which follow from the discussion above. The units are in hartrees.

3. COMPUTATIONAL APPROACHES

3.1. The Art of Locating Transition Structures

Methods for locating transition structures and energy minima on a PES are closely related, because each is a stationary point [4–6,32,33]. As a general rule, it is typically more difficult to locate a transition structure than to find minimum points [6]. Unlike optimization techniques guaranteed to lower a functional value, like the steepest descent or more sophisticated methods, making it possible to locate a minimum, there are no general methods that are certain to find transition structures [5]. There are three strategies for finding energy maximum. These strategies include interpolation between reactant and product [34], systematic variation through a selected reaction coordinate, and employing local information [5,32]. The latter technique, used in a number of research laboratories, requires good approximations to the three-dimensional structure, wave function, and vibrations to determine the transition structure, as described below.

First, it is necessary to determine a three-dimensional geometry near the transition structure. Appropriate starting points are based on good chemical intuition or previous experience with similar types of chemical reactions. An approximate transition structure can be achieved by selecting a few internal coordinates that participate actively in the reaction, such as breaking or forming bonds. The selected internal coordinate values are fixed to those expected, while optimizing the remaining degrees of freedom in the system. For example, in the Diels–Alder reaction, the transition structure for the reaction between butadiene and acrolein can be found by holding the bonds between atoms C1–C5 and C4–C6 fixed at 2.2 Å, and optimizing the remaining degrees of freedom, as shown in Fig. 3.

The distances are known to have this characteristic length for these types of pericyclic reactions [4]. It is critical that the few selected variables make strong contributions to the transition structure for this technique to be successful. As soon as the constrained minimum has been found, all constraints should be released followed by a frequency analysis. The frequency analysis gives a wave function and second derivatives in the three-dimensional approximation to the transition structure. A full transition structure search should then be conducted using the computed structure, wave function and Hessian to determine the true stationary point. Fig. 3 illustrates the transition structure for the gas phase reaction of butadiene and acrolein, which has been verified by a final vibrational analysis resulting in a single imaginary frequency. Table 2 gives a portion of the Gaussian output after a frequency analysis,

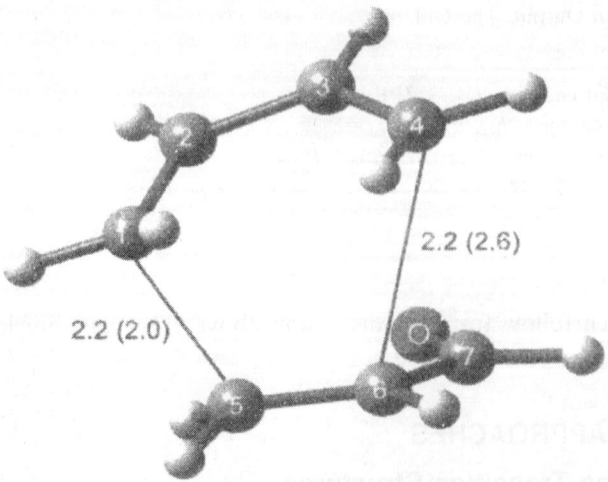

Figure 3 The starting and energy optimized structure (in parentheses) of the reaction between butadiene and acrolein. Distances are measured in angstroms.

showing the lowest three vibrations, including the imaginary frequency at 438 cm^{-1} and atomic displacements for each atom (not shown).

3.2. Computational Approximations

Advances in computational chemistry allow for the determination of stationary points by various approximations to the Schrödinger equation [4,35–43]. Complete discussions and excellent reviews of the different methods can be found in the literature [6,33,44,45]. Over the years, the Diels–Alder reaction between 1,3-butadiene and ethylene has become a prototype reaction to evaluate the accuracy of many different levels of theory. A "level of theory" involves the specific combination of a computational method and basis set. For example, the RHF/3-21G level of theory involves the restricted Hartree–Fock method with the 3-21G basis set. Ken Houk and his research group have pioneered many ideas concerning the fundamental ideas of pericyclic reactions by combining theory and experiment [3,4,37,38,46–48]. For the Diels–Alder

Table 2 Example of Gaussian Output. Single Imaginary Frequency for a TS

	(1)	(2)	(3)
Frequencies	−438.4513	64.8746	76.2705
Red. masses	9.7288	4.3238	3.6629
Frc consts	1.1019	0.0107	0.0126
IR intensity	1.6793	1.4096	0.6294
Raman activity	0.0000	0.0000	0.0000
Depolarization	0.0000	0.0000	0.0000

Imaginary frequencies (negative signs), harmonic frequencies (cm^{-1}), IR intensities (KM/mol), Raman scattering activities (Å/AMU), Raman depolarization ratios, reduced masses (AMU), force constants (mDyn/A), and normal coordinates.

reaction, results from various methods and basis sets (levels of theory) are considered, many of which are summarized in Table 3, to compare with the experimental activation energies ($\Delta E^{\neq}$) and energies of reaction (ΔE_{rxn}).

Two distinct observations from the computations emerge. First, the activation energies are sensitive to the level of theory chosen for the computations. The computed activation energies range from 47.4 to 20.0 kcal/mol, compared with 26.5 ± 2.0 kcal/mol found by experiment [4,53]. Restricted Hartree–Fock (RHF) activation energies are too high, and second-, third- and fourth-order Møller–Plesset values are too low, reinforcing the need to include electron correlation in energy evaluations. In addition, the complete active-space SCF method (CASSCF) overestimates the activation energies, illustrating well-known issues with the selection of the active space orbitals [54]. Gradient-corrected, nonlocal density functional methods, such as the B3LYP (Becke [55] three-parameter exchange functional and the nonlocal correlation functional of Lee et al. [56]), have gained popularity because of computer time efficiency and results that closely match experimental data [37], despite the lack of dispersion forces. Truncated configuration interaction calculations, such as the quadratic configuration interaction method QCISD(T), provide reasonable energies of concerted and stepwise pathways of Diels–Alder activation energies [35,52]. The second observation is that key geometric parameters are relatively insensitive to the level of theory used for the computations. For example, the breaking/forming bond lengths between the C1 and C5 atoms fall within the narrow range of 2.201–2.292 Å, as shown in Table 3. These results suggest that the *geometries* of pericyclic transition structures can be calculated at lower levels of theory, such as RHF/3-21G. More demanding methods that incorporate the effects of electron correlation, such as Møller–Plesset and density functional theory, can be used to evaluate the activation *energies* of the reactions.

Because of the resources required and the limitations on computer size and speed, quantum mechanical (QM) methods typically treat a relatively small number of atoms. The high dependency of the computer time on the number of basis functions makes it

Table 3 Transition Structure Geometries, Activation Energies, and Reaction Energies of the Concerted Diels–Alder Reaction of 1,3-Butadiene and Ethylene

Method	R_{C1-C5}	$\phi_{C2-C1-C5}$	$\Delta E^{\neq}$	ΔE_{rxn}	Ref.
RHF/3-21G	2.210	101.6	35.9	−43.1	[49]
RHF/6-31G*	2.201	102.6	47.4	−36.0	[40,48]
MP2/6-31G*	2.286	101.6	20.0	−45.9	[40,48]
MP4SDTQ/6-31G*[a]	2.286	101.6	22.4	−47.5	[50]
BLYP/6-31G*	2.292	102.6	22.8	29.8	[51]
B3LYP/6-31G*	2.273	102.4	24.8	−36.3	[40]
CASSCF/3-21G	2.217	101.9	37.3		[52]
CASSCF/6-31G*	2.223	103.1	40.7		[52]
UQCISD(T)/6-31G*[b]	2.223	103.1	29.4		[52]
RQCISD(T)/6-31G*[b]	2.223	103.1	25.5		[52]
Experimental			26.5 ± 2.0	−38.4	[4,53]

Energies are reported in kcal/mol, distances in angstroms, and angles in degrees.
[a] Single point on the MP2/6-31G* optimized geometries.
[b] Single point on the CASSCF/6-31G* optimized geometries.
Source: Refs. 4 and 37.

impossible to treat large systems, such as enzymes involving hundreds to thousands of heavy atoms with ab initio techniques, as listed in Table 3. For example, molecular systems with more than six to seven heavy atoms generally cannot be studied at high levels of theory using CCSD(T) (coupled cluster theory, including the coupling between singles and doubles and singles and triples) or CISD (configuration interaction singles and doubles) with large basis sets [6]. Consequently, theoretical studies of large molecules are approximated by smaller model systems using high levels of theory or low levels of theoretical treatments on the full system. Both approaches have obvious limitations and disadvantages. In particular, the truncated model may miss essential electronic or steric influences from missing parts of the real system. Application of a low-level theory—for example, AM1 or PM3—on the full system may give an inaccurate description of the electronics or structure.

Classical computations that use molecular mechanical (MM) force fields have been extremely successful in providing valuable and meaningful insights about intermolecular interactions for large biomolecules in solution [57]. The two independent computational methods, QM and MM, have been combined (known as QM/MM methods) over the last decade to formulate an alternative approach to the ubiquitous size and time-scale problem in computational chemistry. Excellent reviews are available [58–61]. The QM/MM methods have been widely used to study the chemical reactivity in large molecular systems. Even so, applying these methods to large systems studying conformational transitions is often difficult. Many factors contribute, but one important assumption is that there is only one saddle point between any initial and final states. However, in a molecule with a complex energy landscape, there may be a multitude of transition structures and minima.

Another problem lies in the region between QM and MM. Considerable debate has arisen when the partitioning between the two regions cuts across covalent bonds [58,62,63]. The most common method in modeling this middle space is to use the "link atom" method. It consists of adding QM hydrogen atoms in order to fill the free valencies of the QM atoms that are connected to the atoms described by MM. These dummy atoms are explicitly treated during the QM calculations but do not interact with the MM atoms. Whether or not these link atoms should interact via Coulombic interactions is still open to debate.

The QM/MM method has been used to model the Diels–Alder reaction in recent papers [64,65]. The Diels–Alder reaction between the diene cyclopentadiene and dieneophiles, methyl vinyl ketone and isoprene, was carried out in aqueous solution by Furlani and Gao. By employing simulations that combined AM1/TIP3P potential, transition-state stabilization was attributed entirely to hydrophobic effects [64]. The QM/MM method was also carried out with higher-level methods. In the work of Nendel and coworkers [65], the B3LYP/6-3IG* concerted transition structure of 1,2-cyclohexadiene and furan was substituted with bornyl ester, and the ester was optimized with MM2. The induction by a chiral substituent was generally overestimated, seen in similar QM/MM studies [47], hence the relative energy differences were so small that they were no longer chemically significant. Therefore the QM/MM method predicts no stereoselectivity for the 1-bornyl cyclohexadiene carboxylate cycloadditions.

3.3. Assessment of Computed Results

Derivatives are required for energy minimizations [6]. The direction of the first derivative of the energy indicates where the minimum lies, while the second derivative gives the curvature of the function. The first derivative is used to lower the energy of the system by moving each atom in response to the forces acting on it. The second derivative predicts where the function is either a minimum (positive curvature, normal vibration) or maximum (negative curvature, imaginary frequency). The Taylor series expansion of the energy, $V(x)$, about the point x_0 can be used to understand how stationary points are classified. Approximations are commonly made to this series expansion.

$$V(x) = V(x_0) + (x - x_0)V'(x_0) + (x - x_0)^2 V''(x_0)/2 + \cdots \tag{16}$$

Consider how vibrational analysis must be carried out on stationary points, where the first derivatives, $V'(x_0)$, equal zero and the potential at $V(x_0)$ is assumed to be zero. By ignoring higher-order terms, the harmonic approximation results.

$$V(x) = \frac{1}{2} V''(x_0)\Delta(x - x_0)^2 \tag{17}$$

This representation gives a useful way of understanding how the second derivative is related to the force constant (curvature) and how to assign the type of stationary point (sign).

The number of negative eigenvalues resulting after vibrational analysis on the Hessian matrix distinguishes the different types of stationary points. For a ground state, all of the eigenvalues correspond to real vibrations of the molecule, thus are all positive quantities and can be observed by experiment. The transition structure is special because it has a single negative eigenvalue describing the vibration at the transition structure. Of interest is the first-order saddle point, where energy passes through a maximum for movement along the reaction pathway that connects two minima, but is a minimum for displacements in all other directions perpendicular to the path. It is vital to check that the Hessian matrix at any proposed saddle point has the required single negative eigenvalue. The vibration has an imaginary or negative frequency, because, as the molecules move along the reaction path, the energy decreases as the reactants and products are approached. The curvature of such vibrational behavior is negative. Thus the name of negative frequency is commonly used. This is opposite to the energetic behavior of ground state molecules, where displacements from the equilibrium position result in energetic increases. The relationship between the frequency of vibration and force constant is well known. The vibrational frequency, v, is proportional to the square root of the force constant, k, divided by the reduced mass, μ.

$$v = (2\pi)^{-1}(k/\mu)^{1/2} \tag{18}$$

Transition structure calculations provide geometrical and electronic structure details. Determining the geometry of a transition structure and its vibrational force constants allows for the calculation of kinetic isotope effects by applying the equation of Bigelesien and Goeppert-Mayer [66,67]. This is dependent on the Born–Oppenheimer (BO) approximation, which underlies most of the theoretical considerations of isotope effects. The BO approximation allows for the quantum mechanics of molecules to be separated into the electronic (nuclei fixed) and nuclear motion. For the electronic motion, the nuclei only contribute their charge. For the nuclear motion, the nuclei's potential energy is the electronic energy as a function of nuclear configuration. This potential energy is independent of their isotopic masses; however, the kinetic energy expression does contain the masses of the nuclei. Hence the theoretical study of isotope effects on the molecular properties can show how different masses affect motion on the same potential surface [67]. Most importantly, the results can then be compared with the experimentally observed values. In the work of Singleton et al. [3], kinetic isotope effects were determined for all positions on isoprene with methyl vinyl ketone, ethyl acrylate, and acrolein catalyzed by Et_2AlCl. The results supported a highly asynchronous, concerted [4 + 2] transition structure concurrent with calculated values. This method is one of the few means for studying the details of the potential surfaces experimentally. The magnitude of a secondary kinetic isotope effect usually increases as the transition structure changes from reactant-like to product-like [68].

4. PERTURBING THE TRANSITION STATE ENVIRONMENT

4.1. The Diels–Alder Reaction

The Diels–Alder reaction is one of the most powerful carbon–carbon bond forming processes in organic synthesis [69]. Considerable experimental work has been carried out to improve the rate as well as the selectivity of Diels–Alder reactions [69]. Theoretical work in understanding this important reaction is relatively small compared to the huge amount of available experimental data (see references in Ref. 17). As a result, the Diels–Alder reaction is well studied, but not completely understood. From our research efforts accumulated over the last few years, we summarize the differences discovered between the computed transition structures of the Diels–Alder reaction in vacuum, microsolvated environments, and fully solvated systems for one of the simplest Diels–Alder reactions, acrolein, and s-cis butadiene, as schematically illustrated in Fig. 4. Molecular origins leading to the rate enhancement and selectivities are discussed, and then are related to the issues surrounding enzymatic catalysis.

We have recently reported a detailed discussion on solvation effects in this particular reaction [17]. Briefly, the experimental activation energy value at 298 K in the gas phase has been reported to be 19.7 kcal/mol [70]. In toluene, the experimental activation enthalpy was reported at 15.8 ± 1.4 kcal/mol, with an activation entropy of −38 ± 4 cal/mol K [71]. Four possible reaction pathways are possible for the acrolein and s-cis butadiene reaction. Consistent with previous conventions [17,72,73], the transition structures are denoted as NC (*endo, s-cis* acrolein), XC (*exo, s-cis* acrolein), NT (*endo, s-trans* acrolein), and XT (*exo, s-trans* acrolein), as illustrated for the parent reaction in vacuum in Fig. 5.

All ab initio and solvation model calculations were carried out with the Gaussian 98 program [24], by using a 16-node SP IBM RS/6000 supercomputer [74], The discrete, continuum, and discrete-continuum solvation models are used to approximate the effect of solvent on the butadiene and acrolein Diels–Alder reaction. The

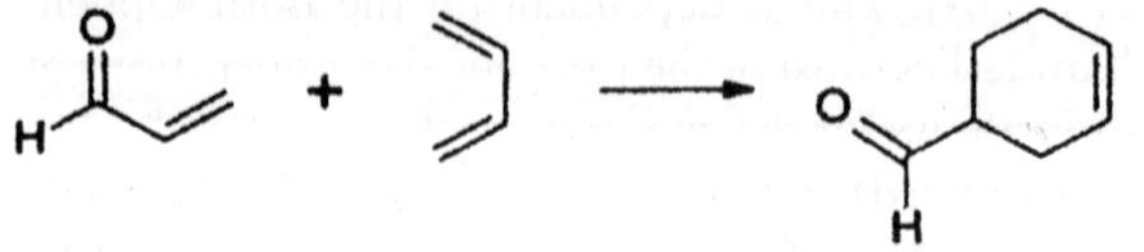

Figure 4 A schematic of the reaction between butadiene and acrolein.

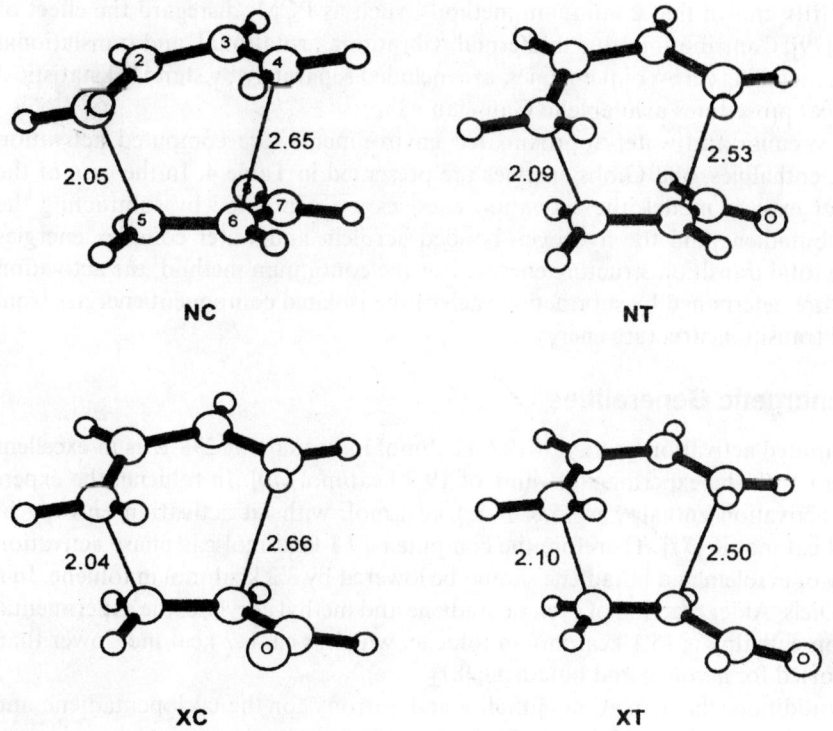

Figure 5 Four different reaction pathways for the Diels–Alder reaction between butadiene and acrolein. The distances are reported in angstroms.

discrete model includes explicit water molecules in the quantum mechanical calculations to satisfy local hydrogen bonding. The Becke three-parameter exchange functional [55] and the nonlocal correlation functional of Lee et al. [56] (B3LYP) with 6-31G* basis set [75] have been employed.

All energy optimizations, frequency analyses, and solvation computations were carried out by using the B3LYP/6-31G* level of theory, which has been shown to produce realistic structures and energies for pericyclic reactions [37,40], and properly treat the butadiene and acrolein Diels–Alder gas phase reaction [17,72,76]. Vibrational frequency calculations at the same level of theory were used to confirm all stationary points as either minima or transition structures, and provide thermodynamic and zero-point energy corrections.

The polarizable continuum model (PCM) by Tomasi and coworkers [77–79] was selected to describe the effects of solvent, because it was used to successfully investigate the effect of solvent upon the energetics and equilibria of other small molecular systems. The PCM method has been described in detail [80]. The solvents and dielectric constants used were benzene ($\varepsilon = 2.25$), methylene chloride ($\varepsilon = 8.93$), methanol ($\varepsilon = 32.0$), and water ($\varepsilon = 78.4$). Full geometry optimizations were carried out for the discrete and PCM models. To simultaneously account for localized hydrogen bonding and bulk solvation effects, PCM single-point energy calculations have been conducted on stationary points of the acrolein and butadiene reaction with two waters explicitly defined. It is known that continuum methods, such as PCM, disregard the effect of entropy [79]. Contributions due to thermal, vibrational, rotational, and translational motions, including zero-point energies, are included separately by standard statistical mechanical procedures available in Gaussian 98.

In vacuum and water-approximated environments, the computed activation energies, enthalpies, and Gibbs energies are presented in Table 4. In the case of the two-water explicit model, the activation energies are computed by subtracting the isolated butadiene and the hydrogen-bonded acrolein and water complex energies from the total transition structure energy. For the continuum method, the activation energies are determined by subtracting each of the isolated component energies from the total transition structure energy.

4.2. Energetic Generalities

The computed activation energy of 19.6 kcal/mol in vacuum at 298 K is in excellent agreement with the experimental value of 19.7 kcal/mol [70]. In toluene, the experimental activation enthalpy is 15.8 ± 1.4 kcal/mol, with an activation entropy of −38 ± 4 cal/mol K [71]. Therefore the computed 19.1 kcal/mol gas phase activation enthalpy of acrolein and butadiene should be lowered by 3.3 kcal/mol in toluene. In a related Diels–Alder reaction of cyclopentadiene and methyl acrylate, the experimental activation enthalpy is 15.1 kcal/mol in toluene, which is ca. 0.7 kcal/mol lower than that reported for acrolein and butadiene [81].

In addition, the activation enthalpy and entropy for the cyclopentadiene and methyl acrylate reaction is 10.2 and −40.9 cal/mol K, respectively, in a methanol/water mixture [81]. Therefore, analogous to the cyclopentadiene and methyl acrylate reaction, the butadiene and acrolein activation enthalpy (adjusted by 0.7 kcal/mol)

Table 4 Activation Energies, Enthalpies, and Gibbs Energies (kcal/mol) of the Reaction Between 1,3-Butadiene and Acrolein in Vacuum, PCM, and Explicit Water Using the B3LYP/6-31G* Level of Theory

TS	$\Delta E_0^{\ddagger}$	$\Delta E_{298}^{\ddagger}$	$\Delta H_{298}^{\ddagger}$	$\Delta G_{298}^{\ddagger}$
Vacuum				
NC	20.1	19.6	19.1	32.2
NT	21.4	20.9	20.3	33.5
XC	20.2	19.7	19.2	32.2
XT	22.0	21.6	21.0	34.1
PCM[a]				
NCP0	16.4	15.9	15.3	28.9
NTP0	17.1	16.6	16.0	29.6
XCP0	17.5	17.0	16.4	30.0
XTP0	18.3	17.7	17.1	31.0
Two water				
NC2W	16.0	15.5	14.9	28.8
NT2W	18.1	17.8	17.2	30.3
XC2W	17.1	16.6	16.0	29.7
XT2W	18.9	18.5	17.9	31.2

[a] The dielectric constant of water 78.39 was used.

is expected to be ca. 10.9 kcal/mol in a methanol/water mixture, which is 8.2 kcal/mol lower than the 19.1 kcal/mol gas phase activation enthalpy. Five different models were used to assess the influence of solvent. The vacuum is not signified, whereas the solvent models are indicated by a two-letter code to reflect the single- (1W), two- (2W), and three-water (3W) explicit models; the PCM (P0) continuum model; and the combined discrete-continuum strategy of two waters with PCM (P2).

An activation enthalpy decreasing with increasing solvent polarity is provided by the PCM method, as shown in Tables 4 and 5. The gas phase NC activation enthalpies decrease by 1.5 and 3.8 kcal/mol in benzene and water, respectively. In benzene (dielectric similar to toluene), the computed activation enthalpy is 17.6 or 1.5 kcal/mol lower than the gas phase activation enthalpy.

Compared to the experimental observation of 15.8 ± 1.4 kcal/mol in toluene [71], the PCM method recovers ca. 45% [1.5 kcal/mol (computed)/3.3 kcal/mol (expected)] of the experimental solvation effect.

In aqueous solution, the PCM method computes an activation enthalpy of 15.3 kcal/mol, which is 46% [3.8 kcal/mol (computed)/8.2 kcal/mol (expected)] of the estimated reduction to a 10.9 kcal/mol activation barrier. In a second comparison, the second-order rate constants between cyclopentadiene and alkyl vinyl ketones have been reported to be 740 times larger in water than in *n*-octane [82]. Overall, the PCM model lowers the Gibbs activation energies of the NC reaction from benzene to water by 2.0 kcal/mol, which correspond to an aqueous rate increase of ca. 30 times. Therefore despite mirroring the experimental trend in activation barrier reduction with increasing solvent polarity, the PCM method does not fully account for the observed rate of aqueous acceleration.

Compared to vacuum, two explicit waters lower the activation enthalpy by 4.2 kcal/mol. Thus in the two-water explicit model, an activation enthalpy of 14.9 kcal/

Table 5 Activation Energies, Enthalpies, and Gibbs Energies (kcal/mol) of the Reaction Between 1,3-Butadiene and Acrolein Computed Using PCM and the B3LYP/6-31G* Level of Theory in Different Solvents

TS	$\Delta E_0^{\ddagger}$	$\Delta E_{298}^{\ddagger}$	$\Delta H_{298}^{\ddagger}$	$\Delta G_{298}^{\ddagger}$
Benzene				
NCP0	18.7	18.2	17.6	30.9
NTP0	19.9	19.4	18.8	32.1
XCP0	19.1	18.7	18.1	31.1
XTP0	20.6	20.2	19.6	32.6
Methylene chloride				
NCP0	18.7	18.1	17.6	31.0
NTP0	19.3	18.9	18.3	31.4
XCP0	19.1	18.7	18.1	31.2
XTP0	20.4	19.9	19.3	32.4
Methanol				
NCP0	17.5	17.1	16.5	30.0
NTP0	18.2	17.7	17.2	30.8
XCP0	18.3	17.9	17.3	30.5
XTP0	19.2	18.8	18.2	31.6

mol is computed, which is 51% [4.2 kcal/mol (computed)/8.2 kcal/mol (expected)] of the estimated activation barrier reduction to 10.9 kcal/mol. In the absence of including all the water explicitly, the computations indicate that the explicit two-water model represents a majority of the possible *local* microscopic phenomena through hydrogen bonding. Consequently, both the implicit and discrete models account for approximately one-half of the observed rate acceleration caused by the aqueous phase.

The PCM and discrete water models are fundamentally different in the description of solvation effects [80]. The question arises if the PCM and discrete water models account for the same portion of the observed acceleration. Thus a combined discrete-continuum (P2) strategy using both PCM and explicit waters was employed. Full transition structure searches using PCM and two explicit waters proved to be problematic, because stationary points could not be located. Thus single-point energy evaluations were carried out on the two-water transition structures using PCM. The results are given in Table 6. To simulate the methanol/water mixture, P2 single-point energy evaluations in methanol were also carried out. Because single-point evaluations are not thermally corrected, the expected activation enthalpy decrease of 8.2 kcal/mol must be adjusted. The average difference between computed $\Delta E_0^{\ddagger}$ and $\Delta H_{298}^{\ddagger}$ values in Table 4 is 1.0 ± 0.2 kcal/mol. Therefore the expected aqueous phase activation energy for the acrolein and butadiene reaction is assumed to be 1 kcal/mol greater than the expected activation enthalpy (8.2 kcal/mol), or 9.2 kcal/mol. The P2 computed NC activation energy lowering of $20.1 - 12.3 = 7.8$ kcal/mol (water) and $20.1 - 11.5 = 8.6$ kcal/mol (methanol/water) recovers 85% and 93% of the expected 9.2 kcal/mol activation energy reduction, respectively [81]. The results indicate that the catalytic effect of water is manifested by a combination of both bulk and local microsolvation phenomena.

To better understand bulk phase effects, the computed activation energy lowering in a methanol/water mixture, as compared to pure water, was examined. PCM single-point evaluations on the two-water transition structure produced solvent corrected activation energies of 12.3 (water) and 11.5 kcal/mol (methanol/water). The computed activation lowering in the water/methanol mixture is consistent with the idea of antihydrophobic effects [83–86]. Essentially, the dielectric of the cavity provided by the continuum interacts with the induced charge polarization of the transition structure caused by the two explicit waters. Hydrophobic regions of the transition structure are better stabilized by the lower dielectric (methanol) provided by

Table 6 Single-Point $\Delta E_0^{\ddagger}$ Activation Energies (kcal/mol) of the Reaction Between 1,3-Butadiene and Acrolein Using PCM and Two Explicit Waters at the B3LYP/6-31G* Level of Theory

TS	Methanol	Water
PCM, Two Water		
NCP2	11.5	12.3
NTP2	16.1	13.2
XCP2	13.7	14.7
XTP2	13.2	13.9

the continuum cavity. The magnitude of the antihydrophobic effect is estimated to be the induced energetic preference of the methanol/water mixture over pure water (12.3 – 11.5 = 0.8 kcal/mol) based upon the P2 computations listed in Table 6, plus the energy to destabilize pure water over methanol (17.7 – 16.4 = 1.1 kcal/mol) by the P0 computations in Table 4. Consequently, 1.9 kcal/mol of the bulk phase effect is composed of an antihydrophobic interaction, which accounts for ca. 21% [1.9 kcal/mol (computed)/9.2 kcal/mol (expected)] of the activation energy lowering. The remaining 4.6– 1.9 = 2.7 kcal/mol or 29% [2.7 kcal/mol (remaining) 9.2 kcal/mol (expected)] is attributed to the enforced hydrophobic effect. The computed 2.7 kcal/mol effect is slightly greater than the simulation work by Jorgensen et al. [87,88], and smaller than the QM/MM value reported by Furlani and Gao [64]. A few significant results should be mentioned. First, it is necessary to induce the charge polarization of the NC transition structure by explicit hydrogen bonds in order to generate the bulk phase effects. Comparison of the activation enthalpies between P0 and P2 continuum models illustrates the effect of explicit hydrogen bonding. Second, local hydrogen bonding accounts for a large portion of the rate enhancement (50%), and the bulk phase composed of enhanced hydrophobic interactions (30%) and antihydrophobic effects (20%) describe the remainder of the rate acceleration observed by experiment.

4.3. Connection with Enzymatic Catalysis

Linus Pauling [89,90] originally proposed that tight binding between a transition structure and enzyme could explain the extraordinary rates in enzymatic catalysis. His postulate defined many research efforts and ideas over the years [8,11,13,16]. Recent research from a number of different experiments show that tight binding of the transition structure alone is insufficient to account for the rate enhancements achieved by enzymes [10,12,14,15]. Other factors have been recently proposed that involve preorganization and orientational effects of transition structure binding, and the role of the enzyme protecting the substrate from solvent. Thus it is clear that to gain a deeper appreciation of enzymatic catalysis, it is essential to understand the properties and structure of the transition structure. How enzymatic catalysis is actually achieved continues to be a matter of debate [8,11,13,16]. Computed information on how the immediate environment surrounding the transition structure affects the rate and selectivity of the Diels–Alder reaction should give clues on how an enzyme affects the enzymatic catalysis.

The microsolvation effect of explicit waters is found to induce a polarization of the transition structure with a lowering of the activation barrier. The microsolvation of hydrogen bonding accounts for approximately one-half of the observed catalytic effect observed by experiment. The macroscopic effects of solvation were studied by the implicit PCM model, which accounts for the remaining 50% of the experimental effect. The explicit hydrogen bonding is necessary to induce charge polarization of the NC transition structure and allow enforced hydrophobic interactions and antihydrophobic effects. The full aqueous acceleration and enhanced *endo/exo* selectivity observed by experiment is only realized when solvation forces are approximated by the discrete-continuum model, taking both local and bulk phase effects into account. The gas phase activation energy is lowered to 11.5 kcal/mol, in excellent agreement with known experimental activation energies of similar Diels–Alder reactions in mixed methanol and water solutions. The computed *endo* preference is enhanced to 2.4 kcal/

mol in aqueous solution, in agreement with experiment. We find that the *endo/exo* selectivity is equally influenced by hydrogen bonding and bulk phase effects. Therefore the catalytic and *endo/exo* selectivity results are consistent with the hypothesis of maximum accumulation of unsaturation.

ACKNOWLEDGMENTS

The authors would like to thank Sue Kong, Amy M. Waligorski, Anne Loccisano, Alba T. Macias, Jason DeChancie, and Tugba Kucukkal for their contributions to this chapter. The Department of Energy, National Energy Technology Laboratory (NETL) is thanked for their support of our research efforts through grant no. DE-FG26-01NT41287. We acknowledge the Department of Defense (DAAH04-96-1-0311 and DAAG55-98-1-0067) for financial support to make this research possible. IBM is acknowledged for their generous contributions toward an IBM 16-node SP supercomputer.

REFERENCES

1. Zewail AH. Femtochemistry: atomic-scale dynamics of the chemical bond using ultrafast lasers (Nobel lecture). Angew Chem Int Ed 2000; 39:2586–2631.
2. Zewail AH. Femtochemistry: recent progress in studies of dynamics and control of reactions and their transition states. J Phys Chem 1996; 100:12701–12724.
3. Singleton DA, Merrigan SR, Beno BR, Houk KN. Isotope effects for Lewis acid catalyzed Diels–Alder reactions. The experimental transition state. Tetrahedron Lett 1999; 40:5817–5821.
4. Houk KN, Li Y, Evanseck JD. Transition structures of hydrocarbon pericyclic reactions. Angew Chem Int Ed Engl 1992; 31:682–708.
5. Jensen F. Transition structure optimization techniques. In: Schleyer PVR, ed. Encyclopedia of Computational Chemistry. New York: John Wiley, 1998:3114–3123.
6. Leach AR. Molecular Modelling: Principles and Applications. 2d ed. Hemel Hempstead, U.K.: Prentice-Hall, 2001.
7. Garrett BC, Truhlar DG. Transition state theory. In: Schleyer, PVR, ed. Encyclopedia of Computational Chemistry. New York: John Wiley, 1998:3094–3104.
8. Schramm VL. Enzymatic transition states and transition state analog design. Annu Rev Biochem 1998; 67:693–720.
9. Bruice TC, Lightstone FC. Ground state and transition state contributions to the rates of intramolecular and enzymatic reactions. Acc Chem Res 1999; 32:127–136.
10. Schramm VL. Transition state variation in enzymatic reactions. Curr Opin Chem Biol 2001; 5:556–563.
11. Sutcliffe M, Scrutton N. Enzyme catalysis: over-the-barrier or through-the-barrier? Trends Biochem Sci 2000; 25:405–408.
12. Kollman PA, Kuhn B, Peräkylä M. Computational studies of enzyme-catalyzed reactions: where are we in predicting mechanisms and in understanding the nature of enzyme catalysis? J Phys Chem B 2002; 106:1537–1542.
13. Bruice TC, Benkovic SJ. Chemical basis for enzyme catalysis. Biochem 2000; 39: 6267–6274.
14. Schramm VL, Shi W. Atomic motion in enzymatic reaction coordinates. Curr Opin Struct Biol 2001; 11:657–665.
15. Warshel A. Perspective on "The energetics of enzymatic reactions." Theor Chem Acc 2000; 103:337–339.

16. Wolfenden R, Snider MJ. The depth of chemical time and the power of enzymes as catalysts. Acc Chem Res 2001; 34:938–945.

17. Kong S, Evanseck JD. Density functional theory study of aqueous-phase rate acceleration and *endo/exo* selectivity of the butadiene and acrolein Diels–Alder reaction. J Am Chem Soc 2000; 122:10418–10427.

18. Lynch BJ, Truhlar DG. How well can hybrid density functional methods predict transition state geometries and barrier heights? J Phys Chem A 2001; 105:2936–2941.

19. Fast PL, Sánchez ML, Truhlar DG. Multi-coefficient Gaussian-3 method for calculating potential energy surfaces. Chem Phys Lett 1999; 306:407–410.

20. Truhlar DG, Garrett BC, Klippenstein SJ. Current status of transition-state theory. J Phys Chem 1996; 100:12771–12800.

21. Albery WJ. Transition-state theory revisited. Adv Phys Org Chem 1993; 28:139–171.

22. Garrett BC. Perspective on "The transition state method." Theor Chem Acc 2000; 103:200–204.

23. Pilling MJ, Seakins PW. Reaction kinetics. New York: Oxford University Press Inc., 1999.

24. Gaussian 98, Frisch MJ, Trucks GW, Schlegel HB, Scuseria GE, Robb MA, Cheeseman JR, Zakrzewski VG, Montgomery JA, Stratmann RE, Burant JC, Dapprich S, Millam JM, Daniels AD, Kudin KN, Strain MC, Farkas O, Tomasi J, Barone V, Cossi M, Cammi R, Mennucci B, Pomelli C, Adamo C, Clifford S, Ochterski J, Peterson GA, Ayala PY, Cui Q, Morokuma K, Malick DK, Rabuck AD, Raghavachari K, Foresman JB, Cioslowski J, Ortiz JV, Stefanov BB, Liu G, Liashenko A, Piskorz P, Komaromi I, Gomperts R, Martin RL, Fox DJ, Keith T, Al-Laham MA, Peng CY, Nanayakkara A, Gonzalez C, Challacombe M, Gill PMW, Johnson BG, Chen W, Wong MW, Andres JL, Head-Gordon M, Replogle ES, Pople JA. Pittsburgh, PA: Gaussian, Inc., 1998.

25. Foresman JB, Frisch A. Exploring chemistry with electronic structure methods. 2d ed. Pittsburgh, PA: Gaussian, Inc., 1996.

26. Ochterski JW. Thermochemistry in *Gaussian*, www.gaussian.com, 1–19, 2000.

27. Del Bene JE, Mettee HD, Frisch MJ, Luke BT, Pople JA. Ab initio computation of the enthalpies of some gas-phase hydration reactions. J Phys Chem 1983; 87:3279–3282.

28. Pople JA, Head-Gordon M, Fox DJ, Raghavachari K, Curtiss LA. Gaussian-1 theory: A general procedure for prediction of molecular energies. J Chem Phys 1989; 90:5622–5629.

29. Curtiss LA, Raghavachari K, Trucks GW, Pople JA. Gaussian-2 theory for molecular energies of first- and second-row compounds. J Chem Phys 1991; 94:7221–7230.

30. Curtiss LA, Raghavachari K, Pople JA. Gaussian-2 theory using reduced Møller–Plesset orders. J Chem Phys 1993; 98:1293–1298.

31. Curtiss LA, Carpenter JE, Raghavachari K, Pople JA. Validity of additivity approximations used in GAUSSIAN-2 theory. J Chem Phys 1992; 96:9030–9034.

32. Schlegel HB. Optimization of equilibrium geometries and transition structures In: Lawley KP, ed. Ab Initio Methods in Quantum Chemistry. New York: 1987.

33. Cramer CJ. Essentials of computational chemistry: theories & models. New York, NY: Wiley, 2002.

34. Hehre WJ, Radom L, Schleyer PvR, Pople JA. Ab initio molecular orbital theory. New York: John Wiley & Sons, 1986.

35. Wiest O. Transition states in organic chemistry: ab initio. In: Schleyer PVR, ed. Encyclopedia of Computational Chemistry. New York: 1998:3104–3114.

36. Liu J, Niwayama S, You Y, Houk KN. Theoretical prediction and experimental tests of conformational switches in transition states of Diels–Alder and 1,3-dipolar cyclo-additions to enol ethers. J Org Chem 1998; 63:1064–1073.

37. Wiest O, Montiel DC, Houk KN. Quantum mechanical methods and the interpretation and prediction of pericyclic reaction mechanisms. J Phys Chem A 1997; 101:8378–8388.

38. Houk KN, Beno BR, Nendel M, Black K, Yoo HY, Wilsey S, Lee JK. Exploration of pericyclic reaction transition structures by quantum mechanical methods: competing

concerted and stepwise mechanisms. J Mol Struct (THEOCHEM) 1997; 398–399:169–179.

39. Barone V, Arnaud R. Diels–Alder reactions: An assessment of quantum chemical procedures. J Chem Phys 1997; 106:8727–8732.

40. Goldstein E, Beno B, Houk KN. Density functional theory prediction of the relative energies and isotope effects for the concerted and stepwise mechanisms of the Diels–Alder reaction of butadiene and ethylene. J Am Chem Soc 1996; 118:6036–6043.

41. Barone V, Arnaud R. Study of prototypical Diels–Alder reactions by a hybrid density functional/Hartree–Fock approach. Chem Phys Lett 1996; 251:393–399.

42. Bernardi F, Bottoni A, Field MJ, Guest MF, Hillier IH, Robb MA, Venturini A. MC-SCF study of the Diels–Alder reaction between ethylene and butadiene. J Am Chem Soc 1988; 110:3050–3055.

43. Bernardi F, Bottoni A, Olivucci M, McDouall JJW, Robb MA, Tonachini G. Potential energy surfaces of cycloaddition reactions. J Mol Struct (THEOCHEM) 1988; 165:341–351.

44. Jensen F. Introduction to Computational Chemistry. New York, NY: Wiley, 1999.

45. Szabo A, Ostlund NS. Modern quantum chemistry introduction to advanced electronic structure theory. 1st ed. Revised. New York, NY: McGraw-Hill, 1989.

46. Houk KN, Gonzalez J, Li Y. Pericyclic reaction transition states. Acc Chem Res 1995; 28:81–90.

47. Eksterowicz JE, Houk KN. Transition-state modeling with empirical force fields. Chem Rev 1993; 93:2439–2461.

48. Storer JW, Raimondi L, Houk KN. Theoretical secondary kinetic isotope effects and the interpretation of transition state geometries. 2. The Diels–Alder reaction transition state geometry. J Am Chem Soc 1994; 116:9675–9683.

49. Houk KN, Lin YT, Brown FK. Evidence for the concerted mechanism of the Diels–Alder reaction of butadiene with ethylene. J Am Chem Soc 1986; 108:554–556.

50. Herges R, Jiao H, Schleyer PVR. Magnetic properties of aromatic transition states: the Diels–Alder reaction. Angew Chem Int Ed 1994; 33:1376–1378.

51. Wiest O, Houk KN, Black KA, Thomas B. IV. Secondary kinetic isotope effects of diastereotopic protons in pericyclic reactions: a new mechanistic probe. J Am Chem Soc 1995; 117:8594–8599.

52. Li Y, Houk KN. Diels–Alder dimerization of 1,3-butadiene: An ab initio CASSCF study of the concerted and stepwise mechanisms and butadiene–ethylene revisited. J Am Chem Soc 1993; 115:7478–7485.

53. Uchiyama M, Tomioka T, Amano A. Thermal decomposition of cyclohexene. J Phys Chem 1964; 68:1878–1881.

54. Schmidt MW, Gordon MS. The construction and interpretation of MCSCF wavefunctions. Annu Rev Phys Chem 1998; 49:233–266.

55. Becke AD. Density-functional thermochemistry: III. The role of exact exchange. J Chem Phys 1993; 98:5648–5652.

56. Lee C, Yang W, Parr RG. Development of the Colle–Salvetti correlation-energy formula into a functional of the electron density. Phys Rev B 1988; 37:785–789.

57. MacKerell ADJ, Bashford D, Bellott M, Dunbrack RL, Evanseck JD, Field MJ, Fischer S, Gao J, Guo H, Ha S, Joseph-McCarthy D, Kuchnir L, Kuczera K, Lau FTK, Mattos C, Michnick S, Ngo T, Nguyen DT, Prodhom B, Reiher WE, Roux B, Schlenkrich M, Smith JC, Stote R, Straub J, Watanabe M, Wiorkiewicz-Kuczera J, Yin D, Karplus M. All-atom empirical potential for molecular modeling and dynamics studies of proteins. J Phys Chem 1998; 102:3586–3616.

58. Gordon MS, Freitag MA, Bandyopadhyay P, Jesen JH, Karirys V, Stevens WJ. The effective fragment potential method: A QM-based MM approach to modeling environmental effects in chemistry. J Phys Chem A 2001; 105:293–307.

59. Monard G, Mertz KMJ. Combined quantum mechanical/molecular mechanical methodologies applied to biomolecular systems. Acc Chem Res 1999; 32:904–911.

60. Hillier IA. Chemical reactivity studied by hybrid QM/MM methods. J Mol Struct (THEOCHEM) 1999; 463:45–52.

61. Dapprich S, Komáromi I, Byun SB, Morokuma KJ, Frisch MJ. A new ONIOM implementation in Gaussian98: Part I. The calculation of energies, gradients, vibrational frequencies and electric field derivatives. J Mol Struct (THEOCHEM) 1999; 461–462:1–21.

62. Sauer J, Sierka M. Combining quantum mechanics and interatomic potential functions in ab initio studies of extended systems. J Comp Chem 2000; 21:1470–1493.

63. Naray-Szabo G. Chemical fragmentation in quantum mechanical methods. Comp Chem 2000; 24:287–294.

64. Furlani TR, Gao J. Hydrophobic and hydrogen-bonding effects on the rate of Diels–Alder reactions in aqueous solution. J Org Chem 1996; 61:5492–5497.

65. Nendel M, Tolbert LM, Herring LE, Islam MN, Houk KN. Strained allenes as dienophiles in the Diels–Alder reaction: An experimental and computational study. J Org Chem 1999; 64:976–983.

66. Bigeleisen J, Mayer MG. J Chem Phys 1947; 15:261.

67. Wolfsberg M. Theoretical evaluation of experimentally observed isotope effects. Acc Chem Res 1972; 5:225–233.

68. Carroll FA. Perspectives on structure and mechanism in organic chemistry. Pacific Grove, CA: Brooks/Cole Publishing Company, 1998.

69. Weinreb SM. Heterodienophile additions to dienes In: Trost BM, ed. Comprehensive Organic Synthesis. Oxford: Pergamon Press, 1991:401.

70. Kistiakowaki GB, Lacher JR. The kinetics of some gaseous Diels–Alder reactions. J Am Chem Soc 1936; 58:123–133.

71. Blankenburg VB, Fiedler H, Hampel M, Hauthal HG, Just G, Kahlert K, Korn J, Müller K-H, Pritzkow W, Reinhold Y, Röllig M, Sauer E, Schnurpfeil D, Zimmermann G. J Prakt Chemie 1974; 316:804–816.

72. García JI, Martinez-Merino V, Mayoral JA, Salvatella L. Density functional theory study of a Lewis acid catalyzed Diels–Alder reaction. The butadiene + acrolein paradigm. J Am Chem Soc 1998; 120:2415–2420.

73. Birney DM, Houk KN. Transition structures of the Lewis Acid catalyzed Diels–Alder reaction of butadiene with acrolein. The origins of selectivity. J Am Chem Soc 1990; 112:4127–4133.

74. The IBM Corporation is thanked for their generous contributions towards an IBM 16-node SP supercomputer.

75. Petersson GA, Al-Laham MA. A complete basis set model chemistry: II. Open-shell systems and the total energies of the first-row atoms. J Chem Phys 1991; 94:6081–6090.

76. Garcia JI, Mayoral JA, Salvatella L. Is it [4 + 2] or [2 + 4]? A new look at Lewis Acid catalyzed Diels–Alder reactions. J Am Chem Soc 1996; 118:11680–11681.

77. Miertus S, Scrocco E, Tomasi J. Electrostatic interaction of a solute with a continuum. A direct utilization of ab initio molecular potentials for the prevision of solvent effects. Chem Phys 1981; 55:117–129.

78. Aguilar MA, Olivares del Valle FJ, Tomasi J. Nonequilibrium solvation: an ab initio quantum-mechanical method in the continuum-cavity-model approximation. J Chem Phys 1993; 98:7375–7384.

79. Cossi M, Barone V, Cammi R, Tomasi J. Ab initio study of solvated molecules: a new implementation of the polarizable continuum model. J Chem Phys Lett 1996; 255:327–335.

80. Tomasi J, Persico M. Molecular interactions in solution: An overview of methods based on continuous distributions of the solvent. Chem Rev 1994; 94:2027.

81. Ruiz-López MF, Assfeld X, Garcia JI, Mayoral JA, Salvatella L. Solvent effects on the mechanism and selectivities of asymmetric Diels–Alder reactions. J Am Chem Soc 1993; 115:8780.

82. Rideout DC, Breslow R. Hydrophobic acceleration of Diels–Alder reactions. J Am Chem Soc 1980; 102:7816–7817.

83. Breslow R, Groves K, Mayer MU. Antihydrophobic cosolvent effects in organic displacement reactions. Org Lett 1999; 1:117–120.

84. Breslow R, Connors R, Zhu Z. Mechanistic studies using antihydrophobic effects. Pure and Appl Chem 1996; 68:1527–1533.

85. Breslow R, Zhu Z. Quantitative antihydrophobic effects as probes for transition state structures. J Am Chem Soc 1995; 117:9923–9924.

86. Breslow R. Hydrophobic and antihydrophobic effects on organic reactions in aqueous solution. Struct React Aq Sol 1994; 568:291–302.

87. Blake JF, Lim D, Jorgensen WL. Enhanced hydrogen bonding of water to Diels–Alder transition states. Ab initio evidence. J Org Chem 1994; 59:803–805.

88. Blake JF, Jorgensen WL. Solvent effects on a Diels–Alder reaction from computer simulations. J Am Chem Soc 1991; 113:7430–7432.

89. Pauling L. Chem Eng News 1946; 24:1375.

90. Pauling L. Nature 1948; 161:707–709.

13

Molecular Similarity, Quantum Topology, and Shape

PAUL G. MEZEY

Memorial University of Newfoundland, St. John's, Newfoundland, Canada

1. INTRODUCTION

Molecular similarity necessarily involves quantum mechanical concepts; hence all similarity approaches applied to molecules must, in some way, be related to the most general aspects of Quantum Similarity proposed by Carbo-Dorca (for a review of Quantum Similarity and its applications, see Ref. 1). Powerful fundamental results of quantum mechanics play an important role, as manifested by the basic result of the Hohenberg–Kohn theorem [2] concerning the all-determining property of nondegenerate ground-state electron density, a quantum-mechanical object. As shown by later developments concerning subsystems, even local parts of such density clouds fully determine all properties of molecules [3–5]. This has been first indicated for an artificial system of a formal, bounded model of molecules with electron densities confined to a closed and finite box [3] and more recently proven by the "Holographic Electron Density Theorem" for actual, boundary-less electron densities of molecules [4,5].

Some recent developments concerning macromolecular quantum chemistry, especially the first linear-scaling method applied successfully for the ab initio quality quantum-chemistry computation of the electron density of proteins, have underlined the importance and the applicability of quantum chemistry-based approaches to molecular similarity. These methods, the linear-scaling numerical Molecular Electron Density Lego Approach (MEDLA) method [6–9] and the more advanced and more generally applicable linear-scaling macromolecular density matrix method called Adjustable Density Matrix Assembler or ADMA method [10,11], have been employed for the calculation of ab initio quality protein electron densities and other

345

properties at a level of detail still inaccessible by experimental methods, such as x-ray crystallography.

However, as Quantum Similarity was formulated originally by Carbo-Dorca, the first approaches relied on a direct comparison of two quantum objects, in the most relevant case, of two molecules, and similarity was expressed by such pair comparisons. A not entirely unrelated but an alternative application of quantum chemistry to the molecular similarity problem was formulated in terms of the intrinsic shape of molecular electron distributions, which resulted in a direct shape representation of the quantum chemical electron density in terms of topological shape codes, exploiting the power of topology in focusing on the essential [12–19]. The origins of this approach can be found in the topological methods first applied to quantum chemistry within the context of multidimensional potential energy hypersurfaces [12]. In this field, besides the quantum mechanical features, some direct and some indirect consequences of the Heisenberg uncertainty relation, and also the large number of internal coordinates and the resulting very high dimensionality of conformational and reactive potential energy hypersurface problems, motivated the application of some of the powerful tools of algebraic and differential topology.

Subsequently, these topological methods have been adopted and modified to the significantly simpler, three-dimensional molecular shape problem, where the shape of the molecule is the quantum mechanical shape of the electron density cloud [13–19]. This has led to the development of the shape group methods, where the ranks of homology groups describing relative convexity domains of the complete set of all isodensity surfaces of the molecule, the so-called Shape Group Betti numbers, provided a detailed, numerical shape code for the quantum chemical electron density [13–19].

Molecular similarity, in turn, was expressed in the next step, directly on the topological representations of shapes, and at this level, quantum similarity is not involved directly. Using differential and algebraic topological representations of molecular shapes had several advantages when compared to the direct Quantum Similarity approach:

1. The algebraic and differential topological similarity measures required much simpler mathematical and computational apparatus than the direct comparisons of the original, complex quantum mechanical objects.
2. Many of the well-established additional methods of algebraic and differential topology could be adapted and employed for the inherently topological molecular problem.
3. When dealing with intrinsic topological shapes, the need to find optimum (or otherwise specified) superposition or mutual arrangement of two complex quantum mechanical objects was no longer required.

Hence while the topological shape analysis and shape similarity measures do not truly represent a "Non-Quantum Similarity" approach, nevertheless, in practical terms, they offer a powerful alternative.

Several additional developments had very positive impact on the advances and applications of the topological shape analysis and similarity methods. Among these are the already mentioned establishment of the holographic property of the electron density clouds of real, boundary-less molecules [4,5] and the extension of many aspects of small-molecule quantum chemistry to macromolecules, such as proteins, by the linear-scaling MEDLA and ADMA methods [6–11].

This has opened the way to a systematic, quantum chemical definition and description of functional groups, to macromolecular shape analysis, to macromolecular force computations, and to shape-code-based macromolecular similarity analysis [20–25].

In the subsequent parts of this chapter, some aspects of these advances are described.

2. A FUNDAMENTAL LAW OF MOLECULAR INFORMATICS: THE HOLOGRAPHIC ELECTRON DENSITY PRINCIPLE

Molecular similarity is a concept based on molecular information, and all information about a molecule is contained within the molecular electron density. The branch of chemistry (indeed, a branch of molecular physics) that deals with the ways molecules represent, store, process, and exhibit information is called molecular informatics.

Evidently, molecular informatics strongly relies on electron density. The central role of electron density in molecular informatics underlines the importance of the Hohenberg–Kohn theorem, referred to above: the nondegenerate ground-state electron density of a molecule determines the molecular energy and, through the Hamiltonian, all other molecular properties [2].

However, even more is true, and a stronger statement may be regarded as the fundamental law of molecular informatics:

> Any nonzero volume part of the nondegenerate ground state electron density cloud of a molecule contains all information about the molecule. (Holographic Electron Density Theorem [4,5])

This result involves some earlier, fundamental developments of density functional theory. Of course, one of the central results of density functional theory, the Hohenberg–Kohn theorem can also be viewed in terms of information [2]: the nondegenerate ground-state electron density determines the energy and, in fact, through the Hamiltonian, all molecular properties, hence the complete molecular information. In addition, for artificial, closed, and bounded systems, Riess and Münch [3] have demonstrated another, information-related, important result: if the complete system is finite, with a closed boundary, then the Hohenberg–Kohn theorem applies to any closed and bounded subsystem of it, as long as the ground-state electron density is nondegenerate. Of course, real molecules are neither bounded nor finite and have no closed boundaries in a rigorous sense. This is an important feature. If one disregards the differences between closed and open systems, one may easily obtain inconclusive or even false results, similarly to "proofs" of $2 + 2 = 5$ involving division by zero. Consequently, the result of Riess and Münch, relying on a closed model, is not applicable to actual, boundary-less molecules, a fact that was well recognized by these authors; nevertheless, their work has motivated various further developments leading to new, fundamental relations [4,5,26,27].

The interpretation of the fundamental results of molecular informatics is rather straightforward if one considers the actual "material" making up the molecules. The atomic nuclei and the fuzzy, boundary-less electron density cloud are the only physical entities contained in molecules, where the electron density contains all information about the location and nature of the nuclei. Consequently, simply on information-theoretical grounds, the result of the Hohenberg–Kohn theorem [2], stating that the nondegenerate ground-state electron density determines all molecular properties, is a very natural idea indeed. Simply stated, there is nothing else in a molecule that could be the carrier of any additional information to determine molecular properties. It appears that on purely information-theoretical grounds, electron density must determine all molecular properties, and the preceding arguments may be regarded as an informal proof of the Hohenberg–Kohn theorem.

Based on the tools employed in the Hohenberg–Kohn theorem, also in part on the result of Riess and Münch, and on a four-dimensional version of the Alexandrov one-point compactification method of topology applied to the complete three-dimensional electron density, it was possible to prove recently that for nondegenerate ground-state electron densities, the Holographic Electron Density Theorem applies: any nonzero volume part of the nondegenerate ground-state electron density cloud contains all information about the molecule [4,5].

It is important to realize that in the proofs of the Hohenberg–Kohn theorem and the Holographic Electron Density Theorem, some very natural properties of molecular electron densities have been assumed. Two of these assumptions are i) the very existence of a ground-state electron density function and ii) the assumption of continuity of this function in the space variable **r**.

It goes beyond the scope of this chapter to discuss the matter of time scale where such a density function appears justified; nevertheless, one should recognize that the effects of zero-point energy and the associated vibrations modify the role of the effective external potential experienced by the actual electron density, which also involves the role of time. Whereas electron density changes and fluctuations can carry information, all such effects on the molecular information are also dependent on the ground-state electron density.

One further development concerning molecular information is relevant to the so-called *latent properties* of molecules, that is, to those properties not directly exhibited by the actual electron density. It turns out that the ground-state electron density may also serve as the source of the information for the study of all latent properties.

Let us regard an isolated molecule A in its nondegenerate electronic ground state. It is natural to assume that many of the molecular properties exhibited by the isolated molecule A are primarily associated with its electronic ground state and nuclear arrangements similar to the most stable one that is typically (but not necessarily for the so-called potential-defying species) an energy minimum on the associated potential energy hypersurface [12]. These properties are regarded as the *exposed properties*, considered to be directly associated with the ground-state electron density. It is probably not at all surprising that all of these properties can be derived from the ground-state electron density.

However, one may consider a much broader family of molecular properties. Evidently, the same molecule also has properties which are not exhibited in the electronic ground state at the most stable conformation of nuclear arrangement. A *latent property* P may be regarded as the response of a molecule A to a specific interaction X. Specifically, a latent property P of the molecule is one that is reproducibly exhibited by the molecule if it is exposed to a specific interaction or to a specific range of interactions. Latent properties include those which are associated with electronic excited states, with highly distorted nuclear arrangements, with different stable conformations, or even with products of dissociation reactions preserving the same overall stoichiometry of the original molecule (where again, the role of the complete potential energy hypersurface is evident [12]).

In fact, an extension of the Holographic Electron Density Theorem applies to latent molecular properties.

We may approach the problem of latent properties the same way as one used the information-theoretical approach in the case of the Hohenberg–Kohn theorem: where is the information stored concerning all the latent properties of a given molecule A? It is evident that many of these latent properties are eventually exhibited in response to some specific interactions, and the interacting partner or partners must have a role in triggering the manifestation of the latent properties. It is also true, however, that a different molecule B has different latent properties and the process involving the same interaction partner or partners leads to a different set of latent properties.

For example, the interaction with a photon of the same energy may, reproducibly, lead to the manifestation of well-defined but different sets of latent properties in two different molecules A and B. The information contribution of the photon of a specific energy may be regarded as minimal; the relevant piece of information are its energy, determined by a single number, and its frequency, which decides which excited state of the given molecule will be manifested. Consequently, most of the information concerning the triggered latent properties must be stored within the molecule itself. In the case of the photon having a specific energy, one may view the information external to the molecule as a simple switch that provides only the selection, deciding which latent property of the molecule (which electronic excited state with all its properties) is going to be exhibited. Consequently, for both molecules A and B, all the essential information concerning the triggered set of latent properties must be stored in the molecules A and B, respectively. The external information of the photon is used merely to select for exhibition of some of the actual latent properties for which the information is already stored in the molecule, that is, to convert some of the latent properties of the molecule into actually exhibited properties.

For a more formal argument, let us regard a molecule A of a nondegenerate ground-state electron density $\rho(\mathbf{r})$ and some latent property P. Being latent, this property P is not directly exhibited by the ground-state electron density $\rho(\mathbf{r})$. However, latent property P is reproducibly exhibited by the molecule A exposed to a specific interaction X. One may regard latent property P as a component of the response of molecule A to the specific interaction X; consequently, all information about this "potential" response must be stored within the molecule A. We also know that all information already present in molecule A must be stored within the ground-state nondegenerate electron density $\rho(\mathbf{r})$. As follows from the Holographic Electron Density Theorem, any small positive volume part of the ground-state electron density $\rho(\mathbf{r})$ must also contain this information. Evidently, an extension of the original Holographic Electron Density Theorem holds that has been called the Holographic Electron Density Theorem for Latent Molecular Properties [26]:

> Any small positive volume part of the nondegenerate ground-state electron density $\rho(\mathbf{r})$ of any molecule A contains all information about any latent property P of the molecule, regarded as a component of a reproducible response of molecule A to a specific interaction X.

By combining this theorem and the original Holographic Electron Density Theorem, one may obtain a single statement:

> Any small, nonzero volume piece of the nondegenerate ground-state electron density cloud of a molecule A contains the complete information on all actual and all latent molecular properties of molecule A.

Based on these considerations, not only the actual properties exhibited by the molecule are fully determined, but also all potential, reproducible responses of the molecule resulting from various interactions which are determined by any small positive volume part of the electron density cloud.

One may focus on the properties exhibited in a given excited electronic state of a molecule and consider it as the molecular response to the interaction with a photon of energy precisely equivalent to the energy of excitation. Evidently, the complete information about this response is already fully encoded within any nonzero volume part of the original electron density $\rho(\mathbf{r})$ of the molecule A, and the role of the photon is only to trigger the manifestation of this information.

Other specific interactions can lead to responses such as a conformational variation or the generation of a transition structure requiring a specific energy of activation. Considering the range of potential responses, one can conclude that for the given overall stoichiometry, all the properties of the potential energy hypersurfaces of all electronic states, all the associated molecular shape properties, the corresponding chemical reactivities, and the biochemical activities are all latent properties of molecules that are fully determined by any small, nonzero volume piece of the nondegenerate ground-state electron density cloud of the molecule.

3. THE TOPOLOGICAL MOLECULAR SHAPE AND SIMILARITY ANALYSIS: THE SHAPE GROUP METHOD

The main tool for a systematic, topological shape and similarity analysis of molecules is the shape group analysis of molecular electron density clouds [13–25]. The shape group methods are not restricted to molecular electron densities; however, in the present context, we shall phrase our brief review of these techniques in terms of electron densities.

In general, the shape groups are algebraic groups, describing in detail various aspects of shape. In a formal, differential, and algebraic topological sense, the shape groups are the homology groups of truncated objects obtained from continuous and twice differentiable functions (such as electron density clouds at most locations), where the truncation is determined by local shape properties of this function. These groups are not directly related to point symmetry groups, although the presence of symmetry may influence the shape groups and the subsequent analysis.

The electron density itself is typically calculated from a molecular wavefunction, for example, one determined by the Hartree–Fock method, and expressed in terms of a density matrix and atomic orbital basis set, as follows. If an LCAO (Linear Combination of Atomic Orbitals) ab initio wavefunction is computed for a molecule A of some fixed conformation K, then the electronic density $\rho(\mathbf{r})$ can be computed in a simple way. If one denotes by n the number of atomic orbitals $\varphi_i(\mathbf{r})$ ($i = 1,2,...,n$), $\mathbf{r}$ is the three-dimensional position vector variable, and $\mathbf{P}$ is the $n \times n$ density matrix, then the electronic density $\rho(\mathbf{r})$ of the molecule can be computed as

$$\rho(\mathbf{r}) = \sum_{i=1}^{n} \sum_{j=1}^{n} \mathbf{P}_{ij} \varphi_i(\mathbf{r}) \varphi_j(\mathbf{r}) \tag{1}$$

This function $\rho(\mathbf{r})$ represents the fuzzy body of the electronic charge cloud that in turn represents the shape of the molecule A.

A related concept is the molecular isodensity contour surface, MIDCO (Molecular IsoDensity Contour) $G(K,a)$, defined for the given nuclear configuration K and density threshold a as

$$G(K, a) = \{\mathbf{r} : \rho(K, \mathbf{r}) = a\} \tag{2}$$

A MIDCO is the collection of all points $\mathbf{r}$ of the three-dimensional space where the electronic density $\rho(K,\mathbf{r})$ is equal to the given threshold value a. We may notice that the $G(K,a)$ collection of points $\mathbf{r}$ fulfilling Eq. (2) is a continuous closed surface, as this follows from the fact that the molecular electronic density $\rho(K,\mathbf{r})$ is a continuous function of the position vector $\mathbf{r}$ (with the possible exception of some pathological cases).

The concept of MIDCO is connected to the concept of approximate molecular body. For an appropriate threshold value $a > a_{min}$, an approximate (open) molecular body is defined as the collection $F(K,a)$ of all those points $\mathbf{r}$ of the three-dimensional space where the electronic density is greater than the threshold a,

$$F(K, a) = \{\mathbf{r} : \rho(K, \mathbf{r}) > a\} \tag{3}$$

This set $F(K,a)$ is a formal level set of the electronic density $\rho(K,\mathbf{r})$ for the threshold level a. Whereas electrons are negatively charged, the density of electrons is either positive or zero, and a negative charge means a positive value for the density function $\rho(K,\mathbf{r})$.

The third related concept is the electronic density domain $DD(K,a)$ of some density threshold a which is a formal (closed) body that includes a MIDCO surface $G(K,a)$ as its boundary, whereas the interior of the density domain $DD(K,a)$ is the level set $F(K,a)$ of the same density threshold value a. The corresponding density domain $DD(K,a)$ is defined [13–25] as

$$DD(K, a) = \{\mathbf{r} : \rho(K, \check{\mathbf{r}}) \geq a\} \tag{4}$$

For a specified electronic state of the molecule A, the shape and size of sets $G(K,a)$, $F(K,a)$, and $DD(K,a)$ depend on both the nuclear arrangement K and on the threshold a.

Density domains include their boundaries, providing a closer analogy with the bodies of ordinary, macroscopic objects the way we see them.

Note that the density domains are not necessarily domains in a mathematical sense; for example, a $DD(K,a)$ may be disconnected. Especially, a formal density domain $DD(K,a)$ of a high-density threshold value a may have several disconnected parts. Often, the individual pieces are also referred to as density domains while being parts of a single $DD(K,a)$.

The density domains provide good insight into chemical bonding and are often much more revealing than the conventional stereochemical "skeletal" bonding patterns of formal single, double, and triple bonds, represented by lines along various directions. The Density Domain Analysis of chemical bonding focuses on the interfacing and mutual interpenetration of local, fuzzy charge density clouds [13–25]. Within the chemically relevant range $[a_{min}, a_{max}]$ of density thresholds, one may consider the associated family of density domains $[DD(K,a_{min}), DD(K,a_{max})]$ and study the topological changes occurring as a function of a change in the threshold value a. This provides a detailed description of the actual pattern of bonding within the molecule and reveals many features not ordinarily appreciated. Whereas at high-density thresholds, only disjoined, local nuclear neighborhoods appear, at intermediate thresholds, the separate pieces of density domains gradually join and change into a series of topologically different bodies, in a series of changes that provides a rigorous basis for quantum chemical definitions of many conventional chemical concepts, such as functional groups. At low threshold values, one finds a single, usually simply connected body, which reveals less and less detail about the molecular shape.

In the most common applications of shape groups, the local shape properties are specified in terms of shape domains: for example, in terms of the locally convex, concave, or saddle-type regions of MIDCOs, relative to some curvature reference parameter b.

For a detailed-enough shape description of most molecules, the use of local shape domains of contour surfaces appears as a practical choice. Unless local shape properties are involved in the definition of the topological objects used for shape characterization, most contour surfaces of small molecules, such as their MIDCO surfaces, are topologically rather simple objects and reveal little about the chemically relevant properties. Within the chemically interesting threshold ranges for the molecular electron density, many of these surfaces are topologically equivalent to a sphere or to a doughnut or to objects topologically equivalent to several doughnuts glued together. Since a direct topological characterization of such simple objects provides only a rather crude shape characterization and may lead to nearly trivial and rather useless results in a similarity analysis, an alternative, more detailed technique has been introduced. It is possible to identify and topologically classify the patterns of local shape domains on such contour surfaces, and then a detailed topological shape description can be obtained. Based on some geometrical or physical conditions (denoted by μ), one can define local shape domains on a contour surface $G(K,a)$, where these local shape domains are denoted by the symbol D_μ indicating that within this domain, the actual condition is fulfilled. For example, by comparing the curvature of a MIDCO surface $G(K,a)$ to a plane, three types of domains are obtained—locally convex, locally concave, and locally saddle-type domains. One may visualize this by assuming that the plane is moved along the MIDCO as a tangent plane, then the local curvature properties of the MIDCO are compared to the zero curvature of the plane. Specifically, each point r of the MIDCO surface is characterized by the local relation between the tangent plane and the actual density domain enclosed by the MIDCO. Locally, the tangent plane may fall on the outside, or on the inside, or it may cut into the given density domain within any small neighborhood of the tangent point r, indicating that at point r, the MIDCO is locally convex, locally concave, or locally saddle-type, respectively. This approach leads to various local curvature domains on the MIDCO surface. If this characterization is extended to all points r of the MIDCO, then one obtains a subdivision; in fact, there is a partition of the molecular contour surface into locally convex, locally concave, and locally saddle-type shape domains, where the symbols D_2, D_0, and D_1 are used for these three domain types, respectively.

However, much more detailed shape description is obtained if the tangent planes are systematically replaced by some other objects. Typically, a MIDCO is compared to a series of tangent spheres of various radii r, but one may find advantageous in direction-dependent problems to use a series of oriented tangent ellipsoids T, especially if a characterization itself involves some reference directions. In the case of oriented tangent ellipsoids, we assume that they can be translated but not rotated as they are brought into tangential contact with the MIDCO surface $G(K,a)$.

In general, the tangent object T may fall locally on the outside, or on the inside, or it may cut into the given density domain $DD(K,a)$ within any small neighborhood of the point of tangent along the surface. Hence the local characterization refers to a close neighborhood surface point $\mathbf{r}$ where the tangential contact occurs, and locality is understood in this context. These differences lead to a family of local shape domains D_2, D_0, and D_1, respectively, relative to the new tangent object T. Of course, if tangent spheres are used, orientation cannot play any role, and one may use the curvature of the sphere, $b = 1/r$, for specification. In the latter case, the local shape domains D_2, D_0, and D_1 represent the local relative convexity domains of the MIDCO, relative to the reference curvature b. Note that $b = 0$ corresponds to the case of the tangent plane.

Consider a specific reference curvature b and the associated local shape domains D_2, D_0, and D_1. These shape domains generate a partitioning of the MIDCO surface $G(K,a)$. In the next step, all D_μ domains of a specified type μ, for example, all the locally convex domains D_2 relative to reference curvature b, are excised from the MIDCO surface $G(K,a)$. This produces a new, topologically more interesting object, a truncated contour surface $G(K,a,\mu)$ that inherits some essential shape information from the original MIDCO surface $G(K,a)$, and also allows a concise formulation of the shape information, where simple topological tools, such as the algebraic homology groups, can be used.

The curvature-based shape analysis of each MIDCO surface $G(K,a)$ can be repeated for a whole range of reference curvature values b, providing a detailed shape characterization of $G(K,a)$. It is important to point out that for the complete range of chemically relevant reference curvature values b, there exist only a finite number of topologically different truncated MIDCOs $G(K,a,\mu)$ obtained from $G(K,a)$. When these truncated surfaces are characterized by their topological invariants, then a numerical shape characterization is obtained.

The topological shape characterization is achieved by computing the homology groups of truncated surfaces $G(K,a,\mu)$. Such homology groups of algebraic topology are topological invariants of $G(K,a,\mu)$, expressing important features of the topological structure of bodies and surfaces in the general case. In a concise characterization, the ranks of these groups, called the Betti numbers, are important topological invariants of each $G(K,a,\mu)$.

In molecular shape analysis, the Shape Groups of the original MIDCO $G(K,a)$ are the homology groups $H_\mu^p(a,b)$ of the truncated surfaces $G(K,a,\mu)$ for each pair of values a and b. The list or table of the corresponding $b_\mu^p(a,b)$ Betti numbers of the $H_\mu^p(a,b)$ shape groups generates numerical shape codes for molecular electron density distributions, and these shape codes are the actual tools for further analysis. Since the curvature patterns on each MIDCO involve points, lines, and surface patches for each reference curvature b and shape domain and truncation pattern μ of a given MIDCO $G(K,a)$ of density threshold a, there are three shape groups, $H_\mu^0(a,b)$, $H_\mu^1(a,b)$, and $H_\mu^2(a,b)$, for each pair of values a and b. Accordingly, the formal dimensions p of these three shape groups are zero, one, and two, collectively expressing aspects of the essential shape information of the MIDCO $G(K,a)$. Consequently, there are three Betti numbers, $b_\mu^0(a,b)$, $b_\mu^1(a,b)$, and $b_\mu^2(a,b)$, for each (a,b) pair of parameters and for each shape domain truncation type μ. For a detailed introduction to shape analysis and topological techniques, for more detailed mathematical derivations of shape groups, and for various examples of actual calculation of the shape groups and the corresponding Betti numbers, the reader may consult Refs. 13–25.

In a more general, although less easily visualizable setting, shape groups can also be defined directly for the entire, three-dimensional electronic charge distribution, where subdivisions of the three-dimensional space are obtained first for a range of parameters $p_1, p_2, \ldots, p_t$, describing the local gradient and second-derivative properties of the charge distribution at each point $\mathbf{r}$. Note that the second derivatives contain the curvature information. As in the more usual approach discussed above, a general range of these parameters is denoted by μ. By excising various three-dimensional domains characterized by some range μ of these parameters, topologically new objects are obtained, subject to a similar analysis. Accordingly, the homology groups of the truncated three-dimensional charge distribution define the shape groups $H_\mu^0(p_1, p_2, \ldots, p_t)$, $H_\mu^1(p_1, p_2, \ldots, p_t)$, $H_\mu^2(p_1, p_2, \ldots, p_t)$, and $H_\mu^3(p_1, p_2, \ldots, p_t)$ of the entire electronic charge density. As before, their Betti numbers $b_\mu^0(p_1, p_2, \ldots, p_t)$, $b_\mu^1(p_1, p_2, \ldots, p_t)$, $b_\mu^2(p_1, p_2, \ldots, p_t)$, and $b_\mu^3(p_1, p_2, \ldots, p_t)$, respectively, provide a concise and detailed shape characterization, as numerical shape codes and topological invariants. Note, however, that for visualization of molecular shape analysis, the shape group method, as applied to individual contour surfaces, appears more informative and conveys more conventional impressions about shapes.

The outlined shape group method of topological shape analysis combines the advantages of geometrical and topological approaches, and it follows the spirit of the GSTE principle: Geometrical Similarity as Topological Equivalence. In particular, the local shape domains and the truncated MIDCOs $G(K, a, \mu)$ are defined in terms of geometrical classification of points of the surfaces, relying on local curvature properties, whereas the truncated surfaces $G(K, a, \mu)$ are characterized topologically by the shape groups and their Betti numbers, leading to shape codes.

An important advantage of the topological shape analysis techniques is the numerical representation of shape information, which can be compared using simple tools. Specifically, the results of a shape group analysis can be represented by a finite family of Betti numbers $b_\mu^p(a, b)$ for all the shape groups which occur for a given molecule, and these numbers form a numerical shape code, represented either as a vector or a matrix. These shape codes can be compared algorithmically, providing a well-defined, nonvisual, numerical measure of molecular shape similarity, and, by a suitable transformation, the shape codes can also be used for generating numerical measure for shape complementarity, important in molecular interactions. The shape-code-based similarity and complementarity measures can be generated automatically by a suitable computer software, eliminating the subjective element of visual shape comparisons and providing reproducibility. This feature of the shape group method is particularly advantageous if large sequences of molecules are to be compared, for example, if a large number of molecules in the data banks of drug companies are the subject of the study. A valuable practical aspect for drug design is the versatility of the method and the fact that it is applicable for both global and local shape analysis.

The most common type of shape groups are those generated for the various ranges of MIDCOs of the electronic density of molecules, using curvature-based shape domains for ranges of curvature parameters b and density threshold values a. According to the usual convention, a positive b value implies that a reference tangent sphere is placed on the exterior side of the MIDCO surface, whereas a negative b value implies that the reference sphere is placed on the interior side of the MIDCO. If the nuclear arrangement K is fixed, then the shape groups of a molecule depend on the electronic density threshold a and on the reference curvature b, that is, on two parameters. If one considers the chemically relevant ranges of these two parameters, a and b, one can define a formal, two-dimensional map, the so-called (a,b) map. A detailed shape characterization of the electronic density of the molecule is obtained using the shape group distribution along this map.

Two approaches are used for shape analysis based on (a,b) maps. In the first approach, the molecular electron density is regarded as a single object, even if in the high-density regions (for high threshold values a), this object consists of several, disjoint pieces, where at such density thresholds, the localized ranges of density domains are typical, belonging to individual surroundings of atomic nuclei. In the second family of (a,b) maps, one regards each disjoint piece of the electronic density at the given threshold a as a separate object, and for each such piece, the actual shape analysis is performed separately.

Even within the first approach, separate (a,b) maps are generated for each of the three types of Betti numbers, $b^0_\mu(a,b)$, $b^1_\mu(a,b)$, and $b^2_\mu(a,b)$, according to the dimension of the shape groups. The set of Betti numbers obtained for a given pair of values of parameters a and b is assigned to the given location of the (a,b) parameter map, so such a map is indeed a map of the distribution of Betti numbers. The chemically most relevant shape information is provided by the important Betti numbers of type $b^1_\mu(a,b)$ of the one-dimensional shape groups.

The typical representation of the (a,b) map of Betti numbers is in a discretized form, given in terms of a grid of a and b values within some interval $[a_{min}, a_{max}]$ of density thresholds and some interval $[b_{min}, b_{max}]$ of reference curvature values, chosen by their chemical relevance. Typically, the condition

$$b_{min} - b_{max} \qquad\qquad (5)$$

is applied for simplicity.

Since the range of these a and b parameters covers several orders of magnitude, it is advantageous to use logarithmic scales. For negative values of the curvature parameters b, the $\log |b|$ values are taken. In an often-employed realization of the (a,b) map technique, a 41×21 grid is used, with range $[0.001, 0.1\ \text{a.u.}]$ (a.u. = atomic unit) for the density threshold values a and range $[-1.0, 1.0]$ for the curvature b, representing the reciprocal of the radii of the test spheres used to characterize the local curvatures of the MIDCOs.

The values of the Betti numbers at the grid points (a,b), or at the points $[\log(a), \log |b|]$ of the logarithmic map, form a matrix, $\mathsf{M}^{(a,b)}$. In either of the direct or the logarithmic representations, this matrix $\mathsf{M}^{(a,b)}$ is a numerical shape code for the fuzzy electronic density cloud of the molecule, representing the actual molecular shape.

If the shape representation is given in matrix form $\mathsf{M}^{(a,b)}$ of the (a,b) map of Betti numbers $b_\mu(^p(a,b)$ for a family of molecules, then numerical similarity measures can be calculated between molecules using these matrices $\mathsf{M}^{(a,b)}$.

Consider, for example, two molecules, A and B, both in some fixed nuclear configuration, and calculate their shape codes in their matrix forms $\mathsf{M}^{(a,b),\text{A}}$ and $\mathsf{M}^{(a,b),\text{B}}$, respectively. Based on these matrices, a numerical shape similarity measure can be defined as follows:

$$s(\text{A}, \text{B}) = m[\mathsf{M}^{(a,b),\text{A}}, \mathsf{M}^{(a,b),\text{B}}]/t, \qquad\qquad (6)$$

where the number $m[\mathsf{M}^{(a,b),\text{A}}, \mathsf{M}^{(a,b),\text{B}}]$ is simply the number of matches between corresponding elements in the two matrices and t is the total number of elements in either matrix. If n_a and n_b are the number of grid divisions for parameters a and b, respectively, then the value of t is

$$t = n_a n_b, \tag{7}$$

If, for the ranges [0.001, 0.1 a.u.] for the density threshold values a and [−1.0, 1.0] for the reference curvature b, a rather natural 41 × 21 grid is used, justified by the logarithmic scales, then the elements of the 41 × 21 matrix $M^{(a,b)}$ can also be stored as an integer vector **C** of 861 compon\ents. Using this alternative, the shape similarity measure $s(A,B)$ can also be expressed as

$$s(A, B) = \sum_{i = 1}^{861} \delta_{j(i),k(i)}/861, \tag{8}$$

where the Kroenecker delta $\delta_{j,k}$ is defined with respect to the indices

$$j(i) = C_i(A), \tag{9}$$

for molecule A and

$$k(i) = C_i(B) \tag{10}$$

for molecule B.

Often, the shape comparisons of local regions of molecules are of interest, which, in many instances, may appear more important than the evaluation of the global similarities of molecules.

In such a case, the electron density fragment additivity principle provides a simple approach for analyzing and evaluating local shape similarity of molecules. The fragment electron densities F are well defined within any LCAO-based quantum chemical electron density, and it is simple to consider the family of MIDCOs, their shape groups, as well as their shape codes $M^{(a,b),F}$ in a manner entirely analogous to the case of complete molecules. Accordingly, a fuzzy density fragment similarity measure can be computed as

$$s(F, F') = m\left[M^{(a,b),F}, M^{(a,b),F'}\right]/t \tag{11}$$

of any two fragments F and F'.

One very useful approach in rational drug design is the study of the local shapes and local shape similarities of molecules showing similar biochemical activities, and the technique of shape groups offers an algorithmic approach to this problem.

Various shape complementarity measures can also be determined based on shape codes. This is an important problem since molecular recognition usually depends on the complementarity of local regions of molecules, where complementarity may refer to electron distributions, polarizability properties, electrostatic potentials, or simply geometric considerations. The powerful topological techniques are suitable for the quantification of the degree of molecular complementarity and can be used as tools for the study of molecular recognition.

The nonvisual shape similarity measures of molecules as well as molecular fragments, using the numerical shape code method, provide the basis for a shape complementarity measure. A simple transformation of the local shape codes generates a representation that is suitable for a direct evaluation of local shape complementarity.

The very same techniques applied for the construction of local shape similarity measures are also applicable for the construction of local shape complementarity measures, which requires only a very simple modification.

If the molecular electron densities are represented by MIDCOs, then shape complementarity implies complementarity of two families of properties of these MIDCOs. The first family of these properties is represented by the density threshold values, whereas the other is represented by the reference curvature parameters. Indeed, shape complementarity of two MIDCOs involves complementary curvatures, as well as complementary values of the charge density contour parameters a. In this context, a locally convex domain relative to a reference curvature b shows some degree of shape complementarity with a locally concave domain relative to a reference curvature of $-b$; hence the sign symmetry of values indeed reflects some aspect of complementarity. In a similar way, shape complementarity between the lower electron density contours of one molecular fragment and the higher electron density contours of another molecular fragment is required, as implied by the partial interpenetration of interacting molecular fragments, which suggests a low–high combination of the density threshold parameter values.

The electron density clouds of interacting molecules penetrate each other only to a limited extent. The stronger the interaction, the more pronounced mutual interpenetration is likely. For each interaction, a formal "contact" density threshold a_0 can be defined as follows. Consider the complete MIDCO sequences of two interacting molecules. Calculate the MIDCOs as they occur for the isolated molecules, and place them according to the mutual nuclear positions of the interacting molecules. Consider a series of common density thresholds for these two families of MIDCOs: at high-density values, the MIDCOs of the two molecules have no contacts, but at low-density thresholds, they intersect each other. Following the changes as they occur for gradually decreasing common threshold values, the threshold where the first contact occurs defines the formal contact density a_0 for this pair of interacting molecules.

Consider the conformations K_1 and K_2 of interacting molecules M_1 and M_2, respectively, and their local shape complementarity, with reference to the MIDCOs $G(K_1, a_0)$ and $G(K_2, a_0)$ for the given interaction and a contact density value a_0. Naturally, shape complementarity is limited to MIDCOs of a single density threshold value a_0, and, generally, one should consider the local shape complementarities of all MIDCO pairs $G(K_1, a_0 - a')$ and $G(K_2, a_0 + a')$ in some narrow density interval surrounding the reference density threshold:

$$[a_0 - \Delta a, a_0 + \Delta a]. \tag{12}$$

Based on the considerations of the mutual interpenetration of electron density clouds, the complementarity of the local shapes of MIDCO pairs of threshold values deviating from the contact density a_0 in the opposite sense is of importance.

On the local level, shape complementarity implies matches between locally concave and locally convex domains, as well as matches between properly placed saddle-type domains, where a directional convex–concave match is important. Replacing the simple $D_\mu(K, a)$ notation, the more elaborate notation $D_{\mu(b), i}(K, a)$ is used sometimes when studying the complementarity of local shape domains, where the notation includes the relative convexity specification $\mu(b)$. This quantity takes values of 2 for convex, 1 for saddle, and 0 for concave domains with respect to reference curvature b, and the shape domains of the given type are specified by a serial index i.

Accordingly, the three types of complementary local matches between curvature domain pairs can be written as

$$D_{0(b),i}(K_1, a_0 - a'), D_{2(-b),i}(K_2, a_0 + a') \tag{13}$$

$$D_{1(b),i}(K_1, a_0 - a'), D_{1(-b),i}(K_2, a_0 + a') \tag{14}$$

and

$$D_{2(b),i}(K_1, a_0 - a'), D_{1(-b),i}(K_2, a_0 + a') \tag{15}$$

respectively.

Shape complementarity, as expressed by the pairings $(b, -b)$ and (a_0-a', a_0+a'), can be represented using a single condition based on the (a,b) parameter maps. With respect to the most informative one-dimensional shape groups for both molecules, we shall consider the shape group $H_\mu^1(a,b)$ with reference to the $\mu = 2$ truncation for molecular fragment F_1 and the shape group $H_{2-\mu}^1(a,b)$ for the complementary $\mu' = 2-\mu = 0$ truncation for molecular fragment F_2. Based on these, one can construct the (a,b) map of the $H_2^1(a,b)$ shape groups of molecular fragment F_1 and the (a,b) map of the $H_0^1(a,b)$ shape groups of molecular fragment F_2 for the subsequent shape complementarity analysis.

Evidently, in these two (a,b) maps, the curvature types for the truncation in the two fragments are complementary. Unfortunately, the above two (a,b) maps cannot be compared directly since if a direct comparison is made by simply overlaying these maps, identical and not complementary, a and b values occur for the two molecular fragments. Nevertheless, a simple transformation of either one of these two maps ensures complementarity of the density threshold and reference curvature values; all one needs to do is to carry out a central inversion of one of these maps. Such a central inversion of the (a,b) parameter map of molecular fragment F_2 with respect to the point $(a_0,0)$ of the map ensures a proper match between complementary parameter values. Consequently, direct comparison of the original (a,b) map of fragment F_1 and the centrally inverted (a,b) map of fragment F_2 is suitable to evaluate shape complementarity.

Evidently, the locally convex domains of the MIDCO $G(K_1, a_0-a')$ of fragment F_1 relative to the reference curvature b are tested for shape complementarity against the locally concave domains of the MIDCO $G(K_2, a_0 + a')$ of fragment F_2 relative to a reference curvature $-b$. This is precisely what is required for both curvature and density threshold.

An additional advantage is the fact that the Centrally Inverted Map Method of molecular shape complementarity analysis [2] relies on the same fundamental method used for similarity measures. In a formal sense, the problem of shape complementarity is replaced by a problem of similarity between the original (a,b) parameter map of shape groups $H_\mu^p(a,b)$ of fragment F_1 and the centrally inverted (a,b) parameter map of the complementary $H_{2-\mu}^p(a,b)$ shape groups of fragment F_2. The approach does not involve much additional computations and can be made a routine part of similarity evaluations of molecular families.

4. MOLECULAR RECOGNITION: UNIQUENESS AND SIMILARITY

Molecular similarity and molecular complementarity are strongly related concepts; in fact, in some sense, similarity and complementarity complement each other. A specific aspect of molecular complementarity is molecular recognition, and the holographic approach provides a rather simple principle for molecular recognition: for each mutual arrangement of each molecule pair, the actual nature of molecular recognition is unique.

Approximately 100 years ago, Emil Fischer's idea of the lock-and-key analogy for molecular recognition and complementarity was, in some sense, the origin of topological considerations in the study of molecular interactions. This more recent topological approach leads not only to rigorous molecular shape descriptors and similarity measures, but also to shape complementarity measures and to algorithmic computer programs for a bias-free, nonvisual evaluation of molecular recognition. In some instances, these approaches are based on local shape features and the associated interactions between local electron density clouds.

This principle has consequences in most biochemical fields and in such applied areas as pharmaceutical drug design, herbicide, pesticide, and fungicide design, medicinal chemistry, and preventive toxicology, as well as fields such as catalysis in petroleum chemistry using zeolites or other catalysts, homogeneous catalysis in the production of fine chemicals, synthesis in natural product chemistry, and design of various supramolecular structures.

The basic idea can be formulated as a formal theorem: the Uniqueness Theorem on Molecular Recognition [27].

The nature of molecular recognition can be described in terms of the electron density deformations due to molecular interactions during the recognition process. Using an optimality condition, phrased in terms of a minimax principle described below, the fundamental holographic properties of molecular electron densities are sufficient to show that molecular recognition is necessarily unique. The theorem is connected to a "Duality Principle of Molecular Recognition," and it also leads to various selectivity measures for molecular recognition.

Many of the principles of molecular recognition are evident and well understood, without invoking any of the fundamental results of quantum chemistry or molecular informatics. The shapes of molecular electron density clouds are of primary importance in molecular recognition; in particular, the degree and specificity of molecular recognition are clearly dependent on the mutual shape conditions, with respect to shape similarity and shape complementarity of molecular electron densities. Nevertheless, a quantum chemistry and molecular informatics approach, involving the Holographic Electron Density Theorem, provides a convenient and sufficiently rigorous approach to a fundamental aspect of molecular recognition.

Since molecular recognition typically involves two or more molecules, it is useful to phrase the problem in terms of a generalization of the Holographic Electron Density Theorem to supermolecular and supramolecular structures involving several interacting, but formally individual molecules. Such a generalization is the Supramolecular Holographic Electron Density Theorem.

In the context of molecular recognition, consider a supramolecular object, for example, the interacting pair ED of an enzyme E and a drug molecule D, where, for simplicity, we assume that for all these molecules, nondegenerate ground-state electron distributions are involved. (If this is not the case, then the extension of the holographic theorem to latent properties can be used in an analogous manner.)

We consider the entire supramolecular object as a single entity that has been produced as a combination of two originally independent molecules. The process of molecular recognition itself can be regarded as a change from the noninteractive states of the two independent molecules to the new, interacting supramolecular entity composed from them.

In the course of the recognition process, the complete electron density changes. Consequently, the information about the recognition itself must be contained in the change of the electron density, as a single supramolecular object is formed from the combination of two originally independent molecules.

In order to apply the Holographic Electron Density Theorem to both the independent molecules and the supramolecular object, consider a nonzero volume part P' of the electron density of independent molecule E. For example, select a spherical volume about a specific nucleus X of molecule E. As what follows from the Holographic Electron Density Theorem, this volume P' contains all the information about the independent molecule E, assumed to be infinitely removed from any other molecule.

In the next step, bring the two molecules into some mutual position where some interaction occurs between them, and consider the same nonzero volume part P in the supramolecular object ED, for example, the spherical volume of the same radius about the same nucleus X. Whereas this volume P was originally specified for the independent molecule E, nevertheless, by applying the Holographic Electron Density Theorem to the entire supramolecular object ED, now, this volume P now contains all the information about the supramolecular object.

This result is a rather trivial consequence of the original Holographic Electron Density Theorem; nevertheless, it can be viewed as a basic aspect of supramolecular chemistry, where the components of the supramolecular structure may retain a sufficient degree of their original, individual autonomy to justify a reference to their original electron density.

The Supramolecular Holographic Electron Density Theorem [27] states the following:

> If P is a nonzero volume primarily associated with a molecular component E in a supramolecular assembly ED, where the electron density of the supramolecular object ED is characterized by a nondegenerate ground state, then the electron density in this volume P contains all information about the entire supramolecular object ED, specifically, all information about all other molecular component(s) D as they occur within the supramolecular object ED.

A Duality Principle of Molecular Recognition appears to affect many aspects of the recognition process. In the most typical case of molecular recognition, the actual recognition is a process characteristic to the given pair of molecules. It is a natural expectation that if a molecule A recognizes another molecule B, either by forming temporarily a complex or by undergoing a specific chemical reaction, then the interaction process involved in the recognition affects both molecules. The two molecules may be affected in different ways and to a different degree. It is well understood that the roles of the two molecules are seldom symmetric; nevertheless, the recognition is mutual. In the two molecules, the changes associated with the recognition process are often not only different but in fact complementary, and the degrees of selectivities concerning the interactions with the actual and some other potential partners can also be markedly different.

However, there are certain regularities. Typically, the process of molecular recognition is characterized by a possibly and often markedly asymmetric *duality*, where the roles of the recognizer molecule A and the recognized molecule B can be interchanged.

It appears useful to consider the changes involved in the recognition process in a way that can be analyzed using the electron density shape analysis methods.

For this purpose, consider the following electron densities:

ρ_A	the electron density of independent molecule A
ρ_B	the electron density of independent molecule B
ρ_{AB}	the electron density of interacting molecule pair AB
$\rho_{A(AB)}$	the electron density of fragment A within interacting molecule pair AB (as obtained by the AFDF density fragmentation process applied to molecule pair AB)
$\rho_{B(AB)}$	the electron density of fragment B within interacting molecule pair AB (as obtained by the AFDF density fragmentation process applied to molecule pair AB)

In terms of a shape group analysis on these densities and the computed shape similarities, the following shape similarities, $s(\rho_A, \rho_{A(AB)})$ and $s(\rho_B, \rho_{B(AB)})$, appear to have special importance.

Depending on the relative magnitudes of these similarity measures, several conclusions can be drawn. If

$$s(\rho_A, \rho_{A(AB)}) < s(\rho_B, \rho_{B(AB)}) \tag{16}$$

then the molecular component A is affected by a greater degree in the recognition process than component B since independent molecule A is less similar to the interacting molecule A, when compared to the change in the case of molecule B. (By reassigning labels A and B, the roles can be reversed.) Of course, these similarities are practically never perfect; consequently, the condition

$$s(\rho_B, \rho_{B(AB)}) = 1 \tag{17}$$

is practically never fulfilled (with the exception of trivial identity). The recognition process usually introduces at least some changes in the electron densities of all molecules involved.

Some asymmetry of recognition as expressed by the inequality Eq. (16) is typical; nevertheless, in the case of a self-recognition process between two identical molecules along a symmetric interaction pathway, and in some other, probably rather exceptional instances, perfect duality is possible, as measured by the shape group similarity measures:

$$s(\rho_A, \rho_{A(AB)}) = s(\rho_B, \rho_{B(AB)}) \tag{18}$$

The selectivity of molecular recognition can also be studied based on the shape group similarity measures. For example, consider a simple family **A** of two molecules,

$$\mathbf{A} = \{A_1, A_2\} \tag{19}$$

taking the formal role of recognizer and another family **B** of k molecules to be recognized:

$$\mathbf{B} = \{B_1, \ldots B_i, \ldots B_k\} \tag{20}$$

With reference to the family B as background information, the selectivity of recognition of molecule B_i by molecule A_1 can be characterized by the quantity $t(A_1, B_i)$

$$t(A_1 B_i) = \min_{k, k \neq i} \left\{ \mathrm{abs} \left[s(\rho_A, \rho_{A(AB_i)}) - s(\rho_{A_1}, \rho_{A(AB_k)}) \right] \right\}. \tag{21}$$

This measure involves a minimax principle. By definition, the greater this number, the greater is the smallest difference between the changes of electron densities caused by the interaction between A_1 and B_i on the one hand, and the next most similar change between A_1 and any of the B_k molecules excluding B_i. In other words, the greater this number $t(A_1, B_i)$, the greater the selectivity of molecule A_1 recognizing molecule B_i as compared to all other molecules from the given family **B**. Hence this measure is, indeed, context-dependent, where the family **B** provides the context.

If the following condition is fulfilled,

$$t(A_1, B_i) > t(A_2, B_i), \tag{22}$$

then molecule A_1 is more selective in recognizing molecule B_i from the molecular family B than molecule A_2. Hence the measure provides a discrimination among the "recognizers."

For a discussion of the uniqueness of molecular recognition, consider now an A and B pair of molecules involved in the recognition process and the electron densities ρ_A, ρ_B, ρ_{AB}, $\rho_{A(AB)}$, and $\rho_{B(AB)}$ discussed above. The associated difference densities $\Delta\rho_{A(AB)}$ and $\Delta\rho_{B(AB)}$ are defined as

$$\Delta\rho_{A(AB)} = \rho_{A(AB)} - \rho_A, \tag{23}$$

and

$$\Delta\rho_{B(AB)} = \rho_{B(AB)} - \rho_B, \tag{24}$$

respectively. The two quantities $\Delta\rho_{A(AB)}$ and $\Delta\rho_{B(AB)}$ can be regarded as the individual electron density responses of molecules A and B to the interaction associated with their mutual recognition process. We know that the holographic theorem applies to densities ρ_A and ρ_B; however, it does not directly apply to fuzzy density fragments $\rho_{A(AB)}$ and $\rho_{B(AB)}$. Nevertheless, the holographic theorem does apply to the supramolecular object AB, that is, to density ρ_{AB}. Consequently, the actual difference densities, ρ_{ABA} and ρ_{ABB}, defined as

$$\Delta\rho_{AB\backslash A} = \rho_{AB} - \rho_A, \tag{25}$$

and

$$\Delta\rho_{AB\backslash B} = \rho_{AB} - \rho_B, \tag{26}$$

respectively, also exhibit the holographic property, provided that molecule A is specified for $\Delta\rho_{ABA}$ and molecule B is specified for $\Delta\rho_{ABB}$, and a mutual arrangement, for example, an optimality condition, is also specified for the relative geometrical placements of the interacting and noninteracting molecules.

Consider the interacting molecule pair AB and select a positive volume P. Since the holographic theorem applies for the corresponding nondegenerate ground electronic states, the difference densities $\Delta\rho_{ABA}$ and $\Delta\rho_{ABB}$, restricted to the volume P, fully determine the complete difference densities $\Delta\rho_{ABA}$ and $\Delta\rho_{ABB}$, respectively. As a consequence, both the local and the global electron density responses involved in A recognizing B and B recognizing A are necessarily unique. This proves the following:

Uniqueness Theorem of Molecular Recognition

> Molecular recognition, as monitored by changes of electron densities in any positive volume P, is necessarily unique, characteristic to the given molecule pair with the given mutual arrangement.

On intuitive grounds, the Uniqueness Theorem of Molecular Recognition is a plausible result; however, the electron density proof presented here provides justification for regarding the detailed analysis of molecular recognition processes as a tool that is sufficient, in principle, for unambiguous identification of molecules, analogous to a formal "molecular fingerprinting."

REFERENCES

1. Carbó R, Arnau M. Molecular Engineering: A General Approach to QSAR. In: de las Heras FG, Vega S, eds. Medicinal Chemistry Advances. Oxford: Pergamon Press, 1981: 72–86.
2. Hohenberg P, Kohn W. Inhomogeneous electron gas. Phys Rev 1964; 136:B864–B865.
3. Riess J, Münch W. The theorem of Hohenberg and Kohn for subdomains of a quantum system. Theor Chim Acta 1981; 58:295–300.
4. Mezey PG. Generalized chirality and symmetry deficiency. J Math Chem 1998; 23:65–84.
5. Mezey PG. The holographic electron density theorem and quantum similarity measures. Mol Phys 1999; 96:169–178.
6. Walker PD, Mezey PG. Molecular electron density Lego approach to molecule building. J Am Chem Soc 1993; 115:12423–12430.
7. Walker PD, Mezey PG. Ab initio quality electron densities for proteins: a MEDLA approach. J Am Chem Soc 1994; 116:12022–12032.
8. Walker PD, Mezey PG. Realistic, detailed images of proteins and tertiary structure elements: ab initio quality electron density calculations for bovine insulin. Can J Chem 1994; 72:2531–2536.
9. Walker PD, Mezey PG. A new computational microscope for molecules: high resolution MEDLA images of taxol and HIV-1 protease, using additive electron density fragmentation principles and fuzzy set methods. J Math Chem 1995; 17:203–234.
10. Mezey PG. Macromolecular density matrices and electron densities with adjustable nuclear geometries. J Math Chem 1995; 18:141–168.
11. Mezey PG. Quantum similarity measures and Löwdin's transform for approximate density matrices and macromolecular forces. Int J Quant Chem 1997; 63:39–48.
12. Mezey PG. Potential Energy Hypersurfaces. Amsterdam: Elsevier, 1987:1–538.
13. Mezey PG. Group theory of electrostatic potentials: a tool for quantum chemical drug design. Int J Quantum Chem, Quantum Biol Symp 1986; 12:113–122.
14. Mezey PG. Tying knots around chiral centres: chirality polynomials and conformational

invariants for molecules. J Am Chem Soc 1986; 108:3976–3984.

15. Mezey PG. The shape of molecular charge distributions: group theory without symmetry. J Comput Chem 1987; 8:462–469.

16. Mezey PG. Group theory of shapes of asymmetric biomolecules. Int J Quantum Chem, Quantum Biol Symp 1987; 14:127–132.

17. Mezey PG. Three-Dimensional Topological Aspects of Molecular Similarity. In: Johnson MA, Maggiora GM, eds. Concepts and Applications of Molecular Similarity. New York: Wiley, 1990:321–368.

18. Mezey PG. Molecular Surfaces. In: Lipkowitz KB, Boyd DB, eds. Reviews in Computational Chemistry. New York: VCH Publ, 1990:265–294.

19. Mezey PG. Shape in Chemistry: an Introduction to Molecular Shape and Topology. New York: VCH Publishers, 1993.

20. Mezey PG. Shape Analysis of Macromolecular Electron Densities. Struct Chem 1995; 6:261–270.

21. Mezey PG. Functional groups in quantum chemistry. Adv Quantum Chem 1996; 27:163–222.

22. Mezey PG. Quantum chemistry of macromolecular shape. Int Rev Phys Chem 1997; 16:361–388.

23. Mezey PG. Combinatorial aspects of biomolecular shape analysis. Bolyai Soc Math Stud 1999; 7:323–332.

24. Mezey PG, Fukui K, Arimoto S, Taylor K. Polyhedral shapes of functional group distributions in biomolecules and related similarity measures. Int J Quantum Chem 1998; 66:99–105.

25. Mezey PG. Molecular similarity and host–guest interactions. Theor Comput Chem 1999; 6:593–612.

26. Mezey PG. The holographic principle for latent molecular properties. J Math Chem 2001; 30:299–303.

27. Mezey PG. A uniqueness theorem on molecular recognition. J Math Chem 2001; 30:305–313.

14

Quantum Similarity and Quantitative Structure–Activity Relationships

RAMON CARBÓ-DORCA and XAVIER GIRONÉS

University of Girona, Campus Montilivi, Girona, Catalonia, Spain

1. INTRODUCTION

Since the middle of the 19th century [1], when several authors started dealing with various kinds of structure–property relationships, up to today the so-called quantitative structure–activity relationships (QSAR) have generated quite a large amount of literature (see for example the contributions of Ref. 2). In the near past, a more general application landscape has naturally emerged from initial QSAR ideas, providing the generic concept of quantitative structure–property relationships (QSPR) (see for example Ref. 3). Even more recently, the particular conceptual and practical use of quantitative structure–toxicity relationships (QSTR) is appearing with some frequency in the current literature (see for example Ref. 4).

The fundamental idea of QSAR consists of the possibility of a relationship between a set of descriptors, which are derived from molecular structure, and a molecular response. Within this scope, several molecular descriptors, which discretely parameterize a given molecular set, have been devised. From the early work of Cross [1], where a relationship between toxicity and water solubility was observed, several other parameters have been proposed, such as Hammett's sigma [5], which accounts for electronic effects due to molecular backbone substitution, or the octanol/water partition coefficient [6], widely used to describe lipophilicity. Later on, other frameworks have been formulated in order to include molecular shape, size, polarizability, and many other structural features, also based on the three-dimensional molecular character. A number of reviews have been published [7–9], concerning the historical development, descriptor generation, and their application into the QSAR field.

In addition, the recent and incredibly fast development of computers, both in architecture and speed, has increased the computing power up to levels that allow the application of quantum theory to fair-sized organic molecules at fairly accurate computational levels (semi-empirical or ab initio with appropriate basis sets) within reasonable time limits and affordable costs. These advances have made possible the development of theoretical molecular descriptors, some of them based on quantum mechanics easier, leading to the concept of Computer-Aided Drug Design.

Within this scope, in the past few years much effort has been devoted to applying the idea of molecular quantum similarity measures (MQSM) to rational drug design [10–25]. Because of its importance, this area of theoretical chemistry has experienced a steady growth. The mathematical background for this new expanding field was formulated some time ago by Carbó et al. [26], who introduced the concept of MQSM. Since then, great progress has been made, not only in basic methodology but also in the formulation of robust computational schemes [27–39] and application to QSAR in both pharmacological [10–13,40–43] and toxicological [44–47] fields.

This chapter is intended to provide the readers with a simple theoretical introduction to the MQSM field, followed with some illustrative examples, where the practical application of MQSM theory can be judged.

2. QUANTUM MECHANICAL BASIS OF MOLECULAR SIMILARITY

2.1. The Density Function Role

As previous chapters have shown, quantum systems such as molecules obey the Schrödinger equation. For the present purposes, we deal with time-independent states, for which we may write:

$$H\Psi = E\Psi$$

Once the hamiltonian operator has been defined for the N-electron system, the Schrödinger equation [48–50] yields both the state energies and the corresponding wavefunctions for the quantum system. From the wavefunction, the density function ρ is obtained as:

$$\rho = |\Psi|^2$$

The first-order density function describes the probability of finding an (or more precisely *any*) electron at a point r, by integrating the wavefunction over all spins and all $N-1$ coordinates:

$$\rho^{(1)}(r_1) = N \int \ldots \int \Psi^*(x_1, \ldots x_N)\Psi(x_1, \ldots x_N)ds_1 \ldots ds_N dr_2 \ldots dr_N$$

It is these first-order density functions that are considered in this work.

These density functions can be used in the calculation of expectation values rather than using the more commonly known formula based on wave functions. For the different observables, one has a formula [51]:

$$\langle\theta\rangle = \int_{x'=x} \Theta\rho(x, x')dx$$

The statistical foundation for the application and interpretation of quantum mechanics, expressed in this way, has to be considered as the basic element of the theory to be used in practical studies, concerning chemical problems. The structure and properties of the density function seem to have given it a secondary role in its application of quantum mechanics within a chemical scope. Maybe this quantum mechanical density function's secondary and ancillary position has been the cause, which has somehow, according to the literature, postponed the possibility of enlarging their application field and perspectives. Another possibility of the low application profile of density functions may be also reflected in the fact that literature trends have usually dealt with chemical systems per se but very seldom *in relationship* with other parent structures. However, the chemical language is full of comparison sentences involving two or more molecules. Even more curious: experimental chemistry, since the initial analysis of the atomic properties, which has led to the construction of the periodic table of the elements, tends to produce information about chemical properties by employing comparative reasoning.

Because the density function of a chemical system, constructed in a precisely described internal energy state, is the recipient of all the observable information, which can be extracted from such a system, it becomes logical to also consider the theoretical possibility of using quantum mechanical density functions in order to develop the procedures, which will finally allow the comparison of two or more microscopic systems.

Density functions can be considered at the same time as *functions and operators*, in the form of projectors (see for example Ref. 51). Thus nothing prevents the extraction of the numerical figures containing the measure of the similarity degree between the compared systems. This can be done in the form of a statistical expectation value, associated with the usual manner to obtain it within a unique system state.

Such is the role, which can be attached, when studying chemical systems, to their quantum mechanical density functions.

2.2. Quantum Similarity Measures

Although we will mainly be interested in molecular species, it is worth introducing the terms "tagged set" and "quantum object." If we consider a collection of objects, such as molecules, we may construct a first set, the object set. The features of the molecules form a second set, the tag set. A tagged set collects both sets, e.g., the identification of the molecule and its tags. A quantum object is defined simply as an element of such a tagged set. In the present application, its object part holds the identity of the molecule involved, together with the state description, while the tags are the density functions for that molecule in its specific state. Once two quantum objects are known, the definition of a quantum similarity measure (QSM) becomes easy to construct. Comparison of two quantum objects can be easily performed using their corresponding density tags. Suppose two quantum objects: $\{\omega_A = (S_A; \rho_A); \omega_B(S_B; \rho_B)\}$, where the first object symbol corresponds to the microscopic system identification and the second to the density function of the involved object. Both densities can be multiplied and integrated over the respective electronic coordinates in a convenient domain, weighted by a positive definite operator $\Omega(\mathbf{r}_1, \mathbf{r}_2)$. That is:

$$z_{AB} = \langle \rho_A | \Omega | \rho_B \rangle = \int\int \rho_A(\mathbf{r}_1)\Omega(\mathbf{r}_1, \mathbf{r}_2)\rho_B(\mathbf{r}_2)d\mathbf{r}_1, d\mathbf{r}_2. \tag{1}$$

The operator Ω can be chosen according to the nature and description of the studied cases. Equation (1), when the operator is chosen as the Dirac's delta function: $\delta(\mathbf{r}_1 - \mathbf{r}_2)$, is customarily called an *overlap* QSM, which permits quite a large set of possible applications [10,52–54] and generalizations [29,55]. Such an operator may be used when the volume of the components of a given molecular system is determinant. Another widely explored possibility of choice [41,42,44] is the use of the Coulomb operator, $|\mathbf{r}_1 - \mathbf{r}_2|^{-1}$, defining a Coulomb QSM. This operator is known to better reflect existent electrostatic interactions. The integral (1), because of the presence of the positive definite operator and density functions, always produces a result that is a positive definite real number. When relating a quantum object by itself, by means of Eq. (1), i.e., when computing Z_{AA}, a norm of the quantum object density function tag is obtained, which can be named a molecular quantum self-similarity measure (MQS-SM).

Given a set of N quantum objects, there is always the possibility of computing the whole array of QSM between quantum object pairs, producing a symmetric $(N \times N)$ matrix: $\mathbf{Z} = \{Z_{IJ}\}$, the so-called similarity matrix (SM) of the quantum object set. Such a matrix is illustrated below. The self-similarity measures are the diagonal elements of this matrix:

$$
\begin{array}{c}
\begin{array}{ccccc} 1 & & \cdots & & N \end{array} \\
\begin{array}{c} 1 \\ \vdots \\ \vdots \\ \vdots \\ N \end{array}
\left[
\begin{array}{cccc}
z_{11} & z_{12} & \cdots & z_{1N} \\
z_{21} & z_{22} & & \vdots \\
\vdots & & \ddots & \vdots \\
z_{N1} & z_{N2} & \cdots & z_{NN}
\end{array}
\right]
\end{array}
$$

Each of the columns or rows of the SM $\mathbf{z}_I$ can be considered the collection of all the QSM between the Ith quantum object and each element of the set, including itself. Consequently, every vector $\mathbf{z}_I$ can be interpreted as a discrete N-dimensional representation of the Ith quantum object and can be related to a projection of the corresponding density function tag, into a subspace generated by the quantum object set density function tags. A new breed of *discrete quantum objects* can be defined, where the objects are the same microscopic systems as before but where the tags are now the columns of the attached SM. As a result, the Ith quantum object can be written with the equivalent tagged set structure: $\Omega_I = (S_I; \rho_I) \leftrightarrow (S_I; \mathbf{z}_I)$ [25,37]. Such a procedure has the advantage in that it can substitute a smeared out density function by an N-dimensional vector, whose elements are positive definite real numbers. Moreover, such collections of newly made quantum object discrete tags, $\{\mathbf{z}_I\}$, can be considered as a set of molecular descriptors by its own nature, when the quantum objects are associated with molecular systems.

However, the SM column collection does not constitute just another set of quantum object descriptors, as those used in general to theoretically describe a given molecule. From the previous discussion, it is easy to attach to such a quantum object description the following properties. Every descriptor $\mathbf{z}_I$ is:

1) *Universal* in the sense that it can be obtained from *any* quantum object set and for any quantum object in the set.
2) *Unbiased* as in the building process there are *no other choices*, but the operator used, than those provided by the knowledge of the involved density function tags and the MQSM as described in Eq. (1).

2.3. Similarity Indices

Molecular quantum similarity measures, like any of the off-diagonal elements of the SM, z_{AB}, involving the QSM between quantum objects A and B, can be easily transformed into a number lying within the interval $[0;1]$, just by using the simple rule:

$$r_{AB} = \frac{z_{AB}}{\sqrt{z_{AA} z_{BB}}},$$

(2)

producing the so-called Carbó similarity index (CSI) [56]. The SI as defined in Eq. (2) corresponds to a cosine of the angle subtended by the involved density functions, considered in turn as vectors. When the index approaches unity, the involved quantum objects can be considered more similar, and on the contrary, the compared objects become more dissimilar with decreasing CSI. The exact unity value is only obtained when both compared objects are identical, corresponding, as commented above, to a MQS-SM.

2.4. Quantum Object Ordering

The Carbó similarity index can be immediately employed to order a given quantum object set as has been discussed and used in many instances in the literature [57]. Suppose, to simplify the arguments and in order to provide a schematic example, that three quantum objects are known and labeled $\{A;B;C\}$, and that their corresponding CSIs are written as: $\{r_{AB}; r_{AC}; r_{BC}\}$. Moreover, suppose that their magnitudes can be related following the ordering sequence: $r_{AB} > r_{BC} > r_{AC}$. From such a relationship, one can conclude that a similarity ordering between the quantum objects can be written using the scheme:

$$A \succ B \succ C$$

(3)

Thus the quantum object set can be ordered at once when the CSIs, relating the quantum objects within the set, are known. Now, the immediate application of this ordering possibility may consist of the following procedure. Suppose that a property is known for quantum objects A and C: $\{\pi_A; \pi_C\}$. One will have either: $\pi_A \geq \pi_C$ or $\pi_A \leq \pi_C$. Any of such relationships permits obtaining an estimate of the supposedly unknown value of the property for the quantum object B, using the relationship (Eq. (3)), in either case as: $\pi_A \geq \pi_B \geq \pi_C$ or $\pi_A \leq \pi_B \leq \pi_C$, respectively. Therefore at least a lower and upper bound for the value of the property for the quantum object B can easily be obtained from the knowledge of the similarity relationship (Eq. (3)).

2.5. Stochastic Transformation

Another possible scaling to be performed on the SM **Z**, other than the previous CSI definition in Eq. (2), can be performed by means of a *stochastic transformation* [58]. Such SM transform can be defined by means of:

$$s_{AB} = z_{AB} \left(\sum_{B=1}^{N} z_{AB} \right)^{-1},$$

(4)

providing a *stochastic* SM, $S = \{s_{AB}\}$, where the elements sum of every row has been used as scale factor. This procedure introduces an alternative uniform nonsymmetric SM, whose columns can be used as descriptors for a given molecular set.

2.6. Fundamental Quantum Quantitative Structure–Activity Relationships Equation

The possibility opened by the SM manipulation over a quantum object set, although appealing, will yield a very limited application of the QSM framework. The praxis of the theoretical findings has conducted the application of the SM, considered as a set of quantum objects descriptors, to QSAR model construction [45]. From the obtained initial results the possible existence of a sound reason for the general applicability of both QSM or the CSI set, in order to obtain quite accurate QSAR models, has been deduced. Soon it was apparent that such a relationship corresponded to the consequence of a simple quantum mechanical application, involving the concept of the expectation value, attached to a general quantum object property. Following quite a lengthy procedure and taking into account the previous definitions, it can be shown that the expectation value of any quantum object property can be written in terms of a linear combination of QSM, related to a parent quantum object set [33].

That is, suppose a quantum object A and the expectation value of a given property for this object: $\langle \pi_A \rangle$, which can be associated with the experimental value of the given property of the object. The following approximate relationship can be found [59]:

$$\langle \pi_A \rangle \approx w_1 z_{1A} + w_2 z_{2A} \cdots + w_A z_{AA} \cdots + w_N z_{NA} = \sum_I w_I z_{IA} = \langle w | z_A \rangle, \qquad (5)$$

where the set, $\{z_{IA}\} = z_A$, is just constituted by the components collection of the QSM N-dimensional descriptor for the quantum object A. In Eq. (5) another N-dimensional vector collection, $\{w_I\} = w$, is also present, which is a set of coefficients to be computed in order to fit optimally all the known property values of the quantum object set. This will be done in the same way as in empirical QSAR model finding. There are no mathematical differences between Eq. (5) and the usual QSAR models, except for the descriptor form. However, the fundamental quantum QSAR (Q^2SAR) Eq. (5) can be deduced from the QSM definitions, as in Eq. (1), plus the definition of the quantum mechanical expectation value concept. In this sense, not only a linear relationship between molecular properties and generally constructed, unbiased molecular descriptors is proven, but also additionally Eq. (5) provides the possible existence of a *causal relationship* between properties and QSM descriptors. Thus even if in order to obtain the coefficients of the vector w, some statistical procedure has to be sought, usually related to a least squares technique or to some connected procedures [60], the final Q^2SAR model described in Eq. (5) contains the seed of a causal connection between the structure, represented by QSM and the properties of any quantum object set. The components of the vector w can be associated with an approximate discrete representation of an unknown operator yielding to the property in Eq. (5) as an expectation value [39].

3. PRACTICAL METHODOLOGY AND APPLICATION EXAMPLES

In this section it is intended to demonstrate the QSAR capabilities of MQSM with three molecular sets, each of them related to different biological activities. The first set is composed of benzenesulfonamides [61], which exhibit binding affinity to carbonic anhydrase. The second series corresponds to 22 benzylamine derivatives [63], which are competitive inhibitors of the proteolytic enzyme trypsin. Finally, the last group is formed by 29 indole derivatives [67], acting as benzodiazepine receptor inverse antagonists. First, a detailed description of each step of the protocol is given, followed then by the application examples.

3.1. Preliminary Considerations

3.1.1. Molecular Modeling

All studied molecules in this work have been drawn and cleaned using WebLab ViewerPro [68]. These initial molecular geometries have been optimized using Ampac 6.55 [69] at the AM1 [70] semi-empirical level. Finally, molecular electronic density functions have been built using the Promolecular Atomic Shell Approximation [34–38], detailed below, using parameters fitted to the 6-311G basis set.

3.1.2. Promolecular Atomic Shell Approximation

Practical computation of the integral (1) becomes computationally expensive when the involved density functions correspond to large molecules or have been calculated at high computational levels. Even concrete applications of MQSM have been carried out at the ab initio level, when several molecules are studied simultaneously, as in QSAR studies, MQSM need to be computed several times, preventing their usage at these stages. In order to overcome this problem, the promolecular atomic shell approximation (PASA) [34–38] has been defined as a model of the true ab initio density, devised as a linear combination of $1S$ functions, and mathematically expressed as:

$$\rho_A^{PASA}(\mathbf{r}) = \frac{1}{P_A} \sum_{a \in A} P_a \rho_a^{PASA}(\mathbf{r} - \mathbf{R}a),$$

(6)

where P_a is the atomic number of each atom present in molecule A and P_A is the total number of electrons. In this way, the molecular density is considered as a simple addition of discrete atomic densities $\rho_a^{PASA}(\mathbf{r}-\mathbf{R}a)$, collapsed at the atomic locations R_a, which in turn are expressed as:

$$\rho_a^{PASA}(\mathbf{r} - \mathbf{R}_a) = \sum_{i \in a} w_i |S_i(\mathbf{r} - \mathbf{R}_a)|^2.$$

(7)

In this way, integral (1) is computed using only $1S$ gaussian functions, thereby decreasing the computational requirements and enlarging the potential application of MQSM. In addition, as it has been proven that calculations done using PASA are within 2% from the true ab initio ones [36], their use is clearly justified.

3.1.3. Molecular Alignment

Molecular quantum similarity measures, as formulated in integral (1), are dependent on the relative position of both studied molecules in space. Consequently, a procedure capable of arranging the molecular coordinates needs to be established. Two methodologies have been implemented to deal with this question: the maximal similarity rule (MSR) [71], which considers that the optimal orientation corresponds to the one that maximizes the value of integral (1); and the topo-geometrical superposition algorithm

(TGSA) [72], based simply on atomic numbers and coordinates, which considers that the optimal alignment corresponds to the pairing of the largest common substructure. In recent studies [39,40], TGSA is currently applied because of its faster execution compared to the MSR. Once the molecules are finally aligned, MQSM are computed, which corresponds to a single calculation within TGSA, whereas an exhaustive, slower, and more expensive search would have been required with the MSR in order to ensure that the maximal value is located. As an illustrative example, a TGSA solution for two benzosulfonamides is presented in Fig. 1, where the common benzosulfonamide group is superposed.

3.1.4. Treatment of Similarity Matrix and Model Building

Common chemometric tools may be applied to deal with similarity matrices. Particularly, partial least squares (PLS) [73,74] stands as an ideal technique for obtaining a generalized regression for modeling the association between the matrices **X** (descriptors) and **Y** (responses). In computational chemistry, its main use is to model the relationship between computed variables, which together characterize the structural variation of a set of N compounds and any property of interest measured on those N substances [75–77]. This variation of the molecular skeleton is condensed into the matrix **X**, whereas the analyzed properties are recorded into **Y**. In PLS, the matrix **X** is commonly built up from nonindependent data, as it usually has more columns than rows; hence it is not called the *independent* matrix, but predictor or descriptor matrix. A good review, as well as its practical application in QSAR, is found in Ref. 78 and a detailed tutorial in Ref. 79.

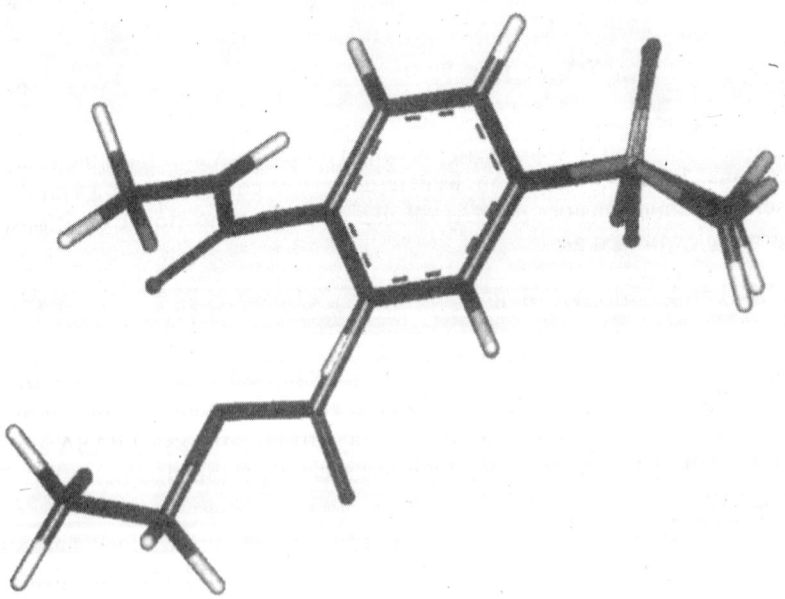

Figure 1   TGSA superposition solution for 4-CONHCH$_3$ and 3-CO$_2$C$_2$H$_5$ benzosulfonamides.

Unlike regression, PLS is not based on the assumption of independent and precise **X** variables but rather on the more realistic assumption that **X** contains more or less collinear and noisy parameters. Partial least squares summarizes these **X** variables by means of a few orthogonal score vectors ($t_a \in \mathbf{T}$), and the matrix **Y** is also resumed in a few score vectors ($u_a \in \mathbf{U}$) which are not orthogonal. Plots of columns from **T** and **U** provide a visual representation of the configuration of the observations in the **X** or **Y** space, respectively. The PLS procedure allows one to derive a number of *factors* and *weights*, which are used to describe the desired properties. Quantitative structure–property relationships models are built up from these factors and weights.

In this work, all obtained models are evaluated by commonly used statistical parameters: goodness-of-fit (r^2) [80], root mean square error between experimental and predicted values (s) [80] by *leave-one-out* [81], and predictive capacity (q^2) [80]. In addition, and in order to avoid chance correlations and excess of parameters, models are submitted to random tests, where the properties are randomly permuted in their positions and the entire modeling procedure is repeated a number of times, a thousand times in our case. If satisfactory correlations are found within the random test, the model obtained should not be trusted, as the methodology used may be potentially capable of correlating any kind of data.

3.2. Inhibition of Human Carbonic Anhydrase by Benzenesulfonamides

As previously introduced in the beginning of this section, a set of 29 benzenesulfonamides with binding affinity to human carbonic anhydrase [61] have been studied. This enzyme is partially responsible for the elimination of CO_2 (the metabolic product formed as the consequence of the use of O_2), which in normal physiological conditions exists as HCO_3 and catalyzes the reaction $HCO_3^- \rightarrow CO_2 + H_2O$; however, although this enzyme is localized in many tissues, it is linked to increased intraocular pressure, a major symptom of glaucoma [62], hence its pharmacological importance and the interest in designing inhibitors. The studied molecular set, whose structures, as well as inhibitor constants (expressed as log K), are listed in Table 1, will be studied using the following methodology:

- Construction and optimization of molecular geometries.
- Calculation of the optimal superposition by molecular pairs using TGSA.
- Computation, in the relative orientation provided by TGSA, of MQSM using the Coulomb Operator for all molecular pairs, thus obtaining a SM.
- Transformation of the SM using the Carbó Index, as in Eq. (2).
- Creation of the QSAR model from the SM and log K using PLS, along with the associated statistical parameters.

Thus, once given the SM, QSAR models have been built using up to 10 PLS factors. The evolution of both correlation and predictive capacity is presented in Fig. 2.

As can be seen, a sharp increase occurs up to the second PLS factor, whereas a moderate increase is found up to the sixth. From here, the predictive capacity slowly decreases, whereas the correlation of the models increases, thus indicating over-parameterized models. In this way, and in order to obtain a representation with an

Table 1 Structures and Inhibitor
Constants for 29 Benzenesulfonamides

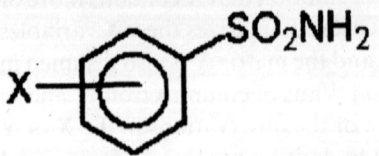

X	log K
H	6.69
4-CH$_3$	7.09
4-C$_2$H$_5$	7.53
4-C$_3$H$_7$	7.77
4-C$_4$H$_9$	8.30
4-C$_5$H$_{11}$	8.86
4-CO$_2$CH$_3$	7.98
4-CO$_2$C$_2$H$_5$	8.50
4-CO$_2$C$_3$H$_7$	8.77
4-CO$_2$C$_4$H$_9$	9.11
4-CO$_2$C$_5$H$_{11}$	9.39
4-CO$_2$C$_6$H$_{13}$	9.39
4-CONHCH$_3$	7.08
4-CONHC$_2$H$_5$	7.53
4-CONHC$_3$H$_7$	8.08
4-CONHC$_4$H$_9$	8.49
4-CONHC$_5$H$_{11}$	8.75
4-CONHC$_6$H$_{13}$	8.88
4-CONHC$_7$H$_{15}$	8.93
3-CO2CH$_3$	5.87
3-CO2C$_2$H$_5$	6.21
3-CO$_2$C$_3$H$_7$	6.44
3-CO$_2$C$_4$H$_9$	6.95
3-CO$_2$C$_5$H$_{11}$	6.86
2-CO$_2$CH$_3$	4.41
2-CO$_2$C$_2$H$_5$	4.80
2-CO$_2$C$_3$H$_7$	5.28
2-CO$_2$C$_4$H$_9$	5.76
2-CO$_2$C$_5$H$_{11}$	6.18

Source: Ref. 61.

optimal balance between correlation and predictive capacity, the one built up from
four PLS factors is taken, yielding the following model:

$$\log K = 1.565 \cdot \mathbf{f}_1 + 4.078 \cdot \mathbf{f}_2 + 1.320 \cdot \mathbf{f}_3 + 4.992 \cdot \mathbf{f}_4.$$

$$r^2 = 0.928 \qquad q^2 = 0.881 \qquad s = 0.477$$

(8)

In the model presented in Eq. (8) a sound relationship, judged from the associated statistical parameters, is obtained between the PLS factors derived from
MQSM and the inhibitor constant. These results are graphically represented in Fig. 3,
where predicted vs. experimental values of log K are plotted.

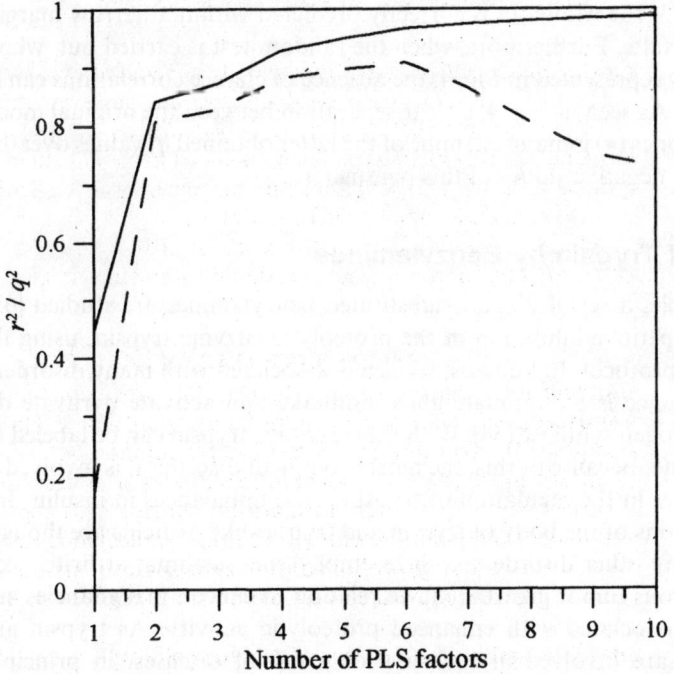

Figure 2 Evolution of r^2 and q^2 vs. the number of PLS factors used in the QSAR model.

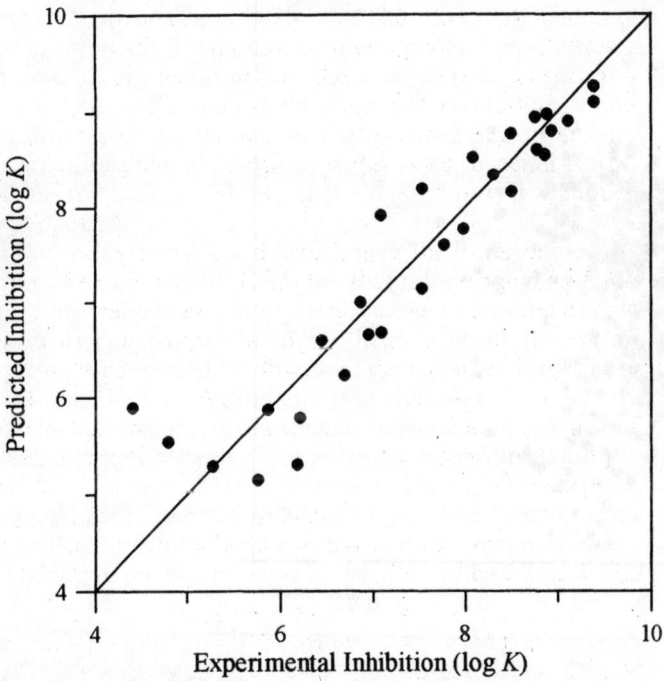

Figure 3 Experimental vs. predicted values of log K for a set of 29 benzenesulfonamides.

As seen in Fig. 3, the whole set is correctly predicted within a narrow margin, obtaining valuable results. Furthermore, when the random test is carried out, whose results are graphically represented in Fig. 4, the absence of chance correlations can be certainly ascertained. As seen in Fig. 4, a clear separation between the original model (**+**) and the random ones (•) is manifest; none of the latter obtained q^2 values over 0.5, and a vast set yielded negative values of this parameter.

3.3. Inhibition of Trypsin by Benzylamines

In this second example, a set of 22 *para*-substituted benzylamines are studied [63], along with their competitive inhibition of the proteolytic enzyme trypsin, using the previously described protocol. In humans, trypsin is associated with many disorders. Its biological significance is to stimulate glucose uptake and activate pyruvate de-hydrogenase and glycogen synthase [64]. With these actions, trypsin can be labeled an insulin mimetic enzyme. Because of this, the most prominent disorder it is involved in is diabetes. Imbalances in the regulation of trypsin cause imbalances in insulin. Imbalances in other regions of the body of trypsin and trypsin-like proteins are thought to play a role in many other disorders such as emphysema, asthma, arthritis, skin disorders, and cancerous tumor growth [65]. In relation to cancer, malignancies and tumor invasion are associated with enhanced proteolytic activity. As trypsin and trypsin-like proteins are involved in such a wide array of diseases, in principle, inhibitors of specific serine proteases could be the cure [66], hence the importance of

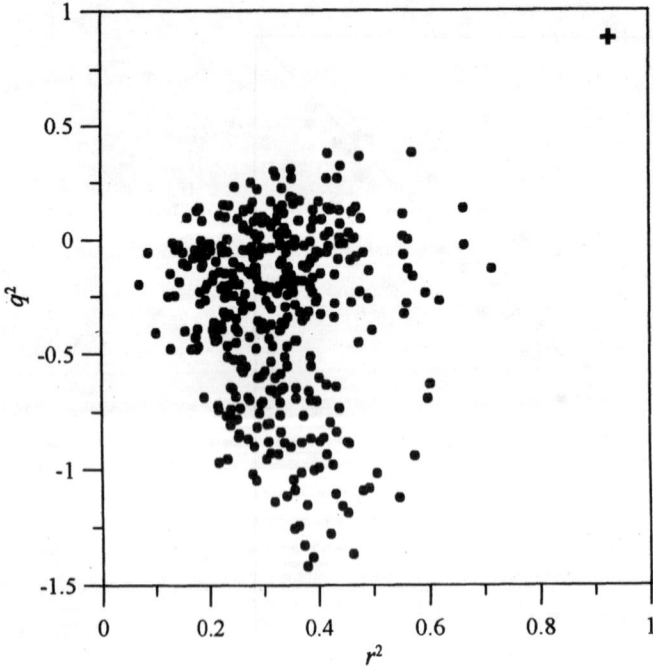

Figure 4   Random test results for a set of 29 benzenesulfonamides. Dots (•) correspond to random models and the cross (**+**) to the original one.

such inhibitors. All data regarding the studied structures and inhibition activity, expressed as pK, are summarized in Table 2.

Similarly to the previous example, and after building models up to 10 PLS factors, an optimal number of four is also chosen, resulting in the following model:

$$pK = 0.066 \cdot \mathbf{f}_1 + 1.351 \cdot \mathbf{f}_2 + 6.300 \cdot \mathbf{f}_3 + 3.375 \cdot \mathbf{f}_4. \tag{9}$$

$$r^2 = 0.799 \qquad q^2 = 0.604 \qquad s = 0.339$$

Even with worse results than the previous example, still a valuable model is obtained with reasonable predictive capacity. These results can be better judged when perusing Fig. 5a, where predicted vs. experimental values of pK are plotted. As seen from the statistical results and the cross-validation plot, it can be considered that the proposed model, as well as the results achieved, is fairly acceptable. In addition, the random test is also fulfilled, as seen in Fig. 5b.

Table 2   Structures and Inhibitor Constants for 22 Benzylamines

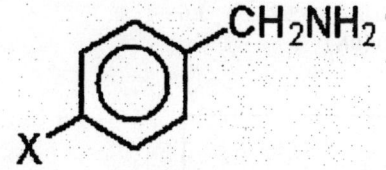

X	$-\log K$
H	0.523
CH$_3$	−0.176
Cl	0.155
OCH$_3$	0.000
OCH$_2$C$_6$H$_5$	0.398
NH$_2$	0.301
COOH	−0.301
COOCH$_3$	−0.362
COOCH$_2$CH$_3$	−0.447
COO(CH$_2$)$_2$CH$_3$	−0.301
COO(CH$_2$)$_3$CH$_3$	−0.041
COO(CH$_2$)$_4$CH$_3$	0.155
COO(CH$_2$)$_5$CH$_3$	0.523
COOCH$_2$C$_6$H$_5$	1.523
COOCH$_2$-p-Cl-C$_6$H$_4$	1.523
COO(CH$_2$)$_2$C$_6$H$_5$	0.222
COO(CH$_2$)$_3$C$_6$H$_5$	0.301
CONH$_2$	−0.398
CONHC$_6$H$_5$	0.699
CONHCH$_2$C$_6$H$_5$	0.398
CONH(CH$_2$)$_2$C$_6$H$_5$	0.523
CONHC$_{10}$H$_7$ (1-naphtyl)	1.000

Source: Ref. 63.

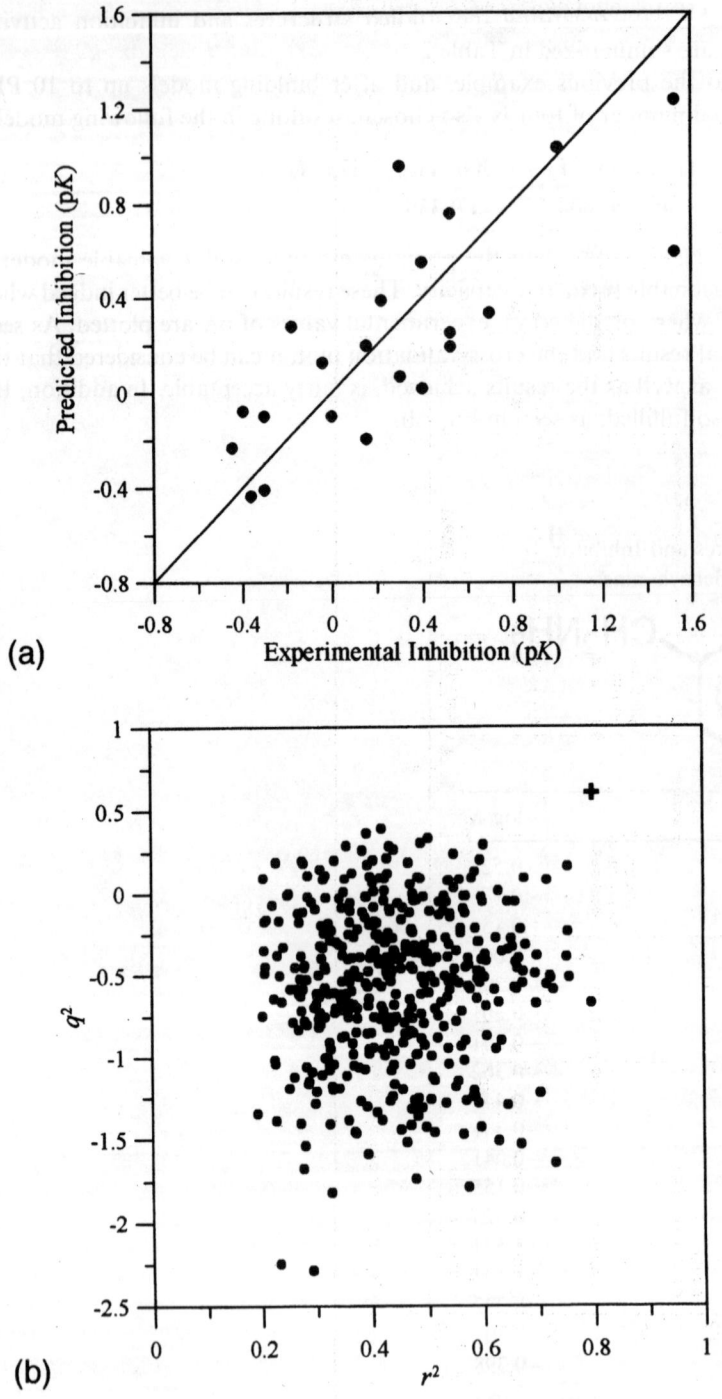

Figure 5 Plots for a set of 22 benzylamines. (a) Experimental vs. predicted values of pK. (b) Random test results.

Table 3 Structures and Inhibitor Constants for 29 Indole Derivatives

R	R_1	R_2	R_3	$-\log K$
H	H	H	H	6.92
Cl	H	H	H	6.31
NO_2	H	H	H	6.93
H	OCH_3	H	H	6.79
Cl	OCH_3	H	H	6.97
NO_2	OCH_3	H	H	7.28
H	H	OCH_3	H	6.54
Cl	H	OCH_3	H	6.79
NO_2	H	OCH_3	H	7.42
H	OCH_3	OCH_3	H	7.03
Cl	OCH_3	OCH_3	H	7.52
NO_2	OCH_3	OCH_3	H	7.96
H	Cl	H	H	7.17
Cl	H	H	Cl	5.59
NO_2	OH	H	H	6.37
H	OH	H	H	6.82
H	OH	H	H	7.92
H	H	OH	H	6.09
Cl	H	OH	H	6.24
NO_2	H	OH	H	7.19
H	OH	OH	H	6.46
Cl	OH	OH	H	6.74
NO_2	OH	OH	H	7.32
H	H	Cl	H	6.52
H	F	H	H	5.82
H	H	H	F	5.77
H	OH	OCH_3	H	6.74
Cl	OH	OCH_3	H	7.04
NO_2	OH	OCH_3	H	7.67

Source: Ref. 67.

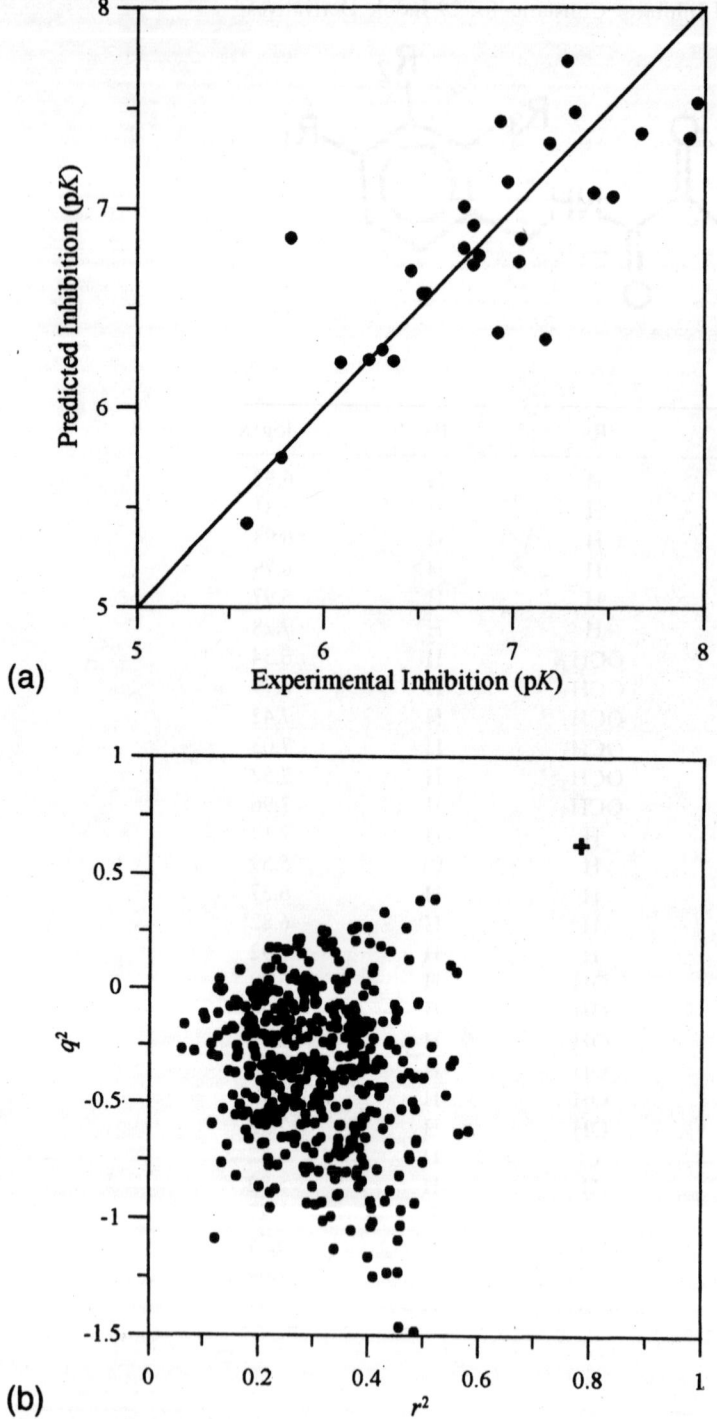

(a)

(b)

Figure 6   Plots for a set of 29 indole derivatives. (a) Experimental vs. predicted values of pK. (b) Random test results.

In this example, it is shown that although predicted results may not be very precise in terms of prediction, the proposed model can still be used for classification purposes by defining regions in the graph sorted by degrees of activity, or inhibition capacity in this case. This type of *discriminant analysis* becomes very useful when the aim of the research is to estimate in which category a molecule belongs, instead of intending to guess its exact activity.

3.4. Indole Derivatives as Antagonists of Benzodiazepine Receptor

In the last example, the protocol is applied to a set of 29 indole derivatives [67], which are benzodiazepine receptor inverse antagonists and are able to displace $[^3H]$fluni-trazepam from binding to bovine cortical membranes. Although in this case the activity refers to animal data, benzodiazepines are therapeutically employed to treat anxiety and convulsions, or to induce hypnotic states and muscle relaxation in humans. However, some people, such as the elderly during surgery, often experience respiratory arrest with such drugs. Therefore antagonists could be employed to reverse the effects of benzodiazepine-induced anesthesia and to indirectly reverse induced central nervous system depression, as well as to shed some light into the whole mechanism of action. All data regarding inhibitor structures and the related activity, expressed in terms of pK, are summarized in Table 3.

Models are built up to 10 PLS factors, finally choosing an optimal number of four, yielding the following model:

$$pK = 1.361 \cdot \mathbf{f}_1 + 4.134 \cdot \mathbf{f}_2 + 4.987 \cdot \mathbf{f}_3 + 6.774 \cdot \mathbf{f}_4.$$

$$r^2 = 0.782 \qquad q^2 = 0.623 \qquad s = 0.357$$

(10)

As seen, similar results as in the previous examples are obtained, both in correlation and predictive capacity. The pK values, as presented in Fig. 6a, are correctly predicted within a reasonable margin. Similarly to the previous example, the proposed model could be used for classification purposes, allowing prediction of the inhibition capacity in absolute terms of *strong* or *weak* inhibitor, instead of pointing to a precise discrete value. The random test, as displayed in Fig. 6b, clearly indicates the absence of chance correlations, just as in the previous cases.

4. CONCLUSIONS

The presented quantum QSAR model building protocol basically consists of MQSM and derived parameters and represents a self-contained theoretical framework, which offers the appropriate universal application, besides an unbiased parameter structure, as well as a causal relationship between structure and activity.

Even if other methodologies may provide better results, it must be stated that the methodology presented in this work, and that includes descriptor generation, similarity matrix transformation, and statistical procedure, has not been altered in any way to take into account the nature of the studied system. In this way, the exposed QSAR protocol is potentially capable of handling and characterizing different molecular biological activities from diverse molecular sets without introducing further information than those provided by quantum similarity, which is based on electronic density functions. Additional refinements or more sophisticated statistical tools may be further applied to each different case in order to improve the results; however, the exposed methodology provides acceptable correlations and may constitute an excellent starting point for subsequent research.

REFERENCES

1. Quoted in: Borman S. New QSAR techniques eyed for environmental assessments. Chem Eng News 1990; 68:20–23.
2. Kubinyi H. 3D QSAR in Drug Design. Theory, Methods and Applications. Leiden: ESCOM Science Publishers, 1993.
3. Charton M. Advances in Quantitative Structure–Property Relationships. 1st ed. London: JAI Press, 1996.
4. Boethling RS, Mackay D. Handbook of Property Estimation Methods for Chemicals. Environmental and Health Sciences. London: Lewis Publishers, 2000.
5. Hammett LP. The effect of structures upon the reactions of organic compounds. Benzene derivatives. J Am Chem Soc 1937; 59:96–103.
6. Hansch C, Fujita T. ρ-σ-π analysis. A method for correlation of biological activity and chemical structure. J Am Chem Soc 1964; 86:5175–5180.
7. Jurs PC. Quantitative Structure–Property Relationships (QSPR). In: Schleyer PvR, Allinger NL, Clark T, Gasteiger J, Kollman PA, Schaefer HF III, Schreiner PR, eds. Encyclopedia of Computational Chemistry. Vol. 4. Chichester, UK: John Wiley and Sons Ltd, 1998:2309–2319.
8. Kubinyi H. Quantitative Structure–Activity Relationships in Drug Design. Schleyer PvR, Allinger NL, Clark T, Gasteiger J, Kollman PA, Schaefer HF III., Schreiner PR, eds. Encyclopedia of Computational Chemistry. Vol. 4. Chichester, UK: John Wiley and Sons Ltd., 1998:2309–2319.
9. Waterbeemd Hvd. Structure–Property Correlations in Drug Research. Austin: Academic, R.G. Landes Company, 1996.
10. Fradera X, Amat L, Besalú E, Carbó-Dorca R. Application of molecular quantum similarity to QSAR. Quant Struct-Act Relatsh 1997; 16:25–32.
11. Lobato M, Amat L, Besalú E, Carbó-Dorca R. Structure–activity relationships of a steroid family using quantum similarity measures and topological quantum similarity indices. Quant Struct-Act Relatsh 1997; 16:465–472.
12. Amat L, Robert D, Besalú E, Carbó-Dorca R. Molecular quantum similarity measures tuned 3D QSAR: an antitumoral family validation study. J Chem Inf Comput Sci 1998; 38:624–631.
13. Robert D, Amat L, Carbó-Dorca R. Three-dimensional quantitative structure–activity relationships from tuned molecular quantum similarity measures: prediction of the corticosteroid-binding globulin binding affinity for a steroid family. J Chem Inf Comput Sci 1999; 39:333–344.
14. Amat L, Carbó-Dorca R, Ponec R. Molecular quantum similarity measures as an alternative to log P values in QSAR studies. J Comput Chem 1998; 19:1575–1583.
15. Ponec R, Amat L, Carbó-Dorca R. Molecular basis of quantitative structure–properties relationships (QSPR): a quantum similarity approach. J Comput-Aided Mol Des 1999; 13:259–270.
16. Ponec R, Amat L, Carbó-Dorca R. Quantum similarity approach to LFER: substituent and solvent effects on the acidities of carboxylic acids. J Phys Org Chem 1999; 12:447–454.
17. Good AC, Hodgkin EE, Richards WG. Similarity screening of molecular data sets. J Comput-Aided Mol Des 1992; 6:513–520.

18. Good AC, So S-S, Richards WG. Structure–activity relationships from molecular similarity matrices. J Med Chem 1993; 36:433–438.

19. Good AC, Peterson SJ, Richards WG. QSAR's from similarity matrices. Technique validation and application in the comparison of different similarity evaluation methods. J Med Chem 1993; 36:2929–2937.

20. Cooper DL, Allan NL. A novel approach to molecular similarity. J Comput-Aided Mol Des 1989; 3:253–259.

21. Measures PT, Mort KA, Allan NL, Cooper DL. Applications of momentum-space similarity. J Comput-Aided Mol Des 1995; 9:331–340.

22. Benigni R, Cotta-Ramusino M, Giorgi F, Gallo G. Molecular similarity matrixes and quantitative structure–activity relationships: a case study with methodological implications. J Med Chem 1995; 38:629–635.

23. Mestres J, Rohrer DC, Maggiora GM. A molecular field-based similarity approach to pharmacophoric pattern recognition. J Mol Graph Model 1997; 15:114–121.

24. Mestres J, Rohrer DC, Maggiora GM. A molecular-field-based similarity study of non-nucleoside HIV-1 reverse transcriptase inhibitors. J Comput Aided-Mol Des 1999; 13:79–93.

25. Carbó-Dorca R, Robert D, Amat L, Gironés X, Besalú E. Molecular Quantum Similarity in QSAR and Drug Design. Lecture Notes in Chemistry 73. Berlin: Springer Verlag, 2000.

26. Carbó R, Arnau J, Leyda L. How similar is a molecule to another? An electron density measure of similarity between two molecular structures. Int J Quant Chem 1980; 17: 1185–1189.

27. Carbó R, Domingo L. LCAO-MO similarity measures and taxonomy. Int J Quant Chem 1987; 23:517–545.

28. Carbó R, Calabuig B. Quantum molecular similarity measures and the n-dimensional representation of a molecular set: Phenyldimethylthiazines. J Mol Struct (THEOCHEM) 1992; 254:517–531.

29. Carbó R, Calabuig B, Vera L, Besalú E. Molecular quantum similarity: theoretical framework, ordering principles, and visualization techniques. Adv Quant Chem 1994; 25: 253–313.

30. Besalú E, Carbó R, Mestres J, Solá M. Foundations and recent developments on molecular quantum similarity. Topics Curr Chem 1995; 173:31–62.

31. Carbó R. Molecular Similarity and Reactivity: from Quantum Chemical to Phenomenological Approaches. Amsterdam: Kluwer Academic, 1995.

32. Carbó-Dorca R, Mezey PG. Advances in Molecular Similarity. Vol. 1 and Vol. 2. Greenwich, CT: JAI Press Inc, 1996 and 1998.

33. Carbó R, Besalú E, Amat L, Fradera X. Quantum molecular similarity measures (QMSM) as a natural way leading towards a theoretical foundation of quantitative structure–properties relationships (QSPR). J Math Chem 1995;18, 237–246.

34. Constans P, Carbó R. Atomic shell approximation: electron density fitting algorithm restricting coefficients to positive values. J Chem Inf Comput Sci 1995; 35:1046–1053.

35. Constans P, Amat L, Fradera X, Carbó-Dorca R. Quantum Molecular Similarity Measures (QMSM) and the Atomic Shell Approximation (ASA). In: Carbó-Dorca R, Mezey PG, eds. Advances in Molecular Similarity. Vol. 1. London: JAI Press, 1996:187–211.

36. Amat L, Carbó-Dorca R. Quantum similarity measures under atomic shell approximation: first-order density fitting using elementary Jacobi rotations. J Comput Chem 1997; 18:2023–2039.

37. Carbó-Dorca R. Quantum Similarity. Carbó-Dorca R, Mezey PG, eds. Advances in Molecular Similarity. Vol. 2. London: JAI Press, 1998:1–42.

38. Amat L, Carbó-Dorca R. Fitted electronic density functions from H to Rn for use in quantum similarity measures: Cis-diamminedichloroplatinum(II) complex as an application example. J Comput Chem 1999; 20:911–920.

39. Carbó-Dorca R, Amat L, Besalú E, Gironés X, Robert D. Quantum Molecular Similarity: theory and Applications to the Evaluation of Molecular Properties, Biological Activities and Toxicity. In: Carbó-Dorca R, Gironés X, Mezey PG, eds. Fundamentals of Molecular Similarity. New York: Kluwer Academic/Plenum Publishers, 2001:187–320.

40. Gironés X, Gallegos A, Carbó-Dorca R. Modeling antimalarial activity: application of kinetic energy density quantum similarity measures as descriptors in QSAR. J Chem Inf Comput Sci 2000; 40:1400–1407.

41. Robert D, Gironés X, Carbó-Dorca R. Quantification of the influence of single point mutations on haloalkane dehalogenase activity: a molecular quantum similarity study. J Chem Inf Comput Sci 2000; 40:839–846.

42. Robert D, Amat L, Carbó-Dorca R. Quantum similarity QSAR: study of inhibitors binding to thrombin, trypsin and factor Xa, including a comparison with CoMFA and CoMSIA methods. Intl J Quantum Chem 2000; 80:265–282.

43. Robert D, Gironés X, Carbó-Dorca R. Facet diagrams for quantum similarity data. J Comput-Aided Mol Des 1999; 13:597–610.

44. Gironés X, Amat L, Robert D, Carbó-Dorca R. Use of electron–electron repulsion energy as a molecular descriptor in QSAR and QSPR studies. J Comput-Aided Mol Des 2000; 14:477–485.

45. Robert D, Gironés X, Carbó-Dorca R. Molecular quantum similarity measures as descriptors for quantum QSAR. Polycycl Aromat Compd 2000; 19:51–71.

46. Robert D, Carbó-Dorca R. Aromatic compounds aquatic toxicity QSAR using quantum similarity measures. SAR QSAR Environ Res 1999; 10:401–422.

47. Gironés X, Amat L, Carbó-Dorca R. Using molecular quantum similarity measures as descriptors in quantitative structure–toxicity relationships. SAR QSAR Environ Res 1999; 10:545–556.

48. von Neumann J. Mathematical Foundations of Quantum Mechanics. Princeton: Princeton University Press, 1955.

49. Born M. Atomic Physics. London: Blackie and Son, 1945.

50. Dirac PAM. The Principles of Quantum Mechanics. Oxford: Clarendon Press, 1983.

51. Mc. Weeny R. Methods of Molecular Quantum Mechanics. London: Academic Press, 1978.

52. Amat L, Carbó-Dorca R, Ponec R. Molecular quantum similarity studies as an alternative to log P values in QSAR studies. J Comput Chem 1998; 14:1575–1583.

53. Ponec R, Amat L, Carbó-Dorca R. Molecular basis of quantitative structure–properties relationships (QSPR): a quantum similarity approach. J Comput-Aided Mol Des 1999; 13:259–270.

54. Mezey PG, Ponec R, Amat L, Carbó-Dorca R. Quantum similarity approach to the characterization of molecular chirality. Enantiomer 1999; 4:371–378.

55. Carbó-Dorca R, Amat L, Besalú E, Lobato M. Quantum Similarity. In: Carbó-Dorca R, Mezey PG, eds. Advances in Molecular Similarity. Vol. 2. London: JAI Press, 1998: 1–42.

56. Carbó R, Besalú E, Amat L, Fradera X. On quantum molecular similarity measures (QMSM) and indices (QMSI). J Math Chem 1996; 19:47–56.

57. Besalú E, Amat L, Fradera X, Carbó R. An Application of the Molecular Quantum Similarity: ordering of Some Properties of the Hexanes. In: Sanz F, Giraldo J, Manaut F, eds. QSAR and Molecular Modeling: Concepts, Computational Tools and Biological Applications. Proceedings of the 10th European Symposium on SAR, QSAR and Molecular Modeling. Barcelona: Prous Science, 1995:396–399.

58. Carbó-Dorca R. Stochastic transformation of quantum similarity matrices and their use in quantum QSAR (QQSAR) models. Intl J Quant Chem 2000; 79:163–177.

59. Carbó-Dorca R. Inward matrix products: extensions and applications to quantum mechanical foundations of QSAR. Theochem 2001; 537:41–54.

60. Neter J, Wasserman W, Kutner MH. Applied Linear Statistical Models. Boston: RD Irwin Inc, 1990.

61. Hansch C, McClarin J, Klein T, Langridge R. A quantitative structure–activity relationship and molecular graphics study of carbonic anhydrase inhibitors. Mol Pharmacol 1985; 27:493–498.

62. Friedenwald JS. The formation of the intraocular fluid. Am J Ophthalmol 1949; 32:9–27.

63. Markwart F, Landmann H, Walsmann P. Comparative studies on the inhibition of trypsin, plasmin, and thrombin by derivatives of benzylamine and benzamidine. Eur J Biochem 1968; 6:502–506.

64. Leef J, Larner J. Insulin-mimetic effect of trypsin on the insulin receptor tyrosine kinase in intact adipocytes. J Biol Chem 1987; 262(30):14837–14842.

65. Tanaka T, McRae B. Mammalia tissue trypsin-like enzymes. J Biol Chem 1983; 258(22): 13552–13557.

66. Bertrand J, Oleksyszyn J. Inhibition of trypsin and thrombin by amino methanephosphonate diphenyl ester derivatives: x-ray structures and molecular models. Biochemistry 1996; 35:3147–3155.

67. Da Settimo A, Primofiore G, Da Settimo F, Marini AM, Novellino E, Greco G, Martini C, Giannaccini G, Lucacchini A. Synthesis, structure–activity relationships, and molecular modeling studies of N-(Indole-3-ylglyoxylyl)benzylamine derivatives acting at the benzodiazepine receptor. J Med Chem 1996; 39:5083–5091.

68. WebLab Viewer Pro 4.0, Molecular Simulations Inc., 2000. A free trial version is available from its website: http://www.msi.com.

69. AMPAC 6.55, Semichem, Inc., 7128 Summit, Shawnee, KS 66216. DA, 2001.

70. Dewar MJS, Zoebisch EG, Healy EF, Stewart JJP. AM1: a new general purpose quantum mechanical molecular model. J Am Chem Soc 1985; 107:3902–3909.

71. Constans P, Amat L, Carbó-Dorca R. Toward a global maximization of the molecular similarity function: superposition of two molecules. J Comput Chem 1997; 18:826–846.

72. Gironés X, Robert D, Carbó-Dorca R. TGSA: a molecular superposition program based on topo-geometrical considerations. J Comput Chem 2000; 22:255–263.

73. Höskuldsson A. Prediction Methods in Science and Technology. Copenhagen: Thor, 1996.

74. Tenenhaus M. Regression de PLS. Paris: Editions, 1997.

75. Wold S, Johansson E, Cocchi M. PLS—Partial Least-Squares Projections to Latent Structures. In: Kubinyi H, ed. 3D QSAR in Drug Design, Theory, Methods and Applications. Leiden: ESCOM Science, 1993:253–550.

76. Wold S. PLS for multivariate linear modelling. In: Waterbeemd Hvd, ed. Methods and Principles in Medicinal Chemistry. Chemometric Methods in Molecular Design. Vol. 2. Weinheim: VCH, 1995:195–218.

77. Cramer RD III., Patterson DE, Bunce JD. Comparative molecular field analysis (CoMFA): 1. Effect on shape on binding of steroids to carrier proteins. J Am Chem Soc 1988; 110:5959–5967.

78. Wold S, Sjöström M, Eriksson L. Partial Least Squares Projections to Latent Structures (PLS). In: Schleyer PvR, Allinger NL, Clark T, Gasteiger J, Kollman PA, Schaefer HF III, Schreiner PR, eds. Encyclopedia of Computational Chemistry. Vol. 4. Chichester, UK: John Wiley and Sons Ltd., 1998:2006–2021.

79. Geladi P, Kowalski BR. Partial least-squares regression: a tutorial. Anal Chim Acta 1986; 185:1–17.

80. Montgomery DC, Peck EA. Introduction to Linear Regression Analysis. New York: Wiley, 1992.

81. Wold S. Cross-validatory estimation of a number of components in factor and principal component models. Technometrics 1978; 20:397–405.

15

Protein Structures: What Good Is Beauty If It Cannot Be Seen?

SANDER B. NABUURS, CHRIS A. E. M. SPRONK, ELMAR KRIEGER, and GERT VRIEND

University of Nijmegen, Nijmegen, The Netherlands

ROB W. W. HOOFT

Bruker Nonius BV, Delft, The Netherlands

The availability of the 3-D coordinates of biological macromolecules has revolutionized scientific fields as diverse as drug design, evolution theory, protein structure prediction, and molecular biology. Since the early 1960s, on the order of 10^5 structures have been solved, and about 18,000 of these are available to the general public via the Protein Data Bank (PDB). The extraordinary importance of these coordinates in designing theoretical and practical experiments in the scientific fields cited above creates a need for quality indicators that are understandable to scientists without crystallographic training.

We report on a project that attempts to design such quality indicators. The described software produces a reader-friendly report on all anomalies, unique features, and errors in PDB files. The magnitude of the coordinate errors can, in many cases, be assessed and, in some cases, these errors can even be corrected without the need of using the experimental structure determination data. Several aspects of structure validation will be discussed and a few examples will be worked out in detail to illustrate the possibilities and difficulties of validation. It is envisaged that the software described here will help the scientists in determining these structures to maintain the quality of their products in an environment that allows for an ever-increasing rate of structure solution.

1. INTRODUCTION

Although it seems likely that the number of unique protein folds is only of the order of 10^3 [1], the total number of possible protein structures is probably many orders of magnitude higher. This means that many aspects of protein structures can be studied by statistical analysis. From the structural analyses that have been performed over the last 40 years, many rules of protein folding have emerged, some of which hold with virtually no exceptions. Knowledge of these rules, either in the human mind, or implemented in computer programs such as Insight [2], O [3], or WHAT IF [4] is a prerequisite for the structure-based design of practical or theoretical experiments on proteins. Today, we are only beginning to understand a few of the factors that govern the structure and function of proteins. The biannual CASP meetings [5–8] on the assessment of the quality of protein structure prediction techniques make us painfully aware that there is still much to be learned. Computational analyses of experimentally determined protein structure coordinates, preferably backed up by molecular biological and biophysical experiments, is at present the only promising route leading to greater insights into the intricacies of protein structures.

1.1. Why?

Structure-based drug design [9] is the simplest way to illustrate the importance of determining the three-dimensional coordinates of proteins to people beyond the boundaries of the scientific community. Despite the fact that structure-based drug design (also called rational drug design [9], a term we consider inappropriate because it suggests that all other methods are irrational) has not lived up to initial expectations, it is now routinely employed in the pharmaceutical industry as one of the many techniques that lead to the design of new medicines.

1.2. The Importance of "Correct" Structures

Immediately after the coordinates of the human immunodeficiency virus (HIV)-1 protease became available [10], they were used to direct the search for inhibitors. However, the initial structure included a mistake in the dimer interface region, which made it difficult to understand the molecular mechanisms of the release of HIV-1 [11]. Blundell and Pearl [11] were the first to suggest that there might be something wrong with the proposed HIV-1 structure based on a comparison with the homologous retroviral Rous sarcoma virus (RSV) protease [12]. The disparity between the two retroviral protease structures was reinforced by a model of the HIV-1 protease, built by using RSV as a template [13]. Finally, the discrepancy was solved by Wlodawer [14], who determined the structure of a synthetic HIV-1 protease, but now with the correct chain tracing of the amino-terminal strands at the dimer interface. The use of HIV protease inhibitors in the fight against the human immunodeficiency virus (HIV) has not only been a major breakthrough in the battle against AIDS, it was also the documentation of a major breakthrough for structure-based drug design.

1.3. Origin of Errors

The main cause of the existence of errors in protein structure coordinates is that x-ray structure determination consists of a long series of complicated steps. Starting with the exposure of imperfect crystals to x-rays, reflections are obtained which, after several steps, including computations and human interpretation, result in a set of three-dimensional coordinates. The resulting structure is so complex that the human brain can only marvel at its beauty without ever fully understanding the underlying interplay of interactions. If atomic resolution (i.e., better than 1.2 Å) can be obtained, it becomes difficult to solve the structure incorrectly. However, as was clearly demonstrated in a large-scale experiment [15], even at 1.0 Å resolution, quite a number of small anomalies can still be observed. Most errors occurred in areas where, even at this resolution, the map was still poor and/or the B-factors were high.

Gerard Kleywegt once deliberately threaded the sequence backward through a 3.0-Å electron density map of CRABP II, thus placing every residue in a wrong position. He refined this wrong structure to an R-factor of 21.4% [16]. Using frequently reported criteria (at that time) such as conventional R-value, root-mean-square (RMS) deviations from ideality of bond lengths and angles and average B-factors, the backward-traced model would have gone undetected. Only more informative quality indicators such as the R-free calculation [17], as well as about every structure validation method ever designed (EU 3D Network, personal communication), detected that this structure was wrong. In this light, it is perhaps worrying that 23% of all x-ray structures in the PDB have an R-factor greater than or equal to 21.4%.

1.4. How to Cope with Errors

If the experimental data is not good enough, the quality of the coordinates will suffer. This is simply unavoidable. If the observed reflections are of high quality, i.e., high resolution and a high signal-to-noise ratio, there are many numerical indicators that can tell the crystallographer whether or not the data was treated optimally. If the crystals are of poor quality, the crystallographer will need to base his judgement of the quality of data handling on intuition rather than on hard numerical determinants. When, after lots of work, the reflections obtained from poor crystals are converted into an interpretable electron density map, the crystallographer faces the problem that it is impossible to avoid all errors. The questions that remain are how to minimize the number of errors and how to optimize the flagging of unavoidable ones. Besides obtaining better crystals, there are three ways to cope with the problem of errors resulting from poor data.

1.   The best thing to do is to improve the data handling. Quite frequently, better data-handling programs are becoming available, either in the form of updates on existing software, or the introduction of totally new methods. The last few years have seen new methods for image processing, multiple isomorphous replacement (MIR) phasing, better signal integration, better data scaling, better detector calibration, and better absorption correction. This list indicates that the structure community is working hard on improving data handling. The weak link in the process lies in the refinement software. A large fraction of the detected errors are not caused by poor x-ray data, but by poor refinement.

2.   The second approach to dealing with poor data is the use of standard conformation libraries [3]. It is clear that the use of "perfect" coordinates in such libraries to help building the model is to be preferred over coordinates with errors. However, there is one danger in the use of standard libraries: circularity. Take, for example, side chain conformation libraries [3,18,19]. Residues positioned in a structure purely based on such a database will not easily be flagged as incorrect by rotamer

validation software, because the validation is based on a similar library. They will end up in the PDB, falsely creating the idea that these rotamers are so common that other ones almost must be wrong. Of course, electron density will reflect reality, but it is not always available or users do not know how to interpret the electron density. Therefore to avoid wrong interpretations, all residues obtained by using standard libraries, rather than by electron density, should be flagged.

3. The third manner of dealing with errors is to accept that certain errors are inevitable, and that these errors become considerably less of a problem if they are properly flagged. If the drug designer or molecular biologist knows that certain residues are likely to be imprecise, the subject can take appropriate action in the design of experiments.

1.5. When to Call an Anomaly an Error

Three conditions must be satisfied before an observation can be considered an error. First, it should be known what the expected value distribution is for that observation. Second, if the distribution is normal, the standard deviation should be known. Third, it should be decided what is the maximally allowed chance that an observation which is called an error is actually correct. Just as a reminder, in Table 1 we give the relation between the Z-score (this is the number of standard deviations that an observation varies from the expected value) and the chance that such an observation is not an error, assuming a normal distribution.

So, if we use a 4σ cutoff, we take a 1 in 10,000 chance that something which is called wrong is actually right. Therefore barring any bugs in the WHAT_CHECK software, the PDBREPORT database (http://www.cmbi.kun.nl/gv/pdbreport) [20], which contained about 10,000,000 error messages in early 2002, should statistically hold about 1000 messages in which something is called an error that is actually correct. A few examples of outliers exceeding the 4σ cutoff are shown in Fig. 1.

1.6. This Chapter

In this chapter, we describe the techniques for finding errors and anomalies observed in protein coordinates. We will show examples of several categories of warnings that the structure validation software WHAT_CHECK can give. For example, we will show how the side chains of asparagine, glutamine, and histidine residues can be corrected by an evaluation of the global hydrogen bonding network. Furthermore, we will list a series of administrative errors. We will also show that sometimes things are perfectly

Table 1 Relation Between the Z-score and the Chance that the Error Message is a False Positive

Z-score	Chance [%]
1	31
2	4.5
3	0.27
4	0.01

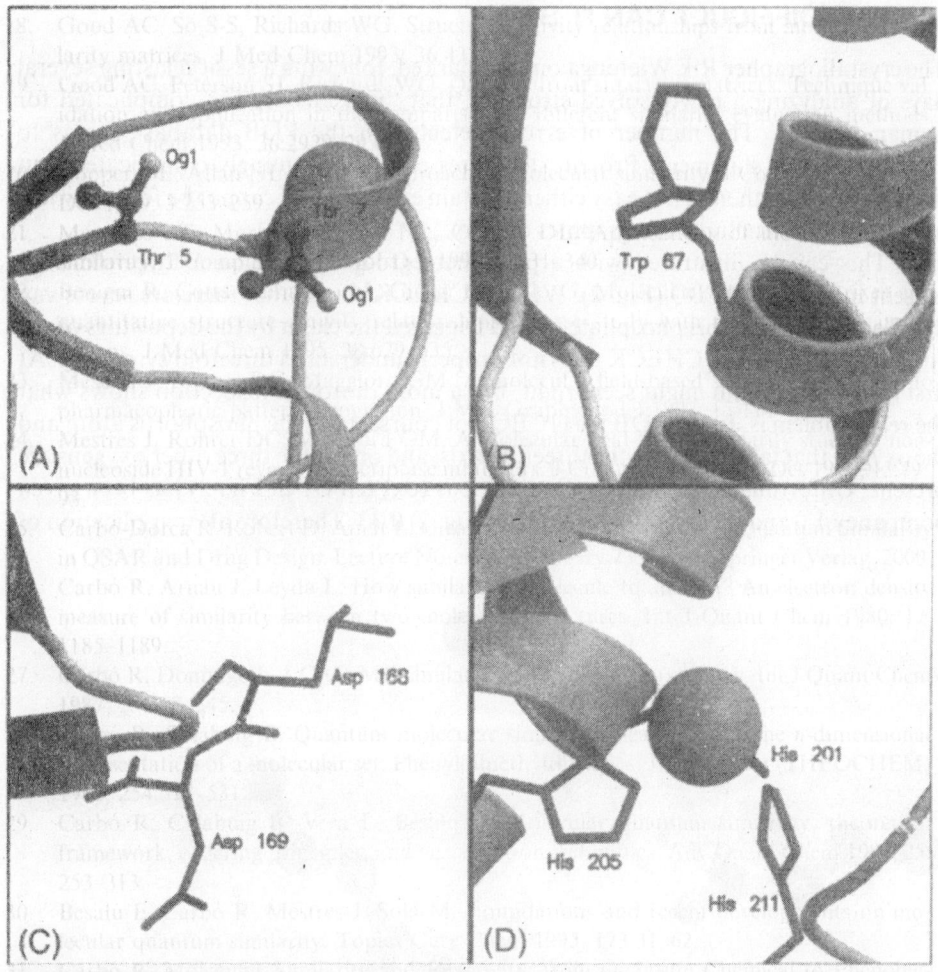

Figure 1 Some structural anomalies. (A) A threonine residue with swapped chirality on the C^β carbon (Thr 5, PDB entry 5RXN). (B) A tryptophan residue with an angle of 106° between the two rings (Trp 67, PDB entry 7GPB). (C) An aspartate residue with a 64σ deviation from planarity (Asp 168, PDB entry 1DLP); an aspartate with correct planarity (Asp 169) is shown for comparison. (D) A histidine residue with a 23σ deviation from planarity (His 211, PDB entry 1BIW). (Figure created using Molscript [37] and Raster3D [38].)

correct from a structure determination viewpoint, but nevertheless need modification before they become useful for a bioscientist mining the PDB. WHAT_CHECK, which is available at no fee to the crystallographic community (academia and industry alike), encourages that structures be validated during the refinement process (more information on the program is available on http://www.cmbi.kun.nl/gv/whatcheck/).

We will end this chapter with a short historical overview. How do the presently solved structures compare in quality to structures published 5 or 10 years ago? Is our improved knowledge of protein structures finding its way into the commonly used structure refinement programs?

2. HOW DIFFICULT CAN IT BE?

The crystallographer Rik Wierenga once remarked, following a session lasting several days of analyzing a newly solved structure, that "proteins are too complicated for human beings." The number of errors detected in the PDB database seems to corroborate this statement. Protein structures can be intriguingly complicated, but locating errors in them is not easy either. Certain errors simply cannot be found by just looking at the structure on a graphics display.

This can be illustrated with a phosphate group located on a threefold axis, present in PDB file 1CBQ (Fig. 2). WHAT_CHECK claims that there are severe van der Waals clashes in this phosphate, which triggered the claim by the depositors of the structure that WHAT_CHECK does not properly understand threefold symmetry. At first glance, this claim might seem right, but a more thorough inspection shows what the real problem is. In the PDB file (1CBQ), of course, only the phosphorus atom and the oxygen that fall exactly on the threefold axis, and one of the three other oxygens are present. Unfortunately, the oxygen on the axis (oxygen O1 in Fig. 2) has been given occupancy 1.0 and the other oxygen (O2 in Fig. 2) 0.33. Therefore after application of

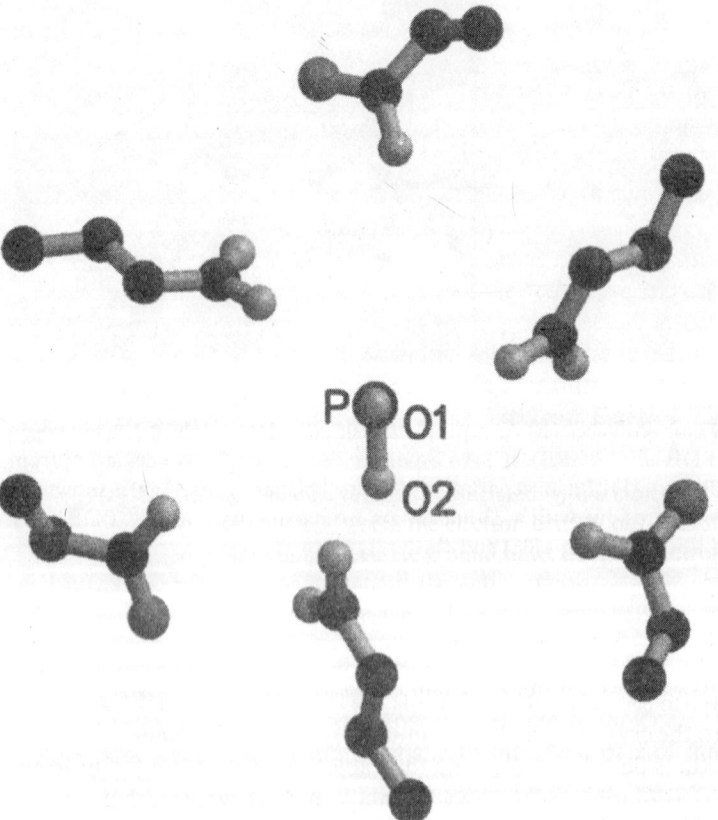

Figure 2   A phosphate on a threefold axis of symmetry (PDB code 1CBQ). The view is along the symmetry axis, with the phosphate and the O1 oxygen lying exactly on the axis. (Figure created using Molscript [37] and Raster3D [38].)

the symmetry matrices, three oxygen atoms lie exactly on top of each other on the symmetry axis, and WHAT_CHECK faithfully reports this as "bumps." In this example, the real origin of the error could only be determined by manual inspection of the PDB entry. The fact that current structure validation software only works with the coordinates from PDB-format files can also lead to a few outliers being flagged as errors, while there might be a justification for them in the header of the PDB file or in the original publication. Unfortunately, this is a choice one has to make at this point, when machine reading of textual information written by humans is far from reliable. So, not only proteins are "too complicated for human beings," but protein structure files are definitely too complicated for computer programs. Some of the errors and outliers reported by WHAT_CHECK are actually interesting findings. An example of this is histidine 21 in one of the KH modules of Vigilin (Fig. 3) [21], which was flagged by several of WHAT_CHECK algorithms as poorly packed. However, all experimental data, NMR restraints in this case, indicated that histidine was placed correctly in the structure. After a follow-up study, it was found that this histidine had an important role in Vigilin [22].

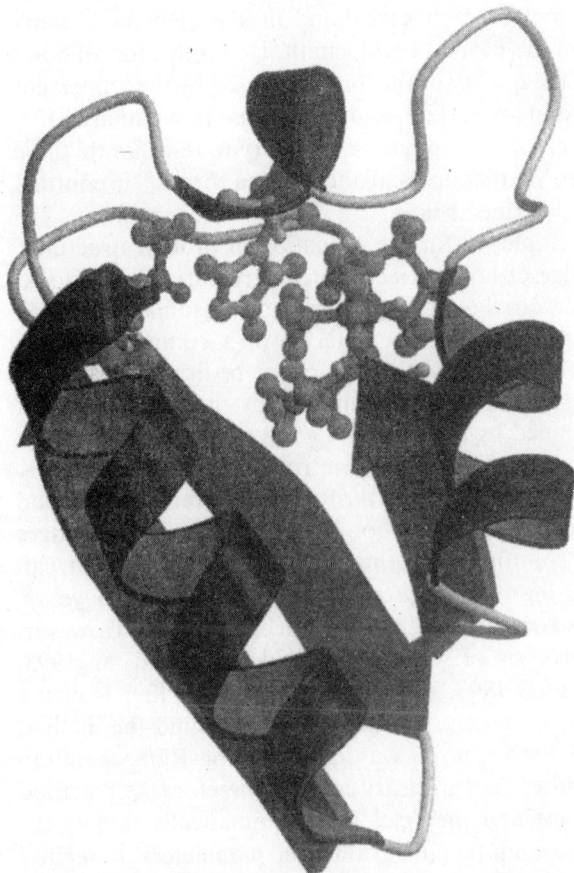

Figure 3 The solution structure of one of the KH modules of Vigilin (PDB code 1VIG) showing the location of the buried histidine. (Courtesy of A. Pastore.)

This last example shows a pleasant side effect of structure validation, namely the detection of interesting outliers. Some outliers are, in effect, indicative of interesting exceptions that often lead to follow-up studies and new knowledge.

3. LIVING CHECKS VERSUS DEAD CHECKS

A few characteristics of protein structures are known with great precision. For example, Engh and Huber [23] determined the ideal parameters for bond lengths and bond angles in biomolecules. They obtained their data from a study on bond lengths and bond angles in peptides and peptide-like molecules in the CSD database [24] of small-molecule x-ray and neutron diffraction structures. These structures are determined with a much greater accuracy than can be obtained for macromolecules. The assumption that these small molecule data can be extrapolated to macromolecules has been proven correct [15]. We therefore know bond lengths and bond angles with such great precision that we can safely assume them to be true. What does it mean to know that, for a certain bond-type the bond length is 1.231 ± 0.020 Å? It means that the natural bond length in a crystal structure for this bond type is 1.231 Å, and that deviations in environmental effects have a root-mean-square deviation of 0.020 Å as their effect. Now, we can determine for each such bond in a protein how many standard deviations it is away from the natural bond length. Doing this for all bond lengths gives a good impression of the quality of the force field used by the refinement software [25]. Bond lengths [26], bond angles [26], and deviations from planarity [27] are known with such great precision that it will not be necessary to re-calibrate these parameters in the foreseeable future. We therefore call them "dead checks," in contrast to the "living checks" that will be described below.

Many techniques have been employed for the validation of protein structures. Still, when one lacks firm knowledge of how perfect protein structures should look, one normally resorts to a statistical analysis of a database of presumably "good" proteins and expresses the quality of structures in terms of a comparison with parameters derived from this database. The scoring can either be firmly based on a proper statistical treatment of the data, or more intuitively, by the introduction of cutoffs, the values of which are defined by an expert. A good example of this latter procedure is the Ramachandran plot [28]. Ever since the introduction of PROCHECK as a validation tool in 1993 [26], it has been clear that the Ramachandran plot is a good indicator of protein structure quality. The introduction of this program, which judges the quality of protein structures by counting the outliers in a Ramachandran plot, can be seen as the dawn of modern-day structure validation. For about 5 years, PROCHECK provided the standard of truth in the validation of structures. However, PROCHECK was based on an analysis of structures considered "good" in 1993. Better structures became available after 1993, and Kleywegt and Jones [29], realizing this, fine-tuned the PROCHECK parameters, leaving the ideas behind the method untouched. Hooft et al. [30] introduced a novel way of evaluating Ramachandran plots by using statistical methods rather than arbitrary cutoffs. Therefore their method does not need input from an expert and they can thus automatically update the database and the resulting Ramachandran plot evaluation parameters at regular intervals, making it a "living" check.

We suggest that all future validation procedures should either be based on elementary laws of physics, or on an objective statistical comparison with a database

Table 2 Quality Indicators for Crambin (1CRN) Determined Late 2000 Versus Those Determined Early 2002

Structure Z-scores, positive is better than average	2000	2002
First-generation packing quality	0.163	0.163
Second-generation packing quality	−0.547	−1.509
Ramachandran plot appearance	−0.230	−0.244
Chi-1/chi-2 rotamer normality	−0.195	−0.738
Backbone conformation	0.470	1.089

The different quality indicators mentioned in this table are explained in the help pages of the PDBREPORT database (http://www.cmbi.kun.nl/gv/pdbreport/checkhelp/).

of structures that is continuously being updated to contain what are currently considered the "best" structures. This continuous updating process will make most structures look poor after time passes. See Table 2, where we compare the quality indicators of the very well-solved structure of Crambin (1CRN) as they were determined using the WHAT_CHECK databases of late 2000 and early 2002, respectively. This could be construed as somewhat dishonest to the crystallographers who deposited their data a long time ago. On the other hand, if quality differences exist between structures, then this should be made known to the users of PDB files, whether the quality differences arose from poor crystals or from the use of old-fashioned refinement software.

4. DATABASE GENERATION

Hooft et al. [27] described a procedure to extract a representative subset of high-quality structures from the PDB. Since this publication, the PDB has grown by almost a factor of three. Consequently, we can now use more strict selection criteria. The criteria for being acceptable to the WHAT_CHECK internal database in 2002 are listed below:

- The keyword "mutant" does not occur in the "Compound" name.
- The structure was solved by using x-ray crystallography.
- The resolution is better than 2.0 Å.
- The R-factor is better than 19%.
- There is at least one water molecule present in the PDB file.
- There are no chain breaks.
- The structure has not been determined to be bad by the previous version of WHAT_CHECK.
- There are no amino acids other than the standard 20.
- There is less than 30% sequence identity to any other protein in the WHAT_CHECK database.

In this study, we concentrate on structures solved by x-ray crystallography. The datasets that represent certain time periods were selected from the PDB by using the PDBFINDER [31] database. We selected protein structures from the database based on the year of publication and the resolution at which the structure was determined.

5. TYPES OF ERRORS

During the course of the structure validation project, we have discovered about one class of errors every 2 weeks. Most of these errors are fully unimportant to the average PDB-file user, but they are errors nonetheless. Consequently, it is impossible even to list all types of errors here. In this chapter, we have to limit ourselves to a discussion of a few classes of errors that, if not detected, could severely hamper the scientist who bases an experiment on a three-dimensional structure. We will discuss "flipping" of asparagine, histidine, and glutamine side-chains, administration of alternate atoms and residues, and the role of water molecules in protein structures. A large class of error types is formed by nomenclature errors. Table 3 lists some nomenclature errors we found in the PDB.

For completeness, we list several other types of errors in Table 4. Some errors are really implausible, so we added one or more PDB files in which the error occurs as "evidence."

For several years, the European Community (EC) has sponsored a structure validation project. The results of this project are available as a WWW-based server http://biotech.ebi.ac.uk:8400/). The server incorporates PROVE [32] by the Wodak group, PROCHECK by the Thornton group, and WHAT_CHECK.

5.1. Flipping Asparagine, Histidine, and Glutamine

Even at very high resolution, it is nearly impossible to see the difference between N, C, and O atoms in an electron density map of a protein structure. Consequently, crystallographers can determine the chi-2 torsion angle of asparagine and histidine and the chi-3 torsion angle of glutamine except for a 180° flip. It is often straightforward to determine the correct values for these torsion angles by looking at the local hydrogen-bonding pattern. Unfortunately, a popular refinement program such as XPLOR [33], by default, has the electrostatic forces switched off for faster convergence. Because not all crystallographers look at all these residues by hand, about 1 in every 6, asparagine, histidine, and glutamine residues in the PDB are placed "the wrong way around" (or "flipped").

It can be harmful if such flipped residues are located in a molecule used as a template for homology modeling, because most modeling programs use residues in the

Table 3 Nomenclature Errors in the PDB

Error	Frequency
$O^{\delta 1}$ and $O^{\delta 2}$ swapped on aspartate	61,388
$O^{\delta 1}$ and $O^{\delta 2}$ swapped on glutamate	75,785
Ring flipped in phenylalanine	49,100
C^{δ} sits on wrong C^{γ} in isoleucine	317[a]
$C^{\delta 1}$ and $C^{\delta 2}$ swapped on leucine	381
C^{γ} and O^{γ} swapped on threonine	385[a]
$N^{\eta 1}$ and $N^{\eta 2}$ swapped on arginine	30,235
Ring swapped in tyrosine	41,951
O^{1} and O^{2} swapped in phosphate group	98,446
Other nomenclature problems	4545

[a] These are actually chemical errors rather than nomenclature errors.

Table 4   Some Examples of Types of Errors Observed

Error	PDB code	Resolution (Å)
Distance between carbon and C-terminal oxygen >2.0 Å.	1PSR	1.05
Groups bound to the same residue have different chain names	1LKK	1.00
Ions with chain name X only contact atoms of chain Y	1ETI	0.90
Different cofactors share the same space	1ABE	1.70[a]
Water atom occurs three times with the same coordinates	1QU9	1.20[b]
Occupancies do not add up to 1.0.	1IFC	1.19[c]
Swapped occupancies on consecutive alternate residues	1BYI	0.97[d]
Threonine with swapped chirality on the C^β	5RXN	1.20
Negated value in the scale matrix	1CRN	1.50
Cluster of water makes no contacts with solute	1NKD	1.07
A water molecule makes zero hydrogen bonds	1GCI	0.78[e]

Four-letter codes refer to PDB files in which examples were found. The resolution at which the structures were also given.
[a] This error often is only administrative. Rather than properly flagging alternate atoms, two different cofactors are given with different occupancies. WHAT_CHECK can only flag such errors if the sum of the occupancies is larger than 1.0.
[b] Mostly the occupancies of the different incarnations of the same water add up to about 1.0.
[c] In 1IFC the occupancies add up to 1.0 for only a few atoms.
[d] The residues 182 and 183 in 1BYI both occur in two conformations, A and B. The occupancies of 182 A and B, and 183 A and B are 0.47, 0.53, 0.51, and 0.49, respectively. Thus the lower occupancy copy of residue 182 is connected to the higher occupancy copy of residue 183.
[e] This error is observed about 105 times.

direct vicinity to determine the optimal side chain conformation for residues that need to be introduced in the model. The presence of flipped residues has particularly negative influence on electrostatic calculations. Nielsen et al. [34,35] calculated the pK_a of active site residues in a series of extensively studied molecules, and calculated these pK_a values again after placing the flipped asparagine, histidine, and glutamine residues in their correct orientation. The pK_a values of these residues, known to be involved in catalytic mechanisms, changed by up to 3 pK_a units as a result of the flip corrections.

5.2.   Alternate Atoms and Residues

Proteins are much more mobile than one would expect by looking at all the very rigid structures stored in the PDB. The development, in recent years, of new crystallization methods and the availability of increasingly intense synchrotron x-ray beams are rapidly increasing the number of structures that can be solved at real atomic resolution (better than 1.2 Å). We see a large number of cases in these very high-resolution structures where atoms, side chains, residues, or even whole loops occupy multiple positions. Although this has not been really been proven, it seems a safe assumption that such alternate states are in a dynamic equilibrium. The fact that they are often seen in atomic resolution structures means that they are undoubtedly also present, but are undetectable, in all structures solved at lower resolution [36].

The validation of alternate atom locations has opened a Pandora's box of its own. The number of errors that crystallographers can make when depositing structures with alternate locations is truly mind-boggling. We only recently started working in WHAT_CHECK on the detection of errors in alternate atom indicators. To our dismay, we found many classes of alternate atom location administration errors that, if undetected, lead to severe misrepresentation of the atomic information. It seems likely that a significant number of errors reported in the PDBREPORT database is the result of errors in the alternate atom indicators, rather than what is reported. Many of these problems will be solved by year 2002, but it cannot be certainly guaranteed that all alternate atom location problems will be detected.

5.3. Water Molecules

Water molecules are an essential aspect of protein structures. Without knowledge of the location of all tightly bound waters, many aspects of the structure, function, and stability of proteins cannot be properly studied. Unfortunately, waters are often abused in crystal structure determination. From the kinds of errors we detect, we assume that some crystallographers, shortly before they manually place waters in the density map, use software that places a water molecule close to each alpha carbon (in a well-determined structure there is about 1 bound water molecule visible per residue). Unfortunately, it regularly occurs that a few waters that are not moved to another position are forgotten, and remain part of the structure.

There are many other water-related problems. The R-factor is determined by how well each peak in the electron density is filled by a positioned atom. Therefore if, at some stage of the refinement procedure, peaks are observed in the electron density map, the R-factor will decrease if waters are placed in these peaks; no matter how nonsensical those waters may be. We found almost 3000 waters or clusters of waters that are not even near anything else but bulk water, and thus surely are the result of ill-advised attempts to lower the R-factor. We also found more than 80,000 water molecules (that is an average of 5 per PDB file) that make no hydrogen bonds whatsoever. A tightly bound water that makes no hydrogen bonds at all is, of course, possible, but it is energetically so unfavorable that we have to assume that a vast majority of these 80,000 waters are not real. Waters are also prone to a series of administrative errors. When waters fall in between crystallographically related proteins, they are present only one time in the PDB file, as they should be. For example, take a molecule that crystallizes as a monomer in space group $P1$. If a water is bound between serine 17 in the central molecule, and asparagine 124 in a translated molecule, then there is also a water touching the asparagine 124 in the central molecule. Only one of the two alternate coordinates for the waters should be given, and it is not common practice to have both present with an occupancy of 0.5. A user who does not have access to proper symmetry generating software will therefore analyze the structure with one water missing. We have built a WWW server that adds these symmetry-related waters (see http://www.cmbi.kun.nl:1100/WIWWWI/). This server should be used with care because, formally speaking, it actually introduces errors; however, if it is clear that these extra waters are not to be used in x-ray refinement, they can be useful to many people.

6. ERRORS OVER THE YEARS

In 1996, we reported that there were more than 1,000,000 outliers in the PDB [20]. These outliers reflect discrepancies with conventions, statistical outliers, and probable errors. At that time, that corresponded to an average of about 1 "error" per amino acid. Many protein crystallographers reacted to that report by saying that all these

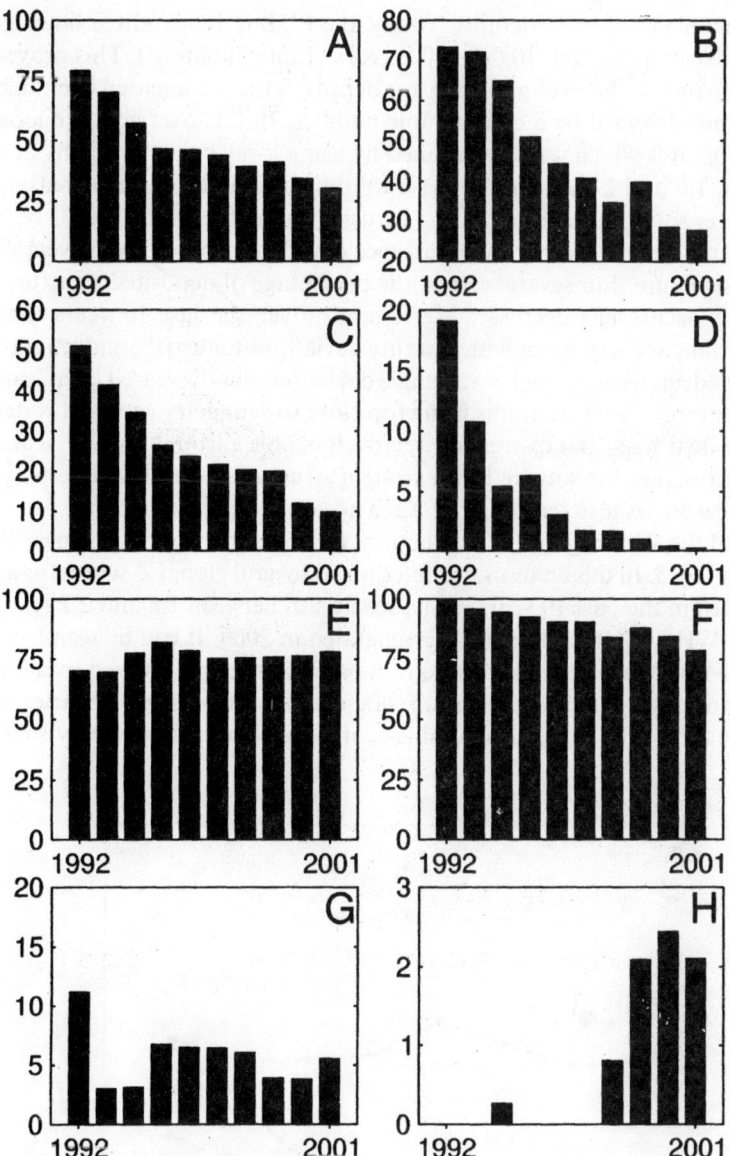

Figure 4 Percentage of deposited structures in a given year containing the following errors: (A) chirality deviations, (B) unusual bond lengths, (C) side chain planarity problems, (D) high bond angle deviations, (E) water molecules without hydrogen bonds, (F) unusual bond angles, (G) atoms too close to a symmetry axis, and (H) chain names not unique. The year of deposition is depicted on the horizontal axis, the percentage of deposited structures containing the error is shown on the vertical axis.

errors were attributable to older structures. Today, the PDB is almost three times as large, and we can detect more than 10,000,000 "errors" (unfortunately!). This proves the proposed explanation to be wrong, as the error density actually increased over time and has certainly not dropped by a considerable number. In this overview, we look only at protein structures which were determined by using x-ray crystallography at a resolution between 1.8 and 2.2 Å. Our analysis will not be biased by the increasing number of structures solved at high resolution by using this subset of the PDB.

Fig. 4 shows the dependency of the occurrence of different errors on the year of deposition of the structure. For several errors, the percentage of deposited structures in which the error occurs has decreased significantly over the last 10 years. For example, the occurrence of structures with chirality deviations, unusual bond lengths, side chain planarity deviations, or high bond angle deviations has decreased by at least a factor of 2. Other errors, such as atoms being too close to symmetry axes and water molecules without hydrogen bonds, remain relatively stable throughout the years. With improved techniques for solving larger protein structures (with multiple polypeptide chains), new errors also tend to arise, such as non-unique chain names.

The course of the WHAT_CHECK structural Z-scores throughout the past 10 years is depicted in Fig. 5. In this analysis, we determined several global Z-scores for all the structures solved in the past 10 years with a resolution between 1.8 and 2.2 Å, by using the internal WHAT_CHECK database generated in 2000. It can be seen from Fig. 5 that the chi-1/chi-2 rotamer distribution, in particular, has improved over the years, which is most likely due to the use of side-chain conformation libraries in structure determination. The other structural Z-scores show an insignificantly small

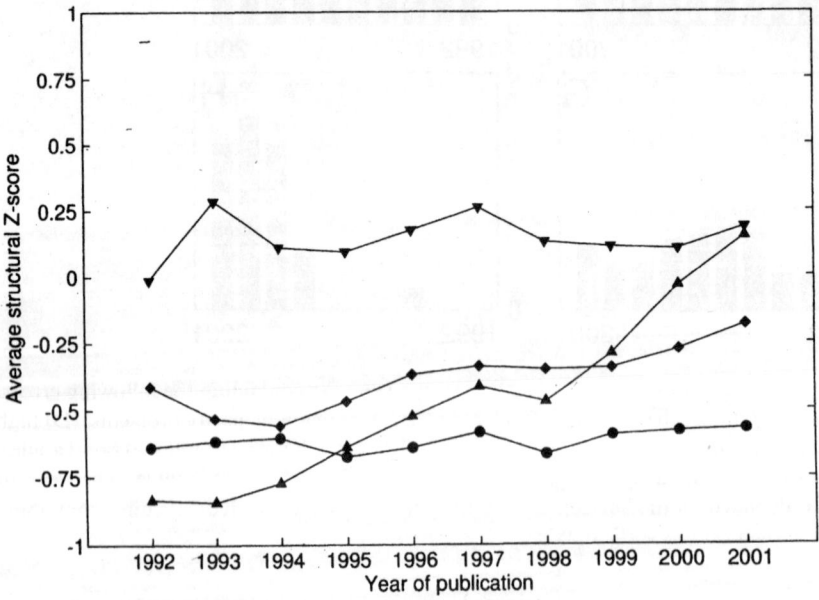

Figure 5   Average WHAT_CHECK structural Z-scores for structures solved at a resolution between 1.8 and 2.2 Å. The average Z-scores for the first-generation packing quality are marked by circles, those for the Ramachandran plot appearance by diamonds, those for the chi-1/chi-2 rotamer distribution by upright triangles and those for the backbone conformation with inverted triangles.

improvement over the years, reflecting the inability of refinement software to improve these properties when experimental data is sparse.

7. CONCLUSION

Everything that could go wrong has gone wrong. For more than a decade, we kept discovering a new class of errors in PDB files every 2 weeks, which makes it impossible to list all types of errors that we have encountered.

As shown in Fig. 4, there are certain errors whose occurrence has decreased over the past 10 years, while the occurrence of others has remained stable or even increased. This shows that a huge amount of work has been carried out to increase the quality of structure determination methods, but it also shows that there still is a lot of work to be performed.

We would like to emphasize the importance of flagging those residues which are very likely not suitable for use in designing structure-based experiments. If bad residues, as well as residues that do not originate from electron density maps but from side chain conformation libraries, would be properly flagged, the experimental scientist would know that the involved coordinates cannot be trusted. Of course, crystallographers are solving increasingly complicated molecules and the availability of powerful synchrotron sources allows poorer crystals to be used, but very many errors in files deposited in the PDB recently have to be attributed to the use of poor refinement software, combined with the unavailability of structure validation software. We hope that this chapter and the freely available software described in it will help to increase the average quality indices to a level that is constrained only by the quality of the experimental data.

After all, protein structures are very beautiful, but what good is beauty if it cannot be seen because of errors?

ACKNOWLEDGMENTS

Throughout this article, we have not put references on the PDB files that we use for illustrating errors. We can only show very few of the detected 10,000,000 outliers, and we do not want to create the false impression that the few crystallographers whose data we used are bad scientists. It is actually the opposite; any crystallographer who deposits data is a better scientist than all crystallographers who do not. We therefore positively acknowledge all crystallographers once, rather than create the impression of the opposite by mentioning a few in the reference list. We would also like to thank K. Henrick, G. Kleywegt and A. Pastore for critical reading of the manuscript. We would like to thank Geerten W. Vuister for stimulating discussions. C.S. acknowledges the Netherlands Organization of Scientific Research (NWO), (grant QLRI-CT-2000-30398), S.N. and E.K. acknowledge the EC (grant QLG2-CT-2000-01313).

REFERENCES

1. Chothia C. Proteins, One thousand families for the molecular biologist. Nature 1992; 357:543–544.
2. Dayringer HE, Tramontano A, Fletterick RJ. Interactive program for visualization and modelling of proteins, nucleic acids and small molecules. J Mol Graph 1986; 4:82–87.

3. Jones TA, Zou JY, Cowan SW, Kjeldgaard M. Improved methods for binding protein models in electron density maps and the location of errors in these models. Acta Crystallogr A 1991; 47:110–119.

4. Vriend G. WHAT IF: a molecular modeling and drug design program. J Mol Graph 1990; 8:52–56 (29).

5. Moult J, Pedersen JT, Judson R, Fidelis K. A large-scale experiment to assess protein structure prediction methods. Proteins 1995; 23:ii–v.

6. Moult J, Hubbard T, Bryant SH, Fidelis K, Pedersen JT. Critical assessment of methods of protein structure prediction (CASP): round II. Proteins Suppl 1997; 1:2–6.

7. Moult J, Hubbard T, Fidelis K, Pedersen JT. Critical assessment of methods of protein structure prediction (CASP): round III. Proteins Suppl 1999; 3:2–6.

8. Moult J, Fidelis K, Zemla A, Hubbard T. Critical assessment of methods of protein structure prediction (CASP): round IV. Proteins 2001; 45:2–7.

9. Hol WGJ. Protein crystallography and computer graphics—towards rational drug design. Angew Chem (Int Ed) 1986; 25:768–778.

10. Navia MA, Fitzgerald PM, McKeever BM, Leu CT, Heimbach JC, Herber WK, Sigal IS, Darke PL, Springer JP. Three-dimensional structure of aspartyl protease from human immunodeficiency virus HIV-1. Nature 1989; 337:615–620.

11. Blundell T, Pearl L. Retroviral proteinases. A second front against AIDS. Nature 1989; 337:596–597.

12. Miller M, Schneider J, Sathyanarayana BK, Toth MV, Marshall GR, Clawson L, Selk L, Kent SB, Wlodawer A. Structure of complex of synthetic HIV-1 protease with a substrate-based inhibitor at 2.3 Å resolution. Science 1989; 246:1149–1152.

13. Weber IT, Miller M, Jaskolski M, Leis J, Skalka AM, Wlodawer A. Molecular modeling of the HIV-1 protease and its substrate binding site. Science 1989; 243:928–931.

14. Wlodawer A, Miller M, Jaskolski M, Sathyanarayana BK, Baldwin E, Weber IT, Selk LM, Clawson L, Schneider J, Kent SB. Conserved folding in retroviral proteases: crystal structure of a synthetic HIV-1 protease. Science 1989; 245:616–621.

15. Who checks the checkers? Four validation tools applied to eight atomic resolution structures. EU 3-D Validation Network. J Mol Biol 1998; 276:417–436.

16. Kleywegt GJ, Jones TA. Where freedom is given, liberties are taken. Structure 1995; 3:535–540.

17. Brunger AT. Free R value: a novel statistical quantity for assessing the accuracy of crystal structures. Nature 1992; 355:472–475.

18. Ponder JW, Richards FM. Tertiary templates for proteins. Use of packing criteria in the enumeration of allowed sequences for different structural classes. J Mol Biol 1987; 193: 775–791.

19. Kleywegt GJ, Jones TA. Databases in protein crystallography. Acta Crystallogr D Biol Crystallogr 1998; 54:1119–1131.

20. Hooft RWW, Vriend G, Sander C, Abola EE. Errors in protein structures. Nature 1996; 381:272.

21. Musco G, Stier G, Joseph C, Castiglione Morelli MA, Nilges M, Gibson TJ, Pastore A. Three-dimensional structure and stability of the KH domain: molecular insights into the fragile X syndrome. Cell 1996; 85:237–245.

22. Fraternali F, Amodeo P, Musco G, Nilges M, Pastore A. Exploring protein interiors: the role of a buried histidine in the KH module fold. Proteins 1999; 34:484–496.

23. Engh RA, Huber R. Accurate bond and angle parameters for X-ray protein structure refinement. Acta Crystallogr A 1991; 47:392–400.

24. Allen FH, Kennard O, Taylor R. Acc Chem Res 1983; 16:146–153.

25. Laskowski RA, Moss DS, Thornton JM. Main-chain bond lengths and bond angles in protein structures. J Mol Biol 1993; 231:1049–1067.

26. Laskowski RA, MacArthur MW, Moss DS, Thornton JM. PROCHECK: a program to

check the stereochemical quality of protein structures. J Appl Crystallogr 1993; 26:283–291.

27. Hooft RWW, Sander C, Vriend G. Verification of protein structures: side-chain planarity. J Appl Crystallogr 1996; 29:714 716.

28. Ramachandran GN, Ramakrishnan C, Sasisekharan V. Stereochemistry of polypeptide chain conformations. J Mol Biol 1963; 7:95–99.

29. Kleywegt GJ, Jones TA. Phi/psi-chology: Ramachandran revisited. Structure 1996; 4: 1395–1400.

30. Hooft RWW, Sander C, Vriend G. Objectively judging the quality of a protein structure from a Ramachandran plot. Comput Appl Biosci 1997; 13:425–430.

31. Hooft RWW, Sander C, Scharf M, Vriend G. The PDBFINDER database: a summary of PDB, DSSP and HSSP information with added value. Comput Appl Biosci 1996; 12: 525–529.

32. Pontius J, Richelle J, Wodak SJ. Deviations from standard atomic volumes as a quality measure for protein crystal structures. J Mol Biol 1996; 264:121–136.

33. Brunger AT. A System for X-ray Crystallography and NMR. New Haven, CT: Yale University Press, 1992.

34. Nielsen JE, Beier L, Otzen D, Borchert TV, Frantzen HB, Andersen KV, Svendsen A. Electrostatics in the active site of an alpha-amylase. Eur J Biochem 1999; 264:816–824.

35. Nielsen JE, Vriend G. Optimizing the hydrogen-bond network in Poisson-Boltzmann equation-based pK(a) calculations. Proteins 2001; 43:403–412.

36. MacArthur MW, Thornton JM. Protein side-chain conformation: a systematic variation of chi 1 mean values with resolution—a consequence of multiple rotameric states? Acta Crystallogr D Biol Crystallogr 1999; 55:994–1004.

37. Kraulis PJ. MOLSCRIPT: a program to produce both detailed and schematic plots of protein structures. J Appl Crystallogr 1991; 24:946–950.

38. Merritt EA, Bacon DJ. Raster3D: photorealistic molecular graphics. Methods Enzymol 1997; 277:505–524.

16

Docking and Scoring

INGO MUEGGE

Boehringer Ingelheim Pharmaceuticals, Inc., Ridgefield, Connecticut, U.S.A.

ISTVAN ENYEDY

Bayer Research Center, West Haven, Connecticut, U.S.A.

1. INTRODUCTION

In the past decade, high-speed synthesis and high-throughput screening (HTS) have revolutionized the lead discovery process in the pharmaceutical industry [1–3]. However, following the excitement of the early years, it became clear that the success of random library design and HTS is limited [4]. Up to 10^6 compounds are screened today in a typical lead discovery effort, which is only a tiny fraction of the total conceivable chemical space for which estimates range between 10^{60} and 10^{100} compounds [5,6]. The main question therefore remains for the medicinal chemists: "What compounds should be made and tested for lead discovery and optimization?"

In parallel to advanced synthesis and screening techniques, the knowledge of structural information about potential target proteins has increased tremendously after the last few years. More than 22,000 protein structures are available today [7]. This sizable body of 3-D structure information at atomic resolution of a variety of relevant drug targets, including enzymes, receptors, and transporter proteins, provides a growing basis for structure-based drug design. Structure-based design has come a long way from its first successes in the seventies, e.g., the identification of hemoglobin ligands [8,9] and the discovery of captopril as an angiotensin-converting enzyme inhibitor [10]. Not surprisingly, structure-based lead optimization has been used increasingly in recent years. Manual and automated molecular docking approaches (computational methods that predict the 3-D structure of a protein–ligand complex) are powerful structure-based design techniques that have been successfully used to discover drug candidates for targets such as HIV protease and human thrombin [11].

Lead optimization is typically performed in a low-throughput mode. However, to keep pace with the ever-increasing speed of biological screening methods, high-throughput computational methods such as protein–ligand docking and scoring had to be developed to evaluate thousands of compounds per day. Consequently, computational docking and scoring techniques have evolved to be used in high-throughput virtual screening protocols for in silico lead identification. As in most approaches, accuracy comes as a trade-in for speed. Sophisticated energy functions, e.g., in first principles methods for protein–ligand affinity prediction [12], are too time-consuming to be used in high-throughput mode. Therefore, so-called scoring functions have been designed that contain significant simplifications to describe protein–ligand interactions; these are fast tools to identify the correct binding geometry of a protein–ligand complex as well as to rank different protein–ligand complexes according to their binding affinities.

In this review, we will describe several docking techniques used for lead identification and optimization. Because scoring functions constitute the Achilles' heel of docking approaches, we will describe the design and performance of fast scoring functions for molecular docking in some detail. Novel attempts to combine different scoring functions to enhance their performance are discussed. Comprehensive reviews on small molecule docking and scoring have appeared in the literature [13–25], and therefore we will emphasize on new trends and describe some applications. Note that we will not discuss the related issue of protein–protein docking; instead, we refer the interested reader to a recent review on the subject by Ehrlich and Wade [26].

2.   DOCKING TECHNIQUES

Protein–ligand docking is a geometric search problem. Protein and ligand conformations, as well as their relative orientations, are the relevant degrees of freedom. Whereas the given protein structure is reasonably well known (although there are many examples of conformational changes that occur upon ligand binding), the protein-bound ligand conformation is usually unknown. Therefore, most docking approaches address ligand flexibility and keep the protein rigid, although examples of methods dealing with protein flexibility will be briefly discussed. The main concepts of docking approaches are outlined below. Table 1 provides a selection of available docking software. Although in some docking programs scoring is not a separate entity but rather intertwined with the sampling algorithms, we will discuss scoring separately in the subsequent chapters. Note that the docking techniques are described here only briefly; a conceptually similar but more comprehensive review has been written by Muegge and Rarey [25].

2.1.   Protein Structure

In order to perform computational protein–ligand docking experiments, a 3-D structure of the target protein at atomic resolution must be available. The most reliable sources are crystal and solution structures provided by the Protein Data Bank (PDB) [7] or from in-house efforts. Homology models [27,28] and pseudoreceptor models [29] are an alternative in the absence of experimental structures. It should be cautioned, however, that the quality of the protein structure is crucial for the success of subsequent docking experiments. Even small changes in structure can drastically alter

Table 1 Selection of Available Protein–Ligand Docking Software

Docking program	Docking/sampling method	Scoring method	Use
GLIDE (www.schrodinger.com)	Rigid protein; multiple-conformation rigid docking; grid-based energy evaluation	Empirical scoring including penalty term for unformed hydrogen bonds; force-field scoring	Single molecule docking; database searching
Liaison (www.schrodinger.com)	Exhaustive sampling, flexible ligand–flexible protein docking	Free energy	Single molecule docking
DOCK (www.cmpharm.ucsf.edu/kuntz/dock.html)	Rigid protein; flexible ligand docking (incremental construction)	Force-field scoring; chemical scoring, contact scoring	Single molecule docking; database searching
FlexX (cartan.gmd.de/FlexX)	Rigid protein; flexible ligand docking (incremental construction)	Empirical scoring intertwined with sampling	Single molecule docking; database searching
DockVision (www.dockvision.com)	Monte Carlo, genetic algorithm	Various force fields	Single molecule docking; database searching
AutoDock (www.scripps.edu/pub/olson-web/doc/autodock/index.html)	Simulated annealing, genetic algorithm	Force-field scoring, free-energy scoring	Single molecule docking
DockIT (www.daylight.com/meetings/emug00/Dixon)	Ligand conformations generated inside binding site spheres using distance geometry	PLP, PMF	Single molecule docking; database searching
FRED (www.eyesopen.com/fred.html)	Exhaustive sampling; rigid protein, multiple-conformation rigid docking	Chemscore, PLP, ScreenScore, and Gaussian shape scoring	Single molecule docking; database searching
LigandFit (www.accelrys.com)	Monte Carlo	LIGSCORE, PLP, PMF, LUDI	Single molecule docking; database searching
Affinity (www.accelrys.com)	Exhaustive sampling; flexible ligand–flexible protein	Molecular mechanics force field	Single molecule docking
Gold (www.ccdc.cam.ac.uk/prods/gold/)	Genetic algorithm	Soft-core vdW potential and hydrogen bond potentials	Single molecule docking; database searching

the outcome of a computational docking experiment [30]. Ideally, the atomic resolution of crystal structures should be below 2.5 Å [31]. On the other hand, the PDB contains a wealth of protein structures of a variety of enzymes and receptors that can be used for homology modeling. It can be expected that reasonable homology models can be built for many proteins coded in the human genome.

2.2. Rigid Docking

Rigid ligand docking is not generally relevant for protein–ligand docking because ligand flexibility, and often also protein flexibility, is crucial. Nevertheless, such simplification is often acceptable for the docking of small fragments or ensembles of conformations and/or molecules. Algorithms such as clique search techniques [32] can be used to search for distance-compatible matches of protein and ligand features [33]. Possible features include, e.g., complementary hydrogen bonding interactions, distances, or volume segments of the receptor site of the protein or the ligand.

Earlier versions of the DOCK program [34] were based on distance-compatible match searches for rigid-body docking (Fig. 1). Spheres are created that map the molecular surface of the protein and fill the receptor site [35–37]. A second set of spheres represents the ligand. Initial orientations of the ligand in the receptor site are generated from sets of up to four distance-compatible matches. The final position of the ligand is reached through optimization and scoring. Since its first introduction in 1982, the DOCK software has undergone several significant changes. Chemical properties are assigned to the matching spheres [38], and the search process has been accelerated by the use of distance bins [39,40]. In DOCK version 4.0 [41], clique-detection algorithms [33] have been introduced as a search algorithm for distance-compatible matches. In addition, a larger spectrum of scoring functions is now available within DOCK [42–46].

Geometric hashing [47] constitutes an alternative to clique searching for the matching of protein–ligand features. The geometric hashing scheme has been developed in computer vision technology for recognizing partially occluded objects in camera scenes. The recognition of partial matches is particularly important in most docking experiments because not all protein features are matched with all ligand features. Parts of the ligand and protein surfaces are often in contact with bulk water. Hashing involves the creation of a key for data entry that can be used as its memory address. Because typically more addresses are available than computer memory, a

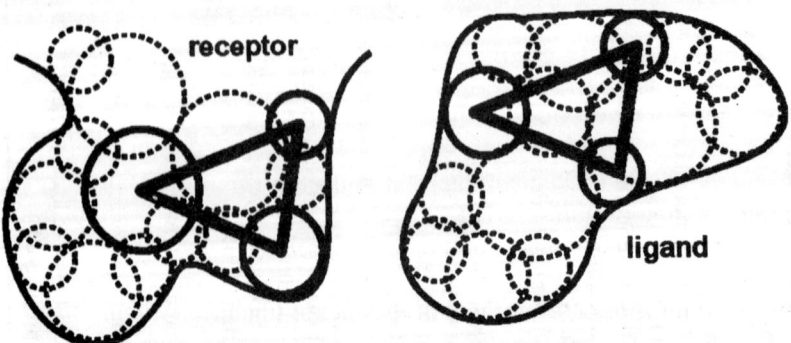

Figure 1 Matching of overlapping sphere triplets in DOCK.

hashing function is applied to map the addresses of the data entry into a smaller address space. In geometric hashing, distance features are used to create the hashing key. As a result, objects with certain geometric features can be accessed very fast through a geometric hashing table. Fischer et al. [48,49] were the first to use geometric hashing to molecular docking using the sphere representation of DOCK. In order to apply fast hashing to the 3-D docking problem, an underdetermined reference frame consisting of only two spheres or atoms has been used.

Pose clustering is another pattern-recognition technique and is used, e.g., in the program FlexX [50,51]. Originally developed to detect objects in 2-D scenes with unknown camera location [52], the algorithm matches each triplet of features of the first object to each triplet of features of the second object. From a match the first object can be located with respect to the second by superimposing the triangles. Locations are stored and clustered. If a cluster grows large, a location with a high number of matching features is found. FlexX uses the LUDI [53,54] representation of molecular interactions as features. Matches are limited by the compatibility of interaction (e.g., a hydrogen bond donor can interact only with an acceptor) and length of triangle edges. A hashing scheme is applied to access and match surface triangles. A complete-linkage algorithm [55] is used to cluster transformations that superimpose two triangles.

2.3. Docking with Flexible Ligands

Energetic differences between alternative ligand conformations are often small compared to the total binding affinity between ligand and target protein. In addition, for flexible ligands it is quite common that the bioactive conformations are different from the minimum energy conformations in solution [56]. Small druglike molecules are typically flexible; 70% of druglike molecules contain between two and eight rotatable bonds [57]. Ligand flexibility is typically handled in docking approaches by combinatorial optimization protocols such as fragmentation, ensembles, genetic algorithms, or simulation techniques.

2.3.1. Fragmentation

In fragmentation approaches, the ligand is dissected into pieces that are either rigid or can be represented by small conformational ensembles. There are two alternative strategies for handling the fragments during the docking process. The first strategy, also called "incremental construction," places the first fragment, which is typically the largest rigid portion of the ligand, in the receptor site and subsequently adds the remaining fragments in a build-up protocol. The second strategy, called "place and join," places all or a subset of the fragments independently and reconnects them in favorable orientations to complete the ligand in the receptor site. While place and join is typically used in de novo design approaches [58,59], incremental construction is more often used as a docking strategy. The reasons for this are twofold. First, if a ligand is divided into fragments, not every fragment has to sit in a minimum-energy position. Second, bond and bond angle distortions in molecules are energetically very costly. Therefore, although some docking algorithms use place and join algorithms [60–63], we will focus on incremental construction algorithms only.

Incremental construction algorithms typically consist of three steps: 1) selection of a set of anchor fragments, 2) placement of the anchor fragments within the active site, and 3) the incremental construction phase. Several fragments of the ligand can serve as the anchor fragment for subsequent construction. Moon and Howe had introduced incremental construction in their peptide design tool GROW [64], and Leach and Kuntz developed the first docking algorithm based on an incremental construction algorithm [65] (Figs. 2 and 3). A single anchor fragment is initially docked into the receptor site using a variant of the DOCK algorithm employing hydrogen-bonding features in the matching phase. A subset of starting placements for the incremental construction procedure is chosen based on several factors, including the number of matched hydrogen bond pairs, a high score for the anchor placement, and low similarity to other placements. After addition of a fragment, steric strain is eliminated and the hydrogen-bonding geometry is optimized. The final placements are then filtered, refined, and scored using a force-field method. Although several manual steps are involved in the procedure, the general applicability of incremental construction for molecular docking could be successfully demonstrated. Fully automated incremental construction algorithms are now available in docking programs such as DOCK [66], FlexX [50], and Hammerhead [67]. The program Hammerhead differs somewhat from DOCK and FlexX by its construction strategy. Instead of forming small fragments by cutting the ligand at each rotatable bond, a set of larger fragments is considered. During the construction phase, the next fragment is added with the connecting atoms or bonds overlapping. Torsion angles of the added fragments are not sampled.

2.3.2. Ensembles of Ligand Conformations

Ligand flexibility can be introduced by evaluating multiple conformations of the ligands in a rigid-body docking algorithm. Because computing time increases linearly with the number of conformations, a balance needs to be sought between computing time and coverage of conformational space.

Flexibase/FLOG is a docking algorithm that uses conformational ensembles [68]. A small set of diverse conformations for each ligand is generated by distance geometry methods [69] and energy minimization. A subset of up to 25 conformations per molecule is selected using root-mean-squared (rms) dissimilarity criteria and then

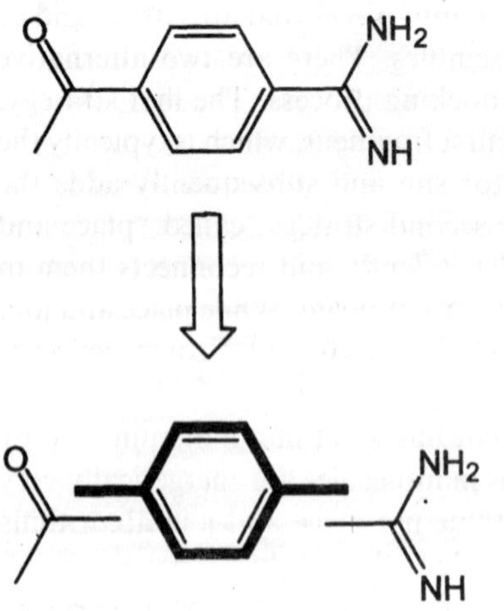

Figure 2   The anchor fragment is selected as the rigid overlapping fragment with the largest number of heavy atoms.

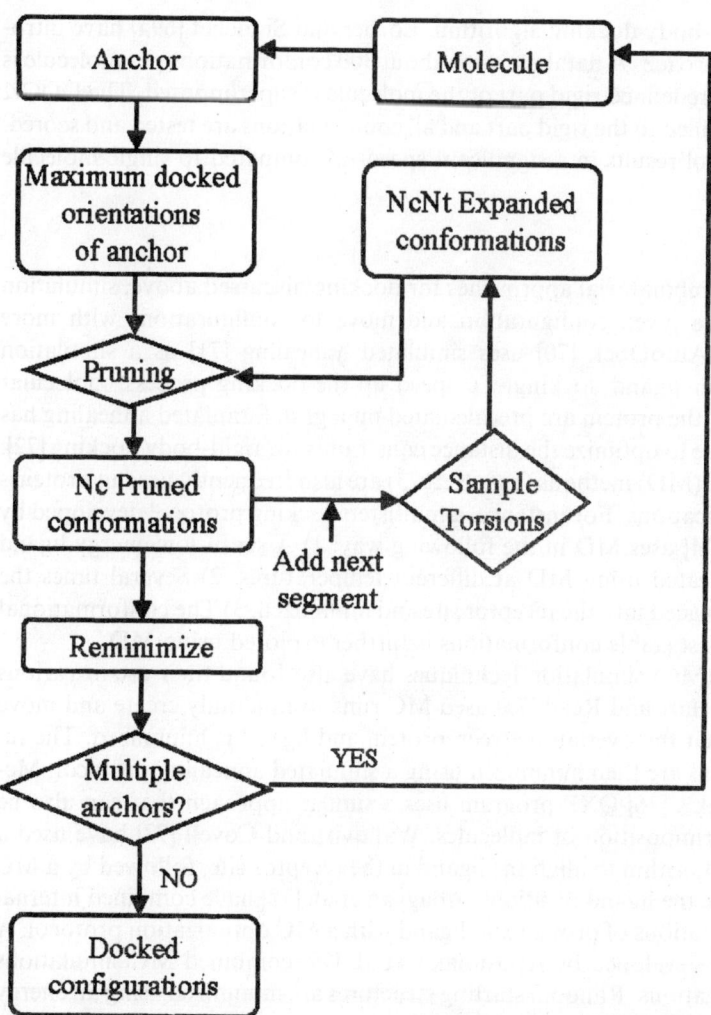

Figure 3 Docking using an incremental construction algorithm. The anchor, the segment with the largest number of heavy atoms among the rigid segments of the molecule, is first placed into the active site of the protein in several orientations. These orientations are pruned according to rank and position to produce the required number of positions per cycle. Pruning attempts to keep the best and most diverse conformations of the ligand. The ensemble of partially built ligands is then expanded by adding a new segment and performing a torsion search on the newly formed bond. Conformations of the complete ligand are reminimized to get the docked conformations. If multiple anchors are used, then the process is repeated using another anchor.

docked using a rigid-body docking algorithm. Lorber and Shoichet [69a] have introduced a different approach. A database with about 300 conformations per molecule is created such that a predefined rigid part of the molecule is superimposed. The DOCK algorithm is then applied to the rigid part and all conformations are tested and scored. This docking protocol results in a significant speed-up compared to single-molecule docking.

2.3.3. Simulation

In contrast to the combinatorial approaches for docking discussed above, simulation methods start with a given configuration and move to configurations with more favorable energies. AutoDock [70] uses simulated annealing [71] as a simulation technique for protein–ligand docking. To speed up the docking process, molecular affinity potentials of the protein are precalculated on a grid. Simulated annealing has also been used by Yue to optimize the distance constraints for rigid-body docking [72]. Molecular dynamics (MD) methods (e.g., Ref. 73) are also frequently used in protein–ligand docking applications. For instance, a multistep docking protocol developed by Given and Gilson [74] uses MD in the following way: 1) A set of low-energy ligand conformations is created using MD at different temperatures. 2) Several times the ligand is randomly placed into the receptor site and minimized. 3) The conformational space around the most stable conformations is further explored using MD.

Monte Carlo (MC) simulation techniques have also found their use in various docking programs. Hart and Read [75] used MC runs to randomly create and move orientations such that the overlap between protein and ligand is minimized. The resulting configurations are then minimized using a simulated annealing protocol. McMartin and Bohacek's [76] QXP program uses a similar approach that can also be applied to the superimposition of molecules. Wallqvist and Covell [77] have used a surface-matching algorithm to align the ligand in the receptor site, followed by a MC protocol to optimize the ligand positions. Abagyan et al. [78] have combined internal coordinate representations of protein and ligand with a MC optimization protocol. A multistep approach developed by Apostolakis et al. [79] combined MC simulations with energy minimizations. Random starting structures are minimized using an energy function consisting of force field terms, hydrophobic solvation terms, and electrostatic solvation terms obtained from Poisson equations. PRODOCK is another example of an application where MC steps are interleaved with minimization steps [80]. It uses AMBER 4.0 [81] or ECEPP/3 force field terms as well as Bezier splines to access the derivatives of the energy function. In LigandFit [82], the ligand is positioned by minimizing the difference between the non-mass-weighted principal moment of inertia (PMI) of the binding site with the non-mass-weighted PMI of the ligand.

$$\text{Fit}_{\text{PMI}} = (\Delta\text{ratio}_{xy}^2 + \Delta\text{ratio}_{xz}^2 + \Delta\text{ratio}_{yz}^2)^{1/2} \tag{1}$$

2.3.4. Other Methods

A variety of other sampling methods are applied in docking programs including genetic algorithms, distance geometry methods, random searching, hybrid methods, and generalized effective potential (GEP) methods. Genetic algorithms have been used in programs such as Gambler [83], AutoDock [70], and GOLD [31]. In GOLD, two bit strings represent a docking configuration. The first string contains the ligand conformations defining the torsion angle of each rotatable bond. The second string contains hydrogen-bond mapping between the relevant protein and ligand atoms. The fitness function takes into account the evaluation of hydrogen bonds, internal energy of the ligand, and the protein–ligand van der Waals energy (vdW). Examples of other genetic algorithm-based applications include a variant of DOCK written by Oshiro et al. [84] and EPDOCK implemented by Gehlhaar et al. [85].

PRO-LEADS uses an alternative search technique called "tabu search" [86]. Starting from a random structure, new structures are created by random moves. A tabu list is maintained during the optimization phase and contains the best and the most recently found binding configurations. Configurations generated that resemble those of the tabu list are rejected except if they are better than the one scoring best. The sampling performance is improved because previously sampled configurations are avoided. Finally, it should be mentioned that multistep hybrid docking procedures have been developed that combine rapid, fragment-based searching with sophisticated MC or MD simulations [87,88].

2.4. Flexible Protein Docking

Most of the current protein–ligand docking programs treat the protein as rigid. This assumption is reasonable in many cases; however, in other cases it has been shown that the protein can adjust its conformation upon ligand binding (e.g., Ref. 89). Therefore, it would be desirable to model protein flexibility in docking studies. In principle, protein flexibility can be introduced through MC or MD. For instance, Luty et al. [90] and Wasserman and Hodge [91] have divided the protein into rigid and flexible parts. During an MD simulation, only flexible receptor site atoms are free to move. The model incorporates implicit or explicit solvation models. Although only a fraction of the protein undergoes MD movements, the procedure is still very slow. Therefore, fragmentation approaches have been developed to speed up the involvement of protein flexibility. Leach [92] developed a docking algorithm that sequentially fixes the degrees of freedom of the protein side-chain atoms. Akin to the idea of introducing ensembles of conformations for ligands to introduce ligand flexibility, Knegtel et al. [93] have extended the idea to protein flexibility by introducing the concept of docking to ensembles of protein structures. Similarly, Broughton [94] reported the use of conformational samples from short protein MD simulation runs. Recently, Lamb et al. [95] introduced the concept of combinatorial docking against multiple targets (note that combinatorial docking as topic has been somewhat neglected here; see, e.g., Refs. 96 and 97 for additional information). Finally, the concept of inverse docking should be mentioned as well. Chen and Zhi recently proposed a method to dock ligands against a database of protein cavities to find potential drug targets of a small molecule [98].

3. SCORING TECHNIQUES

Energy functions derived from first principles are typically used to evaluate binding affinities between proteins and putative ligands. Although some bridging attempts have been made recently [99,100], sophisticated techniques such as free-energy perturbation [12] or linear response theory [101,102] are currently too slow to be of use in molecular docking applications. Fast functions have been developed that incorporate a large number of simplifications and as such do not rigorously describe binding free energies. These functions are usually referred to as scoring functions and have been reviewed recently [21–25]. Here we outline some of the basic scoring techniques and comment on current experiences in consensus scoring methods that try to enhance the performance by combining the outcome of several scoring functions.

Scoring functions can be divided into several classes: 1) Force-field-derived scoring functions typically rely on nonbonded interaction terms [31,103–105]. Solvation terms are sometimes included [45,106]. 2) Regression-based scoring functions are also often used [53,107–113]. 3) Knowledge-based scoring functions have recently appeared and use statistical atom pair potentials to calculate the score [114–120]. 4) Alternative scoring protocols include chemical scores, contact scores, or shape complementary scores [39,56,83,121–125]. Finally, it should be mentioned that scoring functions are currently the Achilles' heel of docking programs, and so far no scoring function has yet been developed that can consistently identify the correct binding mode for any given protein–ligand complex [126].

3.1. Force-Field Scoring

Force-field (FF) scoring has long been the first choice in many successful docking applications including DOCK and AutoDock. When used on precomputed grids, FF scoring is fast and independent of the docking algorithm. Force fields are extensively studied and well understood. However, they typically measure only the potential energy of the system. Important contributions to the binding free energy such as solvation energy and entropy are often ignored. In rigid-body docking of potent protein–ligand complexes, it has been reported that electrostatic contributions are negligible for identifying the correct binding solution [127]. However, for flexible ligand docking, electrostatics often dominates the docking outcome. The overemphasis of electrostatics may help to identify the correct binding mode as the formation of hydrogen bonds and salt bridges is supported. At the same time, however, their contributions to the binding free energy are typically overestimated.

FF scoring most often relies on the nonbonded interaction energy terms of standard force fields, e.g., in vacuo electrostatic terms (sometimes modified by scaling constants that assume the protein to be an electrostatic continuum) and vdW terms [22,105,128,129]. As an example, DOCK and GREEN [130] have implemented the intermolecular terms of the AMBER energy function [103,104] with the exception of an explicit hydrogen-bonding term [42]:

$$E_{\text{non-bond}} = \sum_{i}^{\text{lig}} \sum_{j}^{\text{prot}} \left[\frac{A_{ij}}{r_{ij}^{12}} - \frac{B_{ij}}{r_{ij}^{6}} + 332 \frac{q_i q_j}{D r_{ij}} \right] \qquad (2)$$

where each term is summed up over ligand atoms i and protein atoms j. A_{ij} and B_{ij} are the vdW repulsion and attraction parameters of the 6-12 potential, r_{ij} is the distance between atoms i and j, q is a point charge at each of the atoms, and D is the dielectric constant. Intraligand interactions are added to the score. Up to a hundredfold gain in docking time is achieved by precomputing these terms on a 3-D grid that represents the protein during docking [70,131].

Soft vdW potentials are often used in simulations of whole-ligand docking approaches. For example, Flog uses a 6-9 Lennard–Jones function for vdW interactions, local dielectric constants in a Coulomb representation of the electrostatic interactions, and additional terms for hydrogen bond potentials and hydrophobic potentials [68]. GOLD [31] combines a soft intermolecular Lennard–Jones 4-8 potential with hydrogen bonding terms precalculated using model fragments [132]. Sometimes containment potentials force a ligand to dock in a certain region [133]. In addition, complete force-field energy functions including bonded terms are sometimes used as scoring function [80].

Other than entropy effects, which are often neglected by the assumption that they are similar within a series of ligands, the role of solvation effects cannot be over-emphasized as an important contributor to the final binding free energy. Solvation alone sometimes correlates already significantly with binding affinities [134]. In docking functions, solvation is introduced by surface area terms for nonpolar and electrostatic contributions [79,135], generalized Born/surface area [45,46,136], or by atomic solvation parameters [137–139]. An analysis of the different parameters in FF scoring using the CHARMM energy function was recently given by Vieth et al. [140], who concluded that a soft-core vdW potential is needed for the kinetic accessibility of the binding site.

3.2. Regression-Based Scoring

Regression-based scoring functions (sometimes called empirical scoring functions) are derived from fitting coefficients of 3-D protein–ligand structure-derived terms of a binding energy equation (e.g., hydrogen bonding energy and lipophilic contact energy) to reproduce experimental binding affinities of a training set of known protein–ligand complexes [107–112]. To illustrate the design of regression-based scoring functions, we describe here the ChemScore scoring function as developed by Eldridge et al. [110] and Murray et al. [141] and implemented in the program PRO_LEADS [142]. ChemScore is one of the most successful empirical scoring functions available today [83] and is written as:

$$\Delta G_{\text{binding}} = \Delta G_0 + \Delta G_{\text{hbond}} \sum_{iI} g_1(\Delta r) g_2(\Delta \alpha) + \Delta G_{\text{metal}} \sum_{aM} f(r_{aM})$$
$$+ \Delta G_{\text{lipo}} \sum_{iL} f(r_{iL}) + \Delta G_{\text{rot}} H_{\text{rot}} \tag{3}$$

where the different ΔGs are determined via multiple linear regression. The hydrogen bond term $\sum_{iI} g_1 g_2$ is calculated for all hydrogen bonds between ligand atoms i and protein atoms I:

$$g_1(\Delta r) = \begin{cases} 1 & \text{if} & \Delta r \le 0.25 \text{ Å} \\ 1 - (\Delta r - 0.25)/0.4 & \text{if} & 0.25 Å < \Delta r \le 0.65 \text{ Å} \\ 0 & \text{if} & \Delta 0.65 \text{ Å} \end{cases} \tag{4}$$

$$g_2(\Delta \alpha) = \begin{cases} 1 & \text{if} & \Delta \alpha \le 30° \\ 1 - (\Delta \alpha - 30)/50 & \text{if} & 30° < \Delta \alpha \le 80° \\ 0 & \text{if} & \Delta \alpha > 80° \end{cases} \tag{5}$$

Note that this scoring function does not distinguish between ionic and nonionic hydrogen bonds. Δr is the deviation from the ideal hydrogen bond length of 1.85 Å (HO/N) and $\Delta \alpha$ is the deviation from the ideal angle of 180°. The lipophilic $\Sigma_{iL} f(r_{iL})$ and metal $\Sigma_{aM} f(r_{aM})$ terms are calculated as simple contact terms. ChemScore also introduces a unique way of addressing ligand entropy terms. Frozen rotatable bonds are identified as those of which the atoms on both sides of the bond are in contact with the receptor. Eldridge et al. argued that if one atom is not in contact with the receptor, the rotation may not be completely impaired and the entropy penalty stemming from the rotatable bond should be smaller. The flexibility penalty of ligands for frozen rotatable bonds is calculated as:

$$H_{rot} = 1 + (1 - 1/N_{rot}) \sum_r (P_{nl}(r) + P'_{nl}(r))/2 \qquad (6)$$

where N_{rot} is the number of frozen rotatable bonds and $P_{nl}(r)$ and $P'_{nl}(r)$ are the percentages of nonlipophilic heavy atoms on either side of the rotatable bond, respectively. Trained on 82 protein–ligand complexes taken from the PDB, the following coefficients were obtained: $\Delta G_0 = -5.48$ kJ/mol, $\Delta G_{hbond} = -3.34$ kJ/mol, $\Delta G_{metal} = -6.03$ kJ/mol, $\Delta G_{lipo} = -0.117$ kJ/mol, and $\Delta G_{rot} = 2.56$ kJ/mol. The coefficients derived by Eldridge et al. are similar to those found by Böhm [108]. With the exception of the arbitrary ΔG_0, the coefficients appear to be generally reasonable; for example, the ΔG_{rot} estimation is very similar to the entropy estimates for rotatable bonds ($\sim 1\ k_B T$) by Searle and Williams [143] and Searle et al. [144]. ChemScore has achieved a statistically significant correlation between prediction and experiment of protein–ligand binding affinities with a standard error of 8.68 kJ/mol for the training set of 82 complexes.

Regression-based scoring functions are fast and are easily tailored to the problem at hand. Disadvantages are often a consequence of the fact that the derivation has been based on complexes involving potent ligands only. Unfavorable conformations are often not penalized. Finally, as with all regression-based approaches, it is somewhat unclear to what extent empirical scoring functions can be successfully applied beyond the structural range of the training set.

3.3. Knowledge-Based Scoring

After the early attempts of using knowledge-based functions for the prediction of protein–ligand binding affinities [114,119,120], the technology has matured in the past couple of years toward a promising methodology for molecular docking applications [115,117,118,145,146]. The technology offers the hope that the implicit treatment of all relevant contributions to binding will "automatically" provide the correct balance between solvation, enthalpy contributions, and entropy. Knowledge-based scoring functions were recently reviewed by Gohlke and Klebe [147]. Therefore, we will explain here only briefly the concept based on the example of the PMF scoring function [118].

Knowledge-based functions are based on the derivation of statistical preferences in the form of potentials for protein–ligand atom pair interactions. Similar to potentials derived for protein folding and protein structure evaluation (e.g., Ref. 148), pair potentials akin to potentials of mean force (PMFs) are derived for various protein and ligand atom types using the PDB as a knowledge base. The PMF scoring function [118] is defined as the sum over all protein–ligand atom pair interaction free energies $A_{ij}(r)$ at distance r:

$$\text{PMF\_score} = \sum_{\substack{kl \\ r < r^{ij}_{\text{cut-off}}}} A_{ij}(r) \tag{7}$$

where kl is a ligand–protein atom pair of type ij and $r^{ij}_{\text{cut-off}}$ is the distance at which atom pair interactions are truncated. The terms $A_{ij}(r)$ are calculated as:

$$A_{ij}(r) = -k_{\text{B}}T \, ln \left[f^{j}_{\text{Vol\_corr}}(r) \frac{\rho^{ij}_{\text{seg}}(r)}{\rho^{ij}_{\text{bulk}}} \right] \tag{8}$$

where k_{B} is the Boltzmann factor, T is the absolute temperature, $f^{j}_{\text{Vol\_corr}}(r)$ is a ligand volume correction factor [146], $\rho^{ij}_{\text{seg}}(r)$ is the number density of atom pairs of type ij at a certain atom pair distance r, and ρ^{ij}_{bulk} is the number density of a ligand–protein atom pair of type ij in a reference sphere with a radius of 12 Å [145]. For docking purposes, the PMF score adds a vdW term to account for short-ranged interactions [149]. PMF scoring in docking experiments has been shown to work well on a variety of targets, including the FK506 binding protein [149], neuraminidase [30], and stromelysin [150,151]. Scoring functions similar to PMF score include DrugScore [115] and BLEEP [117,152]. In contrast to PMF and BLEEP, DrugScore contains an additional solvation term derived from the solvent-accessible surface area.

3.4. Complementarity Score

Surface complementarity, e.g., of hydrophobic or hydrophilic nature, can be used for scoring, and sometimes even quantitatively impressive results can be achieved (e.g., Ref. 153). Principal moment of inertia complementarity, as implemented in LigandFit, is very successful for the fast and accurate docking of small molecules. Shape and chemical complementarity scoring are implemented in DOCK. The volume of the binding site is filled with a cubic lattice using atomic contact potentials for polar and apolar contacts in the receptor site [39]. A lattice point scores one for atoms below a certain distance. Close contacts are penalized with negative scores. Ligand scoring is facilitated by mapping ligand atoms on the nearest lattice points and accumulative scoring. Chemical scoring is also available in DOCK; the DOCK spheres can be "colored" with respect to the properties of nearby positioned receptor atoms [33,154]. During the docking procedure, a sphere can be matched only with a complementary ligand atom. In DOCK, chemical "colors" such as hydrogen bond donors, hydrogen bond acceptors, or hydrophobic groups have to be assigned manually. Various other concepts and techniques have been attempted for complementarity scoring, including vdW-like soft-core potentials [155], solvent-accessible surface area [125], volume overlap [121], empirical atom hydrophobicities [43,156], and quadratic shape descriptors [157].

3.5. Comparison of Scoring Functions

The current weaknesses of existing scoring functions are best illustrated by the observation that all available docking programs are using somewhat different scoring functions. Comparison studies of scoring functions have appeared in the recent literature. Here we will separate them into studies that evaluate scoring functions as 1) docking functions (also sometimes called fitness functions) for predicting the correct binding mode and 2) scoring functions for ranking putative protein–ligand complexes according to their experimental binding affinities.

3.5.1. Scoring Functions to Predict Binding Modes

Knegtel et al. [158] used chemical and energy scoring in DOCK and FlexX to predict the correct binding mode of 32 known thrombin inhibitors by flexible ligand docking. Chemical score outperformed both the energy and FlexX scores. With a root-mean-square deviation (rmsd) threshold of 2 Å compared to the crystal coordinates, 10–35% of the native binding conformations could be identified correctly. Gohlke et al. [115] presented a docking study based on two sets of 91 and 68 protein–ligand complexes, respectively. It was found that the DOCK energy score and the DrugScore were similar in their performance, and superior to chemical scoring in identifying the correct binding modes [115]. For a case of 61 biphenyl carboxylic acid inhibitors of stromelysin, it was found that DOCK4/PMF outperformed DOCK4/energy scoring as well as FlexX in identifying the correct binding modes [150]. Cross-docking experiments performed on 11 influenza virus neuraminidase/inhibitor structures showed DOCK4/PMF to perform better than DOCK4/energy scoring in identifying correct binding modes of ligands docked into crystal structures of other cocrystallized ligand analogs [30]. Perez and Ortiz [159] evaluated docking functions in a rigid ligand docking protocol using AMBER energy scoring and PMF scoring. They found that for reproducing crystallographic binding modes within 1.5 Å rmsd, the success rate for 34 protein–ligand structures taken from the PDB was 79% for AMBER scoring compared to 59% for PMF scoring. This compares to success rates of 76–86% as reported for PRO_LEADS [86,160] and 55% for GOLD [31]. A comprehensive study involving 200 protein–ligand complexes from the PDB showed 46% of the highest-ranking binding modes found with FlexX resemble the crystallographic binding modes with an rmsd of less than 2 Å [161]. David et al. [162] reported a success rate of 63% for 27 complexes and an rmsd threshold of 2 Å. Note that extensive tests such as reported for GOLD [31] or FlexX [161] are much more rigorous than those involving smaller test sets mentioned above. Therefore, they cannot be compared on equal footing. Finally, it should be noted that assembling a good set of protein–ligand complexes is essential for further testing of docking approaches; in this respect, Roche et al. have recently assembled a database including consistent binding data available on the World Wide Web [127].

3.5.2. Scoring Functions for Affinity Prediction and Ranking

Ha et al. [150] have shown that DOCK/PMF predicted scores correlate significantly with the measured binding affinities for the highest scoring binding modes of 61 stromelysin/inhibitor complexes [150].

More important than predicting structure–activity relationships by using docking/scoring techniques is the use of docking as a virtual screening tool. In virtual screening, the goal is to identify active compounds in a large pool of biologically inactive molecules [6]. Docking studies have shown that it is possible to identify highly potent compounds from pools of nonbinders (e.g., Refs. 66 and 67). For practical applications in the pharmaceutical industry, however, it is also important to identify weakly active compounds as starting points for lead identification. With this requirement in mind, a comparative attempt to identify very weakly active ligands for FK506 binding protein has been presented recently [149]. A database with approximately 3200 building blocks was used in a docking study involving DOCK/energy score and DOCK/PMF, respectively, to identify 27 weak ligands (K_is between 0.06 and 2 mM) previously detected by NMR techniques. Comparing DOCK/PMF to DOCK/energy scoring, it was found that screening only half of the compounds in the database could have retrieved 90% of the active compounds.

Bissantz et al. [163] evaluated different docking/scoring combinations for the virtual screening against thymidine kinase (TK) and estrogen receptor (ER). The docking programs DOCK [41], FlexX [50], and GOLD [31], in combination with the scoring functions ChemScore [110], DOCK, FlexX, Fresno [164], GOLD, PMF [118], and Score [124], were used. Different scoring routines were accessed using the CScore module in SYBYL [165] to postscore binding conformations found with FlexX. For 10 known TK ligands mixed in a pool with 990 random compounds from the ACD [166], FlexX and PMF scores performed best in identifying the known ligands correctly. However, for 10 ER antagonists both scoring functions performed rather poorly. This is somewhat surprising, as a similar study by Stahl and Rarey suggested FlexX and PMF to perform similar, if not better, compared to DrugScore and PLP [122] in another virtual screening experiment involving ER antagonists [167]. In the same study, it was reported that the scoring functions involved (FlexX, PMF, DrugScore, and PLP) perform quite differently for different targets. For instance, while PMF outperformed the other scoring functions for neuraminidase, it performed quite poorly for gelatinase A and thrombin. FlexX and PLP performed well for thrombin and p38 MAP kinase. The above examples underline the observation that there is currently no single scoring function available that performs consistently better than the others. As such, Bissantz et al. [163] suggested to use a two-step protocol in which a reduced data set containing only a few known ligands is used to derive the best docking/scoring procedure, which is then subsequently applied to a virtual screen run of the entire database [163].

3.6. Consensus Scoring

Because there is no single scoring function available today to reliably score putative protein–ligand complexes according to their binding affinities, researchers have tried to combine different scoring functions to enhance the performance. Typical consensus approaches evaluate the ranking of binding modes measured with different scoring functions and favor those that rank consistently high in several of them. In an attempt to distinguish between docking and scoring functions for affinity prediction and ranking, Wallqvist and Covell used a surface complementarity score in first instance to come up with a limited set of binding modes. A preference-based free energy surface score was then used to calculate the binding affinity [77]. Gohlke et al. [115] have generated a limited number of possible binding modes with FlexX, which were then subjected to DrugScore to find the correct binding mode. In a similar study, Stahl [168] used FlexX to generate protein–ligand conformations, and PMF for postscoring. Often, scoring results are improved using more general filter functions that remove structurally unfavorable molecules [169].

Charifson et al. [83] reported the first comprehensive consensus scoring study involving the docking programs DOCK and Gambler, in combination with 13 scoring functions: LUDI [53], ChemScore [110,141], Score [124], PLP [122], Merck force field [170], DOCK energy score [41,42], DOCK chemical score, Flog [68], strain energy, Poisson Boltzmann [171], buried lipophilic surface area [125], DOCK contact score [39], and volume overlap [121]. The study involved p38 MAP kinase, inosine monophosphate dehydrogenase, and HIV protease. The intersection of the top scoring compounds of each scoring hit list led to a significant reduction in the list of false positives (inactive compounds that have high predicted scores). Active compounds with most of the compounds being weakly active ($1 \mu M < IC_{50} < 30 \mu M$) were found with a hit rate between 2% and 7%, significantly better than the scoring hit rates of 0–3% in typical virtual screens. A comparison of the different scoring functions revealed that ChemScore, PLP, and DOCK energy score performed best as single scoring functions and also in consensus combination. Consensus scoring experiments reported by Bissantz et al. found that docking/consensus scoring performances varied widely among targets [163]. In contrast, Stahl and Rarey suggested that the combinations of FlexX and PLP scores are ideal for consensus scoring for a variety of targets including COX-2, ER, p38 MAP kinase, gyrase, thrombin, gelatinase A, and neuraminidase [167]. It should be mentioned that a consensus scoring spreadsheet, called CScore, is available in the commercial software package SYBYL [165]. It includes FlexX (also used as docking function here), PMF score [118], DOCK energy score [42], and GOLD score [31]. Finally, it should also be mentioned that Terp et al. have recently expanded the concept of consensus scoring in a quantitative statistical manner [172].

4. APPLICATIONS

In this section, we will illustrate the use of docking/scoring techniques in drug discovery. Docking can be applied to lead identification and lead optimization. Docking can help to establish interaction models between ligands and protein to rationalize and expand structure–activity relationships in a lead optimization program, or to identify the structural source of selectivity (example A). Alternatively, docking/scoring can be used as high-throughput virtual (in silico) screening tool for lead identification (example B).

4.1. Docking as a Modeling Tool: Understanding the Selectivity of Thrombin/Matriptase Inhibitors

Matriptase is a novel trypsinlike serine protease that is thought to be involved in cancer invasion, metastasis, and tissue remodeling [173]. To investigate the role of matriptase in cancer, virtual screening studies were conducted to identify inhibitors of the enzyme. Two bis-benzamidine inhibitors have been identified as potent matriptase inhibitors (Table 2), and have shown opposite selectivity against matriptase and thrombin. Here we illustrate how molecular docking has been used to understand the basis of selectivity between the two enzymes for the two bis-benzamidines.

4.1.1. Model Building

A homology model of matriptase was built using human thrombin (PDB ID 1hxe) as a template (34% sequence identity; 53% sequence similarity). Because the sequence

Table 2 K_i Values Obtained for *bis*-Benzamidine Inhibitors of Matriptase and Thrombin

Compound	Structure	K_i (nM)	
		Matriptase	Thrombin
1		924	224
2		208	2670

identity is higher than 30%, the model of matriptase is expected to be close to the real structure [174]. Friedrich et al. [175] confirmed later that the x-ray structure of matriptase has indeed 0.73 Å rmsd for the 212 Cα atoms when compared to human thrombin. It was also found that matriptase has a large 60 insertion loop of the same length and with a similar β-hairpin conformation compared to the corresponding loop in thrombin [175]. Because this insertion loop is not commonly found in trypsinlike serine proteases, it was important to find a template with the proper insertion loop. The structure of matriptase was optimized using molecular dynamics simulation in water to relieve artificial constraints imposed by the homology modeling calculations. This is especially important for loop regions because these regions are very flexible and the uncertainty of their conformation is high.

4.1.2. Thrombin Docking

Compound 1 is a 4-fold more potent inhibitor of thrombin than of matriptase, whereas compound 2 is about 13-fold more potent for matriptase than for thrombin (Table 2). To explain this somewhat surprising selectivity difference, docking calculations were performed on the matriptase and thrombin structures using the DOCK program in standard settings. The proteins were set up for docking as shown in Fig. 4. Depending on the number of minimization steps and cycles used, docking of compound 1 into the active site of thrombin resulted in two different orientations (Fig. 5A). Orientations A and B were obtained using 50 or 100 maximum iterations and 2 or 100 minimization cycles, respectively. Further increasing the number of minimization iterations to 10,000 resulted in orientation B only. This result nicely illustrates why it is necessary to exhaustively sample the configurational and orientational space of the ligand in the receptor site. To identify the most likely binding mode, molecular dynamics simulations with the GEP were performed [176]. The goal of the method is to reduce barrier heights on the potential surface while keeping the locations of the potential energy minima unchanged. This way the ligand can more efficiently sample the potential surface during docking. The generalized effective potential is calculated using Eq. (9). The parameter q controls the reduction of barrier heights. When $q = 1.0$, the generalized effective potential is equal to the potential energy and the simulation becomes regular MD. With q being larger than 1.0, GEP has been shown before to be effective for docking compounds into the C1b domain of PKC [177].

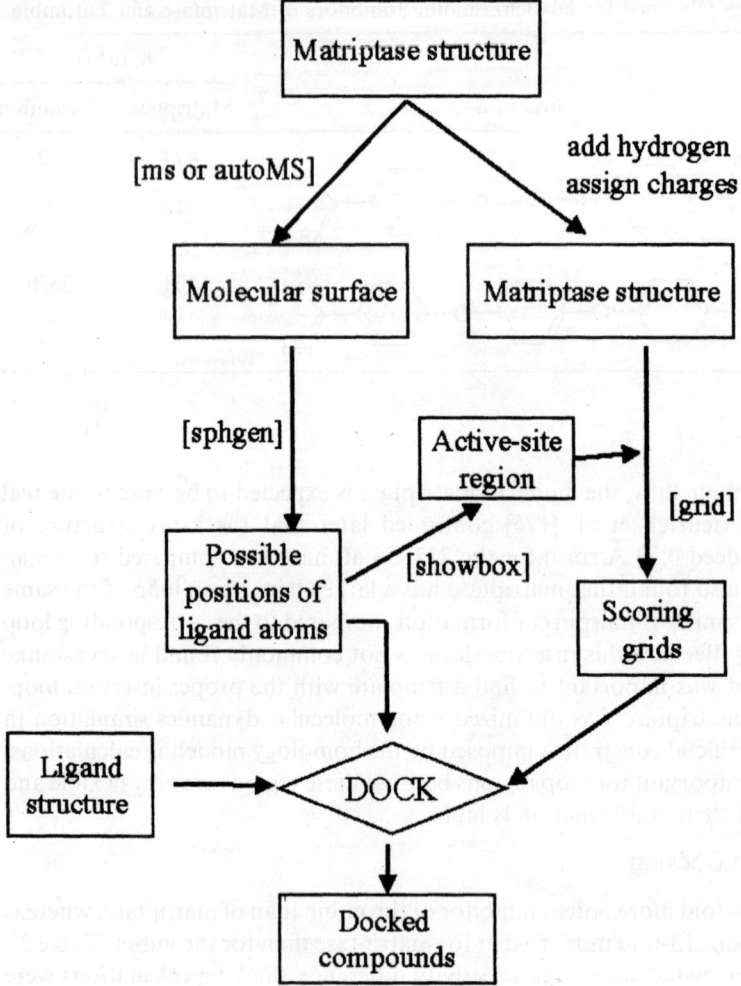

Figure 4 Flowchart for protein–ligand docking with the program DOCK.

$$\overline{U}_q(r^N) = \frac{q}{\beta(q-1)} \ln[1 - (1-q)\beta(U(r^N) + \varepsilon)] \qquad (9)$$

where $\beta = 1/k_B T$ and ε is an arbitrary energy term to guarantee that the term $U(r^N) + \varepsilon$ is always positive.

Parallel simulations were conducted of the complex of thrombin with **1** in orientations A and B. The active site was solvated with explicit water molecules. The maximum q value for calculating the GEP was set to 1.0005. The position of the inhibitor was optimized using an annealing procedure. The final orientation was very similar to orientation B from Fig. 5A, confirming the predicted binding mode from DOCK. The binding mode obtained after docking compound **2** using the same protocol is presented in Fig. 6A. The predicted binding mode of **1** and **2** to thrombin shows that both compounds interact with the S1 site through a salt bridge with Asp189, hydrophobic interactions with Val213, and a hydrogen bond with the

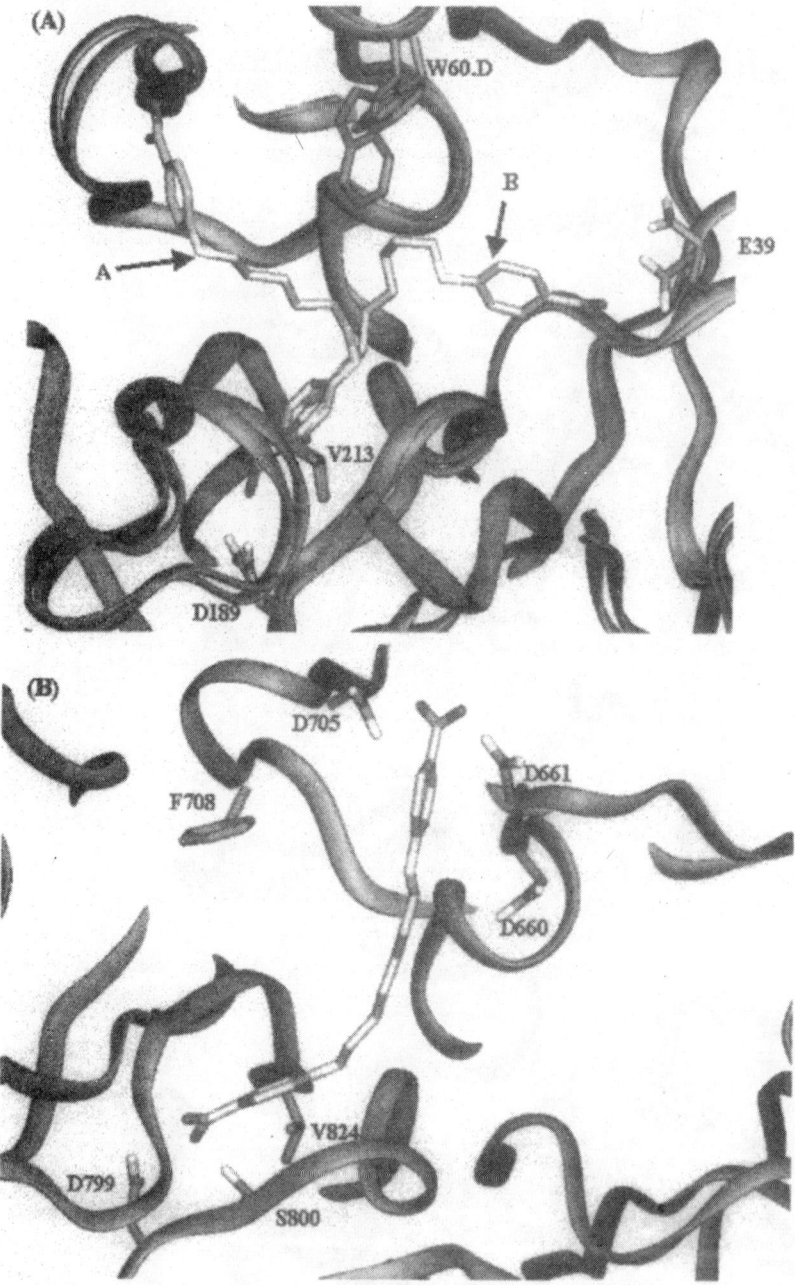

Figure 5 Proposed binding mode of compound **1** in complex with thrombin (A) and matriptase (B) as obtained from docking with DOCK. (See color plate at end of chapter.)

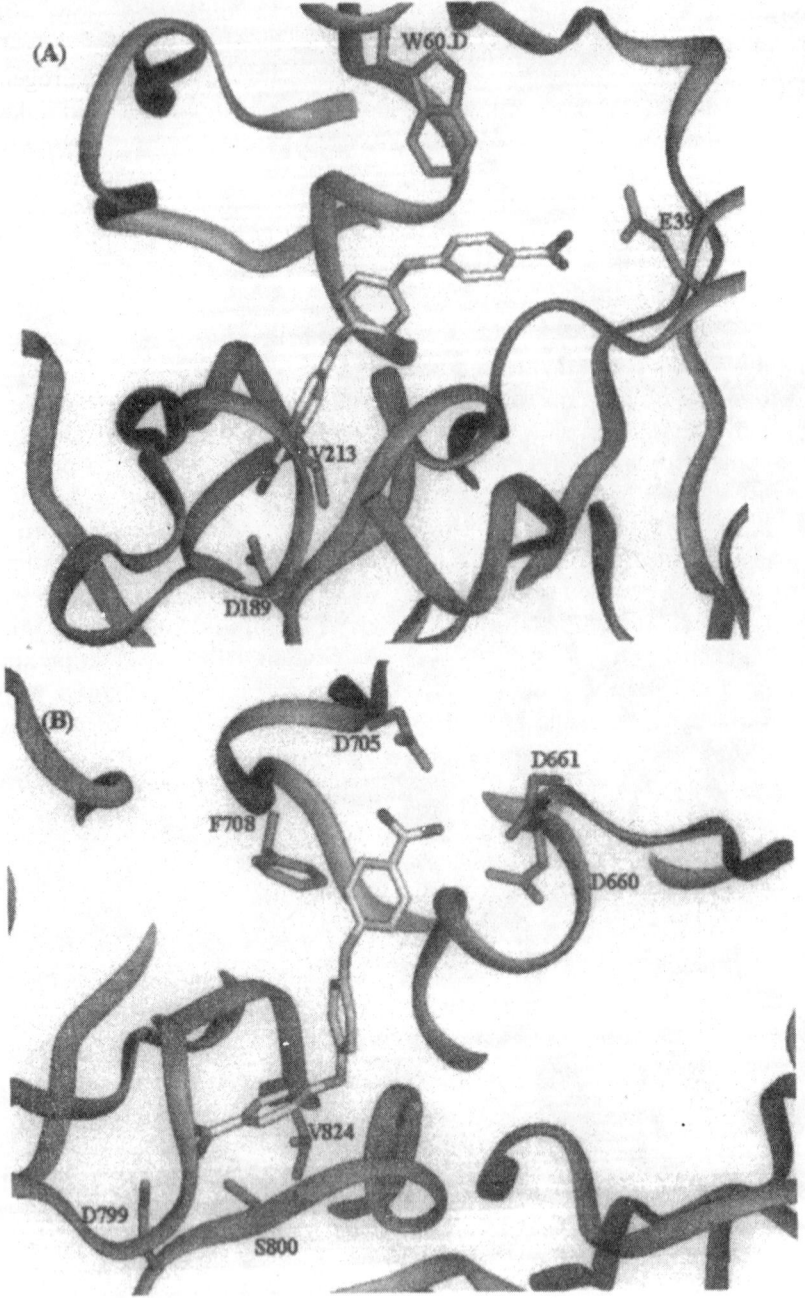

Figure 6 Proposed binding mode of compound **2** in complex with (A) thrombin and (B) matriptase as obtained from docking with DOCK. (See color plate at the end of chapter.)

carbonyl oxygen of Ala190. Both ligands also form hydrophobic interactions with Trp60. These interactions cannot explain the difference in K_i observed experimentally. However, because compound **1** has a flexible linker it could form bidentate hydrogen-bonding interactions with the carboxylate of Glu39. In contrast, **2** has a rigid linker that is also shorter than the linker in **1**; it forms only a monodentate hydrogen-bonding interaction with the carboxylate of Glu39. Bidentate hydrogen-bonding interactions between oppositely charged groups have been shown to be stronger than monodentate interactions for a series of thrombin inhibitors [178]. This finding might provide an explanation why compound **2** is less potent than **1**.

4.1.3. Matriptase Docking

Docking of compounds **1** and **2** into the active site of matriptase using the same protocol led to single orientation of both compounds (Figs. 5B and 6B). A 1-ns molecular dynamics simulation with the GEP protocol did not significantly alter the orientations obtained from DOCK. Predicted binding modes of the two compounds showed that both interact in the same way with the S1 binding site residues Asp799, Ser800, and Val824. Compound **1** forms two monodentate hydrogen-bonding interactions with Asp705 and Asp661, while **2** forms bidentate hydrogen-bonding interactions with Asp705 and monodentate hydrogen-bonding interaction with Asp660. The shorter linker of **2** allows a better interaction of this compound with Asp705. In addition, the linker in **1** is too long to allow the interaction with Phe708, while the shorter linker in **2** enables the interaction of the second benzamidine group with Phe708. Taken together, these findings may provide an explanation of the reverse

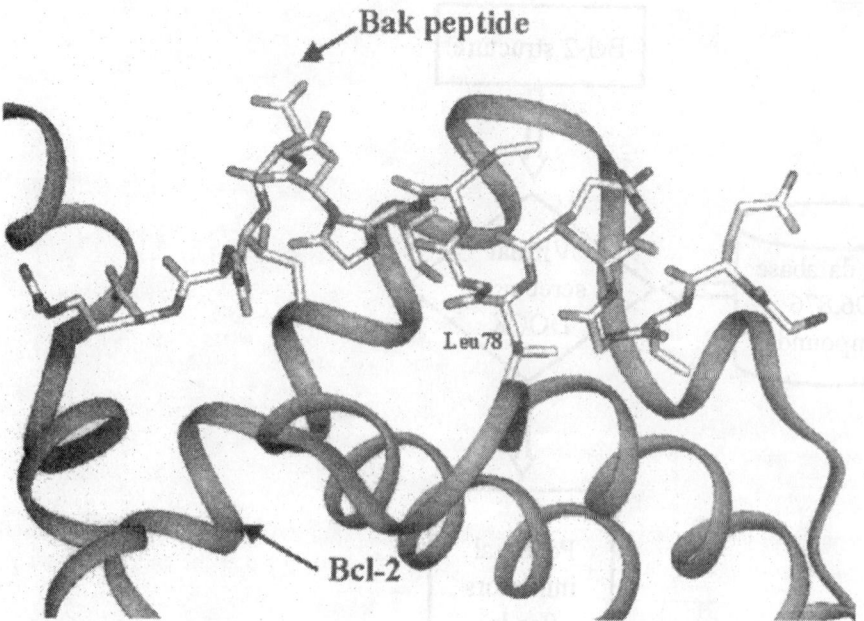

Figure 7 Proposed binding mode of the BAK BH3 peptide to Bcl-2. The binding site where Leu78 from the BAK peptide binds is the most sensible one for Ala mutation. (See color plate at end of chapter.)

potency in inhibiting matriptase versus thrombin. Figs. 5 and 6 indicate that while in matriptase the anionic site is opposite to the S1 site; in thrombin it is positioned to the right of the S1 site. As such, bis-benzamidines with rigid linkers will bind easier to matriptase than to thrombin because the linker is not favorable for optimal binding to thrombin.

4.2. Docking as an In Silico Screening Tool. Discovery of Bcl-2 Inhibitors

Bcl-2 belongs to a family of proteins that regulate apoptosis. Overexpression of Bcl-2 has been observed in several forms of cancer, including breast, prostate, colorectal cancer, and as such this protein provides an attractive target for the development of anticancer therapies. However, the discovery of small molecules that can inhibit protein–protein interactions has been considered difficult because of the large surface through which proteins interact [179]. Fortunately, protein–protein interactions are governed by several key residues crucial for good binding. Alanine scanning of the BAK BH3 peptide has shown that the Leu78Ala mutation decreases the binding

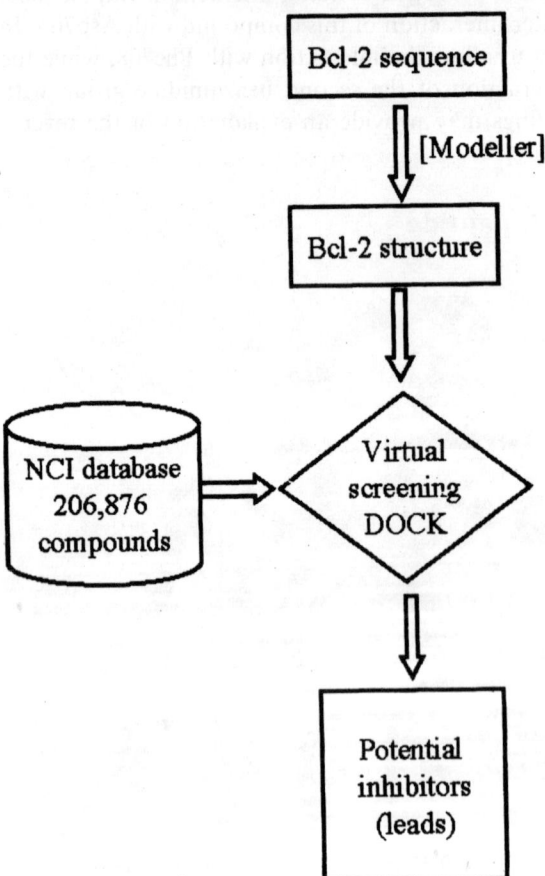

Figure 8   Flowchart for performing a virtual screen using DOCK.

Table 3 Chemical Structures of Selected Active Compounds

Compound	Structure	Binding IC$_{50}$ (µM)	Compound	Structure	Binding IC$_{50}$ (µM)
3		10.4 ± 0.3	6		5.8 ± 2.2
4		1.6 ± 0.1	7		11.7 ± 2.4
5		10.4 ± 1.2	8		7.7 ± 4.5

affinity by about 800-fold [180]. Thus, the binding region of Leu78 to Bcl-2 was a valid target for virtual screening (Fig. 7). Using a model structure of Bcl-2, a query was built based on the binding region of Leu78. The NCI 3-D database containing 206,876 small molecules was searched for compounds fitting into the binding site of Leu78 using the program DOCK. The strategy used for virtual screening is presented in Fig. 8. Molecules were ranked based on their energy score. Compounds with more than 10 flexible bonds, less than 10 heavy atoms, or more than 50 heavy atoms were not considered. The position of the molecules was optimized by considering 50 configurations per ligand building cycle, 100 maximum anchor orientations, two minimization cycles with maximum 10 iterations per cycle. From the 500 top-scoring compounds, 35 were selected for further biochemical testing (the selection was based on structural diversity and availability). Six compounds proved to have an IC_{50} below 12 μM (Table 3). These results suggest that small molecules can indeed inhibit protein–protein interactions and that in silico structure-based 3-D database screening is a successful method in lead identification.

5. CONCLUSIONS

Computational protein–ligand docking has been the subject of intense research in the last decade. Recent successes in structure-based design and virtual screening have prompted an increasing interest in the development of new high-throughput docking methods for in silico screening. The main parts of current docking approaches are the exploration of the geometry and conformational degrees of freedom of the interacting molecules (usually protein and ligand), as well as the evaluation of the complexes according to their binding affinities. The problem of automated sampling of the conformational space can be considered as solved. However, the design of reliable scoring functions, able to identify correctly the binding mode of a putative protein–ligand complex and rank protein–ligand complexes according their binding affinities, still remains the Achilles' heel of current docking routines.

ACKNOWLEDGMENTS

The authors thank Dr. Matthias Rarey for helpful suggestions and discussions.

REFERENCES

1. Gallop MA, Barrett RW, Dower WJ, Fodor SPA, Gordon AM. Applications of combinatorial technologies to drug discovery. 1. Background and peptide combinatorial libraries. J Med Chem 1994; 37:1233–1251.
2. Gordon EM, Barrett RW, Dower WJ, Fodor SPA, Gallop MA. Applications of combinatorial technologies to drug discovery. 2. Combinatorial organic synthesis, library screening strategies, and future directions. J Med Chem 1994; 37:1385–1401.
3. Lutz MW, Menius JA, Choi TD, Laskody RG, Domanico PL, Goetz AS, Saussy DL. Experimental-design for high throughput screening. Drug Discov Today 1996; 1:277–286.
4. Spencer RW. High-throughput screening of historic collections: observations on file size, biological targets, and file diversity. Biotechnol Bioeng 1998; 61:61–67.
5. Martin YC. Challenges and prospects for computational aids to molecular diversity. Perspect Drug Discov Des 1997; 7/8:159–172.

6. Walters WP, Stahl MT, Murcko MA. Virtual screening—an overview. Drug Discov Today 1998; 3:160–178.

7. Berman IIM, Westbrook J, Feng Z, Gilliland G, Baht TN, Weissig H, Shindyalov IN, Bourne PE. The Protein Data Bank. Nucleic Acids Res 2000; 28:235–242.

8. Abraham DJ. The potential role of single crystal x-ray diffraction in medicinal chemistry. Intra-Sci Chem Rep 1974; 8:1–9.

9. Bedell CR, Goodford PJ, Norrington FE. Compounds designed to fit a site of known structure in human hemoglobin. Br J Pharmacol 1976; 57:201–209.

10. Cushman DW, Cheung HS, Sabo EF, Ondetti MA. Design of potent competitive inhibitors of angiotensin-converting enzyme. Carboxyalkanoyl and mercaptoalkanoyl amino acids. Biochemistry 1977; 16:5484–5491.

11. Klebe G. Recent developments in structure-based drug design. J Mol Med 2000; 78:269–281.

12. Kollman P. Free energy calculations—applications to chemical and biological phenomena. Chem Rev 1993; 7:2395–2417.

13. Kuntz ID. Structure-based strategies for drug design and discovery. Science 1992; 257:1078–1082.

14. Blaney JM, Dixon JS. A good ligand is hard to find: automatic docking methods. Perspect Drug Discov Des 1993; 15:301–319.

15. Guida WC. Software for structure-based drug design. Curr Opin Struct Biol 1994; 4:777–781.

16. Colman PM. Structure-based drug design. Curr Opin Struct Biol 1994; 4:868–874.

17. Lybrand TP. Ligand–protein docking and rational drug design. Curr Opin Struct Biol 1995; 5:224–228.

18. Rosenfeld R, Vajda S, Delisi C. Flexible docking and design. Annu Rev Biophys Biomol Struct 1995; 24:677–700.

19. Böhm H. Current computational tools for de novo ligand design. Curr Opin Biotech 1996; 7:433–436.

20. Lengauer T, Rarey M. Computational methods for biomolecular docking. Curr Opin Struct Biol 1996; 6:402–406.

21. Ajay, Murcko MA. Computational methods to predict binding free energy in ligand–receptor complexes. J Med Chem 1995; 38:4953–4967.

22. Holloway MK. A priori prediction of ligand affinity by energy minimization. Perspect Drug Discov Des 1998; 9/10/11:63–84.

23. Oprea TI, Marshall GR. Receptor-based prediction of binding affinities. Perspect Drug Discov Des 1998; 9/10/11:35–61.

24. Tame JRH. Scoring functions: a view from the bench. J Comput-Aided Mol Des 1999; 13:99–108.

25. Muegge I, Rarey M. Small molecule docking and scoring. In: Boyd DB, Lipkowitz KB, eds. Reviews in Computational Chemistry. Vol. 17. New York: Wiley-VCH, 2001:1–60.

26. Ehrlich LP, Wade RC. Protein–protein docking. In: Lipkowitz KB, Boyd DB, eds. Reviews in Computational Chemistry. Vol. 17. New York: Wiley-VCH, 2001:61–97.

27. Sander C, Schneider R. Database of homology-derived protein structures and the structural meaning of sequence alignment. Proteins Struct Funct Genet 1991; 9:56.

28. Blundell TL, Sibanda BL, Sternberg MJE, Thornton JM. Knowledge-based prediction of protein structures and the design of novel molecules. Nature 1987; 326:347.

29. Vedani A, Zbinden P, Snyder JP, Greenidge PA. Pseudoreceptor modeling: the construction of three-dimensional receptor surrogates. J Am Chem Soc 1995;1174987–4994.

30. Muegge I. The effect of small changes in protein structure on predicted binding modes of known inhibitors of influenza virus neuraminidase: PMF-scoring in DOCK4. Med Chem Res 1999; 9:490–500.

31. Jones G, Willett P, Glen RC, Leach AR. Development and validation of a genetic algorithm for flexible docking. J Mol Biol 1997; 267:727–748.

32. Bron C, Kerbosch J. Finding all cliques of an undirected graph. Commun Assoc Comput Mach 1973; 16:575–577.

33. Kuhl FS, Crippen GM, Friesen DK. A combinatorial algorithm for calculating ligand binding. J Comput Chem 1984; 5:24–34.

34. Kuntz ID, Blaney JM, Oatley SJ, Langridge RL. A geometric approach to macro-molecule–ligand interactions. J Mol Biol 1982; 161:269–288.

35. Richards FM. Areas, volumes, packing, and protein structure. Annu Rev Biophys Bioeng 1977; 6:151–176.

36. Connolly ML. Analytical molecular surface calculation. J Appl Cryst 1983; 16:548–558.

37. Connolly ML. Molecular surface triangulation. J Appl Cryst 1985; 18:499–505.

38. Shoichet BK, Stroud RM, Santi DV, Kuntz ID, Perry KM. Structure-based discovery of inhibitors of thymidylate synthase. Science 1993; 259:1445–1450.

39. Shoichet BK, Bodian DL, Kuntz ID. Molecular docking using shape descriptors. J Comput Chem 1992; 13:380–397.

40. Meng EC, Gschwend DA, Blaney JM, Kuntz ID. Orientational sampling and rigid-body minimization in molecular docking. Proteins Struct Funct Genet 1993; 17:266–278.

41. Ewing TJA, Kuntz ID. Critical evaluation of search algorithms for automated molecular docking and database screening. J Comput Chem 1997; 18:1175–1189.

42. Meng EC, Shoichet BK, Kuntz ID. Automated docking with grid-based energy eval-uation. J Comput Chem 1992; 13:505–524.

43. Meng EC, Kuntz ID, Abraham DJ, Kellogg GE. Evaluating docked complexes with the hint exponential function and empirical atomic hydrophobicities. J Comput-Aided Mol Des 1994; 8:299–306.

44. Gschwend DA, Kuntz ID. Orientational sampling and rigid-body minimization in molecular docking revisited—on-the-fly optimization and degeneracy removal. J Com-put-Aided Mol Des 1996; 10:123–132.

45. Shoichet BK, Leach AR, Kuntz ID. Ligand solvation in molecular docking. Proteins Struct Funct Genet 1999; 34:4–16.

46. Zou XQ, Sun YX, Kuntz ID. Inclusion of solvation in ligand binding free energy calculations using the generalized-born model. J Am Chem Soc 1999; 121:8033–8043.

47. Lamdan Y, Wolfson HJ. Geometric Hashing: a general and efficient model-based recog-nition scheme. In: eds. Proceedings of the IEEE International Conference on Computer Vision Vol. New York: IEEE Computer Society Press, 1988:238–249.

48. Fischer D, Norel R, Wolfson H, Nussinov R. Surface motifs by a computer vision technique: searches, detection and implications for protein–ligand recognition. Proteins Struct Funct Genet 1993; 16:278–292.

49. Fischer D, Lin SL, Wolfson HL, Nussinov R. A geometry-based suite of molecular docking processes. J Mol Biol 1995; 248:459–477.

50. Rarey M, Kramer B, Lengauer T, Klebe G. A fast flexible docking method using an in-cremental construction algorithm. J Mol Biol 1996; 261:470–489.

51. Rarey M, Wefing S, Lengauer T. Placement of medium-sized molecular fragments into active sites of proteins. J Comput-Aided Mol Des 1996; 10:41–54.

52. Linnainmaa S, Harwood D, Davis LS. Pose determination of a three-dimensional object using triangle pairs. IEEE Trans Pattern Anal Mach Intell 1988; 10:634–646.

53. Böhm H-J. LUDI: rule-based automatic design of new substituents for enzyme inhibitor leads. J Comput-Aided Mol Des 1992; 6:593–606.

54. Böhm H. The computer program LUDI: a new method for the de novo design of enzyme inhibitors. J Comput-Aided Mol Des 1992; 6:61–78.

55. Duda RO, Hart PE. Pattern Classification and Scene Analysis. New York: John Wiley & Sons, Inc., 1973.

56. Nicklaus MC, Wang S, Driscoll JS, Milne GWA. Conformational changes of small molecules binding to proteins. Bioorg Med Chem 1995; 3:411–428.

57. Oprea TI. Property distribution of drug-related chemical databases. J Comput-Aided Mol Des 2000; 14:251–264.

58. Murcko MA. Recent advances in ligand design methods. In: Lipkowitz KB, Boyd DB, eds. Reviews in Computational Chemistry. Vol. 11. New York: Wiley-VCH, 1997:1–66.

59. Clark DE, Murray CW, Li J. Current issues in de novo molecular design. In: Lipkowitz KB, Boyd DB, eds. Reviews in Computational Chemistry. Vol. 11. New York: Wiley-VCH, 1997:67–125.

60. DesJarlais RL, Sheridan RP, Dixon JS, Kuntz ID, Venkataraghavan R. Docking flexible ligands to macromolecular receptors by molecular shape. J Med Chem 1986; 29:2149–2153.

61. Sandak B, Nussinov R, Wolfson HJ. 3-D flexible docking of molecules. IEEE Workshop on Shape and Pattern Matching in Computational Biology. Calitano A, Rigoutsos I, Wolfson HJ. New York: IEEE Computer Society Press, 1994:41–54.

62. Sandak B, Nussinov R, Wolfson HJ. An automated computer vision and robotics-based technique for 3-D flexible biomolecular docking and matching. Comput Appl Biosci 1995; 11:87–99.

63. Sandak B, Nussinov R, Wolfson HJ. A method for biomolecular structural recognition and docking allowing conformational flexibility. J Comp Biol 1998; 5:631–654.

64. Moon JB, Howe WJ. Computer design of bioactive molecules: a method for receptor-based de novo ligand design. Proteins Struct Funct Genet 1991; 11:314–328.

65. Leach AR, Kuntz ID. Conformational analysis of flexible ligands in macromolecular receptor sites. J Comput Chem 1992; 13:730–748.

66. Makino S, Kuntz ID. Automated flexible ligand docking method and its application for database search. J Comput Chem 1997; 18:1812–1825.

67. Welch W, Ruppert J, Jain AN. Hammerhead—fast, fully automated docking of flexible ligands to protein binding sites. Chem Biol 1996; 3:449–462.

68. Miller MD, Kearsley SK, Underwood DJ, Sheridan RP. Flog—a system to select quasi-flexible ligands complementary to a receptor of known 3-dimensional structure. J Comput-Aided Mol Des 1994; 8:153–174.

69. Havel TF, Kuntz ID, Crippen GM. The theory and practice of distance geometry. Bull Math Biol 1983; 45:665–720.

69a. Lorber DM, Shoichet BK. Flexible ligand docking using conformational ensembles. Protien Sci 1998; 7:938–950.

70. Goodsell DS, Olson AJ. Automated docking of substrates to proteins by simulated annealing. Proteins Struct Funct Genet 1990; 8:195–202.

71. Kirkpatrik S, Gelatt CDJ, Vecchi MP. Optimization by simulated annealing. Science 1983; 220:671–680.

72. Yue S. Distance-constrained molecular docking by simulated annealing. Protein Eng 1990; 4:177–184.

73. Lybrand TP. Computer simulation of biomolecular systems using molecular dynamics and free energy perturbation methods. In: Boyd DB, Lipkowitz KB, eds. Reviews in Computational Chemistry. Vol. 1. New York: VCH Publishers, 1990:295–320.

74. Given JA, Gilson MK. A hierarchical method for generating low-energy conformers of a protein–ligand complex. Proteins Struct Funct Genet 1998; 33:475–495.

75. Hart TN, Read RJ. A multiple-start Monte Carlo docking method. Proteins Struct Funct Genet 1992; 13:206–222.

76. McMartin C, Bohacek RS. QXP: powerful, rapid computer algorithms for structure-based drug design. J Comput-Aided Mol Des 1997; 11:333–344.

77. Wallqvist A, Covell DG. Docking enzyme–inhibitor complexes using a preference-based free-energy surface. Proteins Struct Funct Genet 1996; 25:403–419.

78. Abagyan R, Totrov M, Kuznetsov D. ICM—a new method for protein modeling and design: applications to docking and structure prediction from the distorted native conformation. J Comput Chem 1994; 15:488–506.

79. Apostolakis J, Pluckthun A, Caflisch A. Docking small ligands in flexible binding sites. J Comput Chem 1998; 19:21–37.

80. Trosset JY, Scheraga HA. PRODOCK: software package for protein modeling and docking. J Comput Chem 1999; 20:412–427.

81. Pearlman DA, Case DA, Caldwell JW, Ross WS, Cheatham TEI, DeBolt S, Ferguson D, Seibel G, Kollman P. AMBER, a package of computer programs for applying molecular dynamics, normal mode analysis, molecular dynamics and free energy calculations to simulate the structural and energetic properties of molecules. Comput Phys Commun 1995; 91:1–41.

82. LigandFit is a docking routine in the software package Cerius2 available from Accelrys, San Diego, CA.

83. Charifson PS, Corkery JJ, Murcko MA, Walters WP. Consensus scoring: a method for obtaining improved hit rates from docking databases of three-dimensional structures into proteins. J Med Chem 1999; 42:5100–5109.

84. Oshiro CM, Kuntz ID, Dixon JS. Flexible ligand docking using a genetic algorithm. J Comput-Aided Mol Des 1995; 9:113–130.

85. Gehlhaar DK, Verkhivker GM, Rejto PA, Fogel DB, Fogel LJ, Freer ST. Docking conformationally flexible small molecules into a protein binding site through evolutionary programming. In: McDonnell JR, Reynolds RG, Fogel DB, eds. Proceedings of the Fourth Annual Conference on Evolutionary Programming Vol. Cambridge, MA: MIT Press, 1995:615–627.

86. Baxter CA, Murray CW, Clark DE, Westhead DR, Eldridge MD. Flexible docking using tabu search and an empirical estimate of binding affinity. Proteins Struct Funct Genet 1998; 33:367–382.

87. Wang J, Kollman PA, Kuntz ID. Flexible ligand docking: a multistep strategy approach. Proteins Struct Funct Genet 1999; 36:1–19.

88. Hoffmann D, Kramer B, Washio T, Steinmetzer T, Rarey M, Lengauer T. Two-stage method for protein–ligand docking. J Med Chem 1999; 42:4422–4433.

89. Najmanovich R, Kuttner J, Sobolev V, Edelman M. Side-chain flexibility in proteins upon ligand binding. Prot Struct Funct Genet 2000; 39:261–268.

90. Luty BA, Wasserman ZR, Stouten P, Hodge CN, Zacharias M, McCammon JA. A molecular mechanics/grid method for evaluation of ligand–receptor interactions. J Comput Chem 1995; 16:454–464.

91. Wasserman ZR, Hodge CN. Fitting an inhibitor into the active site of thermolysin: a molecular dynamics case study. Proteins Struct Funct Genet 1996; 24:227–237.

92. Leach AR. Ligand docking to proteins with discrete side-chain flexibility. J Mol Biol 1994; 235:345–356.

93. Knegtel RMA, Kuntz ID, Oshiro CM. Molecular docking to ensembles of protein structures. J Mol Biol 1997; 266:424–440.

94. Broughton HB. A method for including protein flexibility in protein–ligand docking: Improving tools for database mining and virtual screening. J Mol Graph Model 2000; 18:247–257.

95. Lamb ML, Burdick KW, Toba S, Young MM, Skillman AG, Zou X, Arnold JR, Kuntz ID. Design, docking, and evaluation of multiple libraries against multiple targets. Proteins Struct Funct Genet 2001; 42:296–318.

96. Murray CW, Clark DE, Auton TR, Firth MA, Li J, Sykes RA, Waszkowycz B, Westhead DR, Young SC. PRO_SELECT: combining structure-based drug design and combinatorial chemistry for rapid lead discovery. 1. Technology. J Comput-Aid Mol Des 1997; 11:193–207.

97. Sun Y, Ewing TJA, Skillman AG, Kuntz ID. CombiDOCK: structure-based combinatorial docking and library design. J Comput-Aid Mol Des 1998; 12:597–604.

98. Chen YZ, Zhi DG. Ligand–protein inverse docking and its potential use in the computer

search of protein targets of small molecules. Proteins Struct Funct Genet 2001; 43:217–226.

99. Pearlman DA. Free energy grids: a practical qualitative application of free energy perturbation to ligand design using the OWFEG method. J Med Chem 1999; 42:4313–4324.

100. Pearlman DA, Charifson PS. Improving scoring of ligand–protein interactions using OWFEG free energy grids. J Med Chem 2001; 44:502–511.

101. Lee FS, Chu ZT, Warshel A. Microscopic and semimicroscopic calculations of electrostatic energies in proteins by the POLARIS and ENZYMIX programs. J Comput Chem 1993; 14:161–185.

102. Aqvist J, Medina C, Samuelsson JE. New method for predicting binding affinity in computer-aided drug design. Protein Eng 1994; 7:386–391.

103. Weiner SJ, Kollman PA, Case DA, Singh UC, Ghio C, Alagona G, Profeta S Jr, Weiner P. A new force field for molecular mechanical simulation of nucleic acids and proteins. J Am Chem Soc 1984; 106:765.

104. Weiner SJ, Kollman PA, Nguyen DT, Case DA. An all atom force field for simulations of proteins and nucleic Acids. J Comput Chem 1986; 7:230.

105. Holloway MK, Wai JM, Halgren TA, Fitzgerald PMD, Vacca JP, Dorsey BD, Levin RB, Thompson WJ, Chen LJ, deSolms SJ, Gaffin N, Ghosh AK, Giuliani EA, Graham SL, Guare JP, Hungate RW, Lyle TA, Sanders WM, Tucker TJ, Wiggins M, Wiscount CM, Woltersdorf OW, Young SD, Darke PL, Zugay JA. A priori prediction of activity for HIV-1 protease inhibitors employing energy minimization in the active site. J Med Chem 1995; 38:305–317.

106. Zhang C, Vasmatzis V, Cornette J, DeLisi C. Determination of atomic desolvation energies from the structures of crystallized proteins. J Mol Biol 1997; 267:707–726.

107. Horton N, Lewis M. Calculation of the free energy of association for protein complexes. Protein Sci 1992; 1:169–181.

108. Böhm H-J. The development of a simple empirical scoring function to estimate the binding constant for a protein–ligand complex of known three-dimensional structure. J Comput-Aided Mol Design 1994; 8:243–256.

109. Böhm HJ. Prediction of binding constants of protein ligands—a fast method for the prioritization of hits obtained from de novo design or 3D database search programs. J Comput-Aided Mol Design 1998; 12:309–323.

110. Eldridge MD, Murray CW, Auton TR, Paolini GV, Mee RP. Empirical scoring functions: I. The development of a fast empirical scoring function to estimate the binding affinity of ligands in receptor complexes. J Comput-Aided Mol Des 1997; 11:425–445.

111. Head RD, Smythe ML, Oprea TL, Waller CL, Green SM, Marshall GM. VALIDATE: a new method for the receptor-based prediction of binding affinities of novel ligands. J Am Chem Soc 1996; 118:3959–3969.

112. Jain AN. Scoring noncovalent protein–ligand interactions: a continuous differentiable function tuned to compute binding affinities. J Comput-Aided Mol Des 1996; 10:427–440.

113. Alex A, Finn P. Fast and accurate prediction of relative binding energies. J Mol Struct (Theochem) 1997; 398–399:551–554.

114. DeWitte RS, Shakhnovich EI. SMoG: de novo design method based on simple, fast, and accurate free energy estimates. 1. Methodology and supporting evidence. J Am Chem Soc 1996; 118:11733–11744.

115. Gohlke H, Hendlich M, Klebe G. Knowledge-based scoring function to predict protein–ligand interactions. J Mol Biol 2000; 295:337–356.

116. Jernigan RL, Bahar I. Structure-derived potentials and protein simulations. Curr Opin Struct Biol 1996; 6:195–209.

117. Mitchell JBO, Laskowski RA, Alex A, Thornton JM. BLEEP-potential of mean force

describing protein–ligand interactions: I. Generating potential. J Comput Chem 1999; 20: 1165–1176.

118. Muegge I, Martin YC. A general and fast scoring function for protein–ligand interactions: a simplified potential approach. J Med Chem 1999; 42:791–804.

119. Verkhivker G, Appelt K, Freer ST, Villafranca JE. Empirical free energy calculations of ligand–protein crystallographic complexes. I. Knowledge-based ligand–protein interaction potentials applied to the prediction of human immunodeficiency virus 1 protease binding affinity. Protein Eng 1995; 8:677–691.

120. Wallqvist A, Jernigan RL, Covell DG. A preference-based free energy parametrization of enzyme–inhibitor binding. Application to HIV-1-protease inhibitor design. Protein Sci 1995; 4:1881–1903.

121. Stouch TR, Jurs PC. A simple method for the representation, quantification, and comparison of the volumes and shapes of chemical compounds. J Chem Inf Comput Sci 1986; 26:4–12.

122. Gehlhaar DK, Verkhivker GM, Rejto PA, Sherman CJ, Fogel DB, Fogel LJ, Freer ST. Molecular recognition of the inhibitor Ag-1343 by Hiv-1 protease—conformationally flexible docking by evolutionary programming. Chem Biol 1995; 2:317–324.

123. Bostrom J, Norby PO, Liljefors T. Conformational energy penalties of protein bound ligands. J Comput-Aided Mol Des 1998; 12:383–396.

124. Wang RX, Liu L, Lai LH, Tang YQ. SCORE: a new empirical method for estimating the binding affinity of a protein–ligand complex. J Mol Model 1998; 4:379–394.

125. Flower DR. SERF: a program for accessible surface area calculations. J Mol Graph Model 1998; 15:238–244.

126. Dixon JS. Evaluation of the CASP2 docking section. Proteins Struct Funct Genet Suppl 1997; 1:198–204.

127. Roche O, Kiyama R, Brooks CL III. Ligand–protein database: linking protein–ligand complex structures to binding data. J Med Chem 2001; 44:3592–3598.

128. Grootenhuis PDJ, vanGalen PJM. Correlation of binding affinities with non-bonded interaction energies of thrombin–inhibitor complexes. Acta Cryst 1995; D51:560–566.

129. Blom NS, Sygusch J. High resolution fast quantitative docking using Fourier domain correlation techniques. Proteins Struct Funct Genet 1997; 27:493–506.

130. Tomioka N, Itai A. GREEN: a program package for docking studies in rational drug design. J Comput-Aided Mol Des 1994; 8:347–366.

131. Goodford PJ. A computational procedure for determining energetically favorable binding sites on biologically important macromolecules. J Med Chem 1985; 28:849–857.

132. Jones G, Willett P, Glen RC. Molecular recognition of receptor sites using a genetic algorithm with a description of solvation. J Mol Biol 1995; 254:43–53.

133. Liu M, Wang SM. MCDOCK: a Monte Carlo simulation approach to the molecular docking problem. J Comput-Aid Mol Des 1999; 13:435–451.

134. Nauchatel V, Villaverde MC, Sussman F. Solvent accessibility as a predictive tool for the free energy of inhibitor binding to the HIV-1 protease. Protein Sci 1995; 4:1356–1364.

135. Majeux N, Scarsi M, Apostolakis J, Ehrhardt C, Caflisch A. Exhaustive docking of molecular fragments with electrostatic solvation. Proteins Struct Funct Genet 1999; 37:88–105.

136. Qui D, Shenkin PS, Hollinger EP, Still WC. The GB/SA continuum model for solvation. A fast analytical method for the calculation of approximate Born radii. J Phys Chem 1997; 101:3005–3014.

137. Eisenberg D, McLachlan AD. Solvation energy in protein folding and binding. Nature 1986; 319:199–203.

138. Stouten PFW, Frommel C, Nakamura H, Sander C. An effective solvation term based on atomic occupancies for use in protein simulations. Mol Simul 1993; 10:97–120.

139. Vajda S, Weng Z, Rosenfeld R, DeLisi C. Effect of conformational flexibility and solvation on receptor–ligand binding free energies. Biochemistry 1994; 33:13977–13988.

140. Vieth M, Hirst JD, Kolinski A, Brooks CL III. Assessing energy functions for flexible docking. J Comput Chem 1998; 14:1612–1622.

141. Murray CW, Auton TR, Eldridge MD. Empirical scoring functions. II. The testing of an empirical scoring function for the prediction of ligand–receptor binding affinities and the use of Bayesian regression to improve the quality of the model. J Comput-Aid Mol Des 1998; 12:503–519.

142. Westhead DR, Clark DE, Murray CW. A comparison of heuristic search algorithms for molecular docking. J Comput-Aid Mol Des 1997; 11:209–228.

143. Searle MS, Williams DH. The cost of conformational order: entropy changes in molecular associations. J Am Chem Soc 1992; 114:10690–10697.

144. Searle MS, Williams DH, Gerhard U. Partitioning of free energy contributions in the estimate of binding constants: residual motions and consequences for amide–amide hydrogen bond strengths. J Am Chem Soc 1992; 114:10697–10704.

145. Muegge I. A knowledge-based scoring function for protein–ligand interactions: probing the reference state. Perspect Drug Discov Des 2000; 20:99–114.

146. Muegge I. Effect of ligand volume correction on PMF-scoring. J Comput Chem 2001; 22:418–425.

147. Gohlke H, Klebe G. Statistical potentials and scoring functions applied to protein–ligand binding. Curr Opin Struct Biol 2001; 11:231–235.

148. Sippl MJ. Calculation of conformational ensembles from potentials of mean force. J Mol Biol 1990; 213:859–883.

149. Muegge I, Martin YC, Hajduk PJ, Fesik SW. Evaluation of PMF scoring in docking weak ligands to the FK506 binding protein. J Med Chem 1999; 42:2498–2503.

150. Ha S, Andreani R, Robbins A, Muegge I. Evaluation of docking/scoring approaches: a comparative study based on MMP-3 inhibitors. J Comput-Aided Mol Des 2000; 14:435–448.

151. Muegge I, Podlogar B. 3D-quantitative structure activity relationship of biphenyl carboxylic acid MMP-3 inhibitors: exploring automated docking as alignment tool. Quant Struct-Act Relat 2001; 20:215–222.

152. Mitchell JBO, Laskowski RA, Alex A, Forster MJ, Thornton JM. BLEEP-potential of mean force describing protein–ligand interactions: II. Calculation of binding energies and comparison with experimental data. J Comput Chem 1999; 20:1177–1185.

153. Bohacek RS, McMartin C. Definition and display of steric, hydrophobic, and hydrogen-bonding properties of ligand binding sites in proteins using Lee and Richards accessible surface: validation of a high-resolution graphical tool for drug design. J Med Chem 1992; 35:1671–1684.

154. Shoichet BK, Kuntz ID. Matching chemistry and shape in molecular docking. Protein Eng 1993; 6:723–732.

155. Walls PH, Sternberg MJE. New algorithm to model protein–protein recognition based on surface complementarity. J Mol Biol 1992; 228:277–297.

156. Viswanadhan VN, Reddy MR, Wlodawer A, Varney MD, Weinstein JN. An approach to rapid estimation of relative binding affinities of enzyme inhibitors: application to peptidomimetic inhibitors of the human immunodeficiency virus type 1 protease. J Med Chem 1996; 39:705–712.

157. Goldman BB, Wipke WT. QSD quadratic shape descriptors. 2. Molecular docking using quadratic shape descriptors (QSDock). Proteins Struct Funct Genet 2000; 38:79–94.

158. Knegtel RMA, Bayada DM, Engh RA, von der Saal W, van Geerestein VJ, Grootenhuis PDJ. Comparison of two implementations of the incremental construction algorithm in flexible docking of thrombin inhibitors. J Comput-Aided Mol Des 1999; 13: 167–183.

159. Perez C, Ortiz AR. Evaluation of docking functions for protein–ligand docking. J Med Chem 2001; 44:3768–3785.

160. Murray CW, Baxter CA, Frenkel AD. The sensitivity of the results of molecular docking to induced fit effects: application to thrombin, thermolysin and neuraminidase. J Comput-Aided Mol Des 1999; 13:547–562.

161. Kramer B, Rarey M, Lengauer T. Evaluation of the FLEXX incremental construction algorithm for protein–ligand docking. Proteins Struct Funct Genet 1999; 37:228–241.

162. David L, Luo R, Gilson MK. Ligand–receptor docking with the mining minima optimizer. J Comput-Aided Mol Des 2001; 15:157–171.

163. Bissantz C, Folkers G, Rognan D. Protein-based virtual screening of chemical databases. 1. Evaluation of different docking/scoring combinations. J Med Chem 2000; 43:4759–4767.

164. Rognan D, Laumoeller SL, Holm A, Buus S, Tschinke V. Predicting binding affinities of protein ligands from three dimensional coordinates: application to peptide binding to class I major histocompatibility proteins. J Med Chem 1999; 42:4650–4658.

165. SYBYL version 6.6. Tripos Associates, St. Louis, MO. Website www.tripos.com.

166. Available Chemicals Directory is available from MDL Information Systems Inc., San Leandro 94577, CA, and contains specialty bulk chemicals from commercial sources. Website www.mdli.com.

167. Stahl M, Rarey M. Detailed analysis of scoring functions for virtual screening. J Med Chem 2001; 44:1035–1042.

168. Stahl M. Modifications of the scoring function in FlexX for virtual screening applications. Perspect Drug Discov Des 2000; 20:83–98.

169. Stahl M, Böhm HJ. Development of filter functions for protein–ligand docking. J Mol Graph Model 1998; 16:121–132.

170. Halgren TA. Merck molecular force field. II. MMFF94 van der Waals and electrostatic parameters for intermolecular interactions. J Comput Chem 1996; 17:520–552.

171. Honig B, Nicholls A. Classical electrostatics in biology and chemistry. Science 1995; 268:1144–1149.

172. Terp GE, Johansen BN, Christensen IT, Jorgensen FS. A new concept for multidimensional selection of ligand conformations (MultiSelect) and multidimensional scoring (MultiScore) of protein–ligand binding affinities. J Med Chem 2001; 44:2333–2343.

173. Enyedy IJ, Lee S-L, Kuo AH, Dickson RB, Lin C-Y, Wang S. Structure-based approach for the discovery of bis-benzamidines as novel inhibitors of matriptase. J Med Chem 2001; 44:1349–1355.

174. Sali A, Potterton L, Yuan F, van Vlijmen H, Karplus M. Evaluation of comparative protein modeling by MODELLER. Proteins Struct Funct Genet 1995; 23:318–326.

175. Friedrich R, Fuentes-Prior P, Ong E, Coombs G, Hunter M, Oehler R, Pierson D, Gonzales R, Huber R, Bode W, Madison EL. Catalytic domain structures of MT-SP1/matriptase, a matrix-degrading transmembrane serine proteinase. J Biol Chem 2002; 277:2160–2168.

176. Pak Y, Wang S. Application of a molecular dynamics simulation method with a generalized effective potential to the flexible molecular docking problems. J Phys Chem B 2000; 104:354–359.

177. Pak Y, Enyedy IJ, Varady J, Kung JW, Lorenzo PS, Blumberg PM, Wang S. Structural basis of binding of high-affinity ligands to protein kinase C: prediction of the binding modes through a new molecular dynamics method and evaluation by site-directed mutagenesis. J Med Chem 2001; 44:1690–1701.

178. Weber PC, Lee S-L, Lewandowski FA, Schadt MC, Chang C-H, Kettner CA. Kinetic and crystallographic studies of thrombin with Ac-(D)Phe-Pro-boroArg-OH and its lysine, amidine, homolysine, and ornithine analogs. Biochemistry 1995; 34:3750–3757.

179. Enyedy IJ, Ling Y, Nacro K, Tomita Y, Wu X, Cao Y, Guo R, Li B, Zhu X, Huang Y, Long Y-Q, Roller P, Yang D, Wang S. Discovery of small molecule inhibitors of Bcl-2 through structure-based computer screening. J Med Chem 2001; 44:4313–4324.

180. Sattler M, Liang H, Nettesheim D, Meadows RP, Harlan JE, Eberstadt M, Yoon HS, Shuker SB, Chang BS, Minn AJ, Thompson CB, Fesik SW. Structure of Bcl-xL–Bak peptide complex: recognition between regulators of apoptosis. Science 1997; 275:983–986.

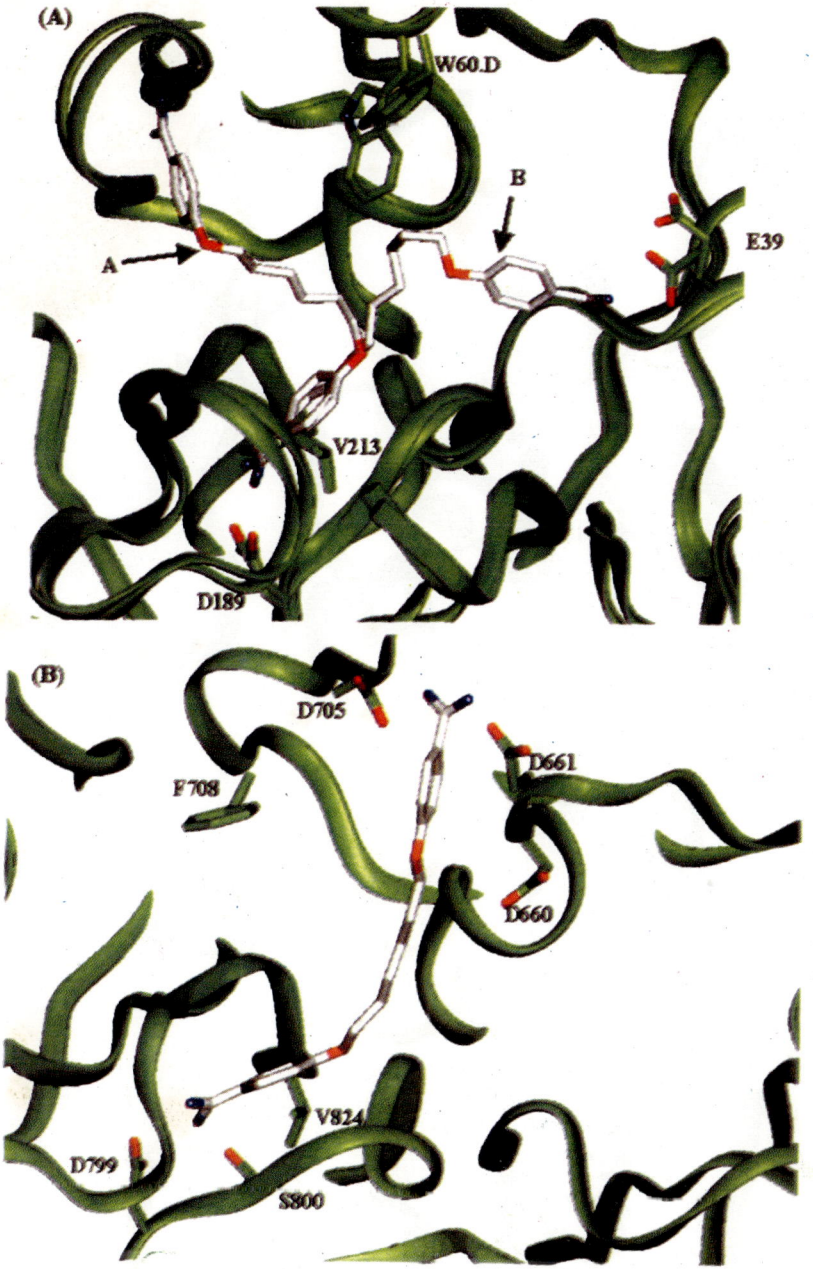

Figure 5 Proposed binding mode of compound **1** in complex with thrombin (A) and matriptase (B) as obtained from docking with DOCK.

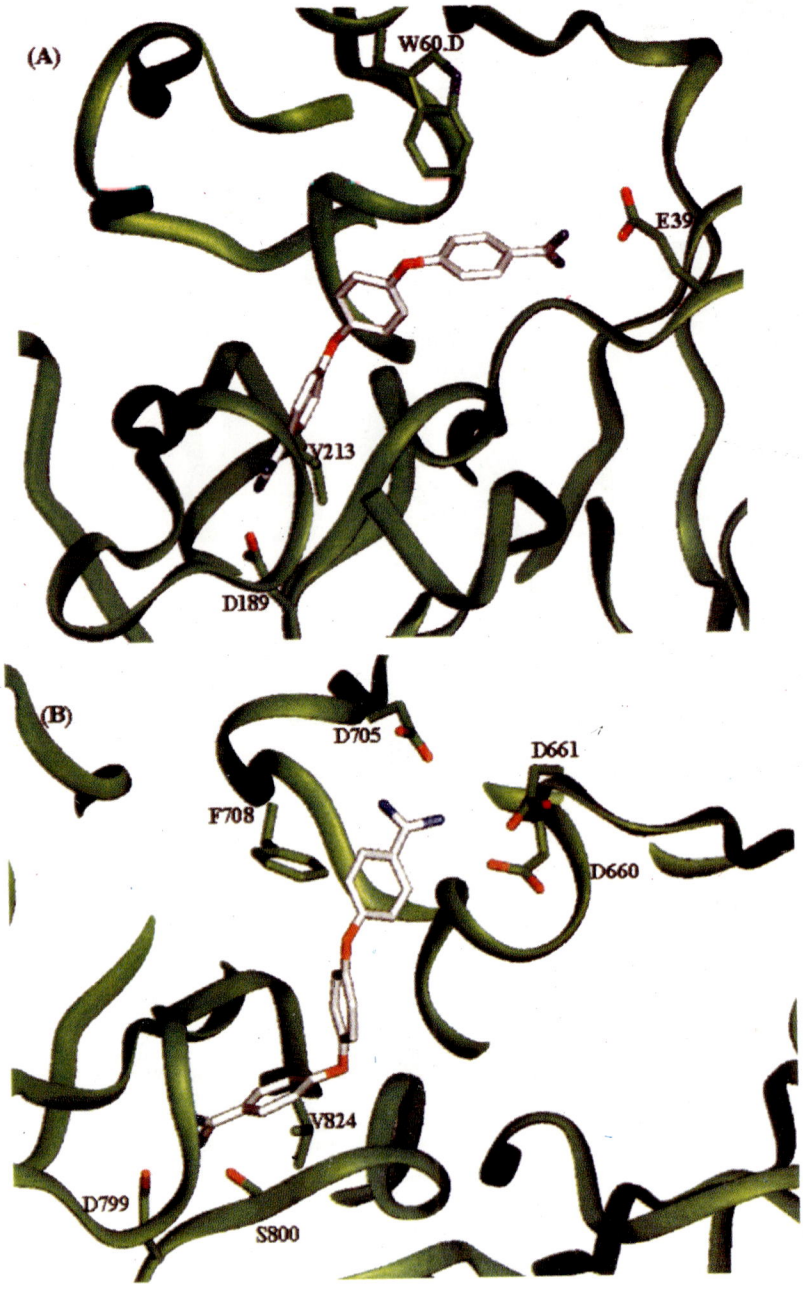

Figure 6 Proposed binding mode of compound **2** in complex with (A) thrombin and (B) matriptase as obtained from docking with DOCK.

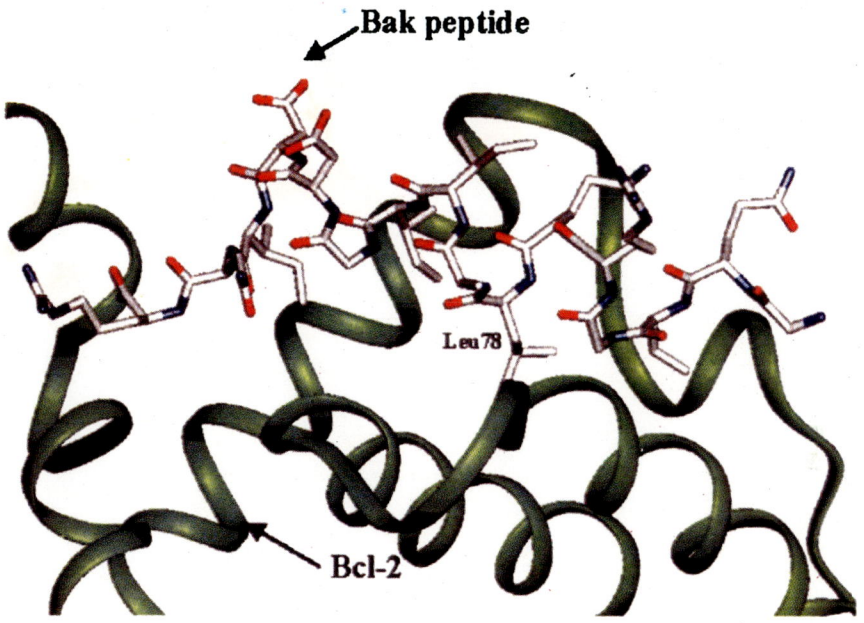

Figure 7 Proposed binding mode of the BAK BH3 peptide to Bcl-2. The binding site where Leu78 from the BAK peptide binds is the most sensible one for Ala mutation.

17

Pharmacophore Discovery: A Critical Review

JOHN H. VAN DRIE

Vertex Pharmaceuticals, Cambridge, Massachusetts, U.S.A.

1. INTRODUCTION

The medicinal chemists' primary challenge is answering well the question, "What molecule should be made next?" Modern synthetic chemistry has advanced sufficiently far that *making* any desired molecule is deemed possible; the real challenge is figuring out *which* molecule to make. Until the emergence of the Hansch method in the 1960s, the medicinal chemists' only guide was the structure–activity relationship (SAR) and the set of molecules already synthesized, along with their biological activities. Clever medicinal chemists are frequently quite skilled at "reading" the SAR, to discern trends.

This fundamental question of medicinal chemistry provides an opportunity for the computational chemist to enter the picture. Yet, the computational chemist is immediately faced with a dilemma: Even if we had the complete quantum mechanical description of a ligand, augmented with its complete statistical–mechanical description in solution, *we could not predict its activity*. The biological activity of any ligand is determined both by *intrinsic* properties, calculable from such fundamental theoretical descriptions, and *extrinsic* properties, the nature of the receptor and how the ligand interacts with that receptor. In the case where the structure of that receptor is determined experimentally, as in structure-based drug design, one may in theory calculate these extrinsic properties. But, to the surprise of many, still at the dawn of the 21st century, the majority of drug targets do not have experimentally determined structures; in these cases, the computational chemist is faced with the problem of *inferring* these extrinsic properties from the structure–activity relationship. The process of computationally inferring these properties in a physically relevant way is called *pharmacophore discovery* (Fig. 1). A pharmacophore is a description of a set of molecules, generally a set of minimal requirements for a molecule to be active. A pharmacophore is typically taken to be the 3D spatial arrangement of features required for activity. Pharmacophore discovery is the process of determining a pharmacophore from the SAR in an "automated" way. By "automated," it is understood that this process requires minimal intervention on the part of the scientist; this helps to ensure that the process is objective and reproducible.

The Hansch method, known as quantitative structure–activity relationships (QSAR), has evolved to embrace a variety of techniques. A glance at the recently published proceedings of the European QSAR Conference [1] shows how much of an impact the methods of pharmacophore discovery have on the computational aspects of medicinal chemistry. Indeed, looking up publications that cite various pharmaco-phore discovery methods papers, it is surprising to see that the total has rapidly accelerated in the past few years, demanding that a review such as this sort through hundreds of papers.

It cannot be expected that this review will cite all these papers. The aim here is to review primarily the methodology papers, and to highlight a selected set of applications that display key aspects of how such methodologies may be applied. In contrast to earlier, less ambitious reviews [2,3], here we aspire to portray a wide-ranging view of the field. In addition, as this has been a field that is growing and maturing, it is important that this review be critical, with an emphasis on "lessons learned"; as such, it is a challenge to distinguish those reports in the literature that are reliable from those that are less so. Furthermore, it is difficult for those outside the field to get a grasp on what innovations have been felicitous, and which have been less so. This inevitably requires numerous judgments on the author's part. The goal of such critiques is to point toward new directions needed for the development of the field of pharmacophore discovery.

One further challenge to performing such a review is that many of the key events driving the evolution of this field were either not published, or published in obscure

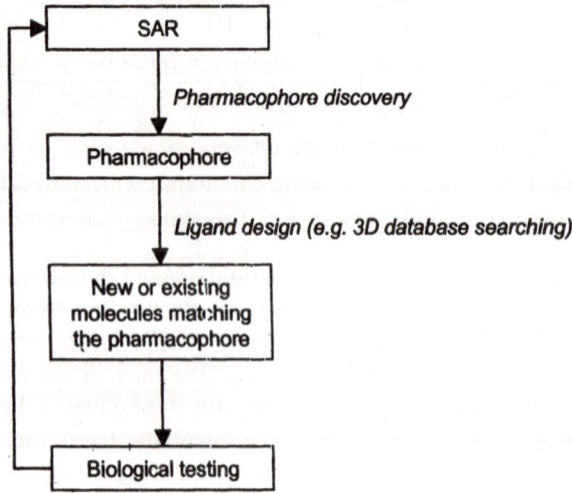

Figure 1   Overview of dataflow in pharmacophore discovery.

venues independent of peer review. The QSAR Gordon Conference, held every 2 years in the USA, has often featured talks on pharmacophore discovery, but no proceedings have ever been published. Furthermore, the rules of that conference prohibit even citing those talks. As a scientist actively developing these methodologies, this author has the advantage of having witnessed and participated in these developments first hand, which brings the disadvantage of having one's own biases. This review will inevitably incorporate both those advantages and disadvantages.

It is also important to clarify what distinguishes methods of pharmacophore discovery from related but distinct computational methods. Pharmacophore discovery is distinct from the methods of 3D-QSAR, such as CoMFA [4], although the alignment rules that CoMFA needs as input may be generated by a pharmacophore discovery method. This review also distinguishes from pharmacophore discovery the various 3D similarity methods: GASP [5] and MIMIC [6,7] both align molecules in 3D space based on the similarity of the molecular fields; Mason et al. [8] and Bradley et al. [9] both identified thousands of pharmacophoric patterns in molecules, and used bitstrings computed from those to build abstract mathematical models for evaluating new ideas for synthesis. The 3D field overlay methods GASP, MIMIC, etc., indeed provide an overlay of molecules, but it is not immediately apparent from those overlays what the actual pharmacophore is (i.e., what are the required features) and what spatial arrangement is required. The methods of Mason et al. and Bradley et al. do not attempt to identify the single best pharmacophore, but instead have cleverly found a way to embrace all possibilities. Although all of these methods have their place in computer-aided drug design, they will be considered as distinct from the methods of pharmacophore discovery, and will not be covered in this review.

The final general aspect of pharmacophore discovery that must be stressed is that, fundamentally, this is a method of *inference*. By contrast, quantum mechanics is deductive, in that one begins with the laws of quantum mechanics to deduce consequences, the correctness of which is guaranteed by those laws. The results of inference cannot be guaranteed to be correct. Their validity can only be determined by their successful prospective application, "prospective" emphasizing that these models must be applied to molecules never before seen by the computational scientist.

Fundamentally, pharmacophore discovery consists of looking for patterns in data. Most of these patterns will be physically irrelevant; only the occasional pattern will be physically meaningful, and will be useful in guiding the medicinal chemist in deciding on which molecule to make next.

This review will first briefly describe the key pharmacophore discovery methods in a linear, temporal fashion to emphasize the evolution of the concepts. Following that, selected methods will be compared and contrasted in detail, and, finally, the lessons learned over the past decade will be summarized.

2. HISTORICAL DEVELOPMENT

Although its flowering occurred all over the world, in large part, pharmacophore discovery germinated from the intellectual efforts of a group of people loosely affiliated in the Midwestern part of the USA, stretching from Washington University in St. Louis through the pharmaceutical research centers near Chicago and onto a mix of institutions in Michigan—a veritable Prairie school of computer-aided drug design.

The ultimate progenitor of all these approaches has been the active analog approach of Marshall et al. [10], out of which emerged the first semiautomated pharmacophore discovery method—the unnamed method of Mayer, Naylor, Motoc, and Marshall [11] (hereafter referred to as the "MNMM" method). The term "semiautomated" refers to the fact that the user is still required to identify those features of the molecules which comprise the pharmacophore; the MNMM method was used to automatically determine a unique proper spatial orientation of these features to account for angiotensin-converting enzyme (ACE) activity, drawing upon over a decade of work on the SAR of ACE, in the successful commercial development of ACE inhibitors (such as captopril and enalapril). Fig. 2 shows their ACE pharmacophore. This MNMM method relied critically on the work of Dammkoehler et al. for the exhaustive conformational analysis of molecules. Most developers of pharmacophore discovery methodology have, consciously or not, rediscovered this MNMM method.

In 1988, a new semiautomated method, APOLLO, was developed by Koehler and Snyder, then at Searle in Chicago [12]. Like the MNMM method, it relied upon the user to select the features for the pharmacophore. APOLLO's main significance was that it served as a progenitor of a new line of thought: the "pseudoreceptor" modeling methods Yak and Prgen, developed by Snyder, Koehler, and Vedani.

In the mid-1980s, this author worked with Martin at Abbott, near Chicago, in part aiming to develop a language, which would provide an objective universal language for describing a pharmacophore, to be used in automated molecular design. This methodology, ALADDIN, was first described in van Drie et al. [13]; its first successful application occurred in 1987—the first successful "virtual screening" application—which played a crucial role in the discovery of a novel lead for a D1 agonist program [14]. The ALADDIN language was widely mimicked, both in 3D database search software [15] as well as in pharmacophore discovery methods.

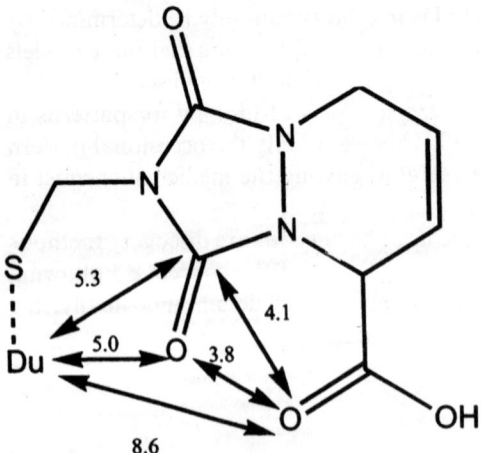

Figure 2   ACE pharmacophore of Mayer et al. All distances are in angstrom. "Du" refers to a "dummy atom" projected from the sulfur atom. Tolerances on these constraints were not given.

After ALADDIN, this author and Martin parted ways, but each played a role in the rise of two distinct pharmacophore discovery methods. In 1990, this author joined, as a founding member and first scientist, a new company in Silicon Valley, BioCAD, dedicated to the development of software to assist drug discovery. Our software, Catalyst, contained three components: conformational analysis, 3D database searching, and "Hypothesis Generation"—a novel approach to pharmacophore discovery. Catalyst represented the first fully automated method for pharmacophore discovery.

Catalyst was first described in the public domain in 1991–1992, but the publications surrounding it did not appear until many years later. Catalyst's Hypothesis Generation ("HypoGen") identifies ALADDIN-style features from a restricted subset of possibilities, and employs a combinatorial optimization algorithm to place those features in space, optimizing a novel function to place a best-fitting regression line through the SAR.

Although Catalyst was not a commercial success (the company BioCAD folded in 1994, and the software was taken over by MSI), it stimulated much interest in fully automated pharmacophore discovery. Martin et al. [16] developed DISCO, borrowing code from ALADDIN to detect features and employing a clique detection algorithm mathematically similar to the MNMM method. DISCO also relies on separate, exhaustive conformational analysis, and, in general, produces many pharmacophores consistent with the SAR.

The original work of Marshall et al. has been carried on, primarily by Beusen and Shands [17a] and Dammkoehler et al. [17b]. Their work has been innovative, primarily in that the pharmacophore discovery algorithm is now integrated with the conformational analysis, a step that should, in principle, significantly improve the quality of the results. The work of SCAMPI [18] is conceptually similar to the work of Beusen and Shands.

Walters and Hinds [19] at the Chicago Medical School devised the first use of a genetic algorithm applied to pharmacophore discovery, with their GERM software. The later work of Pei et al. [20], PARM, was conceptually similar to the Walters and Hinds method. These approaches allow facile integration of both feature detection and 3D analyses.

After developing novel approaches to exhaustive conformational analysis, Crippen [21] at the University of Michigan took a novel approach to pharmacophore discovery, based on Voronoi polyhedra (using hyperplanes to partition space into regions encompassing active molecules). This line of investigation was ultimately abandoned, as Crippen was unable to find a satisfactory resolution to the problem of multiple solutions consistent with the SAR.

The demise of BioCAD spawned two new approaches to pharmacophore discovery: My former colleagues at BioCAD, joined by Barnum, introduced Hiphop [22], a variant of the MNMM approach with a statistical metric added, to alleviate some of the known problems with HypoGen. This author joined Upjohn (now Pharmacia, Kalamazoo, MI), which provided him the opportunity to develop and publish his novel pharmacophore discovery method, DANTE. The two key innovations in DANTE were its use of the principle of selectivity [23,24] to rank possible pharmacophore solutions arising from the MNMM method, and the automatic inference of sterically forbidden regions (another concept that originates from

Marshall's pioneering active analog approach) [25].* DANTE relies on external conformational analysis and produces a ranked list of pharmacophores, the rank reflecting the statistical likelihood that the pattern can arise by chance in the SAR.

Beyond the fertile Midwestern prairies, a number of contributions to pharmacophore discovery appeared, geographically dispersed, stimulated by the work described above and generally as an outgrowth of interests in artificial intelligence (AI). In response to the appearance of Catalyst, APEX-3D was introduced as an extension of earlier 2D work [26]. This is a fully automated method, relying on a standard library of features, external conformational analysis, and produces a large number of possible pharmacophores consistent with the SAR. Another AI outgrowth was COMPASS, developed by a group at Arris Pharmaceuticals in South San Francisco, CA [27]. COMPASS is not a true pharmacophore discovery method, in that it relies on an initial guess for a pharmacophore; it then simultaneously optimizes the molecular alignment and the objective function used to provide estimates for the predicted activity, giving a *local* solution to the pharmacophore. This simultaneous optimization, referred to as "dynamic reposing," is the primary novel aspect implemented in COMPASS. An expert system approach was also adopted by Ting et al. [28]. A final AI outgrowth was the work of Dolata et al. [29], then at Ohio University, who extended his AI-based approach to conformational analysis to develop an AI-based approach to pharmacophore discovery, CLEW.

Finally, a burst of creativity from central Italy should be noted—the work of Pastor et al. [30] in the development of novel descriptors, GRIND, feeding into the novel model-building algorithms of ALMOND [31].

3. ANALYSIS OF DIFFERENT METHODS OF PHARMACOPHORE DISCOVERY

Sec. 2 gave a comprehensive overview of pharmacophore discovery methods, following the conceptual development in roughly a linear, temporal order. We will now study selected methods in more detail, considering similar types of methods as a group. If any of these methods has been mischaracterized, the authors are encouraged to contact this author to ensure that future publications do not propagate the mischaracterizations. Here, we compare and contrast each of these different approaches according to the following criteria:

- Representation. This refers to the manner in which a molecule is converted into objects that are manipulated within the computer. This aspect of computational chemistry often tends to be underappreciated, in that decisions

---

*The experimentation that led to DANTE began with actually an "Easter Egg" that I had hidden in the code of Catalyst version 1.0, which implemented the MNMM method. Applying this to the wide variety of datasets that we encountered at BioCAD, it became apparent that their published ACE example was an anomaly: in general, unique solutions for a pharmacophore do not emerge from a given SAR; multiple pharmacophores are usually consistent with a given SAR. This observation led me to begin experimentation with statistical measures, which would rank pharmacophores by their selectivity, a metric computed from 3D searches. This Easter Egg disappeared by Version 2.0, at which time I was no longer part of the software development group.

about representation are frequently taken for granted, without a conscious analysis of alternatives.

- Conformational analysis. Pharmacophore discovery inherently involves the 3D structure of molecules and, by necessity, goes beyond merely the low-energy conformation.
- Core algorithm, or functional subjected to optimization. This determines how solutions are found.
- Methods to deal with the fundamental problem of inference (i.e., multiple pharmacophores being consistent with a typical SAR).
- Classification vs. activity estimate as output. Some methods aspire to predict the activity of molecules, whereas others merely attempt to classify molecules as "active" or "inactive."
- User intervention required. Most pharmacophore discovery methods are not fully automated, but require some aspects of the solution to be provided by the user.
- Link to design methodologies and types of applications. These methods only prove their worth when used to design new molecules, or to look for novel activities among known molecules.

3.1. Not Quite Pharmacophore Discovery: APOLLO/Yak/Prgen, COMPASS

3.1.1. Representation

Yak/Prgen and COMPASS work fundamentally with the surface of molecules, and derive representations of the predicted binding surface of the receptor ("pseudo-receptor"). APOLLO appears to work with "features"—groups on the molecule that meet specific topological criteria (i.e., which match a specified subgraph isomorphism query); these features must be defined by the user.

3.1.2. Conformational Analysis

APOLLO evidently relies on external programs to perform conformational analysis, and searches for a pharmacophore among them. This pharmacophore is used to select conformations and alignments for further analysis by Yak or Prgen. COMPASS relies primarily on the input conformations and alignments, and makes small tweaks to those.

3.1.3. Core Algorithm, or Functional Subjected to Optimization

APOLLO appears to rely on a variation of the MNMM algorithm to identify a geometry of features common to the active molecules. Yak/Prgen aims to infer a "mini-receptor" or "pseudoreceptor" (a constellation of unconnected sites representing a protein surface) such that the computed binding energies of each ligand in that minireceptor correlate with the experimental binding energy. COMPASS relies on a neural network, with an objective function driving the process toward a regression line that minimizes the root mean squared (rms) deviation between predicted activity and experimental activity, relying on Gaussian functions to transform the input and a sigmoidal function to transform the output.

3.1.4. Method for Handling Multiple Solutions to One SAR

This problem is not treated by these methods. Yak/Prgen assume that the input pharmacophore is correct.

3.1.5. Classification vs. Activity Prediction

The net result of both COMPASS and Yak/Prgen is a prediction of activity.

3.1.6. User Intervention

APOLLO requires the user to identify the important features. Yak/Prgen either take an APOLLO pharmacophore as input, or assume that the user provides such a pharmacophore.

3.1.7. Link to Design Methodologies

Both methods may be used to predict the activity of an unknown molecule (i.e., both may be used in "design mode.") Looking at the literature to see examples of how each of these methods has been used in design mode, no examples of the use of COMPASS have appeared, and with Yak, manual docking, 3D database searching, and de novo design were all applied using a minireceptor model [32].

3.2. Semiautomated Pharmacophore Discovery: Methods Developed by Marshall and Others at Washington University at St. Louis

3.2.1. Representation

These methods follow the original MNMM approach, by which features (i.e., functional groups, each defined by a subgraph isomorphism query) are the primary objects manipulated by the algorithm.

3.2.2. Conformational Analysis

Initially, these methods relied on the exhaustive conformational search algorithms of Dammkoehler et al. [33]. The most recent versions of this methodology, however, employ a Constrained Conformational Search [34], in which interesting regions of space are determined in an initial phase, and subject to more detailed conformational analysis in a later phase.

3.2.3. Core Algorithm, or Functional Subjected to Optimization

Fig. 3 diagrams the fundamental algorithm first utilized in the MNMM method [11] and used in all subsequent versions. Along one axis is the geometrical constraint on one pair of features, along another axis is another geometrical constraint, etc. The algorithm looks at the values of those constraints among all conformations of all active molecules, and determines the intersection region (i.e., the sets of geometries common to all the active molecules). The geometrical constraints for the pharmacophore may be read directly from this intersection region.

3.2.4. Method for Handling Multiple Solutions to One SAR

It is left to the user to sift through the multiple possibilities.

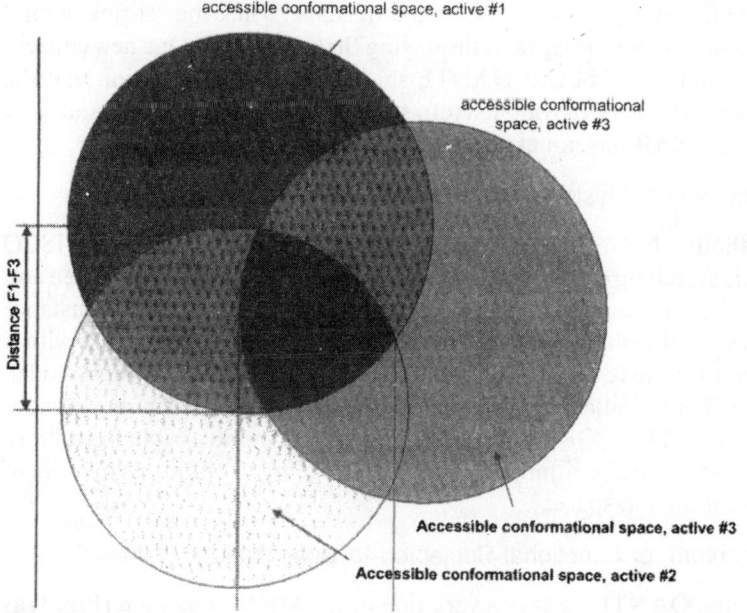

accessible conformational space, active #1

accessible conformational space, active #3

Distance F1-F3

Accessible conformational space, active #3

Accessible conformational space, active #2

Figure 3 Depiction of the algorithm of Mayer, Naylor, Motoc, and Marshall (MNMM).

3.2.5. Classification vs. Activity Prediction

These methods focus strictly on classifying molecules as active or inactive, based on a user-defined threshold to distinguish "active" from "inactive." CoMFA has been used as a postprocessing step to extend this into an activity prediction, as shall be noted in detail in Sec. 4.

3.2.6. User Intervention

The user must both identify relevant features and must decipher which of the pharmacophores consistent with the SAR are most likely physically relevant.

3.2.7. Link to Design Methodologies

Although it would be straightforward to use such pharmacophores as input to a 3D database search, it appears that these pharmacophores have primarily been used to evaluate new ideas one at a time via interactive graphics.

3.3. Fully Automated Pharmacophore Discovery: Catalyst's HypoGen, DISCO, Catalyst's Hiphop, DANTE

3.3.1. Representation

Each of these methods uses ALADDIN-style "features"—functional groups on each molecule. Each relies on a standard library of features, usually those features commonly associated with ligand–receptor interactions (hydrogen bond acceptors and donors, groups having or capable of acquiring a positive charge, hydrophobic groups, and aromatic rings).

DANTE additionally employs a description of shape, using the "shrink–wrap" algorithm for computing, describing, and displaying that shape [25]. One new concept emerged from the application of that DANTE shape algorithm: the notion that the inferred receptor surface can be partitioned into sterically forbidden regions, and terra incognita (regions the SAR has not explored) [24].

3.3.2. Conformational Analysis

Each of these methods relies on external exhaustive conformational analysis. DISCO employs systematic searching or Monte Carlo conformational analysis. HypoGen and Hiphop both use the conformational analysis tools in Catalyst, a variation of distance geometry with the novel technique of "poling" [35]. DANTE has been used with a variety of different exhaustive conformational analysis tools: Catalyst, Macromodel, Omega (written by Stahl, available from OpenEye Software), a fixed-library torsion-driving algorithm, and CONFORT (written by Pearlman, and available from him), and a clever torsion-driving algorithm with adaptive sampling to optimize coverage of important regions of torsion space.

3.3.3. Core Algorithm, or Functional Subjected to Optimization

DISCO, Hiphop, and DANTE all have a variation of the MNMM method (Fig. 3) as their core pharmacophore discovery algorithm. DISCO implements that method via a clique detection algorithm. DANTE implements that method via a quadratic clustering algorithm, with heuristics to ameliorate the problems inherent in a quadratic algorithm. DANTE additionally employs a minimum-volume principle to determine the shape of the putative receptor site [26]. HypoGen is unique in that it uses an MNMM-like method as an initial filter, and then in its final stages applies combinatorial optimization to discover the best fit between predicted and actual activity, where predicted activity is linearly related to the best-fitting conformer of that molecule, and fit is defined as [36]:

$$\text{Fit} = \sum_i w_i \left(1 - \left(\frac{\text{error}_i}{\text{tolerance}_i} \right)^2 \right)$$

where the sum is taken over the features for which the $\text{error}_i < \text{tolerance}_i$, the tolerances and weights w_i are adjustable parameters ("size of the blob and weight of the blob"), and error_i is the distance from the center of the feature on that best-fitting conformer from the Cartesian coordinates of that feature in the pharmacophore (x_i, y_i, z_i), which are also adjustable parameters ("position of the blob"). In essence, Catalyst aims to find locations in 3D space, such that the best-fit rms of each molecule to those points correlates linearly with activity. In addition to this equation being at variance to the fundamental biophysics of ligand–receptor interactions, it also has a pernicious numerical instability, which is probably responsible for the ill behavior noted by a number of workers (e.g., Norinder [37] and Langgard et al. [38]). These types of numerical instabilities make it difficult to obtain identical results on computers produced by different manufacturers.

3.3.4. Method for Handling Multiple Solutions to One SAR

HypoGen ranks all possibilities according to the scoring function above and outputs the top 10. The user is expected to sift through these. Hiphop relies on a statistical measure (which counters HypoGen's trend to employ the most frequently occurring features), but produces many solutions that the user must analyze. DISCO produces many pharmacophores in the general case; many published applications of DISCO have relied upon ad hoc heuristics to deal with this problem. In using DISCO, it is frequently recommended that one reference conformer (i.e., a user-selected bioactive conformation of one molecule) be specified to minimize this problem, but this essentially demands the user to supply the answer. A novel, rigorous solution to this problem in DISCO was proposed by Demeter et al. [39]: Rank each of the DISCO pharmacophore solutions by the quality of the resulting CoMFA model. By contrast, DANTE employs a statistical procedure, *the principle of selectivity*, to rank pharmacophores according to their selectivity S:

$$S = \sum_{k=M}^{N} \binom{N}{k} q^k (1 - q)^{N-k}$$

where N denotes the number of molecules in the dataset, M is the number of molecules in the dataset that match the pharmacophore, $\binom{N}{k}$ represents the number of ways k things may be selected from N things $N!/[(N-k)!k!]$, and q represents the proportion of molecules in a druglike database, which are retrieved from a 3D database search when that pharmacophore is used as a search query [24]. S measures the likelihood that that pharmacophoric pattern could appear by chance in the SAR. Values of S greater than 10^{-5} are usually indicative of a poor pharmacophore or a poor dataset; values of S smaller than 10^{-10} are generally indicative of a good pharmacophore. Depending on the difficulty of the dataset, the best pharmacophore is often the top-ranking one, although the user should investigate the lower-ranked ones as well. DANTE's use of shape also tends to decrease the problem of multiple solutions because only certain pharmacophores will meet the consistency requirements of that shape-computation procedure.

To gain an intuitive understanding of this mathematical expression for selectivity, it is easiest to view it stepwise, in versions of steadily increasing sophistication. The simplest way to rank pharmacophores is by q, the proportion of molecules each pharmacophore returns from a 3D database search. One can do this based on the notion that pharmacophores should mimic receptors, which must be highly selective to perform their function. Low values of q indicate that a pharmacophore is selective. The next step in sophistication is to appreciate that q represents a probability—the probability that a randomly chosen molecule will match a given pharmacophore. Viewed in this way, q^N represents the likelihood of finding a given pharmacophore among N randomly chosen molecules. Thus, ranking by q^N is equivalent to ranking pharmacophores by the likelihood that a given pharmacophoric pattern could have arisen by chance among the dataset. The final sophistication that leads to the equation above is that, in some cases, a pharmacophore matches only M molecules of the N in the dataset. In this case, to recover the conceptual meaning of the selectivity index as measuring the likelihood of finding a pattern at random in the dataset, one must sum up the binomial distribution from M to N, to include the combinatorics of all the ways one can choose M molecules from a larger set N. Note that when $M = N$, the above equation reduces to $S = q^N$.

3.3.5. Classification vs. Activity Prediction

DISCO, Hiphop, and DANTE all perform classification. Catalyst's HypoGen predicts activity according to the linear regression performed, using the scoring function described above.

3.3.6. User Intervention

Befitting the notion that these are fully automated pharmacophore discovery methods, this is minimal, although DISCO, Hiphop, and Catalyst's HypoGen require considerable user intervention to sift through the multiple pharmacophores that emerge. Furthermore, considerable effort is advocated in selecting a "representative dataset" for HypoGen to avoid nonsensical results.

3.3.7. Link to Design Methodologies

All of these methods can easily be coupled to a 3D database search. In particular, the Hiphop and HypoGen pharmacophores can directly be used as input to the Catalyst 3D database search engine. DANTE pharmacophores can be converted into Catalyst 3D search queries, via scripts that convert DANTE output into the Catalyst.chm format. Both DISCO and HypoGen pharmacophores have been used to perform the alignment prior to a CoMFA study.

4. APPLICATIONS OF PHARMACOPHORE DISCOVERY METHODS

Despite the immaturity of these methodologies of pharmacophore discovery, a tremendous variety of apparently successful applications have appeared in the literature. What follows is only a selection of these applications, focusing on applications relevant to drug discovery, categorized by the type of receptor being studied.

4.1. GPCRs

G-protein-coupled receptors remain one of the most important classes of drug targets, and we still await their experimental structure determination. Hence, it is of no surprise that the greatest number of applications of pharmacophore discovery targets GPCRs. The NK-1 receptor has been studied using the Constrained Search method coupled to CoMFA [34] and using DISCO [40], which led to the discovery of a peptide of modest affinity but high selectivity for NK-1. A pharmacophore for the β_2 adrenergic receptor was developed using DANTE [23], and was studied via Yak pseudoreceptor modeling [41]. Catalyst/HypoGen was used to discover pharmacophores for the α_1 and α_2 adrenergic receptors [42].

Serotonin subtypes have been an area of recent interest in the pharmaceutical industry, and that is apparent by the wealth of applications that have appeared against serotonin subtypes: COMPASS was used to study the 5-HT$_{1a}$ receptor [43], as was Yak [32,44]. DANTE was used to develop a never-published pharmacophore for the 5-HT$_{2a}$ receptor, based on public data. That pharmacophore is shown in Fig. 4. Note that this DANTE pharmacophore uses only two features, a basic amine and an aromatic ring, and relies on distance and angular relationships between those features and a sprinkling of sterically forbidden regions. Finally, a pharmacophore for the most recent subtype, 5-HT$_7$, was determined using Catalyst/HypoGen [45].

Pharmacophores for opioid receptors have been determined using a custom implementation of the MNMM method [46], and cannabinoid receptors have been studied using Yak [47]. Pharmacophores for the leukotriene receptors have been determined: one for cysLT(1), developed using Catalyst with manual manipulation, was used to perform the alignment prior to a CoMFA model [48], and one for the

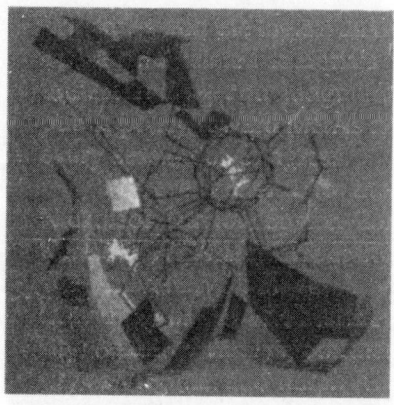
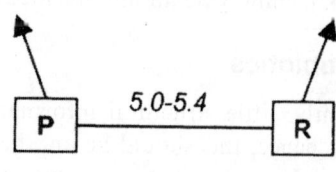

Figure 4 The 5-HT$_{2a}$ pharmacophore produced by DANTE. The selectivity index for this pharmacophore is $10^{-13.7}$. The distance from the center of the aromatic ring "R" to the basic amine "P" is 5.0–5.4 Å. The angle from the center of the ring to the lone pair direction vector on the amine is 49–115°. The angle from the basic amine to the normal of the ring is 75–143°. The torsion angle is 33–101°. Note that, although this pharmacophore has only two features, it is chiral by virtue of the signed torsion angle.

leukotriene D4 was also discovered using Catalyst/HypoGen [49]. The muscarinic receptor was studied by combining protein models with DISCO pharmacophores [50]. Another group used DISCO pharmacophores of the muscarinic M3 receptor, and two potent, novel compounds emerged from a 3D database search [51]. DISCO was used to determine the pharmacophore of the metabotropic glutamate receptor [52], which guided the synthesis of two novel compounds. Yak was used to develop a pharmacophore for the melatonin receptor [53]. Catalyst/HypoGen was used to develop a pharmacophore for the corticotrophin-releasing hormone receptor [54].

4.2. Ion Channels

Ion channels are another important class of integral membrane proteins that frequently serve as drug targets. Catalyst/HypoGen was used to develop a pharmacophore for the 5-HT$_3$ ion channel [55], which evidently played a role in the discovery of a novel series of 5-HT$_3$ ligands [56]. One group determined the pharmacophore of the GABA$_A$ receptor [57,58] using a custom implementation of the MNMM method, which was used in a 3D database search to discover novel GABA$_A$ ligands. An NMDA pharmacophore was determined using DISCO [59], and was used to align molecules for a CoMFA study. The α_7 nicotinic acetylcholine receptor pharmacophore was determined using DISCO [60], and was used to rationalize the SAR of Abbott's program against that receptor.

4.3. Transporters

Transporters are the final class of integral membrane proteins that are important as drug targets. Structurally, they are among the most complex receptors known; hence, they are ideally suited to analysis via pharmacophore discovery. Catalyst/HypoGen was used in the discovery of a pharmacophore for the Na$^+$/bile acid transporter [61]. A stereoselective pharmacophore for the selective serotonin reuptake site [target of selective serotonin reuptake inhibitors (SSRIs), a major class of antidepressants] was reported [62], employing an unpublished implementation of the MNMM method.

4.4. Antibiotics

Until recently, little structural information was available for the targets of most antibiotics; hence, this should be another fertile area for pharmacophore discovery applications. However, it is surprising to find little in the literature on this topic. At Roche (Basel, Switzerland), interactive graphics, coupled with the protein structure, was used in a nonautomated way to create pharmacophores for DNA gyrase, which were then used for LUDI and Catalyst 3D database searching to discover novel inhibitors [63]. At Pharmacia (née Upjohn), linezolid was developed as the first of a novel class of antibiotics, the oxazolidinones [64]. No structural information on the target is available, but it has been extensively studied using DANTE. Fig. 5 shows a DANTE pharmacophore for the oxazolidinones, based on *H. inf.* MICs as the biological endpoint. Because the SAR comprises over 3000 compounds, the sterically forbidden regions cover virtually all of space; the regions of terra incognita have been intentionally obscured in this figure. The details of the distance and angle constraints remain proprietary to Pharmacia.

4.5. Inflammation Targets

The GPIIa/IIIb receptor, target for the RGD (Arg–Gly–Asp) peptides, has been the target for many drug discovery programs. DANTE was used to determine an RGD pharmacophore [23], and Yak was used to study this receptor as well [65]. Interestingly, one of the first successful applications of 3D database searching was to this receptor, where a very coarse manually determined pharmacophore was used to discover the lead for the program of Merck (West Point, PA) [66].

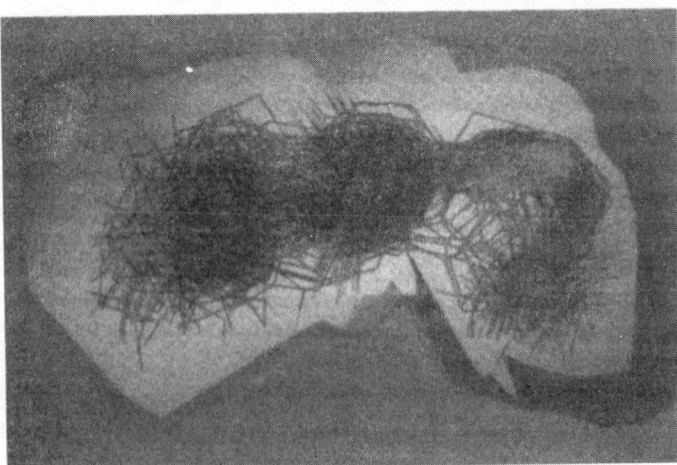

Figure 5 The pharmacophore for the oxazolidinone antibiotics. This is a cutaway view of the sterically forbidden region defining the shape of the binding site.

Catalyst/Hiphop was recently used to discover the pharmacophore for mesangial cell (MC) proliferation inhibition [67]. This pharmacophore was used in a 3D database search to discover 41 novel inhibitors of MC proliferation.

4.6. Miscellaneous

Catalyst/HypoGen was used to discover the pharmacophore for farnesyl transferase (FT) inhibitors [68]. Although one might question the utility of a pharmacophore for which four of five features are hydrophobes, nonetheless, it was successful in a 3D database search of discovering three structurally novel classes of FT inhibitors. Catalyst/HypoGen was also used to determine the pharmacophore for in vitro hepatocyte clearance [69,70]. Finally, Catalyst/HypoGen pharmacophores have been in use for many years in the design of fragrances; part that work has been recently published [71].

5. LESSONS LEARNED; THE FUTURE OF PHARMACOPHORE DISCOVERY

This is an appropriate stage in the development of pharmacophore discovery to look back and summarize what we have learned about what works, what does not work, and where the areas for future work lie.

5.1. A Link to 3D Searching Is Helpful

A facile link between a pharmacophore discovery method and 3D database searching is useful in validating and exploiting a pharmacophore. It is a testament to the power of the notion of a pharmacophore and the notion of pharmacophoric 3D database searching that even weak pharmacophores are successful in retrieving active molecules from databases. The treatment of shape in 3D database searching has always been problematic and must be improved. In the development of ALADDIN, it was thought that any surface, in principle, could be approximated by a large set of spheres. Mathematically, this is true, and the experience of Greenidge et al. [72] shows that, indeed, a protein pocket may be represented to a 3D database search this way, but this approach is unwieldy and makes surface comparisons, in particular, difficult. This shape representation must both be easy to determine, from a pharmacophore discovery algorithm, and easy to apply in a search. It was for this purpose that the "shrink–wrap" algorithm was introduced.

5.2. Better Feature Detection Is Needed

The persistence of the use of a standard library of features (H-bond acceptors/donors, positive/negative charges, aromatic rings, and hydrophobes) is a surprise. We introduced this at BioCAD (this author originally suggested this as a "hack" in 1990 to get us started, never imagining that this offhand suggestion would persist as it has), yet it has been widely copied and little improved upon. This approach appears to exacerbate the problem of multiple pharmacophore solutions to a given SAR. A better approach would be one that uses the SAR to explicitly identify possibilities for features (e.g., if the transformation of an –OH to an –OCH$_3$ destroys activity, it suggests that the –OH is functioning as an H-bond donor). In this author's judgment, the AI-based methods appear superior in this regard, yet this approach has not merged with the algorithmic approaches.

5.3. Inclusion of Sterically Forbidden Regions Is Vital

As a model, a pharmacophore as a spatial arrangement of features is fundamentally a weak model of activity, but the simplest thing that can be done to improve this model is to incorporate sterically forbidden regions. Experience with DANTE suggested that adding these greatly improved the selectivity of a pharmacophore. As is apparent in Sec. 5.2, most workers use the pharmacophore only as alignment rules in preparation for pseudoreceptor modeling, or 3D-QSAR modeling. A variety of methods have recently been introduced for introducing shape into 3D database searching [73], but typically these definitions of shape do not make simultaneous reference to the feature-based pharmacophore.

5.4. How Should One Best Treat Conformational Analysis?

Constrained Search and the related approach of SCAMPI both integrate the conformational analysis closely with the pharmacophore discovery. This has the advantage that the sampling of conformational space can be more focused on key regions. With both Catalyst and DANTE, conformational analysis was explicitly kept separate, in the latter to allow one to take advantage of any innovations in conformational analysis tools. And, indeed, there continues to be a steady flow of new approaches in conformational analysis—pharmacophore discovery is critically dependent on high-quality exhaustive conformational analysis. Based on our experience thus far, we cannot conclude that either approach is superior (integrated vs. external). Furthermore, a consensus has not yet been reached on the optimal manner to perform conformational search as needed by pharmacophore discovery. This will continue to be a fruitful area of research.

Another issue with the exhaustive conformational analysis performed for pharmacophore discovery is that all conformers whose energy is within X kcal/mol of the global minimum are retained, with X ranging from 3 to 20 in various studies. There appears to be no consensus here, and systematic studies of the best value are limited [46,74]. In Catalyst's conformational analysis using their drastically simplified CHARMm force field, this author has tended to use a value of 8 kcal/mol, based on unpublished work comparing open, inactive analogs to cyclic-constrained active analogs. Of course, the appropriate value of X is highly force field-dependent. A related issue is: Should the internal energies be considered as biased against the inclusion of highly strained conformations in the superposition? These are issues that deserve further careful study.

5.5. Retrospective Predictivity Does Not Guarantee Prospective Utility: The Kubinyi Paradox

However, one must be wary of the complete integration of pharmacophore discovery with the process of constructing a full model of activity, as we attempted with HypoGen. For example, many people have been puzzled over why DANTE has stopped short of attempting to incorporate electrostatic fields in addition to steric fields, and has only aimed to classify molecules into active/inactive rather than provide a continuous prediction of activity. For some time, many workers in the field have been

aware of a fundamental limitation in the field of 3D model building, which may be labeled the "Kubinyi paradox" after the observation of Kubinyi et al. [75], who conclude, "in principle, one could expect that good internal predictivities should be indicative for good external predictivities... [in fact] there is no relationship at all."*
The general mathematical problem is that the more complex the models become, the easier it is to find models that are consistent with the SAR but which are meaningless when applied prospectively. This point is more profound than the simple statement "beware of overfitting": If the optimization procedures one pursues are explicitly aiming to find ever-better fits to the data, then such procedures may be explicitly heading in the wrong direction vis-à-vis their prospective utility. The key issue raised by the Kubinyi paradox is: What metric computed on the data is best indicative of the prospective utility of a model? We must understand the mathematical basis of this paradox and answer this last question before we can progress properly on this front.

5.6. Objectivity Is Vital; Computational Controls Are Crucial to Exposing Methodological Weaknesses

As mentioned Sec. 1, a pharmacophore represents an objective description of what is required for activity, and pharmacophore discovery should represent an objective way of determining that pharmacophore. It is difficult to discern how objective this process has been in the applications Sec. 5.5 (i.e., to determine how much user bias was injected into selecting which of the many pharmacophores consistent with the SAR was chosen). User bias is not necessarily bad; —in the hands of a skilled practitioner, this allows judgments to be made based on data from a variety of sources. But, ultimately, this field will best advance if different noncommunicating users applying the same method to the same data achieve the same final result.

One simple way of achieving an objective process of pharmacophore discovery, independent of the method used, is in the rigorous application of computational controls. Biologists, especially cell biologists, are accustomed to relying on controls as they interpret their results of their complex systems. We in computational chemistry would do well to learn from their experiences, in using control experiments to expose artifacts. Pharmacophore discovery is an example of the study of a complex system, where it is easy to make inferences consistent with the data that are meaningless. Computational controls help prevent falling into such traps. Examples of controls applied to DANTE results were:

- Adding random noise of varying size to each atom position in the conformations;
- Renumbering the atoms in each molecule;
- Choosing different subsets of the dataset as input; and
- Choosing totally random datasets as input.

The first control experiment can be especially revealing: noise as small as 0.0001 Å can have major effects on regression lines, which is one reason why some prefer classification over regression. The fundamental expectation is that any reliable method

---

* This paper notes that their observation is not novel, and cites the work of Norinder and Novellino et al., suggesting that, possibly, this paradox should more properly be called the Kubinyi–Norinder–Novellino paradox.

will *not* amplify the noise, but ideally will dampen the noise (i.e., the change in output is smaller than the change in the input). The third control experiment is a simple way to test the effects of selection bias, another easy way for nonobjectivity to creep into a pharmacophore discovery analysis. And the list above is only a start: Other researchers should expand this list of control experiments appropriate to pharmacophore discovery (indeed, how one designs control experiments is, among cell biologists, one of the tests of intellectual rigor and creativity). Finally, control experiments can be done by anyone, even those using commercial software—no access to source code is needed.

The ultimate test of the objectivity of the process of pharmacophore discovery is that these results should conform to the physical reality of the receptor, as determined by independent biophysical experimentation. For example, the standard pharmacophore determined by DANTE for most GPCRs contains a basic amine separated by 5–7 Å from an aromatic ring. These groups interact with the conserved Asp on helix III and conserved aromatic residues on helices VI and VII in most protein models of GPCRs.

5.7. Some Datasets Are Easy; Some Are Hard

Another important lesson learned in over a decade of use of pharmacophore discovery methods is that not all datasets are equal in difficulty. This author has suggested that a rating system should be established for datasets, as canoeists have done for whitewater rivers (Fig. 6). Class I, the calmest of rivers, would correspond to the easiest datasets: those for which many rigid nanomolar ligands are known and all artifacts have been removed. Examples of Class I datasets would be the specific ACE dataset of Mayer et al. [11] and D2 antagonists [23]. Class II, the first hints of whitewater, appears in CCK antagonists [76] and RGD antagonists [23]. One can continue this analogy up to Class VI, the most challenging whitewater suitable only for the most expert canoeists. In general, datasets that are available after the drug has advanced to the clinic are comparatively easy; datasets in the early days of a project are usually more challeng-

Figure 6 A canoeist navigating a Class III rapid. Can we similarly rank SAR datasets by their difficulty?

ing. Yet, it is precisely in this early setting that the need for the application of computational methods is most critical. This author has only recently discovered a way to objectively determine the degree of difficulty for a dataset, based on arguments of SAR nonadditivity (manuscript in preparation). But a scale of difficulty is important as one evaluates different applications of a given methodology and as one evaluates different methodologies.

A new pharmacophore discovery method should be initially tested on Class I datasets; if this is a failure, one should discard that method. If it is a success, that is no great achievement; one must progress to steadily more difficult datasets. When comparing different pharmacophore discovery methodologies, one must take into account the relative differences of difficult of datasets to which they have been applied.

6. THE FUTURE CHALLENGES

The great challenge of the future for pharmacophore discovery is to make the models more sophisticated, yet simultaneously being able to sift through the multiple models consistent with the data to discern those models that will be meaningful prospectively. With DANTE, the models were intentionally kept relatively simple, and yet a sophisticated method—the principle of selectivity—was needed to sift through the multiple possibilities consistent with the SAR. Furthermore, this problem of multiple models consistent with the data gets worse as the datasets are of higher degree of difficulty.

During the 20 years that molecular modeling has been in use in the pharmaceutical industry, computing power has increased by three to four orders of magnitude. It would be nice if we could say that our contributions to pharmaceutical discovery have increased by a similar amount. This discrepancy may be attributed directly to the weaknesses of our methodology—much work is still needed. The problem of pharmacophore discovery has proven more intellectually challenging than anyone would have initially guessed. Hopefully, this review will stimulate major rethinking of overall approaches—simple tweaking of existing methodologies will not suffice. But these initial successes are encouraging, suggesting that we are on the right path, and that the concept of a pharmacophore is a valuable one and that methods to discover pharmacophores are needed. Our goal is that the computer model should always be consulted when the medicinal chemist poses the question, "What molecule should I make next?"

ACKNOWLEDGMENTS

Critical reading of this manuscript by Klaus Gundertofte and my Vertex colleagues Mark Murcko, Paul Charifson, and Brian Goldman has helped improve this manuscript considerably. Hugo Kubinyi has also provided helpful input. Norman Laurin of the Vertex Library has been invaluable in assisting with the electronic literature searching and document retrieval.

REFERENCES

1. Gundertofte K, Jorgensen FS. Molecular Modeling and Prediction of Bioactivity. New York: Kluwer Academic/Plenum Publishers, 2000.

2. Wermuth C-G, Langer T. Pharmacophore identification. In: Kubinyi H, ed. 3D QSAR in Drug Design. Leiden: ESCOM, 1993:117–149.

3. Bures M. Recent techniques and applications in pharmacophore mapping. In: Charifson P, ed. Practical Applications of Computer-Aided Drug Design. New York: Dekker, 1997:39–72.

4. Cramer RD, Patterson DE, Bunce JD. Comparative molecular field analysis (CoMFA): 1. Effect of shape on binding of steroids to carrier proteins. J Am Chem Soc 1988; 110: 5959–5967.

5. Jones G, Willett P, Glen RC. A genetic algorithm for flexible molecular overlay and pharmacophore elucidation. J Comput-Aided Mol Des 1995; 9:532–549.

6. Mestres J, Rohrer DC, Maggiora GM. A molecular field-based similarity approach to pharmacophoric pattern recognition. J Mol Graph Model 1997; 15:114–121.

7. Mestres J, Rohrer DC, Maggiora GM. A molecular-field-based similarity study of non-nucleoside HIV-1 reverse transcriptase inhibitors. J Comput-Aided Mol Des 1999; 13:79–93.

8. Mason JS, Good AC, Martin EJ. 3-D pharmacophores in drug discovery. Curr Pharm Des 2001; 7:567–597.

9. Bradley EK, Beroza P, Penzotti JE, Grootenhuis PD, Spellmeyer DC, Miller JL. A rapid computational method for lead evolution: description and application to alpha(1)-adrenergic antagonists. J Med Chem 2000; 43:2770–2774.

10. Marshall GR, Barry CD, Bosshard HE, Dammkoehler RA, Dunn DA. The conformational parameter in drug design: the active analog approach. In: Olson EC, Christofersen RE, eds. Computer-Assisted Drug Design. Washington: American Chemical Society, 1979:205–226.

11. Mayer D, Naylor CB, Motoc I, Marshall GR. A unique geometry of the active site of ACE consistent with structure–activity studies. J Comput-Aided Mol Des 1987; 1:3–16. This was the very first paper published in this journal.

12. Snyder JP, Rao SN, Koehler KF, Vedani A, Pellicciari R. APOLLO pharmacophores and the pseudoreceptor concept. In: Wermuth C-G, ed. Trends QSAR Molecular Modelling. Leiden: ESCOM (APOLLO was first described at a 1988 ACS meeting by K. Koehler).

13. van Drie JH, Weininger D, Martin YC. ALADDIN: an integrated tool for computer-assisted molecular design and pharmacophore recognition from geometric, steric, and substructure searching of 3D molecular structures. J Comput-Aided Mol Des 1989; 3:225–251.

14. Martin YC. 3D databases in drug design. J Med Chem 1992; 35:2145–2154.

15. van Drie JH. 3D database searching in drug discovery. www.netsci.org/Science/Cheminform/feature06.html.

16. Martin YC, Bures MG, Danaher EA, Delazzer J, Lico I, Pavlik PA. A fast new approach to pharmacophore mapping and its application to dopaminergic and benzodiazepine agonists. J Comput-Aided Mol Des 1993; 7:83–102.

17a. Beusen DD, Shands EFB. Systematic search strategies in conformational analysis. Drug Discov Today 1996; 1:429–437.

17b. Dammkoehler RA, Karasek SF, Shands EFB, Marshall GR. Constrained Search of conformational hyperspace. J Comput-Aided Mol Des 1989; 3:3–21.

18. Chen X, Rusinko A, Tropsha A, Young SS. Automated pharmacophore identification for large chemical data sets. J Chem Inf Comput Sci 1999; 39:887–896.

19. Walters DE, Hinds RM. Genetically evolved receptor models: a computational approach to construction of receptor models. J Med Chem 1994; 37:2527–2536.

20. Pei JF, Zhou JJ, Xie GR, Chen HM, He XF. PARM: a practical utility for drug design. J Mol Graph Model 2001; 19:448–452.

21. Crippen GM. Voronoi binding site models. J Comput Chem 1987; 8:943–955.

22. Barnum D, Greene J, Smellie A, Sprague P. Identification of common functional configurations among molecules. J Chem Inf Comput Sci 1996; 36:563–571.

23. van Drie JH. Strategies for the determination of pharmacophoric 3D database queries. J Comput-Aided Mol Des 1997; 11:39–52.

24. van Drie JH, Nugent RA. Addressing the challenges of combinatorial chemistry: 3D databases, pharmacophore recognition and beyond. SAR QSAR Environ Res 1998; 9:1–21.

25. van Drie JH. "Shrink-wrap" surfaces: a new method for incorporating shape into pharmacophoric 3D database searching. J Chem Inf Comput Sci 1997; 37:38–42.

26. Golender V, Vesterman B, Eliyahu O, Kardash A, Kletzkin M, Shokhen M, Vorpagel E. Knowledge engineering approach to drug design and its implementation in the APEX-3D expert system. In: Sanz F, Giraldo J, Manaut F, eds. QSAR and Molecular Modelling: Concepts, Computational Tools and Biological Applications. Barcelona: Prous Science, 1995:246–251.

27. Jain AN, Koile K, Chapman D. Compass: predicting biological activities from molecular surface properties. Performance comparisons on a steroid benchmark. J Med Chem 1994; 37:2315–2327.

28. Ting A, McGuire R, Johnson AP, Green S. Expert system assisted pharmacophore identification. J Chem Inf Comput Sci 2000; 40:347–353.

29. Dolata DP, Parrill AL, Walters WP. CLEW: the generation of pharmacophore hypotheses through machine learning. SAR QSAR Environ Res 1998; 9:53–81.

30. Pastor M, Cruciani G, McLay I, Pickett S, Clementi S. GRid-INdependent descriptors (GRIND): a novel class of alignment-independent three-dimensional molecular descriptors. J Med Chem 2000; 43:3233–3243.

31. Gnerre C, Thull U, Gaillard P, Carrupt PA, Testa B, Fernandes E, Silva F, Pinto M, Pinto MMM, Wolfender JL, Hostettmann K, Cruciani G. Natural and synthetic xanthones as MAO inhibitors: biological assay and 3D-QSAR. Helv Chim Acta 2001; 84:552–570.

32. Jansen JM, Koehler KF, Hedberg MM, Johansson AM, Hacksell U, Nordvall G, Snyder JP. Molecular design using the minireceptor concept. J Chem Inf Comput Sci 1997; 37:812–818.

33. Dammkoehler RA, Karasek SF, Shands EFB, Marshall GR. Constrained Search of conformational hyperspace. J Comput-Aided Mol Des 1989; 3:3–21.

34. Takeuchi Y, Shands EFB, Beusen DD, Marshall GR. Derivation of a 3D pharmacophore model of substance P antagonists bound to the NK-1 receptor. J Med Chem 1998; 41:3609–3623.

35. Smellie A, Teig SL, Towbin P. Poling—promoting conformational variation. J Comput Chem 1995; 16:171–187.

36. Anonymous. Hypotheses in Catalyst. Mountain View, CA: BioCAD, 1992. This document, never formally published, is the sole documentation on the mathematical basis of Catalyst's Hypothesis Generation.

37. Norinder U. Refinement of Catalyst hypotheses using simplex optimisation. J Comput-Aided Mol Des 2000; 14:545–547.

38. Langgard M, Bjornholm B, Gundertofte K. Pharmacophore modeling by automated methods: possibilities and limitations. In: Guner O, ed. Pharmacophore Perception, Development and Use in Drug Design. La Jolla: International University Line, 2000:237–250.

39. Demeter DA, Weintraub HJR, Knittel JJ. The local minima method of pharmacophore determination: a protocol for predicting the bioactive conformation of small, conformationally flexible molecules. J Chem Inf Comput Sci 1998; 38:1125–1136.

40. Goldstein S, Neuwels M, Moureau F, Berckmans D, Lassoie MA, Differding E, Houssin R, Henichart JP. Bioactive conformations of peptides and mimetics as milestones

in drug design—investigation of NK-1 receptor antagonists. Lett Pept Sci 1995; 2:125–134.

41. Vedani A, Zbinden P, Snyder JP, Greenidge PA. Pseudorecpetor modeling—the construction of 3-dimensional receptor surrogates. J Am Chem Soc 1995; 117:4987–4994.

42. Barbaro R, Betti L, Botta M, Corelli F, Giannaccini G, Maccari L, Manetti F, Strappaghetti G, Corsano S. Synthesis, biological evaluation, and pharmacophore generation of new pyridazinone derivatives with affinity toward alpha(1)- and alpha(2)-adrenoceptors. J Med Chem 2001; 44:2118–2132.

43. Jain AN, Harris NL, Park JY. Quantitative binding-site model generation—COMPASS applied to multiple chemotypes targeting the 5-HT$_{1a}$ receptor. J Med Chem 1995; 38:1295–1308.

44. Hedberg MH, Linnanen T, Jansen JM, Nordvall G, Hjorth S, Unelius L, Johansson AM. 11-Substituted (R)-apomorphines—synthesis, pharmacology, and modeling of D-2A and 5-HT$_{1A}$ receptor interactions. J Med Chem 1996; 39:3503–3513.

45. Lopez-Rodriguez ML, Porras E, Benhamu B, Ramos JA, Morcillo MJ, Lavandera JL. First pharmacophoric hypothesis for 5-HT$_7$ antagonism. Bioorg Med Chem Lett 2000; 10:1097–1100.

46. Filizola M, Villar HO, Loew GH. Molecular determinants of non-specific recognition of delta, mu, and kappa opioid receptors. Bioorg Med Chem 2001; 9:69–76.

47. Zbinden P, Dobler M, Folkers G, Vedani A. PrGen: pseudoreceptor modeling using receptor-mediated ligand alignment and pharmacophore equilibration. QSAR 1998; 17:122–130.

48. Griera R, Armengol M, Reyes A, Alvarez M, Palomer A, Cabre F, Pascual J, Garcia M, Mauleon D. Synthesis and pharmacological evaluation of new cysLT(1) receptor antagonists. Eur J Med Chem 1997; 32:547–570.

49. Palomer A, Pascual J, Cabre F, Garcia ML, Mauleon D. Derivation of pharmacophore and CoMFA models for leukotriene D-4 receptor antagonists of the quinolinyl(bridged) aryl series. J Med Chem 2000; 43:392–400.

50. Toy-Palmer A, Wu H, Liu X. Ligand docking in a muscarinic G protein-coupled receptor model. Med Chem Res 1999; 9:565–578.

51. Marriott DP, Dougall IG, Meghani P, Liu YJ, Flower DR. Lead generation using pharmacophore mapping and three-dimensional database searching: application to muscarinic M-3 receptor antagonists. J Med Chem 1999; 42:3210–3216.

52. Amori L, Costantino G, Marinozzi M, Pellicciari R, Gasparini F, Flor PJ, Kuhn R, Vranesic I. Synthesis, molecular modeling and preliminary biological evaluation of 1-amino-3-phosphono-3-cyclopentene-1-carboxylic acid and 1-amino-3-phosphono-2-cyclopentene-1-carboxylic acid, two novel agonists of metabotropic glutamate receptors of group III. Bioorg Med Chem Lett 2000; 10:1447–1450.

53. Jansen JM, Copinga S, Gruppen G, Molinari EJ, Dubocovich ML, Grol CJ. The high-affinity melatonin binding-site probed with conformationally restricted ligands: 1. Pharmacophore and minireceptor models. Bioorg Med Chem 1996; 4:1321–1332.

54. Keller PA, Bowman M, Dang KH, Garner J, Leach SP, Smith R, McCluskey A. Pharmacophore development for corticotrophin-releasing hormone: new insights into inhibitor activity. J Med Chem 1999; 42:2351–2357.

55. Daveu C, Bureau R, Baglin I, Prunier H, Lancelot LC, Rault S. Definition of a pharmacophore for partial agonists of serotonin 5-HT$_3$ receptors. J Chem Inf Comput Sci 1999; 39:362–369.

56. Baglin I, Daveu C, Lancelot JC, Bureau R, Dauphin F, Pfeiffer B, Renard P, Delagrange P, Rault S. First tricyclic oximino derivatives as 5-HT$_3$ ligands. Bioorg Med Chem Lett 2001; 11:453–457.

57. Harris DL, Loew GH. Development and assessment of a 3D pharmacophore for ligand

recognition of BDZR/GABA(A) receptors initiating the anxiolytic response. Bioorg Med Chem 2000; 8:2527–2538.

58. Filizola M, Harris DL, Loew GH. Benzodiazepine-induced hyperphagia: development and assessment of a 3D pharmacophore by computational methods. J Biomol Struct Dyn 2000; 17:769–778.

59. Kroemer RT, Koutsilieri E, Hecht P, Liedl KR, Riederer P, Kornhuber J. Quantitative analysis of the structural requirements for blockade of the N-methyl-D-aspartate receptor at the phencyclidine binding site. J Med Chem 1998; 41:393–400.

60. Holladay MW, Dart MJ, Lynch JK. Neuronal nicotinic acetylcholine receptors as targets for drug discovery. J Med Chem 1997; 40:4169–4194.

61. Baringhaus KH, Matter H, Stengelin S, Kramer W. Substrate specificity of the ileal and the hepatic Na$^+$/bile acid cotransporters of the rabbit: II. A reliable 3D QSAR pharmacophore model for the ileal Na$^+$/bile acid cotransporter. J Lipid Res 1999; 40: 2158–2168.

62. Gundertofte K, Bogeso KP, Liljefors T. Computer-assisted lead finding and optimization. In: van de Waterbeemd H, Testa B, Folkers G, eds. Current Tools for Medicinal Chemistry. Weinheim: Wiley-VCH, 1997:443–459.

63. Boehm HJ, Boehringer M, Bur D, Gmuender H, Huber W, Klaus W, Kostrewa D, Kuehne H, Luebbers T, MeunierKeller N. Novel inhibitors of DNA gyrase: 3D structure based biased needle screening, hit validation by biophysical methods, and 3D guided optimization. A promising alternative to random screening. J Med Chem 2000; 43:2664–2674.

64. Gadwood RC, Shinabarger DL. Progress in oxazolidinone antibacterials. Ann Rep Med Chem 2000; 35:135–144.

65. Gurrath M, Muller G, Holtje HD. Pseudoreceptor modelling in drug design: applications of Yak and PrGen. Perspect Drug Discov Des 1998; 12:135–157.

66. Hartman GD, Egbertson MS, Halczenko W, Laswell WL, Duggan ME, Smith RL, Naylor AM, Manno PD, Lynch RJ, Zhang G. Non-peptide fibrinogen receptor antagonists: 1. Discovery and design of exosite inhibitors. J Med Chem 1992; 35:4640–4642.

67. Kurogi Y, Miyata K, Okamura T, Hashimoto K, Tsutsumi K, Nasu M, Moriyasu M. Discovery of novel mesangial cell proliferation inhibitors using a three-dimensional database searching method. J Med Chem 2001; 44:2304–2307.

68. Kaminski JJ, Rane DF, Snow ME, Weber L, Rothofsky ML, Anderson SD, Lin SL. Identification of novel farnesyl protein transferase inhibitors using three-dimensional database searching methods. J Med Chem 1997; 40:4103–4112.

69. Ekins S, Obach RS. Three-dimensional quantitative structure activity relationship computational approaches for prediction of human in vitro intrinsic clearance. J Pharmacol Exp Ther 2000; 295:463–473.

70. Ekins S, Bravi G, Binkley S, Gillespie JS, Ring BJ, Wikel JH, Wrighton SA. Three- and four-dimensional quantitative structure activity relationship (3D/4D-QSAR) analyses of CYP2C9 inhibitors. Drug Metab Dispos 2000; 28:994–1002.

71. Frater G, Bajgrowicz JA, Kraft P. Fragrance chemistry. Tetrahedron 1998; 54:7633–7703.

72. Greenidge PA, Carlsson B, Bladh LG, Gillner M. Pharmacophores incorporating numerous excluded volumes defined by x-ray crystallographic structure in three-dimensional database searching: application to the thyroid hormone receptor. J Med Chem 1998; 41:2503–2512.

73. Hahn M. Three-dimensional shape-based searching of conformationally flexible compounds. J Chem Inf Comput Sci 1997; 37:80–86.

74. Gundertofte K, Liljefors T, Norrby PO, Pettersson I. A comparison of conformational energies calculated by several molecular mechanics methods. J Comput Chem 1996; 17:429–449.

75. Kubinyi H, Hamprecht FA, Mietzner T. 3D quantitative similarity–activity relationships from SEAL similarity matrices. J Med Chem 1998; 41:2553–2564. See also Golbraikh T, Tropsha A. Beware of Q^2. J Mol Graph Model 2002; 20: 269–276.

76. Evans BE, Rittle KE, Bock MG, DiPardo RM, Freidinger RM, Whitter WL, Lundell GF, Veber DF, Anderson PS, Chang RS, et al. Methods for drug discovery: development of potent, selective, orally effective CCK antagonists. J Med Chem 1988; 31:2235–2246.

18

Use of 3D Pharmacophore Models in 3D Database Searching[†]

RÉMY D. HOFFMANN and SONJA MEDDEB

Parc Club Orsay Université, Orsay, France

THIERRY LANGER

University of Innsbruck, Innsbruck, Austria

1. INTRODUCTION

The key goal of computer-aided molecular design methods in modern medicinal chemistry is to reduce the overall cost associated to the discovery and development of a new drug by identifying the most promising candidates to focus the experimental efforts on. Very often, many drug discovery projects have reached already a well-advanced stage before detailed structural data on the protein target have become available, although it has been shown that novel methods of molecular biology together with biophysics and computational approaches enhance the likelihood of successfully obtaining detailed atomic structure information. A possible consequence is that often, medicinal chemists develop novel compounds for a target using preliminary structure–activity information, together with the theoretical models of interaction. Only responses that are consistent with the working hypotheses contribute to an evolution of the used models. Within this framework, the pharmacophore approach has proven to be successful, allowing the perception and understanding of key interactions between a receptor and a ligand.

[†] In memory of Anne and Lou-Anne, two little girls who did not have the chance to discover and enjoy all the beauties of this world. Rest in peace.

In the past 15 years, combined advances in computer technology and innovative algorithms development provided the possibility to perform complex computational operations in a reasonable time scale. Therefore these theoretical methods, when used together with modern experimental techniques (combinatorial chemistry and high-throughput screening), are now widely used.

In this chapter, we will start by providing an overview of the evolution of the 3D pharmacophore concept and subsequently show the usefulness of 3D pharmacophore searching in modern lead discovery. The use of this approach for combinatorial library design, compound classification, and molecular diversity analysis is presented. Examples of successful applications reported in the last couple of years are reviewed.

2. THE 3D PHARMACOPHORE CONCEPT

The concept of pharmacophore has first been described by Ehrlich [1] at the beginning of the 20th century as "the molecular framework that carries (*phorein*) the essential features responsible for a drug's (*pharmacon*) biological activity." This definition does not consider any 3D information, but it is interesting to note that it has not been significantly modified since then; only the additional notion of a precise geometry of the pharmacophore elements has been added to it ("...an ensemble of interactive functional groups with a defined geometry") [2]. Pharmacophores are widely used tools in modern drug discovery, and many interesting reviews have been published on this topic (see Ref. 3 for a recent publication). Different problems can be tackled with pharmacophores such as activity evaluation against a biological target [4], analysis of toxicological effects (toxicophores) [5], metabolism [6], and subtype selectivity [7]. Pharmacophores can also be used successfully in other areas such as olfaction research (olfactophores) [8].

Pharmacophores are often derived from ligands, which are, in most cases, small organic molecules (ligand-based approach). However, nowadays, there is an increas-

Table 1   Pharmacophores: Origins and Characteristics

Origin	Number	Generation method	Pharmacophoric groups	Conformation dependency	3D	Nature
Molecule	Single	Manual Automatic	Structural elements Functional elements	"Editing" conformation	Yes	Qualitative
	Multiple	Automatic	Structural elements Functional elements	Conformer generation	Yes	Qualitative Quantitative
Protein	Single	Manual Automatic	Functional elements	Bioactive conformation when co-crystallized ligand	Yes	Qualitative
	Multiple	Automatic	Functional elements	Bioactive conformation when co-crystallized ligand	Yes	Qualitative

ing interest in using 3D structural information from the active site of proteins to derive pharmacophore models; this approach is often referred to as the protein structure-based approach [9,10]. Clark et al. [11] have shown that by combining a number of tools, in casu the programs PRO_PHARMEX and PRO_SCOPE, they could come up with an efficient approach to design new thrombin inhibitors. Their method was based on a pharmacophore extraction based on the protein's active site to search 3D databases, followed by docking the resulting database hits back in the active site.

In Table 1, information about the origin and the characteristics of the different types of pharmacophores that can be generated is summarized. In their great majority, pharmacophore models are qualitative tools, but some methods can associate experimental activity values of the molecules in the building process to derive quantitative pharmacophore models. Examples of these are HASL, APEX, and HypoGen. Conformer generation is often a prerequisite, especially when working with flexible molecules.

The nature of the functional groups that together constitute a pharmacophore has a critical influence on the quality of the results of a database mining experiment. In the early days, pharmacophores most often consisted of structural groups interconnected by geometric constraints in the form of distances and angles. Using these pharmacophore models, only structurally highly related compounds could be retrieved. Today, the focus is no longer on retrieving "me-too" molecules in a hit list, but to explore "unknown regions" and return rather structurally diverse molecules. This can only be achieved by moving away from structural pharmacophoric groups and to expand to the more general concept of chemical complementarity.

The following example illustrates this comment (Fig. 1). In the early 1990s, Hibert et al. [12] published a pharmacophore for 5-HT$_3$ antagonists. This very simple pharmacophore model did consist of three major features: an indole ring, a carbonyl oxygen atom, and a basic nitrogen atom with its lone pair of electrons. Very tight distance constraints were defined to connect these three elements.

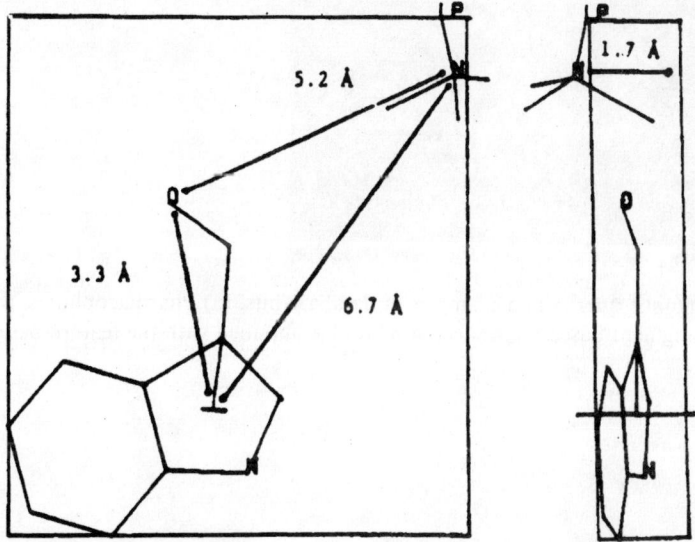

Figure 1   Pharmacophore for 5-HT$_3$ antagonists as described by Hibert et al. [12].

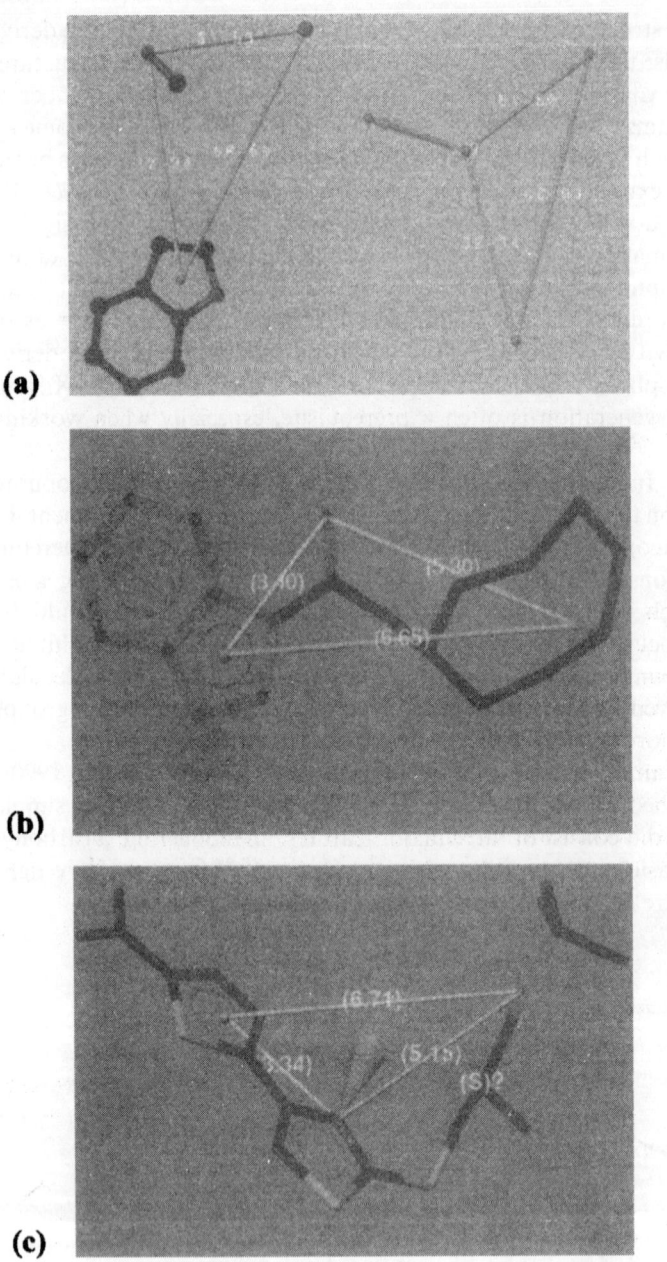

Figure 2   5-HT$_3$ antagonist queries and examples of database hits. (a) Pharmacophores; (b) hit obtained with the fragment-based pharmacophore; (c) hit obtained with the feature-based pharmacophore.

We have rebuilt this pharmacophore and have used the same tight distance constraints between the pharmacophoric points, allowing only a 0.1-Å variation around each distance constraint (Fig. 2a). Searching the Derwent WDI [13] with this model retrieved only two hits (Fig. 2b). Not surprisingly, the molecules retrieved using this pharmacophore model are very similar to the ones used by Hibert et al. to build his pharmacophore for 5-HT$_3$ antagonists. In a second experiment, we replaced the structural elements of the initial pharmacophore by chemical features, namely, a hydrophobic group, a hydrogen bond acceptor, and a basic center, making sure that the same distance constraints were kept. This time, 27 hits were retrieved from the same Derwent WDI [13]. As an example, the molecule shown in Fig. 2c differs significantly from the molecules used in Hibert's training set.

Considering that the majority of drug-like molecules are flexible, there is no unique pharmacophore answer to a given problem. The major challenge will be to design the most specific pharmacophore or set of pharmacophores that can be used to mine a particular database. Questions about the functional groups to consider, the geometric relationship between these functional groups, and the tolerances one should impose to these geometric relationships will affect the quality of the pharmacophore queries. Van Drie [14] has published an excellent paper on different strategies aiming to increase the specificity of pharmacophore models.

3. THREE-DIMENSIONAL MOLECULAR DATABASES AND PHARMACOPHORE SEARCHING

3.1. Molecular Databases

As of today, there are only a few, but nevertheless widely used, molecular database systems that can be searched with 3D pharmacophore queries. Tables 2 and 3 summarize these systems. Although performing a 3D search always requires considering the flexibility of the molecules, the different database systems use different approaches

Table 2 Three-Dimensional Database Systems that can be Queried with 3D Pharmacophores

DB system	Molecule storage	Starting	3D conversion DB building	DB indices	Software vendor
Unity	Single conformation	2D	CONCORD StereoPlex	Fingerprints	Tripos
Catalyst	Single/multiple conformation	2D/3D	catDB	*Substructure	Accelrys
				*Chemical features	
				*Shape	
MACCS-3D	Single/multiple conformation	2D/3D	CONCORD Internal registration system	2D keys 3D keys	MDL
Chem-X	Single conformation	2D		Search key (FP)	Chemical design

Table 3 Comparing the *Pros* and *Contras* of the Different Methods to Store Molecular Conformations

Application	Multiconformer database	Single-conformer database
	Catalyst MACCS-3D	Chem-X Unity MACCS-3D
DB-building step	Slow	Fast
DB-search step	Fast	Slow
DB-size	Large	Small

for compound storage in their databases: single conformation storage for each molecule (Unity, Chem-X, and MACCS-3D) vs. multiple conformations storage for each molecule (Catalyst and MACCS-3D). The first approach, implemented by Unity and Chem-X, requires "on the fly" conformer generation each time the database is searched, but has the advantage that the database is faster to build and is occupying less disk space. In the second approach (Catalyst), the limiting step is the database building time associated with the generation of the molecular conformations, as well as the larger amount of disk space that is required for the storage. This has actually become less and less of a bottleneck with the access to Linux farms to build these multiconformer databases. The advantage, on the other hand, is the faster search performance and the fact that different types of indices can be associated to each molecule in the database to reduce the overall search time even more.

3.2. Searching Molecular Databases

Searching 3D databases with 3D pharmacophore queries originates from some early work by Gund et al. in the late 1970s. They described an experimental program, Molpat, which allowed chemists to specify a pharmacophore and a test molecule and to search for occurrences of the pattern in the molecule [15]. Additional studies, all involving the identification of 3D patterns in relatively small numbers of molecules, have been reported since then [16,17]. To extend this approach to larger collections of molecules (databases), Jakes et al. [18,19] developed methods based on the selection of screens to be used for pharmacophoric pattern searches. These screening techniques used atoms, connectivity information, as well as interatomic distance information. Their goal was to eliminate molecules that could not satisfy these requirements. Further developments, including screens based on aromaticity, hybridization, connectivity, charge, position of lone pair, and center of mass of rings, have been described by Sheridan et al. [20]. Clark et al. [21], and showed that screening for functional groups and geometric searching algorithms could be used for rigid as well as flexible molecules, those flexible molecules being represented using graph-theory methods analogous to those that are used in 2D and rigid 3D substructure searching systems. Further development of this method using smoothed bounded distance matrices minimized the computational requirements of flexible searching [22]. Several search systems using these methods have been developed since that time (CAVEAT, ALADDIN, 3D SEARCH, MACCS-3D, and Chem-X) [20,23–27]. Although this approach was efficient, any molecule missing a particular element was screened out, leading to near misses. To solve the issue of partial mapping of a query, Ho and Marshall developed the programs FOUNDATION [28] and SPLICE [29]. FOUNDATION allows the identification of all molecules containing a user-defined minimum number of matching query elements using a clique detection algorithm, and SPLICE assembles partial query hit molecules into novel ligands. They have illustrated their work by the reconstruction of the known HIV-1 protease pharmacophore.

Retrieving molecules that map partially on a pharmacophore query is critical for the researcher to help him in identifying new structural families that could be made more active by simple chemical modifications (for example, by the addition of an extra functional group). This is important in a pharmaceutical company where molecules stored in the corporate collection may reveal an interesting activity profile against a new target one wants to find original leads. Modern commercial software tools integrate partial mapping of database hits onto pharmacophore queries. The approach differs from one vendor to another. Unity allows real partial mapping, whereas Catalyst uses lists of partial queries ($n - 1$ features) derived from an initial query containing n features. Some pharmacophores are not specific enough and return large and unmanageable hit lists. Molecular similarity calculations on the resulting hit lists have been shown to be useful tools for this purpose [30].

3.3. Molecular Flexibility

Molecular flexibility is another key issue that needs to be addressed. Besides the two above-mentioned classical strategies to deal with molecular flexibility in 3D searching, Güner et al. proposed an original approach. They dealt directly at the query level by creating a flexible query in a stepwise manner [31]. In this type of query, atoms belonging to rigid (or semirigid) parts are fixed, and the flexible parts are anchored to the fixed parts. Distance ranges between rigid and flexible parts are assigned. The first query is then submitted to database search, and further refinements are done by refining the distance ranges, adjusting the tolerance on the fixed atoms, and so forth, until the query is optimized and a search in the corporate or proprietary database can be performed. Another important aspect to consider when searching databases of flexible molecules is the flexibility of saturated alkyl ring systems (five-membered ring and higher). Systematic conformer generation methods generally do not consider the flexibility of these systems. Not considering this element of flexibility in a molecule will certainly result in poor conformational space coverage and will affect the overall performance of the database search as some key functional elements of the molecules will be unable to occupy the proper region of 3D space. This issue has been addressed by Sadowski [32] who used a hybrid approach combining the use of Corina for 2D to 3D conversion [33] and the generation of multiple ring conformations by Unity [34] for the flexible fitting of chain torsions. This approach increases the hit rate of 3D database searches by 10–20% on average [32].

3.4. Pharmacophore Features

The 1990s have also seen the emergence of new types of 3D pharmacophores, in which pharmacophoric points were no longer represented by structural groups or atoms, but rather by chemical functions (donors, acceptors, hydrophobic groups, acidic/basic groups, and so forth). In these pharmacophores, geometric constraints are replaced by location constraints in the form of spatial coordinates. They are generated from a set of ligands (DISCO [35], HipHop [36], HypoGen [37], and Chem-X [27]) or from protein-active sites. Instead of applying a tolerance on the geometric constraints connecting the features, tolerances are applied on the location constraints and as such can be represented by spheres. The search queries can be as simple as one pharmacophoric point (for example, one HB-donor), which can be considered as an equivalent of an extended substructure query, to very complex queries containing a large number of pharmacophoric points. Complex queries often return very little or no hits, and the user has to decide beforehand if it makes sense to generate such complex queries.

3.5. Shape

Shape is another critical component in molecular recognition. Shape can be integrated in 3D pharmacophores. Two major approaches can be used: i) addition of exclusion areas to prevent candidate molecules to occupy some defined regions in space and ii) use of a molecular surface/volume that has to be occupied by potential hits. There has been a growing interest in considering this component as part of a pharmacophore— either alone or combined with other pharmacophoric features. Van Drie [38] has described "Shrink-Wrap" surfaces as a new method for incorporating shape into pharmacophoric searching. Hahn has described another way of describing the shape of a molecule or a set of molecules and search of a 3D database with this query. The shape is constructed from a receptor surface model [39] derived from one or several molecules. The search occurs in three phases: i) database screen using shape-based indices (to find compounds potentially similar in shape), ii) alignment of the candidate molecules and evaluation of shape similarity, and iii) flexible fit of candidate molecules to the receptor surface model [40]. These queries can be used to search Catalyst multiconformer databases that have been previously indexed with the corresponding shape indices. The use of pharmacophore shown in Fig. 3b allowed a significant focus

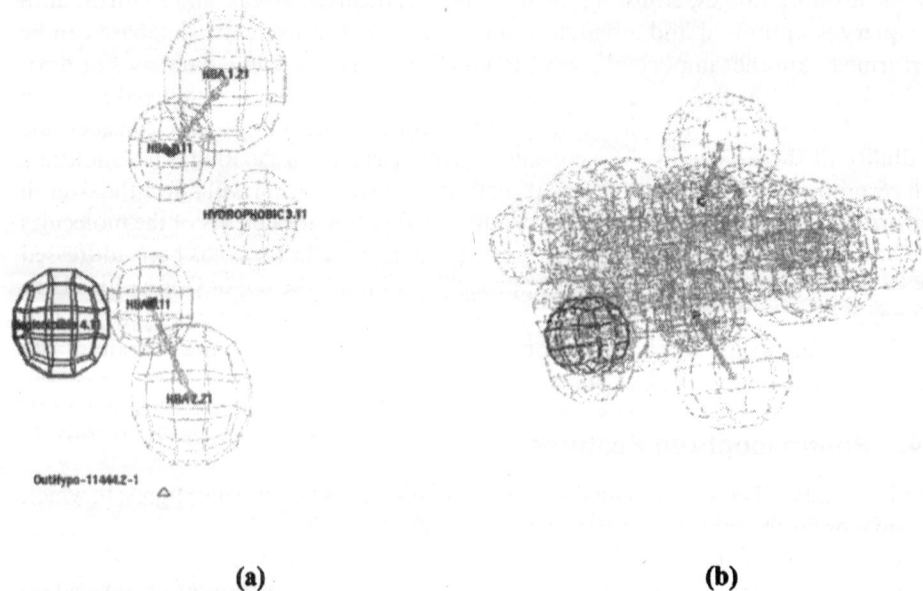

(a)   (b)

Figure 3   Feature-based (a) and combined shape-feature (b) pharmacophores for inhibitors of aldose reductase [41].

of an initial hit list of potential aldose reductase inhibitors found in the Derwent World Drug Index [13] compound collection as compared to the simple feature-based pharmacophore shown in Fig. 3a.

4. SELECTION OF THE MOST PROMISING CANDIDATE MOLECULE(S) FROM A HIT LIST

Since in most cases it is impossible to test every molecule that is returned as a hit from a 3D database search, a selection of candidates for biological evaluation has to be done. Some tools are available in order to help the researcher to decide on which molecule(s) the attention should be focused on first. In addition to the medicinal chemist's knowledge and intuition, tools are available for this task. Some of them are listed below:

- Calculation of drug-likeness parameters (like ADME parameters and Lipinski's rule [43]).
- Calculation of physicochemical parameters and evaluation of the similarity/ diversity against a reference or a set of known compounds with the desired biological activity.
- Ranking of the molecules by calculating some indices describing the quality of the fit on the pharmacophore model used as query (shape similarity if the 3D query contains a shape component, geometric fit value [42], or estimated activity if the query has an associated predictive capability).

Instead of applying additional filtering mechanisms on a hit list, one can evaluate the quality of the pharmacophores to be used as 3D search queries beforehand. The assessment of the quality of a 3D pharmacophore will strictly depend on the main objective(s) of the project (coverage and/or selectivity and/or enrichment). To evaluate one or multiple pharmacophores, one requires a test database that contains i) multiple families of molecules (diverse in structure and type of activity) and ii) a clear activity annotation (if possible with the molecular mechanism of action). Güner and Henry [44] have developed a series of indices that can be calculated for multiple pharmacophores in order to make an informed decision about the one to be used to screen a virtual library or a corporate collection.

Finding a set of known molecules (with the desired activity) in a hit list is always very encouraging and constitutes another quick way of validating a pharmacophore model. However, one should pay attention to the "unknown" molecules in the hit list. In a previously reported study on retinoic acid receptor (RAR) ligands [45], we have manually built a pharmacophore based on the "active" conformation of CD367, a known RAR ligand. This query was used to search the Derwent World Drug Index [13]. Among the hits, bexarotene, a known RAR ligand, was retrieved. Other molecules like haloxyfop (a herbicide with teratogenic activity) or boswellic acid derivatives were picked by the query. Boswellic acid, an anticancer drug originating from *Boswellia serrata*, had previously been described as an effective inducer of cell differentiation in HL-60 cells [46]. As of today, there is no direct evidence about the mechanism of action of this compound, although teratogenic and cell differentiation activities have been observed. The pharmacophore we have used suggests a RAR-mediated mode of action. This, however, remains to be validated experimentally.

5.   USAGE OF 3D PHARMACOPHORES IN LEAD STRUCTURE DISCOVERY

5.1.   Utilization of 3D Pharmacophores for Searching Molecular Databases

A considerable number of studies aimed at the discovery of bioactive compounds using different 3D pharmacophore techniques together with database search, and covering a wide range of pharmacological applications, have been published during the last years [47–57]. Overviews on such investigations are given in recent review articles [58,59]. In most of these papers, the authors describe the different methods to generate pharmacophore models and their use as 3D queries for searching databases containing large numbers of 3D molecular structures. The filtering of such databases or libraries of candidate compounds through the use of computational approaches based on discrimination functions that permit the selection of compounds to be tested for biological activity has been termed virtual screening. Within this context, not only 3D pharmacophores, but also the complete 3D structures of target proteins are used as filtering criteria. Computational efforts in the latter case are, however, much higher, and in addition, the rating of the retrieved ligands according to predicted affinity values still suffers from the lack of appropriate scoring functions, although much progress has been achieved within the last years in this area, especially with the use of consensus scoring [60].

Among the different pharmacophore types described in literature and used for searching molecular databases, the chemical feature-based approach [42] is by far the most successful as indicated by a steadily growing number of publications. In organic molecules, different structural motifs can express a similar chemical behavior and as such the same biological effect. In this respect, the application of chemical feature-based pharmacophore approaches is straightforward to retrieve novel bioactive compounds bearing the necessary interaction functions on molecular frameworks that are different from hitherto known ligands for a certain target. Successful recent applications include 1) the utilization of chemical feature-based pharmacophores derived directly from protein-active sites [9,10,61,62] and 2) ligand-based studies aimed at the discovery of ligands for various enzymes and receptors of which the 3D structures are still unknown [7,41,45,63–83]. A more detailed outlook on a number of these studies is given in Sec. 6 of this chapter.

5.2.   Use of 3D Pharmacophores for Focusing and Profiling Virtual Combinatorial Libraries

Current interest in combinatorial chemistry for lead discovery has enforced the development of new methods for the design and evaluation of the diversity of resultant compound libraries. These methods have had a high impact on the selection of diverse sets of compounds when corporate databases have to be screened. Moreover, such an approach is especially useful when designing new virtual libraries, e.g., using fragment combination-based molecule-building methods [84]. Pickett et al. [85] have proposed pharmacophore-derived queries as a tool for diversity profiling and design. In their paper, they describe a novel methodology for calculating diversity and identifying common features based on the 3D pharmacophores defined by a compound. Using an in-house developed atom-type parameterization scheme, their strategy was to generate pharmacophores consisting of three-point combinations of six pharmacophoric groups. This method of pharmacophore derived queries provides a means for assessing the diversity of compound libraries, utilizing 3D structural information and taking into account the conformational flexibility. Another approach has been developed by Chemical Design Ltd (ChemDiverse),* which is similar in philosophy but implemented differently. The ChemDiverse approach provides detailed molecular information coupled with pharmacophore and whole library data, with the advantage of faster execution times and finer resolution in the pharmacophoric space. Both methods have a wide range of applications beyond designing and profiling combinatorial libraries. The pharmacophore information can be used to partition a large database into diverse subsets covering pharmacophore space. This may be especially useful if several diverse leads are identified during screening, providing a tool for pharmacophore identification by appending the molecular descriptor. The descriptor may also be used in similarity searching of leads against a collection of compounds. This would be useful in the situation that arises frequently, where it is not possible to identify the key pharmacophores, and would allow the identification of molecules with similar pharmacophore profiles, thus giving rise to more detailed structure–activity relationship information during the early stages of a project. Another interesting approach has been proposed by Martin and Hoeffel [86]. With their oriented substituent pharmacophore property space method, the authors claim to overcome limitations in the characterization of large combinatorial libraries when using 3D information and performing calculations on the enumerated library products (product space), rather than just on the substituents (reactant space). A comparison of the efficiency of reagent-based selections vs. product-based combinatorial subsetting in the identification of representative library subsets has been presented [92], indicating that for some descriptors, product-based approaches provide distinct advantages, whereas for others, reactant pools offer comparable results. Brown and Martin [87] have published a study aimed at the evaluation of a variety of structure-based clustering methods. The use of MACCS, Unity, and Daylight 2D descriptors were compared to Unity 3D rigid and flexible descriptors and to two in-house developed parameters based on potential pharmacophore points. The results presented in this paper suggested that 2D descriptors and hierarchical clustering methods, especially MACCS descriptors and Ward's clustering, are well suited for a rapid and low-computing resources demanding classification of molecules into active and inactive ones. Another strategy for the design and comparison of combinatorial libraries using pharmacophore descriptors has been developed recently [88]. The DIVSEL and COMPLIB procedures employ multipharmacophore 3D descriptors in combination with procedures for dissimilarity-based compound selection and library comparison. Moreover, they allow the design of compounds to be performed in product space and library comparison to consider all pairwise intermolecular contributions to the overall diversity. When designing a focused "smart" combinatorial library, the probability of finding a hit (i.e., a bioactive compound) is enhanced if the library is specifically built for a given target, or the complementary pharmacophore, respectively. Ghose et al. [91] have

---

*ChemDiverse was implemented within the Chem-X software package from Chemical Design Ltd., since 1998 part of the Oxford Molecular Group, since 2000 part of Accelrys Inc, 9685 Scranton Road, San Diego, CA 92121–3752, USA.

recently reviewed the use of pharmacophore models for adapting structure-based drug design in the paradigm of combinatorial chemistry and high-throughput screening.

In the past years, much progress has been made in the development of methods for compound selection, classification, design, as well as prediction of their biological activity. For example, cell-based partitioning methods and multiple-point pharmacophore-based approaches have significantly advanced. In addition, a number of studies have already been initiated to evaluate the performance of different molecular descriptors—for detailed reviews, see Refs. 89 and 90. From the results obtained, we can conclude that increasingly complex representations of molecules are not always better, i.e., also relatively simple 2D methods perform remarkably well in a number of applications. Expectedly, the relative performance of methods and molecular descriptors appears to be highly influenced by the specific nature of the problem under investigation. Considering the large range of methods available for virtual screening and database analysis, the concepts based on 3D pharmacophores are among the most exciting ones in this area. This provides opportunities for further development, especially concerning the complementary nature of 2D and 3D methods and the combination of ligand- and protein-based approaches. Finally, it has to be mentioned that nonlinear methods, such as neural networks [93], and evolutionary algorithms, such as genetic algorithms [94], have been successfully proposed for the optimization of chemical libraries in multidimensional space.

6. SELECTED RECENT SUCCESS STORIES

The aim of this section is to review and comment on some success stories in the generation and usage of pharmacophore models as queries for 3D database searches. In a single contribution, it is impossible to provide a complete overview and give full credit to all different interesting applications in the field. Therefore some areas have been, subjectively, emphasized more than others, while others have been omitted. As the 3D searching technology has evolved over the years, it has effectively been used for lead identification and optimization. A number of patent applications for 3D chemical feature-based pharmacophore models have been submitted, including VLA-4 inhibitors [96], Hepatitis C NS3 protease inhibitors [97], and the identification of CYP2D6 inhibitors.* In all these cases, all patents claim compounds that fit the pharmacophore model, although compound structures identified by the 3D database searching are not described.

6.1. Farnesyl Protein Transferase Inhibitors

One of the first success stories published in the context of lead identification using chemical feature-based pharmacophore models was the discovery by Kaminski et al. [64] of farnesyl protein transferase (FPT) inhibitors as potential anticancer drugs in 1997. From a series of known FPT inhibitors, a training set of 35 compounds exhibiting IC_{50} values ranging from 10^{-1} to 10^3 μM was used for pharmacophore hypothesis generation with the Catalyst™ software package.* A model was obtained containing

---

*Catalyst™ software package available from Accelrys Inc., 9685 Scranton Road, San Diego, CA 92121–3752, USA.

four hydrophobic regions and one hydrogen bond acceptor. Using this template, a 3D database search of the corporate database was performed yielding a hit list of 718 compounds. After in vitro FTP inhibitory assay of 330 available compounds, 5 compounds were found with an IC_{50} value below 5 μM (0.69% overall hit rate, 1.5% with respect to the compounds tested). On the other hand, a random screening approach of 84,000 compounds yielded 22 compounds with an IC_{50} value below 5 μM, an overall hit rate of less than 0.03%. After optimization of the hits obtained, a new potent and structurally diverse lead compound was found ($IC_{50} = 0.18$ μM).

6.2. HIV-Integrase Inhibitors

Also in 1997, a series of papers was published on the discovery of novel inhibitors of HIV-1 integrase using 3D pharmacophore generation followed by 3D database searching [51–54]. Since at that time no experimentally determined 3D structure of the target was available, a purely ligand-based approach for the pharmacophore generation was chosen in all cases. In the first case [51], starting from the known HIV-1 integrase inhibitors caffeic acid phenethyl ester and NSC115290, a putative three-point pharmacophore was derived using manual alignment of low-energy molecular structures identified by molecular mechanics. The model derived as such consisted of three H-bond acceptor atoms located at defined distances together with an exclusion volume sphere. A search of the nonproprietary 3D NCI database containing more than 200,000 entries using the ChemDBS-3D software within the Chem-X package [27]* resulted in 267 structures matching the pharmacophore. Out of this list, 60 compounds were selected for testing based on availability in the repository, estimated solubility, and heuristically assessed potential usefulness of the molecule as a drug lead. Finally, 19 compounds were found to inhibit both the 3'-processing and strand transfer of HIV-1 integrase at micromolar concentration. The significance of the pharmacophore model was assessed independently by retrieving compounds from a validation 3D database containing 152 known integrase inhibitors which had no overlap with the group of compounds found in the initial search.

In a subsequent part of their study [52], the generated models were based on the orientation of crucial polar/hydrogen bond interaction sites present in a set of highly active "tetrameric" 4-hydroxycoumarin derivatives, e.g., in compound NSC158393 exhibiting IC_{50} values of 1.5 and 0.8 μM for 3'-processing and strand transfer, respectively. Based upon exhaustive molecular modeling studies comprising Monte Carlo-based conformational analysis, 621 possible conformations of NSC158393 were found that had a potential energy of less than 10 kcal/mol above the calculated global energy minimum. This subset was classified into five different clusters depending on the value of the torsion angles of the rotatable bonds. From these sets, five four-point pharmacophore models were derived using a distance matrix obtained from all interatomic distances between the keto and hydroxyl oxygen atoms. The derived models were checked for their ability to retrieve compounds from the validated 3D database containing 152 HIV-1 integrase inhibitors [51]. Even by using relatively loose distance tolerances of 0.7 Å, the hit rates were found to be disappointingly low, ranging from 0% to 17%, and therefore attempts to define a four-point pharmacophore model were dropped. Based on the conclusion that a three-point pharmacophore, as it is commonly used in enzyme–ligand systems, may be a better choice in the HIV-1 integrase case, consequently, all possible three-point pharmacophores were extracted from the interatomic distance matrices obtained from the five clusters previously defined. From the six interatomic distances, 17 unique possible three-point pharmacophores could be defined. These models were then used as queries for a 3D database search within the 152 compounds library. Expectedly, the hit rates found were significantly higher than before, ranging from 20% to 67%. A cluster analysis of the interatomic distance values in the three-point models revealed that the 17 models could be classified into four families, the first one being the most successful in retrieving hits. In order to obtain a reasonable pharmacophore query for the search in the entire NCI 3D multiconformational database, the multiple sets of distances in each cluster were reduced to a single-distance triad by a weighted averaging of the distances of the original models accordng to their hit rate. The search of the structure database containing more than 200,000 compounds yielded a total of 340 molecules that contained the pharmacophoric pattern in one or more of their conformations. After removal of charged molecules and after clustering according to chemical diversity criteria, a set of 29 compounds was selected and submitted to the integrase bioassay. Interestingly, all 29 compounds exhibited integrase inhibiting activity at a concentration of 100 µg/ml, and 10 of them had IC_{50} values between 1.5 and 350 µM. Four of these 10 inhibitors are particularly potent and may constitute useful lead compounds for the development of anti-AIDS drugs. It is interesting to note that with a very simple pharmacophore model based on three points marking hydrogen-bonding regions starting from a single bioactive molecule, a subset of less than 0.2% of the original database could be retrieved. Furthermore, these molecules are structurally not related to the 4-hydroxycoumarine derivatives used for the model generation.

In the third paper [53], a new structural class of noncatechol-containing integrase inhibitors, bisaroylhydrazines, was identified by modification of hits obtained in the previously presented studies. These compounds may provide valuable avenues for the development of HIV integrase inhibitors. Finally, also natural products, i.e., lichen acids containing depsides and depsidones, and their synthetic derivatives were identified from 3D database search as novel HIV-1 integrase inhibitors with IC_{50} values below 50 µM [54].

6.3. Muscarinic M_3 Receptor Antagonists

Marriott et al. [2] reported the finding of new muscarinic M_3 antagonists using a combination of pharmacophore generation, 3D database searching, and medium-throughput screening. Their interest in M_3 muscarinic antagonists aroused from the observation that a number of potent M_3-selective antagonists (e.g., zamifenacin) exhibited further selectivity between M_3 receptors expressed in different tissues of the same species, thus providing therapeutic advantage as drug candidates in diseases like chronic obstructive airways disease or urinary incontinence [98]. They converted a set of databases (from in-house database to commercial ones in order to get structural diversity) into the Unity format [34]. DISCO [35] was utilized as the pharmacophore generation method. Three series of compounds were used to define the M_3 pharmacophore: one series containing a lung-selective M_3 antagonist and two-related M_1 ligands also showing affinity for the M_3 receptor. One of the two M_1 ligands was used as the reference for the DISCO run. Among the five pharmacophore models generated, visual inspection eliminated three pharmacophore models that did not match the tertiary nitrogen common to the three molecules. The two remaining pharmacophores consisted of an N^+-tertiary nitrogen atom (cationic head), two hydrogen-bond donors, and a hydrogen-bond acceptor. A tolerance of 0.3 Å was applied on the interfeature distances. Searching with the two pharmacophores generated a comparable number of hits from the corporate database, most of them being common to both models, reflecting the similarity in the pharmacophores. Most of these compounds were screened against an M_3 receptor assay in the guinea pig isolated trachea. Three of the selected molecules showed activity with pA_2 values ranging from 4.8 to 6.7. Interestingly, none of these compounds was structurally related to the molecules used to derive the pharmacophore models.

6.4. 3D Pharmacophore for BDZR/GABA$_A$ Receptors

Benzodiazepine receptor (BDZR) ligands belonging to different structural classes bind to specific binding sites on GABA$_A$ receptors and modulate, in an allosteric way, the effects of GABA on chloride flux. Depending on the composition of these heteromeric receptors—several subtypes exist from various isoforms—the ligands exert different biological effects including anxiolytic, sedative, hyperphagic, anticonvulsant, and hyperthermic activity. In the study of Harris and Loew [99], a set of 17 structurally diverse BDZR ligands of the receptor subtypes initiating the anxiolytic response has been studied using the MOLMOD procedure [100], a pharmacophore determination procedure that is, in principle, similar to that used within DISCO [35]. Finally, the five-component 3D recognition pharmacophore query was built in Unity format [34], consisting of two proton acceptors, a hydrophobic group, an aromatic electron accepting ring, and a ring containing polar moieties. The model was validated by searching 3D databases and finding known BDZR ligands active at the anxiolytic endpoint, including 1,4-benzodiazepine derivatives, imidazo derivatives, and β-carboline ligands. In the paper, however, no evidence was given for the retrieval of compounds being novel bioactive BDZR ligands.

6.5. Structure-Based Pharmacophore Model for Selective COX-2 Inhibitors

In a recent paper of Palomer et al. [61], the authors investigate whether pharmacophore models may account for the activity and selectivity of the known cyclooxygenase-2 (COX-2) selective inhibitors of the phenylsulfonyl tricyclic series, i.e., celecoxib and rofecoxib. Moreover, they studied whether transferring structural information onto the frame of a nonsteroidal anti-inflammatory drug (NSAID), known to tightly bind the enzyme active site, might be useful for the design of novel COX-2 selective inhibitors. The authors developed a pharmacophore hypothesis based on the geometric disposition of chemical features in the most favorable conformation of the COX-2 selective inhibitors SC-558 analog of celecoxib, rofecoxib, and more restrained compounds. The pharmacophore model presented in the study contains a sulfonyl S atom and two aromatic rings forming a dihedral angle of 290°. The final disposition of the pharmacophoric groups parallels the geometry of the ligand SC-558 in the known crystal structure of the COX-2 complex [101]. Moreover, the nonconserved residue 523 is known to be important for COX-2 selective inhibition; thus crystallographic information was used to position an area of excluded volume in the pharmacophore, accounting for the space limits imposed by this nonconserved residue. The geometry of the final five-feature pharmacophore was found to be consistent with the crystal structure of the nonselective NSAID indomethacin in the COX-2 complex [101]. This information was subsequently used to design several indomethacin analogs that were shown to exhibit consistent structure–activity relationships leading to a novel potent and selective COX-2 inhibitor (compound LM-1685), which was selected as a promising candidate for further pharmacological evaluation.

7. SUMMARY AND CONCLUSIONS

The pharmacophore concept has proven to be extremely successful not only in rationalizing structure–activity relationships, but also by its large impact in developing the appropriate 3D tools for efficient virtual screening. Profiling of combinatorial libraries and compound classification are other often-used applications of this concept.

As shown in this review, the complexity of pharmacophores can range from very simple objects (two- or three-point pharmacophores) to more sophisticated objects by the addition of more pharmacophoric features, different types of geometric constraints, shape, or excluded regions information. 2D (substructure) as well as 1D (relational data) information can also be added to a 3D pharmacophore. The nature of the pharmacophoric points (feature vs. substructure) will directly affect the overall performance of a database search. In general, an overspecification of the pharmacophoric points will result in hit lists with limited structural diversity. However, the use of pharmacophores is an efficient procedure since it eliminates quickly molecules that do not possess the required features. Unfortunately, all the retrieved hits are not always active as expected since the presence of the pharmacophoric groups is only one of the multiple components that account for the activity of a molecule. Other properties (physicochemical, ADME, and toxicological properties) are other components of the multidimensional approach that is used to turn a hit into a drug.

REFERENCES

1. Ehrlich P. Present status of chemotherapy. Ber Dtsch Chem Ges 1909; 42:17–47.
2. Marriott DP, Dougall IG, Meghani P, Liu YJ, Flower DR. Lead generation using pharmacophore mapping and three-dimensional database searching: application to muscarinic M₃ receptor antagonists. J Med Chem 1999; 42:3210–3216.
3. Mason JS, Good AC, Martin EJ. 3D pharmacophores in drug discovery. Curr Pharm Des 2001; 7:567–597.
4. Doweyko AM. Three-dimensional pharmacophores from binding data. J Med Chem 1994; 37:1769–1778.
5. Kaufman JJ, Koski WS, Hariharan PC, Crawford J, Garmer DM, Chan-Lizardo L. Prediction of toxicology and pharmacology based on model toxicophores and pharmacophores using the new Tox-MATCH–PHARM-MATCH program. Int J Quantum Chem, Quantum Biol Symp 1983; 10:375–416.
6. Ekins S, Bravi G, Binkley S, Gillespie JS, Ring BJ, Wikel JH, Wrighton SA. Three and four dimensional-quantitative structure activity relationship (3D/4D-QSAR) analyses of CYP2D6 inhibitors. Pharmacogenetics 1999; 9:477–489.
7. Bureau R, Daveu C, Lancelot JC, Rault S. Molecular design based on 3D-pharmaco-

phore. Application to 5-HT subtypes receptors. J Chem Inf Comput Sci 2002; 42:429–436.

8. Kraft P, Bajgrowicz JA, Denis C, Frater G. Odds and trends: recent developments in the chemistry of odorants. Angew Chem Int Ed 2000; 39:2980–3010.

9. Hoffrén AM, Murray CM, Hoffmann RD. Structure-based focusing using pharmacophores derived from the active site of 17 β-hydroxysteroid dehydrogenase. Curr Pharm Des 2001; 7:547–566.

10. Greenidge PA, Carlsson B, Bladh LG, Gillner M. Pharmacophores incorporating numerous excluded volumes defined by X-ray crystallographic structure in three-dimensional database searching: application to the thyroid hormone receptor. J Med Chem 1998; 41: 2503–2512.

11. Clark DE, Westhead DR, Sykes RA, Murray CW. Active-site-directed 3D database searching: pharmacophore extraction and validation of hits. J Comput-Aided Mol Des 1996; 10:397–416.

12. Hibert MF, Hoffmann R, Miller RC, Carr AA. Conformation–activity relationship study of 5-HT3 receptor antagonists and a definition of a model for this receptor site. J Med Chem 1990; 33:1594–1600.

13. Derwent World Drug Index, Version 96.2, Derwent Ltd, London, 1996 in Catalyst™ data format, available from Accelrys Inc, 9685 Scranton Road, San Diego, CA 92121-3752, USA.

14. Van Drie JH. Strategies for the determination of pharmacophoric 3D database queries. J Comput-Aided Mol Des 1997; 11:39–52.

15. Gund P. Three-dimensional pharmacophoric pattern searching. Prog Mol Subcell Biol 1977; 5:117–143.

16. Lesk AM. Detection of three-dimensional patterns of atoms in chemical structures. Commun ACM 1979; 22:219–224.

17. Crandell CW, Smith DH. Computer-assisted examination of compounds for common three-dimensional structures. J Chem Inf Comput Sci 1983; 23:186–197.

18. Jakes SE, Willett P. Pharmacophoric pattern matching in files of three-dimensional chemical structures. Selection of interatomic screens. J Mol Graph 1986; 4:12–20.

19. Jakes SE, Watts N, Willett P, Bawden D, Fisher JD. Pharmacophoric pattern matching in files of 3D chemical structures: evaluation of search performance. J Mol Graph 1987; 5:41–48.

20. Sheridan RP, Rusinko A III., Nilakantan R, Venkataraghavan R. Searching for pharmacophores in large coordinate databases and its use in drug design. Proc Natl Acad Sci U S A 1989; 86:8165–8169.

21. Clark DE, Willett P, Kenny PW. Pharmacophoric pattern matching in files of three-dimensional chemical structures; use of bounded distance matrices for the representation and searching of conformationally flexible molecules. J Mol Graph 1992; 10:194–204.

22. Clark DE, Willett P, Kenny PW. Pharmacophoric pattern matching in files of three-dimensional chemical structures: implementation of flexible searching. J Mol Graph 1993; 11:146–156.

23. Barlett PA, Shea GT, Telfer SJ, Waterman S. Roberts SM, ed. Molecular Recognition: Chemical and Biological Problem. London: Royal Society of London, 1989:182–196.

24. Van Drie JH, Weininger D, Martin YC. Alladin: an integrated tool for computer-assisted molecular design and pharmacophore recognition from geometric, steric, and substructure searching of three-dimensional molecular structures. J Comput-Aided Mol Des 1989; 3: 225–251.

25. Christie BD, Henry DR, Güner OF, Moock TE. MACCS-3D: a tool for three-dimensional drug design. In: Raitt DI, ed. Online Information '90, 14th International Online Information Meeting Proceedings. Oxford: Learned Information, 1990:137–161.

26. Moock TE, Christie BD, Henry DR. MACCS-3D: a new database system for three-

dimensional molecular models. In: Mitchell EM, Bawden D, eds. Chemical Information Systems. Chichester: Ellis Horwood, 1990:42–49.

27. Chemical Design Ltd, Oxford OX2 OJB, UK.
28. Ho CMW, Marshall GR. FOUNDATION: a program to retrieve all possible structures containing a user-defined minimum number of matching query elements from three-dimensional databases. J Comput-Aided Mol Des 1993; 7:3–22.
29. Ho CMW, Marshall GR. SPLICE: a program to assemble partial query solutions from three-dimensional database searches into novel ligands. J Comput-Aided Mol Des 1993; 7:623–647.
30. Good AC, Hodgkin EE, Richards WG. Similarity screening of molecular datasets. J Comput-Aided Mol Des 1992; 6:513–520.
31. Güner OF, Henry DR, Pearlman RS. Use of flexible queries for searching conformationally flexible molecules in databases of three-dimensional structures. J Chem Inf Comput Sci 1992; 32:102–109.
32. Sadowski J. A hybrid approach for addressing ring flexibility in 3D database searching. J Comput-Aided Mol Des 1997; 11:53–60.
33. Gasteiger J, Rudolph C, Sadowski J. Automatic generation of 3D atomic coordinates for organic molecules. Tetrahedron Comput Methodol 1990; 3:537–547.
34. UNITY™ is available from Tripos Inc., 1699 South Hanley Rd., St. Louis, MO, 63144, USA.
35. Martin YC, Bures MG, Danaher EA, DeLazzer J, Lico I, Pavlik PA. A fast new approach to pharmacophore mapping and its application to dopaminergic and benzodiazepine agonists. J Comput-Aided Mol Des 1993; 7:83–102.
36. Barnum D, Greene J, Smellie A, Sprague P. Identification of common functional configurations among molecules. J Chem Inf Comput Sci 1996; 36:563–571.
37. Li H, Sutter J, Hoffmann R. HypoGen: an automated system for generating 3D predictive pharmacophore models. Pharmacophore Perception, Development, and Use in Drug Design. La Jolla: Güner, OF, Int. Univ. Line, 2000:171–189.
38. Van Drie JH. 'Shrink-Wrap' surfaces: a new method for incorporating shape into pharmacophoric 3D database searching. J Chem Inf Comput Sci 1997; 37:38–42.
39. Hahn M. Receptor Surface Models, 1. Definition and construction. J Med Chem 1995; 38:2080–2090.
40. Hahn M. Three-dimensional shape-based searching of conformationally flexible compounds. J Chem Inf Comput Sci 1997; 37:80–86.
41. Langer T, Hoffmann RD, Bachmair F, Begle S. Chemical function based pharmacophore models as suitable filters for virtual 3D database screening. J Mol Struct Theochem 2000; 503:59–72.
42. Sprague PW, Hoffmann R. Catalyst pharmacophore models and their utility as queries for searching 3D databases. In: Van de Waterbeemd H, Testa, B, Folkers G, eds. Computer Assisted Lead Finding and Optimization—Current Tools for Medicinal Chemistry. Basel: VHCA, 1997:225–240.
43. Lipinski CA, Lombardo F, Dominy BW, Feeney PJ. Experimental and computational approaches to estimate solubility and permeability in drug discovery and development settings. Adv Drug Deliv Rev 1997; 23:3–25.
44. Güner OF, Henry DR. Metric for analyzing hit lists and pharmacophores. Pharmacophore Perception, Development, and Use in Drug Design. La Jolla: Güner, OF, Int. Univ. Line, 2000:191–211.
45. Bachmair F, Hoffmann RD, Daxenbichler G, Langer T. Studies of structure activity relationship of retinoic acid receptor ligands by means of molecular modeling. Vitam Horm 2000; 59:159–215.
46. Han R. Highlight on the studies of anticancer drugs derived from plants in China. Stem Cells 1994; 12:53–63.

47. Anzini M, Cappelli S, Vomero S, Seeber M, Menziani MC, Langer T, Hagen B, Manzoni C, Bourguignon JJ. Mapping and fitting the peripheral benzodiazepine receptor binding site by carboxamide derivatives related to 1-(2-chlorophenyl)-N-methyl-N-(1-methyl-propyl)-3-isoquinolinecarboxamide (PK11195). Comparison of different approaches to ligand–receptor interaction modelling. J Med Chem 2001; 44:1134–1150.

48. Pineda LF, Liebmann C, Hensellek S, Paegelow I, Steinmetzer T, Schweinitz A, Stürzebecher J, Reissmann S. Novel non-peptide lead structures for bradykinin B_2-receptor antagonists. Lett Pept Sci 2000; 7:69–77.

49. Filizola M, Villar HO, Loew GH. Molecular determinants of non-specific recognition of delta, mu, and kappa opioid receptors. Bioorg Med Chem Lett 2001; 9:69–76.

50. Astles PC, Brown TJ, Handscombe CM, Harper MF, Harris NV, Lewis RA, Lockey PM, McCarthy C, McLay IM, Porter B, Roach AG, Smith C, Walsh RJA. Selective endothelin A receptor ligands. 1. Discovery and structure–activity of 2,4-disubstituted benzoic acid derivatives. Eur J Med Chem 1997; 32:409–423.

51. Nicklaus MC, Neamati N, Hong H, Mazumder A, Sunder S, Chen J, Milne GWA, Pommier Y. HIV-1 integrase pharmacophore: discovery of inhibitors through three-dimensional database searching. J Med Chem 1997; 40:920–929.

52. Hong H, Neamati N, Wang S, Nicklaus MC, Mazumder A, Zhao H, Burke TR Jr, Pommier Y, Milne GWA. Discovery of HIV-1 integrase inhibitors by pharmacophore searching. J Med Chem 1997; 40:930–936.

53. Zhao H, Neamati N, Sunder S, Hong H, Huixiao W, Wang S, Milne GWA, Pommier Y, Burke TR Jr. Hydrazide-containing inhibitors of HIV-1 integrase. J Med Chem 1997; 40: 937–941.

54. Neamati N, Hong H, Mazumder A, Wang S, Sunder S, Nicklaus MC, Milne GWA, Proksa B, Pommier Y. Depsides and depsidones as inhibitors of HIV-1 integrase: discovery of novel inhibitors through 3D database searching. J Med Chem 1997; 40:942–951.

55. Wang S, Sakamuri S, Enyedy IJ, Kozikowski AP, Deschaux O, Bandyopadhyay BC, Tella SR, Zaman WA, Johnson KM. Discovery of a novel dopamine transporter inhibitor, 4-hydroxy-1-methyl-4-(4-methylphenyl)-3-piperidyl 4-methylphenyl ketone, as a potential cocaine antagonist through 3D-database pharmacophore searching. Molecular modeling, structure–activity relationships, and behavioral pharmacological studies. J Med Chem 2000; 43:351–360.

56. Hiramatsu Y, Tsukida T, Nakai Y, Inoue Y, Kondo H. Study on selectin blocker. 8. Lead discovery of a non-sugar antagonist using a 3D-pharmacophore model. J Med Chem 2000; 43:1476–1483.

57. Webb TR, Melman N, Lvovskiy D, Ji X, Jacobson KA. The utilization of a unified pharmacophore query in the discovery of new antagonists of the adenosine receptor family. Bioorg Med Chem Lett 2000; 10:31–34.

58. Langer T, Hoffmann RD. Virtual screening: an effective tool for lead structure discovery? Curr Pharm Des 2001; 7:509–527.

59. Kurogi Y, Güner OF. Pharmacophore modeling and three-dimensional database searching for drug design using catalyst. Curr Med Chem 2001; 8:1035–1055.

60. Charifson PS, Corkery JJ, Murcko MA, Walters WP. Consensus scoring: a method for obtaining improved hit rates from docking databases of three-dimensional structures into proteins. J Med Chem 1999; 42:5100–5109.

61. Palomer A, Cabre F, Pascual J, Campos J, Trujillo MA, Entrena A, Gallo MA, Garcia L, Mauleon D, Espinosa A. Identification of novel cyclooxygenase-2 selective inhibitors using pharmacophore models. J Med Chem 2002; 47:1402–1411.

62. Debnath AK. Pharmacophore mapping of a series of 2,4-diamino-5-deazapteridine inhibitors of *Mycobacterium avium* complex dihydrofolate reductase. J Med Chem 2002; 45:41–53.

63. Orús L, Pérez-Silanesa S, Oficialdegui AM, Martínez-Esparza J, Del Castillo JC, Mourelle M, Langer T, Guccione S, Donzella G, Krovat EM, Poptodorov K, Lasheras B, Ballaz S, Hervías I, Tordera R, Del Río J, Monge A. Synthesis and molecular modeling of new 1-aryl-3-[4-arylpiperazin-1-yl]-1-propane derivatives with high affinity at the serotonin transporter and at 5-HT$_{1A}$ receptors. J Med Chem. In press.

64. Kaminski JJ, Rane DF, Snow ME, Weber L, Rothofsky ML, Anderson SD, Lin SL. Identification of novel farnesyl protein transferase inhibitors using three-dimensional database searching methods. J Med Chem 1997; 40:4103–4112.

65. Bureau R, Daveu C, Lancelot JC, Rault S. Molecular design based on 3D-pharmacophores. Application to 5-HT subtype receptors. J Chem Inf Comput Sci 2001; 41:815–823.

66. Saladino R, Crestini C, Palamara AT, Danti MC, Manetti F, Corelli F, Garaci E, Botta M. Synthesis, biological evaluation, and pharmacophore generation of uracil, 4(3H)-pyrimidinone, and uridine derivatives as potent and selective inhibitors of parainfluenza 1 (Sendai) virus. J Med Chem 2001; 44:4554–4562.

67. Nicolotti O, Pellegrini-Calace M, Carrieri A, Altomare C, Centeno NB, Sanz F, Carotti A. Neuronal nicotinic receptor agonists: a multi-approach development of the pharmacophore. J Comput-Aided Mol Des 2001; 15:859–872.

68. Gritsch S, Guccione S, Hoffmann RD, Cambria A, Raciti G, Langer T. A 3D QSAR study of monoamine oxidase B inhibitors using the chemical function based pharmacophore generation approach. J Enzyme Inhib 2001; 16:199–215.

69. Chen GS, Chang CS, Kang WM, Chang CL, Wang KC, Chern JW. Novel lead generation through hypothetical pharmacophore three-dimensional database searching: discovery of isoflavonoids as nonsteroidal inhibitors of rat 5alpha reductase. J Med Chem 2001; 44: 3759–3763.

70. Barbaro R, Betti L, Botta M, Corelli F, Giannaccini G, Maccari L, Manetti F, Strappaghetti G, Corsano S. Synthesis, biological evaluation, and pharmacophore generation of new pyridazinone derivatives with affinity toward alpha-1 and alpha-2 adrenoceptors. J Med Chem 2001; 44:2118–2132.

71. Bureau R, Daveu C, Baglin I, Sopkova-De Olivera J, Santos JC, Lancelot, Rault S. Association of two 3D QSAR analyses, application to the study of partial agonist serotonin-3 ligands. J Chem Inf Comput Sci 2001; 41:815–823.

72. Greenidge PA, Weiser J. A comparison of methods for pharmacophore generation with the catalyst software and their use for 3D-QSAR: application to a set of 4-aminopyridine thrombin inhibitors. Min Rev Med Chem 2001; 1:79–87.

73. Karki RG, Kulkarni VM. A feature based pharmacophore for *Candida albicans* MyristoylCoA: protein *N*-myristoyltransferase inhibitors. Eur J Med Chem 2001; 36:147–163.

74. Kurogi Y, Miyata K, Okamura T, Hashimoto K, Tsutsumi K, Nasu M, Moriyasu M. Discovery of novel mesangial cell proliferation inhibitors using a three-dimensional database searching method. J Med Chem 2001; 44:2304–2307.

75. Zhu LL, Hou TJ, Chen LR, Xu XJ. 3D QSAR analyses of novel tyrosine kinase inhibitors based on pharmacophore alignment. J Chem Inf Comput Sci 2001; 41:1032–1040.

76. Bajgrowicz JA, Frater G. Chiral recognition of sandalwood odorants. Enantiomer 2000; 5:225–234.

77. Bremner JB, Coban B, Griffith R, Groenwoud KM, Yates BF. Ligand design for alpha1-adrenoceptor subtype selective antagonists. Bioorg Med Chem 2000; 8:201–214.

78. Carlson HA, Masukawa KM, Rubins K, Bushman FD, Jorgenses WL, Lins RD, Briggs JM, McCammon JA. Developing a dynamic pharmacophore model for HIV-1 integrase. J Med Chem 2000; 43:2100–2114.

79. Clement O, Mehl AT. HipHop: pharmacophores based on multiple common-feature

alignments. In: Güner OF, ed. Pharmacophore Perception, Development, and Use in Drug Design. La Jolla: IUL Biotechnology Series, 2000:71–84.

80. Ekins S, Ring BJ, Bravi G, Wikel JH, Wrighton SA. Predicting drug–drug interactions in silico using pharmacophores: paradigm for the next millennium. In: Güner OF, ed. Pharmacophore Perception, Development, and Use in Drug Design. La Jolla: IUL Biotechnology Series, 2000:269–300.

81. Hoffmann RD, Li H, Langer T. Feature-based pharmacophores: application to some biological systems. In: Güner OF, ed. Pharmacophore Perception, Development, and Use in Drug Design. La Jolla: IUL Biotechnology Series, 2000:301–318.

82. Kaminski JJ, Rane DF, Rothofsky ML. Database mining using pharmacophore models to discover novel structural properties. In: Güner OF, ed. Pharmacophore Perception, Development, and Use in Drug Design. La Jolla: IUL Biotechnology Series, 2000:252–268.

83. Kim SG, Yoon CJ, Kim SH, Cho YJ, Kang DI. Building a common feature hypothesis for thymidilate synthase inhibition. Bioorg Med Chem 2000; 8:11–17.

84. Wolber G, Langer T. CombiGen: a novel software package for the rapid generation of virtual combinatorial libraries. In: Höltje HD Sippl W, eds. Rational Approaches to Drug Design. Barcelona: Prous Science, 2001:390–395.

85. Pickett SD, Mason JS, McLay IM. Diversity profiling and design using 3D pharmacophores: pharmacophore-derived queries (PDQ). J Chem Inf Comput Sci 1996; 36:1214–1223.

86. Martin EJ, Hoeffel TJ. Oriented substituent pharmacophore property space (OSPPREYS): a substituent-based calculation that describes combinatorial library products better than the corresponding product-based calculation. J Mol Graphics Mod 2000; 18: 383–403.

87. Brown RD, Martin YC. Use of structure–activity data to compare structure-based clustering methods and descriptors for use in compound selection. J Chem Inf Comput Sci 1996; 36:572–584.

88. Pickett SD, Luttmann C, Guerin V, Laoui A, James E. Divsel and Complib—strategies for the design and comparison of combinatorial libraries using pharmacophoric descriptors. J Chem Inf Comput Sci 1998; 38:144–150.

89. Bajorath J. Selected concepts and investigations in compound classification, molecular descriptor analysis, and virtual screening. J Chem Inf Comput Sci 2001; 41:233–245.

90. Livingstone DJ. The characterization of chemical structures using molecular properties. A survey. J Chem Inf Comput Sci 2000; 40:195–209.

91. Ghose AK, Viswanadhan VN, Wendoloski JJ. Adapting structure-based drug design in the paradigm of combinatorial chemistry and high-throughput screening: an overview and new examples with important caveats for newcomers to combinatorial library design using pharmacophore models or multiple copy simultaneous search fragments. ACS Symp Ser 719 (Rational Drug Design), 1999:226–238.

92. Jamois EA, Hassan M, Waldman M. Evaluation of reagent-based strategies in the design of combinatorial library subsets. J Chem Inf Comput Sci 2000; 40:63–70.

93. Sadowski J. Optimization of chemical libraries by neural networks. Curr Opin Chem Biol 2000; 4:280–282.

94. Gillet VJ, Willet P, Bradshaw J, Green DVS. Selecting combinatorial libraries to optimize diversity and physical properties. J Chem Inf Comput Sci 1999; 39:169–177.

95. Singh J, Zheng Z, Sprague P, Van Vlihmen H, Castro AC, Adams SP. WO 9804913.

96. Hart T, Quibell M. WO 9846630.

97. Ekins S. US 2002/0013372.

98. Wallis RM. Preclinical and clinical-pharmacology of selective muscarinic M_3 receptor antagonists. Life Sci 1995; 56:861–868.

99. Harris DL, Loew G. Development and assessment of a 3D pharmacophore for ligand

recognition of BDZR/GABA$_A$ receptors initiating the anxiolytic response. Bioorg Med Chem 2000; 8:2527–2538.

100. Huang P, Kim S, Loew G. Development of a common 3D pharmacophore for delta-opioid recognition from peptides and non-peptides using a novel computer program. J Comput-Aided Mol Des 1997; 11:21–28.

101. Kurumbail RG, Stevens AM, Gierse JK, McDonald JJ, Stegeman RA, Pak JY, Gildehaus D, Miyashiro JM, Penning TD, Seibert K, Isakson PC, Stallings WC. Structural basis for selective inhibition of cyclooxygenase-2 by anti-inflammatory agents. Nature 1996; 384:644–648.

19

Substructure and Maximal Common Substructure Searching

LINGRAN CHEN

MDL Information Systems, Inc., San Leandro, California, U.S.A.

1. INTRODUCTION

There are three important and closely related methods for the manipulation and searching of chemical structures: structure isomorphism identification, substructure searching (SSS), and maximal common substructure (MCSS) perception [1]. To understand the internal relationships of these three concepts, it is useful to define several terms before further proceeding with the discussion of SSS and MCSS problems. Because the structure of a chemical molecule can be treated as a mathematical graph with atoms corresponding to vertices and bonds to edges, it is convenient to introduce the following definitions [2] according to graph theory.

Definition 1: A graph $G = (V, E)$ is a finite set of vertices (V) and a finite set of edges (E). Each edge (v_i, v_j) consists of an unordered pair of distinct vertices. Two vertices v_i and v_j of a graph are said to be adjacent if (v_i, v_j) is an edge of the graph [3]. If vertices and edges of a graph possess colors, we say it is a colored graph. The colors of vertices and edges of a colored graph separately correspond to the properties of atoms and bonds of a chemical structure.

Definition 2: Two graphs $G_1 = (V_1, E_1)$ and $G_2 = (V_2, E_2)$ are said to be isomorphic if their vertices can be identified in a one-to-one fashion so that, if v_{1i} and v_{1j} are vertices in G_1, and v_{2i} and v_{2j} are the corresponding vertices in G_2, then (v_{1i}, v_{1j}) is an edge of G_1 if and only if (v_{2i}, v_{2j}) is an edge of G_2 [3]. For two colored graphs, the vertex pairs "v_{1i}, v_{2i}" and "v_{1j}, v_{2j}," and edge pairs (v_{1i}, v_{1j}) and (v_{2i}, v_{2j}) must have the same kind of colors, respectively.

Definition 3: Given two graphs $G_1 = (V_1, E_1)$ and $G_2 = (V_2, E_2)$, we say G_1 is a subgraph of G_2 if V_1 is a subset of V_2 and E_1 is a subset of E_2 [3].

Definition 4: Given three graphs G_1, G_2, and G_3, if G_1 is a subgraph of both G_2 and G_3, we say G_1 is a common subgraph of graphs G_2 and G_3. Usually two graphs have more than one common subgraph. Thus we have the following definition:

Definition 5: Of all possible common subgraphs of two given graphs, those that contain the largest number of vertices are called maximal common subgraphs of the two graphs. (Note: The MCSS can also be defined as those common substructures that contain the largest number of edges.)

Some graphs are shown in Fig. 1 to illustrate the above definitions.

For convenience, in the subsequent sections of this chapter, the mathematical terms introduced above, such as graph, subgraph, and maximal common subgraph, are interchangeably used with the corresponding chemical terms, such as structure, substructure, and maximal common substructure.

Brief analysis of the internal relationships between the structure isomorphism, SSS, and MSCC problems can help us understand these problems better. The MCSS problem is that of determining all possible maximal common substructures of two (or more) given structures. If the detected maximal common substructure is isomorphic to the smaller of the two given structures, this kind of MCSS problem is, in fact, a substructure isomorphism problem. If the maximal common substructure found is isomorphic to both of the given structures, such a problem belongs to the structure isomorphism problem. Therefore the structure and substructure isomorphism problems are only two special cases of the more general MCSS problem (see also Fig. 1).

Structure isomorphism problems for general structures are not easy to solve, but substructure isomorphism problems are, however, much more complicated than structure isomorphism problems. In fact, it has been proven that substructure isomorphism problems belong to a set of well-known difficult problems called NP-

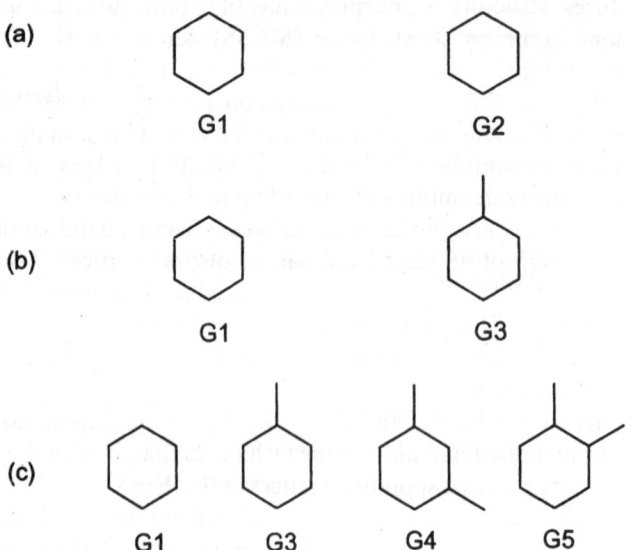

Figure 1 A few graphs to illustrate the related definitions [2]. (a) Graph G_1 is isomorphic to graph G_2. (b) Graph G_1 is a subgraph of graph G_3. (c) Graph G_4 cannot include graph G_5 and vice versa; both graphs G_1 and G_3 are common subgraphs of graphs G_4 and G_5; however, of the two, only graph G_3 is a maximal common subgraph of graphs G_4 and G_5.

complete problems, for which no one has discovered a polynomial time algorithm [3]. The NP-complete problem is a problem that is both NP (solvable in polynomial time by a nondeterministic Turing machine) and NP-hard (any other NP-problem can be translated into this problem).

Although structure and substructure isomorphism problems are difficult to deal with, there are some known definite conditions that can be used to guide the search process. For example, for the structure isomorphism problem, an atom with n attached neighbors in one structure must be matched to an atom with exactly the same number of directly bonded neighbors in the other structure. For the substructure isomorphism problem, an atom with n neighbors in the query structure must be matched to an atom with m ($m \geq n$) direct neighbors in the target structure. Furthermore, the number of atoms and/or bonds of the query structure can be used to decide when the searching can be safely terminated. Applying these conditions greatly enhances the efficiency of the algorithm. However, for the general MCSS problem, none of these conditions can be used. The main difficulty of the general MCSS problem is just here [1].

According to the above analysis, all three types of closely related problems, i.e., the structure isomorphism problem, the substructure isomorphism problem, and the maximal common substructure problem, can be solved by using one well-designed MCSS algorithm as demonstrated by Chen and Robien [1,4,5]. However, because of historical reasons, a variety of different methods have been developed for solving each of the above three problems by different developers for different purposes. For example, the very first algorithm among these methods was developed for finding chemical structures that contain a user-specified substructure query from a molecule database as early as 1957 [6]. Eight years later (1965), Morgan [7] described an algorithm for the generation of a unique machine description for chemical structures. The Morgan algorithm was later extended by Wipke and Dyott [8] to include stereochemistry, leading to the SEMA method (stereochemically extended Morgan algorithm). The SEMA method has been widely used to generating canonical names from 2-D structures and thus makes it possible to perform a very fast duplicate checking of a given chemical structure in a large molecule database without involving time-consuming atom-by-atom searching. The 3-D exact match can be performed by retrieving the 3-D coordinates for each 2-D match found using the SEMA name, and then performing an RMS (root mean square) fit to the query [9]. Later, another approach called hashing code [10] has also been employed to perform fast exact 2-D structure matching, although this method cannot guarantee the correctness of the results. In 1967, the early MCSS algorithm was proposed for analysis of structural changes in chemical reactions [11].

Because of limitations of space, the structure isomorphism problem will not be further discussed. This chapter will focus on the discussion of the major algorithms and methodologies for solving substructure and maximal common substructure problems and their applications.

Historically, substructure search methods were first developed for dealing with 2-D structures, and later these techniques were either extended to become, or incorporated into, 3-D substructure search approaches. The same is true for maximal common substructure searching methods. Two-dimensional methods are not only useful tools themselves for many applications, but also an important foundation for the understanding and development of the 3-D counterparts in many cases. In this chapter, the SSS methods will be introduced first. Then, attention will be focused toward the discussion on MCSS methodologies.

2. SUBSTRUCTURE SEARCHING

As already discussed in the previous section, a chemical structure can be abstracted as a graph in the graph theory, where atoms correspond to vertices and bonds to edges. Here an edge represents only a topological connection between two end vertices. It must be pointed out that such an edge has no meaning of the geometric distance with regard to the two end vertices. This kind of 2-D or constitutional structure can be best represented using the so-called connection table [12], because it can capture topology of chemical structures in tabular forms. This approach directly represents the connectivity of atoms/bonds of a molecule, providing the most valuable representation in chemical structure manipulation. Therefore the connection table has become the predominant form of chemical structure representation in modern chemical computer systems. Given such a representation of chemical structures, the substructure search, i.e., the determination of a query substructure in a target structure, can be performed using the subgraph match algorithm called subgraph-isomorphism algorithm. There are two major types of methods for substructure search: 1) backtracking match algorithms and 2) partitioning–relaxation match algorithms. Other approaches, such as screening techniques and tree-structured search methods, have also been developed for improving the performance of searching large databases.

2.1. Major Substructure Searching Methods

2.1.1. Backtracking-Based Substructure Matching Algorithms

The backtracking algorithm [13] can be viewed as a general control procedure that allows one to explore all possibilities in a search space. In more detail, it is an algorithm for solving problems that may have one or more possible solutions, and that consists of a series of subproblems, each of which may also have multiple possible solutions. Furthermore, selection of a solution for a subproblem may affect the choosing of possible solutions for later subproblems. The algorithm starts with finding a solution to the first subproblem. Then, it tries to find a solution for the next subproblem based on the solution to the first subproblem. This process is recursively carried out until one of the following conditions is met: 1) No solution can be found for the current subproblem. In this case, the algorithm backtracks to the previous subproblem and finds another solution for it, and then continues to solve the next subproblem. 2) The solution to the last subproblem has been found, meaning that a complete solution has been found for the overall problem. In this case, if we want to find a next possible solution for the overall problem, the algorithm backtracks to the previous subproblem and attempts to find another solution to it, and then continues to solve the next subproblem, and so on. The backtracking algorithm terminates when all possible solutions to the first subproblem have been tried.

The Ray–Kirsch Algorithm

The application of the backtracking algorithm to solve the substructure search problem was first described in 1957 by Ray and Kirsch [6]. This approach has since been widely cited, and has been called either the atom-by-atom matching algorithm or backtracking algorithm. The substructure match problem is to find a matching of a query structure onto a target structure. Each subproblem is to find a matching for each query atom onto a target atom. The backtracking match technique is quite similar to the manual method that chemists usually employ to compare two chemical structures. The basic backtracking match procedure can be best described using an example shown in Fig. 2.

First, the backtracking match algorithm selects atom 1 of the query as the starting atom and tries to match it onto atom *a* of the target. But this matching fails because the query atom is a carbon atom, while the target atom is a fluorine atom. Then, the algorithm tries matching query atom 1 onto the next target atom *b*. This time the matching succeeds, as both atoms are carbons. Now the method proceeds, trying to match query atom 1's only direct neighbor (atom 2) onto target atom *b*'s first direct neighbor (atom *a*), but fails as their atom types are not identical. The algorithm then tries to match query atom 2 onto target atom *b*'s second neighbor (atom *c*), but fails as well because the query's bond type connecting atoms 1 and 2 is different from the target's bond type connecting atoms *b* and *c*. Subsequently, the procedure retries to match query atom 1 onto the third target atom *c*, this time with success. It then fails to continue matching query atom 2 onto target atom *b* because of differences in bond types. But the algorithm successfully matches query atom 2 to target atom *d*, and continues to match query atoms 3, 4, and 5 onto target atoms *e, f,* and *g*, respectively [Fig. 2(a)].

During the following steps, the procedure is trying to match query atom 6, connected to query atom 3, onto a target atom that is connected to target atom *e* and that has not been matched to any other query atoms so far. However, this exercise

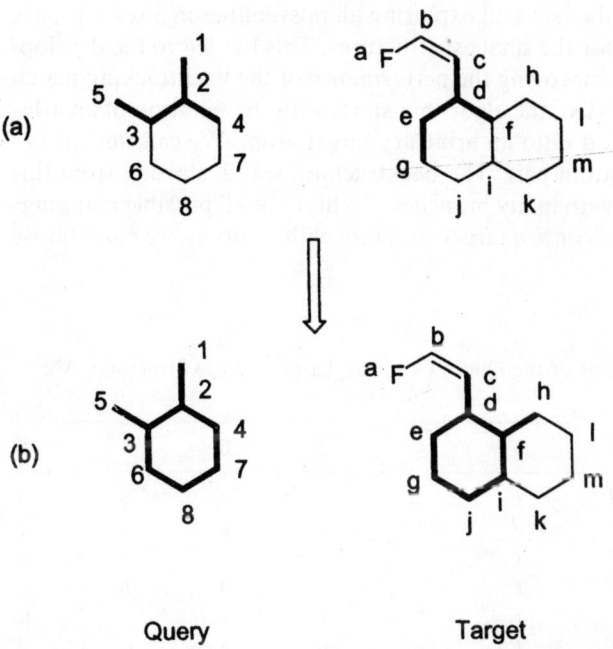

Query   Target

Figure 2 Examples for description of the backtracking match algorithm. (The hydrogen atoms are treated as implicit ones and thus ignored in the procedure.)

fails because no such a target atom exists. Thus the algorithm needs to unmatch query–target atom pair "5-*g*", and backtracks to the last matched query atom 4, and further unmatches query–target atom pairs "4-*f*" and "3-*e*," with backtracking to the last matched query atom 2. Now the algorithm explores another possibility: matches query atom 3 onto target atom *f*, 4 with *e*, 5 onto *h*, 6 with atom *i*, 7 onto *g*, and finally, query atom 8 onto target atom *j*. As such, the algorithm successfully identified a substructure isomorphism as shown in the highlighted parts in Fig. 2(b) and the procedure terminates.

Finding one substructure isomorphism between a given query–target pair is usually enough if the substructure search method is used to identify all compounds in a database that contain the query structure. However, in some other applications, such as prediction of ^{13}C NMR spectra [14], reaction substructure search [15], it is necessary to find out all possible nonoverlapping, or even overlapping mappings of a query onto a target structure. Such a task is not suitable for human vision analysis, even for simple cases such as the one shown in Fig. 2. To identify all mappings of a query–target pair, two additional functionalities must be added into the backtracking match algorithm. First, the algorithm must be able to continue to backtrack (no matter how many substructure mappings have already been identified) until all possible mappings of the query onto the target structure, based on the same query–target starting atom pair, have been tried. Second, the algorithm has to try to match the selected starting atom of the query to all target atoms, one-by-one, no matter how many substructure mappings have already been identified. Such a complete backtracking match algorithm will detect a total of six substructures between the query and target structures shown in Fig. 2. The results are summarized in Table 1.

The basic backtracking match algorithm described above seems not so complicated. The challenge lies in the fact that exploring all possibilities in a search space is too time-consuming except for the smallest structures. This has led to the development of various strategies for improving the performance of the backtracking.match algorithm. As previously discussed, the algorithm starts with the selection of an arbitrary query atom and matches it onto an arbitrary target atom. We call this query–target atom pair the starting atom pair. The backtracking search starting from this atom pair forms a search tree with many branches. To find out all possible mappings of a query consisting of M atoms onto a target structure with N atoms, we may choose

Table 1 A Total of Six Mappings of the Query Onto the Target Whose Structures Are Shown in Fig. 2

Mapping No.	1	2	3	4	5	6	7	8
I	*c*	*d*	*f*	*e*	*h*	*i*	*g*	*j*
II	*h*	*f*	*d*	*i*	*c*	*e*	*j*	*g*
III	*h*	*f*	*i*	*d*	*k*	*j*	*e*	*g*
IV	*k*	*i*	*f*	*j*	*h*	*d*	*g*	*e*
V	*d*	*f*	*i*	*h*	*j*	*k*	*l*	*m*
VI	*j*	*i*	*f*	*k*	*d*	*h*	*m*	*l*

Numbers 1, 2, 3, etc. in the first row are the atom numbers of the query structure. Letters *c*, *d*, *e*, etc. in the remaining rows are the atom labels of the target structure.

any one query starting atom (one is enough) to try matching each of all N target atoms. That is, there are a total of $1 \times N$ query–target starting atom pairs, and as such, the entire search space consists of N search trees. It is thus obvious that the most efficient way to reduce the search space is to cut the fruitless search trees at the very beginning. This can be performed by choosing the most unusual query atom as a starting atom, hoping that this atom cannot be matched to any atoms in the target structure. Such an unusual atom can be a heteroatom or an atom with the largest number of attachments, and so on. For example, in the example shown in Fig. 2, the target structure consists of 13 atoms. If query atom 1 is chosen as the starting atom for the query, the back-tracking algorithm has to explore 12 search trees (query atom 1 cannot be matched to target atom a because of the difference of their atom types, so this search tree can be cut off at the beginning). On the other hand, if query atoms 2 or 3, both of which have the largest number of attachments ($= 3$), are chosen as the respective starting atoms for the query, then the backtracking algorithm needs to explore only 3 out of 13 possible search trees, starting with target atoms d, f, and i, respectively, because only these three target atoms have the same number of attachments as the query starting atom. The symmetry of the target structure is also a way to reduce the number of starting atom pairs [1].

The Ullmann Algorithm

The performance of the original backtracking algorithm can be further improved by pruning-off fruitless branches of a search tree as early as possible. One technique for achieving this was introduced into a backtracking match algorithm by Ullmann in 1976 [16].

Unlike the Ray–Kirsch algorithm that uses basic properties, including atom types, bond orders, etc., for testing the match between query–target atom pairs, the Ullmann's backtracking algorithm [16] incorporates a relaxation procedure [17] to dynamically refine the atom properties of each query–target atom pair by taking into account the neighboring atoms. The refinement results are then used to test whether the matching of each possible query–target atom pair is still valid. If a query–target atom pair is found to no longer match each other, the algorithm backtracks immediately, leading to the cutting-off of this branch of the search tree. Some studies indicate that the application of this relaxation procedure significantly improves the performance of the backtracking match algorithm [18].

A major advantage of the backtracking match algorithm over other approaches (next section) is that this technique, if implemented correctly, can guarantee to reach the correct result for any given query–target pair. Almost all other alternative substructure match algorithms cannot guarantee to obtain correct results in all cases. Therefore some of them resort to a backtracking procedure to resolve problematic cases, while others that do not include a backtracking match procedure as a fallback, may falsely identify some isomorphisms. In the next sections, we discuss some major alternative substructure search methods.

2.1.2. Partitioning-Relaxation-Based Substructure Matching Algorithms

The partitioning-relaxation algorithm for substructure matching, which is also called set reduction algorithm, was first published in 1965 by Sussenguth [19]. The partitioning–relaxation match algorithms are very different from the backtracking match algorithms. Instead of performing atom-by-atom matching between query and target structures from the beginning to the end, the former algorithms first classify atoms of the query and target structures into different sets based on certain criteria, such as node color (atom types), degree (the number of attachments) of each atom, edge color (bond order), order (smallest ring membership). For instance, the eight atoms of the query shown in Fig. 2 will be put into the same set based on the atom type criterion because they are all carbon atoms. However, the 13 target atoms are partitioned into two sets: A first set that contains only atom a, the fluorine atom, and a second set that consists of the remaining 12 carbon atoms. The set membership established for atoms in both query and target structures based on the selected properties (node color, degree, edge color, and order, except connectivity) are summarized in Table 2. Then, a simple the-number-of-set-elements matching is performed to determine whether the substructure matching between the given query–target pair should fail: If the number of elements of a query set is larger than that of the corresponding target set, then the query cannot be matched to the target and the substructure searching immediately terminates, a very efficient termination test.

However, if all query sets can be matched to the corresponding target sets, then the sets are partitioned using logical operation of set intersection to reduce the membership of each set. In other words, this partitioning procedure combines the existing sets to form new sets with fewer elements. The goal is to establish a one-to-one correspondence between query and target atoms. If this goal cannot be fully achieved at this step, the criterion of connectivity is applied to the existing sets to refine the description of each atom, leading to the generation of more selective query–target set pairs by examining the neighbors of each correspondent atom pair of the existing set pairs. The generalization of this method, i.e., the method for enhancement of the description of each atom by iteratively taking into account the properties of the direct neighbor atoms, is called relaxation [17]. At each iteration step, the properties of one more layer of neighbor atoms will be included into the description of the center atom under investigation. After all possible new sets are generated based on the connectivity criterion, another partitioning step is performed again. These partitioning and relaxation steps are repeated until either no substructure isomorphism exists between the

Table 2 Set Membership Established for Atoms in Both Query and Target Structures Based on the Selected Properties (Node Color, Degree, Edge Color, and Order, Except Connectivity), with Reference to Fig. 2.

Partition criteria		Query atom sets	Target atom sets	Set no.
Node color	C	{1,2,3,4,5,6,7,8}	{b,c,d,e,f,g,h,i,j,k,l,m}	1
(atom type)	F	{}	{a}	2
Degree (number of	1	{1,5}	{a}	3
attachments)	2	{4,6,7,8}	{b,c,e,g,h,j,k,l,m}	4
	3	{2,3}	{d,f,i}	5
Edge color	Single	{1,2,3,4,5,6,7,8}	{a,b,c,d,e,f,g,h,i,j,k,l,m}	6
(bond order)	Double	{}	{b,c}	7
Order (smallest ring	6-membered	{2,3,4,6,7,8}	{d,e,f,g,h,i,j,k,l,m}	8
membership)	ring			
	chain	{1,5}	{a,b,c}	9

query–target pair, or until a substructure isomorphism has been found. In both cases, the algorithm immediately terminates. However, the above algorithm contains pitfalls. In some special cases, a third situation occurs when potential correspondences between query–target sets exist, but it is impossible to determine whether there is a substructure isomorphism between query–target pair or not because no further partitioning is possible. For example, if there is more than one mapping between the query–target structure pair (such as the example shown in Fig. 2), this might happen. In this case, a backtracking procedure must be used to resolve this ambiguity.

Other partitioning-relaxation-based algorithms have also been reported, including Figueras's algorithm [20] and Von Scholley's algorithm [21]. One attractive feature of the partitioning-relaxation match algorithms is that they run in polynominal timescale. A common shortcoming of the partitioning-relaxation algorithms is that they cannot guarantee to reach correct solutions for any substructure isomorphism problems. Unlike Sussenguth's algorithm, both Figueras's and Von Scholley's algorithms do not employ a backtracking algorithm as fallback. Therefore both algorithms may falsely identify substructure isomorphisms in some special cases.

2.2. Substructure Matching Techniques for Searching Large Databases

The substructure matching algorithms described in Sec. 2.1 are designed for matching a smaller structure (query) onto a larger structure (target). One of the most important applications of this technique is to search and retrieve compounds from a chemical structure database that all contain a common substructure—the query. However, as previously mentioned, the substructure matching is a very time-consuming task, and as such it is not practical to perform a pure substructure search on a database containing hundreds of thousands of molecules, even with the fastest computers available nowadays. Therefore a great deal of effort has been put into the design of other techniques for speeding up the substructure search of large structure databases, which are the topics of this section.

2.2.1. Screening

Screening is complementary to the substructure matching algorithms described in the previous sections. It is used to quickly screen out those structures that definitely cannot match a given query, and to retrieve the remaining structures from the database as hit candidates that are then passed onto a detailed atom-by-atom matching algorithm. A basic screening system consists of three procedures: 1) selection of keys, 2) generation of inverted key files for all database structures, and 3) retrieval of structure candidates from the database that meet the keys required by a query.

Selection of Keys

The keys, also called screens or fragment codes, are a set of predefined structural features, such as ring sizes, functional groups, and other substructures of various types. They are used to characterize structures in the database. Therefore the selection of proper keys plays a very important role in designing a good screening system. According to Lynch's studies in the 1970s [22], the most useful keys are those structural features that exist in average structures. The substructures that can be found in most of the database structures may provide little discrimination and thus should not be chosen as keys. Similarly, the structural features that exist only in few database structures are also of little use as screening keys.

Generation of Key Files

The screening keys are usually implemented as bit strings. If a screening system predefines M keys, there will be M bit strings. On the other hand, if a database contains N entries (structures, registry numbers, or regnos), each bit string will have the length of N bits. This means that the relationship between keys and regnos of structures in the database can be represented in a bit map. Keys can be arranged in two different ways: sequential keys and inverted keys. In the sequential keys (see Table 3), each row corresponds to one regno. The bit map contains status of each key (0 or 1) for the first regno, status of each key for the second regno, and so on.

During the registration of each structure into a database, all structure features (predefined keys) present in that structure are identified using either a substructure match algorithm as described previously, or a simpler procedure. Then, in the corresponding bit string, all bits corresponding to the keys detected in the structure are set to 1, while the remaining bits that correspond to all other keys that do not exist in the structure are set to 0. The information generated in this way is stored in the sequential key file.

However, the sequential key file is not optimized for searching, because the program must access every regno in the database to determine whether its keys match the query. Therefore after all structures have been registered in the database, the corresponding sequential key file is usually converted into another form called inverted keys (see Table 4). The inverted key file contains regnos that set the first key, regnos that set the second key, and so on. Therefore we say that the structures in the database are indexed by key. Although it is not optimal for registration of structures in the database (the program needs to access the file many times—one time for each key that is set in the structure), the inverted key file is best suitable for searching: The program can immediately find regnos by accessing only the keys that are set in the query.

Performing a Key Search

The purpose of the key search is to quickly screen out structures that cannot match the query. At search time, all keys contained in the query are first identified. Then, each query key is used to retrieve the corresponding key bit string from the inverted

Table 3 The Sequential Key Format. Each Row Corresponds to One Registry Number (regno). The File Contains the Status of Each Key (0 or 1) for Each Regno

		Key				
		1	2	3	4	...
Regno	1	1	1	0	1	
	2	1	1	0	0	
	3	0	1	0	1	
	4	1	0	1	0	
	...					

Table 4 The Inverted Key Format

		Regno			
		1	2	3	4	...
Key	1	1	1	0	1
	2	1	1	1	0
	3	0	0	0	1
	4	1	0	1	0
	...				

key file. Finally, all key bit strings retrieved are combined with the binary AND operations to produce one final bit string that contains the result of the key search: The position of each bit that is set to 1 corresponds to the regno of the hit structure of the key search.

For example, suppose a query containing two keys, 1 and 4, and a database containing four structures with the corresponding inverted key file shown in Table 4. It can be seen from Fig. 3 that in the resulting bit string, only the first bit is set to 1, meaning that among the four database structures, only the first structure contains both keys 1 and 4 as required by the query. Therefore the key searching for this query will return the first database entry as the hit.

In summary, because the inverted key files are pregenerated before searching time, the retrieval of the most likely structure candidates from a large database for a given query using screening techniques is very fast. The most time-consuming atom-by-atom matching needs to be performed for only those structure candidates that are retrieved from the database by key search. Therefore great overall search performance can be achieved by the combination of key screening techniques and substructure matching algorithms.

2.2.2. Tree-Structured Fragment Searching

Although screening techniques based on tree-structured fragments have already been developed in the late 1970s [23] and early 1980s [24], the progress in the development of searching systems in this direction since the middle of the 1980s marks a major advance in substructure searching techniques for large databases.

Hierarchical Tree Substructure Search

The Hierarchical Tree Substructure Search (HTSS) system [25–27] generates a single rooted decision tree from database structures by gradually refining classification of

	1	1	0	1	bit string of key 1
AND	1	0	1	0	bit string of key 4
	1	0	0	0	final bit string

Figure 3 The bit strings retrieved by each query key are combined with the binary AND operator to produce the key search result.

the atoms. All atoms from all structures in a database as a whole are classified first by the number of their attachments, then, each group of atoms is classified by atom type, and finally by bond order. The atoms' positions in chains and rings are also taken into account. Next, the atoms in each subclass are further subdivided using a relaxation procedure, i.e., taking into account the properties of center atom's first-order neighbors, second-order neighbors, and so on, until the furthermost neighbor atoms are reached. Substructure search in this decision tree is a tree-walk from the root through the relevant branches toward the leaves guided by the query. If a leaf can be reached for each query atom during the tree-walk, a substructure isomorphism has been found. If no leaf can be reached for any query atom, then no mapping can be found. It should be stressed that this search process does not rely on atom-by-atom matching and is, as such, very fast. Furthermore, the dependency of the database size on the retrieval time is relatively small: For example, increasing the size of the database from 150,000 to 12,000,000 compounds results in an increase of the retrieval time of only 50% [25]. However, the HTSS system could occasionally lead to false hits. For example, if the query contains ring(s) and the structures retrieved from the search tree contain larger ring systems than those in the query structure, they must be verified [27] using a conventional atom-by-atom match.

The HTSS system has been previously used for substructure searches of the Beilstein database on the Orbit online system, but this service was withdrawn in 1992 and replaced by the Dialog online system [28], which allows one to perform substructure searches of the Beilstein database using another substructure search system called S4.

Softron Substructure Search System

S4 (Softron Substructure Search System) [29–31] was developed by Softron GmbH for searching the Beilstein database. This system is based on Dubois' concept of concentric environments (FREL: Fragment Reduced to an Environment which is Limited) [24]. A FREL is a structural fragment consisting of a central atom and its first and second spheres of neighbor atoms. All FRELs generated from structures of a database are stored in a tree form for searching. S4 extended the FREL concept by taking into account all possible neighbor atoms of a central atom in a structure, sphere-by-sphere, when building an S4's fragment from a structure. In this "full-structure" fragment, all atoms are uniquely numbered with the central atom indexed 1. All atoms spreading from the same root atom are defined as a "bundle" and the center atom is defined as bundle 0. Then, S4 writes all atoms of the fragment as a linear string looking like "b_0, $b_1, \ldots, b_k$', where k is the number of bundles minus one of a particular structure. This bit string is then compressed using the Huffman coding technique, leading to a very compact code. Finally, a 3-byte registry number is added to the string, yielding the final bit string. Therefore such a bit string can be used as an exact hash code for a structure. For a structure with M atoms (not including implicit hydrogen atoms), S4 generates M bit strings: one bit string for each atom. Therefore if a database contains a total of N atoms from all structures, S4 will produce N bit strings. These bit strings are sorted before written to the search file so that the repetitive bytes of successive bit strings as well as identical bit strings generated for symmetric central atoms of the same structure are stored only once.

When doing substructure searching, the best starting atom is first selected in the query in terms of the most selective concentric environment calculated using a heuristic algorithm. Then, bit strings in the search file are decoded and reorganized in a single rooted tree. This is performed as follows: all 0-bundles, i.e., center atoms of the bit strings, matching the starting atom of the query are picked from the search file. Then all 1-bundles matching the corresponding query atoms can be selected and concatenated with the corresponding 0-bundle. This procedure is iterated until one of the two conditions is met: 1) all query atoms have been matched to the corresponding target atoms in the search tree; 2) the 10-bundle is reached (S4 uses only the first 11 bundles during the search). In the first case, if n bit strings have been matched to all query atoms, S4 directly returns n corresponding structures from the database as the hits; that is, no individual atom-by-atom matching is needed for each of the n structures. This leads to a tremendous performance improvement. In the second case, all bit strings whose first 11 bundles have been matched with the first 11 bundles of a large query are to be used to access the corresponding structures from the database. These structures cannot be directly used as the final hits. Rather, each of them must be further verified using an atom-by-atom match algorithm.

An approach very similar to S4 has been adopted in the Bayer AG's in-house ReSy system [32]. The MDL ISIS [33] also incorporated similar ideas.

In summary, when compared with the first generation of substructure searching methods based on the traditional screening technique, tree-structured search approaches such as S4 can be regarded as the second generation of substructure search technology because they have broken new grounds: These systems are able to build such high-discrimination index files for a structure database in the preprocessing stage that at search time the final hitlists can be directly obtained from searching index files for the relatively small queries normally encountered, without the need to perform individual atom-by-atom matching. As a result, superior performance can be achieved [29,30].

2.2.3. Other Techniques for Improving Substructure Searching Performance

Several other techniques have been explored with the goal to improve the performance of the substructure search process. In the first category, some special representations were proposed either to simplify a structure into a reduced graph [34] in which a node consists of a group of atoms, or to build a hyperstructure [35,36] for a set of structures by superimposing them on each other with their common part storing just once. Then, any of the conventional SSS methods previously described can be used to search these special structures. Theoretically, they can be more quickly searched than the corresponding original structures. However, in practice, both methods have not yet been successfully applied so far [32].

In the second category, a variety of parallel processing methods have been explored. For example, Wipke and Rogers [37] described an algorithmic parallel solution for the backtracking method. A well-known example of the application of parallel computing to substructure searching is probably the CAS online system. In this system, the large CAS Registry File containing millions of structures is divided into separate smaller segments, so that it can be searched in parallel on multiple minicomputers using the conventional SSS method. One computer is assigned to search each file segment. This parallel structure search architecture is scalable. As the database grows, more computers can be added, so that the search time can be kept constant for a fixed number of users [38].

2.3. Extensions to Basic 2-D Substructure Searching Methods

As described above, a variety of 2-D substructure matching algorithms have been developed and have become essential tools for the manipulation of 2-D structures in many different areas. These algorithms have been augmented by other techniques, such as key screening and tree-structured organization of database entries, allowing fast substructure searches of large 2-D structure databases. In addition, most of the major 2-D structure database retrieval systems implement certain types of query features (such as generic atom and bond types), allowing users to create more flexible queries. The basic substructure match algorithms can also be extended to handle generic queries and/or generic target structures containing R-groups that are important in combinatorial chemistry [39,40], and even to deal with more general Markush structures that are often encountered in patent databases [41].

All of the substructure match algorithms described so far rely entirely on the topology of 2-D molecular structures. These algorithms can be further extended to compare query–target pairs of structures containing stereochemistry [42]. It is in this area where atom coordinates of 2-D structures play a certain role: The configuration of double bonds and tetrahedral centers can be computed from atom display coordinates and bond symbols [42].

On the other hand, many of the 2-D substructure match methods described above have also been modified to perform 3-D substructure searching (geometric searching), where atom coordinates of 3-D structures play a crucial role. The pioneering work in this important area was first carried out by Gund et al. [43] in the 1970s. Similar to the graph representation of 2-D chemical structures, a geometric representation of the 3-D chemical structure described by its 3-D atom coordinates can also be conveniently abstracted into a graph in which the nodes represent the atoms of a 3-D structure, while the edges of the graph correspond to the geometric distances between atoms (as against the bond types in a 2-D structure). Such a 3-D chemical graph is a complete graph, because it contains not only the distances between chemically bonded atoms, but also all distances between chemically nonbonded atom pairs. With the above graph representation of the 3-D structures established, 3-D substructure matching is analogous to 2-D substructure matching [44] and thus can be carried out using the properly modified versions of the 2-D substructure match algorithms, such as the Sussenguth algorithm [19,45] and Ullman algorithm [16,46]. The performance of the 3-D substructure searching can be significantly improved using 2-D and/or 3-D key screening techniques [47].

3. MAXIMAL COMMON SUBSTRUCTURE SEARCH

As mentioned in Sec. 1, the substructure isomorphism is only a special case of the more general MCSS match problem.

3.1. Early Work in the Development of Maximal Common Substructure Search Algorithms

The interest in the development of MCSS algorithms was originally motivated by the effort to develop methods for automatic recognition of reacting centers of organic reactions suggested by Vleduts [48] in 1963. Early work in this area was carried out by

Armitage et al. [49,50] in 1967. Their MCSS algorithm for the comparison of two given chemical structures was based on an atom-by-atom comparison of derived subsets. The MCSS program was applied to the automatic analysis of structural changes in chemical reactions involving acyclic compounds. Later, they found that their method was "uneconomic" for the analysis of chemical reactions for storage and retrieval and was abandoned [51].

In 1977, Vleduts [52] proposed an algorithm for generating a collection of largest common substructures for the analysis of chemical reactions. His method consisted of an atom-by-atom mapping of one reactant structure onto one product structure. However, the common substructures found by his method were not guaranteed to be maximal. Nevertheless, they would be used to establish reactant-product atom equivalencies, so that the amount of iterative mappings that had to be performed could be reduced. Vleduts' approach for the perception of reaction sites was adopted by Lynch and Willett [53] using a rapid method for detecting certain of the subgraph isomorphisms that are present in the pair of graphs separately representing the set of reactant molecules and the set of product molecules in a reaction. Their algorithm was derived from Sussenguth's set reduction method (see Sec. 2.1.2) [19]. However, their procedure was approximate because nonisomorphic substructures could be identified as being equivalent in a number of cases [54].

3.2. Clique-Detection-Based Maximal Common Substructure Search Algorithms

In 1972, Levi [55] described a new MCSS algorithm based on the analysis of a compatibility table, which establishes the local matching between atoms of two molecules under investigation. Later, Cone et al. [56] described a method for the identification of maximal common substructures based on Levi's compatibility table approach. In their method, chemical structures are represented as binary occurrence vectors. These vectors are used to construct a compatibility table. In the table "K-cover," incompatibility positions and positions not consistent with "neighbor lists" and "degree lists" representing the connectivities of the molecules are cleared, leading to an optimized compatibility table. Each pair of equivalent nodes of the two structures in this table are compared with all other of their equivalent node pairs, to make sure whether the nodes of one molecule are structurally related in the same way as the nodes in another molecule are related. The analysis of the optimized compatibility table thus yields the longest node-string paths, which represent the optimal substructures [56].

The compatibility table is also called a compatibility graph [57,58] or correspondence graph [59]. Barrow and Burstall [60] demonstrated that the MCSSs of two given graphs G_1 and G_2 correspond to the cliques of the correspondence graph of G_1 and G_2. Thus the problem of identification of the MCSSs for a pair of two chemical structures becomes the problem of finding maximum cliques of their corresponding correspondence graphs. To understand the clique detection algorithms better, a few new graph theoretic terms related to the clique concept will be first introduced here.

An undirected graph is said to be complete if each pair of vertices is connected by an edge [61]. The complete graph with n vertices is denoted K_n. Some examples of the complete graphs are shown in Fig. 4.

A complete subgraph of a graph is also called a clique. A clique is maximal if it is not contained in any other clique [62]. A maximum clique is the maximal clique with

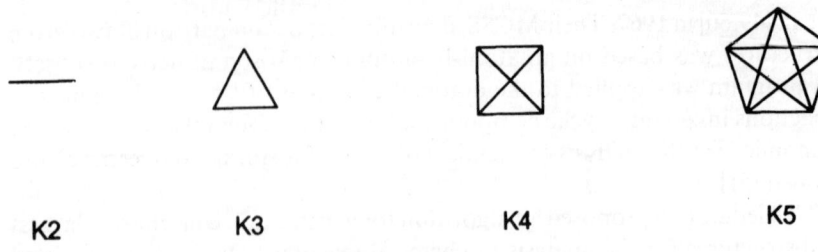

K2   K3   K4   K5

Figure 4   Examples of complete graphs.

the largest cardinality among all cliques. The graph shown in Fig. 5 contains three maximal cliques defined by the subsets {1,2,3}, {3,4,5,6}, and {6,7}. The maximum clique consists of the vertices subset {3,4,5,6}.

Now let us use a simple example to explain the method for finding MCSSs based on the clique-detection algorithm. The procedures for finding the MCSS for two structures 1 and 2 in Fig. 6 involves several stages:

(1)   Converting atom graphs into bond graphs. The graphs 1 and 2 of Fig. 6 are normal chemical structure graphs in which vertices stand for atoms and edges for bonds. For simplicity, we call them atom graphs. These atom graphs must first be converted into the corresponding bond graphs 3 and 4, respectively. A bond graph is composed of a vertex set equal to the bond set of the corresponding atom graph, and an edge set containing those edges that connect two vertices of bonds that are adjacent in the corresponding atom graph. The bond graph is called a line graph or an interchange graph in graph theory.

(2)   Creating the correspondence graph (CG) of two bond graphs, G_1 and G_2. The CG of G_1 and G_2 is built in the following way: 1) if a vertex (v_1) of G_1 is matched (compatible) with a vertex (v_2) of G_2, then the vertex pair (v_1, v_2) is used to create one vertex of the corresponding CG. Thus if G_1 and G_2 have N_1 and N_2 vertices, respectively, the maximal number of vertices of a CG is equal to $N_1 \times N_2$. 2) Two CG vertices $v = (v_1, v_2)$ and $v' = (v_1', v_2')$ are connected if any of the following two conditions is satisfied: (a) v_1 is connected to v_2 in G_1 and v_1' is connected to v_2' in G_2; (b) v_1 is not connected to v_2 in G_1 and v_1' is not connected to v_2' in G_2 either. The CG for the above two bond graphs 3 and 4 is shown as graph 5 in Fig. 6.

(3)   Finding maximum cliques of the correspondence graph. According to the definition of the maximum clique previously discussed, there are six cliques in the correspondence graph 5 of Fig. 6; each of these is composed of two vertices and one edge, e.g., (b_1, b_1')–(b_2, b_2'). In this case, all of them are maximum cliques. Generally, a

2   4   5

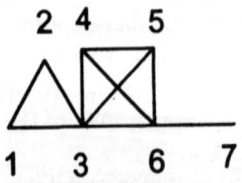

1   3   6   7

Figure 5   Cliques, maximal cliques, and maximum clique.

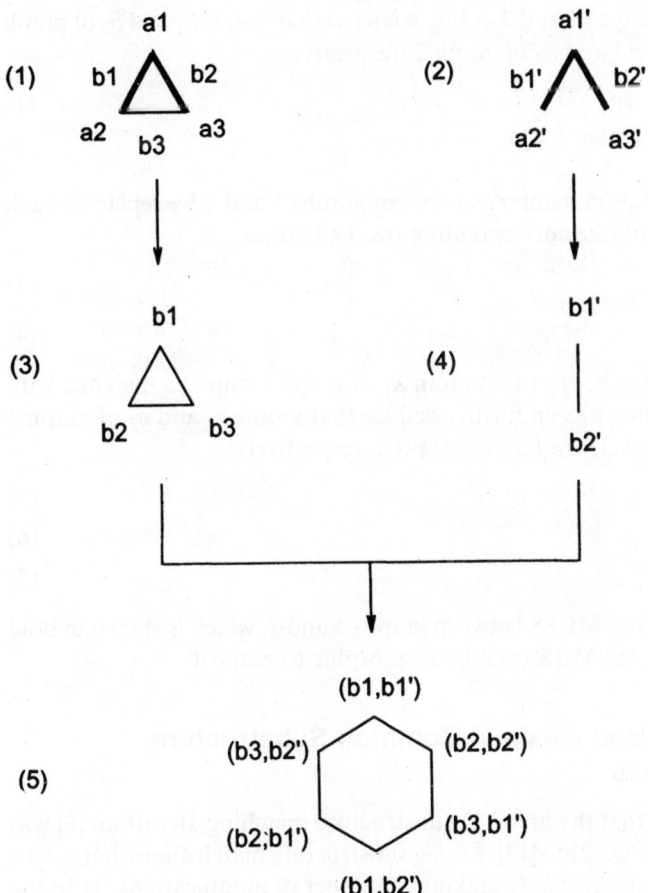

Figure 6   From atom graphs, to bond graphs, to correspondence graph.

CG may be very complicated. In fact, the problem of finding maximum cliques for a graph is NP-complete [63], and many algorithms have been proposed for solving this problem. The best-known clique-detection algorithm in computational chemistry community is probably the Bron–Kerbosch algorithm [64]. This algorithm is based on a backtracking algorithm, using a branch-and-bound technique [65] to cut off branches that cannot lead to a clique. The Bron–Kerbosch algorithm is not only fast, but is able to find all maximal cliques as well. Because of the importance of maximum clique detection algorithms in many areas, new algorithms for solving this problem are still a major area of research. For example, Carraghan and Pardalos [66] have recently developed a very fast algorithm for the maximum clique problem using a partially enumerative procedure. It was found that this new algorithm is several times faster than the Bron–Kerbosch algorithm for dealing with those graphs that have low edge densities [67], although the former algorithm can find only the maximum cliques of a graph.

(4)   Deducing MCSSs from maximum cliques found in step 3. In our example, one maximum clique is induced by $\{(b_1, b_1'), (b_2, b_2')\}$. This maximum clique of the

correspondence graph of graphs 1 and 2 in Fig. 6 tells us that bonds b_1 and b_2 of graph 1 are matched with bonds b_1' and b_2' of graph 2, respectively:

$$b_1 \rightarrow b_2' \tag{1}$$

$$b_2 \rightarrow b_2' \tag{2}$$

Thus we have the possible atom mappings between graphs 1 and 2 by replacing each bond in Eqs. (1) and (2) with the corresponding two end atoms:

$$\{a_1, a_2\} \rightarrow \{a_1', a_2'\} \tag{3}$$

$$\{a_1, a_3\} \rightarrow \{a_1', a_3'\} \tag{4}$$

From Eqs. (3) and (4), it can be seen that atom a_1 of graph 1 must be matched with atom a_1' of graph 2, and thus we can further deduce that atoms a_2 and a_3 of graph 1 must be matched with atoms a_2' and a_3' of graph 2, respectively:

$$a_1 \rightarrow a_2' \tag{5}$$

$$a_2 \rightarrow a_2' \tag{6}$$

$$a_3 \rightarrow a_3' \tag{7}$$

These results correspond to an MCSS between graphs 1 and 2, which is shown in bold lines in Fig. 6. In this case, the MCSS is fully isomorphic to graph 1.

3.3. Backtracking-Based Maximal Common Substructure Search Algorithms

In Sec. 2.1.1, we described that the first 2-D substructure matching algorithm [6] was based on the backtracking algorithm [13]. Such a substructure match algorithm can be extended to handle MCSS problems by making a number of modifications. 1) In the substructure-matching algorithm, a query atom with n neighbors must be matched with a target atom with m neighbors ($m \geq n$). This limitation must be eliminated in the case of MCSS. 2) For the substructure isomorphism problem, the number of atoms and/or bonds of the query can be used as termination condition. That is, if the numbers of atoms and bonds of the common substructure between the query and the target structures are equal to the numbers of atoms and bonds of the query, respectively, then search can be terminated. This termination condition must also be eliminated in the case of MCSS because, in general, we do not know how big the MCSS between the two structures could be. 3) The MCSS searching for two given structures will not terminate until all possible atom pairs have been tried.

3.3.1. The McGregor Algorithm

A nice application of a backtracking algorithm to the MCSS problem was given by McGregor [68] in 1982. In his algorithm, a maximal common subgraph of two graphs was defined to be the common subgraph that contains the largest possible number of arcs (edges), because the algorithm was designed for the identification of reacting centers for a given reaction [54]. To guide the backtracking search in such a way that good solutions will be found as early as possible, the backtrack search is dynamically ordered. However, as McGregor pointed out in his paper [68], "Performing such an operation at every stage in the backtrack search would be extremely expensive in terms of computing time required." Also, the molecules that can be handled by their MCSS program could only contain at most 24 atoms or bonds because the implementation of their algorithm makes extensive use of bit handling procedures and the computer used for their work is based on a word length of 24 bits [54].

3.3.2. The Chen–Robien Algorithm

A very fast MCSS algorithm based on the backtracking technique was described by Chen and Robien [1] in 1992. Their algorithm consists of two parts. The first part is the basic MCSS algorithm for detection of maximal common substructures, and the second part uses the basic algorithm to handle those structures that consist of two or more disconnected fragments. The basic MCSS algorithm consists of the following four steps: 1) quick determination of the initial value for backtracking condition, 2) selection of starting atom pairs, 3) comparison of the two structures, and 4) selection of MCSS candidates. The output contains all possible common substructures that contain the maximal number of bonds. The main procedure of the full Chen–Robien algorithm is as follows:

1. Analysis of structural connectivity. The connectivity of each structure is analyzed to determine whether it consists of two or more disconnected fragments.

2. Calculation of structural environments. Structural environments (EV) for each atom of the two structures are calculated using an improved Morgan algorithm based on the following equations [1]:

$$EV_j(i, 1) = [NB_j(i)]^2 \tag{8}$$

$$EV_j(i, L) = EV_j(i, L - 1) + \sum_{k=1}^{NB_j(i)} EV_j(CT_j(i, k), L - 1) \tag{9a}$$

$$EV_j(i, L) = EV_j(i, L - 1) + \sum_{k=1}^{NB_j(i)} EV_j(CT_j(i, k), L - 1)$$

$$+ ATTP_j(i) + HYBR_j(i) \tag{9b}$$

where L is the Lth iteration cycle; $EV_j(i,1)$ is the initial EV value of the ith atom of the jth structure; $EV_j(i,L)$ is the EV value of the ith atom of the jth structure after L iteration cycles; $NB_j(i)$ is the number of neighbors of the ith atom of the jth structure; $ATTP_j(i)$ is the atom types of the ith atom of the jth structure (e.g., $C=6$, $N=7$, $O=8$, etc.); $HYBR_j(i)$ is the hybridization of the ith atom of the jth structure ($sp^3 = 1$, $sp^2 = 2$, $sp = 3$); $CT_j(i,k)$ is the connectivity table—the ith atom connects to the kth atom within the jth structure. Equations (9a) and (9b) are used for handle topological and colored structure, respectively. The EV values are then used to analyze the symmetry within a structure and the similarities between two given structures.

3. Detection of structure isomorphism. For structure isomorphism problems, the algorithm is used to generate atom–atom correspondences between two chemical structures. In this case, as soon as one MCSS has been detected the search terminates immediately.

4. Perception of substructure isomorphism. After all starting atom pairs that are chosen using method 2 (see *Selection of Starting Atom Pairs* subsection below) have been tried and the number of atoms of the largest MCSS candidate is equal to that of the smaller structure, all MCSSs have been detected and the search terminates.

5. Dealing with general MCSS problem. During this step, all additional starting atom pairs are chosen using method 4 (see *Selection of Starting Atom Pairs* subsection below) to generate all possible MCSS candidates. The results obtained from steps 4 and 5 are combined and used to determine the final results in the next step.

6. Removing duplicate MCSSs. The results obtained from steps 4 and 5 contain all possible MCSSs, and some of them may be duplicate ones. This step then removes all duplicate substructures, leading to the final solution.

7. Handling disconnected structures. If at least one structure consists of two or more disconnected fragments, each fragment is isolated from the original structure representation. Then, all possible combinations of these fragments of the two structures are calculated. According to each combination, each structural fragment from one structure is compared with each fragment of the other structure using the basic MCSS algorithm. All intermediate results are collected and analyzed. One MCSS candidate set that corresponds to one combination and that has the largest total number of bonds is selected as the final solution.

The above algorithm can deal with topological or colored graphs depending on the user's choice.

As previously noted, the MCSS problem for general structures is usually very complicated to solve. For the two given structures containing m and n atoms, respectively, the maximal number of possible atom-by-atom comparisons for the identification of all common substructures containing k atoms is [55]:

$$\frac{m!n!}{(m-k)!(n-k)!k!}$$

It is obvious that trying to explore all possibilities is too time-consuming except for small structures only. The high efficiency of the Chen–Robien MCSS algorithm [1] is achieved by using several new strategies to reduce the search space, which will be described below.

Selection of Starting Atom Pairs

The method for selection of starting atom pairs is extremely important in designing a fast algorithm. A good method should not only choose as few starting pairs as possible, but should also guarantee that all necessary ones are selected. In the Chen–Robien algorithm, several methods for choosing starting atom pairs were designed to deal with a variety of structure pairs under investigation. The most important ones are listed below:

1. For structure isomorphism problems, the atom–atom correspondence of the two isomorphic structures is calculated from a pair of starting atoms. One starting atom is chosen from the first structure in such a way that it shares the fewest identical EV values with other atoms of the same structure. The other starting atom is chosen from the second structure so that

its EV value is equal to that of the already-chosen starting atom of the first structure. These two atoms comprise the unique starting pair for isomorphism problem.

2. For substructure isomorphism problems, the atom having the largest number of neighbors within the query structure is chosen to comprise starting pairs with all those atoms from the other structure, each of which has at least the same number of neighbors as that of the query starting atom.

3. In some applications, we may be more interested in comparing two very similar compounds, for example those that contain the same ring system. The second method is designed to handle such structure pairs. For complicated structures that contain the same number (≥ 3) of rings, the starting atom pairs are those having the largest identical local structural environments [1,2].

4. For general MCSS problems, all atoms having three or more neighbor atoms are selected from the two structures to form starting atom pairs. The symmetry information available is used to reduce the number of possible starting atom pairs.

For colored graph matching problems, all other atom properties, such as atom types and hybridizations, are also taken into account during the selection of starting atom pairs.

Terminating Conditions

Terminating conditions are used to determine whether all possible MCSSs have been found. A good terminating condition allows the search to terminate as soon as all MCSSs have been detected. In contrast, an inappropriate terminating condition may abort the search too early, resulting in loss of some or even all MCSSs. In the Chen–Robien algorithm, several terminating conditions are designed to handle different types of matching problems:

1. For the structure isomorphism problem, the total number of bonds of one structure is used as the terminating condition. The MCSS candidate that contains the same number of bonds as the query structure does represent the final solution.

2. For the substructure isomorphism problem, the number of atoms of the smaller query structure is used as terminating condition.

3. For two very similar structures whose starting atom pair is selected by method 3 described in the *Selection of Starting Atom Pairs* subsection, if the MCSS candidate, which is obtained using this starting atom pair, contains the same parent ring system as the two structures and which represents at least 80% of the bonds of the larger structure under investigation, it is reasonable to believe that these two structures are indeed very similar and thus the MCSS candidate can be used as the final solution.

4. For general MCSS problems, to find all possible MCSSs of the two given structures, the search will not terminate until all possible starting pairs chosen using the methods described in the previous section have been tried.

Reordering of Branches of Search Trees

Within a search tree, the computational time for detecting the largest MCSS candidates depends on the order of its branches. Appropriate rearrangement of the branches may result in earlier detection of larger MCSS candidates. To clearly explain this strategy, first let us briefly discuss the representation of the search process. In the Chen–Robien MCSS algorithm, a Decreased Matrix (DM) is used to describe the search process and a Corresponding Matrix (CM) is used to store the correspondences of atoms within the two structures. Fig. 7 gives an example (atom 2 in structure 1 and atom 7 in structure 2 are chosen as starting atom pair). In the DM and CM, the vertical axis corresponds to the smaller structure, and the horizontal axis to the larger one. As shown in Fig. 7, atoms of each structure are ranked according to their distances from the starting atom.

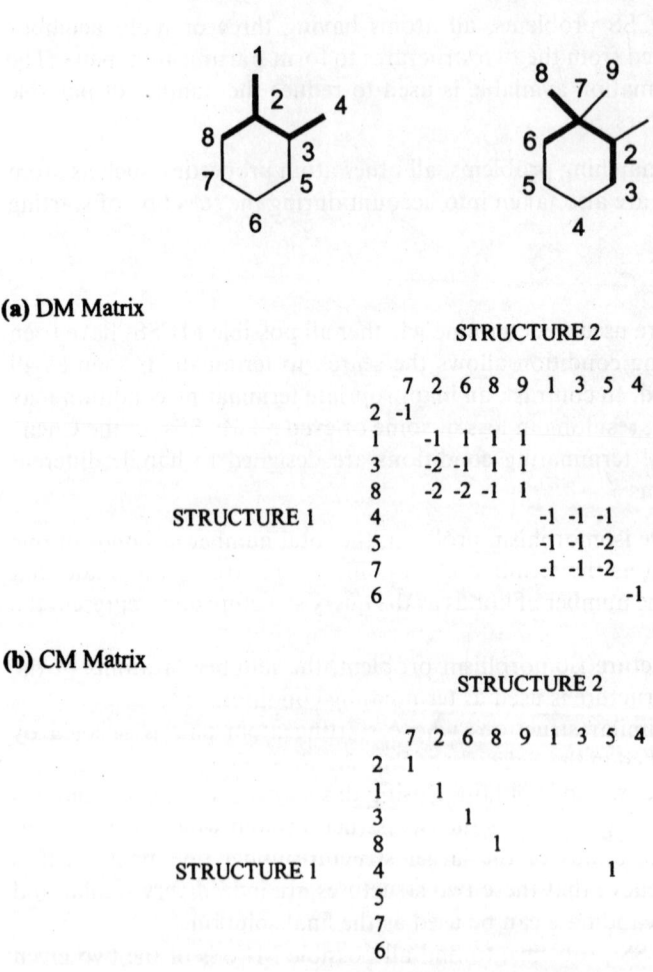

Figure 7 Decreased Matrix and Corresponding Matrix of structures 1 and 2 with original order (starting atom pair: 2–7) [1]. The information shown belongs to one partial mapping between structures 1 and 2, before any backtracking operation is performed and with the common substructures found drawn in bold lines. [From the American Chemical Society (the bold highlighting added and all # signs removed.)]

One basic matching condition of atoms between two structures is that both atoms should have the same distance from their starting atoms, respectively. The original values of DM elements are set according to this condition. If atom i of structure 1 and atom j of structure 2 have the same distance value, it is possible for them to match each other; therefore $DM(i,j)$ is set to 1; otherwise $DM(i,j)$ is set to 0. For example, the selected starting atoms 2 and 7 have the same distance value 0, thus $DM(2,7)$ is initially set to 1. The DM matrix clearly shows that application of this "equal-distance" condition dramatically reduces the search space. There are three basic rules to change the elements of DM and CM matrices during the search process:

1. If atom i of structure 1 has been successfully matched with atom j of structure 2, then the $DM(i,j)$ element is changed from 1 into -1 to indicate that this possibility has been tried. To avoid matching atom j with other previously possible atoms k of structure 1 in the next steps, all elements $DM(k,j)$ are changed from 1 into -2. The matching relationship between atoms i and j is stored in the CM-matrix by setting the $CM(i,j)$ element to 1.
2. If atom i fails to match atom j, $DM(i,j)$ will also be changed from 1 into -1, but $CM(i,j)$ and $DM(k,j)$ remain unchanged.
3. After the search through a branch has been completed, the corresponding lines of the DM and CM matrices will be reinitialized, a necessary preparation for searching a new branch via backtracking operations.

As shown in Fig. 7, atom 1 in structure 1 is first tried to be matched onto atom 2 in structure 2. Following this search branch, a substructure with five atoms and four bonds is found before any backtracking is performed (bold lines in structures 1 and 2 of Fig. 7). The size of the common structure found so far is used as a backtrack condition. With this backtracking condition, no branches were cut off before the next; more extended substructure containing six atoms and five bonds is found. Then, this new "best" number of bonds is used as a new backtrack condition for further search. It can be easily seen from structures 1 and 2 in Fig. 7 that the correct MCSS will not be found until atom 1 in structure 1 is matched against atom 8 or 9 in structure 2. This example reveals that most CPU time is wasted on investigating many useless searches on bad branches within the search tree.

Now let us consider the rearrangement of the order of branches within the same search tree. The method is based on the idea that the mapping of two atoms having more neighbor atoms usually results in an earlier identification of larger substructures. The operations of reordering the branches are performed within each atom group, in which all atoms have the same distance values to the starting atom. The atom having the largest number of directly bonded neighbors is ranked first. This can be clearly illustrated using Fig. 8. The order 1, 3, 8 and 4, 5, 7 of the atoms in structure 1 has been changed into 3, 8, 1 and 5, 7, 4, respectively. Similarly, the order 1, 3, 5 of the atoms in structure 2 has been changed into the new sequence 3, 5, 1. It is interesting to note that this simple procedure of reordering allows immediate detection of the correct MCSS with eight atoms and eight bonds (shown in bold lines in structures 1 and 2), using the same starting atom pair, and even before any backtrack operation is performed.

Thus the backtrack condition immediately obtains its maximal value of 8. Using this value, as long as one atom in structure 1 fails to match any atom in structure 2, the corresponding branch will be immediately cut off.

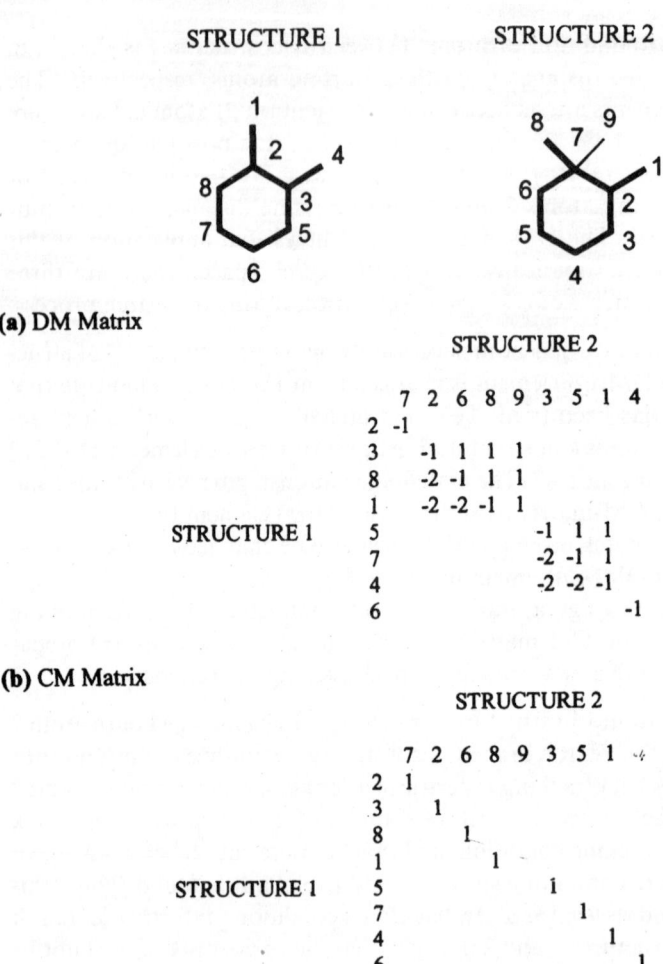

(a) DM Matrix

STRUCTURE 2

	7	2	6	8	9	3	5	1	4
2	-1								
3		-1	1	1	1				
8		-2	-1	1	1				
1		-2	-2	-1	1				
5						-1	1	1	
7						-2	-1	1	
4						-2	-2	-1	
6									-1

STRUCTURE 1

(b) CM Matrix

STRUCTURE 2

	7	2	6	8	9	3	5	1	4
2	1								
3		1							
8			1						
1				1					
5						1			
7							1		
4								1	
6									1

STRUCTURE 1

Figure 8   Decreased Matrix and Corresponding Matrix of structures 1 and 2 with new sequences of atoms (starting pair: 2–7) [1]. The information shown here corresponds to one partial mapping between structures 1 and 2, before any backtracking operation is performed. [From the American Chemical Society (the bold highlighting added and all # signs removed.)]

It should be pointed out that unlike the McGregor algorithm [68], where reordering of search tree branches is performed dynamically and thus very time-consuming (see Sec. 3.3.1), the Chen–Robien algorithm rearranges a search tree only once (before searching begins), and is therefore much faster.

The method described in this section can be used to guide the backtracking search within one search tree in such a way that larger common substructures may be found as early as possible during the search process.

Assigning A Larger Initial Value to the Backtrack Condition

As previously discussed, for general MCSS problems, it is impossible to predict how many atoms and bonds MCSSs will have. Therefore an initial value of 0 is assigned to the backtrack condition. In certain cases, some useless trees may be searched first, and in such cases the backtracking condition will not obtain a larger value until a search tree containing a larger common substructure has been processed. Thus much time is spent on searching many unnecessary trees.

Chen and Robien [1] developed a method for setting the backtracking condition to a larger initial value instead of zero. With this method, it becomes possible to avoid useless searches at a very early stage. The principle of their approach is based on the following findings. In the search forest, every first branch of each tree may contain different common substructures of different sizes. Using the technique for reordering branches described above, larger common substructures, even final MCSSs, may appear in the first branch. In the backtracking search method, most computational time is spent in the backtracking process. However, it usually costs only little time to explore the first branch of each search tree before any backtrack operations occur. Therefore searching each first branch of each tree, in turn, may find a large common substructure. The number of bonds contained in the largest common substructure found in this way can then be used as the initial value for the backtracking condition, leading to very efficiently pruning of the search trees to be explored thereafter.

Atom and Bond Property Conditions

Various atom and/or bond properties can also be used to cut off some bad branches, even to completely cut down some fruitless trees. The Chen–Robien algorithm allows the user to decide whether these additional conditions should be taken into account during the matching process. However, as Chen and Robien [1] have demonstrated, the actual decision on using these conditions depends on the application.

The combination of the new methods for optimizing and pruning search trees described above, and the backtracking technique, makes the Chen–Robien algorithm a very fast MCSS algorithm. It has become a central mainstay of several other commands within the CSEARCH-NMR database system [69].

3.4. Applications of the Maximal Common Substructure Search Method

Finding common structural features among a set of chemical structures is a frequent problem in chemical research. As described in Sec. 3.1, the earliest successful application of the MCSS algorithm in chemistry is to recognize reacting centers of organic chemical reactions, which is of great importance in designing systems for reaction searching, synthesis design as well as reaction prediction. This arena was mainly pioneered by Lynch et al. [49,53,54] and followed by others [58,70,71].

There has been little interest in the MCSS method for the interpretation of spectra of organic compounds until recently. An early work on the application of MCSS approach for unknown mass spectra was reported by Cone et al. [56] in 1977. Chen and Robien developed a novel approach for the automatic deduction of common structural features from a set of structures obtained using ^{13}C-NMR spectral similarity search [72] in 1994. They demonstrated that the detected MCSSs often show the main structural features of the unknown compounds under investigation. This method has recently been adopted by Varmuza et al. [73] for the automatic extraction of common structural features from the hitlist structures obtained using infrared (IR) spectral similarity searches.

In the drug discovery area, the MCSS algorithm has been used for the perception of common structural features for quantitative structure–activity studies [74–76], and for studies on effective components in Chinese medicines as suggested by Wang and

Zhou [77]. Ghose and Crippen [78] described the use of stereochemical information in ligand binding studies with a clique detection algorithm of Kuhl et al. [79]. The Chen–Robien MCSS algorithm [1] has recently been implemented in the MDL Sculpt program [80], a 3-D structure visualization and modeling system for the automatic alignment of different ligands. This alignment method can be used in conjunction with other techniques for studying interactions between ligand and receptor [81], and even for identification of initial specification of pharmacophoric pattern [82]. Other MCSS methods, such as genetic algorithm-based approaches [83], have also been used to derive pharmacophores.

Other applications of the MCSS method include the work of Brown et al. [84] for the construction of hyperstructures from a set of 2-D structures for improving the speed of atom-by-atom searches, the MCSS-based 2-D [85,86] and 3-D [87] similarity searching, as well as identification of tertiary structure resemblance in proteins.[88].

4. CONCLUSION

Historically, 2-D representations of molecular structures have been the foundation for the development of algorithms, methodologies, and the corresponding software in the field of chemoinformatics. Two-dimensional substructure search methods are among the most important and most widely used techniques for dealing with 2-D chemical structures. The basic 2-D substructure matching algorithms and search methods have now been well established. They have been playing a central role in almost all important, 2-D structure related chemical information and expert systems, such as computer-assisted structure elucidation [89], computer-assisted synthesis design [90], chemical structure database management and retrieval systems, and chemical reaction database management and retrieval systems. The 2-D substructure search methods implemented in many important systems are capable of dealing with stereochemistry. However, further improvement in this area is always possible. In recent years, combinatorial chemistry methods have become commonplace in many laboratories. This new technique augmented by the automated synthesis systems allows chemists to construct large libraries of chemical structures in only a few days. The requirement for managing and handling ever-growing combinatorial libraries has led to the development and improvement of methods for representation and searching of these new types of structure libraries [39]. In particular, 2-D substructure searching methods need to be able to deal with not only conventional discrete query vs. discrete structure target matching problems, but also with discrete query vs. Markush structure target, or Markush query vs. discrete structure target, and even Markush query vs. Markush structure target matching problems [39]. On the other hand, with the shift of methods from product-based combinatorial synthesis to reaction-based combinatorial synthesis over the past few years, reaction databases, which are collections of chemical reactions consist of 2-D component structures, have become a more valuable source for the design of new combinatorial libraries. This has sparked the innovation of new reaction substructure searching and indexing methodologies [15].

The 2-D substructure match algorithms were the foundation for the development of early 3-D substructure searching methods. These rigid geometric searching approaches have now widely used for searching large 3-D structure databases, especially in the pharmaceutical and agrochemical industries as well as academic community. More recent 3-D substructure search approaches have been developed in the attempt to tackle the problem of the 3-D structure flexibility, which is the nature of most, if not all, 3-D structures. These methods allow one to search 3-D databases containing single computed conformations of flexible molecules more effectively (i.e., finding more hits) than the rigid geometric methods do. However, the flexible searching is usually slower than the rigid searching. More detailed discussion about 3-D searching methods is given in Chapter 18 on 3-D-pharmacophore searching.

The substructure search problem is only a special case of the more general and more difficult MCSS problem. It is extremely computationally intensive to perform MCSS searches in large structure database. This is because the identification of MCSSs for two given structures is more time consuming than determination of isomorphism between them. Well-established methods for rapidly eliminating candidate structures for substructure searching (such as key screening) are not applicable to the MCSS problem. This is because, in general, the results of an MCSS search are unknown before the searching is completed. More work in this area needs to be performed [91].

The MCSS method has been much less popular than the SSS approach. This is not because of the lack of interest in it, but rather because of the extremely computational demands that have restricted its application. This situation has greatly been changed recently owing to the significant progress in the development of efficient MCSS algorithms as well as the increase of the computer power. Besides its traditional application to the automatic identification of reacting centers of chemical reactions, the MCSS method has successfully been used for the interpretation of spectra of organic compounds, for perception of common structural features for structure–activity studies, for identification of initial specification of pharmacophoric pattern, and for structure similarity searching, and so on. With the increase of the performance of the MCSS searching method, it is to be expected that its application areas can only grow further.

REFERENCES

1. Chen L, Robien W. MCSS: a new algorithm for perception of maximal common substructures and its application to NMR spectral studies. 1. The algorithm. J Chem Inf Comput Sci 1992; 32:501–506.
2. Chen L. Ph.D. Thesis. Vienna, Austria: University of Vienna, 1993:9–10.
3. Tarjan RE. Graph algorithms in chemical computation. Algorithms for Chemical Computations, Washington, DC: American Chemical Society, 1977:1–20.
4. Chen L, Robien W. MCSS: a new algorithm for perception of maximal common substructures and its application to NMR spectral studies. 2. Applications. J Chem Inf Comput Sci 1992; 32:507–510.
5. Chen L, Robien W. Optimized prediction of ^{13}C NMR spectra using increments. Anal Chim Acta 1993; 272:301–308.
6. Ray LC, Kirsch RA. Finding chemical records by digital computers. Science 1957; 126:814–819.
7. Morgan HL. The generation of a unique machine description for chemical structures—A technique developed at Chemical Abstracts Service. J Chem Doc 1965; 5:107–113.
8. Wipke WT, Dyott TM. Stereochemically unique naming algorithm. J Am Chem Soc 1974; 96:4825–4834.
9. Christie BD, Henry DR, Wipke WT, Moock TE. Database structure and searching in MACCS-3D. Tetrahedron Comput Methodol 1990; 3:653–664.

10. Ihlenfeldt W-D, Gasteiger J. Hash codes for the identification and classification of molecular structure elements. J Comput Chem 1994; 15:793–813.
11. Armitage JE, Crowe JE, Evans PN, Lynch MF. Documentation of chemical reactions by computer analysis of structural changes. J Chem Doc 1967; 7:209–215.
12. Barnard JM. Structure representation. In: Schleyer PvR, Allinger NL, Clark T, Gasteiger J, Kollman PA, Schaefer HF, Schreiner PR, eds. The Encyclopedia of Computational Chemistry. Chichester: John Wiley & Sons, 1998:2818–2826.
13. Golomb SW, Baumert LD. Backtrack programming. J Assoc Comput Mach 1965; 12: 516–524.
14. Chen L, Robien W. OPSI: a universal method for prediction of ^{13}C-NMR spectra based on optimized additivity Models. Anal Chem 1993; 65:2282–2287.
15a. Chen L, Nourse JG, Christie BD, Leland BA, Grier DL. From REACCS to reaction data cartridge: evolution of reaction substructure search methods. The 6th International Conference on Chemical Structures. The Netherlands: Noordwijkerhout, June 2–6, 2002.
15b. Chen L, Nourse JG, Christie BD, Leland BA, Grier DL. Over 20 years of chemical structure access systems from MDL: evolution of reaction substructure search methods. J Chem Inf Comput Sci 2002; 42:1296–1310.
16. Ullmann JR. An algorithm for subgraph isomorphism. J Assoc Comput Mach 1976; 23: 31–42.
17. Relaxation is a numeric technique that can be used to find a solution for a problem by iteratively refining the initial "guess" of the solution. The first application of this technique in computational chemistry is probably the well-known Morgan algorithm [7].
18. Downs GM, Lynch MF, Willett P, Manson GA, Wilson GA. Transputer implementations of chemical substructure searching algorithms. Tetrahedron Comput Methodol 1988; 1:207–217.
19. Sussenguth EH Jr. A graph-theoretic algorithm for matching chemical structures. J Chem Doc 1965; 5:36–43.
20. Figueras J. Substructure search by set reduction. J Chem Doc 1972; 12:237–244.
21. Scholley AV. A relaxation algorithm for generic chemical structure screening. J Chem Inf Comput Sci 1984; 24:235–241.
22. Lynch MF. Screening large chemical files. In: Ash JE, Hyde E, eds. Chemical Information Systems. Chichester: Ellis Horwood, 1974:177–194.
23. Feldmann RJ, Milne GWA, Heller SR, Fein A, Miller JA, Koch B. An interactive substructure search system. J Chem Inf Comput Sci 1977; 17:157–163.
24. Attias R. DARC substructure search system: a new approach to chemical information. J Chem Inf Comput Sci 1983; 23:102–108.
25. Bruck P, Nagy MZ, Kozics S. Substructure search on hierarchical trees. Online Information 87; Proceedings of the 11th International Online Information Meeting, London, Dec 8–10, 1987. Oxford: Learned Information, 1987:41–43.
26. Nagy MZ, Kozies S, Veszpremi T, Bruck P. Substructure search on very large files using tree-structured databases. In: Warr WA, ed. Chemical Structures: The International Language of Chemistry; Proceedings of an International Conference at the Leeuwenhorst Congress Center. Noordwijkerhout, The Netherlands, May 31–June 4, 1987. Heidelberg: Springer, 1988:127–130.
27. Nagy ZM. How can parallel algorithms help to find new sequential algorithms? J Chem Inf Comput Sci 1993; 33:542–544.
28. Hartwell IO, Haglund KA. An overview of Dialog. In: Heller SR, ed. The Beilstein Online Database: implementation, content and retrieval. ACS Symposium Series 436. Washington, DC: American Chemical Society, 1990:42–63.
29. Hicks MG, Jochum C. Substructure search systems. 1. Performance comparison of the MACCS, DARC, HTSS, CAS Registry, MVSSS, and S4 substructure search systems. J Chem Inf Comput Sci 1990; 30:191–199.

30. Hicks MG, Jochum C, Maier H. Substructure search systems for large chemical data bases. Anal Chim Acta 1990; 235:87–92.

31. Bartmann A, Maier H, Roth B, Walkowiak D. Substructure search on very large files by using multiple storage techniques. J Chem Inf Comput Sci 1990; 33:539–541.

32. Barnard JM. Substructure searching methods: old and new. J Chem Inf Comput Sci 1993; 33:532–538.

33. Christie BD, Leland BA, Nourse JG. Structure searching in chemical databases by direct lookup methods. J Chem Inf Comput Sci 1993; 33:545–547.

34. Cringean JK, Lynch MF. Subgraphs of reduced chemical graphs as screens for substructure searching of specific chemical structures. J Inf Sci 1989; 15:211–222.

35. Vladutz G, Gould SR. Joint compound/reaction storage and retrieval and possibilities of a hyperstructure-based solution. In: Warr WA, ed. Chemical Structures: The International Language of Chemistry; Proceedings of an International Conference at the Leeuwenhorst Congress Center, Noordwijkerhout, The Netherlands, May 31–June 4, 1987. Heidelberg: Springer, 1988:371–383.

36. Brown RD, Downs GM, Willett P, Cook APF. A hyperstructure model for chemical structure handling: generation and atom-by-atom searching of hyperstructures. J Chem Inf Comput Sci 1992; 32:522–531.

37. Wipke WT, Rogers D. Rapid subgraph search using parallelism. J Chem Inf Comput Sci 1984; 24:255–262.

38. Farmer N, Amoss J, Farel W, Fehribach J, Zeidner CR. The evolution of the CAS parallel structure searching architecture. In: Warr WA, ed. Chemical Structures: The International Language of Chemistry; Proceedings of an International Conference at the Leeuwenhorst Congress Center, Noordwijkerhout, The Netherlands, May 31–June 4, 1987. Heidelberg: Springer, 1988:283–296.

39. Leland BA, Christie BD, Nourse JG, Grier DL, Carhart RE, Maffett T, Welford SM, Smith DH. Managing the combinatorial explosion. J Chem Inf Comput Sci 1997; 37: 62–70.

40. Warr WA. Combinatorial chemistry. In: Schleyer PvR, Allinger NL, Clark T, Gasteiger J, Kollman PA, Schaefer HF, Schreiner PR, eds. The Encyclopedia of Computational Chemistry. Chichester: John Wiley & Sons, 1998:407–417.

41. Berks AH, Barnard JM, O'Hara MP. Markush structure searching in patents. In: Schleyer PvR, Allinger NL, Clark T, Gasteiger J, Kollman PA, Schaefer HF, Schreiner PR, eds. The Encyclopedia of Computational Chemistry. Chichester: John Wiley & Sons, 1998: 1552–1559.

42. Rohde B. Stereochemistry: representation and manipulation. In: Schleyer PvR, Allinger NL, Clark T, Gasteiger J, Kollman PA, Schaefer HF, Schreiner PR, eds. The Encyclopedia of Computational Chemistry. Chichester: John Wiley & Sons, 1998.2726–2737.

43. Gund P, Wipke WT, Langridge R. Computer searching of a molecular structure file for pharmacophic patterns. Proceedings of International Conference On Computers in Chemical Research and Education, Ljubljana, July, 1974:5–21.

44. Gund P. Three-dimensional pharmacophoric pattern searching. Progress in Molecular and Subcellular Biology. Vol. 5. Berlin: Springer-Verlag, 1977:117–143.

45. Jakes SE, Watts N, Willett P, Barden D, Fisher JD. Pharmacophoric pattern matching in files of 3D chemical structures: evaluation of search performance. J Mol Graph 1987; 5:41–48.

46. Brint AT, Willett P. Pharmacophoric pattern matching in files of three-dimensional chemical structures: comparison of geometric searching algorithm. J Mol Graph 1987; 5:49–56.

47. Jakes SE, Willett PJ. Pharmacophoric pattern matching in files of three-dimensional chemical structures. Selection of interatomic distance screens. J Mol Graph 1986; 4:12–20.

48. Vleduts GE. Concerning one system of classification and codification of organic reactions. Inf Storage Retr 1963; 1:117–146.

49. Armitage JE, Crowe JE, Evans PN, Lynch MF. Documentation of chemical reactions by computer analysis of structural changes. J Chem Doc 1967; 7:209–215.

50. Armitage JE, Lynch MF. Automatic detection of structural similarities among chemical compounds. J Chem Soc, C 1967; 7:521–528.

51. Harrison JM, Lynch MF. Computer analysis of chemical reactions for storage and retrieval. J Chem Soc Section C 1970; 10:2082–2087.

52. Vleduts GE. Development of a combined WLN/CTR multilevel approach to algorithmic analysis of chemical reactions in view of their automatic indexing (Report No. 5399). London: British Library Research and Development, 1977.

53. Lynch MF, Willett P. The automatic detection of chemical reaction sites. J Chem Inf Comput Sci 1978; 18:154–159.

54. McGregor JJ, Willett P. Use of a maximal common subgraph algorithm in the automatic identification of the ostensible bond changes occurring in chemical reactions. J Chem Inf Comput Sci 1981; 21:137–140.

55. Levi G. A note on the derivation of maximal common subgraphs of two directed or undirected graphs. Calcolo 1972; 9:341–352.

56. Cone MM, Venkataraghavan R, McLafferty FW. Molecular structure comparison program for the identification of maximal common substructures. J Am Chem Soc 1977; 99:7668–7671.

57. Nicholson V, Tsai C-C, Johnson M, Naim M. A subgraph isomorphism theorem for molecular graphs. Stud Phys Theor Chem 1987; 51:226–230.

58. Tonnelier C, Jauffret P, Hanser T, Kaufmann G. Machine learning of generic reactions: 3. An efficient algorithm for maximal common substructure determination. Tetrahedron Comput Methodol 1990; 3:351–358.

59. Brint AT, Willett P. Algorithms for the identification of three-dimensional maximal common substructures. J Chem Inf Comput Sci 1987; 27:152–158.

60. Barrow HG, Burstall RM. Subgraphs isomorphism, matching relational structures and maximal cliques. Inf Process Lett 1976; 4:83–84.

61. National Institute of Standards and Technology, Dictionary of Algorithms and Data Structures, http://www.nist.gov/dads/HTML/completeGraph.html (Accessed on February 28, 2002).

62. Cavique L, Rego C, Themido I. A scatter search algorithm for the maximum clique problem, *http://hces.bus.olemiss.edu/reports/hces0101.pdf* (Accessed on Feb. 28, 2002).

63. Corno F, Prinetto P, Reorda MS. Using symbolic techniques to find the maximum clique in very large sparse graphs. ED&TC'95: IEEE European Design and Test Conference, Paris, March 1995. *http://www.cad.polito.it/pap/db/edtc95c.pdf*. (Accessed on Feb. 28, 2002).

64. Bron C, Kerbosch J. Algorithm 457. Finding all cliques of an undirected graph. Commun Assoc Comput Mach 1973; 16:575–577.

65. Little JDC, Murty KG, Sweeney DW, Karel C. An algorithm for the traveling salesman problem. Oper Res 1963; 11:972–989.

66. Carraghan R, Pardalos PM. Exact algorithm for the maximum clique problem. Oper Res Lett 1990; 9:375–382.

67. Gardiner EJ, Artymiuk PJ, Willett P. Clique-detection algorithms for matching three-dimensional molecular structures. J Mol Graph Model 1997; 15:245–253.

68. McGregor JJ. Backtrack search algorithms and the maximal common subgraph problem. Softw Prac Exp 1982; 12:23–34.

69. Kalchhauser H, Robien W. CSEARCH: a computer program for identification of organic compounds and fully automated assignment of carbon-13 nuclear magnetic resonance spectra. J Chem Inf Comput Sci 1985; 25:103–108.

70. Moock TE, Nourse JG, Grier D, Hounshell WD. The implementation of AAM and

related reaction features in the reaction access system (REACCS). In: Warr WA, ed. Chemical Structures. Berlin: Springer-Verlag, 1988:303–313.

71. Wipke WT, Rogers D. Tree-structured maximal common subgraph searching. An example of parallel computation with a single sequential processor. Tetrahedron Comput Methodol 1989; 2:177–202.

72. Chen L, Robien W. Application of the maximal common substructure algorithm to automatic interpretation of ^{13}C-NMR spectra. J Chem Inf Comput Sci 1994; 34:934–941.

73. Varmuza K, Penchev PN, Scsibrany H. Maximum common substructures of organic compounds exhibiting similar infrared spectra. J Chem Inf Comput Sci 1998; 38:420–427.

74. Crandell CW, Smith DH. Computer-assisted examination of compounds for common three-dimensional substructures. J Chem Inf Comput Sci 1983; 23:186–197.

75. Golender V, Rosenblit A, eds. Logical and combinatorial algorithms for drug design. Letchworth: Research Studies Press, 1983.

76. Yuan S, Zheng C, Zhao X, Zeng F. Identification of maximal common substructures in structure/activity studies. Anal Chim Acta 1990; 235:239–241.

77. Wang T, Zhou J. EMCSS: a new method for maximal common substructure search. J Chem Inf Comput Sci 1997; 37:828–834.

78. Ghose AK, Crippen GM. Geometrically feasible binding modes of a flexible ligand molecule at the receptor site. J Comput Chem 1986; 6:350–359.

79. Kuhl FS, Crippen GM, Friesen DK. A combinatorial algorithm for calculating ligand binding. J Comput Chem 1984; 5:24–34.

80. Surles M, Richardson J, Richardson D, Brooks F. Sculpting proteins interactively. Protein Sci 1994; 3:198–210. (MDL Sculpt is a Trademarked program developed and marked by MDL Information Systems, Inc., San Diego, California, U.S.A.)

81. Detailed description and examples can be found in the following web page: *http://www.mdl.com/products/predictive/sculpt/tutorials/tutorial4/index.jsp*.

82. Özkabak AG, Miller MA, Henry DR, Güner OF. Development and optimization of property-based pharmacophores. In: Güner OF, ed. Pharmacophore Perception, Development, Use in Drug Design. La Jolla: International University Line, 2000:479–497.

83. Hadschuh S, Gasteiger J. Pharmacophores derived from the 3D substructure perception. In: Guener O, ed. Pharmacophore: Perception, Development and Use in Drug Design. La Jolla, CA: International University Line, 1999:430–453.

84. Brown RD, Downs GM, Willett P, Cook APF. A hyperstructure model for chemical structure handling: generation and atom-by-atom searching of hyperstructures. J Chem Inf Comput Sci 1992; 32:522–531.

85. Hagadone TR. Molecular substructure similarity searching: efficient retrieval in two-dimensional structure databases. J Chem Inf Comput Sci 1992; 32:515–521.

86. Raymond JW, Gardiner EJ, Willett P. Heuristics for similarity searching of chemical graphs using a maximum common edge subgraph algorithm. J Chem Inf Comput Sci 2002; 42:305–316.

87. Artymiuk PJ, Bath PA, Grindley HM, Pepperrell CA, Poirrette AR, Rice DW, Thorner DA, Wild DJ, Willett P. Similarity searching in databases of three-dimensional molecules and macromolecules. J Chem Inf Comput Sci 1992; 32:617–630.

88. Grindley HM, Artymiuk PJ, Rice DW, Willett P. Identification of tertiary structure resemblance in proteins using a maximal common subgraph isomorphism algorithm. J Mol Biol 1993; 229:707–721.

89. Gray NAB. Computer-Assisted Structure Elucidation. New York: John Wiley & Sons, 1986.

90. Wipke WT, Howe WJ, eds. Computer-Assisted Organic Synthesis. Washington DC: American Chemical Society, 1977.

91. Bayada DM, Simpson RW, Johnson AP. An algorithm for the multiple common subgraph problem. J Chem Inf Comput Sci 1992; 32:680–685.

20

Molecular Descriptors

GEOFF M. DOWNS

Barnard Chemical Information Ltd., Stannington, Sheffield, United Kingdom

1. INTRODUCTION

There is great complexity in the nature of molecules and the interactions between them. Chemists use the simplification of molecules to some form of structural representation to provide a basis for analysis and communication of chemical concepts. The computational analysis of chemical structures requires comparisons to be made between them. Consideration of every aspect of a molecule is computationally infeasible, and so, such comparisons rely on the selection of appropriate descriptors (i.e., features of the molecule) that can be derived from the structural representation. In general, molecular structure is a combination of topological, geometrical, and electronic features. Topology deals with the type and connection of atoms in 2D space; geometry deals with the arrangement of atoms in 3D space, in terms of bond lengths, angles, and dihedral angles; and electronic features deal with the electronic distribution resulting from the molecular wave function.

This chapter gives a brief overview of some of the molecular descriptors that have been developed for a variety of structural analyses, and considers ways in which descriptors have been selected and represented for particular applications. The subject area of this review is vast and so only an introduction can be given here. Consequently, this review is intended as a complement to, rather than a substitute for, previous overviews, such as those given in articles by Brown [1], Mason and Pickett [2], and Bajorath [3]. This chapter is also intended to complement other chapters in this book.

The two main applications of molecular descriptors have been for screens in substructure searching [4] and for attributes in similarity calculations [5]. Historically, many descriptors were developed first for substructure searching (see Chapter 19 of this book) and then used in similarity calculations for applications such as similarity analysis, clustering, combinatorial library design, and data mining (see Chapters 14,

23, and 25 of this book). As similarity applications developed, there was increased use of descriptors from the QSAR/QSPR (see Chapters 21 and 22 of this book) and quantum mechanics disciplines. The result is a wide range of potential descriptors, each with its own characteristics.

2. DESCRIPTOR TYPES

There are several ways in which molecular descriptors can be classified. The majority of descriptors are atom-based rather than field-based. The bulk of this review is focused on a discussion of atom-based descriptors. As the name implies, atom-based descriptors are based on individual atoms, with the description extending outward to incorporate information about the atom's environment. The descriptors are typically generated by analysis of 2D or 3D connection tables, and can include 1D, 2D, or 3D information about the molecule. Atom-based descriptors include individual atoms, feature counts, substructural fragments, topological indices, atomic properties, pharmacophores (see Chapters 17 and 18 of this book), and calculated physicochemical properties.

2.1. Atom, Bond, and Feature Counts

Simple presence or counts of individual atoms, bonds, or features, such as rings, have been included in many descriptor sets used for substructure search. The CAS ONLINE dictionary [6], for instance, contains a variety of simple descriptors to represent individual atom counts, degree of connectivity, bond composition, unusual mass, valence, ring count, and type of ring (size and ring connectivity).

The four Lipinski properties [7], namely, molecular weight, the number of H-bond donors and H-bond acceptors, and log P, were originally applied as filters to eliminate molecules unlikely to be druglike. They can equally be used in similarity analyses, and all, except log P, are simple counts or summations.

2.2. Physicochemical Properties

Many calculated physicochemical properties can be used as descriptors. For instance, Chemical Abstracts Service generated 20 global molecular properties for use as descriptors in an experimental searching system [8]. The semiempirical molecular orbital MOPAC program was used to calculate heat of formation; total energy; ionization potential; dipole moment; HOMO, LUMO, and their difference; minimum, maximum, mean, and standard deviation of electron density and charge; and number of filled orbitals. MOPAC was also used to calculate certain local properties, such as atomic electron densities and eigenvalues for molecular orbitals. The SAVOL program was used to calculate van der Waals volume and surface area, and the Ghose and Crippen method was used to calculate log P and molar refractivity (MR; the review by Mannhold and van de Waterbeemd [9] covers the many fragment-based and atom-based methods that have been developed to calculate log P).

2.3. 2D Fragments

The development of efficient and effective substructure search systems has been based on a fast screening search to eliminate compounds that cannot match the query, followed by a more rigorous but slower atom-by-atom matching. The screening search is typically based on the presence/absence of substructural fragments derived from 2D connection tables. Early systems used fragmentation codes, with each fragment corresponding to some feature such as a functional group, ring description, or broad structural classification. Examples include the Central Patents Index (CPI) code [10] and the GREMAS code [11]. These codes were largely based on the chemical features of molecules that chemists use to classify and to distinguish between molecules. They were originally applied manually, and are relatively slow and difficult to generate automatically. Later systems, such as CAS ONLINE [6] and DARC [12], were based on algorithmically generated fragment families. Applications requiring similarity calculations between molecules, such as clustering and QSAR analyses, have used many of the fragment types originally developed for substructure search screening and have also led to the development of further types. Over the past 30 years, a great many different fragment families have been investigated and used. A brief overview of these families is given below.

Research at the University of Sheffield in the 1970s (see Ref. 13 and earlier papers) examined the occurrence distribution of many families of atom-centered, bond-centered, and ring-centered fragments in test datasets of molecules. Examples of these fragment families are given in Fig. 1. The atom-centered fragments begin with a single atom, then extend outward to include its connectivity (coordinated atom), incident bond types (bonded atom), and adjacent atoms (augmented atom). The bond-centered fragments begin with a single bond, then extend outward to include its incident atoms (simple pair), connectivities of the incident atoms (augmented pair), and bond types incident to the incident atoms (bonded pair). The ring-centered

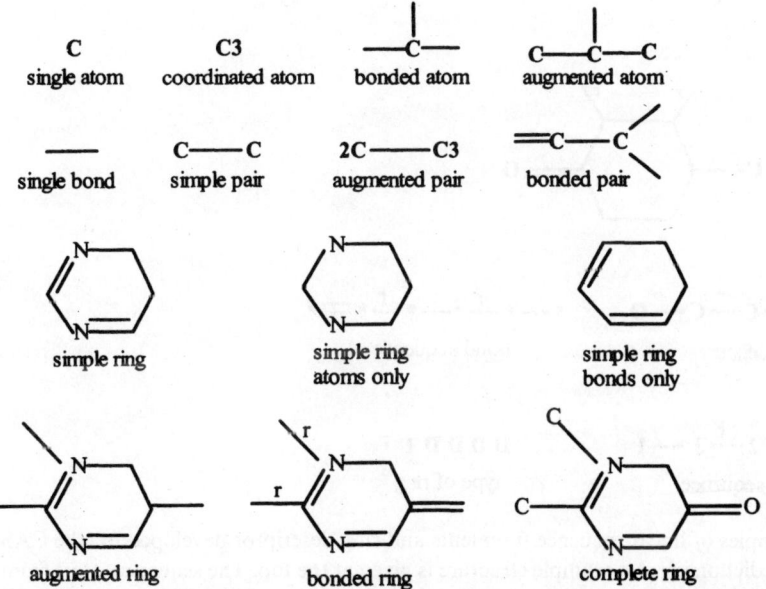

Figure 1 Examples of atom, bond and ring-centred fragments investigated at Sheffield University. Bonds marked 'r' are explicitly ring bonds.

fragments start with the simple ring (with atoms only, bonds only, or both atoms and bonds). The simple ring with both atoms and bonds differentiated then extends outward to include incident bonds (augmented ring if incident bonds are not differentiated, bonded ring if incident bonds are differentiated) and adjacent atoms (complete ring, with incident bonds differentiated). Of these families, the augmented atom family has become a standard type that has been included in many commercial fragment sets, such as the CAS ONLINE screen dictionary. A simple description of ring sizes and heteroatom content was also examined [14], and was used later for screening Markush structures [15].

The CAS ONLINE screen dictionary was developed from the 1960s to the 1980s [6], with major contributions from the Basel Information Center for Chemistry (BASIC) [16,17]. Atom and augmented atom fragments were included. The augmented atoms are supplemented by two variants: hydrogen-augmented atoms (specifying the number of hydrogens attached to the central atom) and twin-augmented atoms (specifying the number of hydrogens attached to the central atom and one of its adjacent atoms). Additional fragment families are based on linear sequences, as shown in Fig. 2. Atom sequences comprise linear sequences of four to six nonhydrogen atoms (with or without the bond types differentiated); bond sequences are atom sequences with the bonds differentiated but the atoms undifferentiated; connectivity sequences are bond sequences (with the bonds undifferentiated or differentiated only as ring or chain) with the atoms replaced by the atom connectivities; and type of ring is the sequence of ring connectivities around the ring [the sequence length gives the ring size, and the connectivities are differentiated into just unfused (D) or fused to another ring (T)]. Within these fragment families, the atom types can be specific, grouped (e.g., halogen), or general (any atom), and the bond types can be specific, grouped (ring or chain), or general (any bond). To support matching free sites on queries, the

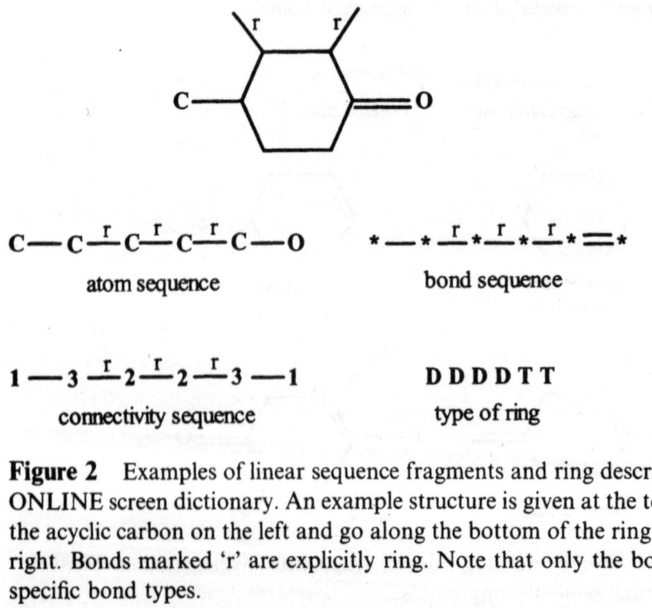

Figure 2   Examples of linear sequence fragments and ring descriptor developed for the CAS ONLINE screen dictionary. An example structure is given at the top. The sequences start from the acyclic carbon on the left and go along the bottom of the ring to the acyclic oxygen on the right. Bonds marked 'r' are explicitly ring. Note that only the bond sequence has completely specific bond types.

augmented atoms can also be reduced such that an augmented atom with three adjacent atoms would also generate augmented atoms with the same central atom and all combinations two and one adjacent atoms.

Another approach to substructure search screens was developed at the Walter Reed Army Institute of Research (WRAIR) [18]. Rather than generating different families of fragments starting from each nonhydrogen atom or bond in the connection table, the WRAIR approach generates all substructures of up to a given number of atoms (11 in the screen set used at WRAIR) starting from each nonhydrogen atom. To limit the number of substructures generated, as substructures are grown, they are compared with a screen dictionary. If the substructure is not present, then that substructure is rejected and any further growth along that path is stopped. The algorithm is similar to those used in game theory where pruning of possible moves is used to eliminate unproductive moves. Their method for selection of substructures for inclusion in the screen set is discussed later in Sec. 3.

In the DARC substructure search system [19–22], the fragments are based on the concepts of an Environment which is Limited, Concentric, and Ordered (ELCO) and a Fragment Reduced to an Environment that is Limited (FREL). The main idea is to generate concentric ordered graphs around a focus. A simple example is an augmented atom in which the central atom is the focus and the incident bonds and adjacent atoms (canonicalized to give the order) form the first concentric level around the focus. An ELCO is a concentric level that does not repeat information contained in concentric levels from which it has been grown. The information contained in each ELCO can take a variety of forms of specificity or generalization. A FREL comprises one or more ELCOs, typically only extending to a maximum of the fourth concentric level (hence, the environment is limited). The concept is a powerful generalization of fragment families such as augmented atoms, and has spawned a rich terminology and a series of acronyms for its various applications [20]. Example simple FRELs around a single atom focus are shown in Fig. 3. The FRELs are designated letters (uppercase for atoms and lowercase for bonds) for each concentric level moving out from the focus. Hence, for a focus that is a single atom, a FREL-a is equivalent to a bonded atom (differentiated incident bonds, undifferentiated adjacent atoms) and a FREL-A is an augmented atom. The focus can be any useful connected subgraph, but is typically an atom, a bond, or a ring. Varying degrees of information and specificity can be represented by different coloring of the nodes and edges within the FREL, giving, for instance, generic FRELs, loose FRELs, and fuzzy FRELs.

At Lederle Laboratories, two new types of descriptors were developed for use in similarity analysis and for prediction of biological activity. The first of these is the atom pair [23], which comprises a pair of nonhydrogen atoms, their separation (i.e., the number of atoms in the shortest path connecting them), and a description of each of the pair of atoms. The description of each atom consists of the atom type, the number of nonhydrogen atoms attached to it, and the number of pi electrons associated with it. The second type of descriptor is the topological torsion fragment, which is a linear sequence of four atoms in which the connecting bonds are unspecified and the atoms are described in a similar manner as the atoms in an atom pair. One difference is that the count of attached nonhydrogen atoms does not include attachments to atoms included in the topological torsion fragment. It can be seen that the atom pair describes more distantly related parts of a molecule (in contrast to the fragments discussed above for use in substructure search screens, which are very localized) and the

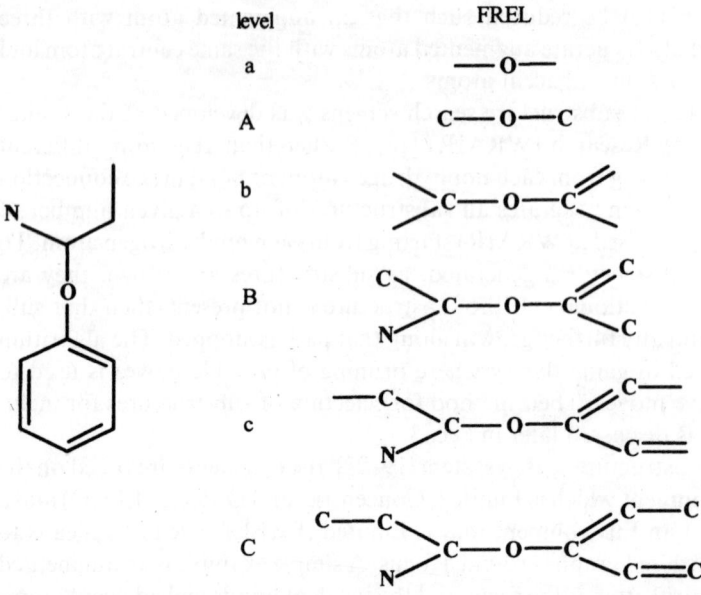

Figure 3 Examples of atom-centred FRELs. The focus atom is the oxygen in the example structure on the left. The FRELs are shown on the right in increasing levels.

topological torsion is a 2D analogue to the torsion (dihedral) angle of a 3D representation. Example atom pair and topological torsion fragments are shown in Fig. 4. Several variants of the atom pair and topological torsion have been developed at Merck Research Laboratories. In one variant [24], the atoms are replaced by physicochemical property descriptions of the atom environment. Three different property descriptions are used: binding property class, atomic log P contribution (hydrophobicity), and partial atomic charge. The binding property pairs and torsions use seven classes: cations, anions, H-bond donors, H-bond acceptors, polar (both donor and acceptor), hydrophobic, and other. For hydrophobic and charge pairs and torsions, the contributions are continuous rather than discrete values, and so a series of seven overlapping bins of values is used to represent ranges. Atom contributions can be assigned to two bins (upper and lower), and so each pair or torsion generates two fragments, one using the lower bins and one using the upper bins. No fragments are generated with a mixture of upper and lower bins. In addition, for charge pair and torsions, polar hydrogens are valid atom types. In another variant [25a], the separation in topological atom pairs and binding property pairs is replaced by the actual distance to give a 3D descriptor. Distances are assigned according to their position in a series of 30 overlapping bins to give a fuzzy contribution that depends on how close the distance is to the center of the bins to which it has been assigned.

At the University of Santiago de Compostela, recent work has developed a family of 2D molecular descriptors based on the local spectral moments of a bond adjacency matrix [25b,26]. Particular attention has been paid to using the bond spectral moments of the bond matrix that correspond to the central bond of a dihedral angle as descriptors for the dihedral angle.

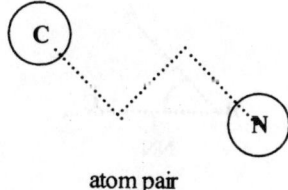

atom pair

(atom environment = atom type + #pi electrons + #non-H connections)

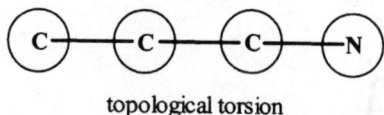

topological torsion

(atom environment = atom type + #pi electrons + #non-H connection not in fragment)

Figure 4 Example atom pair and topological torsion fragments. The information included in the circled atom environments is given below the fragment-type label.

2.4. 3D Fragments

In a manner similar to the 2D fragments algorithmically generated from 2D connection tables discussed above, 3D connection tables can be analyzed to generate a variety of fragment descriptors that can be used for substructure search screens. The additional information available in a 3D environment includes geometrical distance and angles. The simplest and most common descriptors are distance-based screens comprising a pair of (usually) nonhydrogen atoms and a distance range between them. The atoms may have their elemental type specified (AA descriptor), or only one may have its type specified (AX descriptors) [27]. The atoms can have additional information associated with them, such as their connectivity [27,28], aromaticity, number of pi electrons, and charge [28], culminating in the geometrical atom pair descriptors previously mentioned in Sec. 2.3. Research at Sheffield University into angle-based screens included the generalized valence angle [29] and generalized torsion (dihedral) angle [30,31] descriptors. Because they are continuous values, distances and angles are assigned to bins that cover a range of values. A valence angle descriptor comprises three atoms, ABC (where A and C are the apical atoms and B is the vertex atom of the angle), and the angle between them (i.e., between BA and BC) (see Fig. 5). Generalized valence angle screens are classified according to whether the apical atoms are bonded to the vertex atom or not, giving three classes: bonded/bonded (BB), bonded/nonbonded (BN), and nonbonded/nonbonded (NN). The atom types are restricted to three classes: C (carbon), X (nitrogen or oxygen), and Y (any other nonhydrogen). There are thus 18 BB, 27 BN, and 13 BB combinations of atom classes. A torsion angle descriptor comprises four atoms, ABCD (where BC is the central bond of the torsion), and the angle between BA and CD (see Fig. 5). The

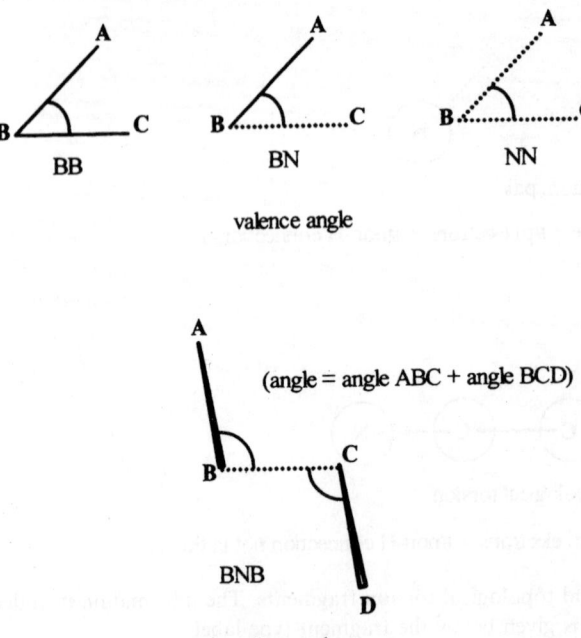

valence angle

(angle = angle ABC + angle BCD)

BNB

torsion angle

Figure 5 Example valence angle and torsion angle fragments. The 'B' and 'N' mean bonded and nonbonded, respectively.

attachments can again be either bonded or nonbonded, with the research investigating four classes of torsion angle descriptor: BBB (bonds joining all atoms of the torsion), BNB (no bond between the central atoms), NBN (only the central atoms are bonded), and NNN (none of the atoms of the torsion is bonded). Generalized BNB torsions are used in the CAVEAT program [32]. Details of nonbonded interactions are given in Chapter 9 of this book. Example distance pair, valence angle, and torsion descriptors are shown in Fig. 5.

Similar research at Chemical Abstracts Service [8,33] has investigated a variety of distance and angle-based screens, some of which are mentioned here. Distance pairs are generalized to include only the atom type pairs any–any, carbon–carbon, carbon–hetero, and hetero–hetero. Distance-based screens are based mainly on atom triangles, with atom types generalized to any, carbon, nitrogen, oxygen, hetero, H-donor, H-acceptor, and "hetero other than nitrogen or oxygen." Atoms can be bonded or nonbonded. The simple atom triangle comprises the three ordered atoms of a triangle. Atom triangle "three-slot" comprises the first two ordered atoms of the triangle along with the binned distance between them. Atom triangle "five-slot" comprises the first two ordered atoms of the triangle along with the three ordered binned distances between atoms in the triangle. The idea of atom triangles has been extended to tetrangles, in which the distances between two bond vectors are used. Each bond vector is described by a base atom and a tip atom. The base atoms are generalized to carbon, hetero, or any, whereas the tip atoms also include hydrogen. The two distances used are the binned distances between the two base atoms and the two tip atoms.

Valence angle and torsion angle type screens include the "three-bonded atoms" and "four-bonded atoms" angle screens in which atom types are generalized to any, carbon, or hetero. These correspond to the BB valence angle and BBB torsion angle descriptors mentioned earlier. In addition, there are "three atoms and one bond vector" and "four atoms and two bond vectors" angle screens in which atom types are generalized to any, carbon, hetero, and hydrogen. These correspond to the BN valence angle and BNB torsion. Examples of these screens are given in Fig. 6.

An important concept in molecular analysis is that of a pharmacophoric pattern (i.e., a geometrical arrangement of structural features that is responsible for a biological effect at a receptor site). The structural features are typically atoms, or collections of atoms and bonds, that can act as pharmacophore points, for instance, H-bond donors and acceptors, charged centers, hydrophobic centers, and ring centers [34]. All geometrically valid combinations of these points can then be considered for use as screens. Typical applications consider pairs, triangles (three-point pharmacophores) [35,36], and quadrangles (four-point pharmacophores). These are related to the distance-based screens mentioned above, but with the atoms replaced by pharmacophore points, and to the geometrical atom pairs and torsions mentioned in an earlier section. The use of pharmacophore screening was pioneered at Lederle Laboratories [28] with their 3DSEARCH program, and at Chemical Design [37] with their Chem-X software, which subsequently formed the basis of screening systems at Rhone-Poulenc Rohrer [38] and Abbott Laboratories [39]. At Rhone-Poulenc Rohrer, six features are used as pharmacophore points: H-bond donor, H-bond acceptor, acid, base, aromatic ring centers, and hydrophobic centers. The distances are assigned to six bins. The same

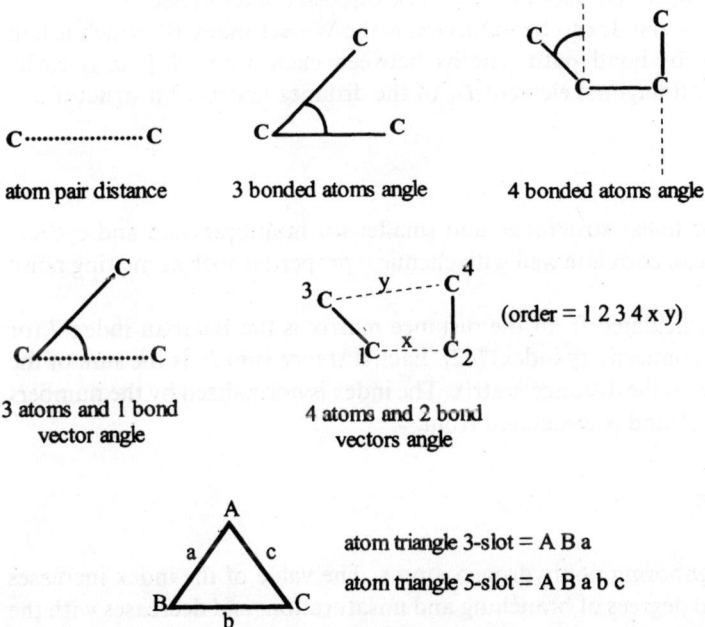

Figure 6 Example types of 3D fragments investigated at Chemical Abstracts Service.

pharmacophore points, with an additional positive charge center, are used within the ChemDiverse module of Chem-X, with distances assigned to a default of 32 bins for three-point pharmacophores. It is also possible to generate all four-point pharmaco-phores [2]; the number of valid pharmacophores reaches around 350 million, but this can be reduced to a more manageable level by using six features and fewer bins and by eliminating impossible distance combinations [40]. At Abbott Laboratories, their 3D-FEATURES program uses rules encoded as Daylight SMARTS patterns to identify atoms that are potential H-bond donors or acceptors, positive or negatively charged, or hydrophobic. The distances are calculated between all such points and then the points and distances are used in two types of descriptors: potential pharmacophore point (PPP) pairs and potential pharmacophore point (PPP) triangles. The five point types give rise to 15 combinations of PPP pairs and 35 combinations of PPP triangles. The assignment of distances to bins can be varied so that the bins are overlapping or nonoverlapping, fixed-width, or based on equifrequent occurrence (as is used in Lederle Laboratories and Chemical Design programs).

2.5. Topological and Topographical Indices

Mathematical invariants such as topological and topographical indices are relatively fast and easy to calculate from a 2D or 3D structural representation, and typically yield a single real or integer value that characterizes the structure. Their greatest use is in QSAR/QSPR analysis, where particular indices show high correlations with observed activities and properties. They are being used increasingly to assist in the discovery of lead compounds, particularly in combinatorial library design and virtual screening. The number of indices and variants that have been developed is huge, but only about a dozen have found general applicability, and so only a few will be men-tioned here. Selection of which ones to use will be discussed later in Sec. 3.

The first topological index to be published is the Wiener index W, which is half the sum of the bond-by-bond path lengths between each atom [41]. It is easily calculated from each offdiagonal element D_{ij} of the distance matrix of a structure:

$$\frac{1}{2} \sum D_{ij}$$

The value is larger for linear structures, and smaller for multibranched and cyclical structures. The index can correlate well with chemical properties such as melting point and viscosity.

Another index calculated from the distance matrix is the Balaban index J (or average distance sum connectivity index) [42]. Each distance sum D_i is the sum of the elements of the ith row of the distance matrix. The index is normalized by the numbers of bonds B and rings C, and is calculated from:

$$\frac{B}{C+1} \sum \frac{1}{\sqrt{D_i D_j}}$$

where i and j are neighboring nonhydrogen atoms. The value of the index increases with structure size and degrees of branching and unsaturation, and decreases with the number of rings. Once the distance matrix is known for a specific molecule, the Wiener index W and the Balaban Index J are quite easily calculated. Consider as an example 2-

methyl butane, with the atom numbering as shown below and hydrogen atoms not displayed:

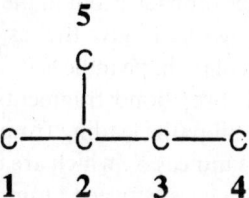

The distance matrix for the carbon atoms is given by:

$$D_i = \sum_{J=1}^{5} D_{ij}$$

	1	2	3	4	5	
1	0	1	2	3	2	8
2	1	0	1	2	1	5
3	2	1	0	1	2	6
4	3	2	1	0	3	9
5	2	1	2	3	0	8

The Wiener index is then easily calculated to be:

$$W = \frac{1}{2}\sum_{i=1}^{5}\sum_{j=i}^{5} D_{ij} = \frac{1}{2}\sum_{i=1}^{5} D_i = \frac{1}{2}(8 + 5 + 6 + 9 + 8) = 18$$

With $B=4$ and $C=1$, one finds for the Balaban index:

$$J = \frac{B}{C+1}\sum_{bonds}\frac{1}{\sqrt{D_i D_j}} = 4\left(\frac{1}{\sqrt{8 \times 5}} + \frac{1}{\sqrt{5 \times 6}} + \frac{1}{\sqrt{6 \times 9}} + \frac{1}{\sqrt{5 \times 8}}\right) = 2.54$$

Randic and Wilkins [43] described the use of path numbers to construct topological sequences, an idea extended by Ruecker and Ruecker [44], among others. All self-avoiding paths are traced through the connection table and the length of each path is assigned to the corresponding element of the sequence. The zeroth element of the sequence indicates the number of atoms in the structure and the last element indicates the longest path length in the structure. Several improvements to the basic path number sequence have been suggested [45], such as normalization of the sequences or weighting in favor of the less frequent path lengths to reduce the effect of different sizes when comparing structures. Weighted path numbers can be used as a sequence, or they can be summed to produce a single-valued topological index.

Some of the most widely used indices are the molecular connectivity indices χ, which were originally formulated by Randic [46] and subsequently generalized and extended by Kier and Hall [47]. The family of indices is extensive and incorporates information about size, branching, unsaturation, heteroatoms, and numbers of rings. Each atom is given two descriptors, δ (the number of adjacent nonhydrogen atoms) and δ_v (the number of valence electrons less the number of adjacent hydrogen atoms). Connectivity indices for a wide variety of path, cluster, path/cluster, and cycle fragments are all calculated by multiplying the descriptor values for each atom in the fragment. The reciprocal square roots are summed over the structure to give the resultant index value. Hall and Kier [48] have also introduced molecular shape indices κ, which are based on counts of all the one-bond, two-bond, and three-bond fragments in a structure; topological state indices T_i, which are based on all paths leading from each atom to every other atom [49]; and electrotopological state indices S_i, which are based on the topological state plus intrinsic electronic state differences between atoms [50].

Other well-established indices include the Zagreb index (the sum of the squared vertex valencies, where valency means number of connections to heavy atoms, regardless of bond order) and the Hosoya index [51]. New indices are published regularly to add to the many hundreds that already exist. Many are highly correlated with others; procedures to reduce redundancy are discussed in Sec. 3.

Topological indices are useful in modelling through-bond interactions but do not consider molecular geometry. Graph theoretical analysis of molecules in 3D space has yielded indices that model through-space interactions; these are referred to as topographical indices. Much of the initial research was started by Randic et al. in the 1980s (see the references in Randic and Razinger [52]). The works by Randic et al. [53,54] on various matrices for topological indices were extended into topographical matrices based on a distance matrix (obtained by embedding the molecular graph into a hexagonal or diamond lattice), [55] or a squared distance matrix [52]. At the same time, the work of Bogdanov et al. [56] led to a 3D version of the Wiener index, based on a 3D distance matrix. Significant contributions from the 1990s to date have come from Estrada and Molina [57] (see references therein). The work by Estrada and Ramirez [58] has included topographical adjacency matrices (built using quantum chemical parameters for the vertex and edge weights) and a 3D bond connectivity index.

The number of topological and topographical indices increases each month, with recent additions including new shape descriptors [59], descriptors based on graph valence shells [60], and a descriptor based on topological distance counts [61]. The latter is shown to be easier to compute while giving results at least as good as the CoMFA and EVA descriptors mentioned below.

2.6. Eigenvalues

In parallel with the work in the QSAR community on matrices, descriptors have been developed based on eigenvalues. In 1989, Burden [62] published a method that generates a 1D descriptor of a molecule by arranging the nonhydrogen atomic numbers along the diagonal of a 2D connectivity matrix and then assigning different values to the offdiagonal elements according to whether the associated pair of atoms is connected or not. The lowest eigenvalue of the resultant matrix represents the molecule. This idea was extended by work at Chemical Abstracts Service and the University of Texas to produce Burden, Cas, University of Texas (BCUT) descriptors [63]. Instead of using atomic numbers, different matrices can be created by putting different information into the diagonal and offdiagonal elements. The standard matrices comprise three classes of information in the diagonal: atomic charge, polarizability, and H-bond ability. Each of these classes can be represented by different values (for instance, atomic charge can be represented by AM1 charges, AM1 densities, Gasteiger–Marsili charges, etc.). The offdiagonal connectivity information can also be supplemented by information on interatomic distances, bond orders, etc. Once the matrices have been created and the offdiagonal elements have been completed, both the highest and lowest eigenvalues are calculated and used as descriptors. Thus, the standard matrices result in a set of six eigenvalue descriptors. However, by introducing different information to the matrix elements, many different BCUT descriptors can be generated. For instance, in work at Rhone-Poulenc Rorer [64], over 60 BCUT descriptors were generated per structure, and at Proctor and Gamble [65], 85 BCUT descriptors have been evaluated.

Another descriptor based on eigenvalues is the EVA descriptor [66,67] (where EVA stands for EigenVAlue). The EVA descriptor is an example of a 3D descriptor that does not require alignment of the 3D structures (i.e., it is invariant to translation or rotation of the structure). It is based on the infrared spectrum of a molecule, which is related to the 3D structure. The molecular vibrations are calculated using a normal coordinate analysis (NCA) of the energy-minimized structure (e.g., using MOPAC AM1), and the eigenvalues from the NCA are used to derive the EVA descriptor (see Ref. 67).

The EVA descriptor contrasts with the research by Benigni et al. [68] who use the infrared spectra more directly as descriptors. The spectra are binned into normalized nonoverlapping ranges and then the bins used as screens. The resultant QSAR models were similar to those obtained using the BME descriptors of Burden [69], but not as good as those obtained using more traditional descriptors (such as log P and polarizability). Burden's BME descriptor is a development of the original eigenvalue method that is based on modified adjacency matrices.

Other whole molecule descriptors that do not require alignment include the Weighted Holistic Invariant Molecular (WHIM) indices developed by Todeschini et al. [70]. These indices are calculated from the 3D coordinates, which are weighted and centered to make them invariant to translation; principal component analysis (PCA) is applied to obtain three principal components. These are used to produce new coordinates, which can be analyzed to obtain a series of 10 descriptors based on eigenvalues and the third-order and fourth-order moments of the three score column vectors. These descriptors are related to molecular size, shape, symmetry, and atom distribution and density.

2.7. Field-Based Descriptors

The descriptors mentioned so far are all essentially atom-based. Some of them represent 3D aspects of molecules, but they do not represent the relative 3D arrangement of the fields responsible for molecular interaction. The alternative is to use field-based descriptors, an area that is receiving increasing interest. A few examples are introduced here.

Projection of molecular features, such as electron density, electrostatic potential, and hydrostatic potential, onto the surface of a sphere has been used successfully by van Geerestein et al. [71]. A small number of surface points are chosen as representative vertices (typically 12 or 20 arranged as an icosahedron or dodecahedron, respectively) that can be used as anchor points for the descriptors. Shape, for instance, can be represented by finding the shortest distance between each vertex and the surface of the molecule.

The most common method of field-based descriptor generation draws a 3D grid around a molecule and then determines the value of the property of interest (steric, electrostatic, or hydrophobic) at each lattice intersection of the grid. This method originally came from work by Carbo et al. [72] to calculate electron density descriptors, and has been extended by others (e.g., Refs. 73 74 75) to include electrostatic potentials and steric fields.

The standard implementation of grid-based 3D QSAR is the Comparative Molecular Field Analysis (CoMFA) method [76]. These approaches can be very time-consuming and approximate values for electrostatic and electron density fields can be obtained using the much faster methods developed by Good et al. [77] and Good and Richards [78], which are based on Gaussian functions. All of the above require molecules to be aligned before they can be compared. Many methods have been explored to find ones that perform alignment efficiently, including those based on the use of field graphs [79] and genetic algorithms [80]. For details about molecular fields, see Chapters 1 and 8 of this book.

2.8. Other Descriptors

The sections above have given a few examples of descriptors of certain categories. No mention has been made of reactivity descriptors because they are covered in Chapter 11, or some of the very specialized descriptors, such as specific chirality descriptors (for instance, the work by Aires-de-Sousa and Gasteiger [81]). There are also other descriptors that fall outside the categories used above, of which a few are mentioned here.

A 1D descriptor has been developed at Pharmacopeia Inc. [82]. Instead of ordering the atoms along the diagonal of a connectivity matrix, the pairwise distances from a 2D or 3D representation are mapped onto 1D coordinates while optimizing the preservation of the distances among the points. Multidimensional Dimensional Scaling is used to perform the mapping. The method is initialized with an approximation of where the 1D coordinates might be, and then coordinates are refined in an iterative manner by minimizing the sum of squared errors. The atoms can be colored by associated information such as elemental type, hybridization state, and connectivity to give a more detailed description of the molecule. Two molecules can be compared by using a sequence alignment method developed from amino acid sequence alignment methods.

Combined descriptors attempt to put more than one type of information into the same descriptor. For instance, the electronic–topological descriptor, used by Katritzky and Gordeeva [83], is an electronic version of the valence connectivity index. The work by Stanton and Jurs [84] has produced over 25 combined descriptors based on charged partial surface areas. Other descriptors are mentioned in the review by Brown [1].

3. DESCRIPTOR SELECTION

As can be seen from the review of molecular descriptors above, there are a great many to choose from. The selection of which descriptors to use, the techniques used for the selection, and the manner in which the descriptors are represented is largely application-dependent.

For substructure search and similarity analyses, descriptors are usually represented as fingerprints or dataprints. A fingerprint is a bitstring indicating the presence or absence of the chosen descriptors (typically 2D or 3D substructural fragments). A dataprint is an array of real numbers indicating the value of the chosen descriptors (typically physicochemical properties or topological/topographical indices).

3.1. Selection for Fingerprints

For a fingerprint, if the descriptors are assigned on the basis of stored screen numbers (bit positions; usually stored in some form of dictionary to give the correspondence between the descriptor and the position in the bitstring), then the fingerprint is "keyed." There is a direct correspondence between the "on" position in the bitstring and a particular descriptor or group of descriptors. Keyed fingerprints enable the selection and assignment of descriptors deemed relevant to particular activities or properties, but if relevant descriptors are omitted from the dictionary, then the desired features are not represented in the fingerprint. The numbers of occurrence of particular descriptors can be indicated in the bitstring by using separate bits to indicate ranges of occurrence. Careful selection of descriptors is crucial to avoid underrepresentation or overrepresentation of desired features. The alternative is not to select particular descriptors but to generate all descriptors of a particular type and then to assign several bit positions on the basis of a hash function applied to the descriptor; these fingerprints are "hashed." Because each descriptor assigns several bits that are overlaid on the bitstring along with bits assigned from other descriptors, there is the possibility of collisions (different descriptors setting the same bit position), there is no direct correspondence between a particular bit position being "on" and a particular descriptor or group of descriptors, and it is difficult to emphasize desired features. The major advantage of hashed fingerprints is that it avoids having to select particular features; all descriptors of the chosen type are represented in the bitstring and so there is no danger of missing ones out.

The best-known keyed fingerprints are those produced from the CAS ONLINE screen dictionary [6] and MDLs MACCS keys. The CAS ONLINE screen dictionary is derived from the BASIC screen dictionary [85,86] and took many years to develop. Twelve fragment types are included as screens, including augmented, hydrogen-augmented, and twin-augmented atoms, and atom, bond, and connectivity sequences. To reduce the number of screens to a reasonable number and to support less specific search queries, the atom and bond types in some fragments are generalized, and some related fragments are assigned to the same screen number. Generalized fragments also help when the screens are used to support similarity analyses. The fragments selected for use as screens were largely chosen on an empirical basis through manual analysis of fragments generated from a test dataset. The result is a dictionary of around 6000 descriptors assigned to 2128 screen numbers. Details about the MACCS keys have not been published, but some information is given in the study by Brown and Martin [39]. The full set of keys consists of 960 keys, of which 166 are defined in the user manual for use in searching.

Selection of descriptors for substructure search can be done on a statistical basis. The aim is to select a set in which the descriptors are independent of each other and roughly equifrequent in distribution. Much of the statistical analysis of fragment distributions was done by Lynch et al. in the 1970s (for a summary, see Ref. 86). The conclusion was that fragment distributions are skewed so that a few fragments occur most of the time and a lot of fragments occur rarely. For screen set selection, the most efficient use of a limited number of bit positions is to choose descriptors that occur neither too frequently nor too infrequently. Frequent descriptors can be eliminated from the selected set, whereas infrequent descriptors can be generalized or grouped so that they become more frequent.

Analysis of the distribution of fragments was used at Upjohn for the production of a screen dictionary for the COUSIN system. This was based on selection from the BASIC dictionary with some additional descriptors added [88], resulting in a dictionary of 1590 screens. Several automated schemes for selection of descriptors for screens have been suggested. Early work was conducted on 2D fragments by Hodes [89], Willett [90], and Gannon and Willett [91], with subsequent work on 3D fragments by Jakes and Willett [27] and Cringean et al. [92]. Selective elimination of fragments based on statistical and relational criteria is used in the Pickfrag program [93].

The best-known hashed fingerprints are the 2D fingerprints produced by the Daylight software, which encode single atom types, augmented atoms, and atom sequences of length two to seven atoms [94]. These are typically hashed onto a fingerprint of length 1024 or 2048 bits. All fragment types are hashed onto the same fingerprint. In contrast, the Tripos Unity 2D fingerprints allocate hydrogen-containing atom sequences, of lengths two to four, and ordinary atom sequences, of lengths four to six, to separate regions of a 988-bit bitstring. 3D descriptors are typically assigned to bins, each of which is assigned bit positions in a similar manner to the 2D descriptors (see Ref. 39 for further details).

3.2. Selection for Dataprints

As mentioned earlier, studies at Chemical Abstracts Service have calculated large numbers of physicochemical properties for use as descriptors. The numbers were reduced subsequently by removing properties that were highly correlated with another property. In a study that used these descriptors as the basis for cluster analysis [95], 29 descriptors were reduced to 13.

In QSAR and QSPR studies, the standard ways of removing redundancy from large numbers of topological and topographical indices include principal component analysis, chi-squared analysis, and multiple regression analysis (MRA). Most QSAR and QSPR applications deal with very small datasets, and so the dimensionality does not cause a problem for PCA or chi-squared analysis. MRA does not impose any restrictions on the type and number of descriptors. The selection process is based on two principles, namely, to cover as much of parametric space as possible (principle of variance) while choosing independent descriptors (principle of orthogonality).

In 1988, Basak et al. [96] published one of the first studies using large numbers of molecules and indices. They generated around 90 topological indices from a 3692 subset of the U.S. Environmental Protection Agency's TSCA database. These indices were chosen to encode aspects size, shape, bonding type, and branching pattern information. Many of these indices were highly correlated; PCA reduced them to 10 principal components that explained over 90% of the variance. These principal components were used as descriptors in similarity searching of the full database.

There are many examples of the use of regression analysis to select descriptors. Dixon et al. [82], for instance, used it to select the best of 84 topological and geometrical descriptors for predicting various physicochemical and biological activities. It is, however, debatable as to whether descriptors used to produce QSAR models are equally valid for use in applications such as similarity analysis [97].

The utility of chi-squared analysis is demonstrated by the "autochoose" algorithm used to select BCUT descriptors [97]. The algorithm essentially takes all combinations of descriptors and evaluates them on the basis of choosing a few that are most orthogonal to each other. The result is typically five or six descriptors. The method is generally applicable, and so descriptors other than BCUTs can be processed to find an orthogonal subset. At Bristol-Myers Squibb [40], reduction of many BCUT descriptors down to five or six, combined with four-point pharmacophore descriptors, has given good results for analyzing large virtual libraries.

In contrast to the above, Taraviras et al. [98] have applied clustering to 240 topological indices to cluster together those that are highly correlated. Seven clustering methods were applied to overcome the limitations of a single clustering method. Descriptors that were found to exist in the same clusters across all seven methods were regarded as being strongly correlated. Selection of one descriptor from each cluster was shown to provide representative sets of orthogonal descriptors for use in QSAR analysis.

The number of elements necessary in a dataprint to provide sufficient discrimination between molecules can be surprisingly few. For instance, the Diverse Property-Derived (DPD) code developed at Rhone-Poulenc Rorer [99] is a combination of six descriptors. An initial 49 descriptors (including topological indices, physicochemical properties, and functional group counts) were analyzed statistically to select a subset that reflected hydrophobicity, polarity, flexibility, shape, hydrogen bonding, and aromatic interactions. The resultant set comprised six descriptors: number of H-bond acceptors and donors, a flexibility index, the normalized sum of squared electrotopological indices, $C \log P$, and aromatic density. Even fewer descriptor families are used by the Chemical Computing Group [100], who use the method of Wildman and Crippen [101] to generate the $S \log P$ and molar refractivity, and the method of Gasteiger and Marsili [102] to calculate partial atomic charges. The values are assigned to 10, 8, and 14 range bins, respectively, to give a total of 32 screens.

The use of genetic algorithms to select descriptors has been investigated by several groups. Xue and Bajorath [103] have linked a genetic algorithm with principal component analysis to produce "minifingerprints" (short-length keyed fingerprints). When selecting 2D descriptors on the basis of their ability to classify compounds according to their biological activity, the result was fingerprints of 50–60 descriptors that performed very well [104]. Yasri and Hartsough [105] use a combination of genetic algorithm and neural network to select descriptors. The number of descriptors required is not fixed beforehand; the genetic algorithm performs selections and the backpropagation neural network works out the fitness score. The method requires a training set to enable the neural network to map descriptors against activity. The combination works particularly well in analyzing nonlinear datasets.

4. SUMMARY

The intention of this brief review has been to give an overall impression of the range and types of descriptors available for representing molecules, with examples of the ones that have general use. The descriptors can contain 2D, 3D, or field information;

can be very localized (as in augmented atoms) or wider-ranging (as in atom pairs); or represent the whole molecule (as in many topological indices). The information can be very specific (giving explicit atom, bond, or connectivity details) or generalized (nonspecific atom or bond types, or features used in pharmacophores). Particular applications may have very specific requirements, especially regarding the amount and type of chemical information contained within the descriptors. The generation of molecular descriptors is followed by selection of those most relevant to a given application and representation in a suitable form. Several methods that assist in the selection of descriptors have been outlined.

REFERENCES

1. Brown RD. Descriptors for diversity analysis. Perspect Drug Discov Des 1997; 7/8:31–49.
2. Mason JS, Pickett SD. Partition-based selection. Perspect Drug Discov Des 1997; 7/8:85–114.
3. Bajorath J. Selected concepts and investigations in compound classification, molecular descriptor analysis, and virtual screening. J Chem Inf Comput Sci 2001; 41:233–245.
4. Welford SM. Substructure search of chemical structure files. In: Ash J, Chubb P, Ward S, Welford S, Willett P, eds. Communication, Storage and Retrieval of Chemical Information. Chichester: Ellis Horwood, 1985:157–181.
5. Johnson MA, Maggiora GM, Lajiness MS, Moon JB, Petke JD, Rohrer DC. Molecular similarity analysis: applications in drug discovery. In: van de Waterbeemd H, ed. Advanced Computer-Assisted Techniques in Drug Discovery. Weinheim: VCH, 1994:89–110.
6. Dittmar PG, Farmer NA, Fisanick W, Haines RC, Mockus J. The CAS ONLINE search system: 1. General system design and selection, generation, and use of search screens. J Chem Inf Comput Sci 1983; 23:93–102.
7. Lipinski CA, Lombardo F, Dominy BW, Feenay PJ. Experimental and computational approaches to estimate solubility and permeability in drug discovery and development settings. Adv Drug Deliv Rev 1997; 23:3–25.
8. Fisanick W, Cross KP III, Rusinko A. Similarity searching on CAS registry substances: 1. Global molecular property and generic atom triangle geometric searching. J Chem Inf Comput Sci 1992; 32:664–674.
9. Mannhold R, van de Waterbeemd H. Substructure and whole molecule approaches for calculating log P. J Comput-Aided Mol Des 2001; 15:337–354.
10. Simmons ES. Central Patents Index chemical code: a user's viewpoint. J Chem Inf Comput Sci 1984; 24:10–15.
11. Rossler S, Kolb A. The system GREMAS, an integral part of the IDC system for chemical documentation. J Chem Doc 1970; 10:128–134.
12. Dubois JE, Sobel Y. DARC system for documentation and artificial intelligence in chemistry. J Chem Inf Comput Sci 1985; 25:326–333.
13. Adamson GW, Clinch VA, Cressey SE, Lynch MF. Distributions of fragment representations in a chemical substructure search screening system. J Chem Doc 1974; 14:72–74.
14. Adamson GW, Cowell J, Lynch MF, Town WG, Yapp AM. Analysis of structural characteristics of chemical compounds in a large computer-based file: Part 4. Cyclic fragments. J Chem Soc Perkin Trans 1973; 1:863–865.
15. Downs GM, Gillet VJ, Holliday JD, Lynch MF. Computer storage and retrieval of

generic chemical structures: 10. Assignment and logical bubble-up of ring screens for structurally explicit generics. J Chem Inf Comput Sci 1989; 29:215–224.

16. Graf W, Kaindl HK, Kniess H, Schmidt B, Warszawski R. Substructure retrieval by means of the BASIC fragment search dictionary based on the CAS Chemical Registry III system. J Chem Inf Comput Sci 1979; 19:51–55.

17. Graf W, Kaindl HK, Kneiss H, Warszawski R. The third BASIC fragment search dictionary. J Chem Inf Comput Sci 1982; 22:177–181.

18. Feldmann A, Hodes L. An efficient design for chemical structure searching: I. The screens. J Chem Inf Comput Sci 1975; 15:147–152.

19. Attias R. DARC substructure search system: a new approach to chemical information. J Chem Inf Comput Sci 1983; 23:102–108.

20. Dubois JE, Panaye A, Attias R. DARC system: notions of defined and generic substructures. Filiation and coding of FREL substructure (SS) classes. J Chem Inf Comput Sci 1987; 27:74–82.

21. Attias R, Dubois J-E. Substructure systems: concepts and classifications. J Chem Inf Comput Sci 1990; 30:2–7.

22. Dubois J-E, Carrier G, Panaye A. DARC topological descriptors for pattern recognition in molecular database management systems and design. J Chem Inf Comput Sci 1991; 31:574–578.

23. Carhart RE, Smith DH, Venkataraghavan R. Atom pairs as molecular features in structure–activity studies: definition and applications. J Chem Inf Comput Sci 1985; 25:64–73.

24. Kearsley SK, Sallamack S, Fluder EM, Andose JD, Mosley RT, Sheridan RP. Chemical similarity using physicochemical property descriptors. J Chem Inf Comput Sci 1996; 36:118–127.

25a. Sheridan RP, Miller MD, Underwood DJ, Kearsley SK. Chemical similarity using geometric atom pair descriptors. J Chem Inf Comput Sci 1996; 36:128–136.

25b. Estrada E, Molina E, Perdomo-Lopez I. Can 3D structural parameters be predicted from 2D (topological) molecular descriptors? J Chem Inf Comput Sci 2001; 41:1015–1021.

26. Estrada E, Molina E. Novel local (fragment-based) topological molecular descriptors for QSPR/QSAR and molecular design. J Mol Graph Model 2001; 20:54–65.

27. Jakes SE, Willett P. Pharmacophoric pattern matching in files of 3D chemical structures: selection of interatomic distance screens. J Mol Graph 1986; 4:12–20.

28. Sheridan RP, Nilakantan R III, Rusinko A, Bauman N, Haraki KS, Venkataraghavan R. 3DSEARCH, a system for three-dimensional substructure searching. J Chem Inf Comput Sci 1989; 29:255–260.

29. Poirette AR, Willett P, Allen FH. Pharmacophoric pattern matching in files of three-dimensional chemical structures: characterisation and use of generalised valence angle screens. J Mol Graph 1991; 9:203–217.

30. Poirrette AR, Willett P. Pharmacophoric pattern matching in files of three-dimensional chemical structures: characterization and use of generalized torsion angle screens. J Mol Graph 1993; 11:2–14.

31. Bath PA, Poirette AR, Willett P. Similarity searching in files of three-dimensional chemical structures: comparison of fragment-based measures of shape similarity. J Chem Inf Comput Sci 1994; 34:141–147.

32. Bartlett PA, Shea GT, Telfer SJ, Waterman S. CAVEAT: a program to facilitate the structure-derived design of biologically active molecules. In: Roberts SM, ed. Molecular Recognition: Chemical and Biochemical Problems. Cambridge, UK: Royal Society of Chemistry, 1990:182–196.

33. Fisanick W, Cross KP, Forman JC III, Rusinko A. An experimental system for similarity and 3D searching of CAS Registry substances: I. 3D substructure searching. J Chem Inf Comput Sci 1993; 33:548–559.

34. Murrall NW, Davies EK. Conformational freedom in 3-D databases: 1. Techniques. J Chem Inf Comput Sci 1990; 30:312–316.

35. Bemis G, Kuntz ID. Fast and efficient method for 2D and 3D molecular shape description. J Comput-Aided Mol Des 1992; 6:607–628.

36. Nilakantan R, Bauman N, Venkataraghavan R. New method for rapid characterization of molecular shape. J Chem Inf Comput Sci 1993; 33:79–85.

37. Murrall NW, Davies EK. Conformational freedom in 3-D databases: 1. Techniques. J Chem Inf Comput Sci 1990; 30:312–316.

38. Pickett SD, Mason JS, McLay IM. Diversity profiling and design using 3D pharmacophores: pharmacophore-derived queries (PDQ). J Chem Inf Comput Sci 1996; 36: 1214–1223.

39. Brown RD, Martin YC. Use of structure–activity data to compare structure-based clustering methods and descriptors for use in compound selection. J Chem Inf Comput Sci 1996; 36:572–584.

40. Mason JS, Beno BR. Library design using BCUT chemistry-space descriptors and multiple four-point pharmacophore fingerprints: simultaneous optimisation and structure-based diversity. J Mol Graph Model 2000; 18:438–451.

41. Wiener H. Structural determination of paraffin boiling point. J Am Chem Soc 1947; 69:17–20.

42. Balaban AT, ed. Chemical Applications of Graph Theory. London: Academic Press, 1976.

43. Randic M, Wilkins CL. Graph theoretical approach to recognition of structural similarity in molecules. J Chem Inf Comput Sci 1979; 19:31–37.

44. Ruecker G, Ruecker C. Counts of all walks as atomic and molecular descriptors. J Chem Inf Comput Sci 1993; 33:683–695.

45. Randic M. Design of molecules with desired properties. In: Johnson MA, Maggiora GM, eds. Concepts and Applications of Molecular Similarity. New York: John Wiley and Sons, 1990:77–145.

46. Randic M. On characterisation of molecular branching. J Am Chem Soc 1975; 97:6609–6615.

47. Kier LB, Hall LH. Molecular Connectivity in Structure–Activity Analysis. New York: John Wiley and Sons, 1986.

48. Hall LH, Kier LB. The molecular connectivity chi indexes and kappa shape indexes in structure–property modelling. Rev Comput Chem 1991; 2:367–422.

49. Hall LH, Kier LB. Determination of topological equivalence in molecular graphs from the topological state. Quant Struct–Act Relat 1990; 9:115–131.

50. Hall LH, Mohney B, Kier LB. The electrotopological state: structure information at the atomic level for molecular graphs. J Chem Inf Comput Sci 1991; 31:76–82.

51. Hosoya H. Topological index: a newly proposed quantity characterizing the topological nature of structural isomers of saturated hydrocarbons. Bull Chem Soc Jpn 1971; 44: 2332–2339.

52. Randic M, Razinger M. Molecular topographic indices. J Chem Inf Comput Sci 1995; 35:140–147.

53. Randic M, Guo X, Oxley T, Krishnapriyan H. Wiener matrix: source of novel graph invariants. J Chem Inf Comput Sci 1993; 33:709–716.

54. Randic M, Kleiner AF, DeAlba LM. Distance/distance matrices. J Chem Inf Comput Sci 1994; 34:277–286.

55. Randic M, Jerman-Blazic B, Trinajstic N. Development of 3-dimensional molecular descriptors. Comput Chem 1990; 14:237–246.

56. Bogdanov B, Nikolic S, Trinajstic N. On the three-dimensional Wiener number, a comment. Math Chem 1989; 3:299–309.

57. Estrada E, Molina E. 3D connectivity indices in QSPR/QSAR studies. J Chem Inf Comput Sci 2001; 41:791–797.

58. Estrada E, Ramirez A. Edge adjacency relationships and molecular topographic descriptors. Definition and QSAR applications. J Chem Inf Comput Sci 1996; 36:837–843.

59. Randic M. Novel shape descriptors for molecular graphs. J Chem Inf Comput Sci 2001; 41:607–613.

60. Randic M. Graph valence shells as molecular descriptors. J Chem Inf Comput Sci 2001; 41:627–630.

61. Baumann K. An alignment-independent versatile structure descriptor for QSAR and QSPR based on the distribution of molecular features. J Chem Inf Comput Sci 2002; 42: 26–35.

62. Burden FR. Molecular identification number for substructure searches. J Chem Inf Comput Sci 1989; 29:225–227.

63. Pearlman RS. Novel software tools for addressing chemical diversity. Network Science (http://www.netsci.org/Science/Combichem/feature08.html).

64. Menard PR, Mason JS, Morize I, Bauerschmidt S. Chemistry space metrics in diversity analysis, library design, and compound selection. J Chem Inf Comput Sci 1998; 38:1204–1213.

65. Stanton DT. Evaluation and use of BCUT descriptors in QSAR and QSPR studies. J Chem Inf Comput Sci 1999; 39:11–20.

66. Turner DB, Willett P, Ferguson AM, Heritage T. Evaluation of a novel infrared range vibration-based descriptor (EVA) for QSAR studies: 1. General application. J Comput-Aided Mol Des 1997; 11:409–422.

67. Turner DB, Willett P. The EVA spectral descriptor. Eur J Med Chem 2000; 35:367–375.

68. Benigni R, Giuliani A, Passerini L. Infrared spectra as chemical descriptors for QSAR models. J Chem Inf Comput Sci 2001; 41:727–730.

69. Burden FR. A chemically intuitive molecular index based on the eigenvalues of a modified adjacency matrix. Quant Struct–Act Relat 1997; 16:309–314.

70. Todeschini R, Lasagni M, Marengo E. New molecular descriptors for 2D and 3D structures. Theory. J Chemom 1994; 8:263–272.

71. van Geerestein V, Perry NC, Grootenhuis PDJ, Haasnoot CA. 3D database searching on the basis of ligand shape using the SPERM prototype method. Tetrahedron Comput Methodol 1990; 3:595–613.

72. Carbo R, Leyda L, Arnau M. An electron density measure of the similarity between two compounds. Int J Quantum Chem 1980; 17:1185–1189.

73. Hodgkin EE, Richards WG. Molecular similarity based on electrostatic potential and electric field. Int J Quantum Chem Quantum Biol Symp 1987; 14:105–110.

74. Kearsley SK, Smith GM. A alternative method for the alignment of molecular structures maximising electrostatic and steric overlap. Tetrahedron Comput Methodol 1990; 3:615–633.

75. Manaut F, Sanz F, Jose J, Milesi M. Automatic search for maximum similarity between molecular electrostatic potential distributions. J Comput-Aided Mol Des 1991; 5:371–380.

76. Cramer RD, Patterson DE, Bunce JD. Comparative molecular field analysis (CoMFA): 1. Effect of shape on binding of steroids to carrier proteins. J Am Chem Soc 1988; 110:5959–5967.

77. Good AC, Hodgkin EE, Richards WG. The utilisation of Gaussian functions for the rapid evaluation of molecular similarity. J Chem Inf Comput Sci 1992; 32:188–191.

78. Good AC, Richards WG. The utilisation of Gaussian functions for the rapid evaluation of molecular similarity. J Chem Inf Comput Sci 1993; 33:112–116.

79. Thorner DA, Willett P, Wright PM. Similarity searching in files of three-dimensional chemical structures: representation and searching of molecular electrostatic potentials using field-graphs. J Comput-Aided Mol Des 1997; 11:163–174.

80. Wild DJ, Willett P. Similarity searching in files of three-dimensional chemical structures. Alignment of molecular electrostatic potential fields with a genetic algorithm. J Chem Inf Comput Sci 1996; 36:159–167.
81. Aires-de-Sousa J, Gasteiger J. Prediction of enantiomeric selectivity in chromatography. Application of conformation-dependent and conformation-independent descriptors of molecular chirality. J Mol Graph Model 2002; 20:373–388.
82. Dixon SL, Merz KM. One-dimensional molecular representations and similarity calculations: methodology and validation. J Med Chem 2001; 44:3795–3809.
83. Katritzky AR, Gordeeva EV. Traditional topological indices vs. electronic, geometrical, and combined molecular descriptors in QSAR/QSPR research. J Chem Inf Comput Sci 1993; 33:835–857.
84. Stanton DT, Jurs PC. Development and use of charged partial surface area structural descriptors in computer-assisted quantitative structure–property relationship studies. Anal Chem 1990; 62:2323–2329.
85. Graf W, Kaindl HK, Kniess H, Schmidt B, Warszawski R. Substructure retrieval by means of the BASIC fragment search dictionary based on the CAS Chemical Registry III system. J Chem Inf Comput Sci 1979; 19:51–55.
86. Graf W, Kaindl HK, Kneiss H, Warszawski R. The third BASIC fragment search dictionary. J Chem Inf Comput Sci 1982; 22:177–181.
87. Adamson GW, Cowell J, Lynch MF, McLure AHW, Town WG, Yapp AM. Strategic considerations in the design of a screening system for substructure searches of chemical structure files. J Chem Doc 1973; 13:153–157.
88. Howe WJ, Hagadone TR. Molecular substructure searching: microcomputer-based query execution. J Chem Inf Comput Sci 1982; 22:182–186.
89. Hodes L. Selection of descriptors according to discrimination and redundancy. Application to chemical structure searching. J Chem Inf Comput Sci 1976; 16:88–93.
90. Willett P. A screen set generation algorithm. J Chem Inf Comput Sci 1979; 19:159–162.
91. Gannon MT, Willett P. Sampling considerations in the selection of fragment screens for chemical substructure search systems. J Chem Inf Comput Sci 1979; 19:251–255.
92. Cringean JK, Pepperrell CA, Poirrette AR, Willett P. Selection of screens for three-dimensional substructure searching. Tetrahedron Comput Methodol 1990; 3:37–46.
93. Barnard JM, Downs GM. Chemical fragment generation and clustering software. J Chem Inf Comput Sci 1997; 37:141–142.
94. http://www.daylight.com/dayhtml/doc/theory/theory.toc.html
95. Downs GM, Willett P, Fisanick W. Similarity searching and clustering of chemical-structure databases using molecular property data. J Chem Inf Comput Sci 1994; 34:1094–1102.
96. Basak SC, Magnuson VR, Niemi GJ, Regal RR. Determining structural similarity of chemicals using graph-theoretic indices. Disc Appl Math 1988; 19:17–44.
97. Pearlman RS, Smith KM. Metric validation and the receptor-relevant subspace concept. J Chem Inf Comput Sci 1999; 39:28–35.
98. Taraviras SL, Ivanciuc O, Cabrol-Bass D. Identification of groupings of graph theoretical descriptors using a hybrid cluster analysis approach. J Chem Inf Comput Sci 2000; 40:1128–1146.
99. Lewis RA, Mason JS, McLay IM. Similarity measures for rational set selection and analysis of combinatorial libraries: the Diverse Property-Derived (DPD) approach. J Chem Inf Comput Sci 1997; 37:599–614.
100. Labute P. A widely applicable set of descriptors. J Mol Graph Model 2000; 18:464–477.
101. Wildman SA, Crippen GM. Prediction of physicochemical parameters by atomic contributions. J Chem Inf Comput Sci 1999; 39:868–873.
102. Gasteiger J, Marsili M. Iterative partial equalisation of orbital electronegativity: a rapid access to atomic charges. Tetrahedron 1980; 36:3219–3228.

103. Xue L, Bajorath J. Molecular descriptors for effective classification of biologically active compounds based on principal component analysis identified by a genetic algorithm. J Chem Inf Comput Sci 2000; 40:801-809.

104. Xue L, Stahura FL, Godden JW, Bajorath J. Mini-fingerprints detect similar activity of receptor ligands previously recognised only by three-dimensional pharmacophore-based methods. J Chem Inf Comput Sci 2001; 41:394-401.

105. Yasri A, Hartsough D. Toward an optimal procedure for variable selection and QSAR model building. J Chem Inf Comput Sci 2001; 41:1218-1227.

21

2D QSAR Models: Hansch and Free–Wilson Analyses

HUGO KUBINYI

BASF AG, Ludwigshafen, Germany

1. QUANTITATIVE STRUCTURE–ACTIVITY RELATIONSHIPS (QSAR) HISTORY

Quantitative structure–activity relationships developed slowly, over a time range of more than a hundred years. During this period, several relationships were observed between the toxicity and/or narcotic activities of organic compounds and their lipophilicity, originally expressed by aqueous solubility and later by oil–water partition coefficients [1–3]. Crum Brown and Fraser, in 1868, observed significantly different pharmacological activities before and after quaternation of the basic nitrogen atom of several organic bases. They proposed Eq. (1) to describe the dependence of "physiological properties" Φ on chemical structures C [4]:

$$\Phi = f(C) \tag{1}$$

Of course, Eq. (1) is only a mathematical formalism. Whereas biological activities can easily be defined by a certain endpoint (e.g., an effective dose, a 50% inhibition constant, a 100% lethal dose, etc.), it was and still is impossible to describe chemical structures in an absolute manner. Only changes in biological activities $\Delta\Phi$ can be correlated with certain changes in chemical structures ΔC (Eq. (2)). Such chemical changes can be quantified either by structural terms (indicator variables, dummy variables, Free–Wilson parameters) or by the resulting change in various physicochemical or other properties:

$$\Delta\Phi = f(\Delta C) \tag{2}$$

At the end of the 19th century, Meyer [5], and Overton [6] independently formulated theories of narcosis, which contributed to the understanding of why narcotic and toxic activities are related to lipophilicity. No significant progress developed for the next 50–60 years. Fühner and Neubauer [7] realized that narcotic activities increase within homologous series in a geometrical progression (i.e., $1:3:3^2:3^3$); this observation gave a first indication of the additivity of group contributions in a logarithmical scale of activities. Ferguson provided a thermodynamic interpretation of nonspecific lipophilicity–activity relationships that also explained the "cutoff" of activities at a certain lipophilicity optimum [8].

A major breakthrough in theoretical organic chemistry resulted from the work of Hammett, around 1935. He derived Eqs. (3) and (4) to describe equilibrium constants K and rate constants k of various aromatic compounds by a certain reaction constant ρ, which depends on the reaction, and by substituent parameters σ, which only depend on the nature of the substituents X of the corresponding aromatic compounds, based on hydrogen as a reference substituent; ρ values are based on the ionization constants of substituted benzoic acids [9]:

$$\log K_{R-X} - \log K_{R-H} = \rho\sigma \tag{3}$$

$$\log k_{R-X} - \log k_{R-H} = \rho\sigma \tag{4}$$

In the following decades, various σ scales were derived for different systems and several attempts were made to derive such relationships also for biological activities of organic compounds. Bruice et al. [10] formulated group contributions to biological activity values in a series of thyroid hormone analogs, which may be considered as a first Free–Wilson-type analysis. Zahradnik and Chvapil [11] and Zahradnik [12,13] tried to apply the concept of the Hammett equation also to biological data (Eq. (5)):

$$\log \tau_i - \log \tau_{Et} = \alpha\beta \tag{5}$$

In this "biological Hammett equation," τ_i stands for the activity value of the ith member of a series, τ_{Et} is the biological activity value of the ethyl compound of the same series, β is a substituent constant (corresponding to the electronic σ parameter in the Hammett equation), and α is a constant characterizing the biological system, which corresponds to the Hammett reaction constant ρ. However, Eq. (5) only applies to nonspecific biological activities, most often within homologous series and only within a certain lipophilicity range.

In 1962, Hansen [14] derived a first Hammett-type relationship between the toxicities of substituted benzoic acids and the electronic σ constants of their substituents. However, later, it turned out that this was a chance correlation that only resulted from a close interrelationship between the Hammett σ parameter and the lipophilicity constant π (Sec. 4; Eqs. (42) and (43)). In the same year, for the very first time, a nonlinear multiparameter equation (Eq. 6) [15] was used to describe biological activity values:

$$\log 1/C = -2.14\pi^2 + 4.08\pi + 2.78\sigma + 3.36 \tag{6}$$

Time was ready for more general formulations (how to treat structure–activity relationships in a quantitative manner). Two independent publications, in 1964, must be considered as the start of modern QSAR methodology: one by Hansch and Fujita [16] on "ρ–σ–π Analysis. A method for the correlation of biological activity and chemical structure," and the other one by Free and Wilson [17] on "A mathematical contribution to structure activity studies." They described two new approaches for quantitative structure–activity relationships, later called Hansch analysis (linear free energy-related approach, extrathermodynamic approach) and Free–Wilson analysis. The real breakthrough in QSAR resulted from the combination of different phys- icochemical parameters in a linear additive manner (cf. Eq. (6), more generally for- mulated as Eq. (7); log $1/C$ is the logarithm of an inverse molar dose that produces or prevents a certain biological response, and log P is the logarithm of the n-octanol/ water partition coefficient P), as applied earlier in theoretical organic chemistry. Further contributions were the definition of a calculated lipophilicity parameter π (Eq. (8)), to be used instead of measured log P values (like Hammett, σ values are used to describe electronic properties), and the formulation of a parabolic equation for the quantitative description of nonlinear lipophilicity–activity relationships (Eq. (9)) [18– 20]:

$$\log 1/C = a \log P + b\sigma + \cdots + \text{const} \tag{7}$$

$$\pi_X = \log P_{R-X} \log P_{R-H} \tag{8}$$

$$\log 1/C = a(\log P)^2 + b \log P + c\sigma + \cdots + \text{const} \tag{9}$$

Considering a significant contribution by Fujita and Ban [21], the Free–Wilson model can be expressed by Eq. (10), where a_{ij} is the group contribution of a substituent X_i in the position j and μ is the (calculated) biological activity value of a reference compound within the series; all group contributions a_{ij} of the different substituents X_i refer to the corresponding substituents (most often being hydrogen) of this reference compound:

$$\log 1/C = \Sigma a_{ij} + \mu \tag{10}$$

Equations (9) and (10) constitute the fundament of all QSAR studies. Since 1964, they have remained essentially unchanged, with the exception of two minor modifi- cations. Improvements resulted from the combination of Hansch equations with indicator variables [22], which may be considered as a mixed Hansch/Free–Wilson model (Eq. (11)) [23], and from the formulation of a theoretically derived nonlinear model for transport and distribution of drugs in a biological system, the bilinear model (Sec. 4; Eq. (30)) [24]:

$$\log 1/C = a(\log P)^2 + b \log P + c\sigma + \cdots + \Sigma a_{ij} + \text{const} \tag{11}$$

For more detailed reviews on QSAR history, see Refs. 1–3,25,26.

2. HANSCH AND FREE–WILSON ANALYSES

The proper application of Hansch and Free–Wilson analyses and the associated problems can best be illustrated with a well-investigated example. The antiadrenergic activities of meta-, para-, and meta,para-disubstituted N,N-dimethyl-α-bromophene- thylamines have been investigated by Hansch and Lien [27], Unger and Hansch [28], Cammarata [29], and Kubinyi and Kehrhahn [30]. Table 1 presents the substituents, experimentally observed activity values, the parameter values for π, σ^+ and E_s^{meta}, as well as biological activity values calculated from Eqs. (14) and (16), respectively. To perform a Hansch analysis, the biological data are taken as Y values and an

Table 1 Antiadrenergic Activities and Physicochemical Properties of *meta*-, *para*-, and *meta,para*-Disubstituted N,N-Dimethyl-α-Bromophenethylamines

meta (X)	*para* (Y)	log $1/C$ observed	π	σ^+	E_s^{meta}	log $1/C$ calculated[a]	log $1/C$ calculated[b]
H	H	7.46	0.00	0.00	1.24	7.82	7.88
H	F	8.16	0.15	−0.07	1.24	8.09	8.17
H	Cl	8.68	0.70	0.11	1.24	8.46	8.60
H	Br	8.89	1.02	0.15	1.24	8.77	8.94
H	I	9.25	1.26	0.14	1.24	9.06	9.26
H	Me	9.30	0.52	−0.31	1.24	8.87	8.98
F	H	7.52	0.13	0.35	0.78	7.45	7.43
Cl	H	8.16	0.76	0.40	0.27	8.11	8.05
Br	H	8.30	0.94	0.41	0.08	8.30	8.22
I	H	8.40	1.15	0.36	−0.16	8.61	8.51
Me	H	8.46	0.51	−0.07	0.00	8.51	8.36
Cl	F	8.19	0.91	0.33	0.27	8.38	8.34
Br	F	8.57	1.09	0.34	0.08	8.57	8.51
Me	F	8.82	0.66	−0.14	0.00	8.78	8.65
Cl	Cl	8.89	1.46	0.51	0.27	8.75	8.77
Br	Cl	8.92	1.64	0.52	0.08	8.94	8.94
Me	Cl	8.96	1.21	0.04	0.00	9.15	9.08
Cl	Br	9.00	1.78	0.55	0.27	9.06	9.11
Br	Br	9.35	1.96	0.56	0.08	9.25	9.29
Me	Br	9.22	1.53	0.08	0.00	9.46	9.43
Me	Me	9.30	1.03	−0.38	0.00	9.56	9.47
Br	Me	9.52	1.46	0.10	0.08	9.35	9.33

π = lipophilicity parameter; σ^+ = Hammett constant for benzyl cations; E_s = Taft's steric parameter.
[a] Calculated by Eq. (14).
[b] Calculated by Eq. (16).
Source: Refs. 28 and 30.

appropriate selection of physicochemical parameters as the X values; then regression analysis is performed, using any standard software.

The very first quantitative model on this data set included a lipophilicity term π and an electronic term σ (Eq. (12)) [27]:

$$\log 1/C = 1.221\pi - 1.587\sigma + 7.888 \qquad (n = 22; \quad r = 0.918; \quad s = 0.238) \quad (12)$$

Cammarata [29] reinvestigated the data and came up with some "better" models (e.g., Eq. (13)):

$$\log 1/C = 0.747(\pm 0.12)\pi_m - 0.911(\pm 0.25)\sigma_m + 1.666(\pm 0.12)r_v^{para}$$
$$+ 5.769 \qquad (n = 22; \quad r = 0.961; \quad s = 0.164) \quad (13)$$

This model was criticized by Unger and Hansch [28]. The r_v^{para} term would indicate that an increase in substituent size, expressed by the van der Waals radius, would increase biological activities, which has no biological meaning at all. The α-bromophenethylamines are unstable at physiological pH values; most probably, they exert their biological action via the formation of cyclic ethyleneiminium ions, a species that is best described by σ^+ (a parameter that was derived earlier for the chemical reactivity of benzyl cations) instead of σ values (Table 1; Eq. (14)) [28]. An improvement of the quantitative description of the biological data could be achieved by a separation of the lipophilic and electronic effects in the *meta-* and *para*-positions (Eq. (15)) [28]:

$$\log 1/C = 1.15\pi - 1.47\sigma^+ + 7.82 \qquad (n = 22; \quad r = 0.994; \quad s = 0.197) \tag{14}$$

$$\log 1/C = 0.83(\pm 0.27)\pi_m + 1.33(\pm 0.20)\pi_p - 0.92(\pm 0.50)\sigma_m^+$$
$$-1.89(\pm 0.57)\sigma_p^+ + 7.80 \qquad (n = 22; \quad r = 0.966; \quad s = 0.164) \tag{15}$$

A further investigation of the same biological data led to Eq. (16) [30]; in this equation, the positive sign of the E_s^{meta} term makes sense because the E_s^{meta} scale is an inverse scale. The steric term indicates that an increase in the size of the *meta*-substituents decreases biological activity, most probably due to steric hindrance:

$$\log 1/C = 1.259(\pm 0.19)\pi - 1.460(\pm 0.34)\sigma^+ + 0.208(\pm 0.17)E_s^{meta}$$
$$+ 7.619 \qquad (n = 22; \quad r = 0.959; \quad s = 0.173) \tag{16}$$

Free Wilson analysis [31,32] is much easier to apply. Biological activity values are correlated with indicator variables, which, for each position of substitution and every substituent, indicate the presence (value 1) or absence (value 0) of the corresponding substituent (Table 2). If there is more than one substituent in a certain position or if symmetrical positions (e.g., *meta,meta'*-disubstituted compounds) are condensed into one variable, numbers of two or higher are used instead of one. Regression analysis leads to Eq. (17) [30–32]:

$$\log 1/C = -0.301(\pm 0.50)[m\text{-}F] + 0.207(\pm 0.29)[m\text{-}Cl]$$
$$+ 0.434(\pm 0.27)[m\text{-}Br] + 0.579(\pm 0.50)[m\text{-}I]$$
$$+ 0.454(\pm 0.27)[m\text{-}Me] + 0.340(\pm 0.30)[p\text{-}F]$$
$$+ 0.768(\pm 0.30)[p\text{-}Cl] + 1.020(\pm 0.30)[p\text{-}Br]$$
$$+ 1.429(\pm 0.50)[p\text{-}I] + 1.256(\pm 0.33)[p\text{-}Me]$$
$$+ 7.821(\pm 0.27) \qquad (n = 22; \quad r = 0.969; \quad s = 0.194; \quad F = 16.99) \tag{17}$$

Equations (12) to (17) already show some of the problems in QSAR studies. First, Hansch models are never unique. Several related models of comparable quality but also meaningless models can be derived. A "good" correlation is never a proof of the validity of a model (cf. the storks and the babies). Which model is really "good" can only be checked by successful predictions of compounds outside the investigated series of compounds. Free–Wilson analysis is straightforward and unique. However, chemical variation must occur in at least two positions of a molecule; otherwise, the algorithm could not be applied. Correspondingly, Free–Wilson analysis wastes a large number of degrees of freedom. Predictions are possible only for new combinations of substituents, not for any other substituents. But also in the case of Hansch analysis, predictions far outside the investigated chemical space are risky. In the example discussed above, extrapolations will be possible for ethyl or trifluoromethyl groups

Table 2 Antiadrenergic Activities of *meta*-, *para*-, and *meta,para*-Disubstituted *N,N*-Dimethyl-α-Bromophenethylamines and Matrix for Free–Wilson Analysis

meta (X)	para (Y)	log 1/C observed	meta- F	Cl	Br	l	Me	para- F	Cl	Br	l	Me	log 1/C calculated
H	H	7.46											7.82
H	F	8.16						1					8.16
H	Cl	8.68							1				8.59
H	Br	8.89								1			8.84
H	I	9.25									1		9.25
H	Me	9.30										1	9.08
F	H	7.52	1										7.52
Cl	H	8.16		1									8.03
Br	H	8.30			1								8.26
I	H	8.40				1							8.40
Me	H	8.46					1						8.28
Cl	F	8.19		1				1					8.37
Br	F	8.57			1			1					8.60
Me	F	8.82					1	1					8.62
Cl	Cl	8.89		1					1				8.80
Br	Cl	8.92			1				1				9.02
Me	Cl	8.96					1		1				9.04
Cl	Br	9.00		1						1			9.05
Br	Br	9.35			1					1			9.28
Me	Br	9.22					1			1			9.30
Me	Me	9.30					1					1	9.53
Br	Me	9.52			1							1	9.51

Source: Refs. 30–32.

because they are chemically related to the substituents included in the investigation; predictions for hydroxyl, amino, or nitro groups will most probably fail.

Free–Wilson analysis can be used for a first inspection of biological activity data [30–32]. The values of the group contributions indicate which physicochemical properties might be responsible for the variations in biological activity values and whether nonlinear lipophilicity–activity relationships are involved. Free–Wilson contributions can be derived from Hansch equations (e.g., by Eq. (18) from Eq. (14), or by Eq. (19) from Eq. (15)) [30]:

$$a_i = 1.15\pi - 1.47\sigma^+ \tag{18}$$

$$a_i = 0.83\pi_{meta} + 1.33\pi_{para} - 0.92\sigma^+_{meta} - 1.89\sigma^+_{para} \tag{19}$$

On the other hand, Free–Wilson analysis can be used to ease the derivation of a Hansch model. An inspection of the coefficients of the different terms in Eq. (17) shows that activity increases with increasing lipophilicity of the halogens, that methyl leads to a higher activity than to be expected from its lipophilicity (corresponds to chlorine), and that *meta*-substituents have smaller activity contributions than the *para*-substituents. This investigation of the Free-Wilson parameters led to the Hansch model presented in Eq. (16) [30].

3. VALIDATION AND SELECTION OF QSAR MODELS

In the context of the reinvestigation of Hansch models for the substituted *N,N*-dimethyl-α-bromophenethylamines, Unger and Hansch [28] formulated recommendations for the proper derivation of extrathermodynamic equations (supplementary comments are given in parentheses):

1. Selection of independent variables. A wide range of different parameters, such as log P or π, σ, MR, and steric parameters, should be tried; molecular orbital (MO) parameters and indicator variables should not be overlooked. The parameters selected for the "best equation" should be essentially independent [i.e., the intercorrelation coefficients r should not be larger than 0.6–0.7; exceptions are combinations of linear and squared terms, such as $(\log P)^2$ and log P, which are usually highly interrelated, with r values > 0.9].

2. Justification of the choice of independent variables. All "reasonable" parameters must be validated by an appropriate statistical procedure (e.g., by stepwise regression analysis). The "best equation" is normally the one with the lowest standard deviation, all terms being significant (indicated by the 95% confidence intervals or by a sequential F test). Alternatively, the equation with the highest overall F value may be selected as the "best" model (nowadays crossvalidation and/or Y-scrambling are recommended as validation tools).

3. Principle of parsimony (Occam's Razor; William of Ockham, 1285–1349/1350, English philosopher and logician). All things being (approximately) equal, one should accept the simplest model.

4. Number of terms. One should have at least five to six data points per variable to avoid chance correlations (this rule applies only to data sets of intermediate size; for small data sets, more parameters may be allowed if they are based on a reasonable model; for large data sets, e.g., $n > 30$, this recommendation leads to equations that include too many variables).

5. Qualitative model. It is important to have a qualitative model that is consistent with the known physical–organic and biomedicinal chemistry of the process under consideration, in order to avoid chance correlations (i.e., "statistical unicorns, beasts that exist on paper but not in reality") [28].

Today, more powerful tools are available for the validation of QSAR models. The most important one is crossvalidation, a procedure in which one object (i.e., compound) or more objects are eliminated from the data set and a model is derived, using only the other compounds. This model is used to predict the biological activity value(s) of the eliminated compound(s). If only one compound is eliminated at a time, in the so-called "leave-one-out" crossvalidation, all objects are eliminated once and only once; a squared crossvalidation correlation coefficient Q^2 (can also have negative values, if the predictivity of the model is worse than "no model", i.e., using y_{mean} values) and a standard deviation s_{PRESS} (PRESS = sum of squared errors of predictions) are calculated, using only the predicted values of the compounds that were not included in the respective models [33]. If crossvalidation is performed in groups, the elimination of groups of compounds is performed many times, in a random manner.

Occasionally, it was observed that models with good internal predictivity did not perform well on external test sets [34,35]. In principle, one would expect that good internal predictivity Q^2 should result in good external predictivity values r^2_{pred} for a test set, which the statistical procedure has never seen. However, there is no relationship at all, as confirmed by systematic investigations [36,37]. High Q^2 values may be associated with very poor r^2_{pred} values and vice versa, a lack of relationship that has recently been called the "Kubinyi paradox" [38]. One outlier, either in the training set or in the test set, can already be responsible for such differences; the only way to avoid such problems seems to be a selection of test and training sets where all compounds cover about the same chemical space, and to eliminate real outliers (compounds that differ from the others in some important feature, not just analogs that do not fit the model for whatever reason).

Another powerful validation procedure is Y-scrambling (i.e., the random shuffling of all biological activity values without any changes in the X variables block; e.g., Ref. 36). If any model resulting from Y-scrambling has a comparable quality of fit, this is a strong indication of a chance correlation for the original data set as well as for the scrambled data set; of course, one should check in such a case whether the scrambled Y values are, by fortune, correlated with the original Y values. Alternatively, one could investigate whether 95% of the models from Y-scrambling are inferior to the original model and accept this as a 95% confidence value for the validity of the model; if a higher percentage of the Y-scrambled models performs worse, the better.

As illustrated in Fig. 1, the final model should include the following information (Eq. (14), recalculated from the original data).

The most important statistical parameters r, s, and F and the 95% confidence intervals of the regression coefficients are calculated by Eqs. (20) to (23) (for details on Eqs. (20) to (23), see Refs. 39 to 42). For more details on linear (multiple) regression analysis and the calculation of different statistical parameters, as well as other validation techniques (e.g., the jackknife method and bootstrapping), see Refs. 33,39–42:

$$r^2 = 1 - \Sigma\Delta^2/S_{yy} \tag{20}$$

$$s^2 = \Sigma\Delta^2/(n - k - 1) \tag{21}$$

$$F = r^2(n - k - 1)/(k(1 - r^2)) \tag{22}$$

$$\text{confidence intervals of } k_i = \pm st\sqrt{c_{ii}} \tag{23}$$

When more X variables are to be investigated, a proper variable selection method has to be applied. Too many variables in the final model and even too many tested variables will significantly increase the risk of mere chance correlations, as already shown by Topliss and Costello [43] and Topliss and Edwards [44]. However, whereas it is desirable to derive a qualitative model first and to test only this one model (good practice in statistics), reality differs from this idealistic view. Thus, the problem arises on how to derive "good" and reliable models from such large sets of different X variables.

Forward selection (i.e., starting from the best single variable regression and adding further variables) by selecting in every step the one that improves the fit to the

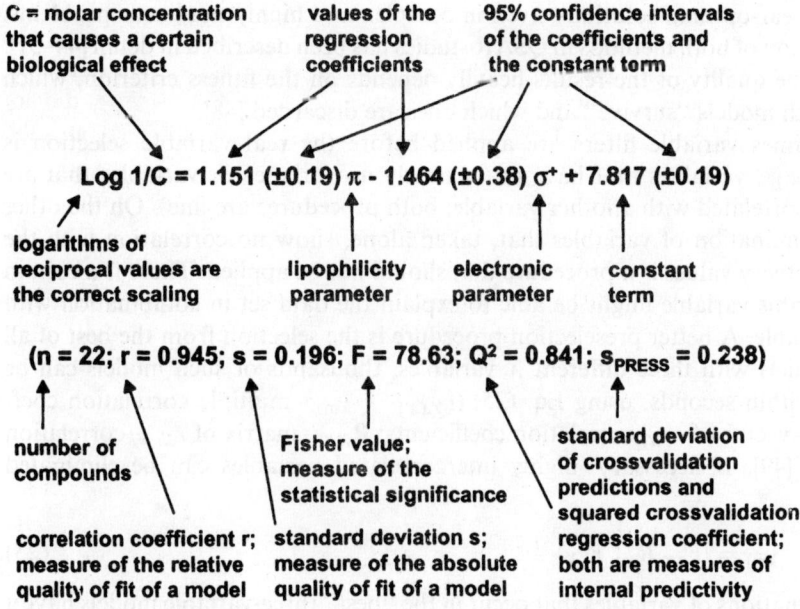

Figure 1 A QSAR equation (cf. Eq. (14); recalculated from original values) should contain 95% confidence intervals of all regression coefficients (preferable to standard deviations of regression coefficients), the number of objects (=compounds) included in the analysis, the correlation coefficient r or a squared correlation coefficient r^2, the standard deviation s, the Fisher F value (optional), the crossvalidation correlation coefficient Q^2, and the crossvalidation standard deviation s_{PRESS} (both optional but recommended).

largest extent, until no further improvement can be obtained (most often checked by a sequential F test, Eq. (24); k_1,r_1 = number of variables and correlation coefficient of the model with the smaller number of variables, k_2,r_2 = number of variables and correlation coefficient of the model with the larger number of variables) does not help too much. If several variables are involved, the procedure most often ends up in a local optimum whereas the "best" model (at least from its statistical parameters) is missed,

$$\text{sequential } F = \frac{(r_2^2 - r_1^2)(n - k_2 - 1)}{(k_2 - k_1)(1 - r_2^2)} \tag{24}$$

Backward elimination starts from all variables and eliminates the one that contributes least to the model (i.e., which produces the lowest partial F test value). This procedure is repeated until no more X variables can be eliminated because the remaining ones are all justified by their sequential F values. However, also this procedure often ends up in a local optimum; in addition, it cannot be applied if the number of tested variables is larger than the number of objects. Stepwise regression avoids some of these problems. It is a forward selection procedure where, after every addition of a new variable, the possible elimination of any other one is checked by a sequential F test [41].

Much better variable selection procedures are genetic algorithms (GAs) and evolutionary algorithms (EAs). In general, GAs and EAs are well suited to find optimal or near-optimal solutions, even in complex and highly nonlinear problems. The application of both methods in QSAR studies has been described in detail [44–51]. Of course, the quality of the results heavily depends on the fitness criterion, which decides which models "survive" and which ones are discarded [48].

Sometimes variable filters are applied before the real variable selection is performed (e.g., variables that have no or nearly no variance or variables that are highly intercorrelated with another variable; both procedures are fine). On the other hand, the elimination of variables that, taken alone, show no correlation with the biological activity values is a procedure that should not be applied. There is a certain chance that this variable might be able to explain the data set in combination with another variable. A better preselection procedure is the selection from the best of all possible models with three different X variables; thousands of such models can be calculated within seconds, using Eq. (25) ($r_{Y,(X1,\ldots,Xm)}$ = multiple correlation coefficient; r_{YX} = vector of r_{YXi} correlation coefficients; $\mathbf{R}_{XX}$ = matrix of $r_{Xi,Xj}$ correlation coefficients) [49]. If necessary, highly intercorrelated variables can be eliminated afterward:

$$r_{Y,(X1,\ldots,Xm)} = (r_{YX}^{T} \, \mathbf{R}_{XX}^{-1} \, r_{YX})^{-1/2} \tag{25}$$

Combinations of variables that occur in the "best" three-variable models have a higher chance to yield explanatory models than those that do not show up [49,50]. The final QSAR model should explain the data set with the smallest number of variables and should yield good predictions for test set compounds.

In the case of many X variables, principal component analysis (PCA) or partial least squares (PLS) analysis (for reviews, see, e.g., Ref. 52) can be used instead of regression analysis, often leading to more stable models. Both methods are discussed in other chapters of this book (see, e.g., Chapters 22 and 25).

4. NONLINEAR QSAR MODELS

Penniston et al. [53] compared the absorption of drugs and its distribution in a biological organism with a "random walk," where the drug has to cross several aqueous and organic phases to arrive from its site of application at a receptor site. Polar drugs will not reach the receptor site due to their inability to permeate lipid membranes, whereas lipophilic drugs will enter the membranes and will not pass through aqueous phases. Only compounds with intermediate lipophilicity cross all these barriers and have a chance to arrive at the binding site of their receptor, in reasonable time and concentration. To model this process, he formulated a parabolic model (Eqs. (9) and (26); Fig. 2) [16,18–20,53], where a linear and a quadratic lipophilicity term may be combined with any other physicochemical terms. The lipophilicity terms describe the transport and distribution, whereas part of the linear lipophilicity term and the other physicochemical terms model the drug–receptor interaction:

$$\log 1/C = a\pi^2 + b\pi + c\sigma + MR + \cdots + const \tag{26}$$

Besides the kinetics of drug transport, also the equilibrium distribution of polar compounds (preferably in the aqueous phases), compounds with intermediate lipophilicity (in both phases), and lipophilic compounds (preferably in the lipid phases)

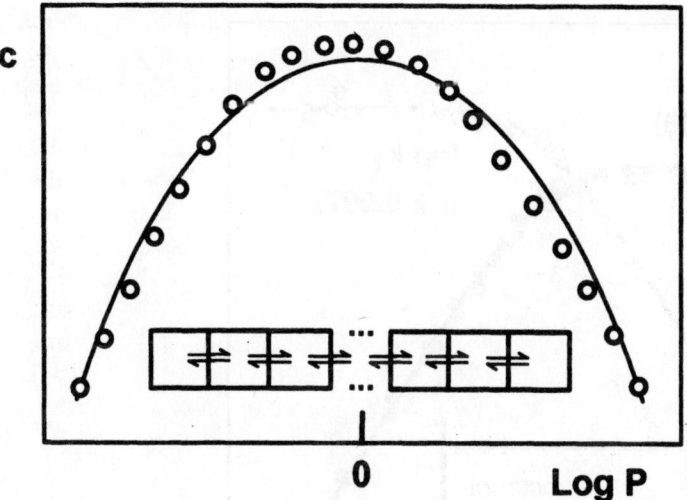

Figure 2 A simulation of drug transport through various aqueous and membrane phases—using the (wrong) assumption that the product of the rate constants of drug transport, k_1 and k_2 (compare Eqs. (28) and (29)), is equal to one—generated a curve that can be approximated by a parabola. (From Ref. 53.)

have been made responsible for nonlinear lipophilicity–activity relationships, as well as steric hindrance at the binding site, allosteric effects, different pharmacokinetics and/or metabolism of lipophilic analogs, poor solubility and/or micelle formation of higher analogs of a series, end product inhibition of enzymes by lipophilic compounds, and, last but not least, the principle of drug receptor occupation that requires that a minimum number of compounds have to interact with receptors to exert a biological effect [20]. Whereas some of these effects may indeed be responsible in certain cases, the kinetics of drug transport is the general reason for nonlinear lipophilicity–activity relationships, as can be seen from the experimental investigation of the rate constants of drug transport by Lippold and Schneider [54,55] and van de Waterbeemd et al. [56,57] (Fig. 3). From the data of Lippold and Schneider, Eq. (27) could be derived for the relationship between k_1, the rate constant of transport from the aqueous phase into the organic phase, and k_2, the rate constant in the reverse direction. Together with the definition of the partition coefficient $P = k_1/k_2$, Eq. (27) led to Eqs. (28) and (29) for the description of the rate constants k_1 and k_2 as functions of the partition coefficient P (Fig. 3) [58]:

$$k_2 = -\beta k_1 + c \tag{27}$$
$$\log k_1 = \log P - \log (\beta P + 1) + c \tag{28}$$
$$\log k_2 = -\log (\beta P + 1) + c \tag{29}$$

The parabolic Hansch model is a good approximation of observed nonlinear structure–activity relationships. However, whereas the left and right sides of a parabola are always nonlinear, many nonlinear lipophilicity relationships show linear left and right sides, as also observed for the function describing the rate constants of

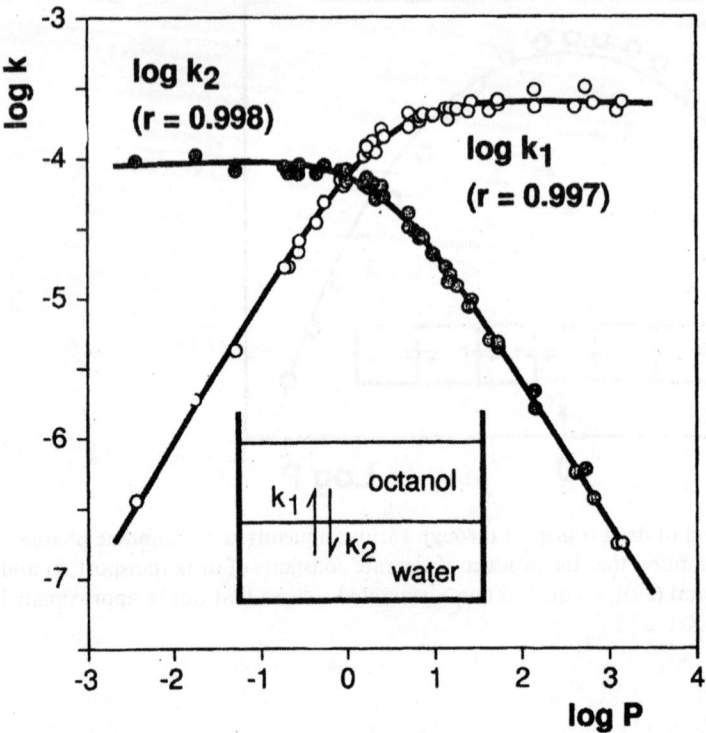

Figure 3 Experimental rate constants of drug transport of 30 sulfonamides and 15 standard compounds, including several drugs. (From Refs. 56 and 57.) The rate constants k_1 (from the aqueous phase into the organic phase) and k_2 (the reverse rate constant) are fitted by Eqs. (28) and (29).

drug transport. In addition, a parabola is always symmetrical whereas many observed relationships are unsymmetrical, some even leading to a plateau for lipophilic analogs. Due to this fact, several empirical models and models based on theoretical considerations were derived as alternatives to the parabolic model (for reviews, see e.g., Refs. 59 and 60). Neither one could describe all different kinds of nonlinear structure–activity relationships. Of all these models, the McFarland [61] model came closest to reality. It started from a consideration of the probability of a drug molecule, located at the aqueous–organic interface, to enter either the aqueous phase or the organic phase. Corresponding to the random walk model, symmetrical curves with linear ascending and descending sides and a nonlinear part at $\log P = 0$ resulted in the transport of a drug from an aqueous phase, through various lipid and aqueous phases, at a distant aqueous phase. If McFarland had added another lipid phase, to consider also a receptor site, an unsymmetrical curve would have resulted.

The bilinear model, in combination with other physicochemical properties (Eq. (30)) [24,62], was the first mathematical expression to describe nonlinear lipophilicity–activity relationships, precisely and in a flexible manner. Besides pharmacokinetic properties, such as absorption, distribution, elimination, and permeation of the blood–brain and blood–placenta barriers, it could also fit unspecific antibacterial, antifungal, hemolytic, narcotic, and toxic activities [60,63,64]:

$$\log 1/C = a\log P - b\log (\beta P + 1) + c\sigma + d\,\mathrm{MR} + \cdots + \mathrm{const} \tag{30}$$

The principal advantage of the bilinear model as compared to the parabolic model can be illustrated with the following example. For the antibacterial activity of homologous amines toward *Rhinocladium beurmannii*, a parabolic relationship (Eq. (31)) is only a rough approximation of the real structure–activity relationship; as often, also the position of the lipophilicity optimum is not well described (Fig. 4). Lien and Wang [65] arrived at a better quantitative model by additionally including a log MW parameter (Eq. (32)). However, Eq. (32) is a chance correlation, as can be seen from the meaningless large coefficients of the individual terms; log P and log MW are highly intercorrelated ($r^2 = 0.969$). A better fit of the data is obtained using the bilinear model, without any additional MW term (Eq. (33)) [60,64]:

$$\log 1/C = -0.199(\pm 0.048)(\log P)^2 + 2.119(\pm 0.42)\log P - 1.382(\pm 0.80)$$

$$\log P \text{ optimum} = 5.31(5.02/5.730) \tag{31}$$

$$(n = 15; \quad r = 0.967; \quad s = 0.354; \quad F = 85.6)$$

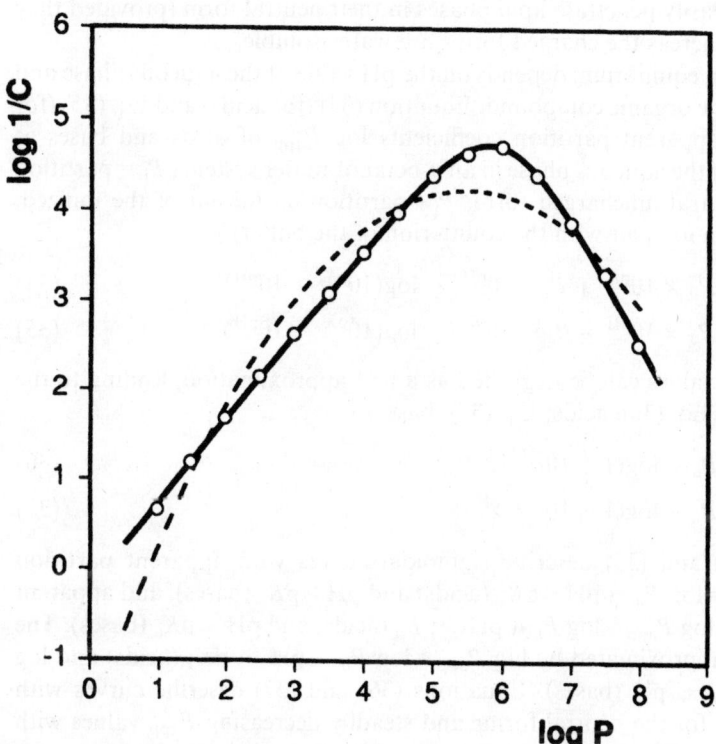

Figure 4 Experimental data for the inhibition of *Rhinocladium beurmanii* by aliphatic amines, fitted by a parabolic model (dotted line; Eq. (31)) (from Ref. 65) and by the bilinear model (Eq. (33)). (From Refs. 60 and 64.)

$$\log 1/C = -0.501(\pm 0.090)(\log P)^2 + 8.049(\pm 1.73)\log P$$
$$- 43.01(\pm 12.5)\log MW + 74.80(\pm 22.1)$$

$\log P$ optimum $= 8.04(7.63/8.35)$

$(n = 15; \quad r = 0.995; \quad s = 0.148; \quad F = 345.1)$

$$(32)$$

$$\log 1/C = 0.944(\pm 0.014)\log P - 2.347(\pm 0.047)\log (\beta P + 1)$$
$$- 0.0534(\pm 0.049) \qquad \log \beta = -5.787$$

$\log P$ optimum $= 5.62 \qquad (n = 15; \quad r = 1.000; \quad s = 0.031; \quad F = 7945)$

$$(33)$$

In addition, bilinear lipophilicity–activity and/or molar refractivity–activity relationships are frequently observed for ligand–enzyme interactions (cf. Sec. 6) [60,64,66–68]. In these relationships, they model the limited size of a lipophilic binding pocket.

5. DISSOCIATION AND IONIZATION

Quantitative structure–activity models for acids and bases are much more complex than models for neutral compounds. Whereas the neutral forms of acids and bases are more or less lipophilic, their charged forms, anions of acids or protonated bases, are more hydrophilic by about three to four orders of magnitude. Correspondingly, many acids and bases can easily penetrate lipid phases in their neutral form (provided they are not too polar), whereas the charged forms are water-soluble.

The dissociation equilibrium depends on the pH value of the aqueous phase and on the pK_a value of the organic compound. Equation (34) (for acids) and Eq. (35) (for bases) describe the apparent partition coefficients $\log P_{app}$ of acids and bases at different pH values of the aqueous phase in an n-octanol/buffer system (P_u = partition coefficient of the neutral, uncharged form; P_i = partition coefficient of the ionized, charged form or of an ion pair with the counterion of the buffer):

$$\log P_{app} = \log(P_u \times 10^{pK_a} + P_i \times 10^{pH}) - \log(10^{pK_a} + 10^{pH}) \qquad (34)$$

$$\log P_{app} = \log(P_u \times 10^{pH} + P_i \times 10^{pK_a}) - \log(10^{pK_a} + 10^{pH}) \qquad (35)$$

Most often, P_i values can be neglected as a first approximation, leading to the simplified equations (Eq. (36), acids; Eq. (37), bases):

$$\log P_{app} = \log P_u - \log(1 + 10^{pH-pK_a}) \qquad (36)$$

$$\log P_{app} = \log P_u - \log(1 + 10^{pK_a-pH}) \qquad (37)$$

Equations (34) and (35) describe sigmoidal curves with apparent partition coefficients $\log P_{app} = \log P_u$ at pH $< pK_a$ (acids) and pH $> pK_a$ (bases), and apparent partition coefficients $\log P_{app} = \log P_i$ at pH $\gg pK_a$ (acids) and pH $\ll pK_a$ (bases). The range in between is approximated by $\log P_{app} \approx \log P_u - pH + pK_a$ (acids) and $\log P_{app} \approx \log P_u - pK_a + pH$ (bases). Equations (36) and (37) describe curves with constant P_{app} values for the neutral forms and steadily decreasing P_{app} values with increasing ionization, at pH $> pK_a$ for acids and pH $< pK_a$ for bases.

Under equilibrium conditions, pH absorption profiles are identical with pH distribution profiles ("pH partitioning hypothesis" [69,70]). However, absorption is most often kinetically controlled. Correspondingly, the pH absorption profiles of lipophilic acids and bases are not any longer identical with their pH distribution profiles. A pH shift (Fig. 5) [71,72] is observed, which results from a rapid dissociation equilibrium close to the aqueous–organic interface that steadily generates neutral species, which are immediately distributed into the organic phase.

In an attempt to formulate a QSAR model for the colonic absorption of aromatic acids, phenols, and other acidic compounds, Lien [73] derived Eq. (38):

$$\log\% \ \text{ABS} = 0.156(\pm 0.08)(pK_a - pH) + 0.366(\pm 0.44)\log P$$

$$+ 0.755 \qquad (n = 10; \quad r = 0.866; \quad s = 0.258) \tag{38}$$

This model is wrong for two reasons. Log % absorption values are used instead of rate constants, and differences pK_a–pH are used to model the pK_a/pH dependence of the colonic absorption (valid only if pH $<$ pK_a). The correct pK_a/pH correction procedure was applied by Scherrer and Howard [74] (Eq. (39)). They performed a nonlinear transformation of the percentage values (log %ABS) to absorption rate constants (log k_{abs}) and converted the log P values into log D values (distribution values between n-octanol and an aqueous buffer of a certain pH value, nowadays most often called log P_{app}) by Eqs. (36) and (37):

$$\log k_{abs} = -0.078(\pm 0.041)(\log P_{app})^2 + 0.265(\pm 0.045)\log P_{app}$$
$$-0.425 \tag{39}$$

$$\text{optimum} \ \log P_{app} = 1.70 \qquad (n = 10; \quad r = 0.984; \quad s = 0.102; \quad F = 105.9)$$

An even better description of the experimental data can be achieved by application of the bilinear model (Eq. (40)) [60,64]:

$$\log k_{abs} = 1.024(\pm 0.31)\log P_{app} - 0.881(\pm 0.36)\log (\beta P_{app} + 1) + 0.935$$
$$\log \beta = 1.600; \qquad \text{no optimum of} \ \log P_{app} \tag{40}$$
$$(n = 10; \quad r = 0.991; \quad s = 0.081; \quad F = 112.9)$$

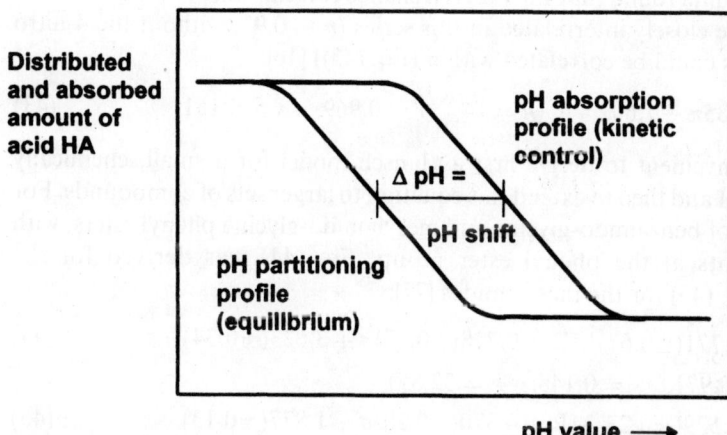

Figure 5 A pH shift is observed for the absorption of lipophilic acids and bases, That is, the pH absorption profile is shifted to the right (for acids, shown in this schematic diagram) or to the left (bases), with respect to the pH–distribution profile. (From Refs. 71 and 72.)

The use of log P_{app} values is appropriate for absorption rate constants but not for binding or equilibrium constants. In such cases, the biological activity parameters have to be corrected for the concentration of either the neutral or the ionized form, dependent on the species that interacts with the protein [75]. An investigation of the inhibition of monoaminoxidase by alcohols and amines shows that the amines ($I = 0$) have much higher activities than the alcohols ($I = 1$) and that the inhibitory activity of the amines decreases with increasing pH values. The corresponding QSAR equation (Eq. (41)) uses pK_a/pH-corrected log $1/K_i$ values [59,60]:

$$\log 1/K_i^{corr} = \log 1/K_i + \log (1 + 10^{K_a - pH}) = 3.130(\pm 0.17)\log P$$
$$- 3.797(\pm 0.32)\log (\beta P + 1) - 3.507(\pm 0.12)I$$
$$+ 3.379(\pm 0.15) \tag{41}$$

$$\log \beta = -1.781; \quad \text{optimum} \log P = 2.45$$

$$(n = 21; \quad r = 0.999; \quad s = 0.118; \quad F = 1737)$$

6. APPLICATIONS OF QSAR ANALYSES: SCOPE AND LIMITATIONS

Of the numerous applications of Free–Wilson and Hansch analyses (the BioByte QSAR Database [76] contains about 9400 data sets on biological applications and another 9000 physicochemical relationships, July 2003) [60,66–68], only a few shall be discussed here to illustrate the most important features and problems.

Equation (42) was derived for the toxicity of substituted benzoic acids against mosquito larvae, after elimination of a 4-nitro analog [14]; a much worse fit ($r = 0.711$) results if the 4-nitro analog is included:

$$\log 1/C = 1.454\sigma + 1.787 \quad (n = 13; \quad r = 0.918; \quad s = 0.243) \tag{42}$$

Hansch and Fujita found that Eq. (42) is a chance correlation, resulting from the fact that π and σ are closely interrelated in this series ($r = 0.91$, without the 4-nitro analog). All analogs could be correlated with π (Eq. (43)) [16]:

$$\log 1/C = 0.535\pi + 1.602 \quad (n = 14; \quad r = 0.969; \quad s = 0.151) \tag{43}$$

Often it is convenient to derive first a Hansch model for a small, chemically closely related subset and then to extend this equation to larger sets of compounds. For the papain binding of benzamido-glycine and mesylamido-glycine phenyl esters, with different substituents at the phenyl ester group, Eq. (44) was derived for the benzamides and Eq. (45) for the mesylamides [77]:

$$\log 1/K_m = 0.771(\pm 0.67)MR + 0.728(\pm 0.37)\sigma + 3.623(\pm 0.34) \tag{44}$$
$$(n = 7; \quad r = 0.971; \quad s = 0.148; \quad F = 32.85)$$
$$\log 1/K_m = 0.529(\pm 0.23)MR + 0.370(\pm 0.20)\sigma + 1.877(\pm 0.13) \tag{45}$$
$$(n = 13; \quad r = 0.935; \quad s = 0.105; \quad F = 34.51)$$

Whereas there are no significant differences in the regression coefficients of the MR and σ terms of Eqs. (44) and (45), the constant terms are quite different, indicating that the benzamides are about two orders of magnitude more potent than the corresponding mesylamides. Indeed, both data sets can be described by Eq. (46), in which an indicator variable I ($I = 0$ for benzamides; $I = 1$ for mesylamides) accounts for the activity differences in both series [77]:

$$\log 1/K_m = 0.569(\pm 0.26)MR + 0.561(\pm 0.19)\sigma - 1.922(\pm 0.15)I$$
$$+ 3.743(\pm 0.17) \tag{46}$$
$$(n = 20; \quad r = 0.990; \quad s = 0.148; \quad F = 272.04)$$

To check whether hydrophobic interactions are responsible for the variation in biological activities at the "amide side" of these compounds, a series of benzamido-glycine esters with different substitution in the benzamide group but no structural variation in the ester part was investigated. The π term of Eq. (47) indeed confirms a hydrophobic ligand–enzyme interaction at this side of the molecule [78]:

$$\log 1/K_m = 1.01(\pm 0.11)\pi + 1.46 \qquad (n = 16; \quad r = 0.981; \quad s = 0.165) \tag{47}$$

The stepwise derivation of Hansch models is further illustrated by equations that were derived for the inhibition of chymotrypsin by thiophosphonates. First, Eq. (48) correlated data where chemical variation was performed only in R_2 of Me-$P(=O)(OR_2)SR_3$; then further analogs with chemical variation in R_3 were included and finally an indicator variable I accounted for charged groups in some residues R_3 (Eq. (49)) [79]:

$$\log K_i = 1.60(\pm 0.22)MR_2 - 3.85(\pm 1.17)\log(\beta \times 10^{MR_2} + 1)$$
$$- 4.76(\pm 0.51)$$
$$\log \beta = -3.86; \text{ optimum } MR_2 = 3.72 \tag{48}$$
$$(n = 19; \quad r = 0.978; \quad s = 0.258)$$

$$\log K_i = 1.47(\pm 0.10)MR_2 - 3.43(\pm 0.74)\log(\beta \times 10^{MR_2} + 1)$$
$$+ 0.34(\pm 0.09)MR_3 + 1.25(\pm 0.19)\sigma_3^* - 1.06(\pm 0.31)I$$
$$- 5.26(\pm 0.38) \tag{49}$$
$$\log \beta = -3.85; \quad \text{optimum } MR_2 = 3.71$$
$$(n = 53; \quad r = 0.985; \quad s = 0.243)$$

In this manner, QSAR models can be derived for very large and heterogeneous data sets (e.g., Eq. (50) for the antimalarial activity of substituted phenanthrenes, quinolines and pyridines) [80]:

$$\log 1/C = 0.576(\pm 0.09)\Sigma\sigma + 0.168(\pm 0.05)\Sigma\pi + 0.105(\pm 0.05)\log P$$
$$- 0.167(\pm 0.07)\log(\beta P + 1) - 0.169(\pm 0.10)c\text{-side}$$
$$+ 0.319(\pm 0.136)CNR_2 - 0.139(\pm 0.06)AB - 0.795(\pm 0.06)$$
$$< 3\text{-cures} + 0.278(\pm 0.11)MR - 4' - Q + 0.252(\pm 0.18) \tag{50}$$
$$\text{Me-6.8} - Q + 0.084(\pm 0.10)2\text{-Pip} + 0.151(\pm 0.19)NBrPy$$
$$- 0.683(\pm 0.22)Q2P378 + 0.267(\pm 0.11)Py + 2.726(\pm 0.15)$$
$$\log \beta = -3.959; \quad \text{optimum } \log P = 4.19$$
$$(n = 646; \quad r = 0.898; \quad s = 0.309)$$

In the early time of QSAR studies, often meaningless combinations of quantum chemical parameters were used to correlate biological data. The proper application of such parameters is, for example, illustrated by Eq. (51), which describes the mutagenic activity of various nitro-substituted aromatic compounds in a *Salmonella typhimurium* TA$_{98}$ strain [81]:

$$\log TA_{98} = 0.65(\pm 0.16)\log P - 2.90(\pm 0.59)\log (\beta P + 1)$$
$$-1.38(\pm 0.25)\varepsilon_{LUMO} + 1.88(\pm 0.39)I_1 - 2.89(\pm 0.81)I_a$$
$$-4.15(\pm 0.58)$$

(51)

$$\log \beta = -5.48; \quad \text{optimum} \log P = 4.93 \qquad (n = 188; \quad r = 0.900; \quad s = 0.886)$$

Most often, it is important to inspect a diagram before deriving a quantitative model. Saxena investigated the interrelationship between molecular connectivity values $^1\chi$ and $^2\chi^v$, molar refractivity MR, and log P for a series of 183 organic molecules, including hydrocarbons, alcohols, ethers, esters, carboxylic acids, amines, and ketones. Saxena observed the following relationships: MR vs. $^1\chi$: ($r = 0.908$; $s = 0.380$; $F = 855.26$); MR vs. $^2\chi^v$: ($r = 0.826$; $s = 0.419$; $F = 389.58$); log P vs. $^1\chi$: ($r = 0.719$; $s = 0.632$; $F = 193.36$); log P vs. $^2\chi^v$: ($r = 0.635$; $s = 0.574$; $F = 122.33$) [82]. From the results, Saxena concluded that connectivity values are more closely related to molar refractivity than they are to lipophilicity. However, a plot of $^1\chi$ against log P (Fig. 6) clearly shows that there are two linear relationships: one for the hydrocarbons

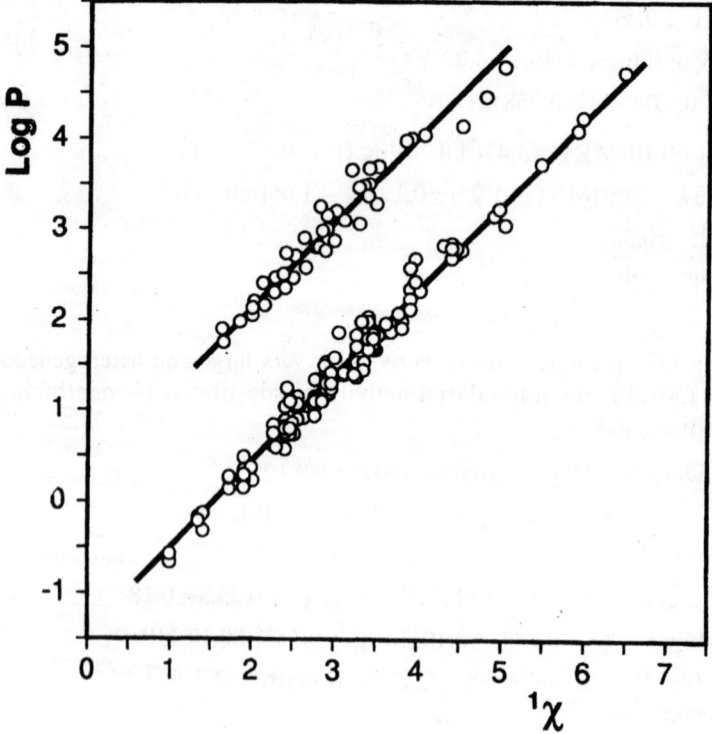

Figure 6 A close relationship exists between lipophilicity and $^1\chi$ connectivity values if hydrocarbons (upper line) and organic compounds with one functional group (lower line) are treated separately (Eq. (52)). (From Ref. 83.)

and one for the other molecules, each one bearing a polar group. Correspondingly, all molecules can be described by Eq. (52), with an indicator variable $I = 0$ for the hydrocarbons and $I = 1$ for all other molecules [83]:

$$\log P = 0.941(\pm 0.02)^1\chi - 1.693(\pm 0.05)I + 0.244(\pm 0.08)$$
$$(n = 183; \quad r = 0.990; \quad s = 0.150; \quad F = 4633) \tag{52}$$

QSAR studies contribute also to the termination of a project, if the quantitative models indicate that no further improvement is to be expected. Equation (53) was derived for the antitumor activity of triazenes [84]. The negative σ term indicates that the largest increase in activity would result from electron-releasing substituents. However, chemical stability of these compounds also depends on σ (Eq. (54)); in aqueous solution, even the 4-OMe analog has such a short half-life time that it is too unstable for practical purposes. In addition, the toxicity of these compounds is correlated with their antitumor activities; thus, it was recommended to stop the project [85]:

$$\log 1/C = -0.042(\log P)^2 + 0.100 \log P - 0.312\Sigma\sigma^+$$
$$-0.178 MR\text{-}2.6 + 0.391 E_s - R + 4.124; \tag{53}$$
$$\text{optimum } \log P = 1.18 \quad (n = 61; \quad r = 0.836; \quad s = 0.191)$$

$$\log k_X/k_H = -4.42\sigma - 0.016 \quad (n = 14; \quad r = 0.995; \quad s = 0.171) \tag{54}$$

Similar conclusions could be derived from an investigation of antihypertensive clonidine analogs, with respect to their peripheral hypertensive side effects. For a series of clonidine-related imidazolines, $\log P_{app}$ values were determined in n-octanol/buffer at pH 7.4, central antihypertensive activities C_{25} (molar doses that cause a 25% blood pressure decrease, intravenous application) in anesthesized rats, peripheral hypertensive activities C_{60} (molar doses that cause a 60-mm Hg blood pressure increase, intravenous application) in pithed rats (i.e., rats without central nervous system control), binding affinities $IC_{50}\alpha_1$ to α_1 receptors (replacement of prazosin), and binding affinities $IC_{50}\alpha_2$ to α_2 receptors (replacement of clonidine) [86,87].

For the peripheral activity, the blood pressure increase that is caused by blood vessel contraction, Eq. 55 was observed, whereas for the central nervous system-mediated activity, the blood pressure decrease, Eq. (56), resulted [60,88]:

$$pC_{60} = 1.16(\pm 0.21)pIC_{50}\alpha_2 - 0.96 \quad (n = 21; \quad r = 0.936; \quad s = 0.317) \tag{55}$$

$$pC_{25} = 1.07(\pm 0.20)pIC_{50}\alpha_2 + 0.81(\pm 0.22)\log P$$
$$- 3.37(\pm 1.02)\log(\beta P + 1) - 1.16 \tag{56}$$
$$\log \beta = -1.99; \quad \log P_{opt} = 1.48 \quad (n = 21; \quad r = 0.971; \quad s = 0.284)$$

The differences between both equations arise from the fact that the compounds have to cross the blood–brain barrier to exert their central nervous system-mediated activity; however, both effects are linearly related to the α_2 binding affinities. Correspondingly, the peripheral hypertensive side effect and the antihypertensive therapeutic activity are related by Eq. (57). No separation of both effects is to be expected in this series of compounds [86–88]:

$$pC_{25} = 0.83(\pm 0.20)pC_{60} + 0.78(\pm 0.26)\log P - 3.69(\pm 1.39)\log(\beta P + 1)$$
$$- 0.19 \tag{57}$$
$$\log \beta = -2.08; \quad \text{optimum } \log P = 1.51 \quad (n = 21; \quad r = 0.954; \quad s = 0.354)$$

In addition to this nonlinear relationship for the blood–brain barrier permeation of imidazolines, many other nonlinear lipophilicity relationships have been derived for buccal and gastrointestinal absorption, skin permeation, as well as blood–brain and blood–placenta barrier permeation (e.g., Eqs. (58) to (62); Fig. 7 [59,60,63,64]:

Barbiturates: permeation through an organic membrane:

$$\log k_{abs} = 0.949(\pm 0.06)\log P - 1.238(\pm 0.11)\log (\beta P + 1)$$
$$- 3.131$$

$$\log \beta = -5.27; \quad \text{optimum } \log P = 1.79$$

$$(n = 23; \quad r = 0.992; \quad s = 0.081; \quad F = 389.66)$$

(58)

Homologous alkyl carbamates: gastric absorption:

$$\log k_{abs} = 0.138(\pm 0.06)\log P - 0.228(\pm 0.16)\log (\beta P + 1)$$
$$- 2.244$$

$$\log \beta = -1.678; \quad \text{optimum } \log P = 1.87$$

$$(n = 8; \quad r = 0.971; \quad s = 0.030; \quad F = 22.14)$$

(59)

Homologous alkyl carbamates: intestinal absorption:

$$\log k_{abs} = 0.234(\pm 0.10)\log P - 0.502(\pm 0.15)\log (\beta P + 1)$$
$$- 0.786$$

$$\log \beta = -0.621; \quad \text{optimum } \log P = 0.56$$

$$(n = 8; \quad r = 0.989; \quad s = 0.031; \quad F = 61.10)$$

(60)

Primary alcohols: neurotoxicity (= permeation of blood–brain barrier):

$$\log 1/C = + 0.892(\pm 0.050)\log P - 1.766(\pm 0.10)\log (\beta P + 1)$$
$$+ 1.586$$

$$\log \beta = -1.933; \quad \text{optimum } \log P = 1.94$$

$$(n = 10; \quad r = 0.998; \quad s = 0.041; \quad F = 637.6)$$

(61)

Various drugs: permeation of blood–placenta barrier:

$$\log TR = 0.354(\pm 0.06)\log P - 0.469(\pm 0.13)\log (\beta P + 1)$$
$$- 0.116$$

$$\log \beta = -0.658; \quad \text{optimum } \log P = 1.15$$

$$(n = 21; \quad r = 0.949; \quad s = 0.106; \quad F = 51.17)$$

(62)

All these models describe small data sets with only limited predictability for other series of compounds. With the increasing importance of combinatorial chemistry and high-throughput screening, it became necessary to formulate more general models for oral bioavailability and blood–brain barrier permeability. Such quantitative models for oral bioavailability are difficult to derive: there is a lack of consistent data and many drugs are either absorbed by an active transport (e.g., by the amino acid or dipeptide transporters), or are eliminated by one of several different efflux pumps.

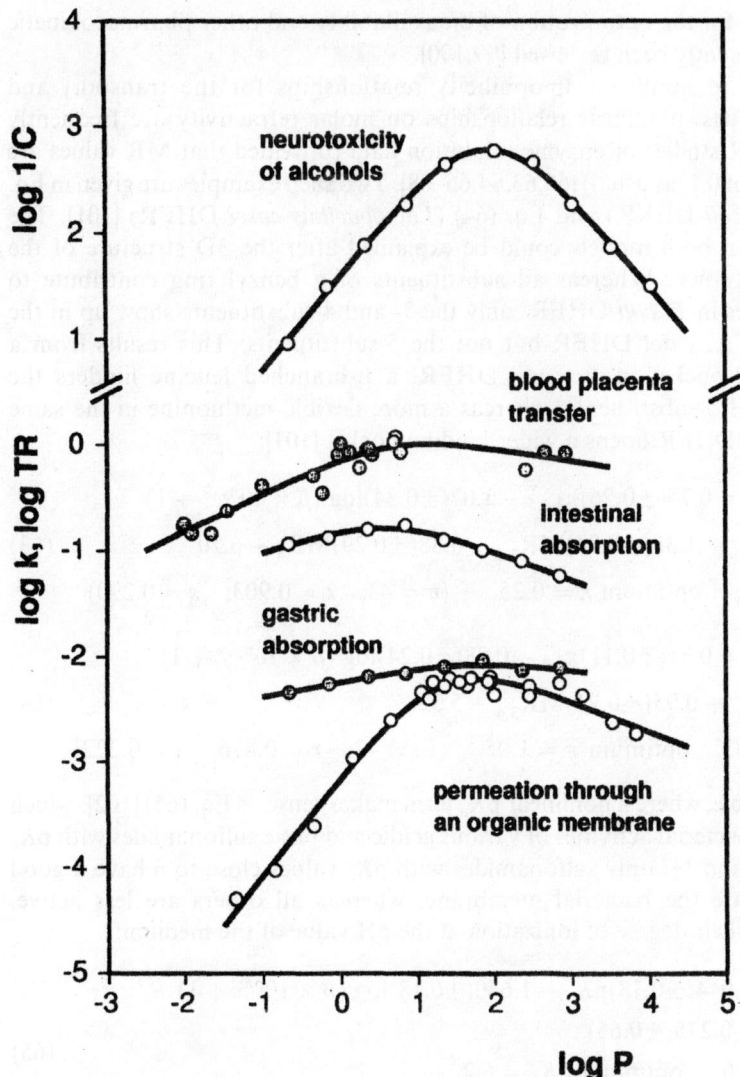

Figure 7 The permeation of barbiturates through an organic membrane, the gastric and intestinal absorption of carbamates, the blood–placenta transfer rate constants of various drugs, and the neurotoxicity of homologous primary alcohols, as a measure of blood–brain barrier permeation, follow nonlinear lipophilicity relationships (Eqs. (58)–(62)). (From Refs. 58,59,62,63.)

Nevertheless, quantitative models for oral bioavailability and blood–brain barrier permeation have been derived from data sets of limited size, using different approaches [89–97]; in most cases, no details of the models are provided. Lipinski's "rule of five" gives the recommendation that a molecular weight > 500, lipophilicity values of log $P > 5$, and more than five hydrogen bond donors and/or more than 10 hydrogen bond acceptors should be avoided to achieve good oral bioavailability [98]. General recommendations for the optimization of bioavailability and other pharmacokinetic properties have recently been reviewed [99,100].

In addition to nonlinear lipophilicity relationships for the transport and distribution of drugs, nonlinear relationships on molar refractivity are frequently observed in QSAR studies of enzyme inhibition data (provided that MR values are scaled by a factor of 0.1, as usual) [60,63,64,66–68]. Two such examples are given in Eq. (63) (*Escherichia coli* DHFR) and Eq. (64) (*Lactobacillus casei* DHFR) [101]. The differences between both models could be explained after the 3D structure of the enzyme became known. Whereas all substituents of a benzyl ring contribute to biological activities in *E. coli* DHFR, only the 3- and 4-substituents show up in the QSAR model for *L. casei* DHFR but not the 5-substituents. This results from a narrower binding pocket in *L. casei* DHFR; a β-branched leucine hinders the accommodation of 5-substituents, whereas a more flexible methionine in the same position of *E. coli* DHFR opens a wider binding pocket [101]:

$$\log 1/K_{i\,\text{app}} = 0.75(\pm0.26)\pi_{3,4,5} - 1.07(\pm0.34)\log\left(\beta \times 10^{\pi_{3,4,5}} + 1\right)$$
$$+ 1.36(\pm0.24)\text{MR}'_{3,5} + 0.88(\pm0.29)\text{MR}'_{4} + 6.20 \tag{63}$$

$$\log \beta = 0.12; \quad \text{optimum } \pi = 0.25 \quad (n = 43; \quad r = 0.903; \quad s = 0.290)$$

$$\log 1/K_{i\,\text{app}} = 0.31(\pm0.11)\pi_{3,4} - 0.88(\pm0.24)\log\left(\beta \times 10^{\pi_{3,4,5}} + 1\right)$$
$$+ 0.95(\pm0.21)\text{MR}'_{3,4} + 5.32 \tag{64}$$

$$\log \beta = -1.33; \quad \text{optimum } \pi = 1.05 \quad (n = 42; \quad r = 0.876; \quad s = 0.222)$$

A rare example, where a nonlinear pK_a term makes sense, is Eq. (65) [102], which describes the antibacterial activities of various acidic and basic sulfonamides with pK_a values between 3 and 11; only sulfonamides with pK_a values close to 6 have a good chance to penetrate the bacterial membrane, whereas all others are less active, corresponding to their degree of ionization at the pH value of the medium:

$$\log 1/C = 1.044(\pm0.13)\text{p}K_a - 1.640(\pm0.18)\log\left(\beta \times 10^{\text{p}K_a} + 1\right)$$
$$+ 0.275(\pm0.65)$$
$$\log \beta = -5.96; \quad \text{optimum p}K_a = 6.22 \tag{65}$$
$$(n = 39; \quad r = 0.956; \quad s = 0.275; \quad F = 124.1)$$

Free–Wilson models are appropriate if certain chemical features are responsible for changes in biological activities. The corticosteroid-binding globulin affinities of steroids are a standard data set for 3D QSAR analyses [103,104]. However, the whole data set can be described with just one Free–Wilson parameter 4.5 > C=C < (value is 1 if a double bond is present in the 4,5-position of ring A; value is 0 if there is a single bond or an aromatic bond; Eq. (66)), with about the same quality of fit and predictivity [105]:

$$\log 1/\text{CBG} = 2.022(\pm0.52)[4.5 > \text{C}=\text{C}<] + 5.186(\pm0.36)$$
$$(n = 21; \quad r = 0.882; \quad s = 0.568; \tag{66}$$
$$F = 66.41; \quad Q^2 = 0.726; \quad s_{\text{PRESS}} = 0.630)$$

For DHFR inhibition by a large group of 2,4-diaminopyrimidines, Hansch derived Eq. (67), which includes six Free Wilson-type indicator variables and one interaction term [106]:

$$\log 1/C = 0.365(\pm 0.12)I\text{-}1 + 1.013(\pm 0.12)I\text{-}8 - 0.784(\pm 0.19)I\text{-}9$$
$$+0.419(\pm 0.20)I\text{-}13 - 0.220(\pm 0.09)I\text{-}15$$
$$+ 0.513(\pm 0.18)I\text{-}20 + 0.674(\pm 0.23)I\text{-}4.1\text{-}8 \tag{67}$$
$$+7.174(\pm 0.07)$$

$$(n = 105; \quad r = 0.903; \quad s = 0.229)$$

The binding affinities of 128 quaternary ammonium compounds, $X\text{-}CH_2\text{-}CH_2\text{-}N^+$ ($R^1R^2R^3$), to a postganglionic acetylcholine receptor [107] show significant differences on chemical variation of X (16 different groups), whereas there is no major influence of the small R groups (eight different groups). Correspondingly, biological activities can be described by 16 Free–Wilson parameters (Eq. (68)) [32,88]:

$$\log K = -2.479[CH_3CH_2O-] - 2.175[CH_3CH_2CH_2-]$$
$$- 1.228[C_6H_5CH_2COO-] - 1.177[C_6H_5CH_2CH_2O-]$$
$$- 0.909[C_6H_5CH_2CH_2CH_2-] - 1.035[C_6H_{11}CH_2COO-]$$
$$- 0.819[C_6H_{11}CH_2CH_2O-] - 0.683[C_6H_{11}CH_2CH_2CH_2-] \tag{68}$$
$$+0.872[(C_6H_5)_2CHCOO-] - 0.070[(C_6H_5)_2CHCH_2O-]$$
$$+0.374[(C_6H_5)_2CHCH_2CH_2-] + 2.035[C_6H_5(C_6H_{11})CHCOO-]$$
$$+2.047[(C_6H_5)_2C(OH)COO-] + 1.467[(C_6H_{11})_2CHCOO-]$$
$$+0.806[(C_6H_{11})_2CHCH_2O-] + 2.975[C_6H_5(C_6H_{11})C(OH)COO-]$$
$$+6.499 \quad (n = 128; \quad r = 0.991; \quad s = 0.231)$$

A closer inspection of the individual group contributions shows that there are some nonlinear effects. The exchange of hydrogen in the X group against phenyl or cyclohexyl or the exchange of phenyl to cyclohexyl and vice versa depends on the groups that are already present in X (Table 3). A much simpler Free–Wilson model

Table 3 Nonadditivities in Free–Wilson Group Contributions

Differences in receptor affinity (logarithmic scale), by changing			
R in group X from	H to C_6H_5	H to C_6H_{11}	C_6H_5 to C_6H_{11}
$R\text{-}CH_2CH_2CH_2-$	1.266	1.492	0.226
$R\text{-}CH_2CH_2O-$	1.302	1.660	0.358
$C_6H_5CH(R)COO-$	2.100	3.263	1.163
$C_6H_{11}CH(R)COO-$	3.070	2.502	-0.568

Source: Refs. 32 and 88.

can be derived if this nonlinear effect is considered by an indicator variable INT (Eq. (69); INT = 1 if X bears both phenyl and cyclohexyl; INT = 0 if this is not the case) [32,88]:

$$\log K = 1.258(\pm 0.09)[\text{PHE} + 1.545(\pm 0.11)[c\text{-HEX}]$$
$$+ 1.609(\pm 0.19)[\text{OH}] + 0.755(\pm 0.17)[\text{COO}]$$
$$+ 0.769(\pm 0.19)\text{INT} + 4.142(\pm 0.12)$$

$$(n = 128; \quad r = 0.983; \quad s = 0.290; \quad F = 711.14)$$

(69)

7. SIMILARITY OF QSAR, HQSAR, BINARY OSAR, AND OTHER APPROACHES

Since decades ago, medicinal chemists have applied the principle of bioisosteric replacement of atoms and/or groups in their search for new powerful drugs. Starting from a lead structure, they perform minor chemical changes to optimize its biological properties (i.e., to generate more active and selective molecules with better oral bioavailability, less toxicity, and fewer side effects). Inherently, they follow the additivity principle of Free–Wilson analysis, that a certain structural change in a certain position adds a constant contribution to biological activity, irrespective of any other modifications of the molecule.

The successful application of topological indices [108,109] and of electrotopological state parameters [110] in drug design is based on the combination of two facts: such artificial parameters are similar for chemically similar compounds and similar compounds most often exert similar biological activities, qualitatively and quantitatively. Although the relationships between chemical similarity and biological activity are not always straightforward [111], molecular similarity of a molecule to a certain reference compound is a valuable parameter for quantitative structure–activity relationships, as first demonstrated by Rum and Herndon [112]. Their approach has been extended to quantum similarity [113,114] and hydrophobic similarity [60,105]. With hydrophobic similarity, even nonlinear lipophilicity–activity relationships can be described [60,105,115].

The use of $N \times N$ 3D similarity matrices in quantitative similarity–activity relationship (QSiAR) studies [36] has several advantages. Instead of a common alignment of all active and inactive molecules to a reference pharmacophore, only pairwise alignments need to be performed. This leads to a matrix with individual "similarity vectors" for each analog. Inactive analogs do not any longer distort the alignment of the active analogs. If SEAL similarity indices [116,117] are used in the generation of the $N \times N$ matrix, not even a grid is needed. A disadvantage of the calculation of similarity coefficients is the fact that no contour maps can be generated from the resulting QSAR model.

Molecular holograms are made up of fingerprints that encode all information about branched and cyclic fragments of a molecule, as well as stereochemistry. Within large data sets, hologram QSAR (HQSAR) identifies those fragments that are responsible for enhancing or reducing biological activities [118–121]. Binary QSAR accepts qualitative biological data (i.e., binary activity measurements; e.g., active vs. inactive). Any set of descriptors and a Bayesian inference technique are used to predict whether a certain compound should be active or inactive [122–125].

In classical QSAR, neural nets do not offer any major advantages, as compared to regression or partial least squares analysis. However, the assignment of "druglike" character to a certain organic molecule corresponds to a pattern recognition problem that can be approached by neural nets. For this purpose, neural nets are fed with training sets of bioactive compounds from a drug database and compounds from a catalog of chemicals, considered to contain "non-druglike" organic compounds [126,127]. About 80% of both categories are correctly assigned to their respective groups. The external predictivity of the trained net is tested with all other compounds from the drug database and the catalog of chemicals. Even in the prediction of the test set compounds, which are structurally unknown to the neural net, 75–80% of the compounds are correctly classified.

Classical QSAR has often been criticized as being applicable only in retrospective studies. The successful application of QSAR in the design of drugs and agrochemicals has been reviewed [128], but it is true that QSAR most often does not have a direct impact on drug design. However, it had an enormous indirect impact: the work of Hansch and his group contributed most to our understanding of the nature of ligand–receptor interactions. About two or three decades ago, most medicinal chemists did not care too much about lipophilicity and ionization. Nowadays, they are aware that there is, in each series, an optimum lipophilicity for oral bioavailability and another one for blood–brain barrier penetration, either to achieve CNS activity or to avoid CNS and unspecific toxic side effects [129]. From the parameters π, σ, and MR, and the steric parameters, we understand lipophilic, electronic, polarizable, and steric contributions of different parts of the molecules under investigation. And we understand why so many drugs are either acids or weak bases. In their neutral form, they penetrate membranes whereas their ionized forms are water-soluble. Thus, they penetrate aqueous and lipid barriers to arrive at their site of action in much larger concentration and in a shorter time than highly polar or lipophilic neutral compounds.

QSAR should be considered as a tool to derive quantitative models that are confirmed or disproved by the syntheses and biological tests of further analogs. Predictions are only a means for the design of new analogs; understanding a structure–activity relationship is the main goal of a QSAR study. In this context, Hansch [130] formulated already in 1977: "the great advantage of the QSAR paradigm lies not in the extrapolations which can be made from known QSAR to fantastically potent new drugs, but in the less spectacular slow development of science in medicinal chemistry."

REFERENCES

1. Purcell WP, Bass GF, Clayton JM. Strategy of Drug Design. A Molecular Guide to Biological Activity. New York: Wiley, 1973:1–20.
2. Tute MS. History and objectives of quantitative drug design. In: Ramsden CA, ed. Quantitative Drug Design. In: Hansch C, Sammes PG, Taylor JB, eds. Comprehensive Medicinal Chemistry. The Rational Design, Mechanistic Study and Therapeutic Application of Chemical Compounds 1990; Vol. 4. Oxford: Pergamon Press, 1990:1–31.
3. Rekker RF. The history of drug research: from Overton to Hansch. Quant Struct–Act Relat 1992; 11:195–199.

4. Crum Brown A, Fraser TR. On the connection between chemical constitution and physiologic action: Part 1. On the physiological action of salts of the ammonium bases, derived from strychnia, brucia, thebia, codeia, morphia and nicotia. Trans R Soc Edinb 1868; 25:151–203.

5. Meyer H. Zur Theorie der Alkoholnarkose. Erste Mitteilung. Welche Eigenschaft der Anästhetica bedingt ihre narkotische Wirkung. Naunyn Schmiedeberg's Arch Exp Path Pharm 1899; 42:109–118; cf. Lipnick RL. Hans Horst Meyer and the lipoid theory of narcosis. Trends Pharmacol Sci 1989; 10:265–269.

6. Overton E. Studien über die Narkose, zugleich ein Beitrag zur allgemeinen Pharmakologie. Jena: G Fischer, 1901; English translation by Lipnick RL, ed. Studies on Narcosis, Charles Ernest Overton. London: Chapman and Hall, 1991; cf. Lipnick RL. Charles Ernest Overton: narcosis studies and a contribution to general pharmacology. Trends Pharmacol Sci 1986; 7:161–164.

7. Führer H, Neubauer E. Hämolyse durch Substanzen homologen Reihen. Arch Exp Path Pharm 1907; 56:333–345.

8. Ferguson J. Use of chemical potentials as indexes of toxicity. Proc Roy Soc (London) 1939; B127:387–404.

9. Hammett LP. Physical Organic Chemistry. Reaction Rates, Equilibria and Mechanism 2nd ed. New York: McGraw-Hill, 1970.

10. Bruice TC, Kharasch N, Winzler RJ. A correlation of thyroxin-like activity and chemical structure. Arch Biochem Biophys 1956; 62:305–317.

11. Zahradnik R, Chvapil M. Study of the relationships between the magnitude of biological activity and the structure of aliphatic compounds. Experientia 1960; 16:511–512.

12. Zahradnik R. Influence of the structure of aliphatic substituents on the magnitude of the biological effect of substances. Arch Int Pharmacodyn Ther 1962; 135:311–329.

13. Zahradnik R. Correlation of the biological activity of organic compounds by means of the linear free energy relationships. Experientia 1962; 18:534–536.

14. Hansen OR. Hammett series with biological activity. Acta Chem Scand 1962; 16:1593–1600.

15. Hansch C, Maloney PP, Fujita T, Muir RM. Correlation of biological activity of phenoxyacetic acids with Hammett substituent constants and partition coefficients. Nature 1962; 194:178–180.

16. Hansch C, Fujita T. ρ–σ–π Analysis. A method for the correlation of biological activity and chemical structure. J Am Chem Soc 1964; 86:1616–1626.

17. Free SM Jr, Wilson JW. A mathematical contribution to structure–activity studies. J Med Chem 1964; 7:395–399.

18. Hansch C. A quantitative approach to biochemical structure–activity relationships. Acc Chem Res 1969; 2:232–239.

19. Hansch C. Quantitative structure–activity relationships in drug design. In: Ariëns EJ, ed. Drug Design 1971; Vol. I. New York: Academic Press, 1971:271–342.

20. Hansch C, Clayton JM. Lipophilic character and biological activity of drugs: II. The parabolic case. J Pharm Sci 1973; 62:1–21.

21. Fujita T, Ban T. Structure–activity study of phenethylamines as substrates of biosynthetic enzymes of sympathetic transmitters. J Med Chem 1971; 14:148–152.

22. Hansch C, Yoshimoto M. Structure–activity relationships in immuno-chemistry: 2. Inhibition of complement by benzamidines. J Med Chem 1974; 17:1160–1167.

23. Kubinyi H. Quantitative structure–activity relationships: 2. A mixed approach, based on Hansch and Free–Wilson analysis. J Med Chem 1976; 19:587–600.

24. Kubinyi H. Quantitative structure–activity relationships: 7. The bilinear model, a new model for nonlinear dependence of biological activity on hydrophobic character. J Med Chem 1977; 20:625–629.

25. van de Waterbeemd H. The history of drug research: from Hansch to the present. Quant Struct–Act Relat 1992; 11:200–204.

26. The QSAR and Modelling Society (www.qsar.org; http://www.ndsu.nodak.edu/qsar_soc/aboutsoc/histlist.htm).
27. Hansch C, Lien EJ. An analysis of the structure–activity relationship in the adrenergic blocking activity of the β-haloalkylamines. Biochem Pharmacol 1968; 17:709–720.
28. Unger SH, Hansch C. On model building in structure–activity relationships. A reexamination of adrenergic blocking activity of β-halo-β-arylalkylamines. J Med Chem 1973; 16:745–749.
29. Cammarata A. Interrelationship of the regression models used for structure–activity analyses. J Med Chem 1972; 15:573–577.
30. Kubinyi H, Kehrhahn OH. Quantitative structure–activity relationships: 1. The modified Free–Wilson approach. J Med Chem 1976; 19:578–586.
31. Kubinyi H. Free–Wilson analysis. Theory, applications and its relationship to Hansch analysis. Quant Struct–Act Relat 1988; 7:121–133.
32. Kubinyi H. The Free–Wilson method and its relationship to the extrathermodynamic approach. In: Ramsden CA, ed. Quantitative Drug DesignHansch C, Sammes PG, Taylor JB, eds. Comprehensive Medicinal Chemistry. The Rational Design, Mechanistic Study and Therapeutic Application of Chemical Compounds 1990; Vol. 4. Oxford: Pergamon Press, 1990:589–643.
33. Cramer RD III, Bunce JD, Patterson DE, Frank IE. Crossvalidation, bootstrapping, and partial least squares compared with multiple regression in conventional QSAR studies. Quant Struct–Act Relat 1988; 7:18–25 Erratum in 1988; 7:91.
34. Norinder U. Single and domain mode variable selection in 3D QSAR applications. J Chemom 1996; 10:95–105.
35. Novellino E, Fattorusso C, Greco G. Use of comparative molecular field analysis and cluster analysis in series design. Pharm Acta Helv 1995; 70:149–154.
36. Kubinyi H, Hamprecht FA, Mietzner T. Three-dimensional quantitative similarity–activity relationships (3D QSiAR) from SEAL similarity matrices. J Med Chem 1998; 41:2553–2564.
37. Golbraikh A, Tropsha A. Beware of q^2. J Mol Graph Model 2002; 20:269–276.
38. van Drie J. Personal communication.
39. Snedecor GW, Cochran WG. Statistical Methods. Ames: The Iowa State University Press, 1973.
40. Daniel C, Wood FS. Fitting Equations to Data. New York: Wiley, 1980.
41. Draper NR, Smith H. Applied Regression Analysis 2nd ed. New York: Wiley, 1981.
42. Tranter RL, ed. Design and Analysis in Chemical Research. Sheffield: Sheffield Academic Press, 2000.
43. Topliss JG, Costello RJ. Chance correlations in structure–activity studies using multiple regression analysis. J Med Chem 1972; 15:1066–1068.
44. Topliss JG, Edwards RP. Chance factors in studies of quantitative structure–activity relationships. J Med Chem 1979; 22:1238–1244.
45. Leardi R, Boggia R, Terrile M. Genetic algorithms as a strategy for feature selection. J Chemom 1992; 6:267–281.
46. Rogers D, Hopfinger AJ. Application of genetic function approximation (GFA) to quantitative structure–activity relationships. J Chem Inf Comput Sci 1994; 34:854–866.
47. Rogers D. Genetic function approximation: a genetic approach to building quantitative structure–activity relationship models. In: Sanz F, ed. Trends in QSAR and Molecular Modelling 94. Barcelona: Prous Science Publishers, 1995:420–426.
48. Kubinyi H. Variable selection in QSAR studies: I. An evolutionary algorithm. Quant Struct–Act Relat 1994; 13:285–294.
49. Kubinyi H. Variable selection in QSAR studies: II. A highly efficient combination of systematic search and evolution. Quant Struct–Act Relat 1994; 13:393–401.
50. Kubinyi H. Evolutionary variable selection in regression and PLS analyses. J Chemom 1996; 10:119–133.

51. So SS. Quantitative structure–activity relationships. In: Clark DE, ed. Evolutionary
     Algorithms in Molecular Design. In: Mannhold R, Kubinyi H, Timmerman H, eds.
     Methods and Principles in Medicinal Chemistry Vol. 8. Weinheim: Wiley-VCH, 2000.
52. van de Waterbeemd H, ed. Chemometric MethodsMannhold R, Krogsgaard-Larsen P,
     Timmerman H, eds. Methods and Principles in Medicinal Chemistry 1995; Vol. 2.
     Weinheim: VCH, 1995.
53. Penniston JT, Beckett L, Bentley DL, Hansch C. Passive permeation of organic compounds
     through biological tissue: a non-steady state theory. Mol Pharmacol 1969; 5:333–341.
54. Lippold BC, Schneider GF. Zur Optimierung der Bioverfügbarkeit homologer quartärer
     Ammoniumverbindungen: 2. Mitteilung: in-vitro-Versuche zur Verteilung von Benzil-
     säureestern homologer Dimethyl-(2-hydroxyäthyl)-alkylammoniumbromide. Arzneim-
     Forsch (Drug Res) 1975; 25:843–852.
55. Lippold BC, Schneider GF. Zur Optimierung der Bioverfügbarkeit homologer quartärer
     Ammoniumverbindungen: 3. Mitteilung: Einfluß von Salzen auf die in-vitro-Verteilung
     von Benzil-säureestern homologer Dimethyl-(2-hydroxyäthyl)-alkylammoniumbromide.
     Arzneim-Forsch (Drug Res) 1975; 25:1683–1686.
56. van de Waterbeemd JTM, van Boekel CCAA, de Sevaux RLFM, Jansen ACA, Gee-
     ritsma KW. Transport in QSAR: IV. The interfacial drug transfer model. Relationships
     between partition coefficients and rate constants of drug partitioning. Pharm Weekbl Sci
     Ed 1981; 3:224–237.
57. van de Waterbeemd H, van Bakel P, Jansen A. Transport in quantitative structure–
     activity relationships: VI. Relationship between transport rate constants and partition
     coefficients. J Pharm Sci 1981; 70:1081–1082.
58. Kubinyi H. Drug partitioning: relationships between the forward and reverse rate
     constants and the partition coefficient. J Pharm Sci 1978; 67:262–263.
59. Kubinyi H. Lipophilicity and drug activity. Prog Drug Res 1979; 23:97–198.
60. Kubinyi H. QSAR: Hansch analysis and related approaches. In: Mannhold R,
     Krogsgaard-Larsen P, Timmerman H, eds. Methods and Principles in Medicinal
     Chemistry 1993; Vol. 1. Weinheim: VCH, 1993.
61. McFarland JW. On the parabolic relationship between drug potency and hydro-
     phobicity. J Med Chem 1970;13, 1192–1196.
62. Kubinyi H. Quantitative structure–activity relationships: 4. Nonlinear dependence of
     biological activity on hydrophobic character: a new model. Arzneim-Forsch (Drug Res)
     1976; 26:1991–1997.
63. Kubinyi H. Lipophilicity and biological activity: drug transport and drug distribution in
     model systems and in biological systems. Arzneim-Forsch (Drug Res) 1979; 29:1067–
     1080.
64. Kubinyi H. Lipophilicity and biological activity: the use of the bilinear model in QSAR.
     In: Kuchar M ed. QSAR in Design of Bioactive Compounds. Proceedings of the First
     International Telesymposium on Medicinal Chemistry. Barcelona: Prous Science
     Publishers, 1984:321–346.
65. Lien EJ, Wang PH. Lipophilicity, molecular weight, and drug action: reexamination of
     parabolic and bilinear models. J Pharm Sci 1980; 69:648–650.
66. Blaney JM, Hansch C, Silipo C, Vittoria A. Structure–activity relationships of
     dihydrofolate reductase inhibitors. Chem Rev 1984; 84:333–407.
67. Gupta SP. QSAR studies on enzyme inhibitors. Chem Rev 1987; 87:1183–1253.
68. Hansch C, Leo A. Exploring QSAR. Fundamentals and Applications in Chemistry and
     Biology. Washington: American Chemical Society, 1995.
69. Shore PA, Brodie BB, Hogben CAM. The gastric secretion of drugs; a pH partition
     hypothesis. J Pharmacol Exp Ther 1957; 119:361–369.
70. Wagner JG. Fundamentals of Clinical Pharmacokinetics. Hamilton: Drug Intelligence
     Publications, 1975.

71. Winne D. Shift of pH–absorption curves. J Pharmacokinet Biopharm 1977; 5:53–94.
72. Tsuji A, Miyamoto E, Hashimoto N, Yamana T. GI absorption of β-lactam antibiotics: II. Deviation from pH-partition hypothesis in penicillin absorption through in situ and in vitro lipoidal barriers. J Pharm Sci 1978; 67:1705–1711.
73. Lien EJ. Structure–absorption–distribution relationships. In: Ariëns EJ, ed. Drug Design 1975; Vol. V. New York: Academic Press, 1975:81–132.
74. Scherrer RA, Howard SM. Use of distribution coefficients in quantitative structure–activity relationships. J Med Chem 1977; 20:53–58.
75. Taylor PJ. Hydrophobic Properties of Drugs. In: Ramsden CA, ed. Quantitative Drug Design. In: Hansch C, Sammes PG, Taylor JB, eds. Comprehensive Medicinal Chemistry The Rational Design, Mechanistic Study and Therapeutic Application of Chemical Compounds Vol. 4. Oxford: Pergamon Press, 1990:241–294.
76. BioByte Corp., Claremont, CA 91711, USA (www.biobyte.com); cf. http://clogp.pomona.edu/chem/qsar-db/index.html.
77. Hansch C, Calef DF. Structure–activity relationships in papain–ligand interactions. J Org Chem 1976; 41:1240–1243.
78. Hansch C, Smith RN, Rockoff A, Calef DF, Jow PYC, Fukunaga JY. Structure–activity relationships in papain and bromelain ligand interactions. Arch Biochem Biophys 1977; 183:383–392.
79. Silipo C, Hansch C, Grieco C, Vittoria A. Inhibition of chymotrypsin by alkyl phosphonates: a quantitative structure–activity analysis. Arch Biochem Biophys 1979; 194: 552–557.
80. Kim KH, Hansch C, Fukunaga JY, Steller EE, Jow PYC, Craig PN, Page J. Quantitative structure–activity relationships in 1-aryl-2-(alkylamino)ethanol antimalarials. J Med Chem 1979; 22:366–391.
81. Debnath AK, Lopez de Compadre RL, Debnath G, Shusterman AJ, Hansch C. Structure–activity relationship of mutagenic aromatic and heteroaromatic nitro compounds. Correlation with molecular orbital energies and hydrophobicity. J Med Chem 1991; 34:786–797.
82. Saxena AK. Physicochemical significance of topological parameters: molecular connectivity index and information content: Part 2. Correlation studies with molar refractivity and lipophilicity. Quant Struct–Act Relat 1995; 14:142–148.
83. Kubinyi H. The physicochemical significance of topological parameters. A rebuttal. Quant Struct–Act Relat 1995; 14:149–150.
84. Hatheway GJ, Hansch C, Kim KH, Milstein SR, Schmidt CL, Smith RN, Quinn FR. Antitumor 1-(X-aryl)-3,3-dialkyltriazenes: 1. Quantitative structure–activity relationships vs L1210 leukemia in mice. J Med Chem 1978; 21:563–574.
85. Hansch C, Hatheway GJ, Quinn FR, Greenberg N. Antitumor 1-(X-aryl)-3,3-dialkyltriazenes: 2. On the role of correlation analysis in decision making in drug modification. Toxicity quantitative structure–activity relationships of 1-(X-phenyl)-3,3-dialkyltriazenes in mice. J Med Chem 1978; 21:574–577.
86. Timmermans PBMWM, de Jonge A, van Meel JCA, Slothorst-Grisdijk FP, Lam E, van Zwieten PA. Characterisation of α-adrenoceptor populations. Quantitative relationships between cardiovascular effects initiated at central and peripheral α-adrenoceptors. J Med Chem 1981; 24:502–507.
87. Timmermans PBMWM, de Jonge A, Thoolen MJMC, Wilffert B, Batink H, van Zwieten PA. Quantitative relationships between α-adrenergic activity and binding affinity of α-adrenoceptor agonists and antagonists. J Med Chem 1984; 27:495–503.
88. Kubinyi H. Current problems in quantitative structure–activity relationships. In: Jochum C, Hicks MG, Sunkel J, eds. Physical Property Prediction in Organic Chemistry. Berlin: Springer-Verlag, 1988:235–247.
89. Wessel MD, Jurs PC, Tolan JW, Muskal SM. Prediction of human intestinal absorption

of drug compounds from molecular structure. J Chem Inf Comput Sci 1998; 38:726–735.

90. Oprea TI, Gottfries J. Toward minimalistic modeling of oral drug absorption. J Mol Graph Model 1999; 17:261–274.

91. Luco JM. Prediction of the brain–blood distribution of a large set of drugs from structurally derived descriptors using partial least-squares (PLS) modelling. J Chem Inf Comput Sci 1999; 39:396–404.

92. Keider J, Grootenhuis PDJ, Bayada DM, Delbressine LPC, Ploemen JP. Polar molecular surface as a dominating determinant for oral absorption and brain penetration of drugs. Pharm Res 1999; 16:1514–1519.

93. Ekins S, Waller CL, Swaan PW, Cruciani G, Wrighton SA, Wikel JH. Progress in predicting human ADME parameters in silico. J Pharmacol Toxicol Methods 2000; 44:251–272.

94. Andrews CW, Bennett L, Yu LX. Predicting human oral bioavailability of a compound: development of a novel quantitative structure–bioavailability relationship. Pharm Res 2000; 17:639–643.

95. Yoshida F, Topliss JG. QSAR model for drug human oral bioavailability. J Med Chem 2000; 43:2575–2585.

96. Egan WJ, Merz KM Jr, Baldwin JJ. Prediction of drug absorption using multivariate statistics. J Med Chem 2000; 43:3867–3877.

97. Zhao YH, Le J, Abraham MH, Hersey A, Eddershaw PJ, Luscombe CN, Boutina D, Beck G, Sherborne B, Cooper I, Platts JAB. Evaluation of human intestinal absorption data and subsequent derivation of a quantitative structure–activity relationship (QSAR) with the Abraham descriptors. J Pharm Sci 2001; 90:749–784.

98. Lipinski CA, Lombardo F, Dominy BW, Feeney PJ. Experimental and computational approaches to estimate solubility and permeability in drug discovery and development settings. Adv Drug Deliv Rev 1997; 23:3–25.

99. Navia MA, Chaturvedi PR. Design principles for orally bioavailable drugs. Drug Discov Today 1996; 1:179–189.

100. van de Waterbeemd H, Smith DA, Beaumont K, Walker DK. Property-based design: optimization of drug absorption and pharmacokinetics. J Med Chem 2001; 44:1313–1333.

101. Hansch C, Li RL, Blaney JM, Langridge R. Comparison of the inhibition of *Escherichia coli* and *Lactobacillus casei* dihydrofolate reductase by 2,4-diamino-5-(substituted-benzyl)pyrimidines: quantitative structure–activity relationships, x-ray crystallography, and computer graphics in structure–activity analysis. J Med Chem 1982; 25:777–784.

102. Silipo C, Vittoria A. A reanalysis of the structure–activity relationships of sulfonamide derivatives. Farmaco Ed Sci 1979; 34:858–868.

103. Cramer RD III, Patterson DE, Bunce JD. Comparative molecular field analysis (CoMFA): I. Effect of shape on binding of steroids to carrier proteins. J Am Chem Soc 1988; 110:5959–5967.

104. Coats EA. The CoMFA steroids as a benchmark dataset for development of 3D QSAR methods. In: Kubinyi H, Folkers G, Martin YC, eds. 3D QSAR in Drug Design: Vol. 3. Recent Advances. Dordrecht: Kluwer/ESCOM, 1998:199–213 also published in Perspect Drug Discov Des 1998; 12–14:199–213.

105. Kubinyi H. A general view on similarity and QSAR studies. In: van de Waterbeemd H, Testa B, Folkers G, eds. Computer-Assisted Lead Finding and Optimization. Proceedings of the 11th European Symposium on Quantitative Structure–Activity Relationships, Lausanne, 1996. Basel: Verlag Helvetica Chimica Acta and VCH, 1997:9–28.

106. Hansch C, Silipo C, Steller EE. Formulation of de novo substituent constants in correlation analysis: inhibition of dihydrofolate reductase by 2,4-diamino-5-(3,4-dichlorophenyl)-6-substituted pyrimidines. J Pharm Sci 1975; 64:1186–1191.

107. Abramson FB, Barlow RB, Mustafa MG, Stephenson RP. Relationships between chemical structure and affinity for acetylcholine receptors. Br J Pharmacol 1969; 37:207–233.
108. Kier LB, Hall LH. Molecular Connectivity in Chemistry and Drug Research. New York: Academic Press, 1976.
109. Kier LB, Hall LH. Molecular Connectivity in Structure–Activity Analysis. New York: Wiley, 1986.
110. Kier LB, Hall LH. Molecular Structure Description: The Electrotopological State. New York: Academic Press, 1999.
111. Kubinyi H. Similarity and dissimilarity—a medicinal chemist's view. In: Kubinyi H, Folkers G, Martin YC, eds. 3D QSAR in Drug Design: Vol. 2. Ligand–Protein Complexes and Molecular Similarity. Dordrecht: Kluwer/ESCOM, 1998:225–252 also published in Perspect Drug Discov Des 1998; 9–11:225–252.
112. Rum G, Herndon WC. Molecular similarity concepts: 5. Analysis of steroid–protein binding constants. J Am Chem Soc 1991; 113:9055–9060.
113. Good AC, Peterson SJ, Richards WG. QSAR's from similarity matrices. Technique validation and application in the comparison of different similarity evaluation methods. J Med Chem 1993; 36:2929–2937.
114. Good AC. 3D Molecular similarity indices and their application in QSAR studies. In: Dean P ed. Molecular Similarity in Drug Design. New York: Chapman and Hall, 1995: 24–56.
115. Martin YC, Lin CT, Hetti C, DeLazzer J. PLS analysis to detect nonlinear relationships between biological potency and molecular properties. J Med Chem 1995; 38:3009–3015.
116. Kearsley SK, Smith GM. An alternative method for the alignment of molecular structures: maximizing electrostatic and steric overlap. Tetrahedron Comput Methodol 1990; 3:615–633.
117. Klebe G, Mietzner T, Weber F. Methodological developments and strategies for a fast flexible superposition of drug-size molecules. J Comput-Aided Mol Des 1999; 13:35–49.
118. Hurst T, Heritage T, HQSAR—a highly predictive QSAR technique based on molecular holograms. 213th ACS National Meeting, San Francisco, CA, 1997, CINF 019.
119. Tong W, Lowis DR, Perkins R, Chen Y, Welsh WJ, Goddette DW, Heritage TW, Sheehan DM. Evaluation of quantitative structure–activity relationship methods for large-scale prediction of chemicals binding to the estrogen receptor. J Chem Inf Comput Sci 1998; 38:669–677.
120. So SS, Karplus M. A comparative study of ligand–receptor complex binding affinity prediction methods based on glycogen phosphorylase inhibitors. J Comput-Aided Mol Des 1999; 13:243–258.
121. Viswanadhan VN, Mueller GA, Basak SC, Weinstein JN. Comparison of a neural net-based QSAR algorithm (PCANN) with hologram- and multiple linear regression-based QSAR approaches: application to 1,4-dihydropyridine-based calcium channel antagonists. J Chem Inf Comput Sci 2001; 41:505–511.
122. Gao H, Bajorath J. Comparison of binary and 2D QSAR analyses using inhibitors of human carbonic anhydrase II as a test case. Mol Divers 1998/1999; 4:115–130.
123. Labute P. Binary QSAR: a new method for the determination of quantitative structure–activity relationships. In: Altman R, Dunker A, Hunder L, Klein T, Lauderdale K, eds. Biocomputing. Proceedings of the 1999 Pacific Symposium. Singapore: World Scientific Publishing, 1999:444–455.
124. Gao H, Williams C, Labute P, Bajorath J. Binary quantitative structure–activity relationship (QSAR) analysis of estrogen receptor ligands. J Chem Inf Comput Sci 1999; 39:164–168.
125. Gao H. Application of BCUT metrics and genetic algorithm in binary QSAR analysis. J Chem Inf Comput Sci 2001; 41:402–407.

126. Ajay, Walters WP, Murcko MA. Can we learn to distinguish between "drug-like" and "nondrug-like" molecules? J Med Chem 1998; 41:3314–3324.
127. Sadowski J, Kubinyi H. A scoring scheme for discriminating between drugs and nondrugs. J Med Chem 1998; 41:3325–3329.
128. Fujita T. Recent success stories leading to commercializable bioactive compounds with the aid of traditional QSAR procedures. Quant Struct–Act Relat 1997; 16:107–112.
129. Hansch C, Björkroth JP, Leo A. Hydrophobicity and central nervous system agents: on the principle of minimal hydrophobicity in drug design. J Pharm Sci 1987; 76:663–687.
130. Hansch C. In: Keverling Buisman JA, ed. Biological Activity and Chemical Structure. Proceedings of the IUPAC–IUPHAR Symposium, Noordwijkerhout, 1977. Pharmacochem Libr 2. Amsterdam: Elsevier, 1977:47–61.

22

3D QSAR Modeling in Drug Design

TUDOR I. OPREA

AstraZeneca R&D Mölndal, Mölndal, Sweden

1. 3D-QSAR HISTORY

Three-dimensional (3-D) quantitative structure–activity relationship (3D-QSAR) has emerged two decades ago as a natural extension to the Hansch [1] and Free-Wilson approaches [2], reviewed by Kubinyi (this volume). The forerunner of 3-D approaches was dynamic lattice-oriented molecular modeling system (DYLOMMS) [3], later known as comparative molecular field analysis (CoMFA) [4]. Combining two existing techniques, GRID [5] and partial least squares (PLS) [6], CoMFA establishes a relationship between the molecular interaction fields (MIFs) of a series of molecules and the target property, via PLS, seeking differences and similarities in the MIFs (steric and electrostatic are default) that can be matched with differences and similarities in the target property values. The interactive graphical analysis [7] of the CoMFA-defined volumes associated with biological activities (via PLS coefficients) and its comparison with molecular structures ushered a new era in understanding structure–activity relationships.

Published in that same year, 1988, the hypothetical active site lattice (HASL) method [8] distributes partial activities at molecular lattice (grid) points, defined within the van der Waals volume of each ligand, which are then correlated to the target property via multilinear regression (MLR). One year later [9], a method using multiple conformers, their conformational energy, and atom-based physicochemical properties to model the binding site, was introduced in the REMOTEDISC [10] program. GRID-based MIFs were introduced to the 3D-QSAR arena in 1993, in the generating optimal linear PLS estimations (GOLPE) program [11]. That same year, a book dedicated to the understanding of 3D-QSAR approaches was published [12]. The methodology itself became widely used, warranting two more volumes dedicated to it [13,14]. An

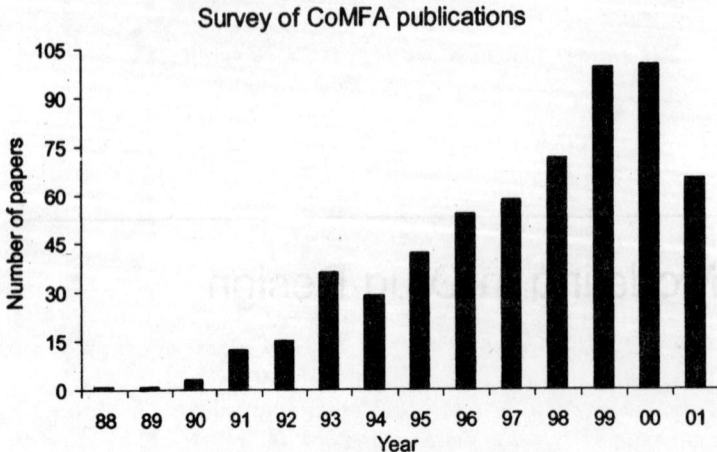

Figure 1 Histogram analysis of CoMFA publications indexed by the Chemical Abstracts Service between 1988 and 2001. Keyword search: "CoMFA" and "comparative molecular field analysis"; search date: April 2, 2002; search conducted using SciFinder 5.1 in the CAPLUS database.

annual survey of CoMFA-related papers (586 papers between 1988 and 2001) is presented in Fig. 1.

Further additions to the 3D-QSAR arsenal include comparative molecular similarity indices analysis (CoMSIA) [15], 4D-QSAR [16], COMPASS [17], receptor surface models [18], the pseudoreceptor approach [19], ComPharm [20], and comparative molecular surface analysis (CoMSA) [21]. 3-D-invariant, alignment-free descriptor systems such as comparative molecular moment analysis (CoMMA) [22], EVA [23], WHIM [24], and ALMOND [25], have also become available. A survey of the 3D-QSAR literature reveals 1154 entries in the Chemical Abstracts Plus database*; of these, 79% are journal publications, 19% are conference proceedings, and four are patents related to, or using, 3D-QSAR models. As the number of potential targets amenable to drug discovery is increasing exponentially, it is likely that 3D-QSAR models and methodologies will continue to be developed in the next decade.

This chapter outlines the premises for attempting a 3D-QSAR model, illustrating some of the difficulties and biases related to obtaining high-quality target property values, as well as some of the caveats related to 3D-QSAR models in the context of computational medicinal chemistry. Introductions to the 3D-QSAR procedures are available for CoMFA [26–28], while 3D-QSAR models and techniques have been recently reviewed critically [29–34]. The reader is kindly referred to those texts for further detail.

---

*Keywords used: "3D QSAR," "three dimensional quantitative structure activity," "4D QSAR," "5D QSAR," "CoMFA," and "comparative molecular field analysis"; search date: April 2, 2002; search conducted using SciFinder 2000.

2. QUANTITATIVE STUCTURE–ACTIVITY RELATIONSHIP: MODEL EVALUATION

Described as a "soft modeling technique" by Stone and Jonathan [35], the QSAR philosophy [36] aims at explaining, and above all, predicting the target property, even though "predictions are always difficult, especially about the future" (Niels Bohr). The target property may include different forms of binding affinity measurements [37], e.g., pA_2, IC_{50}, and/or K_i for antagonists or inhibitors, and EC_{50} and/or pD_2 for agonists or substrates, as well as different forms of physicochemical (e.g., hydrolysis rates and octanol–water partition coefficients) and biological [e.g., passive oral absorption and passive blood–brain barrier (BBB) permeability] property profiles.

The fundamental hypothesis of 3D-QSAR is that macroscopic (e.g., biological or physicochemical) properties are determined by the spatial arrangements (conformations) of a given molecular structure. 3D-QSAR models relate computed atom-based properties (e.g., steric and electrostatic potentials in CoMFA, or fragmental activities in HASL) to macroscopic target properties, and further use the assumption that alterations of these spatial arrangements (and structures) lead to different properties. Most 3D-QSAR methods rely on the linear free-energy formalism [2], which relates ΔG^0_{bind}, the standard free energy of binding, directly to $-\log K_D$, the dissociation constant, at thermodynamic equilibrium concentrations of the ligand, [L], its receptor, [R], and their associated complex, [LR]:

$$\Delta G^0_{bind} = -RT\ln\frac{[L][R]}{[LR]} = -RT\ln K_D = 2.303\ RT\,pK_D \qquad (1)$$

where $R = 8.31451$ J/mol $\times$ K and T is the temperature. K_D represents the molar concentration of the LR complex at which half dissociation is present. Assuming the reaction L + R $\leftrightarrow$ LR at $T = 310$ K, $2.303RT = 5.93624$ kJ/mol, so ΔG^0_{bind} is numerically equal to approximately $6 \times pK_D$ in kJ/mol, or $1.42 \times pK_D$ in kcal/mol. Target property values are routinely transformed as $-\log(\text{activity})$, and expressed as, e.g., pK_i or pIC_{50}.

Stored in a table where columns are descriptors, and rows are compounds (or conformers), QSAR data sets contain separate columns for the measured target property (Y), attributed to the "training set," as well as computed descriptors for (external) reference compounds on which the QSAR model is tested—the "test set." Statistical procedures, e.g., multiple linear regression (MLR), projection to latent structures (PLS), or neural networks (NN) [38], are then used to establish a mathematical "soft" model relating the observed measurement(s) in the Y column(s) with some combination of the properties represented in the subsequent columns. PLS, NN, and AI (artificial intelligence) techniques have been explored by Green and Marshall in the context of 3D-QSAR models [39], and were shown to extract similar information. A problem that may lead to spurious (chance) correlations when using MLR techniques, the colinearity between various descriptors, or cross-correlation, is usually dealt with in PLS [40].

The quality of the resulting QSAR models can be judged by statistical means such as r (the regression coefficient), r^2 [the fraction of explained variance, Eq. (2)] for the training set, and by q^2 [the cross-validated, or predictive r^2, Eq. (3)] for the test set. The fraction of explained variance, r^2, measures the QSAR model's ability to explain the variance in the data; in other words it estimates the goodness-of-fit of the regression model derived from the training set [see Eq. (2)]. The predictive r^2, or q^2, measures the internal robustness of the QSAR model; in other words, it estimates the predictive ability of the model against the test set [see Eq. (3)]. Predictive estimates are obtained either by use of the cross-validation procedure (internal) or by predicting external compounds (previously not used in the model).

$$r^2 = 1 - \frac{\sum_{i=1}^{N}(Y_c - Y_a)^2}{\sum_{i=1}^{N}(Y_m - Y_a)^2} \tag{2}$$

$$q^2 = 1 - \frac{PRESS}{SD} = 1 - \frac{\sum_{i=1}^{N}(Y_a - Y_p)^2}{\sum_{i=1}^{N}(Y_a - Y_m)^2} \tag{3}$$

where Y are the target property values, as follows: Y_a are measured (actual) values; Y_c are calculated values obtained in the fitted models; Y_m is the mean activity of the N given ligands; Y_p are predicted values obtained via, e.g., cross-validation. PRESS is the sum of squared deviations between predicted (Y_p) and measured (Y_a) binding affinity values over N compounds, and SD is the sum of the squared deviations between the measured binding affinity values of the molecules in the test set and the average Y value (Y_m) for all N molecules in the training set. Other statistical indices used in QSAR analyses have been detailed elsewhere [27].

In cross-validation [41], one or more compounds are excluded from the model; the remainder of the training set is used to derive another model, thus evaluating model predictivity. However, when the training set is composed of multiple-member clusters, the procedure of removing just one compound at a time leaves other cluster members in the training set to represent the chemistry, thus increasing the odds of yielding good predictivity. This is more likely to illustrate the redundancy of the training set composition, while masking the (lack of) true predictivity against previously unseen compounds. If the purpose of cross-validation were evaluating model predictivity, then a conceptually correct procedure would be leave-one-cluster out. However, cluster definition and measures are neither metric nor descriptor independent. Therefore, it is equally appropriate to perform successive runs for randomly or systematically eliminated groups (15–50% of the training set) in order to test model significance.

Cross-validation estimates model robustness and predictivity to avoid overfitting in QSAR [27]. In 3D-QSAR models, PLS and NN model complexity are established by testing the significance of adding a new dimension to the current QSAR, i.e., a PLS component or a hidden neuron, respectively. The optimal number of PLS components or hidden neurons is usually chosen from the analysis with the highest q^2 (cross-validated r^2) value, Eq. (3). The most popular cross-validation technique is leave-one-out (LOO), where each compound is left out of the model once and only once, yielding reproducible results. An extremely fast LOO method, SAMPLS [42], which evaluates the covariance matrix only, allows the end user to rapidly estimate the robustness of 3D-QSAR models. Randomly repeated cross-validation rounds using leave 20% out (L5G), or leave 50% out (L2G), are routinely used to check internal model validity. Additional statistical methods and the use of cross-validation were covered by Stone and Jonathan [43].

3. QUANTITATIVE STRUCTURE–ACTIVITY RELATIONSHIP AND MEDICINAL CHEMISTRY

The need for using cross-validation is to ensure that the (3D) QSAR models are not misleading. This is, in part, related to the plethora of (3D) QSAR descriptors [44,45], as it is not easy to decide, a priori, what type or class of (3D) QSAR descriptors is better suited for a particular target property. In fact, choosing the appropriate descriptors leading to directly interpretable (3D) QSAR models is crucial for establishing a dialogue between computational and medicinal chemists. The lack of communication between computational chemists—as QSAR practitioners—and medicinal chemists, is amply discussed by Camille Wermuth [46], who points out that this dialogue, while one of the major goals when deriving QSARs, is often neglected. The need for such dialogue could not be better illustrated than by discussing the computational alert procedure, introduced in Pfizer, and its impact on drug discovery.

Starting from the comparison of early high-throughput screening (HTS) hits at Pfizer (up to 1994), Lipinski et al. [47] analyzed a subset of 2245 drugs from the World Drug Index (WDI) in order to better understand what the common molecular features of orally available drugs are. Using a simplified, yet efficient version of the QSAR paradigm for structure permeability* suggested by Van de Waterbeemd et al. [48], they concluded that poor oral drug absorption or permeation are more likely to occur when

The molecular weight (MW) is over 500.
The calculated [49] octanol/water partition coefficient (CLOGP) is over 5.
There are more than 5 H-bond (hydrogen bond) donors (HDO—expressed as the sum of O–H and N–H groups).
There are more than 10 H-bond acceptors (HAC—expressed as the sum of N and O atoms).

Thus, any pairwise combination of the following conditions: MW > 500, CLOGP > 5, HDO > 5, and HAC > 10, could result in compounds with poor permeability (exceptions are actively transported compounds and peptides). This computational alert procedure, today referred to as the "rule of 5" (RO5) had a major impact on the pharmaceutical industry. Because of its simplicity, it was implemented early on in the drug discovery process in all the pharmaceutical discovery units. Medicinal chemists use, therefore, RO5 criteria in order to avoid compounds with potential permeability problems.

However, RO5 criteria were misused [50]: although established for drugs, they were applied in everyday practice for post-HTS analysis in lead generation and identification. This resulted in compounds that were not easily amenable to further optimization. More restrictive criteria for leadlike compounds had to be formulated [51–54]. Besides oral drug absorption, BBB permeability has recently become of

---

* The QSAR paradigm for structure-permeability expresses the passive permeability as a function of hydrophobicity, molecular size, and hydrogen-bond capacity.

interest to both computational and medicinal chemists. Norinder and Haeberlein recently proposed [55] a simple two-rule system in order to evaluate the potential of a compound to permeate the BBB:

> If the sum of nitrogen and oxygen atoms $(N + O)$ is less than, or equal to five
> If subtracting $(N + O)$ from LogP yields a positive number

then the molecule is likely to have a positive log BB (brain/blood concentration ratio), and permeate the BBB.

While both the RO5 and the Norinder–Haeberlein rules appear to be simple, they were achieved after significant QSAR efforts. They are, however, encouraging in the sense that such simple rules can prove extremely useful for the medicinal chemistry practitioners involved in drug discovery and that they may provide a good starting point for the dialogue between experimentalists and theorists.

4. THE DISTRIBUTION OF ACTIVITIES IN PHYSICOCHEMICAL PROPERTY SPACE

In an effort to highlight model simplicity in deriving QSAR models, Hansch et al. note [56] that more than 60% of the 8500 biological QSAR equations stored in the Pomona College C-QSAR database [57] contain either the LogP [49] term (4614 equations) or the π constant [1] (784 equations). LogP measures the partition of a compound between water and 1-n-octanol, which, in turn, is a model for the propensity of a compound to cross membrane barriers via passive diffusion. LogP is often used to rationalize whether the target property displays significant hydrophobic components. Furthermore, LogP correlates with the difficulty of removing ligands from solvent (e.g., the cavity formed in water), as it docks into a (hydrophobic) target binding site. Often biological activity shows an optimal value for LogP; that is, a parabolic curve can be drawn through a plot of activity vs. LogP. Mathematically, this can be accommodated by including a nonlinear term for LogP by introducing a square term, by using a bilinear model, or by using a neural network [38]. Since all biological systems have lipid compartments, compounds with LogP values above optimal may be sequestered in the lipid phase, and thus be less able to interact with the receptor, whereas compounds with low LogP values are expected to be less prone to bind into hydrophobic target-binding sites.

To evaluate the impact of hydrophobicity and size on biological activities, we examined activity distribution in a physicochemical property space defined by MW, ELogP (LogP estimated with the Kowwin program [58]) and ELogS, the logarithm of the intrinsic water solubility (estimated with the Wskowwin program [59]), on a set of 11,965 medicinal chemistry related compounds, covering 22,763 reported activities and 279 different targets. These compounds were extracted from papers published in mainstream medicinal chemistry journals, mostly in 1997 and 1998].* From the pie

---

* The following publications were surveyed: Journal of Medicinal Chemistry, Vol. 40, 1997 (5203 structures) and Vol. 41, 1998 (5648 structures)—forming 90.69% of the structures surveyed; Quantitative Structure Activity Relationships Vols. 17–19, 1998–2000 (901 structures), and European Journal of Medicinal Chemistry 36, 2001 (188 structures).

Biological Activity Distribution

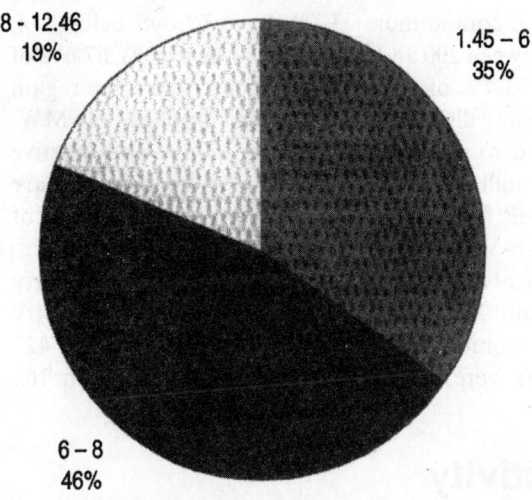

Figure 2 Pie chart illustrating the distribution of biological activities, in the negative log(activity) format, for 11,965 structures with 22,763 reported activities. Inactive compounds (up to 6) are in red; active compounds (above 8) are in yellow; middle-range compounds (between 6 and 8) are in blue. (See color plate at end of chapter.)

chart in Fig. 2, it can be noted that 35% of the compounds exhibit micromolar (or less) activity in at least one instance, whereas 19% of the compounds have activities above 10 nM on a negative log scale. Almost 50% of the reported activities are in the 1 μM–10 nM range (Fig. 2).

The distribution of active molecules for this data set was examined in the physicochemical property space defined by ELogP, ELogS, and MW.* We note that ELogS is related to ELogP and MW by Eq. (4):

$$\text{ElogS (mol/L)} = 0.796 - 0.854E \ \log P° - 0.00728 \ \text{MW} + \text{corrections} \qquad (4)$$

where ELogP° indicates that ELogP values are computed for de novo structures, but measured LogP values are used by the Wskowwin program whenever available. Correction terms are applied to 15 structure types (e.g. alcohols, acids, selected phenols, nitros, amines, alkyl pyridines, amino acids, multinitrogen types, etc) depending on the available melting points, as described by Meylan et al. [59].

---

* The accuracy of ELogP and ELogS estimates on this data set was monitored by comparing ELogP and ELogS with measured values: For LogP, $r^2 = 0.96$ ($N = 292$); for LogS, $r^2 = 0.88$ ($N = 123$). All compounds were considered in neutral form, except for quaternary amines. The structures were computed "as published," i.e., no tautomers were investigated.

From the scatterplot in Fig. 3, it can be noted that there are no apparent clusters for the highly active (yellow) or inactive (red) compounds, although over 50% of the high-activity compounds exhibit MW above 425, ELogP above 4.25, and ELogS below −4.75 [60]. Of the highly active compounds, 73.2% have ELogS below −4, ELogP between −2 and 8, and MW between 200 and 700 daltons (see Fig. 4); 67.6% of the medium-activity compounds and 53.4% of the inactives occupy the same region (data not shown). The lack of significant distribution differences implies that MW, ELogP, and ELogS cannot be utilized to discriminate between active and inactive compounds on a global scale. Even though it appears that more active compounds are located in the lower solubility, mid-LogP, and mid-MW range, this is just an artifact of the trend to synthesize high-LogP, high-MW compounds [47,51,60].

The relationship between biological activities and the physicochemical property space defined by ELogP, ELogS, and MW for this set of medicinal chemistry compounds was further examined by monitoring the regression coefficient (r) in 423 data sets. A total of 1269 regressions were derived on 14,022 activities from 162

Properties vs. Activity

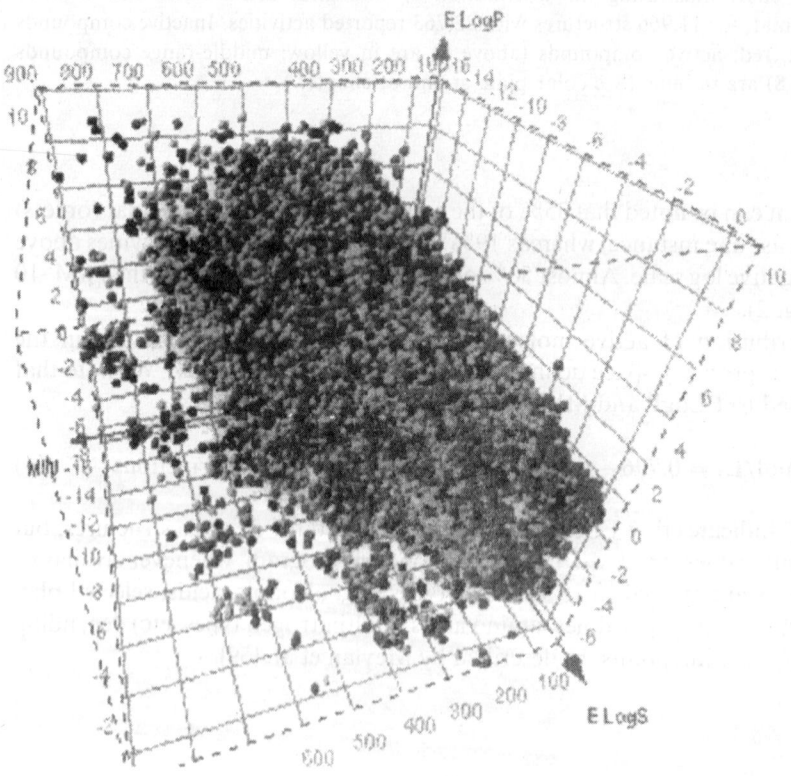

Figure 3   Three-dimensional scatterplot illustrating the distribution of biological activities, in −log (activity) format, for 11,965 structures with 22,763 reported activities, against molecular weight, MW, estimated LogP, ElogP, and estimated aqueous solubility, ELogS. The color code for activity distribution is the same as in Fig. 2. (See color plate at end of chapter.)

Distribution of Actives

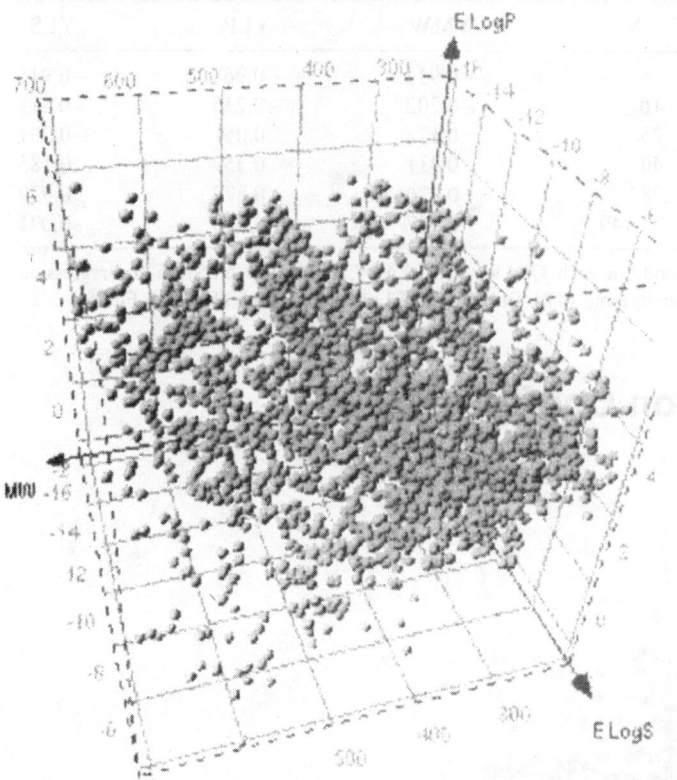

Figure 4 Three-dimensional scatterplot illustrating the distribution of 3111 compounds with activities higher than 10 nM, against molecular weight, MW, estimated LogP, ElogP, and estimated aqueous solubility, ELogS.

different targets. From the descriptive statistics of the regression coefficient (r) distributions summarized in Table 1, it can be observed that the median QSAR data set size is 25 compounds, although N ranges from 5 to 228. More than 50% of the regression coefficients between −log(activity) and MW, ELogP, and ELogS − YMW, YLP, and YLS, respectively, are below the significance level (i.e., below 0.707107 in absolute value). The distribution of YMW, YLP, and YLS for these 1269 regressions is illustrated in Fig. 5, color-coded by activity type (see color plate at end of chapter). No systematic relationships between any particular target, or any activity type, and MW, ELogP, or ELogS, respectively, can be detected.

By applying a minimum level of statistical significance ($r^2 \geq 0.5$, or $|r| > 0.707$), we found 59 data sets from 46 series, where a linear relationship between biological activity and MW, ELogP, and ELogS, respectively, could be established. Regression coefficients, as well as the targets and references, are given in Tables 2 and 3. Although Table 2 is focused on YMW, we note that 7 of the data sets contain significant regressions for MW, ELogP, and ELogS, simultaneously. Three other data sets show

Table 1 Descriptive Statistics for the Regression Coefficient (r) Distributions Derived from 423 Data Set (1269) Regressions, Using 14022 Activities

	N	YMW	YLP	YLS
Minimum	5	−0.914	−0.968	−0.916
Quartile 25%	16	−0.202	−0.235	−0.378
Median	25	0.175	0.051	−0.091
Quartile 75%	40	0.411	0.359	0.185
Maximum	228	0.926	0.879	0.929
Average	33.149	0.097	0.056	−0.071

N is the number of compounds in each QSAR series. YMW, YLP, and YLS are the regression coefficients between −log (activity) and MW, ELogP, and ELogS, respectively. See also Fig. 5.

Regression Coefficients

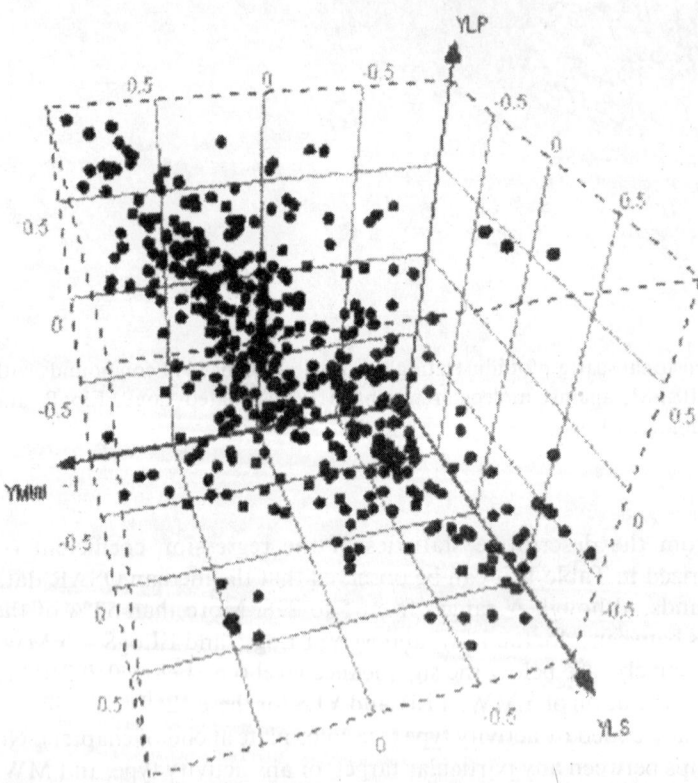

Figure 5 Three-dimensional scatterplot illustrating the distribution of regression coefficients between biological activity, in −log (activity) format, and molecular weight, YMW, estimated LogP, YLP, and estimated aqueous solubility, YLS, respectively. Data are shown for 423 data sets, encompassing 14,022 activities. The color code for activity type is as follows: blue for inhibition estimates (A_2, K_i, IC_{50}), black for agonism/substrate estimates (EC_{50}, D_2), and green for binding affinity estimates (K_D and RBA). See Tables 1–3 and text for details. (See color plate at the end of this chapter.)

Table 2 Regression Coefficients (r Values) That Are Above the 50% Explained Variance Level ($r^2 \geq 0.5$) for MW (37 Data sets from 28 Series)

Series	Receptor	Activity type	N	YMW	YLP	YLS	Source
1	5alpha-R1	K_i	15	0.78	0.62	−0.76	J Med Chem 1997; 40:1293–1315
2	5-HT$_{1A}$	IC$_{50}$	30	−0.75	−0.65	0.64	J Med Chem 1997; 40:1808–1819
3	5-HT$_{1A}$	K_i	28	0.72	0.68	−0.84	J Med Chem 1997; 40:300–312
2	5-HT$_{1B}$	IC$_{50}$	25	−0.71	−0.51	0.50	J Med Chem 1997; 40:1808–1819
2	5-HT$_{1D}$	IC$_{50}$	15	−0.77	−0.48	0.51	J Med Chem 1997; 40:1808–1819
4	5-HT$_{2A}$	K_i	30	0.78	0.37	−0.63	Quant Struct-Act Relat 1999; 18: 548–560
2	5-HT$_{2C}$	IC$_{50}$	19	−0.84	−0.63	0.66	J Med Chem 1997; 40:1808–1819
5	A$_{2a}$	K_i	43	0.84	0.72	−0.76	J Med Chem 1997; 40:4396–4405
6	AChE	IC$_{50}$	19	−0.84	−0.37	0.61	J Med Chem 1997; 40:3516–3523
7	α$_1$-adrenergic	IC$_{50}$	12	0.75	0.52	−0.56	J Med Chem 1998; 41:4165–4170
8	α$_{1D}$-adrenergic	K_i	28	−0.80	−0.52	0.63	J Med Chem 1998; 41:2643–2650
9	Ca-channel	K_i	22	−0.74	0.04	0.75	J Med Chem 1998; 41:5393–5401
3	**D$_2$**	K_i	**26**	**0.72**	**0.86**	**0.86**	**J Med Chem 1997; 40:300–312**
10	DAT	IC$_{50}$	15	−0.72	−0.47	0.70	J Med Chem 1998; 41:2380–2389
11	ENOS	IC$_{50}$	16	−0.77	−0.64	0.72	J Med Chem 1998; 41:3675–3683
12	**ET-A**	**IC$_{50}$**	**30**	**0.82**	**0.84**	**−0.86**	**J Med Chem 1998; 40:322–330**
13	FXa	K_i	36	0.74	−0.43	−0.09	J Med Chem 1998; 41:3551–3556
14	**HIV-1 P**	**IC$_{50}$**	**61**	**0.75**	**0.73**	**0.74**	**J Med Chem 1997; 40:3781–3792**
15	HIV-1 Tat	IC50	8	0.78	0.78	0.57	J Med Chem 1998; 41:2994–3000
16	m1	K_i	24	0.88	0.69	−0.67	J Med Chem 1998; 41:3220–3231
16	m2	K_i	25	0.90	−0.32	0.13	J Med Chem 1998; 41:3220–3231
16	m3	K_i	25	0.93	−0.14	−0.05	J Med Chem 1998; 41:3220–3231
17	MGluR5	K_i	14	0.81	0.18	0.70	J Med Chem 1997; 40:3645–3650

Table 2 Continued

Series	Receptor	Activity type	N	YMW	YLP	YLS	Source
18	MMP-1	K_i	9	−0.74	0.77	0.46	J Med Chem 1998; 41:1745–1748
19	**MMP-9**	**IC$_{50}$**	**6**	**−0.72**	**−0.71**	**0.77**	**J Med Chem 1998; 41:3568–3571**
20	MR	IC$_{50}$	13	−0.75	−0.70	0.53	J Med Chem 1998; 41:291–302
21	NEP	IC$_{50}$	8	0.82	−0.29	0.07	J Med Chem 1997; 40:1570–1577
22	NMDA	IC$_{50}$	13	0.74	0.34	−0.43	J Med Chem 1997; 40:3679–3686
23	**NPY Y$_1$**	**K_i**	**9**	**0.90**	**0.88**	**−0.92**	**J Med Chem 1997; 40:3712–3714**
24	NPY Y$_1$	K_i	38	0.81	0.31	−0.54	J Med Chem 1998; 41:2709–2719
25	**PDE2**	**IC$_{50}$**	**9**	**0.78**	**0.73**	**−0.76**	**J Med Chem 1997; 40:2196–2210**
26	PLA$_2$	IC$_{50}$	37	0.75	0.47	−0.52	J Med Chem 1997; 40:2694–2705
27	RAR-β	K_i	9	−0.91	−0.38	0.93	J Med Chem 1997; 40:4222–4234
27	RXR-α	K_i	9	−0.81	−0.29	0.75	J Med Chem 1997; 40:4222–4234
27	RXR-β	K_i	9	−0.85	−0.35	0.81	J Med Chem 1997; 40:4222–4234
27	RXR-γ	K_i	9	−0.79	−0.27	0.74	J Med Chem 1997; 40:4222–4234
28	**Thrombin**	**IC$_{50}$**	**31**	**−0.76**	**−0.84**	**0.83**	**J Med Chem 1998; 41:401–406**

Series with $r^2 \geq 0.5$ for MW, LogP and LogSw are highlighted in **bold**; series with $r^2 \geq 0.5$ for MW and LogSw are highlighted in light gray; series with $r^2 \geq 0.5$ for MW and LogP are highlighted in dark gray. See also Table 1.

significant YMW and YLP values, whereas 10 data sets show significant YMW and YLS values. Among the 22 data sets summarized in Table 3, all of which show significant YLP values (except one), 10 show significant YLP and YLS values. This is not surprising, since ELogP is a major component of the ELogS estimate [Eq. (4)].

From these data sets, it can be concluded that YLP (the correlation between LogP and biological activity) is significant in 31 data sets, to which perhaps a further 11 data sets for YLS could be added. What becomes immediately apparent, however, is the remarkable data set dependence of these relationships. For example, the same nine compounds in series **27** (Table 2) show significant relationships between MW and ELogS, on one hand, and four biological activities for retinoic acid receptors (RAR and RXR) inhibition on the other hand. This result confirms, indirectly, 2D- and 3D-QSAR results derived on larger series on RAR agonists [61]. Two similarly sized series, **2** and **3** (Table 2), show significant YMW and marginally significant YLP and YLS values on the same target (5HT$_{1A}$ antagonists); however, the regressions are of

Table 3 Regression Coefficients (r Values) That Are Above the 50% Explained Variance Level ($r^2 \geq 0.5$) for LogP (21 Data Sets from 18 Series) and LogSw (1 Data Set, **Bold**)

Series	Receptor	Activity type	N	YMW	YLP	YLS	Source
29	5-HTT	IC_{50}	20	0.66	0.81	−0.66	J Med Chem 1997; 40:2525–2532
30	A_1	App K_i	19	0.70	0.81	−0.81	J Med Chem 1997; 40:3765–3772
31	A_1	K_i	63	0.43	−0.78	0.64	J Med Chem 1997; 40:2156–2163
32	ADH sigma	K_i	23	0.53	−0.73	0.36	J Med Chem 1998; 41:1696–1701
33	α_1-adrenergic	IC_{50}	8	0.66	0.73	−0.77	J Med Chem 1997; 40:4146–4153
34	AR	IC_{50}	15	−0.37	−0.88	0.68	J Med Chem 1998; 41:291–302
34	AR	K_i	14	−0.53	−0.97	0.81	J Med Chem 1998; 41:291–302
35	CCK-A	ED_{50}	26	0.55	−0.72	0.59	J Med Chem 1997; 40:2706–2725
35	CCK-B	IC_{50}	24	0.53	−0.72	0.62	J Med Chem 1997; 40:2706–2725
36	DA uptake	IC_{50}	16	−0.16	0.74	−0.72	J Med Chem 1998; 41:4973–4982
36	DA uptake	K_i	16	−0.16	0.74	−0.72	J Med Chem 1998; 41:4973–4982
36	**DAT**	**K_i**	**16**	**0.02**	**0.70**	**−0.71**	**J Med Chem 1998; 41:4973–4982**
37	ENOS	IC_{50}	10	−0.70	−0.71	0.72	J Med Chem 1998; 41:1361–1366
38	ER	IC_{50}	10	−0.49	−0.82	0.77	J Med Chem 1998; 41:2928–2931
39	ER	RBA	14	0.24	0.77	−0.82	J Med Chem 1998; 41:1272–1283
40	GTPase	EC_{50}	20	−0.18	0.79	−0.58	J Med Chem 1997; 40:3130–3139
41	MMP-3	K_i	9	−0.61	0.76	0.35	J Med Chem 1998; 41:1745–1748
42	MMP-9	IC_{50}	15	0.56	0.71	−0.63	J Med Chem 1998; 41:1209–1217
41	MMP-9	K_i	9	−0.54	0.80	0.29	J Med Chem 1998; 41:1745–1748
43	PDE1	IC_{50}	47	0.67	0.78	−0.78	J Med Chem 1997; 40:2196–2210
44	PDGFR	IC_{50}	11	0.52	0.71	−0.70	J Med Chem 1998; 41:1752–1763
45	PR	EC_{50}	21	−0.38	−0.75	0.38	J Med Chem 1998; 41:4354–4359
46	Sigma1	IC_{50}	10	−0.61	−0.81	0.80	J Med Chem 1998; 41:468–477

Series with $r^2 \geq 0.5$ for MW and LogP are excluded. Series with $r^2 \geq 0.5$ for LogP and LogSw are highlighted in gray. See also Tables 1 and 2.

opposite sign. The same situation occurs for series **38** and **39** (Table 3)—estrogen receptor (ER) antagonists—where YLP and YLS have opposite signs. Such findings should prompt further analysis and comparison across series in order to establish if any of these trends is more than just a chance correlation [62,63]. They also indicate that QSAR results are not only descriptor and statistics dependent, but also data set dependent.

To visualize the potential impact that data set dependence may have on the practice of computational medicinal chemistry, imagine a list of names in the phone book: with a large enough sample size, one could find a significantly large number of entries where the names are listed not only alphabetically, but also in an increasing order of the phone number; conversely, one could find a similar list where the phone numbers are listed in decreasing order, but still alphabetically. Both lists are true, in the sense that they are a partial reflection of reality, and yet the conclusions from analyzing them can be equally misleading.

The above survey, probing the impact of ELogP, ELogS, and MW on the distribution of biological activities, performed on nearly 12,000 medicinal-chemistry-related compounds, illustrates beyond doubt that there is an increasing need for global understanding of chemical and biological interactions [56]. Unlike the curated efforts of Hansch et al. [57], this survey was intended to reflect real-life situations, where raw, unfiltered data sets are the only data a medicinal and computational chemist have. More often than not, the entire data set is not amenable to QSAR modeling, and further processing of the input data that may require removing compounds or descriptors is needed before one can derive useful hypotheses and conclusions about the biological and chemical processes. Size, hydrophobicity, and solubility estimates are likely to play an important role, but perhaps not as significant as 60%. In the above survey, less than 15% of the data sets show significant YMW, YLP, or YLS. Even these instances warrant further analyses: whenever a target binding site can be hypothesized, 3D-QSAR models are likely to prove useful in understanding not only the transfer-related components (e.g., LogP), but also the molecular determinants responsible for biological activity.

5.   ASSUMPTIONS IN 3D-QSAR. THE BIOACTIVE CONFORMATION AND BIOLOGICAL ACTIVITY

3D-QSAR methods rely on a set of assumptions [64], i.e.,

> The modeled ligand, not its metabolite(s) or any of its derivatives, produces the observed effect.
>
> The ligand is modeled in a single (bioactive) conformation that exerts the binding effects. The dynamic nature of this process, as shown for lactate dehydrogenase that assumes different conformational states at the binding site [65], is typically ignored.
>
> The geometry of the receptor binding site is, with few exceptions, considered rigid.
>
> The loss of translational and rotational entropy upon binding [66] is assumed to follow a similar pattern for all compounds, even though thermodynamic data suggest otherwise [67].

The entropic cost for freezing nonterminal single-bond rotors [68] is frequently estimated only by counting the number of rotatable bonds.

The protein binding site is the same for all modeled ligands.

The on–off rate is similar for modeled compounds, i.e. the system is considered to be at equilibrium, and kinetic aspects are usually not considered.

Solvent effects, temperature, diffusion, transport, pH, salt concentrations, and other factors that contribute to the overall ΔG_{bind}^{LR}, the ligand–receptor free energy of binding, are not considered.

For 3D-QSAR methods based on molecular mechanics force fields, ΔG_{bind}^{LR} is largely explained by the enthalpic component (the internal energy derived from force field calculations) that is prone to inherent force field errors.

By their very nature, 3D-QSAR models need to address the issue of conformational flexibility and multiple binding modes. Alternative binding modes in CoMFA, e.g., have been explored for a diverse set of HIV-1 protease inhibitors [69] using a flexible docking procedure, NewPred [70], that have shown a 3 log unit variability in predicted affinity for 12 conformers of the same compound. The proposed solution is often not unique, as shown by experimentally determined alignments [71]. In the case of flexible molecules, many conformers can match a particular pharmacophoric pattern, and thus may appear to be consistent with a 3D-QSAR model. Therefore, the rationale for choosing one (the "alignment rule") is usually done on an energetic basis. When the choice of the alignment has no reference to experimentally determined structures, results should be treated with caution [72]: other conformers may in fact bind to the receptor, while the correlations obtained from the proposed alignment may be fortuitous, perhaps compensating for the inadequacies in the estimation of entropic effects. Multiple, alternate conformers are also considered in REMOTEDISC [10] and COMPASS [17], among other methods. One formalism, related to biological activity adjustment, is outlined below.

Biological activity is the result of sampling multiple conformational substates L_i, $i = 1, 2, \ldots, n$, of which only one (of very few) fit in the receptor binding site. Any solvent-accessible conformer can, in principle, transition into the active one during this process. Some ligands may already be in their lowest energy conformation when binding to the receptor, while others may have to transition to the active conformation, lower in energy. As proposed by Janssen [73], these ligands exhibit on the average higher energies compared to the (global) minimum. All ligands having internal energies higher than the minimum must have a higher affinity than molecules occurring in the minimum energy conformation, independent of the energy level of the receptor bound conformation [73,74]. It follows that the concentration C_F corresponding to the bioactive conformer, L_F, represents only a fraction, α_F, of the total ligand concentration, C_T [see Eq. (5)]:

$$C_F = \alpha_F C_T \tag{5}$$

One could, in principle, adjust [74] the biological activity Y_{exp} of a ligand by taking into consideration the concentration of its bioactive conformer, L_F (6):

$$Y_{F,adj} = Y_{exp} - \log \alpha_F \tag{6}$$

The relative amount, α_F, of the bioactive conformer L_F can be estimated according to Boltzmann statistics [75], using Eq. (7):

$$\alpha_F = \frac{g_F \cdot e^{-U_F/RT}}{\sum_{i=1}^{n} g_i \cdot e^{-U_i/RT}} \tag{7}$$

where U_i are relative conformational energies, calculated by standard conformational analyses and g_i are degeneracy degrees of the conformational energy levels due to the appearance of certain symmetric conformers (i.e., *gauche* forms for example). U_F and g_F represent the corresponding properties for the bioactive conformation. If there is more than a single bioactive conformation, the corresponding sum is performed in the denominator. Adjusted biological activities have been used to select bioactive conformers for a series of 25 acetylcholinesterase (AChE) substrates in the absence of any knowledge of the binding site, and were shown to fit well [74] into the active site of a crystallographic structure of AChE [76], thereby indicating the potential utility of this approach.

In a similar approach, Lukáčová and Baláž [77] have considered the association constant, K_i, as a sum of association constants for individual conformers, K_{ij}, for which individual CoMFA parameters could be computed (8):

$$K_i = e^{c_0} \times \sum_{j=1}^{m} e^{\sum_{k=1}^{p} c_k X_{ijk}} \tag{8}$$

where $\sum_{k=1}^{p} c_k \times X_{ijk}$ is the sum of association constant models derived using c_k adjustable parameters and X_{ijk} interaction energies for a p number of binding points for the ith conformer in the jth binding mode. In this approach, $1 \ldots m$ conformers are considered; once optimized, the contributions of individual binding modes to the overall binding (K_{ij}/K_i) can be estimated. This is conceptually equivalent to the $Y_{F,adj}$ approach in Eq. (6), but has the advantage of considering multiple combinations of the conformational poise for each ligand.

Regardless of the approach taken to model the active conformers, one has to ascertain that no rate-limiting steps occur during intermediate stages, and that non-specific binding does not obscure the experimental binding affinity. All the parameters that cannot be directly measured remain hidden (e.g., receptor-induced conformational changes of the ligand, geometric variations of the binding site, etc.). QSAR methods use a time-sliced (frozen) model, i.e., the system is at equilibrium and time independent [74]. Kinetic bottlenecks in the intermediate steps may occur, and ΔG_{bind}^{LR} remains thermodynamic in nature (not kinetic).

6. CoMFA. THE ALIGNMENT PROBLEM

Quite frequently, ΔG_{bind}^{LR} is expressed as the sum of the free energy components, conceptually shown [37] in the "master equation" (9):

$$\Delta G_{bind}^{LR} = \Delta G_{sol} + \Delta G_{conf} + \Delta G_{int} + \Delta G_{motion} \tag{9}$$

which accounts for free energy terms due to solvent (ΔG_{sol}), to conformational changes in both ligand and protein (ΔG_{conf}), to ligand–protein interactions (ΔG_{int}), and to the motion in the ligand and protein once they are at close range (ΔG_{motion}). This can also be expressed [27] as (10):

$$\Delta G_{bind}^{LR} = \Delta G_{sol} + \Delta U_{vac} - T\Delta S_{vac} \tag{10}$$

where ΔG_{bind}^{LR} is separated, at equilibrium, into solvation effects (ΔG_{sol}) and two components for the process in vacuo: the internal energy, ΔU_{vac}, and the entropic contribution, $T\Delta S_{vac}$. ΔG_{sol} can be calculated with a variety of methods [37], whereas $T\Delta S_{vac}$ is often related to the number of nonterminal single bonds [83–85]. Both ΔG_{sol} and $T\Delta S_{vac}$ are assumed to have similar values for congeneric series; hence, Eq. (6) is widely used in QSAR studies by expanding only the internal energy term (11)

$$\Delta U_{vac} = \Delta U_{vdW}^{LR} + \Delta U_{coul}^{LR} + \Delta U_{distort}^{L} + \Delta U_{distort}^{R} + \Delta U_{conf}^{R} \tag{11}$$

which includes steric (vdW) and electrostatic (coul) aspects of the ligand–receptor interaction (ΔU_{vdW}^{LR} and ΔU_{coul}^{LR}), distortions (distort) induced by the L–R interaction ($\Delta U_{distort}^{L}$ and $\Delta U_{distort}^{R}$), and ligand-induced conformational changes of the receptor (ΔU_{conf}^{R}). ΔU_{conf}^{R} represents agonist-induced conformational rearrangements of the receptor that may be an important component of signal transduction and are not considered to occur upon antagonist binding to the same receptor [66].

CoMFA [4] computes the steric, ΔU_{vdW}^{LR}, and electrostatic, ΔU_{coul}^{LR}, interactions on a uniform grid around each ligand, using hypothetical probes (Csp3 for steric, H+ for electrostatic) that mimic receptor atoms (see Fig. 6). These grid-point calculated interaction energies are tabulated for each molecule (row) in the series. The resulting matrix is analyzed with multivariate statistics [6], yielding Eq. (12) that relates CoMFA fields to ΔU_{bind}^{LR}:

$$\Delta U_{bind}^{LR} = \sum_{x,m}^{X,M} A_x U_{vdW}(x, L_m) + \sum_{x,m}^{X,M} B_x U_{coul}(x, L_m) \tag{12}$$

where X grid-based CoMFA probes x interact with M ligand atoms L_m, and A_x and B_x are PLS (pseudo)coefficients.

Thus, CoMFA compares the steric and electrostatic MIFs within a series, searching to match differences in biological with differences in the interaction fields. In CoMFA, the receptor is approximated by a rigid grid, therefore $\Delta U_{distort}^{R}$ and ΔU_{conf}^{R} are not computed. However, $\Delta U_{distort}^{L}$ is indirectly included since only a particular conformer is usually chosen for the alignment. Other 3D-QSAR methods, while yielding models of similar statistical significance, appear to extract quite different information features from the same data set [81,82].

Ten years after the original CoMFA report [4], more than 100 CoMFA models have been reported on enzyme binding affinity and more than 100 on receptor binding affinity, respectively [31]. An alphabetical list of the 383 CoMFA papers published between 1993 and 1997 is available [83]. Over 200 papers illustrate the use of CoMFA and other 3D-QSAR methods in estimating ΔG_{bind}^{LR}, with an average prediction error of 0.6–0.7 log units (0.85–1 kcal) for external sets of compounds [31]. This estimate, however, is unlikely to reflect the predictive power of CoMFA models for novel classes of compounds. Care needs to be exercised that the prediction step uses interpolation,

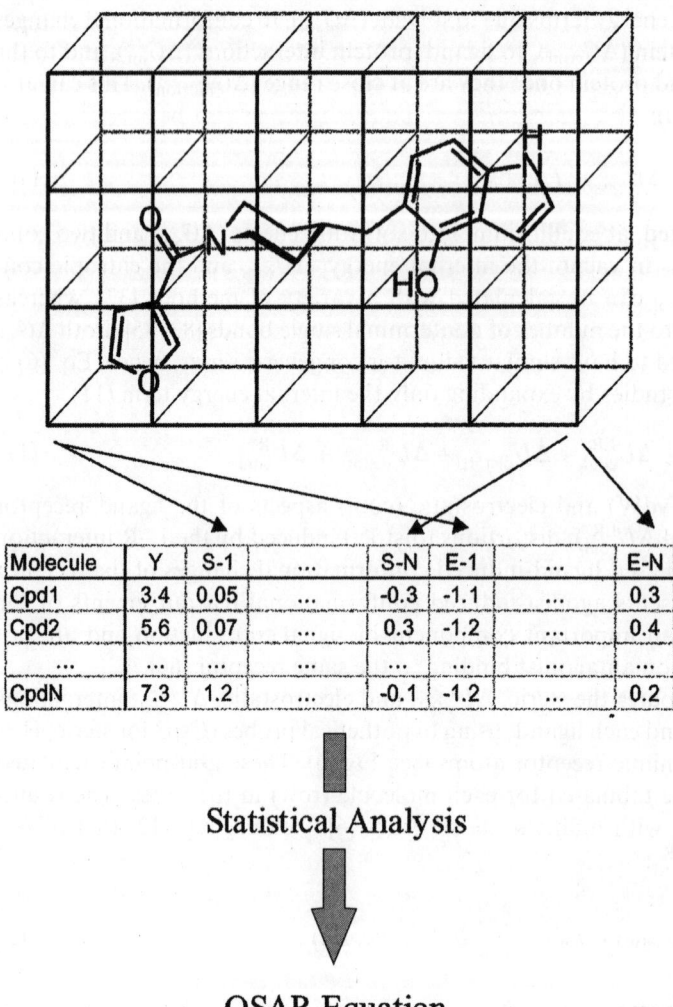

Molecule	Y	S-1	...	S-N	E-1	...	E-N
Cpd1	3.4	0.05	...	-0.3	-1.1	...	0.3
Cpd2	5.6	0.07	...	0.3	-1.2	...	0.4
...	...	...	...	...	...	...	...
CpdN	7.3	1.2	...	-0.1	-1.2	...	0.2

Statistical Analysis

QSAR Equation

Figure 6 An overview of the CoMFA method. See text for details. (Modified from Cramer et al. [4].)

rather than extrapolation [27]. Furthermore, the use of consistent alignment rules for external (test) compounds that are frequently dissimilar from those in the training set remains key among the caveats of alignment-based 3D-QSAR methods. In fact, it has been clearly recognized that the definition of proper alignment rules, i.e., the need to identify a bioactive conformation matching a user-defined pharmacophore, is one of the most important sources of wrong conclusions and errors in all CoMFA studies [84].

The basis of the pharmacophore concept is that all ligands, regardless of chemical structure, bind in conformations that present similar steric and electrostatic features to the receptor—features that are recognized at the receptor site and are responsible for the biological activity [85]. Key among the criteria for valid pharmacophore models [46] is that the training set should include at least one rigid representative, which should be used as a template structure. Even if rigid structures are available, it is possible that seemingly related analogues bind in a different way. Whenever available, experimentally determined alignments should be strongly preferred in 3D-QSAR.

Figure 7 illustrates the indeterminant problem in 3D-QSAR. The first case, of flexible molecules, applies in particular to the common (core) moiety: if the core moiety can adopt multiple conformations, it is likely that if valid QSARs can be established for one conformer, then others may lead to equally relevant QSARs. For example, if our data set consists of tetrapeptides, then the backbone could adopt *all-trans*, α-helical, 3_{10}-helical, and β-turn conformations. The 3D-QSAR models would be equally valid because the biological activity differences would be explained by differences in chemical structure, i.e., by side-chain variations. The relationship between the

1. Flexible molecules: which conformer(s)?

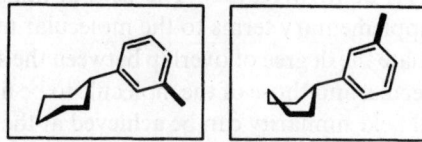

2. Achiral molecules: eight possible alignments

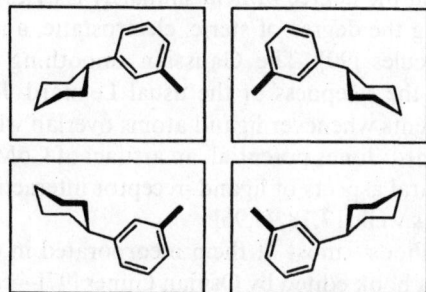

3. Steric, electrostatic or hydrophobic: which one dominates?

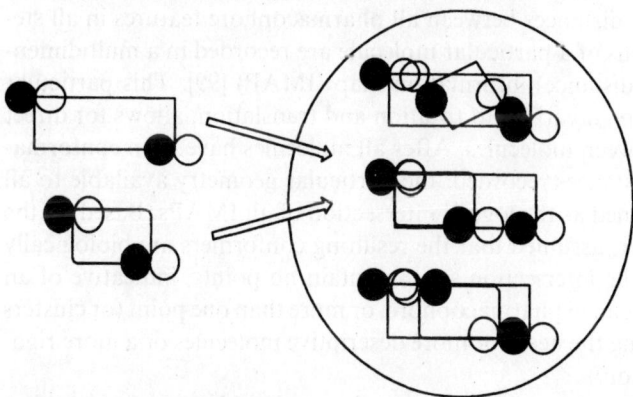

Figure 7 The indeterminant alignment problem in 3D-QSAR.

various substituents could lead to different design hypotheses, in particular if various conformations of the core structure lead to different spatial relationships among them. Thus, using rigid compounds is always helpful. It comes as no surprise that the benchmark for 3D-QSAR method developments is a rigid (31 steroids) data set, which is also illustrative of how errors propagate in the QSAR literature [86].

The second case (Fig. 7), for achiral molecules, relates to the existence of symmetry operators: eight possible alignments can be generated in XYZ coordinates—all of each equally valid except when adding a chiral molecule to the data set. Since receptors are chiral, any pertinent information about the binding site, or chiral ligands, should be incorporated in the alignment procedure(s). The third case illustrates the crux of the alignment problem: what weight should be applied to the different steric, electrostatic, and hydrophobic features when overlapping molecules? More often than not, we are preinclined to consider the steric parameter [87,88] as being dominant. Most targets are hydrophobic in nature [64,79], yet others have a significant section that is exposed to solvent, e.g., (some) serine proteases and monoclonal antibodies.

In the absence of rigid structures, one could use field-fit alignment [89] procedures, e.g., SEAL [90], CoMSIA [15,91] or MIMIC [92]. The field-fit minimization available in CoMFA [89] adds two supplementary terms to the molecular mechanics force field calculation, in order to evaluate the degree of overlap between the steric and electrostatic fields of the template molecule, and those of the molecule to be fitted. The fitted molecule being flexible, maximal field similarity can be achieved at the expense, however, of the internal energy. In SEAL, the fitted molecule is randomly superimposed to the template molecule, and the degree of field similarity is assessed using atom-based penalty functions, scoring the degree of steric, electrostatic, and hydrophobic similarity between the molecules [90]. The Gaussian smoothing function introduced in CoMSIA [15] reduces the steepness of the usual Lennard–Jones potential, yielding significant improvements whenever ligand atoms overlap with probe atoms [93]. The steepness of the Lennard–Jones potential, an artifact of CoMFA that is not likely to be reflected in the natural aspects of ligand–receptor interactions [94], has been addressed by other groups as well [17,18,95,96].

Pharmacophore perception methods—most of them incorporated in commercially available software discussed in a book edited by Ösman Güner [97]—can be exemplified by Marshall et al.'s active analog approach [98]. This approach uses systematic conformational searching to identify all available conformations for the molecules under study. The distances between all pharmacophore features in all sterically allowed conformations of a particular molecule are recorded in a multidimensional (one dimension per distance) information map (IMAP) [99]. This particular representation of conformers, invariant to rotation and translation, allows for direct comparisons of IMAPs between molecules. After all molecules have been conformationally searched and all distances recorded, the particular geometry available to all molecules is readily determined as the logical intersection of all IMAPs. Based on the pharmacophore concept, it is assumed that the resulting conformers are biologically relevant. It is possible for the intersection set to contain no points, indicative of an overconstrained or poorly defined pharmacophore, or more than one point (or clusters of points), possibly indicating the need for more descriptive molecules or a more rigorously defined pharmacophore.

Properly designed alignments and pharmacophores should not only identify a pattern for active compounds, but also account for the lack of activity for weak or inactive molecules. Proper use of negative information in 3D-QSAR is equally important, and deriving models that can also account for true negatives can have a positive influence in the drug discovery decision-making process [100].

7.   ALMOND. ALIGNMENT INDEPENDENCE

To avoid the issue of alignments in 3D-QSAR, several autocorrelation methods have been proposed by the groups of Broto et al. [101], Wagener et al. [102], and Clementi et al. [103]. Except for Broto et al.'s method (used on 2-D or 3-D structures), these methods require a 3-D molecular structure, i.e., they are based on the choice of a conformer. The ALMOND program [25], built on initial work from Clementi et al. [103], is described below because of its convenient use of graphical outputs that relate autocorrelation vectors to MIFs, as well as to molecular structures.

In its standard implementation, ALMOND starts by computing GRID [5] MIFs for three probes: the DRY probe (hydrophobic), the carbonyl O probe (hydrogen bond acceptor) and the amide nitrogen, N1, probe (hydrogen bond donor). Its fundamental assumption is that within these interaction fields, there is pertinent information available, related to the receptor binding site. In other words, spatial regions defined by these MIFs encompass receptor atoms that are involved in the ligand–receptor interaction. ALMOND is based on the virtual receptor site (VRS) concept. The VRS is believed to overlap partially, at least for the bioactive conformer, with the actual receptor binding site, so a subset of the VRS should be relevant for estimating the binding properties of the ligand. To achieve statistical significance, the choice of the 3-D coordinates for the ligand series should be consistent or related to the bioactive conformation as much as possible. ALMOND descriptors are a reduced set of variables representing the geometrical relationships between relevant regions of the VRS, and as such are independent of the spatial framework (grid) where the MIFs are computed. As formulated by Pastor et al. [25], ALMOND descriptors "represent the VRS in the same way that the measures a tailor obtains in order to make tailor-made clothes represent a person." To add to the metaphor, the best-fitting clothes (3D-QSAR models) are obtained when the person (molecule) is in a relaxed (binding-mode) posture (conformation).

The procedure for obtaining ALMOND descriptors requires three steps: 1) compute the MIFs, 2) filter the fields to extract the most relevant regions that define the VRS, and 3) encode the VRS into ALMOND descriptors via autocorrelation. The first step is basically the derivation of GRID fields using the DRY, O, and N1 probes. The second step starts by identifying regions of interest (favorable interaction energies) using an optimization algorithm, selecting from each MIF a fixed number of GRID points (nodes) optimizing a scoring scheme. The scoring scheme involves the optimization of two variables: the energy level of the MIF at a node and the internode distance between two nodes. Therefore, the method extracts from each field a number of nodes (in the order of hundreds) that represent independent, favorable probe–ligand interaction regions. The filtering procedure, based on a Fedorov-like optimization algorithm [104], is conceptually illustrated in Fig. 8. The ensemble of these filtered nodes for all relevant probes defines the VRS. The filtering procedure yields a balanced selection for neutral compounds with a reasonable balance of polar/nonpolar groups, but the exact cut-off for interaction energies needs to be fine-tuned for charged, or unusually polar, molecules. The number of extracted nodes can become

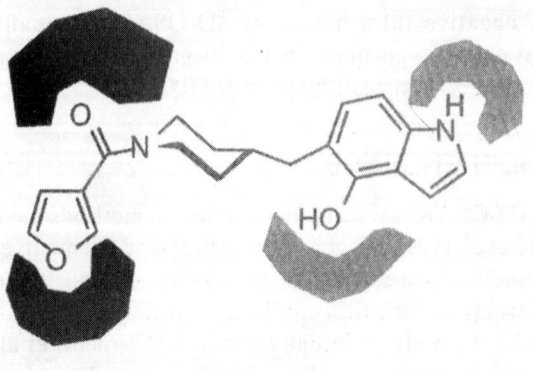

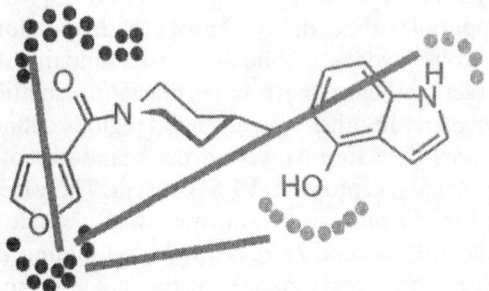

Figure 8 Two-dimensional sketch illustrating the MIF filtering step in ALMOND. MIFs for the N1 probe (hydrogen bond donor, green) and the O probe (hydrogen bond acceptor, red), top, are processed at the same energy level, e.g., -2 kcal/mol, to yield a few hundred representative nodes, bottom. The grey lines illustrate distances used in the auto- (N1–N1) and cross- (N1–O) correlograms by the MACC-2 encoding algorithm. (See color plate at end of chapter.)

relevant when attempting to compare different data sets, because the number of nodes should be identical across all series. Finally, the autocorrelogram is generated in the third step. The autocorrelogram encodes the geometrical relationships between the VRS regions in a spatially independent manner, using the maximum auto- and cross-correlation (MACC)-2 method [105].

The pairwise product of interaction energies between all two-node pairs is managed according to the internode distance. Rather than summing all computed terms, as in traditional autocorrelation methods, only the highest products are stored via MACC-2, while the rest is ignored. The results are managed according to each MIF category. This is significant for the "reversibility" properties of ALMOND: while a sum cannot be reverted to all its terms, the nodes producing the maximum product can be stored in the computer memory, then traced back as necessary during the interpretation step. MACC-2-derived results can be viewed directly in correlogram plots, where the products of the internode energies are plotted against internode distances. One single energy value is obtained for each of the categories considered, representing a small distance range. Fig. 9 shows correlograms obtained for the furoyl-piperidinyl-methyl-indole depicted in Fig. 8 with the O and N1 probes. Every peak in

ALMOND Profile

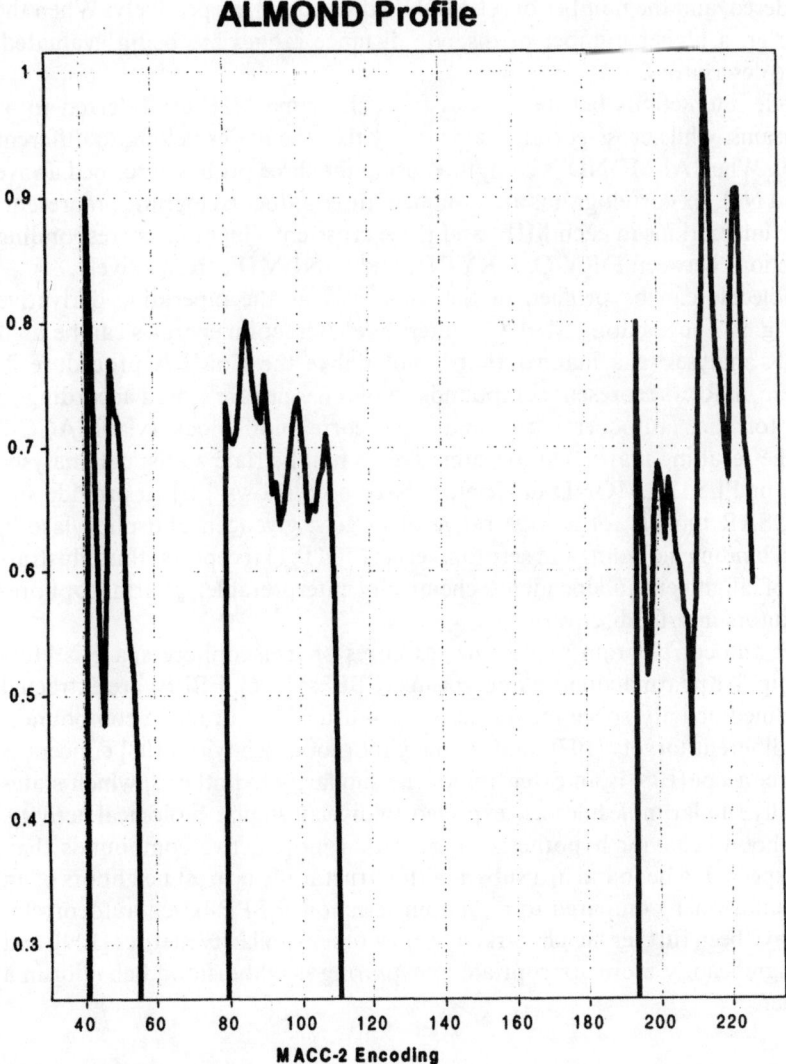

Figure 9 Auto- and cross-correlogram profile obtained for the furoyl-piperidinyl-methyl-indol sketched in Fig. 8, using a hydrogen bond donor (N1) and hydrogen bond acceptor (O) probe. The O–O autocorrelogram is on the left, the N1–N1 autocorrelogram in the middle, and the N1–O cross-correlogram on the right.

the correlogram indicates that the VRS contains two regions separated by a distance corresponding to the abscissa of the peak (short distances are on the left-hand side, longer distances on the right-hand side). The height of the peak expresses the product of the intensity of the field on both nodes. The shape of the peak is also relevant. Chemical moieties producing intense interactions are represented, after filtering, by many contiguous nodes, thus producing wider peaks. Conversely, narrow peaks are associated with weaker interactions. There is a direct relationship between the size of the GRID lattice, defining the maximum internode distance and the size of the distance ranges considered, and the number of ALMOND descriptors, respectively. When the range is shorter, a higher number of discrete distance ranges are being evaluated, yielding more descriptors.

Internode interactions between nodes from the same MIF are referred to as autocorrelograms, while cross-correlograms imply that the nodes belong to different MIFs (Fig. 9). When ALMOND is computed using the three probes described above (DRY, O, and N1), six correlograms are obtained: three autocorrelograms, representing internode interactions in each MIF, and three cross-correlograms corresponding to the interactions between DRY-O, DRY-N1, and O-N1 MIFs, respectively.

Any molecule can be profiled in the same way as the piperidine derivative depicted in Fig. 9. The resulting MACC-2 internode interaction energies can be used for 3D-QSAR analyses in a manner that is not unlike the CoMFA procedure illustrated in Fig. 6. Rows represent compounds, while energies are stored according to the distance for each autocorrelation and cross-correlation block (via MACC-2 encoding). The resulting matrix can be subjected to multivariate statistical analysis, such as PCA or PLS. ALMOND descriptors have been shown [25] to provide significant 3D-QSAR models for a wide range of targets: glycogen phosphorylase b, corticosteroid binding globulin, and serotoninergic ($5\text{-}HT_{2A}$) receptors, thus illustrating the use of alignment-independent, chemically interpretable, pharmacophore-related descriptors in drug discovery.

Another autocorrelogram metric that explores pharmacophore space is Horvath's fuzzy bipolar pharmacophore fingerprints (FBPFs) [106]. FBPFs are extracted from a predefined activity space measured for 584 drugs and druglike compounds, tested on 42 different targets [107], and the neighborhood behavior [108] concept is applied. This concept [108] is an extension of the similarity hypothesis, which states that structurally similar molecules are expected to display similar biological activity. The neighborhood behavior hypothesis suggests that more active compounds than random are expected to be found in a subset of the structurally nearest neighbors of an active compound, when compared to a random selection. FBPF-based autocorrelograms [106] have been further benchmarked against other similarity metrics [109], and found to be significantly more appropriate in capturing neighborhood behavior in a general manner.

8. 2D- AND 3D-QSAR: IN BALANCE

As indicated in Sec. 4, significant information about the data set can be extracted by using 2-D-based QSAR descriptors, in particular those related to size (MW), hydrophobicity (LogP, LogS), and electronic effects. 3D-QSAR methods become relevant if one finds the appropriate bioactive conformation, perhaps by considering multiple conformers and aligning the molecules via a pharmacophore model. As shown by Brown and Martin, simple substructure keys (2-D-based) are more successful in grouping diverse active compounds, compared to more elaborate keys based on 3-D structures [110]. Two-dimensional based descriptors have been found to be more useful in predicting some physicochemical properties, e.g., logP and pK_a, when compared to 3-D descriptors [111]. The balance between 2D- and 3D-QSAR models was further discussed by Yvonne Martin [29].

To understand the information overlap between 2D and 3D descriptors, we report here the results of comparing PCA scores derived from SaSA [112] and

ALMOND [113], on a set of 5998 compounds.* SaSA calculates 72 descriptors starting from the 2-D representation of the molecule: size-related descriptors included MW, the number of heavy atoms, the number of carbons, and the calculated molecular refractivity (CMR) [114]. Polarizability is estimated by CMR and by an atom-based polarizability scheme [115]. Flexibility and rigidity are estimated by counting the total number of bonds and rings (RNG), the number of rotatable bonds (RTB), and the number of rigid bonds (RGB) [80], and by several topological indices that estimate other properties [116] as well. The Wiener, Balaban, Randić and Motoc indices, as well as the Kier and Hall suite of topological descriptors [117] are used in SaSA. Hydrogen-bonding capacity is estimated using HYBOT [118] descriptors. Furthermore, SaSA uses simple counts for oxygen, nitrogen, HDOs, and HACs, positive and negative ionization centers, as well as the maximum positive and negative charge, as calculated using the Gasteiger-Marsili method [119]. Hydrophobicity and solubility were estimated for this data set using two logP methods: CLOGP [120] and ELogP [58] (Sec. 4), as well as the aqueous solubility, ELogS [59]. Thus, overall, SaSA includes a representative set of 2-D-based descriptors computed at the molecular level.

The 5998 2-D structures were processed in SaSA and HYBOT, then multivariate analyses using block scaling for SaSA descriptors were performed in SIMCA [121]. The corresponding 5998 3-D structures were automatically generated (this volume, chapter by Gasteiger et al.) using CORINA [122]. ALMOND descriptor generation and multivariate analysis for the 3-D data set, using blockwise normalization, were performed in ALMOND [113]. Six PCA scores per compound were extracted from both SaSA and ALMOND and further analyzed in SIMCA [121]. To evaluate the degree of information overlap between SaSA and ALMOND, the PCA scores from SaSA were used as X descriptors to explain the PCA scores from ALMOND (Table 4, top). For reciprocity, ALMOND descriptors were used as the X block to model the PCA scores from SaSA (Table 4, bottom). Since the data were PCA scores, no centering, scaling, or normalization was performed in this case. Model overviews, highlighted in yellow, illustrate the fact that there is an approximately 40% overlap between ALMOND and SaSA descriptors—as modeled by the fraction of explained variance (r^2) and by seven-groups cross-validation (L7G q^2) in a two-PLS components model.

The relationship between individual components appears to indicate that the first and second SaSA dimensions can be related to the first and third ALMOND dimensions, with an emphasis on the first component. The first SaSA component is dominated by size-related descriptors, e.g., the total number of heavy atoms, MW, CMR, and polarizability. These properties explain approximately 60% of the first ALMOND component, suggesting that large internode distances and high internode interaction energies dominate this component. The third ALMOND component has a weaker relationship (25%) to hydrophobicity descriptors such as CLOGP, ELogP, nonpolar surface area, and the number of nonpolar atoms [123]. Only the fourth ALMOND component appears to be related (under 25%) with hydrogen-bond-related, e.g., HYBOT descriptors, whereas the second, fifth, and sixth ALMOND

---

* This data set consists, literally, of the first half of the medicinal chemistry related data set discussed in Sec. 4. See also footnote on p. 576.

Table 4 Information Content Overlap Between SaSA and ALMOND
Principal Component Scores for 5998 Compounds, Broken Down on
Individual PCs

Model type	No. PC	r^2	L7G q^2
SaSA/X, ALMOND/Y	1	0.35	0.35
	2	0.40	0.40
t1AL/Y	1	0.59	0.59
	2	0.62	0.61
t3AL/Y	1	0.02	0.02
	2	0.25	0.25
ALMOND/X, SaSA/Y	1	0.32	0.32
	2	0.40	0.40
t1SA/Y	1	0.57	0.57
	2	0.59	0.59
t2SA/Y	1	0.06	0.06
	2	0.39	0.39

In gray are highlighted model overviews. Insignificant r^2 and q^2 values were omitted. See
text and Figs. 10–11 for details.

components did not display any significant relationships to SaSA, 2-D-based,
descriptors. Similar observations can be derived by mapping the r^2 values of the
pairwise correlations among ALMOND and SaSA components (see Fig. 10). Furthermore, the size and shape of the clusters in PCA space differ significantly when
comparing ALMOND and SaSA scores, as illustrated in Fig. 11. We obtained similar
results when comparing VolSurf-derived descriptors [124] to SaSA, since only the first

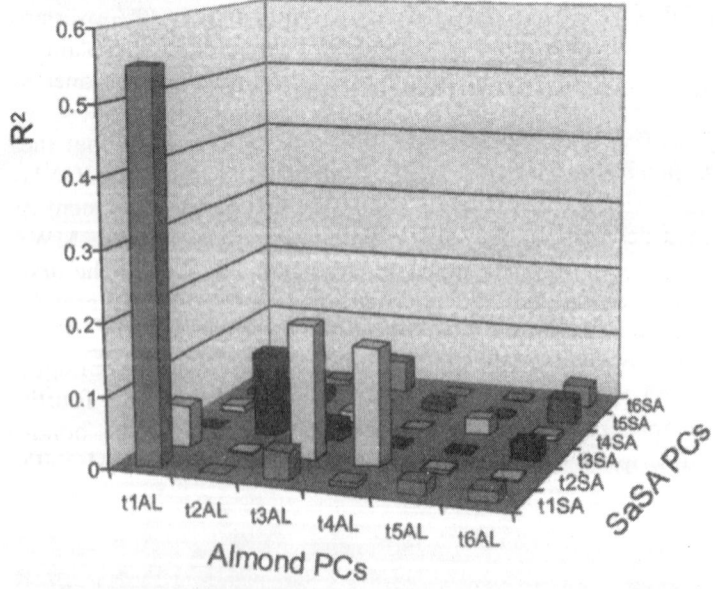

Figure 10 Pairwise correlation between PCA scores derived from SaSA descriptors (2D)
and Almond descriptors (3D). (See color plate at end of chapter.)

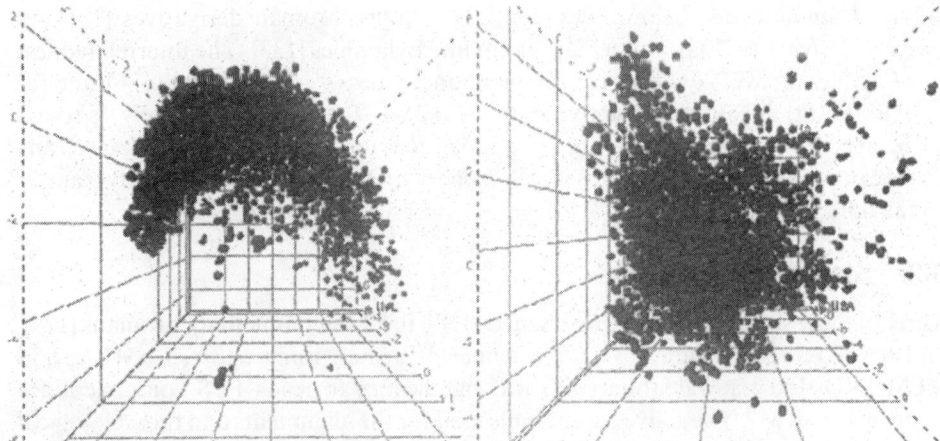

Figure 11 PCA Scores from ALMOND (left) and SaSA (right) for 5998 structurally diverse compounds. The first three principal components are shown.

principal components appeared to be related at a significant level [125], a reflection of the importance of size-related descriptors.

From the above, it can be concluded that while capturing similar information with respect to size, hydrophobicity, polarizability, and perhaps hydrogen bonding, the 2-D and 3-D descriptor systems are likely to differ significantly in the type of information that can be extracted from QSAR modeling. Since the information redundancy is under 40% overall, it is advisable to employ both 2D-QSAR and 3D-QSAR methods when modeling receptor-mediated target properties. This combination of QSAR methods is illustrated in the next section.

9. 3D-QSAR MODELING: AN EXAMPLE

One of the emerging techniques in QSAR is comparative QSAR, or lateral validation, first formulated by Hansch [126]. In this approach, the choice of parameters, their sign, and the size of their coefficients are compared with those from other ("classical") QSARs. An extension to 3-D-based methods was proposed by examining the latent variables [127] and by graphical inspection [27]. For example, even though two independent data sets, **23** and **24** in Table 2 (Sec. 4) indicate that antagonism at the neuropeptide Y Y1 receptor is related to increased size and, perhaps, increased hydrophobicity and reduced solubility, similar information can be inferred, e.g., for the next two targets in Table 2, phosphodiesterase 2 (PDE2) and phospholipase A_2 (PLA_2), as shown by data sets **25** and **26**, respectively. One may expect that medicinal chemists would be less than happy with this kind of general, bland conclusions.

9.1. Biological Activities

In this section we examine biological activities for three types of receptor antagonism: against serotoninergic $5HT_{1A}$, against alpha$_1$-adrenergic (α_1-AR), and against D_2 dopaminergic receptors. An initial survey of the 11,965 medicinal-chemistry-related compounds (see footnote 3) identified 279 compounds, from six different publications, with activities in this area: 48 phenylpiperazines [128], 62 2-(aminomethyl)chromans

[129], 44 aminomethyl-benzamides [130], 37 6-fluorochroman derivatives [131], 19 tricyclic derivatives [132], and 69 2-pyridinylmethylamines [133]. The interrelatedness of pK_i values for $5HT_{1A}$ and alpha$_1$ adrenergic activities ($r^2 = 0.347$, $N = 176$), and for α_1-AR and D_2 dopaminergic activities ($r^2 = 0.254$, $N = 133$), respectively, is found to be very low. This indicates that the three activities are orthogonal; hence, any significant (3-D) QSAR model observed on one activity should not necessarily transfer to the others.

9.2. Statistical Methods

Three of these series, the phenylpiperazines [128], the 2-(aminomethyl)chromans [129], and the 6-fluorochromans [131] were modeled further, using SaSA, CoMFA, and ALMOND. Indices evaluating statistical significance for each PLS component are listed in Tables 5–7. The active compounds used for the alignments and their biological activities are given in Fig. 12. CoMFA alignments were performed using the maximum

Table 5 Statistical Summary for the Phenylpiperazines

	PC1	PC2	PC3	PC4
pK_i $5HT_{1A}$ (N = 38)				
L5G q^2 CoMFA	0.44	0.63	0.73	**0.76**
L5G SEP CoMFA	0.92	0.75	0.65	**0.63**
r^2 CoMFA				**0.91**
SEE CoMFA				**0.39**
Steric Contribution				**0.37**
Electrostatic Contribution				**0.63**
L5G q^2 ALMOND	0.17	0.48	**0.54**	
L5G SDEP ALMOND	1.09	0.86	**0.81**	
r^2 ALMOND	0.36	0.69	**0.77**	
SDEC ALMOND	0.95	0.66	**0.57**	
L7G q^2 SaSA	0.43	0.45	**0.52**	
L7G SEP SaSA	1.12	0.92	**0.74**	
r^2 SA	0.56	0.63	**0.81**	
pK_i α_1-AR (N = 32)				
L5G q^2 CoMFA	0.42	0.77	**0.81**	
L5G SEP CoMFA	1.04	0.67	**0.62**	
r^2 CoMFA			**0.92**	
SEE CoMFA			**0.41**	
Steric contribution			**0.34**	
Electrostatic contribution			**0.66**	
L5G q^2 ALMOND	0.21	0.69	**0.84**	
L5G SDEP ALMOND	1.18	0.74	**0.54**	
r^2 ALMOND	0.35	0.84	**0.95**	
SDEC ALMOND	1.07	0.54	**0.30**	
L7G q^2 SaSA	0.63	**0.75**		
L7G SEP SaSA	1.06	**0.79**		
r^2 SaSA	0.67	**0.86**		

Source: Ref. 128.

Biological Activity Distribution

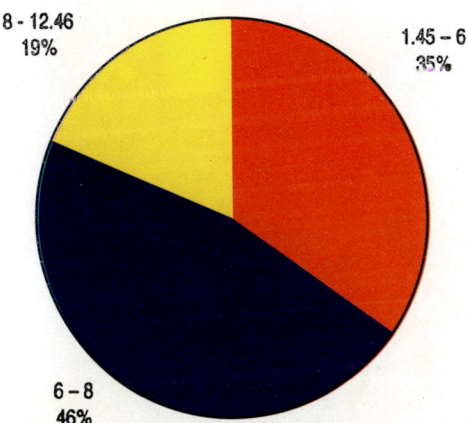

Figure 2 Pie chart illustrating the distribution of biological activities, in the negative log(activity) format, for 11,965 structures with 22,763 reported activities. Inactive compounds (up to 6) are in red; active compounds (above 8) are in yellow; middle-range compounds (between 6 and 8) are in blue.

Properties vs. Activity

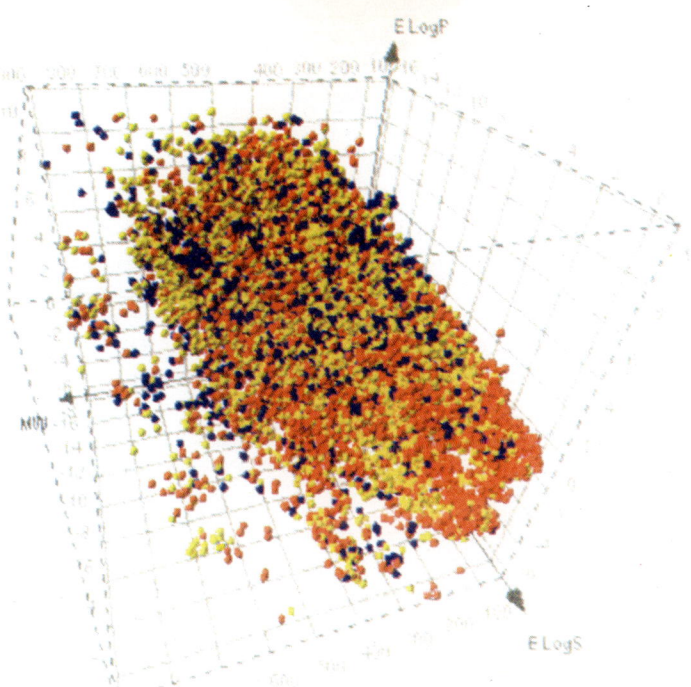

Figure 3 Three-dimensional scatterplot illustrating the distribution of biological activities, in −log (activity) format, for 11,965 structures with 22,763 reported activities, against molecular weight, MW, estimated LogP, ElogP, and estimated aqueous solubility, ELogS. The color code for activity distribution is the same as in Fig. 2.

Regression Coefficients

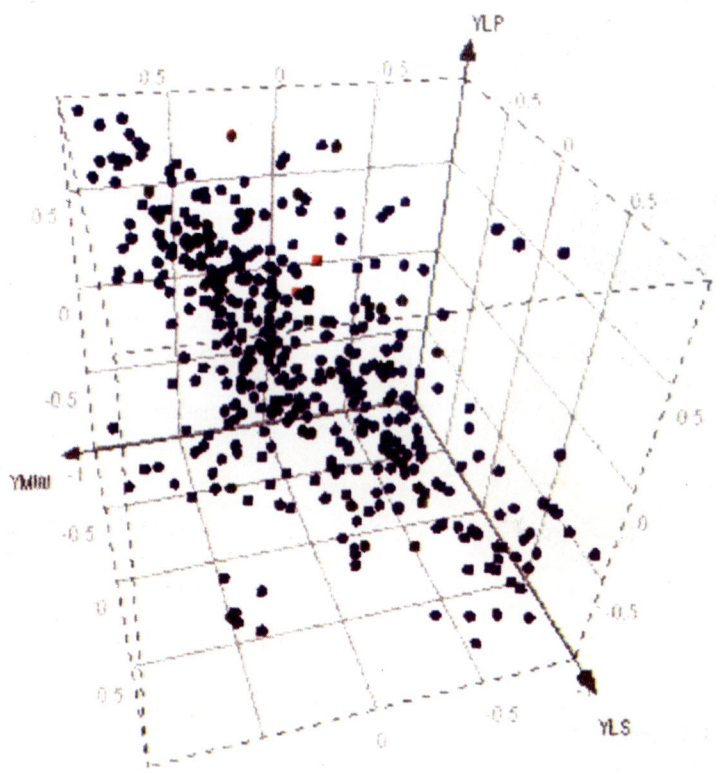

Figure 5 Three-dimensional scatterplot illustrating the distribution of regression coefficients between biological activity, in −log (activity) format, and molecular weight, YMW, estimated LogP, YLP, and estimated aqueous solubility, YLS, respectively. Data are shown for 423 data sets, encompassing 14,022 activities. The color code for activity type is as follows: blue for inhibition estimates (A_2, K_i, IC_{50}), black for agonism/substrate estimates (EC_{50}, D_2), and green for binding affinity estimates (K_D and RBA). See Tables 1–3 and text for details.

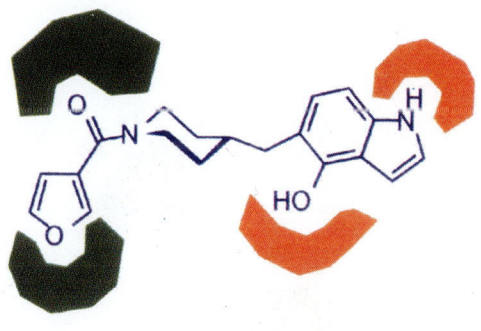

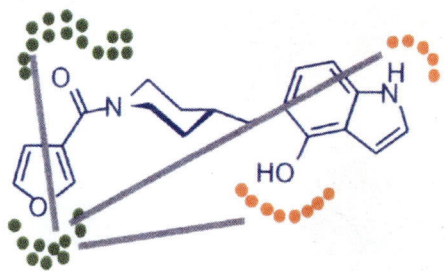

Figure 8 Two-dimensional sketch illustrating the MIF filtering step in ALMOND. MIFs for the N1 probe (hydrogen bond donor, green) and the O probe (hydrogen bond acceptor, red), top, are processed at the same energy level, e.g., −2 kcal/mol, to yield a few hundred representative nodes, bottom. The grey lines illustrate distances used in the auto- (N1–N1) and cross- (N1–O) correlograms by the MACC-2 encoding algorithm.

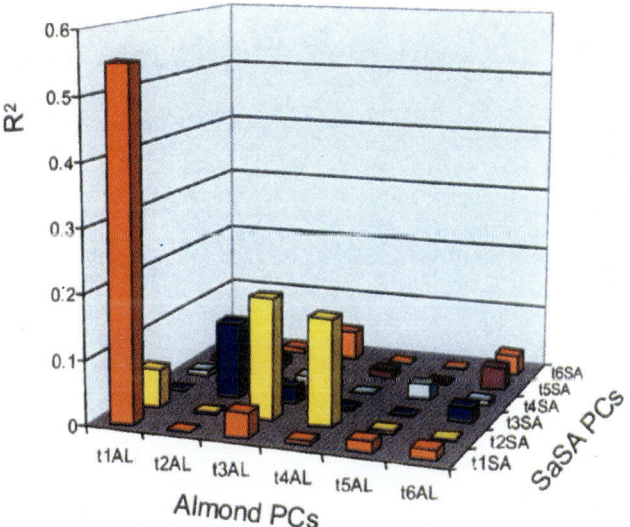

Figure 10 Pairwise correlation between PCA scores derived from SaSA descriptors (2D) and Almond descriptors (3D).

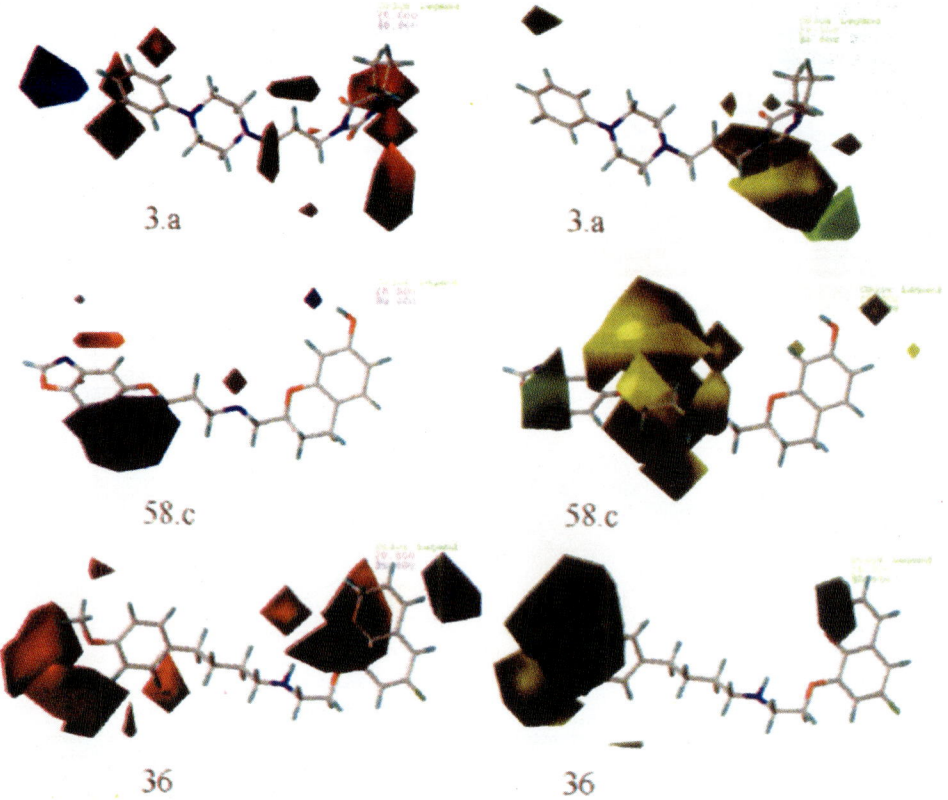

Figure 15 Comparison of stdev*coeff CoMFA fields for CoMFA models derived on 5-HT$_{1A}$ activity (compound numbers as in Fig. 12). Electrostatic fields are on the left, steric fields on the right.

Table 6 Statistical Summary for the 2-(Aminomethyl)chromans

	PC1	PC2	PC3	PC4
pK$_i$ 5HT$_{1A}$ (N=48)				
L5G q^2 CoMFA	**0.51**	0.39	0.50	0.55
L5G SEP CoMFA	**0.72**	0.81	0.75	0.71
r^2 CoMFA	**0.60**			
SEE CoMFA	**0.65**			
Steric contribution	**0.64**			
Electrostatic contribution	**0.36**			
L5G q^2 ALMOND	0.40	**0.52**	0.47	0.37
L5G SDEP ALMOND	0.78	**0.70**	0.74	0.80
r^2 ALMOND	0.48	**0.69**	0.75	0.84
SDEC ALMOND	0.73	**0.56**	0.50	0.41
L7G q^2 SaSA	0.51	**0.58**		
L7G RMSP SaSA	0.73	**0.67**		
r^2 SaSA	0.53	**0.70**		
RMSE SaSA	0.72	**0.55**		
pK$_i$ α$_1$-AR (N=48)				
L5G q^2 CoMFA	**0.71**	0.74	0.75	0.76
L5G SEP CoMFA	**0.42**	0.40	0.39	0.39
r^2 CoMFA	**0.74**			
SEE CoMFA	**0.39**			
Steric contribution	**0.65**			
Electrostatic contribution	**0.35**			
L5G q^2 ALMOND	0.53	**0.66**	0.62	0.51
L5G SDEP ALMOND	0.51	**0.43**	0.45	0.52
r^2 ALMOND	0.58	**0.80**	0.83	0.90
SDEC ALMOND	0.48	**0.34**	0.31	0.24
L7G q^2 SaSA	0.50	**0.70**		
L7G SEP SaSA	0.54	**0.46**		
r^2 SaSA	0.55	**0.77**		
pK$_i$ D$_2$ high (N=53)				
L5G q^2 CoMFA	0.15	0.46	0.59	**0.64**
L5G SEP CoMFA	0.88	0.71	0.62	**0.59**
r^2 CoMFA				**0.91**
SEE CoMFA				**0.30**
Steric contribution				**0.51**
Electrostatic contribution				**0.49**
L5G q^2 ALMOND	−0.05	−0.19		
L5G SDEP ALMOND	0.94	1.00		
r^2 ALMOND	0.11	0.30		
SDEC ALMOND	0.86	0.76		
L7G q^2 SaSA	0.12	0.19	0.29	
L7G SEP SaSA	0.89	0.86	0.75	
r^2 SaSA	0.17	0.33	0.50	

Table 6 Continued

	PC1	PC2	PC3	PC4
pK_i D_2 low (N=51)				
L5G q^2 CoMFA	0.23	0.51	**0.58**	0.54
L5G SEP CoMFA	0.73	0.59	**0.55**	0.58
r^2 CoMFA			**0.85**	
SEE CoMFA			**0.33**	
Steric contribution			**0.53**	
Electrostatic contribution			**0.47**	
L5G q^2 ALMOND	−0.05	−0.18		
L5G SDEP ALMOND	1.02	1.09		
r^2 ALMOND	0.10	0.29		
SDEC ALMOND	0.94	0.83		
L7G q^2 SaSA	0.24	0.32		
L7G SEP SaSA	0.73	0.67		
r^2 SaSA	0.28	0.41		

Source: Ref. 129.

common substructure (MCS) search algorithm, as implemented in the Distill module of SYBYL [134]. However, compounds had to be manually aligned in each series, as the MCS search cannot handle fuzzy atom queries (e.g., piperidine vs. piperazine). The alignments were not biased by pharmacophore patterns for any of these receptor subtypes. Rather, a general alignment, independent of biology-related information, was attempted in order to monitor the influence of biological activity on the overall QSAR results. For consistency, the same (aligned) conformers were used in both CoMFA and ALMOND. PCA and PLS analyses were performed in SIMCA for SaSA, in SYBYL for CoMFA, and in ALMOND, respectively. All the 3-D structures were minimized with the MMFF94 force field, using MMFF94 partial charges, as implemented in SYBYL [134]. Block scaling was applied in ALMOND, while the default options for steric and electrostatic fields were used in CoMFA.

Tables 5–7 provide the model overview (with "best" models highlighted in bold) for each of the QSARs models, as follows: five-groups randomized cross-validation (L5G) q^2 and SEP (standard error of prediction) are reported for CoMFA, while SDEP (standard deviation error of prediction) is being reported for ALMOND. The seven-groups randomized cross-validated q^2 (cumulative) and the standard error of the predicted Y value, SEP [121], are reported for SaSA. In a similar manner, r^2, SEE (standard error of estimate), and SDEC (standard deviation error of calculation), as well as the steric and electrostatic contributions, are given for the fitted models. When default options in CoMFA and/or ALMOND could not yield significant QSARs, variable selection was applied: region focusing (from a marginally significant fitted PLS model) in CoMFA, and fractional factorial design (with the option to remove uncertain variables, using L5G at 2 components), was used in ALMOND.

9.3. Summary of Statistical Results

A survey of Tables 5–7 shows that each technique behaves differently: there is no single data set where uniform conclusions can be drawn. The phenylpiperazines (Table 5) appear to be one of the "easy" series in the sense defined by John Van Drie [135], since

Table 7 Statistical Summary for the 6-Fluorochromans

	PC1	PC2	PC3	PC4
pK$_i$ 5HT$_{1A}$ (N = 34)				
L5G q^2 CoMFA	0.41	0.54	**0.60**	0.60
L5G SEP CoMFA	0.69	0.62	**0.59**	0.59
r^2 CoMFA			**0.88**	
SEE CoMFA			**0.32**	
Steric contribution			**0.44**	
Electrostatic contribution			**0.56**	
L5G q^2 FFD ALMOND	0.36	0.54	**0.60**	
L5G SDEP FFD ALMOND	0.70	0.59	**0.55**	
r^2 FFD ALMOND	0.65	0.77	**0.85**	
SDEC FFD ALMOND	0.52	0.42	**0.34**	
L7G q^2 SaSA	−0.10	0.07	0.24	
L7G SEP SaSA	0.78	0.65	0.56	
r^2 SaSA	0.25	0.49	0.63	
pK$_i$ α$_1$-AR (N = 35)				
L5G q^2 CoMFA	0.24	0.04	0.08	0.18
L5G SEP CoMFA	0.85	0.98	0.97	0.93
r^2 CoMFA	0.59			
SEE CoMFA	0.63			
L5G q^2 focus, no prazosin	0.44	0.42	0.39	
SEP focus, no prazosin	0.63	0.66	0.68	
r^2 Focus, no prazosin	0.59			
SEE focus, no prazosin	0.54			
Steric contribution	0.28			
Electrostatic contribution	0.72			
L5G q^2 FFD ALMOND	0.54	**0.66**	0.64	
L5G SDEP FFD ALMOND	0.65	**0.55**	0.57	
r^2 FFD ALMOND	0.77	**0.79**	0.80	
SDEC FFD ALMOND	0.46	**0.43**	0.42	
L7G q^2 SaSA	−0.79			
L7G SEP SaSA	0.27			
r^2 SaSA	0.47			
pK$_i$ D$_2$ (N = 31)				
L5G q^2 CoMFA	**0.52**	0.52	0.51	**0.53**
L5G SEP CoMFA	**0.70**	0.72	0.74	**0.74**
r^2 CoMFA	**0.69**			**0.94**
SEE CoMFA	**0.57**			**0.26**
Steric contribution	**0.22**			**0.33**
Electrostatic contribution	**0.78**			**0.67**
L5G q^2 FFD ALMOND	0.33	0.58	**0.60**	
L5G SDEP FFD ALMOND	0.81	0.64	**0.62**	
r^2 FFD ALMOND	0.76	0.85	**0.92**	
SDEC FFD ALMOND	0.49	0.38	**0.28**	
L7G q^2 SaSA	−0.13			
L7G SEP SaSA	1.01			
r^2 SaSA	0.55			

Source: Ref. 131.

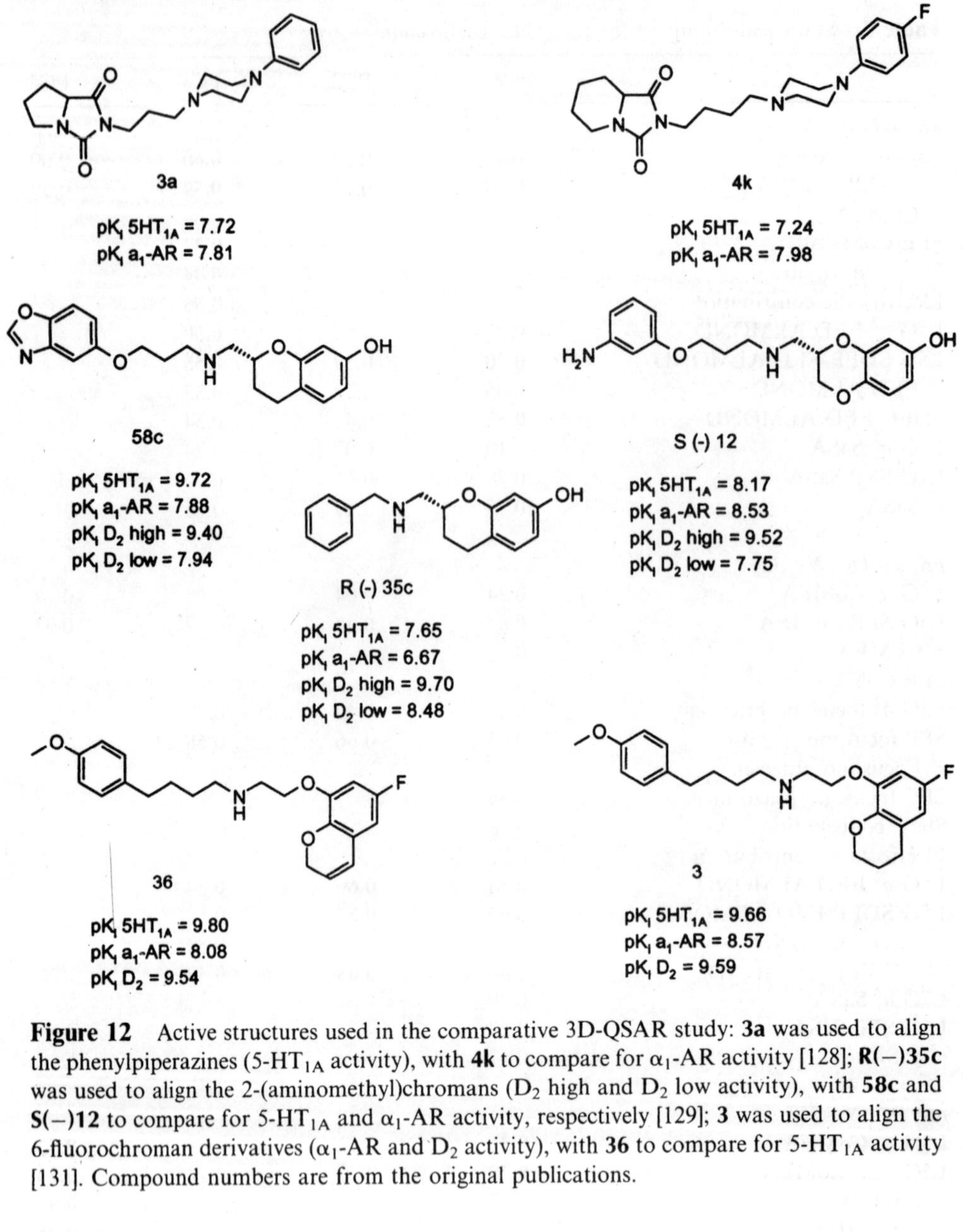

Figure 12 Active structures used in the comparative 3D-QSAR study: **3a** was used to align the phenylpiperazines (5-HT$_{1A}$ activity), with **4k** to compare for α_1-AR activity [128]; **R(−)35c** was used to align the 2-(aminomethyl)chromans (D$_2$ high and D$_2$ low activity), with **58c** and **S(−)12** to compare for 5-HT$_{1A}$ and α_1-AR activity, respectively [129]; **3** was used to align the 6-fluorochroman derivatives (α_1-AR and D$_2$ activity), with **36** to compare for 5-HT$_{1A}$ activity [131]. Compound numbers are from the original publications.

each method can provide statistically significant QSARs. There is, however, no consensus as to what number of components is significant: given that ALMOND and SaSA converge at three PLS components for 5HT$_{1A}$, with q^2 around 0.53, it is not unreasonable to suggest that the three-component CoMFA model should be considered. CoMFA and ALMOND converge at three PLS components for α_1-AR, with q^2 above 0.8, whereas 2-D-based descriptors suggest a lower complexity, marginally less significant model. In fact, model complexity reported here is at variance with the CoMFA models reported in the original publication [128]: an eight-PLS-component model was reported for 5HT$_{1A}$, with a 0.54 contribution from electrostatic fields

($N = 48$), and a seven-PLS-component model was reported for α_1-AR, with a 0.6 contribution from electrostatic fields ($N = 42$). This difference can be traced back to the 10 compounds removed from the initial data set while using the MCS alignment, but the models were perhaps overfitted in this particular case. We further notice that the electrostatic CoMFA field contributions (Table 5), 0.63 for $5HT_{1A}$, and 0.66 for α_1-AR, respectively, are in qualitative agreement with the previously reported higher component models [128]. Based on the comparative QSAR approach, the two- or three-PLS component models are warranted for both $5HT_{1A}$ and α_1-AR in this data set.

The 2-(aminomethyl)chromans are, compared to the phenylpiperazines, a more difficult series: except for CoMFA, no significant QSAR could be achieved for D_2 high (agonism) and D_2 low (antagonism) activity (see Table 6). However, all three methods yield good models for $5HT_{1A}$ and α_1-AR activities, respectively. CoMFA results peak at 1 PLS component for $5HT_{1A}$, whereas ALMOND has a clear peak at 2 components (Table 6) for both $5HT_{1A}$ and α_1-AR. ALMOND and SaSA converge at 2 PLS components for $5HT_{1A}$ (q^2 around 0.55), and for α_1-AR (q^2 around 0.68). Since CoMFA appears to yield significant models at one component, it is reasonable to suggest that the one-component CoMFA models capture the pharmacophore. Given that the same conformers were used in CoMFA and ALMOND, these two QSARs should be compared with respect to the understanding and interpretation of pharmacophoric patterns. Neither ALMOND nor SaSA descriptors could be used to derive significant QSARs for this activity, even after applying variable selection procedures

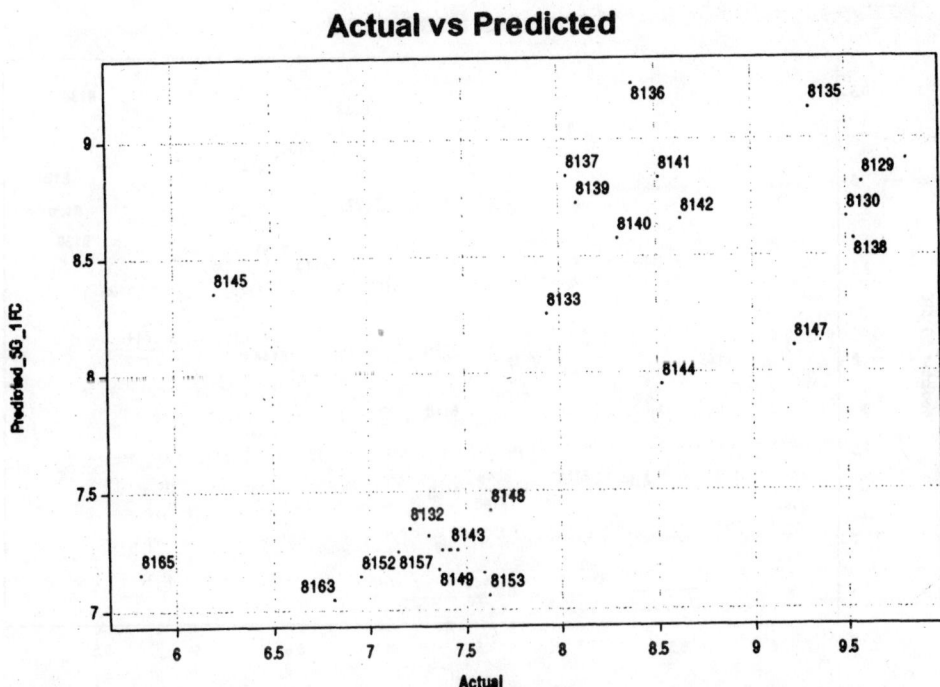

Figure 13 CoMFA plot for actual vs. predicted pK_i activities for D_2, using the L5G PC1 model of the 6-fluorochroman derivatives. See also Table 7. (From Ref. 131.)

(data not shown). These results indicate that at least for this series, CoMFA is the method of choice.

The 6-fluorochromans are, compared to the phenylpiperazines and the 2-(aminomethyl)chromans, a series of intermediate difficulty for 3D-QSAR and of high difficulty for 2D-QSAR. None of the activities for this series is amenable to 2D modeling via SaSA, whereas CoMFA and ALMOND provide good models for $5HT_{1A}$ and D_2 activities, respectively. No significant QSAR could be derived for α_1-AR activity, except when using ALMOND with variable selection (see Table 7). For this series, the pharmacophore does not appear to be correctly captured by the conformational choice, at least for α_1-AR: even after removing prazosin, a clear chemical outlier, and performing variable selection via region focusing, CoMFA modeling remains ineffective (see Table 7). Furthermore, 2D-QSAR methods prove less effective compared to 3D-QSAR for all three biological activities in this series. This is likely to be a data-set-dependent effect for $5HT_{1A}$ and α_1-AR antagonism, since valid QSARs for these activities were derived from the other two series (Tables 5 and 6). However, no D_2 activity was amenable to 2D-QSAR modeling by SaSA descriptors (Tables 6 and 7), even though significant regressions between D_2 K_i values and MW, ELogP, and ELogS could be obtained for series 3 in Table 2.

CoMFA results peak at three PLS components for $5HT_{1A}$, in agreement with ALMOND after FFD selection (Table 7). A one-component and a four-component model for D_2 antagonism are illustrated for CoMFA with respect to cross-validated predictivity (see Figs. 13 and 14). Both plots indicate a clear separation between active

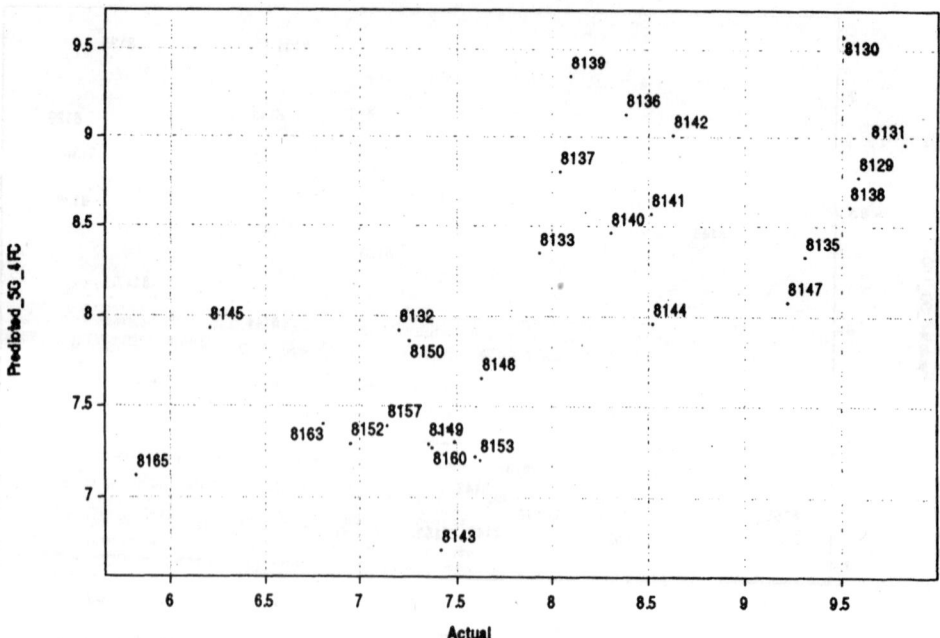

Figure 14 CoMFA plot for actual vs. predicted pK_i activities for D_2, using the L5G PC4 model of the fluorochroman derivatives. See also Table 7. (From Ref. 131.)

compounds (pK_i higher then 8), compared to inactive ones (pK_i lower than 8). This explains why the one- and four-PLS-component models are significant, illustrating a clear case where an artificial (spurious) separation between actives and inactives is at play. This separation was not observed for the three-component ALMOND model (data not shown). Such behavior suggests that while pharmacophoric elements can be detected in ALMOND, this particular alignment, though consistent with $5HT_{1A}$ activity, cannot be utilized for either α_1-AR or D_2 activities. The above results further indicate that for this series, ALMOND is the method of choice.

9.4. Graphical Analyses of CoMFA Plots

Figure 15 illustrates the graphical summary of the stdev*coeff CoMFA fields for 5-HT_{1A} activity. By convention, green regions are associated with steric bulk tolerance, whereas yellow regions are sterically hindered; red regions are associated with negative charges, whereas blue regions have preference for positive charges. A detailed interpretation of CoMFA fields is available [27]. From the comparative QSARs

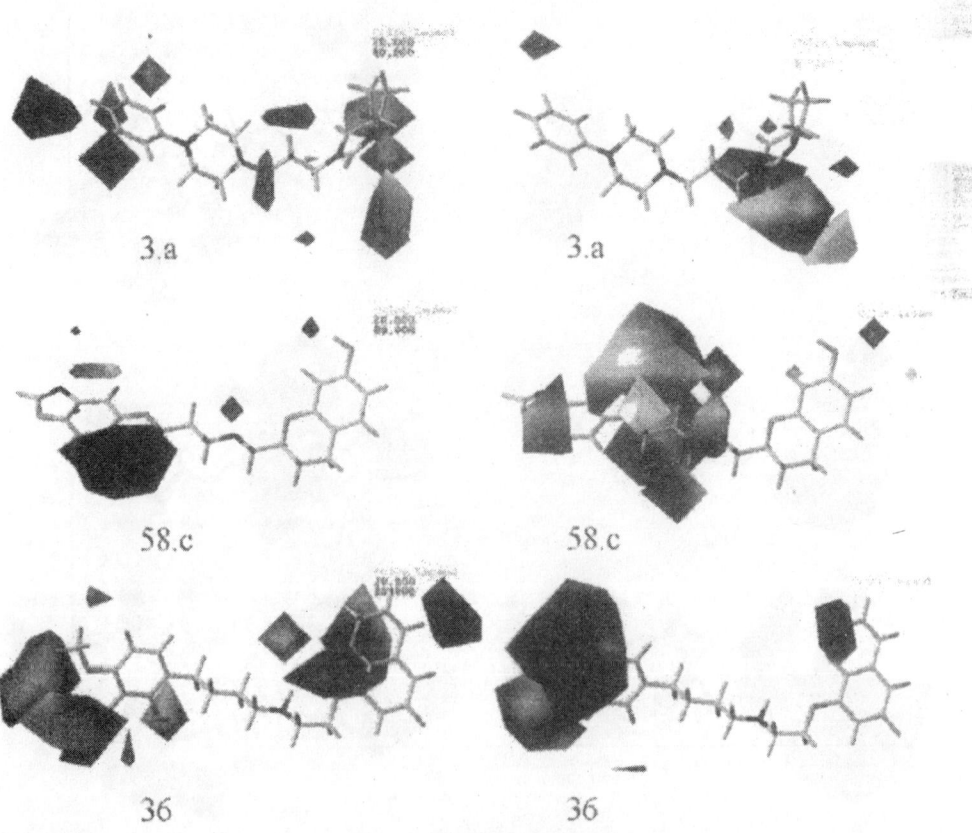

Figure 15 Comparison of stdev*coeff CoMFA fields for CoMFA models derived on 5-HT_{1A} activity (compound numbers as in Fig. 12). Electrostatic fields are on the left, steric fields on the right. (See color plate at end of chapter.)

illustrated in Figs. 15–17, it can be noticed that no two CoMFA fields are the same, even when comparing results for the same biological activity. It can be suggested, however, that the fields presented by the fluorochroman **36** could be aligned to the fields of the aminochroman **58c**, whereas the phenylpiperazine **3a** should be rotated 180°.

Figure 16 illustrates the graphical summary of the stdev*coeff CoMFA fields for α_1-AR activity (see color plate at end of chapter). The CoMFA models for $5HT_{1A}$ and α_1-AR are quite similar when comparing both the phenylpiperazines **3a** and **4k**, and the aminochromans **58c** and **S(−)12**. Further chemical variation should be applied to both series before deriving any meaningful conclusions for drug design.

Figure 17 illustrates the comparative QSAR between D_2 high and D_2 low activities for the same (active) compound **R(−)35** (see color plate at end of chapter). "D_2 high" activity refers to the high affinity state of D_2 receptors that binds agonists, whereas "D_2 low" activity relates to the low affinity state of D_2 receptors, which binds antagonists [129]. By comparing the CoMFA stdev*coeff plots in Fig. 17, one can notice that the region contours which are significant for agonism (top) appear to be more restrictive, compared to the more diffuse contours for antagonism (middle and

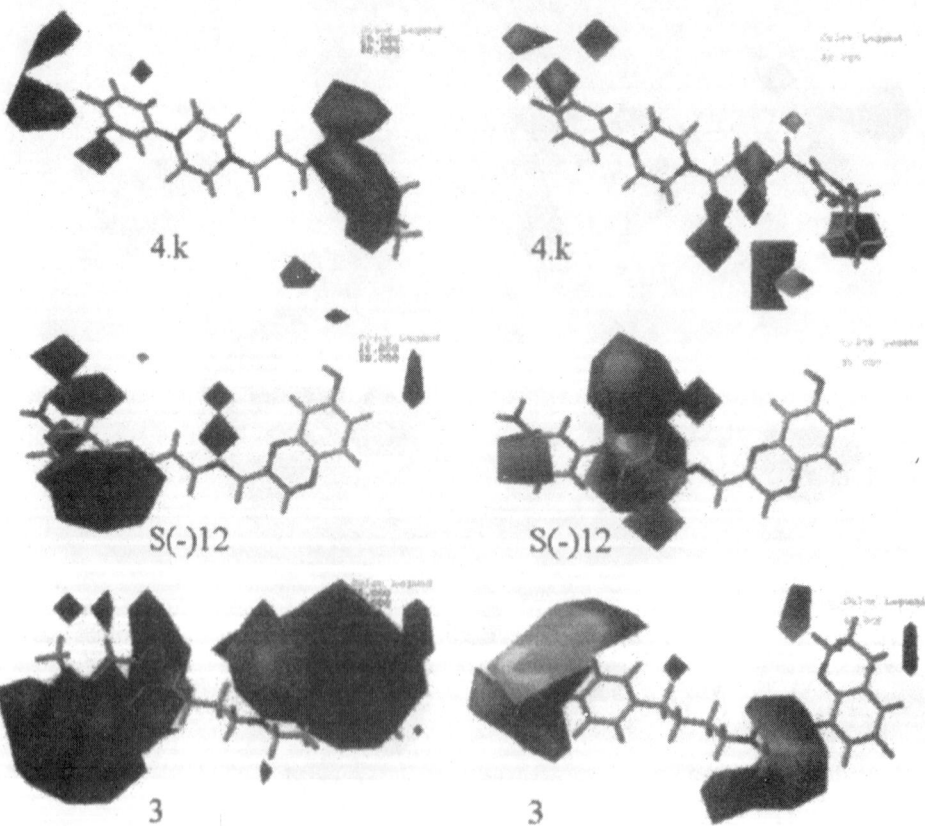

Figure 16 Comparison of stdev*coeff CoMFA fields for CoMFA models derived on α_1-AR activity (compound numbers as in Fig. 12). Electrostatic fields are on the left, steric fields on the right. (See color plate at end of chapter.)

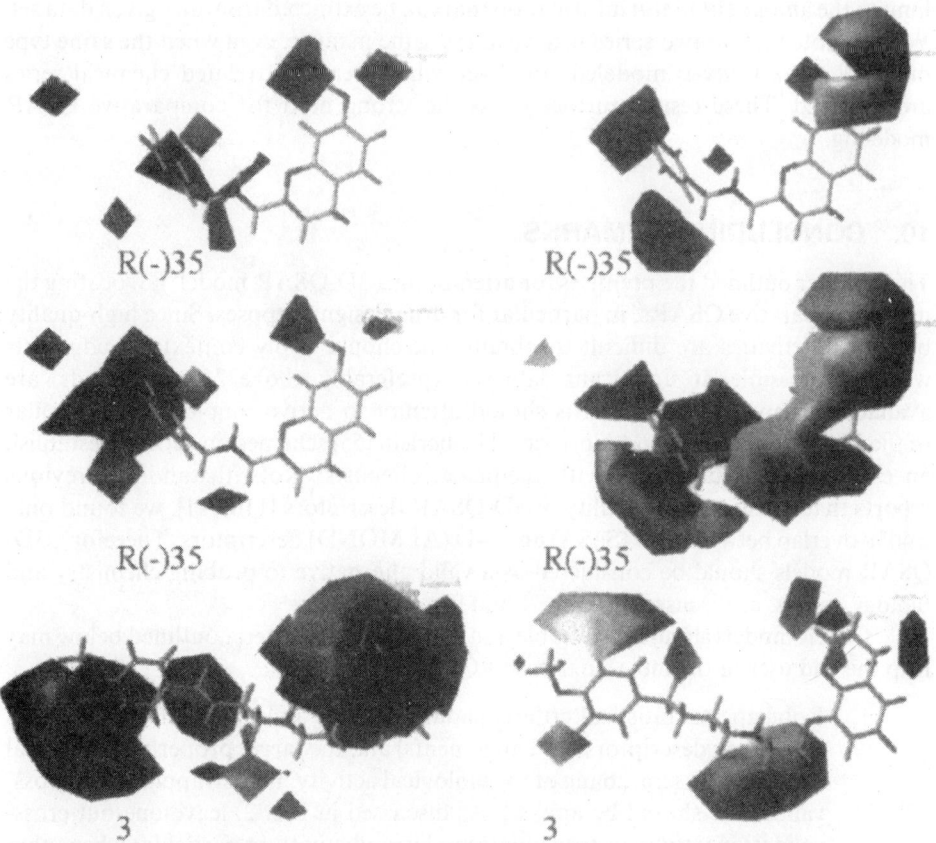

Figure 17 Comparison of stdev*coeff CoMFA fields for CoMFA models derived on $D_{2'}$ high (top) and D_2 low (middle) activity, and D_2 antagonism (low-like) activity (bottom). Compounds are numbered as in Fig. 12. Electrostatic fields are on the left, steric fields on the right. (See color plate at end of chapter.)

bottom), for both the steric and electrostatic fields. This observation is consistent with the general knowledge that agonists are, by definition, much more restrictive in terms of conformational and indeed chemical space, compared to the more loosely defined, entropy-driven antagonists, at least for G-protein-coupled receptors. We also note that the 51-49 split in terms of steric and electrostatic fields for D_2 high activity (Table 6) is the most balanced contribution among all nine CoMFA models reported here.

The pattern of D_2 antagonism presented at the bottom of Fig. 17 cannot be easily overlapped to the middle one. A similar observation can be made from Figs. 15 (middle vs. bottom) and 16 (middle vs. bottom). It can be inferred that the chroman moiety, present in these two series of ligands, does not bind in the same receptor regions. This further suggests that in this particular case, the maximum common substructure algorithm in Distill should not be applied when attempting to merge the two series.

To summarize, the three classes of QSAR models appear to yield comparable results in terms of errors and q^2 (except for D_2 antagonism), indicating that there is a limit to the amount of useful information that can be extracted from any given data set. We also note that no two series behave in the same manner, even when the same type of biological activity is modeled, and even when seemingly related chemical series are explored. These results further stress the strong need for comparative QSAR modeling.

10. CONCLUDING REMARKS

This chapter outlined the premises for attempting a 3D-QSAR model, advocating the use of comparative QSARs, in particular for drug design purposes. Since high-quality biological activities are difficult to obtain, one should apply contextual judgments whenever possible. If significant data sets (preferably above 25 compounds) are available, computational chemists should attempt to derive simple criteria, similar to the Lipinski et al. [47] and Norinder–Haeberlein [55] schemes, in order to establish an effective communication with medicinal chemists. Notwithstanding previous reports that questioned the utility of 3D-QSAR descriptors [110,111], we found only a 40% overlap between 2-D (SaSA) and 3-D (ALMOND) descriptors. Therefore, 3D-QSAR models should be considered as a valid alternative to probing chemistry and biology spaces, as discussed in Secs. 4 and 9.

QSAR models should be testable and verifiable. The criteria outlined below may help the end user in the meta-analyses of QSARs:

1. Robustness: statistical criteria should indicate a clear correlation between the chosen descriptors (and alignments) and the target property. Additional tests, such as scrambling of the biological activity, bootstrapping, and cross-validation, should be applied. As discussed in Sec. 2, leave-one-out cross-validation is unlikely to evaluate model predictivity, in particular when other cluster members in the training set have similar chemical and biological properties. Therefore, successive cross-validation runs for randomly or systematically eliminated groups should be used to test model robustness.

2. Predictive power: this is the main goal of QSAR. Internal prediction is evaluated during cross-validation, whereas external, "true" prediction requires a test set. Using chronological criteria to separate training and test sets has a clear bearing on real-life situations, since novel compounds are tested and evaluated in later stages of the drug discovery process. Regardless of how the test set is chosen, its distribution of Y values should not differ significantly from that of the training set, else statistical tools for evaluating model predictivity might prove inadequate [136].

3. Explanatory power: less important in the initial stages of QSAR model developing, the explanatory power analysis becomes instrumental in understanding the mode of action for active compounds originating from different data sets. Interactive analyses of ALMOND profiles and MIFs or graphical analyses of CoMFA stdev*coeff fields should offer insight into the mechanism of action of individual ligands. Whenever structural information is available, relevant contributions of the ALMOND and CoMFA MIFs should be validated with receptor information. Alignment rules should be experimentally validated by identifying active, and rigid compounds, or by comparing the binding modes via 3-D structural studies.

4. Relevance: closely associated with explanatory power is the pharmacological (mechanistic) relevance of the descriptor fields used in a 3D-QSAR model. As additional MIFs (e.g., extracted from quantum mechanical calculations) become available as 3D-QSAR descriptors, one must exercise caution in the application of a particular field to the problem at hand.

5. Simplicity: descriptors that are easier to interpret should be used whenever trying to understand the modeled system, as discussed in Sec. 3. 2D-QSARs are, by their very nature, simpler than 3D-QSARs, as all issues related to molecular alignment are removed. However, one should always look beyond size and hydrophobicity, as discussed in Secs. 6 and 9.

6. Uniqueness: there is no rigorous way to demonstrate that 3D-QSAR models are unique. Even if the alignments are experimentally determined, the use of different partial atomic charges or the use of different grid probes could lead to different, valid, models. Alternative binding modes, while tolerated by some targets [137], may not be amenable to straightforward CoMFA modeling [138]. Therefore, the validity of 3D-QSAR models should be questioned experimentally in various manners. As discussed in Sec. 9, seemingly related compound series tested on the same targets may lead to different, and sometimes difficult to compare, QSARs.

Since data-set-dependent biases may cloud the analysis, additional sources of information, e.g., pharmacophore models, 3-D structural data for related targets, scientific and patent literature, can and should be used in a comparative manner. Whether or not meaningful alignments can be established, 2-D descriptors and alignment independent methods such as ALMOND should be used to enhance our understanding of the molecular basis for structure–activity relationships.

ACKNOWLEDGMENT

Drs. Jin Li (EST Chemical Computing) and Vladimir Sherbukhin (Medicinal Chemistry), both from AstraZeneca R&D Mölndal, Sweden, are gratefully acknowledged for valuable input. Drs. Maria Mracec and Magdalena Banda, from the Romanian Academy Institute of Chemistry, and Drs. Marius Olah and Zeno Simon, from the University of West, both in Timişoara, Romania, are acknowledged for assistance with the literature survey.

REFERENCES

1. Hansch C, Fujita T. ρ–σ–π Analysis. A method for the correlation of biological activity and chemical structure. J Am Chem Soc 1964; 86:1616–1626.

2. Free SM Jr, Wilson JW. A mathematical contribution to structure–activity studies. J Med Chem 1964; 7:395–399.

3. Cramer RD III, Bunce JD. The DYLOMMS method: initial results from a comparative study of approaches to 3D QSAR. In: Hadzi D, Jerman-Blazic B, eds. QSAR in Drug Design and Toxicology. Amsterdam: Elsevier, 1987:3–12.

4. Cramer RD III, Patterson DE, Bunce JD. Comparative molecular field analysis (CoMFA). 1. Effect of shape on binding of steroids to carrier proteins. J Am Chem Soc 1988; 110:5959–5967.

5. Goodford PJ. Computational procedure for determining energetically favourable

binding sites on biologically important macromolecules. J Med Chem 1985; 28:849–857.

6. Wold S, Johansson E, Cocchi M. PLS—partial least-squares projections to latent structures. In: Kubinyi H, ed. 3D QSAR in Drug Design: Theory, Methods and Applications. Leiden: ESCOM, 1993:523–550.

7. Cramer RD III, Wold SB. Comparative molecular field analysis (CoMFA). U.S. patent US 5025388 A 19910618, 1991, 22 pp.

8. Doweyko A. The hypothetical active site lattice: An approach to modelling active sites from data on inhibitor molecules. J Med Chem 1988; 31:1396–1406.

9. Ghose A, Crippen G, Revankar G, McKernan P, Smee D, Robbins R. Analysis of the in vitro activity of certain ribonucleosides against parainfluenza virus using a novel computer-aided molecular modeling procedure. J Med Chem 1989; 32:746–756.

10. Ghose AK, Crippen GM. Modeling the benzodiazepine receptor binding site by the general three-dimensional structure-directed quantitative structure–activity relationship method REMOTEDISC. Mol Pharmacol 1990; 37:725–734.

11. Baroni M, Costantino G, Cruciani G, Riganelli D, Valigi R, Clementi S. Generating optimal linear PLS estimations (GOLPE): an advanced chemometric tool for handling 3D-QSAR problems. Quant Struct-Act Relat 1993; 12:9–20.

12. Kubinyi H, ed. 3D QSAR in Drug Design: Theory, Methods and Applications. Leiden: ESCOM, 1993.

13. Kubinyi H, Folkers G, Martin YC, eds. 3D QSAR in Drug Design. Vol. 2. Ligand Protein Interactions and Molecular Similarity. Dordrecht: Kluwer/ESCOM, 1998.

14. Kubinyi H, Folkers G, Martin YC, eds. 3D QSAR in Drug Design. Vol. 3. Recent Advances. Dordrecht: Kluwer/ESCOM, 1998.

15. Klebe G. Comparative molecular similarity indices analysis: CoMSIA. In: Kubinyi H, Folkers G, Martin YC, eds. 3D QSAR in Drug Design. Vol. 3. Recent Advances. Dordrecht: Kluwer/ESCOM, 1998:87–104.

16. Dunn WJ, Hopfinger AJ. 3D QSAR of flexible molecules using tensor representation. In: Kubinyi H, Folkers G, Martin YC, eds. 3D QSAR in Drug Design. Vol. 3. Recent Advances. Dordrecht: Kluwer/ESCOM, 1998:167–182.

17. Jain AN, Koile K, Chapman D. Compass: Predicting biological activities from molecular surface properties. Performance comparisons on a steroid benchmark. J Med Chem 1994; 37:2315–2327.

18. Hahn M, Rogers D. Receptor surface models. In: Kubinyi H, Folkers G, Martin YC, eds. 3D QSAR in Drug Design. Vol. 3. Recent Advances. Dordrecht: Kluwer/ESCOM, 1998: 117–133.

19. Gurrath M, Müller G, Höltje HD. Pseudoreceptor modelling in drug design: Applications of Yak and PrGen. In: Kubinyi H, Folkers G, Martin YC, eds. 3D QSAR in Drug Design. Vol. 3. Recent Advances. Dordrecht: Kluwer/ESCOM, 1998:135–157.

20. Horvath D. ComPharm—automated comparative analysis of pharmacophoric patterns and derived QSAR approaches, novel tools in high-throughput drug discovery. A proof-of-concept study applied to farnesyl protein transferase inhibitor design. In: Diudea MV ed. QSPR/QSAR Studies by Molecular Descriptors. Huntington, NY: Nova Science Publishers, 2001:389–433.

21. Polanski J, Gieleciak R, Bak A. The comparative molecular surface analysis (COMSA)—a nongrid 3D QSAR method by a coupled neural network and PLS system: Predicting pKa values of benzoic and alkanoic acids. J Chem Inf Comp Sci 2002; 42:184–191.

22. Silverman DB, Platt DE, Pitman M, Rigoutsos I. Comparative molecular moment analysis (CoMMA). In: Kubinyi H, Folkers G, Martin YC, eds. 3D QSAR in Drug Design. Vol. 3. Recent Advances. Dordrecht: Kluwer/ESCOM, 1998:183–196.

23. Heritage TW, Ferguson AM, Turner DB, Willett P. EVA: A novel theoretical descriptor for QSAR studies. In: Kubinyi H, Folkers G, Martin YC, eds. 3D QSAR in Drug Design.

Vol. 2. Ligand Protein Interactions and Molecular Similarity. Dordrecht: Kluwer/ESCOM, 1998:381–398.

24 Todeschini R, Gramatica P. New 3D molecular descriptors: The WHIM theory and QSAR applications. In: Kubinyi H, Folkers G, Martin YC, eds. 3D QSAR in Drug Design. Vol. 2. Ligand Protein Interactions and Molecular Similarity. Dordrecht: Kluwer/ESCOM, 1998:355–380.

25. Pastor M, Cruciani G, McLay I, Pickett S, Clementi S. GRID-independent descriptors (GRIND): A novel class of alignment-independent three-dimensional molecular descriptors. J Med Chem 2000; 43:3233–3243.

26. Kim KH. Comparative molecular field analysis (CoMFA). In: Dean PM ed. Molecular Similarity in Drug Design. Glasgow. Blackie, 1995:291–331.

27. Oprea TI, Waller CL. Theoretical and practical aspects of three dimensional quantitative structure–activity relationships. Lipkowitz KB, Boyd DB, eds. Reviews in Computational Chemistry. Vol. 11. New York: Wiley, 1997:127–182.

28. Greco G, Novellino E, Martin YC. Approaches to three-dimensional quantitative structure–activity relationships. Lipkowitz KB, Boyd DB, eds. Reviews in Computational Chemistry. Vol. 11. New York: Wiley, 1997:183–240.

29. Martin YC. 3D QSAR: Current state, scope, and limitations. In: Kubinyi H, Folkers G, Martin YC, eds. 3D QSAR in Drug Design. Vol. 3. Recent Advances. Dordrecht: Kluwer/ESCOM, 1998:3–23.

30. Norinder U. Recent progress in CoMFA methodology and related techniques. In: Kubinyi H, Folkers G, Martin YC, eds. 3D QSAR in Drug Design. Vol. 3. Recent Advances. Dordrecht: Kluwer/ESCOM, 1998:25–39.

31. Kim KH, Greco G, Novellino E. A critical review of recent CoMFA applications. In: Kubinyi H, Folkers G, Martin YC, eds. 3D QSAR in Drug Design. Vol. 3. Recent Advances. Dordrecht: Kluwer/ESCOM, 1998:257–315.

32. Hasegawa K, Funatsu K. Partial least squares modeling and genetic algorithm optimization in quantitative structure–activity relationships. SAR QSAR Environ Res 2000; 11:189–209.

33. Debnath AK. Quantitative structure–activity relationship (QSAR) paradigm—Hansch era to new millennium. Mini-Rev Med Chem 2001; 1:187–195.

34. Ivanciuc O. 3D QSAR models. In: Diudea MV ed. QSPR/QSAR Studies by Molecular Descriptors. Huntington NY: Nova Science Publishers, 2001:233–280.

35. Stone M, Jonathan P. Statistical thinking and technique for QSAR and related studies. Part II: Specific methods. J Chemom 1994; 8:1–20.

36. Hansch C, Leo A. Exploring QSAR. Fundamentals and Applications in Chemistry and Biology. Washington, DC: ACS Publishers, 1995.

37. Ajay, Murcko M. Computational methods to predict binding free energy in ligand–receptor complexes. J Med Chem 1995, 38:4953–4967.

38. Anzali S, Gasteiger J, Holzgrabe U, Polanski J, Sadowski J, Teckentrup A, Wagener M. The use of self-organizing neural networks in drug design. In: Kubinyi H, Folkers G, Martin YC, eds. 3D QSAR in Drug Design. Vol. 2. Ligand Protein Interactions and Molecular Similarity. Dordrecht: Kluwer/ESCOM, 1998:273–299.

39. Green SM, Marshall GR. 3D QSAR: a current perspective. Trends Pharmacol Sci 1995; 16:285–291.

40. Wold S, Ruhe A, Wold H, Dunn WJ. The collinearity problem in linear regression. The partial least squares approach to generalised inverses. J Sci Stat Comp 1984; 5:735–743.

41. Wold S. Cross-validatory estimation of the number of components in factor and principal components models. Technometrics 1978; 20:397–405.

42. Bush B, Nachbar RB. Sample-distance partial least squares: PLS optimized for many variables, with application to CoMFA. J Comput-Aided Mol Design 1993; 7:587–619.

43. Stone M, Jonathan P. Statistical thinking and technique for QSAR and related studies. Part I: General theory. J Chemom 1993; 7:455–475.

44. Todeschini R, Consonni V. Handbook of Molecular Descriptors. Weinheim: Wiley-VCH, 2000.

45. Livingstone DJ. The characterization of chemical structures using molecular properties. A survey. J Chem Inf Comput Sci 2000; 40:195–209.

46. Wermuth CG. The impact of QSAR and CADD methods in drug discovery. In: Höltje HD, Sippl W, eds. Rational Approaches to Drug Design. Barcelona: Prous Science, 2001:3–20.

47. Lipinski CA, Lombardo F, Dominy BW, Feeney PJ. Experimental and computational approaches to estimate solubility and permeability in drug discovery and development settings. Adv Drug Deliv Rev 1997; 23:3–25.

48. Van de Waterbeemd H, Camenisch G, Folkers G, Raevsky OA. Estimation of Caco-2 cell permeability using calculated molecular descriptors. Quant Struct-Act Relat 1996; 15: 480–490.

49. Leo A. Estimating $LogP_{oct}$ from structures. Chem Rev 1993; 5:1281–1306.

50. Oprea TI. Virtual screening in lead discovery: a viewpoint. Molecules 2002; 7:55–64.

51. Teague SJ, Davis AM, Leeson PD, Oprea TI. The design of leadlike combinatorial libraries. Angew Chem Int Ed 1999; 38:3743–3748.

52. Hann MM, Leach AR, Harper G. Molecular complexity and its impact on the probability of finding leads for drug discovery. J Chem Inf Comput Sci 2001; 41:856–864.

53. Sneader W. Drug Prototypes and Their Exploitation. Chichester: John Wiley & Sons Ltd, 1996.

54. Oprea TI, Davis AM, Teague SJ, Leeson PD. Is there a difference between leads and drugs? A historical perspective. J Chem Inf Comput Sci 2001; 41:1308–1315.

55. Norinder U, Haeberlein M. Computational approaches to the prediction of the blood–brain distribution. Adv Drug Deliv Rev 2002; 54:291–313.

56. Hansch C, Hoekman D, Leo A, Weininger D, Selassie CD. Chem-bioinformatics: Comparative QSAR at the interface between chemistry and biology. Chem Rev 2002; 102:783–812.

57. Hansch C, Hoekman D, Leo A, Weininger D, Selassie CD, C-QSAR database. Available from the BioByte Corporation, 201 West 4th St. Suite 204, Claremont, CA 91711.

58. Meylan WM, Howard PH. Atom/fragment contribution method for estimating octanol–water partition coefficients. J Pharm Sci 1995; 84:83–92.

59. Meylan WM, Howard PH, Boethling RS. Improved method for estimating water solubility from octanol/water partition coefficient. Environ Toxicol Chem 1996; 15:100–106.

60. Oprea TI. Lead structure searching: Are we looking at the appropriate property? J Comput Aided Mol Design 2002; 16:325–334.

61. Douguet D, Thoreau E, Grassy G. Quantitative structure–activity relationship studies of RAR a, b, g retinoid agonists. Quant Struct-Act Relat 1999; 18::107–123.

62. Topliss J, Edwards R. Chance factors in studies of QSAR. J Med Chem 1979; 22:1238–1244.

63. Clark M, Cramer RD III. The probability of chance correlation using PLS. Quant Struct-Act Relat 1993; 12:137–145.

64. Oprea TI, Zamora I, Svensson P. Qvo vadis, scoring functions? Toward an integrated pharmacokinetic and binding affinity prediction framework. In: Ghose AK, Viswanadhan VN, eds. Combinatorial Library Design and Evaluation for Drug Design. New York: Marcel Dekker Inc., 2001:233–266.

65. Xue Q, Yeung ES. Differences in the chemical reactivity of individual molecules of an enzyme. Nature 1995; 373:681–683.

66. Searle MS, Williams DH. The cost of conformational order: Entropy changes in molecular associations. J Am Chem Soc 1992; 114:10690–10697.

67. Davies TG, Hubbard RE, Tame JR. Relating structure to thermodynamics: the crystal structures and binding affinity of eight OppA–peptide complexes. Protein Sci 1999; 8:1432–1444.

68. Williams DH, Cox JPL, Doig AJ, Gardner M, Gerhard U, Kaye PT, Lal AR, Nicholls IA, Salter CJ, Mitchell RC. Toward the semiquantitative estimation of binding constants. Guides for peptide–peptide binding in aqueous solution. J Am Chem Soc 1991; 113:7020–7030.

69. Waller CL, Oprea TI, Giolitti A, Marshall GR. Three-dimensional QSAR of human immunodeficiency virus (I) protease inhibitors. 1. A CoMFA study employing experimentally-determined alignment rules. J Med Chem 1993; 36:4152–4160.

70. Oprea TI, Waller CL, Marshall GR. Three-dimensional quantitative structure–activity relationship of human immunodeficiency virus (I) protease inhibitors. 2. Predictive power using limited exploration of alternate binding modes. J Med Chem 1994; 37:2206–2215.

71. Diana GD, Kowalczyck P, Treasurywala AM, Oglesby RC, Peavar DC, Dutko FJ. CoMFA analysis of the interactions of antipicornavirus compounds in the binding pocket of human rhinovirus-14. J Med Chem 1992; 35:1002–1008.

72. Klebe G, Abraham U. On the prediction of binding properties of drug molecules by comparative molecular field analysis. J Med Chem 1993; 36:70–80.

73. Janssen LHM. Conformational flexibility and receptor interaction. Bioorg Med Chem 1998; 6:785–788.

74. Sulea T, Kurunczi L, Oprea TI, Simon Z. MTD-ADJ: a multiconformational minimal topologic difference for determining bioactive conformers using adjusted biological activities. J Comput-Aided Mol Design 1998; 12:133–146.

75. McClelland BJ. Statistical Thermodynamics. London: Chapman and Hall, 1973.

76. Sussman JL, Harel M, Frolow F, Oefner C, Goldman A, Toker L, Silman I. Atomic structure of acetylcholinesterase from *Torpedo californica*: a prototypic acetylcholine-binding protein. Science 1991; 253:872–879.

77. Lukáčová V, Baláž Š. Incorporation of multiple binding modes into 3D-QSAR methods. In: Höltje HD, Sippl W, eds. Rational Approaches to Drug Design. Barcelona: Prous Science, 2001:354–358.

78. Krystek S, Stouch T, Novotny J. Affinity and specificity of serine endopeptidase–protein inhibitor interactions. Empirical free energy calculations based on X-ray crystallographic structures. J Mol Biol 1993; 234:661–679.

79. Head RD, Smythe ML, Oprea TI, Waller CL, Greene SM, Marshall GR. VALIDATE: a new method for the receptor-based prediction of binding affinities of novel ligands. J Am Chem Soc 1996; 118:3959–3969.

80. Oprea TI. Property distribution of drug-related chemical databases. J Comput-Aided Mol Design 2000; 14:251–264.

81. Oprea TI, Ciubotariu D, Sulea T, Simon Z. Comparison of the minimal steric difference (MTD) and comparative molecular field analysis (CoMFA) methods for analysis of binding of steroids to carrier proteins. Quant Struct-Act Relat 1993; 12:21–26.

82. Woolfrey JR, Avery MA, Doweyko AM. Comparison of 3D quantitative structure–activity relationship methods: analysis of the in vitro antimalarial activity of 154 artemisinin analogs by hypothetical active-site lattice and comparative molecular field analysis. J Comput-Aided Mol Des 1998; 12:165–181.

83. Kim KH. List of CoMFA references, 1993–1997. In: Kubinyi H, Folkers G, Martin YC, eds. 3D QSAR in Drug Design. Vol. 3. Recent Advances. Dordrecht: Kluwer/ESCOM, 1998:317–338.

84. Kubinyi H. Comparative molecular field analysis (CoMFA). In: Von Ragué Schleyer P, Allinger NL, Clark T, Gasteiger J, Kollman PA, Schaefer HF III, eds. Encyclopedia of Computational Chemistry. Vol. 1. New York: Wiley, 1998:448–460.

85. Gund P. Three-dimensional pharmacophoric pattern searching. Hahn FE ed. Progress in Molecular and Subcellular Biology. Vol. 5. Berlin: Springer Verlag, 1977:117–143.

86. Coats EA. The CoMFA steroids as a benchmark data set for development of 3D-QSAR methods. In: Kubinyi H, Folkers G, Martin YC, eds. 3D QSAR in Drug Design. Vol. 3. Recent Advances. Dordrecht: Kluwer/ESCOM, 1998:199–213.

87. Verloop A, Hoogenstraaten W, Tipker J. Development and application of new steric substituent parameters in drug design. In: Ariens EJ, ed. Drug Design. Vol. 7. New York: Academic Press, 1976:165–207.

88. Simon Z, Chiriac A, Holban S, Ciubotariu D, Mihalas GI. Minimum Steric Difference. The MTD-Method for QSAR Studies. Letchworth: Research Studies Press, 1984.

89. Clark M, Cramer RD III, Jones D, Patterson DE, Simeroth P. Comparative molecular field analysis (CoMFA). 2. Toward its use with. Tetrahedron Comput Methodol 1990; 3:47–59.

90. Kearsley S, Smith G. An alternative method for the alignment of molecular structures: Maximizing electrostatic and steric overlap. Tetrahedron Comput Methodol 1990; 3:615–633.

91. Klebe G, Abraham U, Mietzner T. Molecular similarity indices in a comparative analysis (CoMSIA) of drug molecules to correlate and predict their biological activity. J Med Chem 1994; 37:4130–4146.

92. Mestres J, Rohrer DC, Maggiora GM. MIMIC: A molecular field matching program. Exploiting applicability of molecular similarity approaches. J Comput Chem 1997; 18:934–954.

93. Klebe G, Mietzner T, Weber F. Different approaches toward an automatic structural alignment of drug molecules: Applications to sterol mimics, thrombin and thermolyisin inhibitors. J Comput-Aided Mol Des 1994; 8:751–778.

94. Simon Z. Comparative molecular field analysis. Critical comments. Rev Roum Chim 1992; 37:323–325.

95. Perkins TDJ, Mills JEJ, Dean PM. Molecular surface-volume and property matching to superpose flexible dissimilar molecules. J Comput-Aided Mol Des 1995; 9:479–490.

96. Sulea T, Oprea TI, Muresan S, Chan SL. A different method for steric field evaluation in CoMFA improves model robustness. J Chem Inf Comput Sci 1997; 37:1162–1170.

97. Güner O, ed. Pharmacophore Perception, Development and Use in Drug Design. La Jolla: International University Line, 2000.

98. Marshall GR, Barry CD, Bosshard HE, Dammkoehler RA, Dunn DA. The conformational parameter in drug design: The active analog approach. Olson EC, Christoffersen RE, eds. Computer-Assisted Drug Design. American Chemical Society Symposium Series 1979; Vol. 112. Washington, DC: ACS, 1979:205–226.

99. Beusen DD, Marshall GR. Pharmacophore definition using the active analog approach. In: Güner O, ed. Pharmacophore Perception, Development and Use in Drug Design. La Jolla: International University Line, 2000:21–45.

100. Olsson T, Oprea TI. Cheminformatics: A tool for decision makers in drug discovery. Curr Opin Drug Discov Dev 2001; 4:308–313.

101. Broto P, Moreau G, Vandycke C. Molecular structures: perception, autocorrelation descriptor and SAR studies. Autocorrelation descriptor. Eur J Med Chem 1984; 19:66–70.

102. Wagener M, Sadowski J, Gasteiger J. Autocorrelation of molecular surface properties for modeling corticosteroid binding globulin and cytosolic Ah receptor activity by neural networks. J Am Chem Soc 1995; 117:7769–7775.

103. Clementi S, Cruciani G, Riganelli D, Valigi R, Costantino G, Baroni M, Wold S. Autocorrelation as a tool for a congruent description of molecules in 3D-QSAR studies. Pharm Pharmacol Lett 1993; 3:5–8.

104. Fedorov VV. Theory of Optimal Experiments. New York: Academic Press, 1972.

105. Clementi M, Clementi S, Clementi S, Cruciani G, Pastor M. Chemometric detection of

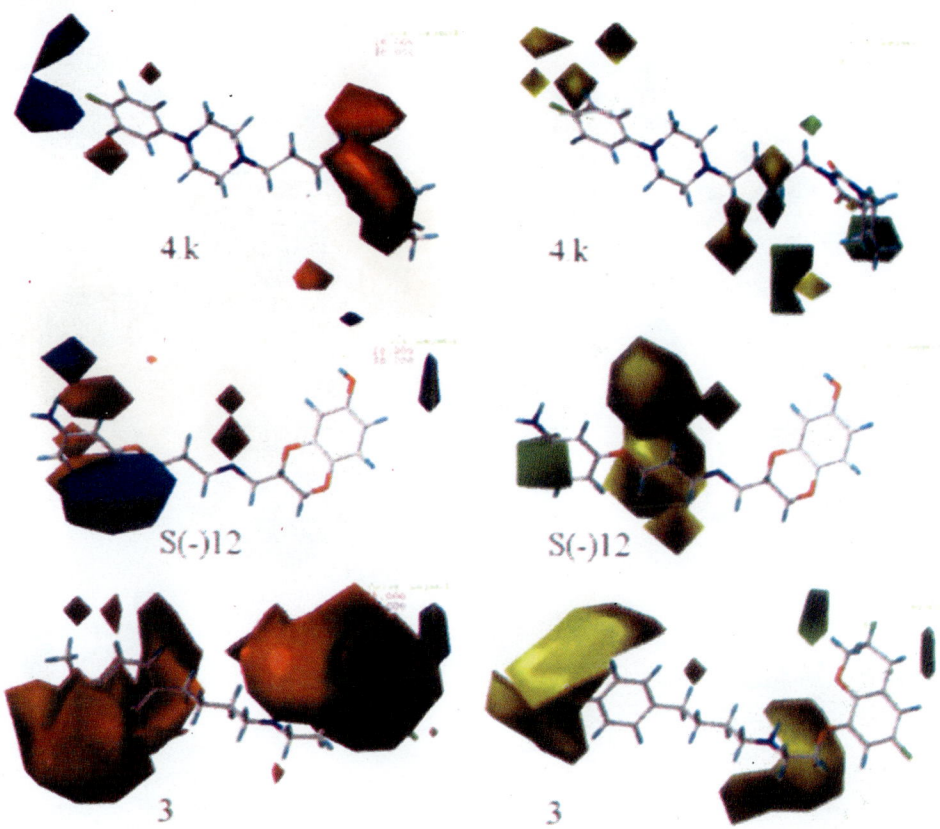

Figure 16 Comparison of stdev*coeff CoMFA fields for CoMFA models derived on α_1-AR activity (compound numbers as in Fig. 12). Electrostatic fields are on the left, steric fields on the right.

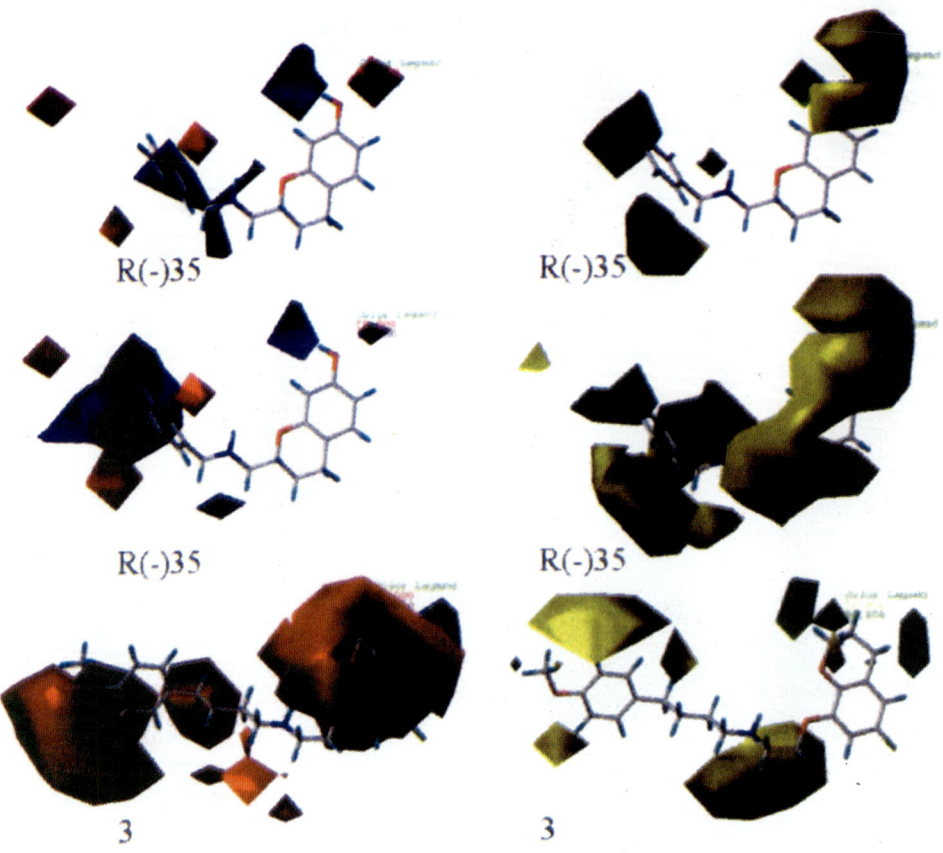

Figure 17 Comparison of stdev*coeff CoMFA fields for CoMFA models derived on D_2 high (top) and D_2 low (middle) activity, and D_2 antagonism (low-like) activity (bottom). Compounds are numbered as in Fig. 12. Electrostatic fields are on the left, steric fields on the right.

Figure 3 The rate-determining transition state in peptide hydrolysis by thermolysin from an AM1/AMBER QM/MM computation. The sticks and balls and sticks part correspond to the QM fragment.

binding sites of 7TM receptors. Gundertofte K, Jørgensen FS, eds. Molecular Modeling and Prediction of Bioactivity. New York: Kluwer Academic/Plenum Publishers, 1996: 207–212.

106. Horvath D. High throughput conformational sampling & fuzzy similarity metrics: a novel approach to similarity searching and focused combinatorial library design and its role in the drug discovery laboratory. In: Ghose AK, Viswanadhan VN, eds. Combinatorial Library Design and Evaluation for Drug Design. New York: Marcel Dekker, 2001:429–472.

107. Horvath D, Jeandenans C. Neighborhood behavior of in silico structural spaces with respect to in vitro activity spaces—a novel understanding of the molecular similarity principle in the context of multiple receptor binding profiles. J Chem Inf Comput Sci 2003; 43:680–690.

108. Patterson DE, Cramer RD III, Ferguson AM, Clark RD, Weinberger LE. Neighborhood behavior: a useful concept for validation of "molecular diversity" descriptors. J Med Chem 1996; 39:3049–3059.

109. Horvath D, Jeandenans C. Neighborhood behavior of in silico structural spaces with respect to in vitro activity spaces—a benchmark for neighborhood behavior assessment of different in silico similarity metrics. J Chem Inf Comput Sci 2003; 43:691–698.

110. Brown RD, Martin YC. Use of structure–activity data to compare structure-based clustering methods and descriptors for use in compound selection. J Chem Inf Comput Sci 1997; 36:572–584.

111. Brown RD, Martin YC. The information content of 2D and 3D structural descriptors relevant to ligand–receptor binding. J Chem Inf Comput Sci 1997; 37:1–9.

112. Olsson T, Sherbukhin V, Synthesis and Structure Administration (SaSA). © AstraZeneca R&D Mölndal 1997–2001.

113. ALMOND 2.0, 2002. Available from Multivariate Informetric Analysis srl, Perugia, Italy.

114. Leo A, Weininger D. CMR3 Reference Manual. 1995. CMR3 is available from Daylight Chemical Information Systems, Santa Fe, New Mexico.

115. Glen RC. A fast empirical method for the calculation of molecular polarizability. J Comput-Aided Mol Des 1994; 8:457–466.

116. Basak SC, Balaban AT, Grunwald GD, Gute BD. Topological indices: Their nature and mutual relatedness. J Chem Inf Comput Sci 2000; 40:891–898.

117. Balaban AT. Topological and stereochemical molecular descriptors for databases useful in QSAR similarity/dissimilarity and drug design. SAR QSAR Environ Res 1998; 8: 1–21.

118. Raevsky OA, Grigor'ev VYu, Kireev D, Zefirov NS. Complete thermodynamic description of H-Bonding in the framework of multiplicative approach. Quant Struct-Act Relat 1992; 11:49–64.

119. Gasteiger J, Marsili M. Iterative partial equalization of orbital electronegativity: A rapid access to atomic charges. Tetrahedron 1980; 36:3219–3222.

120. CLOGP 4.0, 2001. CLOGP is available from Biobyte Inc., Claremont, California.

121. SIMCA 9.0P, 2001. Available from Umetrics AB, Umeå, Sweden.

122. Sadowski J, Gasteiger J, Corina 1.8, 2000. Available from Molecular Networks, Erlangen, Germany.

123. Oprea TI. Rapid estimation of hydrophobicity for virtual combinatorial library analysis. SAR QSAR Environ Res 2001; 12:129–141.

124. Cruciani G, Crivori P, Carrupt PA, Testa B. Molecular fields in quantitative structure–permeation relationships: The VolSurf approach. J Mol Struct (THEOCHEM) 2000; 503:17–30.

125. Oprea TI, Zamora I, Ungell AL. Pharmacokinetically based mapping device for chemical space navigation. J Comb Chem 2002; 4:258–266.

126. Hansch C. Quantitative structure activity relationships and the unnamed science. Acc Chem Res 1993; 26:147–153.

127. Kim KH. Comparison of classical QSAR and comparative molecular field analysis: Toward lateral validations. In: Hansch C, Fujita T, eds. Classical and Three-Dimensional QSAR In Agro-Chemistry, ACS Symposium series. Vol. 606. Washington, DC: American Chemical Society, 1995;302–317.

128. Lopez-Rodriguez ML, Rosado ML, Benhamu B, Morcillo MJ, Fernandez E, Schaper KJ. Synthesis and structure-activity relationships of a new model of arylpiperazines. 2. 3D-QSAR of hydantoin-phenylpiperazine derivatives with affinity for 5-HT1A and alpha1 receptors. A comparison of CoMFA models. J Med Chem 1997; 40:1648–1656.

129. Mewshaw RE, Kavanagh J, Stack G, Marquis KL, Shi X, Kagan MZ, Webb MB, Katz AH, Park A, Kang YH, Abou-Gharbia M, Scerni R, Wasik T, Cortes-Burgos L, Spangler T, Brennan JA, Piesla M, Mazandarani H, Cockett MI, Ochalski R, Coupet J, Andree TH. New generation dopaminergic agents. 1. Discovery of a novel scaffold which embraces the D2 agonist pharmacophore. Structure–activity relationships of a series of 2-(aminomethyl)chromans. J Med Chem 1997; 40:4235–4256.

130. Reitz AB, Baxter EW, Codd EE, Davis CB, Jordan AD, Maryanoff BE, Maryanoff CA, McDonnell ME, Powell ET, Renzi MJ, Schott MR, Scott MK, Shank RP, Vaught JL. Orally active benzamide antipsychotic agents with affinity for dopamine D2, serotonin 5-HT1A, and adrenergic alpha1 receptors. J Med Chem 1998; 41:1997–2009.

131. Yasunaga T, Kimura T, Naito R, Kontani T, Wanibuchi F, Yamashita H, Nomura T, Tsukamoto S, Yamaguchi T, Mase T. Synthesis and pharmacological characterization of novel 6-fluorochroman derivatives as potential 5-HT1A receptor antagonists. J Med Chem 1998; 41:2765–2778.

132. Chern JW, Tao PL, Wang KC, Gutcait A, Liu SW, Yen MH, Chien SL, Rong JK. Studies on quinazolines and 1,2,4-benzothiadiazine 1,1-dioxides. 8.Synthesis and pharmacological evaluation of tricyclic fused quinazolines and 1,2,4-benzothiadiazine 1,1-dioxides as potential alfa1-adrenoceptor antagonists. J Med Chem 1998; 41:3128–3141.

133. Vacher B, Bonnaud B, Funes P, Jubault N, Koek W, Assie MB, Cosi C. Design and synthesis of a series of 6-substituted-2-pyridinylmethylamine derivatives as novel, high-affinity, selective agonists at 5-HT1A receptors. J Med Chem 1998; 41:5070–5083.

134. SYBYL 6.8, 2001. Available from Tripos, Inc., St. Louis, Missouri.

135. Van Drie JH. Future directions in pharmacophore discovery. In: Güner O, ed. Pharmacophore Perception, Development and Use in Drug Design. La Jolla: International University Line, 2000:517–530.

136. Oprea TI, García AE. Three-dimensional quantitative structure activity relationships of steroid aromatase inhibitors. J Comput-Aided Mol Des 1996; 10:186–200.

137. Arevalo JH, Hassig CA, Stura EA, Sims MJ, Taussig MJ, Wilson IA. Structural analysis of antibody specificity. Detailed comparison of five Fab′–steroid complexes. J Mol Biol 1994; 241:663–690.

138. Oprea TI, Head RD, Marshall GR. The basis of cross-reactivity for a series of steroids binding to a monoclonal antibody against progesterone (DB3). A molecular modeling and QSAR study. In: Sanz F, Giraldo J, Manaut F, eds. QSAR and Molecular Modelling; Concepts, Computational Tools and Biological Applications. Barcelona: JR Prous Publishers, 1995:451–455.

23

Computational Aspects of Library Design and Combinatorial Chemistry

VALERIE J. GILLET

University of Sheffield, Sheffield, United Kingdom

1. INTRODUCTION

The techniques of high-throughput screening and combinatorial synthesis have revolutionized the drug discovery process during the last decade. High-throughput screening is an automated process whereby large numbers of compounds (10^4–10^5) are rapidly screened for biological activity [1]. The related technique of combinatorial synthesis refers to the synthesis of large numbers of compounds in parallel where product molecules are formed as combinations of the available reagents or building blocks. For example, a combinatorial synthesis involving three positions of variability with N_A, N_B, and N_C possible reagents at each substitution position, respectively, will result in $N_A \times N_B \times N_C$ product molecules. Thus, combinatorial synthesis represents a source of novel compounds that can be screened using high-throughput screening.

The techniques began to be widely adopted during the early to mid 1990s when early opinion was that: simply increasing throughput would be sufficient to improve the chances of finding novel bioactive compounds. However, it soon became apparent that a "make-and-test-all" approach was neither practical nor possible. For example, Cramer et al. [2] describe a three-component diamine library for which there are 10^{12} potential products using commercially available reagents extracted from the Available Chemicals Directory. At a rate of testing of 10^5 compounds per day, it would take 30,000 years to test this number of molecules. Clearly, it is not possible to make and test everything.

Considerable effort has been devoted to the development of computational techniques that can be used to choose compounds to screen and compounds to make via combinatorial synthesis. Virtual screening [3–5] (also referred to as in silico screening) refers to the use of computational techniques to select compounds, either from existing libraries such as in-house databases or external suppliers, or from virtual libraries that represent the compounds that could potentially be made via combinatorial synthesis. Many techniques have also been developed to assist in combinatorial library design.

In the early days of combinatorial chemistry, there was an emphasis on building very large libraries; however, these early approaches were disappointing, with libraries either failing to produce the hit rates expected or resulting in hits that had undesirable physicochemical properties to be considered as lead compounds [6]. More recently, the focus has shifted toward more rationally designed smaller libraries [7].

The criterion used for selecting compounds depends on the use for which the library is intended. When a library is to be screened against a range of targets, then the criterion is usually diversity. The basis for diverse subsets lies in the similar property principle [8], which makes the assumption that structurally similar molecules are likely to have similar properties. In terms of a screening experiment, compounds that are closely related structurally are likely to represent redundant information as far as structure–activity relationship studies are concerned. Hence, the focus has been on structurally diverse subsets that will provide good coverage of bioactive space while minimizing redundancy.

When information is available about the therapeutic target, either through known active compounds or when the 3D structure of the receptor site is known, then the emphasis is usually on focused or targeted libraries. For example, compounds can be selected on the basis of similarity to a known active or actives, on predicted activity according to a quantitative structure–activity relationship (QSAR), or on their predicted ability to be able to bind to a receptor.

In both diverse and focused libraries, increasing use is being made of computational filters prior to subset selection. The filters are used to remove undesirable compounds, such as ones containing toxic groups, compounds with functional groups that are likely to interfere with the assay, etc. More sophisticated filtering techniques that attempt to rank libraries of compounds according to their likelihood of exhibiting druglike properties have been developed.

The chapter begins with a discussion of similarity and diversity measures and how they can be applied in a virtual screening context. The various computational filters in use are also discussed. The rest of the chapter is concerned with different approaches to combinatorial library design, beginning with reagent-based methods followed by product-based approaches of cherry picking and combinatorial subset selection. Finally, approaches to designing libraries optimized on multiple properties simultaneously are discussed.

2. VIRTUAL SCREENING

Virtual screening techniques require the definition of a chemistry space in order that the similarity (and distance) between compounds within the space can be quantified. Once a space has been defined, a diverse subset is one that covers the chemistry space well, whereas a focused subset is one that is restricted to a localized region within the space. A chemistry space is defined through the use of numerical descriptors, which can be calculated for molecules, as shown schematically in Fig. 1. The similarity (and

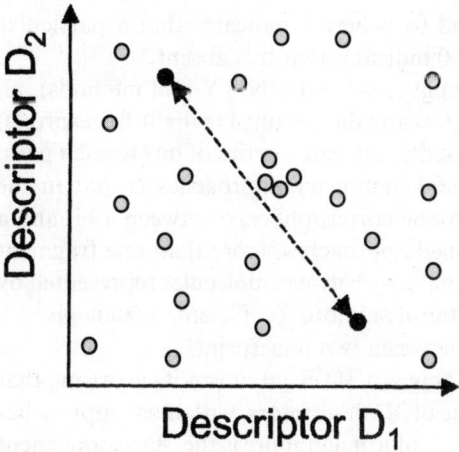

Figure 1 Numerical descriptors are used to define a chemistry space, and the similarity (or distance) between molecules is determined using a similarity coefficient applied to the descriptor representations of the molecules.

distance) between pairs of molecules in the space is measured by applying a similarity coefficient to the numerical descriptor representations of the molecules.

Many different descriptors have been developed in an attempt to define chemistry spaces that are relevant for bioactivity. Aside from being relevant to biological activity, to be useful for combinatorial chemistry, descriptors should also be relatively easy to calculate in order that the methods can be applied to large datasets. Some descriptors commonly used in virtual screening are described below. For more detail on descriptors, see the chapter by Downs and Barnard and the earlier review by Brown [9]. The most commonly used similarity coefficients are the Tanimoto coefficient and Euclidean distance [10]. A weighting scheme may also be applied to assign relative weights to different descriptors.

2.1. Descriptors

Descriptors can be divided into: whole molecule properties, descriptors that can be calculated from the 2D graph of a structure, and descriptors that are based on 3D properties of molecules.

Whole molecule descriptors include physicochemical properties such as molecular weight, log P, molar refractivity, etc. Descriptors that can be calculated from the 2D graph representation of structures include topological indices and 2D fingerprints. A topological index is a single-valued number that represents the shape or connectivity of a molecule. Many different topological indices have been developed (e.g., see Molconn-Z) [11]. Typically, a large number of topological indices are reduced to a smaller number of orthogonal descriptors using a data reduction technique such as principal components analysis.

The fingerprint methods can be divided into dictionary-based and hashed-based methods. In the dictionary-based methods, such as the MDL MACCS keys [12] and BCI fingerprints [13], a binary fingerprint is defined in which each bit represents a particular substructural fragment contained in a fragment dictionary. The fingerprint of a molecule consists of a series of 1s and 0s, where 1 indicates that a particular substructure is present in the molecule and 0 indicates that it is absent.

In the hashed methods (e.g., the Daylight [14] and UNITY [15] methods), all atom paths up to a predefined length (e.g., 0–7 for the default Daylight fingerprints) are determined and a hashing algorithm is used to allocate a series of bits to each path. The main difference between the hashed and dictionary approaches is that in the dictionary-based methods there is a one-to-one correspondence between a bit and a substructural fragment, whereas in the hashed approaches, more than one fragment can result in a particular bit being set. The similarity between molecules represented by fingerprints is usually determined using the Tanimoto coefficient, which gives a measure of the number of bits in common between two fingerprints.

Given the fact that drug–receptor binding is a 3D event, it is not surprising that there has been considerable interest in the use of 3D descriptors, with most approaches being based on the pharmacophore concept. A pharmacophore is the 3D arrangement of functional groups presumed to be relevant for receptor binding. Pharmacophore techniques were originally developed for 3D database searching [16], where the 3D conformations of a series of active compounds are compared to deduce a potential pharmacophore, which is then used to search a database of 3D structures in an attempt to find other potentially active compounds.

The most commonly used 3D descriptors in virtual screening and library design are multipoint pharmacophores [17–19]. A three-point pharmacophore is a set of three features (typically acids, bases, hydrogen bond donors, hydrogen bond acceptors, aromatic centers, and hydrophobes) and the distances between them. A molecule is represented by a bitstring or fingerprint, analogous to the 2D fingerprint, where each position in the fingerprint represents one triangle of features with one set of distances separating the features (the distances are typically binned). Pharmacophoric fingerprints can be used in pairwise molecular similarity calculations in the same way as the 2D fingerprint; however, typically, they are much sparser than 2D fingerprints and small molecular differences can lead to large differences in fingerprints because the number of pharmacophores present in a molecule varies approximately with the cube of the number of pharmacophoric features [20].

A limitation of the three-point pharmacophore is that it is planar and, hence, has limited ability to describe shape and chirality. Recently, there has been interest in four-point pharmacophores, which are based on four features and their associated distances. Four-point pharmacophores provide increased resolution over three-point pharmacophores and can provide a better representation of shape and chirality, which is important in ligand–receptor interactions [21]. However, the number of potential pharmacophores increases enormously in going from three-point to four-point pharmacophores. For example, when distances are divided into seven bins and there are six features, then there are over 9000 possible three-point pharmacophores and 2.3 million four-point pharmacophores [19].

A difficulty associated with 3D descriptors such as multipoint pharmacophores is the handling of conformational flexibility. Ligands are known to bind to receptors in conformations other than their lowest energy conformations, and there may be many hundreds of accessible conformations for a single molecule. A pharmacophore fingerprint should include all pharmacophore points found in each distinct conformer. This can have both a time implication in generating the conformations and an accuracy implication because the bioactive conformation is typically unknown.

BCUT descriptors are recently developed descriptors that can encode 2D and 3D properties [22,23]. Matrices are constructed in which the diagonal matrix elements are based on calculated properties that are believed to relate to ligand–receptor binding, including atomic charge, atomic polarizability, and hydrogen bond donor and acceptor ability. The off-diagonal elements encode either 2D topology or 3D inter-atomic distances. The BCUT descriptors are then the highest or lowest eigenvalues of these matrices. Typically, many different BCUTs are calculated and a subset is chosen to represent a low-dimensional space into which molecules can be mapped. In recent work, Pearlman and Smith have focused on identifying subsets of BCUT descriptors that define receptor relevant spaces (i.e., chemistry spaces that are relevant to ligand–receptor binding).

2.2. Selecting Diverse Subsets

In general, there are:

$$\frac{N!}{n!(N-n)!}$$

different subsets of size n contained within a library of N compounds and, because in a typical library design, $n \ll N$, it is not possible to enumerate all subsets and compare them directly. Thus, approximate methods for selecting subsets must be used. There are four main approaches to selecting diverse subsets. These are clustering, dissimilarity-based compound selection (DBCS), partitioning or cell-based methods, and optimization techniques.

Clustering has been used for many years for generating diverse compound sets for screening [24,25]. It is a two-step process where, firstly, the dataset is divided into clusters so that molecules in the same cluster are similar whereas molecules in different clusters are dissimilar. Secondly, a diverse, or representative, set of compounds is chosen by selecting one or more compounds from each cluster. Thus, clustering is based on calculating the pairwise similarities between all molecules in the dataset. Many different clustering methods have been developed; the most commonly used for chemical applications are the Jarvis–Patrick and Ward's methods [25].

Clustering is typically used with 2D fingerprints where pairwise similarity is quantified using the Tanimoto coefficient. The sparse nature of pharmacophoric fingerprints makes them unsuitable for clustering. Clustering is a computationally expensive process and thus the size of the datasets that can be handled is limited.

Dissimilarity-based compound selection [26,27] is also based on calculating pairwise similarities; however, in this case, compounds are selected directly, rather than via the two-stage process described for clustering. A compound is chosen to seed the subset then an iterative procedure is begun where, in each iteration, the next compound to be added to the subset is the one remaining in the dataset that is most dissimilar to those already included in the subset.

Several different DBCS algorithms have been described and they differ in the way the seed compound is chosen and the way in which the dissimilarity of one compound to a set of compounds is measured [28]. For example, in the MaxMin method, the subset is chosen to maximize the minimum distance between all pairs of molecules in the subset [29], whereas in the MaxSum method, the subset that maximizes the sum of pairwise dissimilarities in the subset is chosen [28]. The basic

DBCS algorithm has an expected time complexity of $O(n^2 N)$ for the selection of a subset of size n from a dataset of size N. A fast $O(N)$ implementation of the MaxSum method has been described [30]; however, it has been shown that the method can result in subsets that contain pairs of closely related molecules [31,32]. The MaxMin function is slower than MaxSum, although it has been shown to be more effective in identifying subsets that exhibit a range of different biological activities [33]. An $O(n\log N)$ implementation of MaxMin that can be used with low-dimensional data has been developed [31]. DBCS approaches are usually used with high-dimensional descriptors such as fingerprints. The approach is illustrated in Fig. 2 for the selection of a subset of five compounds.

Mount et al. [32] describe a DBCS method that is based on a minimum spanning tree. A spanning tree is a set of edges that connect a set of objects. The objects in this method are the molecules in the subset, and each edge is labeled by the dissimilarity between the two molecules it connects. A minimum spanning tree is the spanning tree that connects all molecules in the subset with the minimum sum of pairwise dissimilarities; thus, the diversity is the sum of just some of the intermolecular similarities rather than all of them as in MaxSum. A similar function has also been developed by Brown et al. [34].

Like clustering, the partitioning or cell-based method is also a two-step process. It involves, firstly, defining a low-dimensional chemistry space (e.g., an early approach involved the use of simple physicochemical properties such as molecular weight, log P, numbers of hydrogen bond donors, etc. [35,36]; more recent approaches have used BCUT descriptors) [23]. Each property, or descriptor, forms an axis in a multidimensional space. Each axis is then divided into a series of bins and the combination of all bins forms a set of cells that define a chemistry space. Molecules can then be mapped onto the space according to the particular properties they possess. It is then trivial to select a diverse subset of compounds by taking one or more from each occupied cell. The partitioning approach is illustrated in Fig. 3.

The cell-based method has been referred to as an absolute method because it involves the definition of a chemical space that is independent of the molecules that are

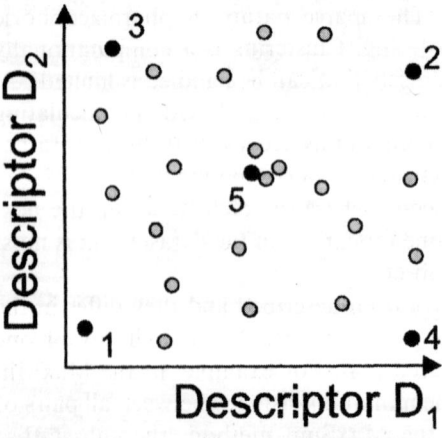

Figure 2   The selection of a subset of five diverse compounds using DBCS.

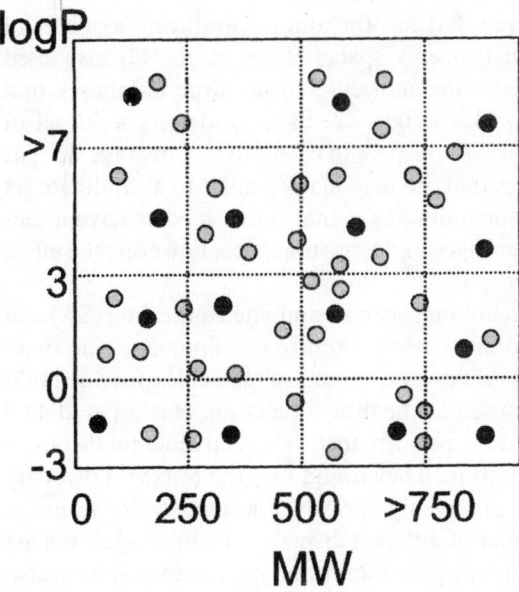

Figure 3 The selection of a diverse subset of compounds using a cell-based partitioning scheme.

subsequently mapped onto it. Absolute measures have several advantages over methods such as clustering and DBCS, which are based on calculating pairwise intermolecular similarities. For example, they allow the diversity or similarity of different datasets to be compared simply by looking at the overlap in occupied cells, and voids in a dataset can be identified easily as the cells that are underoccupied or that contain no compounds. This can be useful when aiming to enhance a corporate collection with externally available compounds.

Multipoint pharmacophore fingerprints have also been used to compare libraries. For example, Pickett et al. [17] have represented libraries by the union of the individual molecular fingerprints and were able to identify regions of multipoint pharmacophore space that were not covered or that were underrepresented. McGregor and Muskal [37,38] developed a similar approach that is based on a low-dimensional pharmacophore space obtained by applying principal components analysis to the three-point pharmacophore representations of the compounds.

A disadvantage of the partitioning approach is that it can only be used with low-dimensional descriptors so that it cannot be used with fingerprints, for example. This is because the number of cells rises exponentially with the number of dimensions. A further disadvantage of cell-based methods is that the cell boundaries are somewhat arbitrary and can result in compounds that are very similar, being assigned to different cells. Given the sensitivity of the method to the location of cell boundaries, some effort has been devoted to finding optimal bin sizes in partitioning schemes [39,40].

The final category of subset selection algorithms is termed optimization techniques and it includes a number of different approaches. For example, Martin et al. [41] have described a subset selection technique based on D-optimal design, which is a statistical experimental design technique that has the aim of producing a subset of compounds that are evenly spread in property space. Higgs et al. [42] also used experimental design algorithms for selecting molecules from large databases that approximate spread and coverage. Spread designs are used to identify a subset of molecules that are maximally dissimilar with respect to each other. Coverage designs are used to identify subsets of molecules that are maximally similar to a candidate set of molecules. A disadvantage of statistical designs is that linear models have a tendency to select molecules that are at the edges of descriptor space; however, the effect can be improved by using higher-order designs.

Other approaches are based on techniques such as simulated annealing (SA) and genetic algorithms (GAs), which are designed to maximize or minimize some functions. These techniques can be used to select diverse subsets of molecules provided that an appropriate measure of diversity is used as the fitness function. Hassan et al. [29] developed an SA method and compared the performance of several different distance-based diversity functions for subset selection. They found that the MaxMin diversity function was most effective in producing evenly spread molecules. More recently, Brown et al. [34] have compared a number of different diversity functions against a list of requirements for an ideal diversity function. Lobanov and Agrafiotis [43] have also developed an SA method for maximizing diversity, their method employs a user-defined objective function and can therefore be tailored to encode different selection criteria.

2.3. Evaluation of Methods

Given the large variety of molecular descriptors that exist, several approaches have been taken to evaluate which descriptors are most suitable for virtual screening although, to some extent, the choice of descriptors and subset selection method is interlinked. For example, as already mentioned, cell-based approaches are only appropriate for low-dimensional chemistry space whereas both DBCS and clustering can be used with high-dimensional data such as fingerprints.

Early evaluations, such as the studies conducted by Brown and Martin [44,45], tended to suggest that 2D descriptors are more effective than 3D descriptors at identifying molecules that exhibit similar bioactivity. Ward's clustering was used as the compound selection method. The 2D descriptors they tested were originally developed for performing substructure searching and, hence, their effectiveness for similarity searching is somewhat surprising. More recently, the limitations of 2D fingerprints for similarity searching have been noted [46] (e.g., they have been shown to be most suited to finding structural analogues and are less effective at finding structural diverse compounds that exhibit similar activities) [47].

In a later study, three-point and four-point pharmacophore fingerprints have been shown to perform better than 2D Daylight fingerprints in some circumstances [48]. More recently, the mini-fingerprints (MFPs), based on a small number of 2D descriptors, were specifically designed for similarity searching and were shown to have comparable performance to the 3D pharmacophores [49].

Some recent approaches have involved using combinations of 2D and 3D descriptors. For example, BCUT descriptors have been used in combination with four-point pharmacophores [50], and field-based methods have been combined with 2D Daylight fingerprints [47].

Patterson et al. [51] have validated descriptors using the concept of neighborhood behavior. The differences between calculated similarities of pairs of molecules were compared with their biological activities. A descriptor is said to exhibit neighborhood behavior if small changes in structural similarity correlate with small changes in biological activity.

2.4. Designing Focused Libraries

Similarity searching and clustering have been used to select compounds for screening for many years [52]. For example, given an active compound, a dataset can be ranked in order of decreasing similarity to the active and, according to the similar property principle, compounds near the top of the ranked list should have an increased probability of also being active and hence should be screened first. Similarly, in clustering approaches, compounds in clusters that contain known actives are more likely to be active than randomly selected compounds. A similar argument can be extended to cell-based approaches. These techniques are typically used when the 3D structure of the target enzyme is unknown.

When the 3D structure of the target enzyme is available, then ligand docking methods can be used to filter out compounds that are unlikely to be able to bind to the receptor [53]. Docking programs attempt to predict how a ligand will react with a receptor. They consist of two components: an algorithm to orient the ligand within the receptor pocket and a scoring function to estimate the binding affinity of the docked ligand in order that the best orientation is found. There is currently a great deal of interest in the use of docking for virtual screening and for docking of combinatorial libraries [54–58]; however, the main difficulty lies in the accurate prediction of binding affinities, and many different algorithms have been developed and compared [59]. Recently, it has been found that accuracy can be improved if several scoring functions are used simultaneously in what is known as a consensus approach [60,61]. The most widely used docking programs are FlexX [62], DOCK [63], and GOLD [64]. Docking is discussed in more detail in the chapter by Muegge.

3. COMPUTATIONAL FILTERING

Given the large number of compounds that are available for screening, the limited numbers that can be handled, and the high costs associated with developing potential drug candidates, most approaches to virtual screening aim to discard unlikely candidates as early as possible in the drug discovery process, based on the "fail fast, fail cheap" scenario. Currently, there is much interest in developing predictive models that can be used to optimize Absorption, Distribution, Metabolism, Excretion (ADME) properties as early as possible in the drug discovery pipeline (see Clark and Pickett [65] and Beresford et al. [66] for reviews on predicting ADME properties).

The simplest filtering technique is to use substructure searching to remove compounds that contain reactive functional groups that are likely to interfere with the biological assay and compounds that contain toxic functional groups [67].

A widely implemented approach to predicting oral absorption is to apply the "Rule-of-Five" by Lipinski et al. [68]. The rules were derived following an analysis of over 2000 drugs believed to have entered Phase II clinical trials, and predict that any compound that satisfies two of the following rules is unlikely to be absorbed following oral administration:

- Molecular mass >500 Da
- Greater than five hydrogen bond donors
- Greater than 10 hydrogen bond acceptors
- $C \log P > 5.0$.

More recently, a number of groups have developed more sophisticated methods that aim to predict druglikeness. The methods use different techniques to compare the properties of known active compounds with (presumed) inactive compounds. They attempt to classify compounds as druglike or non-druglike, or to rank the compounds in a dataset based on their probability of being druglike. For example, Gillet et al. [69] developed a GA-based approach that compares two datasets according to the following physicochemical properties: molecular weight, hydrogen bond donors, hydrogen bond acceptors, rotatable bonds, aromatic rings, and a shape index (the kappa alpha 2 index). The algorithm was trained using a sample of the SPRESI database to represent non-druglike compounds and a sample of World Drugs Index (WDI) to represent druglike compounds. The method was found to be surprisingly effective at distinguishing between the two classes of compounds and has been used to filter compounds prior to high-throughput screening [70].

Other related approaches are based on the use of neural networks and recursive partitioning [71–74].

Recently, work by Oprea et al. [75] has shown that lead compounds are generally less complex than drug molecules. In fact, the lead optimization process usually involves adding functionality to molecules in an attempt to increase their potency. Thus, if the complexity of the molecules being screened is already of the order of known drugs, then there is little scope for adjusting activity. These findings are supported by the work of Hann et al. [76] and suggest that screening libraries should contain molecules that have leadlike properties rather than druglike properties.

A recent approach to computational filtering has focused on identifying "frequent hitters" (i.e., compounds that show up as hits in many different biological assays covering a range of targets) [77]. This can happen because they show nonspecific activity or because they interfere with the assay itself. In both cases, these molecules are likely to be poor starting points for lead optimization and their removal can increase efficiency in screening.

4. COMBINATORIAL LIBRARY DESIGN

So far, the chapter has been concerned with selecting subsets of compounds for screening (e.g., from in-house databases or from external suppliers). Here the focus is on how computational techniques can be used to assist in the design of combinatorial libraries. As described in Sec. 1, a combinatorial library consists of all products resulting from the combinatorial joining of all possible reagents at each position of variability. Typically, there are many more reagents available than can be handled practically and, hence, careful design of combinatorial libraries is required through the use of reagent selection techniques.

There are two different approaches that can be taken to reagent selection. In reagent-based selection, each pool of reagents is handled independently. An alternative approach is to enumerate the full virtual combinatorial library from all possible reagents. This is illustrated in Fig. 4 where a two-component combinatorial library is represented by a 2D array, with the rows representing the reagents available at one substitution position and the columns representing the reagents available at the second substitution position. Product-based selection techniques are then applied to the virtual library. This can either be done by cherry picking (a cherry-picked subset of nine compounds is shown by the shaded elements of Fig. 4a) or by taking direct account of the combinatorial constraint to select a combinatorial subset directly (shown by the shaded elements in Fig. 4b, which represents a 3 × 3 combinatorial subset of the virtual library).

These approaches are discussed in detail in the following sections.

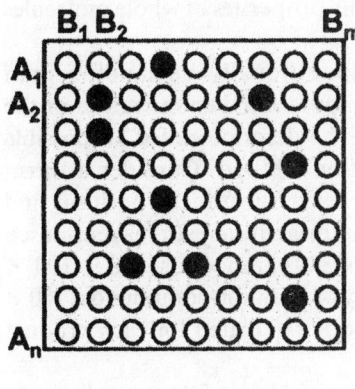

(a)

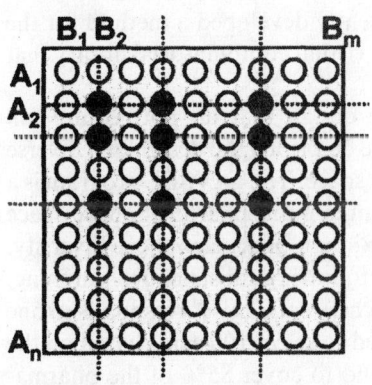

(b)

Figure 4 A two-component combinatorial library is represented by a 2D array. (a) A cherry-picked subset of nine compounds is highlighted. (b) A 3×3 combinatorial subset is highlighted.

4.1. Reagent-Based Selection

Any of the techniques described previously for subset selection can be applied to select subsets of reagents; hence, they are not discussed further here.

4.2. Product-Based Selection

Selecting reagents without considering the product molecules makes the assumption that the sites of variation on the core scaffold are independent; however, experience suggests that this is not the case. For example, Gillet et al. [78] and Gillet and Nicolotti [79] have shown that more diverse libraries result if selection is performed in product space rather than in reagent space. This result was shown to hold for a number of different libraries, descriptors, and diversity indices, although some related studies have also shown that in some circumstances, reagent-based selection may be comparable in performance to product-based selection [80,81]. Product-based designs are preferred in focused library design because it is generally properties of whole molecules that are to be optimized.

Product-based selection is much more computationally demanding than reagent-based selection. Typically, it requires the computational enumeration of the full virtual combinatorial library and calculation of the descriptors for all possible products, prior to the application of a subset selection method. Consider a three-component reaction with 100 reagents available at each substituent position and assume that the aim is to build a $10 \times 10 \times 10$ combinatorial library. In reagent-based selection, this requires the calculation of descriptors for 300 compounds ($100 + 100 + 100$). In product-based design, however, the full library of 1 million compounds ($100 \times 100 \times 100$) must be enumerated and descriptors must be calculated for each product molecule.

The size of a virtual library can be reduced by applying filters to eliminate reagents that are known to be undesirable [67]. However, in some cases, the virtual library may still be too large to allow full enumeration, and thus full product-based design is infeasible. (Although the need for full enumeration may not be necessary in the future, for example, Barnard et al. [82] have recently developed a method for the rapid calculation of descriptors for the products in a virtual combinatorial library that avoids the need for enumeration.)

Several product-based approaches to library design that do not require full enumeration have been developed. Pickett et al. have described the design of a diverse amide library where diversity is measured in product space. The DIVSEL program is a DBCS method where dissimilarity is measured in three-point pharmacophore space [83]. Initially, 11 amines were selected based on maximum pharmacophore diversity. Then a total of 1100 carboxylic acids were identified following substructure searching. A set of 1100 pharmacophores keys was generated, where each key corresponds to one acid combined with the 11 amines. DIVSEL was used to select 100 acids based on the diversity of the products. The final library was found to cover 85% of the pharmacophores represented by the entire 12,100 virtual libraries.

Lobanov and Agrafiotis [84] have developed a similar approach to focused library design where each substitution position is considered in turn. They describe an iterative procedure where, in each iteration, the best substituents at one position are determined whereas the substituents at the other positions are held fixed, with this process being repeated for each substituent position in turn. The fitness of the final library is then compared and if it is an improvement over the previous cycle, the process is repeated starting with the first substituent; if not, the algorithm terminates.

Several groups have used cherry picking techniques to select promising products. Cherry picking does not take direct account of the combinatorial constraint required in combinatorial synthesis. Hence, these methods suffer from the problem of synthetic inefficiency from the viewpoint of combinatorial synthesis. For example, synthesizing the nine products highlighted in Fig. 4a using combinatorial synthesis would require seven reagents from pool A and six reagents from pool B, leading to a combinatorial library of 42 (7 × 6) compounds. Thus, these approaches typically involve a second step in which combinatorial libraries are designed using reagents that occur frequently in the selected products.

Sheridan and Kearsley [85] and Sheridan et al. [86] developed a product-based approach to focused library design that does not require full enumeration. They used a GA to optimize a population of molecules in product space where each molecule is similar to a target. The final population of molecules is then decomposed into the constituent reagents and a combinatorial library can be designed based on reagents that occur frequently within the selected products. The method was tested on the construction of tripeptoid libraries where there are three positions of variability, with 2507 amines available for two of the substitution positions and 3312 for the third position. This represents a virtual library of ~20 billion possible tripeptoids. The GA was able to find molecules that were very similar to given target molecules after exploring a very small fraction of the total search space. The approach is based on optimizing individual molecules and is valid when the fitness function involves pairwise molecular comparisons, for example, with a target molecule. It is not appropriate when the optimization is based on library properties such as diversity or distributions of physicochemical property profiles.

A similar approach has also been developed in the program Focus-2D [87,88], where molecules are described using topological descriptors and are evolved to be similar to a known target compound, or the predicted activity is maximized based on a precomputed QSAR. Both a GA and simulated annealing have been implemented as optimization techniques. Combinatorial libraries are designed following Monomer Frequency Analysis of the product molecules obtained with monomers or reagents that occur frequently in products being considered for the combinatorial synthesis.

4.3. Combinatorial Subset Selection

Taking account of the combinatorial constraint requires methods that are able to select a combinatorial subset directly from within product space, as illustrated in Fig. 4b. Combinatorial subset selection represents an enormous search space and is typically implemented using an optimization technique such as a GA or SA. For example, there are:

$$\prod_{i=1}^{R} \frac{N_i!}{n_i!(N_i - n_i)!}$$

possible combinatorial subsets for a combinatorial reaction consisting of R components with n_i reagents to be selected from N_i available in the ith reagent pool.

The SELECT program [89] for combinatorial subset selection is based on a GA where each chromosome of the GA encodes one possible combinatorial subset. Assume a two-component combinatorial synthesis in which n_1 of a possible N_1 first reagents are to be reacted with n_2 of a possible N_2 second reagents. The chromosome of the GA contains $n_1 + n_2$ elements, with each element specifying one possible reagent and the crossproduct of the two sets of reagents specifying one of the possible $n_1 n_2$ combinatorial libraries that could be made. The fitness function scores the quality of the sublibrary encoded in a chromosome and the GA evolves new subsets in an attempt to maximize quality. Initially, SELECT was designed to optimize distance-based diversity measures such as the sum of pairwise dissimilarities using 2D fingerprints; more recent versions include other diversity measures such as a cell-based method.

Other approaches to product-based library design include the GALOPED program developed by Brown and Martin [90] that is also based on a GA. Their method was developed for the design of diverse combinatorial libraries synthesized as mixtures, where several compounds are synthesized and screened in the same vessel. The synthesis of mixtures allows much higher throughputs to be achieved than the more commonly used one-compound-per-vessel approach. If activity is seen in a vessel, the mixture must be deconvoluted into individual compounds. Deconvolution can be achieved using mass spectroscopy techniques, where the amount of resynthesis and testing is minimized by reducing the redundancy in molecular weights. GALOPED was thus designed to optimize mixtures based on their diversity simultaneously with ease of deconvolution. Each chromosome encodes a combinatorial subset as a binary string, with the number of bits in a chromosome equal to the sum of reagents available in each reagent pool (i.e., $N_1 + N_2$). Thus, a virtual library of 1000×1000 potential products will require chromosomes with 2000 bits. A bit value of "1" indicates that a reagent is included in the combinatorial subset and a value of "0" indicates that the reagent has not been selected. The size of the subset selected can vary according to the number of bits set to "1"; thus, minimum and maximum thresholds are set by the user and libraries outside the desired size range are penalized in the fitness function.

Lewis and Good [91] describe both GA and SA versions of the program Rpick for designing libraries maximized on pharmacophore coverage. Methods for maximizing pharmacophore coverage have a tendency to select highly functionalized, flexible, and high-molecular-weight reagents because these tend to form products with large numbers of pharmacophores. Therefore, it is usually desirable to include additional constraints in the optimization procedure. Subsequently, the HARPick program [20] was developed to design libraries based on a number of terms including pharmacophore coverage, flexibility of molecules, and optimization to distributions of pharmacophores so that the method can be used for void filling or filling out underrepresented pharmacophores.

Beno and Mason [19] describe a product-based method based on simulated annealing that simultaneously optimizes four-point pharmacophore coverage and BCUT diversity. The virtual library is preenumerated; however, the four-point pharmacophores are calculated on-the-fly, that is, during the optimization itself. The approach was used to select 20 carboxylic acids and 20 amines from a virtual library of 86,140 amines (292 acids and 295 amines). The library was optimized on pharmacophore coverage simultaneously with diversity in BCUT space. They found a 20–23% increase in BCUT cell coverage and a 1.8- to 2.6-fold increase in the number of pharmacophores covered compared with randomly selected reagents.

Multipoint pharmacophore approaches to library design are computationally demanding because they require the exploration of conformational space for each product molecule prior to calculating the pharmacophore keys. Thus, Beno and Mason [19] report that the four-point pharmacophore method is currently limited to selecting subsets of a few hundred compounds.

Mason et al. [21] have adapted the four-point pharmacophore methods to take account of "privileged" substructures. Privileged substructures are substructures able to provide high-affinity ligands for more than one type of receptor or enzyme. One of the four pharmacophoric features is forced to be a special feature associated with the privileged substructure itself. For example, a dummy atom is assigned as the centroid of the substructure and all pharmacophoric patterns must include this dummy atom. Thus, similarity and diversity measures can be focused around the privileged substructure.

Other examples of product-based approaches to library design that are based on SA include the PICCOLO program [92] and the method developed by Brown et al. [34,93].

An alternative approach to product-based library design has been developed in the PLUMS program for the design of focused libraries [94]. A library of compounds is first screened to identify privileged structures (e.g., structures that have properties within a given range or that fit a 3D pharmacophore). The starting point for PLUMS is then the combinatorial library that contains all the virtual hits. Monomers are removed from the library iteratively, with the worst monomer being removed in each iteration. The worst monomer is the one that adds least value to the library. The resulting library is optimized according to a user-defined balance between effectiveness and efficiency. Effectiveness is determined by the number of compounds in the library that are virtual hits, and efficiency is the ratio of the number of privileged molecules to the total size of the library. Related approaches have also been described by Stanton et al. [95] and Pickett et al. [96].

4.4. Multiobjective Library Design

It is now recognized that combinatorial library design is a multiobjective optimization problem [e.g., when designing libraries that are diverse or focused, it is also important to take into account other criteria such as the physicochemical properties of the libraries, the ease of synthesis, and the availability (or cost) of reagents]. The importance of optimizing several properties was highlighted by the disappointing results seen for early combinatorial libraries where many of the hits had physicochemical properties that make them undesirable as drug candidates.

It is relatively easy to include additional library properties such as druglike physicochemical property profiles in the product-based approaches to combinatorial subset selection. For example, several groups have handled multiobjectives through the use of weighted-sum fitness functions. In the SELECT program [89], multiple objectives are handled via a fitness function such as the one shown below:

$$f(n) = w_1\text{diversity} + w_2\text{cost} + w_3\text{property1} + w_4\text{property2} + \cdots$$

Using such a function, SELECT can be configured to design to a library consisting of a diverse set of druglike compounds that are relatively cheap to make. Diversity is typically measured via a distance-based or a cell-based diversity function. The physicochemical property profiles could include properties such as molecular weight or $C \log P$, and they are optimized by minimizing the difference between the distribution of a property in the library and some reference distribution (e.g., the distribution of the property in a collection of known drugs). Each of the properties is normalized and relative weights are defined by the user at run time.

The benefits of performing multiobjective library design are shown in Fig. 5 where SELECT was configured to design 30×30 combinatorial subsets from a two-component amide library consisting of 100 amines and 100 carboxylic acids (representing a virtual library of 10,000 amides). Two runs of SELECT were performed: in the first, the combinatorial subset (Subset 1) was optimized on diversity alone; in the second run, the subset (Subset 2) was optimized on both diversity and molecular weight profile. It can be seen that a more druglike molecular weight profile is achieved for Subset 2, albeit at the expense of a small amount of diversity.

A similar approach to multiobjective library design has been implemented in several other library design programs [20,92,93,96].

Recently, however, the limitations of using a weighted sum to handle multiple objectives in library design have been highlighted. These are summarized as follows:

- The setting of appropriate weights is often nonintuitive; in the SELECT program, it is often done by trial and error [94].
- When the objectives to be optimized are of different types (e.g., diversity and cost), it is not obvious how they should be combined.
- When there are more than two objectives, it is difficult to monitor the progress of the search.
- The objectives to be optimized are often conflicting and a weighted-sum fitness function results in a single solution that represents one compromise in the objectives.

Some of these limitations are illustrated in Fig. 6, which shows the results of a number of runs of SELECT for the amide library optimized on diversity and molecule

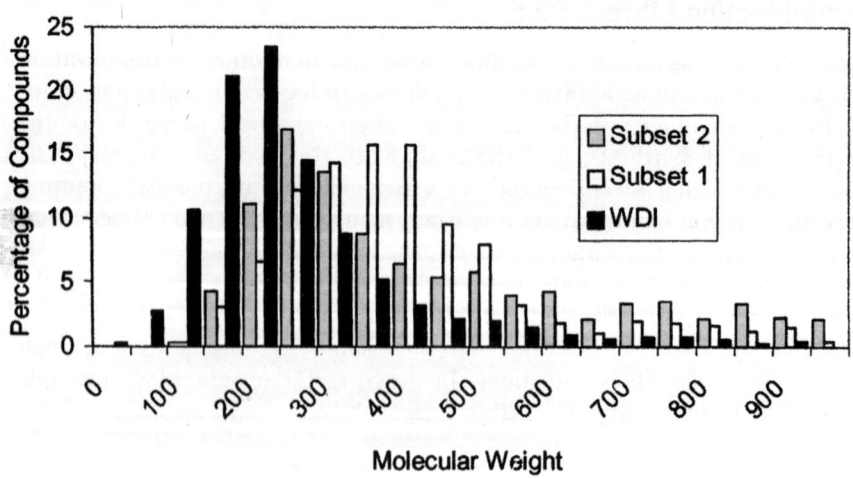

Figure 5   The molecular weight profile of a library optimized on diversity alone (Subset 1) is compared with the profiles found in a library simultaneously optimized on diversity and molecular weight profile (Subset 2) and the World Drugs Index.

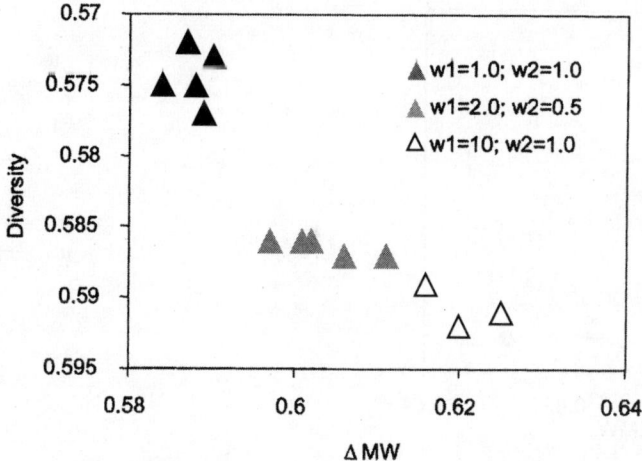

Figure 6 Diversity is plotted on the *y*-axis with the normal direction of the axis reversed, and the difference in the molecular weight profile of the library and the profile found in the WDI is plotted on the *x*-axis. Thus, the direction of improvement of both objectives is toward the bottom left-hand corner.

weight profile, where the relative weights of the two objectives are varied. Three series of runs were performed with: equal weights (black triangles); with $w_1 = 2.0$ and $w_2 = 0.5$ (grey triangles); and with $w_1 = 10$ and $w_2 = 1.0$ (white triangles). The runs show that as the relative weight given to diversity increases, there is a tendency for SELECT to find more diverse libraries, but that this is achieved at the expense of the molecular weight profile. So it can be seen that the two objectives are in competition and that, in fact, a family of solutions exists. A single run of SELECT will find one solution whose position in the objective space depends on the relative weights assigned to the properties being optimized. The conflicting nature of objectives in library design has also been highlighted for pharmacophore methods, where highly flexible molecules are less desirable as lead compounds (less druglike) they lead to better pharmacophore coverage [91].

The MoSELECT program [98–100] is a recent development of SELECT in which the GA is replace by a MultiObjective Genetic Algorithm (MOGA) [101]. The MOGA exploits the population nature of the GA and searches for multiple solutions in parallel. Each objective is handled independently without summation and without the need to choose relative weights, and a set of nondominated solutions is sought rather than a single solution. A nondominated solution is one where an improvement in one objective results in the deterioration in one or more of the other objectives when compared with the other solutions in the population. Thus, one solution dominates another if it is either equivalent or better in all the objectives and, strictly, it is better in at least one objective. Fig. 7 shows the results of a MoSELECT run for the same amide library design problem in Fig. 6, where it can be seen that MoSELECT is able to find an entire family of equivalent solutions in a single run. The library designer can then make an informed choice on what is an appropriate compromise solution. MoSELECT has been applied to the design of both diverse and focused libraries.

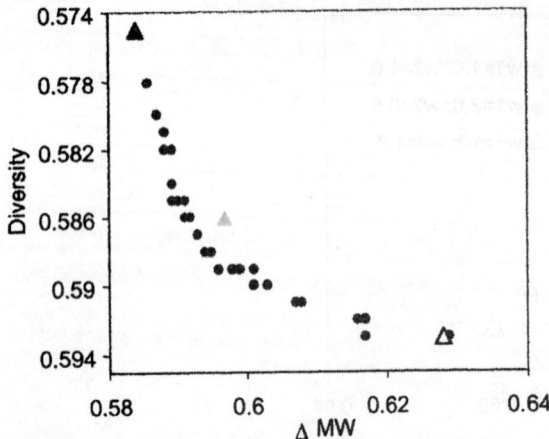

Figure 7   A single run of MoSELECT finds a family of different compromise solutions.

5. SUMMARY

The chapter has examined a variety of computational approaches to virtual screening and combinatorial library design. In particular, reagent-based and product-based approaches to combinatorial library design have been compared and it has been noted that, despite the higher computational cost, product-based approaches can be more effective for the design of diverse libraries. Further advantages of product-based approaches are that additional library-based properties such as physicochemical property profiles can be optimized by allowing multiobjective library design, and that whole molecule properties, which are appropriate when designing focused libraries, can be optimized. Recent efforts have focused on ways of classifying druglike and leadlike compounds, and these methods are now being incorporated within library design. Although diverse libraries are still appropriate when little is known about the biological target, or when libraries are to be screened against a range of targets, the current trend is moving away from very large, diverse libraries toward smaller, more focused libraries. Future trends are likely to be in the development of more accurate ADME models that can be used to eliminate compounds with undesirable physico-chemical properties as early in the design process as possible, and in the further integration of combinatorial chemistry with structure-based drug design techniques, which are currently limited by the computational cost of exploring conformational space.

REFERENCES

1.   Eglen RM, Schneider G, Bohm H-J. High-throughput screening and virtual screening: entry points to drug discovery. In: Böhm H-.J, Schneider G, eds. Virtual Screening for Bioactive Molecules. Weinheim: Wiley-VCH, 2000.
2.   Cramer RD, Patterson DE, Clark RD, Soltanshahi F, Lawless MS. Virtual compound libraries: a new approach to decision making in molecular diversity research. J Chem Inf Comput Sci 1998; 38:1010–1023.

3. Walters WP, Stahl MT, Murcko MA. Virtual screening—an overview. Drug Discov Today 1998; 3:160–178.

4. Böhm H-.J, Schneider G, eds. Virtual Screening for Bioactive Molecules. Weinheim: Wiley-VCH, 2000.

5. Leach AR, Hann MM. The in silico world of virtual libraries. Drug Discov Today 2000; 5:326–336.

6. Martin EJ, Critchlow RE. Beyond mere diversity: tailoring combinatorial libraries for drug discovery. J Com Chem 1999; 1:32–45.

7. Valler MJ, Green D. Diversity screening versus focussed screening in drug discovery. Drug Discov Today 2000; 5:286–293.

8. Johnson MA, Maggiora GM, eds. Concepts and Applications of Molecular Similarity. New York: Wiley, 1990.

9. Brown RD. Descriptors for diversity analysis. Perspect Drug Discov Des 1997; 7/8:31–49.

10. Barnard JM, Downs GM, Willett P. Chemical similarity searching. J Chem Inf Comput Sci 1998; 38:983–996.

11. Molconn-Z. Massachusetts: Hall Associates.

12. MACCS II. San Leandro, CA: Molecular Design Ltd.

13. Barnard Chemical Information. 46 Uppergate Road, Stannington, Sheffield S6 6BX, UK.

14. Mission Viejo, CA, USA: Daylight Chemical Information Systems, Inc.

15. Tripos Inc., 1699 Hanley Road, St. Louis, MO 63144, USA: UNITY Chemical Information Software.

16. Willett P. Searching for pharmacophoric patterns in databases of three-dimensional chemical structures. J Mol Recognit 1995; 8:290–303.

17. Pickett SD, Mason JS, McLay IM. Diversity profiling and design using 3-D pharmacophores: pharmacophore-derived queries (PDQ). J Chem Inf Comput Sci 1996; 36: 1214–1223.

18. Mason JS, Pickett SD. Partition-based selection. Perspect Drug Discov Des 1997; 7/8:85–114.

19. Beno BR, Mason JS. The design of combinatorial libraries using properties and 3D pharmacophore fingerprints. Drug Discov Today 2001; 6:251–258.

20. Good AC, Lewis RA. New methodology for profiling combinatorial libraries and screening sets: cleaning up the design with HARPick. J Med Chem 1997; 40:3926–3936.

21. Mason JS, Morize I, Menard PR, Cheney DL, Hulme C, Labaudiniere RF. New 4-point pharmacophore method for molecular similarity and diversity applications: overview of the method and applications, including a novel approach to the design of combinatorial libraries containing privileged substructures. J Med Chem 1999; 42:3251–3264.

22. Pearlman RS, Smith KM. Metric validation and the receptor-relevant subspace concept. J Chem Inf Comput Sci 1999; 39:28–35.

23. Pearlman RS, Smith KM. Novel software tools for chemical diversity. Perspect Drug Discov Des 1998; 9–11:339–353.

24. Dunbar JB Jr. Cluster-based selection. Perspect Drug Discov Des 1997; 7/8:51–63.

25. Downs GM, Barnard JM. Clustering methods and their uses in computational chemistry. In: Lipkowitz KB, Boyd DB, eds. Reviews in Computational Chemistry Volume 18. New York, VCH Publishers, 2002:1–40.

26. Lajiness MS. Dissimilarity-based compound selection techniques. Perspect Drug Discov Des 1997; 7/8:65–84.

27. Gillet VJ, Willett P. Dissimilarity-based compound selection for library design. In: Ghose AK, Viswanadhan VN, eds. Principles, Software Tools and Applications in Drug Discovery. New York: Marcel Dekker, 2001:379–398.

28. Holliday JD, Willett P. Definitions of "dissimilarity" for dissimilarity-based compound selection. J Biomol Screen 1996; 1:145–151.
29. Hassan M, Bielawski JP, Hempel JC, Waldman M. Optimization and visualization of molecular diversity of combinatorial libraries. Mol Divers 1996; 2:64–74.
30. Holliday JD, Ranade SS, Willett P. A fast algorithm for selecting sets of dissimilar structures from large chemical databases. QSAR 1996; 15:285–289.
31. Agrafiotis DK, Lobanov VS. An efficient implementation of distance-based diversity measures based on k–d trees. J Chem Inf Comput Sci 1999; 39:51–58.
32. Mount J, Ruppert J, Welch W, Jain AN. IcePick: a flexible surface-based system for molecular diversity. J Med Chem 1999; 42:60–66.
33. Snarey M, Terrett NK, Willett P, Wilton DJ. Comparison of algorithms for dissimilarity-based compound selection. J Mol Graph Model 1997; 15:372–385.
34. Brown RD, Hassan M, Waldman M. Tools for designing diverse, druglike, cost-effective combinatorial libraries. In: Ghose AK, Viswanadhan VN, eds. Principles, Software Tools and Applications in Drug Discovery. New York: Marcel Dekker, 2001:301–335.
35. Mason JS, McLay IM, Lewis RA. Applications of computer-aided drug design techniques to lead generation. In: Dean PM, Jolles G, Newton CG, eds. New Perspectives in Drug Design. London: Academic Press, 1994:225–253.
36. Lewis RA, Mason JS, McLay IM. Similarity measures for rational set selection and analysis of combinatorial libraries: the diverse property-derived (DPD) approach. J Chem Inf Comput Sci 1997; 37:599–614.
37. McGregor MJ, Muskal SM. Pharmacophore fingerprinting. 1. Application to QSAR and focused library design. J Chem Inf Comput Sci 1999; 39:569–574.
38. McGregor MJ, Muskal SM. Pharmacophore fingerprinting. 2. Application to primary library design. J Chem Inf Comput Sci 2000; 40:117–125.
39. Bayley MJ, Willett P. Binning schemes for partition-based compound selection. J Mol Graph Model 1999; 17:10–18.
40. Agrafiotis DK, Rassokhin DN. A fractal approach for selecting an appropriate bin size for cell-based diversity estimation. J Chem Inf Comput Sci 2002; 42:117–122.
41. Martin EJ, Blaney JM, Siani MS, Spellmeyer DC, Wong AK, Moos WH. Measuring diversity—experimental design of combinatorial libraries for drug discovery. J Med Chem 1995; 38:1431–1436.
42. Higgs RE, Bemis KG, Watson IA, Wikel JH. Experimental designs for selecting molecules from large chemical databases. J Chem Inf Comput Sci 1997; 37:861–870.
43. Lobanov VS, Agrafiotis DK. Stochastic similarity selections from large combinatorial libraries. J Chem Inf Comput Sci 2000; 40:460–470.
44. Brown RD, Martin YC. Use of structure–activity data to compare structure-based clustering methods and descriptors for use in compound selection. J Chem Inf Comput Sci 1996; 36:572–584.
45. Brown RD, Martin YC. The information content of 2D and 3D structural descriptors relevant to ligand–receptor binding. J Chem Inf Comput Sci 1997; 37:1–9.
46. Flower DR. On the properties of bit string-based measures of chemical similarity. J Chem Inf Comput Sci 1998; 38:379–386.
47. Schuffenhauer A, Gillet VJ, Willett P. Similarity searching in files of three-dimensional chemical structures: analysis of the BIOSTER·database using two-dimensional fingerprints and molecular field descriptors. J Chem Inf Comput Sci 2000; 40:295–307.
48. Bradley EK, Beroza P, Penzotti JE, Grootenhuis PDJ, Spellmeyer DC, Miller JL. A rapid computational method for lead evolution: description and application to α_1-adrenergic antagonists. J Med Chem 2000; 43:2770–2774.
49. Xue L, Stahura FL, Godden JW, Bajorath J. Mini-fingerprints detect similar activity of receptor ligands previously recognized only by three-dimensional pharmacophore-based methods. J Chem Inf Comput Sci 2001; 41:394–401.

50. Mason JS, Beno BR. Library design using BCUT chemistry-space descriptors and multiple four-point pharmacophore fingerprints: simultaneous optimisation and structure-based diversity. J Mol Graph Model 2000; 18:438–451.

51. Patterson DE, Cramer RD, Ferguson AM, Clark RD, Weinberger LE. Neighborhood behavior: a useful concept for validation of molecular diversity descriptors. J Med Chem 1996; 39:3049–3059.

52. Downs GM, Willett P. Similarity searching in databases of chemical structures. In: Lipkowitz KB, Boyd DB, eds. Reviews in Computational Chemistry. Vol. 7. New York: Wiley-VCH, 1995:1–66.

53. Schneider G, Bohm H-J. Virtual screening and fast automated docking methods. Drug Discov Today 2002; 7:64–70.

54. Makino S, Ewing TJ, Kuntz ID. Dream + +: flexible docking program for virtual combinatorial libraries. J Comput-Aided Mol Des 1999; 13:513–532.

55. Lamb ML, Burdick KW, Toba S, Young MM, Skillman AG, Zou X, Arnold JR, Kuntz ID. Design, docking, and evaluation of multiple libraries against multiple targets. Proteins 2001; 42:296–318.

56. Sun Y, Ewing TJ, Skillman AG, Kuntz ID. CombiDOCK: structure-based combinatorial docking and library design. J Comput-Aided Mol Des 1998; 12:597–604.

57. Böhm H-J, Banner DW, Weber L. Combinatorial docking and combinatorial chemistry: design of potent non-peptide thrombin inhibitors. J Comput-Aided Mol Des 1999; 13:51–56.

58. Böhm H-J, Boehringer M, Bur D, Gmuender H, Huber W, Klaus W, Kostrewa D, Kuehne H, Luebbers T, Meunier-Keller N, Mueller F. Novel inhibitors of DNA gyrase: 3D structure based biased needle screening, hit validation by biophysical methods, and 3D guided optimization. A promising alternative to random screening. J Med Chem 2000; 43:2664–2674.

59. Bissantz C, Folkers G, Rognan D. Protein-based virtual screening of chemical databases: 1. Evaluation of different docking/scoring combinations. J Med Chem 2000; 43:4759–4767.

60. Stahl M, Rarey M. Detailed analysis of scoring functions for virtual screening. J Med Chem 2001; 44:1035–1042.

61. Charifson PS, Corkery JJ, Murcko MA, Walters WP. Consensus scoring: a method for obtaining improved hit rates from docking databases of three-dimensional structures into proteins. J Med Chem 1999; 42:5100–5109.

62. Rarey M, Kramer B, Lengauer T, Klebe G. A fast flexible docking method using an incremental construction algorithm. J Mol Biol 1996; 261:470–489.

63. Kuntz ID, Blaney JM, Oatley SJ, Langridge R, Ferrin TE. A geometric approach to macromolecule–ligand interactions. J Mol Biol 1982; 161:269–288.

64. Jones G, Willett P, Glen RC, Leach AR, Taylor R. Development and validation of a genetic algorithm for flexible docking. J Mol Biol 1997; 267:727–748.

65. Clark DE, Pickett SD. Computational methods for the prediction of "druglikeness." Drug Discov Today 2000; 5:49–58.

66. Beresford AP, Selick HE, Tarbit MH. The emerging importance of predictive ADME simulation in drug discovery. Drug Discov Today 2002; 7:109–116.

67. Leach AR, Bradshaw J, Green DVS, Hann MM, Delany JJ III. Implementation of a system for reagent selection, library enumeration, profiling and design. J Chem Inf Comput Sci 1999; 39:1161–1172.

68. Lipinski CA, Lombardo F, Dominy BW, Feeney PJ. Experimental and computational approaches to estimate solubility and permeability in drug discovery and development settings. Adv Drug Deliv Rev 1997; 23:3–25.

69. Gillet VJ, Willett P, Bradshaw J. Identification of biological activity profiles using substructural analysis and genetic algorithms. J Chem Inf Comput Sci 1998; 38:165–179.

70. Hann M, Hudson B, Lewell X, Lifely R, Miller L, Ramsden N. Strategic pooling of compounds for high-throughput screening. J Chem Inf Comput Sci 1999; 39:897–902.

71. Ajay, Walter WP, Murcko MA. Can we learn to distinguish between "drug-like" and "non-drug-like" molecules? J Med Chem 1998; 41:3314–3324.

72. Sadowski J, Kubinyi H. A scoring scheme for discriminating between drugs and non-drugs. J Med Chem 1998; 41:3325–3329.

73. Frimurer TM, Bywater R, Nærum L, Lauriten LN, Brunak S. Improving the odds in discriminating "drug-like" from "non drug-like" compounds. J Chem Inf Comput Sci 2000; 40:1315–1324.

74. Wang J, Ramnarayan K. Toward designing drug-like libraries: a novel computational approach for prediction of drug feasibility of compounds. J Com Chem 1999; 1:524–533.

75. Oprea TI, Davis AM, Teague SD, Leeson PD. Is there a difference between leads and drugs? A historical perspective. J Chem Inf Comput Sci 2001; 41:1308–1315.

76. Hann MM, Leach AR, Harper G. Molecular complexity and its impact on the probability of finding leads for drug discovery. J Chem Inf Comput Sci 2001; 41:856–864.

77. Roche O, Schneider P, Zuegge J, Guba W, Kansy M, Alanine A, Bleicher K, Danel F, Gutknecht EM, Rogers-Evans M, Neidhart W, Stalder H, Dillon M, Sjogren E, Fotouhi N, Gillespie P, Goodnow R, Harris W, Jones P, Taniguchi M, Tsujii S, von der Saal W, Zimmermann G, Schneider G. Development of a virtual screening method for identification of "frequent hitters" in compound libraries. J Med Chem 2002; 45:137–142.

78. Gillet VJ, Willett P, Bradshaw J. The effectiveness of reactant pools for generating structurally diverse combinatorial libraries. J Chem Inf Comput Sci 1997; 37:731–740.

79. Gillet VJ, Nicolotti O. New algorithms for compound selection and library design. Perspect Drug Discov Des 2000; 20:265–287.

80. Jamois EA, Hassan M, Waldman M. Evaluation of reagent-based and product-based strategies in the design of combinatorial library subsets. J Chem Inf Comput Sci 2000; 40:63–70.

81. Linussen A, Gottfries J, Lindgren F, Wold S. Statistical molecular design of building blocks for combinatorial chemistry. J Med Chem 2000; 43:1320–1328.

82. Barnard JM, Downs GM, von Scholley-Pfab A, Brown R. Use of Markush structure analysis techniques for descriptor generation and clustering of large combinatorial libraries. J Mol Graph Model 2000; 18:452–463.

83. Pickett SD, Luttmann C, Guerin V, Laoui A, James E. DIVSEL and COMPLIB—strategies for the design and comparison of combinatorial libraries using pharmacophoric descriptors. J Chem Inf Comput Sci 1998; 38:144–150.

84. Lobanov VS, Agrafiotis DK. Stochastic similarity selections from large combinatorial libraries. J Chem Inf Comput Sci 2000; 40:460–470.

85. Sheridan RP, Kearsley SK. Using a genetic algorithm to suggest combinatorial libraries. J Chem Inf Comput Sci 1995; 35:310–320.

86. Sheridan RP, SanFeliciano SG, Kearsley SK. Designing targeted libraries with genetic algorithms. J Mol Graph Model 2000; 18:320–334.

87. Zheng W, Cho SJ, Tropsha A. Rational combinatorial library design: 1. Focus-2D A new approach to the design of targeted combinatorial chemical libraries. J Chem Inf Comput Sci 1998; 38:251–258.

88. Cho SJ, Zheng W, Tropsha A. Rational combinatorial library design. 2. Rational design of targeted combinatorial peptide libraries using chemical similarity probe and the inverse QSAR approaches. J Chem Inf Comput Sci 1998; 38:259–268.

89. Gillet VJ, Willett P, Bradshaw J, Green DVS. Selecting combinatorial libraries to optimise diversity and physical properties. J Chem Inf Comput Sci 1999; 39:167–177.

90. Brown RD, Martin YC. Designing combinatorial library mixtures using a genetic algorithm. J Med Chem 1997; 40:2304–2313.

91. Lewis RA, Good AC. Quantification of molecular similarity and its application to combinatorial chemistry. In: van de Waterbeemd H, Testa B, Folkers G, eds. Computer-Assisted Lead Finding and Optimization. Zurich: Wiley-VCH, 1997:137–156.

92. Zheng W, Hung ST, Saunders JT, Seibel GL. PICCOLO: a tool for combinatorial library design via multicriterion optimization. In: Atlman RB, Dunkar AK, Hunter L, Lauderdale K, Klein TE, eds. Pacific Symposium on Biocomputing 2000. Singapore: World Scientific, 2000:588–599.

93. Brown JD, Hassan M, Waldman M. Combinatorial library design for diversity, cost efficiency, and drug-like character. J Mol Graph Model 2000; 18:427–437.

94. Bravi G, Green DVS, Hann MM, Leach AR. PLUMS: a program for the rapid optimization of focused libraries. J Chem Inf Comput Sci 2000; 40:1441–1448.

95. Stanton RV, Mount J, Miller JL. Combinatorial library design: maximizing model-fitting compounds within matrix synthesis constraints. J Chem Inf Comput Sci 2000; 40:701–705.

96. Pickett SD, McLay IM, Clark DE. Enhancing the hit-to-lead properties of lead optimization libraries. J Chem Inf Comput Sci 2000; 40:263–272.

97. Rassokhin DN, Agrafiotis DK. Kolmogorov–Smirnov statistic and its application in library design. J Mol Graph Model 2000; 18:427–437.

98. Gillet VJ, Khatib W, Willett P, Fleming PJ, Green DVS. Combinatorial library design using a MultiObjective Genetic Algorithm. J Chem Inf Comput Sci. 2002; 42:375–385.

99. Gillet VJ, Willett P, Fleming PJ, Green DVS. Designing focused libraries using MoSELECT. J Mol Graph Model 2002; 20:491–498.

100. International Patent Application PCT/GB01/05347.

101. Fonseca CM, Fleming PJ. An overview of evolutionary algorithms in multiobjective optimization. In: De Jong K, ed. Evolutionary Computation. Vol. 3, No. 1. Boston: Massachusetts Institute of Technology, 1995:1–16.

24

Quantum-Chemical Descriptors in QSAR

MATI KARELSON

University of Tartu, Tartu, Estonia

1. INTRODUCTION

The quantitative structure–activity and structure–property relationships (QSARs/QSPRs) have become efficient tools to study complex chemical and biochemical systems. In principle, once a correlation between the molecular structure and activity/property is found, any number of compounds, including those not yet synthesized, can be readily screened on the computer in order to predict the structures with the desired properties. Subsequently, the respective chemical compounds can be synthesized and tested in the laboratory. Thus the QSAR/QSPR approach conserves resources and accelerates the process of development of new drugs, materials, etc. In addition to the structure predictions, the QSAR/QSPR models often help to understand the physical nature of the processes and interactions behind the property or activity studied.

The predictive power, robustness, and reliability of the QSAR/QSPR models depend critically on the use of appropriate molecular descriptors. A myriad of descriptors, either empirical or those calculated on the basis of the molecular structure alone ("theoretical" descriptors), have been developed both for the predictive and cognitive purposes [1,2]. Many of those descriptors are based directly on the results of quantum-mechanical calculations or can be derived from the electronic wave function or electrostatic field of the molecule. It is the purpose of the present chapter to give an overview of such molecular descriptors, together with some key applications.

The modern computer hardware power and software enables one to calculate realistic quantum-chemical molecular characteristics in a relatively short computational time. Depending on the size of the molecular system studied, either the semiempirical or the ab initio level of algorithms would be applicable. In general, the ab initio model Hamiltonians represent all nonrelativistic interactions between the nuclei and electrons in a molecule. The computer time is proportional to a high exponential (N^4, N^5) of the number of electrons in the molecule, N, that is usually limiting the calculations to short series of relatively small molecules (10–20 atoms). The Hartree–Fock method employed in basic ab initio calculations is expected to provide better results the larger the basis set used. However, according to the variational principle, this is strictly valid only for the total electronic energy of the molecule. Other molecular properties, particularly those depending on the electron distribution in the molecule (electrical moments and partial charges on atoms), are related to the size of the basis set more vaguely. In the case of such properties, special attention has to be paid to the balance of the basis set used [3]. Various ab initio methods beyond Hartree–Fock [4,5] that account for electron correlation in a molecule [configuration inter-action (CI), multiconfigurational self-consistent field (MC SCF), correlated pair many-electron theory (CPMET) and its various coupled-cluster approximations, perturbation theory, etc.] are much more demanding in terms of computer time and memory. Therefore those would be impractical for the calculation of extended sets of relatively large molecules required in most QSAR model developments.

As an alternative to ab initio methods, the semi-empirical quantum-chemical methods are fast and applicable for the calculation of molecular descriptors of long series of structurally complex and large molecules. Most of these methods have been developed within the mathematical framework of the molecular orbital theory (SCF MO), but use a number of simplifications and approximations in the computational procedure that reduce dramatically the computer time [6]. The most popular semi-empirical methods are Austin Model 1 (AM1) [7] and Parametric Model 3 (PM3) [8]. The results produced by different semi-empirical methods are generally not compa-rable, but they often do reproduce similar trends. For example, the electronic net charges calculated by the AM1, MNDO (modified neglect of diatomic overlap), and INDO (intermediate neglect of diatomic overlap) methods were found to be quite different in their absolute values, but were consistent in their trends. Intermediate between the ab initio and semi-empirical methods in terms of the demand in computa-tional resources are algorithms based on density functional theory (DFT) [9].

2. QUANTUM-CHEMICAL MOLECULAR DESCRIPTORS

2.1. Energy-Related Descriptors

The calculated total energy of the molecule and its different partitionings can be used as theoretical molecular descriptors. The quantum-chemically calculated total energy of the molecule (E_{tot}) refers usually to the quantum mechanical standard state for the energy (isolated electrons and nuclei at 0 K). Within the Born–Oppenheimer approx-imation, it can be divided into total electronic energy, E_{el}, and the nuclear–nuclear electrostatic repulsion energy.

$$E_{tot} = E_{el} + \sum_{A \neq B} \frac{Z_A Z_B}{R_{AB}}$$

Proceeding from the mathematical structure of the molecular Hamiltonian, various further partitionings of both the total energy and the total electronic energy of a molecule are possible. The electron–electron repulsion energy and the nuclear–

electron attraction energy for a given atomic species (atom A) in the molecule can be calculated within the MO method using the elements of the density matrix, the electron repulsion integrals $\{\mu\nu|\lambda\sigma\}$, and the nuclear–electron attraction integrals $\langle\mu|Z_B/R_{iB}|\nu\rangle$ on the given atomic basis $\{\mu\nu\lambda\sigma\}$ (cf. Table 1). Those energies may describe the electron–electron repulsion or nuclear–electron attraction-driven processes in the molecule and may be related to the conformational changes (rotation and inversion) or atomic reactivity in the molecule. As related to a given atom, it may specify the site of a particular chemical activity or conformational change in the molecule. The atomic valence state energies for a given atomic species in a molecule and its various partitionings could also be useful molecular descriptors. The valence state energy characterizes the magnitude of the perturbation experienced by an atom in the molecular environment as compared to the isolated atom. In the framework of transferable atom equivalent (TAE) method [10], two formulations have been proposed for the electron kinetic energy density (cf. Table 1).

Within the LCAO MO (linear combination of atomic orbitals molecular orbit) theory, the electronic exchange energy can be calculated between two given atoms (A and B) in the molecule that reflects the change in the Fermi correlation energy between the two electrons localized on atoms A and B, respectively. It can also be related to the conformational changes of the molecule. In the framework of the semi-empirical quantum-chemical theories, the resonance energy between two given atomic species in the molecule can also be defined using the resonance integrals on the atomic basis, $\beta_{\mu\nu}$. Furthermore, the energy of intramolecular electron–electron repulsion, the nuclear–electron attraction energy, or the nuclear repulsion energy between two given atoms (atoms A and B) in the molecule can be calculated and applied as molecular descriptors. Those descriptors relate to a certain chemical bond (in the case of bonded atoms) or to intramolecular nonbonded interactions. The extreme, i.e., the maximum or the minimum values of such energies for a given atomic species or for a pair of given atomic species, can be considered as global descriptors for a compound.

The heat of formation of the molecule (ΔH_f°) that refers to the energy of the molecule in the thermodynamic standard scale (chemical elements in ideal gas state at 298.15 K and 101,325 Pa) can also be calculated and used as a molecular descriptor. Analogously, the calculated reaction energies or activation energies serve also as specific chemical reactivity-oriented theoretical descriptors. For instance, the energy of protonation has been related to the strength of the hydrogen-bonding acceptance of a compound. Proceeding from the results of the modeling of the molecular potential energy surface, it is possible to calculate the vibrational and rotational energies of the molecule or its fragments. This information also enables one to calculate the total partition function of the molecule Q, its electronic, translational, rotational, and vibrational components and the respective thermodynamic functions [11]. A special descriptor, called EVA (normal coordinate eigenvalues), has been proposed as a characteristic of the standardized vibrational frequencies in the molecule [12].

2.2. Electrostatic Descriptors

Many molecular properties, especially those measured in condensed media, depend on intermolecular interactions. The main term of such interactions is, in most cases, the electrostatic attraction or repulsion between the molecules or their different parts. Consequently, the respective electrostatic descriptors characterizing a molecule would

Table 1 Quantum Chemical Molecular Description

Descriptor	Definition	Reference
Energy-related descriptors		
Total energy	$E_t = E_{el} + \sum_{A \neq B} Z_A Z_B / R_{AB}$ E_{el}—total electronic energy of the molecule Z_A, Z_B—nuclear charges of atoms A and B R_{AB}—distance between nuclei A and B	57
Electronic energy	$E_{el} = 2\text{Tr}(RF) - \text{Tr}(RG)$ R—first-order density matrix F—matrix representation of the Hartree–Fock operator G—matrix representation of the electron repulsion energy	57
Binding energy	$E_b = E_t - \sum_i^N E_{A_i}$ E_{A_i}—energies of atoms A_i	57
Heat of formation	$\Delta H_f^0 = H_f - \sum_A H_f^A$ H_f—quantum-chemically calculated heat of formation of the molecule H_f^A—quantum-chemically calculated heats of formation of isolated atoms, A	58
Ionization energy	$\text{IE} = E_t(A^+) - E_T(A)$	22
Electron affinity	$\text{EA} = E_t(A) - E_T(A^-)$	22
Energy of protonation	$\Delta E = E_t(BH^+) - E_T(B)$	59
Difference in the heat of formation between acid AH and the corresponding anion A^-	$\Delta(\Delta H_f^0) = \Delta H_f^0(AH) - \Delta H_f^0(A^-)$	60
Electron–electron repulsion energy for a given atomic species, A ($B \neq A$)	$E_{ee}(A) = \sum_{B \neq A} \sum_{\mu,\nu \in A} \sum_{\lambda,\sigma \in B} P_{\mu\nu} P_{\lambda\sigma} \langle \mu\nu | \lambda\sigma \rangle P_{\mu\nu}$ $P_{\lambda\sigma}$—density matrix elements over atomic basis $\{\mu\nu\lambda\sigma\}$ $\langle \mu\nu | \lambda\sigma \rangle$—electron repulsion integrals on atomic basis $\{\mu\nu\lambda\sigma\}$	61
Nuclear–electron attraction energy for a given atomic species, A ($B \neq A$)	$E_{ne}(A) = \sum_B \sum_{\mu,\nu \in A} P_{\mu\nu} \left\langle \mu \left| \frac{Z_B}{R_{iB}} \right| \nu \right\rangle$ $\left\langle \mu \left| \frac{Z_B}{R_{iB}} \right| \nu \right\rangle$—electron–nuclear attraction integrals on atomic basis $\{\mu\nu\}$	61
Electron–electron repulsion between two given atoms, A and B	$E_{ee}(AB) = \sum_{\mu,\nu \in A} \sum_{\lambda,\sigma \in B} P_{\mu\nu} P_{\lambda\sigma} \langle \mu\nu | \lambda\sigma \rangle$	61
Nuclear–electron attraction between two given atoms, A and B	$E_{ne}(AB) = \sum_B \sum_{\mu,\nu \in A} P_{\mu\nu} \left\langle \mu \left| \frac{Z_B}{R_{iB}} \right| \nu \right\rangle$	61
Nuclear repulsion energy between two given atoms	$E_{nn}(AB) = \frac{Z_A Z_B}{R_{AB}}$	61

Table 1 Continued

Descriptor	Definition	Reference
Electronic exchange energy between two given atoms, A and B	$E_{\mathrm{exc}}(AB) = \sum\limits_{\mu,\nu \in A} \sum\limits_{\lambda,\sigma \in B} P_{\mu\lambda} P_{\nu\sigma} \langle \mu\lambda \vert \nu\sigma \rangle$	61
Resonance energy between two given atomic species, A and B	$E_{\mathrm{R}}(AB) = \sum\limits_{\mu \in A} \sum\limits_{\nu \in B} P_{\mu\nu} \beta_{\mu\nu}$ $\beta_{\mu\nu}$—resonance integrals on atomic basis $\{\mu\nu\}$	61
Total electrostatic interaction energy between two given atomic species, A and B	$E_{\mathrm{C}}(AB) = E_{\mathrm{ee}}(AB) + E_{\mathrm{ne}}(AB) + E_{\mathrm{nn}}(AB)$	61
Total interaction energy between two given atomic species, A and B	$E_{\mathrm{tot}}(AB) = E_{\mathrm{C}}(AB) + E_{\mathrm{exc}}(AB)$	61
Total molecular one-center electron–electron repulsion energy	$E_{\mathrm{ee}}(\mathrm{tot}) = \sum\limits_{A} E_{\mathrm{ee}}(A)$	61
Total molecular one-center electron–nuclear attraction energy	$E_{\mathrm{ne}}(\mathrm{tot}) = \sum\limits_{A} E_{\mathrm{ne}}(A)$	61
Total intramolecular electrostatic interaction energy	$E_{\mathrm{C}}(\mathrm{tot}) = \dfrac{1}{2} \sum\limits_{A} E_{\mathrm{C}}(A)$	61
Electron kinetic energy density	$K = -\dfrac{N}{4} \int \left(\Psi^* \nabla^2 \Psi + \Psi \nabla^2 \Psi^* \right) \mathrm{d}_{r'}^{r}$ $G = \dfrac{N}{2} \int \nabla\Psi^* \nabla\Psi \mathrm{d}_{r'}^{r}$ N—number of electrons in the molecule Ψ—electronic wave function of the molecule	62
EVA (normal coordinate eigenvalues)	$\mathrm{EVA}_x = \sum\limits_{i=1}^{3N-6} \dfrac{1}{\sigma\sqrt{2\pi}} \exp\left[-\dfrac{(x-f_i)^2}{2\sigma^2} \right]$ f_i—the vibrational frequencies of the molecule σ—the fixed standard deviation for all Gaussian functions characterizing the shape of the vibrational peak	63
Electrostatic descriptors		
Net atomic charge on atom A (Mulliken charge)	$Q_A = Z_A - \sum P_{\mathrm{kl}}$ Z_A—atomic nuclear charge P_{kl}—atomic population matrix elements	57
Net charges of the most negative and most positive atoms	$Q_{A,\mathrm{min}}, \, Q_{A,\mathrm{max}}$	64, 65
Net group charge on atoms A and B	Q_{AB}	66, 67
Sum of squared charge densities on atoms of type A	$\sum Q_A^2$	54
π- and σ-electron densities of the atom A	$q_{A,\sigma}, \, q_{A,\pi}$	19, 68
Sum of absolute values of the charges of all the atoms in a given molecule or functional group	QT, QA	20, 69

Table 1 Continued

Descriptor	Definition	Reference
Sum of squares of the charges of all the atoms in a given molecule or functional group	QT2, QA2	20
Mean absolute atom charge (i.e., the average of the absolute values of the charges on all atoms)	$Q_m = \sum_A \dfrac{Q_A}{N}$	64
Relative positive charge	$RPCG = \dfrac{Q_{max}^+}{\sum_A Q_A} \quad A \in \{\delta_A > 0\}$ Q_{max}^+—maximum atomic positive charge in the molecule	28, 29
Relative negative charge	$RNCG = \dfrac{Q_{min}^-}{\sum_A Q_A} \quad A \in \{\delta_A < 0\}$ Q_{min}^-—minimum atomic negative charge in the molecule	28, 29
Molecular dipole moment	μ	22
Charge and hybridization dipole moments	μ_{char}, μ_{hybr}	70
Square of the molecular dipole moment	μ^2	69, 71
Components of dipole moment along axes of inertia	D_X, D_Y, D_Z	69
Submolecular polarity parameter (largest difference in electron charges between two atoms)	$\Delta = Q_{max} - Q_{min}$	26
Topological electronic index	$T^E = \sum_{ij, i \neq j} \dfrac{|Q_i - Q_j|}{r_{ij}^2}$ r_{ij}—interatomic distances	27
Local dipole index, sum over all connected pairs of atoms	$D = \sum_{A,B} |Q_A - Q_B| / N_{AB}$	19, 64
Quadrupole moment tensor	τ	72
Self-atom polarizabilities π_{AA} and atom–atom polarizabilities π_{AB}	$\pi_{AA}, \pi_{AB} = 4 \sum_i \sum_a \sum_p \sum_r \dfrac{C_{pi}^A C_{pa}^A C_{ri}^B C_{ra}^B}{\varepsilon_i - \varepsilon_a}$ summation over MOs (i,a) and over valence AOs (p,r)	44
Sum of self-atom polarizabilities	$\sum \pi_{AA}$	44
Molecular polarizability	α	22, 70
Mean polarizability of the molecule	$\alpha = \dfrac{1}{3}(\alpha_{xx} + \alpha_{yy} + \alpha_{zz})$	69
Anisotropy of the polarizability	$\beta^2 = \dfrac{1}{2}\left[(\alpha_{xx} - \alpha_{yy})^2 + (\alpha_{yy} - \alpha_{zz})^2 + (\alpha_{zz} - \alpha_{xx})^2\right]$	69
Polarization of molecule, sum of net atomic charges over all atoms in a molecule	$P = \sum_{A=1}^{N} |Q_A| / N$	19
Polarizability tensor	α	72

Table 1 Continued

Descriptor	Definition	Reference
Partial positively charged surface area	$PPSA1 = \sum_A S_A \quad A \in \{\delta_A > 0\}$ S_A—positively charged solvent-accessible atomic surface area	28, 29
Total charge weighted partial positively charged surface area	$PPSA2 = \sum_A Q_A \cdot \sum_A S_A \quad A \in \{\delta_A > 0\}$	28, 29
Atomic charge weighted partial positively charged surface area	$PPSA3 = \sum_A Q_A \cdot S_A \quad A \in \{\delta_A > 0\}$	28, 29
Partial negatively charged surface area	$PNSA1 = \sum_A S_A \quad A \in \{\delta_A < 0\}$ S_A—negatively charged solvent-accessible atomic surface area	28, 29
Total charge weighted partial negatively charged surface area	$PNSA2 = \sum_A Q_A \cdot \sum_A S_A \quad A \in \{\delta_A < 0\}$	28, 29
Atomic charge weighted partial negatively charged surface area	$PNSA3 = \sum_A q_A \cdot S_A \quad A \in \{\delta_A < 0\}$	28, 29
Difference between partial positively 1 and negatively charged surface areas	$DPSA1 = PPSA1 - PNSA1$	28, 29
Difference between total charge weighted partial positively and negatively charged surface areas	$DPSA2 = PPSA2 - PNSA2$	28, 29
Difference between atomic charge weighted partial positively and negatively charged surface areas	$DPSA3 = PPSA3 - PNSA3$	28, 29
Fractional partial positive surface area	$FPSA1 = \dfrac{PPSA1}{TMSA}$ TMSA—total molecular surface area	28, 29
Fractional total charge weighted partial positive surface area	$FPSA2 = \dfrac{PPSA2}{TMSA}$	28, 29
Fractional atomic charge weighted partial positive surface area	$FPSA3 = \dfrac{PPSA3}{TMSA}$	28, 29
Fractional partial negative surface area	$FNSA1 = \dfrac{PNSA1}{TMSA}$	28, 29
Fractional total charge weighted partial negative surface area	$FNSA2 = \dfrac{PNSA2}{TMSA}$	28, 29
Fractional atomic charge weighted partial negative surface area	$FNSA3 = \dfrac{PNSA3}{TMSA}$	28, 29
Surface weighted partial positive charged surface area WPSA1	$WPSA1 = \dfrac{PPSA1 \times TMSA}{1000}$	28, 29
Surface weighted partial positive charged surface area WPSA2	$WPSA2 = \dfrac{PPSA2 \times TMSA}{1000}$	28, 29
Surface weighted partial positive charged surface area WPSA3	$WPSA3 = \dfrac{PPSA3 \times TMSA}{1000}$	28, 29
Surface weighted partial negative charged surface area WNSA1	$WNSA1 = \dfrac{PNSA1 \times TMSA}{1000}$	28, 29
Surface weighted partial negative charged surface area WNSA2	$WNSA2 = \dfrac{PNSA2 \times TMSA}{1000}$	28, 29

Table 1 Continued

Descriptor	Definition	Reference
Surface weighted partial negative charged surface area WNSA3	$\text{WNSA3} = \dfrac{\text{PNSA3} \times \text{TMSA}}{1000}$	28, 29
Average ionization energy	$\bar{I}(r) = \dfrac{\sum_i \rho_i(r)|\varepsilon_i|}{\rho(r)}$ $\rho(\mathbf{r})$—electron density of the ith molecular orbital at the point $\mathbf{r}$ ε_i—ith molecular orbital energy	32
Minimum electrostatic potential at the molecular surface	$V_{S,\min} = \min[V(r)] = \min\left[\sum_A \dfrac{Z_A}{|\mathbf{R}_A - \mathbf{r}|} - \int \dfrac{\rho(\mathbf{r}')d\mathbf{r}'}{|\mathbf{r}' - \mathbf{r}|}\right]$ Z_A—charge on atomic nucleus A at point $\mathbf{R}_A$ $\rho(\mathbf{r})$—total electron density of the molecule	31, 73, 74
Maximum electrostatic potential at the molecular surface	$V_{S,\max} = \max\left[\sum_A \dfrac{Z_A}{|\mathbf{R}_A - \mathbf{r}|} - \int \dfrac{\rho(\mathbf{r}')d\mathbf{r}'}{|\mathbf{r}' - \mathbf{r}|}\right]$	31, 73, 74
Local polarity of molecule	$\Pi = \dfrac{1}{A}\int_S |V(r) - \bar{V}_S|\,dS \approx \dfrac{1}{n}\sum_{i=1}^n |V_i(r) - \bar{V}_S|$ A—molecular surface area $\bar{V}_S$—average value of the electrostatic potential in the molecule $V(\mathbf{r})$—electrostatic potential in the molecule n—number of integration points	31, 73, 74
Total variance of the surface electrostatic potential	$\sigma_{\text{tot}}^2 = \sigma_+^2 - \sigma_-^2 = \dfrac{1}{m}\sum_{i=1}^m \left[V^+(\mathbf{r}_i) - \bar{V}_S^+\right]^2$ $+ \dfrac{1}{n}\sum_{i=1}^n \left[V^-(\mathbf{r}_i) - \bar{V}_S^-\right]^2$ $\bar{V}_S^+, \bar{V}_S^-$—average value of the positive and negative electrostatic potentials in the molecule, respectively $V(\mathbf{r}_i)^+, V(\mathbf{r}_i)^-$—positive and negative electrostatic potentials in the molecule m, n—number of integration points	31, 73, 74
Electrostatic balance parameter	$\nu = \dfrac{\sigma_+^2 \sigma_-^2}{[\sigma_{\text{tot}}^2]^2}$ σ_+^2, σ_-^2—variances of the positive and negative electrostatic potentials in the molecule σ_{tot}^2—total variance of the electrostatic potential in the molecule	31, 73, 74
Electrostatic potential distribution shape factors	$\upsilon_i = \dfrac{\lambda_i}{\sum\limits_{i=1}^3 \lambda_i} \quad i = 1, 2, 3$ λ_i—G-WHIM weighted covariance matrix eigenvalues for electrostatic potential in the molecule	37, 38
Linear dimension of molecular electrostatic potential	$T = \lambda_1 + \lambda_2 + \lambda_3$ (G-WHIM)	37, 38

Table 1 Continued

Descriptor	Definition	Reference
Quadratic dimension of molecular electrostatic potential	$A = \lambda_1\lambda_2 + \lambda_1\lambda_3 + \lambda_2\lambda_3$ (G-WHIM)	37, 38
Total volume of molecular electrostatic potential	$V = \prod_{m=1}^{3}(1 + \lambda_m) - 1 = T + A + \lambda_1\lambda_2\lambda_3$ (G-WHIM)	37, 38
Global shape factor of molecular electrostatic potential	$K = \dfrac{3\sum_{i=1}^{3}\left\|\dfrac{\lambda_i}{\sum_i \lambda_i} - \dfrac{1}{3}\right\|}{4}$ (G – WHIM)	37, 38
Global density factor of molecular electrostatic potential	$D = \dfrac{\sum_{i=1}^{3}\lambda_i}{3}$ (G-WHIM)	37, 38
Integrated molecular transform	$FT_m = \sqrt{\int_1^{3l} I^2(s)\,ds}$ $I(s)$—intensity of the scattered radiation on nuclei	45, 46
Integrated electronic transform	$FT_e = \sqrt{\int_1^{3l} I^2(s)_\rho\,ds}$ $I(s)$—intensity of the scattered radiation for electron density ρ	45, 46
Integrated charge transform	$FT_e = \sqrt{\int_1^{3l} I^2(s)_q\,ds}$ $I(s)$—intensity of the scattered radiation for atomic charges q_i	45, 46
Normalized molecular moment	$M_n = \left(\dfrac{1}{W_m}\right)\sum_{i=1}^{n} A_i\left[\left[\begin{matrix}x_i\\y_i\\z_i\end{matrix}\right]^2 - \left[\begin{matrix}x_0\\y_0\\z_0\end{matrix}\right]^2\right]^{1/2}$ W_m—molecular mass A_i—atomic masses x_i, y_i, z_i—atomic coordinates x_0, y_0, z_0—coordinates of mass center	45, 46
Normalized electronic moment	$M_n = \left(\dfrac{1}{W_m}\right)\sum_{i=1}^{n} \rho_i\left[\left[\begin{matrix}x_i\\y_i\\z_i\end{matrix}\right]^2 - \left[\begin{matrix}x_0\\y_0\\z_0\end{matrix}\right]^2\right]^{1/2}\left(\sum_{i=1}^{n}\rho_i\right)^{-1}$ ρ_i—electron charge on ith atom	45, 46
Normalized charge moment	$M_n = \left(\dfrac{1}{W_m}\right)\sum_{i=1}^{n} q_i\left[\left[\begin{matrix}x_i\\y_i\\z_i\end{matrix}\right]^2 - \left[\begin{matrix}x_0\\y_0\\z_0\end{matrix}\right]^2\right]^{1/2}$ q_i—atomic charges	45, 46

MO-related descriptors

Descriptor	Definition	Reference
Energies of the highest occupied (HOMO) and lowest unoccupied (LUMO) molecular orbitals	$\varepsilon_{HOMO}, \varepsilon_{LUMO}$	57, 75, 76
Fraction of HOMO–LUMO energies arising from the atomic orbitals of the atom A	$\varepsilon_{HOMO,A}, \varepsilon_{LUMO,A}$	77
HOMO and LUMO orbital energies difference	$\varepsilon_{LUMO} - \varepsilon_{HOMO}$	20, 77

Table 1 Continued

Descriptor	Definition	Reference
Energy of singly occupied MO (SOMO)	ε_{SOMO}	78
Absolute hardness	$\eta = (\varepsilon_{LUMO} - \varepsilon_{HOMO})/2$	40, 41, 79
Activation hardness, R and T stand for reactant and transition state, respectively	$\Delta\eta = \eta_R - \eta_T$	40
HOMO–LUMO electron densities on the atom A	$Q_{A,HOMO}, Q_{A,LUMO}$	80
Electrophilic atomic frontier electron densities	$f_r^E = \sum(C_{HOMO,n})^2$ $C_{HOMO,n}$ are the coefficients of the nth atomic orbital in the HOMO	70, 81
Nucleophilic atomic frontier electron densities	$f_r^N = \sum(C_{LUMO,n})^2$ $C_{LUMO,n}$ are the coefficients of the nth atomic orbital in the LUMO	43, 70
Indices of frontier electron density	$F_r^E = f_r^E / \varepsilon_{HOMO}$ $F_r^N = f_r^N / \varepsilon_{LUMO}$	82
Electrophilic superdelocalizability	$S_{E,A} = 2 \sum_j \sum_{m=1}^{N_A} (C_{jm}^A)^2 / \varepsilon_j$ summation over occupied or MOs (j) and over the valence AOs in the atom A (m)	19, 42
Nucleophilic superdelocalizability	$S_{N,A} = 2 \sum_j \sum_{m=1}^{N_A} (C_{jm}^A)^2 / \varepsilon_j$ summation over unoccupied MOs (j) and over the valence AOs in the atom A (m)	19, 42
Sum of electrophilic superdelocalizabilities	$\sum S_{E,A}$	69
Sum of nucleophilic superdelocalizabilities	$\sum S_{N,A}$	69

be of significant importance in determining its properties. The charge distribution in the molecule can be calculated using various empirical schemes based mostly on the principle of the equalization of electronegativities of bonded atoms [13]. The charge distribution and the partial charges on atoms in the molecule can also be obtained from quantum-chemical calculations. A standard output of almost any quantum-chemical program gives the Mulliken atomic partial charges. For a series of structurally related compounds (congeneric series), they may give a semiquantitative relative ordering of atomic charges. A more reliable charge distribution in the molecule can be obtained using Shannon's information theory [14] or by fitting the predicted interaction energies [15], dipole moments, or electrostatic potential values inside and around the molecule [16,17].

Commonly, the minimum (most negative) and maximum (most positive) partial charges in the molecule or the minimum and maximum partial charges for particular types of atoms (e.g., C, O, etc.) have been used as electrostatic descriptors in the development of QSAR/QSPR equations. Also, a polarity parameter can be defined as the difference between the values of the most positive and the most negative charge in the molecule (cf. Table 1). Atomic partial charges have been used as static chemical reactivity indices [18,19]. The calculated σ- and π-electron densities on a particular atom characterize the possible direction of the chemical reactions (directional reactivity indices) [19]. Various sums of absolute or squared values of partial charges have been used to describe intermolecular interactions, e.g., solute–solvent interactions [20,21]. Other common charge-based descriptors are the averages of the absolute values of atomic partial charges.

Proceeding from the quantum-mechanical wave function, it is possible to calculate electrical moments and their components of a molecule. The polarization of a molecule by an external electric field can be described in terms of respective susceptibility tensors of the molecule [22]. The first-order term that is referred to as the polarizability of the molecule, α, represents the constant of proportionality between the induced dipole moment μ' and the strength of the external field E. At higher field strengths, the higher-order polarizabilities, called superpolarizabilities, have to be accounted for (β, γ, etc.). One of the most significant properties of the molecular polarizability is the close relation to the molecular bulk or molar volume. The polarizability values have also been shown to be related to hydrophobicity and thus to the biological activity of compounds [23,24]. Furthermore, the electronic polarizability of molecules shares common features with electrophilic superdelocalizability [24]. The first-order polarizability tensor contains also information about possible inductive interactions in the molecule [25]. The total anisotropy of the polarizability (the second-order term) characterizes the properties of a molecule as an electron acceptor. Local polarities can be represented by the local dipole moments, calculated for a fragment of a molecule, but these are conceptually difficult to define. First approximations of these quantities can be obtained by considering the atomic charges in the localized regions of the molecule. The local dipole index [19], the differences between net charges on atoms [26], and the topological electronic index [27] have been applied as charge-based polarity indices. However, such tensors depend on the choice of the coordinate system, and therefore the orientation of the congeneric molecular fragment must be the same for all molecules in the series.

A special class of electrostatic descriptors [charged partial surface area (abbreviated as CPSA) descriptors] has been proposed by Jurs et al. [28–30] in terms of the surface area of the whole molecule or its fragments and in terms of the charge distribution in the molecule. These descriptors, listed in the table, should account for the polar interactions between molecules.

The general interaction properties function (GIPF) descriptors describe the charge distribution in the molecule and its surface [31]. In contrast to descriptors that are intended to measure certain elements of an interaction, such as the hydrogen-bonding ability or polarizability, the GIPF descriptors identify quantities that allow to characterize most effectively the electrostatic potential over an entire molecular surface. For instance, the average ionization energy, $\bar{I}(r)$ [32], can be interpreted as the average energy needed to remove an electron from any given point in the space of the molecule. The plot of this descriptor on the molecular surface gives an indication of the most sensitive region toward electrophilic attack. The minimum and the maximum values of the electrostatic potential at the molecular surface, $V_{S,min}$ and $V_{S,max}$, respectively, have been found suitable for the analysis of the behavior of molecules in electrophilic and nucleophilic processes and also for determining the acidity and basicity of the compound [33,34]. The GIPF descriptor Π is defined as a measure of the local polarity in the molecule describing the amount, by which the electrostatic potential on the molecular surface deviates from its average value [35]. The Π descriptor correlates well with various empirical polarity or polarizability scales and with the dielectric constant of the bulk compound. The total variance of the surface electrostatic potential, σ_{tot}^2, is designed to reflect the molecule's ability to participate in noncovalent electrostatic interactions, whereas the electrostatic "balance" parameter, v, describes the ability of a molecule to act simultaneously as a hydrogen bond donor and a hydrogen bond acceptor [36].

The recently developed grid-weighted holistic invariant molecular (G-WHIM) descriptors are independent of molecule alignment and summarize all the information of the whole distribution in terms of dimension and shape indices [36–38]. The G-WHIM descriptors involving the molecular electrostatic potential include the directional and global descriptors of a molecule. The molecular electrostatic potential is calculated at selected grid points between the van der Waals and the threshold surface and the weighted covariance matrix of the grid points created by weighting each point with its potential value. Principal component analysis (PCA) of the weighted covariance matrix provides thus the directions of maximum property variance, and different G-WHIM descriptors can be calculated from the obtained eigenvalues and scores of the PCA. The global G-WHIM descriptors have been defined for the whole distribution of the molecular electrostatic potential. As the molecular electrostatic potential may have both negative and positive values, the respective two sets of G-WHIM descriptors can be obtained for any given molecule.

2.3. MO-Related Descriptors

A number of useful molecular descriptors related to the physical properties and chemical reactivity of molecules can be derived on the basis of the information available within the molecular orbital formalism. The energies of the highest occupied molecular orbital and the lowest unoccupied molecular orbital (HOMO and LUMO energies) belong to the most popular quantum-chemical descriptors [39]. Indeed, in many cases, these orbitals determine the chemical reactivity of a compound and the possible mechanism of a chemical reaction. The difference between the HOMO and LUMO energies has been related to the electronic band gaps in solids and the transition frequencies in the electronic spectra of compounds. The HOMO–LUMO gap has also been related to the chemical stability of compounds. The concept of chemical hardness has been derived on the basis of the HOMO–LUMO energy gap [40,41]. The activation hardness is expected to be useful in distinguishing between the reaction rates at different sites in the molecule and thus is relevant for predicting orientation effects. The energy localization of frontier molecular orbitals is also found to be important for the description of molecular charge–transfer complexes.

A variety of molecular descriptors have been defined and used proceeding from frontier molecular orbital theory (FMO) of chemical reactivity [42]. This theory is based on the concept of the superdelocalizability, an index characterizing the affinity of occupied and unoccupied orbitals in chemical reactions. A distinction has been made between the electrophilic and the nucleophilic superdelocalizability (or acceptor and donor superdelocalizability), respectively. The former describes the interaction of a compound with the electrophilic center at another reagent. The nucleophilic super-delocalizability characterizes the interaction of a compound with the nucleophilic center at the other reactant. The extreme (maximum and minimum) values of simplified atomic nucleophilic (N'_A), electrophilic (E'_A), and one-electron (R'_A) reactivity indices for a given atomic species in the molecule have often been used as descriptors of molecular reactivity.

The electron densities on frontier orbitals on atoms can be used for the description of donor–acceptor interactions between molecules [43]. According to this approach, the HOMO or nucleophilic electron density, f_r^N, of the donor molecule and the LUMO or electrophilic electron density, f_r^E, of an acceptor molecule are responsible for the charge transfer. However, the frontier electron densities can strictly be used only to describe the reactivity of different atoms in the same molecule. To compare the reactivity of different molecules, frontier electron densities have been normalized by the energy of the corresponding orbitals.

The self-atom and atom–atom polarizabilities (π_{AA}, π_{AB}) defined using pertur-bation theory have been also employed to describe chemical reactivity [44]. These quantities represent the effect of an electric field perturbation at one atom on the electronic charge at the same (π_{AA}) or another atom (π_{AB}), respectively.

For the characterization of the charge distribution in a molecule, the numerically unitary integrated molecular transform (FT_m), its analogous electronic (FT_e) and charge (FT_c) transforms, and the normalized molecular moment (M_n), its analogous electronic (M_e) and charge (M_c) moment, have been developed as molecular structure descriptors [45,46]. Those descriptors have been applied successfully for the develop-ment of QSAR models for various physicochemical, pharmacological, and thermody-namic properties of compounds.

2.4. Quantum-Chemical Modeling of Empirical Descriptors

Numerous molecular descriptors have been developed using the results from exper-imental measurements (empirical descriptors). The experimental limitations (chemical stability, solubility, etc.) often restrict the derivation of such descriptors from the results of direct measurements. However, the empirical descriptors have been very useful in the description of chemical reactivity and biological activity in many systems. Thus in order to expand the predictive power of the respective QSAR models, it would be beneficial to calculate the empirical descriptors using some quantum-chemical characteristics of molecules. For instance, it was found long ago [47] that the value of Hammett σ constant was linearly related with π-electron densities. Also, the Hammett σ and Taft σ^0 constants were examined using the semi-empirical quantum-chemical characteristics of a series of benzoic acids and benzoate anions. The most significant correlation for σ was established with the calculated electronic charge on the oxygen of the anion (q_{O-}). A nearly equally good linear correlation was observed for the reciprocal value of E_{HOMO} of the anion [48]. These results suggested that the equilibrium constant for the ionization of benzoic acid is controlled by the structure of anion. Significant correlations were also found between the Hammett σ constants and GIPF descriptors [49].

A large group of empirical molecular descriptors involves solvent effects on various chemical or physical processes. Because of the complexity of solvent effects, the respective QSAR/QSPR correlation equations are usually multiparametric, involving the descriptors reflecting the polarity and the polarizability of the solvent, its ability to act as an acceptor or a donor in a hydrogen bond, and the short-range dispersion and repulsion interactions [1]. Also, the hydrophobicity parameter log P can be referred to as a solvational characteristic since it is directly related to the change of the free energy of solvation of a solute in two solvents (water and octanol). The linear solvatochromic relation descriptors have been shown to be successful in correlating a wide range of chemical and physical properties involving solute–solvent interactions as well as biological activities of compounds [50,51]. In order to extend those parameters for wider selection of solvents, the so-called theoretical linear solvation energy relationship (TLSER) descriptors have been derived [52,53]. The general form of a TLSER is as follows

$$\log(\gamma) = c_0 + c_1 V_{mc} + c_2 \pi^* + c_3 \varepsilon_a + c_4 \varepsilon_b + c_5 q^+ + c_6 q^-$$

where V_{mc} is the molecular van der Waals volume and the polarizability term π^* is derived from the polarization volume of a compound. The covalent contribution to Lewis basicity, ε_b, is calculated as the difference in energy between the lowest unoccupied molecular orbital (E_{LUMO}) of water and the highest occupied molecular orbital (E_{HOMO}) of the solute. The electrostatic basicity contribution, denoted as q^-, is simply the most negative atomic charge in the solute molecule. Analogously, the hydrogen-bonding donating ability is divided into two components: ε_a is the energy difference between the E_{HOMO} of water and E_{LUMO} of solute, whereas q^+ is the most positive charge of a hydrogen atom in the solute molecule.

Various procedures have been proposed for calculating partition coefficients, log P, from the molecular structure. Different quantum-chemically calculated characteristics of a molecule have been used for this purpose. A good multilinear model for the prediction of log P has been developed using atomic charge densities [54]. In another approach, more quantum-chemically calculated molecular descriptors were used including the calculated dipole moment, the sums of absolute values of atomic charges, and the charge dispersions [20]. More sophisticated theoretical estimations of the partition coefficients involve detailed description of the molecular charge distribution and electrostatic potential [55,56].

3. QUANTUM QSAR IN BIOLOGICAL SYSTEMS

Quantum-chemical molecular descriptors have been actively used in the quantitative structure–activity relationship studies of biological activities [1,2,72]. In the following, examples of QSARs involving quantum-chemical descriptors and applied on the enzymatic reactivity, pharmacological activity, and toxicity of compounds are discussed.

3.1. Enzymatic Reactions

In QSAR of enzyme inhibition reactions, quantum-chemically calculated electrostatic or MO-related descriptors have been widely used. The former are expected to describe the complex formation between enzyme and the substrate, whereas the latter reflect the chemical reactivity of the substrate at the site. Already in 1967, Klopman and Hudson [83] developed a polyelectronic perturbation theory, according to which the drug–receptor interactions can be under either charge or orbital control. Thus the net atomic charges may be considered as the characteristics of electrostatic interactions, while the superdelocalizability or other MO-related characteristics characterize the covalent component of the interaction.

Thus good correlations were obtained between the CNDO/2 calculated total net $(\sigma + \pi)$ atomic and group charges in the heterocyclic sulfonamides and their anhydrase inhibition activity [84].

$$\log\varPi_{50} = 37.84 q_{SO_2NH_2} + 8.78$$

$$n = 28 \quad R = 0.909 \quad s = 0.336 \quad F = 123.2$$

where $q_{SO_2NH_2}$ is the charge of the $-SO_2NH_2$ group. In this and the subsequent equations, n denotes the number of samples, R is the regression correlation coefficient, s is the standard deviation of the regression, and F denotes the Fisher's F-value. More recently, various models of electrostatic interactions have been applied and tested for several leucine aminopeptidase (LAP) inhibitors interacting with the enzyme's active site [85]. The results indicate that atomic multipoles up to quadrupole moment as well as the electrostatic potential derived from different charge calculation schemes [16,86,87] reproduce reasonably well the ab initio electrostatic interaction energies and the expectation values of the molecular electrostatic potential. These electrostatic models together with CHELP (CHarges from ELectrostatic Potentials) atomic point charges yielded also a satisfactory correlation of the electrostatic interaction energy with the experimental activities of the inhibitors, in contrast to the results obtained from atomic Mulliken charges and atomic dipoles.

The electrostatic descriptors have been used in QSAR analysis together with other quantum-chemically derived descriptors. In such cases, both the substrate or inhibitor reactivity and binding are assumed to be important for the prediction of overall activity of compounds. For example, the inhibitory activity of a large group of benzenesulfonamides containing both a primary and secondary sulfonamide moiety has been measured towards several isozymes of carbonic anhydrase. This activity was found to depend on semi-empirical AM1-calculated electrostatic potential-based charges on the atoms of sulfonamide groups, HOMO and LUMO energies, dipole moments, and lipophilicities [88]. In a parallel work by the same authors, the activity of 1,3,4-thiadiazole- and 1,3,4-thiadiazoline disulfonamides was described using similar quantum-chemical descriptors [89]. A significant correlation has been obtained between the inhibition potency of indanone-benzylpiperidine inhibitors of acetylcholinesterase and the MNDO HOMO energy [90]:

$$-\log(IC_{50}) = -757.52 + 2.21 C_4 - 162.9 E_{HOMO} - 8.85 E_{HOMO}^2 - 6.65\mu + 1.18\mu^2$$

$$R^2 = 0.882 \quad s = 0.25 \quad n = 16 \quad F = 14.8$$

where C_4 is the HOMO out-of-plane π orbital coefficient of the ring carbon atom and μ is the total dipole moment.

Quantum mechanical molecular electrostatic potentials have been combined with artificial neural networks to predict the binding energy of bioactive molecules with enzyme targets and to identify the quantum mechanical features of inhibitory molecules that contribute to binding [91]. It was demonstrated that quantum neural networks could help in the identification of critical areas of inhibitor potential surfaces involved in binding and predict with quantitative accuracy the binding strength of new inhibitors. This conclusion was reached by examining three enzyme systems, i.e., adenosine monophosphate nucleosidase, adenosine deaminase, and cytidine deaminase.

In another study, it was also found that neural networks using the optimum descriptors from multiple linear regression analysis improved the correlations between the descriptors and the activities, implying that the relationship between the biological activity and descriptors is nonlinear. Comparative quantitative structure–activity relationship (QSAR) studies were carried out for flavonoid derivatives as cytochrome P450 1A2 inhibitors [92]. The results by both methods indicated that apart from the Hammett constant, the highest occupied molecular orbital energy (HOMO), the nonoverlap steric volume, the partial charge of the C_3 carbon atom, and the HOMO coefficients of C_3, C_3 and C_4 carbon atoms of flavonoids play an important role in inhibitory activity.

3.2. Pharmacological Activity

The quantum-chemical molecular descriptors have been widely used in the development of quantitative structure–activity relationships for various pharmacological activities of compounds. Again, most of the QSARs developed include the electrostatic and/or MO-related descriptors.

The first of them can describe the long-distance intermolecular electrostatic interactions in biological environment in vivo. For instance, the electrostatic descriptors may reflect the drug–receptor interactions at the docking site. However, an alternative interpretation can be based on the influence of electrostatic interactions on the physicochemical properties determining the bioavailability of compounds (solubility, permeability, distribution between phases, etc.) [93]. Thus if such data are available for a particular set of compounds, it would be useful to develop the respective quantitative structure–property relationship (QSPR) and compare this with the QSAR of the pharmacological activity. The presence of the same or similar descriptors in the comparative equations would give further information about the possible mechanism of the pharmacological action.

Apart from reflecting directly the reactivity between receptor and pharmacological agent, the MO-related descriptors may be related to the intermolecular donor–acceptor interactions responsible for bioavailability of compounds. Once again, the search for analogous correlations for the properties like solubility or distribution coefficients could be useful for determining the mechanism of biological action.

In quantum QSAR of pharmacological properties, several examples involve the description of the *antitumor activity* of compounds. For instance, both the hydrophobicity and the LUMO energy were found to determine the activity of a series of alkyl-substituted phenols against Chinese hamster V76 tumor cells according to the following quadratic equation [94]

$$\log\left(\frac{1}{ED_{50}}\right) = (0.818 \pm 0.062) + (0.278 \pm 0.132)\log P - (0.017 \pm 0.001)(\log P)^2$$

$$- (3.485 \pm 0.458)E_{LUMO}$$

$$R^2 = 0.714 \quad s = 0.227 \quad n = 29 \quad F = 20.74$$

The presence of E_{LUMO} in this equation was interpreted as showing that the activity of these compounds does not only depend on their bioavailability (ability to penetrate the cell membrane), but also upon their ability to participate as electron acceptor in the interaction with receptor. Another antitumor activity, the toxicity of substituted phenols against L1210 leukemia cells, has been related to the HOMO–LUMO energy gap, ΔE_{FMO} [95]:

$$\log\left(\frac{1}{IC_{50}}\right) = (26.58 \pm 3.30) + (0.25 \pm 0.05)\log P - (2.50 \pm 0.37)\Delta E_{FMO}$$

$$R^2 = 0.903 \quad s = 0.176 \quad n = 26 \quad R_{cv}^2 = 0.874$$

According to this equation, the activity increases in parallel with the decrease in the HOMO–LUMO energy gap. Such trend is expected as the smaller gap of frontier orbital energies is usually related to higher radical reactivity. For a comprehensive set of phenols, it was established that the inhibition of growth in murine leukemia cells correlates, together with log P, with both the HOMO–LUMO energy gap and the quantum-chemically calculated homolytic OH bond dissociation energy [96].

In another study, the electrophilic superdelocalizability at the 6-position of purine derivatives, S_6^E, was established as the main factor determining the activity of these compounds against murine solid adenocarcinoma CA 755. The respective QSAR equation [97]:

$$\log\left(\frac{1}{C}\right) = (3.69 \pm 0.14) + (0.51 \pm 0.14)S_6^E + (0.24 \pm 0.14)\pi_6$$

$$R^2 = 0.846 \quad s = 0.265 \quad n = 17 \quad F = 39.67$$

involves the hydrophobic constant for the substituent at 6-position of purine (π_6). It was concluded from this result that the charge transfer from this position to the biomacromolecule is an important electronic process related to the activity of compounds. Alternatively, the antitumor activity of some 3,5-disubstituted N-formylheteroaromatic thiosemicarbazones has been correlated with the nucleophilic superdelocalizability on carbon atoms as a single parameter determining the reactivity [98].

$$\log\left(\frac{1}{IC_{50}}\right) = -2.5 + 18.6S_C^N$$

$$R^2 = 0.835 \quad n = 10$$

In this case, the biological target was ribonucleoside diphosphate reductase and it was proposed on the basis of this result that before the compound interacts with the enzyme, it should form a complex with the Fe(II) ion.

The quantum-chemical descriptors have also shown their usefulness in the development of QSARs for *antiviral activities*. The antirhinoviral activity of 9-benzylpurines has been correlated with Hückel MO-generated electronic parameters and empirical substituent constants [99]. The respective QSAR equation included the LUMO energy and the total π-electron energy (E_π^T) of the compounds as quantum-chemical descriptors:

$$-\log(IC_{50}) = 6.044 + 2.056R_2 + 0.873F_4 - 0.289\pi_4 - 0.094E_\pi^T - 2.323E_{LUMO}$$

$$R^2 = 0.684 \quad s = 0.503 \quad n = 50 \quad F = 19.0$$

where R and F are the Swain–Lupton resonance and field parameters [100], respectively, and π is the hydrophobicity substituent constant at a given position in the purine ring. However, it was established that various serotypes of rhinovirus behave differently in terms of the electronic parameters that inhibit their action.

The QSAR of anti-HIV drugs continue to be of large interest. Principal component and hierarchical cluster studies showed that the semi-empirical molecular orbital method PM3-calculated LUMO energy, electronegativity χ, and charges on certain "active" atoms are related to the anti-HIV activity of flavonoid compounds [101]. The general interaction properties function (GIPF) approach has been used to develop analytical representations for the anti-HIV-1 potencies of two groups of reverse transcriptase inhibitors [102]. The molecular surface electrostatic potentials were calculated using the HF/STO-5G*//HF/STO-3G* level theory. The compounds examined were the derivatives of two main heterocyclic structures, TIBO (tetrahydro-imidazo[4,5,1-jk][1,4]-benzodiazepin-2(1H)-thione) and HEPT (1-(2-hydroxyethoxymethyl)-6-(phenylthio)-thymine). The following best four- and three-parameter correlation equations were obtained for the anti-HIV potency for TIBO derivatives.

$$\log(10^6/C_{50}) = 54.37\upsilon - 6.341\upsilon\Pi + 0.0138A_S + 4.81 \times 10^{-4}A_S^+\sigma_+^2 - 2.695$$

$$R = 0.930 \qquad s = 0.597$$

$$\log(10^6/C_{50}) = 152.4\upsilon^2 - 6.715\upsilon\Pi + 3.995 \times 10^{-8}\left(A_S^+\sigma_+^2\right)^2 - 8.023$$

$$R = 0.922 \qquad s = 0.618$$

where υ, Π, A_S, A_S^+, and σ_+^2 are the respective GIPF descriptors (cf. Table 1). For the HEPT derivatives, analogous equations were obtained, involving four and three parameters, respectively.

$$\log(10^6/C_{50}) = 0.4588\sigma_+^2 + 6.505 \times 10^{-2}A_S + 7.460 \times 10^{-3}A_S^- V_S^-$$
$$- 2.311 \times 10^{-3}\left(V_{S,max} - V_{S,min}\right)^2 - 4.754$$

$$R = 0.952 \quad s = 0.371$$

$$\log(10^6/C_{50}) = 0.3417\sigma_+^2 + 6.078 \times 10^{-2}A_S + 9.394 \times 10^{-3}A_S^- V_S^- - 9.079$$

$$R = 0.939 \quad s = 0.404$$

Other pharmacological activities have also been correlated with quantumchemically derived descriptors. For instance, the quantitative structure–activity relationship developed for the *antibacterial activity* of a series of monocyclic β-lactam antibiotics included the atomic charges, the bond orders, the dipole moment, and the first excitation energy of the compound [103]. The *fungicidal activity* of Δ^3-1,2,4-thiadiazolines has been correlated with an index of frontier orbital electron density derived from semi-empirical PM3 molecular orbital calculations [104].

$$pEC_{50} = 0.42R(1) + 2.04$$

$$R = 0.94 \quad s = 0.17 \quad n = 7 \quad F = 39.46$$

$$pEC_{50} = 2.14R(1) - 0.18R(1)^2 - 1.91$$

$$R = 0.88 \quad s = 0.21 \quad n = 17 \quad F = 24.76$$

The index $R(1) = f_r(1)/-E_{HOMO} \times 10^2$ was derived from the HOMO electron density at the sulfur atom $[f_r(1)]$ and the HOMO energy (E_{HOMO}), which is equivalent to the ionization potential of the molecule. The *anticonvulsant activity* of a set of structurally diverse compounds has been correlated with various quantum-chemical descriptors Interestingly, the best correlation was obtained using a single parameter—the LUMO energy [105].

$$\log(\text{ED}_{50}) = -11.669 E_{\text{LUMO}} + 1.206$$

$$R = 0.931 \quad s = 0.213 \quad n = 11 \quad F = 58.46$$

This relationship was traditionally explained in terms of an acceptor–donor interaction involving electronic transfer to the ligand.

The Integrated Molecular Transform and the Normalized Molecular Moment Structure Descriptors [46] have been used for the description of various physicochemical, thermodynamic, and pharmacological properties of compounds. The first included the polarizability, octanol/water partition coefficients ($\log P$), pK_a in aqueous solutions and in gas phase, organic magnetic susceptibility, peptide distribution coefficients ($\log P'$), transition frequencies in UV spectra, gas chromatographic retention indices, Hammett constants, dipole moments, and heats of formation of compounds. Several pharmacological properties were modeled by one-parameter QSAR equations. The best correlation of data on minimum blocking concentration for local anesthesia was obtained with integrated molecular transform (FT_m)

$$\log \text{MBC} = -0.013 \text{FT}_m + 3.387$$

$$R = 0.978 \quad s = 0.446 \quad n = 36 \quad F = 745$$

The same descriptor correlated well also with the acetylcholinesterase (AChE) and butyrylcholinesterase (BuChE) enzyme inhibition activity of organophosphorus compounds. The best one-parameter correlation for the toxicity of organophosphorus compounds was achieved by using the normalized electronic moment (M_e) [46].

3.3. Toxicity

Quantum-chemical descriptors have been extensively used in the development of QSARs of various toxic activities of compounds. In principle, the possible interactions that determine toxicity coincide with those determining the pharmacological activity of compounds. Therefore the descriptors may either reflect the direct interaction of toxic agents with the biological targets or they may be related to the bioavailability of such agents.

The best correlation for the mutagenicity of quinolines was obtained with the AM1-calculated net atomic charges on a carbon atom (q_2) and the hydrophobic parameter ($\log P$) [106].

$$\ln(\text{TA100}) = -5.39 - 45.76 q_2 + 1.14 \log P$$

$$R^2 = 0.726 \quad s = 0.565 \quad n = 21 \quad F = 11.9$$

Notably, the HOMO and LUMO energies and electron densities were also correlating this property. The involvement of the net atomic charge on carbon atom in the 2-position (q_2) suggests that this might be the site for activity. In addition, linear correlations have been established between the calculated HOMO or LUMO energy and the mutagenicity of aromatic and heteroaromatic nitro-compounds [107–109], aromatic and heteroaromatic amines [110], and aryltriazenes and heterocyclic triazenes [111].

The toxicity of compounds has often been related to the polarizability of compounds. This descriptor is related to the intermolecular interactions in biological environments and can be ascribed both to the drug–receptor interactions as well as to the properties determining the bioavailability of a compound [112]. Thus it was shown that even the CNDO/2 calculated molecular polarizability (α) can be successfully correlated with the acute toxicity in a series of 20 nitriles [113]:

$$-\log LD_{50} = -0.03\alpha + 0.43$$

$$R = 0.87 \quad s = 0.199 \quad n = 13 \quad F = 42.1$$

$$-\log LD_{50} = -1.69\alpha/\Delta E + 0.47$$

$$R = 0.87 \quad s = 0.199 \quad n = 13 \quad F = 42.4$$

The $\alpha/\Delta E$ parameter, where ΔE is the difference in HOMO and LUMO energies, is an orbital energy-weighted polarizability term. The last equation implies that the acute toxicity of nitriles is a function of molecular size/polarity and electronic activation energy. As the dipole moment of the compounds did not correlate with the activity for this series, it was suggested that the enzyme–substrate interaction might be of secondary importance.

The importance of molecular orbital-related descriptors in the QSARs related to biotransformation and toxicity has been reviewed recently [114,115]. For example, significant correlations were found between Ames TA100 mutagenicity and the AM1-calculated electron affinity or LUMO energy (i.e., the stability of the corresponding anion radical) of chlorofuranones but also with the frontier electron density of the LUMO at the α-carbon [43,116,117]. The correlations observed suggest a reaction mechanism in which chlorofuranones act as electron acceptors in the interaction with DNA. In general, the participation of frontier orbitals in mutagenic activity seems to be essential, even if it is masked almost entirely by the hydrophobicity.

The acute toxicity of soft electrophiles such as substituted benzenes, phenols, and anilines has been correlated with MNDO-calculated descriptors [118]:

$$\log(1/LC_{50}) = -1.49 + 0.56\log P + 13.7 S_{av}^{N}$$

$$R^2 = 0.81 \quad s^2 = 0.19 \quad n = 114 \quad F = 238.7$$

The average acceptor superdelocalizability S_{av}^{N} is the average of S_i^{N} over the atoms (i) involved in the π bonds. The hydrophobicity ($\log P$) and soft electrophilicity descriptors were shown to be orthogonal for the 114 compounds studied.

The proliferation toxicity toward the algae *Scenedesmus vacuolatus* in a 24-hr one-generation reproduction assay has been correlated with hydrophobicity ($\log K_{ow}$) and various quantum-chemical descriptors of molecular reactivity using AM1 parameterization [119]. The possible mechanism of the toxic action has been proposed in view of the strong correlations with the LUMO and SOMO (singly occupied molecular orbital for radicals) energies. Also, the molecular hardness and softness parameters have been employed to describe the genotoxicity of chlorinated hydrocarbons [120].

In another recent study, QSAR models were developed using quantum-chemical descriptors to describe the toxic influence of polychlorinated organic compounds on the rainbow trout (*Oncorhynchus mykiss*). The logarithm of the bioconcentration factor (BCF) was best correlated with the AM1-calculated α-polarizability, energies of the frontier orbitals, and the core–core repulsion energy (CCR), as follows [121]:

$$\log \text{BCF} = 10.4678(\pm 0.8129) + 0.0033(\pm 0.0014)\alpha + 0.7415(\pm 0.0751)E_{\text{HOMO}}$$

$$+ 0.05696(\pm 0.0788)E_{\text{LUMO}} + 0.0266(\pm 0.0102)\text{CCR}$$

$$R^2 = 0.8613 \quad s = 0.2254 \quad n = 31$$

The presence of the orbital energies in the QSAR equation was interpreted as reflecting the donor–acceptor interactions between the tested compounds and biotarget molecule in the fish. The positive correlation of the log BCF with CCR is ascribed to the relationship between this descriptor and the partition coefficient log K_{ow}.

An attempt has been made to determine which descriptor parameterizes the best the electrophilicity of aromatic compounds with regard to their acute toxicity [122]. To achieve this, toxicity data for 203 substituted aromatic compounds containing nitro or cyano groups were evaluated in the 40-hr *Tetrahymena pyriformis* population growth impairment assay. The quantitative structure–activity relationships (QSARs) relating the toxic potency involved hydrophobicity quantified by the 1-octanol-water partition coefficient (log P) and electrophilic reactivity quantified by the molecular orbital parameters, either by the energy of the lowest unoccupied molecular orbital (E_{LUMO}) or the maximum acceptor superdelocalizability [A(max)]. For the full data set, E_{LUMO} and A(max) were found to be almost collinear ($R = 0.87$). The results, however, indicated that A(max) would be the superior descriptor of electrophilicity for the purpose of toxicological QSARs for aromatic compounds. Development of QSARs using partial least-squares yielded similar conclusions.

In most cases, descriptors calculated using some semi-empirical quantum-chemical parameterization (MNDO, AM1, PM3, etc.) have been used in the development of QSARs on biological activities. However, little attention has been paid to the quality of these data. Notably, a comparative analysis of the quality of descriptors obtained by using different quantum-chemical methods has been carried out as applied to the toxicity data [123]. It was demonstrated that the performance of AM1 in deriving QSARs for toxicity could be improved by employing ab initio Hartree–Fock, density functional theory B3LYP, and MP2 perturbation methods together with a split-valence basis set with polarization functions. The Hartree–Fock method with a minimal basis set did not perform well, and it was suggested to avoid it in descriptor calculations.

Karelson et al. [124] had also carried out a comparative analysis of the molecular descriptors calculated for the isolated molecules (gas phase) and for the molecules embedded into a dielectric continuum corresponding to aqueous solution. The self-consistent reaction field method [125] was used for the latter calculations. The results indicated that, in general, the quantum-chemically derived descriptors are rather insensitive towards the change in the environment surrounding the molecule. However, the most influenced are the polarizability and several other MO-related descriptors, often used in biological QSARs. Thus one should be cautious when using the descriptors calculated for the isolated molecules.

REFERENCES

1. Karelson M. Molecular Descriptors in QSAR/QSPR. New York: J Wiley & Sons, 2000: 436.
2. Kubinyi H, ed. 3-D QSAR in Drug Design. Vol. 1. Leiden: ESCOM, 1993.
3. Wilson S, Diercksen GHF, eds. Problem Solving in Computational Molecular Science. Dordrecht: Kluwer Academic Publishers, 1997:416.
4. McWeeny R. Methods of Molecular Quantum Mechanics. 2d ed. London: Academic Press, 1992:573.
5. Atkins PW, Friedman RS. Molecular Quantum Mechanics. 3rd ed. Oxford: Oxford University Press, 1999:562.
6. Pople JA, Beveridge DL. Approximate Molecular Orbital Theory. New York: McGraw-Hill, 1970:234.
7. Dewar MJS, Zoebisch EG, Healy EF, Stewart JJP. AM1: a new general purpose quantum mechanical model. J Am Chem Soc 1985; 107:3902–3909.
8. Stewart JJP. Optimization of parameters for semi-empirical methods. I—Method. J Comp Chem 1989; 10:209–220.
9. Koch W, Holthausen M. A Chemist's Guide to Density Functional Theory. Weinheim: Wiley-VCH Wiley & Sons, 2000:300.
10. Breneman CM, Martinov M. The use of electrostatic potential fields in QSAR and QSPR. In: Murray JS, Sen K, eds. Molecular Electrostatic Potentials: Concepts and Applications, Theoretical and Computational Chemistry. Vol. 3. Amsterdam: Elsevier Science BV, 1996.
11. Akhiezer AI, Peltminskii SV. Methods of Statistical Physics. Oxford: Pergamon Press, 1981.
12. Heritage TW, Ferguson AM, Turner DB, Willett P. EVA: a novel theoretical descriptor for QSAR studies. Perspect Drug Discov Des 1998; 9–11:381–398.
13. Gasteiger J, Marsili M. A new model for calculating atomic charges in molecules. Tetrahedron Lett, 1978:3181–3184.
14. Bader RF. Atoms in Molecules. A Quantum Theory. London: Oxford Science Publications, Clarendon Press, 1990.
15. Price SL, Stone AJ. The electrostatic interactions in van der Waals complexes involving aromatic molecules. J Chem Phys 1987; 86:2859–2868.
16. Bayly CI, Cieplak P, Cornell WD, Kollman PA. A well-behaved electrostatic potential based method using charge restraints for determining atom-centered charges: the RESP model. J Phys Chem 1993; 97:10269–10280.
17. Francl MM, Chirlian LE. The pluses and minuses of mapping atomic charges to electrostatic potentials. Rev Comput Chem 2000; 14:1–31.
18. Franke R. Theoretical Drug Design Methods. Amsterdam: Elsevier, 1984.
19. Kikuchi O. Systematic QSAR procedures with quantum chemical descriptors. Quant Struct-Act Relatsh 1987; 6:179–184.
20. Bodor N, Gabanyi Z, Wong C-K. A new method for the estimation of partition coefficient. J Am Chem Soc 1989; 111:3783–3786.
21. Klopman G. In: Klopman G, ed. Chemical Reactivity and Reaction Paths. New York: John Wiley & Sons, 1974:55–165.
22. Atkins PW. Quanta. 2d ed. Oxford: Oxford University Press, 1991: 434.
23. Hansch C, Coats E. Chymotrypsin: a case study of substituent constants and regression analysis in enzymic structure–activity relationships. J Pharm Sci 1970; 59:731–743.
24. Lewis DVF. The calculation of molar polarizabilities by the CNDO/2 method: correlation with the hydrophobic parameter, logP. J Comput Chem 1989; 10:145–151.

25. Takahata Y, Gaudio AC, Korolkovas A. Quantitative structure–activity relationships for calcium antagonist 1,4-dihydropyridine (nifedipine analogues) derivatives: a quantum chemical/classical approach. J Pharm Sci 1994; 83:1110–1115.

26. Osmialowski K, Halkiewicz J, Radecki A, Kaliszan R. Quantum chemical parameters in correlation analysis of gas–liquid chromatographic retention indices of amines. J Chromatogr A 1985; 346:53–60.

27. Osmialowski K, Halkiewicz J, Kaliszan R. Quantum chemical parameters in correlation analysis of gas–liquid chromatographic retention indices of amines; II. Topological electronic index. J Chromatogr A 1986; 361:63–69.

28. Stanton DT, Jurs PC. Development and use of charged partial surface area structural descriptors in computer-assisted quantitative structure–property relationship studies. Anal Chem 1990; 62:2323–2329.

29. Stanton DT, Jurs PC. Computer-assisted study of the relationship between molecular structure and surface tension of organic compounds. J Chem Inf Comp Sci 1992; 32:109–115.

30. Wessel MD, Jurs PC. Prediction of normal boiling points for a diverse set of industrially important organic compounds from molecular structure. J Chem Inf Comput Sci 1995; 35:841–850.

31. Murray JS, Politzer P. In: Politzer P, Murray JS, eds. Quantitative Treatments of Solute/Solvent Interactions. Amsterdam: Elsevier, 1994:243–289.

32. Brinck T, Murray JS, Politzer P. Molecular surface electrostatic potentials and local ionization energies of group {V–VII} hydrides and their anions: relationships for aqueous and gas-phase acidities. Int J Quant Chem 1993; 48:73–88.

33. Gross KC, Seybold PG, Peralta-Inga Z, Murray JS, Politzer P. Comparison of quantum chemical parameters and Hammett constants in correlating pK_a values of substituted anilines. J Org Chem 2001; 66:6919–6925.

34. Politzer P, Murray JS. Molecular Electrostatic Potentials and Chemical Reactivity. In: Lipkowitz KB, Boyd DB, eds. Reviews in Computational Chemistry. Vol. 2. New York: VCH Publishers, 1991:273–312.

35. Murray JS, Politzer P. In: von P, Schleyer R, eds. Encyclopedia of Computational Chemistry. Vol. 2. New York: J Wiley & Sons, 1998:912–920.

36. Todeschini R, Gramatica P. 3D-modelling and prediction by WHIM descriptors. Part 5. Theory development and chemical meaning of the WHIM descriptors. Quant Struct-Act Relat 1997; 16:113–119.

37. Todeschini R, Gramatica P. The WHIM theory: New 3D-molecular descriptors for QSAR in environmental modelling. SAR QSAR Environ Res 1997; 7:89–115.

38. Todeschini R, Gramatica P. New 3D-molecular descriptors: the WHIM theory and QSAR applications. In: Kubinyi H, Folkers G, Martin YC, eds. 3D QSAR in Drug Design. Vol. 2. Dordrecht: Kluwer Escom, 1998:355–380.

39. Clare BW. Frontier orbital energies in quantitative structure–activity relationships: a comparison of quantum chemical methods. Theor Chim Acta 1994; 87:415–430.

40. Zhou Z, Parr RG. Activation hardness. New index for describing the orientation of electrophilic aromatic substitution. J Am Chem Soc 1990; 112:5720–5724.

41. Pearson RG. Absolute electronegativity and hardness: applications to organic chemistry. J Org Chem 1989; 54:1423–1432.

42. Fukui K. Theory of Orientation and Stereoselection. New York: Springer-Verlag, 1975.

43. Tuppurainen K, Lötjönen S, Laatikainen R, Vartiainen T, Maran U, Strandberg M, Tamm T. About the mutagenicity of chlorine-substituted furanones and halopropenals. A QSAR study using molecular orbital indexes. Mutat Res 1991; 247:97–102.

44. Coulson CA, Longuet-Higgins HC. The electronic structure of conjugated systems. I. General theory. Proc R Soc Lond Ser A 1947; 191:39–60.

45. Molnar SP, King JW. Molecular structural index control in property-directed clustering and correlation. Int J Quantum Chem 2000; 80:1164–1171.

46. Molnar SP, King JW. Theory and applications of the integrated molecular transform and the normalized molecular moment structure descriptors: QSAR and QSPR paradigms. Int J Quantum Chem 2001; 85:662–675.

47. Jaffe HH. A reexamination of the Hammett equation. Chem Rev 1953; 53:191–261.

48. Gilliom RD, Beck JP, Purcell WP. An MNDO treatment of sigma values. J Comput Chem 1985; 6:437–440.

49. Murray JS, Brinck T, Politzer P. Applications of calculated local surface ionization energies to chemical reactivity. J Mol Struct Theochem 1992; 255:271–281.

50. Kamlet MJ, Taft RW, Abboud J-LM. Regarding the generalized scale of solvent polarities. J Am Chem Soc 1977; 99:8325–8327.

51. Reichardt Chr. Solvents and Solvent Effects in Organic Chemistry. 2d ed. New York: VCH Publishers, 1990.

52. Famini GR, Wilson LY. Using Theoretical Descriptors in Linear Solvation Energy Relationships. In: Politzer P, Murray JS, eds. Quantitative Treatments of Solute/Solvent Interactions. Amsterdam: Elsevier, Amsterdam 1994:213–242.

53. Murray JS, Politzer P., Famini G. Theoretical alternatives to linear solvation energy relationships. J Mol Struct Theochem 1998; 454(2–3):299–306.

54. Klopman G, Iroff LD. Calculation of partition coefficients by the charge density method. J Comput Chem 1981; 2:157–160.

55. Brinck T, Murray JS, Politzer P. Octanol/water partition coefficients expressed in terms of their molecular surface areas and electrostatic potentials. J Org Chem 1993; 58: 7070.

56. Essex JW, Reynolds CA, Richards WG. Theoretical determination of partition co-efficients. J Am Chem Soc 1992; 114, 3634–3639.

57. Csizmadia IG. Theory and Practice of MO Calculations on Organic Molecules. Amsterdam: Elsevier, 1976.

58. Atkins PW. Physical Chemistry. 3rd ed. Oxford: Oxford University Press, 1988.

59. Major DT, Halbfinger E, Fischer B. Molecular recognition of modified adenine nucleotides by the P2Y(1)-receptor. 2. A computational approach. J Med Chem 1999; 42:5338–5347.

60. Sotomatsu T, Murata Y, Fujita T. Correlation analysis of substituent effects on the acidity of benzoic acids by the AM1 method. J Comput Chem 1989; 10:94–98.

61. Clementi E. Computational Aspects of Large Chemical Systems. New York: Springer Verlag, 1980.

62. Breneman CM, Rhem M. A QSPR analysis of HPLC column capacity factors for a set of high-energy materials using electronic van der Waals surface property descriptors computed by the transferable atom equivalent method. J Comput Chem 1997; 18:182–197.

63. Ferguson AM, Heritage T, Jonathon P, Pack SE, Phillips L, Rogan J, Snaith PJ. EVA—a new theoretically based molecular descriptor for use in QSAR/QSPR analysis. J Comput-Aided Mol Des 1997; 11:143–152.

64. Clare BW, Supuran CT. Semi-empirical atomic charges and dipole moments in hyper-valent sulfonamide molecules: descriptors in QSAR studies. J Mol Struct Theochem 1998; 428:109–121.

65. Katritzky AR, Sild S, Lobanov V, Karelson M. Quantitative structure–property relationship (QSPR) correlation of glass transition temperatures of high molecular weight polymers. J Chem Inf Comput Sci 1998; 38:300–304.

66. DeBenedetti PG. Electrostatics in quantitative structure–activity relationship analysis. J Mol Struct Theochem 1992; 256:231–248.

67. Fleming I. Frontier Orbitals and Organic Chemical Reactions. New York: J Wiley & Sons, 1976.

68. Cosentino U, Moro G, Quintero MG, Giraldo E, Rizzi CA, Schiavi GB, Turconi M. The role of electronic and conformational properties in the activity of 5-HT3 receptor antagonists. J Mol Struct Theochem 1993; 286:275–291.

69. Cocchi M, Menziani MC, De Benedetti PG, Cruciani G. Theoretical versus empirical

molecular descriptors in monosubstituted benzenes; a chemometric study. Chemometr Intell Lab Syst 1992; 14:209–224.

70. Gaudio AC, Korolkovas A, Takahata Y. Quantitative structure–activity relationships for 1,4-dihydropyridine calcium channel antagonists (nifedipine analogs): a quantum/classical approach. J Pharm Sci 1994; 83:1110–1115.

71. Buydens L, Geerlings P, Massart DL. Prediction of gas chromatographic retention indices with topological, physicochemical and quantum chemical parameters. Anal Chem 1983; 55:738–744.

72. Karelson M, Lobanov VS, Katritzky AR. Quantum-chemical descriptors in QSAR/QSPR studies. Chem Rev 1996; 96:1027–1043.

73. Murray JS, Peralta-Inga Z, Politzer P. Conformational dependence of molecular surface electrostatic potentials. Int J Quant Chem 1999; 75:267–273.

74. Politzer P, Murray JS, Peralta-Inga Z. Molecular surface electrostatic potentials in relation to noncovalent interactions in biological systems. Int J Quantum Chem 2001; 85:676–684.

75. Cardozo MG, Iimura Y, Sugimoto H, Yamanishi Y, Hopfinger AJ. QSAR analysis of the substituted inanone and benzylpiperidine rings of a series of indanone-benzylpiperidine inhibitors of acetylcholinesterase. J Med Chem 1992; 35:584–589.

76. Debnath AK, Compadre RLL, Debnath D, Shusterman AJ, Hansch C. Structure–activity relationship of mutagenic aromatic and heteroaromatic nitro compounds. Correlation with molecular orbital energies and hydrophobicity. J Med Chem 1991; 34:786–797.

77. Cartier A, Rivail J-L. Electronic descriptors in quantitative structure–activity relationships. Chemometr Intell Lab Syst 1987; 1:335–347.

78. Schmitt H, Altenburger R, Jastorff B, Schüürmann G. Quantitative structure–activity analysis of the algae toxicity of nitroaromatic compounds. Chem Res Toxicol 2000; 13:441–450.

79. Pearson RG. Chemical Hardness. Weinheim: Wiley-VCH, 1997.

80. Tuppurainen K, Lötjönen S, Laatikainen R, Vartiainen T. Structural and electronic properties of MX compounds related to TA100 mutagenicity: a semiempirical molecular orbital QSAR study. Mutat Res 1992; 266:181–188.

81. Langenaeker W, Demel K, Geerlings P. Quantum-chemical study of the Fukui function as a reactivity index. Part 2. Electrophilic substitution on mono-substituted benzenes. J Mol Struct Theochem 1991; 234:329–342.

82. Ishikawa Y, Kishi K. Molecular orbital approach to possible discrimination of musk odor intensity. Int J Quantum Chem 2000; 79:109–119.

83. Klopman G, Hudson RF. Polyelectronic perturbation treatment of chemical reactivity. Theor Chim Acta 1967; 8:165–174.

84. De Benedetti PG, Menziani MC, Cocchi M, Frassineti C. A quantum chemical QSAR analysis of carbonic anhydrase inhibition by heterocyclic sulfonamides. Sulfonamide carbonic anhydrase inhibitors: Quantum chemical QSAR. Quant Struct-Act Relat 1987; 6:51–53.

85. Grembecka J, Kêdzierski P, Sokalski WA, Leszczyñski J. Electrostatic models of inhibitory activity. Int J Quant Chem 2001; 83, 180–192.

86. Besler BH, Merz JKM, Kollman PA. Atomic charges derived from semiempirical methods. J Comput Chem 1990; 11:431–439.

87. Francl MM, Carey C, Chirlian LE, Gange D. Charges fit to electrostatic potentials II: can atomic charges be unambiguously fit to electrostatic potentials? J Comput Chem 1996; 17:367–383.

88. Clare BW, Supuran CT. Carbonic anhydrase inhibitors. Part 61. Quantum chemical QSAR of a group of benzenedisulfonamides. Eur J Med Chem 1999; 34:463–474.

89. Clare BW, Supuran CT. Carbonic anhydrase inhibitors. Part 57: quantum chemical QSAR of a group of 1,3,4-thiadiazole- and 1,3,4-thiadiazoline disulfonamides with carbonic anhydrase inhibitory properties. Eur J Med Chem 1999; 34:41–50.

90. Cardozo MG, Iimura Y, Sugimoto H, Yamanishi Y, Hopfinger AJ. QSAR analyses of the substituted indanone and benzylpiperidine rings of a series of indanone-benzyl-piperidine inhibitors of acetylcholinesterase. J Med Chem 1992; 35:584–589.

91. Braunheim BB, Bagdassarian CK, Schramm VL, Schwartz SD. Quantum neural networks can predict binding free energies for enzymatic inhibitors. Int J Quant Chem 2000; 78:195–204.

92. Moon T, Chi MH, Kim D-H, Yoon CN, Choi Y-S. Quantitative structure–activity relationships (QSAR) study of flavonoid derivatives for inhibition of cytochrome P450 1A2. Quant Struct-Act Relat 2000; 19:257–263.

93. Katritzky AR, Fara DC, Petrukhin R, Tatham DB, Maran U, Lomaka A, Karelson M. The present utility and future potential for medicinal chemistry of QSAR/QSPR with whole molecule descriptors. Curr Top Med Chem 2002; 2:1333–1356.

94. Itokawa H, Totsuka N, Nakahara K, Maezuru M, Takeya K, Kondo M, Inamatsu M, Morita H. A quantitative structure–activity relationship for antitumor activity of long-chain phenols from *Ginkgo biloba* L. Chem Pharm Bull 1989; 37:1619–1621.

95. Zhang L, Gao H, Hansch C, Selassie CD. Molecular orbital parameters and comparative QSAR in the analysis of phenol toxicity to leukemia cells. J Chem Soc Perkin Trans 1998; 2:2553–2556.

96. Selassie CD, Shusterman AJ, Kapur S, Verma RP, Zhang L, Hansch C. On the toxicity of phenols to fast growing cells. A QSAR model for a radical-based toxicity. J Chem Soc Perkin Trans 1999; 2:2729–2733.

97. Mekenyan OG, Bonchev D, Rouvray DH, Petichev D, Bangov I. Modeling the interaction of small molecules with biomacromolecules. IV. The in-vivo interaction of substituted purines with murine tumor adenocarcinoma CA 755. Eur J Med Chem 1991; 26:305–312.

98. Miertuš S, Miertušova J, Filipovic P. In: Tichy M, ed. QSAR in Toxicology and Xenobiochemistry. Amsterdam: Elsevier, 1985:143.

99. Prabhakar YS. Quantum QSAR of the antirhinoviral activity of 9-benzylpurines. Drug Des Deliv 1991; 7:227–239.

100. Swain CG, Lupton EC. Field and resonance components of substituent effects. J Am Chem Soc 1968; 90:4328–4337.

101. Alves CN, Pinheiro JC, Camargo AJ, de Souza AJ, Carvalho RB, da Silva ABF. A quantum chemical and statistical study of flavonoid compounds with anti-HIV activity. J Mol Struct Theochem 1999; 491:123–131.

102. Gonzalez OG, Murray JS, Peralta-Inga Z, Politzer P. Computed molecular surface electrostatic potentials of two groups of reverse transcriptase inhibitors: relationships to anti-HIV-1 activities. Int J Quant Chem 2001; 83:115–121.

103. Li L, Maoshuang, Zhao K, Tian A. Semi-empirical quantum-chemical study of structure–activity relationship in monocyclic β-lactam antibiotics. J Mol Struct Theochem 2001; 545:1–5.

104. Nakayama A, Hagiwara K, Hashimoto S, Shimoda S. QSAR of fungicidal Δ3-1,2,4-thiadiazolines. Reactivity-activity correlation of sulfhydryl inhibitors. Quant Struct-Act Relat 1993; 12:251–255.

105. Tasso SM, Bruno-Blanch LE, Moon SC, Estiu GL. Pharmacophore searching and QSAR analysis in the design of anticonvulsant drugs. J Mol Struct Theochem 2000; 504: 229–240.

106. Debnath AK, Lopez de Compadre RL, Hansch C. Mutagenicity of quinolines in *Salmonella typhimurium* TA100. A QSAR study based on hydrophobicity and molecular orbital determinants. Mutat Res 1992; 280:55–65.

107. Debnath AK, Lopez de Compadre RL, Shusterman AJ, Hansch C. Quantitative structure–activity relationship investigation of the role of hydrophobicity in regulating mutagenicity in the Ames Test: 2. Mutagenicity of aromatic and heteroaromatic nitro compounds in *Salmonella typhimurium* TA100. Environ Mol Mutagen 1992; 19:53–70.

108. Debnath AK, Lopez de Compadre RL, Debnath G, Shusterman AJ, Hansch C. The structure–activity relationship of mutagenic aromatic nitro compounds. Correlation with molecular orbital energies and hydrophobicity. J Med Chem 1991; 34:786–797.

109. Debnath AK, Hansch C. Structure–activity relationship of genotoxic polycyclic aromatic nitro compounds: further evidence for the importance of hydrophobicity and molecular orbital energies in genetic toxicity. Environ Mol Mutagen 1992; 20:140–144.

110. Debnath AK, Debnath G, Shusterman AJ, Hansch C. A QSAR Investigation of the role of hydrophobicity in regulating mutagenicity in the Ames test: 1. Mutagenicity of aromatic and heteroaromatic amines in *Salmonella typhimurium* TA98 and TA100. Environ Mol Mutagen 1992; 19:37–52.

111. Debnath AK, Shusterman AJ, Lopez de Compadre RL, Hansch C. The importance of the hydrophobic interaction in the mutagenicity of organic compounds. Mutat Res 1994; 305:63–72.

112. Romanelli GP, Cafferata LFR, Castro EA. An improved QSAR study of toxicity of saturated alcohols. J Mol Struct Theochem 2000; 504:261–265.

113. Lewis DFV, Ioannides C, Parke DV. Interaction of a series of nitriles with the alcohol-inducible isoform of P450: computer analysis of structure–activity relationships. Xenobiotica 1994; 24:401–408.

114. Schultz TW, Seward JR. Health-effects related structure–toxicity relationships: a paradigm for the first decade of the new millennium. Sci Total Environ 2000; 249:73–84.

115. Soffers AEMF, Boersma MG, Vaes WHJ, Vervoort J, Tyrakowska B, Hermens JLM, Rietjens IMCM. Computer-modeling-based QSARs for analyzing experimental data on biotransformation and toxicity. Toxicol In Vitro 2001; 15:539–551.

116. Tuppurainen K, Lötjönen S, Laatikainen R, Vartiainen T. Structural and electronic properties of MX compounds related to TA100 mutagenicity: a semiempirical molecular orbital QSAR study. Mutat Res 1992; 266:181–188.

117. Tuppurainen K. QSAR approach to molecular mutagenicity. A survey and a case study: MX compounds. J Mol Struct Theochem 1994; 306:49–56.

118. Veith GD, Mekenyan OG. A QSAR approach for estimating the aquatic toxicity of soft electrophiles [QSAR for electrophiles]. Quant Struct-Act Relat 1993; 12:349–356.

119. Schmitt H, Altenburger R, Jastorff B, Schüürmann G. Quantitative structure–activity analysis of the algae toxicity of nitroaromatic compounds. Chem Res Toxicol 2000; 13:441–450.

120. Baeten A, Tafazoli M, Kirsch-Volders M, Geerlings P. Use of the HSAB principle in quantitative structure–activity relationships in toxicological research: application to the genotoxicity of chlorinated hydrocarbons. Int J Quant Chem 1999; 74:351–355.

121. Wei D, Zhang A, Wu C, Han S, Wang L. Progressive study and robustness test of QSAR model based on quantum chemical parameters for predicting BCF of selected poly-chlorinated organic compounds (PCOCs). Chemosphere 2001; 44:1421–1428.

122. Cronin MTD, Manga N, Seward JR, Sinks GD, Schultz TW. Parametrization of electrophilicity for the prediction of the toxicity of aromatic compounds. Chem Res Toxicol 2001; 14:1498–1505.

123. Trohalaki S, Gifford E, Pachter R. Improved QSARs for predictive toxicology of halogenated hydrocarbons. Comput Chem 2000; 24:421–427.

124. Karelson M, Sild S, Maran U. Non-linear QSAR treatment of genotoxicity. Mol Simul 2000; 24:229–242.

125. Karelson MM, Zerner MC. A theoretical treatment of solvent effects on spectroscopy. J Phys Chem 1992; 96:6949–6957.

25

Data Mining Applications in Drug Discovery

MICHAEL F. M. ENGELS and THEO H. REIJMERS

Johnson & Johnson Pharmaceutical Research and Development,
A Division of Janssen Pharmaceutica N.V., Beerse, Belgium

1. INTRODUCTION

The search for lead compounds in the pharmaceutical industry has historically followed an inherently iterative process of synthesis and testing (see Fig. 1). Recent developments, however, fueled by the revelation of the human genome [1,2] and the widespread implementation of high throughput technologies are about to challenge the classical synthesis-and-testing paradigm. These new types of technologies are highly miniaturized and operate on a massively parallel processing mode. As a consequence, the number of generated data points grew exponentially over the last decade. On first sight, these large amounts of data seem to provide drug discovery research with an unprecedented number of opportunities. However, the increase in compound and particular data flow has led to a paradigm shift, diverting the focus of attention from pure synthesis and testing activities, toward the handling and analysis of the produced data. The extraction of knowledge became a bottleneck in modern data driven drug discovery, endangering the informative interplay of between synthesis and testing.

This chapter is reviewing the data mining approach to data analysis. It is a quite novel approach which focuses on the analysis of large data sets [3]. Given this quality, data mining has the potential to become an important tool in modern drug discovery research.

Each attempt to cover data mining and its applications in drug discovery is bound to be incomplete. Therefore we restrict our discussion to those areas of drug discovery that are relevant to the field of medicinal chemistry. The chapter is divided in

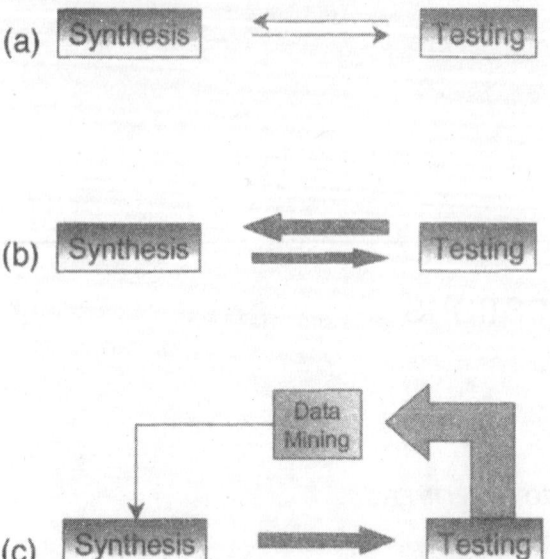

Figure 1   Change of the synthesis-and-testing paradigm in drug discovery research. (a) Classical situation where compound (from left to right) and data flows (right to left) are balanced. (b) Increase in synthesis and testing capacities increased data and compound flows. (c) Further increase of data flow due to increased profiling activities and introduction of data mining as a tool to extract information and knowledge.

two major sections. The first section is quite theoretical and discusses general principles in data mining. The second section provides examples of applications within drug discovery.

2. PRINCIPLES OF DATA MINING

2.1. Definition

Data mining is a quite new discipline representing the confluence of ideas that originated from several well-established scientific disciplines such as statistics, machine learning, pattern recognition, and database technology. There are many definitions of data mining, one of the early statements by Fayyad et al. [4] as "a nontrivial process of identifying valid, novel, potentially useful and ultimately understandable patterns or models in data" describes best the scope of this novel approach to data analysis.

Its main objective is to enable and maximize the extraction of useful and interesting knowledge from large data sets in an efficient and timely manner. This is achieved by the novel concept of computational-driven exploratory data analysis that automates many of the analytical tasks by using one or several computing programs. Data mining puts much emphasis on the utilization of prior or context-specific information of the subject of interest. This context-specific knowledge does not necessarily need to be part of the data but can come from the expert user knowledgeable in the field of interest, the so-called domain expert. Although integration of this context-specific knowledge in the early analysis process might decrease the objectivity toward the presented data, practice has shown that it becomes much easier for the domain expert to identify those patterns in the flood of facts that are most interesting or valuable. In that respect, data mining may be defined as a data-driven analysis process in which the algorithm generates a novel description of the data and it is up to the prepared mind of the domain expert to translate the information, inherent to the description, in useful and novel knowledge [5]. It goes without saying that the simplicity of the algorithm-generated description of the data is important in conveying information. This is the reason why visualization of data is important in data mining [6].

Besides this novel knowledge discovery concept, the emphasis on large data sets brings about further requirements. Scalability of the analysis algorithms, i.e., the adaptation of the algorithms to larger data sets, as well as an improved link to the data under consideration, such as by data warehouses or data marts, became integrated in the knowledge discovery strategy [3]. An understanding of all these factors, the "mathematical modeling" view, the "computational algorithm" view, and the "database management" view are essential in the science of data mining.

Although most of the terms will be revisited later in the chapter, it is important to perceive data mining as a process that involves the interactive exploration of large data sets with the intention to identify novel, interesting, or useful patterns or models. In this respect, data mining shows large resemblance with the recently introduced "global approach" to data analysis [7]. It should be noted that the statement above also corresponds to the definition of the process named knowledge discovery in databases (KDD). However, the term data mining has become more popular, and is therefore used in this chapter.

2.2. Model and Pattern

Data mining refers to extracting or "mining" knowledge from large amounts of data [3]. This knowledge is typically manifested in the form of a model or as patterns [8]. Both are based on the discovery of structures or signals in the data and the relationship of these structures or signals to the existing knowledge. A model can be considered as a global description or overall summary of the data encapsulating and rendering its main aspects and trends. This is in a way the standard statistical approach. One can speak of a neural network regression model, a cluster model, a decision tree model, and so on. Knowledge derived from such structures represents a large-scale summary of a mass of data. In contrast, a pattern is a local structure referring to only a relatively small number of objects in the data set. Patterns are of interest because they represent departures from the general run of the data. Outliers are typical manifestations of patterns because they may suggest deviations from the general course within the data.

It should be emphasized that within data mining, the patterns and models are clearly the primary products of the data. Data mining is data-driven, indicating that the patterns and models in the data give rise to hypotheses. This is in contrast to situations where hypotheses are generated from theoretical arguments about underlying mechanisms, which are then confirmed or invalidated by posterior analysis of the data. Hand et al. [8] provide several examples around this topic and discuss implications of this important distinction.

2.3. The Data Mining Process

The extraction and discovery of knowledge in data mining involves an interactive and iterative process of several stages, each again consisting of a sequence of steps. In the first stage, the so-called exploration stage, the data are prepared to be handled by the data mining method of choice. The way data are collected and preprocessed is at least as important as the application of the data mining methods. Important elements in *the exploration stage* are:

1. Integration of data from different primary data sources into a target database. Different sources of data are often stored in databases having their own unique properties (e.g., the total number of records/objects in the database, the way how these objects are labeled, the way how the data is actually stored for each object, the sequence of the records in the database). Therefore, it is important that the integration of the data into the target database is carried out carefully so that the records in the first data source are correctly linked with the records in the second database, etc.

2. Selection of the data in the target database. The data that are stored in the primary source databases are often collected by different users using different automated methods and business rules. As a consequence, the quality of the data is not the same for all the records in the database and data will be contaminated. Depending on the goal of the data mining process and method, the data in the target database should be cleaned first and obvious inconsistencies between data points should be resolved.

3. Preprocessing/transformation of the data in the target database. The result of many data mining methods is largely affected by the way the data are represented in the target database. Many techniques can be applied to the data to enhance the outcome of the mining techniques. For a description of how molecules are represented in databases, the reader is referred to the paragraph "handling representations and descriptors."

Additional steps in the exploration stage include the splitting of the data set in different representative subsets (see below), and the reduction of the dimensionality of the data both for the number of records/objects (sampling) and the number of features/variables (feature selection).

In the second stage, the *model building stage*, models are built and patterns identified. Again, this stage can be subdivided into several important elements:

1. Mining of the data. In this step, methods are applied for the enumeration of patterns or compilation of models. Many different data mining methods can be used to extract information from the preprocessed target database. The most suitable method is determined in the next step of the process.

2. Validation and verification. After patterns and models have been obtained from the data, it is important to validate these by applying the detected patterns and models to new subsets of data.

3. Evaluation and interpretation. A very important step in the data mining process is to evaluate and interpret the obtained patterns and models. This process is also known as the translation of the information from the data into knowledge. This knowledge can eventually guide the search for better and new patterns and/or models.

In the last stage, the *deployment stage*, the validated and interpreted patterns and models are applied to new data. One way to do this is by means of consolidation and incorporation of the discovered knowledge into a decision support system.

Fig. 2 provides an illustration of the basic flow in a data mining study. The large interdependence of steps and the numerous decision points in the process gain such a degree of complexity that favors informed human intervention over complete automation. In that respect, the success of a data mining process largely on the domain expert and only secondary on the computational aids or tools. Therefore it is necessary to develop a deeper understanding of the application domain and to properly comprehend the goals of the end user.

2.3.1. An Illustrative Example of a Data Mining Process

The now seminal study by Lipinski et al. [9] represents an excellent showcase to demonstrate the workflow in data mining. Here we will focus on the procedure; interested readers are referred to Chap. 22 by Tudor Oprea, which covers the scientific aspects of this study.

Table 1 summarizes the sequence of actions taken in the course of this study. The *exploration stage* included extensive cleaning and removal of records from the primary data source, the World Drug Index (WDI) [10]. The cleaning was primarily based on 3 out of more than 15 attributes available in the WDI annotating each record. It should be noted that the selection of the three attributes is by far not obvious and based on a deeper understanding of the content of the WDI and the domain of research. The elimination of records from the original data source resulted in the USAN library a truncated target database, which included less than 5% of the original number of

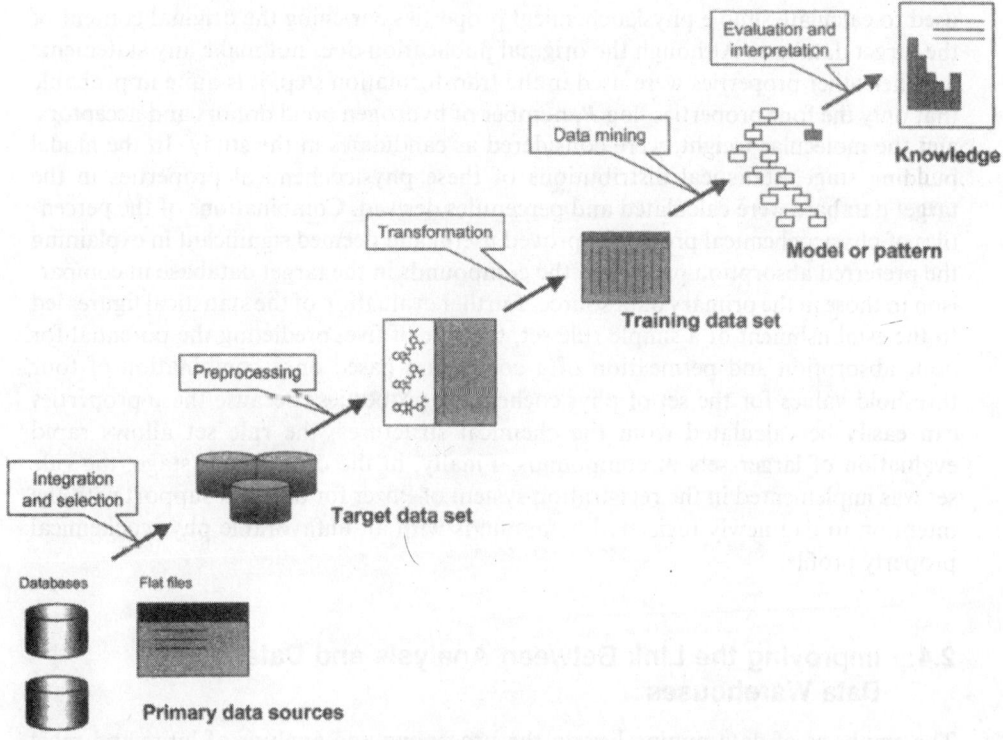

Figure 2 Work flow in data mining.

Table 1   The Data Mining Process Employed by Lipinski et al. [9]

Generic Steps in Data Mining	Corresponding Actions in Ref. 9
Selection	Selection criteria applied to WDI compounds for the definition of the USAN library • Presence in phase II studies • Clinical exposure • Absence of four different types of substructures
Preprocessing	Not applied
Transformation	Calculation of physicochemical properties from chemical structure information • AlogP (ClogP) • molecular weight • O, N, NH, and OH counts
Data mining	• Percentile analysis of individual properties • Percentile analysis of combinations of properties
Evaluation and interpretation	• Comparison of derived percentiles of the target database with those of the complete WDI database • Derivation of simple rule set
Consolidation of knowledge	Implementation of rule sets as alerting system in Pfizer registration system

records in the WDI database. The chemical information in the target database was used to calculate simple physicochemical properties enriching the original content of the target database. Although the original publication does not make any statements whether other properties were used in the transformation step, it is quite improbable that only the four properties, log P, number of hydrogen bond donors and acceptors, and the molecular weight, were considered as candidates in the study. In the model building stage, statistical distributions of these physicochemical properties in the target database were calculated and percentiles derived. Combinations of the percentiles of physicochemical properties proved useful and deemed significant in explaining the preferred absorption profiles of the compounds in the target database in comparison to those in the primary data source. Further evaluation of the statistical figures led to the establishment of a simple rule set, the rule of five, predicting the potential for poor absorption and permeation of a compound based on a combination of four threshold values for the set of physicochemical properties. Because these properties can easily be calculated from the chemical structures, the rule set allows rapid evaluation of larger sets of compounds. Finally, in the deployment stage, the rule set was implemented in the registration system of Pfizer for decision support with the intention to flag newly registered compounds with an unfavorable physicochemical property profile.

2.4. Improving the Link Between Analysis and Data: Data Warehouses

The emphasis of data mining lies on the processing and analysis of large and most often heterogeneous data sets. Therefore improving the link between the relevant data sources and the computational platform that performs data mining seems a logical and necessary requirement for the timely and successful extraction of knowledge [3]. However, today's database infrastructure is often not aligned with the demands of data mining. Organizational databases that are used to conduct the daily operations and that are tuned to answer well-defined and repetitive queries do not meet the demands of complex retrieval. In addition, relevant data for decision making are most often distributed across multiple organizational databases. These are either not connected or the connections are not enabled for data-intensive and/or query-diverse communication [8].

Data warehouses have been proposed as a novel form of storage organization focusing on the information retrieval on a large scale (Fig. 3). Data warehouses have been defined as an integrated collection of data that stores data from various operational databases [3]. A typical example in drug discovery research is the combination of biological and chemical data for structure–activity relationship analysis in one integrated system. However, a data warehouse is not just the sum of the individual operational databases. Due to the specific requirements of flexible and fast querying of large data sets, restrictions have to be imposed on what to store in a data warehouse and how. This is the reason why, in a complex and domain expert-driven selection process, one has to select those properties that might be relevant for the subject of interest. Relevant data items might be augmented by computer-derived properties that are computationally expensive. An example is the annotation of sequence databases with literature data, or augmentation of protein sequence data with predicted properties such as secondary structure or protein domain assignments. Examples in the chemical area would be the storage of computationally expensive quantum mechanical properties [11] or pharmacophore fingerprints.

Examples of data warehouses published in literature are the SPINE [12] and CerBeruS [13] systems. While the SPINE system is focusing on the support and mining of protein crystallization data, the CerBeruS system links data relevant for the SAR analysis of larger data sets such as HTS data.

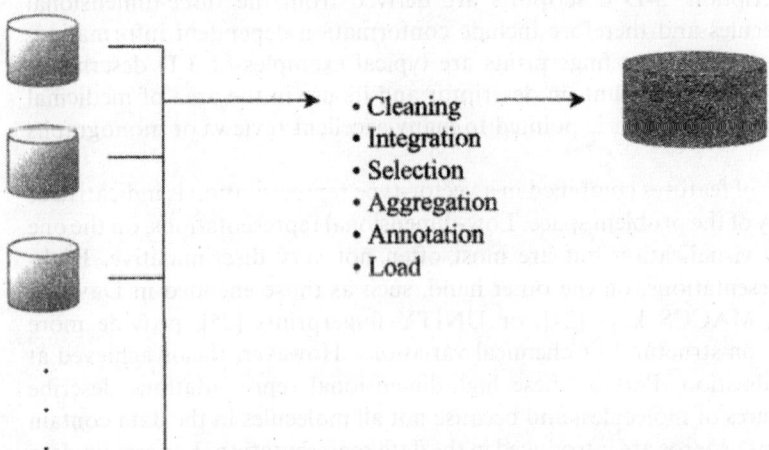

- Cleaning
- Integration
- Selection
- Aggregation
- Annotation
- Load

Figure 3 Differentiating between organizational databases (left) and a data warehouse (right).

2.5. Representations and Descriptors

A general concern in data mining is the representation of objects. Molecules, text documents, images, nucleic acid, or protein sequences all represent nonnumerical objects. However, all data mining methods require the transformation of objects into an algebraic, i.e., numerical, representation.

The most common representation in data mining is the *propositional* or *vector representation*. In that representation, each object is described by a set of properties summarized in a vector of fixed length. This vector representation is also the preferred form for the characterization of molecules (see Fig. 4). Each molecule is thereby characterized by the same set of features or properties. This arrangement of features is characteristic for each molecule and is often referred to as fingerprint. Features in such a vector representation can be encoded in binary form indicating its presence (bit on) or absence (bit off), by a series of integer counts indicating its frequency of the occurrence, by real numbers, or by a combination of these different forms. Next to this propositional representation, *relational representations* are becoming more practiced in data mining [14]. In a relational representation, molecules are characterized by a set of relations instead of a series of features, resulting in a much more compact description of objects in the data. These relations can be quite complex [15]. Here, we will focus on vector-based representations because most data mining methods still require that the feature space is presented in a vector representation.

The type of information that is collected in a vector-type representation depends on the type of descriptor(s) used during the transformation of a molecular object into a set of numerical features. Descriptors provide characteristic views on particular properties of a molecule. One roughly distinguishes 1-D, 2-D, or 3-D type of descriptors. 1-D descriptors characterize a global property of a molecule. Examples are physicochemical properties such as molecular weight, dipole moment, log P, or biological properties such as pIC_{50}, pK_i, or ED_{50}. 2-D descriptors are derived from the chemical graph of a molecule, and characterize topological, fragment, atom-type or path-related properties. Topological indices [16], atom-pair descriptors [17], topological torsions [18], or keys of substructures [19] are typical examples of this type of description. 3-D descriptors are derived from the three-dimensional structure of molecules and therefore include conformation-dependent information. For example, pharmacophore fingerprints are typical examples of 3-D descriptors [20]. For a more detailed account on descriptor and its use in the area of medicinal chemistry, the interested reader is pointed to many excellent reviews or monographs [19,21,22].

The number of features combined in a vector-type representation is indicative of the dimensionality of the problem space. Low-dimensional representations, on the one hand, allow easy visualization but are most often not very discriminative. High-dimensional representations, on the other hand, such as those encoded in Daylight fingerprints [23], MACCS keys [24], or UNITY fingerprints [25], provide more detailed accounts on structural or chemical variations. However, this is achieved at the cost of visualization. Part of these high-dimensional representations describe specific local features of molecules, and because not all molecules in the data contain these features, gaps or zeros are introduced in the data representation. For certain data mining methods, this could be problematic. In many cases, dimensionality reduction procedures are applied to reduce the complexity of the representation. The reduction of the dimensionality is accomplished by means of 1) variable selection procedures, 2)

	CC-NC	OC-CN	...	C(1,2)-6-C(3,1)	N(2,0)-5-C(3,1)	C(2,0)-4-C(2,1)
(structure 1)	1	0	...	1	0	1
(structure 2)	1	1	...	0	0	0
(structure 3)	0	0	...	0	1	1
(structure 4)	1	0	...	0	0	1

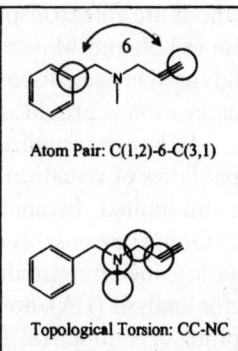

Atom Pair: C(1,2)-6-C(3,1)

Topological Torsion: CC-NC

Figure 4 Typical vector representation employed for representing chemical objects in data mining. Each object is represented by a set of vector elements that encode chemical, topological, biological, or structural properties of the chemical object. The illustration shows paragyline, an MAO inhibitor, and three of its analogs characterized by bit string representations indicating the presence ("1") or absence ("0") of the so-called topological torsions and atom pairs. Topological torsions [18] and atom pairs [17] are descriptors that are generated from the topological (2-D) representation of a molecular structure capturing local substructural environment or more global-path-related characteristics, respectively (see inlet for examples). The atom code C(1,2) in one of the atom pair descriptors indicates a carbon atom with one connection to nonhydrogen substituents and two π-electrons, i.e., sp-hybridized carbon. (From Ref. 53.)

mapping and/or projection, or 3) designing low-dimensional descriptors that contain enough information to be discriminative in future studies. Examples of this latter type of descriptors are BCUT descriptors [26] and the recently introduced 1-D similarity descriptors by Dixon and Merz [27].

2.6. Tasks in Data Mining

It is convenient to categorize data mining into types of tasks corresponding to the different objectives. The categorization below is not unique and underlines only the most dominant tasks encountered in drug discovery applications.

2.6.1. Descriptive Data Mining

In descriptive data mining, the aim is to present the derived patterns or models in a concise and convenient form providing either a novel, simpler, or comprehensive perspective on the data. It is essentially a summary of the data (for models) or a compilation of the most interesting structures in the data (for patterns), permitting to study the most important aspects of the data without being obscured by the sheer size of the data set [5,8]. In that respect, visualization remains, such as in statistics, to play a very important role in the area of descriptive data mining [6]. The strength of visualization lies in the fact that it represents the data in an unbiased fashion, making it ideal for obtaining a global summary of the data. In addition, visualization approaches are also helpful in the interpretation of the derived computational models [28]. Next to classical visualization methods such as scatter plots, bar charts, or histograms modern visualization methods such as parallel coordinate representations, trellis plots, brushing, or dimensional stacking, become more popular [29]. These methods are able to display and dynamically link more than three variables. Ladd and Kenner [29] and Meyer and Cook [30] provide brief overviews on visualization methods used in drug discovery. A more elaborated compilation of the art of information visualization is provided by Ref. [31].

Although visualization has proven to be an important tool in data mining, the capabilities of visualization techniques of data sets in high-dimensional feature space are still limited. Estimates are that human cognification is limited to 20 dimensions [32]. Going beyond this limit, algorithmic methods reducing the dimensionality of the data set, including multidimensional scaling, principal component analysis (PCA), factor analysis (FA), nonlinear mapping methods (NLM), or clustering, can be aids to explore or summarize the higher-dimensional space [4]. Also, statistical summary counts such as correlation or association measures, probability measures, or simple rule sets can provide means to obtain a first grip on the data. Many of these techniques have found their way in drug discovery and are usually applied in the area of medicinal chemistry.

2.6.2. Predictive Data Mining

Predictive data mining is aiming at predicting a certain property Y, the so-called target property, and its relationship to the properties X. The goal of predictive data mining is to estimate a mapping or a function. Two types of predictive data mining approaches can be distinguished. If the target property is binary, such as in "TRUE" or "FALSE," or categorical, such as for the categories "active", "inactive", and "equivocal," the data mining task is referred to as *classification*. Data mining methods such as decision trees, artificial neural networks (ANN), support vector machines (SVM), discriminant analysis, or logistic regression fall in this type of approach. If the target property is interval scaled, i.e., being continuous, the task is called *regression*. Here methods such as regression trees, artificial neural networks (ANN), or partial least squares regression (PLS) are often applied.

In typical data mining problems, very little is known in advance about the functional form of the mapping between X and Y. The most important factor in favoring one certain functional form over the other is by comparing the predictive performance of the derived models. The performance of a regression model is primarily judged based on the difference between the actual and predicted values for the response variable, while the performance of a classification model is judged by the misclassification rate, which refers to the percentage of the incorrectly classified observations. However, both measures provide only a limited view on the validity of the model. For example, if the diversity of the training data set is restricted, only limited conclusions on the validity of the model can be drawn. Therefore the validity of a predictive data mining model should always be challenged by several methods. Examples of validation methods are external hold-out test sets, cross-validation, bootstrapping, or randomization. The choice of the method depends on the amount of data that is available and on the dimensionality of the problem. Validation methods such as hold-out test sets or randomization tests are computationally inexpensive and provide a first impression on the validity of the model. More expensive methods such as cross-validation or bootstrapping have the advantage that they can be useful in determining error significance or confidence ranges.

2.6.3. Comparative Data Mining

In comparative data mining, one distinguishes between overlay analysis and the retrieval of patterns from one or several data sets given a set of patterns of interest. The latter task is also known as retrieval by content [8]. Typical manifestations of comparative data mining in drug discovery are for example similarity analysis of chemical compounds [33], or the comparison of large chemical libraries [34].

The notion of distance or similarity plays an important role in comparative data mining. The type of data mining methods used in comparative data mining tasks depends on whether the focus is on retrieval of content or overlay analysis. While the emphasis for retrieval of content is on the speed with which similar patterns can be identified in different data sets, overlay analysis focuses on the degree of coarseness and summarization of the comparison. A typical example of a data mining method used for retrieval by content is latent semantic structure indexing [35]. Overlay or superposition analyses in chemical library analysis are performed by methods such as clustering [36], mapping [37], or principal component analysis [34]. As in descriptive data mining, visualization of the compared data sets can provide interesting information on the scope of data and potential outliers.

2.7. Components of Data Mining Methods

In an attempt to simplify the broad range of data mining methods, Hand et al. [8] identified four primary components commonly found in all data mining methods.

The first component is the structure of a data mining method, also referred to as the *model representation* or *functional form*. It determines the boundaries of what can be approximated or learned by the data mining method. For example, decision trees and artificial neural networks are both methods that are used for classification tasks. However, the underlying structure of both methods significantly differs. While a decision tree model partitions the attribute space in one-dimensional splits parallel along the axes of the chosen attributes, a neural network assumes nonlinear dependencies over the whole attribute space. As a consequence, neural networks are able to learn any type of functional relationship, while decision trees will have problems to learn simple relationships such as $X = Y$ from a set of data. The set of structures that are commonly found in current data mining methods is quite extensive. Typical structures are linear regressions, hierarchical dendrograms, decision trees, Kernel density distributions, nonlinear functions, and association rules [4]. Each of these functional forms has its own limitations and it is up to the user to decide which functional form tackles the problem best. Because this decision is most often not known ahead of time, many commercial data mining products, such as SAS Enterprise Miner [38] or Clementine [39], implement several data mining methods. They leave it up to the user to decide which functional form is best suited to describe the problem.

The *scoring function** is the second component of a data mining method. The scoring function judges the quality of the fitted model or pattern. Typical scoring functions in predictive data mining are misclassification rate for classification tasks and the sum of squared error for regression tasks. In descriptive data mining, the portfolio of scoring functions is much broader and differs from algorithm to algorithm. For example, in clustering the sum of squared errors within each cluster is a typical measure. Other measures include the "support" in association rule mining, or probabilistic measures in Kernel density distributions. Typical scoring functions in comparative data mining are related to the distance between two data sets. Euclidean distances or Tanimoto indices are as such typical examples. During learning or run time, the value of the scoring function must be optimized. This is carried out by fitting parameters to the model or pattern. Therefore, it is important that the scoring function reflects the relative practical utility of different parameterizations of the model or pattern structures. The scoring function is critical for learning and generalization.

The *search* or *optimization method*, the third component, describes the computational procedure to search over parameters and structures. Issues here include the computational methods used to optimize the scoring function and to search related parameters such as the maximum number of iterations or convergence specifications for iterative algorithms. Typical search methods are greedy search, gradient-dependent search methods, or breadth search methods [4]. One distinguishes between searches that involve only the optimization of the parameters in fixed structures, and the optimization of structures and parameters for data mining methods that include searches over parameter and structure.

The final component in any data mining method is related to the way in which data are *stored* and *accessed*. Most well-known data analysis algorithms in statistics and machine learning have been developed under the assumption that all individual data points can be quickly and efficiently accessed in random-access memory (RAM). However, many massive data sets will not fit in available RAM and will, therefore, still reside largely on disk. This limitation has driven the development of methods that optimize data access and data processing. Three main strategies can be identified. The first strategy attempts to reduce the search space by aggregating and cross-linking only those data that are relevant to the subject or domain of interest. The second strategy is sampling. The last strategy is concerned with the development of improved indexing strategies. Indexing is an well-applied technology in databases. In recent years, development has begun on techniques that support the "primitive" data access operations necessary to implement efficient versions of data mining algorithms. For example, tree-structured indexing systems have been used to retrieve the neighbors of a point in multiple dimensions [8].

---

*Hand et al. [8] originally refers to it as the score function.

2.8. Tools and Methods in Data Mining

This part provides a short theoretical introduction to some tools and methods that have recently been used in drug discovery data mining studies. More detailed accounts have been compiled by Hand et al. [8], Fayyad et al. [4], and others [3,40].

2.8.1. Cluster Analysis

Clustering is a classical tool in computational medicinal chemistry and chemical information [41–44]. Cluster analysis primarily aims at identifying natural groups of similar objects, the so-called clusters, in a data set of interest. Since clustering results in partitioning objects into a smaller number of groups, cluster analysis is helpful in reducing the complexity of a data set. Alternatively, clustering may serve as a preprocessing step for other data mining tasks which then operate on the detected clusters. Newer forms of clustering developed in the field of machine learning go beyond the simple identification of like groups. Conceptual clustering, for example, combines cluster-driven grouping of molecules and chemical characterization of the obtained clusters. An example of a conceptual clustering approach is the method implemented in the Leadpharmer program by Bioreason [45]. The Leadpharmer program attempts to characterize active compounds from a structure–activity data set by combining cluster analysis and maximum common substructure analysis [46].

Clustering algorithms can be classified into four major approaches: hierarchical methods, partitioning-based methods, density-based methods, and grid-based methods. Here, we will focus on the hierarchical cluster approach because it is often used in the context of structure–activity analysis. Recent research has suggested that hierarchical methods perform better than the more commonly used nonhierarchical methods in separating known actives and inactives [41].

In hierarchical clustering, a so-called dendrogram, logically representing the relationship between the objects of the data set, is developed 1) by splitting the data objects into groups to maximize intercluster dissimilarity (divisive clustering), or 2) by linking data objects incrementally based on intracluster similarity (agglomerative algorithms). One of the primary features distinguishing hierarchical techniques from other techniques is that the allocation of an object to a cluster is irrevocable; that is, once an object joins a cluster, it is never again removed or fused with other objects belonging to some other cluster.

Fig. 5 provides an example of a hierarchical clustering. Typical hierarchical clustering methods are Ward clustering, Guenoche, or average linkage clustering. Hierarchical cluster methods are computationally very demanding. This is the reason for further developments in this field. Parallelization of classical hierarchical cluster algorithms is one option to cope with the growing average sizes of data sets [47]. In addition, novel methods such as BIRCH [48], Cure [49], or CHAMELEON [50], all methods developed in the field of information studies and data management, provide new opportunities in this field [3].

2.8.2. Self-Organizing Maps

Self-organizing maps (SOMs) are one manifestation of neural network approaches to clustering. They have been extensively used in many fields of computational medicinal chemistry [51]. SOMs consist of a grid of neural elements, each containing a vector of a certain dimension. The map is trained by presenting a series of new data objects to the

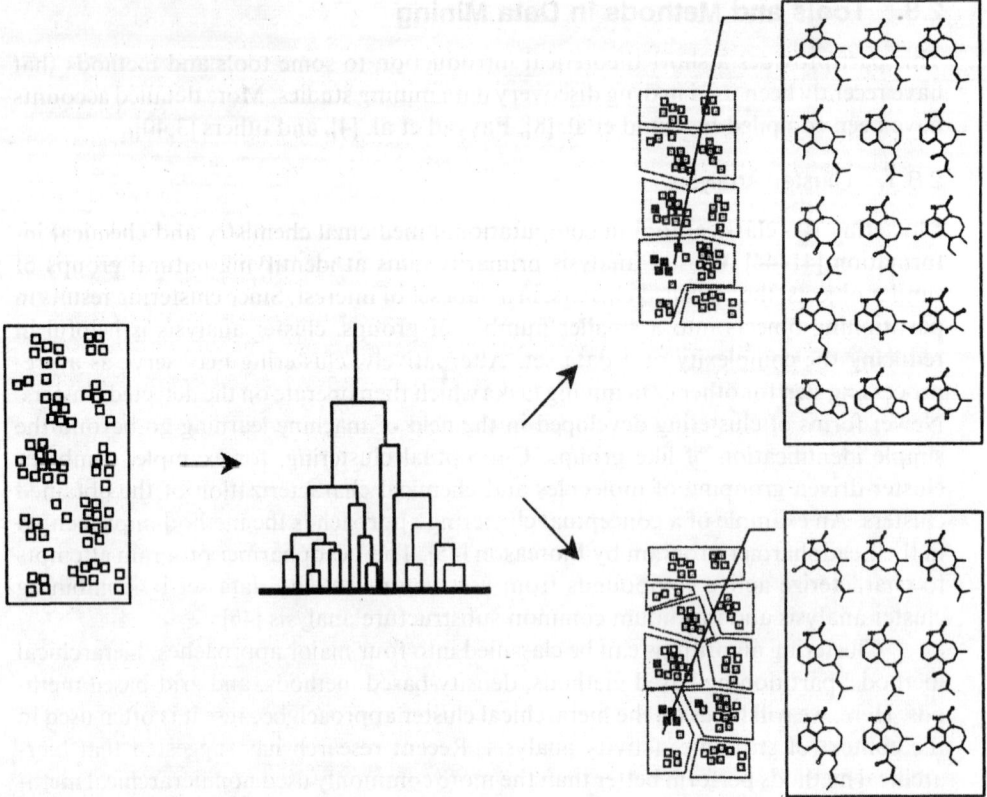

Figure 5 Hierarchical cluster analysis. The dendrogram represents the structural relationships between the molecules of a library. Cluster ensembles of different size and homogeneity can be derived from the dendrogram.

map, and allowing neurons to change their vector in response to the new information. Specifically, when a data vector x is presented, the neuron with a value m closest to x is located in the grid. The value m of this winning neuron is adjusted to reflect the new data. In addition, the neurons in the neighborhood of this winning neuron also have their values adjusted by an amount that decreases with distance from the winner. It is this training of neighboring neurons that gives the map its "self-organizing" property. After the map has been trained, neurons close together in the grid will have similar vectors. Any data belonging to a cluster will therefore appear in some neighborhood of the grid. This neural network approach to clustering has strong theoretical links with actual brain processing [51].

Self-organizing maps are very often used to project concepts on the calculated maps that were not part of the original data set. Fig. 6, for example, displays a typical two-dimensional SOM derived from the structures of more than 40,000 anticancer agents [52]. The spatial relationship between the clusters is indicated by a color-coding scheme projected onto the map so that close and far neighbors are separated by dark and light blue colors, respectively. The size of the brown hexagons provides an indication of the degree of population within each cluster.

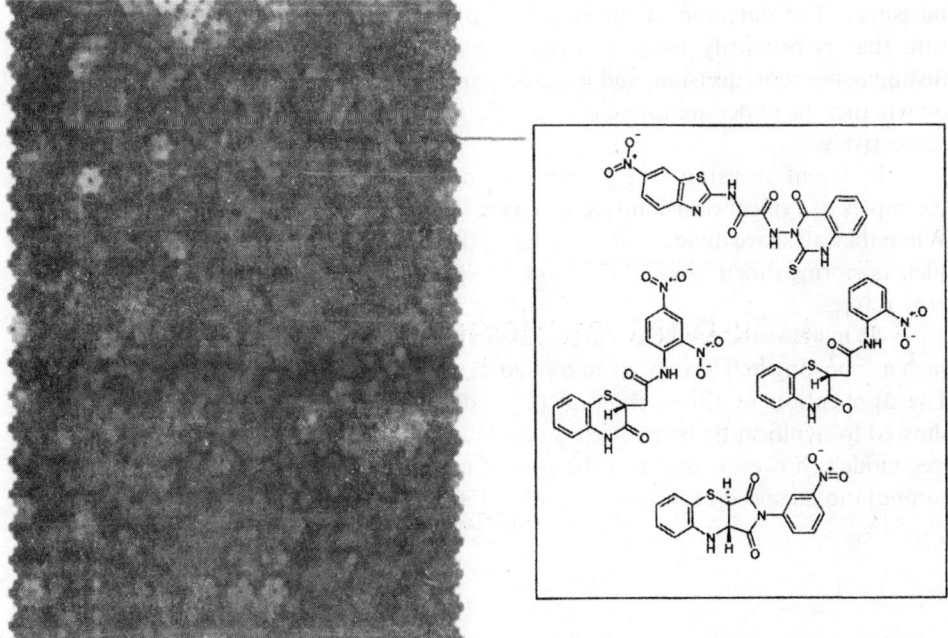

Figure 6 Example of a complex SOM visualizing the structural relationships of more than 40,000 chemical structures of the August 1999 release of the NCI anticancer database [52]. The SOM is partitioned into a hexagonal array of 966 clusters. Distances between clusters are indicated by the colors between clusters (red, close; black, intermediate; purple, far). Close and far neighbors are separated by dark and light blue colors, respectively. As an example, compounds in hexagon 9–23 are highlighted. (Courtesy of Drs. Rabow and Covell. See color plate at end of chapter.)

Self-organizing maps represent an alternative to the more classical use of dendrograms for displaying cluster results. It has proven particularly useful in those cases where the data set does not appear to lend itself to hierarchical organization.

2.8.3. Decision Trees

Decision trees are one of the most versatile tools in data mining. They have been employed for the extraction of patterns in large structure–activity [53–55] or structure–property data sets [56], and for the development of classification models [56,57].

Decision tree learning is based on a quite simple algorithm. It attempts to partition the feature space and at the same time maximizing a score of class purity so that the majority of points in each cell of the partitioning belong to one class. The partitioning of the feature space is recursively carried out; that is, each of these cells is subsequently split into two more pure cells. This process, also called divide-and-conquer, is repeated as many times as is necessary to reach a predefined level of purity. To split a given cell, a search is performed over each possible threshold for each variable to find the threshold split that leads to the greatest improvement in a specified score function. Typical split functions use either entropy-related criteria or χ^2

measures. The outcome of the recursive procedure is a flow-chart-like tree structure that is primarily used in producing classification models (see Fig. 7). One distinguishes root, decision, and leaf nodes in the flow-chart-like structure, dependent on whether the nodes are found on the top, in the middle, or at the bottom of the tree, respectively.

Different decision tree programs have been used in data mining studies. Examples of typical decision tree programs are C4.5 [58], C5.0 [59], or CART [60]. While they all share divide-and-conquer as their search strategy, they differ in aspects such as scoring functions, evaluation of the split, and postprocessing or pruning of the grown tree.

To increase the predictivity of decision tree classification models, statistical tools such as boosting [62] have been employed in the context of decision tree classification. The application of this technique in predicting structure–property relationships showed to significantly increase the accuracy and robustness of the obtained decision tree models; however, this is at the cost of comprehensiveness of the model and the computational speed of model generation [56].

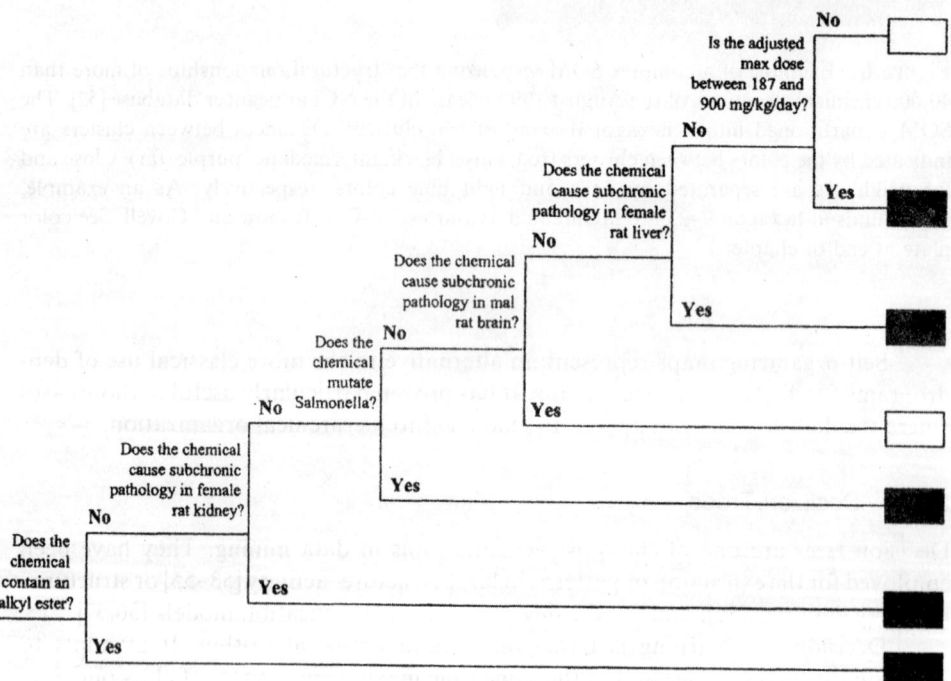

Figure 7 Illustration of a decision tree generated with C4.5 obtained for the classification of chemical carcinogens in rodents [61]. The classification of 122 chemicals was based on a molecular feature representation that included structural alerts, biological activities in different assays and pathological indicators. The classification leaves at the right hand side of the decision tree are visualized graphically by filled or open boxes. A filled box indicates that the chemicals in that box are classified as rodent carcinogenic; open boxes indicate chemicals without rodent carcinogenicity. Note the strong imbalance of the decision tree.

2.8.4. Association Rules

Association rules are among the most popular representations for local structures or patterns in data mining [3]. These patterns are inferred without prior knowledge of predefined classes. A rule consists of a left-hand side proposition called the antecedent and a right-hand side called the consequent. A typical example rule is shown below [14]:

> If a compound shows no activity in the cytotoxicity assay and if the compound contains a sulfide group, **then** the compound is not carcinogenic, with a confidence of 86% and a support of 6%.

This example joins three items or attributes in one "itemset." Because the number of items in this rule is three—cytotoxicity yes/no; sulfide carrying yes/no; carcinogenic yes/no—it is referred to as three-itemset. The frequency of the co-occurrence of the three itemsets is described by the "support," which corresponds to the joint probability of finding these three items in the data set. The "confidence" is referring to the conditional probability that, given no activity in the cytotoxicity assay and given the occurrence of a sulfide group in the molecule, the consequent "not carcinogenic" will happen. This rule structure is quite simple and interpretable, which helps explain the general appeal of that method.

How are rules or patterns extracted from a data set? The typical strategy in association rule mining is based on the identification of itemsets whose frequency corresponds at least to a predefined minimum support count. Based on these frequent itemsets, strong association rules are created that satisfy minimum support and minimum confidence. The Apriori algorithm [63] was the first of its kind that addressed the detection of frequent itemsets in a very efficient way. Because as the number of combinations increases the more items are combined, the search space exponentially grows. Apriori overcomes this problem by employing an iterative approach known as a level-wise search in which k-itemsets are used to explore $(k+1)$-itemsets. At each level, candidate $(k+1)$-itemsets are generated by joining k-itemsets of the previous level. To minimize the otherwise exponentially growing number of candidates, only those $(k+1)$ candidate itemsets are examined that are derived from frequent itemsets of smaller size. This method leads to the elimination of a very large number of itemsets that otherwise had to be examined in an exhaustive procedure. The algorithm finally stops if either a predefined number of levels has been achieved or if no itemsets could be detected to fulfill the minimum support criterion.

Several different flavors of the Apriori algorithm have been implemented. An example of an association rule mining program that uses this type of strategy is the Warmr program [14].

2.8.5. Multilayer Perceptrons

Multilayer perceptrons belong to the large and important family of artificial neural networks (ANNs) [64]. Artificial neural networks are of a class of highly parameterized statistical models that have attracted considerable attention in data mining applications in drug discovery and are particularly used in the context of predictive data mining. Artificial neural networks work by forming a linear combination of the input variables and transforming this linear combination via a nonlinear transfer function. Multilayer perceptrons, also referred to as feedforward ANN, adopt this as the basic element. However, instead of using just one such element, they use multiple layers of many such elements. The outputs from one layer—the transformed linear combinations from each basic element—serve as inputs to the next layer. In this next layer, the inputs are combined in exactly the same way—each element forms a weighted sum that is then nonlinearly transformed.

There is no limit to the number of layers that can be used, although practicality and the fear for overfitting leads to the rule of thumb to reduce the number of layers to a bare minimum. The strength of a multilayer-perceptron, i.e., the approximation of any functional form, is also one of its weaknesses. To correctly learn the underlying functional form of a data set, a massive data set has to be used. This makes it the privileged tool in data-driven data analysis applications. However, neural network can be rather slow to train. This is due to the large number of parameters that needs to be optimized during the training stage. Therefore the learning time can limit the applicability of ANNs in data mining problems involving large data sets.

2.8.6. Sampling Methods

Although many data mining methods were selected for their computational efficiency, it is easy to predict that the pace with which databases are growing will outperform the scalability of current data mining algorithms. Also, hardware improvements may not always be an option because it is linked to rather expensive investments. Sampling, although not a data mining method itself, is a very efficient way to decrease the dimension of the input data set by this reducing the time for developing a data mining model [65]. The advantage of sampling for data reduction is that the costs of obtaining a sample or subset is proportional to the size of the sample or subset, n, as opposed to N, the data set size. Hence the complexity is potentially sublinear to the size of the data, making it an attractive and cheap alternative to the problem of data reduction. Many data mining methods implement sampling methods in their algorithm; for example, the decision and regression tree programs within the SAS Enterprise Miner software suite [38] use sampling techniques to reduce the number of data points brought into computer memory.

The most prominent sampling method is simple random sampling without replacement. A random sample of the total data set is created by simply drawing n out of N objects; the probability of drawing an object is equally likely. Next to simple random sampling without replacement is random sampling with replacement, i.e., an object is drawn from a data set, recorded and then replaced back, so that it can be drawn again. Cluster sampling is a very popular method in chemoinformatics applications for diversity analysis of large compound libraries or as tool in sequential screening [13,36,66]. A cluster sample is generated by clustering the data set in disjoint clusters from which one or a few objects are drawn. Stratified sampling is another method that is applied to cases where a higher concept forces the data to split in disjoint parts called strata. A stratified sample is generated by obtaining a simple random sampling at each stratum. The sampling frequency can be different in the different strata. This sampling method is especially applied when the data are skewed. If the data are highly skewed, biased sampling methods such as the rare-event modeling method [57] (see Fig. 8) seem better suited, in particular if a neighborhood behavior around a certain class of object should be maintained.

Sampling assumes that relationships and features appearing in the whole data set will be tractable as long as they are sufficiently represented in the sample. This makes it an ideal tool for model searching and building. However, it is not suited in situations in

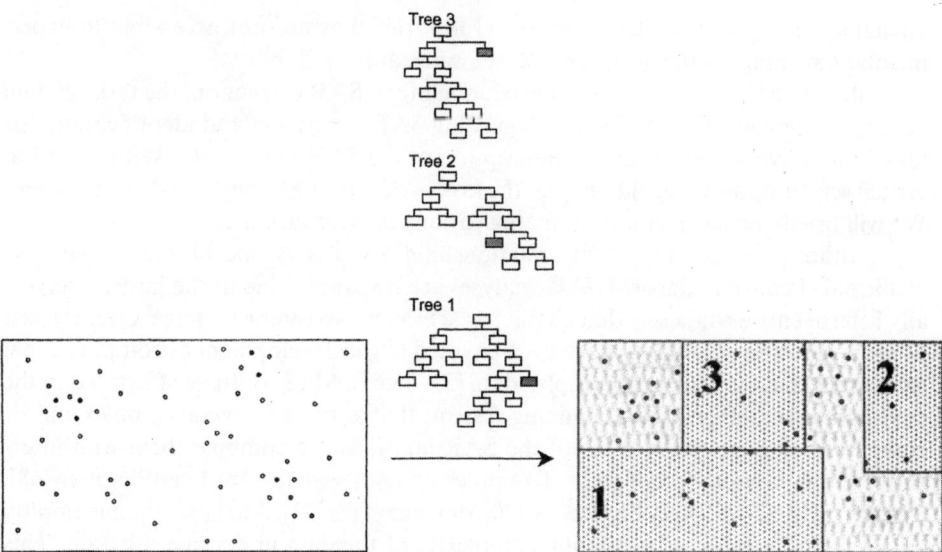

Figure 8 Principles of the "rare event" modeling procedures as implemented in the SAS Enterprise Miner software [57]. Irrelevant parts of the problem space are identified (1,2,3) by decision tree learning and removed from the problem space. The remaining part of the problem space is presented to another learning method for deriving the last model. All models together form the final classification model.

which the data mining search strategy is focusing on the discovery of local structures or on not very well-represented patterns. In these instances, sampling will probably not pick up all relevant objects needed for the maturation of such patterns.

3. APPLICATIONS OF DATA MINING IN DRUG DISCOVERY

3.1. General Comments

In this section, we will review several large-scale applications of data mining methods in the fields of structure–activity and activity–activity relationships (SAR) analyses, and in the areas of absorption, distribution, metabolism, excretion (ADME), and toxicity.

3.1.1. Data Mining of Screening Data Sets

The mining of large screening data sets has become routine in modern pharmaceutical drug discovery. Screening data are produced by the automated testing of many compounds against a biological target. Extraction of useful or relevant structure–activity information from this data set, on the one hand, has become difficult because of the size and complexity of the chemical information. On the other hand, these data sets form valuable knowledge databases that can be used for the generation of predictive in silico models. These in silico models are then applied in electronic or virtual screening of even larger chemical libraries that are not accessible to experimental screening on the grounds of costs, time, and availability.

Based on the emphasis on in silico modeling or SAR extraction, the tasks of data mining in these areas are different. Regarding SAR extraction and identification, the task is to derive a comprehensive description of the SAR in the data. With regard to virtual screening and model building, the task is clearly to establish predictive models. We will briefly review methods and applications in both task areas.

Although structure–activity relationships analysis is one of the domains of medicinal chemistry, classical SAR analyses are not applicable to the large, structurally heterogeneous data sets that characterize modern screening systems. Over the last 14 years, considerable efforts have been devoted to the development of computational techniques to cope with such complex data [53–55,67]. Many of these efforts led to the development of integrated data mining systems that automate several complex and/or computationally intensive tasks of the SAR analysis and combine them with interactive graphics and visualization. Examples of such systems are LeadPharmer [68], CerBeruS [13], Distill [69], Leadscope [70], or ChemTree [71]. All these systems employ data reduction to limit the size and complexity of the data under investigation. This reduction is achieved by cluster analysis techniques such as hierarchical clustering or Kohonen mapping, or classification methods such as recursive partitioning or decision trees. It is interesting to note that although the algorithms seem different, the final SAR is always presented in tree-like or hierarchical representations. In several studies using public available screening data sets, it has been shown that these types of data mining systems can successfully assist in the accelerated identification of structural families and SAR rules [53,72,73].

As mentioned above, the second task is to deal with the derivation of predictive in silico models from the existing screening data. Because of the wealth of information, screening data form a valuable knowledge base for activity prediction. Because predictive models enable rapid scoring of molecules, they can be used in virtual screening campaigns for profiling massive data sets either to support, e.g., the design of combinatorial libraries, the acquisition of external compounds, or the selection of compounds in sequential screening experiments. Preferred data mining methods that have been used for learning in silico models include techniques such as artificial neural networks [57,74], nearest neighbor [75], recursive partitioning [76], and decision trees [57]. Also, nonsupervised methods such as clustering or nonlinear mapping procedures have been used for the identification of similar compounds [66,77]. When applying these models to external data sets, improvements of hit rates by about fourfold to tenfold vs. random picking are common [66,76]. However, the predictive performance of these methods is deteriorated in those situations where the distribution between active and inactive compounds is highly unbalanced. This situation is often found in modern high throughput screening campaigns where hit rates vary between 0.1% and 1% [78]. In some cases, hits are so scarce that they can be characterized as rare events. Strategies for modeling rare events have been recently discussed [57].

While the derived data mining models or patterns can be used for screening large collections of compounds in silico, predictive data mining models also gain more acceptance in the quality control of the screening data. Because the biochemical screening of large compound libraries represents a significant expenditure in terms of resources such as proteins and reagents, the number of measurements that is performed on a compound basis is most often restricted to a minimum. For this reason, compounds are usually tested without replicates in the first stage of an HTS campaign. This has significant implications on the quality of the produced experimental data. Therefore it is of pivotal importance to establish mechanisms that help to identify misleading or inadequate information within the experimental data. Recently, an outlier mining approach has been published to identify outliers in primary screening data [79]. The method is based on the development of a SAR description of the data using logistic regression analysis and comparing it with the actually measured biological activities. Strong inconsistencies between the SAR description and the measured data are indicative of potential outliers. Prospective case studies on in-house HTS campaigns seem to prove the validity of this approach [79].

This brief survey highlighted some of the data mining approaches that are now being applied to the analysis of the SAR present in large screening data sets such as that coming from HTS systems. Reviews by Gedeck and Willett [80] or by Young et al. [81] provide excellent starting points to explore further this very dynamic application area of data mining in drug discovery.

3.1.2. Mining Arrays of Biological Assays

Arrays of biological assays are becoming increasingly important to pharmaceutical research and development. Next to the classical way of screening, the recent development of DNA microarrays permits the simultaneous measurement at the expression level of thousands of genes. Information encoded by the output of such screening can be used to gain essential insights in the mechanism of action of drug response and also on alternative targets and modulators. In addition, the activity pattern of several compounds derived from such screening can be used to organize these compounds into families based on their activity.

In a recent analysis by Rabow et al. [52], structure–activity analyses has been combined with the mining of activity patterns to identify compounds with similar activities against cancer cell lines, which thereby facilitate discoveries of potentially new drug leads and new molecular targets. The study used the extensive screening database of anticancer compounds at the National Cancer Institute [82]. This database contains measures of several responses of cell toxicity for over 100,000 compounds tested in various subsets of 60–100 cancer cell lines. A 20k subset of this database was selected and by SOM mapped onto a 41×26 hexagonal array using 80 different and measured cell activities. From this 1066 cluster map (see Fig. 9), 50 regions could be defined that group individual clusters with the most similar response profiles. Interestingly, several compact areas could be assigned corresponding to functional classes of cellular activities: mitosis (M), nucleic acid synthesis (S), membrane transport and integrity (N), and phosphatase-and kinase-mediated cell cycle regulation (P). The assignment of these classes was obtained by projecting 171 clinically evaluated anticancer agents onto the cluster map. Because for most of these compounds, the mechanism of action is known, assignment of activity pattern to cellular activity was possible. Also interestingly, the analysis of the activity pattern showed that there is strong correspondence between chemical chemotype and type of cellular activity.

In a further large-scale analysis on the same database, gene expression and biological screening data were used to identify a correlation between gene expression and cell sensitivity to compounds [83]. Sixty cancer cell lines were exposed to numerous compounds at the National Cancer Institute, and were determined to be either sensitive or resistant to each compound. Using a Bayesian statistical classifier, Staunton et al.

Figure 9 Self-organizing map of more than 20,000 compounds tested in the NCI's tumor cell screen [52]. The map consists of 966 clusters. Each compound in this data set is characterized by a set of more than 60 biological properties. Color bar at lower right indicates the distance between clusters (red, close; black, intermediate; purple, far). Fifty regions have been defined on this map that group together individual clusters with the most similar response profiles. These regions are assigned to six functional categories according to their apparent cellular activity: M, S, N, P. Regions Q and R have not been assigned to an activity class (see text for further clarification). (Courtesy of Drs. Rabow and Covell. See color plate at end of chapter.)

[83] showed that for at least one third of the tested compounds, cell sensitivity can be predicted with the gene expression pattern of untreated cells.

3.2. Data Mining in the Context of Absorption, Distribution, Metabolism, and Excretion

A recent survey [84,85] shows that failures in clinical testing of new molecular entities can be attributed to issues related to the absorption, distribution, metabolism, and excretion (ADME) of compounds. Researchers are nowadays quite aware of the fact that poor ADME properties are at least as unwanted as lack of efficacy or selectivity. Therefore it is desirable to identify indicators that link the chemical and ADME property space; these indicators can guide the synthesis of potent compounds with improved ADME profiles. As mentioned before, one of the first attempts to correlate physicochemical properties with the absorption behavior of a compound on a large scale was performed by Lipinski et al. [9]. More complex data mining studies focused on the development of models for predicting general CNS activity [86], blood–brain barrier penetration [87], oral bioavailability [88], and human intestinal absorption [89]. Databases such as the WDI [10], the CMC [90], or the MDDR [91] are preferred primary sources to define target databases for the subject of interest. Although these databases are excellent primary data sources, the quality of the data in these databases should be questioned. They gather data from different sources that use slightly different conditions or preparations in the testing. These deviations can have a drastic impact on the modeling result, and it is up to the domain expert to decide whether these deviations will affect the outcome of the data mining study.

Next to ADME phenomena, recent data mining studies also focused on the development or improvement of models predicting physicochemical properties relevant to the field of ADME. Examples are Henry's law constant [92], polar surface area [93], and log P [94]. These models try to overcome limitations of already existing models, see for example SlogP [94] vs. Clogp [95], or aqueous solubility [96]. The latter study used more than 2000 compounds selected from the AQUASOL [97] and PHYSOPROP [98] databases. Comparison with a multilinear regression showed clear preference for the neural network.

While these studies were focusing on well-specified biological or physicochemical effects, recent efforts in the data mining of ADME phenomena are focusing on the determination of structural and/or physicochemical properties to discriminate drug-like and nondrug-like compounds [56,99,100] or drug-and lead-like compounds [101]. However, it is difficult to evaluate the value of these models with respect to their accuracy and hence usefulness. These models are subject to great variability such as the choice of training set or the set of descriptors. In addition, the type of property that is modeled is not always reproducible. Nonetheless, given the virtually unlimited sources of small molecules and the limited capability to test these molecules, a filter-like mechanism eliminating or flagging nondrug-like molecules in early drug discovery can lead to the saving of significant amounts of time and resources.

3.3. Data Mining of Structure–Toxicity Relationships

While computational approaches to toxicity prediction relied in the past mainly on expert systems comprising rules derived from human knowledge or QSAR approaches [102], the availability of large-scale toxicity databases such as RTECS [103] or TOXSYS [104] enables an information-intensive approach to toxicology. Several data mining studies have been published utilizing the data residing in these databases. Wang et al. [105], e.g., have taken an excellent pragmatic approach to modeling and extracting information from the RTECS toxicity database. They have sorted compounds with LD_{50} (dose at which 50% of the population dies) data into different categories and then identified structural patterns that are associated with low LD_{50} values, i.e., high toxicity. This approach may provide a way forward to mine large and variable data collections. By exemplifying their strategy on one reference compound, they identified different toxicity end points in rats. Using similarity searching, compounds similar to the reference compounds were identified from the RTECS database. These analogs had also a wide variety of toxic effects. The set of similar compounds was then used for the generation of independent and selective 3-D-QSAR

Comparative Molecular Field Analysis (CoMFA) models. The same authors used in a slightly similar approach the same database for cataloging structural frameworks associated with different organ specific toxicological end points [106]. This approach was originally introduced by Bemis and Murcko [107] for cataloging drug molecule frameworks, partitioning molecules in smaller moieties such as cyclic systems, functional groups, etc. Rule-based similarity analysis based on these moieties indicates considerable specificity when used for evaluating toxicity end points.

Other studies focus on the extraction of structure–toxicity relationship rules. For example, King et al. [108] could identify a number of features associated with the mutagenicity of heterocyclic aromatic amines, which were explainable as hydrophobic, electronic, and steric. In another experiment, King and Srinivasan [109] identified structural alerts for carcinogenicity; many of them were similar to those identified by Ashby and Paton [110] with their expert knowledge. Sometimes, the results are unexpected and counterintuitive. Lee et al. [111] found that negative results in certain genotoxicity assays are an indication of carcinogenicity, which conflicts with the hypothesis that genetic damage leads to cancer. Such findings require further investigations and may lead to the formulation of new hypotheses. Recently, King et al. [14] showed that the relation between structure and activity in carcinogenesis is bound to a minimum number of atom-bond conditions putting a lower boundary on the complexity of the relationship between chemical structure and carcinogenicity. For a good overview on the current state of the art in this field, confer the reviews by Helma et al. [15] and Durham and Pearl [112].

Despite this progress, the extraction of structure–toxicity knowledge from toxicological data sets remains very challenging. This is mainly because of the complexity of the toxicological science and our limited understanding of the potential mechanisms that are involved in the different toxicological end points. To come to an objective view on the accuracy of computational toxicity prediction, several open competitions have been organized in the past in which attendees could submit predictive models based on several available training data sets. The models are evaluated later on an unknown test set that was withheld by the organizing committee. The outcome of these competitions have been published [113].

4. SUMMARY

In this chapter we described the underlying theoretical concept of data mining and reviewed recent applications of this technology in drug discovery research. Data mining, a technology originally applied in the banking and retailing business, has received much attention in drug discovery because of its conceptual approach to data analysis and a large and versatile set of tools. What has been summarized is the first wave of data mining applications in drug discovery. However, it is quite easy to predict that data mining will become crucial in capitalizing on the huge volumes of data that characterize modern pharmaceutical research. An interesting token might be the fact that data mining has been proposed as a standard tool for data retrieval and analysis by the database management community [114]. Similar to the introduction of the Structured Query Language (SQL) to the database world more than 25 years ago, data-mining tools are about to become more integrated in the information technologies of the future.

ACKNOWLEDGMENT

The authors would like to thank Aziz Yasri, Rudi Verbeeck, and Luc Wouters, for reading the manuscript and providing valuable comments. We also would like to thank the editors for the invitation to write this chapter.

REFERENCES

1. Venter JC, et al. The sequence of the human genome. Science 2001; 291:1304–1351.
2. International Human Genome Sequencing Consortium. Initial sequencing and analysis of the human genome. Nature 2001; 409:860–921.
3. Han J, Kamber M. Data Mining—Concepts and Techniques. San Francisco: Morgan Kaufmann Publishers, 2001.
4. Fayyad UM, Piatetsky-Shapiro G, Smyth P, Uthurusamy R. Advances in Knowledge Discovery and Data Mining. Cambridge: The MIT Press, 1996.
5. Hand DJ, Blunt G, Kelly MG, Adams NM. Data Mining for Fun and Profit. Stat Sci 2000; 15:111–131.
6. Fayyad U, Grinstein GG, Wierse A. Information Visualization in Data Mining and Knowledge Discovery. San Francisco: Morgan Kaufmann Publishers, 2002.
7. Root DE, Kelley BP, Stockwell BR. Global analysis of large-scale chemical and biological experiments. Curr Opin Drug Disc Dev 2002; 5:355–360.
8. Hand D, Mannila H, Smyth P. Principles of Data Mining. Cambridge: The MIT Press, 2001.
9. Lipinski CA, Lombardo F, Dominy BW, Feeney PJ. Experimental and computational approaches to estimate solubility and permeability in drug discovery and development settings. Adv Drug Deliv Rev 1997; 23:3–25.
10. World Drug Index, version 2/96; Derwent Information: London, U.K., 1996.
11. Clark T. Quantum cheminformatics: an oxymoron? In: Hoeltje H-D, Sippl W, eds. "Rational Approaches to Drug Design." Barcelona, Spain: Prous Science, 2001:29–40.
12. Bertone P, Kluger Y, Lan N, Zheng D, Christendat D, Yee A, Edwards AM, Arrowsmith CH, Montelione GT, Gerstein M. SPINE: an integrated tracking database and data mining approach for identifying feasible targets in high-throughput structural proteomics. Nucleic Acids Res 2001; 29:2884–2898.
13. Engels MFM, Thielemans T, Verbinnen D, Tollenaere JP, Verbeeck R. CerBeruS: a system supporting the sequential screening process. J Chem Inf Comput Sci 2000; 40:241–245.
14. King RD, Srinivasan A, Dehaespe L. Warmr: a data mining tool for chemical data. J Comput-Aided Mol Des 2001; 15:173–181.
15. Helma C, Gottmann E, Kramer S, Knowledge discovery and data mining in toxicology. Stat Methods Med Res 2000; 9:329–358.
16. Estrada E, Uriarte E. Recent advances on the role of topological indices in drug discovery research. Curr Med Chem 2001; 8:1573–1588.
17. Carhart RE, Smith DH, Venkataraghavan R. Atom pairs as molecular features in structure–activity studies: definition and applications. J Chem Inf Comput Sci 1985; 25:64–73.
18. Nilakantan R, Bauman N, Dixon JS, Venkataraghavan R. Topological torsion: a new molecular descriptor for SAR applications. Comparison with other descriptors. J Chem Inf Comput Sci 1987; 27:82–85.
19. Merlot C, Domine D, Church DJ. Fragment analysis in small molecule discovery. Curr Opin Drug Discov Dev 2002; 5:391–399.
20. McGregor MJ, Muskal SM. Pharmacophore fingerprinting. 1. Application to QSAR and focused library design. J Chem Inf Comput Sci 1999; 39:569–574.

21. Mannhold R, Kubinyi H, Timmerman H, Todeschini R, Consonni V. Handbook of Molecular Descriptors. Weinheim, Germany: VCH, 1999.

22. Devillers J, Balaban AT. Topological Indices and Related Descriptors in QSAR and QSPR. Amsterdam, NL: Gordon and Breach Science Publishers, 1999.

23. Daylight Chemical Information System Inc, Mission Viejo, CA, U.S.A. *http://www.daylight.com.*

24. MDL Information Systems Inc, San Leandro, CA, U.S.A., *http://www.mdli.com.*

25. Tripos Inc, St Louis, MO, U.S.A. *http://www.tripos.com.*

26. Pearlman RS, Smith KM. Novel software tools for chemical diversity. Perspect Drug Discov Des 1998; 9:339–353.

27. Dixon SL, Merz KM. One-dimensional molecular representations and similarity calculations: methodology and validation. J Med Chem 2001; 44:3795–3809.

28. Thearling K, Becker B, DeCoste D, Mawby WD, Pilote M, Sommerfield D. Visualizing data mining models. In: Fayyd U, Grinstein GG, Wierse A, eds. "Information Visualization in Data Mining and Knowledge Discovery". San Francisco: Morgan Kaufmann Publishers, 2002.

29. Ladd B, Kenner S. Information visualization and analytical data mining in pharmaceutical R&D. Curr Opin Drug Discov Dev 2000; 3:280–291.

30. Meyer RD, Cook D. Visualization of data. Curr Opin Biotechnol 2000; 11:89–96.

31. Card SK, Mackinlay JD, Shneiderman B. Readings in Information Visualization: Using Vision to Think. San Francisco, CA: Morgan Kaufmann Publishers, 1999.

32. Mihalsisin TW. Multidimensional education: visual and algorithmic data mining domains and symbiosis. In: Fayyd U, Grinstein GG, Wierse A, eds. "Information Visualization in Data Mining and Knowledge Discovery". San Francisco: Morgan Kaufmann Publishers, 2002.

33. Willett P, Barnard JM, Downs GM. Chemical similarity searching. J Chem Inf Comput Sci 1998; 38:983–996.

34. McGregor MJ, Muskal SM. Pharmacophore fingerprinting. 2. Application to primary library design. J Chem Inf Comput Sci 2000; 40:117–125.

35. Hull RD, Singh SB, Nachbar RB, Sheridan RP, Kearsley SK, Fluder EM. Latent semantic structure indexing (LaSSI) for defining chemical similarity. J Med Chem 2001; 44:1177–1184.

36. Dunbar JB. Cluster-based selection. Perspect Drug Discov Des 1997; 7/8:51–63.

37. Bernard P, Golbraikh A, Kireev D, Chretien JR, Rozhkova N. Comparison of chemical databases: analysis of molecular diversity with self organising maps (SOM). Analusis 1998; 26:333–341.

38. Enterprise Miner Reference Help, Release 4.1, 2000, SAS Institute Inc., Cary, NC, U.S.A.

39. Clementine, http://www.spssscience.com/clementine/index.cfm.

40. Mitchell TM. Machine Learning. Boston, MA: McGraw-Hill, 1997.

41. Brown RD, Martin YC. Use of structure–activity data to compare structure-based clustering methods and descriptors for use in compound selection. J Chem Inf Comput Sci 1996; 36:572–584.

42. Willett P. Similarity and clustering in chemical information systems. Letchword: Research Studies Press, 1987.

43. Barnard JM, Downs GM. Clustering of chemical structures on the basis of 2-D similarity measures. J Chem Inf Comput Sci 1992; 32:644–649.

44. Downs GM, Willett P. Clustering of chemical structure databases for compound selection. In: van de Waterbeemd H, ed. Advanced Computer-Assisted Techniques in Drug Discovery. Vol 3. Weinheim, Germany: VCH, 1994.

45. Bioreason Inc. (Nicolaou C, Kelley BP, Nutt RF, Bassett SI): Method and system for artificial intelligence directed lead discovery through multidomain clustering. WO-00049530, 2000.

46. Nicolaou CA, Tamura SY, Kelley BP, Bassett SI, Nutt RF. Analysis of large screening data sets via adaptively grown phylogenetic-like trees. J Chem Inf Comput Sci, 2002. ASAP Article.

47. Hierarchical Agglomerative Clustering Package. Barnard Chemical Information Ltd. http://www.bci.gb.com.

48. Zhang T, Ramakrishnan R, Livny M. BIRCH: an efficient data clustering method for very large databases. Proc. 1996 ACM-SIGMOD Int. Conf. Management of Data (SIGMOD'96). Montreal, Canada, June 1996:103–114.

49. Guha S, Rasttogi R, Shim K. Cure: an efficient clustering algorithm for large databases. Proc. 1998 ACM-SIGMOD Int. Conf. Management of Data (SIGMOD'98). Seattle, WA, June 1998:73–84.

50. Karypis G, Han J, Kumar V. CHAMELEON: A hierarchical clustering algorithm using dynamic modeling. Computer 1999; 32:68–75.

51. Zupan J, Gasteiger J. Neural Networks in Chemistry and Drug Design. Second Edition. Weinheim, GE: Wiley-VCH, 1999.

52. Rabow AA, Shoemaker RH, Sausville EA, Covell DG. Mining the National Cancer Institute's tumor-screening database: identification of compounds with similar cellular activities. J Med Chem 2002; 45:818–840.

53. Hawkins DM, Young SS, Rusinko A. Analysis of a large structure–activity data set using recursive partitioning. Quant Struct-Act Relat 1997; 16:296–302.

54. Chen X, Rusinko A, Young SS. Recursive partitioning analysis of a large structure–activity data set using three-dimensional descriptors. J Chem Inf Comput Sci 1998; 38:1054–1062.

55. Engels MFM, De Winter H, Tollenaere JP. A decision tree learning approach for the classification and analysis of high-throughput screening data. In: Gundertofte K, Jorgensen FS, eds. Molecular Modeling and Prediction of Bioactivity. New York, NY, U.S.A.: Kluwer Academic, 2000.

56. Wagener M, van Geerestein VJ. Potential drugs and nondrugs: prediction and identification of important structural features. J Chem Inf Comput Sci 2000; 40:280–292.

57. Engels MFM, Knapen K, Tollenaere JP. Approaches for mining high-throughput screening data sets. In: Sippl, Hoeltje, eds. Rational Approaches to Drug Design. Barcelona, Spain: Prous Science, 2000.

58. Quinlan JR. C4.5 Programs for Machine Learning. San Mateo, CA, U.S.A.: Morgan Kaufmann Publishers, 1993.

59. C5.0, release 1.08; RuleQuest Research Pty Ltd. St Ives NSW, Australia (http://www.rulequest.com).

60. Breiman L, Friedman JH, Ohlson RA, Stone CJ. Classification and Regression Trees. Belmont, CA: Wadsworth, 1984.

61. Bahler D, Bristol DW. The induction of rules for predicting chemical carcinogenesis in rodents. ISMB 1993; 1:29–37.

62. Freund Y, Schapire RE. A decision-theoretic generalization of on-line learning and an application to boosting. J Comput Syst Sci 1997; 55:119–139.

63. Agrawal R, Imaelinski T, Swami A. Mining association rules between sets of items in large databases. Proc. Of the 1993 ACM SIGMOD Conference, Washington DC, U.S.A., May 1993.

64. Bishop CM. Neural Networks for Pattern Recognition. New York, NY, U.S.A.: Oxford University Press, 1995.

65. Cochran WG. Sampling Techniques. New York: Wiley, 1977.

66. Engels MFM, Venkatarangan P. Smart screening: approaches to efficient HTS. Curr Opin Drug Discov Dev 2001; 4:275–283.

67. Cosgrove DA, Willett P. SLASH: a program for analysing the functional groups in molecules. J Mol Graph Model 1998; 16:19–32.

68. Leadpharmer, Bioreason Inc., Santa Fe, NM; http://www.bioreason.com.
69. Distill, Tripos, St Louis, MS, U.S.A. *http://www.tripos.com.*
70. LeadScope. LeadScope Inc. Columbus, Ohio; http://www.leadscope.com.
71. ChemTree, GoldenHelix Inc. Bozeman, MT, U.S.A., http://www.goldenhelix.com.
72. Tamura SY, Bacha PA, Gruver HS, Nutt RF. Data analysis of high-throughput screening results: application of multidomain clustering to the NCI anti-HIV data set. J Med Chem 2002; 45:3082–3093.
73. Roberts G, Myatt GJ, Johnson WP, Cross KP, Blower PE. LeadScope: software for exploring large sets of screening data. J Chem Inf Comput Sci 2000; 40:1302–1314.
74. Ajay, Bemis GW, Murcko MA. Designing libraries with CNS activity. J Med Chem 1999; 42:4942–4951.
75. Stanton DT, Morris TW, Roychoudhury S, Parker CN. Application of nearest-neighbor and cluster analyses in pharmaceutical lead discovery. J Chem Inf Comput Sci 1997; 39:21–27.
76. Jones-Hertzog DK, Mukhopadhyay P, Keefer CE, Young SS. Use of recursive partitioning in the sequential screening of G-protein-coupled receptors. J Pharmacol Toxicol 1999; 42:207–215.
77. Agrafiotis DK, Cedeño W. Feature selection for structure–activity correlation using binary particle swarms. J Med Chem 2002; 45:1098–1107.
78. Spencer RW. Diversity analysis in high throughput screening. J Biomol Screen 1997; 2:69–70.
79. Engels MFM, Wouters L, Verbeeck R, Vanhoof G. Outlier mining in high throughput screening experiments. J Biomol Screen 2002; 7:341–353.
80. Gedeck P, Willett P. Visual and computational analysis of structure–activity relationships in high-throughput screening data. Curr Opin Chem Biol 2001; 5:389–395.
81. Young SS, Lam RLH, Welch WJ. Initial compound selection for sequential screening. Curr Opin Drug Discov Dev 2002; 5:422–427.
82. The NCI data set is available at *http://www.dtp.nih.gov.*
83. Staunton JE, Slonim DK, Coller HA, Tamayo P, Angelo MJ, Park J, Scherf U, Lee JK, Reinhold WO, Weinstein JN, Mesirov JP, Lander ES, Golub TR. Chemosensitivity prediction by transcriptional profiling. Proc Natl Acad Sci USA 2001; 98:10787–10792.
84. Beresford AP, Selick HE, Tabit MH. The emerging importance of predictive ADME simulation in drug discovery. Drug Discov Today 2002; 7:109–116.
85. Caldwell J, Gardner I, Swales N. An introduction to drug disposition: the basic principles of absorption. Toxicol Pathol 1995; 23:102–109.
86. Ajay, Bemis GW, Murcko MA. Designing libraries with CNS activity. J Med Chem 1999; 42:4942–4951.
87. Platts JA, Abraham MH, Zhao YH, Hersey A, Ijaz L, Butina D. Correlation and prediction of a large blood–brain barrier distribution data set—an LFER study. Eur J Med Chem 2001; 36:719–730.
88. Yoshida F, Topliss JG. QSAR model for drug human oral bioavailability. J Med Chem 2000; 43:2575–2585.
89. Zhao YH, Le J, Abraham MH, Hersey A, Eddershaw PJ, Luscombe CN, Boutina D, Beck G, Sherborne, Cooper I, Platts JA. Evaluation of human intestinal absorption data and subsequent derivation of a quantitative structure–activity relationship (QSAR) with Abraham descriptors. J Pharm Sci 2001; 90:749–784.
90. Comprehensive Medicinal Chemistry database is available from MDL Information Systems Inc., San Leandro, CA 94577.
91. Molecular Drug Data Report. Available form Molecular DDR.
92. English NJ, Carroll DG. Prediction of Henry's law constants by a quantitative structure property relationship and neural networks. J Chem Inf Comput Sci 2001; 41:1150–1161.

93. Ertl P, Rohde B, Selzer P. Fast calculation of molecular polar surface area as a sum of fragment-based contributions and its application to the prediction of drug transport properties. J Med Chem 2000; 45:1–18.

94. Wildman SA, Crippen GM. Prediction of physicochemical parameters by atomic contributions. J Chem Inf Comput Sci 1999; 39:868–873.

95. ClogP available from Daylight Chemical Information Systems, http://www.daylight. com.

96. Huuskonen J. Estimation of aqueous solubility for a diverse set of organic compounds based on molecular topology. J Chem Inf Comput Sci 2000; 40:773–777.

97. Yalkowsky S, Dannelfelser RM. The ARIZONA dATAbASE of Aqueous Solubility, College of Pharmacy, University of Arizona, Tucson, AZ, U.S.A.

98. Syracuse Research Corporation. Physical Chemical Property database (PHYSOPROP); SRC Environmental Science center: Syracuse, NY.

99. Ajay, Walters WP, Murcko MA. Can we learn to distinguish between "drug-like" and "nondrug-like" molecules? J Med Chem 1998; 141:3314–3324.

100. Sadowski J, Kubinyi H. Scoring scheme for discriminating between drugs and nondrugs. J Med Chem 1998; 41:3325–3329.

101. Oprea TI, Davis AM, Teague SJ, Leeson PD. Is there a difference between leads and drugs? A historical perspective. J Chem Inf Comput Sci 2001; 41:1308–1315.

102. Reiss C, Parvez S, Labbe G, Parvez H. Advances in Molecular Toxicology. The Netherlands: VSP Zeist, 1998.

103. RTECS C2(96-4); National Institute for Occupational Safety and Health (NIOSH), US Department of Health and Human Services: Washington DC (http://www.ccohs.ca).

104. TOXSYS database. http://www.scivision.com/ToxSys.html.

105. Wang J, Lai L, Tang Y. Data mining of toxic chemicals: structure patterns and QSAR. J Mol Model 1999; 5:252–262.

106. Wang J, Lai L, Tang Y. Structural features of toxic chemicals for specific toxicity. J Chem Inf Comput Sci 1999; 39:1173–1189.

107. Bemis GW, Murcko MA. The properties of known drugs: 1. Molecular frameworks. J Med Chem 1996; 39:2887–2893.

108. King RD, Muggleton SH, Srinivasan A, Sternberg MJE. Structure–activity relationships derived by machine learning: the use of atoms and their bond connectivities to predict mutagenicity by inductive logic programming. Proc Natl Acad Sci USA 1996; 93:438–442.

109. King RD, Srinivasan A. The discovery of indicator variables for QSAR using inductive logic programming. J Comput-Aided Mol Des 1997; 11:571–580.

110. Ashby J, Paton D. The influence of chemical structure on the extent and sites of carcinogenesis for 522 rodent carcinogens and 55 different human carcinogens. Mutat Res 1993; 286:3–74.

111. Lee Y, Buchanan BG, Rosenkranz HS. Carcinogenicity predictions for a group of 30 chemicals undergoing rodent cancer bioassays based on rules derived from subchronic organ toxicities. Environ Health Perspect 1996; 104S(5):1059–1063.

112. Durham SK, Pearl GM. Computational methods to predict drug safety liabilities. Curr Opin Drug Discov Dev 2001; 4:110–115.

113. Benigni R. The first US National Toxicology Program exercise on the prediction of rodent carcinogenicity: definitive results. Mutat Res 1997; 387:35–45.

114. OLE DB for Data Mining, Draft Specification, version 0.9, Microsoft Corporation, February 2000.

26

Vibrational Circular Dichroism Spectroscopy: A New Tool for the Stereochemical Characterization of Chiral Molecules

PHILIP J. STEPHENS

University of Southern California, Los Angeles, California, U.S.A.

1. INTRODUCTION

Pharmaceutical compounds—drugs—are generally small to medium-sized organic molecules. Many are chiral. When this is the case, either a pure enantiomer or a racemic mixture may be used in therapy. Generally, the physiological effects of enantiomeric and racemic forms of a pharmaceutical compound are different. Drug development therefore involves the testing of both enantiomeric and racemic forms of candidate compounds. Before this can be carried out, the enantiomers of the compound of interest must be obtained and their stereochemistry defined.

In this chapter, we discuss a relatively new methodology for the structural characterization of chiral organic molecules, namely, vibrational circular dichroism (VCD) spectroscopy [1]. To date, very few studies of the VCD spectra of pharmaceutical compounds have been reported. Our goal will therefore be to describe the general techniques and applications of VCD spectroscopy in order to illuminate its potential value to specific problems of pharmaceutical chemistry.

In the following sections we discuss in turn the experimental measurement of VCD spectra, the theoretical calculation of VCD spectra, and the application of VCD spectroscopy to the determination of molecular stereochemistry.

2. EXPERIMENT

Circular dichroism (CD) is the differential absorption of right and left circularly polarized light. Using absorbance, A, as a measure of a sample's absorption, its CD is

$$\Delta A = A_L - A_R \tag{1}$$

where L and R refer to left and right circular polarizations, respectively. For dilute solutions of an absorbing solute in a nonabsorbing solvent, to which Beer's Law applies,

$$A = \varepsilon \, cl \tag{2}$$

where ε is the molar extinction coefficient of the solute, c is its molarity, and l is the sample path length in cm. Then

$$\Delta A = (\Delta \varepsilon)cl \tag{3}$$

where $\Delta \varepsilon = \varepsilon_L - \varepsilon_R$. When the absorption originates in electronic excitations, electronic CD (ECD) is observed, generally in the visible–ultraviolet spectral region. When the absorption originates in vibrational excitations, vibrational CD (VCD) is observed, generally in the infra-red (IR) spectral region.

We restrict consideration here to isotropic solutions, those in which the absorption of linearly polarized light is independent of the direction of propagation and the plane of polarization. For unpolarized radiation, $A \equiv \bar{A}$ and $\varepsilon \equiv \bar{\varepsilon}$. A useful parameter in CD spectroscopy is the anisotropy ratio, g, defined by $g = \Delta A / \bar{A} = \Delta \varepsilon / \bar{\varepsilon}$. This ratio quantifies the magnitude of the circular polarization of the absorption. In ECD spectroscopy, g factors of 10^{-3}–10^{-1} are common. In VCD spectroscopy, g factors are generally much smaller, typically in the range 10^{-5}–10^{-3}.

Circular dichroism is exhibited by a solute molecule if and only if the molecule is chiral. Achiral molecules exhibit zero CD. Enantiomeric forms of a chiral molecule exhibit "mirror image" CD spectra. That is, if the two enantiomers are E_1 and E_2,

$$\Delta A(E_1) = -\Delta A(E_2); \qquad \Delta \varepsilon(E_1) = -\Delta \varepsilon(E_2) \tag{4}$$

at every frequency/wavelength. Multiplication of the CD spectrum of E_1 by -1 gives the CD spectrum of E_2. In contrast, E_1 and E_2 exhibit identical unpolarized absorption spectra,

$$\bar{A}\,(E_1) = \bar{A}\,(E_2); \qquad \bar{\varepsilon}(E_1) = \bar{\varepsilon}(E_2) \tag{5}$$

The mirror-image relationship of the CD spectra of enantiomers is the key to the determination of the absolute configuration of chiral molecules using CD spectroscopy.

Circular dichroism is almost universally measured using modulation spectroscopy. Light from a source is phase modulated at a frequency ω_M, so that the polarization of the light oscillates between R and L circular polarizations, the intensity remaining constant. After passing through the sample, if nonzero CD exists, an intensity fluctuation is created of frequency ω_M. Measurement of this intensity fluctuation using phase-sensitive detection electronics yields the CD, ΔA, of the sample. Modulation spectroscopy was introduced into the measurement of ECD ca. 1960 [2]; phase modulation was accomplished using electrooptic modulators. Subsequently, photoelastic modulators (PEMs) were developed [3] and have since become the modulators of choice. Photoelastic modulators use isotropic optical elements and can thus be constructed from many materials. The construction of PEMs from IR-transmitting materials made possible the first measurements of VCD, carried out in the 1970s [4].

The earliest measurements of VCD spectra were carried out using dispersive spectrometers. Subsequently, the methods of modulation spectroscopy were adapted to Fourier transform (FT) spectrometers [5], permitting the advantages of the FT methodology to be exploited. Over the last quarter century, enormous progress has been made in improving the frequency range, sensitivity, and resolution of instrumentation for measuring VCD. Very efficient dispersive spectrometers, dedicated to measurements in specific, relatively narrow, frequency ranges have been constructed [6]. At the same time, FT spectrometers, capable of broad band performance, have achieved high levels of sensitivity [5b,7]. Further, the increasing applicability of VCD spectroscopy to the study of molecular stereochemistry, consequent on theoretical developments described below, has led to the commercialization of VCD instrumentation. Bomem (BioTools), Bruker, Bio-Rad, and Jasco all now market FT VCD instruments. These instruments all use ZnSe PEMs and HgCdTe detectors. Their lower frequency limits are $>700 \text{ cm}^{-1}$.

In Fig. 1, we show the unpolarized absorption (infrared, "IR") and VCD spectrum of a chiral molecule, camphor, in CCl_4 solution in the mid-infrared spectral region. The IR spectrum was recorded with a conventional FT IR spectrometer at 1 cm^{-1} resolution. The spectrum is identical for $(+)$-, $(-)$-, and $(\pm)$-camphor. The VCD spectra of $(+)$- and $(-)$-camphor were recorded using a commercial FT VCD instrument [Bomem (BioTools) ChiralIR] at 4 cm^{-1} resolution. $(+)$- and $(-)$-camphor exhibit "mirror-image" VCD spectra. The largest peak unpolarized absorbances, $\bar{A}$, of the vibrational transitions in this frequency region lie in the range 0.5–1.0. The largest peak CD intensities, ΔA, are $\sim 1 \times 10^{-4}$. Anisotropy ratios vary widely; mostly they are in the range $1 \times 10^{-5} - 1 \times 10^{-3}$. The IR and VCD spectra of camphor are typical of the spectra of organic molecules.

To achieve an optimum signal-to-noise (S/N) ratio in measuring the VCD of a vibrational transition a number of factors are of importance. First, the unpolarized absorbance should be optimized, peak absorbance lying in the range 0.5–1.0. This defines the optimum concentration–pathlength (cl) product required. Second, solvent absorption should be minimized, both by choosing an optimally transmitting solvent and as short a cell pathlength as practicable. Third, spectral resolution should be chosen to be as low as possible, consistent with full resolution of the transition. Optimum S/N ratios cannot be achieved simultaneously for all vibrational transitions. Measurement of a VCD spectrum over a broad spectral range often requires spectra to be obtained over a range of pathlengths and using several solvents. Generally, solution concentrations of 0.01–1 M and pathlengths of 10–1000 μm are used. A spectral resolution of 4 cm^{-1} is a typical choice in the mid-IR spectral region.

The quality of VCD spectra is also determined by the magnitude of CD artifacts, which are signals which arise from phenomena other than CD but which look like CD to the CD instrument. For example, a birefringent plate followed by a linear polarizer generates CD artifacts; in fact, this provides a basis for calibrating the CD spectrometer [4d,4e,8]. Circular dichroism artifacts can arise from a variety of causes, including strain birefringence in cell windows, polarization sensitivity in the detector, excessive sample absorbance, and so on. When CD artifacts are present, it is generally assumed

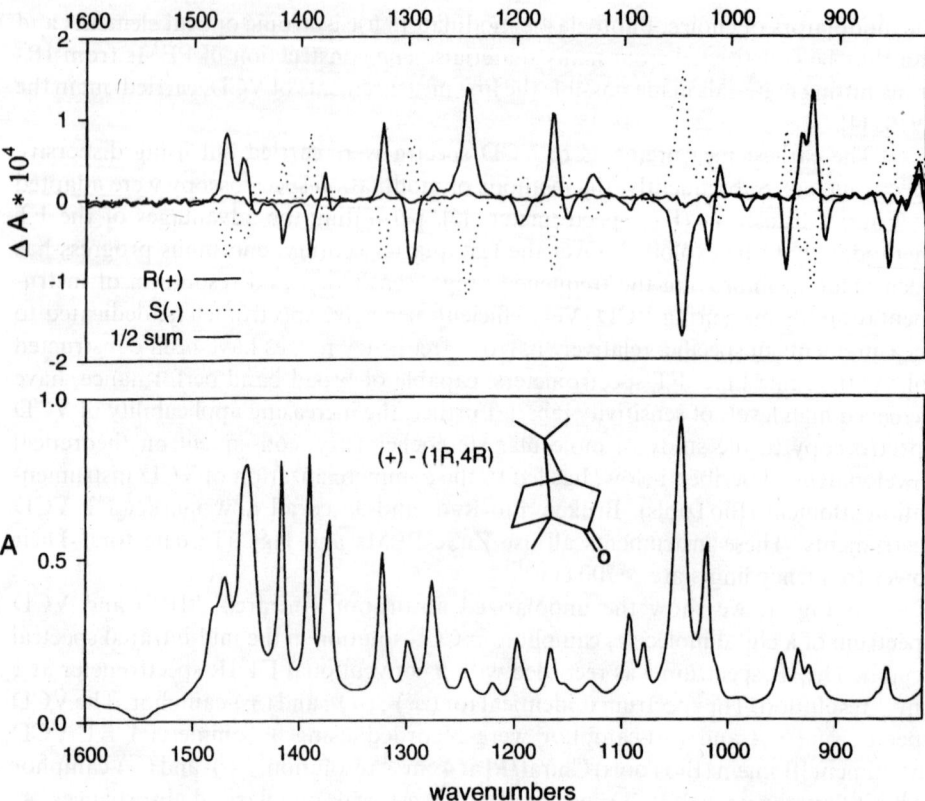

Figure 1 Unpolarized absorption spectrum and VCD spectrum of camphor in CCl$_4$ solution; concentration 0.6 M; pathlength 151 μm; resolution 4 cm^{-1}; VCD scan time 1 hr.

that they are additive with the true CD signal; subtraction of the spectrum of the racemic sample from that of an enantiomer then permits the true CD spectrum to be obtained.

3. THEORY

For a dilute solution of a chiral absorbing molecule in a nonabsorbing solvent [9],

$$\bar{\varepsilon}\,(v) = \left(\frac{8\pi^3 N}{(2.303)3000hc}\right) v \sum_j D_j f_j(v_j, \gamma_j)$$

$$\Delta\varepsilon\,(v) = \left(\frac{32\pi^3 N}{(2.303)3000hc}\right) v \sum_j R_j f_j(v_j, \gamma_j)$$

(6)

where D_j and R_j are the dipole and rotational strengths of excitation $g \rightarrow j$ at frequency v_j, respectively. $f_j(v_j, \gamma_j)$ is the normalized bandshape, γ_j specifying the bandwidth. For the jth excitation, the anisotropy ratio is $g_j = 4R_j/D_j$. D_j and R_j are given by:

$$D_j = \left|\left\langle g\left|\vec{\mu}_{\text{el}}\right|j\right\rangle\right|^2$$

(7)

$$R_j = \text{Im}\left[\left\langle g\left|\vec{\mu}_{el}\right|j\right\rangle \cdot \left\langle j\left|\vec{\mu}_{mag}\right|g\right\rangle\right]$$

where $\vec{\mu}_{el}$ and $\vec{\mu}_{mag}$ are the electric dipole and magnetic dipole moment operators, respectively.

In the case of vibrational excitations, we invoke the harmonic approximation, when the potential energy of the electronic ground state can be written

$$W_G = W_G^o + \frac{1}{2}\sum_{i=1}^{3N-6} k_i Q_i^2 \tag{8}$$

where (for an N-atom molecule) Q_i are the $3N-6$ normal coordinates. The vibrational eigenstates are then characterized by $3N-6$ vibrational quantum numbers, $v_i = 0,1,2\ldots$, and are of energy

$$E(v_1 \ldots v_{3N-6}) = \sum_i \left(v_i + \frac{1}{2}\right)hv_i \tag{9}$$

where v_i is the frequency of the ith normal mode, given by

$$v_i = \frac{1}{2\pi}\sqrt{k_i} \tag{10}$$

Within the harmonic approximation, only fundamental excitations—$\Delta v_i = +1, \Delta v_j = 0$ ($j \neq i$)—of excitation energy hv_i are allowed.

In calculating harmonic vibrational frequencies and normal coordinates via quantum-mechanical methods, the standard procedure begins with the calculation of the Hessian, $[\partial^2 W_G / \partial X_{\lambda\alpha} \, \partial X_{\lambda'\alpha'}]_o$, where $X_{\lambda\alpha}$ are the $3N$ Cartesian displacement coordinates (λ = nucleus; $\alpha = x, y,$ or z), defined relative to the equilibrium positions of the N nuclei, $\vec{R}_{\lambda_0}$, and the derivatives are calculated at the equilibrium geometry. The linear transformation

$$X_{\lambda\alpha} = \sum_i S_{\lambda\alpha,i} \, Q_i \tag{11}$$

then accomplishes the conversion of W_G to the diagonal form of Eq. (8).

The electric dipole transition moment of the fundamental excitation of mode i is given by

$$\left\langle 0\left|(\mu_{el})_\beta\right|1\right\rangle_i = \left(\frac{\hbar}{4\pi v_i}\right)^{1/2}\left[\frac{\partial\left(\mu_{el}^G\right)_\beta}{\partial Q_i}\right]_o \tag{12}$$

where $\vec{\mu}_{el}^G$ is the electric dipole moment of the electronic ground state G and its derivative is evaluated at the equilibrium geometry. In terms of Cartesian displacement coordinates, Eq. (12) can be re-written

$$\left\langle 0\left|(\mu_{el})_\beta\right|1\right\rangle_i = \left(\frac{\hbar}{4\pi v_i}\right)^{1/2}\sum_{\lambda\alpha} S_{\lambda\alpha,i} P_{\alpha\beta}^\lambda \tag{13}$$

where $P_{\alpha\beta}^\lambda = [\partial(\mu_{el}^G)_\beta / \partial X_{\lambda\alpha}]_o$ is referred to as the atomic polar tensor (APT) of nucleus λ [10]. Calculation of harmonic electric dipole transition moments, in addition to harmonic vibrational frequencies and normal coordinates, thus requires the additional calculation of the APTs. Since $\vec{\mu}_{el}^{G} = \langle \Psi_G \mid \vec{\mu}_{el} \mid \Psi_G \rangle$, where Ψ_G is the electronic ground state wave function, $P_{\alpha\beta}^{\lambda}$ can be written

$$P_{\alpha\beta}^{\lambda} = 2\left\langle \left[\frac{\partial \psi_G}{\partial X_{\lambda\alpha}}\right]_0 \middle| (\mu_{el}^e)_\beta \middle| \psi_G^0 \right\rangle + Z_\lambda e \delta_{\alpha\beta} \tag{14}$$

where $\vec{\mu}_{el}^e$ is the electronic contribution to $\vec{\mu}_{el}$ and $Z_\lambda e$ is the charge of nucleus λ. Calculation of $P_{\alpha\beta}^{\lambda}$ thus requires calculation of Ψ_G and $\partial \Psi_G / \partial X_{\lambda\alpha}$ at the equilibrium geometry.

The magnetic dipole transition moment of the fundamental excitation of mode i is given analogously by [11]

$$\left\langle 0 \middle| (\mu_{mag})_\beta \middle| 1 \right\rangle_i = -(4\pi\hbar^3 \nu_i)^{1/2} \sum_{\lambda\alpha} S_{\lambda\alpha,i} M_{\alpha\beta}^{\lambda} \tag{15}$$

where $M_{\alpha\beta}^{\lambda}$ is referred to as the atomic axial tensor (AAT) of nucleus λ. $M_{\alpha\beta}^{\lambda}$ is given by

$$M_{\alpha\beta}^{\lambda} = I_{\alpha\beta}^{\lambda} + J_{\alpha\beta}^{\lambda}$$

$$I_{\alpha\beta}^{\lambda} = \left\langle \left(\frac{\partial \psi_G}{\partial X_{\lambda\alpha}}\right)_0 \middle| \left(\frac{\partial \psi_G}{\partial H_\beta}\right)_0 \right\rangle \tag{16}$$

$$J_{\alpha\beta}^{\lambda} = \frac{i}{4\hbar c} \sum_{\gamma} \varepsilon_{\alpha\beta\gamma} (Z_\lambda e) R_{\lambda\gamma}^{0}$$

$\partial \Psi_G / \partial H_\beta$ is the derivative of Ψ_G with respect to H_β when the molecule is perturbed by the uniform magnetic field perturbation:

$$H' = -\left(\mu_{mag}^e\right)_\beta H_\beta \tag{17}$$

R_λ^0 is the equilibrium position of nucleus λ; $\vec{\mu}_{mag}^e$ is the electronic contribution to $\vec{\mu}_{mag}$. Calculation of harmonic magnetic dipole transition moments, in addition to harmonic vibrational frequencies and normal coordinates, thus requires calculation of the AATs. Calculation of AATs requires calculation of $\partial \Psi_G / \partial X_{\lambda\alpha}$ and $\partial \Psi_G / \partial H_\beta$. Thus, over and above the calculation of APTs, calculation of AATs requires only the calculation of $\partial \Psi_G / \partial H_\beta$.

Substitution of Eqs. (13) and (15) into Eq. (7) gives

$$D_i(0 \rightarrow 1) = \left(\frac{\hbar}{4\pi\nu_i}\right) \sum_{\beta} \left[\sum_{\lambda\alpha} P_{\alpha\beta}^{\lambda} S_{\lambda\alpha,i}\right]^2$$

$$R_i(0 \rightarrow 1) = -(\hbar^2) \mathrm{Im} \sum_{\beta} \left[\sum_{\lambda\alpha} P_{\alpha\beta}^{\lambda} S_{\lambda\alpha,i}\right] \cdot \left[\sum_{\lambda'\alpha'} M_{\alpha'\beta'}^{\lambda'} S_{\lambda'\alpha',i}\right] \tag{18}$$

Writing

$$\left(p_i^{\lambda}\right)_\beta = \left(\frac{\hbar}{4\pi\nu_i}\right)^{1/2} \sum_{\alpha} P_{\alpha\beta}^{\lambda} S_{\lambda\alpha,i}$$

$$\left(m_i^{\lambda}\right)_\beta = -(4\pi\hbar^3 \nu_i)^{1/2} \sum_{\alpha} M_{\alpha\beta}^{\lambda} S_{\lambda\alpha,i} \tag{19}$$

leads to

$$D_i = \sum_{\lambda,\lambda'} \vec{p}_i^{\lambda} \cdot \vec{p}_i^{\lambda'}$$

(20)

$$R_i = \text{Im}\left[\sum_{\lambda,\lambda'} \vec{p}_i^{\lambda} \cdot \vec{m}_i^{\lambda'}\right]$$

$\vec{p}_i^{\lambda}$ and $\vec{m}_i^{\lambda}$ are the contributions to the electric and magnetic dipole transition moments of the motions of nucleus λ in the ith normal mode. The dependencies of D_i and R_i on these atomic contributions are explicitly displayed in Eq. (20).

To obtain accurate predictions of vibrational dipole and rotational strengths ab initio, quantum-mechanical methods must be employed in calculating Hessians, APTs, and AATs. Efficient calculation of these properties requires analytical deriv- ative (AD) methods [12] together with perturbation-dependent (PD) basis sets. Analytical derivative methods calculate derivatives of wave functions and energies analytically (as opposed to numerically, using finite-difference methods). Perturba- tion-dependent basis sets are those basis sets which are explicitly dependent on the variable (the perturbation) with respect to which derivatives are being calculated. Thus, in calculating derivatives with respect to nuclear displacements $X_{\lambda\alpha}$, PD basis sets are nuclear position dependent. In calculating derivatives with respect to the magnetic field H_{β}, PD basis sets are magnetic field dependent. Such basis sets are well- known in quantum chemistry. The standard choice for nuclear position dependent basis sets are nucleus-centered atomic orbitals. The standard choice for magnetic field dependent basis sets are the atomic orbitals referred to either as London orbitals or gauge-invariant (including) atomic orbitals (GIAOs) [13]. Analytical derivative methods using PD basis sets have been implemented for the calculation of Hessians, APTs, and AATs using Hartree–Fock (HF) theory and density functional theory (DFT) [14]. The accuracy of DFT is substantially higher than that of HF theory, while being not much more computationally demanding. At the present time DFT is the most cost-effective methodology. It is available in a widely distributed ab initio quantum chemistry package, GAUSSIAN [15].

Density functional theory calculations require a density functional. By now, a very large number of functionals have been introduced into the literature. The earliest were so-called "local" functionals. Subsequently, so-called "nonlocal" functionals were developed, otherwise known as "gradient-corrected" functionals. Most recently, a class of functionals referred to as "hybrid" functionals have been developed [16]. At the present time, hybrid functionals are the most accurate functionals available for a wide range of properties. The earliest hybrid functional is known as B3PW91 [16]. A very popular hybrid functional is B3LYP [17].

Given a density functional, DFT calculations also require the choice of an atomic orbital basis set. The larger the basis set the closer calculations approach the complete basis set limit and the smaller the basis set error. At the same time, computational demands increase rapidly with increasing basis set size. The optimum choice of basis set is that which provides the optimum compromise of accuracy and computational effort.

We illustrate the variation in predicted IR and VCD spectra resulting from variation in functional and basis set in Figs. 2–5, where spectra predicted for the small rigid chiral molecule propylene oxide (methyloxirane), **1**, are shown. In Figs. 2 and 4, the basis set is varied with the functional held constant; in Figs. 3 and 5, the basis set is fixed and the functional is varied. Over the range of basis sets employed one observes convergence to the complete basis set limit. The cc-pVTZ basis set provides results

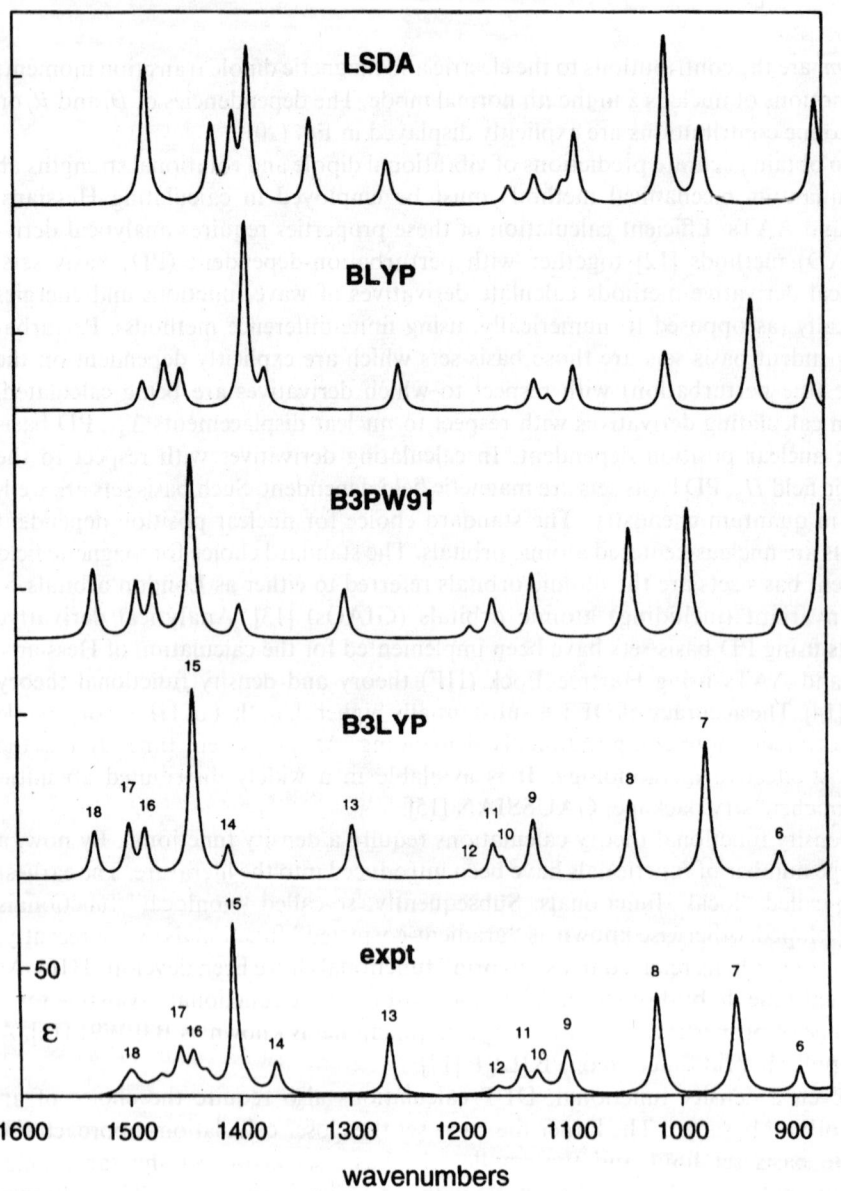

Figure 2 Mid-IR absorption spectra of **1**. The experimental spectrum is in CCl$_4$ solution. Density functional theory spectra are calculated using the cc-pVTZ basis set and a range of functionals. Band shapes are Lorentzian ($\gamma = 4.0$ cm^{-1}). Fundamentals are numbered.

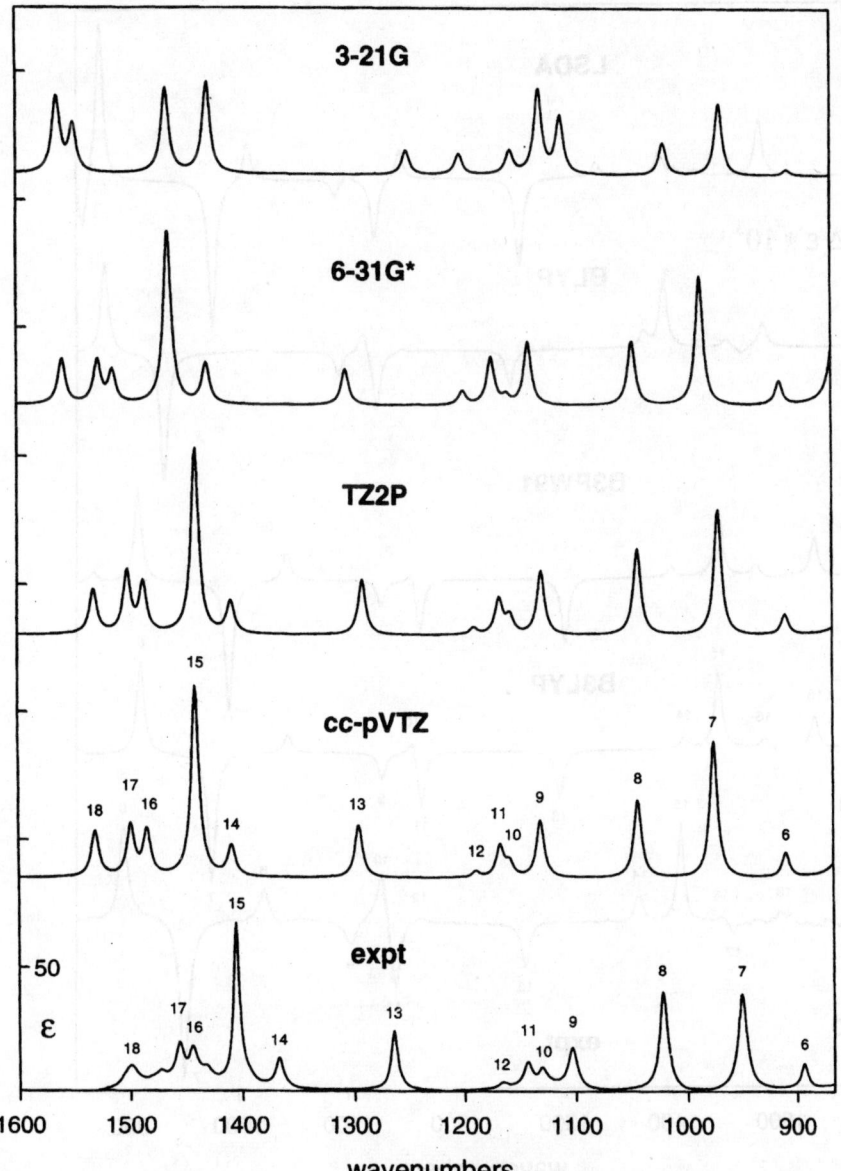

Figure 3 Mid-IR absorption spectra of **1**. The experimental spectrum is as in Fig. 2. Density functional theory spectra are calculated using the B3LYP functional and a range of basis sets. Band shapes are Lorentzian ($\gamma = 4.0$ cm^{-1}). Fundamentals are numbered.

essentially free of basis set error. Over the range of functionals employed, local, nonlocal, and hybrid functionals give qualitatively different spectra; variation from one hybrid functional to another is smaller.

The accuracies of predicted IR and VCD spectra can be gauged by comparison to the experimental spectra, also shown in Figs. 2–5. Clearly, as may be anticipated, hybrid functionals together with large basis sets give the best agreement with

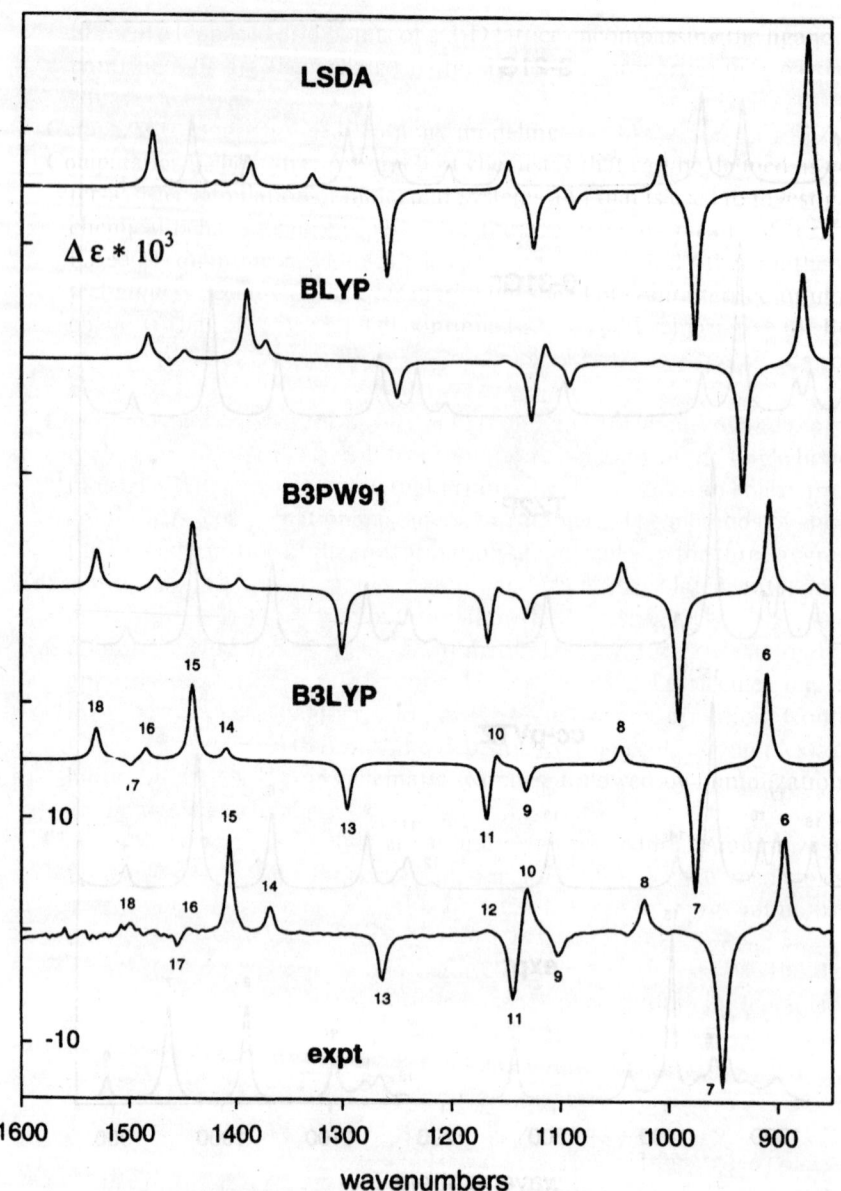

Figure 4 Mid-IR VCD spectra of (+)-*R*-**1**. The experimental spectrum is in CCl₄ solution. Density functional theory spectra are calculated using the cc-pVTZ basis set and a range of functionals. Band shapes are Lorentzian (γ = 4.0 cm⁻¹). Fundamentals are numbered.

experiment. The quantitative accuracies of predicted dipole and rotational strengths can be gauged by comparison to values obtained from the experimental spectra by Lorentzian fitting [18]. As shown in Fig. 6 for the rotational strengths, the most accurate calculations give results in excellent agreement with experiment. The variation in accuracy with variation in basis set and functional is documented in Table 1.

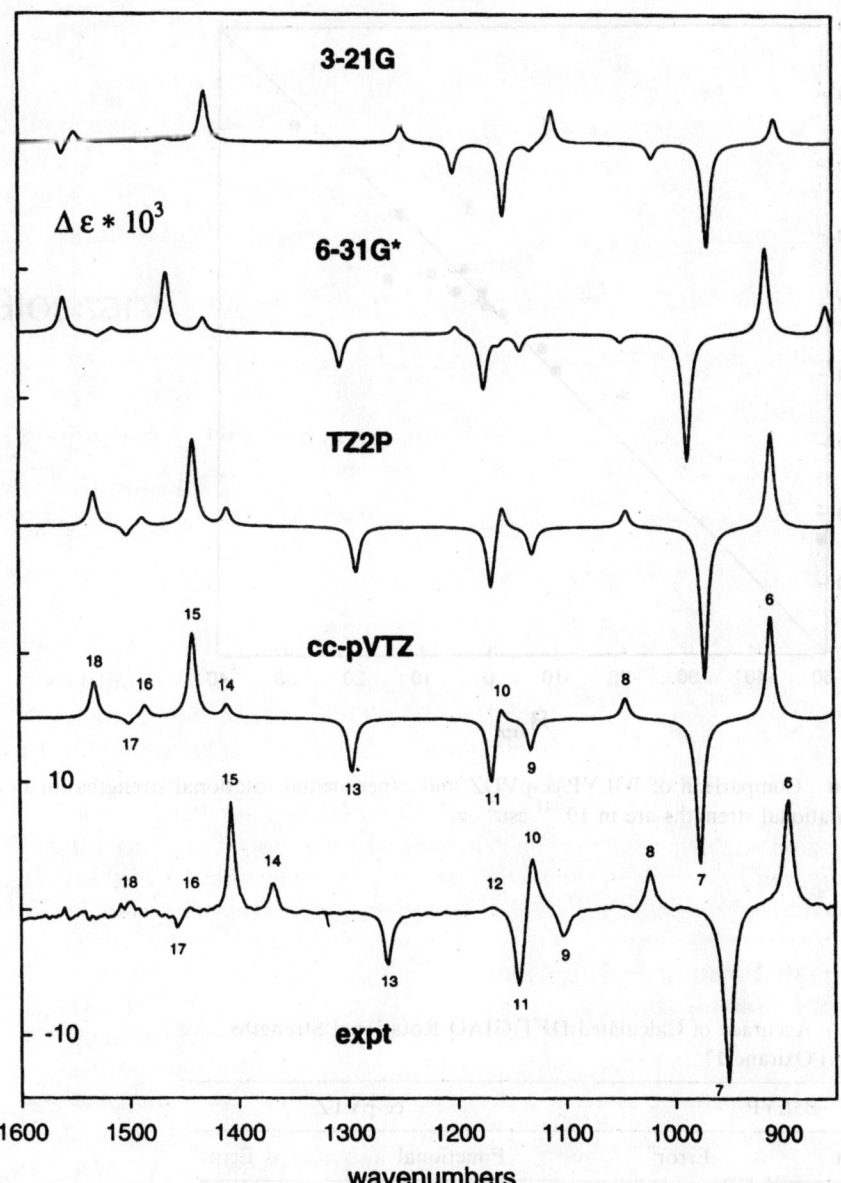

Figure 5 Mid-IR VCD spectra of (+)-*R*-**1**. The experimental spectrum is as in Fig. 4. Density functional theory spectra are calculated using the B3LYP functional and a range of basis sets. Band shapes are Lorentzian (γ = 4.0 cm^{-1}). Fundamentals are numbered.

The methodology we have described above ignores anharmonicity. As a result, calculated vibrational frequencies are a few percent higher than experimental frequencies. However, more importantly, as illustrated by the comparison of calculated and experimental spectra for propylene oxide, in the mid-IR spectral region the effects of anharmonicity on the dipole and rotational strengths of fundamental transitions are not large and, to a good approximation, can be ignored.

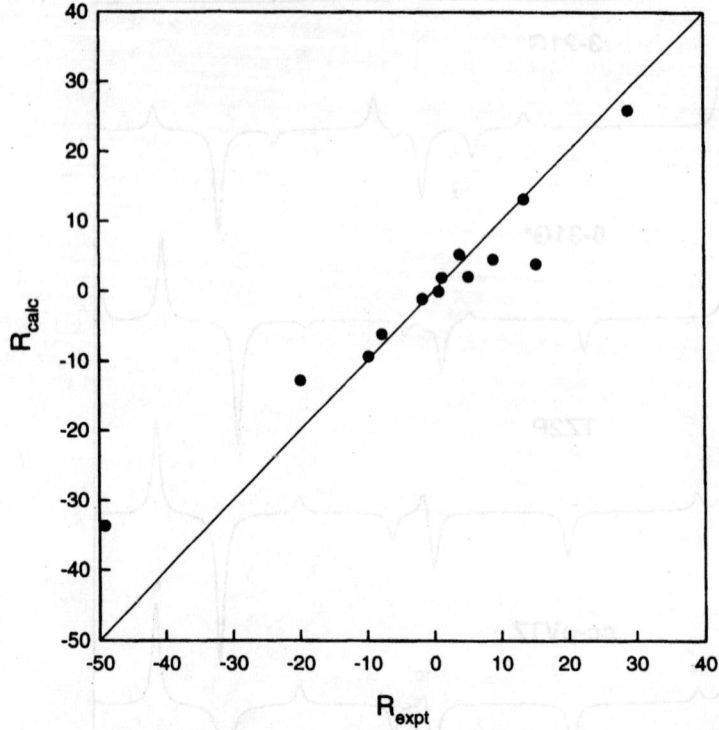

Figure 6 Comparison of B3LYP/cc-pVDZ and experimental rotational strengths for (+)-*R*-**1**. Rotational strengths are in 10^{-44} esu^2 cm^2.

Table 1 Accuracy of Calculated DFT/GIAO Rotational Strengths of Methyl Oxirane **1**[a]

B3LYP		cc-pVTZ	
Basis set	Error[b]	Functional	Error[b]
3-21G	10.5	LSDA	6.7
6-31G*	6.5	BLYP	3.9
6-31G**	6.2	BH and H	9.2
cc-pVDZ	6.4	BH and HLYP	6.0
TZ2P	3.9	B3LYP	3.9
cc-pVTZ	3.9	B3PW91	4.3
VD3P	3.4	B3P86	4.3
cc-pVQZ	3.7	PBE1PBE	4.7

[a] Rotational strengths in 10^{-44} esu^2 cm^2.
[b] Error is average absolute deviation of calculated and experimental rotational strengths for fundamentals 6–18. Experimental rotational strengths were obtained by Lorentzian fitting to the VCD spectrum of **1** in CCl$_4$ solution.

This conclusion does not apply to the C–H stretching region, where Fermi resonance is apparent [19] and harmonic calculations are of insufficient accuracy to account for the observed spectra, either qualitatively or quantitatively. Our methodology also ignores solvent effects. This is a good approximation when solute–solvent interactions are minimal, as in the case of CCl_4 solutions of propylene oxide, but not when large, specific interactions exist.

Comparison of predicted and experimental IR and VCD spectra for a range of functionals and basis sets have also been reported for a number of rigid molecules larger than propylene oxide, including camphor [18a], fenchone [18a], α-pinene [18b], 6,8-dioxabicyclo[3.2.1] octane [20], and phenyl oxirane [21].

4. APPLICATIONS

The IR and VCD spectra of a molecule are exquisitely sensitive to its three-dimensional structure, i.e., its stereochemistry. In addition, enantiomers exhibit mirror image VCD spectra. Consequently, IR and VCD spectra can be used to elucidate the conformational structure(s) of a molecule and VCD can, in addition, be used to determine its absolute configuration (AC). In this section, we discuss the protocol by which conformational and configurational analysis is carried out, using the ab initio DFT calculational methodology described in the previous section.

We cannot deduce molecular structure directly from molecular spectra. We therefore proceed indirectly, first postulating a structure and then evaluating its reliability by comparison of spectra predicted for that structure to experimental spectra. Step 1 is therefore to predict the stable structure(s) of the molecule. This is most accurately carried out using ab initio methods; when IR and VCD spectra are to be calculated using DFT, self-consistency requires the use of DFT. When the molecule is rigid, i.e., only one conformation exists, geometry optimization is straightforward. When the molecule is flexible, and more than one stable conformation exists, the potential energy surface (PES) must first be scanned in order to locate the minima corresponding to these structures. Subsequent geometry optimizations then yield each conformational structure. Potential energy surface scanning becomes increasingly time-consuming as the flexibility of the molecule increases. At this time, scans with respect to one or two degrees of freedom are generally practicable. When the molecule is too flexible to permit a complete PES scan to be carried out, other methods for finding the stable conformations must be used (e.g., semi-empirical methods or molecular mechanics), followed by DFT geometry optimizations. In such cases, one must trust that the conformational structures predicted by the methods used are qualitatively identical to those predicted by DFT, i.e., that these methods reliably predict the shape of the PES. This will not always be the case.

Given the DFT geometries and energies of the stable molecular conformations, step 2 begins, namely, the prediction of their IR and VCD spectra. (Note that this must be carried out using the same functional and basis set used in geometry optimization.) Harmonic frequencies and dipole strengths yield IR spectra; frequencies and rotational strengths yield VCD spectra. This is followed by conformational averaging of the spectra: the spectra of the individual conformations, weighted by their fractional populations, are summed to give the spectra predicted for the equilibrium conformational mixture. Populations reflect the free-energy differences of the conformations. These are often approximated by the calculated DFT energy differences.

Given predicted IR and VCD spectra, analysis of the experimental IR and VCD spectra begins: step 3. Because the frequency range and the S/N ratio of the experimental IR spectrum are invariably greater than those of the VCD spectrum, it is best to commence with the analysis of the IR spectrum. This proceeds in two stages: first, qualitative assignment; second, quantitative analysis. In the first stage, bands of the experimental spectrum are assigned by comparison to the predicted spectrum. In the second stage, the experimental spectrum is deconvoluted using Lorentzian fitting to give the experimental frequencies and dipole strengths [18]. Comparison of these experimental parameters to the predicted ones then defines the agreement of theory and experiment. If this is good, the predicted conformational structures and energies must be correct. If agreement is not good, the predictions are not correct and must be re-assessed.

If, and only if, the predicted and experimental IR spectra are in good agreement, one can proceed to the analysis of the VCD spectrum. Given the successful assignment of the IR spectrum, the VCD spectrum is automatically and simultaneously assigned (as all bands in the VCD spectrum must correspond in frequency to bands in the IR spectrum). It remains to obtain the experimental rotational strengths from the VCD spectrum via Lorentzian fitting and to compare them to the calculated values. Assuming that the experimental VCD spectrum has been normalized to 100% enantiomeric excess (ee), the experimental rotational strengths should be in agreement with those calculated for the enantiomer present in excess. If the latter is unknown, comparison of the experimental rotational strengths to those calculated for both enantiomers establishes which it is: that is, the AC is determined.

The protocol described above can be illuminated by a specific example. The analysis of the cyclic sulfoxide, 1-thiochroman-4-one S-oxide (**2**), will now be described [22]. This molecule is bicyclic, one ring (the phenyl ring) being planar, the other puckered. In order to establish the stable conformations of **2**, a B3LYP/6-31G* PES scan was carried out, varying simultaneously the two dihedral angles C8C9SC1 and C5C4C3C2 (see Fig. 7 for the atom numbering). The results, in the form of a contour plot, are shown in Fig. 7. Two wells exist in the PES. Geometry optimization starting from geometries of minimum energy within these wells leads to two stable conformations, **a** and **b**. Since we will calculate the IR and VCD spectra of **a** and **b** at the B3LYP/TZ2P level (in order to obtain spectra of higher accuracy than obtained at the B3LYP/6-31G* level), geometry optimization is also carried out at the B3LYP/TZ2P level. The structures obtained are shown in Fig. 8. Atoms C2C3C4C5C6C7C8C9S are essentially coplanar. C1 and O1 deviate substantially from this plane. In **a**, C1 and O1 are on the same side of the plane; in **b** they are on opposite sides. The energy difference of **a** and **b** is predicted to be 0.42 kcal/mol, **a** being lower in energy.

Experimental IR and VCD spectra of **2** in the mid-IR spectral region are displayed in Fig. 9. Analysis focuses on the range 800–1500 cm^{-1}. The IR spectra over this range, predicted at the B3LYP/TZ2P level, of **a** and **b**, and the conformationally averaged IR spectrum are shown in Fig. 10. The spectra of **a** and **b** are substantially different. As a result, the conformationally averaged spectrum is substantially different from those of both **a** and **b** individually. In particular, note the presence of many more bands. The experimental IR spectrum is compared to the predicted spectrum in Fig. 10. Agreement of theory and experiment is good, and

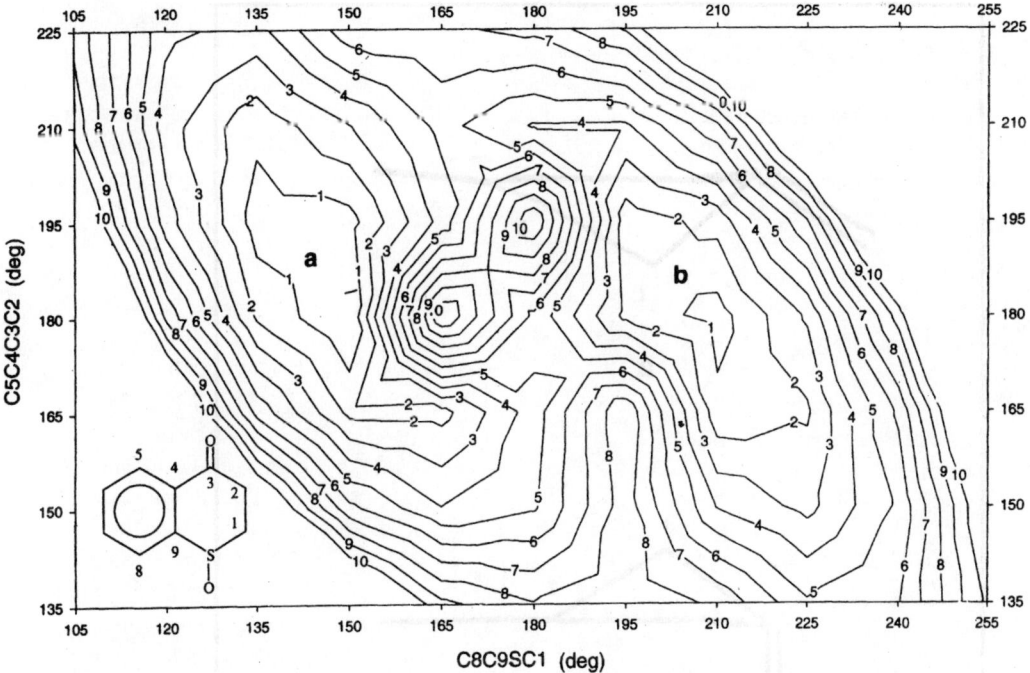

Figure 7 The B3LYP/6-31G* PES of *S*-2. The dihedral angles C5C4C3C2 and C8C9SC1 were varied in 15° steps. Contours are shown at 1 kcal/mol intervals.

assignment of the experimental spectrum is straightforward, as documented in Fig. 10. In particular, many bands are clearly identifiable in the experimental spectrum which are assignable to fundamentals of either **a** or **b**. That is, "conformational splittings" are well resolved for a number of modes. The experimental spectrum thus clearly demonstrates the presence of the two conformations, **a** and **b**, predicted by DFT. Quantitative comparison of predicted and experimental frequencies and dipole strengths shows agreement typical of B3LYP/TZ2P level calculations [18,20,21], further confirming the reliability of the theoretical calculations on which the analysis is based.

The VCD spectra over the range 800–1500 cm^{-1}, predicted at the B3LYP/TZ2P level, of **a** and **b** for the S enantiomer of **2** are shown in Fig. 11, together with the conformationally averaged spectrum. The VCD spectra of **a** and **b** are very different. As a result, the conformationally averaged spectrum differs substantially from those of both **a** and **b**. Again, more bands are present than in the spectra of **a** and **b** individually. The experimental VCD spectrum of the (+) enantiomer is compared to the predicted spectrum in Fig. 11. The assignment of the experimental VCD spectrum follows automatically from the assignment of the IR spectrum. Comparison of the experimental VCD spectrum to the calculated spectrum shows good agreement between theory and experiment. As with the IR spectrum, bands in the VCD spectrum assignable to fundamentals of either **a** or **b** are identifiable, further confirming the presence of the two conformations. Quantitative comparison of predicted and experimental rotational strengths (Fig. 12) shows agreement typical of B3LYP/

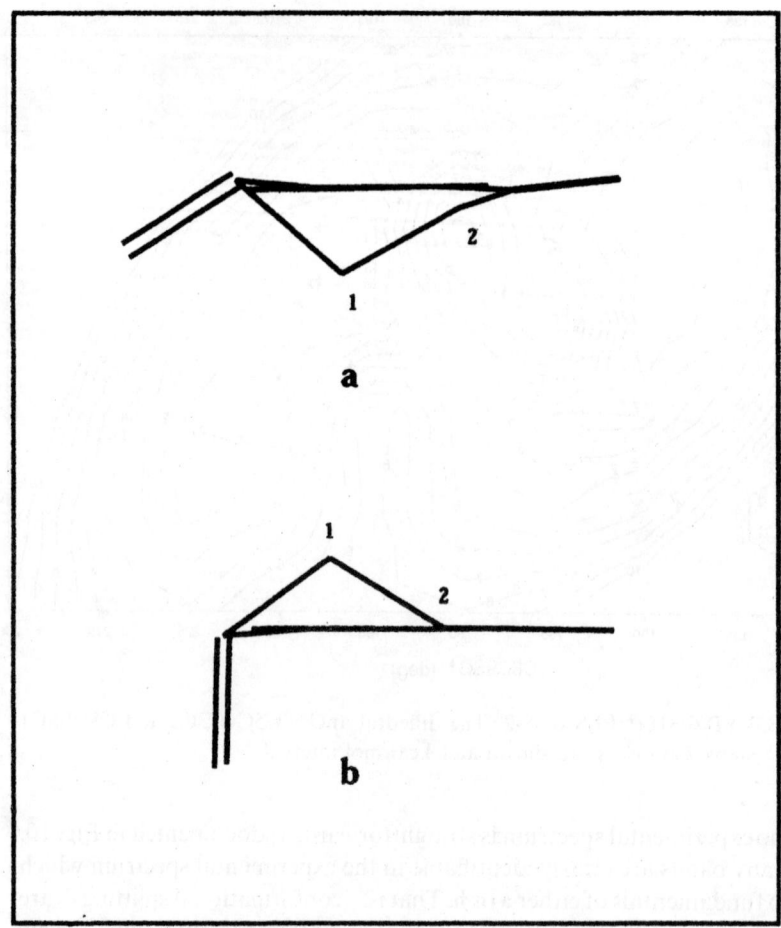

Figure 8 The B3LYP/TZ2P structures of conformations **a** and **b** of *S*-**2**. H atoms are not shown. The perspective demonstrates the near-planarity of the C2C3C4C5C6C7C8C9S moiety.

TZ2P calculations [18,20,21]. Most importantly, the agreement of the predicted VCD spectrum for the S enantiomer with the experimental VCD spectrum for the (+) enantiomer leads to the assignment of the AC of **2** as $S(+)/R(-)$.

By now, IR and VCD spectra of a fairly large number of chiral organic molecules have been analyzed (with varying degrees of thoroughness) using the DFT-based methodology described above. These molecules are listed in Table 2.

5. DISCUSSION

We have described a methodology by means of which the IR and VCD spectra of a chiral molecule can be analyzed and molecular stereochemistry elucidated. Density

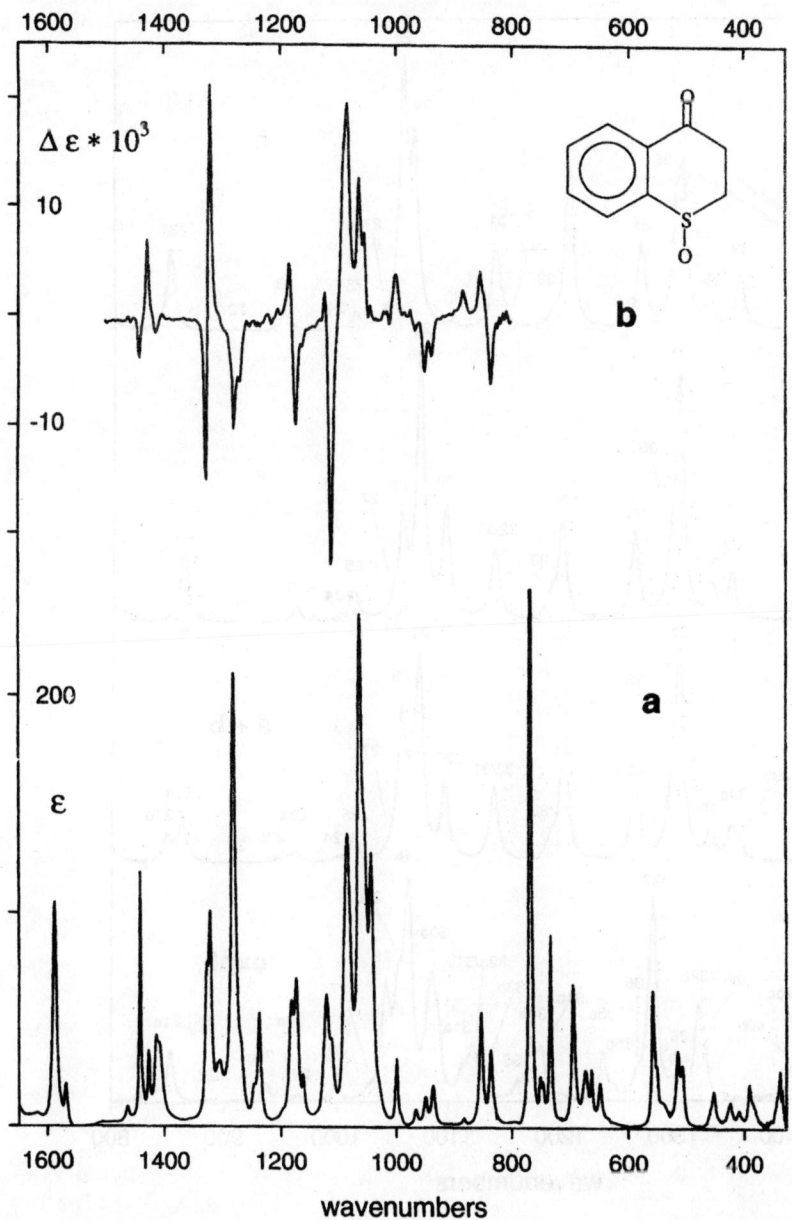

Figure 9 Experimental IR, **a**, and VCD, **b**, spectra of **2**. **a**: 325–715 and 900–1650 cm^{-1}, 0.12M in CCl$_4$, 597 µm path; 715–825 cm^{-1}, 0.05 M in CS$_2$, 597 µm path; 825–900 cm^{-1}, 0.12 M in CCl$_4$, 239 µm path. **b**: 800–842 cm^{-1}, 0.05 M in CS$_2$, 597 µm path; 842–1050 and 1130–1500 cm^{-1}, 0.12 M in CCl$_4$, 597 µm path; 1050–1130 cm^{-1}, 0.12 M in CCl$_4$, 239 µm path. The IR spectrum is for (+)-**1**; the VCD spectrum is the "half-difference" spectrum, [Δε(+)−Δε(−)]/2.

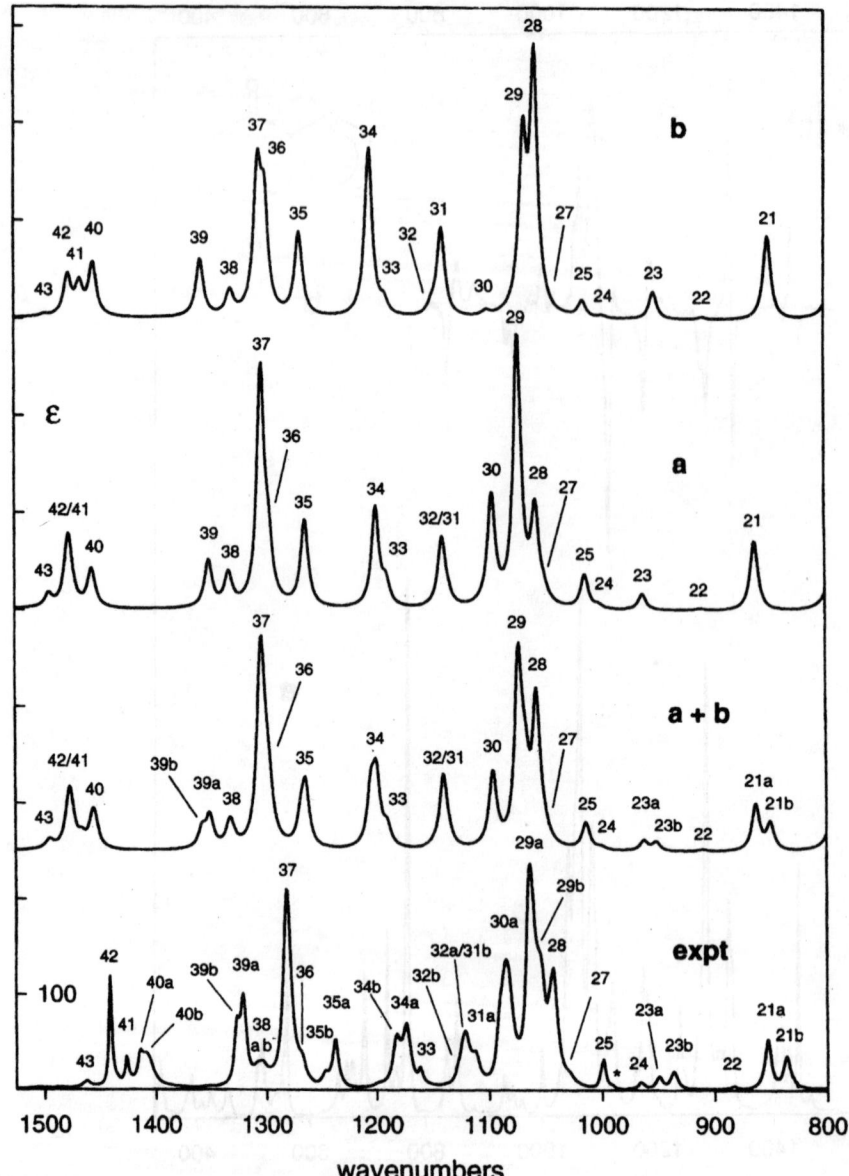

Figure 10 Calculated and experimental IR spectra of **2**. Spectra of conformations **a** and **b** are calculated at the B3LYP/TZ2P level. Lorentzian band shapes are used; $\gamma = 4.0$ cm^{-1}. The spectrum of the equilibrium mixture of **a** and **b** is obtained using populations calculated from the B3LYP/TZ2P energy difference of **a** and **b**. The experimental spectrum is from Fig. 9. The numbers indicate fundamental vibrational modes. Where fundamentals of **a** and **b** are not resolved only the number is shown. The asterisk indicates a band not assigned to fundamentals of **2**.

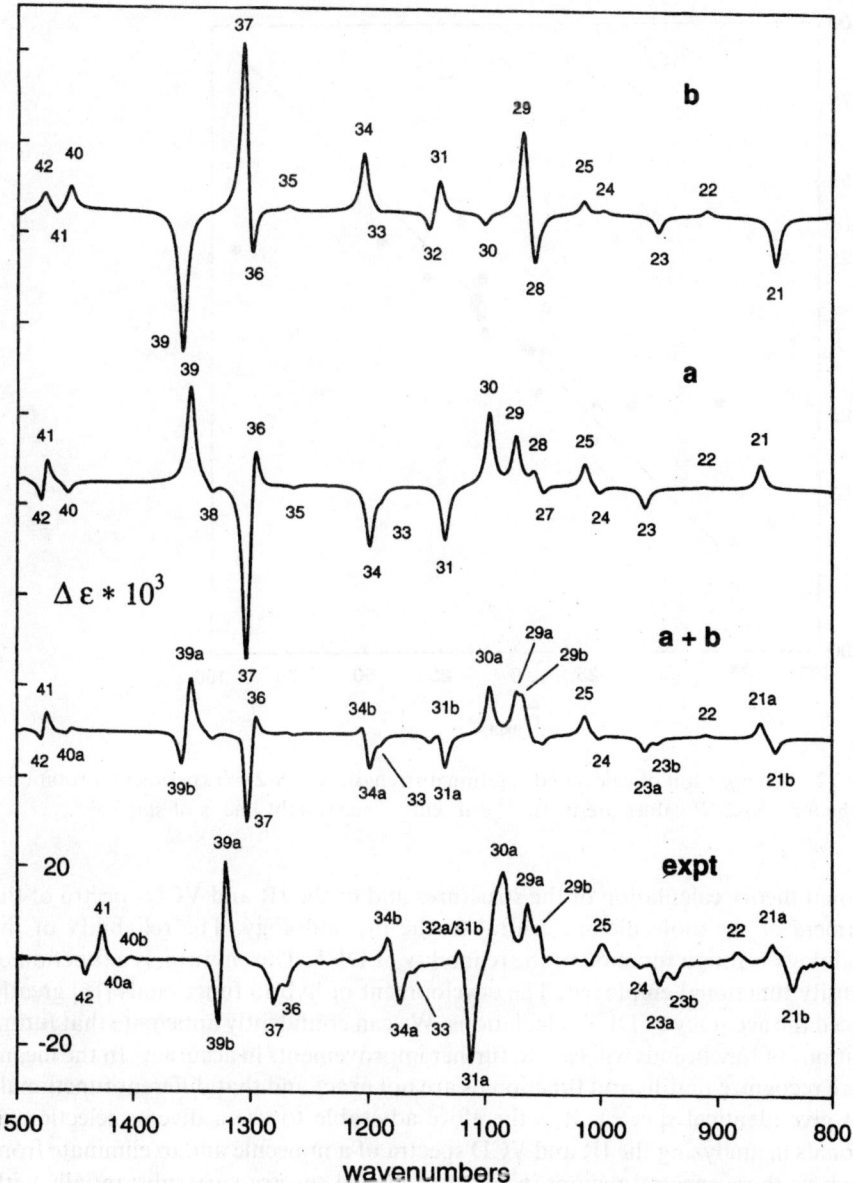

Figure 11 Calculated and experimental VCD spectra of **2**. Spectra of conformations **a** and **b** are calculated at the B3LYP/TZ2P level for *S*-**2**. Lorentzian band shapes are used; $\gamma = 4.0$ cm^{-1}. The spectrum of the equilibrium mixture of **a** and **b** is obtained using populations calculated from the B3LYP/TZ2P energy difference of **a** and **b**. The experimental spectrum is from Fig. 9. The numbers indicate fundamental vibrational modes. Where fundamentals of **a** and **b** are not resolved only the number is shown.

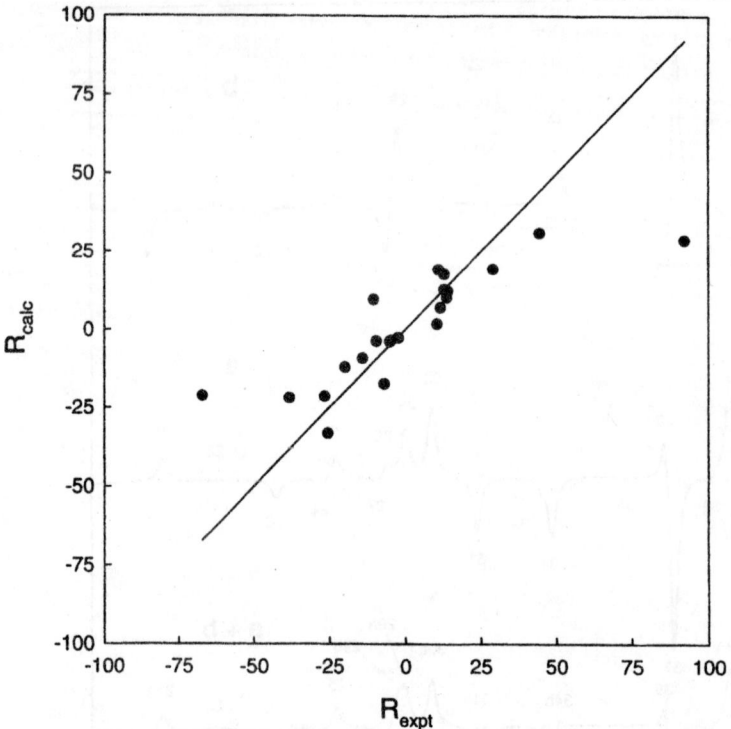

Figure 12   Comparison of calculated rotational strengths for *S*-2 to experimental rotational strengths for (+)-2. *R* values are in 10^{-44} esu^2 cm^2. The straight line is of slope +1.

functional theory calculation of the structures and of the IR and VCD spectra of the conformers of the molecule are central to the methodology. The reliability of the methodology is thus a function of the reliability of DFT. This, in turn, is a function of the density functional employed. The development of hybrid functionals [16] greatly enhanced the accuracy of DFT calculations. We can confidently anticipate that future generations of functionals will lead to further improvements in accuracy. In the meantime, we recognize that hybrid functionals are not exact and that different functionals do not give identical spectra. It is therefore advisable to use a diverse selection of functionals in analyzing the IR and VCD spectra of a molecule and to eliminate from the analysis those spectral regions in which predicted spectra vary substantially with the choice of functional. This increases computational time but, at the same time, increases the reliability of the analysis. Minimally, at least two unrelated functionals of comparable overall accuracy should be used. B3LYP and B3PW91 have been found to be a useful choice [18,20,21,22b].

As with all ab initio calculations, the accuracy of DFT calculations is a function of basis set size and completeness. The larger the basis set the more accurate the results. It is always advisable to examine the basis set convergence of predicted spectra. Minimally, a medium-sized basis set and a substantially larger basis set should be examined. The basis sets 6-31G* and TZ2P constitute a useful choice [18]. 6-31G* is the smallest basis set which gives adequately reliable results. TZ2P is roughly twice as large as 6-31G* and is a good approximation to a complete basis set when predicting

Table 2 Molecules Whose VCD Spectra Have Been Analyzed Using the DFT Methodology

Reference	Molecules
[14c]	*trans*-2,3d$_2$-oxirane
[18a]	camphor
[18a]	fenchone
[18b]	α-pinene
[20]	6,8-dioxabicyclo[3.2.1]octane [DBO]
[23]	exo-7-methyl-DBO
[23]	endo-7-methyl-DBO
[23]	exo-5,7-dimethyl-DBO
[23]	endo-5,7-dimethyl-DBO
[23,24]	1,5-dimethyl-DBO (frontalin)
[25]	2d$_1$-cyclohexanone
[25]	*trans*-2,6d$_2$-cyclohexanone
[26]	3-methylcyclohexanone
[21]	phenyl oxirane
[27]	4,4a,5,6,7,8-hexahydro-4a-methyl-2(3H)-naphthalenone
[27]	3,4,8,8a-tetrahydro-8a-methyl-1,6(2H,7H)-naphthalenedione
[28]	Tröger's Base
[29]	spiropentyl carboxylic acid methyl ester
[29]	spiropentyl acetate
[30]	*tert*-butyl methyl sulfoxide
[31]	1-(2-methyl-naphthyl) methyl sulfoxide
[22b,32]	1-thiochroman *S*-oxide
[22]	1-thiochromanone *S*-oxide
[22b]	1-thiaindan *S*-oxide
[33]	2-butanol
[34]	3-butyn-2-ol
[35]	2,3-butane diol
[36]	epichlorohydrin
[37]	2,5-dimethylthiolane
[37]	2,5-dimethyl sulfolane
[38]	*n*-butyl *tert*-butyl sulfoxide
[39]	*tert*-butyl phenyl phosphine oxide
[40]	1,2,2,2-tetrafluoroethyl methyl ether
[41]	desflurane
[42]	enflurane
[43]	2,2'-di(R$_1$)-3-R$_2$-binaphthyl (R$_1$,R$_2$ = H, H; OH, H; NH$_2$, H; OSO$_2$CF$_3$, H; OH, COOH; OH, COOCH$_3$; $^1/_2$ PO$_4$H, H)
[44]	1,4-oxazin-2-one
[45]	"oxathiane **4**"

IR and VCD spectra. Of course, the computational demands of the larger basis set may exceed the available resources; when only 6-31G* is practicable, the limitations of this basis set must be kept in mind.

The computational demands of DFT calculations increase with increasing molecular size. Eventually, despite the power of the computational resources, DFT calculations with a basis set such as 6-31G* become impractical. At the present time, at this basis set level molecules with >100 atoms present a substantial challenge. However, given the continuous increase in computing power, the threshold is continually increasing and is likely to be much larger in the near future.

There are two major approximations in the theory of VCD which constitutes the foundation for the DFT calculations. One is the harmonic approximation, within which anharmonicity is ignored. As a result, all fundamental excitations occur at harmonic frequencies and intensities of all overtone and combination transitions are predicted to be zero. In the mid-IR spectral region, this has proved to be a very good approximation. In particular, very few bands are observed in IR and VCD spectra which cannot be assigned to fundamental excitations. At higher frequencies, however, this is no longer the case. In particular, the fundamental hydrogenic stretching region of organic molecules is almost always more complex than predicted by harmonic calculations. As a result, analysis using harmonic calculations is not feasible. Clearly, the inclusion of anharmonicity is important. This has been done for one molecule: *trans*-2,3d$_2$-oxirane [46]. Hopefully, this work will be considerably extended in the near future.

The other major approximation is the neglect of solvent effects. To date, most VCD spectroscopy has been carried out using solvents such as CCl_4, CS_2, $CHCl_3$, CH_2Cl_2, and so on, in which, for most organic molecules, solute–solvent interactions are relatively innocuous. Here the neglect of solvent effects has not appeared to be a substantial limitation. However, for many solute–solvent combinations interactions are much stronger and can be expected to cause substantial perturbations to IR and VCD spectra. For such systems, it is clearly important to include solvent effects in the theoretical methodology. There have been major advances recently in the treatment of solvent effects on IR spectra using continuum dielectric solvent models [47]; their extension to VCD spectra is obviously desirable.

In the case of conformationally flexible molecules, IR and VCD spectra are superpositions of the spectra of individual conformers, weighted by the corresponding conformational populations. The latter depend on conformational free-energy differences and on the temperature. The former can be determined experimentally or predicted theoretically. In general, however, experimental values will not be already available, and their determination will be nontrivial. Conformational free-energy differences are most commonly determined using NMR. However, the use of NMR presupposes sufficiently high barriers to conformational interconversion to permit NMR spectra of individual conformers to be observed at accessible temperatures. For many molecules, this is not the situation and NMR cannot be used. Vibrational spectroscopy provides an alternative approach to determining conformational free-energy differences [26], which does not suffer from the limitations of NMR, because the vibrational time scale is always much faster than conformational interconversion times. However, the use of vibrational spectroscopy becomes increasingly difficult as the number of conformations increases. Thus, in practice, one normally relies on calculated free-energy differences to obtain conformational populations. Often free-energy differences are approximated by simple equilibrium energy differences, neglecting zero-point energy and entropic contributions. Generally, solvent effects are ignored. The magnitudes of the errors incurred are not easily quantitated and undoubtedly vary from molecule to molecule. It is always advisable to examine the range of variation in predicted IR and VCD spectra consequent to varying conformational free-energy differences over a realistic range.

The routine use of VCD spectroscopy is limited to liquid solutions. Vibrational circular dichroism is an intrinsically weak phenomenon (g values are very small) and its measurement requires optimum experimental conditions, in addition to state-of-the-art instrumentation. In general, VCD spectra are measured at fairly high concentrations—in the range 0.01–1.0M—in solvents with good mid-IR transmission at fairly short pathlengths (≤ 1 mm). The accessibility of such conditions depends on the solubilities of the molecules to be studied in available solvents. Compounds only soluble to significant extent in water are generally not easily studied.

Probably the most important application of VCD spectroscopy is the determination of absolute configuration (AC). While the stereochemist already has available a variety of tools for determining AC, VCD spectroscopy possesses a number of significant advantages. First and foremost, it employs liquid solutions and does not depend on the availability of single crystal samples, as does x-ray crystallography. Secondly, calculations of VCD spectra are most easily carried out for compounds containing only first-row atoms; in contrast, x-ray crystallography generally requires a "heavy" atom to be present. Thirdly, the high degree of resolution of vibrational spectra allows the study over the experimentally accessible mid-IR region of the VCD intensities of a large number of transitions, allowing the reliability of predicted VCD intensities to be examined with good statistics. This constitutes a substantial advantage over electronic CD spectroscopy, where usually a smaller number of bands are accessible, band widths are much larger, and spectra are much less well resolved.

6. CONCLUSION

At this time, VCD spectroscopy is a viable technique for elucidating the stereochemistry of chiral organic molecules. Most importantly, it provides a powerful new alternative approach to the determination of absolute configuration. Both the experimental instrumentation and the computational software required for the application of VCD are now commercially available. As a result, the VCD technique is ripe for exploitation by stereochemists. It should be of value, inter alia, in stereochemical studies of candidate pharmaceutical compounds.

ACKNOWLEDGMENTS

I gratefully acknowledge the assistance of Dr. Frank J. Devlin in preparing this paper, and the financial support of the National Science Foundation (NSF grant CHE-9902832).

REFERENCES

1a. Stephens PJ, Lowe MA. Vibrational circular dichroism. Ann Rev Phys Chem 1985; 36:213–241.

1b.   Stephens PJ, Devlin FJ. Determination of the structure of chiral molecules using ab initio vibrational circular dichroism spectroscopy. Chirality 2000; 12:172–179.

1c.   Stephens PJ, Devlin FJ, Aamouche A. Determination of the structures of chiral molecules using vibrational circular dichroism spectroscopy. In: Hicks J, ed. Chirality: Physical Chemistry, ACS Symposium Series, 2002; Vol. 810:18–33.

1d.   Nafie LA. Infrared and Raman vibrational optical activity: theoretical and experimental aspects. Ann Rev Phys Chem 1997; 48:357–386.

2.   Velluz L, Legrand M, Grosjean M. Optical Circular Dichroism. Principles, Measurements and Applications. Verlag Chemie/Academic Press, 1965.

3a.   Billardon M, Badoz J. Modulateur de birefringence. CR Acad Sci Paris 1966; 262B:1672–1675.

3b.   Kemp JC. Piezo-optical birefringence modulators: new use for a long-known effect. J Opt Soc Am 1969; 59:950–954.

3c.   Jasperson SN, Schnatterly SE. An improved method for high reflectivity ellipsometry based on a new polarization modulation technique. Rev Sci Instrum 1969; 40:761–767.

3d.   Mollenauer LF, Downie D, Engstrom H, Grant WB. Stress plate optical modulator for circular dichroism measurements. Appl Opt 1969; 8:661–665.

4.   Hsu EC, Holzwarth G. Vibrational circular dichroism observed in crystalline α-$NiSO_4 \cdot 6H_2O$ and α-$ZnSeO_4 \cdot 6H_2O$ between 1900 and 5000 cm^{-1}. J Chem Phys 1973; 59:4678–4685.

4b.   Holzwarth G, Hsu EC, Mosher HS, Faulkner TR, Moscowitz A. Infrared circular dichroism of carbon–hydrogen and carbon–deuterium stretching modes. Observations. J Am Chem Soc 1974; 96:251–252.

4c.   Nafie LA, Cheng JC, Stephens PJ. Vibrational circular dichroism of 2,2,2-Trifluoro-1-Phenylethanol. J Am Chem Soc 1975; 97:3842–3843.

4d.   Nafie LA, Keiderling TA, Stephens PJ. Vibrational circular dichroism. J Am Chem Soc 1976; 98:2715–2723.

4e.   Stephens PJ, Clark R. Vibrational circular dichroism: the experimental viewpoint. In: Mason SF, ed. Optical Activity and Chiral Discrimination. Reidel, 1979:263–287.

5a.   Nafie LA, Diem M, Vidrine DW. Fourier transform infrared vibrational circular dichroism. J Am Chem Soc 1979; 101:496–498.

5b.   Nafie LA. Polarization modulation FTIR spectroscopy. In: Mackenzie NW, ed. Advances in Applied Fourier Transform Infrared Spectroscopy. Wiley, 1988:67–104.

6a.   Diem M, Roberts GM, Lee O, Barlow A. Design and performance of an optimized dispersive infrared dichrograph. Appl Spectrosc 1988; 42:20–27.

6b.   Diem M. Advances in instrumentation for the observation of vibrational optical activity. Vib Spectra Struct 1991; 19:1–54.

6c.   Xie P, Diem M. Measurement of dispersive vibrational circular dichroism: signal optimization and artifact reduction. Appl Spectrosc 1996; 50:675–680.

7.   Keiderling TA. Vibrational circular dichroism. Comparison of techniques and practical considerations. Practical Fourier Transform Infrared Spectroscopy. New York: Academic Press, 1990:203–284.

8.   Osborne GA, Cheng JC, Stephens PJ. A near-infrared circular dichroism and magnetic circular dichroism instrument. Rev Sci Instrum 1973; 44:10–15.

9.   Schellman JA. Circular dichroism and optical rotation. Chem Rev 1975; 75:323–331.

10.   Person WB, Newton JH. Dipole moment derivatives and infrared intensities: I. Polar tensors. J Chem Phys 1974; 61:1040–1049.

11a. Stephens PJ. Theory of vibrational circular dichroism. J Phys Chem 1985; 89:748–752.

11b. Stephens PJ. Gauge dependence of vibrational magnetic dipole transition moments and rotational strengths. J Phys Chem 1987; 91:1712–1715.

12a. Amos RD. Molecular property derivatives. Adv Chem Phys 1987; 67:99–153.

12b.   Pulay P. Analytic derivative methods. Adv Chem Phys 1987; 69:241.

12c.   Yamaguchi Y, Osamura Y, Goddard JD, Schaefer HF. A new dimension to quantum chemistry: analytic derivative methods in ab initio quantum chemistry. OUP, 1994.

13a.   London F. J Phys Radium 1937; 8:397.

13b.   Ditchfield R. Self-consistent perturbation theory of diamagnetism: I. A gauge-invariant LCAO method for NMR chemical shifts. Mol Phys 1974; 27:789–807.

14a.   Bak KL, Jørgensen P, Helgaker T, Ruud K, Jensen HJA. Gauge-origin independent multiconfigurational self-consistent-field theory for vibrational circular dichroism. J Chem Phys 1993; 98:8873–8883.

14b.   Bak KL, Jørgensen P, Helgaker T, Ruud K, Jensen HJA. Basis set convergence of atomic axial tensors obtained from self-consistent field calculations using London atomic orbitals. J Chem Phys 1994; 100:6620–6627.

15.   Cheeseman JR, Frisch MJ, Devlin FJ, Stephens PJ. Ab initio calculation of atomic axial tensors and vibrational rotational strengths using density functional theory. Chem Phys Lett 1996; 252:211–220.

15c.   Frisch MJ, et al. Gaussian 98. Pittsburgh: Gaussian Inc., 1998.

16a.   Becke AD. A new mixing of Hartree–Fock and local density functional theories. J Chem Phys 1993, 98:1372–1377.

16b.   Becke AD. Density functional thermochemistry: III. The role of exact exchange. J Chem Phys 1993; 98:5648–5652.

17.   Stephens PJ, Devlin FJ, Chabalowski CF, Frisch MJ. Ab initio calculation of vibrational circular dichroism spectra using density functional force fields. J Phys Chem 1994; 98:11623–11627.

18a.   Devlin FJ, Stephens PJ, Cheeseman JR, Frisch MJ. Ab initio prediction of vibrational absorption and circular dichroism spectra of chiral natural products using density functional theory: camphor and fenchone. J Phys Chem 1997; 101:6322–6333.

18b.   Devlin FJ, Stephens PJ, Cheeseman JR, Frisch MJ. Ab initio prediction of vibrational absorption and circular dichroism spectra of chiral natural products using density functional theory: α-pinene. J Phys Chem 1997; 101:9912–9924.

19a.   Kawiecki RW, Devlin FJ, Stephens PJ, Amos RD. Vibrational circular dichroism of propylene oxide. J Phys Chem 1991; 95:9817–9831.

19b.   Stephens PJ, Chabalowski CF, Jalkanen KJ, Devlin FJ. Ab initio calculation of vibrational circular dichroism spectra using large basis set MP2 force fields. Chem Phys Lett 1994; 225:247–257.

20.   Stephens PJ, Cheeseman JR, Frisch MJ, Ashvar CS, Devlin FJ. Ab initio calculation of atomic axial tensors and vibrational rotational strengths using density functional theory. Mol Phys 1996; 89:579–594.

21.   Ashvar CS, Devlin FJ, Stephens PJ. Molecular structure in solution: an ab initio vibrational spectroscopy study of phenyloxirane. J Am Chem Soc 1999; 121:2836–2849.

22a.   Devlin FJ, Stephens PJ, Scafato P, Superchi S, Rosini C. Determination of absolute configuration using vibrational circular dichroism spectroscopy: the chiral sulfoxide 1-thiochromanone-S-oxide. Chirality 2002; 14:400–408.

22b.   Devlin FJ, Stephens PJ, Scafato P, Superchi S, Rosini C. Conformational analysis using IR and VCD spectroscopies: the chiral cyclic sulfoxides 1-thiochromanone-S-oxide, 1-thiaindan S-oxide and 1-thiochroman-S-oxide. J Phys Chem A, 2002; 106:10510–10524.

23.   Ashvar CS, Devlin FJ, Stephens PJ, Bak KL, Eggimann T, Wieser H. Vibrational absorption and circular dichroism of mono- and di-methyl derivatives of 6,8-dioxabicyclo [3.2.1] octane. J Phys Chem 1998; 102:6842–6857.

24.   Ashvar CS, Stephens PJ, Eggimann T, Wieser H. Vibrational circular dichroism spectroscopy of chiral pheromones: frontalin (1,5-dimethyl-6,8-dioxabicyclo [3.2.1] octane). Tetrahedron: Asymmetry 1998; 9:1107–1110.

25.   Devlin FJ, Stephens PJ. Ab initio density functional theory study of the structure and

vibrational spectra of cyclohexanone and its isotopomers. J Phys Chem 1999; 103:527–538.

26. Devlin FJ, Stephens PJ. Conformational analysis using ab initio vibrational spectroscopy: 3-methyl-cyclohexanone. J Am Chem Soc 1999; 121:7413–7414.

27. Aamouche A, Devlin FJ, Stephens PJ. Molecular structure of chiral molecules in solution: ab initio vibrational absorption and circular dichroism studies of 4, 4a, 5, 6, 7, 8-hexa hydro-4a-methyl-2(3H)naphthalenone, and 3, 4, 8, 8a, -Tetra Hydro-8a-methyl-1,6(2H,7H)-naphthalenedione. J Am Chem Soc 2000; 122:7358–7367.

28a. Aamouche A, Devlin FJ, Stephens PJ. Determination of absolute configuration using circular dichroism: Tröger's base revisited using vibrational circular dichroism. J Chem Soc Chem Commun, 1999, 361–362.

28b. Aamouche A, Devlin FJ, Stephens PJ. Structure, vibrational absorption and circular dichroism spectra and absolute configuration of Tröger's base. J Am Chem Soc 2000; 122:2346–2354.

29. Devlin FJ, Stephens PJ, Oesterle C, Wiberg KB, Cheeseman JR, Frisch MJ. Configurational and conformational analysis of chiral molecules using IR and VCD spectroscopies: spiropentylcarboxylic acid methyl ester and spiropentylacetate. J Org Chem 2002; 67:8090–8096.

30. Aamouche A, Devlin FJ, Stephens PJ, Drabowicz J, Bujnicki B, Mikolajczyk M. Vibrational circular dichroism and absolute configuration of chiral sulfoxides: *tert*-butyl methyl sulfoxide. Chemistry: A European Journal 2000; 6:4479–4486.

31. Stephens PJ, Aamouche A, Devlin FJ, Superchi S, Donnoli MI, Rosini C. Determination of absolute configuration using vibrational circular dichroism spectroscopy: 1-(2-methylnaphthyl) methyl sulfoxide. J Org Chem 2001; 66:3671–3677.

32. Devlin FJ, Stephens PJ, Scafato P, Superchi S, Rosini C. Determination of absolute configuration using vibrational circular dichroism spectroscopy: the chiral sulfoxide 1-thiochroman-S-oxide. Tetrahedron Asymmetry 2001; 12:1551–1558.

33. Wang F, Polavarapu PL. Vibrational circular dichroism: predominant conformations and intermolecular interactions in R-(−)-2-butanol. J Phys Chem A 2000; 104:10683–10687.

34. Wang F, Polavarapu PL. Vibrational circular dichroism, predominant conformations, and hydrogen bonding in (S)-(−)-3-butyn-2-ol. J Phys Chem A 2000; 104:1822–1826.

35. Wang F, Polavarapu PL. Predominant conformations of ($2R,3R$)-(−)-2,3-butanediol. J Phys Chem A 2001; 105:6991–6997.

36. Wang F, Polavarapu PL. Conformational stability of (+)-epichlorohydrin. J Phys Chem A 2000; 104:6189–6196.

37. Wang F, Wang H, Polavarapu PL, Rizzo CJ. Absolute configuration and conformational stability of (+)-2,5-dimethylthiolane and (−)-2,5-dimethylsulfolane. J Org Chem 2001; 66:3507–3512.

38. Drabowicz J, Dudzinski B, Bogdan M, Mikolajczyk M, Wang F, Dehlavi A, Goring J, Park M, Rizzo CJ, Polavarapu PL, Biscarini P, Wieczorek MW, Majzner WR. Absolute configuration, predominant conformations, and vibrational circular dichroism spectra of enantiomers of n-butyl *tert*-butyl sulfoxide. J Org Chem 2001; 66:1122–1129.

39. Wang F, Polavarapu PL, Drabowicz J, Mikolajczyk M. Absolute configurations, predominant conformations and tautomeric structures of enantiomeric *tert*-butylphenylphosphine oxides. J Org Chem 2000; 65:7561–7565.

40. Polavarapu PL, Zhao C, Ramig K. Vibrational circular dichroism, absolute configuration and predominant conformations of volatile anesthetics: 1,2,2,2-tetrafluoroethyl methyl ether. Tetrahedron Asymmetry 1999; 10:1099–1106.

41. Polavarapu PL, Zhao C, Cholli AL, Vernice GG. Vibrational circular dichroism, absolute configuration and predominant conformations of volatile anesthetics: desflurane. J Phys Chem B 1999; 103:6127–6132.

42. Zhao C, Polavarapu PL, Grosenick H, Schurig V. Vibrational circular dichroism, absolute configuration and predominant conformations of volatile anesthetics: enflurane. J Mol Struct 2000; 550–551:105–115.

43. Setnicka V, Urbanova M, Bour P, Kral V, Volka K. Vibrational circular dichroism of 1,1′-binaphthyl derivatives: experimental and theoretical study. J Phys Chem A 2001; 105:8931–8938.

44. Solladié-Cavallo A, Sedy O, Salisova M, Biba M, Welch CJ, Nafie L, Freedman T. A chiral 1,4-oxazin-2-one: asymmetric synthesis versus resolution, structure, conformation and VCD absolute configuration. Tetrahedron Asymmetry 2001; 12:2703–2707.

45. Solladié-Cavallo A, Balaz M, Salisova M, Suteu C, Nafie LA, Cao X, Freedman TB. A new chiral oxathiane: synthesis, resolution and absolute configuration determination by vibrational circular dichroism. Tetrahedron: Asymmetry 2001; 12:2605–2611.

46. Bak KL, Bludsky O, Jørgensen P. Ab initio calculations of anharmonic vibrational circular dichroism intensities of *trans*-2,3-dideuterio-oxirane. J Chem Phys 1995; 103:10548–10555.

47a. Mennucci B, Cammi R, Tomasi J. Analytical free energy second derivatives with respect to nuclear coordinates: complete formulation for electrostatic continuum solvation models. J Chem Phys 1999; 110:6858–6870.

47b. Cammi R, Cappelli C, Corni S, Tomasi J. On the calculation of infrared intensities in solution within the polarizable continuum model. J Phys Chem A 2000; 104:9874–9879.

47c. Cappelli C, Corni S, Cammi R, Mennucci B, Tomasi J. Nonequilibrium formulation of infrared frequencies and intensities in solution: analytical evaluation within the polarizable continuum model. J Chem Phys 2000; 113:11270–11279.

27

Sialidases: Targets for Rational Drug Design

JEFFREY C. DYASON, JENNIFER C. WILSON, and MARK VON ITZSTEIN

Institute for Glycomics, Griffith University (Gold Coast Campus), Bundall, Queensland, Australia

1. INTRODUCTION

Sialidases (also known as neuraminidases, EC 3.2.1.18) are glycohydrolases that cleave terminal sialic acid residues from a range of glycoconjugates and therefore play an important role in the regulation of the distribution of various sialic acids in biological systems [1]. Sialidases have been identified in various organisms, such as bacteria and viruses, and are widely distributed throughout mammalian tissues, where they may be membrane-associated, cytosolic, or extracellular. Pathogenic organisms such as bacteria utilize sialidases either for nutrition (*Salmonella typhimurium*) or as an aid to pathogenesis (*Vibrio cholerae*). Parasitic organisms, such as *Trypanosoma cruzi* (Chagas disease) and *Trypanosoma brucei* (African sleeping sickness), exploit sialidase and/or *trans*-sialidase (hydrolysis and transfer of sialic acid) activities for their virulence. The viral pathogens orthomyxoviruses (influenza) and paramyxoviruses (parainfluenza, mumps, Newcastle disease) also have sialidase enzymes on their cell surfaces [2].

Influenza virus sialidase, one of two major antigenic glycoproteins (the other is hemagglutinin) on the surface of influenza virus, is undoubtedly the most comprehensively studied sialidase to date [3–5]. Influenza is a highly contagious upper respiratory tract infection that, in addition to humans, infects mammals such as seals, whales, horses, and pigs, and many bird species [6]. There are three distinct types of

---

The manuscript for this chapter was submitted in February 2002.

influenza virus, A, B, and C, with only A and B being infectious to human populations; however, type A is also infectious to some animal and bird species. Type A influenza is further divided into subtypes, according to how their hemagglutinin and sialidase are recognized by antibodies. To date, there are 15 known hemagglutinins, designated as H1–H15, and 9 sialidase N1–N9 antigens. Each of these subtypes can combine by genetic reassortment to produce a different strain of the virus. Interest in the design of drugs against influenza arises from the fact that influenza virus infection can have devastating consequences in epidemic or pandemic situations leading to significant loss of life [7,8]. There were four pandemics in the twentieth century, in 1918, 1957, 1968, and 1977 [9]. The 1918 Spanish influenza pandemic resulted in the death of at least 20 million people worldwide [10,11]. An outbreak of a new virulent strain (designated H5N1) in Hong Kong in 1997, originated in birds and infected 18 people, of which 6 died [12–14]. Major catastrophe was averted by a mass culling of infected birds and, fortunately, human-to-human transmission was not observed. This outbreak served as a timely reminder of the severity of this disease, which has been present since ancient times. The threat of influenza is attributable to the rapid antigenic variation in the two major surface antigens (hemagglutinin and sialidase) that extensively decorate the surface of the virus [6]. Major antigenic shifts occur in influenza Type A periodically, due to a reassortment of the hemagglutinin of one source with the sialidase from another, which then appears in human populations [6]. The virus also undergoes antigenic drift resulting from mutations in the RNA coding for hemagglutinin and sialidase [6]. Vaccine development against newly emergent strains of the virus is largely ineffective because the process can usually take approximately several months to produce a suitable product. In fact, until as recently as the mid-1990s, the only drugs available to treat influenza were amantadine and rimantidine. These drugs exert their therapeutic effect by targeting the M2 ion channel protein of the influenza virus [15]. As therapies, these drugs are not completely satisfactory; however, because they are ineffective against influenza B, which lacks an M2 ion channel, and produce undesirable side effects in some patients. Moreover, drug-resistant mutant viruses were found to develop quickly, rendering these drugs ineffective for long-term usage [16,17].

The breakthrough in the development of a new generation of potent drugs against influenza disease was facilitated by the determination of the x-ray crystal structure of the sialidase from influenza virus [18]. Although antigenic drift contributes to the changing of surface amino acids of influenza virus sialidase, careful alignment of the amino acid sequences of sialidases from several influenza strains clearly demonstrated that there were a number of highly conserved regions across all strains of the virus. The x-ray crystal structure of the influenza virus sialidase revealed that these conserved regions were located around the active site pocket. In particular, those amino acid residues that line the active site and are in contact with the sialic acid, and therefore intimately involved in the hydrolysis process, are strictly conserved [18,19]. Invariant important residues in the active site of the sialidase structure are the chink in the armor of the influenza virus pathogenicity, and makes influenza virus sialidase an excellent target for drug discovery.

Influenza virus sialidase recognizes and cleaves terminal α-ketosidically linked sialic acids from glycoconjugates, and is therefore termed an *exo*-glycohydrolase [20–22]. The role of the sialidase in the infective life cycle of influenza virus appears varied and includes functions such as promoting the release of the viral progeny from the host cell surface by destroying receptors for hemagglutinin allowing their elution away from infected cells [23,24]. The receptor-destroying properties are also useful in assisting the virus to penetrate and move through respiratory mucins that are also rich in sialic acids. Another important role for sialidase arises because progeny virions have newly synthesized sialidase and hemagglutinin proteins on their surface that contain oligosaccharide chains terminating with sialic acid residues. Hemagglutinin of neighboring virions can recognize and bind to these sialic acid residues, leading to aggregation of the viral progeny. Sialidase removes terminal sialic acid residues from these glycoproteins, preventing self-recognition and reducing the propensity of the virus particles to self-aggregate.

Electron micrograph experiments of influenza virus particles reveal that sialidase appears as numerous mushroom-shaped spikes on the surface of the virus [25,26]. The box-shaped head of the mushroom ($100 \times 100 \times 60$ Å) is composed of a tetramer of identical sialidase subunits, centrally attached to a long, thin stalk with a hydrophobic region that embeds itself into the viral membrane. Each monomeric unit has one active site, located in pockets that occur on the upper surface of the mushroom-head tetramer and viewed as a distinct depression on the surface.

(1) R = NHAc

The x-ray crystal structure of influenza virus sialidase was first determined by Colman et al. [18] in the early 1980s. Since then, many high-resolution structures of both influenza A and influenza B sialidase complexed with N-acetylneuraminic acid (Neu5Ac, **1**), modified sialic acids, and various small molecule inhibitors have been determined [3,18,28–43]. Each monomeric unit has a topology of six four-stranded antiparallel β-sheets arranged as if on the blades of a propeller (Fig. 1). The active site lies close to the center of a sixfold pseudo-symmetry axis that passes through the center of the monomer and provides a large, highly charged environment for a sialic acid to reside. Key interactions of the conserved active site residues with the natural ligand sialic acid include charge–charge, hydrogen bonding, and hydrophobic contacts (Fig. 2). Not only are those residues that have direct contact with a sialic acid conserved, but a number of amino acids which provide structural support similar to a scaffolding for the active site are also conserved in both influenza virus A and B sialidase. A cluster of three arginine residues (Arg118, Arg292, and Arg371) orients the sialic acid, which binds as its α-anomer in a distorted half-chair conformation, within the active site. The positively charged residues form charge–charge interactions with the negatively charged carboxylic acid of the sialic acid to firmly position this ligand within the active site. Arg371 is arranged in a planar salt bridge with the carboxylate group. A glutamic acid Glu277 stabilizes the positioning of Arg292, one of the arginine residues of the tri-arginyl cluster. A tyrosine residue, Tyr406, is located below the sialic acid ring. Other interactions that assist in the correct positioning of the sialic acid in the catalytic pocket involve the N-acetyl group of the sialic acid. The carbonyl oxygen of

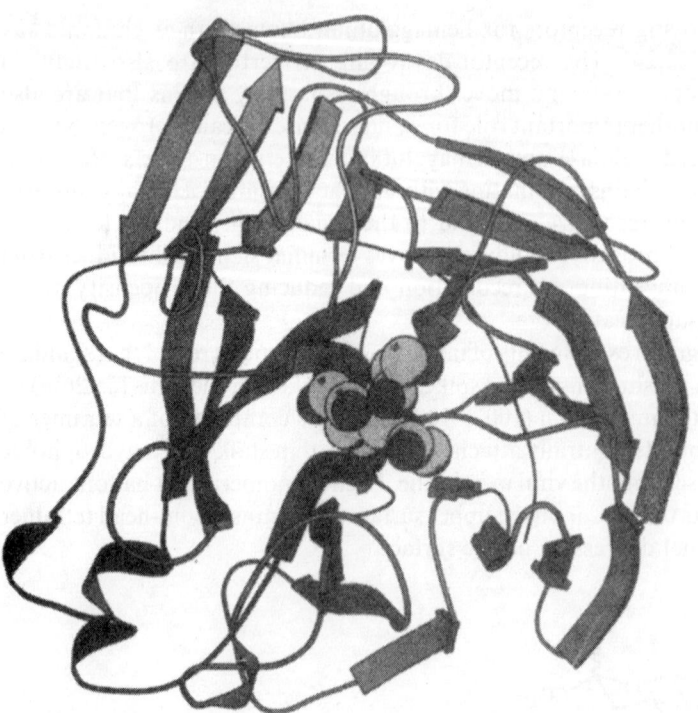

Figure 1 Molscript [27] diagram of an influenza virus sialidase monomer with Neu5Ac (**1**) bound in the active site.

this group hydrogen bonds to Arg152 and a buried water molecule, while the terminal methyl group makes important hydrophobic contacts to the residues that form a small hydrophobic pocket, namely, Trp178 and Ile122, as well as the side chain of Arg152. The glycerol side chain of the sialic acid also assists in maintaining the correct positioning of the sialic acid within the catalytic pocket. The hydroxyl groups of C8 and C9 form a bidentate hydrogen bond to the carboxylate of Glu276. Additionally, the C4-hydroxyl is directed toward the carboxylate oxygen of Glu119. Interestingly, the atoms of the sialic acid ring do not appear to be involved in any strong interactions with any of the active site residues.

There is only about 30% sequence identity between influenza A and B sialidase, yet those amino acids that bind to the sialic acid in the active site of the sialidases associated with each type are strictly conserved. Furthermore, superimposition of the conserved residues of the active site of sialidases (A/Tern/Australia/G70c/75, subtype N9, B/Beijing/1/87, and B/Lee/40) reveals close agreement in both main chain and side-chain positioning. However, there are some subtle differences in the active sites of influenza A and B sialidase that have been shown to affect the selectivity of the enzymes toward some types of inhibitors [33]. These differences are significant around residue 405 and the conserved residue 224. Residue 405 is a tryptophan in influenza virus B sialidase but a glycine in influenza virus A sialidase; this leads to congestion on the floor of the active site in influenza virus B sialidase. A conserved residue Arg224 hydrogen bonds to Ser250 in influenza virus B sialidase, while this interaction is not

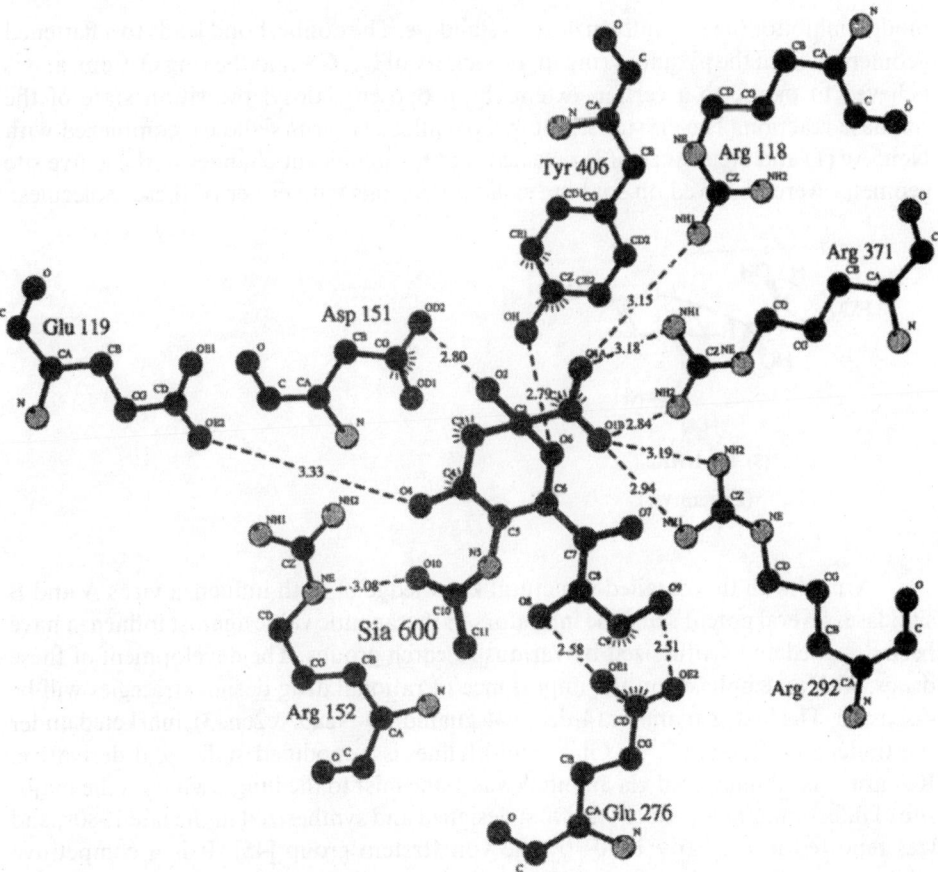

Figure 2 Ligplot [44] diagram showing the important interactions between influenza virus sialidase and Neu5Ac (**1**).

possible in influenza virus A sialidase as the equivalent residue (to Ser250) is an alanine. The side-chain conformation of the acidic oxygen atoms of Glu276 are arranged differently in influenza virus A and B sialidase because of the differences in the second sphere of residues surrounding the active site. These differences lead to a hydrophilic environment around the side-chain of Glu276 in influenza virus A sialidase, whereas there is a hydrophobic environment around this side chain in influenza virus B sialidase.

(2) R = NHAc

The proposed transition state mimetic of the sialidase hydrolysis reaction, 2-deoxy-2,3-didehydro-D-*N*-acetylneuraminic acid (Neu5Ac2en, **2**), was found to be a modest inhibitor (µM) of influenza virus sialidase. The double bond leads to a flattened geometry within the pyranose ring in the vicinity of C2, C3, and the ring O atom, and is believed to mimic to a certain extent the proposed sialosyl transition state of the sialidase reaction. The crystal structures of influenza virus sialidase complexed with Neu5Ac (**1**) and Neu5Ac2en (**2**) revealed that no significant changes in the active site geometry were observed on soaking sialidase crystals with either of these molecules.

(**3**) R = NHAc

(Zanamivir)

Armed with this detailed structural knowledge of both influenza virus A and B sialidase, several potent sialidase inhibitors of therapeutic value against influenza have been designed and synthesized by various research groups. The development of these drugs, with an emphasis on the importance of rational drug design strategies will be discussed. The first, Zanamivir (4-deoxy-4-guanidino-Neu5Ac2en, **3**), marketed under the tradename Relenza™ by GlaxoSmithKline, is a modified sialic acid derivative. Relenza™ is administered via an inhaler as a fine mist to the lungs, which is the major site of infection. Zanamivir (**3**) was first designed and synthesized in the late 1980s, and was reported in the early 1990s by the von Itzstein group [45]. It is a competitive inhibitor of influenza virus sialidase and displays exceptionally potent in vitro and in vivo inhibition of viral replication in both influenza A and B strains [45–48]. Since the pioneering research that led to the development of Relenza™, other influenza sialidase inhibitors have been developed.

(**4a**) R = CH$_2$CH$_3$
(**4b**) R = H

Oseltamivir (GS 4104, **4a**), developed by Gilead/Hoffmann-LaRoche and marketed as Tamiflu™, is the first commercially available orally active, noncarbohydrate mimetic of Zanamivir (**3**) [49–51]. It is based on a cyclohexene framework with a lipophilic moiety replacing the glycerol side chain of Zanamivir (**3**). It is orally administered as an ethyl ester prodrug GS 4104 (**4a**), which—upon absorption in the gastrointestinal tract—undergoes rapid enzymatic conversion by the action of esterases in plasma and tissue [52] to the parent form of the drug GS 4071 (**4b**). Like Zanamivir (**3**), the liberated parent compound, is a slow binding inhibitor of sialidase from both influenza A and B [53] with similar inhibitory levels in both types.

(5)
(BCX-1812)

Another class of potent and selective influenza virus sialidase inhibitors which are orally active against influenza A and B, based on a cyclopentane framework, have also been described [54]. Although not yet commercially available, BCX-1812 (RWJ-270201, **5**) has been subjected to Phase III clinical trials. BCX-1812 (**5**) is a nanomolar inhibitor of both influenza virus A and B sialidase and inhibits growth of influenza virus in tissue culture [55] and has demonstrated efficacy in a mouse influenza model [56].

2. COMPUTATIONAL TECHNIQUES

As mentioned in the introductory remarks, influenza virus sialidase has been extensively characterized in terms of both its structure and function. There have been a number of different approaches adopted, using molecular modeling techniques to design potential inhibitors for this enzyme. Subsequently, other molecular modeling techniques have been used to explain the observed trends in the inhibitory activity of these compounds against this enzyme. These techniques are enumerated herein and an explanation of their relevance in the overall structure-based drug design process is given. The significance of these techniques and their importance in the drug discovery process is highlighted by the availability of two commercially available anti-influenza drugs, Relenza™ and Tamiflu™, with at least two more compounds either in clinical trials (BCX-1812, **5**), or showing great promise ABT-675 (**6**) [57].

(6)
ABT-675

2.1. Visualization

The information gleaned from being able to interactively view the structure of an enzyme, especially its active site, is often overlooked. This is especially so if the structure of the enzyme also contains either a substrate or a known inhibitor, as it is then possible to deduce not only the important interactions between the substrate and the enzyme, but it is also possible to infer the mechanism of action of the enzyme and obtain more information for the drug design process. Chong et al. [58] used the crystal structure of the complex between Neu5Ac (**1**) and influenza virus sialidase to explain the kinetic results in terms of a detailed proposed mechanism. This structure was detailed by Varghese et al. [40], where Neu5Ac (**1**) was shown to bind in a distorted geometry to maximize the interaction between the carboxylic acid group of Neu5Ac (**1**) and the tri-arginyl cluster of the sialidase (Fig. 3).

The visualization of the results produced by GRID calculations [45,59] also proved useful in the design of Relenza™, with the most favorable interactions sites within the active site being displayed along with either Neu5Ac (**1**) or the transition state inhibitor Neu5Ac2en (**2**) as a template from which to design new compounds. Furthermore, visualization has also had an important impact in the design of a wide range of influenza virus sialidase inhibitors, e.g., Oseltamivir (**4a**) [49], BCX-1812 (**5**) [54,60], benzoic acid-based compounds [61–63], and the pyrrolidine-based ABT-675 (**6**) [57,64].

2.2. GRID

The program GRID [65], developed by Peter Goodford, proved to be an invaluable tool in the design of inhibitors of influenza virus sialidase [45,59]. This program was applied to the crystal structure coordinates of influenza virus sialidase in order to calculate a contour map of the most energetically favourable interaction sites within the active site of the enzyme for a range of different probe groups (e.g., carboxylic acid, amine, hydroxyl, etc.). Comparing each map of energetically favourable sites for each probe with the crystal structure of influenza virus sialidase complexed with either Neu5Ac (**1**) or Neu5Ac2en (**2**) provided a guide as to where modifications to these templates could be made to optimize interactions with the sialidase and improve inhibitor binding affinity. This approach proved to be extremely successful. Studying the interaction energy map for the carboxylate probe provided validation of the GRID technique and demonstrated that this strategy was capable of providing useful information for the drug design process. The carboxylate probe map predicted the exact location where the acidic group of Neu5Ac (**1**) and Neu5Ac2en (**2**) binds, surrounded by the tri-arginyl cluster. Interesting results were also obtained for the amine probe, which showed three main locations of favourable interaction energy within the active site pocket. The most accessible and, subsequently, most valuable, in inhibitor design was within a sizable pocket situated adjacent to the C4-hydroxyl of Neu5Ac (**1**) or Neu5Ac2en (**2**) (Fig. 4).

(**7**) R = NHAc

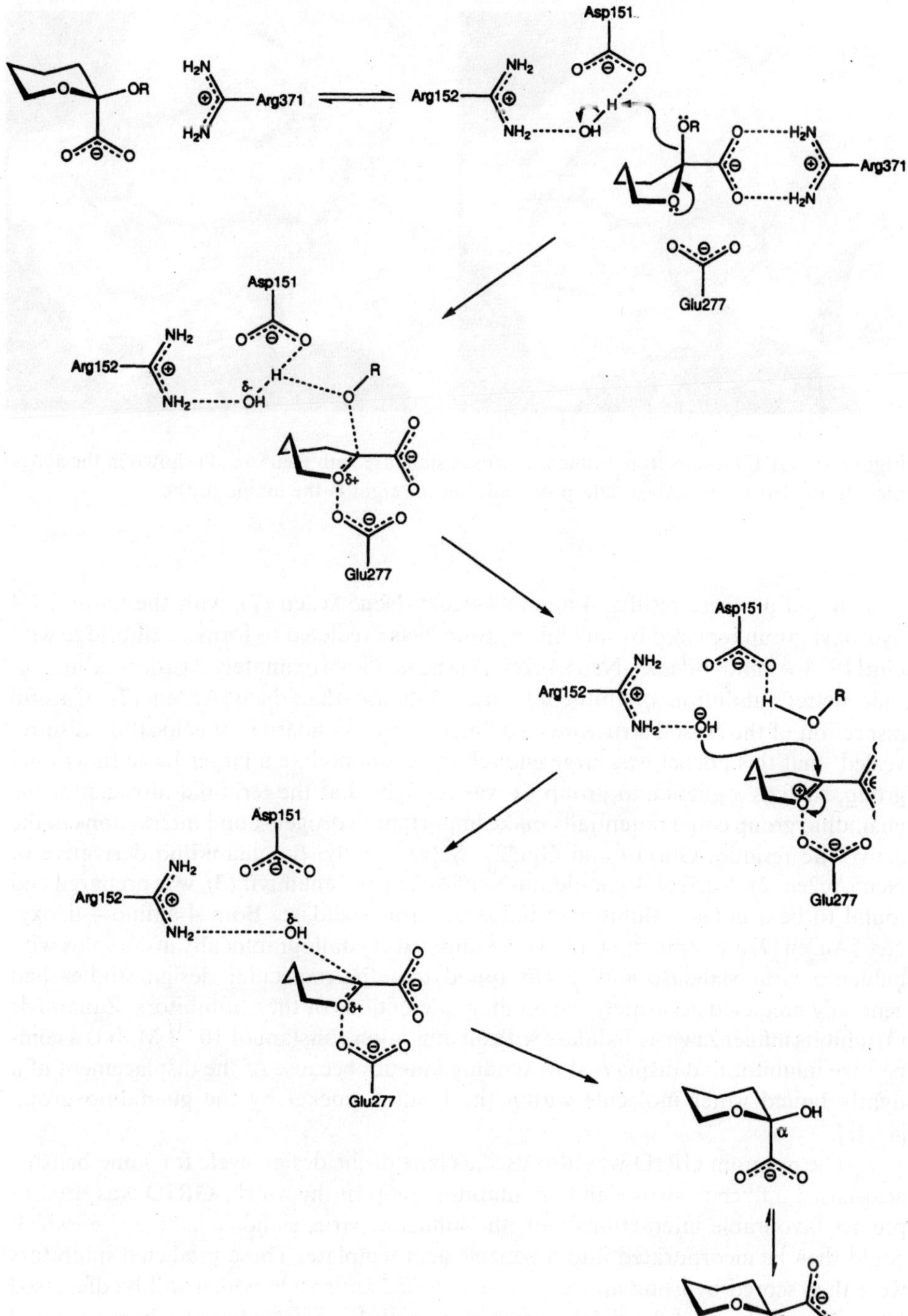

Figure 3 Proposed mechanism of action of influenza virus sialidase.

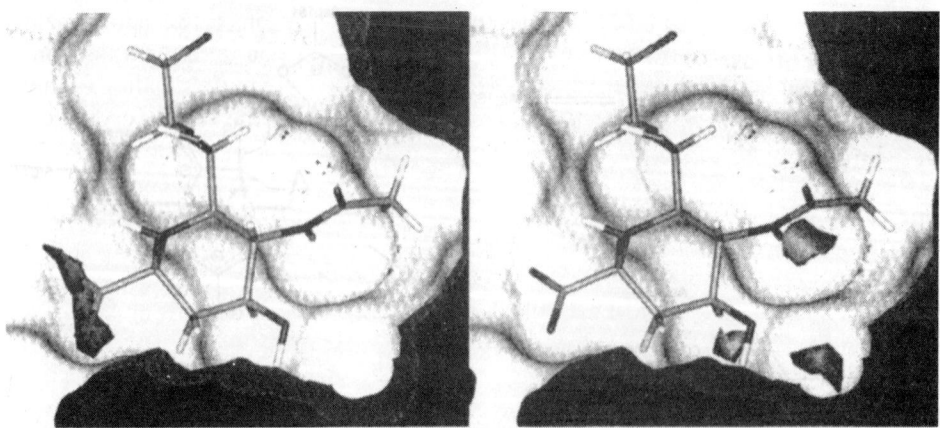

Figure 4   GRID results from influenza virus A sialidasc with Neu5Ac (**1**) shown in the active site. On the left is the carboxylate probe and on the right is the amine probe.

Based on these results, 4-amino-4-deoxy-Neu5Ac2en (**7**), with the natural C4 hydroxyl group replaced by an amino group, was predicted to form a salt bridge with Glu119. 4-Amino-4-deoxy-Neu5Ac2en (**7**) showed approximately 2 orders of magnitude better inhibition of influenza virus sialidase than Neu5Ac2en (**2**). Careful inspection of the crystal structure of influenza virus sialidase in this location also revealed that this pocket was large enough to accommodate a larger basic functional group, such as a guanidino group. It was thought that the terminal nitrogen of the guanidino group could potentially make important hydrogen-bond interactions to the active site residues Glu119 and Glu227. Subsequently, the guanidino derivative of Neu5Ac2en (**2**), 4-deoxy-4-guanidino-Neu5Ac2en or Zanamivir (**3**), was prepared and found to be a potent inhibitor to influenza virus sialidase. Both 4-amino-4-deoxy-Neu5Ac2en (**7**) and Zanamivir (**3**) were examined crystallographically in complex with influenza virus sialidase and it was found that the molecular design studies had generally predicted accurately the binding interactions of these inhibitors. Zanamivir (**3**) inhibits influenza virus sialidase with an inhibition constant of 10^{-11} M. It is a competitive inhibitor and displays slow binding kinetics because of the displacement of a tightly bound water molecule within the binding pocket by the guanidino group [47,48].

The program GRID was also used as part of the design cycle for some benzoic acid-based influenza virus sialidase inhibitors [66]. In this work, GRID was used to predict favourable interactions with the influenza virus sialidase active site, which could then be incorporated into a benzoic acid template. These predicted inhibitors were then scored by using an electrostatics-based approach, which will be discussed later. The active site of the sialidase associated with *V. cholerae* has also been explored [67] using GRID with the major interaction site of interest being in the C5 pocket.

2.3.   Molecular Mechanics and Molecular Dynamics

Molecular mechanics and dynamics calculations are probably the most frequently reported computational techniques used in the study of influenza virus sialidase. The earliest reports [58,68] used the techniques to study the interactions between influenza virus sialidase and not only the natural substrate Neu5Ac (**1**), but also the transition state-based inhibitor Neu5Ac2en (**2**). This led to a more detailed understanding of the important interactions with the active site including the role of the various crystallographic water molecules and assisted in elucidating the mechanism of action.

These studies were then extended to calculate the relative binding energies of a number of C4-substituted Neu5Ac2en-based compounds [69] The binding energies were computed by using molecular-mechanics-derived interactions as the sum of pairwise atomic nonbonded energies. The force field used was the Consistent Valence Force Field (CVFF) as supplied with the Discover simulation package from Accelrys [70], and instead of simply using a single conformation for the calculation, a number of conformations were generated by using a novel molecular dynamics approach. An overall trend was observed between the calculated binding energy and the observed log K_i; however, there were three outliers, including Zanamivir (**3**). A possible explanation for the aberrant behavior of Zanamivir (**3**) involves the displacement of a crystallographic water molecule as part of a slow binding process [47].

The novel molecular dynamics approach mentioned above was adopted to explain the different effects observed between a number of hydrophobic carboxamide derivatives of 4-amino-4-deoxy-Neu5Ac2en (**7**) and the sialidases from both influenza A and B [32,33]. The carboxamide compounds have one or two sterically bulky hydrophobic groups in the vicinity where the glycerol side chain of Neu5Ac2en (**2**) is normally positioned. These compounds showed a difference in observed inhibition of influenza virus A and B sialidase of approximately 2–3 orders of magnitude. The crystal structures of both influenza virus A and B sialidase with these derivatives showed that Glu276 rearranges to produce a salt bridge interaction with Arg224, producing a small hydrophobic pocket. The molecular dynamics calculations show that this rearrangement process is energetically more favourable for influenza virus A sialidase than for the influenza virus B sialidase, which provides a plausible explanation for the observed difference in the inhibition constants between the two types.

(**8**) R = NHAc   (**9**) R = NHAc

Smith et al. [71] used molecular mechanics calculations in conjunction with crystallographic data to study the binding of several Neu5Ac2en-based derivatives to influenza virus A sialidase. The derivatives were 4-amino-4-deoxy-Neu5Ac2en (**7**), 9-amino-9-deoxy-Neu5Ac2en (**8**), and 4,9-diamino-4,9-dideoxy-Neu5Ac2en (**9**), with x-ray crystallography showing that they bound as expected into the active site. The calculations measured the distance between the C4 substituent and the nearest O of Glu119, an important residue in the pocket found adjacent to the C4 group. Interestingly, the distances measured from the minimized structures agreed best when the charge state of Glu119 was treated as neutral rather than the expected negative charge.

The above-mentioned studies all utilized the CVFF. While to date there have been no detailed Free Energy Perturbation (FEP) studies on influenza virus sialidase, there has been one report of the Linear Interaction Energy (LIE) method introduced by Åqvist et al. [72]. This study [73] used the OPLS force field and Monte Carlo simulations were carried out using the MCPRO package [74]. The results, including a detailed statistical treatment, produced a reasonable fit between the calculated and observed ΔG, with a q^2 of 0.74, and contained van der Waals and electrostatic energy terms. One of the original benefits proposed for the LIE method was the ability to fit a relatively simply equation relating electrostatic and van der Waals terms to the ΔG of binding. This study has shown that, while the LIE method can produce reasonable predictions of binding free energies, it seems that there is no unique formula which can be applied to all such calculations.

2.4. Electrostatics

The next most popular choice of computational technique to study the binding of inhibitors to influenza virus sialidase has been the use of electrostatic-based calculations. The first study to report these type of calculations was performed by Jedrzejas et al. [66], who used the program Delphi [75] to perform Finite Difference Poisson–Boltzmann (FDPB) calculations to estimate the binding energy of a number of inhibitors, predominantly benzoic acid-based, with influenza virus sialidase. The results were quite promising for their small training set of molecules; however, prediction of the K_i of several new benzoic acid-based compounds of greater than 10^{-10} M appears to be optimistic, with none of the predicted strong binding compounds living up to their promise.

In addition to their work on Molecular Mechanics Interaction (MMI) energy (mentioned previously), Taylor and von Itztein [69] also used a Continuum Electrostatics (CE) approach in the study of a range of C4-substituted Neu5Ac2en-based compounds. The binding energies were calculated by using the DelPhi [75] program, from the 15 minimized structures obtained for each compound from a novel molecular dynamics protocol. These calculations provided the same overall trend as seen in the case of the MMI calculations but the fit between observed binding affinity and the calculated binding energy was better, although again the same three outliers, including Zanamivir (**3**), did not behave as predicted. The CE approach provides more information on the change in solvation upon binding of the inhibitors than the forcefield-based method, and therefore it is not surprising that it provides a closer fit to what is believed to be actually happening.

Smith et al. [71] followed the molecular mechanics study (mentioned previously) of the following derivatives, 4-amino-4-deoxy-Neu5Ac2en (**7**), 9-amino-9-deoxy-Neu5Ac2en (**8**), and 4,9-diamino-4,9-dideoxy-Neu5Ac2en (**9**), by calculating the binding energy associated with each of the above compounds and influenza virus A sialidase. The calculation of the binding energy involved a modified Delphi [75] approach, where the individual contribution from various desolvation parameters was easier to assign to physical events. In summary, the addition of the 9-amino group to the Neu5Ac2en-based template and 4-amino-4-deoxy-Neu5Ac2en-based template was calculated to improve the solvent-screened interaction energy markedly, as would be predicted a priori from viewing the compounds docked in the active site. However, the partial desolvation energy of the ligand and the protein is greater than the gains in the interaction energy, leading to a lower overall binding energy as observed from the measured inhibition constants.

Recently, Woods et al. [76] recycled the Monte Carlo dynamics trajectories they had collected to study the LIE method [73], as applied to influenza virus sialidase inhibitors and used them to study the configurational dependence of binding free energy calculations. The program UHBD [77] was used for all the FDPB electrostatic calculations. The authors were more interested in studying the configurational dependence of the calculations than actually in developing a predictive model of inhibition of influenza virus sialidase. This study provided some very important conclusions about the use of FDPB energies as a means of scoring the interaction between inhibitors and the influenza virus sialidase. Specifically, they state the following—" just as molecular mechanics energies are very sensitive to configuration, and single-structure values are typically not used to score binding free energies, single FDPB energies should be treated with the same caution." One interpretation of this study is that if FDPB is used in the calculation of the binding free energy of a protein ligand complex, then checking to determine if the calculated energy is dependent on conformation is very important. If it is, then a protocol will need to be devised to calculate the binding energy over an ensemble of structures to produce a statistically meaningful result.

2.5. Combine

The most comprehensive QSAR analysis of the influenza virus sialidase system to date was performed by Wang and Wade [78], using 43 complexes containing 29 different inhibitors with two different subtypes of influenza virus A sialidase. The complexes used included both crystallographically determined structures as well as complexes where the inhibitor was docked into the active site using AUTODOCK3.0 [79]. The COMBINE [80] method was used to correlate the pIC_{50} of the receptor–ligand complexes with a set of selected interaction energy components. The final results showed that the inhibitory activity for the set of inhibitors was predominantly determined by interactions with 12 active site residues and 1 bound water molecule. Strong inhibitors should have the following features: a negatively charged group at the C1-pocket, a positively charged group at the C4-pocket, a hydrogen-bond acceptor and a small hydrophobic group at the C5-pocket, a large hydrophobic group at the C6-pocket, and a hydrogen-bond donor at the water position. An interesting extension to this work would be to determine whether it could be extrapolated to influenza virus B sialidase. Furthermore, when modeling an inhibitor without knowing the crystal structure, it is currently necessary to choose the appropriate conformation of Glu276 so as to either hydrogen bond to the C6-pocket substituent [e.g., Zanamivir (3)], or to form a salt bridge with Arg224, providing a hydrophobic pocket for the C6-pocket substituent [e.g., Oseltamivir (4a)].

2.6. Docking

Muegge [81] has used DOCK4 with a fast knowledge-based scoring function (PMF) [82] to study 11 different crystallographically determined complexes between influenza virus sialidase and various inhibitors. The 11 inhibitors could be correctly docked into the protein structure from which they were extracted; however, only two of the protein structures were able to successfully have all 11 inhibitors docked in the correct orientation. Overall, the DOCK4/PMF method was considerably better than using DOCK4 with the standard force field (FF) scoring module. This study has shown that small differences in protein structure can have a very large effect on the result obtained, with Zanamivir (3) being docked in a flipped orientation (the 4-guanidino group pointing toward the standard C6-pocket) in seven of the eleven structures using DOCK4/PMF; however, DOCK4/PMF correctly positioned Zanamivir (3) in nine of the eleven structures.

More recently, the same author studied the effect of ligand volume correction on PMF scoring [83], using a range of protein–ligand complexes. In this study, the same influenza virus sialidase–ligand complexes were used as described above, and the results showed a slight improvement in the correlation between the experimentally determined inhibition constants and the DOCK4/PMF score.

3. CONCLUSIONS

Computational chemistry techniques have been extensively used in the design of new, potent inhibitors of influenza virus sialidase, which have become commercially available for the treatment of influenza. These same techniques have led to a much better understanding of the mechanism of action of the enzyme, as well as the important interactions within the active site. Developing a predictive model to enable the virtual screening of new inhibitors has inherent difficulties for influenza virus sialidase for a number of reasons. First, the highly charged nature of the active site leads to the electrostatic terms far outweighing the van der Waals contributions, which has implications in every calculation from simply minimizing the x-ray crystal structure of a complex to using docking techniques to predict the bound conformation of a new sialidase–inhibitor complex. Second, although the conservation of active site residues between influenza A and B sialidase is very high, there are some subtle differences, especially in the glycerol side-chain binding pocket, which can lead to large differences in the measured sialidase inhibition constant, between influenza types for the same compound. Notably, the conformational change in Glu276 observed when a hydrophobic residue is introduced into this same pocket, leading to a predominantly hydrophobic interaction, rather than the predominantly hydrogen-bonded interactions seen with the natural substrate and the Neu5Ac2en-based inhibitors, including Zanamivir (3). Furthermore, there are a number of very important water molecules which can be seen in the x-ray crystallography studies, and they form vital hydrogen bonds with many of the observed inhibitors. The displacement of a relatively tightly bound water molecule back into the bulk solvent leads to the additional slow binding component of the interaction with Zanamivir (3) as entropic gains produce a tighter binding inhibitor.

Despite these difficulties, computational chemistry and structure-based drug design have shown their value as part of an integrated drug design strategy.

REFERENCES

1. Varki A. Diversity in the sialic acids. Glycobiology 1992; 2:25–40.
2. Taylor G, Crennell S, Thompson C, Chuenkova M. Sialidsases. In: Ernst B, Hart GW, Sinay P, eds. Carbohydrates in Chemistry and Biology. Weinheim: Wiley-VCH, 2000:485–495.

3. Varghese JN, Colman PM, van Donkelaar A, Blick TJ, Sahasrabudhe A, McKimm-Breschkin JL. Structural evidence for a second sialic acid binding site in avian influenza virus neuraminidases. Proc Natl Acad Sci USA 1997; 94:11808–11812.

4. Colman PM Influenza virus neuraminidase: structure, antibodies, and inhibitors. Protein Sci 1994; 3:1687–1696.

5. Varghese JN. Development of neuraminidase inhibitors as anti-influenza virus drugs. Drug Dev Res 1999; 46:176–196.

6. Laver WG, Bischofberger N, Webster RG. The origin and control of pandemic influenza. Perspect Biol Med 2000; 43:173–192.

7. Kuszewski K, Brydak L. The epidemiology and history of influenza. Biomed Pharmacother 2000; 54:188–195.

8. Webster RG, Shortridge KF, Kawaoka Y. Influenza: interspecies transmission and emergence of new pandemics. FEMS Immunol Med Microbiol 1997; 18:275–279.

9. Oxford JS. Influenza A pandemics of the 20th century with special reference to 1918: virology, pathology and epidemiology. Rev Med Virol 2000; 10:119–133.

10. Taubenberger JK, Reid AH, Fanning TG. The 1918 influenza virus: a killer comes into view. Virology 2000; 274:241–245.

11. Webster RG. 1918 Spanish influenza: the secret remains elusive. Proc Natl Acad Sci USA 1999; 96:1164–1166.

12. Subbarao K, Kilimov A, Katz J, Regnery H, Lim W, Hall H, Perdue M, Swayne D, Bender C, Huang J. Characterisation of an avian influenza A (H5N1) virus isolated from a child with fatal respiratory illness. Science 1998; 279:393–396.

13. de-Jong JC, Claas EC, Osterhaus AD, Webster RG, Lim WL. A pandemic warning? Nature 1997; 389:554.

14. Yuen KY, Chan PK, Peiris M, Tsang DN, Que TL, Shortridge KF, Cheung PT, To WK, Ho ET, Sung R, Cheng AF. Clinical features and rapid viral diagnosis of human disease associated with avian influenza A H5N1 virus. Lancet 1998; 351, 467–471.

15. Pinto LH, Holsinger LJ, Lamb RA. Influenza virus M2 protein has ion channel activity. Cell 1992; 69:517–528.

16. Hayden FG. Amantidine and rimantidine—clinical aspects. In: Richman DD, ed. Antiviral Drug Resistance. Chichester: Wiley, 1996:59–77.

17. Hayden FG. Update on antiviral agents and viral drug resistance. In: Mandell GL, Douglas RGJ, Bennett JE, eds. Principles and Practice of Infectious Disease. New York: Churchill Livingston, 1993:3–15.

18. Colman PM, Varghese JN, Laver WG. Structure of the catalytic and antigenic sites in influenza virus neuraminidase. Nature 1983; 303:41–44.

19. Colman PM. A novel approach to antiviral therapy for influenza. J Antimicrob Chemother 1999; 44(suppl B):17–22.

20. Klenk E, Faillard H, Lempfrid H. Uber die enzymatishe Wirkung von Influenza. Z Physiol Chem 1955; 301·235–246.

21. Gottschalk A. Neuraminidase: the specific enzyme of influenza virus and *Vibrio cholerae*. Biochem Biophys Acta 1957; 23:645–646.

22. Gottschalk A. Neuraminidase: its substrate and mode of action. Adv Enzymol 1958; 20: 135–145.

23. Palese P, Tobita K, Ueda M, Compans RW. Characterisation of temperature sensitive influenza virus mutants defective in neuraminidase. Virology 1974; 61:397–410.

24. Liu C, Eichelberger MC, Compans RW, Air GM. Influenza type A virus neuraminidase does not play a role in viral entry, replication, assembly or budding. J Virol 1995; 69:1099–1106.

25. Laver WG, Valentine RC. Morphology of isolated hemagglutinin and neuraminidase subunits of influenza virus. Virology 1969; 38:105–119.

26. Bucher DJ, Palese P. The biologically active proteins of influenza virus neuraminidase. In:

Kilbourne ED, ed. Influenza Virus and Influenza. New York: Academic Press, 1975:83–125.

27. Kraulis P. MOLSCRIPT: a program to produce both detailed and schematic plots of protein structures. J Appl Crystallogr 1991; 24:946–950.

28. Baker AT, Varghese JN, Laver WG, Air GM, Colman PM. Three-dimensional structure of neuraminidase of subtype N9 from an avian influenza virus. Proteins 1987; 2:111–117.

29. Blick TJ, Tiong T, Sahasrabudhe A, Varghese JN, Colman PM, Hart GJ, Bethell RC, McKimm-Breschkin JL. Generation and characterization of an influenza virus neuraminidase variant with decreased sensitivity to the neuraminidase-specific inhibitor 4-guanidino-Neu5Ac2en. Virology 1995; 214:475–484.

30. Colman PM, Laver WG, Varghese JN, Baker AT, Tulloch PA, Air GM, Webster RG. Three-dimensional structure of a complex of antibody with influenza virus neuraminidase. Nature 1987; 326:358–363.

31. Colman PM, Tulip WR, Varghese JN, Tulloch PA, Baker AT, Laver WG, Air GM, Webster RG. Three-dimensional structures of influenza virus neuraminidase-antibody complexes. Philos Trans R Soc Lond, B Biol Sci 1989; 323:511–518.

32. Smith PW, Sollis SL, Howes PD, Cherry PC, Cobley KN, Taylor H, Whittington HR, Bethell RC, Taylor N, Varghese JN, Colman PM, Singh O, Slkarzynski T, Cleasby A, Wonacott AJ. Novel inhibitors of sialidases related to GG167. Structure–Activity, crystallography, and molecular dynamics studies with 4-H-pyran-2-carboxylic acid 6-carboxamides. Biorg Med Chem Lett 1996; 6:2931–2936.

33. Taylor NR, Cleasby A, Singh O, Skarzynski T, Wonacott AJ, Smith PW, Sollis SL, Howes PD, Cherry PC, Bethell R, Colman P, Varghese J. Dihydropyrancarboxamides related to zanamivir: a new series of inhibitors of influenza virus sialidases: 2. Crystallographic and molecular modeling study of complexes of 4-amino-4*H*-pyran-6-carboxamides and sialidase from influenza virus types A and B. J Med Chem 1998; 41:798–807.

34. Tulip WR, Varghese JN, Webster RG, Air GM, Laver WG, Colman PM. Crystal structures of neuraminidase-antibody complexes. Cold Spring Harbor Symp Quant Biol 1989; 54:257–263.

35. Tulip WR, Varghese JN, Baker AT, van Donkelaar A, Laver WG, Webster RG, Colman PM. Refined atomic structures of N9 subtype influenza virus neuraminidase and escape mutants. J Mol Biol 1991; 221:487–497.

36. Tulip WR, Varghese JN, Webster RG, Laver WG, Colman PM. Crystal structures of two mutant neuraminidase–antibody complexes with amino acid substitutions in the interface. J Mol Biol 1992; 227:149–159.

37. Tulip WR, Varghese JN, Laver WG, Webster RG, Colman PM. Refined crystal structure of the influenza virus N9 neuraminidase–NC41 Fab complex. J Mol Biol 1992; 227:122–148.

38. Varghese JN, Webster RG, Laver WG, Colman PM. Structure of an escape mutant of glycoprotein N2 neuraminidase of influenza virus A/Tokyo/3/67 at 3 A. J Mol Biol 1988; 200:201–203.

39. Varghese JN, Colman PM. Three-dimensional structure of the neuraminidase of influenza virus A/Tokyo/3/67 at 2.2 Å resolution. J Mol Biol 1991; 221:473–486.

40. Varghese JN, McKimm-Breschkin JL, Caldwell JB, Kortt AA, Colman PM. The structure of the complex between influenza virus neuraminidase and sialic acid, the viral receptor. Proteins 1992; 14:327–332.

41. Varghese JN, Laver WG, Colman PM. Structure of the influenza virus glycoprotein antigen neuraminidase at 2.9 A resolution. Nature 1993; 303:35–40.

42. Varghese JN, Epa VC, Colman PM. Three-dimensional structure of the complex of 4-guanidino-Neu5Ac2en and influenza virus neuraminidase. Protein Sci 1995; 4:1081–1087.

43. Varghese JN, Smith PW, Sollis SL, Blick TJ, Sahasrabudhe A, McKimmBreschkin JL, Colman PM. Drug design against a shifting target: a structural basis for resistance to inhibitors in a variant of influenza virus neuraminidase. Structure 1998; 6:735–746.

44. Wallace AC, Laskowski RA, Thornton JM. LIGPLOT: a program to generate schematic diagrams of protein–ligand interactions. Protein Eng 1995; 8:127–134.

45. von Itzstein M, Wu WY, Kok GB, Pegg MS, Dyason JC, Jin B, V Phan T, Smythe ML, White HF, Oliver SW, Colman PM, Varghese JN, Ryan DM, Woods JM, Bethell RC, Hotham VJ, Cameron JM, Penn CR. Rational design of potent sialidase-based inhibitors of influenza virus replication. Nature 1993; 363:418–423.

46. Holzer CT, von Itzstein M, Jin B, Pegg MS, Stewart WP, Wu WY. Inhibition of sialidases from viral, bacterial and mammalian source by analogues of 2-deoxy-2,3-didehydro-N-acetylneuraminic acid modified at the C-4 position. Glycoconj J 1993; 10:40–44.

47. Pegg MS, von Itzstein M. Slow-binding inhibition of sialidase from influenza virus. Biochem Mol Biol Int 1994; 32:851–858.

48. Hart GJ, Bethell RC. -2,3-Didehydro-2,4-dideoxy-4-guanidino-N-acetyl-D-neuraminic acid (4-guanidino-Neu5Ac2en) is a slow-binding inhibitor of sialidase from both influenza A virus and influenza B virus. Biochem Mol Biol Int 1995; 36:695–703.

49. Lew W, Chen X, Kim CU. Discovery and development of GS 4104 (oseltamivir): an orally active influenza neuraminidase inhibitor. Curr Med Chem 2000; 7:663–672.

50. Li W, Escarpe PA, Eisenberg EJ, Cundy KC, Sweet C, Jakeman KJ, Merson J, Lew W, Williams M, Zhang L, Kim CU, Bischofberger N, Chen MS, Mendel DB. Identification of GS 4104 as an orally bioavailable prodrug of the influenza virus neuraminidase inhibitor GS 4071. Antimicrob Agents Chemother 1998; 42:647–653.

51. Mendel DB, Tai CY, Escarpe PA, Li W, Sidwell RW, Huffman JH, Sweet C, Jakeman KJ, Merson J, Lacy SA, Lew W, Williams MA, Zhang L, Chen MS, Bischofberger N, Kim CU. Oral administration of a prodrug of the influenza virus neuraminidase inhibitor GS 4071 protects mice and ferrets against influenza infection. Antimicrob Agents Chemother 1998; 42:640–646.

52. Stella VJ, Charman WNA, Naringrekar VH. Prodrugs: do they have advantages in clinical practice? Drugs 1985; 29:455–473.

53. Kati WM, Saldivar AS, Mohamadi F, Sham HL, Laver WG, Kohlbrenner WE. GS4071 is a slow-binding inhibitor of influenza neuraminidase from both A and B strains. Biochem Biophys Res Commun 1998; 244:408–413.

54. Babu YS, Chand P, Bantia S, Kotian P, Dehghani A, El Kattan Y, Lin TH, Hutchison TL, Elliott AJ, Parker CD, Ananth SL, Horn LL, Laver GW, Montgomery JA. BCX-1812 (RWJ-270201): discovery of a novel, highly potent, orally active, and selective influenza neuraminidase inhibitor through structure-based drug design. J Med Chem 2000; 43:3482–3486.

55. Smee DF, Huffman JH, Morrison AC, Barnard DL, Sidwell RW. Cyclopentane neuraminidase inhibitors with potent in vitro anti-influenza virus activities. Antimicrob Agents Chemother 2001; 45:743–748.

56. Bantia S, Parker CD, Ananth SL, Horn LL, Andries K, Chand P, Kotian PL, Dehghani A, El Kattan Y, Lin T, Hutchison TL, Montgomery JA, Kellog DL, Babu YS. Comparison of the anti-influenza virus activity of RWJ-270201 with those of oseltamivir and zanamivir. Antimicrob Agents Chemother 2001; 45:1162–1167.

57. Maring C, McDaniel K, Krueger A, Zhao C, Sun M, Madigan D, DeGoey D, Chen H-J, Yeung MC, Flosi W, Grampovnik D, Kati W, Klein L, Stewart K, Stoll V, Saldivar A, Montgomery D, Carrick R, Steffy K, Kempf D, Molla A, Kohlbrenner W, Kennedy A, Herrin T, Xu Y, Laver WG. SAR studies of novel pyrrolidine influenza neuraminidase inhibitors: identification of ABT-675 a potent and broad spectrum inhibitor. Antivir Res 2001; 50:A77.

58. Chong AK, Pegg MS, Taylor NR, von Itzstein M. Evidence for a sialosyl cation transition-

state complex in the reaction of sialidase from influenza virus. Eur J Biochem 1992; 207, 335–343.

59. von Itzstein M, Dyason JC, Oliver SW, White HF, Wu WY, Kok GB, Pegg MS. A study of the active site of influenza virus sialidase: an approach to the rational design of novel anti-influenza drugs. J Med Chem 1996; 39:388–391.

60. Chand P, Kotian PL, Dehghani A, El Kattan Y, Lin TH, Hutchison TL, Babu YS, Bantia S, Elliott AJ, Montgomery JA. Systematic structure-based design and stereoselective synthesis of novel multisubstituted cyclopentane derivatives with potent antiinfluenza activity. J Med Chem 2001; 44:4379–4392.

61. Finley JB, Atigadda VR, Duarte F, Zhao JJ, Brouillette WJ, Air GM, Luo M. Novel aromatic inhibitors of influenza virus neuraminidase make selective interactions with conserved residues and water molecules in the active site. J Mol Biol 1999; 293:1107–1119.

62. Chand P, Babu YS, Bantia S, Chu NM, Cole LB, Kotian PL, Laver WG, Montgomery JA, Pathak VP, Petty SL, Shrout DP, Walsh DA, Walsh GW. Design and synthesis of benzoic acid derivatives as influenza neuraminidase inhibitors using structure-based drug design. J Med Chem 1997; 40:4030–4052.

63. Atigadda VR, Brouillette WJ, Duarte F, Babu YS, Bantia S, Chand P, Chu NM, Montgomery JA, Walsh DA, Sudbeck E, Finley J, Air GM, Luo M, Laver GW. Hydrophobic benzoic acids as inhibitors of influenza neuraminidase. Bioorg Med Chem 1999; 7:2487–2497.

64. Wang GT, Chen Y, Wang S, Gentles R, Sowin T, Kati W, Muchmore S, Giranda V, Stewart K, Sham H, Kempf D, Laver WG. Design, synthesis, and structural analysis of influenza neuraminidase inhibitors containing pyrrolidine cores. J Med Chem 2001; 44: 1192–1201.

65. Goodford PJ. A computational procedure for determining energetically favorable binding sites on biologically important macromolecules. J Med Chem 1985; 28:849–857.

66. Jedrzejas MJ, Singh S, Brouillette WJ, Air GM, Luo M. A strategy for theoretical binding constant, K_i, calculations for neuraminidase aromatic inhibitors designed on the basis of the active site structure of influenza virus neuraminidase. Proteins 1995; 23:264–277.

67. Wilson JC, Thomson RJ, Dyason JC, Florio P, Quelch KJ, Abo S, von Itzstein M. The design, synthesis and biological evaluation of neuraminic acid-based probes of Vibrio cholerae sialidase. Tetrahedron Asymmetry 2000; 11:53–73.

68. Taylor NR, von Itzstein M. Molecular modeling studies on ligand binding to sialidase from influenza virus and the mechanism of catalysis. J Med Chem 1994; 37:616–624.

69. Taylor NR, von Itzstein M. A structural and energetics analysis of the binding of a series of N-acetylneuraminic-acid-based inhibitors to influenza virus sialidase. J Comput-Aided Mol Des 1996; 10:233–246.

70. Discover; Accelrys Inc: San Diego.

71. Smith BJ, Colman PM, von Itzstein M, Danylec B, Varghese JN. Analysis of inhibitor binding in influenza virus neuraminidase. Protein Sci 2001; 10:689–696.

72. Aqvist J, Medina C, Samuelsson JE. New method for predicting binding-affinity in computer-aided drug design. Protein Eng 1994; 7:385–391.

73. Wall ID, Leach AR, Salt DW, Ford MG, Essex JW. Binding constants of neuraminidase inhibitors: an investigation of the linear interaction energy method. J Med Chem 1999; 42:5142–5152.

74. MCPRO 1.4; Yale University, New Haven, CT.

75. Delphi; Accelrys Inc: San Diego.

76. Woods CJ, King MA, Essex JW. The configurational dependence of binding free energies: a Poisson–Boltzmann study of neuraminidase inhibitors. J Comput-Aided Mol Des 2001; 15:129–144.

77. Davis ME, Madura JD, Luty BA, McCammon JA. Electrostatics and diffusion of

molecules in solution: simulations with the University of Houston Brownian dynamics program. Comput Phys Commun 1991; 62:187–197.

78. Wang T, Wade RC. Comparative binding energy (COMBINE) analysis of influenza neuraminidase–inhibitor complexes. J Med Chem 2001; 44:961–971.

79. Morris GM, Goodsell DS, Halliday RS, Huey R, Hart WE, Belew RK, Olson AJ. Automated docking using a Lamarckian genetic algorithm and an empirical binding free energy function. J Comp Chem 1998; 19:1639–1662.

80. Ortiz AR, Pisabarro MT, Gago F, Wade RC. Prediction of drug binding affinities by comparative binding energy analysis. J Med Chem 1995; 38:2681–2691.

81. Muegge I. The effect of small changes in protein structure on predicted binding modes of known inhibitors of influenza virus neuraminidase: PMF-scoring in DOCK4. Med Chem Res 1999; 9:490–500.

82. Muegge I, Martin YC. A general and fast scoring function for protein–ligand interactions: a simplified potential approach. J Med Chem 1999; 42:791–804.

83. Muegge I. Effect of ligand volume correction on PMF scoring. J Comp Chem 2001; 22:418–425.

Glossary

ED E. MORET and JAN P. TOLLENAERE

Utrecht University, Utrecht, The Netherlands

Ab initio: A quantum mechanical nonparametrized molecular orbital treatment (Latin: from "first principles") for the description of chemical behavior taking into account nuclei and all electrons. In principle, it is the most accurate of the three computational methodologies: ab initio, semi-empirical all-valence electron methods, and molecular mechanics.

Active analogue approach: In the absence of information regarding the receptor a medicinal chemist may modify known active structures from which one or several pharmacophoric patterns can be deduced. Then a set of possible (low energy) conformations for each compound known to activate the receptor is calculated. For each allowed conformation, the pharmacophoric pattern is determined. The intersection of all generated pharmacophoric patterns may then yield the pharmacophore embedded in all compounds of the set of active analogues.

Adiabatic searching: Adiabatic (Greek: not passing through) conformational searching in which no strain energy enters or leaves the molecule because at each step during the rotation around a bond all molecular strain energy is relaxed by minimizing all bond stretches and bond angles.

All valence electron methods: In contrast to ab initio methods, the semi-empirical molecular orbital methods only consider the valence electrons for the construction of the atomic orbitals. Well-known semi-empirical methods are EHT, CNDO, MNDO, PCILO, AM1, and PM3. These methods are orders of magnitude faster than ab initio calculations.

AM1: Austin Model 1. A semi-empirical quantum chemical Hamiltonian originating from the M.J.S. Dewar group. The quality of the AM1 results in many cases is beyond the simpler ab initio results and is superior to the MNDO

Reprinted in part, with permission from J. P. Tollenaere, Glossary of Terminology, in Guidebook on Molecular Modeling in Drug Design, N. C. Cohen, ed. Academic Press, San Diego, 1996.

method especially in the description of hydrogen bonds, when compared to other all valence electron methods.

Artificial neural networks: A machine or program for supervised or unsupervised learning based on a layered network of neurons. Normally, a network is trained to best describe a biological or chemical system, in order to classify new systems. Used for pattern recognition in cheminformatics, QSAR, and bioinformatics.

Bayesian statistics: Bayesian inference is a variant of statistics where prior information is allowed to influence the posterior probability of an event via application of Bayes' rule. Complex problems of cheminformatics and bioinformatics often benefit from Bayesian models. A schism divides statisticians from Bayesians.

Bioactive conformation: The bioactive conformation or the biologically relevant conformation can be defined either as the conformation a molecule must adopt in order to be recognized by the receptor or as the conformation of the ligand at the receptor site after binding. The dual interpretation of bioactive conformation stems from the fact that the environment of the ligand at the stage of recognition of the receptor or when it is fulfilling its biological role is not well understood.

Bioinformatics: The use of computers and algorithms to store, generate, and analyze the exploding amount of genomic and proteomic data. Applications range from gene finding, via functional annotation of proteins to the description of interaction networks in entire cells and organisms. Sequence alignment (see, e.g., BLAST) is the core technique.

Bioisosterism: Isosteric groups are chemical groups with the same mass or number of electrons, e.g., OH, NH_2, and CH_3. It was postulated that replacement of these groups did not change the physicochemical properties of the molecules much, thereby assuming same size, as well as same polarity and electronegativity. Friedman expanded this concept to bioisosteres, groups of atoms that could replace one another without changing the biological activity of the molecules. This concept has been used extensively in medicinal chemistry.

BLAST: Basic Local Alignment Search Tool is the most popular nucleotide and protein similarity searching tool in bioinformatics. It is web-accessible, free, fast, and can search many databases at the same time. It can detect isolated sequence patterns, especially if used with the more recent position-specific iterative method PSI-BLAST. Similarity of residues can be defined via mutation or physicochemical substitution matrices.

Boltzmann factor: In any system of molecules at equilibrium, the number of systems possessing an energy E is proportional to the Boltzmann factor $\exp(-E/kT)$, where k and T are the Boltzmann constant and the absolute temperature, respectively. The sum of all the Boltzmann factors for all the energy levels E_i is the partition function of that system.

Born–Oppenheimer approximation: The Born–Oppenheimer approximation consists of separating the motion of nuclei from the electronic motion. An often used physical picture is that the nuclei being so much heavier than electrons may be treated as stationary as the electrons move around them. The Schrödinger equation can then be solved for the electrons alone at a definite internuclear separation. The Born–Oppenheimer approximation is quite good for the calculation of the quantum chemical behavior of molecules in the ground state.

BSSE: Basis Set Superposition Error. The BSSE in ab initio quantum chemical calculations of intermolecular interactions arises from a minor imbalance between the description given for the complex and its individual constituents. When two molecules approach each other, the description of a given molecule is energetically better within the complex than for the free monomer because orbitals of the partner molecule also become partly available leading to over-estimated stabilization energies of weakly bonded complexes.

Buckingham potential: The Buckingham potential is an alternative to describe van der Waals interactions where the 12th power of the Lennard–Jones potential is replaced by an exponential function which is an alternative description of the repulsive forces arising from overlapping electron clouds. The Buckingham potential has the advantage of being softer than the Lennard–Jones potential at short distances.

CADD: Computer-Aided Drug Design in the broadest sense is the science and art of finding molecules of potential therapeutic value that satisfy a whole range of quantitative criteria such as, for example, high potency, high specificity, minimal toxic effects, and good bioavailability. Computer-aided drug design relies on computers, information science, statistics, mathematics, chemistry, physics, biology, and medicine. In a more narrow sense, CADD implies the use of computer graphics to visualize, manipulate chemical structures, to synthesize "in computro" new molecules, to determine their conformation, to assess the similarities and dissimilarities between series of molecules. Computer-aided drug design further involves the calculation or scoring of the interaction energetics between drug molecules and hypothetical or experimentally determined macromolecular structures. It should be noted that CADD only helps in designing ligands, whereas it takes much more disciplines to make a drug. Computer-aided drug design leads to insight in molecular recognition processes and above all stimulates the creativity of all those involved in drug research.

Chem(o)informatics: The use of computers and algorithms to store, generate, and analyze data from combinatorial chemistry, virtual libraries, and high through-put screening. The methods of cheminformatics are the same as those of QSAR, but the field got its name after the data revolution caused by high throughput screening and synthesis, just as bioinformatics became eminent after the revolution of molecular biology.

CNDO: Complete Neglect of Differential Overlap. One of the first semi-empirical all-valence electron methods formulated by J.A. Pople et al. in the 1960s. Because of the drastic simplifications dictated by the speed of the computers in those days, CNDO methods are superseded by more elaborate semi-empirical quantum chemical calculations such as AM1 and PM3.

Comparative Molecular Field Analysis: The basic idea of Comparative Molecular Field Analysis developed by R.D. Cramer et al. (Tripos CoMFA®) is that a suitable sampling of the steric and electrostatic field around a ligand molecule may provide all the information necessary for explaining its biological property in a 3D-QSAR. The steric and electrostatic contributions to the total interaction energy between the ligand and a chosen probe are calculated at regularly spaced grid points of a 3-D lattice encompassing the ligand. These contributions are then related to the biological properties in a partial least squares analysis.

Comparative modeling: see homology modeling.

Computational chemistry: A branch of chemistry that can be defined as computer-assisted simulation of molecular systems and that is used to investigate the chemical behavior and properties of these systems by means of formalisms based on quantum mechanics, classical mechanics, and other mathematical techniques. Because of the ever increasing speed of computers computational chemistry has become and will continue to be a viable alternative to chemical experimentation in cases where experiment is either unfeasible, too dangerous, or too costly.

Conformational analysis: The study of the configuration of atoms and the relative molecular energies that result from rotation about any of the single bonds in a molecule. The possible individual arrangements of atoms in space are called conformers, conformational isomers, or rotamers. The methods of choice for the characterization of the conformation of molecules in the three aggregation states viz. solid (crystalline), dissolved, and gaseous (isolated state) are X-ray diffraction, NMR, and computational methods, respectively.

Conformational searching: Theoretical methods of conformational analysis are applied to explore the conformational energy surface of molecules, e.g., to find the minimal energy conformation or the bioactive conformation. Conformational changes are based on time-dependence (molecular dynamics), probability (Monte Carlo), or systematic searches, followed by minimization steps or simulated annealing.

Conjugate gradients: A mathematical first-order procedure to minimize a function such as a potential energy function used in molecular mechanics. The conjugate gradients method is the method of choice to energy minimize large molecular systems.

Connectivity index: The molecular connectivity index is a term used to describe molecular structure in terms of the adjacency of each atom in the molecule. A well-known molecular connectivity index is the χ-index of Kier and Hall which basically reflects a weighted count (based on the connectivity of each atom) of bonds and connected sets of bonds in a molecule.

Connolly surface: The Connolly surface is the molecular surface related to the solvent accessible surface area but traced by the inward-facing part of a solvent probe model, represented by a sphere with a given radius, free to touch but not to penetrate the solute when the probe is rolled over its van der Waals surface. The surface combines the contact surface of a solute atom and the probe and the reentrant surface when the probe is in contact with more than one atom.

Constraint: A constraint in a target function such as the energy function in molecular mechanics is defined as a degree of freedom that is fixed or not allowed to vary during the molecular simulation. This reduces the amount of space to be searched in conformational searching.

Continuum electrostatics: A simplification of molecular electrostatics by using the same values throughout one or a range of molecules for computational efficiency. Continuum electrostatics is often used to mimic the properties of bulk water, not treating every single water molecule or atom as a separate entity, in order to compute molecular solvation implicitly rather than explicitly.

Coulomb interaction: The Coulomb or charge–charge interaction arises from the attraction or repulsion of two charges and is inversely proportional to the distance separating the two charges. Because of this $1/r$ proportion Coulomb interactions are long-range interactions and therefore are one of the major driving forces governing the recognition process between a ligand and its receptor. The interaction energy of two unit charges at a separation of 10 Å in a dielectric medium of $\varepsilon = 1$ amounts to about -332 kcal/mol.

CPK: Corey–Pauling–Koltun or space-filling representation of a molecule in which each atom is represented by a sphere, the radius of which is proportional to the van der Waals radius of that atom.

Cross-terms: Cross or off-diagonal terms in a second- or third-generation force field account for the fact that bonds and bond angles in a molecule are interdependent because the energy for a given stretch or bend depends on the actual value of neighboring bond lengths and bond angles. Cross-terms usually increase the accuracy of a force field and may enhance the transferability of the diagonal terms because these are no longer contaminated by these cross-term effects.

CSD: The Cambridge Structural Database produced by the Cambridge Crystallographic Data Centre contains bibliographic, chemical, and numerical data of x-ray structures. This machine-readable file is a comprehensive compendium of molecular geometries of organic and organometallic compounds.

Cut-off distance: In order to improve the computational efficiency in force field calculations nonbonded interaction energy contributions for pairs of atoms separated by distances larger than a predetermined value are neglected. As van der Waals and electrostatic interactions are significant up to 15 Å and for large systems account for more than 90% of the total computational time, a given cut-off distance is always a compromise between computational efficiency and accuracy of the calculation.

3-D Builders: Expert system techniques are employed, using tables of standard bond lengths and bond angles in conjunction with a simplified force field, to build a 3-D conformation from a 2-D structure representation. Programs such as CONCORD and CORINA accept SMILES strings as input file formats and can quickly generate a structure database of thousands of molecules, which can be used as a virtual library for molecular diversity analysis and pharmacophore searching.

3-D Fingerprints: A 3-D fingerprint is a description of the 3-D topology or pharmacophore of a molecule, binning the distances between atoms or functional groups. This enables rapid prescreening in 3-D fragment searching, which greatly increases the efficiency especially if conformational flexibility is allowed.

3-D Fragment Search: Having converted the traditional 2-D databases of chemical structures to a 3-D database with 3-D builders, 3-D fragment searching is used to find all molecules in that database that contain a specific pharmacophore or other fragment. Conformationally flexible searching addresses the problem of finding molecules with a conformation different from that which is stored in the primary 3-D database, by using 3-D fingerprints.

3-D QSAR: An extension of QSAR to 3-D properties of molecules. Initially, QSARs consisted of 1-D molecular descriptors only, although the method of computation often included 3-D structure, like in quantum-chemical descriptors. 3-D QSAR better reflects the lock-and-key concept by coupling the molecular properties to the molecular coordinates. In all 3-D QSAR models (e.g., Tripos CoMFA®, HASSL), the choice of conformation and alignment of a series of molecules is crucial for the accuracy of the model.

De novo design: De novo design is a ligand design strategy in which the availability of a 3-D structure of a therapeutic target (an enzyme or protein) is used to design and predict the affinity of novel ligands. In principle, all de novo design methodologies identify interaction sites within the target followed by various strategies to create molecular fragments that fit on the interaction sites and finally propose molecular links between the fragments to form real ligands that are ranked according to affinity using a scoring function.

Density functional theory: The DF approach is a calculational procedure according to which all of the electronic properties of a chemical system, including the energy, can be derived from the electronic density. Local DF theory which is steadily gaining popularity in the chemical computational community takes into account electron correlation. It requires considerably less computer time and disk space than ab initio calculations making it feasible to deal with much larger atoms and molecular systems.

Diagonal terms: Diagonal terms in a force field refer to the terms representing the bond stretch and bond angle deformations, torsion angle, and out-of-plane bending contributions. The bond stretching term describes the molecular potential energy change as a bond stretches or contracts relative to an equilibrium bond length. The bond angle term describes the molecular potential energy change as a bond angle deviates from an ideal equilibrium bond angle value. In classical force fields the bond stretching and bond angle term are represented by a harmonic function. The torsional potential is represented by an n-fold Fourier series expansion usually truncated up to $n = 3$. The displacement of a trigonal atom above and below the molecular plane is a mode of motion distinguishable from the bond stretching, angle bending, and torsional motions. This out-of-plane coordinate is often called improper torsion because it treats the four atoms in the plane as if they were bonded in the usual way as in a proper torsional angle. The Urey–Bradley (UB) term in a force field may account for the repulsion between two atoms bonded to a common atom. In essence the UB term takes into account the 1–3 interaction term and is similar to the bond–bond and bond-angle term. Diagonal force fields do not contain cross-terms (off-diagonal terms), as in second-generation force fields.

Dipole–dipole force: The dipole–dipole force, also called the Keesom force, arises from the interaction of the permanent dipoles of two interacting molecules. The interaction energy is inversely proportional to the sixth power of the distance between the two dipoles. Dipole–dipole interactions are temperature dependent as thermal motion of the molecules competes with the tendency toward favorable dipole orientations. The energy of two interacting dipoles of $\mu = 2$ Debye at a distance of 5 Å in vacuum is of the order of -0.25 kcal/mol.

Dipole-induced dipole force: The dipole-induced dipole force, also called the induction or Debye force, arises when a permanent dipole induces a redistribution of electron density in another polarizable molecule, leading to an induced dipole. This type of interaction is inversely proportional to the sixth power of the distance between the two dipoles and is temperature independent. The average dipole-induced dipole interaction energy of a molecule of $\mu = 1$ Debye with, e.g., benzene is about -0.2 kcal/mol at a separation of 3 Å.

Dispersion force: The dispersion or London force arises from the instantaneous transient dipoles that all molecules possess as a result of the changes in the instantaneous positions of electrons. The dispersion force which in fact is an induced dipole-induced dipole interaction depends on the polarizability of the interacting molecules and is inversely proportional to the sixth power of separation. In the case of, e.g., two CH_4 molecules at a separation of 3 Å, the dispersion interaction energy is of the order of -1.1 kcal/mol.

Distance geometry: Distance geometry pioneered by G.M. Crippen is a method for converting a set of distance bounds into a set of coordinates that are consistent with these bounds. In applying distance geometry to conformationally flexible structures the upper and lower bounds to the distance between each pair of points (atoms) are used. This approach is useful for molecular model-building and conformational analysis and has been extended to find a common pharmacophore from a set of biologically active molecules.

Distance-dependent dielectric constant: In computing the Coulomb interaction between two point charges, the dielectric constant is set to the value of the dielectric medium. In an attempt to implicitly simulate water, the dielectric constant should be 80 at long distances, but 1 at close range. Replacing the dielectric constant by r makes the dielectric effect distance-dependent at negligible computational cost.

Docking: An operation in which one molecule is brought into the vicinity of another while calculating the interaction energies of the many mutual orientations and conformations of the two interacting species. A docking procedure is used as a guide to identify the preferred orientation of one molecule relative to the other. In docking, the interaction energy is generally calculated by computing the van der Waals and the Coulombic energy contributions between all atoms of the two molecules, but other scoring methods have been applied as well.

EHT: Extended Hückel Theory. One of the first semi-empirical all-valence electron methods formulated by R. Hoffmann in the mid-1960s.

Electrostatic potential-derived charges: While net atomic charges q_i are not rigorously defined quantum mechanical properties they can be derived by fitting the classical electrostatic potential due to the charges q_i to the rigorously defined quantum mechanical electrostatic potential.

EMBL Data Library: The main role of the European Molecular Biology Laboratory Data Library, currently known as EBI, is to maintain and distribute a database of nucleotide sequences. This work is a collaborative effort with GenBank® and DNA Database of Japan (DDBJ) where each participating group collects a portion of the total reported sequence data.

Ensemble: When treating systems of interacting particles in a molecular dynamics simulation it is useful to introduce the concept of ensemble which basically means "collection." Taking a closed system with a given volume V, composition N, and temperature T and replicating it n times constitute a canonical ensemble (NVT) in which all the identical closed systems are regarded as being in thermal contact with each other and having the same temperature. In the microcanonical ensemble (NVE) the condition of constant temperature is replaced by the requirement that all the systems should have the same energy E. Other ensembles are, e.g., the isobaric-isoenthalpic NPH ensemble and the isobaric-isothermal NPT ensemble. Depending on the molecular dynamics simulation experiment an appropriate choice of ensemble has to be made. For example, the NVT ensemble is the appropriate choice when conformational searching of molecules is carried out in a vacuum and no periodic boundary conditions (PBC) are used.

Enthalpy–entropy compensation: The free energy of equilibria can be decomposed into an enthalpic and entropic component. In biochemical equilibria, there often is a linear relationship between the enthalpy and the entropy, most intuitively explained by the loss of conformational entropy as binding enthalpy increases: the stronger the binding the more rigid the complex. There is still controversy on this topic, however, because it is difficult to measure enthalpy and entropy independently and it is even more difficult to explain thermodynamic phenomena with atomistic models.

Excluded volume: Excluded volume is the union of volumes of a set of active ligands that is available to the ligands interacting with the receptor. Subtraction of the volume in common with the volume of the active and inactive ligands from the volume of the inactive ligand leads to the receptor essential volume, i.e., the volume required by the receptor.

Fold recognition: Protein structure can be predicted with fold recognition. A protein sequence and its predicted secondary structure are compared to the sequences of a library of experimentally known folds. If a structure has more than 50% sequence homology, homology modeling is used instead. A similar procedure is threading.

Force field: A force field is a set of equations and parameters which, when evaluated for a molecular system, yields an energy as a function of the atomic coordinates. This energy expresses the cost of structural deviation from ideal values. Force fields used in molecular mechanics consider the molecular system as a collection of classical masses held together by classical forces. The contributions to the molecular energy include the diagonal terms bond stretching, angle bending, and dihedral angle deformations, as well as the nonbonded van der Waals and electrostatic interactions. Many force fields have been parametrized for biomolecules, such as Gromos and Amber. Other force fields have also been parametrized for hetero-groups such as cofactors and drugs. Some force fields use a united atom model for computational efficiency; others have an additional angle-dependent hydrogen bond term or the explicit use of lone pairs to more closely reproduce hydrogen bonding.

Free energy perturbation: FEP. A statistical mechanical method for deriving the free energy difference between two states a and b from an ensemble average of a potential energy difference ($\Delta V = V\text{b-}V\text{a}$) that can be evaluated using molecular dynamics. In the FEP approach the free energy difference between two states of a system is computed by transforming one state into the other by changing a coupling parameter λ in small increments such that the system is in equilibrium at all values of λ. As λ increases from $\lambda = 0$ (state a) to $\lambda = 1$ the system is transformed into the b state. The free energy difference between the two states a and b is then calculated as the sum of free energy differences between the closely spaced λ states.

Free-Wilson model: The Free-Wilson model is a mathematical approach for *QSAR* and is based on the hypothesis that the biological activity within a series of molecules arises from the constant and additive contributions of the various substituents, without determining their physicochemical basis.

Frontier orbital: Frontier electron theory is based on the idea that a reaction should occur at the position of largest electron density in the frontier orbitals. In the case of an electrophilic reaction, the frontier orbital is the HOMO, and the LUMO in the case of a nucleophilic reaction.

GB/SA: Generalized-Born/Surface-Area. A method for simulating solvation implicitly, developed by W.C. Still's group at Columbia University. The solute–solvent electrostatic polarization is computed using the Generalized-Born equation. Nonpolar solvation effects such as solvent–solvent cavity formation and solute–solvent van der Waals interactions are computed using atomic solvation parameters, which are based on the solvent accessible surface area. Both water and chloroform solvation can be emulated.

GenBank®: GenBank® is the National Institute of Health database of all known nucleotide and protein sequences. Entries in the database include a description of the sequence, scientific name, and taxonomy of the source organism. Collaboration with the EMBL Data Library and the DNA Database of Japan enables shared data collection and sequence information.

Genetic algorithms: Genetic algorithms (GAs) are optimization methods loosely based on Darwinian evolution and are used for a wide range of global optimization problems having to do with high-dimensional spaces. As a conformational search method GAs consist of successively transforming one generation of a series of conformers into the next using the operations of selection (conformers with lower energy are "fitter" than those with higher energy), crossover, and mutation. Because the selection process is biased toward conformations with lower energy, the GA method leads to a collection of low-energy conformers.

GRID: P. Goodford was the first to compute discontinuous atom affinity maps on a 3-D lattice around proteins to aid structure-based design. The use of pre-computed affinity maps for atoms and charges increased the efficiency of docking and de novo design algorithms.

Hamiltonian: An operator which when operating on the wave function of a quantum chemical system returns the energy of that system. The classical Hamiltonian function $H = T + V$ is the sum of the kinetic energy function T and the potential energy function V representing the total energy E of a system.

Hansch analysis: A QSAR method based on extra-thermodynamic principles which expresses the biological activity of a congeneric series of molecules in terms of additive physical quantities as, e.g., hydrophobicity ($\log P$, π), electronic effects (pK_a, σ), and steric effects (E_s of Taft).

Hessian matrix: The Hessian matrix or the force constant matrix is the second derivative of the energy with respect to the atomic coordinates of a molecular system. Diagonalization of the Hessian matrix pertaining to a minimum energy conformation leads to all positive eigenvalues. A transition state structure is characterized by one negative eigenvalue; all the others being positive.

Hidden Markov model: A probabilistic model that is often used as a prediction engine in bioinformatics and cheminformatics. The probability of transition between states is known although the states remain hidden.

HOMO: Highest Occupied Molecular Orbital. A molecular orbital calculation yields a set of eigenvalues or energy levels in which all the available electrons are accommodated. The highest filled energy level is called the HOMO. The next higher energy level which is unoccupied because no more electrons are available is the LUMO or Lowest Unoccupied Molecular Orbital. On the basis of Koopman's theorem the HOMO and LUMO of a molecule can be approximated as its ionization and electron affinity, respectively.

Homology modeling: Homology modeling is the art of building a protein structure knowing only its amino acid sequence and the complete 3-D structure of at least one other reference protein. Protein homology building is based on the fact that there are structurally conserved regions in proteins of a particular family that have nearly identical structure. In homology modeling, sequence alignment methods are used in determining which regions of the reference protein(s) and the unknown protein are conserved. Because homologous sequences have the same ancestor, comparative modeling better reflects this procedure in case no common ancestry has been proven. Related prediction techniques are fold recognition and threading.

Hybrid QM/MM: Hybrid QM/MM is the combination of quantum mechanical (QM) and molecular mechanics (MM) methodologies in Monte Carlo and molecular dynamics calculations where the solute or chemically reacting part of the total system is treated quantum mechanically, whereas the rest of the system is treated in the MM approximation.

Hydrogen bond: A hydrogen bond involves the stabilizing interaction, either inter- or intramolecular, between two moieties XH and Y. It is commonly assumed that for a hydrogen bond to be formed both X and Y should be electronegative elements. Evidence is accumulating that hydrogen bonds can also be formed between, e.g., CH...O and OH... π-bonded systems. Hydrogen bonds have specific geometric directionality and properties and therefore give rise to geometrically well-organized structures in biological systems such as DNA and proteins.

Hydrophobicity: The preference of a solute to dissolve in apolar solvents or molecules over polar solvents. A common measure of hydrophobicity is the partition or distribution coefficient in the n-octanol/water system, often used to predict pharmacokinetics and pharmacodynamics in QSAR and Hansch analysis.

Implicit solvation: To increase the efficiency of computational chemistry explicit water molecules are seldom used in simulations. Instead, implicit solvation models have been defined, which are either based on atomic solvation parameters, continuum electrostatics, or a combination of the two, e.g., in the Generalized Born/Surface Area method.

Induced fit: D. Koshland Jr. introduced the induced fit concept to explain the conformational adjustment of enzyme and substrate, as an improvement over the more rigid lock-and-key model. Neither model can exclusively explain all interactions, though.

Internal coordinates: The internal coordinates of a molecule define its 3-D structure in terms of bond lengths, bond angles, and torsion angles.

Lennard–Jones potential: As two atoms approach one another there is the attraction due to London dispersion forces and eventually a van der Waals repulsion as the interatomic distance r gets smaller than the equilibrium distance. A well-known potential energy function to describe this behavior is the Lennard–Jones (6-12) potential (LJ). The LJ (6-12) potential represents the attractive part as r^{-6}-dependent whereas the repulsive part is represented by an r^{-12} term. Another often used nonbonded interaction potential is the Buckingham potential which uses a similar distance dependence for the attractive part as the LJ (6-12) potential but where the repulsive part is represented by an exponential function.

Lock and key: In 1894, E. Fischer proposed the lock-and-key mechanism for an enzyme that spliced glycoside substrates in a configurational manner: "Um ein Bild zu gebrauchen, will ich sagen, dass Enzym und Glucosid wie Schloss und Schlüssel zu einander passen müssen, um eine chemische Wirkung auf einander ausüben zu können." This intuitive concept was readily accepted and later extended to the concept of pharmacophore and induced fit.

MINDO/3: Modified Intermediate Neglect of Differential Overlap. The MINDO/3 technique representing the third version of MINDO is a semi-empirical all-valence electron self-consistent field molecular orbital approach. MINDO/3 calculations provide fairly accurate values of molecular properties on medium to large organic molecules.

Minimal basis set: A minimal basis set in quantum chemical calculations is the smallest possible set of orbitals consisting of only that number of functions (Gaussian) necessary to accommodate all the electrons of an atom. The minimal basis set for an atom such as carbon is $1s$, $2s$, $2p_x$, $2p_y$, $2p_z$. Minimal basis sets cannot adequately describe nonspherical molecular electron distributions.

Minimization: Minimization of the energy of a molecule is a procedure to find configurations for which the molecular energy is a minimum, i.e., finding a point in configuration space where all the forces acting on the atoms are balanced. As there exist several points in large molecules where the atomic forces are balanced, finding the point of the absolute minimum energy is often not a trivial problem. Different minimization algorithms (e.g., steepest descents, conjugate gradients, Newton–Raphson) and procedures for conformational searching are used to find the minimum energy conformation of a molecule.

Minimum energy conformation: MEC. The MEC is that point in configurational space where the energy of the molecule is an absolute minimum and where all the first derivatives of the energy with respect to the coordinates are zero and the second derivative matrix (Hessian matrix) is positive definite.

MNDO: Modified Neglect of Diatomic Overlap. A semi-empirical all-valence electron quantum chemical method pioneered by M.J.S. Dewar and coworkers. For the molecular properties investigated such as heats of formation, ionization potentials, bond lengths, and dipole moments MNDO values are quite close to the experimental ones and are superior to the MINDO/3 results, particularly for nitrogen-containing compounds. Taking into account that the computational effort for MNDO is only about 20% greater than for a MINDO/3 calculation, MNDO is considered to be a significant improvement over MINDO/3.

Molecular descriptors: Molecular descriptors are physicochemical properties used in QSAR analysis and for computing molecular similarity. The simplest constitutional descriptors depend on the composition of the molecule such as molecular weight, number of atoms, bonds, and rings. Topological descriptors of molecular structure depend on the molecular topology or connectivity and branching of the molecule. Well-known topological descriptors are the Wiener index, the Randic index, and the Kier and Hall connectivity index. The geometric or topographic descriptors reflect the 3-D properties of molecules such as molecular volume, solvent accessible surface area, STERIMOL steric parameters, principal moments of inertia, and torsion angles. The electrostatic descriptors are based on the electronic and electrostatic structure of a molecule such as partial atomic charges, the electronegativity of the atoms, polarizability, and molecular electrostatic potential. The electrotopological index of Hall and Kier combines both the electronic character and the topological characteristics of each atom in a molecule. Quantum chemically derived molecular descriptors are attracting much attention nowadays (see Chapter 24).

Molecular diversity: Synthetic and virtual molecule libraries used in combinatorial chemistry and high throughput screening can only be efficient if they contain as much information as possible. The molecular diversity should be comparable to the diversity of natural compounds. The first combinatorial libraries were biased during screening toward large, hydrophobic molecules, which led to poor bioavailability and the rule-of-5.

Molecular dynamics: MD. Taking the negative gradient of the potential energy as evaluated from the molecular mechanics force field yields the force. Using this force and the mass for each atom, Newton's equation of motion ($F = ma$) can be numerically integrated to compute the positions of the atoms after a short time interval (typically of the order of 1 fsec, 10^{-15} sec). By taking successive time steps, a time-dependent trajectory of all the atomic motions can be constructed.

Molecular electrostatic potential: The molecular electrostatic potential (MEP) associated with a molecule arises from the distribution of electrical charges of the nuclei and electrons of a molecule. The MEP is quantum mechanically defined in terms of the spatial coordinates of the charges on the nuclei and the electronic density function $\rho(r)$ of the molecule. As the MEP is the net result of the opposing effects of the nuclei and the electrons, electrophiles will be guided to the regions of a molecule where the MEP is most negative. The MEP is a useful quantity in the study of molecular recognition processes.

Molecular mechanics: MM. Molecular mechanics is an attempt to formulate a force field that can serve as a computational model for evaluating the potential energy for all degrees of freedom of a molecule. Molecular mechanics calculations are very popular because large structures containing many thousands of atoms can be fully energy minimized at reasonable computational costs. Molecular mechanics methods, however, depend heavily on the parametrization of the force field. Molecular mechanics is not appropriate for simulating situations where electronic effects such as orbital interactions and bond breaking are predominant.

Molecular modeling: Molecular modeling of a molecule consists of a computer graphics visualization and representation of the geometry of a molecule. In addition, it involves the manipulation and modification of molecular structures. In combination with X-ray crystallographic or NMR data, molecular modeling implies the use of theoretical methods such as ab initio, semi-empirical, or molecular mechanics to evaluate and predict the minimum energy conformation and other physical and chemical properties of the molecule. Molecular modeling has become an essential tool for structural molecular biology with applications in *CADD*, protein engineering, and molecular recognition.

Molecular recognition: The interaction between two molecules, e.g., ligand and receptor, is mainly dependent on the ligand being able to sterically fit into the active site of the receptor and on the electrostatic complementarity between drug and receptor. The ligand and receptor are at the molecular recognition state when they are separated by more than two van der Waals radii. The contributing interactions for recognition are electrostatic (Coulomb, dipole–dipole), hydrogen bonding, van der Waals (dispersion), and hydrophobic in nature.

Molecular similarity: The degree of similarity between molecules, although quantitatively measurable, very much depends on what molecular features are used to establish the degree of similarity. One of the many comparators is the electron density of a pair of molecules. Other comparators include electrostatic potentials, reactivity indices, hydrophobicity potentials, molecular geometry such as distances and angles between key atoms, solvent accessible surface area, etc. It is an open question as to how much or what part(s) of the molecular structure is to be compared. The Tanimoto coefficient which compares dissimilarity to similarity is often used in molecular diversity analysis.

Monte Carlo: Straightforward scanning of the complete configuration space of a molecular system containing many degrees of freedom is impossible. In that case, an ensemble of configurations can be generated by the Monte Carlo (MC) method which makes use of random sampling and Boltzmann factors. Given a starting configuration, a new configuration is generated by randomly displacing one or more atoms. The newly generated configuration is either accepted or rejected using an energy criterion involving the change of the potential energy (ΔE) relative to the previous configuration. The current configuration is accepted only if its potential energy is lower than or equal to the previous one ($\Delta E \leq 0$) or for $\Delta E > 0$ if the Boltzmann factor $\exp(-\Delta E/kT)$ is larger than a random number taken from a uniform distribution over the $(0,1)$ interval.

Morse potential: A Morse potential is often used for the bond stretching term in a force field. Instead of the quadratic dependence of a harmonic bond stretching term, a Morse potential describes the bond stretching mode as an exponential function. When a molecule is in a high energy state due to sterically overlapping atoms or at a high temperature in a molecular dynamics simulation, the Morse function may allow the bonded atoms to stretch to unrealistic bond lengths.

Neural nets: see artificial neural networks.

Newton–Raphson: A mathematical technique used for the optimization of a function. In contrast to steepest descents and conjugate gradients methods, where the first derivative or gradient of the function is used, Newton–Raphson (NR) methods also use second derivative information to predict where along the gradient the function will change directions. As the second partial derivative matrix of the energy function (Hessian matrix) is calculated, the NR method is much more time consuming than the steepest descents and conjugate gradients methods. NR minimization becomes unstable when a structure is far from the minimum where the forces are large and the second derivative (the curvature) is small. Because storage requirements scale as $3N^2$ (N the number of atoms), NR methods are not suitable for large structures such as proteins.

Parametrization of force fields: The reliability of a molecular mechanics calculation depends on the potential energy equations and the numerical values of the parameters. One obstacle is the small amount of experimental data available for parametrizing and testing a force field. The energy, first and second derivatives of the energy with respect to the Cartesian coordinate of a molecule obtained from high quality ab initio calculations are used to optimize force-field parameters by adjusting the parameters to fit the energy and the energy derivatives by least squares methods.

Partial least squares: Partial least squares (PLS) is a statistical technique often applied to relate physicochemical properties to one or several measurements of biological activity. The PLS results consist of two sets of computed factors which are, on the one hand, linear combinations of the chemical descriptors and, on the other hand, linear combinations of the biological activities. Partial least squares finds many applications in chemometrics and, e.g., in the Tripos CoMFA® approach. Normally used in conjunction with cross-validation.

Partition function: The partition function Q is the summation of the Boltzmann factors $\exp[-E_i/kT]$ over the energy levels E_i of a molecule. A large value of Q will result when the energy levels E_i are closely spaced. The partition function is a measure of the number of available translational, rotational, vibrational, and electronic energy levels. Its value depends among others on the molecular weight, the temperature, the molecular volume, the internuclear distances, the molecular motions, and the intermolecular forces. Although often the calculation of the energy-level pattern is impossible, reasoning in terms of partition functions may provide a more concrete understanding of the free energy of drug–receptor interactions.

Pattern recognition: Pattern recognition is a branch of artificial intelligence that provides an approach to solving the problem of recognizing an obscure property in a collection of objects from measurements made on the objects.

Pattern recognition techniques can be divided into display, preprocessing, supervised, and unsupervised learning. Pattern recognition methods are used among others in the search for correlations between sequence, structure, and biological activity in cheminformatics and bioinformatics.

PBC: Periodic boundary conditions. The term periodic boundary conditions refers to the simulation of molecular systems in a periodic 3-D lattice of identical replicates of the molecular system under consideration. Using PBC allows to simulate the influence of bulk solvent in such a way as to minimize edge effects such as diffusion of a solute toward a surface or the evaporation of solvent molecules.

PCILO: The Perturbative Configuration Interaction using Localized Orbitals method is a semi-empirical, all-valence electron quantum chemical method where, in addition to the ground state, singly and doubly excited configurations are taken into account. The wave function and the ground-state energy are determined by the Rayleigh–Schrödinger perturbation treatment up to the third order. Because of this summation treatment, PCILO is much faster than the Self Consistent Field methods such as MNDO, AM1, and PM3.

PDB: The Protein Data Bank, originally compiled at Brookhaven National Laboratory and currently distributed from http://www.rcsb.org, contains X-ray diffraction and NMR-based structural data of macromolecular structures such as proteins, nucleic acids, and entire viruses. The PDB is the primary structural databank for the 3-D coordinates of macromolecules and freely accessible on-line.

Pharmacophore: A pharmacophore is the spatial mutual orientation of atoms or groups of atoms assumed to be recognized by and interact with a receptor or the active site of a receptor. In conjunction with the receptor concept, the notion of a pharmacophore relates directly to the lock-and-key theory proposed by E. Fischer and P. Ehrlich around the beginning of the 20th century (Corpora non agunt nisi fixata).

PM3: Parametrized Model 3. PM3 is a version of the AM1 method reparametrized by J.J. Stewart. On the whole PM3 gives better estimates of the heat of formation than AM1.

Poisson–Boltzmann model: The PB approximation is one of the more elaborate continuum solvation models that take into account not only the charge density of the solute but also the mobile charge density within the surrounding continuum. Continuum models of solvation are in general capable of calculating absolute free energies of solvation.

Potential of mean force: The thermodynamic quantity needed to estimate equilibrium constants is the $\Delta G°$ between reactants and products. By sampling a reaction coordinate r a potential of mean force (pmf) can be obtained. From the frequency of occurrence of different r values a distribution function $g(r)$ is calculated that is related to $w(r)$, the relative free energy, or the pmf, by $w(r) = -kT \ln g(r)$. By using an additional constraining or biasing potential (umbrella) a system can be forced to sample a reaction coordinate region which would be infrequently sampled in the absence of the umbrella potential because of high barriers in $w(r)$.

Protein folding: One of the most challenging problems in structural biology is the prediction of the 3-D tertiary structure of a protein from its primary structure.

Despite many years of experimental and theoretical studies devoted to it, the protein folding problem remains essentially unsolved because there are too many conformations that can occur in both the unfolded and the folded structure to be searched. The problem of protein folding is further compounded by transient disulfide bonds, solvent, and environmental effects in general that may play an important role in stabilizing particular folded states such as, e.g., the α- or the 3_{10}-helix.

Quantitative structure–activity relationships: QSAR. The QSAR approach pioneered by Hansch and co-workers relates biological data of congeneric structures to physical properties such as hydrophobicity, electronic, and steric effects using linear regression techniques to estimate the relative importance of each of those effects contributing to the biological effect. The molecular descriptors used can be 1-D or 3-D (3D-QSAR). A statistically sound QSAR regression equation can be used for lead optimization.

Quantum-chemical descriptors: The quantum-chemical molecular descriptors are derived from the eigenvalues and eigenvectors. Descriptors based on the eigenvalues are the HOMO and LUMO. Atomic charges, dipole moment, bond orders, and frontier orbital indices are derived from the coefficients of the eigenvectors of the atomic orbitals. The superdelocalizability index is based on both the values of the eigenvalues and eigenvectors.

Radial distribution function: Radial distribution function (RDF) is a term often utilized in analyzing the results of Monte Carlo or MD calculations. The RDF $g(r)$ gives the probability of occurrence of an atom of type a at a distance r from an atom of type b. Peaks in the $g(r)$ vs. r plots can be associated with solvation shells or specific neighbors and can be integrated to yield coordination numbers.

Ramachandran plot: A Ramachandran plot is the conformational energy distribution as a function of the conventional $\varphi(CNC_\alpha C)$ and $\psi(NC_\alpha CN)$ rotational angles in peptides. In high-quality X-ray structures almost all amino acids, apart from glycine and proline, have their φ-, ψ-angles in the regions of the Ramachandran plot indicative of the secondary structure elements α-helix and β-sheet.

Rational drug design: The majority of drugs on the market today for treating disorders in humans, animals, and plants were discovered either by chance observation or by systematic screening of large series of synthetic and natural substances. This traditional method of drug discovery is now supplemented by methods exploiting the increasing knowledge of the molecular targets assumed to participate in some disorder, computer technology, and the physical principles underlying drug–target interactions. Rational drug design— traditional methods were or are not irrational—or better "structure-based ligand design" continues to increase in importance in the endeavor of promoting a biologically active ligand toward the status of a useful drug.

Receptor: A receptor can be envisioned as a macromolecular structure such as a protein being an integral part of the complex molecular structure of the cellular membrane in which it is anchored or associated with. The recognition elements or receptor sites are oriented in such a way that recognition of and interaction with ligands can take place, leading to a pharmacological effect.

Receptor mapping: Receptor mapping is the topographical feature representation of a receptor based on the SAR and conformational aspects of active and inactive analogs of rigid and flexible molecules all putatively acting on that receptor. Inferences as to a pharmacophore on the basis of molecular interactions such as ionic and hydrogen bonding, dipolar effects, π-π stacking interactions, and hydrophobic interactions can be used to construct a hypothetical model of the receptor in which the accessible parts of the amino acids of the receptor protein are delineated.

Restraint: A restraint biases or forces a target function such as the energy function in molecular mechanics toward a specific value for a degree of freedom. Various restraints are in common use: torsional restraints, distance restraints, and tethering. A constraint is the most restrictive version of restraint.

Rule-of-5: Based on a survey of the molecular descriptors of the most successful drugs on the market, C. Lipinski postulated the empirical rule-of-5 for drug-likeness. The molecular weight should be under 500, the log P (hydrophobicity) should be under 5, the number of hydrogen bond donors should be 5 or less, and the number of hydrogen bond acceptors should be 10 or less. All values are multiples of 5, hence the name of the rule. A double violation of this rule gives a warning that the compound might not have drug-like properties, especially unwanted pharmacokinetics.

Scoring: Scoring is the theoretical prediction of ligand–protein affinity and therefore reflects our current knowledge of molecular recognition at the molecular level. De novo design and docking algorithms completely rely on scoring functions to discriminate ligands and docked states. Most scoring functions evaluate polar interactions, such as hydrogen bonds and salt bridges, apolar interactions such as buried apolar surface area, and rotational freedom.

Second-generation force field: The original force field terms were diagonal. In order to increase the realism of simulations, many cross-terms were introduced in the second generation of force fields. The bond-angle term is a cross-term used in second-generation force fields accounting for the mutual influence of bond stretching and bond-angle deformation of a bond angle and a bond centered on the same atom. The bond–bond term is a cross-term used in second-generation force fields to account for the fact that bonds are not isolated but do interact during the vibration around their equilibrium value. The bond-torsion term is a cross-term used in second-generation force fields to account for the fact that the torsional movement around a bond influences the bond length of that bond.

Semi-empirical: see all-valence electron methods.

Semi-ab initio method: SAM1. The major difference between SAM1 and AM1 involves the repulsion integrals which are calculated using an STO-3G basis set and then scaled to account for electron correlation.

Sequence alignment: The search for similarity of nucleotide and amino acid sequences, as, e.g., in BLAST, is essential to bioinformatics. Aligning two or multiple sequences can be performed with or without allowing gaps or residue substitutions. A score should reflect the alignment quality as well as the probability of finding the alignment by chance alone.

Sequence databanks: Sequence databanks hold the sequence information of DNA, RNA, protein translations, or verified protein sequences. These databanks range from simple data depositories such as the GenBank® to well-curated, -checked and -annotated databanks such as SWISS-PROT. Thanks to web technology, these exponentially growing databanks can still be managed.

Sequence patterns: Protein sequence is not well conserved, because protein function is often attributed to only a small set of residues, such as the catalytic triad Serine–Histidine–Aspartate in serine protease enzymes. Special algorithms have been developed to find these patterns or motifs, of which the constituting parts are most often discontinuous parts of a protein sequence.

SHAKE: One approach to reduce the computer time of computationally expensive MD calculations is to increase the time step Δt used for the numerical integration of Newton's equations. For reasons of numerical stability Δt must be small compared to the period of the highest frequency of motions viz. bond stretching vibrations. SHAKE is an algorithm that can constrain bonds to a fixed length during an MD calculation thereby allowing somewhat larger Δt values.

Simulated annealing: Simulated annealing (SA) is a conformational searching technique used in locating the global minimum energy conformation of polypeptides and proteins. Simulated annealing uses a Monte Carlo search of conformational space starting at high temperature where large changes in conformational energies are allowed. As the temperature is lowered with an appropriate cooling schedule the system is (possibly) trapped into a conformation of lowest energy.

Slow-growth: The slow-growth method for free energy calculations is a free energy perturbation or a thermodynamic integration approach under the assumption that the spacings $d\lambda$ of the coupling parameter λ are so small that one needs to sample only one point at any window. This reduces the ensemble average to a single value and allows the derivative to be approximated by a finite difference.

SMILES: Simplified Molecular Input Line Entry System (SMILES) is a chemical notation system based on the principles of molecular graph theory and denotes a molecular structure as a two-dimensional graph familiar to chemists. It allows a rigorous and unambiguous structure specification representing molecular structures by a linear string of symbols. SMILES is used for chemical structure storage, structural display, and substructure searching.

Solvent accessible surface: The solvent accessible surface is the loci of the center of a solvent probe model, represented by a sphere with a given radius, free to touch but not to penetrate the solute when the probe is rolled over its van der Waals surface. This surface can be regarded as a surface based on expanded van der Waals radii.

SPC: In view of the importance of the water–protein interactions it is of utmost interest to have available intermolecular potential functions for the water dimer that yield a good model for liquid water. The simple point charge (SPC) is a three-point charge (on the hydrogen and oxygen positions) model for water with a (6-12) Lennard–Jones potential on the oxygen atom and a charge of 0.41 and −0.82 on the hydrogen and oxygen atoms, respectively.

Steepest descents: The steepest descents method is a minimization algorithm in which the line search direction is taken as the gradient of the function to be minimized. The steepest descents method is very robust in situations where configurations are far from the minimum but converge slowly near the minimum (where the gradient approaches zero).

Stochastic dynamics: The stochastic dynamics (SD) method is a further extension of the original molecular dynamics method. A space–time trajectory of a molecular system is generated by integration of the stochastic Langevin equation which differs from the simple molecular dynamics equation by the addition of a stochastic force R and a frictional force proportional to a friction coefficient g. The SD approach is useful for the description of slow processes such as diffusion, the simulation of electrolyte solutions, and various solvent effects.

Strain energy: Although the first strain theory was advanced by von Bayer in 1885 there is no generally accepted and unique definition of strain energy. The basic qualitative idea is that simple strainless molecules exist and that larger molecules are strainless if their heats of formation are equal to the summation of the bond energies and other increments from the small strainless molecules. The energy calculated by molecular mechanics is strain energy because the deformation energy occurring in a molecule is equal to the energy of minimized structure relative to the hypothetical reference structure.

Structural genomics: The proteins encoded by the genome perform all biological functions. Because the function is related to structure rather than to amino acid sequence, protein structure prediction is one of the most important tools of bioinformatics. Homology modeling, fold recognition and threading all rely on the experimentally determined structures of the Protein Data Bank, and the structural genomics effort aims at fully automated crystallization and structure elucidation of large amounts of proteins to rapidly extend this knowledge base.

Structure alignment: The superimposition or fitting of the 3-D structures of molecules can be based on electron density, atom positions, molecular electrostatic fields, or secondary structures. This is a crucial step for 3D-QSAR and x-ray structure elucidation.

Structure databanks: Structure databanks are repositories of the 3-D coordinates of molecules. These databanks can contain experimental structures of small (CSD) and large (PDB) molecules, or structures made by 3-D builders in virtual libraries, such as the proprietary databanks of most pharmaceutical companies.

Structure-based design: The revolution of molecular biology (started in the early 1970s) has led to many 3-D structures of biomacromolecules and their complexes with ligands and drugs. From the protein structure, preferential sites and atom affinities can be deduced, using programs such as GRID. Ligands can be docked to explore active sites and de novo design is employed for lead generation and lead improvement. (Protein) structure-based design is portrayed as the ultimate rational drug design.

SWISS-PROT: SWISS-PROT is an annotated protein sequence database maintained by the Swiss Institute of Bioinformatics and the EBI. The SWISS-

PROT database distinguishes itself from other protein databases by (i) the generous annotation information; (ii) a minimal redundancy for a given protein sequence; and (iii) the cross-reference with many other biomolecular databases.

Switching function: In order to avoid discontinuities in derivatives and energies during minimization calculations a switching function is used in conjunction with a cut-off algorithm ensuring nonbonded interactions to be smoothly reduced from full strength to zero over a predefined interatomic distance range.

Systematic conformational search: To a first approximation the conformation of a molecule is defined by the torsional angles about the single bonds of a molecule. The systematic search consists of generating all combinations of the torsion angles through 360°. As the number n of rotatable bonds increases and the angular increment Δa decreases, the total number of conformations $N = (360°/\Delta a)^n$ fairly rapidly leads to a combinatorial explosion.

Template forcing: Template forcing is a type of restraint useful in the identification of possible biologically relevant conformations of a conformationally flexible molecule. By selecting atoms or groups of atoms belonging to the possible pharmacophoric pattern common to two molecules, the atoms of the flexible molecule are forced to superimpose onto the atoms of the rigid or template molecule. The energy expenditure to force the flexible molecule onto the template molecule is a measure of the similarity between the two molecules.

Thermodynamic cycle: The thermodynamic cycle approach used to calculate relative free energies or binding constants of, e.g., drug–receptor interactions is based on the fact that the free energy is a thermodynamic function of state. Thus as long as a system is changed reversibly the change in free energy is independent of the path, and therefore nonchemical processes (paths) can be calculated such as the conversion of one type of atom into another (computational alchemy!).

Thermodynamic integration: TI. An approach to free energy calculations is thermodynamic integration consisting of numerically integrating the ensemble average of the derivative of the potential energy of a given configuration with respect to a coupling parameter λ. Because the free energy is evaluated directly from the ensemble average and not as the logarithm of the average of an exponential function as in the free energy perturbation (FEP), TI is not subject to certain systematic errors inherent to FEP calculations.

Threading: If the sequence of a protein is completely dissimilar from the proteins of which the structure is known, structure prediction is often still possible by threading the sequence through a library of protein folds. All amino acids are evaluated and scored at all positions of this fold library and a best fold is chosen. It is estimated that 35% of all protein folds are currently known, but this percentage should rapidly increase because of structural genomics efforts.

Time correlation function: Time correlation functions are of great value for the analysis of dynamical processes in condensed phases. A time correlation function $C(t)$ is obtained when a time-dependent quantity $A(t)$ is multiplied by itself (auto-correlation) or by another time-dependent quantity $B(t')$ evaluated at time t' (cross-correlation) and the product is averaged over some equilibrium ensemble. For example, the self-diffusion coefficient can be obtained from the velocity auto-correlated function for the molecular center of mass motion.

TIP: Transferable intermolecular potential. The TIP family of potentials is used for simulating liquid water. The TIP4P potential for water involves a rigid water monomer composed of three charge centers and one Lennard–Jones center. Two charge centers ($Q = 0.52$) are placed on the hydrogen site 0.9572 Å away from the oxygen atom. The third charge center ($Q = -1.04$) is placed 0.15 Å away from the oxygen atom along the bisector of the HOH angle (104.52°). A Lennard–Jones center is placed on the oxygen atom. The model yields reasonable geometric and energetic results for a linear water dimer and is therefore used in simulations of aqueous solutions.

Topliss tree: The Topliss tree is an empirical decision scheme for a stepwise aromatic substituent selection and lead optimization that is guided by the supposed influence on potency due to the hydrophobic, electronic, and steric effects of the substituents.

Transition state isostere: Enzymes catalyze reactions by lowering the energy barrier that separates the substrate from the product. L. Pauling and others postulated that a substrate intermediate, the transition state, fits the enzyme better than both the substrate and the product, thereby favoring the conversion. This concept led to the design of enzyme inhibitors that resemble this transition state (bioisosteres) in order to better compete with the substrate, because of their higher affinity for the enzyme.

United atom model: For the sake of speeding up an energy calculation the total number of atoms is artificially reduced by lumping together all nonpolar hydrogen atoms into the heavy atoms (C atoms) to which they are bonded. Although this approximation may speed up the calculation several-fold, an all-hydrogen atom model is preferable for accurate calculations.

Van der Waals forces: The term van der Waals forces denotes the short-range interactions between closed-shell molecules. Van der Waals forces include attractive forces arising from interactions between the partial electric charges and repulsive forces arising from the Pauli exclusion principle and the exclusion of electrons in overlapping orbitals. A very commonly used potential is the so-called Lennard–Jones (6-12) potential to describe the attractive and repulsive components of van der Waals forces.

Virtual library: Virtual library is a database of structures and properties of molecules that may not even have been synthesized. Such a library is used to ensure high molecular diversity in the design of a synthetic combinatorial library or can even be used for virtual high throughput screening by applying QSAR analysis.

X-ray structure: Single-crystal X-ray diffraction analysis yields the 3-D structure of a molecule in the crystalline state. An X-ray structure is likely to be a structure in a minimum energy conformational state or close to an energy minimum. An X-ray structure therefore may or may not be the biologically relevant conformation. Inspection of the molecular packing arrangement may yield valuable information about intermolecular contacts and sites of intermolecular hydrogen bonds. Atomic coordinates based on X-ray diffraction data may serve as the primary input data for theoretical conformational analysis calculations, as well as structure-based design, and are normally stored in structure databanks such as the CSD and the PDB.

Z-matrix: The Z-matrix provides a description of each atom of a molecule in terms of its atomic number, bond length, bond angle, and dihedral angle, the so-called internal coordinates. The information from the Z-matrix is used to calculate the Cartesian (X, Y, Z) coordinates of the atoms.

Zero-point energy: The zero-point energy is the residual vibrational energy of a harmonic oscillator at the lowest vibrational state. It arises from the fact that the position of a particle is uncertain and therefore its momentum and hence its kinetic energy cannot be exactly zero.

ZINDO: A semi-empirical quantum-chemical method developed in the group of M. Zerner. ZINDO is particularly adapted to the calculation of transition metals and lanthanide containing compounds.

Related Sources

Tollenaere JP. Glossary of terminology. Chapter 8 in Guidebook on Molecular Modelling in Drug Design. Cohen NC, ed. Academic Press 1996:337–356.

van de Waterbeemd H, Carter G, Grassy RE, Kubinyi H, Martin YC, Tute MS, Willett P. Glossary of terms used in computational drug design. Pure Appl Chem 1997; 69:1137–1152. IUPAC recommendations 1997.

Minkin VI. Glossary of terms used in theoretical organic chemistry. Pure Appl Chem 1999; 71:1919–1981. IUPAC recommendations 1999.

OSC's Computational chemistry glossary, last visited 12 August 2003. A list of computational programs with hyperlinks to their homepages. *http://oscinfo.osc.edu/chemistry/glossary.html.*

Cambridge Healthtech Institute, last visited 12 August 2003. Glossaries of terms used in Bioinformatics, Chem(o)informatics, Drug discovery and development, Molecular modeling and more. *http://www.genomicglossaries.com.*

Tollenaere JP, Moret EE. Hyperglossary linked to bibliography and programmes, last visited 12 August 2003. *http://wwwcmc.pharm.uu.nl/webcmc/glossary.html.*

Index